This text is dedicated to our mentors—John Waterbury, Richard Losick, Thomas Bott, Hank Heath, Pete Magee, Lou Rigley, Irv Snyder, and R. Balfour Sartor. And to our students.

—Joanne M. Willey
—Linda M. Sherwood
—Christopher J. Woolverton

Brief Contents

Contents

Contents   **ix**

Preface

Prescott, Harley, and Klein's Microbiology has acquired the reputation of covering the broad discipline of microbiology at a depth not found in any other textbook. The seventh edition introduces a new author team. As new authors, we were faced with the daunting task of making a superior textbook even better. We bring over 40 years of combined research and teaching experience. Our keen interest in teaching has been fostered by our involvement in workshops and conferences designed to explore, implement, and assess various pedagogical approaches. Thus one of our goals for this edition was to make the book more accessible to students. To accomplish this we focused on three specific areas: readability, artwork, and the integration of several key themes throughout the text.

Our Strengths

Readability

We have retained the relatively simple and direct writing style used in previous editions of *Prescott, Harley, and Klein's Microbiology*. However, for the seventh edition, we have added style elements designed to further engage students. For example, we have intro-

duced the use of the first person to describe the flow of information (e.g., see chapter openers) and we pose questions within the text, prompting students to reflect on the matter at hand. Each chapter is divided into numbered section headings and organized in an outline format. Some chapters have been significantly reorganized to present the material in a more logical format (e.g., chapters 12, 28, and 39). As in previous editions, key terminology is boldfaced and clearly defined. In addition, some words are now highlighted in red font: these include names of scientists with whom the students should be acquainted, as well as names of techniques and microbes. Every term in the extensive glossary, which includes over 200 new and revised entries, includes a page reference.

Artwork

To engage today's students, a textbook must do more than offer text and images that just adequately describe the topic at hand. Our goal is to make the students *want* to read the text because they find the material interesting and appealing. The seventh edition brings a new art program that features three-dimensional renditions and bright, attractive colors. However, not only have existing figures been updated, over 200 new figures have been added. The updated art program also includes new pedagogical features such as concept maps (*see figures 8.1, 12.1, and 31.1*) and annotation of key pathways and processes (*see figures 9.9 and 11.17*).

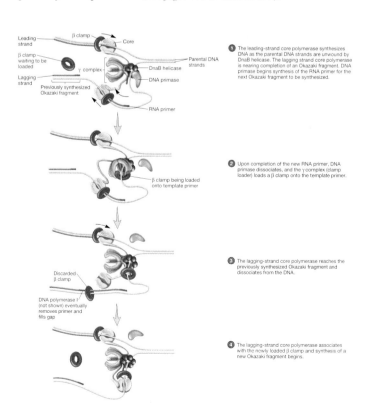

Thematic Integration

With the advent of genomics and the increased reach of cell biology, the divisions among microbiology subdisciplines have become blurred; for instance, the microbial ecologist must also be well-versed in microbial physiology, evolution, and the principles and practices of molecular biology. In addition, the microbiologist must be acquainted with all major groups of microorganisms: viruses, bacteria, archaea, protists, and fungi. Students new to microbiology are asked to assimilate vocabulary, facts, and most importantly, concepts, from a seemingly vast array of subjects. The challenge to the professor of microbiology is to integrate essential concepts throughout the presentation of material while conveying the beauty of microbes and excitement of this dynamic field.

While previous editions of *Microbiology* excelled in incorporating genetics and metabolism throughout the text, in this edition we have attempted to bring the diversity of the microbial world into each chapter. Of course this was most easily done in those chapters devoted to microbial evolution, diversity, and ecology (chapters 19 to 30), but we challenged ourselves to bring microbial diversity into chapters that are traditionally *E. coli*-based. So, although the chapters on genetics (chapters 11 to 13) principally review processes as they are revealed in *E. coli*, we also explore other systems as well, such as the regulation of sporulation in *Bacillus subtilis* and quorum sensing in *V. fischerii* (*see figures 12.19 through 12.21*).

We also thought it was important to weave the thread of evolution throughout the text. We start in the first chapter with a discussion of the universal tree of life (*see figure 1.1*), with various renditions of "the big tree" appearing in later chapters. Importantly, we remind students that structures and processes *evolved* to their current state; that natural selection is always at work (e.g., the title and the tone of chapter 13—now called *Microbial Genetics: Mechanisms of Genetic Variations*—have been changed). Finally, the seventh edition of *Microbiology* explores theories regarding the origin of life at a depth not seen in other microbiology texts (chapter 19).

Indeed, depth of coverage has been one of the mainstays of *Prescott, Harley, and Klein's Microbiology*. The text was founded on two fundamental principles: (1) students need an introduction to the whole of microbiology before concentrating on specialized areas, and (2) this introduction should provide the level of understanding required for students to grasp the conceptual underpinning of facts. We remain committed to this approach. Thus the seventh edition continues to provide a balanced and thorough introduction to all major areas of microbiology. This book is suitable for courses with orientations ranging from basic microbiology to medical and applied microbiology. Students preparing for careers in medicine, dentistry, nursing, and allied health professions will find the text as useful as will those aiming for careers in research, teaching, and industry. While two courses each of biology and chemistry are assumed, we provide a strong overview of the relevant chemistry in appendix I.

CHANGES TO THE SEVENTH EDITION

The seventh edition of *Prescott, Harley, and Klein's Microbiology* is the result of extensive review and analysis of previous editions, the input from reviewers, and casual discussions with our colleagues. As a new author team, we were committed to keeping the in-depth coverage that *Microbiology* is known for, while at the same time bringing a fresh perspective not only to specific topics but to the overall presentation as well.

Up-to-Date Coverage

Each year exciting advances are made in microbiology. While we understand that not all of these are appropriate for discussion in an introductory textbook, we have incorporated the most up-to-date information and exciting, recent discoveries to maintain accurate descriptions of structures and processes and to illustrate essential points. A few specific examples include a current description of the structure and function of DNA polymerase III, the role of viruses in marine ecosystems, the ubiquitous nature of type III secretion systems, an updated coverage of the inflammatory response, and the current understanding of HIV origins and avian influenza epidemiology.

Increased Emphasis on Microbial Evolution and Diversity

Microbial evolution, diversity, and ecology are no longer subdisciplines to be ignored by those interested in microbial genetics, physiology, or pathogenesis. For example, within the last 10 years, polymicrobial diseases, intercellular communication, and biofilms have been recognized as important microbial processes that closely tie evolution to genetics, ecology to physiology, and ecology to pathogenesis. The seventh edition strives to integrate these themes throughout the text. We begin chapter 1 with a discussion of the universal tree of life and whenever possible, bring diverse microbial species into discussions so that students can begin to appreciate the tremendous variation in the microbial world. Chapter 19 now covers microbial evolution in greater depth than other texts. It has been retitled *Microbial Evolution, Taxonomy, and Diversity* and the content significantly revised so that microbial evolution is presented as a key component of microbiology. We also introduce and frequently remind students of the enormity of microbial diversity. Like previous editions, the seventh edition features specific chapters that review the members of the microbial world. The chapters that are specifically devoted to ecology (chapters 27 through 29) have undergone significant revisions. We continue to use the classification scheme set forth in the second edition of *Bergey's Manual of Systematic Bacteriology;* in addition, we have introduced the Baltimore System of virus classification and the International Society of Protistologists' new classification scheme for eucaryotes in chapters 18 and 25, respectively.

Chapter 39—Reorganized and updated to reflect disease transmission routes (similar to chapters 37 and 38); new sections on cyclospora and microsporidia.

Part XI

Chapter 40—Expanded discussion of lactic acid bacteria, probiotics: chocolate fermentation now featured in a Techniques & Applications box.

Chapter 41—Revised to include water purification and wastewater treatment. New section on nanotechnology; expanded section on the biochemistry of bioremediation.

TOOLS FOR LEARNING

Chapter Preview

- Each chapter begins with a preview—a list of important concepts discussed in the chapter.

Cross-Referenced Notes

- In-text notes in blue type refer students to other parts of the book to review.

Review and Reflection Questions within Narrative

- Review questions throughout each chapter assist students in mastering section concepts before moving on to other topics.

the color they fluoresce after treatment with a special mixture of stains (**figure 2.13a**). Thus the microorganisms can be viewed and directly counted in a relatively undisturbed ecological niche. Identification of microorganisms from specimens: Immunologic techniques (section 35.2).

1. List the parts of a light microscope and describe their functions.
2. Define resolution, numerical aperture, working distance, and fluorochrome.
3. If a specimen is viewed using a 5X objective in a microscope with a 15X eyepiece, how many times has the image been magnified?
4. How does resolution depend on the wavelength of light, refractive index, and numerical aperture? How are resolution and magnification related?
5. What is the function of immersion oil?
6. Why don't light microscopes use 30X ocular lenses for greater magnification?
7. Briefly describe how dark-field, phase-contrast, differential interference contrast, and epifluorescence microscopes work and the kind of image provided by each. Give a specific use for each type.

2.3   PREPARATION AND STAINING OF SPECIMENS

Although living microorganisms can be directly examined with the light microscope, they often must be fixed and stained to increase visibility, accentuate specific morphological features, and preserve them for future study.

Fixation

The stained cells seen in a microscope should resemble living cells as closely as possible. **Fixation** is the process by which the internal and external structures of cells and microorganisms are preserved and fixed in position. It inactivates enzymes that might disrupt cell morphology and toughens cell structures so that they do not change during staining and observation. A microorganism usually is killed and attached firmly to the microscope slide during fixation.

There are two fundamentally different types of fixation. **Heat fixation** is routinely used to observe procaryotes. Typically, a film of cells (a smear) is gently heated as a slide is passed

1

The History and Scope of Microbiology

Louis Pasteur, one of the greatest scientists of the nineteenth century, maintained that "Science knows no country, because knowledge belongs to humanity, and is a torch which illuminates the world."

PREVIEW

- Microbiology is defined not only by the size of its subjects but the techniques it uses to study them.
- Microorganisms include acellular entities (e.g., viruses), procaryotic cells, and eucaryotic cells. Cellular microorganisms are found in all three domains of life: *Bacteria, Archaea, Eucarya*.
- The development of microbiology as a scientific discipline has depended on the availability of the microscope and the ability to isolate and grow pure cultures of microorganisms. The development of these techniques in large part grew out of studies disproving the Theory of Spontaneous Generation and others establishing that microorganisms can cause disease.
- Microbiology is a large discipline; it has had and will continue to have a great impact on other areas of biology and general human welfare.

The importance of microorganisms can't be overemphasized. In terms of sheer number and mass—it is estimated that microbes contain 50% of the biological carbon and 90% of the biological nitrogen on Earth—they greatly exceed every other group of organisms on the planet. Furthermore, they are found everywhere from geothermal vents in the ocean depths to the coldest arctic ice, to every person's skin. They are major contributors to the functioning of the biosphere, being indispensable for the cycling of the elements essential for life. They also are a source of nutrients at the base of all ecological food chains and webs. Most importantly, certain microorganisms carry out photosynthesis, rivaling plants in their role of capturing carbon dioxide and releasing oxygen into the atmosphere. Those microbes that inhabit humans also play important roles, including helping the body digest food and producing vitamins B and K. In addition, society in general benefits from microorganisms, as

they are necessary for the production of bread, cheese, beer, antibiotics, vaccines, vitamins, enzymes, and many other important products. Indeed, modern biotechnology rests upon a microbiological foundation.

Although the majority of microorganisms play beneficial or benign roles, some harm humans and have disrupted society over the millennia. Microbial diseases undoubtedly played a major role in historical events such as the decline of the Roman Empire and the conquest of the New World. In 1347, plague or black death, an arthropod-borne disease, struck Europe with brutal force, killing 1/3 of the population (about 25 million people) within four years. Over the next 80 years, the disease struck again and again, eventually wiping out 75% of the European population. The plague's effect was so great that some historians believe it changed European culture and prepared the way for the Renaissance. Today the struggle by microbiologists and others against killers like AIDS and malaria continues.

In this introductory chapter, we introduce the microbial world to provide a general idea of the organisms and agents that microbiologists study. Then we describe the historical development of the science of microbiology and its relationship to medicine and other areas of biology. Finally, we discuss the scope, relevance, and future of modern microbiology.

Dans les champs de l'observation, le hasard ne favorise que les esprits préparés.
(In the field of observation, chance favors only prepared minds.)

—*Lou*

1.1   MEMBERS OF THE MICROBIAL WO

Microbiology often has been defined as the study and agents too small to be seen clearly by the unai is, the study of **microorganisms**. Because objects le one millimeter in diameter cannot be seen clearly

Concept Maps

- Many chapters include a concept map, new to this edition, that outlines critical themes.

Figure 8.1   Overview of Metabolism. The cell structures of organisms are assembled from various macromolecules (e.g., nucleic acids and proteins). Macromolecules are synthesized from monomers and other building blocks (e.g., nucleotides and amino acids), which are the products of biochemical pathways that begin with precursor metabolites (e.g., pyruvate and α-ketoglutarate). In autotrophs, the precursor metabolites arise from CO₂ fixation pathways and related pathways; in heterotrophs, they arise from reactions of the central metabolic pathways. Reducing power and ATP are consumed in many metabolic pathways. All organisms can be defined metabolically in terms of their energy source, carbon source, and electron source. In the case of chemoorganotrophs, the energy source is an organic molecule that is also the source of carbon and electrons. For chemolithotrophs, the energy source ... the carbon source can be either CO₂ (autotrophs) or an organic molecule (heter... bon source can be CO₂ or organic molecules, and the electron source can be wa... such as hydrogen sulfide (anoxygenic phototrophs).

BACTERIA

Transcription — Gene
Genetic regulatory proteins can bind to the DNA and control whether or not transcription begins.

Attenuation: Transcription can terminate very early after it has begun due to the formation of a transcriptional terminator.

Binding of a metabolite to a riboswitch in mRNA can cause premature termination of transcription.

Translation — mRNA
Translational repressor proteins can bind to the mRNA and prevent translation from starting.

Antisense RNA can bind to mRNA and control whether or not translation begins.

Binding of a metabolite to a riboswitch in mRNA can block translation.

Posttranslation — Protein
Small molecules can bind (noncovalently) to a protein and affect its function. An example is feedback inhibition, in which the product of a metabolic pathway inhibits the first enzyme in the pathway.

The structure and function of a protein can be altered by covalent changes to the protein. These can be reversible (e.g., phosphorylation/dephosphorylation) or irreversible (e.g., removal of amino acid residues). These are called posttranslational modifications.

Functional protein

(a)

ARCHAEA

Transcription — Gene
Genetic regulatory proteins can bind to the DNA and control whether or not transcription begins.

The compaction level of chromatin may influence transcription.

Translation — mRNA
Antisense RNA can bind to mRNA and control whether or not translation begins.

Posttranslation — Protein
Feedback inhibition and covalent modifications (reversible and irreversible) may regulate protein function.

Functional protein

(b)

Figure 12.1   Gene Expression and Common Regulatory Mechanisms in the Three Domains of Life.

① Isolate DNA to be cloned.

② Use a restriction enzyme or PCR to generate fragments of DNA.

③ Generate a recombinant molecule by inserting DNA fragments into a cloning vector.

Vector   Linear vector

④ Introduce recombinant molecule into new host.

New host

Figure 14.1   Steps in Cloning a Gene. Each step shown in this overview is discussed in more detail in Chapter 14.

Special Interest Essays

- Interesting essays on relevant topics are included in each chapter. Readings are organized into these topics: Historical Highlights, Techniques & Applications, Microbial Diversity & Ecology, Disease, and Microbial Tidbits.

Microbial Diversity & Ecology

18.1   SARS: Evolution of a Virus

In November 2002, a mysterious pneumonia was seen in the Guangdong Province of China, but the first case of this new type of pneumonia was not reported until February 2003. Thanks to the ease of global travel, it took only a couple of months for the pneumonia to spread to more than 25 countries in Asia, Europe, and North and South America. This newly emergent pneumonia was labeled Severe Acute Respiratory Syndrome (SARS) and its causative agent was identified as a previously unrecognized member of the coronavirus family, the SARS-CoV. Almost 10% of the roughly 8,000 people with SARS died. However, once the epidemic was contained, the virus appeared to "die out," and with the exception of a few mild, sporadic cases in 2004, no additional cases have been identified. From where does a newly emergent virus come? What does it mean when a virus "dies out"?

We can answer these questions thanks to the availability of the complete SARS-CoV genome sequence and the power of molecular modeling. Coronaviruses are large, enveloped viruses with positive-strand RNA genomes. They are known to infect a variety of mammals and birds. Researchers suspected that SARS-CoV had "jumped" from its animal host to humans, so samples of animals at open markets in Guangdong were taken for nucleotide sequencing. These studies revealed that cat-like animals called masked palm civits (*Paguma larvata*) harbored variants of the SARS-CoV. Although thousands of civits were then slaughtered, further studies failed to find widespread infection of domestic or wild civits. In addition, experimental infection of civits with human SARS-CoV strains made these animals ill, making the civit an unlikely candidate for the reservoir species. Such a species would be expected to harbor SARS-CoV without symptoms so that it could efficiently spread the virus.

Bats are reservoir hosts of several zoonotic viruses (viruses spread from animals to people) including the emerging Hendra and Nipah viruses that have been found in Australia and East Asia, respectively. Thus it was perhaps not too surprising when in 2005, two groups of international scientists independently demonstrated that Chinese horseshoe bats (genus *Rhinolophus*) are the natural reservoir of a SARS-like coronavirus. When the genomes of the human and bat SARS-CoV are aligned, 92% of the nucleotides are identical. More revealing is that 96 to 100% identical for all proteins except the receptor-binding spike proteins, which are only 64% identical. The SARS-CoV spike protein mediates both host cell surface attachment and membrane fusion. Thus a mutation of the spike protein allowed the virus to "jump" from bat host cells to those of another species. It is not clear if the SARS-

Receptor Activity

(a) Good   (b) Poor   (c) Poor

Host Range of SARS-CoV Is Determined by Several Amino Acid Residues in the Spike Protien.   (a) The spike protien of the SARS-CoV that caused the SARS epidemic in 2002–2003 fits tightly to the human host cell receptor ACE2. (b) The civit SARS-CoV has two different amino acids at positions 479 and 487. This spike protein binds very poorly to human ACE2, thus the receptor is only weakly variant. (c) The spike protein on the human SARS-CoV that was isolated from patients in 2003 and 2004 also differs from that seen in the epidemic-causing SARS-CoV virus. This SARS-CoV variant caused only mild, sporadic cases.

within the RBD, only four differ between civit and human. Two of these amino acids appear to be critical. As shown in the **Box figure** compared to the spike RBD in the SARS-CoV that caused 2002–2003 epidemic, the civit spike has a serine (S) substituted a threonine (T) at position 487 (T487S) and a lysine (K) at position 479 instead of asparagine (N), N479K. This causes a 1,000-fold decrease in the capacity of the virus to bind to human ACE2. Furthermore, the spike found in SARS-CoV isolated from patients in 2003 and 2004 also has a serine at position 487 as well as a proline (P) leucine (L) substitution at position 472 (L472P). These amino acids substitutions could be responsible for the reduced virulence of the virus found in these more recent infections. In other words, these mutations could be the reason the SARS virus "died out."

Meanwhile a SARS vaccine based on the virulent 2002–2003 strain is being tested. This raises additional questions. Does original virulent SARS-CoV strain still exist? Will the most

---

Microbial Tidbits

35.2   Biosensors: The Future Is Now

The 120-plus-year-old pathogen detection systems based on culture and biochemical phenotyping are being challenged. Fueled by the release of anthrax spores in the U.S. postal system, government agencies have been calling for newer technologies for the near-immediate detection and identification of microbes. In the past, detection technologies have traded speed for cost and complexity. The agar plate technique, refined by Robert Koch and his contemporaries in the 1880s, is a trusted and highly efficient method for the isolation of bacteria into pure cultures. Subsequent phenotyping biochemical methods, often using differential media in a manner similar to that used in the isolation step, then identifies common bacterial pathogens. Unfortunately, reliable results from this process often take several days. More rapid versions of the phenotyping systems can be very efficient, yet still require pure culture inoculations. The rapid immunological tests offer faster detection responses but may sacrifice sensitivity. Even DNA sequence comparisons, which are extremely accurate, may require significant time for DNA amplification and significant cost for reagents and sensitive readers. As usual, necessity has begat invention.

The more recent microbial detection systems, many of which are still untested in the clinical arena, sound like science fiction gizmos, yet promise a new age for near-immediate detection and identification of pathogens. These technologies are collectively referred to as "biosensors," and if the biosensor is integrated with a computer microchip for information management, it is then called a "biochip." Biosensors should ideally be capable of highly specific recognition so as to discriminate between nearest relatives, and

"communicate" detection through some type of transducing system. Biosensors that detect specific DNA sequences, expressed proteins, and metabolic products have been developed that use optical (mostly fluorescence), electrochemical, or even mass displacement, to report detection. The high degree of recognition required to reduce false-positive results has demanded the uniquely specific, receptor-like capture that is associated with nucleic acid hybridization and antibody binding. Several microbial biosensors employ single-stranded DNA or RNA sequences, or antibody, for the detection component. The transducing or sensing component of biosensors may be markedly different, however. For example, microcantilever systems detect the increased mass of the receptor-bound ligand; the surface acoustic wave device detects change in specific gravity; the bulk quartz resonator monitors fluid density and viscosity; the quartz crystal microbalance measures frequency change in proportion to the mass of material deposited on the crystal; the micromirror sensor uses an optical fiber waveguide that changes reflectivity; and the liquid crystal-based system reports the reorientation of polarized light. Thus the specific capture of a ligand is reflected in the net change measured by each system and results in a signal that announces the initial capture event. Microchip control of the primary and subsequent secondary signals has resulted in automation of the detection process. The reliable detection of pathogens in complex specimens will be the real test as each of these technologies continues to compete for a place in the clinical laboratory.

Techniques & Applications

40.3   Chocolate: The Sweet Side of Fermentation

Chocolate could be characterized as the "world's favorite food," and yet few people realize that fermentation is an essential part of chocolate production. The Aztecs were the first to develop chocolate fermentation, serving a chocolate drink made from the seeds of the chocolate tree, *Theobroma cocao* [Greek *theos*, god and *broma*, food, or "food of the gods"]. Chocolate trees now grow in West Africa as well as South America.

The process of chocolate fermentation has changed very little over the past 500 years. Each tree produces a large pod that contains 30 to 40 seeds in a sticky pulp (see **Box Figure**). Ripe pods are harvested and slashed open to release the pulp and seeds. The sooner the fermentation begins, the better the product, so fermentation occurs on the farm where the trees are grown. The seeds and pulp are placed in "sweat boxes" or in heaps in the ground and covered, usually with banana leaves.

Like most fermentations, this process involves a succession of microbes. First, a community of yeasts, including *Candida rugosa* and *Kluyveromyces marxianus*, hydrolyze the pectin that covers the seeds and ferment the sugars to release ethyl alcohol and CO_2. As the temperature and the alcohol concentration increase, the yeasts are inhibited and lactic acid bacteria increase in number. The mixture is stirred to aerate the microbes and ensure an even temperature distribution. Lactic acid production drives the pH down; this encourages the growth of bacteria that produce acetic acid as a fermentation end product. Acetic acid is critical to the production of fine chocolate because it kills the sprout inside the seed and releases enzymes that cause further degradation of proteins and carbohydrates, contributing to the overall taste of the chocolate. In addition, acetate esters, derived from acetic acid, are important for the development of good flavor. Fermentation takes five to seven days. An experienced chocolate grower will know when the fermentation is complete—if it is stopped too soon the chocolate will be bitter and astringent. On the other hand, if fermentation lasts too long, microbes start growing on the seeds instead of in the pulp. "Off-tastes" arise when the gram-positive bacterium *Bacillus* and the filamentous fungi *Aspergillis*, Penicillium, and *Mucor* hydrolyze lipids in the seeds to release short-chain fatty acids. As the pH begins to rise, the bacteria of the genera *Pseudomonas*, *Enterobacter*, and *Escherichia* also contribute to bad tastes and odor.

After fermentation, the seeds, now called beans, are spread out to dry. Ideally this is done in the sun, although drying ovens are also used. The oven-drying method is considered inferior because the beans can acquire a smoky taste. The dried beans are brown and lack the pulp. They are bagged and sold to chocolate manufacturers, who first roast the beans to further reduce the bitter taste and kill most of the microbes (some *Bacillus* spores may remain). The beans are then ground and the nibs—the inner part of each bean—are removed. The nibs are crushed into a thick paste called a chocolate liquor, which contains cocoa solids and cocoa butter, but no alcohol. Cocoa solids are brown and have a rich flavor, and cocoa butter has a high fat con-

---

Historical Highlights

5.1   The Discovery of Agar as a Solidifying Agent and the Isolation of Pure Cultures

The earliest culture media were liquid, which made the isolation of bacteria to prepare pure cultures extremely difficult. In practice, a mixture of bacteria was diluted successively until only one organism, as an average, was present in a culture vessel. If everything went well, the individual bacterium thus isolated would reproduce to give a pure culture. This approach was tedious, gave variable results, and was plagued by contamination problems. Progress in isolating pathogenic bacteria understandably was slow.

The development of techniques for growing microorganisms on solid media and efficiently obtaining pure cultures was due to the efforts of the German bacteriologist Robert Koch and his associates. In 1881 Koch published an article describing the use of boiled potatoes, sliced with a flame-sterilized knife, in culturing bacteria. The surface of a sterile slice of potato was inoculated with bacteria from a needle tip, and then the bacteria were streaked out over the surface so that a few individual cells would be separated from the remainder. The slices were incubated beneath bell jars to prevent airborne contamination, and the isolated cells developed into pure colonies. Unfortunately many bacteria would not grow well on potato slices.

At about the same time, Frederick Loeffler, an associate of Koch, developed a meat extract peptone medium for cultivating

pathogenic bacteria. Koch decided to try solidifying this medium. Koch was an amateur photographer—he was the first to take photomicrographs of bacteria—and was experienced in preparing his own photographic plates from silver salts and gelatin. Precisely the same approach was employed for preparing solid media. He spread a mixture of Loeffler's medium and gelatin over a glass plate, allowed it to harden, and was experienced in the same way he had inoculated his sliced potatoes. The new solid medium worked well, but it could not be incubated at 37°C (the best temperature for most human bacterial pathogens) because the gelatin would melt. Furthermore, some bacteria digested the gelatin.

About a year later, in 1882, agar was first used as a solidifying agent. It had been discovered by a Japanese innkeeper, Minora Tarazaemon. The story goes that he threw out some extra seaweed and discovered the next day that it had jelled during the cold winter night. Agar had been used by the East Indies Dutch to make jellies and jams. Fannie Eilshemius Hesse (see figure 1.7), the New Jersey-born wife of Walther Hesse, one of Koch's assistants, had learned of agar from a Dutch acquaintance and suggested its use when she heard of the difficulties with gelatin. Agar-solidified medium was an instant success and continues to be essential in all areas of microbiology.

Disease

1.2   Koch's Molecular Postulates

Although the criteria that Koch developed for proving a causal relationship between a microorganism and a specific disease have been of great importance in medical microbiology, it is not always possible to apply them in studying human diseases. For example, some pathogens cannot be grown in pure culture outside the host; because other pathogens grow only in humans, their study would require experimentation on people. The identification, isolation, and cloning of genes responsible for pathogen virulence have made possible a new molecular form of Koch's postulates that resolves some of these difficulties. The emphasis is on the virulence genes present in the infectious agent rather than on the agent itself. The molecular postulates can be briefly summarized as follows:

1. The virulence trait under study should be associated much more with pathogenic strains of the species than with nonpathogenic strains.

2. Inactivation of the gene or genes associated with the suspected virulence trait should substantially decrease pathogenicity.
3. Replacement of the mutated gene with the normal wild-type gene should fully restore pathogenicity.
4. The gene should be expressed at some point during the infection and disease process.
5. Antibodies or immune system cells directed against the gene products should protect the host.

The molecular approach cannot always be applied because of problems such as the lack of an appropriate animal system. It is also difficult to employ the molecular postulates when the pathogen is not well characterized genetically.

---

has been used to study the interactions between the *E. coli* GroES and GroEL chaperone proteins, to map plasmids by locating restriction enzymes bound to specific sites, to follow the behavior of living bacteria and other cells, and to visualize membrane proteins (**figure 2.29**).

1. How does a confocal microscope operate? Why does it provide better images of thick specimens than does the standard compound microscope?
2. Briefly describe the scanning probe microscope and compare and contrast its most popular versions—the scanning tunneling microscope and the atomic force microscope. What are these microscopes used for?

Summary

2.1 Lenses and the Bending of Light

a. A light ray moving from air to glass, or vice versa, is bent in a process known as refraction.

b. Lenses focus light rays at a focal point and magnify images (**figure 2.2**).

2.2 The Light Microscope

a. In a compound microscope like the bright-field microscope, the primary image is formed by an objective lens and enlarged by the eyepiece or ocular lens to yield the final image (**figure 2.3**).

b. A substage condenser focuses a cone of light on the specimen.

c. Microscope resolution increases as the wavelength of radiation used to illuminate the specimen decreases. The maximum resolution of a light microscope is about 0.2 μm.

d. The dark-field microscope uses only refracted light to form an image (**figure 2.7**), and objects glow against a black background.

e. The phase-contrast microscope converts variations in the refractive index and density of cells into changes in light intensity and thus makes colorless, unstained cells visible (**figure 2.9**).

f. The differential interference contrast microscope uses two beams of light to create high-contrast, three-dimensional images of live specimens.

g. The fluorescence microscope illuminates a fluorochrome-labeled specimen and forms an image from its fluorescence (**figure 2.12**).

2.3 Preparation and Staining of Specimens

a. Specimens usually must be fixed and stained before viewing them in the bright-field microscope.

b. Most dyes are either positively charged basic dyes or negative acidic dyes and bind to ionized parts of cells.

c. In simple staining a single dye is used to stain microorganisms.

d. Differential staining procedures like the Gram stain and acid-fast stain distinguish between microbial groups by staining them differently (**figure 2.15**).

e. Some staining techniques are specific for particular structures like bacterial capsules, flagella, and endospores (**figure 2.14**).

2.4 Electron Microscopy

a. The transmission electron microscope uses magnetic lenses to form an image from electrons that have passed through a very thin section of a specimen (**figure 2.19**). Resolution is high because the wavelength of electrons is very short.

b. Thin section contrast can be increased by treatment with solutions of heavy metals like osmium tetroxide, uranium, and lead.

c. Specimens are also prepared for the TEM by negative staining, shadowing with metal, or freeze-etching.

d. The scanning electron microscope (**figure 2.23**) is used to study external surface features of microorganisms.

2.5 Newer Techniques in Microscopy

a. The confocal scanning laser microscope (**figure 2.25**) is used to study thick, complex specimens.

b. Scanning probe microscopes reach very high magnifications that allow scientists to observe biological molecules (**figures 2.27** and **2.29**).

Key Terms

acidic dyes 26
acid-fast staining 26
atomic force microscope 36
basic dyes 26
bright-field microscope 18
capsule staining 26
chemical fixation 26
chromophore groups 26
confocal scanning laser microscope (CSLM) 34
dark-field microscope 21
differential interference contrast (DIC) microscope 23

differential staining 26
endospore staining 26
eyepieces 18
fixation 25
flagella staining 28
fluorescence microscope 23
fluorescent light 23
fluorochromes 24
focal length 18
focal point 18
freeze-etching 30

Gram stain 26
heat fixation 25
mordant 26
negative staining 26
numerical aperture 19
objective lenses 18
ocular lenses 18
parfocal 18
phase-contrast microscope 21
refraction 17
refractive index 17

resolution 18
scanning electron microscope (SEM) 30
scanning probe microscope 35
scanning tunneling microscope 35
shadowing 29
simple staining 26
substage condenser 18
transmission electron microscope (TEM) 29
working distance 20

Chapter Summaries

- End-of-chapter summaries are organized by numbered headings and provide a snapshot of important chapter concepts.

End-of-Chapter Material

- *Key Terms* highlight chapter terminology and list term location in the chapter.
- *Critical Thinking Questions* supplement the questions for review and reflection found throughout each chapter; they are designed to stimulate analytical problem solving skills.
- *Learn More* includes a short list of recent and relevant papers for the interested student and professor. Additional references can be found at the Prescott website at *www.mhhe. com/prescott7*.

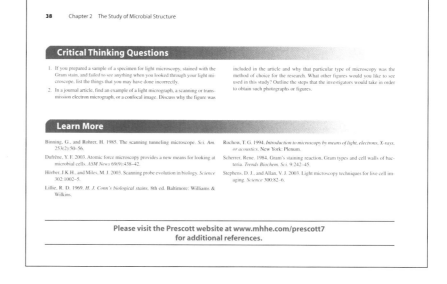

Critical Thinking Questions

1. If you prepared a sample of a specimen for light microscopy, stained with the Gram stain, and failed to see anything when you looked through your light microscope, list the things that you may have done incorrectly.

2. In a journal article, find an example of a light micrograph, a scanning or transmission electron micrograph, or a confocal image. Discuss why the figure was included in the article and why that particular type of microscopy was the method of choice for the research. What other figures would you like to see used in this study? Outline the steps that the investigators would take in order to obtain such photographs or figures.

Learn More

Binning, G., and Rohrer, H. 1985. The scanning tunneling microscope. *Sci. Am.* 253(2):50–56.

Dufrêne, Y. F. 2003. Atomic force microscopy provides a new means for looking at microbial cells. *ASM News* 69(9):438–42.

Hörber, J.K.H., and Miles, M. J. 2003. Scanning probe evolution in biology. *Science* 302:1002–5.

Lillie, R. D. 1969. *H. J. Conn's biological stains*, 8th ed. Baltimore: Williams & Wilkins.

Rochow, T. G. 1994. *Introduction to microscopy by means of light, electrons, X-rays, or acoustics.* New York: Plenum.

Scherrer, Rene. 1984. Gram's staining reaction, Gram types and cell walls of bacteria. *Trends Biochem. Sci.* 9:242–45.

Stephens, D. J., and Allan, V. J. 2003. Light microscopy techniques for live cell imaging. *Science* 300:82–6.

Please visit the Prescott website at www.mhhe.com/prescott7
for additional references.

STUDENT RESOURCES

Student Study Guide

The **Student Study Guide** is a valuable resource that provides learning objectives, study outlines, learning activities, and self-testing material to help students master course content.

Laboratory Exercises in Microbiology

The seventh edition of *Laboratory Exercises in Microbiology* by John P. Harley has been prepared to accompany the text. Like the text, the laboratory manual provides a balanced introduction in each area of microbiology. The class-tested exercises are modular and short so that an instructor can easily choose those exercises that fit his or her course.

ARIS

McGraw-Hill's ARIS—Assessment, Review, and Instruction System for *Prescott, Harley, and Klein's Microbiology*, *www.mhhe.com/prescott7*. This online resource provides helpful study materials that support each chapter in the book. Features include:

Self-quizzes
Animations (with quizzing)
Flashcards
Clinical case studies
Additional course content and more!

INSTRUCTOR RESOURCES

ARIS (www.mhhe.com/prescott7)

McGraw-Hill's ARIS—Assessment, Review, and Instruction System for *Prescott, Harley, and Klein's Microbiology* is a complete, online tutorial, electronic homework, and course management system. Instructors can create and share course materials and assignments with colleagues with a few clicks of the mouse. All PowerPoint lectures, assignments, quizzes, and tutorials are directly tied to text-specific materials. Instructors can also edit questions, import their own content, and create announcements and due dates for assignments. ARIS has automatic grading and reporting of easy-to-assign homework, quizzing, and testing. All student activity within McGraw-Hill's ARIS is automatically recorded and available to the instructor through a fully integrated grade book that can be downloaded to Excel. Contact your local McGraw-Hill Publisher's representative for more information on getting started with ARIS.

ARIS Presentation Center

Build instructional materials wherever, whenever, and however you want!
ARIS Presentation Center is an online digital library containing assets such as photos, artwork, animations, PowerPoints, and other media types that can be used to create customized lectures, visually enhanced tests and quizzes, compelling course websites, or attractive printed support materials.

Access to your book, access to all books!
The Presentation Center library includes thousands of assets from many McGraw-Hill titles. This ever-growing resource gives instructors the power to utilize assets specific to an adopted textbook as well as content from all other books in the library.

Nothing could be easier!
Accessed from the instructor side of your textbook's ARIS website, Presentation Center's dynamic search engine allows you to explore by discipline, course, textbook chapter, asset type, or keyword. Simply browse, select, and download the files you need to build engaging course materials. All assets are copyright McGraw-Hill Higher Education but can be used by instructors for classroom purposes.

- **Art Library**—Color-enhanced, digital files of all illustrations in the book can be readily incorporated into lecture presentations, exams, or custom-made classroom materials. The large, bolded labels make the images appropriate for use in large lecture halls.
- **TextEdit Art Library**—Every line art piece is placed into a PowerPoint presentation that allows the user to revise, move, or delete labels as desired for creation of customized presentations or for testing purposes.
- **Photo Library**—Like the Art Library, digital files of all the photographs from the book are available.
- **Table Library**—Every table that appears in the book is provided in electronic form.
- **Animations Library**—Full-color presentations involving key process figures in the book have been brought to life via animations. These animations offer flexibility for instructors and were designed to be used in lecture or for self-study. Instructors can pause, rewind, fast forward, and turn audio off/on to create dynamic lecture presentations.
- **PowerPoint Lecture Outlines**—These ready-made presentations combine art and lecture notes for each of the 41 chapters of the book. The presentations can be used as they are, or they can be customized to reflect your preferred lecture topics and organization.
- **PowerPoint Outlines**—The art, photos, and tables for each chapter are inserted into blank PowerPoint presentations to which you can add your own notes.

Instructor Testing and Resource CD-ROM

This cross-platform CD contains the *Instructor's Manual and Test Bank*, both available in Word and PDF formats. The *Instructor's Manual* contains chapter overviews, objectives, and answer guidelines for Critical Thinking Questions. The Test Bank provides questions that can be used for homework assignments or the preparation of exams. The computerized test bank allows the user to quickly create customized exams. This user-friendly program allows instructors to search for questions by topic, format, or difficulty level; edit existing questions or add new ones; and scramble questions and answer keys for multiple versions of the same test.

Transparencies

A set of 250 full-color acetate transparencies is available to supplement classroom lectures. These have been enhanced for projection and are available to adopters of the seventh edition.

ACKNOWLEGMENTS

Focus Group Participants

Jeffrey Isaacson, *Nebraska Wesleyan University*
Janice Knepper, *Villanova University*
Donald Lehman, *University of Delaware*
Susan Lovett, *Brandeis University*
Anne Morris Hooke, *Miami University of Ohio*
Mark Schneegurt, *Wichita State University*
Daniel Smith, *Seattle University*
Michael Troyan, *Pennsylvania State University*
Russell Vreeland, *West Chester University*
Stephen Wagner, *Stephen F. Austin State University*
Darla Wise, *Concord University*

Reviewers

Phillip M. Achey, *University of Florida*
Shivanti Anandan, *Drexel University*
Cynthia Anderson, *Mt. San Antonio College*
Michelle L. Badon, *University of Texas*
Larry L. Barton, *University of New Mexico*
Mary Burke, *Oregon State University*
Frank B. Dazzo, *Michigan State University*
Johnny El-Rady, *University of South Florida*
Paul G. Engelkirk, *Central Texas College*
Robert H. Findlay, *University of Alabama*
Steven Foley, *University of Central Arkansas*
Bernard Lee Frye, *University of Texas at Arlington*
Amy M. Grunden, *North Carolina State University*
Janet L. Haynes, *Long Island University*
Diane Herson, *University of Delaware*
D. Mack Ivey, *University of Arkansas*

Colin R. Jackson, *Southeastern Louisiana University*
Shubha Kale Ireland, *Xavier University of Louisiana*
Judith Kandel, *California State University–Fullerton*
James Kettering, *Loma Linda University*
Madhukar B. Khetmalas, *Texas Tech University*
Peter J. King, *Stephen F. Austin State University*
Tina M. Knox, *University of Illinois–Urbana*
Duncan Krause, *University of Georgia*
Don Lehman, *University of Delaware*
Rita B. Moyes, *Texas A&M University*
Ronald D. Porter, *Pennsylvania State University*
Sabine Rech, *San Jose State University*
Pratibha Saxena, *University of Texas at Austin*
Geoffrey B. Smith, *New Mexico State University*
Fred Stutzenberger, *Clemson University*
Karen Sullivan, *Louisiana State University*
Jennifer R. Walker, *University of Georgia*
William Whalen, *Xavier University of Louisiana*
John M. Zamora, *Middle Tennessee State University*

International Reviewers

Judy Gnarpe, *University of Alberta*
Shaun Heaphy, *University of Leicester*
Edward E. Ishiguro, *University of Victoria*
Kuo-Kau Lee, *National Taiwan Ocean University*
Jong-Kang Liu, *National Sun Yat-sen University*
Cheryl L. Patten, *University of New Brunswick*
Clive Sweet, *University of Birmingham*
Chris Upton, *University of Victoria*
Fanus Venter, *University of Pretonia*
Shang-Shyng Yang, *National Taiwan University*
Guang-yu Zheng, *Beijing Normal University*

As a new group of authors, we encountered a very steep learning curve that we could not have overcome without the help of our editors, Lisa Bruflodt, Jayne Klein, Janice Roerig-Blong, and Colin Wheatley. We would also like to thank our photo editor Mary Reeg and the tremendous talent and patience displayed by the artists. We are also very grateful to Professor Norman Pace for his helpful discussions, and the many reviewers who provided helpful criticism and analysis. Finally, we thank our spouses and children who provided support and tolerated our absences (mental if not physical) while we completed this demanding project.

1

The History and Scope of Microbiology

Louis Pasteur, one of the greatest scientists of the nineteenth century, maintained that "Science knows no country, because knowledge belongs to humanity, and is a torch which illuminates the world."

PREVIEW

- Microbiology is defined not only by the size of its subjects but the techniques it uses to study them.

- Microorganisms include acellular entities (e.g., viruses), procarytic cells, and eucaryotic cells. Cellular microorganisms are found in all three domains of life: *Bacteria, Archaea, Eucarya*.

- The development of microbiology as a scientific discipline has depended on the availability of the microscope and the ability to isolate and grow pure cultures of microorganisms. The development of these techniques in large part grew out of studies disproving the Theory of Spontaneous Generation and others establishing that microorganisms can cause disease.

- Microbiology is a large discipline; it has had and will continue to have a great impact on other areas of biology and general human welfare.

The importance of microorganisms can't be overemphasized. In terms of sheer number and mass—it is estimated that microbes contain 50% of the biological carbon and 90% of the biological nitrogen on Earth—they greatly exceed every other group of organisms on the planet. Furthermore, they are found everywhere: from geothermal vents in the ocean depths to the coldest arctic ice, to every person's skin. They are major contributors to the functioning of the biosphere, being indispensable for the cycling of the elements essential for life. They also are a source of nutrients at the base of all ecological food chains and webs. Most importantly, certain microorganisms carry out photosynthesis, rivaling plants in their role of capturing carbon dioxide and releasing oxygen into the atmosphere. Those microbes that inhabit humans also play important roles, including helping the body digest food and producing vitamins B and K. In addition, society in general benefits from microorganisms, as they are necessary for the production of bread, cheese, beer, antibiotics, vaccines, vitamins, enzymes, and many other important products. Indeed, modern biotechnology rests upon a microbiological foundation.

Although the majority of microorganisms play beneficial or benign roles, some harm humans and have disrupted society over the millennia. Microbial diseases undoubtedly played a major role in historical events such as the decline of the Roman Empire and the conquest of the New World. In 1347, plague or black death, an arthropod-borne disease, struck Europe with brutal force, killing 1/3 of the population (about 25 million people) within four years. Over the next 80 years, the disease struck again and again, eventually wiping out 75% of the European population. The plague's effect was so great that some historians believe it changed European culture and prepared the way for the Renaissance. Today the struggle by microbiologists and others against killers like AIDS and malaria continues.

In this introductory chapter, we introduce the microbial world to provide a general idea of the organisms and agents that microbiologists study. Then we describe the historical development of the science of microbiology and its relationship to medicine and other areas of biology. Finally, we discuss the scope, relevance, and future of modern microbiology.

1.1 MEMBERS OF THE MICROBIAL WORLD

Microbiology often has been defined as the study of organisms and agents too small to be seen clearly by the unaided eye—that is, the study of **microorganisms.** Because objects less than about one millimeter in diameter cannot be seen clearly and must be

Dans les champs de l'observation, le hasard ne favorise que les esprits préparés.
(In the field of observation, chance favors only prepared minds.)

—*Louis Pasteur*

examined with a microscope, microbiology is concerned primarily with organisms and agents this small and smaller. However, some microorganisms, particularly some eucaryotic microbes, are visible without microscopes. For example, bread molds and filamentous algae are studied by microbiologists, yet are visible to the naked eye, as are the two bacteria *Thiomargarita* and *Epulopiscium*.   Microbial Diversity & Ecology 3.1: Monstrous Microbes

The difficulty in setting the boundaries of microbiology has led to the suggestion of other criteria for defining the field. For instance, an important characteristic of microorganisms, even those that are large and multicellular, is that they are relatively simple in their construction, lacking highly differentiated cells and distinct tissues. Another suggestion, made by Roger Stanier, is that the field also be defined in terms of its techniques. Microbiologists usually first isolate a specific microorganism from a population and then culture it. Thus microbiology employs techniques—such as sterilization and the use of culture media—that are necessary for successful isolation and growth of microorganisms.

Microorganisms are diverse, and their classification has always been a challenge for microbial taxonomists. Their early descriptions as either plants or animals were too simple. For instance, some microbes are motile like animals, but also have cell walls and are photosynthetic like plants. Such microbes cannot be placed easily into one kingdom or another. Another important factor in classifying microorganisms is that some are composed of procaryotic cells and others of eucaryotic cells. **Procaryotic cells** [Greek *pro,* before, and *karyon,* nut or kernel; organisms with a primordial nucleus] have a much simpler morphology than eucaryotic cells and lack a true membrane-delimited nucleus. In contrast, **eucaryotic cells** [Greek, *eu,* true, and *karyon,* nut or kernel] have a membrane-enclosed nucleus; they are more complex morphologically and are usually larger than procaryotes. These observations eventually led to the development of a classification scheme that divided organisms into five kingdoms: the *Monera, Protista, Fungi, Animalia,* and *Plantae.* Microorganisms (except for viruses, which are acellular and have their own classification system) were placed in the first three kingdoms.

In the last few decades, great progress has been made in three areas that profoundly affect microbial classification. First, much has been learned about the detailed structure of microbial cells from the use of electron microscopy. Second, microbiologists have determined the biochemical and physiological characteristics of many different microorganisms. Third, the sequences of nucleic acids and proteins from a wide variety of organisms have been compared. The comparison of ribosomal RNA (rRNA), begun by Carl Woese in the 1970s, was instrumental in demonstrating that there are two very different groups of procaryotic organisms: *Bacteria* and *Archaea,* which had been classified together as *Monera* in the five-kingdom system. Later, studies based on rRNA comparisons suggested that *Protista* was not a cohesive taxonomic unit and that it should be divided into three or more kingdoms. These studies and others have led many taxonomists to conclude that the five-kingdom system is too simple. A number of alternatives have been suggested, but currently, most

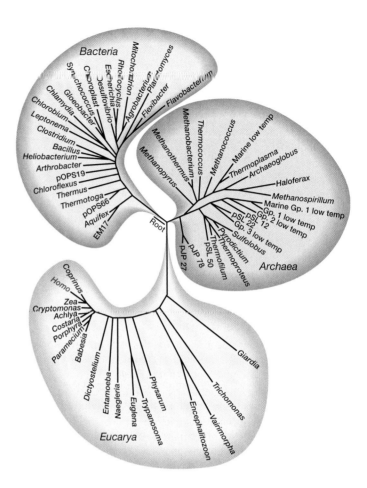

Figure 1.1   Universal Phylogenetic Tree.   These evolutionary relationships are based on rRNA sequence comparisons. Man (Homo) is highlighted in red.

microbiologists believe that organisms should be divided among three domains: *Bacteria* (the true bacteria or eubacteria), *Archaea,*[1] and *Eucarya* (all eucaryotic organisms) (**figure 1.1**). This system, which we shall use here, and the results leading to it are discussed in chapter 19. A brief description of the three domains and of the microorganisms placed in them follows.

Bacteria[2] are procaryotes that are usually single-celled organisms. Most have cell walls that contain the structural molecule peptidoglycan. They are abundant in soil, water, and air and are also major inhabitants of our skin, mouth, and intestines. Some bacteria live in environments that have extreme temperatures,

---

[1] Although this will be discussed further in chapter 19, it should be noted here that several names have been used for the *Archaea.* The two most important are archaeobacteria and archaebacteria. In this text, we shall use only the name *Archaea.*
[2] In this text, the term bacteria (s., bacterium) will be used to refer to procaryotes that belong to domain *Bacteria,* and the term archaea (s., archaeon) will be used to refer to procaryotes that belong to domain *Archaea.* It should be noted that in some publications, the term bacteria is used to refer to all procaryotes. That is not the case in this text.

pH, or salinity. Although some bacteria cause disease, many play more beneficial roles such as cycling elements in the biosphere, breaking down dead plant and animal material, and producing vitamins. Cyanobacteria produce significant amounts of oxygen through the process of photosynthesis.

Archaea are procaryotes that are distinguished from *Bacteria* by many features, most notably their unique ribosomal RNA sequences. They also lack peptidoglycan in their cell walls and have unique membrane lipids. Some have unusual metabolic characteristics, such as the methanogens, which generate methane gas. Many archaea are found in extreme environments. Pathogenic archaea have not yet been identified.

Domain *Eucarya* includes microorganisms classified as protists or *Fungi*. Animals and plants are also placed in this domain. **Protists** are generally larger than procaryotes and include unicellular algae, protozoa, slime molds, and water molds. **Algae** are photosynthetic protists that together with the cyanobacteria produce about 75% of the planet's oxygen. They are also the foundation of aquatic food chains. **Protozoa** are unicellular, animal-like protists that are usually motile. Many free-living protozoa function as the principal hunters and grazers of the microbial world. They obtain nutrients by ingesting organic matter and other microbes. They can be found in many different environments and some are normal inhabitants of the intestinal tracts of animals, where they aid in digestion of complex materials such as cellulose. A few cause disease in humans and other animals. **Slime molds** are protists that are like protozoa in one stage of their life cycle, but are like fungi in another. In the protozoan phase, they hunt for and engulf food particles, consuming decaying vegetation and other microbes. **Water molds,** as their name implies, are found in the surface water of freshwater sources and moist soil. They feed on decaying vegetation such as logs and mulch. Some water molds have produced devastating plant infections, including the Great Potato Famine of 1846–1847. *Fungi* are a diverse group of microorganisms that range from unicellular forms (yeasts) to molds and mushrooms. Molds and mushrooms are multicellular fungi that form thin, threadlike structures called hyphae. They absorb nutrients from their environment, including the organic molecules that they use as a source of carbon and energy. Because of their metabolic capabilities, many fungi play beneficial roles, including making bread rise, producing antibiotics, and decomposing dead organisms. Other fungi cause plant diseases and diseases in humans and other animals.

Viruses are acellular entities that must invade a host cell in order to replicate. They are the smallest of all microbes (the smallest is 10,000 times smaller than a typical bacterium), but their small size belies their power—they cause many animal and plant diseases and have caused epidemics that have shaped human history. The diseases they cause include smallpox, rabies, influenza, AIDS, the common cold, and some cancers.

The development of microbiology as a science is described in sections 1.2 to 1.5. **Figure 1.2** presents a summary of some of the major events in this process and their relationship to other historical landmarks.

1. Describe the field of microbiology in terms of the size of its subject material and the nature of its techniques.
2. Describe and contrast procaryotic and eucaryotic cells.
3. Describe and contrast the five-kingdom classification system with the three-domain system. Why do you think viruses are not included in either system?

1.2   THE DISCOVERY OF MICROORGANISMS

Even before microorganisms were seen, some investigators suspected their existence and responsibility for disease. Among others, the Roman philosopher Lucretius (about 98–55 B.C.) and the physician Girolamo Fracastoro (1478–1553) suggested that disease was caused by invisible living creatures. The earliest microscopic observations appear to have been made between 1625 and 1630 on bees and weevils by the Italian Francesco Stelluti, using a microscope probably supplied by Galileo. In 1665, the first drawing of a microorganism was published in Robert Hooke's *Micrographia*. However, the first person to publish extensive, accurate observations of microorganisms was the amateur microscopist Antony van Leeuwenhoek (1632–1723) of Delft, The Netherlands (**figure 1.3a**). Leeuwenhoek earned his living as a draper and haberdasher (a dealer in men's clothing and accessories), but spent much of his spare time constructing simple microscopes composed of double convex glass lenses held between two silver plates (figure 1.3b). His microscopes could magnify around 50 to 300 times, and he may have illuminated his liquid specimens by placing them between two pieces of glass and shining light on them at a 45° angle to the specimen plane. This would have provided a form of dark-field illumination in which the organisms appeared as bright objects against a dark background and made bacteria clearly visible (figure 1.3c). Beginning in 1673, Leeuwenhoek sent detailed letters describing his discoveries to the Royal Society of London. It is clear from his descriptions that he saw both bacteria and protozoa.

As important as Leeuwenhoek's observations were, the development of microbiology essentially languished for the next 200 years. Little progress was made primarily because microscopic observations of microorganisms do not provide sufficient information to understand their biology. For the discipline to develop, techniques for isolating and culturing microbes in the laboratory were needed. Many of these techniques began to be developed as scientists grappled with the conflict over the Theory of Spontaneous Generation. This conflict and the subsequent studies on the role played by microorganisms in causing disease ultimately led to what is now called the Golden Age of Microbiology.

1. Give some examples of the kind of information you think can be provided by microscopic observations of microorganisms.
2. Give some examples of the kind of information you think can be provided by isolating microorganisms from their natural environment and culturing them in the laboratory.

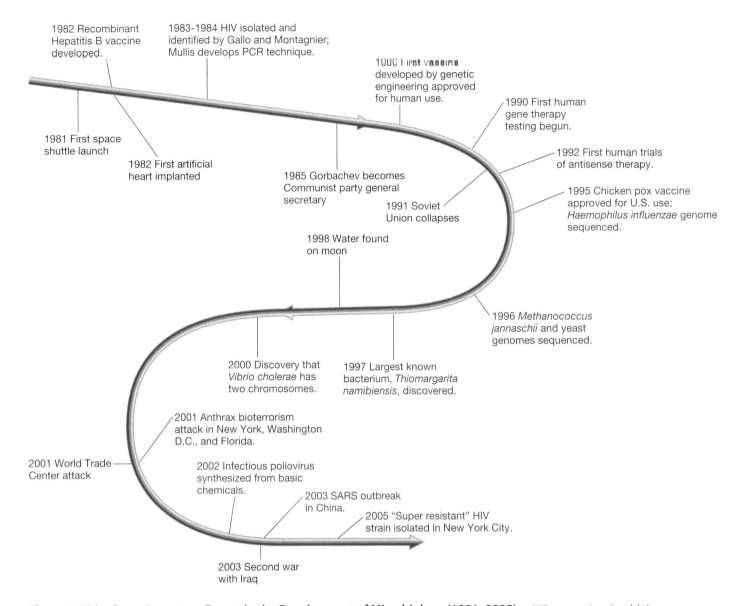

1982 Recombinant Hepatitis B vaccine developed.

1983-1984 HIV isolated and identified by Gallo and Montagnier; Mullis develops PCR technique.

1986 First vaccine developed by genetic engineering approved for human use.

1990 First human gene therapy testing begun.

1981 First space shuttle launch

1982 First artificial heart implanted

1985 Gorbachev becomes Communist party general secretary

1991 Soviet Union collapses

1992 First human trials of antisense therapy.

1995 Chicken pox vaccine approved for U.S. use; *Haemophilus influenzae* genome sequenced.

1998 Water found on moon

1996 *Methanococcus jannaschii* and yeast genomes sequenced.

2000 Discovery that *Vibrio cholerae* has two chromosomes.

1997 Largest known bacterium, *Thiomargarita namibiensis*, discovered.

2001 Anthrax bioterrorism attack in New York, Washington D.C., and Florida.

2001 World Trade Center attack

2002 Infectious poliovirus synthesized from basic chemicals.

2003 SARS outbreak in China.

2005 "Super resistant" HIV strain isolated in New York City.

2003 Second war with Iraq

Figure 1.2(c)   Some Important Events in the Development of Microbiology (1981–2005).   Milestones in microbiology are marked in red; other historical events are in black.

1.3   THE CONFLICT OVER SPONTANEOUS GENERATION

From earliest times, people had believed in **spontaneous generation**—that living organisms could develop from nonliving matter. Even Aristotle (384–322 B.C.) thought some of the simpler invertebrates could arise by spontaneous generation. This view finally was challenged by the Italian physician Francesco Redi (1626–1697), who carried out a series of experiments on decaying meat and its ability to produce maggots spontaneously. Redi placed meat in three containers. One was uncovered, a second was covered with paper, and the third was covered with a fine gauze that would exclude flies. Flies laid their eggs on the uncovered meat and maggots developed. The other two pieces of meat did not

produce maggots spontaneously. However, flies were attracted to the gauze-covered container and laid their eggs on the gauze; these eggs produced maggots. Thus the generation of maggots by decaying meat resulted from the presence of fly eggs, and meat did not spontaneously generate maggots as previously believed. Similar experiments by others helped discredit the theory for larger organisms.

Leeuwenhoek's discovery of microorganisms renewed the controversy. Some proposed that microorganisms arose by spontaneous generation even though larger organisms did not. They pointed out that boiled extracts of hay or meat would give rise to microorganisms after sitting for a while. In 1748, the English priest John Needham (1713–1781) reported the results of his experiments on spontaneous generation. Needham boiled mutton broth and then tightly stoppered the flasks. Eventually many of the flasks became cloudy and contained microorganisms. He thought organic matter contained a vital force that could confer the properties of life on nonliving matter. A few years later, the Italian priest and naturalist Lazzaro Spallanzani (1729–1799) improved on Needham's experimental design by first sealing glass flasks that contained water and seeds. If the sealed flasks were placed in boiling water for 3/4 of an hour, no growth took place as long as the flasks remained sealed. He proposed that air carried germs to the culture medium, but also commented that the external air might be required for growth of animals already in the medium. The supporters of spontaneous generation maintained that heating the air in sealed flasks destroyed its ability to support life.

Several investigators attempted to counter such arguments. Theodore Schwann (1810–1882) allowed air to enter a flask containing a sterile nutrient solution after the air had passed through a red-hot tube. The flask remained sterile. Subsequently Georg Friedrich Schroder and Theodor von Dusch allowed air to enter a flask of heat-sterilized medium after it had passed through sterile cotton wool. No growth occurred in the medium even though the air had not been heated. Despite these experiments the French naturalist Felix Pouchet claimed in 1859 to have carried out experiments conclusively proving that microbial growth could occur without air contamination. This claim provoked Louis Pasteur (1822–1895) to settle the matter once and for all. Pasteur (**figure 1.4**) first filtered air through cotton and found that objects resembling plant spores had been trapped. If a piece of the cotton was placed in sterile medium after air had been filtered through it, microbial growth occurred. Next he placed nutrient solutions in flasks, heated their necks in a flame, and drew them out into a variety of curves, while keeping the ends of the necks open to the atmosphere (**figure 1.5**). Pasteur then boiled the solutions for a few minutes and allowed them to cool. No growth took place even though the contents of the flasks were exposed to the air. Pasteur pointed out that no growth occurred because dust and germs had been trapped on the walls of the curved necks. If the necks were broken, growth commenced immediately. Pasteur had not only resolved the controversy by 1861 but also had shown how to keep solutions sterile.

Lens
Specimen holder
Focus screw
Handle

(b)

(a)

(c)

Figure 1.3   Antony van Leeuwenhoek.   (a) An oil painting of Leeuwenhoek (1632–1723). **(b)** A brass replica of the Leeuwenhoek microscope. Inset photo shows how it is held. **(c)** Leeuwenhoek's drawings of bacteria from the human mouth.

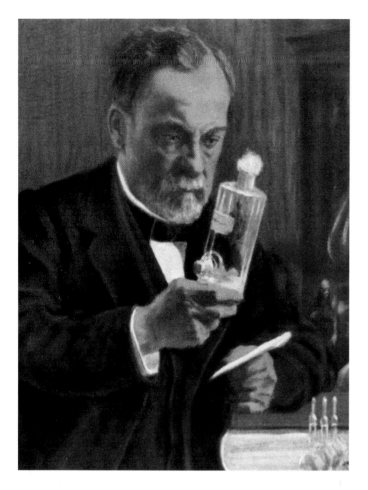

Figure 1.4   Louis Pasteur.   Pasteur (1822–1895) working in his laboratory.

Figure 1.5   The Spontaneous Generation Experiment.
Pasteur's swan neck flasks used in his experiments on the spontaneous generation of microorganisms. *Source: Annales Sciences Naturelle, 4th Series, Vol. 16, pp. 1–98, Pasteur, L., 1861, "Mémoire sur les Corpuscules Organisés Qui Existent Dans L'Atmosphère: Examen de la Doctrine des Générations Spontanées."*

The English physicist John Tyndall (1820–1893) dealt a final blow to spontaneous generation in 1877 by demonstrating that dust did indeed carry germs and that if dust was absent, broth remained sterile even if directly exposed to air. During the course of his studies, Tyndall provided evidence for the existence of exceptionally heat-resistant forms of bacteria. Working independently, the German botanist Ferdinand Cohn (1828–1898) discovered the existence of heat-resistant bacterial endospores. The bacterial endospore (section 3.11)

---

1. How did Pasteur and Tyndall finally settle the spontaneous generation controversy?
2. Why was the belief in spontaneous generation an obstacle to the development of microbiology as a scientific discipline?

1.4   The Golden Age of Microbiology

Pasteur's work with swan neck flasks ushered in the Golden Age of Microbiology. Within 60 years (1857–1914), a number of disease-causing microbes were discovered, great strides in understanding microbial metabolism were made, and techniques for isolating and characterizing microbes were improved. Scientists also identified the role of immunity in preventing disease and controlling microbes, developed vaccines, and introduced techniques used to prevent infection during surgery.

Recognition of the Relationship between Microorganisms and Disease

Although Fracastoro and a few others had suggested that invisible organisms produced disease, most believed that disease was due to causes such as supernatural forces, poisonous vapors called miasmas, and imbalances among the four humors thought to be present in the body. The role of the four humors (blood, phlegm, yellow bile [choler], and black bile [melancholy]) in disease had been widely accepted since the time of the Greek physician Galen (129–199). Support for the idea that microorganisms cause disease—that is, the germ theory of disease—began to accumulate in the early nineteenth century. Agostino Bassi (1773–1856) first showed a microorganism could cause disease when he demonstrated in 1835 that a silkworm disease was due to a fungal infection. He also suggested that many diseases were due to microbial infections. In 1845, M. J. Berkeley proved that the great Potato Blight of Ireland was caused by a water mold, and in 1853, Heinrich de Bary showed that smut and rust fungi caused cereal crop diseases. Following his successes with the study of fermentation, Pasteur was asked by the French government to investigate the pèbrine disease of silkworms that was disrupting the silk industry. After several years of work, he showed that the disease was due to a protozoan parasite. The disease was controlled by raising caterpillars from eggs produced by healthy moths.

Indirect evidence for the germ theory of disease came from the work of the English surgeon Joseph Lister (1827–1912) on the prevention of wound infections. Lister, impressed with Pasteur's studies on the involvement of microorganisms in fermentation and putrefaction, developed a system of antiseptic surgery designed to prevent microorganisms from entering wounds. Instruments were heat sterilized, and phenol was used on surgical dressings and at times sprayed over the surgical area. The approach was remarkably successful and transformed surgery after Lister published his findings in 1867. It also provided strong indirect evidence for the role of microorganisms in disease because phenol, which kills bacteria, also prevented wound infections.

Koch's Postulates

The first direct demonstration of the role of bacteria in causing disease came from the study of anthrax by the German physician Robert Koch (1843–1910). Koch (**figure 1.6**) used the criteria proposed by his former teacher, Jacob Henle (1809–1885), to establish the relationship between *Bacillus anthracis* and anthrax, and published his findings in 1876 (**Techniques & Applications 1.1** briefly discusses the scientific method). Koch injected healthy mice with material from diseased animals, and the mice became ill. After transferring anthrax by inoculation through a series of 20 mice, he incubated a piece of spleen containing the anthrax bacillus in beef serum. The bacilli grew, reproduced, and produced endospores. When the isolated bacilli or their spores were injected into mice, anthrax developed. His criteria for proving the causal relationship between a microorganism and a specific disease are known as **Koch's postulates** (**table 1.1**). Koch's proof that *B. anthracis* caused anthrax was independently confirmed by Pasteur and his coworkers. They discovered that after burial of dead animals, anthrax spores survived and were brought to the surface by earthworms. Healthy animals then ingested the spores and became ill.

Although Koch used the general approach described in the postulates during his anthrax studies, he did not outline them fully until his work on the cause of tuberculosis (table 1.1). In 1884, he reported that this disease was caused by a rod-shaped bacterium, *Mycobacterium tuberculosis;* he was awarded the Nobel Prize in Physiology or Medicine in 1905 for his work. Koch's postulates quickly became the cornerstone of connecting many diseases to their causative agent. However, their use is at times not feasible (**Disease 1.2**). For in-

Figure 1.6   Robert Koch.   Koch (1843–1910) examining a specimen in his laboratory.

stance, some organisms, like *Mycobacterium leprae,* the causative agent of leprosy, cannot be isolated in pure culture.

The Development of Techniques for Studying Microbial Pathogens

During Koch's studies on bacterial diseases, it became necessary to isolate suspected bacterial pathogens in pure culture—a culture containing only one type of microorganism. At first Koch cultured bacteria on the sterile surfaces of cut, boiled potatoes, but this was unsatisfactory because the bacteria would not always grow well. Eventually he developed culture media using meat extracts and protein digests because of their similarity to body fluids. He first tried to solidify the media by adding gelatin. Separate bacterial colonies developed after the surface of the solidified medium had been streaked with a bacterial sample. The sample could also be mixed with liquefied gelatin medium.

Table 1.1	Koch's Application of His Postulates to Demonstrate that *Mycobacterium tuberculosis* is the Causative Agent of Tuberculosis.
Postulate	**Experimentation**
1. The microorganism must be present in every case of the disease but absent from healthy organisms.	Koch developed a staining technique to examine human tissue. *M. tuberculosis* cells could be identified in diseased tissue.
2. The suspected microorganisms must be isolated and grown in a pure culture.	Koch grew *M. tuberculosis* in pure culture on coagulated blood serum.
3. The same disease must result when the isolated microorganism is inoculated into a healthy host.	Koch injected cells from the pure culture of *M.tuberculosis* into guinea pigs. The guinea pigs subsequently died of tuberculosis.
4. The same microorganism must be isolated again from the diseased host.	Koch isolated *M. tuberculosis* from the dead guinea pigs and was able to again culture the microbe in pure culture on coagulated blood serum.

Techniques & Applications

1.1   The Scientific Method

Although biologists employ a variety of approaches in conducting research, microbiologists and other experimentally oriented biologists often use the general approach known as the scientific method. They first gather observations of the process to be studied and then develop a tentative hypothesis—an educated guess—to explain the observations (see **Box figure**). This step often is inductive and creative because there is no detailed, automatic technique for generating hypotheses. Next they decide what information is required to test the hypothesis and collect this information through observation or carefully designed experiments. After the information has been collected, they decide whether the hypothesis has been supported or falsified. If it has failed to pass the test, the hypothesis is rejected, and a new explanation or hypothesis is constructed. If the hypothesis passes the test, it is subjected to more severe testing. The procedure often is made more efficient by constructing and testing alternative hypotheses and then refining the hypothesis that survives testing. This general approach is often called the hypothetico-deductive method. One deduces predictions from the currently accepted hypothesis and tests them. In deduction the conclusion about specific cases follows logically from a general premise ("if . . ., then . . ." reasoning). Induction is the opposite. A general conclusion is reached after considering many specific examples. Both types of reasoning are used by scientists.

When carrying out an experiment, it is essential to use a control group as well as an experimental group. The control group is treated precisely the same as the experimental group except that the experimental manipulation is not performed on it. In this way one can be sure that any changes in the experimental group are due to the experimental manipulation rather than to some other factor not taken into account.

If a hypothesis continues to survive testing, it may be accepted as a valid theory. A theory is a set of propositions and concepts that provides a reliable, systematic, and rigorous account of an aspect of nature. It is important to note that hypotheses and theories are never absolutely proven. Scientists simply gain more and more confidence in their accuracy as they continue to survive testing, fit with new observations and experiments, and satisfactorily explain the observed phenomena. Ultimately, if the support for a hypothesis or theory becomes very strong, it is considered to be a scientific law. Examples include the laws of thermodynamics discussed in section 8.3.

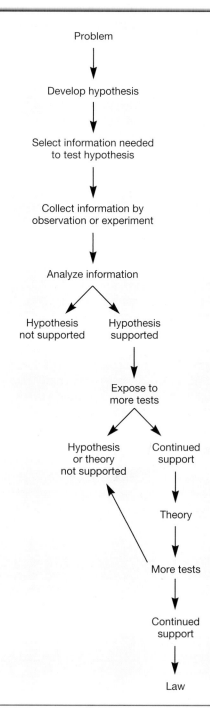

When the gelatin medium hardened, individual bacteria produced separate colonies. Despite its advantages, gelatin was not an ideal solidifying agent because it can be digested by many bacteria and melts at temperatures above 28°C. A better alternative was provided by Fannie Eilshemius Hesse, the wife of Walther Hesse, one of Koch's assistants (**figure 1.7**). She suggested the use of agar as a solidifying agent—she had been using it successfully to make jellies for some time. Agar was not attacked by most bacteria and did not melt until reaching a temperature of 100°C. Furthermore, once melted, it did not solidify until it reached a temperature of 50°C, eliminating the need to handle boiling liquid and providing time for manipulation of the medium. Some of the media developed by Koch and his associates, such as nutrient broth and nutrient agar, are still widely

Disease

1.2   Koch's Molecular Postulates

Although the criteria that Koch developed for proving a causal relationship between a microorganism and a specific disease have been of great importance in medical microbiology, it is not always possible to apply them in studying human diseases. For example, some pathogens cannot be grown in pure culture outside the host; because other pathogens grow only in humans, their study would require experimentation on people. The identification, isolation, and cloning of genes responsible for pathogen virulence have made possible a new molecular form of Koch's postulates that resolves some of these difficulties. The emphasis is on the virulence genes present in the infectious agent rather than on the agent itself. The molecular postulates can be briefly summarized as follows:

1. The virulence trait under study should be associated much more with pathogenic strains of the species than with nonpathogenic strains.

2. Inactivation of the gene or genes associated with the suspected virulence trait should substantially decrease pathogenicity.
3. Replacement of the mutated gene with the normal wild-type gene should fully restore pathogenicity.
4. The gene should be expressed at some point during the infection and disease process.
5. Antibodies or immune system cells directed against the gene products should protect the host.

The molecular approach cannot always be applied because of problems such as the lack of an appropriate animal system. It also is difficult to employ the molecular postulates when the pathogen is not well characterized genetically.

Figure 1.7   Fannie Eilshemius (1850–1934) and Walther Hesse (1846–1911).   Fannie Hesse suggested to her husband Walther (a physican and bacteriologist) that he should try using agar in his culture medium when more typical media failed to meet his needs.

used. Another important tool developed in Koch's laboratory was a container for holding solidified media—the petri dish (plate), named after Richard Petri, who devised it. These developments directly stimulated progress in all areas of bacteriology. Culture media (section 5.7); Isolation of pure cultures (section 5.8)

Viral pathogens were also studied during this time. The discovery of viruses and their role in disease was made possible when Charles Chamberland (1851–1908), one of Pasteur's associates, constructed a porcelain bacterial filter in 1884. Dimitri Ivanowski and Martinus Beijerinck (pronounced "by-a-rink") used the filter to study tobacco mosaic disease. They found that plant extracts and sap from diseased plants were infectious, even after being filtered with Chamberland's filter. Because the infec-

tious agent passed through a filter that was designed to trap bacterial cells, the agent must be something smaller than a bacterium. Beijerinck proposed that the agent was a "filterable virus." Eventually viruses were shown to be tiny, acellular infectious agents.   Early development of virology (section 16.1)

Immunological Studies

In this period progress also was made in determining how animals resisted disease and in developing techniques for protecting humans and livestock against pathogens. During studies on chicken cholera, Pasteur and Roux discovered that incubating their cultures for long intervals between transfers would attenuate the bacteria, which meant they had lost their ability to cause the disease. If the chickens were injected with these attenuated cultures, they remained healthy but developed the ability to resist the disease. He called the attenuated culture a *vaccine* [Latin *vacca*, cow] in honor of Edward Jenner because, many years earlier, Jenner had used material from cowpox lesions to protect people against smallpox. Shortly after this, Pasteur and Chamberland developed an attenuated anthrax vaccine in two ways: by treating cultures with potassium bichromate and by incubating the bacteria at 42 to 43°C. Control of epidemics: Vaccines and immunizations (section 36.8)

Pasteur next prepared rabies vaccine by a different approach. The pathogen was attenuated by growing it in an abnormal host, the rabbit. After infected rabbits had died, their brains and spinal cords were removed and dried. During the course of these studies, Joseph Meister, a nine-year-old boy who had been bitten by a rabid dog, was brought to Pasteur. Since the boy's death was certain in the absence of treatment, Pasteur agreed to try vaccination. Joseph was injected 13 times over the next 10 days with increasingly virulent preparations of the attenuated virus. He survived.

In gratitude for Pasteur's development of vaccines, people from around the world contributed to the construction of the

Pasteur Institute in Paris, France. One of the initial tasks of the Institute was vaccine production.

After the discovery that the diphtheria bacillus produced a toxin, Emil von Behring (1854–1917) and Shibasaburo Kitasato (1852–1931) injected inactivated toxin into rabbits, inducing them to produce an antitoxin, a substance in the blood that would inactivate the toxin and protect against the disease. A tetanus antitoxin was then prepared and both antitoxins were used in the treatment of people.

The antitoxin work provided evidence that immunity could result from soluble substances in the blood, now known to be antibodies (humoral immunity). It became clear that blood cells were also important in immunity (cellular immunity) when Elie Metchnikoff (1845–1916) discovered that some blood leukocytes could engulf disease-causing bacteria (**figure 1.8**). He called these cells phagocytes and the process phagocytosis [Greek *phagein*, eating].

1. Discuss the contributions of Lister, Pasteur, and Koch to the germ theory of disease and to the treatment or prevention of diseases.
2. What other contributions did Koch make to microbiology?
3. Describe Koch's postulates. What is a pure culture? Why are pure cultures important to Koch's postulates?
4. Would microbiology have developed more slowly if Fannie Hesse had not suggested the use of agar? Give your reasoning.
5. What are Koch's molecular postulates? Why are they important?
6. Some individuals can be infected by a pathogen yet not develop disease. In fact, some become chronic carriers of the pathogen. How does this observation impact Koch's postulates? How might the postulates be modified to account for the existence of chronic carriers?
7. Describe the scientific method in your own words. How does a theory differ from a hypothesis? Why is it important to have a control group?
8. How did von Behring and Metchnikoff contribute to the development of immunology?

1.5 THE DEVELOPMENT OF INDUSTRIAL MICROBIOLOGY AND MICROBIAL ECOLOGY

Although humans had unknowingly exploited microbes for thousands of years, industrial microbiology developed in large part from the work of Louis Pasteur and others on the alcoholic fermentations that yielded wine and other alcoholic beverages. In 1837, when Theodore Schwann and others proposed that yeast cells were responsible for the conversion of sugars to alcohol, the leading chemists of the time believed microorganisms were not involved. They were convinced that fermentation was due to a chemical instability that degraded the sugars to alcohol. Pasteur did not agree; he believed that fermentations were carried out by living organisms. In 1856 M. Bigo, an industrialist in Lille, France, where Pasteur worked, requested Pasteur's assistance. His business produced ethanol from the fermentation of beet sugars, and the alcohol yields had recently declined and the product had become sour. Pasteur discovered that the fermentation was failing because the yeast normally responsible for alcohol formation had been replaced by microorganisms that produced lactic acid rather than ethanol. In solving this practical problem, Pasteur demonstrated

Figure 1.8 Elie Metchnikoff. Metchnikoff (1845–1916) shown here at work in his laboratory.

that all fermentations were due to the activities of specific yeasts and bacteria, and he published several papers on fermentation between 1857 and 1860. His success led to a study of wine diseases and the development of pasteurization to preserve wine during storage. Pasteur's studies on fermentation continued for almost 20 years. One of his most important discoveries was that some fermentative microorganisms were anaerobic and could live only in the absence of oxygen, whereas others were able to live either aerobically or anaerobically.   Controlling food spoilage (section 40.3)

Microbial ecology developed when a few of the early microbiologists chose to investigate the ecological role of microorganisms. In particular they studied microbial involvement in the carbon, nitrogen, and sulfur cycles taking place in soil and aquatic habitats. The Russian microbiologist Sergei Winogradsky (1856–1953) made many contributions to soil microbiology. He discovered that soil bacteria could oxidize iron, sulfur, and ammonia to obtain energy, and that many bacteria could incorporate CO_2 into organic matter much like photosynthetic organisms do. Winogradsky also isolated anaerobic nitrogen-fixing soil bacteria and studied the decomposition of cellulose. Martinus Beijerinck (1851–1931) was one of the great general microbiologists who made fundamental contributions to microbial ecology and many other fields. He isolated the aerobic nitrogen-fixing bacterium *Azotobacter,* a root nodule bacterium also capable of fixing nitrogen (later named *Rhizobium*), and sulfate-reducing bacteria. Beijerinck and Winogradsky also developed the enrichment-culture technique and the use of selective media, which have been of such great importance in microbiology.   Biogeochemical cycling (section 27.2); Culture media (section 5.7)

1. Briefly describe Pasteur's work on microbial fermentations.
2. How did Winogradsky and Beijerinck contribute to the study of microbial ecology?

3. Leeuwenhoek is often referred to as the Father of Microbiology. However, many historians feel that Louis Pasteur, Robert Koch, or perhaps both, deserve that honor. Who do you think is the Father of Microbiology? Why?

4. Consider the discoveries described in sections 1.2 to 1.5. Which do you think were the most important to the development of microbiology? Why?

1.6 THE SCOPE AND RELEVANCE OF MICROBIOLOGY

As the late scientist-writer Steven Jay Gould emphasized, we live in the Age of *Bacteria*. They were the first living organisms on our planet and live virtually everywhere life is possible. Furthermore, the whole biosphere depends on their activities, and they influence human society in countless ways. Because microorganisms play such diverse roles, modern microbiology is a large discipline with many different specialties; it has a great impact on fields such as medicine, agricultural and food sciences, ecology, genetics, biochemistry, and molecular biology. One indication of the importance of microbiology is the Nobel Prize given for work in physiology or medicine. About one-third of these have been awarded to scientists working on microbiological problems (*see inside front cover*).

Microbiology has both basic and applied aspects (**figure 1.9**). The basic aspects are concerned with the biology of microorganisms themselves and include such fields as bacteriology, virology, mycology (study of fungi), phycology or algology (study of algae), protozoology, microbial cytology and physiology, microbial genetics and molecular biology, microbial ecology, and microbial taxonomy. The applied aspects are concerned with practical problems such as disease, water and wastewater treatment, food spoilage and food production, and industrial uses of microbes. It is important to note that the basic and applied aspects of microbiology are intertwined. Basic research is often conducted in applied fields and applications often arise out of basic research. A discussion of some of the major fields of microbiology and the occupations they provide follows.

One of the most active and important fields in microbiology is medical microbiology, which deals with diseases of humans and animals. Medical microbiologists identify the agents causing infectious diseases and plan measures for their control and elimination. Frequently they are involved in tracking down new, unidentified pathogens such as the agent that causes variant Creutzfeldt-Jakob disease, (the human version of "mad cow disease") the hantavirus, the West Nile virus, and the virus responsible for SARS. These microbiologists also study the ways in which microorganisms cause disease. Arthropod-borne viral diseases (section 37.2); Microbial Diversity & Ecology 18.1: SARS: Evolution of a virus

Public health microbiology is closely related to medical microbiology. Public health microbiologists try to identify and control the spread of communicable diseases. They often monitor community food establishments and water supplies in an attempt to keep them safe and free from infectious disease agents.

Immunology is concerned with how the immune system protects the body from pathogens and the response of infectious agents. It is one of the fastest growing areas in science; for example, techniques for the production and use of monoclonal antibodies have developed extremely rapidly. Immunology also deals with practical health problems such as the nature and treatment of allergies and autoimmune diseases like rheumatoid arthritis. Techniques & Applications 32.2: Monoclonal Antibody Technology

Agricultural microbiology is concerned with the impact of microorganisms on agriculture. Agricultural microbiologists try to combat plant diseases that attack important food crops, work on methods to increase soil fertility and crop yields, and study the role of microorganisms living in the digestive tracts of ruminants such as cattle. Currently there is great interest in using bacterial and viral insect pathogens as substitutes for chemical pesticides.

Microbial ecology is concerned with the relationships between microorganisms and the components of their living and nonliving habitats. Microbial ecologists study the global and local contributions of microorganisms to the carbon, nitrogen, and sulfur cycles. The study of pollution effects on microorganisms also is important because of the impact these organisms have on the environment. Microbial ecologists are employing microorganisms in bioremediation to reduce pollution.

Scientists working in food and dairy microbiology try to prevent microbial spoilage of food and the transmission of foodborne diseases such as botulism and salmonellosis. They also use microorganisms to make foods such as cheeses, yogurts, pickles, and beer. In the future, microorganisms themselves may become a more important nutrient source for livestock and humans. Microbiology of food (chapter 40)

In 1929, Alexander Fleming discovered that the fungus *Penicillium* produced what he called penicillin, the first antibiotic that could successfully control bacterial infections. Although it took World War II for scientists to learn how to mass produce it, scientists soon found other microorganisms capable of producing additional antibiotics as well as compounds such as citric acid, vitamin B_{12}, and monosodium glutamate. Today, industrial microbiologists use microorganisms to make products such as antibiotics, vaccines, steroids, alcohols and other solvents, vitamins, amino acids, and enzymes. Industrial microbiologists identify microbes of use to industry. They also engineer microbes with desirable traits and devise systems for culturing them and isolating the products they make.

Microbiologists working in microbial physiology and biochemistry study many aspects of the biology of microorganisms. They may study the synthesis of antibiotics and toxins, microbial energy production, the ways in which microorganisms survive harsh environmental conditions, microbial nitrogen fixation, and the effects of chemical and physical agents on microbial growth and survival.

Microbial genetics and molecular biology focus on the nature of genetic information and how it regulates the development and function of cells and organisms. The use of microorganisms has been very helpful in understanding gene structure and function. Microbial geneticists play an important role in applied microbiology because they develop techniques that are useful in agricultural microbiology, industrial microbiology, food and dairy microbiology, and medicine.

---

1. Briefly describe the major subdisciplines in microbiology.
2. Why do you think microorganisms are so useful to biologists as experimental models?
3. List all the activities or businesses you can think of in your community that are directly dependent on microbiology.

(a) Rita Colwell

(b) R. G. E. Murray

(c) Stanley Falkow

(d) Martha Howe

(e) Frederick Neidhardt

(f) Jean Brenchley

Figure 1.9   Important Contributors to Microbiology.   (a) Rita Colwell has studied the genetics and ecology of marine bacteria such as *Vibrio cholerae* and helped establish the field of marine biotechnology. **(b)** R. G. E. Murray has contributed greatly to the understanding of bacterial cell envelopes and bacterial taxonomy. **(c)** Stanley Falkow has advanced our understanding of how bacterial pathogens cause disease. **(d)** Martha Howe has made fundamental contributions to our knowledge of the bacteriophage Mu. **(e)** Frederick Neidhardt has contributed to microbiology through his work on the regulation of *E. coli* physiology and metabolism, and by coauthoring advanced textbooks. **(f)** Jean Brenchley has studied the regulation of glutamate and glutamine metabolism, helped found the Pennsylvania State University Biotechnology Institute, and is now finding biotechnological uses for psychrophilic (cold-loving) microorganisms.

1.7   THE FUTURE OF MICROBIOLOGY

As the preceding sections have shown, microbiology has had a profound influence on society. What of the future? Science writer Bernard Dixon is very optimistic about microbiology's future for two reasons. First, microbiology has a clearer mission than do many other scientific disciplines. Second, microbiology has great practical significance. Dixon notes that microbiology is required both to face the threat of new and reemerging human infectious diseases and to develop industrial technologies that are more efficient and environmentally friendly.

What are some of the most promising areas for future microbiological research and their potential practical impacts? What kinds of challenges do microbiologists face? A discussion of some aspects of the future of microbiology follows.

Medical microbiology, public health microbiology, and immunology will continue to be areas of intense research. New infectious diseases are continually arising and old diseases are once again becoming widespread and destructive. AIDS, SARS, hemorrhagic fevers, and tuberculosis are excellent examples of new and reemerging infectious diseases. Microbiologists will have to respond to these threats, many of them presently unknown. They will also need to find ways to stop the spread of established infectious diseases, as well as the spread of multiple antibiotic resistance, which can render a pathogen resistant to current medical treatment. Microbiologists will also be called upon to create new drugs and vaccines, to study the association between infectious agents and chronic disease (e.g., autoimmune and cardiovascular diseases), and to further our understanding of host defenses and how pathogens interact with host cells. It will be necessary to use techniques in molecular biology and recombinant DNA technology to solve many of these problems.

Industrial microbiology and environmental microbiology also face many challenges and opportunities. Microorganisms are increasingly important in industry and environmental control, and we must learn how to use them in a variety of new ways. For example, microorganisms can serve as sources of high-quality food and other practical products such as enzymes for industrial applications. They may also be used to degrade pollutants and toxic wastes and as vectors to treat diseases and enhance agricultural productivity. There also is a continuing need to protect food and crops from microbial damage.

The development of techniques, especially DNA-based techniques, that allow the study of microorganisms in their natural environment has greatly stimulated research in microbial ecology. Several areas of research will continue to be important. Understanding microbial diversity is one area that requires further research. It is estimated that less than 1% of Earth's microbes have been cultured. Greater efforts to grow previously uncultivated microbes will be required. Much work also needs to be done on microorganisms living in extreme environments. The discovery of new and unusual microorganisms may well lead to further advances in the development of new antimicrobial agents, industrial processes, and bioremediation. Another area of increasing interest to microbial ecologists is biofilms. Microbes often form biofilms on surfaces, and in doing so exhibit a physiology that differs from that observed when they live freely or planktonically. For instance, microbes in a biofilm are often more resistant to killing agents than they are when not in a biofilm. Biofilms are not only of interest to microbial ecologists; they can form on human tissues, on indwelling catheters, and on other man-made medical devices. In fact, microbial ecologists and medical microbiologists now understand that microorganisms are essential partners with higher organisms. Greater knowledge of the nature of these symbiotic relationships can help improve our appreciation of the living world. It also will lead to new approaches in treating infectious diseases in livestock and in humans.

The fields of genomics and proteomics have and will continue to have a tremendous impact on microbiology. The genomes of many microorganisms have already have been sequenced and many more will be determined in the coming years. These sequences are ideal for learning how the genome is related to cell structure and function and for providing insights into fundamental questions in biology, such as how complex cellular structures develop and how cells communicate with one another and respond to the environment. Analysis of the genome and its activity will require continuing advances in the field of bioinformatics and the use of computers to investigate biological problems.

Perhaps the biggest challenge facing microbiologists will be to assess the implications of new discoveries and technological developments. The pace of these discoveries and developments is very rapid, and sometimes it is difficult for nonscientists to follow and assess them. Microbiologists will need to communicate a balanced view of both the positive and the negative long-term impacts of these developments on society.

Clearly, the future of microbiology is bright. The microbiologist René Dubos has summarized well the excitement and promise of microbiology:

> How extraordinary that, all over the world, microbiologists are now involved in activities as different as the study of gene structure, the control of disease, and the industrial processes based on the phenomenal ability of microorganisms to decompose and synthesize complex organic molecules. Microbiology is one of the most rewarding of professions because it gives its practitioners the opportunity to be in contact with all the other natural sciences and thus to contribute in many different ways to the betterment of human life.

---

1. What do you think are the five most important research areas to pursue in microbiology? Give reasons for your choices.

Summary

1.1 Members of the Microbial World

a. Microbiology studies microscopic organisms that are often unicellular, or if multicellular, do not have highly differentiated tissues. The discipline is also defined by the techniques it uses—in particular, those used to isolate and culture microorganisms.

b. Procaryotic cells differ from eucaryotic cells in lacking a membrane-delimited nucleus, and in other ways as well.

c. Microbiologists divide organisms into three domains: *Bacteria, Archaea,* and *Eucarya.*

d. Domains *Bacteria* and *Archaea* consist of procaryotic microorganisms. The eucaryotic microbes (protists and fungi) are placed in *Eucarya.* Viruses are acellular entities that are not placed in any of the domains but are classified by a separate system.

1.2 The Discovery of Microorganisms

a. Antony van Leeuwenhoek was the first person to extensively describe microorganisms.

1.3 The Conflict Over Spontaneous Generation

a. Experiments by Redi and others disproved the theory of spontaneous generation in regard to larger organisms.

b. The spontaneous generation of microorganisms was disproved by Spallanzani, Pasteur, Tyndall, and others.

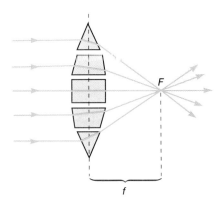

Figure 2.2   Lens Function.   A lens functions somewhat like a collection of prisms. Light rays from a distant source are focused at the focal point *F*. The focal point lies a distance *f*, the focal length, from the lens center.

Table 2.1	Common Units of Measurement	
Unit	**Abbreviation**	**Value**
1 centimeter	cm	10^{-2} meter or 0.394 inches
1 millimeter	mm	10^{-3} meter
1 micrometer	μm	10^{-6} meter
1 nanometer	nm	10^{-9} meter
1 Angstrom	Å	10^{-10} meter

index, it accelerates and is bent away from the normal. Thus a prism bends light because glass has a different refractive index from air, and the light strikes its surface at an angle.

Lenses act like a collection of prisms operating as a unit. When the light source is distant so that parallel rays of light strike the lens, a convex lens will focus these rays at a specific point, the **focal point** (*F* in **figure 2.2**). The distance between the center of the lens and the focal point is called the **focal length** (*f* in figure 2.2).

Our eyes cannot focus on objects nearer than about 25 cm or 10 inches (**table 2.1**). This limitation may be overcome by using a convex lens as a simple magnifier (or microscope) and holding it close to an object. A magnifying glass provides a clear image at much closer range, and the object appears larger. Lens strength is related to focal length; a lens with a short focal length will magnify an object more than a weaker lens having a longer focal length.

1. Define refraction, refractive index, focal point, and focal length.
2. Describe the path of a light ray through a prism or lens.
3. How is lens strength related to focal length?

2.2   THE LIGHT MICROSCOPE

Microbiologists currently employ a variety of light microscopes in their work; bright-field, dark-field, phase-contrast, and fluorescence microscopes are most commonly used. Modern micro-

scopes are all compound microscopes. That is, the magnified image formed by the objective lens is further enlarged by one or more additional lenses.

The Bright-Field Microscope

The ordinary microscope is called a **bright-field microscope** because it forms a dark image against a brighter background. The microscope consists of a sturdy metal body or stand composed of a base and an arm to which the remaining parts are attached (**figure 2.3**). A light source, either a mirror or an electric illuminator, is located in the base. Two focusing knobs, the fine and coarse adjustment knobs, are located on the arm and can move either the stage or the nosepiece to focus the image.

The stage is positioned about halfway up the arm and holds microscope slides by either simple slide clips or a mechanical stage clip. A mechanical stage allows the operator to move a slide around smoothly during viewing by use of stage control knobs. The **substage condenser** is mounted within or beneath the stage and focuses a cone of light on the slide. Its position often is fixed in simpler microscopes but can be adjusted vertically in more advanced models.

The curved upper part of the arm holds the body assembly, to which a nosepiece and one or more **eyepieces** or **ocular lenses** are attached. More advanced microscopes have eyepieces for both eyes and are called binocular microscopes. The body assembly itself contains a series of mirrors and prisms so that the barrel holding the eyepiece may be tilted for ease in viewing (**figure 2.4**). The nosepiece holds three to five **objective lenses** of differing magnifying power and can be rotated to position any objective beneath the body assembly. Ideally a microscope should be **parfocal**—that is, the image should remain in focus when objectives are changed.

The image one sees when viewing a specimen with a compound microscope is created by the objective and ocular lenses working together. Light from the illuminated specimen is focused by the objective lens, creating an enlarged image within the microscope (figure 2.4). The ocular lens further magnifies this primary image. The total magnification is calculated by multiplying the objective and eyepiece magnifications together. For example, if a 45× objective is used with a 10× eyepiece, the overall magnification of the specimen will be 450×.

Microscope Resolution

The most important part of the microscope is the objective, which must produce a clear image, not just a magnified one. Thus resolution is extremely important. **Resolution** is the ability of a lens to separate or distinguish between small objects that are close together.

Resolution is described mathematically by an equation developed in the 1870s by Ernst Abbé, a German physicist responsible for much of the optical theory underlying microscope design. The Abbé equation states that the minimal distance (*d*) between two

Figure 2.3   A Bright-Field Microscope.   The parts of a modern bright-field microscope. The microscope pictured is somewhat more sophisticated than those found in many student laboratories. For example, it is binocular (has two eyepieces) and has a mechanical stage, an adjustable substage condenser, and a built-in illuminator.

Figure 2.4   A Microscope's Light Path.   The light path in an advanced bright-field microscope (see also figure 2.19).

objects that reveals them as separate entities depends on the wavelength of light (λ) used to illuminate the specimen and on the **numerical aperture** of the lens ($n \sin \theta$), which is the ability of the lens to gather light.

$$d = \frac{0.5\lambda}{n \sin \theta}$$

As d becomes smaller, the resolution increases, and finer detail can be discerned in a specimen; d becomes smaller as the wavelength of light used decreases and as the numerical aperture (NA) increases. Thus the greatest resolution is obtained using a lens with the largest possible NA and light of the shortest wavelength, light at the blue end of the visible spectrum (in the range of 450 to 500 nm; *see figure 6.25*).

The numerical aperture ($n \sin \theta$) of a lens is a complex concept that can be difficult to understand. It is defined by two components: n is the refractive index of the medium in which the lens works (e.g., air) and θ is 1/2 the angle of the cone of light entering an objective (**figure 2.5**). When this cone has a narrow angle and tapers to a sharp point, it does not spread out much after leaving the slide and therefore does not adequately separate images of closely packed objects. If the cone of light has a very wide angle and spreads out rapidly after passing through a specimen, closely packed objects appear widely separated and are resolved. The angle of the cone of light that can enter a lens depends on the refractive index (n) of the medium in which the lens works, as well as upon the objective itself. The refractive index for air is 1.00 and sin θ cannot be greater than 1 (the maximum θ is 90° and sin 90° is 1.00). Therefore no lens working in air can have a numerical aperture greater than 1.00. The only practical way to raise the numerical aperture above 1.00, and therefore achieve higher resolution, is to increase the refractive index with immersion oil, a colorless liquid with the same refractive index as glass (**table 2.2**). If air is replaced with immersion oil, many light rays that did not enter the objective due to reflection and refraction at the surfaces of the objective lens and slide will now do so (**figure 2.6**). An increase in numerical aperture and resolution results.

Numerical aperture is related to another characteristic of an objective lens, the working distance. The **working distance** of an objective is the distance between the front surface of the lens and the surface of the cover glass (if one is used) or the specimen when it is in sharp focus. Objectives with large numerical apertures and great resolving power have short working distances (table 2.2).

The preceding discussion has focused on the resolving power of the objective lens. The resolution of an entire microscope must take into account the numerical aperture of its condenser as is evident from the equation below.

$$d_{microscope} = \frac{\lambda}{(NA_{objective} + NA_{condenser})}$$

The condenser is a large, light-gathering lens used to project a wide cone of light through the slide and into the objective lens. Most microscopes have a condenser with a numerical aperture between 1.2 and 1.4. However, the condenser numerical aperture will not be much above about 0.9 unless the top of the condenser is oiled to the bottom of the slide. During routine microscope operation, the condenser usually is not oiled and this limits the overall resolution, even with an oil immersion objective.

Although the resolution of the microscope must consider both the condenser and the objective lens, in most cases the limit of resolution of a light microscope is calculated using the Abbé equation, which considers the objective lens only. The maximum theoretical resolving power of a microscope with an oil immersion objective (numerical aperture of 1.25) and blue-green light is approximately 0.2 μm.

$$d = \frac{(0.5)(530 \text{ nm})}{1.25} = 212 \text{ nm or } 0.2 \mu m$$

At best, a bright-field microscope can distinguish between two dots about 0.2 μm apart (the same size as a very small bacterium).

Figure 2.5 Numerical Aperture in Microscopy. The angular aperture θ is 1/2 the angle of the cone of light that enters a lens from a specimen, and the numerical aperture is n sin θ. In the right-hand illustration the lens has larger angular and numerical apertures; its resolution is greater and its working distance smaller.

Figure 2.6 The Oil Immersion Objective. An oil immersion objective operating in air and with immersion oil.

Table 2.2	The Properties of Microscope Objectives			
	Objective			
Property	**Scanning**	**Low Power**	**High Power**	**Oil Immersion**
Magnification	4×	10×	40–45×	90–100×
Numerical aperture	0.10	0.25	0.55–0.65	1.25–1.4
Approximate focal length (f)	40 mm	16 mm	4 mm	1.8–2.0 mm
Working distance	17-20 mm	4-8 mm	0.5–0.7 mm	0.1 mm
Approximate resolving power with light of 450 nm (blue light)	2.3 μm	0.9 μm	0.35 μm	0.18 μm

Given the limit of resolution of a light microscope, the largest useful magnification—the level of magnification needed to increase the size of the smallest resolvable object to be visible with the light microscope—can be determined. Our eye can just detect a speck 0.2 mm in diameter, and consequently the useful limit of magnification is about 1,000 times the numerical aperture of the objective lens. Most standard microscopes come with 10× eyepieces and have an upper limit of about 1,000× with oil immersion. A 15× eyepiece may be used with good objectives to achieve a useful magnification of 1,500×. Any further magnification does not enable a person to see more detail. Indeed, a light microscope can be built to yield a final magnification of 10,000×, but it would simply be magnifying a blur. Only the electron microscope provides sufficient resolution to make higher magnifications useful.

The Dark-Field Microscope

The **dark-field microscope** allows a viewer to observe living, unstained cells and organisms by simply changing the way in which they are illuminated. A hollow cone of light is focused on the specimen in such a way that unreflected and unrefracted rays do not enter the objective. Only light that has been reflected or refracted by the specimen forms an image (**figure 2.7**). The field surrounding a specimen appears black, while the object itself is brightly illuminated (**figure 2.8a,b**). The dark-field microscope can reveal considerable internal structure in larger eucaryotic microorganisms (figure 2.8b). It also is used to identify certain bacteria like the thin and distinctively shaped *Treponema pallidum* (figure 2.8a), the causative agent of syphilis.

The Phase-Contrast Microscope

Unpigmented living cells are not clearly visible in the bright-field microscope because there is little difference in contrast between the cells and water. As will be discussed in section 2.3, one solution to this problem is to kill and stain cells before observation to increase contrast and create variations in color between cell structures. But what if an investigator must view living cells in order to observe a dynamic process such as movement or phagocytosis? Phase-contrast microscopy can be used in this situation. A **phase-contrast microscope** converts slight differences in refractive index and cell density into easily detected variations in light intensity and is an excellent way to observe living cells (figure 2.8c–e).

(a)

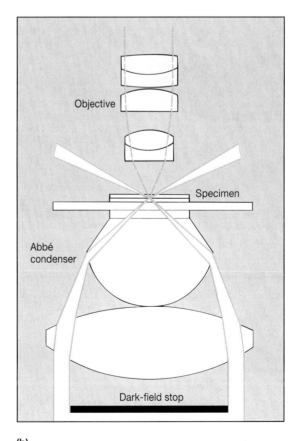

(b)

Figure 2.7 Dark-Field Microscopy.   The simplest way to convert a microscope to dark-field microscopy is to place **(a)** a dark-field stop underneath **(b)** the condenser lens system. The condenser then produces a hollow cone of light so that the only light entering the objective comes from the specimen.

(a) *T. pallidum:* dark-field microscopy

(b) *Volvox* and *Spirogyra:* dark-field microscopy

(c) *Pseudomonas:* phase-contrast microscopy

(d) *Desulfotomaculum:* phase-contrast microscopy

(e) *Paramecium:* phase-contrast microscopy

Figure 2.8   Examples of Dark-Field and Phase-Contrast Microscopy.   (a) *Treponema pallidum,* the spirochete that causes syphilis; dark-field microscopy. **(b)** *Volvox and Spirogyra;* dark-field microscopy (×175). Note daughter colonies within the mature *Volvox* colony (center) and the spiral chloroplasts of *Spirogyra* (left and right). **(c)** A phase-contrast micrograph of *Pseudomonas* cells, which range from 1–3 μm in length. **(d)** *Desulfotomaculum acetoxidans* with endospores; phase contrast (×2,000). **(e)** *Paramecium* stained to show a large central macronucleus with a small spherical micronucleus at its side; phase-contrast microscopy (×100).

and undeviated waves will be about 1/2 wavelength out of phase and will cancel each other when they come together to form an image (**figure 2.10**). The background, formed by undeviated light, is bright, while the unstained object appears dark and well-defined. This type of microscopy is called dark-phase-contrast microscopy. Color filters often are used to improve the image (figure 2.8*d*).

Phase-contrast microscopy is especially useful for studying microbial motility, determining the shape of living cells, and detecting bacterial components such as endospores and inclusion bodies that contain poly-β-hydroxyalkanoates (e.g., poly-β-hydroxybutyrate), polymetaphosphate, sulfur, or other substances. These are clearly visible (figure 2.8*d*) because they have refractive indices markedly different from that of water. Phase-contrast microscopes also are widely used in studying eucaryotic cells. The cytoplasmic matrix: Inclusion bodies (section 3.3)

The Differential Interference Contrast Microscope

The **differential interference contrast (DIC) microscope** is similar to the phase-contrast microscope in that it creates an image by detecting differences in refractive indices and thickness. Two beams of plane-polarized light at right angles to each other are generated by prisms. In one design, the object beam passes through the specimen, while the reference beam passes through a clear area of the slide. After passing through the specimen, the two beams are combined and interfere with each other to form an image. A live, unstained specimen appears brightly colored and three-dimensional (**figure 2.11**). Structures such as cell walls, endospores, granules, vacuoles, and eucaryotic nuclei are clearly visible.

The Fluorescence Microscope

The microscopes thus far considered produce an image from light that passes through a specimen. An object also can be seen because it actually emits light, and this is the basis of fluorescence microscopy. When some molecules absorb radiant energy, they become excited and later release much of their trapped energy as light. Any light emitted by an excited molecule will have a longer wavelength (or be of lower energy) than the radiation originally absorbed. **Fluorescent light** is emitted very quickly by the excited molecule as it gives up its trapped energy and returns to a more stable state.

The **fluorescence microscope** exposes a specimen to ultraviolet, violet, or blue light and forms an image of the object with the resulting fluorescent light. The most commonly used fluorescence microscopy is epifluorescence microscopy, also called incident light or reflected light fluorescence microscopy. Epifluorescence microscopes employ an objective lens that also acts as a condenser (**figure 2.12**). A mercury vapor arc lamp or other source produces an intense beam of light that passes through an exciter filter. The exciter filter transmits only the desired wavelength of excitation light. The excitation light is directed down the microscope by a special mirror called the dichromatic mirror. This mirror reflects light of shorter wavelengths (i.e., the excitation light),

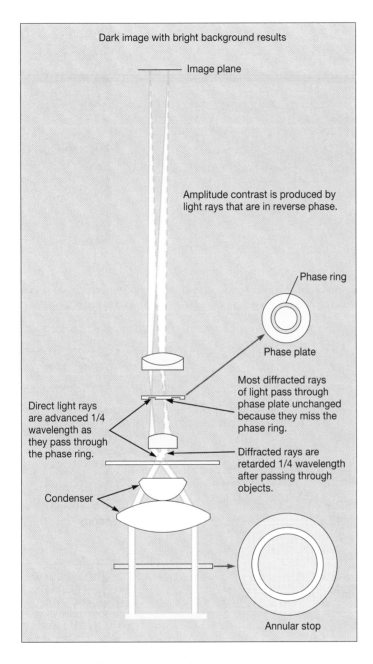

Figure 2.9 Phase-Contrast Microscopy. The optics of a dark-phase-contrast microscope.

The condenser of a phase-contrast microscope has an annular stop, an opaque disk with a thin transparent ring, which produces a hollow cone of light (**figure 2.9**). As this cone passes through a cell, some light rays are bent due to variations in density and refractive index within the specimen and are retarded by about 1/4 wavelength. The deviated light is focused to form an image of the object. Undeviated light rays strike a phase ring in the phase plate, a special optical disk located in the objective, while the deviated rays miss the ring and pass through the rest of the plate. If the phase ring is constructed in such a way that the undeviated light passing through it is advanced by 1/4 wavelength, the deviated

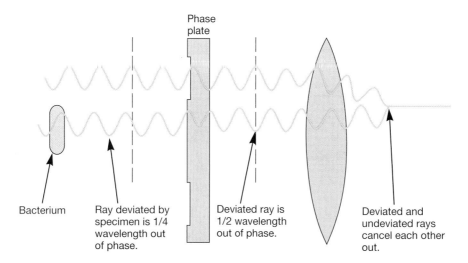

Bacterium | Ray deviated by specimen is 1/4 wavelength out of phase. | Deviated ray is 1/2 wavelength out of phase. | Deviated and undeviated rays cancel each other out.

Figure 2.10 The Production of Contrast in Phase Microscopy. The behavior of deviated and undeviated or undiffracted light rays in the dark-phase-contrast microscope. Because the light rays tend to cancel each other out, the image of the specimen will be dark against a brighter background.

Figure 2.11 Differential Interference Contrast Microscopy. A micrograph of the protozoan *Amoeba proteus.* The three-dimensional image contains considerable detail and is artificially colored ($\times$160).

Figure 2.12 Epifluorescence Microscopy. The principles of operation of an epifluorescence microscope.

but allows light of longer wavelengths to pass through. The excitation light continues down, passing through the objective lens to the specimen, which is usually stained with special dye molecules called **fluorochromes (table 2.3).** The fluorochrome absorbs light energy from the excitation light and fluoresces brightly. The emitted fluorescent light travels up through the objective lens into the microscope. Because the emitted fluorescent light has a longer wavelength, it passes through the dichromatic mirror to a barrier filter, which blocks out any residual excitation light. Finally, the emitted light passes through the barrier filter to the eyepieces.

The fluorescence microscope has become an essential tool in medical microbiology and microbial ecology. Bacterial

pathogens (e.g., *Mycobacterium tuberculosis,* the cause of tuberculosis) can be identified after staining them with fluorochromes or specifically labeling them with fluorescent antibodies using immunofluorescence procedures. In ecological studies the fluorescence microscope is used to observe microorganisms stained with fluorochrome-labeled probes or fluorochromes that bind specific cell constituents (table 2.3). In addition, microbial ecologists use epifluorescence microscopy to visualize photosynthetic microbes, as their pigments naturally fluoresce when excited by light of specific wavelengths. It is even possible to distinguish live bacteria from dead bacteria by

Table 2.3	Commonly Used Fluorochromes	

Fluorochrome	Uses
Acridine orange	Stains DNA; fluoresces orange
Diamidino-2-phenyl indole (DAPI)	Stains DNA; fluoresces green
Fluorescein isothiocyanate (FITC)	Often attached to antibodies that bind specific cellular components or to DNA probes; fluoresces green
Tetramethyl rhodamine isothiocyanate (TRITC or rhodamine)	Often attached to antibodies that bind specific cellular components; fluoresces red

(a)   10 μm   **(b)**   10 μm   **(c)**   10 μm

Figure 2.13   Fluorescent Dyes and Tags.   (a) Dyes that cause live cells to fluoresce green and dead ones red; **(b)** Auramine is used to stain *Mycobacterium* species in a modification of the acid-fast technique; **(c)** Fluorescent antibodies tag specific molecules. In this case, the antibody binds to a molecule that is unique to *Streptococcus pyogenes*.

the color they fluoresce after treatment with a special mixture of stains (**figure 2.13a**). Thus the microorganisms can be viewed and directly counted in a relatively undisturbed ecological niche. Identification of microorganisms from specimens: Immunologic techniques (section 35.2)

1. List the parts of a light microscope and describe their functions.
2. Define resolution, numerical aperture, working distance, and fluorochrome.
3. If a specimen is viewed using a 5X objective in a microscope with a 15X eye-piece, how many times has the image been magnified?
4. How does resolution depend on the wavelength of light, refractive index, and numerical aperture? How are resolution and magnification related?
5. What is the function of immersion oil?
6. Why don't most light microscopes use 30X ocular lenses for greater magnification?
7. Briefly describe how dark-field, phase-contrast, differential interference contrast, and epifluorescence microscopes work and the kind of image provided by each. Give a specific use for each type.

2.3   PREPARATION AND STAINING OF SPECIMENS

Although living microorganisms can be directly examined with the light microscope, they often must be fixed and stained to increase visibility, accentuate specific morphological features, and preserve them for future study.

Fixation

The stained cells seen in a microscope should resemble living cells as closely as possible. **Fixation** is the process by which the internal and external structures of cells and microorganisms are preserved and fixed in position. It inactivates enzymes that might disrupt cell morphology and toughens cell structures so that they do not change during staining and observation. A microorganism usually is killed and attached firmly to the microscope slide during fixation.

There are two fundamentally different types of fixation. **Heat fixation** is routinely used to observe procaryotes. Typically, a film of cells (a smear) is gently heated as a slide is passed through a flame. Heat fixation preserves overall morphology but not structures within cells. **Chemical fixation** is used to protect fine cellular substructure and the morphology of larger, more delicate microorganisms. Chemical fixatives penetrate cells and react with cellular components, usually proteins and lipids, to render them inactive, insoluble, and immobile. Common fixative mixtures contain such components as ethanol, acetic acid, mercuric chloride, formaldehyde, and glutaraldehyde.

Dyes and Simple Staining

The many types of dyes used to stain microorganisms have two features in common: they have **chromophore groups,** groups with conjugated double bonds that give the dye its color, and they can bind with cells by ionic, covalent, or hydrophobic bonding. Most dyes are used to directly stain the cell or object of interest, but some dyes (e.g., India ink and nigrosin) are used in **negative staining,** where the background but not the cell is stained; the unstained cells appear as bright objects against a dark background.

Dyes that bind cells by ionic interactions are probably the most commonly used dyes. These ionizable dyes may be divided into two general classes based on the nature of their charged group.

1. **Basic dyes**—methylene blue, basic fuchsin, crystal violet, safranin, malachite green—have positively charged groups (usually some form of pentavalent nitrogen) and are generally sold as chloride salts. Basic dyes bind to negatively charged molecules like nucleic acids, many proteins, and the surfaces of procaryotic cells.
2. **Acidic dyes**—eosin, rose bengal, and acid fuchsin—possess negatively charged groups such as carboxyls (—COOH) and phenolic hydroxyls (—OH). Acidic dyes, because of their negative charge, bind to positively charged cell structures.

The staining effectiveness of ionizable dyes may be altered by pH, since the nature and degree of the charge on cell components change with pH. Thus acidic dyes stain best under acidic conditions when proteins and many other molecules carry a positive charge; basic dyes are most effective at higher pHs.

Dyes that bind through covalent bonds or because of their solubility characteristics are also useful. For instance, DNA can be stained by the Feulgen procedure in which the staining compound (Schiff's reagent) is covalently attached to its deoxyribose sugars. Sudan III (Sudan Black) selectively stains lipids because it is lipid soluble but will not dissolve in aqueous portions of the cell.

Microorganisms often can be stained very satisfactorily by **simple staining,** in which a single dye is used (**figure 2.14a,b**). Simple staining's value lies in its simplicity and ease of use. One [the] fixed smear with stain for a short period of time, [wash]es stain off with water, and blots the slide dry. Ba[sic crystal] violet, methylene blue, and carbolfuchsin are [used in simple] staining to determine the size, shape, [and arrangement of] cells.

Differential Staining

The **Gram stain,** developed in 1884 by the Danish physician Christian Gram, is the most widely employed staining method in bacteriology. It is an example of **differential staining**—procedures that are used to distinguish organisms based on their staining properties. Use of the Gram stain divides *Bacteria* into two classes—gram negative and gram positive.

The Gram-staining procedure is illustrated in **figure 2.15.** In the first step, the smear is stained with the basic dye crystal violet, the primary stain. This is followed by treatment with an iodine solution functioning as a **mordant.** The iodine increases the interaction between the cell and the dye so that the cell is stained more strongly. The smear is next decolorized by washing with ethanol or acetone. This step generates the differential aspect of the Gram stain; gram-positive bacteria retain the crystal violet, whereas gram-negative bacteria lose their crystal violet and become colorless. Finally, the smear is counterstained with a simple, basic dye different in color from crystal violet. Safranin, the most common counterstain, colors gram-negative bacteria pink to red and leaves gram-positive bacteria dark purple (figures 2.14c and 2.15b).   The bacterial cell wall (section 3.6)

Acid-fast staining is another important differential staining procedure. It is most commonly used to identify *Mycobacterium tuberculosis* and *M. leprae* (figure 2.14d), the pathogens responsible for tuberculosis and leprosy, respectively. These bacteria have cell walls with high lipid content; in particular, mycolic acids—a group of branched-chain hydroxy lipids, which prevent dyes from readily binding to the cells. However, *M. tuberculosis* and *M. leprae* can be stained by harsh procedures such as the Ziehl-Neelsen method, which uses heat and phenol to drive basic fuchsin into the cells. Once basic fuchsin has penetrated, *M. tuberculosis* and *M. leprae* are not easily decolorized by acidified alcohol (acid-alcohol), and thus are said to be acid-fast. Non-acid-fast bacteria are decolorized by acid-alcohol and thus are stained blue by methylene blue counterstain.

Staining Specific Structures

Many special staining procedures have been developed to study specific structures with the light microscope. One of the simplest is **capsule staining** (figure 2.14f), a technique that reveals the presence of capsules, a network usually made of polysaccharides that surrounds many bacteria and some fungi. Cells are mixed with India ink or nigrosin dye and spread out in a thin film on a slide. After air-drying, the cells appear as lighter bodies in the midst of a blue-black background because ink and dye particles cannot penetrate either the cell or its capsule. Thus capsule staining is an example of **negative staining.** The extent of the light region is determined by the size of the capsule and of the cell itself. There is little distortion of cell shape, and the cell can be counterstained for even greater visibility.   Components external to the cell wall: Capsules, slime layers, and S-layers (section 3.9)

Endospore staining, like acid-fast staining, also requires harsh treatment to drive dye into a target, in this case an endospore. An endospore is an exceptionally resistant structure produced by some bacterial genera (e.g., *Bacillus* and *Clostrid-*

Simple Stains   **Differential Stains**   **Special Stains**

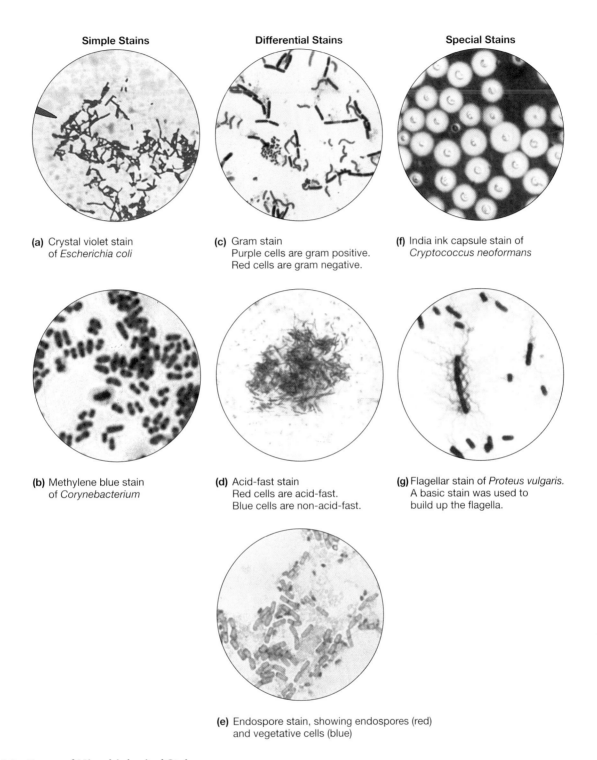

(a) Crystal violet stain
of *Escherichia coli*

(c) Gram stain
Purple cells are gram positive.
Red cells are gram negative.

(f) India ink capsule stain of
Cryptococcus neoformans

(b) Methylene blue stain
of *Corynebacterium*

(d) Acid-fast stain
Red cells are acid-fast.
Blue cells are non-acid-fast.

(g) Flagellar stain of *Proteus vulgaris*.
A basic stain was used to
build up the flagella.

(e) Endospore stain, showing endospores (red)
and vegetative cells (blue)

Figure 2.14   Types of Microbiological Stains.

ium). It is capable of surviving for long periods in an unfavorable environment and is called an endospore because it develops within the parent bacterial cell. Endospore morphology and location vary with species and often are valuable in identification; endospores may be spherical to elliptical and either smaller or larger than the diameter of the parent bacterium. Endospores are not stained well by most dyes, but once stained, they strongly resist decolorization. This property is the basis of most endospore staining methods (figure 2.14*e*). In the Schaeffer-Fulton procedure, endospores are first stained by heating bacteria with malachite green, which is a very strong stain that can penetrate endospores. After malachite green treatment, the rest of the cell

Figure 2.15   Gram Stain.   (a) Steps in the Gram stain procedure. **(b)** Results of a Gram stain. The Gram-positive cells (purple) are *Staphylococcus aureus;* the Gram-negative cells (reddish-pink) are *Escherichia coli.*

is washed free of dye with water and is counterstained with safranin. This technique yields a green endospore resting in a pink to red cell.   The bacterial endospore (section 3.11); Class *Clostridia* (section 23.4); and Class *Bacilli* (section 23.5)

Flagella staining provides taxonomically valuable information about the presence and distribution pattern of flagella on procaryotic cells (figure 2.14*g: see also figure 3.39*). Procaryotic flagella are fine, threadlike organelles of locomotion that are so slender (about 10 to 30 nm in diameter) they can only be seen directly using the electron microscope. To observe them with the light microscope, the thickness of flagella is increased by coating them with mordants like tannic acid and potassium alum, and then staining with pararosaniline (Leifson method) or basic fuchsin (Gray method).   Components external to the cell wall: Flagella and motility (section 3.9)

1. Define fixation, dye, chromophore, basic dye, acidic dye, simple staining, differential staining, mordant, negative staining, and acid-fast staining.
2. Describe the two general types of fixation. Which would you normally use for procaryotes? For protozoa?
3. Why would one expect basic dyes to be more effective under alkaline conditions?
4. Describe the Gram stain procedure and explain how it works. What step in the procedure could be omitted without losing the ability to distinguish between gram-positive and gram-negative bacteria? Why?
5. How would you visualize capsules, endospores, and flagella?

2.4   ELECTRON MICROSCOPY

For centuries the light microscope has been the most important instrument for studying microorganisms. However, even the very best light microscopes have a resolution limit of about 0.2 μm, which greatly compromises their usefulness for detailed studies of many microorganisms. Viruses, for example, are too small to be seen with light microscopes. Procaryotes can be observed, but because they are usually only 1 μm to 2 μm in diameter, just their general shape and major morphological features are visible. The detailed internal structure of larger microorganisms also cannot be effectively studied by light microscopy. These limitations arise from the nature of visible light waves, not from any inadequacy of the light microscope itself. Electron microscopes have much greater resolution. They have transformed microbiology and added immeasurably to our knowledge. The nature of the electron microscope and the ways in which specimens are prepared for observation are reviewed briefly in this section.

The Transmission Electron Microscope

Electron microscopes use a beam of electrons to illuminate and create magnified images of specimens. Recall that the resolution of a light microscope increases with a decrease in the wavelength of the light it uses for illumination. Electrons replace light as the

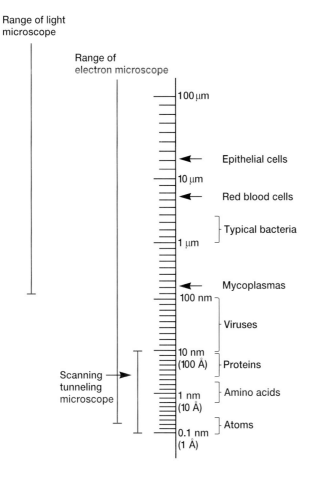

Range of light
microscope

Range of
electron microscope

100 μm

Epithelial cells

10 μm

Red blood cells

Typical bacteria

1 μm

Mycoplasmas

100 nm

Viruses

10 nm
(100 Å)   Proteins

Scanning
tunneling
microscope

1 nm
(10 Å)   Amino acids

Atoms

0.1 nm
(1 Å)

Figure 2.16   The Limits of Microscopic Resolution.   Dimensions are indicated with a logarithmic scale (each major division represents a tenfold change in size). To the right side of the scale are the approximate sizes of cells, bacteria, viruses, molecules, and atoms.

are deflected by collisions with air molecules. The specimen scatters some electrons, but those that pass through are used to form an enlarged image of the specimen on a fluorescent screen. A denser region in the specimen scatters more electrons and therefore appears darker in the image since fewer electrons strike that area of the screen; these regions are said to be "electron dense." In contrast, electron-transparent regions are brighter. The image can also be captured on photographic film as a permanent record.

Table 2.4 compares some of the important features of light and transmission electron microscopes. The TEM has distinctive features that place harsh restrictions on the nature of samples that can be viewed and the means by which those samples must be prepared. Since electrons are deflected by air molecules and are easily absorbed and scattered by solid matter, only extremely thin slices (20 to 100 nm) of a microbial specimen can be viewed in the average TEM. Such a thin slice cannot be cut unless the specimen has support of some kind; the necessary support is provided by plastic. After fixation with chemicals like glutaraldehyde or osmium tetroxide to stabilize cell structure, the specimen is dehydrated with organic solvents (e.g., acetone or ethanol). Complete dehydration is essential because most plastics used for embedding are not water soluble. Next the specimen is soaked in unpolymerized, liquid epoxy plastic until it is completely permeated, and then the plastic is hardened to form a solid block. Thin sections are cut from this block with a glass or diamond knife using a special instrument called an ultramicrotome.

As with bright-field microscopy, cells usually must be stained before they can be seen clearly. The probability of electron scattering is determined by the density (atomic number) of the specimen atoms. Biological molecules are composed primarily of atoms with low atomic numbers (H, C, N, and O), and electron scattering is fairly constant throughout the unstained cell. Therefore specimens are prepared for observation by soaking thin sections with solutions of heavy metal salts like lead citrate and uranyl acetate. The lead and uranium ions bind to cell structures and make them more electron opaque, thus increasing contrast in the material. Heavy osmium atoms from the osmium tetroxide fixative also "stain" cells and increase their contrast. The stained thin sections are then mounted on tiny copper grids and viewed.

Two other important techniques for preparing specimens are negative staining and shadowing. In negative staining, the specimen is spread out in a thin film with either phosphotungstic acid or uranyl acetate. Just as in negative staining for light microscopy, heavy metals do not penetrate the specimen but render the background dark, whereas the specimen appears bright in photographs. Negative staining is an excellent way to study the structure of viruses, bacterial gas vacuoles, and other similar objects (figure 2.17c). In **shadowing,** a specimen is coated with a thin film of platinum or other heavy metal by evaporation at an angle of about 45° from horizontal so that the metal strikes the microorganism on only one side. In one commonly used imaging method, the area coated with metal appears dark in photographs, whereas the uncoated side and the shadow region created by the object is light (**figure 2.20**). This technique is particularly useful in studying virus morphology, procaryotic flagella, and DNA.

illuminating beam. They can be focused, much as light is in a light microscope, but their wavelength is around 0.005 nm, approximately 100,000 times shorter than that of visible light. Therefore, electron microscopes have a practical resolution roughly 1,000 times better than the light microscope; with many electron microscopes, points closer than 0.5 nm can be distinguished, and the useful magnification is well over 100,000× (**figure 2.16**). The value of the electron microscope is evident on comparison of the photographs in **figure 2.17,** microbial morphology can now be studied in great detail.

A modern **transmission electron microscope (TEM)** is complex and sophisticated (**figure 2.18**), but the basic principles behind its operation can be readily understood. A heated tungsten filament in the electron gun generates a beam of electrons that is then focused on the specimen by the condenser (**figure 2.19**). Since electrons cannot pass through a glass lens, doughnut-shaped electromagnets called magnetic lenses are used to focus the beam. The column containing the lenses and specimen must be under high vacuum to obtain a clear image because electrons

(a) (b) (c)

Figure 2.17 Light and Electron Microscopy. A comparison of light and electron microscopic resolution. **(a)** *Rhodospirillum rubrum* in phase-contrast light microscope (×600). **(b)** A thin section of *R. rubrum* in transmission electron microscope (×100,000). **(c)** A transmission electron micrograph of a negatively stained T4 bacteriophage.

Figure 2.18 A Transmission Electron Microscope. The electron gun is at the top of the central column, and the magnetic lenses are within the column. The image on the fluorescent screen may be viewed through a magnifier positioned over the viewing window. The camera is in a compartment below the screen.

The TEM will also disclose the shape of organelles within microorganisms if specimens are prepared by the **freeze-etching** procedure. First, cells are rapidly frozen in liquid nitrogen and then warmed to −100°C in a vacuum chamber. Next a knife that has been precooled with liquid nitrogen (−196°C) frac-

tures the frozen cells, which are very brittle and break along lines of greatest weakness, usually down the middle of internal membranes (**figure 2.21**). The specimen is left in the high vacuum for a minute or more so that some of the ice can sublimate away and uncover more structural detail. Finally, the exposed surfaces are shadowed and coated with layers of platinum and carbon to form a replica of the surface. After the specimen has been removed chemically, this replica is studied in the TEM and provides a detailed, three-dimensional view of intracellular structure (**figure 2.22**). An advantage of freeze-etching is that it minimizes the danger of artifacts because the cells are frozen quickly rather than being subjected to chemical fixation, dehydration, and plastic embedding.

The Scanning Electron Microscope

Transmission electron microscopes form an image from radiation that has passed through a specimen. The **scanning electron microscope (SEM)** works in a different manner. It produces an image from electrons released from atoms on an object's surface. The SEM has been used to examine the surfaces of microorganisms in great detail; many SEMs have a resolution of 7 nm or less.

Specimen preparation for SEM is relatively easy, and in some cases air-dried material can be examined directly. Most often, however, microorganisms must first be fixed, dehydrated, and dried to preserve surface structure and prevent collapse of the cells when they are exposed to the SEM's high vacuum. Before viewing, dried samples are mounted and coated with a thin layer of metal to prevent the buildup of an electrical charge on the surface and to give a better image.

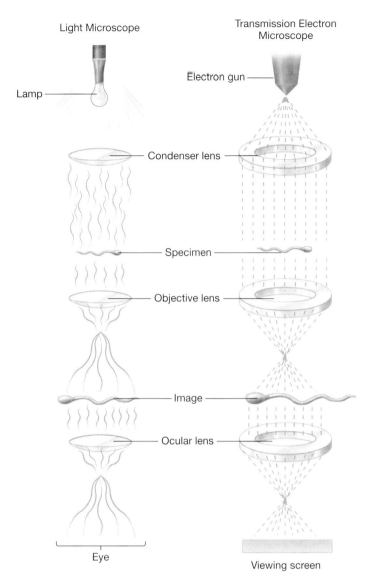

Light Microscope

Transmission Electron Microscope

Lamp

Electron gun

Condenser lens

Specimen

Objective lens

Image

Ocular lens

Eye

Viewing screen

Figure 2.19   Transmission Electron Microscope Operation.
An overview of TEM operation and a comparison of the operation
of light and transmission electron microscopes.

To create an image, the SEM scans a narrow, tapered electron beam back and forth over the specimen (**figure 2.23**). When the beam strikes a particular area, surface atoms discharge a tiny shower of electrons called secondary electrons, and these are trapped by a special detector. Secondary electrons entering the detector strike a scintillator causing it to emit light flashes that a photomultiplier converts to an electrical current and amplifies. The signal is sent to a cathode-ray tube and produces an image like a television picture, which can be viewed or photographed.

The number of secondary electrons reaching the detector depends on the nature of the specimen's surface. When the electron beam strikes a raised area, a large number of secondary electrons enter the detector; in contrast, fewer electrons escape a depression in the surface and reach the detector. Thus raised areas appear lighter on the screen and depressions are darker. A realistic three-dimensional image of the microorganism's surface results (**figure 2.24**). The actual in situ location of microorganisms in ecological niches such as the human skin and the lining of the gut also can be examined.

1. Why does the transmission electron microscope have much greater resolution than the light microscope?
2. Describe in general terms how the TEM functions. Why must the TEM use a high vacuum and very thin sections?
3. Material is often embedded in paraffin before sectioning for light microscopy. Why can't this approach be used when preparing a specimen for the TEM?
4. Under what circumstances would it be desirable to prepare specimens for the TEM by use of negative staining? Shadowing? Freeze-etching?
5. How does the scanning electron microscope operate and in what way does its function differ from that of the TEM? The SEM is used to study which aspects of morphology?

2.5   Newer Techniques in Microscopy

Confocal Microscopy

Like the large and small beads illustrated in **figure 2.25a,** biological specimens are three-dimensional. When three-dimensional objects are viewed with traditional light microscopes, light from

Table 2.4	Characteristics of Light and Transmission Electron Microscopes	
Feature	**Light Microscope**	**Transmission Electron Microscope**
Highest practical magnification	About 1,000–1,500	Over 100,000
Best resolution[a]	0.2 μm	0.5 nm
Radiation source	Visible light	Electron beam
Medium of travel	Air	High vacuum
Type of lens	Glass	Electromagnet
Source of contrast	Differential light absorption	Scattering of electrons
Focusing mechanism	Adjust lens position mechanically	Adjust current to the magnetic lens
Method of changing magnification	Switch the objective lens or eyepiece	Adjust current to the magnetic lens
Specimen mount	Glass slide	Metal grid (usually copper)

[a]The resolution limit of a human eye is about 0.2 mm.

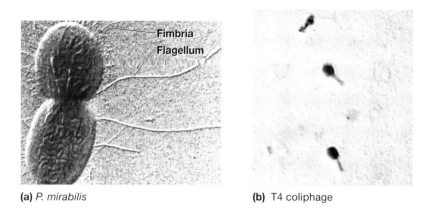

(a) *P. mirabilis* **(b)** T4 coliphage

Figure 2.20 Specimen Shadowing for the TEM. Examples of specimens viewed in the TEM after shadowing with uranium metal. **(a)** *Proteus mirabilis* ($\times$42,750); note flagella and fimbriae. **(b)** T4 coliphage ($\times$72,000).

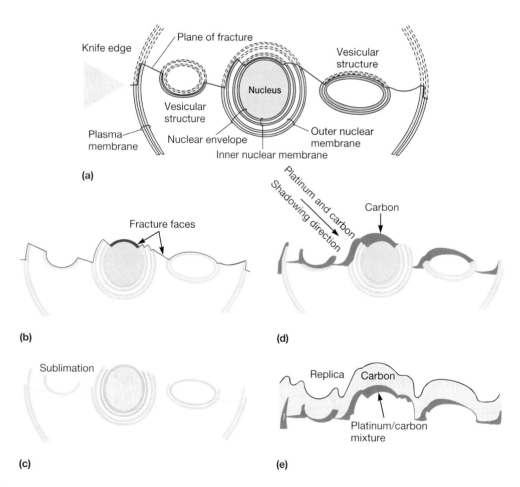

Figure 2.21 The Freeze-Etching Technique. In steps **(a)** and **(b)**, a frozen eucaryotic cell is fractured with a cold knife. Etching by sublimation is depicted in **(c)**. Shadowing with platinum plus carbon and replica formation are shown in **(d)** and **(e)**. See text for details.

Figure 2.22 Example of Freeze-Etching. A freeze-etched preparation of the bacterium *Thiobacillus kabobis*. Note the differences in structure between the outer surface, S; the outer membrane of the cell wall, OM; the cytoplasmic membrane, CM; and the cytoplasm, C. Bar = 0.1 μm.

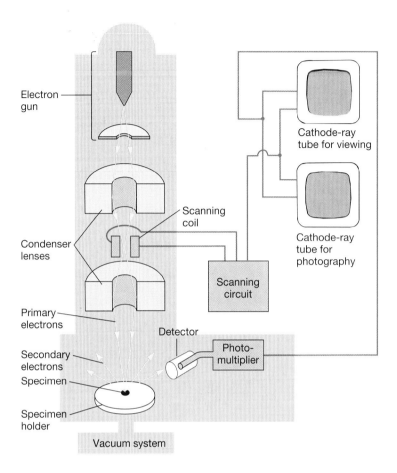

Figure 2.23 The Scanning Electron Microscope. See text for explanation.

(a) *S. aureus*

(b) *Cristispira*

Figure 2.24 Scanning Electron Micrographs of Bacteria. (a) *Staphylococcus aureus* (×32,000). (b) *Cristispira,* a spirochete from the crystalline style of the oyster, *Ostrea virginica.* The axial fibrils or periplasmic flagella are visible around the protoplasmic cylinder (×6,000).

| Schematic diagram of CSLM planes | Light microscopy | CSLM composite of all sections | CSLM 3-D reconstruction |

A x-z plane, x-y planes **B** **C** **D**

E z cross-section, z cross-section **F**

| CSLM: Three different views | CSLM 3-D reconstruction of *P. a.* biofilm |

Figure 2.25 **Light and Confocal Microscopy.** Two beads examined by light and confocal microscopy. Light microscope images are generated from light emanating from many areas of a three-dimensional object. Confocal images are created from light emanating from only a single plane of focus. Multiple planes within the object can be examined and used to construct clear, finely detailed images. **(a)** The planes observable by confocal microscopy. **(b)** The light microscope image of the two beads shown in (a). Note that neither bead is clear and that the smaller bead is difficult to recognize as a bead. **(c)** A computer connected to a confocal microscope can make a composite image of the two beads using digitized information collected from multiple planes within the beads. The result is a much clearer and more detailed image. **(d)** The computer can also use digitized information collected from multiple planes within the beads to generate a three-dimensional reconstruction of the beads. **(e)** Computer generated views of a specimen: the top left panel is the image of a single x-y plane (i.e., looking down from the top of the specimen). The two lines represent the two x-z planes imaged in the other two panels. The vertical line indicates the x-z plane shown in the top right panel (i.e., a view from the right side of the specimen) and the horizontal line indicates the x-z plane shown in the bottom panel (i.e., a view from the front face of the specimen). **(f)** A three-dimensional reconstruction of a *Pseudomonas aeruginosa* biofilm. The biofilm was exposed to an antibacterial agent and then stained with dyes that distinguish living (green) from dead (red) cells. The cells on the surface of the biofilm have been killed, but those in the lower layers of the biofilm are still alive.

all areas of the object, not just the plane of focus, enter the microscope and are used to create an image. The resulting image is murky and fuzzy (figure 2.25*b*). This problem has been solved by the development of the **confocal scanning laser microscope (CSLM),** or simply, confocal microscope. The confocal microscope uses a laser beam to illuminate a specimen, usually one that has been fluorescently stained. A major component of the confocal microscope is an aperture placed above the objective lens, which eliminates stray light from parts of the specimen that lie above and below the plane of focus (**figure 2.26**). Thus the only light used to create the image is from the plane of focus, and a much clearer sharp image is formed.

Computers are integral to the process of creating confocal images. A computer attached to the confocal microscope receives digitized information from each plane in the specimen that is examined. This information can be used to create a composite image that is very clear and detailed (figure 2.25*c*) or to create a three-dimensional reconstruction of the specimen (figure 2.25*d*). Images of x-z plane cross-sections of the specimen can also be generated, giving the observer views of the specimen from three perspectives (figure 2.25*e*). Confocal microscopy has numerous applications, including the study of biofilms, which can form on many different types of surfaces including in-dwelling medical devices such as hip joint replacements. As shown in figure 2.25*f*, it is difficult to kill all cells in a biofilm. This makes them a particular concern to the medical field because formation of biofilms on medical devices can result in infections that are difficult to treat. Microbial growth in natural environments: Biofilms (section 6.6)

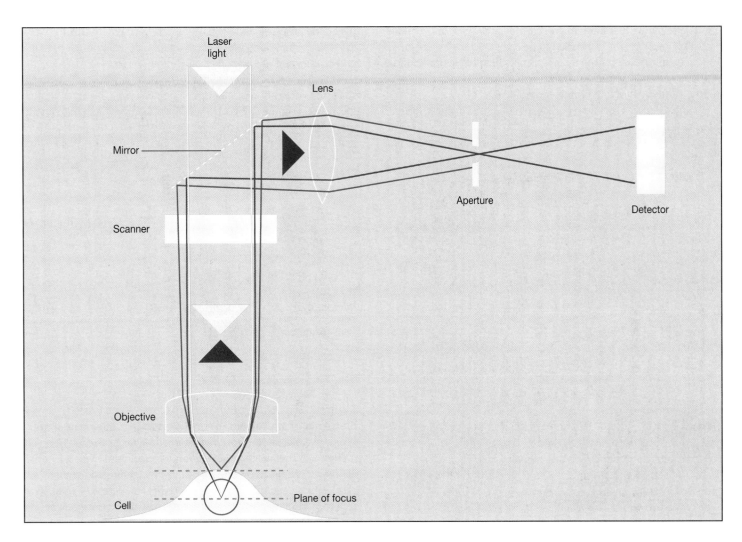

Figure 2.26   A Ray Diagram of a Confocal Laser Scanning Microscope.   The yellow lines represent laser light used for illumination. Red lines symbolize the light arising from the plane of focus, and the blue lines stand for light from parts of the specimen above and below the focal plane. See text for explanation.

Scanning Probe Microscopy

Although light and electron microscopes have become quite sophisticated and have reached an advanced state of development, powerful new microscopes are still being created. A new class of microscopes, called **scanning probe microscopes,** measure surface features by moving a sharp probe over the object's surface. The **scanning tunneling microscope,** invented in 1980, is an excellent example of a scanning probe microscope. It can achieve magnifications of 100 million and allow scientists to view atoms on the surface of a solid. The electrons surrounding surface atoms tunnel or project out from the surface boundary a very short distance. The scanning tunneling microscope has a needle-like probe with a point so sharp that often there is only one atom at its tip. The probe is lowered toward the specimen surface until its electron cloud just touches that of the surface atoms. If a small voltage is applied between the tip and specimen, electrons flow through a narrow channel in the electron clouds. This tun-

neling current, as it is called, is extraordinarily sensitive to distance and will decrease about a thousandfold if the probe is moved away from the surface by a distance equivalent to the diameter of an atom.

The arrangement of atoms on the specimen surface is determined by moving the probe tip back and forth over the surface while keeping it at a constant height above the specimen by adjusting the probe distance to maintain a steady tunneling current. As the tip moves up and down while following the surface contours, its motion is recorded and analyzed by a computer to create an accurate three-dimensional image of the surface atoms. The surface map can be displayed on a computer screen or plotted on paper. The resolution is so great that individual atoms are observed easily. The microscope's inventors, Gerd Binnig and Heinrich Rohrer, shared the 1986 Nobel Prize in Physics for their work, together with Ernst Ruska, the designer of the first transmission electron microscope.

The scanning tunneling microscope is already having a major impact in biology. It can be used to directly view DNA and other biological molecules (**figure 2.27**). Since the microscope can examine objects when they are immersed in water, it may be particularly useful in studying biological molecules.

More recently a second type of scanning probe microscope has been developed. The **atomic force microscope** moves a sharp probe over the specimen surface while keeping the distance be-

tween the probe tip and the surface constant. It does this by exerting a very small amount of force on the tip, just enough to maintain a constant distance but not enough force to damage the surface. The vertical motion of the tip usually is followed by measuring the deflection of a laser beam that strikes the lever holding the probe (**figure 2.28**). Unlike the scanning tunneling microscope, the atomic force microscope can be used to study surfaces that do not conduct electricity well. The atomic force microscope

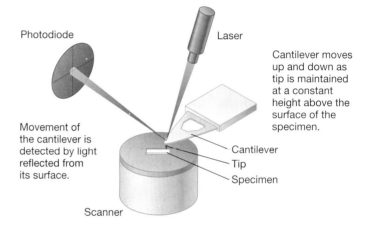

Figure 2.28 Atomic Force Microscopy—The Basic Elements of an Atomic Force Microscope. The tip used to probe the specimen is attached to a cantilever. As the probe passes over the "hills and valleys" of the specimen's surface, the cantilever is deflected vertically. A laser beam directed at the cantilever is used to monitor these vertical movements. Light reflected from the cantilever is detected by the photodiode and used to generate an image of the specimen.

Figure 2.27 Scanning Tunneling Microscopy of DNA. The DNA double helix with approximately three turns shown (false color; ×2,000,000).

(a)

(b)

Figure 2.29 The Membrane Protein Aquaporin Visualized by Atomic Force Micrscopy. Aquaporin is a membrane-spanning protein that allows water to move across the membrane. **(a)** Each circular structure represents the surface view of a single aquaporin protein. **(b)** A single aquaporin molecule observed in more detail and at higher magnification.

has been used to study the interactions between the *E. coli* GroES and GroEL chaperone proteins, to map plasmids by locating restriction enzymes bound to specific sites, to follow the behavior of living bacteria and other cells, and to visualize membrane proteins (**figure 2.29**).

1. How does a confocal microscope operate? Why does it provide better images of thick specimens than does the standard compound microscope?
2. Briefly describe the scanning probe microscope and compare and contrast its most popular versions—the scanning tunneling microscope and the atomic force microscope. What are these microscopes used for?

Summary

2.1 Lenses and the Bending of Light

a. A light ray moving from air to glass, or vice versa, is bent in a process known as refraction.

b. Lenses focus light rays at a focal point and magnify images (**figure 2.2**).

2.2 The Light Microscope

a. In a compound microscope like the bright-field microscope, the primary image is formed by an objective lens and enlarged by the eyepiece or ocular lens to yield the final image (**figure 2.3**).

b. A substage condenser focuses a cone of light on the specimen.

c. Microscope resolution increases as the wavelength of radiation used to illuminate the specimen decreases. The maximum resolution of a light microscope is about 0.2 μm.

d. The dark-field microscope uses only refracted light to form an image (**figure 2.7**), and objects glow against a black background.

e. The phase-contrast microscope converts variations in the refractive index and density of cells into changes in light intensity and thus makes colorless, unstained cells visible (**figure 2.9**).

f. The differential interference contrast microscope uses two beams of light to create high-contrast, three-dimensional images of live specimens.

g. The fluorescence microscope illuminates a fluorochrome-labeled specimen and forms an image from its fluorescence (**figure 2.12**).

2.3 Preparation and Staining of Specimens

a. Specimens usually must be fixed and stained before viewing them in the bright-field microscope.

b. Most dyes are either positively charged basic dyes or negative acidic dyes and bind to ionized parts of cells.

c. In simple staining a single dye is used to stain microorganisms.

d. Differential staining procedures like the Gram stain and acid-fast stain distinguish between microbial groups by staining them differently (**figure 2.15**).

e. Some staining techniques are specific for particular structures like bacterial capsules, flagella, and endospores (**figure 2.14**).

2.4 Electron Microscopy

a. The transmission electron microscope uses magnetic lenses to form an image from electrons that have passed through a very thin section of a specimen (**figure 2.19**). Resolution is high because the wavelength of electrons is very short.

b. Thin section contrast can be increased by treatment with solutions of heavy metals like osmium tetroxide, uranium, and lead.

c. Specimens are also prepared for the TEM by negative staining, shadowing with metal, or freeze-etching.

d. The scanning electron microscope (**figure 2.23**) is used to study external surface features of microorganisms.

2.5 Newer Techniques in Microscopy

a. The confocal scanning laser microscope (**figure 2.25**) is used to study thick, complex specimens.

b. Scanning probe microscopes reach very high magnifications that allow scientists to observe biological molecules (**figures 2.27** and **2.29**).

Key Terms

acidic dyes 26
acid-fast staining 26
atomic force microscope 36
basic dyes 26
bright-field microscope 18
capsule staining 26
chemical fixation 26
chromophore groups 26
confocal scanning laser microscope (CSLM) 34
dark-field microscope 21
differential interference contrast (DIC) microscope 23

differential staining 26
endospore staining 26
eyepieces 18
fixation 25
flagella staining 28
fluorescence microscope 23
fluorescent light 23
fluorochromes 24
focal length 18
focal point 18
freeze-etching 30

Gram stain 26
heat fixation 25
mordant 26
negative staining 26
numerical aperture 19
objective lenses 18
ocular lenses 18
parfocal 18
phase-contrast microscope 21
refraction 17
refractive index 17

resolution 18
scanning electron microscope (SEM) 30
scanning probe microscope 35
scanning tunneling microscope 35
shadowing 29
simple staining 26
substage condenser 18
transmission electron microscope (TEM) 29
working distance 20

Critical Thinking Questions

1. If you prepared a sample of a specimen for light microscopy, stained with the Gram stain, and failed to see anything when you looked through your light microscope, list the things that you may have done incorrectly.

2. In a journal article, find an example of a light micrograph, a scanning or transmission electron micrograph, or a confocal image. Discuss why the figure was included in the article and why that particular type of microscopy was the method of choice for the research. What other figures would you like to see used in this study? Outline the steps that the investigators would take in order to obtain such photographs or figures.

Learn More

Binning, G., and Rohrer, H. 1985. The scanning tunneling microscope. *Sci. Am.* 253(2):50–56.

Dufrêne, Y. F. 2003. Atomic force microscopy provides a new means for looking at microbial cells. *ASM News* 69(9):438–42.

Hörber, J.K.H., and Miles, M. J. 2003. Scanning probe evolution in biology. *Science* 302:1002–5.

Lillie, R. D. 1969. *H. J. Conn's biological stains,* 8th ed. Baltimore: Williams & Wilkins.

Rochow, T. G. 1994. *Introduction to microscopy by means of light, electrons, X-rays, or acoustics.* New York: Plenum.

Scherrer, Rene. 1984. Gram's staining reaction, Gram types and cell walls of bacteria. *Trends Biochem. Sci.* 9:242–45.

Stephens, D. J., and Allan, V. J. 2003. Light microscopy techniques for live cell imaging. *Science* 300:82–6.

**Please visit the Prescott website at www.mhhe.com/prescott7
for additional references.**

3

Procaryotic Cell Structure and Function

Bacterial species may differ in their patterns of flagella distribution. These *Pseudomonas* cells have a single polar flagellum used for locomotion.

PREVIEW

- Procaryotes can be distinguished from eucaryotes in terms of their size, cell structure, and molecular make-up. Most procaryotes lack extensive, complex, internal membrane systems.

- Although some cell structures are observed in both eucaryotic and procaryotic cells, some structures are unique to procaryotes.

- Procaryotes can be divided into two major groups: *Bacteria* and *Archaea*. Although similar in overall structure, *Bacteria* and *Archaea* exhibit important differences in their cell walls and membranes.

- Most bacteria can be divided into two broad groups based on cell wall structure; the differences in cell wall structure correlate with the reaction to the Gram staining procedure.

- Many procaryotes are motile; several mechanisms for motility have been identified.

- Some bacteria form resistant endospores to survive harsh environmental conditions in a dormant state.

Even a superficial examination of the microbial world shows that procaryotes are one of the most important groups by any criterion: numbers of organisms, general ecological importance, or practical importance for humans. Indeed, much of our understanding of phenomena in biochemistry and molecular biology comes from research on procaryotes. Although considerable space in this text is devoted to eucaryotic microorganisms, the major focus is on procaryotes. Therefore the unit on microbial morphology begins with the structure of procaryotes. As mentioned in chapter 1, there are two quite different groups of procaryotes: *Bacteria* and *Archaea*. Although considerably less is known about archaeal cell structure and biochemistry, certain features distinguish the two domains. Whenever possible, these distinctions will be noted. A more detailed discussion of the *Archaea* is provided in chapter 20. A comment about nomenclature is necessary to avoid confusion. The word procaryote will be used in a general sense to include both the *Bacteria* and *Archaea;* the term bacterium will refer specifically to a member of the *Bacteria* and archaeon to a member of the *Archaea.* Members of the microbial world (section 1.1)

3.1 AN OVERVIEW OF PROCARYOTIC CELL STRUCTURE

Because much of this chapter is devoted to a discussion of individual cell components, a preliminary overview of the procaryotic cell as a whole is in order.

Shape, Arrangement, and Size

One might expect that small, relatively simple organisms like procaryotes would be uniform in shape and size. This is not the case, as the microbial world offers almost endless variety in terms of morphology (**figures 3.1** and **3.2**). However, most commonly encountered procaryotes have one of two shapes. **Cocci** (s., **coccus**) are roughly spherical cells. They can exist as individual cells, but also are associated in characteristic arrangements that are frequently useful in their identification. **Diplococci** (s., **diplococcus**) arise when cocci divide and remain together to form pairs. Long chains of cocci result when cells adhere after repeated divisions in one plane; this pattern is seen in the genera *Streptococcus, Enterococcus,* and *Lactococcus* (figure 3.1*b*). *Staphylococcus* divides in random planes to generate irregular grapelike clumps (figure 3.1*a*). Divisions in two or three planes

The era in which workers tended to look at bacteria as very small bags of enzymes has long passed.

—*Howard J. Rogers*

(a) *S. aureus*

(b) *S. agalactiae*

(c) *B. megaterium*

(d) *R. rubrum*

(e) *V. cholerae*

Figure 3.1   Common Procaryotic Cell Shapes.   **(a)** *Staphylococcus aureus* cocci arranged in clusters; color-enhanced scanning electron micrograph; average cell diameter is about 1 μm. **(b)** *Streptococcus agalactiae,* the cause of Group B streptococcal infections; cocci arranged in chains; color-enhanced scanning electron micrograph (×4,800). **(c)** *Bacillus megaterium,* a rod-shaped bacterium arranged in chains, Gram stain (×600). **(d)** *Rhodospirillum rubrum,* phase contrast (×500). **(e)** *Vibrio cholera,* curved rods with polar flagella; scanning electron micrograph.

can produce symmetrical clusters of cocci. Members of the genus *Micrococcus* often divide in two planes to form square groups of four cells called tetrads. In the genus *Sarcina,* cocci divide in three planes producing cubical packets of eight cells.

The other common shape is that of a **rod,** sometimes called a **bacillus** (pl., **bacilli**). *Bacillus megaterium* is a typical example of a bacterium with a rod shape (figure 3.1*c*). Bacilli differ considerably in their length-to-width ratio, the coccobacilli being so short and wide that they resemble cocci. The shape of the rod's end often varies between species and may be flat, rounded, cigar-shaped, or bifurcated. Although many rods occur singly, some re-

main together after division to form pairs or chains (e.g., *Bacillus megaterium* is found in long chains).

Although procaryotes are often simple spheres or rods, other cell shapes and arrangements are not uncommon. **Vibrios** most closely resemble rods, as they are comma-shaped (figure 3.1*e*). Spiral-shaped procaryotes can be either classified as **spirilla,** which usually have tufts of flagella at one or both ends of the cell (figure 3.1*d* and 3.2*c*), or spirochetes. **Spirochetes** are more flexible and have a unique, internal flagellar arrangement. Actinomycetes typically form long filaments called hyphae that may branch to produce a network called a **mycelium** (figure 3.2a). In this sense, they are

(a) *Actinomyces*

(b) *M. pneumoniae*

(c) *Spiroplasma*

Bud

Hypha

Hypha

(d) *Hyphomicrobium*

2 μ

(e) Walsby's square archaeon

(f) *G. ferruginea*

Figure 3.2 Unusually Shaped Procaryotes. Examples of procaryotes with shapes quite different from bacillus and coccus types. **(a)** *Actinomyces,* SEM (×21,000). **(b)** *Mycoplasma pneumoniae,* SEM (×62,000). **(c)** *Spiroplasma,* SEM (×13,000). **(d)** *Hyphomicrobium* with hyphae and bud, electron micrograph with negative staining. **(e)** Walsby's square archaeon. **(f)** *Gallionella ferruginea* with stalk.

similar to filamentous fungi, a group of eucaryotic microbes. The oval- to pear-shaped *Hyphomicrobium* (figure 3.2*d*) produces a bud at the end of a long hypha. Other bacteria such as *Gallionella* produce nonliving stalks (figure 3.2*f*). A few procaryotes actually are flat. For example, Anthony Walsby has discovered square archaea living in salt ponds (figure 3.2*e*). They are shaped like flat, square-to-rectangular boxes about 2 μm by 2 to 4 μm, and only 0.25 μm thick. Finally, some procaryotes are variable in shape and lack a single, characteristic form (figure 3.2*b*). These are called **pleomorphic** even though they may, like *Corynebacterium,* have a generally rod-like form. Phylum *Spirochaetes* (section 21.6)

Bacteria vary in size as much as in shape (**figure 3.3**). *Escherichia coli* is a rod of about average size, 1.1 to 1.5 μm wide by 2.0 to 6.0 μm long. Near the small end of the size continuum

are members of the genus *Mycoplasma,* an interesting group of bacteria that lack cell walls. For many years, it was thought that they were the smallest procaryotes at about 0.3 μm in diameter, approximately the size of the poxviruses. However, even smaller procaryotes have been discovered. Nanobacteria range from around 0.2 μm to less than 0.05 μm in diameter. Only a few strains have been cultured, and these appear to be very small, bacteria-like organisms. The discovery of nanobacteria was quite surprising because theoretical calculations predicted that the smallest cells were about 0.14 to 0.2 μm in diameter. At the other end of the continuum are bacteria such as the spirochaetes, which can reach 500 μm in length, and the photosynthetic bacterium *Oscillatoria,* which is about 7 μm in diameter (the same diameter as a red blood cell). A huge bacterium lives in the intestine of

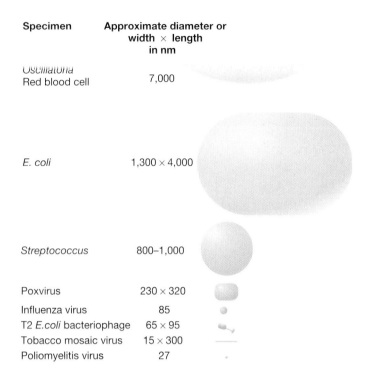

Specimen	Approximate diameter or width × length in nm
Oscillatoria Red blood cell	7,000
E. coli	1,300 × 4,000
Streptococcus	800–1,000
Poxvirus	230 × 320
Influenza virus	85
T2 E.coli bacteriophage	65 × 95
Tobacco mosaic virus	15 × 300
Poliomyelitis virus	27

Figure 3.3   Sizes of Procaryotes and Viruses.   The sizes of selected bacteria relative to a red blood cell and viruses.

the brown surgeonfish, *Acanthurus nigrofuscus. Epulopiscium fishelsoni* grows as large as 600 by 80 μm, a little smaller than a printed hyphen. More recently an even larger bacterium, *Thiomargarita namibiensis,* has been discovered in ocean sediment (**Microbial Diversity & Ecology 3.1**). Thus a few bacteria are much larger than the average eucaryotic cell (typical plant and animal cells are around 10 to 50 μm in diameter).

Procaryotic Cell Organization

Procaryotic cells are morphologically simpler than eucaryotic cells, but they are not just simpler versions of eucaryotes. Although many structures are common to both cell types, some are unique to procaryotes. The major procaryotic structures and their functions are summarized and illustrated in **table 3.1** and **figure 3.4,** respectively. Note that no single procaryote possesses all of these structures at all times. Some are found only in certain cells in certain conditions or in certain phases of the life cycle. However, despite these variations procaryotes are consistent in their fundamental structure and most important components.

Procaryotic cells almost always are bounded by a chemically complex cell wall. Interior to this wall lies the plasma membrane. This membrane can be invaginated to form simple internal membranous structures such as the light-harvesting membrane of some photosynthetic bacteria. Since the procaryotic cell does not contain internal membrane-bound organelles, its interior appears morphologically simple. The genetic material is localized in a discrete region, the nucleoid, and usually is not separated from the

surrounding cytoplasm by membranes. Ribosomes and larger masses called inclusion bodies are scattered about the cytoplasmic matrix. Many procaryotes use flagella for locomotion. In addition, many are surrounded by a capsule or slime layer external to the cell wall.

In the remaining sections of this chapter we describe the major procaryotic structures in more detail. We begin with the plasma membrane, a structure that defines all cells. We then proceed inward to consider structures located within the cytoplasm. Then the discussion moves outward, first to the cell wall and then to structures outside the cell wall. Finally, we consider a structure unique to bacteria, the bacterial endospore.

1. What characteristic shapes can bacteria assume? Describe the ways in which bacterial cells cluster together.
2. Draw a bacterial cell and label all important structures.

3.2   PROCARYOTIC CELL MEMBRANES

Membranes are an absolute requirement for all living organisms. Cells must interact in a selective fashion with their environment, whether it is the internal environment of a multicellular organism or a less protected and more variable external environment. Cells must not only be able to acquire nutrients and eliminate wastes, but they also have to maintain their interior in a constant, highly organized state in the face of external changes.

The **plasma membrane** encompasses the cytoplasm of both procaryotic and eucaryotic cells. It is the chief point of contact with the cell's environment and thus is responsible for much of its relationship with the outside world. The plasma membranes of procaryotic cells are particularly important because they must fill an incredible variety of roles. In addition to retaining the cytoplasm, the plasma membrane also serves as a selectively permeable barrier: it allows particular ions and molecules to pass, either into or out of the cell, while preventing the movement of others. Thus the membrane prevents the loss of essential components through leakage while allowing the movement of other molecules. Because many substances cannot cross the plasma membrane without assistance, it must aid such movement when necessary. Transport systems are used for such tasks as nutrient uptake, waste excretion, and protein secretion. The procaryotic plasma membrane also is the location of a variety of crucial metabolic processes: respiration, photosynthesis, and the synthesis of lipids and cell wall constituents. Finally, the membrane contains special receptor molecules that help procaryotes detect and respond to chemicals in their surroundings. Clearly the plasma membrane is essential to the survival of microorganisms. Uptake of nutrients by the cell (section 5.6)

As will be evident in the following discussion, all membranes apparently have a common, basic design. However, procaryotic membranes can differ dramatically in terms of the lipids they contain. Indeed, membrane chemistry can be used to identify particular bacterial species. To understand these chemical differences and to understand the many functions of the

Microbial Diversity & Ecology

3.1   Monstrous Microbes

Biologists often have distinguished between procaryotes and eucaryotes based in part on cell size. Generally, procaryotic cells are supposed to be smaller than eucaryotic cells. Procaryotes grow extremely rapidly compared to most eucaryotes and lack the complex vesicular transport systems of eucaryotic cells described in chapter 4. It has been assumed that they must be small because of the slowness of nutrient diffusion and the need for a large surface-to-volume ratio. Thus when Fishelson, Montgomery, and Myrberg discovered a large, cigar-shaped microorganism in the intestinal tract of the Red Sea brown surgeonfish, *Acanthurus nigrofuscus,* they suggested in their 1985 publication that it was a protist. It seemed too large to be anything else. In 1993 Esther Angert, Kendall Clemens, and Norman Pace used rRNA sequence comparisons to identify the microorganism, now called *Epulopiscium fishelsoni,* as a procaryote related to the gram-positive bacterial genus *Clostridium.*

E. fishelsoni [Latin, *epulum,* a feast or banquet, and *piscium,* fish] can reach a size of 80 μm by 600 μm, and normally ranges from 200 to 500 μm in length (see **Box figure**). It is about a million times larger in volume than *Escherichia coli.* Despite its huge size the organism has a procaryotic cell structure. It is motile and swims at about two body lengths a second (approximately 2.4 cm/min) using the flagella that cover its surface. The cytoplasm contains large nucleoids and many ribosomes, as would be required for such a large cell. *Epulopiscium* appears to overcome the size limits set by diffusion by having a highly convoluted plasma membrane. This increases the cell's surface area and aids in nutrient transport.

It appears that *Epulopiscium* is transmitted between hosts through fecal contamination of the fish's food. The bacterium can be eliminated by starving the surgeonfish for a few days. If juvenile fish that lack the bacterium are placed with infected hosts, they are reinoculated. Interestingly this does not work with uninfected adult surgeonfish.

In 1997, Heidi Schulz discovered an even larger procaryote in the ocean sediment off the coast of Namibia. *Thiomargarita namibiensis* is a spherical bacterium, between 100 and 750 μm in diameter, that often forms chains of cells enclosed in slime sheaths. It is over 100 times larger in volume than *E. fishelsoni.* A vacuole occupies about 98% of the cell and contains fluid rich in nitrate; it is surrounded by a 0.5 to 2.0 μm layer of cytoplasm filled with sulfur granules. The cytoplasmic layer is the same thickness as most bacteria and sufficiently thin for adequate diffusion rates. It uses sulfur as an energy source and nitrate as the electron acceptor for the electrons released when sulfur is oxidized in energy-conserving processes.

The discovery of these procaryotes greatly weakens the distinction between procaryotes and eucaryotes based on cell size. They are certainly larger than a normal eucaryotic cell. The size distinction between procaryotes and eucaryotes has been further weakened by the discovery of eucaryotic cells that are smaller than previously thought possible. The best example is *Nanochlorum eukaryotum. Nanochlorum* is only about 1 to 2 μm in diameter, yet is truly eucaryotic and has a nucleus, a chloroplast, and a mitochondrion. Our understanding of the factors limiting procaryotic cell size must be reevaluated. It is no longer safe to assume that large cells are eucaryotic and small cells are procaryotic.

(a)

(b)

Giant Bacteria.   **(a)** This photograph, taken with pseudo dark-field illumination, shows *Epulopiscium fishelsoni* at the top of the figure dwarfing the paramecia at the bottom (×200). **(b)** A chain of *Thiomargarita namibiensis* cells as viewed with the light microscope. Note the external mucous sheath and the internal sulfur globules.

Sources: Angert, E. R., Clements, K. D., and Pace, N. R. 1993 The largest bacterium. Nature 362:239–41; and Shulz, H. N., Brinkhoff, T., Ferdelman, T. G., Mariné, M. H., Teske, A., and Jorgensen, B.B. 1999. Dense populations of a giant sulfur bacterium in Namibian shelf sediments. Science 284:493–95.

Table 3.1	Functions of Procaryotic Structures
Plasma membrane	Selectively permeable barrier, mechanical boundary of cell, nutrient and waste transport, location of many metabolic processes (respiration, photosynthesis), detection of environmental cues for chemotaxis
Gas vacuole	Buoyancy for floating in aquatic environments
Ribosomes	Protein synthesis
Inclusion bodies	Storage of carbon, phosphate, and other substances
Nucleoid	Localization of genetic material (DNA)
Periplasmic space	Contains hydrolytic enzymes and binding proteins for nutrient processing and uptake
Cell wall	Gives procaryotes shape and protection from osmotic stress
Capsules and slime layers	Resistance to phagocytosis, adherence to surfaces
Fimbriae and pili	Attachment to surfaces, bacterial mating
Flagella	Movement
Endospore	Survival under harsh environmental conditions

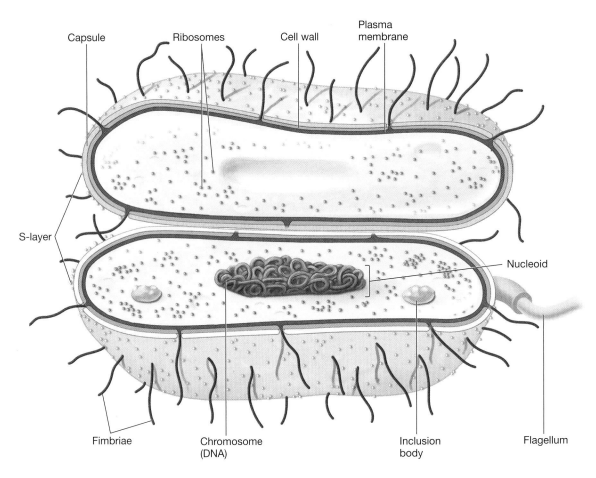

Figure 3.4 Morphology of a Procaryotic Cell.

plasma membrane, it is necessary to become familiar with membrane structure. In this section, the common basic design of all membranes is discussed. This is followed by a consideration of the significant differences between bacterial and archaeal membranes.

The Fluid Mosaic Model of Membrane Structure

The most widely accepted model for membrane structure is the **fluid mosaic model** of Singer and Nicholson (**figure 3.5**), which proposes that membranes are lipid bilayers within which proteins float. The model is based on studies of eucaryotic and bac-

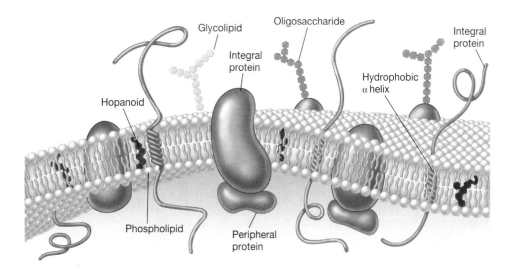

Figure 3.5 Bacterial Plasma Membrane Structure. This diagram of the fluid mosaic model of bacterial membrane structure shows the integral proteins (blue) floating in a lipid bilayer. Peripheral proteins (purple) are associated loosely with the inner membrane surface. Small spheres represent the hydrophilic ends of membrane phospholipids and wiggly tails, the hydrophobic fatty acid chains. Other membrane lipids such as hopanoids (red) may be present. For the sake of clarity, phospholipids are shown in proportionately much larger size than in real membranes.

terial membranes, and a variety of experimental approaches were used to establish it. Transmission electron microscopy (TEM) studies were particularly important. When membranes are stained and examined by TEM, it can be seen that cell membranes are very thin structures, about 5 to 10 nm thick, and that they appear as two dark lines on either side of a nonstained interior. This characteristic appearance has been interpreted to mean that the membrane lipid is organized in two sheets of molecules arranged end-to-end (figure 3.5). When membranes are cleaved by the freeze-etching technique, they can be split down the center of the lipid bilayer, exposing the complex internal structure. Within the lipid bilayer, small globular particles are visible; these have been suggested to be membrane proteins lying within the membrane lipid bilayer. The use of atomic force microscopy has provided powerful images to support this interpretation. Electron microscopy (section 2.4); Newer techniques in microscopy: Scanning probe microscopy (section 2.5)

The chemical nature of membrane lipids is critical to their ability to form bilayers. Most membrane-associated lipids are structurally asymmetric, with polar and nonpolar ends (**figure 3.6**) and are called **amphipathic.** The polar ends interact with water and are **hydrophilic;** the nonpolar **hydrophobic** ends are insoluble in water and tend to associate with one another. In aqueous environments, amphipathic lipids can interact to form a bilayer. The outer surfaces of the bilayer membrane are hydrophilic, whereas hydrophobic ends are buried in the interior away from the surrounding water (figure 3.5). Lipids (appendix I)

Two types of membrane proteins have been identified based on their ability to be separated from the membrane. **Peripheral proteins** are loosely connected to the membrane and can be easily removed (figure 3.5). They are soluble in aqueous solutions

Figure 3.6 The Structure of a Polar Membrane Lipid. Phosphatidylethanolamine, an amphipathic phospholipid often found in bacterial membranes. The R groups are long, nonpolar fatty acid chains.

and make up about 20 to 30% of total membrane protein. About 70 to 80% of membrane proteins are **integral proteins.** These are not easily extracted from membranes and are insoluble in aqueous solutions when freed of lipids. Integral proteins, like membrane lipids, are amphipathic; their hydrophobic regions are buried in the lipid while the hydrophilic portions project from the membrane surface (figure 3.5). Integral proteins can diffuse laterally in the membrane to new locations, but do not flip-flop or rotate through the lipid layer. Carbohydrates often are attached to the outer surface of plasma membrane proteins, where they have important functions.   Proteins (appendix I)

Bacterial Membranes

Bacterial membranes are similar to eucaryotic membranes in that many of their amphipathic lipids are phospholipids (figure 3.6), but they usually differ from eucaryotic membranes in lacking sterols (steroid-containing lipids) such as cholesterol (**figure 3.7a**). However, many bacterial membranes contain sterol-like molecules called **hopanoids** (figure 3.7b). Hopanoids are synthesized from the same precursors as steroids, and like the sterols in eucaryotic membranes, they probably stabilize the membrane. Hopanoids are also of interest to ecologists and geologists: it has been estimated that the total mass of hopanoids in sediments is around 10^{11-12} tons—about as much as the total mass of organic carbon in all living organisms (10^{12} tons)—and there is evidence that hopanoids have contributed significantly to the formation of petroleum.

The emerging picture of bacterial plasma membranes is one of a highly organized and asymmetric system that also is flexible and dynamic. Numerous studies have demonstrated that lipids are not homogeneously distributed in the plasma membrane. Rather, there are domains in which particular lipids are concentrated. It has also been demonstrated that the lipid composition of bacterial membranes varies with environmental temperature in such a way that the membrane remains fluid during growth. For example, bacteria growing at lower temperatures will have fatty acids with lower melting points in their membrane phospholipids. The influence of environmental factors on growth: Temperature (section 6.5)

Although procaryotes do not contain complex membranous organelles like mitochondria or chloroplasts, internal membranous structures can be observed in some bacteria (**figure 3.8**).

(a) Cholesterol (a steroid) is found in eucaryotes

(b) A bacteriohopanetetrol (a hopanoid), as found in bacteria

Figure 3.7   Membrane Steroids and Hopanoids.   Common examples.

Plasma membrane infoldings are common in many bacteria and can become extensive and complex in photosynthetic bacteria such as the cyanobacteria and purple bacteria or in bacteria with very high respiratory activity, like the nitrifying bacteria. These internal membranous structures may be aggregates of spherical vesicles, flattened vesicles, or tubular membranes. Their function may be to provide a larger membrane surface for greater metabolic activity. One membranous structure sometimes reported in bacteria is the mesosome. Mesosomes appear to be invaginations of the plasma membrane in the shape of vesicles, tubules, or lamellae. Although a variety of functions have been ascribed to

(a)

(b)

Figure 3.8   Internal Bacterial Membranes.   Membranes of nitrifying and photosynthetic bacteria. **(a)** *Nitrocystis oceanus* with parallel membranes traversing the whole cell. Note nucleoplasm (n) with fibrillar structure. **(b)** *Ectothiorhodospira mobilis* with an extensive intracytoplasmic membrane system ($\times 60,000$).

mesosomes, many bacteriologists believe that they are artifacts generated during the chemical fixation of bacteria for electron microscopy.

Archaeal Membranes

One of the most distinctive features of the *Archaea* is the nature of their membrane lipids. They differ from both *Bacteria* and *Eucarya* in having branched chain hydrocarbons attached to glycerol by ether links rather than fatty acids connected by ester links (**figure 3.9**). Sometimes two glycerol groups are linked to form an extremely long tetraether. Usually the diether hydrocarbon chains are 20 carbons in length, and the tetraether chains are 40 carbons. Cells can adjust the overall length of the tetraethers by cyclizing the chains to form pentacyclic rings (figure 3.9).

Phosphate-, sulfur- and sugar-containing groups can be attached to the third carbons of the diethers and tetraethers, making them polar lipids. These predominate in the membrane, and 70 to 93% of the membrane lipids are polar. The remaining lipids are nonpolar and are usually derivatives of squalene (**figure 3.10**).

Despite these significant differences in membrane lipids, the basic design of archaeal membranes is similar to that of *Bacteria* and eucaryotes—there are two hydrophilic surfaces and a hydrophobic core. When C_{20} diethers are used, a regular bilayer membrane is formed (**figure 3.11a**). When the membrane is constructed of C_{40} tetraethers, a monolayer membrane with much more rigidity is formed (figure 3.11b). As might be expected from

Figure 3.9 Archaeal Membrane Lipids. An illustration of the difference between archaeal lipids and those of *Bacteria*. Archaeal lipids are derivatives of isopranyl glycerol ethers rather than the glycerol fatty acid esters in *Bacteria*. Three examples of common archaeal glycerolipids are given.

Figure 3.10 Nonpolar Lipids of *Archaea*. Two examples of the most predominant nonpolar lipids are the C_{30} isoprenoid squalene and one of its hydroisoprenoid derivatives, tetrahydrosqualene.

Figure 3.11 Examples of Archaeal Membranes. **(a)** A membrane composed of integral proteins and a bilayer of C_{20} diethers. **(b)** A rigid monolayer composed of integral proteins and C_{40} tetraethers.

their need for stability, the membranes of extreme thermophiles such as *Thermoplasma* and *Sulfolobus,* which grow best at temperatures over 85°C, are almost completely tetraether monolayers. Archaea that live in moderately hot environments have a mixed membrane containing some regions with monolayers and some with bilayers. Phylum *Euryarchaeota:* Thermoplasms (section 20.3); Phylum *Crenarchaeota: Sulfolobus* (section 20.2)

---

1. List the functions of the procaryotic plasma membrane.
2. Describe in words and with a labeled diagram the fluid mosaic model for cell membranes.
3. Compare and contrast bacterial and archaeal membranes.
4. Discuss the ways bacteria and archaea adjust the lipid content of their membranes in response to environmental conditions.

3.3   THE CYTOPLASMIC MATRIX

The **cytoplasmic matrix** is the substance in which the nucleoid, ribosomes, and inclusion bodies are suspended. It lacks organelles bound by lipid bilayers (often called unit membranes), and is largely water (about 70% of bacterial mass is water). Until recently, it was thought to lack a cytoskeleton. The plasma membrane and everything within is called the **protoplast;** thus the cytoplasmic matrix is a major part of the protoplast.

The Procaryotic Cytoskeleton

When examined with the electron microscope, the cytoplasmic matrix of procaryotes is packed with ribosomes. For many years it was thought that procaryotes lacked the high level of cytoplasmic organization present in eucaryotic cells because they lacked a cytoskeleton. Recently homologs of all three eucaryotic cytoskeletal elements (microfilaments, intermediate filaments, and microtubules) have been identified in bacteria, and one has been identified in archaea (**table 3.2**). The cytoskeletal filaments of procaryotes are structurally similar to their eucaryotic counterparts and carry out similar functions: they participate in cell division, localize proteins to certain sites in the cell, and determine cell shape (table 3.2 and **figure 3.12**). The procaryotic cell cycle: Cytokinesis (section 6.1)

Inclusion Bodies

Inclusion bodies, granules of organic or inorganic material that often are clearly visible in a light microscope, are present in the cytoplasmic matrix. These bodies usually are used for storage (e.g., carbon compounds, inorganic substances, and energy), and also reduce osmotic pressure by tying up molecules in particulate form. Some inclusion bodies lie free in the cytoplasm—for example, polyphosphate granules, cyanophycin granules, and some glycogen granules. Other inclusion bodies are enclosed by a shell about 2.0 to 4.0 nm thick, which is single-layered and may consist of proteins or a membranous structure composed of proteins and phospholipids. Examples of enclosed inclusion bodies are poly-β-hydroxybutyrate granules, some glycogen and sulfur granules, carboxysomes, and gas vacuoles. Many inclusion bodies are used for storage; their quantity will vary with the nutritional status of the cell. For example, polyphosphate granules will

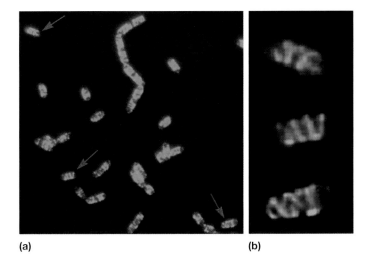

(a)   (b)

Figure 3.12   The Procaryotic Cytoskeleton.   Visualization of the MreB-like cytoskeletal protein (Mbl) of *Bacillus subtilis.* The Mbl protein has been fused with green fluorescent protein and live cells have been examined by fluorescence microscopy. **(a)** Arrows point to the helical cytoskeletal cables that extend the length of the cells. **(b)** Three of the cells from (a) are shown at a higher magnification.

Table 3.2	Procaryotic Cytoskeletal Proteins		
Procaryotic Protein (Eucaryotic Counterpart)	**Function**		**Comments**
FtsZ (tubulin)	Cell division		Widely observed in *Bacteria* and *Archaea*
MreB (actin)	Cell shape		Observed in many rod-shaped bacteria; in *Bacillus subtilis* is called Mbl
Crescentin (intermediate filament proteins)	Cell shape		Discovered in *Caulobacter crescentus*

be depleted in freshwater habitats that are phosphate limited. A brief description of several important inclusion bodies follows.

Organic inclusion bodies usually contain either glycogen or poly-β-hydroxyalkanoates (e.g., poly-β-hydroxybutyrate). **Glycogen** is a polymer of glucose units composed of long chains formed by $\alpha(1\rightarrow4)$ glycosidic bonds and branching chains connected to them by $\alpha(1\rightarrow6)$ glycosidic bonds. **Poly-β-hydroxybutyrate (PHB)** contains β-hydroxybutyrate molecules joined by ester bonds between the carboxyl and hydroxyl groups of adjacent molecules. Usually only one of these polymers is found in a species, but some photosynthetic bacteria have both glycogen and PHB. Poly-β-hydroxybutyrate accumulates in distinct bodies, around 0.2 to 0.7 μm in diameter, that are readily stained with Sudan black for light microscopy and are seen as empty "holes" in the electron microscope (**figure 3.13a**). This is because the solvents used to prepare specimens for electron microscopy dissolve these hydrophobic inclusion bodies. Glycogen is dispersed more evenly throughout the matrix as small granules (about 20 to 100 nm in diameter) and often can be seen only with the electron microscope. If cells contain a large amount of glycogen, staining with an iodine solution will turn them reddish-brown. Glycogen and PHB inclusion bodies are carbon storage reservoirs providing material for energy and biosynthesis. Many bacteria also store carbon as lipid droplets. Carbohydrates (appendix I)

Cyanobacteria, a group of photosynthetic bacteria, have two distinctive organic inclusion bodies. **Cyanophycin granules** (figure 3.13b) are composed of large polypeptides containing approximately equal amounts of the amino acids arginine and aspartic acid. The granules often are large enough to be visible in the light microscope and store extra nitrogen for the bacteria. **Carboxysomes** are present in many cyanobacteria and other CO_2-fixing bacteria. They are polyhedral, about 100 nm in diameter, and contain the enzyme ribulose-1, 5-bisphosphate carboxylase, called Rubisco. Rubisco is the critical enzyme for CO_2 fixation, the process of converting CO_2 from the atmosphere into sugar. The enzyme assumes a paracrystalline arrangement in the carboxysome, which serves as a reserve of the enzyme. Carboxysomes also may be a site of CO_2 fixation. The fixation of CO_2 by autotrophs (section 10.3)

(a)

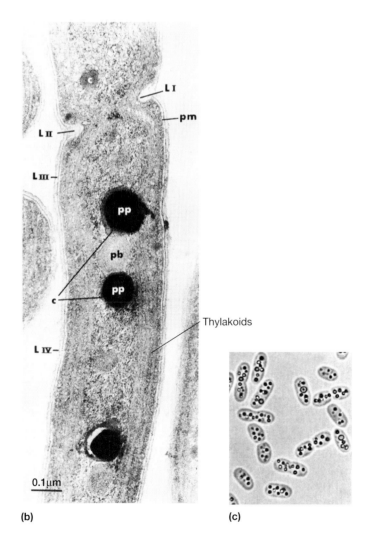

(b)   (c)

Figure 3.13   Inclusion Bodies in Bacteria.   (a) Electron micrograph of *Bacillus megaterium* ($\times30{,}500$). Poly-β-hydroxybutyrate inclusion body, PHB; cell wall, CW; nucleoid, N; plasma membrane, PM; "mesosome," M; and ribosomes, R. **(b)** Ultrastructure of the cyanobacterium *Anacystis nidulans*. The bacterium is dividing and a septum is partially formed, LI and LII. Several structural features can be seen, including cell wall layers, LIII and LIV; polyphosphate granules, pp; a polyhedral body, pb; cyanophycin material, c; and plasma membrane, pm. Thylakoids run along the length of the cell. **(c)** *Chromatium vinosum*, a purple sulfur bacterium, with intracellular sulfur granules, bright-field microscopy($\times2{,}000$).

A most remarkable organic inclusion body is the **gas vacuole,** a structure that provides buoyancy to some aquatic procaryotes. Gas vacuoles are present in many photosynthetic bacteria and a few other aquatic procaryotes such as *Halobacterium* (a salt-loving archaeon) and *Thiothrix* (a filamentous bacterium). Gas vacuoles are aggregates of enormous numbers of small, hollow, cylindrical structures called **gas vesicles** (**figure 3.14**). Gas vesicle walls are composed entirely of a single small protein. These protein subunits assemble to form a rigid enclosed cylinder that is hollow and impermeable to water but freely permeable to atmospheric gases. Procaryotes with gas vacuoles can regulate their buoyancy to float at the depth necessary for proper light intensity, oxygen concentration, and nutrient levels. They descend by simply collapsing vesicles and float upward when new ones are constructed.

Two major types of inorganic inclusion bodies are seen in procaryotes: polyphosphate granules and sulfur granules. Many bacteria store phosphate as **polyphosphate granules** or **volutin granules** (figure 3.13*b*). Polyphosphate is a linear polymer of orthophosphates joined by ester bonds. Thus volutin granules function as storage reservoirs for phosphate, an important component of cell constituents such as nucleic acids. In some cells they act as an energy reserve, and polyphosphate can serve as an energy source in reactions. These granules are sometimes called **metachromatic granules** because they show the metachromatic effect; that is, they appear red or a different shade of blue when stained with the blue dyes methylene blue or toluidine blue. Sulfur granules are used by some procaryotes to store sulfur temporarily (figure 3.13*c*). For example, photosynthetic bacteria can use hydrogen sulfide as a photosynthetic electron donor and accumulate the resulting sulfur in either the periplasmic space or in special cytoplasmic globules. Phototrophy: The light reaction in anoxygenic photosynthesis (section 9.12)

Inorganic inclusion bodies can be used for purposes other than storage. An excellent example is the **magnetosome,** which is used by some bacteria to orient in the Earth's magnetic field. Many of these inclusion bodies contain iron in the form of magnetite (**Microbial Diversity & Ecology 3.2**).

Ribosomes

As mentioned earlier, the cytoplasmic matrix often is packed with **ribosomes;** they also may be loosely attached to the plasma membrane. Ribosomes are very complex structures made of both protein and ribonucleic acid (RNA). They are the site of protein synthesis; cytoplasmic ribosomes synthesize proteins destined to remain within the cell, whereas plasma membrane ribosomes make proteins for transport to the outside. The newly formed polypeptide folds into its final shape either as it is synthesized by the ribosome or shortly after completion of protein synthesis. The shape of each protein is determined by its amino acid sequence. Special proteins called molecular chaperones, or simply chaperones, aid the polypeptide in folding to its proper shape. Protein synthesis, including a detailed treatment of ribosomes and chaperones, is discussed at considerable length in chapter 11.

Procaryotic ribosomes are smaller than the cytoplasmic or endoplasmic reticulum-associated ribosomes of eucaryotic cells. Procaryotic ribosomes are called 70S ribosomes (as opposed to 80S in eucaryotes), have dimensions of about 14 to 15 nm by 20 nm, a molecular weight of approximately 2.7 million, and are constructed of a 50S and a 30S subunit (**figure 3.15**). The S in 70S and similar values stands for **Svedberg unit.** This is the unit of the sedimentation coefficient, a measure of the sedimentation velocity in a centrifuge; the faster a particle travels when centrifuged, the greater its Svedberg value or sedimentation coefficient. The sedimentation coefficient is a function of a particle's molecular weight, volume, and shape (*see figure 16.19*). Heavier and more compact particles normally have larger Svedberg numbers or sediment faster.

Figure 3.14 Gas Vesicles and Vacuoles. A freeze-fracture preparation of *Anabaena flosaquae* (×89,000). Clusters of the cigar-shaped vesicles form gas vacuoles. Both longitudinal and cross-sectional views of gas vesicles can be seen (arrows).

1. Briefly describe the nature and function of the cytoplasmic matrix.
2. List and describe the functions of cytoskeletal proteins, inclusion bodies, and ribosomes.
3. List the most common kinds of inclusion bodies.
4. Relate the structure of a gas vacuole to its function.

Microbial Diversity & Ecology

3.2   Living Magnets

Bacteria can respond to environmental factors other than chemicals. A fascinating example is that of the aquatic magnetotactic bacteria that orient themselves in the Earth's magnetic field. Most of these bacteria have intracellular chains of magnetite (Fe_3O_4) particles that are called magnetosomes. Magnetosomes are around 35 to 125 nm in diameter and are bounded by a lipid bilayer (see **Box figure**). Some species from sulfidic habitats have magnetosomes containing greigite (Fe_3S_4) and pyrite (FeS_2). Since each iron particle is a tiny magnet, the Northern Hemisphere bacteria use their magnetosome chain to determine northward and downward directions, and swim down to nutrient-rich sediments or locate the optimum depth in freshwater and marine habitats. Magnetotactic bacteria in the Southern Hemisphere generally orient southward and downward, with the same result. Magnetosomes also are present in the heads of birds, tuna, dolphins, green turtles, and other animals, presumably to aid navigation. Animals and bacteria share more in common behaviorally than previously imagined.

(a)

(b)

(c)

Magnetotactic Bacteria. (**a**) Transmission electron micrograph of the magnetotactic bacterium *Aquaspirillum magnetotacticum* (×123,000). Note the long chain of electron-dense magnetite particles, MP. Other structures: OM, outer membrane; P, periplasmic space; CM, cytoplasmic membrane. (**b**) Isolated magnetosomes (×140,000). (**c**) Bacteria migrating in waves when exposed to a magnetic field.

30S subunit   50S subunit

Figure 3.15   Procaryotic Ribosome.   The two subunits of a bacterial ribosome are shown. The 50S subunit includes 23S rRNA (gray) and 5S rRNA (light blue), while 16S rRNA (cyan) is found in the 30S subunit. A molecule of tRNA (gold) is shown in the A site. To generate this ribbon diagram, crystals of purified bacterial ribosomes were grown, exposed to X rays, and the resulting diffraction pattern analyzed.

3.4   THE NUCLEOID

Probably the most striking difference between procaryotes and eucaryotes is the way in which their genetic material is packaged. Eucaryotic cells have two or more chromosomes contained within a membrane-delimited organelle, the nucleus. In contrast, procaryotes lack a membrane-delimited nucleus. The procaryotic chromosome is located in an irregularly shaped region called the **nucleoid** (other names are also used: the nuclear body, chromatin body, nuclear region) (**figure 3.16**). Usually procaryotes contain a single circle of double-stranded **deoxyribonucleic acid (DNA),** but some have a linear DNA chromosome and some, such as *Vibrio cholerae* and *Borrelia burgdorferi* (the causative agents of cholera and Lyme disease, respectively), have more than one chromosome.

Both electron and light microscopic studies have been important for understanding nucleoid structure and function, especially during active cell growth and division. The nucleoid has a fibrous appearance in electron micrographs; the fibers are probably DNA. In actively growing cells, the nucleoid has projections that extend into the cytoplasmic matrix. Presumably these projections contain DNA that is being actively transcribed to produce mRNA. Other studies have shown that more than one nucleoid can be observed within a single cell when genetic material has been duplicated but cell division has not yet occurred (figure 3.16*a*).

(a)

0.5 μm

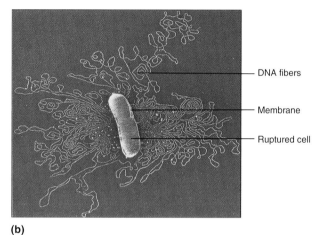

DNA fibers

Membrane

Ruptured cell

(b)

Figure 3.16   Procaryotic Nucleoids and Chromosomes. Procaryotic chromosomes are located in the nucleoid, an area in the cytoplasm. **(a)** A color-enhanced transmission electron micrograph of a thin section of a dividing *E. coli* cell. The red areas are the nucleoids present in the two daughter cells. **(b)** Chromosome released from a gently lysed *E. coli* cell. Note how tightly packaged the DNA must be inside the cell.

It is possible to isolate pure nucleoids. Chemical analysis of purified nucleoids reveals that they are composed of about 60% DNA, 30% RNA, and 10% protein by weight. In *Escherichia coli,* the closed DNA circle measures approximately 1,400 μm or about 230–700 times longer than the cell (figure 3.16*b*). Obviously it must be very efficiently packaged to fit within the nucleoid. The DNA is looped and coiled extensively (*see figure 11.8*), probably with the aid of RNA and a variety of nucleoid proteins. These include condensing proteins, which are conserved in both *Bacteria* and *Archaea*. Unlike the eucaryotes and some archaea, *Bacteria* do not use histone proteins to package their DNA.

There are a few exceptions to the preceding picture. Membrane-bound DNA-containing regions are present in two genera of the unusual bacterial phylum *Planctomycetes* (*see figure 21.12*). *Pirellula* has a single membrane that surrounds a region, the pirellulosome, which contains a fibrillar nucleoid and ribosome-like particles. The nuclear body of *Gemmata obscuriglobus* is bounded by two membranes. More work will be required to determine the functions of these membranes and how widespread this phenomenon is. Phylum *Planctomycetes* (section 21.4)

---

1. Describe the structure and function of the nucleoid and the DNA it contains.
2. List three genera that are exceptional in terms of their chromosome or nucleoid structure. Suggest how the differences observed in these genera might impact how they function.

3.5 PLASMIDS

In addition to the genetic material present in the nucleoid, many procaryotes (and some yeasts and other fungi) contain extrachromosomal DNA molecules called plasmids. Indeed, most of the bacterial and archaeal genomes sequenced thus far include plasmids. In some cases, numerous different plasmids within a single species have been identified. For instance, *B. burgdorferi,* carries 12 linear and 9 circular plasmids. Plasmids play many important roles in the lives of the organisms that have them. They also have proved invaluable to microbiologists and molecular geneticists in constructing and transferring new genetic combinations and in cloning genes, as described in chapter 14. This section discusses the different types of procaryotic plasmids.

Plasmids are small, double-stranded DNA molecules that can exist independently of the chromosome. Both circular and linear plasmids have been documented, but most known plasmids are circular. Linear plasmids possess special structures or sequences at their ends to prevent their degradation and to permit their replication. Plasmids have relatively few genes, generally less than 30. Their genetic information is not essential to the host, and cells that lack them usually function normally. However, many plasmids carry genes that confer a selective advantage to their hosts in certain environments.

Plasmids are able to replicate autonomously. Single-copy plasmids produce only one copy per host cell. Multicopy plasmids may be present at concentrations of 40 or more per cell. Some plasmids are able to integrate into the chromosome and are thus replicated with the chromosome. Such plasmids are called **episomes.** Plasmids are inherited stably during cell division, but they are not always equally apportioned into daughter cells and sometimes are lost. The loss of a plasmid is called **curing.** It can occur spontaneously or be induced by treatments that inhibit plasmid replication but not host cell reproduction. Some commonly used curing treatments are acridine mutagens, UV and ionizing radiation, thymine starvation, antibiotics, and growth above optimal temperatures.

Plasmids may be classified in terms of their mode of existence, spread, and function. A brief summary of the types of bacterial plasmids and their properties is given in **table 3.3. Conjugative plasmids** are of particular note. They have genes for the construction of hairlike structures called pili and can transfer copies of themselves to other bacteria during conjugation. Perhaps the best-studied conjugative plasmid is the **F factor** (fertility factor or F plasmid) of *E. coli,* which was the first conjugative factor to be described. The F factor contains genes that direct the formation of sex pili that attach an F$^+$ cell (a cell containing an F plasmid) to an F$^-$ cell (a cell lacking an F plasmid). Other plasmid-encoded gene products aid DNA transfer from the F$^+$ cell to the F$^-$ cell. The F factor also has several segments called insertion sequences that enable it to integrate into the host cell chromosome. Thus the F factor is an episome. Transposable elements (section 13.5); Bacterial conjugation (section 13.7)

Resistance factors (R factors, R plasmids) are another group of important plasmids. They confer antibiotic resistance on the cells that contain them. R factors typically have genes that code for enzymes capable of destroying or modifying antibiotics. Some R plasmids have only a single resistance gene, whereas others have as many as eight. Often the resistance genes are within mobile genetic elements called transposons, and thus it is possible for multiple-resistance plasmids to evolve. R factors usually are not integrated into the host chromosome.

R factors are of major concern to public health officials because they can spread rapidly throughout a population of cells. This is possible for several reasons. One is that many R factors also are conjugative plasmids. However, a nonconjugative R factor can be spread to other cells if it is present in a cell that also contains a conjugative plasmid. In such a cell, the R factor can sometimes be transferred when the conjugative plasmid is transferred—that is, it is "mobilized." Even more troubling is the fact that some R factors are readily transferred *between* species. When humans and other animals consume antibiotics, the growth of host bacteria with R factors is promoted. The R factors then can be transferred to more pathogenic genera such as *Salmonella* or *Shigella,* causing even greater public health problems. Drug resistance (section 34.6)

Several other important types of plasmids have been discovered. These include bacteriocin-encoding plasmids, virulence plasmids, and metabolic plasmids. Bacteriocin-encoding plasmids may give the bacteria that harbor them a competitive advantage in the microbial world. **Bacteriocins** are bacterial proteins that destroy other bacteria. They usually act only against closely related strains. Some bacteriocins kill cells by forming channels in the plasma membrane, thus breaching the critical selective permeability required for cell viability. They also may degrade DNA and RNA or attack peptidoglycan and weaken the cell wall. **Col plasmids** contain genes for the synthesis of bacteriocins known as colicins, which are directed against *E. coli.* Other plasmids carry genes for bacteriocins against other species. For example, cloacins kill *Enterobacter* species. Some Col plasmids are conjugative and carry resistance genes. It should be noted that not all bacteriocin genes are on plasmids. For example, the bacteriocin genes of *Pseudomonas aeruginosa,* which code for proteins called pyocins, are located on the chromosome. Bacteriocins produced by the normal flora of humans (and other animals) also are

Figure 3.18 Peptidoglycan Subunit Composition. The peptidoglycan subunit of *E. coli,* most other gram-negative bacteria, and many gram-positive bacteria. NAG is *N*-acetylglucosamine. NAM is *N*-acetylmuramic acid (NAG with lactic acid attached by an ether linkage). The tetrapeptide side chain is composed of alternating D- and L-amino acids since *meso*-diaminopimelic acid is connected through its L-carbon. NAM and the tetrapeptide chain attached to it are shown in different shades of color for clarity.

Figure 3.19 Diaminoacids Present in Peptidoglycan. (a) L-Lysine. (b) *meso*-Diaminopimelic acid.

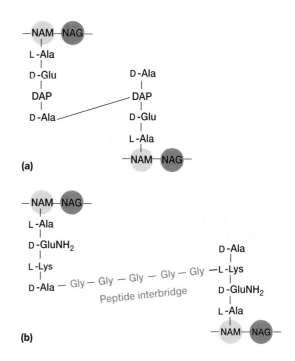

Figure 3.20 Peptidoglycan Cross-Links. (a) *E. coli* peptidoglycan with direct cross-linking, typical of many gram-negative bacteria. (b) *Staphylococcus aureus* peptidoglycan. *S. aureus* is a gram-positive bacterium. NAM is *N*-acetylmuramic acid. NAG is *N*-acetylglucosamine. Gly is glycine. Although the polysaccharide chains are drawn opposite each other for the sake of clarity, two chains lying side-by-side may be linked together (see figure 3.21).

polymer is composed of alternating *N*-acetylglucosamine and *N*-acetylmuramic acid residues. A peptide chain of four alternating D- and L-amino acids is connected to the carboxyl group of *N*-acetylmuramic acid. Many bacteria replace *meso*-diaminopimelic acid with another diaminoacid, usually L-lysine (**figure 3.19**). Carbohydrates (appendix I); Peptidoglycan and endospore structure (section 23.3); Proteins (appendix I)

In order to make a strong, meshlike polymer, chains of linked peptidoglycan subunits must be joined by cross-links between the peptides. Often the carboxyl group of the terminal D-alanine is connected directly to the amino group of diaminopimelic acid,

but a **peptide interbridge** may be used instead (**figure 3.20**). Most gram-negative cell wall peptidoglycan lacks the peptide interbridge. This cross-linking results in an enormous peptidoglycan sac that is actually one dense, interconnected network (**figure 3.21**). These sacs have been isolated from gram-positive bacteria and are strong enough to retain their shape and integrity (**figure 3.22**), yet they are relatively porous, elastic, and somewhat stretchable.

Gram-Positive Cell Walls

Gram-positive bacteria normally have cell walls that are thick and composed primarily of peptidoglycan. Peptidoglycan in gram-positive bacteria often contains a peptide interbridge (figure 3.21 and **figure 3.23**). In addition, gram-positive cell walls usually contain large amounts of **teichoic acids,** polymers of glycerol or ribitol joined by phosphate groups (figure 3.23 and **figure 3.24**). Amino acids such as D-alanine or sugars like glucose are attached to the glycerol and ribitol groups. The teichoic acids are covalently connected to either the peptidoglycan itself or to plasma membrane lipids; in the latter case they are called lipoteichoic acids. Teichoic acids appear to extend to the surface of the peptidoglycan, and, because they are negatively charged, help give the gram-positive cell

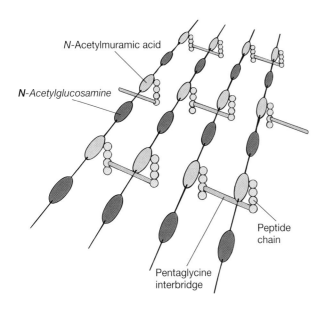

Figure 3.21   Peptidoglycan Structure.   A schematic diagram of one model of peptidoglycan. Shown are the polysaccharide chains, tetrapeptide side chains, and peptide interbridges.

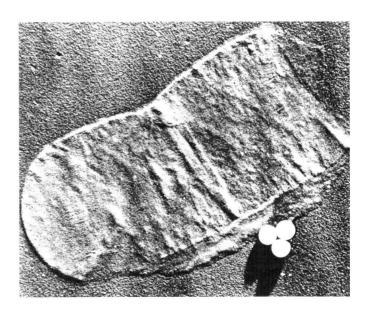

Figure 3.22   Isolated Gram-Positive Cell Wall.   The peptidoglycan wall from *Bacillus megaterium,* a gram-positive bacterium. The latex spheres have a diameter of 0.25 μm.

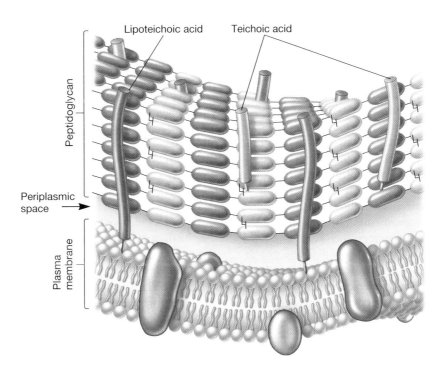

Figure 3.23   The Gram-Positive Envelope.

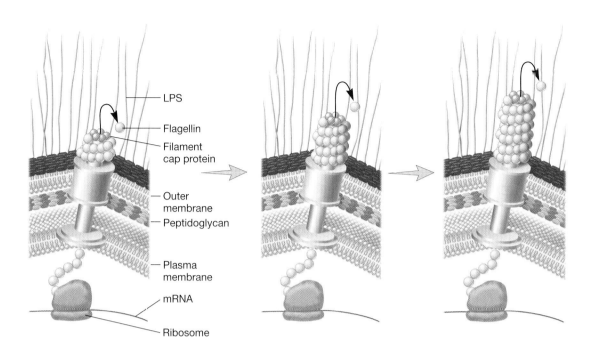

Figure 3.40   Growth of Flagellar Filaments.   Flagellin subunits travel through the flagellar core and attach to the growing tip. Their attachment is directed by the filament cap protein.

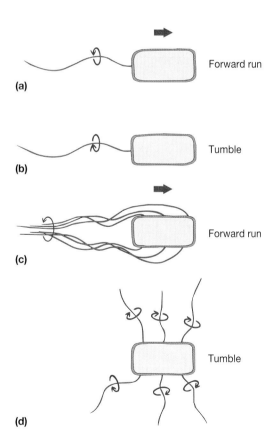

Figure 3.41   Flagellar Motility.   The relationship of flagellar rotation to bacterial movement. Parts **(a)** and **(b)** describe the motion of monotrichous, polar bacteria. Parts **(c)** and **(d)** illustrate the movements of peritrichous organisms.

must bore through the water with its corkscrew-shaped flagella, and if flagellar activity ceases, it stops almost instantly. Despite such environmental resistance to movement, bacteria can swim from 20 to almost 90 μm/second. This is equivalent to traveling from 2 to over 100 cell lengths per second. In contrast, an exceptionally fast 6-ft human might be able to run around 5 body lengths per second.

Bacteria can move by mechanisms other than flagellar rotation. Spirochetes are helical bacteria that travel through viscous substances such as mucus or mud by flexing and spinning movements caused by a special **axial filament** composed of periplasmic flagella. The swimming motility of the helical bacterium *Spiroplasma* is accomplished by the formation of kinks in the cell body that travel the length of the bacterium. A very different type of motility, **gliding motility,** is employed by many bacteria: cyanobacteria, myxobacteria and cytophagas, and some mycoplasmas. Although there are no visible external structures associated with gliding motility, it enables movement along solid surfaces at rates up to 3 μm/second.   Microbial Diversity & Ecology 21.1: The mechanism of gliding motility; Phylum *Spirochaetes* (section 21.6); Photosynthetic bacteria (section 21.3); Class *Deltaproteobacteria: Order Myxococcales* (section 22.4); Class *Mollicutes* (the Mycoplasmas) (section 23.2)

1. Briefly describe capsules, slime layers, glycocalyxes, and S-layers. What are their functions?
2. Distinguish between fimbriae and sex pili, and give the function of each.
3. Be able to discuss the following: flagella distribution patterns, flagella structure and synthesis, and the way in which flagella operate to move a bacterium.
4. What is self-assembly? Why does it make sense that the flagellar filament is assembled in this way?

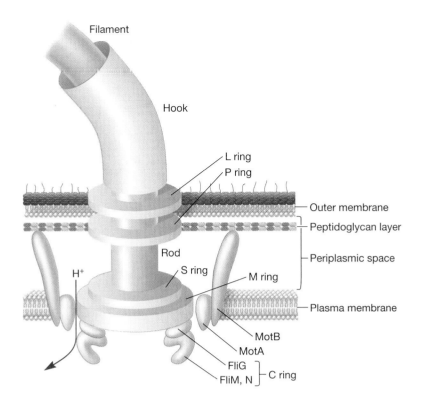

Filament

Hook

L ring

P ring

Outer membrane

Peptidoglycan layer

Periplasmic space

Rod

S ring

M ring

H⁺

Plasma membrane

MotB

MotA

FliG

FliM, N ─ C ring

Figure 3.42   Mechanism of Flagellar Movement.
This diagram of a gram-negative flagellum shows some of the more important components and the flow of protons that drives rotation. Five of the many flagellar proteins are labeled (MotA, MotB, FliG, FliM, FliN).

3.10   CHEMOTAXIS

Bacteria do not always move aimlessly but are attracted by such nutrients as sugars and amino acids, and are repelled by many harmful substances and bacterial waste products. Bacteria also can respond to other environmental cues such as temperature (thermotaxis), light (phototaxis), oxygen (aerotaxis), osmotic pressure (osmotaxis), and gravity; (Microbial Diversity & Ecology 3.2.) Movement toward chemical attractants and away from repellents is known as **chemotaxis.** Such behavior is of obvious advantage to bacteria.

Chemotaxis may be demonstrated by observing bacteria in the chemical gradient produced when a thin capillary tube is filled with an attractant and lowered into a bacterial suspension. As the attractant diffuses from the end of the capillary, bacteria collect and swim up the tube. The number of bacteria within the capillary after a short length of time reflects the strength of attraction and rate of chemotaxis. Positive and negative chemotaxis also can be studied with petri dish cultures (**figure 3.43**). If bacteria are placed in the center of a dish of semisolid agar containing an attractant, the bacteria will exhaust the local supply and then swim outward following the attractant gradient they have created. The result is an expanding ring of bacteria. When a disk of repellent is placed in a petri dish of semisolid agar and bacteria, the bacteria will swim away from the repellent, creating a clear zone around the disk (**figure 3.44**).

Bacteria can respond to very low levels of attractants (about 10^{-8} M for some sugars), the magnitude of their response increasing with attractant concentration. Usually they sense repellents only at higher concentrations. If an attractant and a repellent are present together, the bacterium will compare both signals and respond to the chemical with the most effective concentration.

Attractants and repellents are detected by **chemoreceptors,** special proteins that bind chemicals and transmit signals to the other components of the chemosensing system. About 20 attractant chemoreceptors and 10 chemoreceptors for repellents have been discovered thus far. These chemoreceptor proteins may be located in the periplasmic space or the plasma membrane. Some receptors participate in the initial stages of sugar transport into the cell.

The chemotactic behavior of bacteria has been studied using the tracking microscope, a microscope with a moving stage that automatically keeps an individual bacterium in view. In the absence of a chemical gradient, *E. coli* and other bacteria move randomly. For a few seconds, the bacterium will travel in a straight or slightly curved line called a **run.** When a bacterium is running, its flagella are organized into a coordinated, corkscrew-shaped bundle (figure 3.41c). Then the flagella "fly apart" and the bacterium will stop and **tumble.** The tumble results in the random reorientation of the bacterium so that it often is facing in a different direction. Therefore when it begins the next run, it usually goes in a different direction (**figure 3.45a**). In contrast, when the bacterium is exposed to an attractant, it tumbles less frequently (or has longer runs) when traveling towards the attractant. Although the tumbles can still orient the bacterium away from the attractant, over time, the bacterium gets closer and closer to the attractant (figure 3.45b). The opposite response occurs with a repellent. Tumbling frequency decreases (the run time lengthens) when the bacterium moves away from the repellent.

Figure 3.43 Positive Bacterial Chemotaxis. Chemotaxis can be demonstrated on an agar plate that contains various nutrients. Positive chemotaxis by *E. coli* on the left. The outer ring is composed of bacteria consuming serine. The second ring was formed by *E. coli* consuming aspartate, a less powerful attractant. The upper right colony is composed of motile, but nonchemotactic mutants. The bottom right colony is formed by nonmotile bacteria.

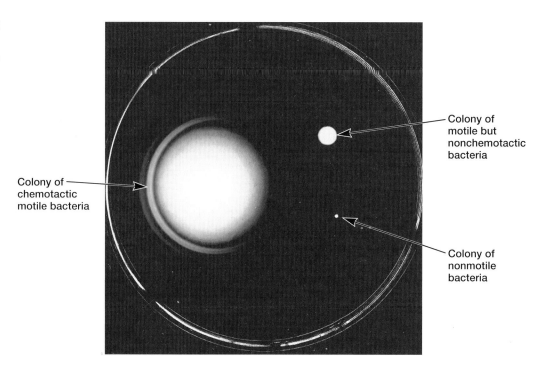

Colony of chemotactic motile bacteria

Colony of motile but nonchemotactic bacteria

Colony of nonmotile bacteria

Figure 3.44 Negative Bacterial Chemotaxis. Negative chemotaxis by *E. coli* in response to the repellent acetate. The bright disks are plugs of concentrated agar containing acetate that have been placed in dilute agar inoculated with *E. coli*. Acetate concentration increases from zero at the top right to 3 M at top left. Note the increasing size of bacteria-free zones with increasing acetate. The bacteria have migrated for 30 minutes.

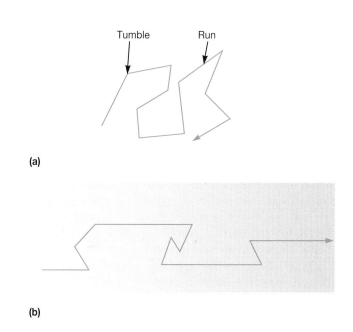

Tumble   Run

(a)

(b)

Figure 3.45 Directed Movement in Bacteria. **(a)** Random movement of a bacterium in the absence of a concentration gradient. Tumbling frequency is fairly constant. **(b)** Movement in an attractant gradient. Tumbling frequency is reduced when the bacterium is moving up the gradient. Therefore, runs in the direction of increasing attractant are longer.

Clearly, the bacterium must have some mechanism for sensing that it is getting closer to the attractant (or is moving away from the repellent). The behavior of the bacterium is shaped by temporal changes in chemical concentration. The bacterium moves toward the attractant because it senses that the concentration of the attractant is increasing. Likewise, it moves away from a repellent because it senses that the concentration of the repellent is decreasing. The bacterium's chemoreceptors play a critical role in this process. The molecular events that enable bacterial cells to sense a chemical gradient and respond appropriately are presented in chapter 8.

---

1. Define chemotaxis, run, and tumble.
2. Explain in a general way how bacteria move toward substances like nutrients and away from toxic materials.

3.11 THE BACTERIAL ENDOSPORE

A number of gram-positive bacteria can form a special resistant, dormant structure called an **endospore.** Endospores develop within vegetative bacterial cells of several genera: *Bacillus* and *Clostridium* (rods), *Sporosarcina* (cocci), and others. These structures are extraordinarily resistant to environmental stresses such as heat, ultraviolet radiation, gamma radiation, chemical disinfectants, and desiccation. In fact, some endospores have remained viable for around 100,000 years. Because of their resistance and the fact that several species of endospore-forming bacteria are dangerous pathogens, endospores are of great practical importance in food, industrial, and medical microbiology. This is because it is essential to be able to sterilize solutions and solid objects. Endospores often survive boiling for an hour or more; therefore autoclaves must be used to sterilize many materials. Endospores are also of considerable theoretical interest. Because bacteria manufacture these intricate structures in a very organized fashion over a period of a few hours, spore formation is well suited for research on the construction of complex biological structures. In the environment, endospores aid in survival when moisture or nutrients are scarce. The use of physical methods in control: Heat (section 7.4)

Endospores can be examined with both light and electron microscopes. Because endospores are impermeable to most stains, they often are seen as colorless areas in bacteria treated with methylene blue and other simple stains; special endospore stains are used to make them clearly visible. Endospore position in the mother cell (**sporangium**) frequently differs among species, making it of considerable value in identification. Endospores may be centrally located, close to one end (subterminal), or definitely terminal (**figure 3.46**). Sometimes an endospore is so large that it swells the sporangium. Preparation and staining of specimens (section 2.3)

Electron micrographs show that endospore structure is complex (**figure 3.47**). The spore often is surrounded by a thin, delicate covering called the **exosporium.** A **spore coat** lies beneath the exosporium, is composed of several protein layers, and may be fairly thick. It is impermeable to many toxic molecules and is responsible for the spore's resistance to chemicals. The coat also is thought to contain enzymes that are involved in germination.

The **cortex,** which may occupy as much as half the spore volume, rests beneath the spore coat. It is made of a peptidoglycan that is less cross-linked than that in vegetative cells. The **spore cell wall** (or core wall) is inside the cortex and surrounds the protoplast or **spore core.** The core has normal cell structures such as ribosomes and a nucleoid, but is metabolically inactive.

It is still not known precisely why the endospore is so resistant to heat and other lethal agents. As much as 15% of the spore's dry weight consists of dipicolinic acid complexed with calcium ions (**figure 3.48**), which is located in the core. It has long been thought that dipicolinic acid was directly involved in

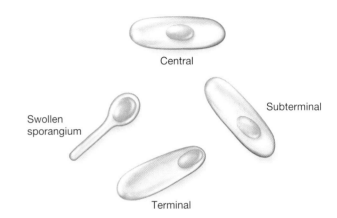

Central

Subterminal

Swollen sporangium

Terminal

Figure 3.46 Examples of Endospore Location and Size.

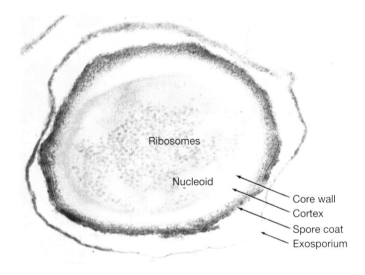

Ribosomes

Nucleoid

Core wall
Cortex
Spore coat
Exosporium

Figure 3.47 Endospore Structure. *Bacillus anthracis* endospore ($\times 151,000$).

HOOC COOH

Figure 3.48 Dipicolinic Acid.

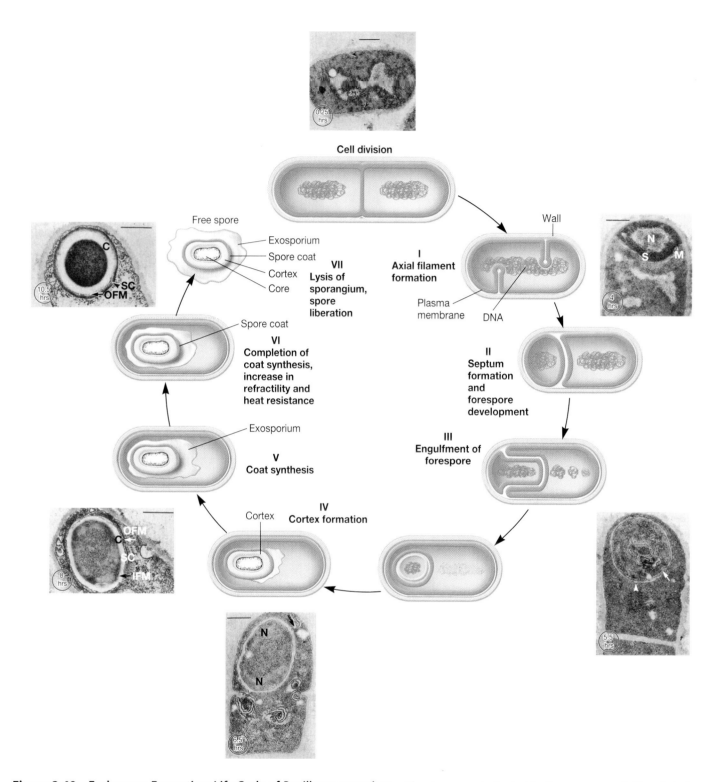

Figure 3.49   Endospore Formation: Life Cycle of *Bacillus megaterium*.   The stages are indicated by Roman numerals. The circled numbers in the photographs refer to the hours from the end of the logarithmic phase of growth: 0.25 h—a typical vegetative cell; 4 h–stage II cell, septation; 5.5 h–stage III cell, engulfment; 6.5 h–stage IV cell, cortex formation; 8 h–stage V cell, coat formation; 10.5 h–stage VI cell, mature spore in sporangium. Abbreviations used: C, cortex; IFM and OFM, inner and outer forespore membranes; M, mesosome; N, nucleoid; S, septum; SC, spore coats. Bars = 0.5 μm.

heat resistance, but heat-resistant mutants lacking dipicolinic acid have been isolated. Calcium does aid in resistance to wet heat, oxidizing agents, and sometimes dry heat. It may be that calcium-dipicolinate stabilizes the spore's nucleic acids. In addition, specialized *small, acid-soluble DNA-binding proteins* (SASPs), are found in the endospore. They saturate spore DNA and protect it from heat, radiation, dessication, and chemicals. Dehydration of the protoplast appears to be very important in heat resistance. The cortex may osmotically remove water from the protoplast, thereby protecting it from both heat and radiation damage. The spore coat also seems to protect against enzymes and chemicals such as hydrogen peroxide. Finally, spores contain some DNA repair enzymes. DNA is repaired once the spore germinates and the cell becomes active again. In summary, endospore heat resistance probably is due to several factors: calcium-dipicolinate and acid-soluble protein stabilization of DNA, protoplast dehydration, the spore coat, DNA repair, the greater stability of cell proteins in bacteria adapted to growth at high temperatures, and others.

Endospore formation, also called **sporogenesis** or **sporulation,** normally commences when growth ceases due to lack of nutrients. It is a complex process and may be divided into seven stages (**figure 3.49**). An axial filament of nuclear material forms (stage I), followed by an inward folding of the cell membrane to enclose part of the DNA and produce the forespore septum (stage II). The membrane continues to grow and engulfs the immature endospore in a second membrane (stage III). Next, cortex is laid down in the space between the two membranes, and both calcium and dipicolinic acid are accumulated (stage IV). Protein coats then are formed around the cortex (stage V), and maturation of the endospore occurs (stage VI). Finally, lytic enzymes destroy the sporangium releasing the spore (stage VII). Sporulation requires about 10 hours in *Bacillus megaterium.* Global regulatory systems: Sporulation in *Bacillus subtilis* (section 12.5)

The transformation of dormant spores into active vegetative cells seems almost as complex a process as sporogenesis. It occurs in three stages: (1) activation, (2) germination, and (3) outgrowth (**figure 3.50**). Often a spore will not germinate successfully, even in a nutrient-rich medium, unless it has been activated. **Activation**

0.5 µm

Figure 3.50   Endospore Germination.   *Clostridium pectinovorum* emerging from the spore during germination.

is a process that prepares spores for germination and usually results from treatments like heating. It is followed by **germination,** the breaking of the spore's dormant state. This process is characterized by spore swelling, rupture or absorption of the spore coat, loss of resistance to heat and other stresses, loss of refractility, release of spore components, and increase in metabolic activity. Many normal metabolites or nutrients (e.g., amino acids and sugars) can trigger germination after activation. Germination is followed by the third stage, **outgrowth.** The spore protoplast makes new components, emerges from the remains of the spore coat, and develops again into an active bacterium.

1. Describe the structure of the bacterial endospore using a labeled diagram.
2. Briefly describe endospore formation and germination. What is the importance of the endospore? What might account for its heat resistance?
3. How might one go about showing that a bacterium forms true endospores?
4. Why do you think dehydration of the protoplast is an important factor in the ability of endospores to resist environmental stress?

Summary

3.1   An Overview of Procaryotic Cell Structure

a. Procaryotes may be spherical (cocci), rod-shaped (bacilli), spiral, or filamentous; they may form buds and stalks; or they may even have no characteristic shape at all (pleomorphic) (**figure 3.1** and **3.2**).

b. Procaryotic cells can remain together after division to form pairs, chains, and clusters of various sizes and shapes.

c. Procaryotes are much simpler structurally than eucaryotes, but they do have unique structures. **Table 3.1** summarizes the major functions of procaryotic cell structures.

3.2   Procaryotic Cell Membranes

a. The plasma membrane fulfills many roles, including acting as a semipermeable barrier, carrying out respiration and photosynthesis, and detecting and responding to chemicals in the environment.

b. The fluid mosaic model proposes that cell membranes are lipid bilayers in which integral proteins are buried (**figure 3.5**). Peripheral proteins are loosely associated with the membrane.

c. Bacterial membranes are composed of phospholipids constructed of fatty acids connected to glycerol by ester linkages (**figure 3.6**). Bacterial membranes usually lack sterols, but often contain hopanoids (**figure 3.7**).

d. The plasma membrane of some bacteria invaginates to form simple membrane systems containing photosynthetic and respiratory assemblies. Other bacteria, like the cyanobacteria, have internal membranes (**figure 3.8**).

e. Archaeal membranes are composed of glycerol diether and diglycerol tetraether lipids (**figure 3.9**). Membranes composed of glycerol diether are lipid bilayers. Membranes composed of diglycerol tetraethers are lipid monolayers (**figure 3.11**). The overall structure of a monolayer membrane is similar to that of the bilayer membrane in that the membrane has a hydrophobic core and its surfaces are hydrophilic.

3.3   The Cytoplasmic Matrix

a. The cytoplasm of procaryotes contains proteins that are similar in structure and function to the cytoskeletal proteins observed in eucaryotes.

b. The cytoplasmic matrix of procaryotes contains inclusion bodies. Most are used for storage (glycogen inclusions, PHB inclusions, cyanophycin granules, carboxysomes, and polyphosphate granules), but others are used for other purposes (magnetosomes and gas vacuoles).

c. The cytoplasm of procaryotes is packed with 70S ribosomes (**figure 3.15**).

3.4   The Nucleoid

a. Procaryotic genetic material is located in an area called the nucleoid and is not usually enclosed by a membrane (**figure 3.16**).

b. In most procaryotes, the nucleoid contains a single chromosome. The chromsosome consists of a double-stranded, covalently closed, circular DNA molecule.

3.5   Plasmids

a. Plasmids are extrachromosomal DNA molecules. They are found in many procaryotes.

b. Although plasmids are not required for survival in most conditions, they can encode traits that confer selective advantage in some environments.

c. Episomes are plasmids that are able to exist freely in the cytoplasm or can be integrated into the chromosome.

d. Conjugative plasmids encode genes that promote their transfer from one cell to another.

e. Resistance factors are plasmids that have genes conferring resistance to antibiotics.

f. Col plasmids contain genes for the synthesis of colicins, proteins that kill *E. coli*. Other plasmids encode virulence factors or metabolic capabilities.

3.6   The Bacterial Cell Wall

a. The vast majority of procaryotes have a cell wall outside the plasma membrane to give them shape and protect them from osmotic stress.

b. Bacterial walls are chemically complex and usually contain peptidoglycan (**figures 3.17–3.21**).

c. Bacteria often are classified as either gram-positive or gram-negative based on differences in cell wall structure and their response to Gram staining.

d. Gram-positive walls have thick, homogeneous layers of peptidoglycan and teichoic acids (**figure 3.23**). Gram-negative bacteria have a thin peptidoglycan layer surrounded by a complex outer membrane containing lipopolysaccharides (LPSs) and other components (**figure 3.25**).

e. The mechanism of the Gram stain is thought to depend on the peptidoglycan, which binds crystal violet tightly, preventing the loss of crystal violet during the ethanol wash.

3.7   Archaeal Cell Walls

a. Archaeal cell walls do not contain peptidoglycan (**figure 3.30**).

b. Archaea exhibit great diversity in their cell wall make-up. Some archaeal cell walls are composed of heteropolysaccharides, some are composed of glycoprotein, and some are composed of protein.

3.8   Protein Secretion in Procaryotes

a. The Sec-dependent protein secretion pathway (**figure 3.32**) has been observed in all domains of life. It transports proteins across or into the cytoplasmic membrane.

b. Gram-negative bacteria have additional protein secretion systems that allow them to move proteins from the cytoplasm, across both the cytoplasmic and outer membranes, to the outside of the cell (**figure 3.33**). Some of these systems work with the Sec-dependent pathway to accomplish this (Type II, Type V, and usually Type IV). Some pathways function alone to move proteins across both membranes (Types I and III).

c. ABC transporters (Type I protein secretion system) are used by all procaryotes for protein translocation.

3.9   Components External to the Cell Wall

a. Capsules, slime layers, and glycocalyxes are layers of material lying outside the cell wall. They can protect procaryotes from certain environmental conditions, allow procaryotes to attach to surfaces, and protect pathogenic bacteria from host defenses (**figures 3.34** and **3.35**).

b. S-layers are observed in some bacteria and many archaea. They are composed of proteins or glycoprotein and have a characteristic geometric shape. In many archaea the S-layer serves as the cell wall (**figure 3.36**).

c. Pili and fimbriae are hairlike appendages. Fimbriae function primarily in attachment to surfaces, but some types of bacterial fimbriae are involved in a twitching motility. Sex pili participate in the transfer of DNA from one bacterium to another (**figure 3.37**).

d. Many procaryotes are motile, usually by means of threadlike, locomotory organelles called flagella (**figure 3.38**).

e. Bacterial species differ in the number and distribution of their flagella.

f. In bacteria, the flagellar filament is a rigid helix that rotates like a propeller to push the bacterium through water (**figure 3.41**).

3.10 Chemotaxis

a. Motile procaryotes can respond to gradients of attractants and repellents, a phenomenon known as chemotaxis.

b. A bacterium accomplishes movement toward an attractant by increasing the length of time it spends moving toward the attractant, shortening the time it spends tumbling. Conversely, a bacterium increases its run time when it moves away from a repellent.

3.11 The Bacterial Endospore

a. Some bacteria survive adverse environmental conditions by forming endospores, dormant structures resistant to heat, desiccation, and many chemicals (**figure 3.47**).

b. Both endospore formation and germination are complex processes that begin in response to certain environmental signals and involve numerous stages (**figures 3.49** and **3.50**).

Key Terms

ABC protein secretion pathway 65	fimbriae 66	mycelium 40	run 71
activation 75	flagellar filament 67	nucleoid 52	Sec-dependent pathway 63
amphipathic 45	flagellar hook 67	O antigen 60	self-assembly 68
amphitrichous 67	flagellin 67	O side chain 60	sex pili 67
axial filament 70	flagellum 67	outer membrane 55	signal peptide 63
bacillus 40	fluid mosaic model 44	outgrowth 75	S-layer 66
bacteriocin 53	gas vacuole 50	penicillin 61	slime layer 65
basal body 67	gas vesicles 50	peptide interbridge 56	spheroplast 61
capsule 65	germination 75	peptidoglycan 55	spirilla 40
carboxysomes 49	gliding motility 70	peripheral proteins 45	spirochete 40
cell envelope 55	glycocalyx 65	periplasm 55	sporangium 73
chemoreceptors 71	glycogen 49	periplasmic space 55	spore cell wall 73
chemotaxis 71	hopanoids 46	peritrichous 67	spore coat 73
coccus 39	hydrophilic 45	plasma membrane 42	spore core 73
Col plasmid 53	hydrophobic 45	plasmid 53	sporogenesis 75
conjugative plasmid 53	inclusion body 48	plasmolysis 61	sporulation 75
core polysaccharide 58	integral proteins 45	pleomorphic 41	Svedberg unit 50
cortex 73	lipid A 58	polar flagellum 67	teichoic acid 57
curing 53	lipopolysaccharides (LPSs) 58	poly-β-hydroxybutyrate (PHB) 49	tumble 71
cyanophycin granules 49	lophotrichous 67	polyphosphate granules 50	type I protein secretion pathway 65
cytoplasmic matrix 48	lysis 61	porin proteins 60	type II protein secretion pathway 65
deoxyribonucleic acid (DNA) 52	lysozyme 61	protoplast 48	type III protein secretion pathway 65
diplococcus 39	magnetosomes 50	pseudomurein 62	type IV protein secretion pathway 65
endospore 73	metabolic plasmid 54	resistance factor (R factor,	type V protein secretion pathway 65
episome 53	metachromatic granules 50	R plasmid) 53	vibrio 40
exoenzyme 58	monotrichous 67	ribosome 50	virulence plasmid 54
exosporium 73	murein 55	rod 40	volutin granules 50
F factor 53			

Critical Thinking Questions

1. Propose a model for the assembly of a flagellum in a gram-positive cell envelope. How would that model need to be modified for the assembly of a flagellum in a gram-negative cell envelope?

2. If you could not use a microscope, how would you determine whether a cell is procaryotic or eucaryotic? Assume the organism can be cultured easily in the laboratory.

3. The peptidoglycan of bacteria has been compared with the chain mail worn beneath a medieval knight's suit of armor. It provides both protection and flexibility. Can you describe other structures in biology that have an analogous function? How are they replaced or modified to accommodate the growth of the inhabitant?

Learn More

Cannon, G. C.; Bradburne, C. E.; Aldrich, H. C.; Baker, S. H.; Heinhorst, S.; and Shively, J. M. 2001. Microcompartments in prokaryotes: Carboxysomes and related polyhedra. *Appl. Env. Microbiol.* 67(12):5351–61.

Drews, G. 1992. Intracytoplasmic membranes in bacterial cells: Organization, function and biosynthesis. In *Prokaryotic structure and function,* S. Mohan, C. Dow, and J. A. Coles, editors, 249–74. New York: Cambridge University Press.

Frankel, R. B., and Bazylinski, D. A. 2004. Magnetosome mysteries. *ASM News* 70(4):176–83.

Ghuysen, J.-M., and Hekenbeck, R., editors. 1994. *Bacterial cell wall.* New York: Elsevier.

Gital, Z. 2005. The new bacterial cell biology: Moving parts and subcellular architecture. *Cell* 120:577–86.

Harshey, R. M. 2003. Bacterial motility on a surface: Many ways to a common goal. *Annu. Rev. Microbiol.* 57:249–73.

Henderson, I. R.; Navarro-Garcia, F.; Desvaux, M.; Fernandez, R. C.; and Ala'Aldeen, D. 2004. Type V protein secretion pathway: The autotransporter story. *Microbiol. Mol. Biol. Rev.* 68(4):692–744.

Hoppert, M., and Mayer, F. 1999. Prokaryotes. *American Scientist* 87:518–25.

Kerfeld, C. A.; Sawaya, M. R.; Tanaka, A.; Nguyen, C. V.; Phillips, M.; Beeby, M.; and Yeates, T. O. 2005. Protein structures forming the shell of primitive bacterial organelles. *Science* 309:936–38.

Kostakioti, M.; Newman, C. L.; Thanassi, D. G.; and Stathopoulos, C. 2005. Mechanisms of protein export across the bacterial outer membrane. *J. Bacteriol.* 187(13):4306–14.

Macnab, R. M. 2003. How bacteria assemble flagella. *Annu. Rev. Microbiol.* 57:77–100.

Mattick, J. S. 2002. Type IV pili and twitching motility. *Annu. Rev. Microbiol.* 56:289–314.

Nicholson, W. L.; Munakata, N.; Horneck, G.; Melosh, H. J.; and Setlow, P. 2000. Resistance of *Bacillus* endospores to extreme terrestrial and extraterrestrial environments. *Microbiol. Mol. Biol. Rev.* 64(3):548–72.

Nikaido, H. 2003. Molecular basis of bacterial outer membrane permeability revisited. *Microbiol. Mol. Biol. Rev.* 67(4):593–656.

Parkinson, J. S. 2004. Signal amplification in bacterial chemotaxis through receptor teamwork. *ASM News* 70(12):575–82.

Robinow, C., and Kellenberger, E. 1994. The bacterial nucleoid revisited. *Microbiol. Rev.* 58(2):211–32.

Sára, M., and Sleytr, U. B. 2000. S-layer proteins. *J. Bacteriol.* 182(4):859–68.

Scherrer, R. 1984. Gram's staining reaction, Gram types and cell walls of bacteria. *Trends Biochem. Sci.* 9:242–45.

Schulz, H. N., and Jorgensen, B. B. 2001. Big bacteria. *Annu. Rev. Microbiol.* 55:105–37.

Trun, N. J., and Marko, J. F. 1998. Architecture of a bacterial chromosome. *ASM News* 64(5):276–83.

Walsby, A. E. 1994. Gas vesicles. *Microbiol. Rev.* 58(1):94–144.

Walsby, A. E. 2005. Archaea with square cells. *Trends Microbiol.* 13(5):193–95.

Wätermann, M., and Steinbüchel, A. 2005. Neutral lipid bodies in prokaryotes: Recent insights into structure, formation, and relationship to eukaryotic lipid depots. *J. Bacteriol.* 187(11):3607–19.

Please visit the Prescott website at www.mhhe.com/prescott7 for additional references.

4

Eucaryotic Cell Structure and Function

Often we emphasize procaryotes and viruses, but eucaryotic microorganisms also have major impacts on human welfare. For example, the protozoan parasite *Trypanosoma brucei gambiense* is a cause of African sleeping sickness. The organism invades the nervous system and the victim frequently dies after suffering several years from symptoms such as weakness, headache, apathy, emaciation, sleepiness, and coma.

PREVIEW

- Eucaryotic cells differ most obviously from procaryotic cells in having a variety of complex membranous organelles in the cytoplasmic matrix and the majority of their genetic material within membrane-delimited nuclei. Each organelle has a distinctive structure directly related to specific functions.

- A cytoskeleton composed of microtubules, microfilaments, and intermediate filaments helps give eucaryotic cells shape; the cytoskeleton is also involved in cell movements, intracellular transport, and reproduction.

- When eucaryotes reproduce, genetic material is distributed between cells by the highly organized, complex processes called mitosis and meiosis.

- Despite great differences between eucaryotes and procaryotes with respect to such things as morphology, they are similar on the biochemical level.

In chapter 3 considerable attention is devoted to procaryotic cell structure and function because procaryotes are immensely important in microbiology and have occupied a large portion of microbiologists' attention in the past. Nevertheless, protists and fungi also are microorganisms and have been extensively studied. These eucaryotes often are extraordinarily complex, interesting in their own right, and prominent members of ecosystems (**figure 4.1**). In addition, many protists and fungi are important model organisms, as well as being exceptionally useful in industrial microbiology. A number of protists and fungi are also major human pathogens; one only need think of candidiasis, malaria, or African sleeping sickness to appreciate the significance of eucaryotes in medical microbiology. So although this text emphasizes procaryotes, eucaryotic microorganisms also demand attention and are briefly discussed in this chapter.

Chapter 4 focuses on eucaryotic cell structure and its relationship to cell function. Because many valuable studies on eucaryotic cell ultrastructure have used organisms other than microorganisms, some work on nonmicrobial cells is presented. At the end of the chapter, procaryotic and eucaryotic cells are compared in some depth.

4.1 AN OVERVIEW OF EUCARYOTIC CELL STRUCTURE

The most obvious difference between eucaryotic and procaryotic cells is in their use of membranes. Eucaryotic cells have membrane-delimited nuclei, and membranes also play a prominent part in the structure of many other organelles (**figures 4.2** and **4.3**). **Organelles** are intracellular structures that perform specific functions in cells analogous to the functions of organs in the body. The name organelle (little organ) was coined because biologists saw a parallel between the relationship of organelles to a cell and that of organs to the whole body. It is not satisfactory to define organelles as membrane-bound structures because this would exclude such components as ribosomes and bacterial flagella. A comparison of figures 4.2 and 4.3 with figures 3.4 and 3.13*a* shows how structurally complex the eucaryotic cell is. This complexity is due chiefly to the use of internal membranes for several purposes. The partitioning of the eucaryotic cell interior by membranes makes possible the placement of different biochemical and physiological functions in separate compartments so that they can more easily take place simultaneously under independent control and proper coordination. Large membrane surfaces

The key to every biological problem must finally be sought in the cell.

—E. B. Wilson

(a) *Paramecium*

(b) Diatom frustules

(c) *Penicillium*

(d) *Penicillium*

(e) *Stentor*

(f) *Amanita muscaria*

Figure 4.1   Representative Examples of Eucaryotic Microorganisms.
(a) *Paramecium* as seen with interference-contrast microscopy (×115). **(b)** Mixed diatom frustules (×100). **(c)** *Penicillium* colonies, and **(d)** a microscopic view of the mold's hyphae and conidia (×220). **(e)** *Stentor.* The ciliated protozoa are extended and actively feeding, dark-field microscopy (×100). **(f)** *Amanita muscaria,* a large poisonous mushroom (×5).

make possible greater respiratory and photosynthetic activity because these processes are located exclusively in membranes. The intracytoplasmic membrane complex also serves as a transport system to move materials between different cell locations. Thus abundant membrane systems probably are necessary in eucaryotic cells because of their large volume and the need for adequate regulation, metabolic activity, and transport.

Figures 4.2 and 4.3 illustrate most of the organelles to be discussed here. **Table 4.1** briefly summarizes the functions of the major eucaryotic organelles. Our detailed discussion of eucaryotic cell structure begins with the eucaryotic membrane. We then proceed to organelles within the cytoplasm, and finally to components outside the membrane.

1. What is an organelle?
2. Why is the compartmentalization of the cell interior advantageous to eucaryotic cells?

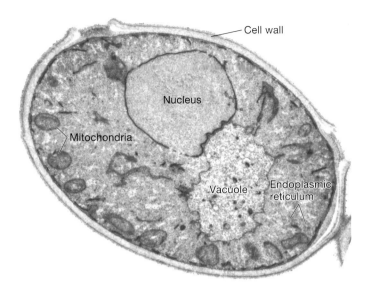

Figure 4.2   Eucaryotic Cell Ultrastructure.   The yeast *Saccharomyces* (×7,200). Note the nucleus, mitochondrion, vacuole, endoplasmic reticulum, and cell wall.

4.2   THE PLASMA MEMBRANE AND MEMBRANE STRUCTURE

As discussed in chapter 3, the fluid mosaic model of membrane structure is based largely on studies of eucaryotic membranes. In eucaryotes, the major membrane lipids are phosphoglycerides, sphingolipids, and cholesterol (**figure 4.4**). The distribution of these lipids is asymmetric. Lipids in the outer monolayer differ from those of the inner monolayer. Although most lipids in individual monolayers mix freely with each other, there are microdomains that differ in lipid and protein composition. One such microdomain is the **lipid raft,** which is enriched in cholesterol and lipids with many saturated fatty acids including some sphingolipids. The lipid raft spans the membrane bilayer, and lipids in the adjacent monolayers interact. These lipid rafts appear to participate in a variety of cellular processes (e.g., cell movement and signal transduction). They also may be involved in the entrance of some viruses into their host cells and the assembly of some viruses before they are released from their host cells.

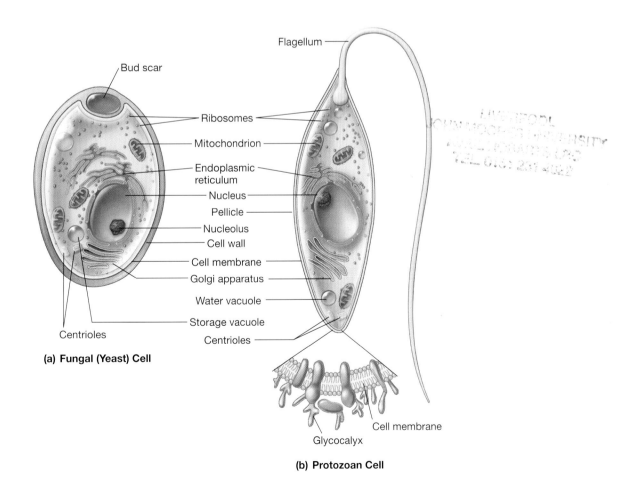

(a) Fungal (Yeast) Cell

(b) Protozoan Cell

Figure 4.3   The Structure of Two Representative Eucaryotic Cells.   Illustrations of a yeast cell (fungus) **(a)** and the flagellated protozoan *Peranema* **(b)**.

Table 4.1	Functions of Eucaryotic Organelles
Plasma membrane	Mechanical cell boundary, selectively permeable barrier with transport systems, mediates cell-cell interactions and adhesion to surfaces, secretion
Cytoplasmic matrix	Environment for other organelles, location of many metabolic processes
Microfilaments, intermediate filaments, and microtubules	Cell structure and movements, form the cytoskeleton
Endoplasmic reticulum	Transport of materials, protein and lipid synthesis
Ribosomes	Protein synthesis
Golgi apparatus	Packaging and secretion of materials for various purposes, lysosome formation
Lysosomes	Intracellular digestion
Mitochondria	Energy production through use of the tricarboxylic acid cycle, electron transport, oxidative phosphorylation, and other pathways
Chloroplasts	Photosynthesis—trapping light energy and formation of carbohydrate from CO_2 and water
Nucleus	Repository for genetic information, control center for cell
Nucleolus	Ribosomal RNA synthesis, ribosome construction
Cell wall and pellicle	Strengthen and give shape to the cell
Cilia and flagella	Cell movement
Vacuole	Temporary storage and transport, digestion (food vacuoles), water balance (contractile vacuole)

Figure 4.4   Examples of Eucaryotic Membrane Lipids.   (a) Phosphatidylcholine, a phosphoglyceride. **(b)** Sphingomyelin, a sphingolipid. **(c)** Cholesterol, a sterol.

4.3   THE CYTOPLASMIC MATRIX, MICROFILAMENTS, INTERMEDIATE FILAMENTS, AND MICROTUBULES

The many organelles of eucaryotic cells lie in the **cytoplasmic matrix.** The matrix is one of the most important and complex parts of the cell. It is the "environment" of the organelles and the location of many important biochemical processes. Several physical changes seen in cells—viscosity changes, cytoplasmic streaming, and others—also are due to matrix activity.

A major component of the cytoplasmic matrix is a vast network of interconnected filaments called the **cytoskeleton.** The cytoskeleton plays a role in both cell shape and movement. Three types of filaments form the cytoskeleton: microfilaments, microtubules, and intermediate filaments (**figure 4.5**). **Microfilaments** are minute protein filaments, 4 to 7 nm in diameter, that may be either scattered within the cytoplasmic matrix or organized into networks and parallel arrays. Microfilaments are composed of an actin protein that is similar to the actin contractile protein of muscle tissue. Microfilaments are involved in cell motion and shape changes such as the motion of pigment granules, amoeboid movement, and protoplasmic streaming in slime molds. Interestingly, some pathogens use the actin proteins of their eucaryotic hosts to move rapidly through the host cell and to propel themselves into new host cells (**Disease 4.1: Getting Around**). Protist classification: *Eumycetozoa* and *Stramenopiles* (section 25.6)

Microtubules are shaped like thin cylinders about 25 nm in diameter. They are complex structures constructed of two spherical protein subunits—α-tubulin and β-tubulin. The two proteins are the same molecular weight and differ only slightly in terms of their amino acid sequence and tertiary structure. Each tubulin is approximately 4 to 5 nm in diameter. These subunits are assembled in a helical arrangement to form a cylinder with an average of 13 subunits in one turn or circumference (figure 4.5).

Microtubules serve at least three purposes: (1) they help maintain cell shape, (2) are involved with microfilaments in cell movements, and (3) participate in intracellular transport processes. Microtubules are found in long, thin cell structures requiring support such as the axopodia (long, slender, rigid pseudopodia) of protists (**figure 4.6**). Microtubules also are present in structures that participate in cell or organelle movements—the mitotic spindle, cilia, and flagella.

Intermediate filaments are heterogeneous elements of the cytoskeleton. They are about 10 nm in diameter and are assembled from a group of proteins that can be divided into several classes. Intermediate filaments having different functions are assembled from one or more of these classes of proteins. The role

Figure 4.5   The Eucaryotic Cytoplasmic Matrix and Cytoskeleton.   The cytoplasmic matrix of eucaryotic cells contains many important organelles. The cytoskeleton helps form a framework within which the organelles lie. The cytoskeleton is composed of three elements: microfilaments, microtubules, and intermediate filaments.

Disease

4.1   Getting Around

Listeria monocytogenes is a gram-positive, rod-shaped bacterium responsible for the disease listeriosis. Listeriosis is a food-borne infection that is usually mild but can cause serious disease (meningitis, sepsis, and stillbirth) in immunocompromised individuals and pregnant women. *L. monocytogenes* is an intracellular pathogen that has a number of important virulence factors. One virulence factor is the protein ActA, which the bacterium releases after entering a host cell. ActA causes actin proteins to polymerize into filaments that form a tail at one end of the bacterium (see **Box figure**). As more and more of the actin proteins are polymerized, the growing tail pushes the bacterium through the host cell at rates up to 11 μm/minute. The bacterium can even be propelled through the cell surface and into neighboring cells.

***Listeria* Motility and Actin Filaments.**   A *Listeria* cell is propelled through the cell surface by a bundle of actin filaments.

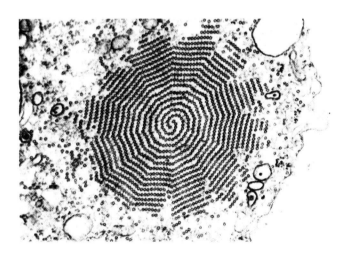

Figure 4.6   Cytoplasmic Microtubules.   Electron micrograph of a transverse section through the axopodium of a protist known as a heliozoan (×48,000). Note the parallel array of microtubules organized in a spiral pattern.

of intermediate filaments, if any, in eucaryotic microorganisms is unclear. Thus far, they have been identified and studied only in animals: some intermediate filaments have been shown to form the nuclear lamina, a structure that provides support for the nuclear envelope (see p. 91); and other intermediate filaments help link cells together to form tissues.

1. Compare the membranes of *Eucarya*, *Bacteria*, and *Archaea*. How are they similar? How do they differ?
2. What are lipid rafts? What roles do they play in eucaryotic cells?
3. Define cytoplasmic matrix, microfilament, microtubule, and tubulin. Discuss the roles of microfilaments, intermediate filaments, and microtubules.
4. Describe the cytoskeleton. What are its functions?

4.4   ORGANELLES OF THE BIOSYNTHETIC-SECRETORY AND ENDOCYTIC PATHWAYS

In addition to the cytoskeleton, the cytoplasmic matrix is permeated with an intricate complex of membranous organelles and vesicles that move materials into the cell from the outside (endocytic pathway), and from the inside of the cell out, as well as from location to location within the cell (biosynthetic-secretory pathway). In this section, some of these organelles are described. This is followed by a summary of how the organelles function in the biosynthetic-secretory and endocytic pathways.

The Endoplasmic Reticulum

The **endoplasmic reticulum (ER)** (figure 4.3 and **figure 4.7**) is an irregular network of branching and fusing membranous tubules, around 40 to 70 nm in diameter, and many flattened sacs called **cisternae** (s., **cisterna**). The nature of the ER varies with the functional and physiological status of the cell. In cells synthesizing a great deal of protein for purposes such as secretion, a large part of the ER is studded on its outer surface with ribosomes and is called **rough endoplasmic reticulum (RER)**. Other cells, such as those producing large quantities of lipids, have ER that lacks ribosomes. This is **smooth ER (SER)**.

The endoplasmic reticulum has many important functions. Not only does it transport proteins, lipids, and other materials through the cell, it is also involved in the synthesis of many of the materials it transports. Lipids and proteins are synthesized by ER-associated enzymes and ribosomes. Polypeptide chains synthesized on RER-bound ribosomes may be inserted either into the ER membrane or into its lumen for transport elsewhere. The ER is also a major site of cell membrane synthesis.

The Golgi Apparatus

The **Golgi apparatus** is composed of flattened, saclike cisternae stacked on each other (**figure 4.8**). These membranes, like the smooth ER, lack bound ribosomes. There are usually around 4 to 8 cisternae in a stack, although there may be many more. Each is 15 to 20 nm thick and separated from other cisternae by 20 to 30 nm. A complex network of tubules and vesicles (20 to 100 nm in diameter) is located at the edges of the cisternae. The stack of cisternae has a definite polarity because there are two faces that are quite different from one another. The sacs on the cis or forming face often are associated with the ER and differ from the sacs on the trans or maturing face in thickness, enzyme content, and degree of vesicle formation.

The Golgi apparatus is present in most eucaryotic cells, but many fungi and ciliate protozoa lack a well-formed structure. Sometimes the Golgi consists of a single stack of cisternae; however, many cells may contain up to 20, and sometimes more, separate stacks. These stacks of cisternae, often called **dictyosomes,** can be clustered in one region or scattered about the cell.

The Golgi apparatus packages materials and prepares them for secretion, the exact nature of its role varying with the organism. For instance, the surface scales of some flagellated photosynthetic and radiolarian protists appear to be constructed within the Golgi apparatus and then transported to the surface in vesicles. The Golgi often participates in the development of cell membranes

Figure 4.7   The Endoplasmic Reticulum.   A transmission electron micrograph of the corpus luteum in a human ovary showing structural variations in eucaryotic endoplasmic reticulum. Note the presence of both rough endoplasmic reticulum lined with ribosomes and smooth endoplasmic reticulum without ribosomes (×26,500).

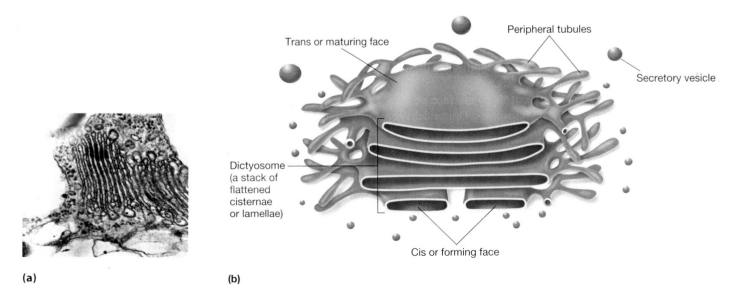

(a)   (b)

Figure 4.8   Golgi Apparatus Structure.   Golgi apparatus of *Euglena gracilis*. Cisternal stacks are shown in the electron micrograph (×165,000) in **(a)** and diagrammatically in **(b).**

and i
hyph
the w

Lysc

Lyso
euca
mals
mem
50 nt
gesti
macr
hydr
cond
acidi

The

The
to ly:
men
unde
and i
     I
creti
dopl
quer
throi
bud
ofter
glyc
     T
Golg
secri
face
the (
proc
Som
catio
ditic
     '
Golg
prote
vesi
stitu
then
the |
are :
serv
imn
leas
the s
and

gans

(a) Bloom of cyanobacteria (photolithoautotrophic bacteria)

(b) Purple sulfur bacteria (photoheterotrophs)

(c) Purple sulfur bacteria

Figure 5.1   Phototrophic Bacteria.   Phototrophic bacteria play important roles in aquatic ecosystems, where they can cause blooms. **(a)** A cyanobacterial and an algal bloom in a eutrophic pond. **(b)** Purple sulfur bacteria growing in a bog. **(c)** A bloom of purple sulfur bacteria in a sewage lagoon.

Internal membrane system used for oxidation of nitrite

0.25 U

(a) *Nitrobacter winogradskyi*, a chemolithoautotroph

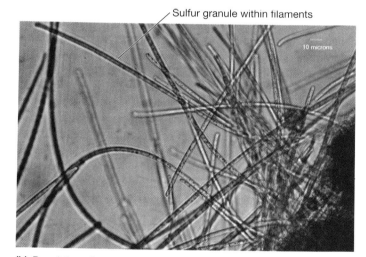

Sulfur granule within filaments

10 microns

(b) *Beggiatoa alba*, a chemolithoheterotroph (mixotroph)

Figure 5.2   Chemolithotrophic Bacteria.   **(a)** Transmission electron micrograph of *Nitrobacter winogradskyi*, an organism that uses nitrite as its source of energy ($\times 213,000$). **(b)** Light micrograph of *Beggiatoa alba*, an organism that uses hydrogen sulfide as its energy source and organic molecules as carbon sources. The dark spots within the filaments are granules of elemental sulfur produced when hydrogen sulfide is oxidized.

5.4   REQUIREMENTS FOR NITROGEN, PHOSPHORUS, AND SULFUR

To grow, a microorganism must be able to incorporate large quantities of nitrogen, phosphorus, and sulfur. Although these elements may be acquired from the same nutrients that supply carbon, microorganisms usually employ inorganic sources as well.

Nitrogen is needed for the synthesis of amino acids, purines, pyrimidines, some carbohydrates and lipids, enzyme cofactors, and other substances. Many microorganisms can use the nitrogen in amino acids. Others can incorporate ammonia directly through the action of enzymes such as glutamate dehydrogenase or glutamine synthetase and glutamate synthase (*see figures 10.11 and 10.12*). Most phototrophs and many chemotrophic microorganisms reduce nitrate to ammonia and incorporate the ammonia in a process known as assimilatory nitrate reduction (*see p. 235*). A variety of bacteria (e.g., many cyanobacteria and the symbiotic bacterium *Rhizobium*) can assimilate atmospheric nitrogen (N_2) by reducing it to ammonium (NH_4^+). This is called nitrogen fixation.   Synthesis of amino acids (section 10.5)

Phosphorus is present in nucleic acids, phospholipids, nucleotides like ATP, several cofactors, some proteins, and other cell components. Almost all microorganisms use inorganic phosphate as their phosphorus source and incorporate it directly. Low phosphate levels actually limit microbial growth in many aquatic environments. Some microbes, such as *Escherichia coli,* can use both organic and inorganic phosphate. Some organophosphates such as hexose 6-phosphates can be taken up directly by the cell. Other organophosphates are hydrolyzed in the periplasm by the enzyme alkaline phosphatase to produce inorganic phosphate, which then is transported across the plasma membrane.   Synthesis of purines, pyrimidines, and nucleotides (section 10.6)

Sulfur is needed for the synthesis of substances like the amino acids cysteine and methionine, some carbohydrates, biotin, and thiamine. Most microorganisms use sulfate as a source of sulfur and reduce it by assimilatory sulfate reduction; a few microorganisms require a reduced form of sulfur such as cysteine.

---

1. Briefly describe how microorganisms use the various forms of nitrogen, phosphorus, and sulfur.
2. Why do you think ammonia (NH_3) can be directly incorporated into amino acids while other forms of combined nitrogen (e.g., NO_2^- and NO_3^-) are not?

5.5   Growth Factors

Some microorganisms have the enzymes and biochemical pathways needed to synthesize all cell components using minerals and sources of energy, carbon, nitrogen, phosphorus, and sulfur. Other microorganisms lack one or more of the enzymes needed to manufacture indispensable constituents. Therefore they must obtain these constituents or their precursors from the environment. Organic compounds that are essential cell components or precursors of such components but cannot be synthesized by the organism are called **growth factors.** There are three major classes of growth factors: (1) amino acids, (2) purines and pyrimidines, and (3) vitamins. Amino acids are needed for protein synthesis; purines and pyrimidines for nucleic acid synthesis. **Vitamins** are small organic molecules that usually make up all or part of enzyme cofactors and are needed in only very small amounts to sustain growth. The functions of selected vitamins, and examples of microorganisms requiring them, are given in **table 5.3.** Some microorganisms require many vitamins; for ex-

ample, *Enterococcus faecalis* needs eight different vitamins for growth. Other growth factors are also seen; heme (from hemoglobin or cytochromes) is required by *Haemophilus influenzae,* and some mycoplasmas need cholesterol.   Enzymes (section 8.7)

Understanding the growth factor requirements of microbes has important practical applications. Both microbes with known, specific requirements and those that produce large quantities of a substance (e.g., vitamins) are useful. Microbes with a specific growth factor requirement can be used in bioassays for the factor they need. A typical assay is a growth-response assay, which allows the amount of growth factor in a solution to be determined. These assays are based on the observation that the amount of growth in a culture is related to the amount of growth factor present. Ideally, the amount of growth is directly proportional to the amount of growth factor; if the growth factor concentration doubles the amount of microbial growth doubles. For example, species from the bacterial genera *Lactobacillus* and *Streptococcus* can be used in microbiological assays of most vitamins and amino acids. The appropriate bacterium is grown in a series of culture vessels, each containing medium with an excess amount of all required components except the growth factor to be assayed. A different amount of growth factor is added to each vessel. The standard curve is prepared by plotting the growth factor quantity or concentration against the total extent of bacterial growth. The quantity of the growth factor in a test sample is determined by comparing the extent of growth caused by the unknown sample with that resulting from the standards. Microbiological assays are specific, sensitive, and simple. They still are used in the assay of substances like vitamin B_{12} and biotin, despite advances in chemical assay techniques.

On the other hand, those microorganisms able to synthesize large quantities of vitamins can be used to manufacture these compounds for human use. Several water-soluble and fat-soluble vitamins are produced partly or completely using industrial fermentations. Good examples of such vitamins and the microorganisms that synthesize them are riboflavin (*Clostridium, Candida, Ashbya, Eremothecium*), coenzyme A (*Brevibacterium*), vitamin B_{12} (*Streptomyces, Propionibacterium, Pseudomonas*), vitamin C (*Gluconobacter, Erwinia, Corynebacterium*), β-carotene (*Dunaliella*), and vitamin D (*Saccharomyces*). Current research focuses on improving yields and finding microorganisms that can produce large quantities of other vitamins.

---

1. What are growth factors? What are vitamins?
2. How can humans put to use a microbe with a specific growth factor requirement?
3. List the growth factors that microorganisms produce industrially.
4. Why do you think amino acids, purines, and pyrimidines are often growth factors, whereas glucose is not?

5.6   Uptake of Nutrients by the Cell

The first step in nutrient use is uptake of the required nutrients by the microbial cell. Uptake mechanisms must be specific—that is, the necessary substances, and not others, must be acquired. It

Table 5.3	Functions of Some Common Vitamins in Microorganisms	
Vitamin	**Functions**	**Examples of Microorganisms Requiring Vitamin[a]**
Biotin	Carboxylation (CO_2 fixation) One-carbon metabolism	*Leuconostoc mesenteroides* (B) *Saccharomyces cerevisiae* (F) *Ochromonas malhamensis* (P) *Acanthamoeba castellanii* (P)
Cyanocobalamin (B_{12})	Molecular rearrangements One-carbon metabolism—carries methyl groups	*Lactobacillus* spp. (B) *Euglena gracilis* (P) Diatoms (P) *Acanthamoeba castellanii* (P)
Folic acid	One-carbon metabolism	*Enterococcus faecalis* (B) *Tetrahymena pyriformis* (P)
Lipoic acid	Transfer of acyl groups	*Lactobacillus casei* (B) *Tetrahymena* spp. (P)
Pantothenic acid	Precursor of coenzyme A—carries acyl groups (pyruvate oxidation, fatty acid metabolism)	*Proteus morganii* (B) *Hanseniaspora* spp. (F) *Paramecium* spp. (P)
Pyridoxine (B_6)	Amino acid metabolism (e.g., transamination)	*Lactobacillus* spp. (B) *Tetrahymena pyriformis* (P)
Niacin (nicotinic acid)	Precursor of NAD and NADP—carry electrons and hydrogen atoms	*Brucella abortus, Haemophilus influenzae* (B) *Blastocladia pringsheimii* (F) *Crithidia fasciculata* (P)
Riboflavin (B_2)	Precursor of FAD and FMN—carry electrons or hydrogen atoms	*Caulobacter vibrioides* (B) *Dictyostelium* spp. (P) *Tetrahymena pyriformis* (P)
Thiamine (B_1)	Aldehyde group transfer (pyruvate decarboxylation, α-keto acid oxidation)	*Bacillus anthracis* (B) *Phycomyces blakesleeanus* (F) *Ochromonas malhamensis* (P) *Colpidium campylum* (P)

[a] The representative microorganisms are members of the following groups: *Bacteria* (B), *Fungi* (F), and protists (P).

does a cell no good to take in a substance that it cannot use. Because microorganisms often live in nutrient-poor habitats, they must be able to transport nutrients from dilute solutions into the cell against a concentration gradient. Finally, nutrient molecules must pass through a selectively permeable plasma membrane that prevents the free passage of most substances. In view of the enormous variety of nutrients and the complexity of the task, it is not surprising that microorganisms make use of several different transport mechanisms. The most important of these are facilitated diffusion, active transport, and group translocation. Eucaryotic microorganisms do not appear to employ group translocation but take up nutrients by the process of endocytosis. Organelles of the biosynthetic-secretory and endocytic pathways (section 4.4)

Passive Diffusion

A few substances, such as glycerol, can cross the plasma membrane by **passive diffusion.** Passive diffusion, often called diffusion or simple diffusion, is the process in which molecules move from a region of higher concentration to one of lower concentration. The rate of passive diffusion is dependent on the size of the concentration gradient between a cell's exterior and its interior (**figure 5.3**). A fairly large concentration gradient is required for adequate nutrient uptake by passive diffusion (i.e., the external nutrient concentration must be high while the internal concentration is low), and the rate of uptake decreases as more nutrient is acquired unless it is used immediately. Very small molecules such as H_2O, O_2, and CO_2 often move across membranes by passive diffusion. Larger molecules, ions, and polar substances must enter the cell by other mechanisms.

Facilitated Diffusion

The rate of diffusion across selectively permeable membranes is greatly increased by using carrier proteins, sometimes called **permeases,** which are embedded in the plasma membrane. Diffusion involving carrier proteins is called **facilitated diffusion.** The rate of facilitated diffusion increases with the concentration gradient

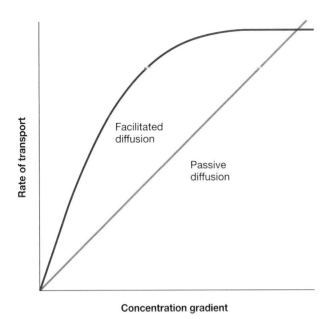

Figure 5.3 Passive and Facilitated Diffusion. The dependence of diffusion rate on the size of the solute's concentration gradient (the ratio of the extracellular concentration to the intracellular concentration). Note the saturation effect or plateau above a specific gradient value when a facilitated diffusion carrier is operating. This saturation effect is seen whenever a carrier protein is involved in transport.

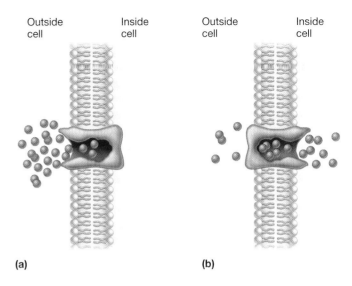

Figure 5.4 A Model of Facilitated Diffusion. The membrane carrier **(a)** can change conformation after binding an external molecule and subsequently release the molecule on the cell interior. **(b)** It then returns to the outward oriented position and is ready to bind another solute molecule. Because there is no energy input, molecules will continue to enter only as long as their concentration is greater on the outside.

much more rapidly and at lower concentrations of the diffusing molecule than that of passive diffusion (figure 5.3). Note that the diffusion rate levels off or reaches a plateau above a specific gradient value because the carrier is saturated—that is, the carrier protein is binding and transporting as many solute molecules as possible. The resulting curve resembles an enzyme-substrate curve (*see figure 8.18*) and is different from the linear response seen with passive diffusion. Carrier proteins also resemble enzymes in their specificity for the substance to be transported; each carrier is selective and will transport only closely related solutes. Although a carrier protein is involved, facilitated diffusion is truly diffusion. A concentration gradient spanning the membrane drives the movement of molecules, and no metabolic energy input is required. If the concentration gradient disappears, net inward movement ceases. The gradient can be maintained by transforming the transported nutrient to another compound. Once the nutrient is inside a eucaryotic cell, the gradient can be maintained by moving the nutrient to another membranous compartment. Some permeases are related to the major intrinsic protein (MIP) family of proteins. MIPs facilitate diffusion of small polar molecules. They are observed in virtually all organisms. The two most widespread MIP channels in bacteria are aquaporins (*see figure 2.29*), which transport water. Other important MIPs are the glycerol facilitators, which aid glycerol diffusion.

Although much work has been done on the mechanism of facilitated diffusion, the process is not yet understood completely. It appears that the carrier protein complex spans the membrane (**fig-**

ure 5.4). After the solute molecule binds to the outside, the carrier may change conformation and release the molecule on the cell interior. The carrier subsequently changes back to its original shape and is ready to pick up another molecule. The net effect is that a hydrophilic molecule can enter the cell in response to its concentration gradient. Remember that the mechanism is driven by concentration gradients and therefore is reversible. If the solute's concentration is greater inside the cell, it will move outward. Because the cell metabolizes nutrients upon entry, influx is favored.

Although glycerol is transported by facilitated diffusion in many bacteria, facilitated diffusion does not seem to be the major uptake mechanism. This is because nutrient concentrations often are lower outside the cell. Facilitated diffusion is much more prominent in eucaryotic cells where it is used to transport a variety of sugars and amino acids.

Active Transport

Because facilitated diffusion can efficiently move molecules to the interior only when the solute concentration is higher on the outside of the cell, microbes must have transport mechanisms that can move solutes against a concentration gradient. This is important because microorganisms often live in habitats characterized by very dilute nutrient sources. Microbes use two important transport processes in such situations: active transport and group translocation. Both are energy-dependent processes.

Active transport is the transport of solute molecules to higher concentrations, or against a concentration gradient, with the input of metabolic energy. Because active transport involves permeases, it resembles facilitated diffusion in some ways. The

Table 5.6	Some Common Complex Media
Nutrient Broth	**Amount (g/liter)**
Peptone (gelatin hydrolysate)	5
Beef extract	3
Tryptic Soy Broth	
Tryptone (pancreatic digest of casein)	17
Peptone (soybean digest)	3
Glucose	2.5
Sodium chloride	5
Dipotassium phosphate	2.5
MacConkey Agar	
Pancreatic digest of gelatin	17.0
Pancreatic digest of casein	1.5
Peptic digest of animal tissue	1.5
Lactose	10.0
Bile salts	1.5
Sodium chloride	5.0
Neutral red	0.03
Crystal violet	0.001
Agar	13.5

in the absence of organic substances and to determine carbon sources for heterotrophic bacteria by supplementing the medium with various organic compounds.

Functional Types of Media

Media such as tryptic soy broth and tryptic soy agar are called general purpose media or **supportive media** because they sustain the growth of many microorganisms. Blood and other special nutrients may be added to general purpose media to encourage the growth of fastidious microbes. These specially fortified media (e.g., blood agar) are called **enriched media (figure 5.9)**.

Selective media favor the growth of particular microorganisms (**table 5.7**). Bile salts or dyes like basic fuchsin and crystal violet favor the growth of gram-negative bacteria by inhibiting the growth of gram-positive bacteria; the dyes have no effect on gram-negative organisms. Endo agar, eosin methylene blue agar, and MacConkey agar (tables 5.6 and 5.7) are three media widely used for the detection of *E. coli* and related bacteria in water supplies and elsewhere. These media contain dyes that suppress gram-positive bacterial growth. MacConkey agar also contains bile salts. Bacteria also may be selected by incubation with nutrients that they specifically can use. A medium containing only cellulose as a carbon and energy source is quite effective in the isolation of cellulose-digesting bacteria. The possibilities for selection are endless, and there are dozens of special selective media in use.

Historical Highlights

5.1   The Discovery of Agar as a Solidifying Agent and the Isolation of Pure Cultures

The earliest culture media were liquid, which made the isolation of bacteria to prepare pure cultures extremely difficult. In practice, a mixture of bacteria was diluted successively until only one organism, as an average, was present in a culture vessel. If everything went well, the individual bacterium thus isolated would reproduce to give a pure culture. This approach was tedious, gave variable results, and was plagued by contamination problems. Progress in isolating pathogenic bacteria understandably was slow.

The development of techniques for growing microorganisms on solid media and efficiently obtaining pure cultures was due to the efforts of the German bacteriologist Robert Koch and his associates. In 1881 Koch published an article describing the use of boiled potatoes, sliced with a flame-sterilized knife, in culturing bacteria. The surface of a sterile slice of potato was inoculated with bacteria from a needle tip, and then the bacteria were streaked out over the surface so that a few individual cells would be separated from the remainder. The slices were incubated beneath bell jars to prevent airborne contamination, and the isolated cells developed into pure colonies. Unfortunately many bacteria would not grow well on potato slices.

At about the same time, Frederick Loeffler, an associate of Koch, developed a meat extract peptone medium for cultivating

pathogenic bacteria. Koch decided to try solidifying this medium. Koch was an amateur photographer—he was the first to take photomicrographs of bacteria—and was experienced in preparing his own photographic plates from silver salts and gelatin. Precisely the same approach was employed for preparing solid media. He spread a mixture of Loeffler's medium and gelatin over a glass plate, allowed it to harden, and inoculated the surface in the same way he had inoculated his sliced potatoes. The new solid medium worked well, but it could not be incubated at 37°C (the best temperature for most human bacterial pathogens) because the gelatin would melt. Furthermore, some bacteria digested the gelatin.

About a year later, in 1882, agar was first used as a solidifying agent. It had been discovered by a Japanese innkeeper, Minora Tarazaemon. The story goes that he threw out extra seaweed soup and discovered the next day that it had jelled during the cold winter night. Agar had been used by the East Indies Dutch to make jellies and jams. Fannie Eilshemius Hesse (*see figure 1.7*), the New Jersey-born wife of Walther Hesse, one of Koch's assistants, had learned of agar from a Dutch acquaintance and suggested its use when she heard of the difficulties with gelatin. Agar-solidified medium was an instant success and continues to be essential in all areas of microbiology.

(a)

(b)

Figure 5.9 Enriched Media. **(a)** Blood agar culture of bacteria from the human throat. **(b)** Chocolate agar, an enriched medium used to grow fastidious organisms such as *Neisseria gonorrhoeae.* The brown color is the result of heating red blood cells and lysing them before adding them to the medium. It is called chocolate agar because of its chocolate brown color.

Differential media are media that distinguish among different groups of microbes and even permit tentative identification of microorganisms based on their biological characteristics. Blood agar is both a differential medium and an enriched one. It distinguishes between hemolytic and non-hemolytic bacteria. Hemolytic bacteria (e.g., many streptococci and staphylococci isolated from throats) produce clear zones around their colonies because of red blood cell destruction (figure 5.9*a*). MacConkey

agar is both differential and selective. Since it contains lactose and neutral red dye, lactose-fermenting colonies appear pink to red in color and are easily distinguished from colonies of nonfermenters.

---

1. Describe the following kinds of media and their uses: defined media, complex media, supportive media, enriched media, selective media, and differential media. Give an example of each kind.
2. What are peptones, yeast extract, beef extract, and agar? Why are they used in media?

5.8 ISOLATION OF PURE CULTURES

In natural habitats microorganisms usually grow in complex, mixed populations with many species. This presents a problem for microbiologists because a single type of microorganism cannot be studied adequately in a mixed culture. One needs a **pure culture,** a population of cells arising from a single cell, to characterize an individual species. Pure cultures are so important that the development of pure culture techniques by the German bacteriologist Robert Koch transformed microbiology. Within about 20 years after the development of pure culture techniques most pathogens responsible for the major human bacterial diseases had been isolated (*see figure 1.2*). There are several ways to prepare pure cultures; a few of the more common approaches are reviewed here.

The Spread Plate and Streak Plate

If a mixture of cells is spread out on an agar surface at a relatively low density, every cell grows into a completely separate **colony,** a macroscopically visible growth or cluster of microorganisms on a solid medium. Because each colony arises from a single cell, each colony represents a pure culture. The **spread plate** is an easy, direct way of achieving this result. A small volume of dilute microbial mixture containing around 30 to 300 cells is transferred to the center of an agar plate and spread evenly over the surface with a sterile bent-glass rod (**figure 5.10**). The dispersed cells develop into isolated colonies. Because the number of colonies should equal the number of viable organisms in the sample, spread plates can be used to count the microbial population.

Pure colonies also can be obtained from **streak plates.** The microbial mixture is transferred to the edge of an agar plate with an inoculating loop or swab and then streaked out over the surface in one of several patterns (**figure 5.11**). After the first sector is streaked, the inoculating loop is sterilized and an inoculum for the second sector is obtained from the first sector. A similar process is followed for streaking the third sector, except that the inoculum is from the second sector. Thus this is essentially a dilution process. Eventually, very few cells will be on the loop, and single cells will drop from it as it is rubbed along the agar surface. These develop into separate colonies. In both spread-plate and streak-plate techniques, successful isolation depends on spatial separation of single cells.

Table 5.7	Mechanisms of Action of Selective and Differential Media	
Medium	**Functional Type**	**Mechanism of Action**
Blood agar	Enriched and differential	Blood agar supports the growth of many fastidious bacteria. These can be differentiated based on their ability to produce hemolysins—proteins that lyse red blood cells. Hemolysis appears as a clear zone around the colony (β-hemolysis) or as a greenish halo around the colony (α-hemolysis) (e.g., *Streptococcus pyogenes,* a β-hemolytic streptococcus).
Eosin methylene blue (EMB) agar	Selective and differential	Two dyes, eosin Y and methylene blue, inhibit the growth of gram-positive bacteria. They also react with acidic products released by certain gram-negative bacteria when they use lactose or sucrose as carbon and energy sources. Colonies of gram-negative bacteria that produce large amounts of acidic products have a green, metallic sheen (e.g., fecal bacteria such as *E. coli*).
MacConkey (MAC) agar	Selective and differential	The selective components in MAC are bile salts and crystal violet, which inhibit the growth of gram-positive bacteria. The presence of lactose and neutral red, a pH indicator, allows the differentiation of gram-negative bacteria based on the products released when they use lactose as a carbon and energy source. The colonies of those that release acidic products are red (e.g., *E. coli*).
Mannitol salt agar	Selective and differential	A concentration of 7.5% NaCl selects for the growth of staphylococci. Pathogenic staphylococci can be differentiated based on the release of acidic products when they use mannitol as a carbon and energy source. The acidic products cause a pH indicator (phenol red) to turn yellow (e.g., *Staphylococcus aureus*).

(a)

(b)

Figure 5.10   Spread-Plate Technique.   (a) The preparation of a spread plate. (*1*) Pipette a small sample onto the center of an agar medium plate. (*2*) Dip a glass spreader into a beaker of ethanol. (*3*) Briefly flame the ethanol-soaked spreader and allow it to cool. (*4*) Spread the sample evenly over the agar surface with the sterilized spreader. Incubate. **(b)** Typical result of spread-plate technique.

Note: This method only works if the spreading tool (usually an inoculating loop) is resterilized after each of steps 1–4.

1   2   3   4   5

(a) Steps in a Streak Plate

(b)

Figure 5.11   Streak-Plate Technique.   A typical streaking pattern is shown **(a)** as well as an example of a streak plate **(b).**

1.0 ml   1.0 ml   1.0 ml   1.0 ml

Original sample   9 ml H_2O (10^{-1} dilution)   9 ml H_2O (10^{-2} dilution)   9 ml H_2O (10^{-3} dilution)   9 ml H_2O (10^{-4} dilution)

Mix with warm agar and pour.   1.0 ml   1.0 ml

Figure 5.12   The Pour-Plate Technique.   The original sample is diluted several times to thin out the population sufficiently. The most diluted samples are then mixed with warm agar and poured into petri dishes. Isolated cells grow into colonies and can be used to establish pure cultures. The surface colonies are circular; subsurface colonies are lenticular (lens shaped).

The Pour Plate

Extensively used with procaryotes and fungi, a **pour plate** also can yield isolated colonies. The original sample is diluted several times to reduce the microbial population sufficiently to obtain separate colonies when plating (**figure 5.12**). Then small volumes of several diluted samples are mixed with liquid agar that has been cooled to about 45°C, and the mixtures are poured immediately into sterile culture dishes. Most bacteria and fungi are not killed by a brief exposure to the warm agar. After the agar has hardened, each cell is fixed in place and forms an individual colony. Like the spread plate, the pour plate can be used to determine the number of cells in a population. Plates containing between 30 and 300 colonies are counted. The total number of colonies equals the number of viable microorganisms in the sample that are capable of growing in the medium used. Colonies growing on the surface also can be used to inoculate fresh medium and prepare pure cultures (**Techniques & Applications 5.2**).

6. Suppose the generation time of a bacterium is 90 minutes and the initial number of cells in a culture is 10^3 cells at the start of the log phase. How many bacteria will there be after 8 hours of exponential growth?

7. What effect does increasing a limiting nutrient have on the yield of cells and the growth rate?

8. Contrast and compare the viable but nonculturable status of microbes with that of programmed cell death as a means of responding to starvation.

6.3 MEASUREMENT OF MICROBIAL GROWTH

There are many ways to measure microbial growth to determine growth rates and generation times. Either population number or mass may be followed because growth leads to increases in both. Here the most commonly employed techniques for growth measurement are examined briefly and the advantages and disadvantages of each noted. No single technique is always best; the most appropriate approach will depend on the experimental situation.

Measurement of Cell Numbers

The most obvious way to determine microbial numbers is through direct counting. Using a counting chamber is easy, inexpensive, and relatively quick; it also gives information about the size and morphology of microorganisms. Petroff-Hausser counting chambers can be used for counting procaryotes; hemocytometers can be used for both procaryotes and eucaryotes. These specially designed slides have chambers of known depth with an etched grid on the chamber bottom (**figure 6.12**). Procaryotes are more easily counted in these chambers if they are stained, or when a phase-contrast or a fluorescence microscope is employed. The number of microorganisms in a sample can be calculated by taking into account the chamber's volume and any sample dilutions required. One disadvantage to the technique is that the microbial population must be fairly large for accuracy because such a small volume is sampled.

Larger microorganisms such as protists and yeasts can be directly counted with electronic counters such as the Coulter Counter, although more recently the flow cytometer is increasingly used. The microbial suspension is forced through a small hole or orifice in the Coulter Counter. An electrical current flows through the hole, and electrodes placed on both sides of the orifice measure its electrical resistance. Every time a microbial cell passes through the orifice, electrical resistance increases (or the conductivity drops) and the cell is counted. The Coulter Counter gives accurate results with larger cells and is extensively used in hospital laboratories to count red and white blood cells. It is not as useful in counting bacteria because of interference by small debris particles, the formation of filaments, and other problems. Identification of microorganisms from specimens (section 35.2)

The number of bacteria in aquatic samples is frequently determined from direct counts after the bacteria have been trapped on special membrane filters. In the membrane filter technique, the sample is first filtered through a black polycarbonate membrane filter. Then the bacteria are stained with a fluorescent dye such as acridine orange or the DNA stain DAPI, and observed microscopically. The stained cells are easily observed against the black

(a)

(b)

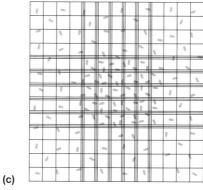

(c)

Figure 6.12 The Petroff-Hausser Counting Chamber.
(a) Side view of the chamber showing the cover glass and the space beneath it that holds a bacterial suspension. (b) A top view of the chamber. The grid is located in the center of the slide. (c) An enlarged view of the grid. The bacteria in several of the central squares are counted, usually at ×400 to ×500 magnification. The average number of bacteria in these squares is used to calculate the concentration of cells in the original sample. Since there are 25 squares covering an area of 1 mm^2, the total number of bacteria in 1 mm^2 of the chamber is (number/square)(25 squares). The chamber is 0.02 mm deep and therefore,

$$\text{bacteria/mm}^3 = (\text{bacteria/square})(25 \text{ squares})(50).$$

The number of bacteria per cm^3 is 10^3 times this value. For example, suppose the average count per square is 28 bacteria:

$$\text{bacteria/cm}^3 = (28 \text{ bacteria})(25 \text{ squares})(50)(10^3) = 3.5 \times 10^7.$$

background of the membrane filter and can be counted when viewed with an epifluorescence microscope. The light microscope: The fluorescence microscope (section 2.2)

Traditional methods for directly counting microbes in a sample usually yield cell densities that are much higher than the plating methods described next because direct counting procedures do not distinguish dead cells from live cells. Newer methods for direct counts avoid this problem. Commercial kits that use fluorescent reagents to stain live and dead cells differently are now available, making it possible to directly count the number of live and dead microorganisms in a sample (*see figures 2.13a and 27.16*).

Several plating methods can be used to determine the number of viable microbes in a sample. These are referred to as viable counting methods because they count only those cells that are alive and able to reproduce. Two commonly used procedures are the spread-plate technique and the pour-plate technique. In both of these methods, a diluted sample of bacteria or other microorganisms is dispersed over a solid agar surface. Each microorganism or group of microorganisms develops into a distinct colony. The original number of viable microorganisms in the sample can be calculated from the number of colonies formed and the sample dilution. For example, if 1.0 ml of a 1×10^{-6} dilution yielded 150 colonies, the original sample contained around 1.5×10^{8}

cells per ml. Usually the count is made more accurate by use of a special colony counter. In this way the spread-plate and pour-plate techniques may be used to find the number of microorganisms in a sample. Isolation of pure cultures: The spread plate and streak plate: The pour plate (section 5.8)

Another commonly used plating method first traps bacteria in aquatic samples on a membrane filter. The filter is then placed on an agar medium or on a pad soaked with liquid media (**figure 6.13**) and incubated until each cell forms a separate colony. A colony count gives the number of microorganisms in the filtered sample, and special media can be used to select for specific microorganisms (**figure 6.14**). This technique is especially useful in analyzing water purity. Water purification and sanitary analysis (section 41.1)

Plating techniques are simple, sensitive, and widely used for viable counts of bacteria and other microorganisms in samples of food, water, and soil. Several problems, however, can lead to inaccurate counts. Low counts will result if clumps of cells are not

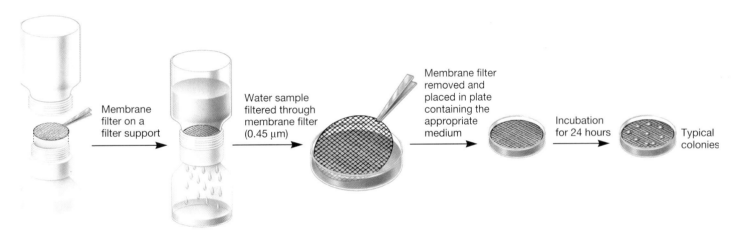

Figure 6.13   The Membrane Filtration Procedure.   Membranes with different pore sizes are used to trap different microorganisms. Incubation times for membranes also vary with the medium and microorganism.

(a)   (b)   (c)

Figure 6.14   Colonies on Membrane Filters.   Membrane-filtered samples grown on a variety of media. **(a)** Standard nutrient media for a total bacterial count. An indicator colors colonies red for easy counting. **(b)** Fecal coliform medium for detecting fecal coliforms that form blue colonies. **(c)** Wort agar for the culture of yeasts and molds.

broken up and the microorganisms well dispersed. Because it is not possible to be absolutely certain that each colony arose from an individual cell, the results are often expressed in terms of **colony forming units (CFU)** rather than the number of microorganisms. The samples should yield between 30 and 300 colonies for most accurate counting. Of course the counts will also be low if the agar medium employed cannot support growth of all the viable microorganisms present. The hot agar used in the pour-plate technique may injure or kill sensitive cells; thus spread plates sometimes give higher counts than pour plates.

Measurement of Cell Mass

Increases in the total cell mass, as well as in cell numbers, accompany population growth. Therefore techniques for measuring changes in cell mass can be used in following growth. The most direct approach is the determination of microbial dry weight. Cells growing in liquid medium are collected by centrifugation, washed, dried in an oven, and weighed. This is an especially useful technique for measuring the growth of filamentous fungi. It is time-consuming, however, and not very sensitive. Because bacteria weigh so little, it may be necessary to centrifuge several hundred milliliters of culture to collect a sufficient quantity.

Spectrophotometry can also be used to measure cell mass. These methods are more rapid and sensitive. They depend on the fact that microbial cells scatter light that strikes them. Because microbial cells in a population are of roughly constant size, the amount of scattering is directly proportional to the biomass of cells present and indirectly related to cell number. When the concentration of bacteria reaches about 10 million cells (10^7) per ml, the medium appears slightly cloudy or turbid. Further increases in concentration result in greater turbidity and less light is transmitted through the medium. The extent of light scattering can be measured by a spectrophotometer and is almost linearly related to cell concentration at low absorbance levels (**figure 6.15**). Thus population growth can be easily measured as long as the population is high enough to give detectable turbidity.

If the amount of a substance in each cell is constant, the total quantity of that cell constituent is directly related to the total microbial cell mass. For example, a sample of washed cells collected from a known volume of medium can be analyzed for total protein or nitrogen. An increase in the microbial population will be reflected in higher total protein levels. Similarly, chlorophyll determinations can be used to measure algal and cyanobacterial populations, and the quantity of ATP can be used to estimate the amount of living microbial mass.

Figure 6.15   Turbidity and Microbial Mass Measurement.   Determination of microbial mass by measurement of light absorption. As the population and turbidity increase, more light is scattered and the absorbance reading given by the spectrophotometer increases. The spectrophotometer meter has two scales. The bottom scale displays absorbance and the top scale, percent transmittance. Absorbance increases as percent transmittance decreases.

1. Briefly describe each technique by which microbial population numbers may be determined and give its advantages and disadvantages.
2. When using direct cell counts to follow the growth of a culture, it may be difficult to tell when the culture enters the phase of senescence and death. Why?
3. Why are plate count results expressed as colony forming units?

6.4 THE CONTINUOUS CULTURE OF MICROORGANISMS

Up to this point the focus has been on closed systems called batch cultures in which nutrient supplies are not renewed nor wastes removed. Exponential growth lasts for only a few generations and soon the stationary phase is reached. However, it is possible to grow microorganisms in an open system, a system with constant environmental conditions maintained through continual provision of nutrients and removal of wastes. These conditions are met in the laboratory by a **continuous culture system.** A microbial population can be maintained in the exponential growth phase and at a constant biomass concentration for extended periods in a continuous culture system.

The Chemostat

Two major types of continuous culture systems commonly are used: (1) chemostats and (2) turbidostats. A **chemostat** is constructed so that sterile medium is fed into the culture vessel at the same rate as the media containing microorganisms is removed (**figure 6.16**). The culture medium for a chemostat possesses an essential nutrient (e.g., an amino acid) in limiting quantities. Because one nutrient is limiting, the growth rate is determined by the rate at which new medium is fed into the growth chamber, and the final cell density depends on the concentration of the limiting nutrient. The rate of nutrient exchange is expressed as the dilution rate (D), the rate at which medium flows through the culture vessel relative to the vessel volume, where f is the flow rate (ml/hr) and V is the vessel volume (ml).

$$D = f/V$$

For example, if f is 30 ml/hr and V is 100 ml, the dilution rate is 0.30 hr^{-1}.

Both the microbial population level and the generation time are related to the dilution rate (**figure 6.17**). The microbial population density remains unchanged over a wide range of dilution rates. The generation time decreases (i.e., the rate of growth increases) as the dilution rate increases. The limiting nutrient will be almost completely depleted under these balanced conditions. If the dilution rate rises too high, the microorganisms can actually be washed out of the culture vessel before reproducing because the dilution rate is greater than the maximum growth rate. This occurs because fewer microorganisms are present to consume the limiting nutrient.

Figure 6.16 A Continuous Culture System: The Chemostat. Schematic diagram of the system. The fresh medium contains a limiting amount of an essential nutrient. Growth rate is determined by the rate of flow of medium through the culture vessel.

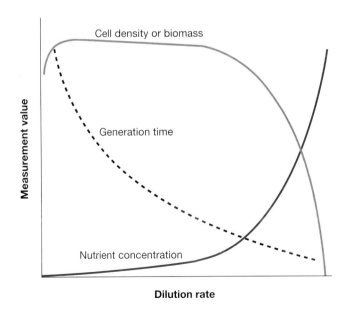

Figure 6.17 Chemostat Dilution Rate and Microbial Growth. The effects of changing the dilution rate in a chemostat.

At very low dilution rates, an increase in D causes a rise in both cell density and the growth rate. This is because of the effect of nutrient concentration on the growth rate, sometimes called the Monod relationship (figure 6.1b). Only a limited supply of nutrient is available at low dilution rates. Much of the available energy must be used for cell maintenance, not for growth and reproduction. As the dilution rate increases, the amount of nutrients and the resulting cell density rise because energy is available for both maintenance and reproduction. The growth rate increases when the total available energy exceeds the maintenance energy.

The Turbidostat

The second type of continuous culture system, the **turbidostat,** has a photocell that measures the absorbance or turbidity of the culture in the growth vessel. The flow rate of media through the vessel is automatically regulated to maintain a predetermined turbidity or cell density. The turbidostat differs from the chemostat in several ways. The dilution rate in a turbidostat varies rather than remaining constant, and its culture medium contains all nutrients in excess. That is, none of the nutrients is limiting. The turbidostat operates best at high dilution rates; the chemostat is most stable and effective at lower dilution rates.

Continuous culture systems are very useful because they provide a constant supply of cells in exponential phase and growing at a known rate. They make possible the study of microbial growth at very low nutrient levels, concentrations close to those present in natural environments. These systems are essential for research in many areas—for example, in studies on interactions between microbial species under environmental conditions resembling those in a freshwater lake or pond. Continuous systems also are used in food and industrial microbiology (chapters 40 and 41, respectively).

---

1. How does an open system differ from a closed culture system or batch culture?
2. Describe how the two different kinds of continuous culture systems, the chemostat and turbidostat, operate.
3. What is the dilution rate? What is maintenance energy?
4. How is the rate of growth of a microbial population controlled in a chemostat? In a turbidostat?

6.5 THE INFLUENCE OF ENVIRONMENTAL FACTORS ON GROWTH

As we have seen, microorganisms must be able to respond to variations in nutrient levels, and particularly to nutrient limitation. The growth of microorganisms also is greatly affected by the chemical and physical nature of their surroundings. An understanding of environmental influences aids in the control of microbial growth and the study of the ecological distribution of microorganisms.

The ability of some microorganisms to adapt to extreme and inhospitable environments is truly remarkable. Procaryotes are present anywhere life can exist. Many habitats in which procary-

otes thrive would kill most other organisms. Procaryotes such as *Bacillus infernus* are even able to live over 1.5 miles below the Earth's surface, without oxygen and at temperatures above 60°C. Microorganisms that grow in such harsh conditions are often called **extremophiles.**

In this section we shall briefly review how some of the most important environmental factors affect microbial growth. Major emphasis will be given to solutes and water activity, pH, temperature, oxygen level, pressure, and radiation. **Table 6.3** summarizes the way in which microorganisms are categorized in terms of their response to these factors. It is important to note that for most environmental factors, a range of levels supports growth of a microbe. For example, a microbe might exhibit optimum growth at pH 7, but will grow, though not optimally, at pH values down to pH 6 (its pH minimum) and up to pH 8 (its pH maximum). Furthermore, outside this range, the microbe might cease reproducing but will remain viable for some time. Clearly, each microbe must possess adaptations that allow it to adjust its physiology within its preferred range, and it may also have adaptations that protect it in environments outside this range. These adaptations will also be discussed in this section.

Solutes and Water Activity

Because a selectively permeable plasma membrane separates microorganisms from their environment, they can be affected by changes in the osmotic concentration of their surroundings. If a microorganism is placed in a hypotonic solution (one with a lower osmotic concentration), water will enter the cell and cause it to burst unless something is done to prevent the influx. Conversely if it is placed in a hypertonic solution (one with a higher osmotic concentration), water will flow out of the cell. In microbes that have cell walls (i.e., most procaryotes, fungi, and algae), the membrane shrinks away from the cell wall—a process called plasmolysis. Dehydration of the cell in hypertonic environments may damage the cell membrane and cause the cell to become metabolically inactive.

It is important, then, that microbes be able to respond to changes in the osmotic concentrations of their environment. For instance, microbes in hypotonic environments can reduce the osmotic concentration of their cytoplasm. This can be achieved by the use of inclusion bodies. Some bacteria and archaea also have mechanosensitive (MS) channels in their plasma membrane. In a hypotonic environment, the membrane stretches due to an increase in hydrostatic pressure and cellular swelling. MS channels then open and allow solutes to leave. Thus they can act as escape valves to protect cells from bursting. Since many protists do not have a cell wall, they must use contractile vacuoles (*see figures 25.5 and 25.17b*) to expel excess water. Many microorganisms, whether in hypotonic or hypertonic environments, keep the osmotic concentration of their protoplasm somewhat above that of the habitat by the use of compatible solutes, so that the plasma membrane is always pressed firmly against their cell wall. **Compatible solutes** are solutes that do not interfere with metabolism and growth when at high intracellular concentrations. Most procaryotes increase their

Table 6.3	Microbial Responses to Environmental Factors	
Descriptive Term	**Definition**	**Representative Microorganisms**
Solute and Water Activity		
Osmotolerant	Able to grow over wide ranges of water activity or osmotic concentration	*Staphylococcus aureus, Saccharomyces rouxii*
Halophile	Requires high levels of sodium chloride, usually above about 0.2 M, to grow	*Halobacterium, Dunaliella, Ectothiorhodospira*
pH		
Acidophile	Growth optimum between pH 0 and 5.5	*Sulfolobus, Picrophilus, Ferroplasma, Acontium, Cyanidium caldarium*
Neutrophile	Growth optimum between pH 5.5 and 8.0	*Escherichia, Euglena, Paramecium*
Alkalophile	Growth optimum between pH 8.0 and 11.5	*Bacillus alcalophilus, Natronobacterium*
Temperature		
Psychrophile	Grows well at 0°C and has an optimum growth temperature of 15°C or lower	*Bacillus psychrophilus, Chlamydomonas nivalis*
Psychrotroph	Can grow at 0–7°C; has an optimum between 20 and 30°C and a maximum around 35°C	*Listeria monocytogenes, Pseudomonas fluorescens*
Mesophile	Has growth optimum around 20–45°C	*Escherichia coli, Neisseria gonorrhoeae, Trichomonas vaginalis*
Thermophile	Can grow at 55°C or higher; optimum often between 55 and 65°C	*Geobacillus stearothermophilus, Thermus aquaticus, Cyanidium caldarium, Chaetomium thermophile*
Hyperthermophile	Has an optimum between 80 and about 113°C	*Sulfolobus, Pyrococcus, Pyrodictium*
Oxygen Concentration		
Obligate aerobe	Completely dependent on atmospheric O_2 for growth	*Micrococcus luteus, Pseudomonas, Mycobacterium;* most protists and fungi
Facultative anaerobe	Does not require O_2 for growth, but grows better in its presence	*Escherichia, Enterococcus, Saccharomyces cerevisiae*
Aerotolerant anaerobe	Grows equally well in presence or absence of O_2	*Streptococcus pyogenes*
Obligate anaerobe	Does not tolerate O_2 and dies in its presence	*Clostridium, Bacteroides, Methanobacterium, Trepomonas agilis*
Microaerophile	Requires O_2 levels below 2–10% for growth and is damaged by atmospheric O_2 levels (20%)	*Campylobacter, Spirillum volutans, Treponema pallidum*
Pressure		
Barophilic	Growth more rapid at high hydrostatic pressures	*Photobacterium profundum, Shewanella benthica, Methanocaldococcus jannaschii*

internal osmotic concentration in a hypertonic environment through the synthesis or uptake of choline, betaine, proline, glutamic acid, and other amino acids; elevated levels of potassium ions are also involved to some extent. Photosynthetic protists and fungi employ sucrose and polyols—for example, arabitol, glycerol, and mannitol—for the same purpose. Polyols and amino acids are ideal solutes for this function because they normally do not disrupt enzyme structure and function. The cytoplasmic matrix: Inclusion bodies (section 3.3)

Some microbes are adapted to extreme hypertonic environments. **Halophiles** grow optimally in the presence of NaCl or other salts at a concentration above about 0.2 M (**figure 6.18**). Extreme halophiles have adapted so completely to hypertonic, saline conditions that they require high levels of sodium chloride to grow—concentrations between about 2 M and saturation (about 6.2 M). The archaeon *Halobacterium* can be isolated from the

Dead Sea (a salt lake between Israel and Jordan and the lowest lake in the world), the Great Salt Lake in Utah, and other aquatic habitats with salt concentrations approaching saturation. *Halobacterium* and other extremely halophilic procaryotes accumulate enormous quantities of potassium in order to remain hypertonic to their environment; the internal potassium concentration may reach 4 to 7 M. Furthermore, their enzymes, ribosomes, and transport proteins require high potassium levels for stability and activity. In addition, the plasma membrane and cell wall of *Halobacterium* are stabilized by high concentrations of sodium ion. If the sodium concentration decreases too much, the wall and plasma membrane disintegrate. Extreme halophiles have successfully adapted to environmental conditions that would destroy most organisms. In the process they have become so specialized that they have lost ecological flexibility and can prosper only in a few extreme habitats. Phylum Euryarchaeota: The Halobacteria (section 20.3)

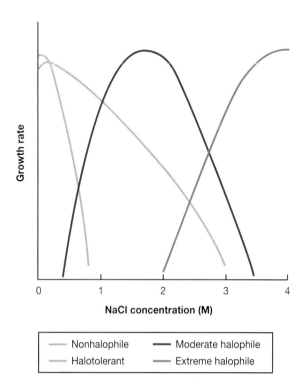

Figure 6.18 The Effects of Sodium Chloride on Microbial Growth. Four different patterns of microbial dependence on NaCl concentration are depicted. The curves are only illustrative and are not meant to provide precise shapes or salt concentrations required for growth.

Because the osmotic concentration of a habitat has such profound effects on microorganisms, it is useful to be able to express quantitatively the degree of water availability. Microbiologists generally use **water activity (a$_w$)** for this purpose (water availability also may be expressed as water potential, which is related to a$_w$). The water activity of a solution is 1/100 the relative humidity of the solution (when expressed as a percent). It is also equivalent to the ratio of the solution's vapor pressure (P_{soln}) to that of pure water (P_{water}).

$$a_w = \frac{P_{soln}}{P_{water}}$$

The water activity of a solution or solid can be determined by sealing it in a chamber and measuring the relative humidity after the system has come to equilibrium. Suppose after a sample is treated in this way, the air above it is 95% saturated—that is, the air contains 95% of the moisture it would have when equilibrated at the same temperature with a sample of pure water. The relative humidity would be 95% and the sample's water activity, 0.95. Water activity is inversely related to osmotic pressure; if a solution has high osmotic pressure, it's a$_w$ is low.

Microorganisms differ greatly in their ability to adapt to habitats with low water activity (**table 6.4**). A microorganism must expend extra effort to grow in a habitat with a low a$_w$ value be-

cause it must maintain a high internal solute concentration to retain water. Some microorganisms can do this and are **osmotolerant;** they will grow over wide ranges of water activity or osmotic concentration. For example, *Staphylococcus aureus* is halotolerant (figure 6.18) and can be cultured in media containing sodium chloride concentration up to about 3 M. It is well adapted for growth on the skin. The yeast *Saccharomyces rouxii* will grow in sugar solutions with a$_w$ values as low as 0.6. The photosynthetic protist *Dunaliella viridis* tolerates sodium chloride concentrations from 1.7 M to a saturated solution.

Although a few microorganisms are truly osmotolerant, most only grow well at water activities around 0.98 (the approximate a$_w$ for seawater) or higher. This is why drying food or adding large quantities of salt and sugar is so effective in preventing food spoilage. As table 6.4 shows, many fungi are osmotolerant and thus particularly important in the spoilage of salted or dried foods. Controlling food spoilage (section 40.3)

1. How do microorganisms adapt to hypotonic and hypertonic environments? What is plasmolysis?
2. Define water activity and briefly describe how it can be determined.
3. Why is it difficult for microorganisms to grow at low a$_w$ values?
4. What are halophiles and why does *Halobacterium* require sodium and potassium ions?

pH

pH is a measure of the hydrogen ion activity of a solution and is defined as the negative logarithm of the hydrogen ion concentration (expressed in terms of molarity).

$$pH = -\log [H^+] = \log(1/[H^+])$$

The pH scale extends from pH 0.0 (1.0 M H$^+$) to pH 14.0 (1.0 × 10^{-14} M H$^+$), and each pH unit represents a tenfold change in hydrogen ion concentration. **Figure 6.19** shows that the habitats in which microorganisms grow vary widely—from pH 0 to 2 at the acidic end to alkaline lakes and soil that may have pH values between 9 and 10.

It is not surprising that pH dramatically affects microbial growth. Each species has a definite pH growth range and pH growth optimum. **Acidophiles** have their growth optimum between pH 0 and 5.5; **neutrophiles,** between pH 5.5 and 8.0; and **alkalophiles** prefer the pH range of 8.0 to 11.5. Extreme alkalophiles have growth optima at pH 10 or higher. In general, different microbial groups have characteristic pH preferences. Most bacteria and protists are neutrophiles. Most fungi prefer more acidic surroundings, about pH 4 to 6; photosynthetic protists also seem to favor slight acidity. Many archaea are acidophiles. For example, the archaeon *Sulfolobus acidocaldarius* is a common inhabitant of acidic hot springs; it grows well around pH 1 to 3 and at high temperatures. The archaea *Ferroplasma acidarmanus* and *Picrophilus oshimae* can actually grow at pH 0, or very close to it.

Although microorganisms will often grow over wide ranges of pH and far from their optima, there are limits to their tolerance.

Table 6.4	Approximate Lower a_w Limits for Microbial Growth			
Water Activity	**Environment**	**Procaryotes**	**Fungi**	**Photosynthetic protists**
1.00—Pure water	Blood · Vegetables, Plant wilt · meat, fruit Seawater	Most gram-negative bacteria and other nonhalophiles		
0.95	Bread	Most gram-positive rods	*Basidiomycetes*	Most genera
0.90	Ham	Most cocci, *Bacillus*	*Fusarium* *Mucor, Rhizopus* Ascomycetous yeasts	
0.85	Salami	*Staphylococcus*	*Saccharomyces rouxii* (in salt)	
0.80	Preserves		*Penicillium*	
0.75	Salt lakes Salted fish	*Halobacterium* *Actinospora*	*Aspergillus*	*Dunaliella*
0.70	 Cereals, candy, dried fruit		*Aspergillus*	
0.60	 Chocolate Honey Dried milk		*Saccharomyces rouxii* (in sugars) *Xeromyces bisporus*	
0.55—DNA disordered				

Adapted from A. D. Brown, "Microbial Water Stress," in *Bacteriological Reviews*, 40(4):803–846 1976. Copyright © 1976 by the American Society for Microbiology. Reprinted by permission.

Drastic variations in cytoplasmic pH can harm microorganisms by disrupting the plasma membrane or inhibiting the activity of enzymes and membrane transport proteins. Most procaryotes die if the internal pH drops much below 5.0 to 5.5. Changes in the external pH also might alter the ionization of nutrient molecules and thus reduce their availability to the organism.

Microorganisms respond to external pH changes using mechanisms that maintain a neutral cytoplasmic pH. Several mechanisms for adjusting to small changes in external pH have been proposed. The plasma membrane is impermeable to protons. Neutrophiles appear to exchange potassium for protons using an antiport transport system. Extreme alkalophiles like *Bacillus alcalophilus* maintain their internal pH closer to neutrality by exchanging internal sodium ions for external protons. Internal buffering also may contribute to pH homeostasis. However, if the external pH becomes too acidic, other mechanisms come into play. When the pH drops below about 5.5 to 6.0, *Salmonella enterica* serovar Typhimurium and *E. coli* synthesize an array of new proteins as part of what has been called their acidic tolerance response. A proton-translocating ATPase contributes to this protective response, either by making more ATP or by pumping protons out of the cell. If the external pH decreases to 4.5 or lower, chaperone proteins such as acid shock proteins and heat shock proteins are synthesized. These prevent the acid denaturation of proteins and aid in the refolding of denatured proteins. Uptake of nutrients by the cell (section 5.6); Translation: Protein folding and molecular chaperones (section 11.8).

Microorganisms frequently change the pH of their own habitat by producing acidic or basic metabolic waste products. Fermentative microorganisms form organic acids from carbohydrates, whereas chemolithotrophs like *Thiobacillus* oxidize reduced sulfur components to sulfuric acid. Other microorganisms make their environment more alkaline by generating ammonia through amino acid degradation. Fermentation (section 9.7); Chemolithotrophy (section 9.11).

Because microorganisms change the pH of their surroundings, buffers often are included in media to prevent growth inhibition by large pH changes. Phosphate is a commonly used buffer and a good example of buffering by a weak acid ($H_2PO_4^-$) and its conjugate base (HPO_4^{2-}).

$$H^+ + HPO_4^{2-} \longrightarrow H_2PO_4^-$$

$$OH^- + H_2PO_4^- \longrightarrow HPO_4^{2-} + HOH$$

If protons are added to the mixture, they combine with the salt form to yield a weak acid. An increase in alkalinity is resisted because the weak acid will neutralize hydroxyl ions through proton donation to give water. Peptides and amino acids in complex media also have a strong buffering effect.

1. Define pH, acidophile, neutrophile, and alkalophile.
2. Classify each of the following organisms as an alkalophile, a neutrophile, or an acidophile: *Staphylococcus aureus, Microcycstis aeruginosa, Sulfolobus acidocaldarius,* and *Pseudomonas aeruginosa.* Which might be pathogens? Explain your choices.

Figure 6.19 The pH Scale. The pH scale and examples of substances with different pH values. The microorganisms are placed at their growth optima.

3. Describe the mechanisms microbes use to maintain a neutral pH. Explain how extreme pH values might harm microbes.

4. How do microorganisms change the pH of their environment? How does the microbiologist minimize this effect when culturing microbes in the lab?

Temperature

Environmental temperature profoundly affects microorganisms, like all other organisms. Indeed, microorganisms are particularly susceptible because their temperature varies with that of the external environment. A most important factor influencing the effect of temperature on growth is the temperature sensitivity of enzyme-catalyzed reactions. Each enzyme has a temperature at which it functions optimally (*see figure 8.19b*). At some temperature below the optimum, it ceases to be catalytic. As the temperature rises from this low temperature, the rate of catalysis increases to that observed for the optimal temperature. The ve-

locity of the reaction will roughly double for every 10°C rise in temperature. When all enzymes in a microbe are considered together, as the rate of each reaction increases, metabolism as a whole becomes more active, and the microorganism grows faster. However, beyond a certain point, further increases actually slow growth, and sufficiently high temperatures are lethal. High temperatures damage microorganisms by denaturing enzymes, transport carriers, and other proteins. Temperature also has a significant effect on microbial membranes. At very low temperatures, membranes solidify. At high temperatures, the lipid bilayer simply melts and disintegrates. In summary, when organisms are above their optimum temperature, both function and cell structure are affected. If temperatures are very low, function is affected but not necessarily cell chemical composition and structure.

Because of these opposing temperature influences, microbial growth has a fairly characteristic temperature dependence with distinct **cardinal temperatures**—minimum, optimum, and max-

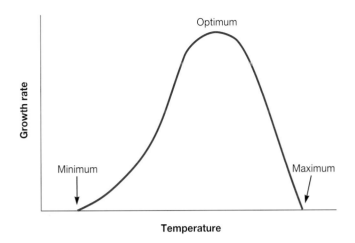

Figure 6.20   Temperature and Growth. The effect of temperature on growth rate.

imum growth temperatures (**figure 6.20**). Although the shape of the temperature dependence curve can vary, the temperature optimum is always closer to the maximum than to the minimum. The cardinal temperatures for a particular species are not rigidly fixed but often depend to some extent on other environmental factors such as pH and the available nutrients. For example, *Crithidia fasciculate,* a flagellated protist living in the gut of mosquitoes, will grow in a simple medium at 22 to 27°C. However, it cannot be cultured at 33 to 34°C without the addition of extra metals, amino acids, vitamins, and lipids.

The cardinal temperatures vary greatly between microorganisms (**table 6.5**). Optima usually range from 0°C to 75°C, whereas microbial growth occurs at temperatures extending from less than −20°C to over 120°C. Some archaea can even grow at 121°C (250°F), the temperature normally used in autoclaves (**Microbial Diversity and Ecology 6.1**). The major factor determining this growth range seems to be water. Even at the most extreme temperatures, microorganisms need liquid water to grow. The growth temperature range for a particular microorganism usually spans about 30 degrees. Some species (e.g., *Neisseria gonorrhoeae*) have a small range; others, like *Enterococcus faecalis,* will grow over a wide range of temperatures. The major microbial groups differ from one another regarding their maximum growth temperatures. The upper limit for protists is around 50°C. Some fungi can grow at temperatures as high as 55 to 60°C. Procaryotes can grow at much higher temperatures than eucaryotes. It has been suggested that eucaryotes are not able to manufacture organellar membranes that are stable and functional at temperatures above 60°C. The photosynthetic apparatus also appears to be relatively unstable because photosynthetic organisms are not found growing at very high temperatures.

Microorganisms such as those listed in table 6.5 can be placed in one of five classes based on their temperature ranges for growth (**figure 6.21**).

1. **Psychrophiles** grow well at 0°C and have an optimum growth temperature of 15°C or lower; the maximum is around

Table 6.5	Temperature Ranges for Microbial Growth		
	Cardinal Temperatures (°C)		
Microorganism	**Minimum**	**Optimum**	**Maximum**
Nonphotosynthetic Procaryotes			
Bacillus psychrophilus	−10	23–24	28–30
Micrococcus cryophilus	−4	10	24
Pseudomonas fluorescens	4	25–30	40
Staphylococcus aureus	6.5	30–37	46
Enterococcus caecalis	0	37	44
Escherichia coli	10	37	45
Neisseria gonorrhoeae	30	35–36	38
Thermoplasma acidophilum	45	59	62
Bacillus stearothermophilus	30	60–65	75
Thermus aquaticus	40	70–72	79
Sulfolobus acidocaldarius	60	80	85
Pyrococcus abyssi	67	96	102
Pyrodictium occultum	82	105	110
Pyrolobus fumarii	0	106	113
Photosynthetic Bacteria			
Rhodospirillum rubrum	ND[a]	30–35	ND
Anabaena variabilis	ND	35	ND
Oscillatoria tenuis	ND	ND	45–47
Synechococcus eximius	70	79	84
Protists			
Chlamydomonas nivalis	−36	0	4
Fragilaria sublinearis	−2	5–6	8–9
Amoeba proteus	4–6	22	35
Euglena gracilis	ND	23	ND
Skeletonema costatum	6	16–26	>28
Naegleria fowleri	20–25	35	40
Trichomonas vaginalis	25	32–39	42
Paramecium caudatum		25	28–30
Tetrahymena pyriformis	6–7	20–25	33
Cyclidium citrullus	18	43	47
Cyanidium caldarium	30–34	45–50	56
Fungi			
Candida scotti	0	4–15	15
Saccharomyces cerevisiae	1–3	28	40
Mucor pusillus	21–23	45–50	50–58

[a]ND, no data.

10°C. They are readily isolated from Arctic and Antarctic habitats; because 90% of the ocean is 5°C or colder, it constitutes an enormous habitat for psychrophiles. The psychrophilic protist *Chlamydomonas nivalis* can actually turn a snowfield or glacier pink with its bright red spores. Psychrophiles are widespread among bacterial taxa and are found in such genera as *Pseudomonas, Vibrio, Alcaligenes, Bacillus, Arthrobacter,*

Microbial Diversity & Ecology

6.1   Life above 100°C

Until recently the highest reported temperature for procaryotic growth was 105°C. It seemed that the upper temperature limit for life was about 100°C, the boiling point of water. Now thermophilic procaryotes have been reported growing in sulfide chimneys or "black smokers," located along rifts and ridges on the ocean floor, that spew sulfide-rich super-heated vent water with temperatures above 350°C (see **Box figure**). Evidence has been presented that these microbes can grow and reproduce at 121°C and can survive temperatures to 130°C for up to 2 hours. The pressure present in their habitat is sufficient to keep water liquid (at 265 atm; seawater doesn't boil until 460°C).

The implications of this discovery are many. The proteins, membranes, and nucleic acids of these procaryotes are remarkably temperature stable and provide ideal subjects for studying the ways in which macromolecules and membranes are stabilized. In the future it may be possible to design enzymes that can operate at very high temperatures. Some thermostable enzymes from these organisms have important industrial and scientific uses. For example, the Taq polymerase from the thermophile *Thermus aquaticus* is used extensively in the polymerase chain reaction. The polymerase chain reaction (section 14.3)

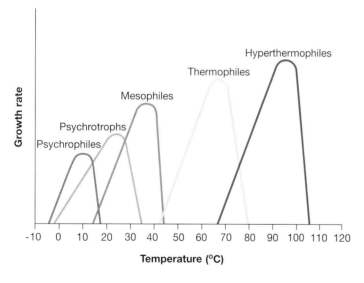

Figure 6.21   Temperature Ranges for Microbial Growth.
Microorganisms can be placed in different classes based on their temperature ranges for growth. They are ranked in order of increasing growth temperature range as psychrophiles, psychrotrophs, mesophiles, thermophiles, and hyperthermophiles. Representative ranges and optima for these five types are illustrated here.

Moritella, Photobacterium, and *Shewanella.* A psychrophilic archaeon, *Methanogenium,* has been isolated from Ace Lake in Antarctica. Psychrophilic microorganisms have adapted to their environment in several ways. Their enzymes, transport systems, and protein synthetic mechanisms function well at low temperatures. The cell membranes of psychrophilic microorganisms have high levels of unsaturated fatty acids and remain semifluid when cold. Indeed, many psychrophiles begin to leak cellular constituents at temperatures higher than 20°C because of cell membrane disruption.

2. Many species can grow at 0 to 7°C even though they have optima between 20 and 30°C, and maxima at about 35°C. These are called **psychrotrophs** or **facultative psychrophiles.** Psychrotrophic bacteria and fungi are major factors in the spoilage of refrigerated foods as described in chapter 40.

3. **Mesophiles** are microorganisms with growth optima around 20 to 45°C; they often have a temperature minimum of 15 to 20°C. Their maximum is about 45°C or lower. Most microorganisms probably fall within this category. Almost all human pathogens are mesophiles, as might be expected because their environment is a fairly constant 37°C.

4. Some microorganisms are **thermophiles;** they can grow at temperatures of 55°C or higher. Their growth minimum is usually around 45°C and they often have optima between 55 and 65°C. The vast majority are procaryotes although a few photosynthetic protists and fungi are thermophilic (table 6.5). These organisms flourish in many habitats including composts, self-heating hay stacks, hot water lines, and hot springs.

Thermophiles differ from mesophiles in many ways. They have more heat-stable enzymes and protein synthesis systems, which function properly at high temperatures. These proteins are stable for a variety of reasons. Heat-stable proteins have highly organized, hydrophobic interiors; more hydrogen bonds and other noncovalent bonds strengthen the structure. Larger quantities of amino acids such as proline also make the polypeptide chain less flexible. In addition, the proteins are stabilized and aided in folding by special chaperone proteins. There is evidence that in thermophilic bacteria, DNA is stabilized by special histonelike proteins. Their membrane lipids are also quite temperature stable. They tend to be more saturated, more branched, and of higher molecular weight. This increases the melting points of membrane lipids. Archaeal thermophiles have membrane lipids with ether linkages, which protect the lipids from hydrolysis at high temperatures. Sometimes archaeal lipids actually span the membrane to form a rigid, stable monolayer. Proteins (appendix I); Procaryotic cell membranes (section 3.2)

5. As mentioned previously, a few thermophiles can grow at 90°C or above and some have maxima above 100°C. Procaryotes that have growth optima between 80°C and about 113°C are called **hyperthermophiles.** They usually do not grow well below 55°C. *Pyrococcus abyssi* and *Pyrodictium occultum* are examples of marine hyperthermophiles found in hot areas of the seafloor.

1. What are cardinal temperatures?
2. Why does the growth rate rise with increasing temperature and then fall again at higher temperatures?
3. Define psychrophile, psychrotroph, mesophile, thermophile, and hyperthermophile.
4. What metabolic and structural adaptations for extreme temperatures do psychrophiles and thermophiles have?

Oxygen Concentration

The importance of oxygen to the growth of an organism correlates with its metabolism—in particular, with the processes it uses to conserve the energy supplied by its energy source. Almost all energy-conserving metabolic processes involve the movement of electrons through an electron transport system. For chemotrophs, an externally supplied terminal electron acceptor is critical to the functioning of the electron transport system. The nature of the terminal electron acceptor is related to an organism's oxygen requirement.

An organism able to grow in the presence of atmospheric O_2 is an **aerobe,** whereas one that can grow in its absence is an **anaerobe.** Almost all multicellular organisms are completely dependent on atmospheric O_2 for growth—that is, they are **obligate aerobes** (table 6.3). Oxygen serves as the terminal electron acceptor for the electron-transport chain in aerobic respiration. In addition, aerobic eucaryotes employ O_2 in the synthesis of sterols and unsaturated fatty acids. **Facultative anaerobes** do not require O_2 for growth but grow better in its presence. In the presence of oxygen they use aerobic respiration. **Aerotolerant anaerobes** such as *Enterococcus faecalis* simply ignore O_2 and grow equally well whether it is present or not. In contrast, **strict** or **obligate anaerobes** (e.g., *Bacteroides, Fusobacterium, Clostridium pasteurianum, Methanococcus, Neocallimastix*) do not tolerate O_2 at all and die in its presence. Aerotolerant and strict anaerobes cannot generate energy through aerobic respiration and must employ fermentation or anaerobic respiration for this purpose. Finally, there are aerobes such as *Campylobacter,* called **microaerophiles,** that are damaged by the normal atmospheric level of O_2 (20%) and require O_2 levels below the range of 2 to 10% for growth. The nature of bacterial O_2 responses can be readily determined by growing the bacteria in culture tubes filled with a solid culture medium or a special medium like thioglycollate broth, which contains a reducing agent to lower O_2 levels (**figure 6.22**). Oxidation-reduction reactions, electron carriers, and electron transport systems (section 8.6); Aerobic respiration (section 9.2); Anaerobic respiration (section 9.6); Fermentation (section 9.7)

A microbial group may show more than one type of relationship to O_2. All five types are found among the procaryotes and protozoa. Fungi are normally aerobic, but a number of species—particularly among the yeasts—are facultative anaerobes. Photosynthetic protists are almost always obligate aerobes. It should be noted that the ability to grow in both oxic and anoxic environments provides considerable flexibility and is an ecological advantage.

Although obligate anaerobes are killed by O_2, they may be recovered from habitats that appear to be oxic. In such cases they associate with facultative anaerobes that use up the available O_2 and thus make the growth of strict anaerobes possible. For example, the strict anaerobe *Bacteroides gingivalis* lives in the mouth where it grows in the anoxic crevices around the teeth.

Figure 6.22 Oxygen and Bacterial Growth. Each dot represents an individual bacterial colony within the agar or on its surface. The surface, which is directly exposed to atmospheric oxygen, will be oxic. The oxygen content of the medium decreases with depth until the medium becomes anoxic toward the bottom of the tube. The presence and absence of the enzymes superoxide dismutase (SOD) and catalase for each type are shown.

These different relationships with O_2 are due to several factors, including the inactivation of proteins and the effect of toxic O_2 derivatives. Enzymes can be inactivated when sensitive groups like sulfhydryls are oxidized. A notable example is the nitrogen-fixation enzyme nitrogenase, which is very oxygen sensitive. *Synthesis of amino acids: Nitrogen assimilation (section 10.5)*

Oxygen accepts electrons and is readily reduced because its two outer orbital electrons are unpaired. Flavoproteins, which function in electron transport, several other cell constituents, and radiation promote oxygen reduction. The result is usually some combination of the reduction products **superoxide radical, hydrogen peroxide,** and **hydroxyl radical.**

$$O_2 = e^- \rightarrow O_2\cdot \text{ (superoxide radical)}$$

$$O_2\cdot + e^- + 2H^+ \rightarrow H_2O_2 \text{ (hydrogen peroxide)}$$

$$H_2O_2 + e^- + H^+ \rightarrow H_2O + OH\cdot \text{ (hydroxyl radical)}$$

These products of oxygen reduction are extremely toxic because they oxidize and rapidly destroy cellular constituents. A microorganism must be able to protect itself against such oxygen products or it will be killed. Indeed, neutrophils and macrophages, two important immune system cells, use these toxic oxygen products to destroy invading pathogens. *Phagocytosis (section 31.3)*

Many microorganisms possess enzymes that afford protection against toxic O_2 products (figure 6.22). Obligate aerobes and facultative anaerobes usually contain the enzymes **superoxide dismutase (SOD)** and **catalase,** which catalyze the destruction of superoxide radical and hydrogen peroxide, respectively. Peroxidase also can be used to destroy hydrogen peroxide.

$$2O_2\cdot + 2H^+ \xrightarrow{\text{superoxide dismutase}} O_2 + H_2O_2$$

$$2H_2O_2 \xrightarrow{\text{catalase}} 2H_2O + O_2$$

$$H_2O_2 + NADH + H^+ \xrightarrow{\text{peroxidase}} 2H_2O + NAD^+$$

Aerotolerant microorganisms may lack catalase but almost always have superoxide dismutase. The aerotolerant *Lactobacillus plantarum* uses manganous ions instead of superoxide dismutase to destroy the superoxide radical. All strict anaerobes lack both enzymes or have them in very low concentrations and therefore cannot tolerate O_2.

Because aerobes need O_2 and anaerobes are killed by it, radically different approaches must be used when growing the two types of microorganisms. When large volumes of aerobic microorganisms are cultured, either the culture vessel is shaken to aerate the medium or sterile air must be pumped through the culture vessel. Precisely the opposite problem arises with anaerobes; all O_2 must be excluded. This can be accomplished in several ways. (1) Special anaerobic media containing reducing agents such as thioglycollate or cysteine may be used. The medium is boiled during preparation to dissolve its components; boiling also drives off oxygen very effectively. The reducing agents will eliminate any dissolved O_2 remaining within the medium so that anaerobes can grow beneath its surface. (2) Oxygen also may be eliminated from an anaerobic system by removing air with a vacuum pump and

Figure 6.23 An Anaerobic Work Chamber and Incubator. This anaerobic system contains an oxygen-free work area and an incubator. The interchange compartment on the right of the work area allows materials to be transferred inside without exposing the interior to oxygen. The anaerobic atmosphere is maintained largely with a vacuum pump and nitrogen purges. The remaining oxygen is removed by a palladium catalyst and hydrogen. The oxygen reacts with hydrogen to form water, which is absorbed by desiccant.

flushing out residual O_2 with nitrogen gas (**figure 6.23**). Often CO_2 as well as nitrogen is added to the chamber since many anaerobes require a small amount of CO_2 for best growth. (3) One of the most popular ways of culturing small numbers of anaerobes is by use of a GasPak jar (**figure 6.24**). In this procedure the environment is made anoxic by using hydrogen and a palladium catalyst to remove O_2 through the formation of water. The reducing agents in anaerobic agar also remove oxygen, as mentioned previously. (4) Plastic bags or pouches make convenient containers when only a few samples are to be incubated anaerobically. These have a catalyst and calcium carbonate to produce an anoxic, carbon-dioxide-rich atmosphere. A special solution is added to the pouch's reagent compartment; petri dishes or other containers are placed in the pouch; it then is clamped shut and placed in an incubator. A laboratory may make use of all these techniques since each is best suited for different purposes.

1. Describe the five types of O_2 relationships seen in microorganisms.
2. How do chemotrophic aerobes use O_2?
3. What are the toxic effects of O_2? How do aerobes and other oxygen-tolerant microbes protect themselves from these effects?
4. Describe four ways in which anaerobes may be cultured.

Pressure

Organisms that spend their lives on land or on the surface of water are always subjected to a pressure of 1 atmosphere (atm), and are never affected significantly by pressure. Yet many procaryotes live in the deep sea (ocean of 1,000 m or more in depth)

Figure 6.24   The GasPak Anaerobic System.   Hydrogen and carbon dioxide are generated by a GasPak envelope. The palladium catalyst in the chamber lid catalyzes the formation of water from hydrogen and oxygen, thereby removing oxygen from the sealed chamber.

where the hydrostatic pressure can reach 600 to 1,100 atm and the temperature is about 2 to 3°C. Many of these procaryotes are **baro-tolerant:** increased pressure adversely affects them but not as much as it does nontolerant microbes. Some procaryotes in the gut of deep-sea invertebrates such as amphipods (shrimplike crustaceans) and holothurians (sea cucumbers) are truly **barophilic**— they grow more rapidly at high pressures. These microbes may play an important role in nutrient recycling in the deep sea. A barophile recovered from the Mariana trench near the Philippines (depth about 10,500 m) is actually unable to grow at pressures below about 400 to 500 atm when incubated at 2°C. Thus far, barophiles have been found among several bacterial genera (e.g., *Photobacterium, Shewanella, Colwellia*). Some archaea are thermobarophiles (e.g., *Pyrococcus* spp., *Methanocaldococcus jannaschii*).   Microorganisms in marine environments (section 28.3)

Radiation

Our world is bombarded with electromagnetic radiation of various types (**figure 6.25**). This radiation often behaves as if it were composed of waves moving through space like waves traveling on the surface of water. The distance between two wave crests or troughs is the wavelength. As the wavelength of electromagnetic radiation decreases, the energy of the radiation increases—gamma rays and X rays are much more energetic than visible light or infrared waves. Electromagnetic radiation also acts like a stream of energy packets called photons, each photon having a quantum of energy whose value will depend on the wavelength of the radiation.

Sunlight is the major source of radiation on the Earth. It includes visible light, ultraviolet (UV) radiation, infrared rays, and radio waves. Visible light is a most conspicuous and important aspect of our environment: most life is dependent on the ability of photosynthetic organisms to trap the light energy of the sun. Almost 60% of the sun's radiation is in the infrared region rather than the visible portion of the spectrum. Infrared is the major source of the Earth's heat. At sea level, one finds very little ultraviolet radiation below about 290 to 300 nm. UV radiation of

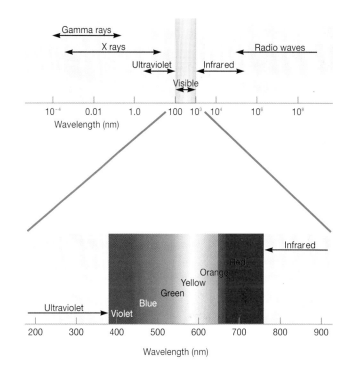

Figure 6.25   The Electromagnetic Spectrum.   A portion of the spectrum is expanded at the bottom of the figure.

wavelengths shorter than 287 nm is absorbed by O_2 in the Earth's atmosphere; this process forms a layer of ozone between 25 and 30 miles above the Earth's surface. The ozone layer then absorbs somewhat longer UV rays and reforms O_2. The fairly even distribution of sunlight throughout the visible spectrum accounts for the fact that sunlight is generally "white."   Phototrophy (section 9.12)

Many forms of electromagnetic radiation are very harmful to microorganisms. This is particularly true of **ionizing radiation,** radiation of very short wavelength and high energy, which can cause atoms to lose electrons (ionize). Two major forms of ionizing radiation are (1) X rays, which are artificially produced, and (2) gamma rays, which are emitted during radioisotope decay. Low levels of ionizing radiation will produce mutations and may indirectly result in death, whereas higher levels are directly lethal. Although microorganisms are more resistant to ionizing radiation than larger organisms, they will still be destroyed by a sufficiently large dose. Ionizing radiation can be used to sterilize items. Some procaryotes (e.g., *Deinococcus radiodurans*) and bacterial endospores can survive large doses of ionizing radiation. The use of physical methods in control: Radiation (section 7.4); *Deinococcus-Thermus* (section 21.2)

A variety of changes in cells are due to ionizing radiation; it breaks hydrogen bonds, oxidizes double bonds, destroys ring structures, and polymerizes some molecules. Oxygen enhances these destructive effects, probably through the generation of hydroxyl radicals (OH·). Although many types of constituents can be affected, it is reasonable to suppose that destruction of DNA is the most important cause of death.

Ultraviolet (UV) radiation can kill all kinds of microorganisms due to its short wavelength (approximately from 10 to 400 nm) and high energy. The most lethal UV radiation has a wavelength of 260 nm, the wavelength most effectively absorbed by DNA. The primary mechanism of UV damage is the formation of thymine dimers in DNA. Two adjacent thymines in a DNA strand are covalently joined to inhibit DNA replication and function. Microbes are protected from shorter wavelengths of UV light because they are absorbed by oxygen, as described previously. The damage caused by UV light that reaches Earth's surface can be repaired by several DNA repair mechanisms, which are discussed in chapter 13. Excessive exposure to UV light outstrips the organism's ability to repair the damage and death results. Longer wavelengths of UV light (near-UV radiation; 325 to 400 nm) are not absorbed by oxygen and so reach the Earth's surface. They can also harm microorganisms because they induce the breakdown of tryptophan to toxic photoproducts. It appears that these toxic tryptophan photoproducts plus the near-UV radiation itself produce breaks in DNA strands. The precise mechanism is not known, although it is different from that seen with 260 nm UV. Mutations and their chemical basis (section 13.1)

Visible light is immensely beneficial because it is the source of energy for photosynthesis. Yet even visible light, when present in sufficient intensity, can damage or kill microbial cells. Usually pigments called photosensitizers and O$_2$ are required. All microorganisms possess pigments like chlorophyll, bacteriochlorophyll, cytochromes, and flavins, which can absorb light energy, become excited or activated, and act as photosensitizers. The excited photosensitizer (P) transfers its energy to O$_2$ generating **singlet oxygen** (^{1}O$_2$).

$$P \xrightarrow{light} P \text{ (activated)}$$
$$P \text{ (activated)} + O_2 \longrightarrow P + {}^1O_2$$

Singlet oxygen is a very reactive, powerful oxidizing agent that will quickly destroy a cell. It is probably the major agent employed by phagocytes to destroy engulfed bacteria. Phagocytosis (section 31.3)

Many microorganisms that are airborne or live on exposed surfaces use carotenoid pigments for protection against photooxidation. Carotenoids effectively quench singlet oxygen—that is,

they absorb energy from singlet oxygen and convert it back into the unexcited ground state. Both photosynthetic and nonphotosynthetic microorganisms employ pigments in this way.

1. What are barotolerant and barophilic bacteria? Where would you expect to find them?
2. List the types of electromagnetic radiation in the order of decreasing energy or increasing wavelength.
3. Why is it so important that the Earth receives an adequate supply of sunlight? What is the importance of ozone formation?
4. How do ionizing radiation, ultraviolet radiation, and visible light harm microorganisms? How do microorganisms protect themselves against damage from UV and visible light?

6.6 MICROBIAL GROWTH IN NATURAL ENVIRONMENTS

Section 6.5 surveys the effects on microbial growth of individual environmental factors such as water availability, pH, and temperature. Although microbial ecology is introduced in more detail in chapters 27 to 30, we now briefly consider the effect of the environment as a whole on microbial growth.

Growth Limitation by Environmental Factors

The microbial environment is complex and constantly changing. It often contains low nutrient concentrations (**oligotrophic environment**) and exposes microbes to many overlapping gradients of nutrients and other environmental factors. The growth of microorganisms depends on both the nutrient supply and their tolerance of the environmental conditions present in their habitat at any particular time. Two laws clarify this dependence. Liebig's law of the minimum states that the total biomass of an organism will be determined by the nutrient present in the lowest concentration relative to the organism's requirements. This law applies in both the laboratory (figure 6.7) and in terrestrial and aquatic environments. An increase in a limiting essential nutrient such as phosphate will result in an increase in the microbial population until some other nutrient becomes limiting. If a specific nutrient is limiting, changes in other nutrients will have no effect. Shelford's law of tolerance states that there are limits to environmental factors below and above which a microorganism cannot survive and grow, regardless of the nutrient supply. This can readily be seen for temperature as shown in figure 6.21. Each microorganism has a specific temperature range in which it can grow. The same rule applies to other factors such as pH, oxygen level, and hydrostatic pressure in the marine environment. Inhibitory substances in the environment can also limit microbial growth. For instance, rapid, unlimited growth ensues if a microorganism is exposed to excess nutrients. Such growth quickly depletes nutrients and often results in the release of toxic products. Both nutrient depletion and the toxic products limit further growth. Another example is seen with microbes growing in nutrient-poor or oligotrophic environments, where the growth of microbes can be directly inhibited by a variety of natural substances including phenolics, tannins, ammonia, ethylene, and volatile sulfur compounds.

(a)   (b)

Figure 6.26   Morphology and Nutrient Absorption. Microorganisms can change their morphology in response to starvation and different limiting factors to improve their ability to survive. **(a)** *Caulobacter* has relatively short stalks when phosphorous is plentiful. **(b)** The stalks are extremely long under phosphorus-limited conditions.

In response to oligotrophic environments and intense competition, many microorganisms become more competitive in nutrient capture and exploitation of available resources. Often the organism's morphology will change in order to increase its surface area and ability to absorb nutrients. This can involve conversion of rod-shaped procaryotes to "mini" and "ultramicro" cells or changes in the morphology of prosthecate or stalked bacteria, in response to starvation. Nutrient deprivation induces many other changes as discussed previously (**figure 6.26**). For example, microorganisms can undergo a step-by-step shutdown of metabolism except for housekeeping maintenance genes.

Many factors can alter nutrient levels in oligotrophic environments. Microorganisms may sequester critical limiting nutrients, such as iron, making them less available to competitors. The atmosphere can contribute essential nutrients and support microbial growth. This is seen in the laboratory as well as natural environments. Airborne organic substances have been found to stimulate microbial growth in dilute media, and enrichment of growth media by airborne organic matter can allow significant populations of microorganisms to develop. Even distilled water, which contains traces of organic matter, can absorb one-carbon compounds from the atmosphere and grow microorganisms. The presence of such airborne nutrients and microbial growth, if not detected, can affect experiments in biochemistry and molecular biology, as well as studies of microorganisms growing in oligotrophic environments.

Counting and Identifying Microorganisms in Natural Environments

Microbial ecologists ask two important questions: What microbes are in a microbial habitat, and how many there are? Although microbiologists have developed numerous techniques for identifying and counting microbes, these questions are not easily answered. There are two reasons for this. First, many identification and counting methods rely on the ability of a microbe to form colonies. This presupposes that the microbiologist knows how to construct a growth medium and create environmental conditions that will support all the microbes in a habitat. Unfortunately, this knowledge eludes microbiologists, and it is estimated that only about 1% of the microbes in natural environments have been cultured. Increasingly, molecular methods are being used to analyze the diversity of microbial populations. The second reason is related to the "stress"

microbes experience in natural environments. John Postgate of the University of Sussex in England was one of the first to note that microorganisms stressed by survival in natural habitats—or in many selective laboratory media—were particularly sensitive to secondary stresses. Such stresses can produce viable microorganisms that have lost the ability to grow on media normally used for their cultivation. To determine the growth potential of such microorganisms, Postgate developed what is now called the Postgate microviability assay, in which microorganisms are cultured in a thin agar film under a coverslip. The ability of a cell to change its morphology, even if it does not grow beyond the single-cell stage, indicates that the microorganism does show "life signs."

Since that time many workers have developed additional sensitive microscopic, isotopic, and molecular genetic procedures to evaluate the presence and significance of these viable but nonculturable procaryotes (VBNC) in both lab and field. The new field of environmental genomics, or metagenomics, is discussed in chapter 15. In a more routine approach, levels of fluorescent antibody and acridine orange-stained cells often are compared with population counts obtained by the most probable number (MPN) method and plate counts using selective and nonselective media. The release of radioactive-labeled cell materials also is used to monitor stress effects on microorganisms. Despite these advances, the estimation of substrate-responsive viable cells by Postgate's method is still important. These studies show that even when pathogenic bacteria such as *Escherichia coli, Vibrio cholerae, Klebsiella pneumoniae, Enterobacter aerogenes,* and *Enterococcus faecalis* have lost their ability to grow on conventional laboratory media using standard cultural techniques, they still might be able to play a role in infectious disease. Microbial ecology and its methods: An overview (section 27.4); Water purification and sanitary analysis (section 41.1)

Biofilms

Although scientists observed as early as the 1940s that more microbes in aquatic environments were found attached to surfaces (sessile) rather than were free-floating (planktonic), only relatively recently has this fact gained the attention of microbiologists. These attached microbes are members of complex, slime-encased communities called **biofilms.** Biofilms are ubiquitous in nature. There they are most often seen as layers of slime on rocks or other objects in water (**figure 6.27a**). When they form on the hulls of

(a)

(b)

Figure 6.27   Examples of Biofilms.   Biofilms form on almost any surface exposed to microorganisms. **(a)** Biofilm on the surface of a stromatolite in Walker Lake (Nevada, USA), an alkaline lake. The biofilm consists primarily of the cyanobacterium *Calothrix*. **(b)** Photograph taken during surgery to remove a biofilm-coated artificial joint. The white material is composed of pus, bacterial and fungal cells, and the patient's white blood cells.

boats and ships, they cause corrosion, which limits the life of the ships and results in economic losses. Of major concern is the formation of biofilms on medical devices such as hip and knee implants (figure 6.27*b*). These biofilms often cause serious illness and failure of the medical device. Biofilm formation is apparently an ancient ability among the procaryotes, as evidence for biofilms can be found in the fossil record from about 3.4 billion years ago.

Biofilms can form on virtually any surface, once it has been conditioned by proteins and other molecules present in the environment (**figure 6.28**). Microbes reversibly attach to the conditioned surface and eventually begin releasing polysaccharides, proteins, and DNA. These polymers allow the microbes to stick more stably to the surface. As the biofilm thickens and matures, the microbes reproduce and secrete additional polymers. The end result is a complex, dynamic community of microorganisms. The microbes interact in a variety of ways. For instance, the waste products of one microbe may be the energy source for another microbe. The cells also communicate with each other as described next. Finally, the presence of DNA in the extracellular slime can be taken up by members of the biofilm community. Thus genes can be transferred from one cell (or species) to another.

While in the biofilm, microbes are protected from numerous harmful agents such as UV light, antibiotics, and other antimicrobial agents. This is due in part to the extracellular matrix in which they are embedded, but it also is due to physiological changes. Indeed, numerous proteins synthesized or activated in biofilm cells are not observed in planktonic cells and vice versa. The resistance of biofilm cells to antimicrobial agents has serious consequences. When biofilms form on a medical device such as a hip implant (figure 6.27*b*), they are difficult to kill and can cause serious illness. Often the only way to treat patients in this

situation is by removing the implant. Another problem with biofilms is that cells are regularly sloughed off (figure 6.28). This can have many consequences. For instance, biofilms in a city's water distribution pipes can serve as a source of contamination after the water leaves a water treatment facility.

Cell-Cell Communication Within Microbial Populations

For decades, microbiologists tended to think of bacterial populations as collections of individual cells growing and behaving independently. But about 30 years ago, it was discovered that the marine luminescent bacterium *Vibrio fischeri* controls its ability to glow by producing a small, diffusible substance called autoinducer. The autoinducer molecule was later identified as an **acylhomoserine lactone (AHL).** It is now known that many gram-negative bacteria make AHL molecular signals that vary in length and substitution at the third position of the acyl side chain (**figure 6.29**). In many of these species, the AHL is freely diffusible across the plasma membrane. Thus at a low cell density it diffuses out of the cell. However, when the cell population increases and AHL accumulates outside the cell, the diffusion gradient is reversed so that the AHL enters the cell. Because the influx of AHL is cell-density-dependent, it enables individual cells to assess population density. This is referred to as **quorum sensing;** a quorum usually refers to the minimum number of members in an organization, such as a legislative body, needed to conduct business. When AHL reaches a threshold level inside the cell, it serves to induce the expression of target genes that regulate a number of functions, depending on the microbe. These functions are most effective only if a large number of microbes are present. For instance, the light produced by one cell

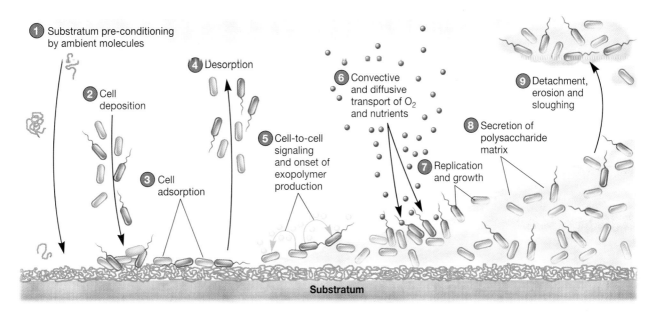

Figure 6.28   Biofilm Formation.

Signal and Structure	Representative Organism	Function Regulated
N-acyl homoserine lactone (AHL)	Vibrio fischeri	Bioluminescence
	Agrobacterium tumefaciens	Plasmid transfer
	Erwinia carotovora	Virulence and antibiotic production
	Pseudomonas aeruginosa	Virulence and biofilm formation
	Burkholderia cepacia	Virulence
Furanosylborate (AI-2)	Vibrio harveyi	Virulence
Cyclic thiolactone (AIP-II)	Staphylococcus aureus	Virulence
Hydroxy-palmitic acid methyl ester (PAME)	Ralstonia solanacearum	Virulence
Methyl dodecenoic acid (DSF)	Xanthomonas campestris	Virulence
Farnesoic acid	Candida albicans	Dimorphic transition and virulence

Figure 6.29   Representative Cell-Cell Communication Molecules.

(a) *E. scolopes*, the bobtail squid

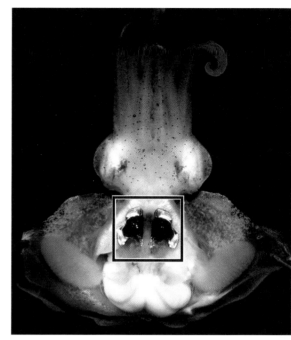

(b) Light organ

Figure 6.30 *Euprymna scolopes.* **(a)** *E. scolopes* is a warm-water squid that remains buried in sand during the day and feeds at night. **(b)** When feeding it uses its light organ (boxed, located on its ventral surface) to provide camouflage by projecting light downward. Thus the outline of the squid appears as bright as the water's surface to potential predators looking up through the water column. The light organ is colonized by a large number of *Vibrio fischeri* so autoinducer accumulates to a threshold concentration, triggering light production.

is not visible, but cell densities within the light organ of marine fish and squid reach 10^{10} cells per milliliter. This provides the animal with a flashlight effect while the microbes have a safe and nutrient-enriched habitat (**figure 6.30**). In fact, many of the processes regulated by quorum sensing involve host-microbe interactions such as symbioses and pathogenicity.   Global regulatory systems: Quorum sensing (section 12.5)

Many different bacteria use AHL signals. In addition to *V. fischeri* bioluminescence, the opportunistic pathogens *Burkholderia cepacia* and *Pseudomonas aeruginosa* use AHLs to regulate the expression of virulence factors (figure 6.29). These gram-negative bacteria cause debilitating pneumonia in people who are immunocompromised, and are important pathogens in cystic fibrosis patients. The plant pathogens *Agrobacterium tumefaciens* will not infect a host plant and *Erwinia carotorvora* will not produce antibiotics without AHL signaling. Finally, *B. cepacia*, *P. aeruginosa*, as well as *Vibrio cholerae* use AHL intercellular communication to control biofilm formation, an important strategy to evade the host's immune system.

The discovery of additional molecular signals made by a variety of microbes underscores the importance of cell-cell communication in regulating procaryotic processes. For instance, while only gram-negative bacteria are known to make AHLs, both gram-negative and gram-positive bacteria make autoinducer-2 (AI-2). Gram-positive bacteria usually exchange short peptides called oligopeptides instead of autoinducer-like molecules. Examples include *Enterococcus faecalis*, whose oligopeptide signal is used to

determine the best time to conjugate (transfer genes). Oligopeptide communication by *Staphylococcus aureus* and *Bacillus subtilis* is used to trigger the uptake of DNA from the environment. The soil microbe *Streptomyces griseus* produces a gamma-butyrolactone known as A-factor. This small molecule regulates both morphological differentiation and the production of the antibiotic streptomycin. Eucaryotic microbes also rely on cell-cell communication to coordinate key activities within a population. For example, the pathogenic fungus *Candida albicans* secretes farnesoic acid to govern morphology and virulence.

These examples of cell-cell communication demonstrate what might be called multicellular behavior in that many individual cells communicate and coordinate their activities to act as a unit. Other examples of such complex behavior is pattern formation in colonies and fruiting body formation in the myxobacteria. Isolation of pure cultures: Microbial growth on agar surfaces (section 5.8); Class *Deltaproteobacteria*: Order *Myxococcales* (section 22.4)

1. How are Liebig's law of the minimum and Shelford's law of tolerance related? Why are generation times in nature usually much longer than in culture?
2. Describe how microorganisms respond to oligotrophic environments.
3. Briefly discuss the Postgate microviability assay and other ways in which viable but nonculturable microorganisms can be counted or studied.
4. What is a biofilm? Why might life in a biofilm be advantageous for microbes?
5. What is quorum sensing? Describe how it occurs and briefly discuss its importance to microorganisms.

Summary

Growth is an increase in cellular constituents and results in an increase in cell size, cell number, or both.

6.1 The Procaryotic Cell Cycle

a. Most procaryotes reproduce by binary fission, a process in which the cell elongates and the chromosome is replicated and segregates to opposite poles of the cell prior to the formation of a septum, which divides the cell into two progeny cells (**figures 6.1** and **6.3**).

b. Two overlapping pathways function during the procaryotic cell cycle: the pathway for chromosome replication and segregation and the pathway for septum formation (**figure 6.2**). Both are complex and poorly understood. The partitioning of the progeny chromosomes may involve homologues of eucaryotic cytoskeletal proteins.

c. In rapidly dividing cells, initiation of DNA synthesis may occur before the previous round of synthesis is completed. This allows the cells to shorten the time needed for completing the cell cycle.

6.2 The Growth Curve

a. When microorganisms are grown in a closed system or batch culture, the resulting growth curve usually has four phases: the lag, exponential or log, stationary, and death phases (**figure 6.6**).

b. In the exponential phase, the population number of cells undergoing binary fission doubles at a constant interval called the doubling or generation time (**figure 6.10**). The mean growth rate constant (k) is the reciprocal of the generation time.

c. Exponential growth is balanced growth, cell components are synthesized at constant rates relative to one another. Changes in culture conditions (e.g., in shift-up and shift-down experiments) lead to unbalanced growth. A portion of the available nutrients is used to supply maintenance energy.

6.3 Measurement of Microbial Growth

a. Microbial populations can be counted directly with counting chambers, electronic counters, or fluorescence microscopy. Viable counting techniques such as the spread plate, the pour plate, or the membrane filter can be employed (**figures 6.12** and **6.14**).

b. Population changes also can be followed by determining variations in microbial mass through the measurement of dry weight, turbidity, or the amount of a cell component (**figure 6.15**).

6.4 The Continuous Culture of Microorganisms

a. Microorganisms can be grown in an open system in which nutrients are constantly provided and wastes removed.

b. A continuous culture system is an open system that can maintain a microbial population in the log phase. There are two types of these systems: chemostats and turbidostats (**figure 6.16**).

6.5 The Influence of Environmental Factors on Growth

a. Most bacteria, photosynthetic protists, and fungi have rigid cell walls and are hypertonic to the habitat because of solutes such as amino acids, polyols, and potassium ions. The amount of water actually available to microorganisms is expressed in terms of the water activity (a_w).

b. Although most microorganisms will not grow well at water activities below 0.98 due to plasmolysis and associated effects, osmotolerant organisms survive and even flourish at low a_w values. Halophiles actually require high sodium chloride concentrations for growth (**figure 6.18** and **table 6.3**).

c. Each species of microorganism has an optimum pH for growth and can be classified as an acidophile, neutrophile, or alkalophile (**figure 6.19**).

d. Microorganisms can alter the pH of their surroundings, and most culture media must be buffered to stabilize the pH.

e. Microorganisms have distinct temperature ranges for growth with minima, maxima, and optima—the cardinal temperatures. These ranges are determined by the effects of temperature on the rates of catalysis, protein denaturation, and membrane disruption (**figure 6.20**).

f. There are five major classes of microorganisms with respect to temperature preferences: (1) psychrophiles, (2) facultative psychrophiles or psychrotrophs, (3) mesophiles, (4) thermophiles, and (5) hyperthermophiles (**figure 6.21** and **table 6.3**).

g. Microorganisms can be placed into at least five different categories based on their response to the presence of O_2: obligate aerobes, facultative anaerobes, aerotolerant anaerobes, strict or obligate anaerobes, and microaerophiles (**figure 6.22** and **table 6.3**).

h. Oxygen can become toxic because of the production of hydrogen peroxide, superoxide radical, and hydroxyl radical. These are destroyed by the enzymes superoxide dismutase, catalase, and peroxidase.

i. Most deep-sea microorganisms are barotolerant, but some are barophilic and require high pressure for optimal growth.

j. High-energy or short-wavelength radiation harms organisms in several ways. Ionizing radiation—X rays and gamma rays—ionizes molecules and destroys DNA and other cell components. Ultraviolet (UV) radiation induces the formation of thymine dimers and strand breaks in DNA.

k. Visible light can provide energy for the formation of reactive singlet oxygen, which will destroy cells.

6.6 Microbial Growth in Natural Environments

a. Microbial growth in natural environments is profoundly affected by nutrient limitations and other adverse factors. Some microorganisms can be viable but unculturable and must be studied with special techniques.

b. Many microbes form biofilms, aggregations of microbes growing on surfaces and held together by extracellular polysaccharides (**figure 6.28**). Life in a biofilm has several advantages, including protection from harmful agents.

c. Often, bacteria will communicate with one another in a density-dependent way and carry out a particular activity only when a certain population density is reached. This phenomenon is called quorum sensing (**figures 6.29** and **6.30**).

Key Terms

acidophile 134
acylhomoserine lactone (AHL) 144
aerobe 139
aerotolerant anaerobe 139
alkalophile 134
anaerobe 139
balanced growth 123
barophilic 141
barotolerant 141
batch culture 123
binary fission 119
biofilm 143
cardinal temperatures 136
catalase 140
cell cycle 119
chemostat 131
coenocytic 119

colony forming units (CFU) 130
compatible solutes 132
continuous culture system 131
cytokinesis 121
doubling time 126
exponential phase 123
extremophiles 132
facultative anaerobe 139
facultative psychrophiles 138
FtsZ protein 122
generation time 126
growth 119
halophile 133
hydrogen peroxide 140
hydroxyl radical 140
hyperthermophile 139
ionizing radiation 141

lag phase 123
log phase 123
mean generation time 126
mean growth rate constant (k) 126
mesophile 138
microaerophile 139
MreB protein 120
neutrophile 134
obligate aerobe 139
obligate anaerobe 139
oligotrophic environment 142
origin of replication 120
osmotolerant 134
programmed cell death 125
psychrophile 137
psychrotroph 138
quorum sensing 144

replisome 120
septation 121
singlet oxygen 142
starvation proteins 124
stationary phase 124
strict anaerobe 139
superoxide dismutase (SOD) 140
superoxide radical 140
terminus 120
thermophile 138
turbidostat 132
ultraviolet (UV) radiation 142
unbalanced growth 123
viable but nonculturable (VBNC) 125
water activity (a_w) 134
Z ring 122

Critical Thinking Questions

1. As an alternative to diffusable signals, suggest another mechanism by which bacteria can quorum sense.

2. Design an "enrichment" culture medium and a protocol for the isolation and purification of a soil bacterium (e.g., *Bacillus subtilis*) from a sample of soil. Note possible contaminants and competitors. How will you adjust conditions of growth and what conditions will be adjusted to differentially enhance the growth of the *Bacillus?*

3. Design an experiment to determine if a slow-growing microbial culture is just exiting lag phase or is in exponential phase.

4. Why do you think the cardinal temperatures of some microbes change depending on other environmental conditions (e.g., pH)? Suggest one specific mechanism underlying such change.

Learn More

Atlas, R. M., and Bartha, R. 1997. *Microbial ecology: Fundamentals and applications,* 4th ed. Menlo Park, CA: Benjamin/Cummings.

Bartlett, D. H., and Roberts, M. F. 2000. Osmotic stress. In *Encyclopedia of microbiology,* 2d ed., vol. 3, J. Lederberg, editor-in-chief, 502–15. San Diego: Academic Press.

Cavicchioli, R., and Thomas, T. 2000. Extremophiles. In *Encyclopedia of microbiology,* 2d ed., vol. 2, J. Lederburg, editor-in-chief, 317–37. San Diego: Academic Press.

Cotter, P. D., and Hill, C. 2003. Surviving the acid test: Responses of gram-positive bacteria to low pH. *Microbiol. Mol. Biol. Rev.* 67(3):429–53.

Gitai, Z.; Thanbichler, M.; and Shapiro, L. 2005. The choreographed dynamics of bacterial chromosomes. *Trends Microbiol.* 13(5):221–28.

Hall-Stoodley, L.; Costerton, J. W.; and Stoodley, P. 2004. Bacterial biofilms: From the natural environment to infectious diseases. *Nature Rev. Microbiol.* 2:95–108.

Hoskisson, P. A., and Hobbs, G. 2005. Continuous culture—making a comeback? *Microbiology* 151:3153–59.

Kashefi, K., and Lovley, D. R. 2003. Extending the upper temperature limit for life. *Science* 301:934.

Krieg, N. R., and Hoffman, P. S. 1986. Microaerophily and oxygen toxicity. *Annu. Rev. Microbiol.* 40:107–30.

Krulwich, T. A., and Guffanti, A. A. 1989. Alkalophilic bacteria. *Annu. Rev. Microbiol.* 43:435–63.

Marr, A. G. 2000. Growth kinetics, bacterial. In *Encyclopedia of microbiology,* 2d ed., vol. 2, J. Lederberg, editor-in-chief, 584–89. San Diego: Academic Press.

Morita, R. Y. 2000. Low-temperature environments. In *Encyclopedia of microbiology,* 2d ed., vol. 3, J. Lederberg, editor-in-chief, 93–98. San Diego: Academic Press.

Nyström, T. 2004. Stationary-phase physiology. *Annu. Rev. Microbiol.* 58:161–81.

Visick, K. L., and Fuqua, C. 2005. Decoding microbial chatter: Cell-cell communication in bacteria. *J. Bact.* 187(16):5507–19.

Weiss, D. S. 2004. Bacterial cell division and the septal ring. *Molec. Microbiol.* 54(3):588–97.

Please visit the Prescott website at www.mhhe.com/prescott7
for additional references.

7

Control of Microorganisms by Physical and Chemical Agents

Bacteria are trapped on the surface of a membrane filter used to remove microorganisms from fluids.

PREVIEW

- Microbial population death is exponential, and the effectiveness of an agent is not fixed but influenced by many environmental factors.

- Solid objects can be sterilized by physical agents such as heat and radiation; liquids and gases are sterilized by heat, radiation, and filtration.

- Most chemical agents do not readily destroy bacterial endospores and therefore cannot sterilize objects; they are used as disinfectants, sanitizers, and antiseptics. Objects can be sterilized by gases like ethylene oxide and vaporized hydrogen peroxide that destroy endospores.

- Chemotherapeutic agents are chemicals used to kill or inhibit the growth of microorganisms within host tissues.

Chapters 5 and 6 are concerned with microbial nutrition and growth. In this chapter we address the subject of the control and destruction of microorganisms, a topic of immense practical importance. Although most microorganisms are beneficial and necessary for human well-being, microbial activities may have undesirable consequences, such as food spoilage and disease. Therefore it is essential to be able to kill a wide variety of microorganisms or inhibit their growth to minimize their destructive effects. The goal is twofold: (1) to destroy pathogens and prevent their transmission, and (2) to reduce or eliminate microorganisms responsible for the contamination of water, food, and other substances.

From the beginning of recorded history, people have practiced disinfection and sterilization, even though the existence of microorganisms was unknown. The Egyptians used fire to sterilize infectious material and disinfectants to embalm bodies, and the Greeks burned sulfur to fumigate buildings. Mosaic law commanded the Hebrews to burn any clothing suspected of being contaminated with leprosy. Today the ability to destroy micro-organisms is no less important: it makes possible the aseptic techniques used in microbiological research, the preservation of food, and the treatment and prevention of disease. The techniques described in this chapter are also essential to personal safety in both the laboratory and hospital (**Techniques & Applications 7.1**).

This chapter focuses on the control of microorganisms by physical and chemical agents, including chemotherapeutic agents, which are discussed in more detail in chapter 35. However, microbes can be controlled by many mechanisms that will not be considered in this chapter. For instance, the manipulation of environmental parameters is used extensively in the food industry to preserve foods. Increased solutes, such as salt and sugar, preserve meats, jams, and jellies. Microbial fermentations of milk and vegetables decrease the pH of these foods, creating new foods such as yogurt, cheese, and pickles—all of which have a longer shelf life than the milk and vegetables from which they are made. Heat and the generation of anoxic conditions are important in the preservation of canned foods, and ionizing radiation is used to extend the shelf life of seafood, fruits, and vegetables. The use of these control measures is described in more detail in chapter 40.

7.1 DEFINITIONS OF FREQUENTLY USED TERMS

Terminology is especially important when the control of microorganisms is discussed because words like disinfectant and antiseptic often are used loosely. The situation is even more confusing because a particular treatment can either inhibit growth or kill depending on the conditions. The types of control agents and their uses are outlined in **figure 7.1**.

We all labour against our own cure, for death is the cure of all diseases.

—Sir Thomas Browne

Techniques & Applications

7.1   Safety in the Microbiology Laboratory

Personnel safety should be of major concern in all microbiology laboratories. It has been estimated that thousands of infections have been acquired in the laboratory, and many persons have died because of such infections. The two most common laboratory-acquired bacterial diseases are typhoid fever and brucellosis. Most deaths have come from typhoid fever (20 deaths) and Rocky Mountain spotted fever (13 deaths). Infections by fungi (histoplasmosis) and viruses (Venezuelan equine encephalitis and hepatitis B virus from monkeys) are also not uncommon. Hepatitis is the most frequently reported laboratory-acquired viral infection, especially in people working in clinical laboratories and with blood. In a survey of 426 U.S. hospital workers, 40% of those in clinical chemistry and 21% in microbiology had antibodies to the hepatitis B virus, indicating their previous exposure (though only about 19% of these had disease symptoms).

Efforts have been made to determine the causes of these infections in order to enhance the development of better preventive measures. Although often it is not possible to determine the direct cause of infection, some major potential hazards are clear. One of the most

frequent causes of disease is the inhalation of an infectious aerosol. An aerosol is a gaseous suspension of liquid or solid particles that may be generated by accidents and laboratory operations such as spills, centrifuge accidents, removal of closures from shaken culture tubes, and plunging of contaminated loops into a flame. Accidents with hypodermic syringes and needles, such as self-inoculation and spraying solutions from the needle, also are common. Hypodermics should be employed only when necessary and then with care. Pipette accidents involving the mouth are another major source of infection; pipettes should be filled with the use of pipette aids and operated in such a way as to avoid creating aerosols.

People must exercise care and common sense when working with microorganisms. Operations that might generate infectious aerosols should be carried out in a biological safety cabinet. Bench tops and incubators should be disinfected regularly. Autoclaves must be maintained and operated properly to ensure adequate sterilization. Laboratory personnel should wash their hands thoroughly before and after finishing work.

Figure 7.1   Microbial Control Methods.

The ability to control microbial populations on inanimate objects, like eating utensils and surgical instruments, is of considerable practical importance. Sometimes it is necessary to eliminate all microorganisms from an object, whereas only partial destruction of the microbial population may be required in other situations. **Sterilization** [Latin *sterilis,* unable to produce offspring or barren] is the process by which all living cells, spores, and acellular entities (e.g., viruses, viroids, and prions) are either destroyed or removed from an object or habitat. A sterile object is totally free of viable microorganisms, spores, and other infectious agents. When sterilization is achieved by a chemical agent, the chemical is called a sterilant. In contrast, **disinfection** is the killing, inhibition, or removal of microorganisms that may cause disease. The primary goal is to destroy potential pathogens, but disinfection also substantially reduces the total microbial population. **Disinfectants** are agents, usually chemical, used to carry out disinfection and are normally used only on inanimate objects. A disinfectant does not necessarily sterilize an object because viable spores and a few microorganisms may remain. **Sanitization** is closely related to disinfection. In sanitization, the microbial population is reduced to levels that are considered safe by public health standards. The inanimate object is usually cleaned as well as partially disinfected. For example, sanitizers are used to clean eating utensils in restaurants. Prions (section 18.10); Viroids and virusoids (section 18.9)

It also is frequently necessary to control microorganisms on or in living tissue with chemical agents. **Antisepsis** [Greek *anti,* against, and *sepsis,* putrefaction] is the prevention of infection or sepsis and is accomplished with **antiseptics.** These are chemical agents applied to tissue to prevent infection by killing or inhibiting pathogen growth; they also reduce the total microbial population. Because they must not destroy too much host tissue, antiseptics are generally not as toxic as disinfectants. **Chemotherapy** is the use of chemical agents to kill or inhibit the growth of microorganisms within host tissue.

A suffix can be employed to denote the type of antimicrobial agent. Substances that kill organisms often have the suffix -cide [Latin *cida,* to kill]; a **germicide** kills pathogens (and many nonpathogens) but not necessarily endospores. A disinfectant or antiseptic can be particularly effective against a specific group, in which case it may be called a **bactericide, fungicide, algicide,** or **viricide.** Other chemicals do not kill, but they do prevent growth. If these agents are removed, growth will resume. Their names end in -static [Greek *statikos,* causing to stand or stopping]—for example, **bacteriostatic** and **fungistatic.**

Although these agents have been described in terms of their effects on pathogens, it should be noted that they also kill or inhibit the growth of nonpathogens as well. Their ability to reduce the total microbial population, not just to affect pathogen levels, is quite important in many situations.

---

1. Define the following terms: sterilization, sterilant, disinfection, disinfectant, sanitization, antisepsis, antiseptic, chemotherapy, germicide, bactericide, bacteriostatic.

7.2 THE PATTERN OF MICROBIAL DEATH

A microbial population is not killed instantly when exposed to a lethal agent. Population death, like population growth, is generally exponential or logarithmic—that is, the population will be reduced by the same fraction at constant intervals (**table 7.1**). If the logarithm of the population number remaining is plotted against the time of exposure of the microorganism to the agent, a straight-line plot will result (**figure 7.2**). When the population has been greatly reduced, the rate of killing may slow due to the survival of a more resistant strain of the microorganism.

To study the effectiveness of a lethal agent, one must be able to decide when microorganisms are dead, a task by no means as easy as with macroorganisms. It is hardly possible to take a bacterium's pulse. A bacterium is often defined as dead if it does not grow and reproduce when inoculated into culture medium that would normally support its growth. In like manner, an inactive virus cannot infect a suitable host. This definition has flaws, however. It has been demonstrated that when bacteria are exposed to certain conditions, they can remain alive but are temporarily unable to reproduce. When in this state, they are referred to as viable but nonculturable (VBNC) *(see figure 6.8).* In conventional tests to demonstrate killing by an antimicrobial agent, VBNC bacteria would be thought to be dead. This is a serious problem because

Table 7.1	A Theoretical Microbial Heat-Killing Experiment			
Minute	Microbial Number at Start of Minute[a]	Microorganisms Killed in 1 Minute (90% of total)[a]	Microorganisms at End of 1 Minute	Log$_{10}$ of Survivors
1	10^6	9×10^5	10^5	5
2	10^5	9×10^4	10^4	4
3	10^4	9×10^3	10^3	3
4	10^3	9×10^2	10^2	2
5	10^2	9×10^1	10	1
6	10^1	9	1	0
7	1	0.9	0.1	−1

[a]Assume that the initial sample contains 10^6 vegetative microorganisms per ml and that 90% of the organisms are killed during each minute of exposure. The temperature is 121°C.

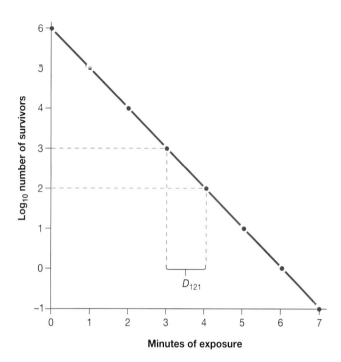

Figure 7.2   The Pattern of Microbial Death.   An exponential plot of the survivors versus the minutes of exposure to heating at 121°C. In this example the D_{121} value is 1 minute. The data are from table 7.1.

after a period of recovery, the bacteria may regain their ability to reproduce and cause infection.   The growth curve: Senescence and death (section 6.2)

---

1. Describe the pattern of microbial death and how one decides whether microorganisms are actually dead.

7.3   CONDITIONS INFLUENCING THE EFFECTIVENESS OF ANTIMICROBIAL AGENTS

Destruction of microorganisms and inhibition of microbial growth are not simple matters because the efficiency of an **antimicrobial agent** (an agent that kills microorganisms or inhibits their growth) is affected by at least six factors.

1. Population size. Because an equal fraction of a microbial population is killed during each interval, a larger population requires a longer time to die than a smaller one. This can be seen in the theoretical heat-killing experiment shown in table 7.1 and figure 7.2. The same principle applies to chemical antimicrobial agents.
2. Population composition. The effectiveness of an agent varies greatly with the nature of the organisms being treated because microorganisms differ markedly in susceptibility. Bacterial endospores are much more resistant to most antimicrobial agents than are vegetative forms, and younger cells are usually more readily destroyed than mature organisms. Some species are able to withstand adverse conditions better than

others. For instance, *Mycobacterium tuberculosis,* which causes tuberculosis, is much more resistant to antimicrobial agents than most other bacteria.

3. Concentration or intensity of an antimicrobial agent. Often, but not always, the more concentrated a chemical agent or intense a physical agent, the more rapidly microorganisms are destroyed. However, agent effectiveness usually is not directly related to concentration or intensity. Over a short range a small increase in concentration leads to an exponential rise in effectiveness; beyond a certain point, increases may not raise the killing rate much at all. Sometimes an agent is more effective at lower concentrations. For example, 70% ethanol is more effective than 95% ethanol because its activity is enhanced by the presence of water.
4. Duration of exposure. The longer a population is exposed to a microbicidal agent, the more organisms are killed (figure 7.2). To achieve sterilization, an exposure duration sufficient to reduce the probability of survival to 10^{-6} or less should be used.
5. Temperature. An increase in the temperature at which a chemical acts often enhances its activity. Frequently a lower concentration of disinfectant or sterilizing agent can be used at a higher temperature.
6. Local environment. The population to be controlled is not isolated but surrounded by environmental factors that may either offer protection or aid in its destruction. For example, because heat kills more readily at an acidic pH, acidic foods and beverages such as fruits and tomatoes are easier to pasteurize than foods with higher pHs like milk. A second important environmental factor is organic matter, which can protect microorganisms against heating and chemical disinfectants. Biofilms are a good example. The organic matter in a biofilm protects the biofilm's microorganisms, and the biofilm and its microbes often are hard to remove. Furthermore, it has been clearly documented that bacteria in biofilms are altered physiologically, and this makes them less susceptible to many antimicrobial agents. Because of the impact of organic matter, it may be necessary to clean objects, especially syringes and medical or dental equipment, before they are disinfected or sterilized. The same care must be taken when pathogens are destroyed during the preparation of drinking water. When a city's water supply has a high content of organic material, steps are taken to decrease the organic matter or to add more chlorine.   Microbial growth in natural environments: Biofilms (section 6.6)

---

1. Briefly explain how the effectiveness of antimicrobial agents varies with population size, population composition, concentration or intensity of the agent, treatment duration, temperature, and local environmental conditions.
2. How does being in a biofilm affect an organism's susceptibility to antimicrobial agents?
3. Suppose hospital custodians have been assigned the task of cleaning all showerheads in patient rooms in order to prevent the spread of infectious disease. What two factors would have the greatest impact on the effectiveness of the disinfectant the custodians use? Explain what that impact would be.

7.4 THE USE OF PHYSICAL METHODS IN CONTROL

Heat and other physical agents are normally used to control microbial growth and sterilize objects, as can be seen from the continual operation of the autoclave in every microbiology laboratory. The four most frequently employed physical agents are heat, low temperatures, filtration, and radiation.

Heat

Fire and boiling water have been used for sterilization and disinfection since the time of the Greeks, and heating is still one of the most popular ways to destroy microorganisms. Either moist or dry heat may be applied.

Moist heat readily kills viruses, bacteria, and fungi (**table 7.2**). Moist heat is thought to kill by degrading nucleic acids and by denaturing enzymes and other essential proteins. It may also disrupt cell membranes. Exposure to boiling water for 10 minutes is sufficient to destroy vegetative cells and eucaryotic spores. Unfortunately the temperature of boiling water (100°C or 212°F at sea level) is not high enough to destroy bacterial endospores, which may survive hours of boiling. Therefore boiling can be used for disinfection of drinking water and objects not harmed by water, but boiling does not sterilize.

In order to destroy bacterial endospores, moist heat sterilization must be carried out at temperatures above 100°C, and this requires the use of saturated steam under pressure. Steam sterilization is carried out with an **autoclave** (**figure 7.3**), a device somewhat like a fancy pressure cooker. The development of the autoclave by Chamberland in 1884 tremendously stimulated the growth of microbiology. Water is boiled to produce steam, which is released through the jacket and into the autoclave's chamber (figure 7.3b). The air initially present in the chamber is forced out until the chamber is filled with saturated steam and the outlets are closed. Hot, saturated steam continues to enter until the chamber reaches the desired temperature and pressure, usually 121°C and 15 pounds of pressure. At this temperature saturated steam destroys all vegetative cells and endospores in a small volume of liquid within 10 to 12 minutes. Treatment is continued for at least 15 minutes to provide a margin of safety. Of course, larger containers of liquid such as flasks and carboys require much longer treatment times.

Table 7.2	Approximate Conditions for Moist Heat Killing	
Organism	**Vegetative Cells**	**Spores**
Yeasts	5 minutes at 50–60°C	5 minutes at 70–80°C
Molds	30 minutes at 62°C	30 minutes at 80°C
Bacteria[a]	10 minutes at 60–70°C	2 to over 800 minutes at 100°C
		0.5–12 minutes at 121°C
Viruses	30 minutes at 60°C	

[a]Conditions for mesophilic bacteria.

Autoclaving must be carried out properly or the processed materials will not be sterile. If all air has not been flushed out of the chamber, it will not reach 121°C even though it may reach a pressure of 15 pounds. The chamber should not be packed too tightly because the steam needs to circulate freely and contact everything in the autoclave. Bacterial endospores will be killed only if they are kept at 121°C for 10 to 12 minutes. When a large volume of liquid must be sterilized, an extended sterilization time is needed because it takes longer for the center of the liquid to reach 121°C; 5 liters of liquid may require about 70 minutes. In view of these potential difficulties, a biological indicator is often autoclaved along with other material. This indicator commonly consists of a culture tube containing a sterile ampule of medium and a paper strip covered with spores of *Geobacillus stearothermophilus*. After autoclaving, the ampule is aseptically broken and the culture incubated for several days. If the test bacterium does not grow in the medium, the sterilization run has been successful. Sometimes either special tape that spells out the word *sterile* or a paper indicator strip that changes color upon sufficient heating is autoclaved with a load of material. If the word appears on the tape or if the color changes after autoclaving, the material is supposed to be sterile. These approaches are convenient and save time but are not as reliable as the use of bacterial endospores.

Many substances, such as milk, are treated with controlled heating at temperatures well below boiling, a process known as **pasteurization** in honor of its developer Louis Pasteur. In the 1860s the French wine industry was plagued by the problem of wine spoilage, which made wine storage and shipping difficult. Pasteur examined spoiled wine under the microscope and detected microorganisms that looked like the bacteria responsible for lactic acid and acetic acid fermentations. He then discovered that a brief heating at 55 to 60°C would destroy these microorganisms and preserve wine for long periods. In 1886 the German chemists V. H. and F. Soxhlet adapted the technique for preserving milk and reducing milk-transmissible diseases. Milk pasteurization was introduced into the United States in 1889. Milk, beer, and many other beverages are now pasteurized. Pasteurization does not sterilize a beverage, but it does kill any pathogens present and drastically slows spoilage by reducing the level of nonpathogenic spoilage microorganisms.

Many objects are best sterilized in the absence of water by **dry heat sterilization.** Some items are sterilized by incineration. For instance, inoculating loops, which are used routinely in the laboratory, can be sterilized in a small, bench-top incinerator (**figure 7.4**). Other items are sterilized in an oven at 160 to 170°C for 2 to 3 hours. Microbial death apparently results from the oxidation of cell constituents and denaturation of proteins. Dry air heat is less effective than moist heat. The spores of *Clostridium botulinum*, the cause of botulism, are killed in 5 minutes at 121°C by moist heat but only after 2 hours at 160°C with dry heat. However, dry heat has some definite advantages. It does not corrode glassware and metal instruments as moist heat does, and it can be used to sterilize powders, oils, and similar items. Most laboratories sterilize glassware and pipettes with dry heat. Despite these advantages, dry heat sterilization is slow and not suitable for heat-sensitive materials like many plastic and rubber items.

Figure 7.3 The Autoclave or Steam Sterilizer. **(a)** A modern, automatically controlled autoclave or sterilizer. **(b)** Longitudinal cross section of a typical autoclave showing some of its parts and the pathway of steam. From John J. Perkins, *Principles and Methods of Sterilization in Health Science,* 2nd edition, 1969. Courtesy of Charles C. Thomas, Publisher, Springfield, Illinois.

Because heat is so useful in controlling microorganisms, it is essential to have a precise measure of the heat-killing efficiency. Initially effectiveness was expressed in terms of thermal death point (TDP), the lowest temperature at which a microbial suspension is killed in 10 minutes. Because TDP implies that a certain temperature is immediately lethal despite the conditions, **thermal death time (TDT)** is now more commonly used. This is the shortest time needed to kill all organisms in a microbial suspension at a specific temperature and under defined conditions. However, such destruction is logarithmic, and it is theoretically not possible to completely destroy microorganisms in a sample, even with extended heating. Therefore an even more precise figure, the **decimal reduction time (D)** or **D value** has gained wide

acceptance. The decimal reduction time is the time required to kill 90% of the microorganisms or spores in a sample at a specified temperature. In a semilogarithmic plot of the population remaining versus the time of heating, the D value is the time required for the line to drop by one log cycle or tenfold (figure 7.2). The D value is usually written with a subscript, indicating the temperature for which it applies. D values are used to estimate the relative resistance of a microorganism to different temperatures through calculation of the **z value.** The z value is the increase in temperature required to reduce D to 1/10 its value or to reduce it by one log cycle when log D is plotted against temperature (**figure 7.5**). Another way to describe heating effectiveness is with the F value. The **F value** is the time in minutes at a specific tem-

Figure 7.4　Dry Heat Incineration.　Bench-top incinerators are routinely used to sterilize inoculating loops used in microbiology laboratories.

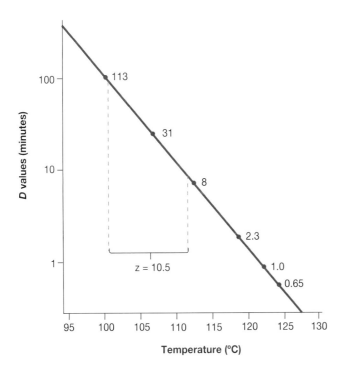

Figure 7.5　z Value Calculation.　The z value used in calculation of time-temperature relationships for survival of a test microorganism, based on D value responses at various temperatures. The z value is the increase in temperature needed to reduce the decimal reduction time (D) to 10% of the original value. For this homogeneous sample of a test microorganism the z value is 10.5°. The D values are plotted on a logarithmic scale.

perature (usually 250°F or 121.1°C) needed to kill a population of cells or spores.

The food processing industry makes extensive use of D and z values. After a food has been canned, it must be heated to eliminate the risk of botulism arising from *Clostridium botulinum* spores. Heat treatment is carried out long enough to reduce a population of 10^{12} *C. botulinum* spores to 10^0 (one spore); thus there is a very small chance of any can having a viable spore. The D value for these spores at 121°C is 0.204 minute. Therefore it would take 12D or 2.5 minutes to reduce 10^{12} spores to one spore by heating at 121°C. The z value for *C. botulinum* spores is 10°C—that is, it takes a 10°C change in temperature to alter the D value tenfold. If the cans were to be processed at 111°C rather than at 121°C, the D value would increase by tenfold to 2.04 minutes and the 12D value to 24.5 minutes. D values and z values for some common food-borne pathogens are given in **table 7.3.** Three D values are included for *Staphylococcus aureus* to illustrate the variation of killing rate with environment and the protective effect of organic material.　Controlling food spoilage (section 40.3)

Low Temperatures

Although our emphasis is on the destruction of microorganisms, often the most convenient control technique is to inhibit their growth and reproduction by the use of either freezing or refrigeration. This approach is particularly important in food microbiology. Freezing items at −20°C or lower stops microbial growth because of the low temperature and the absence of liquid water. Some microorganisms will be killed by ice crystal disruption of cell membranes, but freezing does not destroy all contaminating microbes. In fact, freezing is a very good method for long-term storage of microbial samples when carried out properly, and many laboratories have a low-temperature freezer for culture storage at −30 or −70°C. Because frozen food can contain many microorganisms, it should be thawed in a refrigerator and consumed promptly in order to avoid spoilage and pathogen growth.　The influence of environmental factors on growth: Temperature (section 6.5)

Refrigeration greatly slows microbial growth and reproduction, but does not halt it completely. Fortunately most pathogens are mesophilic and do not grow well at temperatures around 4°C. Refrigerated items may be ruined by growth of psychrophilic and psychrotrophic microorganisms, particularly if water is present. Thus refrigeration is a good technique only for shorter-term storage of food and other items.

1. Describe how an autoclave works. What conditions are required for sterilization by moist heat? What three things must one do when operating an autoclave to help ensure success?
2. In the past, spoiled milk was responsible for a significant proportion of infant deaths. Why is untreated milk easily spoiled? Why is boiling milk over prolonged periods not a desirable method for controlling spoilage and spread of milk-borne pathogens?
3. Define thermal death point (TDP), thermal death time (TDT), decimal reduction time (D) or D value, z value, and the F value.

Table 7.3	*D* Values and *z* Values for Some Food-Borne Pathogens		
Organism	**Substrate**	**D Value (°C) in Minutes**	**z Value (°C)**
Clostridium botulinum	Phosphate buffer	$D_{121} = 0.204$	10
Clostridium perfringens (heat-resistant strain)	Culture media	$D_{90} = 3–5$	6–8
Salmonella spp.	Chicken à la king	$D_{60} = 0.39–0.40$	4.9–5.1
Staphylococcus aureus	Chicken à la king	$D_{60} = 5.17–5.37$	5.2–5.8
	Turkey stuffing	$D_{60} = 15.4$	6.8
	0.5% NaCl	$D_{60} = 2.0–2.5$	5.6

Values taken from F. L. Bryan, 1979, "Processes That Affect Survival and Growth of Microorganisms," *Time-Temperature Control of Foodborne Pathogens,* Atlanta: Centers for Disease Control and Prevention, Atlanta, GA.

4. How can the *D* value be used to estimate the time required for sterilization? Suppose that you wanted to eliminate the risk of salmonellosis by heating your food ($D_{60} = 0.4$ minute, *z* value = 5.0). Calculate the 12*D* value at 60°C. How long would it take to achieve the same results by heating at 50, 55, and 65°C?

5. In table 7.3, why is the *D* value so different for the three conditions in which *S. aureus* might be found?

6. How can low temperatures be used to control microorganisms? Compare the control goal for using heat with that for using low temperatures.

Filtration

Filtration is an excellent way to reduce the microbial population in solutions of heat-sensitive material, and sometimes it can be used to sterilize solutions. Rather than directly destroying contaminating microorganisms, the filter simply removes them. There are two types of filters. **Depth filters** consist of fibrous or granular materials that have been bonded into a thick layer filled with twisting channels of small diameter. The solution containing microorganisms is sucked through this layer under vacuum, and microbial cells are removed by physical screening or entrapment and also by adsorption to the surface of the filter material. Depth filters are made of diatomaceous earth (Berkefield filters), unglazed porcelain (Chamberlain filters), asbestos, or other similar materials.

Membrane filters have replaced depth filters for many purposes. These circular filters are porous membranes, a little over 0.1 mm thick, made of cellulose acetate, cellulose nitrate, polycarbonate, polyvinylidene fluoride, or other synthetic materials. Although a wide variety of pore sizes are available, membranes with pores about 0.2 µm in diameter are used to remove most vegetative cells, but not viruses, from solutions ranging in volume from 1 ml to many liters. The membranes are held in special holders (**figure 7.6**) and are often preceded by depth filters made of glass fibers to remove larger particles that might clog the membrane filter. The solution is pulled or forced through the filter with a vacuum or with pressure from a syringe, peristaltic pump, or nitrogen gas bottle, and collected in previously sterilized containers. Membrane filters remove microorganisms by screening them out much as a sieve

separates large sand particles from small ones (**figure 7.7**). These filters are used to sterilize pharmaceuticals, ophthalmic solutions, culture media, oils, antibiotics, and other heat-sensitive solutions.

Air also can be sterilized by filtration. Two common examples are surgical masks and cotton plugs on culture vessels that let air in but keep microorganisms out. Other important examples are **laminar flow biological safety cabinets,** which employ **high-efficiency particulate air (HEPA) filters** (a type of depth filter) to remove 99.97% of 0.3 µm particles. Laminar flow biological safety cabinets or hoods force air through HEPA filters, then project a vertical curtain of sterile air across the cabinet opening. This protects a worker from microorganisms being handled within the cabinet and prevents contamination of the room (**figure 7.8**). A person uses these cabinets when working with dangerous agents such as *Mycobacterium tuberculosis* and tumor viruses. They are also employed in research labs and industries, such as the pharmaceutical industry, when a sterile working surface is needed for conducting assays, preparing media, examining tissue cultures, and the like.

Radiation

In chapter 6, the types of radiation and the ways in which radiation damages or destroys microorganisms were discussed. Microbiologists take advantage of the effects of ultraviolet and ionizing radiation to sterilize or disinfect objects.

Ultraviolet (UV) radiation around 260 nm (*see figure 6.25*) is quite lethal but does not penetrate glass, dirt films, water, and other substances very effectively. Because of this disadvantage, UV radiation is used as a sterilizing agent only in a few specific situations. UV lamps are sometimes placed on the ceilings of rooms or in biological safety cabinets to sterilize the air and any exposed surfaces. Because UV radiation burns the skin and damages eyes, people working in such areas must be certain the UV lamps are off when the areas are in use. Commercial UV units are available for water treatment (**figure 7.9**). Pathogens and other microorganisms are destroyed when a thin layer of water is passed under the lamps.

Ionizing radiation is an excellent sterilizing agent and penetrates deep into objects. It will destroy bacterial endospores and vegetative cells, both procaryotic and eucaryotic; however, ion-

(a)

(c)

1/4" stepped hose connector

Vent

Durapore filters

Filter support

Filling bell

Bell cap

Cross section
Millipak-40 filter unit

Liquid

Filter

Pore

Filter

Sterilized fluid

Vacuum pump suction

(b)

Figure 7.6   Membrane Filter Sterilization.   The liquid to be sterilized is pumped through a membrane filter and into a sterile container. **(a)** A complete filtering setup. The nonsterile solution is in the Erlenmeyer flask, *1*. A peristaltic pump, *2*, forces the solution through the membrane filter unit, *3*. **(b)** Schematic representation of a membrane filtration setup that uses a vacuum pump to force liquid through the filter. The inset shows a cross section of the filter and its pores, which are too small for microbes to pass through. **(c)** Cross section of a membrane filtration unit. Several membranes are used to increase its capacity.

izing radiation is not always effective against viruses. Gamma radiation from a cobalt 60 source is used in the cold sterilization of antibiotics, hormones, sutures, and plastic disposable supplies such as syringes. Gamma radiation has also been used to sterilize and "pasteurize" meat and other food (**figure 7.10**). Irradiation can eliminate the threat of such pathogens as *Escherichia coli*

O157:H7, *Staphylococcus aureus,* and *Campylobacter jejuni.* Based on the results of numerous studies, both the Food and Drug Administration and the World Health Organization have approved food irradiation and declared it safe. Currently irradiation is being used to treat poultry, beef, pork, veal, lamb, fruits, vegetables, and spices.

(a) *B. megaterium*

(b) *E. faecalis*

Figure 7.7 Membrane Filter Types. **(a)** *Bacillus megaterium* on an Ultipor nylon membrane with a bacterial removal rating of 0.2 μm (×2,000). **(b)** *Enterococcus faecalis* resting on a polycarbonate membrane filter with 0.4 μm pores (×5,900).

(a)

(b)

Figure 7.8 A Laminar Flow Biological Safety Cabinet. **(a)** A technician pipetting potentially hazardous material in a safety cabinet. **(b)** A schematic diagram showing the airflow pattern.

---

1. What are depth filters and membrane filters, and how are they used to sterilize liquids? Describe the operation of a biological safety cabinet.
2. Give the advantages and disadvantages of ultraviolet light and ionizing radiation as sterilizing agents. Provide a few examples of how each is used for this purpose.

7.5 THE USE OF CHEMICAL AGENTS IN CONTROL

Physical agents are generally used to sterilize objects. Chemicals, on the other hand, are more often employed in disinfection and antisepsis. The proper use of chemical agents is essential to lab-

oratory and hospital safety (**Techniques & Applications 7.2**). Chemicals also are employed to prevent microbial growth in food, and certain chemicals are used to treat infectious disease. Techniques & Applications 35.1: Standard microbiological practices

Many different chemicals are available for use as disinfectants, and each has its own advantages and disadvantages. In selecting an agent, it is important to keep in mind the characteristics of a desirable disinfectant. Ideally the disinfectant must be effective against a wide variety of infectious agents (gram-positive and gram-negative bacteria, acid-fast bacteria, bacterial endospores, fungi, and viruses) at low concentrations and in the presence of organic matter. Although the chemical must be toxic for infectious agents, it should not be toxic to people or corrosive for common materials. In practice, this balance between effectiveness and low toxicity for animals is hard to achieve. Some chemicals are used despite their low effectiveness because they are relatively nontoxic. The ideal disinfectant should be stable upon storage, odorless or with a pleasant odor, soluble in water and lipids for penetration into microorganisms, have a low surface tension so that it can enter cracks in surfaces, and be relatively inexpensive.

One potentially serious problem is the overuse of antiseptics. For instance, the antibacterial agent triclosan is found in products such as deodorants, mouthwashes, soaps, cutting boards, and baby toys. Unfortunately, the emergence of triclosan-resistant bacteria has become a problem. For example, *Pseudomonas aeruginosa* ac-

tively pumps the antiseptic out of the cell. There is now evidence that extensive use of triclosan also increases the frequency of bacterial resistance to antibiotics. Thus overuse of antiseptics can have unintended harmful consequences. Drug resistance (section 34.6)

The properties and uses of several groups of common disinfectants and antiseptics are surveyed next. Chemotherapeutic agents are briefly introduced at the end of this section. Many of the characteristics of disinfectants and antiseptics are summarized in **tables 7.4** and **7.5**. Structures of some common agents are given in **figure 7.11.**

Phenolics

Phenol was the first widely used antiseptic and disinfectant. In 1867 Joseph Lister employed it to reduce the risk of infection during surgery. Today phenol and phenolics (phenol derivatives) such as cresols, xylenols, and orthophenylphenol are used as disinfectants in laboratories and hospitals. The commercial disinfectant Lysol is made of a mixture of phenolics. Phenolics act by denaturing proteins and disrupting cell membranes. They have some real advantages as disinfectants: phenolics are tuberculocidal, effective in the presence of organic material, and remain active on surfaces long after application. However, they have a disagreeable odor and can cause skin irritation.

Hexachlorophene (figure 7.11) has been one of the most popular antiseptics because once applied it persists on the skin and reduces skin bacteria for long periods. However, it can cause brain damage and is now used in hospital nurseries only in response to a staphylococcal outbreak.

Figure 7.9   Ultraviolet (UV) Treatment System for Disinfection of Water.   Water flows through racks of UV lamps and is exposed to 254 nm UV radiation. This system has a capacity of several million gallons per day and can be used as an alternative to chlorination.

Alcohols

Alcohols are among the most widely used disinfectants and antiseptics. They are bactericidal and fungicidal but not sporicidal; some lipid-containing viruses are also destroyed. The two most popular alcohol germicides are ethanol and isopropanol, usually used in

Radiation room
Chamber with radiation shield
Conveyor system with pallets of sterilized materials
Radioactive source

(a)

(b)

Figure 7.10   Sterilization with Ionizing Radiation.   (a) An irradiation machine that uses radioactive cobalt 60 as a gamma radiation source to sterilize fruits, vegetables, meats, fish, and spices. **(b)** The universal symbol for irradiation that must be affixed to all irradiated materials.

Figure 7.11 Disinfectants and Antiseptics. The structures of some frequently used disinfectants and antiseptics.

sometimes has been used successfully as an alternative to chlorination in Europe and Canada.

Chlorine is also an excellent disinfectant for individual use because it is effective, inexpensive, and easy to employ. Small quantities of drinking water can be disinfected with halazone tablets. Halazone (parasulfone dichloramidobenzoic acid) slowly releases chloride when added to water and disinfects it in about a half hour. It is frequently used by campers lacking access to uncontaminated drinking water.

Chlorine solutions make very effective laboratory and household disinfectants. An excellent disinfectant-detergent combination can be prepared if a 1/40 dilution of household bleach is combined with a nonionic detergent, such as a dishwashing detergent, to give a 0.8% detergent concentration. This mixture will remove both dirt and bacteria.

Heavy Metals

For many years the ions of heavy metals such as mercury, silver, arsenic, zinc, and copper were used as germicides. These have now been superseded by other less toxic and more effective germicides (many heavy metals are more bacteriostatic than bactericidal). There are a few exceptions. In some hospitals, a 1% solution of silver nitrate is added to the eyes of infants to prevent ophthalmic gonorrhea. Silver sulfadiazine is used on burns. Copper sulfate is an effective algicide in lakes and swimming pools.

Heavy metals combine with proteins, often with their sulfhydryl groups, and inactivate them. They may also precipitate cell proteins.

Quaternary Ammonium Compounds

Quaternary ammonium compounds are detergents that have antimicrobial activity and are effective disinfectants. **Detergents** [Latin *detergere*, to wipe away] are organic cleansing agents that are amphipathic, having both polar hydrophilic and nonpolar hydrophobic components. The hydrophilic portion of a quaternary ammonium compound is a positively charged quaternary nitrogen; thus quaternary ammonium compounds are cationic detergents. Their antimicrobial activity is the result of their ability to disrupt microbial membranes; they may also denature proteins.

Cationic detergents like benzalkonium chloride and cetylpyridinium chloride kill most bacteria but not *M. tuberculosis* or endospores. They have the advantages of being stable and nontoxic but they are inactivated by hard water and soap. Cationic detergents are often used as disinfectants for food utensils and small instruments and as skin antiseptics. Several brands are on the market. Zephiran contains benzalkonium chloride and Ceepryn, cetylpyridinium chloride.

Aldehydes

Both of the commonly used aldehydes, formaldehyde and glutaraldehyde (figure 7.11), are highly reactive molecules that combine with nucleic acids and proteins and inactivate them, probably by cross-linking and alkylating molecules (**figure 7.12**). They are sporicidal and can be used as chemical sterilants. Formaldehyde is usually dissolved in water or alcohol before use.

A 2% buffered solution of glutaraldehyde is an effective disinfectant. It is less irritating than formaldehyde and is used to disinfect hospital and laboratory equipment. Glutaraldehyde usually disinfects objects within about 10 minutes but may require as long as 12 hours to destroy all spores.

Sterilizing Gases

Many heat-sensitive items such as disposable plastic petri dishes and syringes, heart-lung machine components, sutures, and catheters are sterilized with ethylene oxide gas (figure 7.11). Ethylene oxide (EtO) is both microbicidal and sporicidal and kills by combining with cell proteins. It is a particularly effective sterilizing agent because it rapidly penetrates packing materials, even plastic wraps.

Sterilization is carried out in a special ethylene oxide sterilizer, very much resembling an autoclave in appearance, that controls the EtO concentration, temperature, and humidity (**figure 7.13**). Because pure EtO is explosive, it is usually supplied in a 10 to 20% concentration mixed with either CO_2 or dichlorodifluoromethane. The ethylene oxide concentration, humidity, and temperature influence the rate of sterilization. A clean object can be sterilized if treated for 5 to 8 hours at 38°C or 3 to 4 hours at 54°C when the relative humidity is maintained

Figure 7.12   Effects of Glutaraldehyde. Glutaraldehyde polymerizes and then interacts with amino acids in proteins (left) or in peptidoglycan (right). As a result, the proteins are alkylated and cross-linked to other proteins, which inactivates them. The amino groups in peptidoglycan are also alkylated and cross-linked, which prevents them from participating in other chemical reactions such as those involved in peptidoglycan synthesis.

Figure 8.13 The Structure of Heme. Heme is composed of a porphyrin ring and an attached iron atom. It is the nonprotein component of many cytochromes. The iron atom alternatively accepts and releases an electron.

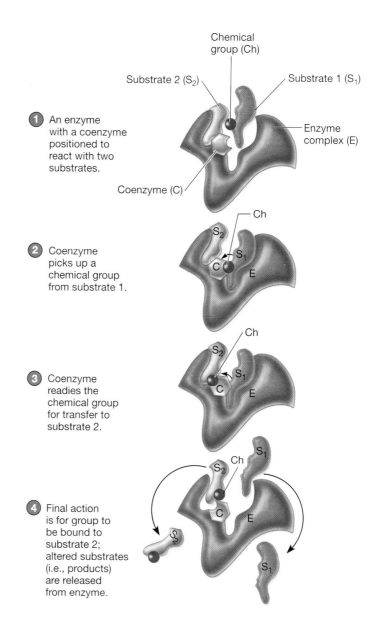

1. An enzyme with a coenzyme positioned to react with two substrates.

2. Coenzyme picks up a chemical group from substrate 1.

3. Coenzyme readies the chemical group for transfer to substrate 2.

4. Final action is for group to be bound to substrate 2; altered substrates (i.e., products) are released from enzyme.

Figure 8.14 Coenzymes as Carriers.

their component monosaccharides is exergonic and will occur spontaneously. However, an organic chemist would have to carry out this reaction in 6 M HCl and at 100°C for several hours to get it to go to completion. A cell, on the other hand, can accomplish the same reaction at neutral pH, at a much lower temperature, and in just fractions of a second. How are cells able to do this? They can do so because they manufacture proteins called enzymes that speed up chemical reactions. Enzymes are critically important to cells, since most biological reactions occur very slowly without them. Indeed, enzymes make life possible.

Structure and Classification of Enzymes

Enzymes may be defined as protein catalysts that have great specificity for the reaction catalyzed and the molecules acted on. A **catalyst** is a substance that increases the rate of a chemical reaction without being permanently altered itself. Thus enzymes speed up cellular reactions. The reacting molecules are called **substrates,** and the substances formed are the **products.** Proteins (appendix I)

Many enzymes are composed only of proteins. However, some enzymes consist of a protein, the **apoenzyme,** and a nonprotein component, a **cofactor,** required for catalytic activity. The complete enzyme consisting of the apoenzyme and its cofactor is called the **holoenzyme.** If the cofactor is firmly attached to the apoenzyme it is a **prosthetic group.** If the cofactor is loosely attached to the apoenzyme and can dissociate from the protein after products have been formed, it is called a **coenzyme.** Many coenzymes can carry one of the products to another enzyme (**figure 8.14**). For example, NAD^+ is a coenzyme that carries electrons within the cell. Many vitamins that humans require serve as

coenzymes or as their precursors. Niacin is incorporated into NAD^+ and riboflavin into FAD. Metal ions may also be bound to apoenzymes and act as cofactors.

Despite the large number and bewildering diversity of enzymes present in cells, they may be placed in one of six general classes (**table 8.2**). Enzymes usually are named in terms of the substrates they act on and the type of reaction catalyzed. For example, lactate dehydrogenase (LDH) removes hydrogens from lactate.

$$\text{Lactate} + NAD^+ \xrightleftharpoons{\text{LDH}} \text{pyruvate} + NADH + H^+$$

Lactate dehydrogenase can also be given a more complete and detailed name, L-lactate:NAD oxidoreductase. This name describes the substrates and reaction type with even more precision.

Table 8.2	Enzyme Classification	
Type of Enzyme	**Reaction Catalyzed by Enzyme**	**Example of Reaction**
Oxidoreductase	Oxidation-reduction reactions	Lactate dehydrogenase: Pyruvate + NADH + H $\rightleftharpoons$ lactate + NAD$^+$
Transferase	Reactions involving the transfer of groups between molecules	Aspartate carbamoyltransferase: Aspartate + carbamoylphosphate $\rightleftharpoons$ carbamoylaspartate + phosphate
Hydrolase	Hydrolysis of molecules	Glucose-6-phosphatase: Glucose-6-phosphate + H$_2$O → glucose + P$_i$
Lyase	Removal of groups to form double bonds or addition of groups to double bonds	Fumarate hydratase: L-malate $\rightleftharpoons$ fumarate + H$_2$O
Isomerase	Reactions involving isomerizations	Alanine racemase: L-alanine $\rightleftharpoons$ D-alanine
Ligase	Joining of two molecules using ATP energy (or that of other nucleoside triphosphates)	Glutamine synthetase: Glutamate + NH$_3$ + ATP → glutamine + ADP + P$_i$

$$\underset{/}{\overset{\backslash}{c}} = \underset{\backslash}{\overset{/}{c}} + x - y \rightleftharpoons - \overset{\overset{x}{|}}{\underset{|}{c}} - \overset{\overset{y}{|}}{\underset{|}{c}} -$$

The Mechanism of Enzyme Reactions

It is important to keep in mind that enzymes increase the rates of re-actions but do not alter their equilibrium constants. If a reaction is endergonic, the presence of an enzyme will not shift its equilibrium so that more products can be formed. Enzymes simply speed up the rate at which a reaction proceeds toward its final equilibrium.

How do enzymes catalyze reactions? Although a complete answer would be long and complex, some understanding of the mechanism can be gained by considering the course of a simple exergonic chemical reaction.

$$A + B \rightleftharpoons C + D$$

When molecules A and B approach each other to react, they form a **transition-state complex,** which resembles both the sub-strates and the products (**figure 8.15**). **Activation energy** is re-quired to bring the reacting molecules together in the correct way to reach the transition state. The transition-state complex can then resolve to yield the products C and D. The difference in free energy level between reactants and products is $\Delta G^{\circ\prime}$. Thus the equilibrium in our example will lie toward the products because $\Delta G^{\circ\prime}$ is negative (i.e., the products are at a lower energy level than the substrates).

As seen in figure 8.15, A and B will not be converted to C and D if they are not supplied with an amount of energy equivalent to the activation energy. Enzymes accelerate reactions by lowering the activation energy; therefore more substrate molecules will have sufficient energy to come together and form products. Even though the equilibrium constant (or $\Delta G^{\circ\prime}$) is unchanged, equilib-rium will be reached more rapidly in the presence of an enzyme because of this decrease in the activation energy.

Researchers have worked hard to discover how enzymes lower the activation energy of reactions, and the process is be-coming clearer. Enzymes bring substrates together at a specific

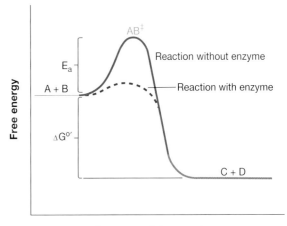

Figure 8.15   Enzymes Lower the Energy of Activation.
This figure traces the course of a chemical reaction in which A and B are converted to C and D. The transition-state complex is represented by AB‡, and the activation energy required to reach it, by E$_a$. The red line represents the course of the reaction in the presence of an enzyme. Note that the activation energy is much lower in the enzyme-catalyzed reaction.

place on their surface called the **active site** or **catalytic site** to form an **enzyme-substrate complex (figures 8.16, 8.17;** *see also appendix figure AI.19*). An enzyme can interact with its substrate in two general ways. It may be rigid and shaped to precisely fit the substrate so that the correct substrate binds specifically and is positioned properly for reaction. This mechanism is referred to as the lock-and-key model (figure 8.16). An enzyme also may change shape when it binds the substrate so that the active site surrounds and precisely fits the substrate. This has been called the

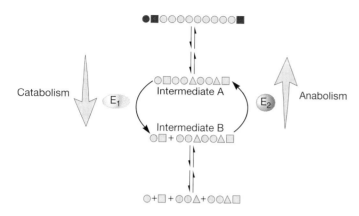

Figure 9.4 Amphibolic Pathway. A simplified diagram of an amphibolic pathway, such as glycolysis. Note that the interconversion of intermediates A and B is catalyzed by two separate enzymes, E_1 operating in the catabolic direction, and E_2 in the anabolic.

greatly increases metabolic efficiency by avoiding the need for a large number of less metabolically flexible pathways.

Although biosynthesis is the topic of chapter 10, it is important to point out that many catabolic pathways also are important in anabolism. They supply materials needed for biosynthesis, including precursor metabolites and reducing power. Precursor metabolites serve as the starting molecules for biosynthetic pathways. Reducing power is used to reduce the carbon skeletons provided by the precursor metabolites as they are transformed into amino acids, nucleotides, and the other small molecules needed for synthesis of macromolecules. Pathways that function both catabolically and anabolically are called **amphibolic pathways** [Greek *amphi*, on both sides]. Three of the most important amphibolic pathways are the Embden-Meyerhof pathway, the pentose phosphate pathway, and the tricarboxylic acid (TCA) cycle. Many of the reactions of the Embden-Meyerhof pathway and the TCA cycle are freely reversible and can be used to synthesize or degrade molecules depending on the nutrients available and the needs of the microbe. The few irreversible catabolic steps are bypassed in biosynthesis with alternate enzymes that catalyze the reverse reaction (**figure 9.4**). For example, the enzyme fructose bisphosphatase reverses the phosphofructokinase step when glucose is synthesized from pyruvate. The presence of two separate enzymes, one catalyzing the reversal of the other's reaction, permits independent regulation of the catabolic and anabolic functions of these amphibolic pathways. The precursor metabolites (section 10.2)

---

1. Compare and contrast fermentation and respiration. Give examples of the types of electron acceptors used by each process. What is the difference between aerobic respiration and anaerobic respiration?
2. Why is it to the cell's advantage to catabolize diverse organic energy sources by funneling them into a few common pathways?
3. What is an amphibolic pathway? Why are amphibolic pathways important?

9.3 THE BREAKDOWN OF GLUCOSE TO PYRUVATE

Microorganisms employ several metabolic pathways to catabolize glucose and other sugars. Because of this metabolic diversity, their metabolism is often confusing. To avoid confusion as much as possible, the ways in which microorganisms degrade sugars to pyruvate and similar intermediates are introduced by focusing on only three routes: (1) the Embden-Meyerhof pathway, (2) the pentose phosphate pathway, and (3) the Entner-Doudoroff pathway. In this text, these three pathways will be referred to collectively as **glycolytic pathways** or as **glycolysis** [Greek *glyco*, sweet, and *lysis*, a loosening]. However, in other texts, the term glycolysis is often reserved for use in reference only to the Embden-Meyerhof pathway. For the sake of simplicity, the detailed structures of metabolic intermediates are not used in pathway diagrams. Common metabolic pathways (appendix II)

The Embden-Meyerhof Pathway

The **Embden-Meyerhof pathway** is undoubtedly the most common pathway for glucose degradation to pyruvate in stage two of aerobic respiration. It is found in all major groups of microorganisms and functions in the presence or absence of O_2. As noted earlier, it is also an important amphibolic pathway and provides several precursor metabolites. The Embden-Meyerhof pathway occurs in the cytoplasmic matrix of procaryotes and eucaryotes.

The pathway as a whole may be divided into two parts (**figure 9.5** and *appendix II*). In the initial six-carbon phase, energy is consumed as glucose is phosphorylated twice, and is converted to fructose 1,6-bisphosphate. This preliminary phase consumes two ATP molecules for each glucose and "primes the pump" by adding phosphates to each end of the sugar. In essence, the organism invests some of its ATP so that more can be made later in the pathway.

The three-carbon, energy-conserving phase begins when the enzyme fructose 1,6-bisphosphate aldolase catalyzes the cleavage of fructose 1,6-bisphosphate into two halves, each with a phosphate group. One of the products, dihydroxyacetone phosphate, is immediately converted to glyceraldehyde 3-phosphate. This yields two molecules of glyceraldhyde 3-phosphate, which are then converted to pyruvate in a five-step process. Because dihydroxyacetone phosphate can be easily changed to glyceraldehyde 3-phosphate, both halves of fructose 1,6-bisphosphate are used in the three-carbon phase. First, glyceraldehyde 3-phosphate is oxidized with NAD^+ as the electron acceptor (to form NADH), and a phosphate (P_i) is simultaneously incorporated to give a high-energy molecule called 1,3-bisphosphoglycerate. The high-energy phosphate on carbon one is subsequently donated to ADP to produce ATP. This synthesis of ATP is called **substrate-level phosphorylation** because ADP phosphorylation is coupled with the exergonic breakdown of a high-energy bond. The role of ATP in metabolism (section 8.5)

A somewhat similar process generates a second ATP by substrate-level phosphorylation. The phosphate group on 3-phosphoglycerate shifts to carbon two, and 2-phosphoglycerate is

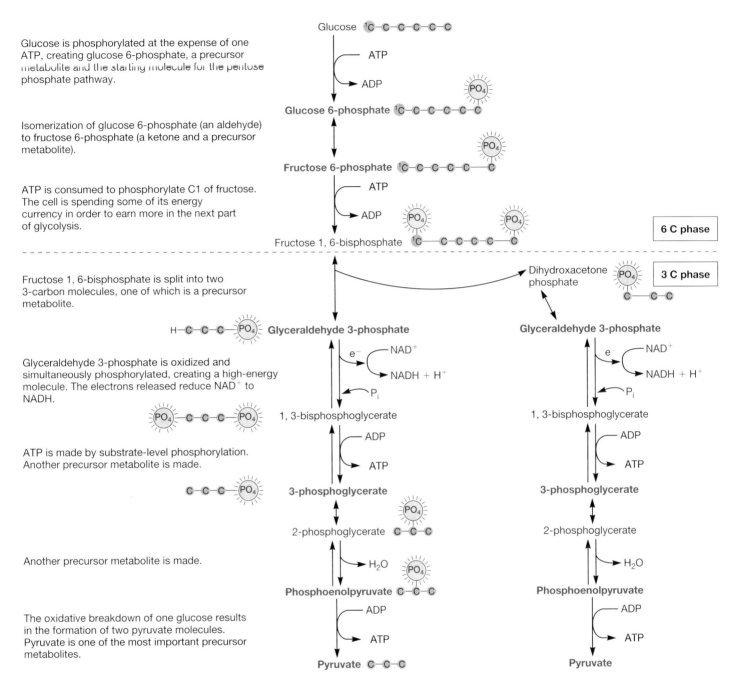

Glucose is phosphorylated at the expense of one ATP, creating glucose 6-phosphate, a precursor metabolite and the starting molecule for the pentose phosphate pathway.

Isomerization of glucose 6-phosphate (an aldehyde) to fructose 6-phosphate (a ketone and a precursor metabolite).

ATP is consumed to phosphorylate C1 of fructose. The cell is spending some of its energy currency in order to earn more in the next part of glycolysis.

Fructose 1, 6-bisphosphate is split into two 3-carbon molecules, one of which is a precursor metabolite.

Glyceraldehyde 3-phosphate is oxidized and simultaneously phosphorylated, creating a high-energy molecule. The electrons released reduce NAD^+ to NADH.

ATP is made by substrate-level phosphorylation. Another precursor metabolite is made.

Another precursor metabolite is made.

The oxidative breakdown of one glucose results in the formation of two pyruvate molecules. Pyruvate is one of the most important precursor metabolites.

Figure 9.5   Embden-Meyerhof Pathway.   This is one of three glycolytic pathways used to catabolize glucose to pyruvate, and it can function during aerobic respiration, anaerobic respiration, and fermentation. When used during a respiratory process, the electrons accepted by NAD^+ are transferred to an electron transport chain and are ultimately accepted by an exogenous electron acceptor. When used during fermentation, the electrons accepted by NAD^+ are donated to an endogenous electron acceptor (e.g., pyruvate). The Embden-Meyerhof pathway is also an important amphibolic pathway, as it generates several precursor metabolites (shown in blue).

dehydrated to form a second high-energy molecule, phosphoenolpyruvate. This molecule donates its phosphate to ADP forming a second ATP and pyruvate, the final product of the pathway.

The Embden-Meyerhof pathway degrades one glucose to two pyruvates by the sequence of reactions just outlined and shown in figure 9.5. ATP and NADH are also produced. The yields of ATP and NADH may be calculated by considering the two phases separately. In the six-carbon phase, two ATPs are used to form fructose 1,6-bisphosphate. For each glyceraldehyde 3-phosphate transformed into pyruvate, one NADH and two ATPs are formed.

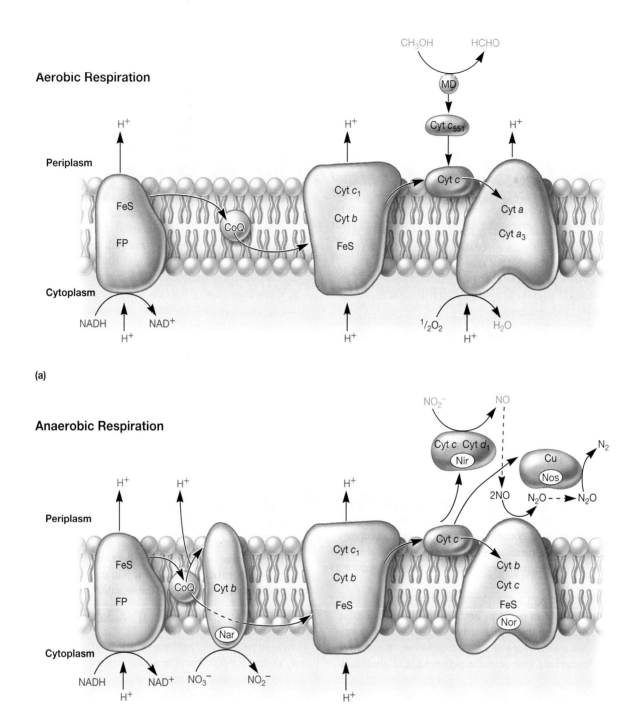

Aerobic Respiration

(a)

Anaerobic Respiration

(b)

Figure 9.16 *Paracoccus denitrificans* **Electron Transport Chains.** **(a)** The aerobic transport chain resembles a mitochondrial electron transport chain and uses oxygen as its acceptor. Methanol and methylamine can contribute electrons at the cytochrome *c* level. **(b)** The highly branched anaerobic chain is made of both membrane and periplasmic proteins. Nitrate is reduced to diatomic nitrogen by the collective action of four different reductases that receive electrons from CoQ and cytochrome *c*. Locations of proton movement are shown, but the number of protons involved has not been indicated. Abbreviations used: flavoprotein (FP), methanol dehydrogenase (MD), nitrate reductase (Nar), nitrite reductase (Nir), nitric oxide reductase (Nor), and nitrous oxide reductase (Nos).

Paracoccus and *Pseudomonas aeruginosa*), and the other is a copper-containing protein (e.g., *Alcaligenes*). Nitrite reductase seems to be periplasmic in gram-negative bacteria. Nitric oxide reductase catalyzes the formation of nitrous oxide from NO and is a membrane-bound cytochrome *bc* complex. In *P. denitrificans*, the nitrate reductase and nitric oxide reductase are membrane-bound, whereas nitrite reductase and nitrous oxide reductase are periplasmic (figure 9.16*b*).

In addition to *P. denitrificans*, some members of the genera *Pseudomonas* and *Bacillus* carry out denitrification. All three genera use denitrification as an alternative to aerobic respiration and may be considered facultative anaerobes. Indeed, if O_2 is present, these bacteria use aerobic respiration, which is far more efficient in capturing energy. In fact, the synthesis of nitrate reductase is repressed by O_2. Denitrification in anoxic soil results in the loss of soil nitrogen and adversely affects soil fertility. Biogeochemical cycling: Nitrogen cycle (section 27.2)

Not all microbes employ anaerobic respiration facultatively. Some are obligate anaerobes that can carry out only anaerobic respiration. The methanogens are an example. These archaea use CO_2 or carbonate as a terminal electron acceptor. They are called methanogens because the electron acceptor is reduced to methane. Bacteria such as *Desulfovibrio* are another example. They donate eight electrons to sulfate, reducing it to sulfide (S_2^- or H_2S).

$$SO_4^{2-} + 8e^- + 8H^+ \rightarrow S^{2-} + 4H_2O$$

It should be noted that both methanogens and *Desulfovibrio* are able to function as chemolithotrophs, using H_2 as an energy source (section 9.11).

As we saw for denitrification, anaerobic respiration using sulfate or CO_2 as the terminal electron acceptors is not as efficient in ATP synthesis as is aerobic respiration. Reduction in ATP yield arises from the fact that these alternate electron acceptors have less positive reduction potentials than O_2 (*see table 8.1*). The difference in standard reduction potential between a donor like NADH and nitrate is smaller than the difference between NADH and O_2. Because energy yield is directly related to the magnitude of the reduction potential difference, less energy is available to make ATP in anaerobic respiration. Nevertheless, anaerobic respiration is useful because it allows ATP synthesis by electron transport and oxidative phosphorylation in the absence of O_2. Anaerobic respiration is prevalent in oxygen-depleted soils and sediments.

The ability of microbes to use a variety of electron acceptors has ecological consequences. Often one sees a succession of microorganisms in an environment when several electron acceptors are present. For example, if O_2, nitrate, manganese ion, ferric ion, sulfate, and CO_2 are available in a particular environment, a predictable sequence of electron acceptor use takes place when an oxidizable substrate is available to the microbial population. Oxygen is employed as an electron acceptor first because it inhibits nitrate

use by microorganisms capable of respiration with either O_2 or nitrate. While O_2 is available, sulfate reducers and methanogens are inhibited because these groups are obligate anaerobes.

Once the O_2 and nitrate are exhausted and fermentation products (section 9.7), including hydrogen, have accumulated, competition for use of other electron acceptors begins. Manganese and iron are used first, followed by competition between sulfate reducers and methanogens. This competition is influenced by the greater energy yield obtained with sulfate as an electron acceptor. Differences in enzymatic affinity for hydrogen, an important energy and electron source used by both groups, also are important. The sulfate reducer *Desulfovibrio* grows rapidly and uses the available hydrogen at a faster rate than *Methanobacterium*. When the sulfate is exhausted, *Desulfovibrio* no longer oxidizes hydrogen, and the hydrogen concentration rises. The methanogens finally dominate the habitat and reduce CO_2 to methane. The subsurface biosphere (section 29.7)

---

1. Describe the process of anaerobic respiration. Is as much ATP produced in anaerobic respiration as in aerobic respiration? Why or why not?
2. What is denitrification? Why do farmers dislike this process?
3. *E. coli* can use O_2, fumarate^{2-}, or nitrate as a terminal electron acceptor under different conditions. What is the order of energy yield from highest to lowest for these electron acceptors? Explain your answer in thermodynamic terms.

9.7 FERMENTATIONS

Despite the tremendous ATP yield obtained by oxidative phosphorylation, some chemoorganotrophic microbes do not respire because either they lack electron transport chains or they repress the synthesis of electron transport chain components under anoxic conditions, making anaerobic respiration impossible. Yet NADH produced by the Embden-Meyerhof pathway reactions during glycolysis (figure 9.5) must still be oxidized back to NAD$^+$. If NAD$^+$ is not regenerated, the oxidation of glyceraldehyde 3-phosphate will cease and glycolysis will stop. Many microorganisms solve this problem by slowing or stopping pyruvate dehydrogenase activity and using pyruvate or one of its derivatives as an electron acceptor for the reoxidation of NADH in a **fermentation** process (**figure 9.17**). There are many kinds of fermentations, and they often are characteristic of particular microbial groups (**figure 9.18**). A few of the more common fermentations are introduced here, and several others are discussed at later points. Three unifying themes should be kept in mind when microbial fermentations are examined: (1) NADH is oxidized to NAD$^+$, (2) the electron acceptor is often either pyruvate or a pyruvate derivative, and (3) oxidative phosphorylation cannot operate, reducing the ATP yield per glucose significantly. In fermentation, the substrate is only partially oxidized, ATP is formed exclusively by substrate-level phosphorylation, and oxygen is not needed.

Glutamine synthetase reaction

COOH
|
CH$_2$
|
CH$_2$ + NH$_3$ + ATP ⟶
|
CH—NH$_2$
|
COOH

Glutamic acid

O
‖
C—NH$_2$
|
CH$_2$
|
CH$_2$ + ADP + P$_i$
|
CH — NH$_2$
|
COOH

Glutamine

Glutamate synthase reaction

COOH   COOH
| |
C=O   CH—NH$_2$
| |
CH$_2$ +   CH$_2$ + NADPH + H$^+$ ⟶
| | or
CH$_2$ CH$_2$ Fd$_{reduced}$
| |
COOH   C—NH$_2$
            ‖
            O

α-Ketoglutaric   Glutamine
acid

COOH   COOH
| |
CH—NH$_2$ CH—NH$_2$
| |
CH$_2$ +   CH$_2$ + NADP$^+$
| | or
CH$_2$ CH$_2$ Fd$_{oxidized}$
| |
COOH   COOH

Two glutamic acids

Figure 10.17   Glutamine Synthetase and Glutamate Synthase.   The glutamine synthetase and glutamate synthase reactions involved in ammonia assimilation. Some glutamine synthases use NADPH as an electron source; others use reduced ferredoxin (Fd). The nitrogen being incorporated and transferred is shown in turquoise.

Figure 10.18   Ammonia Incorporation Using Glutamine Synthetase and Glutamate Synthase.   This route is effective at low ammonia concentrations.

Assimilatory nitrate reduction takes place in the cytoplasm in bacteria. The first step in nitrate assimilation is its reduction to nitrite by **nitrate reductase,** an enzyme that contains both FAD and molybdenum (**figure 10.19**). NADPH is the electron source.

$$NO_3^- + NADPH + H^+ \longrightarrow NO_2^- + NADP^+ + H_2O$$

Nitrite is next reduced to ammonia with a series of two electron additions catalyzed by **nitrite reductase** and possibly other enzymes. The ammonia is then incorporated into amino acids by the routes already described.

Nitrogen Fixation

The reduction of atmospheric gaseous nitrogen to ammonia is called **nitrogen fixation.** Because ammonia and nitrate levels often are low and only a few bacteria and archaea can carry out nitrogen fixation (eucaryotic cells completely lack this ability), the rate of this process limits plant growth in many situations. Nitrogen fixation occurs in (1) free-living bacteria and archaea (e.g., *Azotobacter, Klebsiella, Clostridium,* and *Methanococcus*), (2) bacteria living in symbiotic association with plants such as legumes (*Rhizobium*), and (3) cyanobacteria (*Nostoc, Anabaena,* and *Trichodesmia*). The biological aspects of nitrogen fixation

Figure 10.19   Assimilatory Nitrate Reduction.   This sequence is thought to operate in bacteria that can reduce and assimilate nitrate nitrogen. See text for details.

Figure 10.20   Nitrogen Reduction.   A hypothetical sequence of nitrogen reduction by nitrogenase.

Figure 10.21   Structure of the Nitrogenase Fe Protein. The Fe protein's two subunits are arranged like a pair of butterfly wings with the iron sulfur cluster between the wings and at the "head" of the butterfly. The iron sulfur cluster is very exposed, which helps account for nitrogenase's sensitivity to oxygen. The oxygen can readily attack the exposed iron atoms.

are discussed in chapters 28 and 29. The biochemistry of nitrogen fixation is the focus of this section.

The reduction of nitrogen to ammonia is catalyzed by the enzyme **nitrogenase.** Although the enzyme-bound intermediates in this process are still unknown, it is believed that nitrogen is reduced by two-electron additions in a way similar to that illustrated in **figure 10.20.** The reduction of molecular nitrogen to ammonia is quite exergonic, but the reaction has a high activation energy because molecular nitrogen is an unreactive gas with a triple bond between the two nitrogen atoms. Therefore nitrogen reduction is expensive and requires a large ATP expenditure. At least 8 electrons and 16 ATP molecules, 4 ATPs per pair of electrons, are required.

$$N_2 + 8H^+ + 8e^- + 16ATP \longrightarrow$$
$$2NH_3 + H_2 + 16ADP + 16P_i$$

The electrons come from ferredoxin that has been reduced in a variety of ways: by photosynthesis in cyanobacteria, respiratory processes in aerobic nitrogen fixers, or fermentations in anaerobic bacteria. For example, *Clostridium pasteurianum* (an anaerobic bacterium) reduces ferredoxin during pyruvate oxidation, whereas the aerobic *Azotobacter* uses electrons from NADPH to reduce ferredoxin.

Nitrogenase is a complex system consisting of two major protein components, a MoFe protein (MW 220,000) joined with one or two Fe proteins (MW 64,000). The MoFe protein contains 2 atoms of molybdenum and 28 to 32 atoms of iron; the Fe protein has 4 iron atoms (**figure 10.21**). Fe protein is first reduced by ferredoxin, then it binds ATP (**figure 10.22**). ATP binding changes the conformation of the Fe protein and lowers its reduction potential, enabling it to reduce the MoFe protein. ATP is hydrolyzed when this electron transfer occurs. Finally, reduced MoFe protein donates electrons to atomic nitrogen. Nitrogenase is quite sensi-

tive to O_2 and must be protected from O_2 inactivation within the cell. In many cyanobacteria, this protection against oxygen is provided by a special structure called the heterocyst (*see figure 21.9*).

The reduction of N_2 to NH_3 occurs in three steps, each of which requires an electron pair (figures 10.20 and 10.22). Six electron transfers take place, and this requires a total 12 ATPs per N_2 reduced. The overall process actually requires at least 8 electrons and 16 ATPs because nitrogenase also reduces protons to H_2. The H_2 reacts with diimine (HN = NH) to form N_2 and H_2. This futile cycle produces some N_2 even under favorable conditions and makes nitrogen fixation even more expensive. Symbiotic

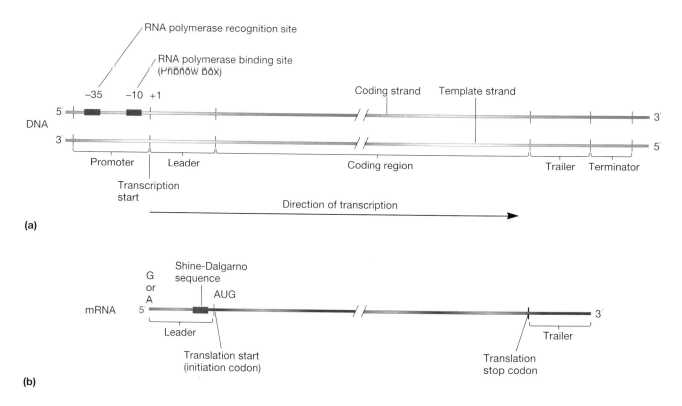

Figure 11.23 A Bacterial Structural Gene and Its mRNA Product. **(a)** The organization of a typical structural gene in bacteria. Leader and trailer sequences are included even though some genes lack one or both. Transcription begins at the +1 position in DNA and proceeds to the right as shown. The template is read in the 3′ to 5′ direction. **(b)** Messenger RNA product of the gene shown in part **a**. The first nucleotide incorporated into mRNA is usually GMP or AMP. Translation of the mRNA begins with the AUG initiation codon. Regulatory sites are not shown.

Immediately next to (and downstream of) the leader is the most important part of the gene, the **coding region** (figure 11.23). In genes that direct the synthesis of proteins, the coding region typically begins with the template DNA sequence 3′-TAC-5′. This produces the codon 5′-AUG-3′, which in bacteria codes for *N*-formylmethionine, a specially modified amino acid used to initiate protein synthesis. The remainder of the coding region consists of a sequence of codons that specifies the sequence of amino acids for that particular protein. The coding region ends with a special codon called the **stop codon,** which signals the end of the protein and stops the ribosome during translation. The stop codon is immediately followed by the **trailer sequence** (figure 11.23), which is needed for proper expression of the coding region of the gene. The stop codon is not recognized by RNA polymerase during transcription. Instead, a **terminator sequence** is used to stop transcription by dislodging the RNA polymerase from the template DNA.

Besides these basic components—the promoter, leader, coding region, trailer, and terminator—many bacterial genes have a variety of regulatory sites. These are locations where DNA-recognizing regulatory proteins bind to stimulate or prevent gene

expression. Regulatory sites often are associated with promoter function, and some consider them to be parts of special promoters. Two such sites, operator and activator binding sites, are discussed in sections 12.2 and 12.5. Certainly everything is not known about genes and their structure. With the ready availability of cloned genes and DNA sequencing technology, major discoveries continue to be made in this area.

Genes That Code for tRNA and rRNA

The DNA segments that code for tRNA and rRNA also are considered genes, although they give rise to important RNA rather than protein. In *E. coli* the genes for tRNA are fairly typical, consisting of a promoter and transcribed leader and trailer sequences that are removed during the process of tRNA maturation (**figure 11.24a**). The precise function of the leader is not clear; however, the trailer is required for transcription termination. Genes coding for tRNA may code for more than a single tRNA molecule or type of tRNA (figure 11.24a). The segments coding for tRNAs are separated by short spacer sequences that are removed after transcription by special ribonu-

Figure 11.24   tRNA and rRNA Genes.   (a) A tRNA precursor from *E. coli* that contains two tRNA molecules. The spacer and extra nucleotides at both ends are removed during processing. **(b)** The *E. coli* ribosomal RNA gene codes for a large transcription product that is cleaved into three rRNAs and one to three tRNAs. The 16S, 23S, and 5S rRNA segments are represented by blue lines, and tRNA sequences are placed in brackets. The seven copies of this gene vary in the number and kind of tRNA sequences.

cleases, at least one of which contains catalytic RNA. RNA molecules with catalytic activity are called **ribozymes (Microbial Tidbits 11.2)**.

The genes for rRNA also are similar in organization to genes coding for proteins because they have promoters, trailers, and terminators (figure 11.24*b*). Interestingly all the rRNAs are transcribed as a single, large precursor molecule that is cut up by ribonucleases after transcription to yield the final rRNA products. *E. coli* pre-rRNA spacer and trailer regions even contain tRNA genes. Thus the synthesis of tRNA and rRNA involve

posttranscriptional modification, a relatively rare process in procaryotes.

1. Define or describe the following: gene, template and coding strands, promoter, leader, coding region, reading frame, trailer, and terminator.
2. How do the genes of procaryotes and eucaryotes usually differ from each other?
3. Briefly discuss the general organization of tRNA and rRNA genes. How does their expression differ from that of structural genes with respect to posttranscriptional modification of the gene product?

Microbial Tidbits

11.2   Catalytic RNA (Ribozymes)

Biologists once thought that all cellular reactions were catalyzed by proteins called enzymes (*see section 8.7*). The discovery during 1981–1984 by Thomas Cech and Sidney Altman that RNA also can sometimes catalyze reactions has transformed our way of thinking about topics as diverse as catalysis and the origin of life. It is now clear that some RNA molecules, called ribozymes, catalyze reactions that alter either their own structure or that of other RNAs.

This discovery has stimulated scientists to hypothesize that the early Earth was an "RNA world" in which RNA acted as both the genetic material and a reaction catalyst. Experiments showing that introns from *Tetrahymena thermophila* can catalyze the formation of polycytidylic acid under certain circumstances have further encouraged such speculations. Some have suggested that RNA viruses are "living fossils" of the original RNA world.   The first self-replicating entity: The RNA world (section 19.1)

The best-studied ribozyme activity is the self-splicing of RNA. This process is widespread and occurs in *Tetrahymena* pre-rRNA; the mitochondrial rRNA and mRNA of yeast and other fungi; chloroplast tRNA, rRNA, and mRNA; in mRNA from some bacteriophages (e.g., the T4 phage of *E. coli*); and in the hepatitis delta virusoid. The 413-nucleotide rRNA intron of *T. thermophila* provides a good example of the self-splicing reaction. The reaction occurs in three steps and requires the presence of guanosine (see **Box figure**). First, the 3′-OH group of guanosine attacks the intron's 5′-phosphate group and cleaves the phosphodiester bond. Second, the new 3′-hydroxyl on the left exon attacks the 5′-phosphate of the right exon. This joins the two exons and releases the intron. Finally, the intron's 3′-hydroxyl attacks the phosphate bond of the nucleotide 15 residues from its end. This releases a terminal fragment and cyclizes the intron. Self-splicing of this rRNA occurs about 10 billion times faster than spontaneous RNA hydrolysis. Just as with enzyme proteins, the RNA's shape is essential to catalytic efficiency. The ribozyme even has Michaelis-Menten kinetics (*see figure 8.18*). The ribozyme from the hepatitis delta virusoid catalyzes RNA cleavage that is involved in its replication. It is unusual in that the same RNA can fold into two shapes with quite different catalytic activities: the regular RNA cleavage activity and an RNA ligation reaction.

The discovery of ribozymes has many potentially important practical consequences. Ribozymes act as "molecular scissors" and will enable researchers to manipulate RNA easily in laboratory experiments. It also might be possible to protect hosts by specifically removing RNA from pathogenic viruses, bacteria, and fungi. For example, ribozymes are being tested against the AIDS, herpes, and tobacco mosaic viruses.

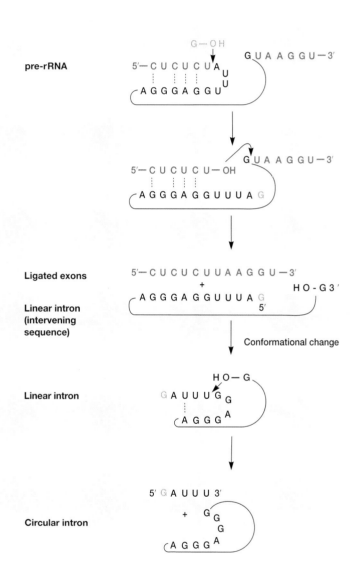

Ribozyme Action.   The mechanism of *Tetrahymena thermophila* pre-rRNA self-splicing. See text for details.

11.6 TRANSCRIPTION

As mentioned earlier, synthesis of RNA under the direction of DNA is called transcription. The RNA product has a sequence complementary to the DNA template directing its synthesis (**table 11.2**). Thymine is not normally found in mRNA and rRNA. Although adenine directs the incorporation of thymine during DNA replication, it usually codes for uracil during RNA synthesis. Transcription generates three kinds of RNA. Messenger RNA (mRNA) bears the message for protein synthesis. In *Bacteria* and *Archaea,* the mRNA often bears coding information transcribed from adjacent genes. Therefore it is said to be polygenic or polycistronic (**figure 11.25**). Eucaryotic mRNAs, on the other hand, are usually monocistronic, containing infor-

Table 11.2	RNA Bases Coded for by DNA
DNA Base	**Purine or Pyrimidine Incorporated into RNA**
Adenine	Uracil
Guanine	Cytosine
Cytosine	Guanine
Thymine	Adenine

Figure 11.25 A Polycistronic Bacterial Messenger RNA.

mation for a single polypeptide. **Transfer RNA (tRNA)** carries amino acids during protein synthesis, and **ribosomal RNA (rRNA)** molecules are components of ribosomes. The synthesis of bacterial mRNA is described first.

Transcription in *Bacteria*

RNA is synthesized under the direction of DNA by the enzyme **RNA polymerase.** The reaction is quite similar to that catalyzed by DNA polymerase (figure 11.14). ATP, GTP, CTP, and UTP are used to produce an RNA complementary to the DNA template. As mentioned earlier, these nucleotides contain ribose rather than deoxyribose (figure 11.15).

$$n[\text{ATP, GTP, CTP, UTP}] \xrightarrow[\text{DNA template}]{\text{RNA polymerase}} \text{RNA} + n\text{PP}_i$$

RNA synthesis, like DNA synthesis, proceeds in a 5′ to 3′ direction with new nucleotides being added to the 3′ end of the growing chain at a rate of about 40 nucleotides per second at 37°C (**figure 11.26**). In both DNA and RNA polymerase reactions, pyrophosphate (PP_i) is produced. It is then hydrolyzed to orthophosphate in a reaction catalyzed by the pyrophosphatase enzyme. Hydrolysis of the pyrophosphate product makes DNA and RNA synthesis irreversible. If the pyrophosphate level were too high, DNA and RNA would be degraded by a reversal of the polymerase reactions.

Most bacterial RNA polymerases contain five types of polypeptide chains: α, β, β′, ω, and σ (**figure 11.27**). The **core enzyme** is composed of five chains (α₂, β, β′, and ω) and catalyzes RNA synthesis. The **sigma factor** (σ) has no catalytic activity but helps the core enzyme recognize the start of genes. When sigma is bound to core enzyme, the six-subunit complex is termed **RNA polymerase holoenzyme.** Only holoenzyme can begin transcription, but as we will see, core enzyme completes RNA synthesis once it has been initiated. The precise functions of the α, β, β′, and ω polypeptides are not yet clear. The α subunits

seem to be involved in the assembly of the core enzyme, recognition of promoters, and interaction with some regulatory factors. The binding site for DNA is on β′, and the ω subunit seems to be involved in stabilizing the conformation of the β′ subunit. The β subunit binds ribonucleotide substrates. Rifampin, an RNA polymerase inhibitor, binds to the β subunit.

Recently the atomic structure of RNA polymerase from *Thermus aquaticus* has been determined (figure 11.27a,b). In this bacterium, the core enzyme is composed of four different subunits (α₂, β, β′, and ω) and is complexed with the sigma factor (σ). The enzyme is claw-shaped with a clamp domain that closes on an internal channel, which contains an essential magnesium and the active site. Sigma interacts extensively with the core enzyme and specifically binds to elements of the promoter. It also may widen the channel so that DNA can enter the interior of the polymerase complex.

Transcription involves three separate processes: initiation, elongation, and termination. Only a relatively short segment of DNA is transcribed (unlike replication in which the entire chromosome must be copied), and initiation begins when the RNA polymerase binds to the promoter for the gene. RNA polymerase core enzyme is not able to bind DNA tightly or specifically. This situation is drastically changed when sigma is bound to core to make the holoenzyme, which binds the promoter tightly. Recall that the promoter serves only as a target for the binding of the RNA polymerase and is not transcribed. Bacterial promoters have two characteristic features: a sequence of six bases (often TTGACA) about 35 bases pairs before the transcription starting point and a TATAAT sequence, or **Pribnow box,** usually about 10 base pairs upstream of the transcriptional start site (**figure 11.28;** also 11.23). These regions are called the −35 and −10 sites, respectively, while the first nucleotide to be transcribed is referred to as the +1 site. As noted previously, RNA polymerase holoenzyme recognizes the specific sequences at the −10 and −35 sites of promoters. Because the sites must be similar in all promoters, they are called **consensus sequences.**

Once bound to the promoter site, RNA polymerase is able to unwind the DNA without the aid of helicases (**figure 11.29**). The −10 site is rich in adenines and thymines, making it easier to break the hydrogen bonds that keep the DNA double stranded; when the DNA is unwound at this region, it is called **open complex.** A region of unwound DNA equivalent to about two turns of the helix (about 16–20 bases pairs) becomes the "transcription bubble," which moves with the RNA polymerase as it proceeds to transcribe mRNA from the template DNA strand during elongation (**figure 11.30**). Within the transcription bubble, a temporary RNA:DNA hybrid is formed. As the RNA polymerase progresses in the 3′ to 5′ direction along the DNA template, the sigma factor soon dissociates from core RNA polymerase and is available to aid another unit of core enzyme initiate transcription. The mRNA is made in the 5′ to 3′ direction so it is complementary and antiparallel to the template DNA. As elongation of the mRNA continues, single-stranded mRNA is released and the two strands of DNA behind the transcription bubble resume their double helical structure. As shown in figure 11.26, RNA polymerase is a remarkable

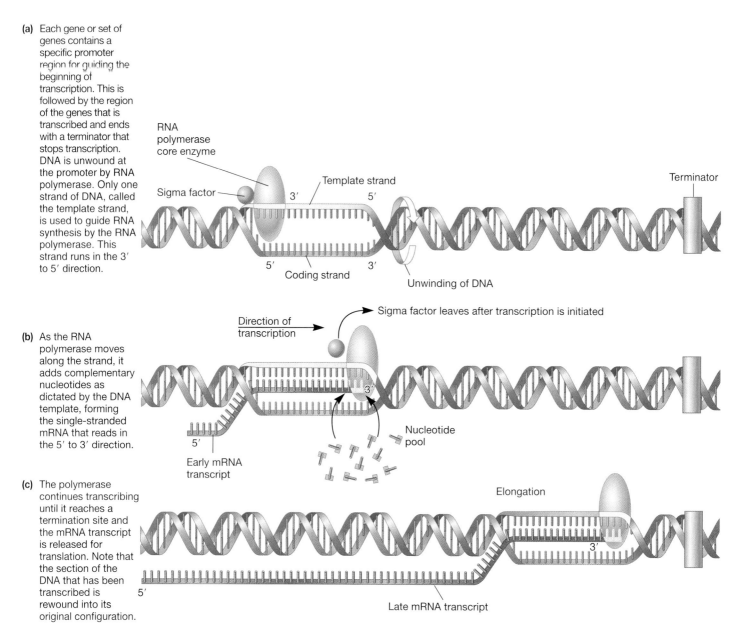

(a) Each gene or set of genes contains a specific promoter region for guiding the beginning of transcription. This is followed by the region of the genes that is transcribed and ends with a terminator that stops transcription. DNA is unwound at the promoter by RNA polymerase. Only one strand of DNA, called the template strand, is used to guide RNA synthesis by the RNA polymerase. This strand runs in the 3′ to 5′ direction.

RNA polymerase core enzyme

Sigma factor

Template strand

3′   5′

5′   3′

Coding strand

Terminator

Unwinding of DNA

(b) As the RNA polymerase moves along the strand, it adds complementary nucleotides as dictated by the DNA template, forming the single-stranded mRNA that reads in the 5′ to 3′ direction.

Direction of transcription

Sigma factor leaves after transcription is initiated

3′

5′

Early mRNA transcript

Nucleotide pool

(c) The polymerase continues transcribing until it reaches a termination site and the mRNA transcript is released for translation. Note that the section of the DNA that has been transcribed is rewound into its original configuration.

Elongation

3′

5′

Late mRNA transcript

Figure 11.26 The Major Events in Transcription.

enzyme capable of several activities, including unwinding the DNA, moving along the template, and synthesizing RNA.

Termination of transcription occurs when the core RNA polymerase dissociates from the template DNA. The end of a gene or group of genes is marked by DNA sequences in the trailer (which is transcribed but not translated) and the terminator. The sequences within procaryotic terminators often contain nucleotides that, when transcribed into RNA, form hydrogen bonds within the single-stranded RNA. This intrastrand base pairing creates a hairpin-shaped loop-and-stem structure. This structure appears to cause the RNA polymerase to pause or stop transcribing DNA. There are two kinds of terminators. The first type causes intrinsic or rho-independent termination (**figure 11.31**). It features the

mRNA hairpin followed by a stretch of about six uridine residues. Once the RNA polymerase has paused at the hairpin loop, the A-U base pairs in the uracil-rich region are too weak to hold the RNA:DNA duplex together and the RNA polymerase falls off. The second kind of terminator lacks a poly-U region, and often the hairpin; it requires the aid of a special protein, the **rho factor** (ρ). This terminator causes rho-dependent termination. It is thought that rho binds to mRNA and moves along the molecule until it reaches the RNA polymerase that has halted at a terminator (**figure 11.32**). The rho factor, which has hybrid RNA:DNA helicase activity, then causes the polymerase to dissociate from the mRNA, probably by unwinding the mRNA-DNA complex.

(a)

(b)

Upstream ←   Downstream →

Extended
−10 element

−35
element

−40   −30   −20

σ₂

σ₄

5′nt

−10
element

3′nt

σ₃   5′t

β′

αI

β

(c)

Zipper

Wall

Clamp
head

Metal

Bridge

(d)

Transcribed DNA
(upstream)

Lid

Exit

Clamp

Rudder

Wall

Entering DNA
(downstream)

Mg²⁺   Pore

Bridge

Jaw

α Amanitin

Funnel

NTPs

Transcription →

Figure 11.27 RNA Polymerase Structure. The atomic structures of RNA polymerase from the bacterium *Thermus aquaticus* (**a** and **b**) and yeast RNA polymerase II (**c** and **d**) are presented here. (**a**) The *Taq* RNA polymerase holoenzyme is shown with the σ subunit depicted as an α-carbon backbone with cylinders for α-helices. Two of the three σ factor domains are labeled. (**b**) The holoenzyme-DNA complex with the σ surface rendered slightly transparent to show the α-carbon backbone inside. Protein surfaces that contact the DNA are in green and are located on the σ factor. The −10 and −35 elements in the promoter are in yellow. The internal active site is covered by the β subunit in this view. (**c**) Yeast RNA polymerase II transcribing complex with some peptide chains removed to show the DNA. The active site metal is a red sphere. A short stretch of DNA-RNA hybrid (blue and red) lies above the metal. (**d**) A cutaway side view of the polymerase II transcribing complex with the pathway of the nucleic acids and some of the more important parts shown. The enzyme is moving from left to right and the DNA template strand is in blue. A protein "wall" forces the DNA into a right-angle turn and aids in the attachment of nucleoside triphosphates to the growing 3′ end of the RNA. The newly synthesized RNA (red) is separated from the DNA template strand and exits beneath the rudder and lid of the polymerase protein complex. The binding site of the inhibitor α-amanitin also is shown.

Figure 11.28 The conventional numbering system of promoters. The first nucleotide that acts as a template for transcription is designated +1. The numbering of nucleotides to the left of this spot is in a negative direction, while the numbering to the right is in a positive direction. For example, the nucleotide that is immediately to the left of the +1 nucleotide is numbered −1, and the nucleotide to the right of the +1 nucleotide is numbered +2. There is no zero nucleotide in this numbering system. In many bacterial promoters, sequence elements at the −35 and −10 regions play a key role in promoting transcription.

Transcription in Eucaryotes

Transcriptional processes in eucaryotic microorganisms (and in other eucaryotic cells) differ in several ways from bacterial transcription. There are three major RNA polymerases, not one as in *Bacteria*. RNA polymerase II, associated with chromatin in the nuclear matrix, is responsible for mRNA synthesis. Polymerases I and III synthesize rRNA and tRNA, respectively (**table 11.3**). The eucaryotic RNA polymerase II is a large aggregate, at least 500,000 daltons in size, with about 10 or more subunits. The atomic structure of the 10-subunit yeast RNA polymerase II associated with DNA and RNA has been determined (figure 11.27c,d). The entering DNA is held in a clamp that closes down on it. A magnesium ion is located at the active site and the 9 base pair DNA-RNA hybrid in the transcription bubble is bound in a cleft formed by the two large polymerase subunits. The newly synthesized RNA exits the polymerase beneath the rudder and lid regions. The substrate nucleoside triphosphates probably reach the active site through a pore in the complex. Unlike bacterial polymerase, RNA polymerase II requires extra transcription factors to recognize its promoters (**figure 11.33**). The polymerase binds near the start point; the transcription factors bind to the rest of the promoter. Eucaryotic promoters also differ from those in *Bacteria*. They have combinations of several elements. Three of the most common are the TATA box (located about 30 base pairs before or upstream of the start point), and the GC and CAAT boxes located between 50 to 100 base pairs upstream of the start site (**figure 11.34**). The TFIID transcription factor (figure 11.33) plays an important role in transcription initiation in eucaryotes. This multi-

Figure 11.29 The Initiation of Transcription in *Bacteria*. The sigma factor of the RNA polymerase holoenzyme is responsible for positioning the core enzyme properly at the promoter. Sigma factor recognizes two regions in the promoter, one centered at −35 and the other centered at −10. Once positioned properly, the DNA at the −10 region unwinds to form an open complex. The sigma factor dissociates from the core enzyme as it begins transcribing the gene.

protein complex contains the TATA-binding protein (TBP). TBP has been shown to sharply bend the DNA on attachment. This makes the DNA more accessible to other initiation factors. A variety of general transcription factors, promoter specific factors, and promoter elements have been discovered in different eucaryotic cells. Each eucaryotic gene seems to be regulated differently, and more research will be required to fully understand the regulation of eucaryotic gene transcription.

Unlike bacterial mRNA, eucaryotic mRNA arises from **post-transcriptional modification** of large RNA precursors, about 5,000 to 50,000 nucleotides long, sometimes called heteroge-

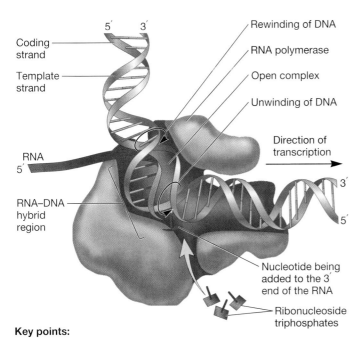

Key points:

- RNA polymerase slides along the DNA, creating an open complex as it moves.

- The DNA strand known as the template strand is used to make a complementary copy of RNA as an RNA–DNA hybrid.

- The RNA is synthesized in a 5′ to 3′ direction using ribonucleoside triphosphates as precursors. Pyrophosphate is released (not shown).

- The complementarity rule is the same as the AT/GC rule except that U is substituted for T in the RNA.

Figure 11.30   The "Transcription Bubble."

Figure 11.31   Intrinsic Termination of Transcription.   This type of terminator contains a U-rich sequence downstream from a stretch of nucleotides that can form a stem-loop and stem structure. Formation of the stem loop in the newly synthesized RNA causes RNA polymerase to pause. This pausing is stabilized by the NusA protein. The U-A bonds in the uracil-rich region are not strong enough to hold the RNA and DNA together. Therefore, the RNA, DNA, and RNA polymerase dissociate and transcription stops.

neous nuclear RNA (hnRNA) (**figure 11.35**). As hnRNA is synthesized, a 5′ cap is added. After synthesis is completed, the precursor RNA is modified by the addition of a 3′ poly-A tail. It is also processed, if necessary, to remove any introns. The 5′ cap is the unusual nucleotide 7-methylguanosine. It is attached to the 5′-hydroxyl of the hnRNA by a triphosphate linkage (**figure 11.36**). Addition of a poly-A tail is initiated by an endonuclease that shortens the hnRNA and generates a 3′-OH group. The enzyme polyadenylate polymerase then catalyzes the addition of adenylic acid to the 3′ end of hnRNA to produce a poly-A sequence about 200 nucleotides long. The functions of the 5′ cap and poly-A tail are not completely clear. The 5′ cap on eucaryotic mRNA may promote the initial binding of ribosomes to the mRNA. It also may protect the mRNA from enzymatic attack. Poly-A protects mRNA from rapid enzymatic degradation. The poly-A tail must be shortened to about 10 nucleotides before mRNA can be degraded. Poly-A also seems to aid in mRNA translation.

As noted earlier, many eucaryotic genes are split or interrupted, which leads to the final type of posttranscriptional pro-

cessing. **Split** or **interrupted genes** have **exons** (*ex*pressed sequences), regions coding for RNA that end up in the mRNA. Exons are separated from one another by **introns** (*int*ervening sequences), sequences coding for RNA that is never translated into protein (**figure 11.37**). The initial RNA transcript contains both exon and intron sequences. Genes coding for rRNA and tRNA may also be interrupted. Except for cyanobacteria and *Archaea* (*see chapters 20 and 21*), interrupted genes have not been found in procaryotes.

Introns are removed from the initial RNA transcript (also called pre-mRNA or primary transcript) by a process called **RNA splicing** (figure 11.37). The intron's borders are clearly marked for accurate removal. Exon-intron junctions have a GU sequence at the intron's 5′ boundary and an AG sequence at its 3′ end. These two sequences define the splice junctions and are recognized by special RNA molecules. The nucleus contains several **small nuclear RNA (snRNA)** molecules, about 60 to 300 nucleotides long. These complex with proteins to form *s*mall *n*uclear *ri*bonucleoprotein particles called snRNPs or snurps. Some of the snRNPs recognize splice junctions and ensure splicing accuracy. For example, U1-snRNP recognizes the 5′ splice junction, and U5-snRNP recognizes

Table 11.3	Eucaryotic RNA Polymerases	
Enzyme	**Location**	**Product**
RNA polymerase I	Nucleolus	rRNA (5.8S, 18S, 28S)
RNA polymerase II	Chromatin, nuclear matrix	mRNA
RNA polymerase III	Chromatin, nuclear matrix	tRNA, 5S rRNA

the 3′ junction. Splicing of pre-mRNA occurs in a large complex called a **spliceosome** that contains the pre-mRNA, at least five kinds of snRNPs, and non-snRNP splicing factors. Sometimes a pre-mRNA will be spliced so that different patterns of exons remain. This alternative splicing allows a single gene to code for more than one protein. The splice pattern determines which protein will be synthesized. Splice patterns can be cell-type specific or determined by the needs of the cell. The importance of alternative splicing in multicellular eucaryotes was emphasized when it was discovered that the human genome has only about 20,000 genes rather than the anticipated 100,000. It is thought that alternative splicing is one mechanism by which human cells produce such a vast array of proteins.

As just mentioned, a few rRNA genes also have introns. Some of these pre-rRNA molecules are self-splicing, that is, the pre-rRNA is a ribozyme (Microbial Tidbits 11.2). Thomas Cech first discovered that pre-rRNA from the ciliate protozoan *Tetrahymena thermophila* is self-splicing. Sidney Altman then showed that ribonuclease P, which cleaves a fragment from one end of pre-tRNA, contains a piece of RNA that catalyzes the reaction. Several other self-splicing rRNA introns have since been discovered. Cech and Altman received the 1989 Nobel Prize in chemistry for these discoveries. Microbial evolution (section 19.1)

Transcription in the *Archaea*

Transcription in the *Archaea* is similar to and distinct from what is observed in *Bacteria* and eucaryotes. Each archaeon has a single RNA polymerase responsible for transcribing all genes in the cell (as in *Bacteria*). However, the RNA polymerase is larger and contains more subunits, many of which are similar to subunits in RNA polymerase II of eucaryotes. The promoters of archaeal genes are similar to those of eucaryotes in having a TATA box; binding of the archaeal RNA polymerase to its promoter requires a TATA-binding protein, just as in eucaryotes. Like the eucaryotic counterpart, the archaeal RNA polymerase also needs several additional transcription factors to function properly. Furthermore, some archaeal genes have introns, which must be removed by posttranscriptional processing. Finally, the mRNA molecules produced by transcription in *Archaea* are usually polycistronic, as in *Bacteria*. This intriguing mixture of bacterial and eucaryotic features in the *Archaea* has fueled a great deal of speculation about the evolution of all three domains of life. Introduction to the *Archaea*: Genetics and molecular biology (section 20.1)

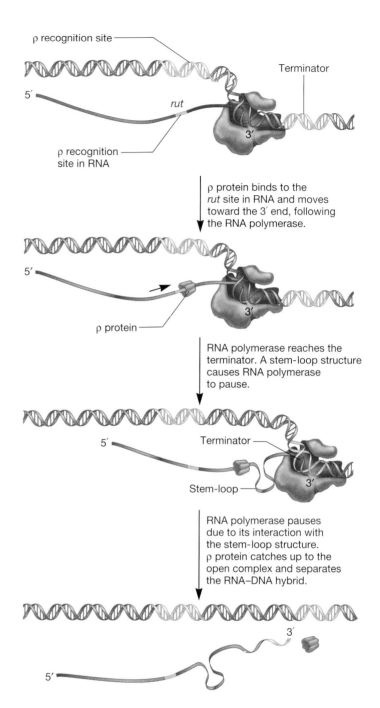

Figure 11.32   Rho-Factor (ρ)-Dependent Termination of Transcription.   The *rut* site stands for *rho utilization* site.

1. Define the following terms: polygenic mRNA, RNA polymerase core enzyme, sigma factor, RNA polymerase holoenzyme, and rho factor.
2. Define or describe posttranscriptional modification, heterogeneous nuclear RNA, 3′ poly-A sequence, 5′ capping, split or interrupted genes, exon, intron, RNA splicing, snRNA, spliceosome, and ribozyme.
3. Describe how RNA polymerase transcribes bacterial DNA. How does the polymerase know when to begin and end transcription?
4. How do bacterial RNA polymerases and promoters differ from those of *Archaea* and eucaryotes?

Figure 11.33   Initiation of Transcription in Eucaryotes.
The TATA box is a major component of eucaryotic promoters. *TATA-binding protein* (TBP), which is a component of a complex of proteins called TFIID (*transcription factor IID*), binds the TATA box. Note that numerous other transcription factors are required for initiation of transcription, unlike *Bacteria* where only the sigma factor is needed. Initiation of transcription in *Archaea* is similar to that seen in eucaryotes.

11.7   THE GENETIC CODE

The final step in the expression of genes that encode proteins is translation. The mRNA nucleotide sequence is translated into the amino acid sequence of a polypeptide chain. Protein synthesis is called translation because it is a decoding process. The information encoded in the language of nucleic acids must be rewritten in the language of proteins. Therefore, before we discuss protein synthesis, we will examine the nature of the genetic code.

Establishment of the Genetic Code

The realization that DNA is the genetic material triggered efforts to understand how genetic instructions are stored and organized in the DNA molecule. Early studies on the nature of the genetic code showed that the DNA base sequence corresponds to the amino acid sequence of the polypeptide specified by the gene. That is, the nucleotide and amino acid sequences are colinear. It also became evident that many mutations are the result of changes of single amino acids in a polypeptide chain. However, the exact nature of the code was still unclear.

Theoretical considerations directed much of the early work on deciphering the code. Scientists reasoned that because only 20 amino acids normally are present in proteins, there must be at least 20 different code words in DNA. Therefore the code must be contained in some sequence of the four nucleotides commonly found in DNA.

If the code words were two nucleotides in length, there would be only 16 possible combinations (4^2) of the four nucleotides and this would not be enough to code for all 20 amino acids. Therefore a code word, or codon, had to consist of at least nucleotide triplets even though this would give 64 possible combinations (4^3), many more than the minimum of 20 needed to specify the common amino acids. Research eventually confirmed this and the code was deciphered in the early 1960s by Marshall Nirenberg, Heinrich Matthaei, Philip Leder, and Har Gobind Khorana. In 1968 Nirenberg and Khorana shared the Nobel Prize with Robert Holley, the first person to sequence a nucleic acid (phenylalanyl-tRNA).

Organization of the Code

The genetic code, presented in RNA form, is summarized in **table 11.4.** Note that there is **code degeneracy.** That is, there are up to six different codons for a given amino acid. Only 61 codons, the **sense codons,** direct amino acid incorporation into protein. The remaining three codons (UGA, UAG, and UAA) are involved in the termination of translation and are called stop or **nonsense codons.** Despite the existence of 61 sense codons, there are not 61 different tRNAs, one for each codon. The 5′ nucleotide in the anticodon can vary, but generally, if the nucleotides in the second and third anticodon positions complement the first two bases of the mRNA codon, an aminoacyl-tRNA with the proper amino acid will bind to the mRNA-ribosome complex. This pattern is evident on inspection of changes in the amino acid specified

Table 11.4	The Genetic Code

First Position (5′ End)[a]	Second Position				Third Position (3′ End)
	U	**C**	**A**	**G**	
U	UUU } Phe UUC UUA } Leu UUG	UCU } UCC } Ser UCA UCG	UAU } Tyr UAC UAA } STOP UAG	UGU } Cys UGC UGA STOP UGG Trp	U C A G
C	CUU } CUC } Leu CUA CUG	CCU } CCC } Pro CCA CCG	CAU } His CAC CAA } Gln CAG	CGU } CGC } Arg CGA CGG	U C A G
A	AUU } AUC } Ile AUA AUG Met	ACU } ACC } Thr ACA ACG	AAU } Asn AAC AAA } Lys AAG	AGU } Ser AGC AGA } Arg AGG	U C A G
G	GUU } GUC } Val GUA GUG	GCU } GCC } Ala GCA GCG	GAU } Asp GAC GAA } Glu GAG	GGU } GGC } Gly GGA GGG	U C A G

[a]The code is presented in the RNA form. Codons run in the 5′ to 3′ direction. See text for details.

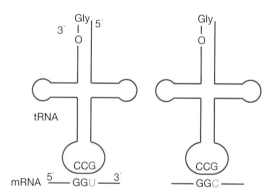

(a) Base pairing of one glycine tRNA with two codons due to wobble

Glycine mRNA codons: GGU, GGC, GGA, GGG (5′ ⟶ 3′)

Glycine tRNA anticodons: CCG, CCU, CCC (3′ ⟶ 5′)

(b) Glycine codons and anticodons

Figure 11.38 Wobble and Coding. The use of wobble in coding for the amino acid glycine. **(a)** Because of wobble, G in the 5′ position of the anticodon can pair with either C or U in the 3′ position of the codon. Thus two codons can be recognized by the same tRNA. **(b)** Because of wobble, only three tRNA anticodons are needed to translate the four glycine (Gly) codons.

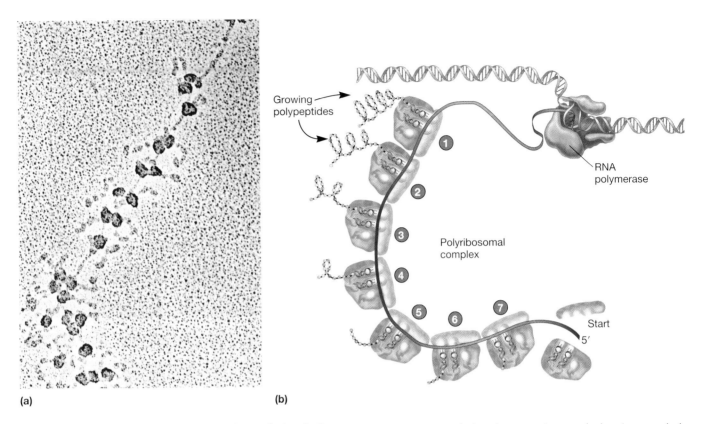

Figure 11.39   Coupled Transcription and Translation in Procaryotes.   (a) A transmission electron micrograph showing coupled transcription and translation. **(b)** A schematic representation of coupled transcription and translation. As the DNA is transcribed, ribosomes bind the free 5′ end of the mRNA. Thus translation is started before transcription is completed. Note that there are multiple ribosomes bound to the mRNA, forming a polyribosome.

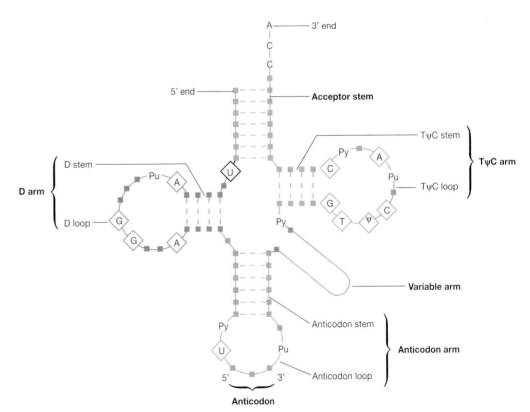

Figure 11.40   tRNA Structure.
The cloverleaf structure for tRNA in procaryotes and eucaryotes. Bases found in all tRNAs are in diamonds; purine and pyrimidine positions in all tRNAs are labeled Pu and Py respectively.

amino acid corresponding to the proper mRNA codon is added to the C-terminal end of the polypeptide chain. The bacterial elongation cycle is described next.

The ribosome has three sites for binding tRNAs: (1) the **peptidyl** or **donor site** (the **P site**), (2) the **aminoacyl** or **acceptor site** (the **A site**), and (3) the **exit site** (the **E site**). At the beginning of an elongation cycle, the peptidyl site is filled with either *N*-formylmethionyl-tRNA^fMet or peptidyl-tRNA, and the aminoacyl and exit sites are empty (**figure 11.48**). Messenger RNA is bound to the ribosome in such a way that the proper codon interacts with the P site tRNA (e.g., an AUG codon for fMet-tRNA). The next codon is located within the A site and is ready to accept an aminoacyl-tRNA.

The first phase of the elongation cycle is the aminoacyl-tRNA binding phase. The aminoacyl-tRNA corresponding to the codon in the A site is inserted so its anticodon is aligned with the codon on the mRNA. GTP and the elongation factor EF-Tu, which carries the aminoacyl-tRNA to the ribosome, are required for this insertion. When GTP is bound to EF-Tu, the protein is in its active state and delivers aminoacyl-tRNA to the A site. This is followed by GTP hydrolysis, and the EF-Tu · GDP complex leaves the ribosome. EF-Tu · GDP is converted to EF-Tu · GTP with the aid of a second elongation factor, EF-Ts. Subsequently another aminoacyl-tRNA binds to EF-Tu · GTP (figure 11.48).

Aminoacyl-tRNA binding to the A site initiates the second phase of the elongation cycle, the transpeptidation reaction (figure 11.48 and **figure 11.49**). This is catalyzed by the 23S rRNA ribozyme activity called **peptidyl transferase,** located on the 50S subunit. The α-amino group of the A site amino acid nucleophilically attacks the α-carboxyl group of the C-terminal amino acid on the P site tRNA (figure 11.49). The peptide chain attached to the tRNA in the P site is transferred to the A site as a peptide bond is formed between the chain and the incoming amino acid. No extra energy source is required for peptide bond formation because the bond linking an amino acid to tRNA is high in energy (figure 11.43). Evidence strongly suggests that 23S rRNA contains the peptidyl transferase function, and is therefore a ribozyme. Almost all protein can be removed from the 50S subunit, leaving the 23S rRNA and protein fragments and the remaining complex still has peptidyl transferase activity. The high-resolution structure of the large subunit has been obtained by X-ray crystallography. There is no protein in the active site region. A specific adenine base seems to participate in catalyzing peptide bond formation. Thus the 23S rRNA appears to be the major component of the peptidyl transferase and contributes to both A and P site functions.

The final phase in the elongation cycle is **translocation.** Three things happen simultaneously: (1) the peptidyl-tRNA moves about 20 Å from the A site to the P site; (2) the ribosome moves one codon along mRNA so that a new codon is positioned in the A site; and (3) the empty tRNA leaves the P site. Instead of immediately being ejected from the ribosome when the ribosome moves along the mRNA, the empty tRNA is moved from the P site to the E site and then leaves the ribosome. Ribosomal proteins are involved in these tRNA movements. The intricate process also requires the participation of the EF-G or translocase

protein and GTP hydrolysis. The ribosome changes shape as it moves down the mRNA in the 5′ to 3′ direction.

Termination of Protein Synthesis

Protein synthesis stops when the ribosome reaches one of three nonsense codons—UAA, UAG, and UGA (**figure 11.50**). The nonsense (stop) codon is found on the mRNA immediately before the trailer region. Three **release factors** (RF-1, RF-2, and RF-3) aid the ribosome in recognizing these codons. Because there is no cognate tRNA for a nonsense codon, the ribosome stops. The peptidyl transferase hydrolyzes the peptide free from the tRNA in the P site, and the empty tRNA is released. GTP hydrolysis seems to be required during this sequence, although it may not be needed for termination in *Bacteria*. Next the ribosome dissociates from its mRNA and separates into 30S and 50S subunits. IF-3 binds to the 30S subunit to prevent it from reassociating with the 50S subunit until the proper stage in initiation is reached. Thus ribosomal subunits associate during protein synthesis and separate afterward. The termination of eucaryotic protein synthesis is similar except that only one release factor appears to be active.

Protein synthesis is a very expensive process. Three GTP molecules probably are used during each elongation cycle, and two ATP high-energy bonds are required for amino acid activation (ATP is converted to AMP rather than to ADP). Therefore five high-energy bonds are required to add one amino acid to a growing polypeptide chain. GTP also is used in initiation and termination of protein synthesis (figures 11.47 and 11.50). Presumably this large energy expenditure is required to ensure the fidelity of protein synthesis. Very few mistakes can be tolerated.

Although the mechanism of protein synthesis is similar in *Bacteria* and eucaryotes, bacterial ribosomes differ substantially from those in eucaryotes. This explains the effectiveness of many important antibacterial agents. Either the 30S or the 50S subunit may be affected. For example, streptomycin binding to the 30S ribosomal subunit inhibits protein synthesis and causes mRNA misreading. Erythromycin binds to the 50S subunit and inhibits peptide chain elongation. Antibacterial drugs (section 34.4)

Protein Folding and Molecular Chaperones

For many years it was believed that polypeptides spontaneously folded into their final native shape, either as they were synthesized by ribosomes or shortly after completion of protein synthesis. Although the amino acid sequence of a polypeptide does determine its final conformation, it is now clear that special helper proteins aid the newly formed or nascent polypeptide in folding to its proper functional shape. These proteins, called **molecular chaperones** or chaperones, recognize only unfolded polypeptides or partly denatured proteins and do not bind to normal, functional proteins. Their role is essential because the cytoplasmic matrix is filled with nascent polypeptide chains and proteins. Under such conditions it is quite likely that new polypeptide chains often will fold improperly and aggregate to form nonfunctional complexes. Molecular chaperones suppress incorrect folding and may reverse

Figure 11.48   Elongation Cycle.
The elongation cycle of protein synthesis. The ribosome possesses three sites, a peptidyl or donor site (P site), an aminoacyl or acceptor site (A site), and an exit site (E site). The arrow below the ribosome in the translocation step shows the direction of mRNA movement. See text for details.

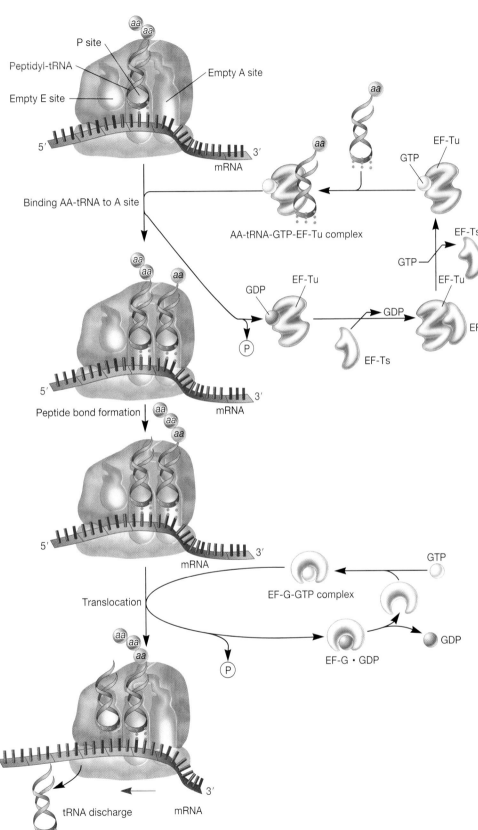

Critical Thinking Questions

1. Many scientists say that RNA was the first of the information molecules (i.e., RNA, DNA, protein) to arise during evolution. Given the information in this chapter, what evidence is there to support this hypothesis?

2. *Streptomyces coelicolor* has a linear chromosome. Interestingly, there are no genes that encode essential proteins near the ends of the chromosome in this bacterium. Why do you think this is the case?

3. You have isolated several *E. coli* mutants:

   Mutant #1 has a mutation in the -10 region of the promoter of a structural gene encoding an enzyme needed for synthesis of the amino acid serine.

   Mutant #2 has a mutation in the -35 region in the promoter of the same gene.

   Mutant #3 is a double mutant with mutations in both the -10 and -35 region of the promoter of the same gene.

   Only Mutant #3 is unable to make serine. Why do you think this is so?

4. Suppose that you have isolated a microorganism from a soil sample. Describe how you would go about determining the nature of its genetic material.

Learn More

Bao, K., and Cohen, S. N. 2004. Reverse transcriptase activity innate to DNA polymerase I and DNA topoisomerse I proteins of *Streptomyces* telomere complex. *Proc. Natl. Acad. Sci.* 101(40):14361–66.

Borukhov, S., and Nudler, E. 2003. RNA polymerase holoenzyme: Structure, function and biological implications. *Current Opinion Microbiol.* 6:93–100.

Brooker, R. J. 2005. *Genetics: Analysis and principles,* 2d ed. New York: McGraw-Hill.

Gogarten, J. P.; Senejani, A. G.; Zhaxybayeva, O.; Olendzenski, L.; and Hilario, E. 2002. Inteins: Structure, function, and evolution. *Annu. Rev. Microbiol.* 56:263–87.

Grabowski, B., and Kelman, Z. 2003. Archaeal DNA replication: Eukaryal proteins in a bacterial context. *Annu. Rev. Microbiol.* 57:487–516.

Hartl, F. J., and Hayer-Hartl, M. 2002. Molecular chaperones in the cytosol: From nascent chain to folded protein. *Science* 295:1852–58.

Johnson, A., and O'Donnell, M. 2005. Cellular DNA replicases: Components and dynamics at the replication fork. *Annu. Rev. Biochem.* 74:283–315.

Kelman, L. M., and Kelman, Z. 2004. Multiple origins of replication in archaea. *Trends Microbiol.* 12(9):399–401.

Laursen, B. S.; Søjorensen, H. P.; Mortensen, K. K.; and Sperling-Petersen, H. U. 2005. Initiation of protein synthesis in bacteria. *Microbiol. Mol. Biol. Rev.* 69(1):101–23.

McKee, T., and McKee, J. R. 2003. *Biochemistry: The molecular basis of life,* 3d ed. Dubuque, Iowa: McGraw-Hill.

Narberhaus, F. 2002. Crystallin-type heatshock proteins: Socializing minichaperones in the context of a multichaperone network. *Microbiol. Mol. Biol. Rev.* 66(1):64–93.

Nelson, D. L., and Cox, M. M. 2005. *Lehninger: Principles of biochemistry,* 4th ed. New York: W. H. Freeman.

Neylon, C.; Kralicek, A. V.; Hill, T. M.; and Dixon, N. E. 2005. Replication termination in *Escherichia coli:* Structure and antihelicase activity of the Tus-*Ter* complex. *Microbiol. Mol. Biol. Rev.* 69(3):501–26.

Paul, B. J.; Ross, W.; Gaal, T.; and Gourse, R. L. 2004. rRNA transcription in *Escherichia coli. Annu. Rev. Genet.* 38:749–70.

Reeve, J. N. 2003. Archaeal chromatin and transcription. *Mol. Microbiol.* 48(3):587–98.

Reeve, J. N.; Bailey, K. A.; Li, W-T.; Marc, F.; Sandman, K.; and Soares, D. J. 2004. Archaeal histones: Structures, stability and DNA binding. *Biochem. Soc. Trans.* 32 (part 2):227–30.

Smogorzewska, A., and deLange, T. 2004. Regulation of telomerase by telomeric proteins. *Annu. Rev. Biochem.* 73:177–208.

Yarus, M.; Caporaso, J. G.; and Knight, R. 2005. Origins of the genetic code: The escaped triplet theory. *Annu. Rev. Biochem.* 74:179–98.

12

Microbial Genetics:
Regulation of
Gene Expression

Lactose operon activity is under the control of repressor and activator proteins. The *lac* repressor (pink) and catabolite activator protein (blue) are bound to the *lac* operon. The repressor blocks transcription when bound to the operators (red).

PREVIEW

- The long-term regulation of metabolism, behavior, and morphology is brought about by control of gene expression. This can occur at many levels including transcription initiation, transcription elongation, translation, and posttranslation.

- In *Bacteria*, control at the level of transcription initiation is often achieved by regulatory proteins. Some block transcription and others promote transcription. Furthermore, transcription can be terminated prematurely by a process called attenuation, in which ribosome behavior affects RNA polymerase activity. Translation in *Bacteria* can be regulated by small molecules that bind the leader of mRNA. The conformation changes that result block ribosome binding. Translation can also be blocked by antisense RNA molecules.

- Microorganisms must be able to respond rapidly to changing environmental conditions. Their responses often involve many genes or operons. The simultaneous control of many genes or operons is called global control. Important examples of bacterial global control systems are catabolite repression, quorum sensing, and endospore formation.

- Eucaryotic gene expression involves more steps than does bacterial gene expression; thus there are more points in the process where regulation can occur.

- Although archaeal genome organization is similar to that seen in *Bacteria*, the machinery used by the *Archaea* during information flow is more like that of the *Eucarya*. Thus archaeal regulatory mechanisms may be similar to those observed in the other domains of life.

The gram-positive soil bacterium *Bacillus subtilis* senses that the nutrient levels in its environment are decreasing, and it must determine if it should initiate sporulation. An *Escherichia coli* cell is in an environment rich in carbon and energy sources, and it must determine which to use and when to use them. A pathogen is transmitted from a stream to the intestinal tract of its animal host, and it must adjust to the warmer temperature, increased nutrient supply, and defenses of the host. These are just a few examples of situations to which microbes must respond. To make the most efficient use of the resources in the current environment and their own cellular machinery, microbes must respond to changes by altering physiological and behavioral processes. How is this accomplished?

The control of cellular processes by regulation of the activity of enzymes and other proteins is a fine-tuning mechanism: it acts rapidly to adjust metabolic activity from moment to moment. Microorganisms also are able to control the expression of their genome, although over longer intervals. For example, the *E. coli* chromosome can code for about 4500 polypeptides, yet not all proteins are produced at the same time. Regulation of gene expression serves to conserve energy and raw materials, to maintain balance between the amounts of various cell proteins, and to adapt to long-term environmental change. Thus control of gene expression complements the regulation of enzyme activity. (Control of enzyme activity (section 8.10)

The particular field which excites my interest is the division between the living and the non-living, as typified by, say, proteins, viruses, bacteria and the structure of chromosomes. The eventual goal, which is somewhat remote, is the description of these activities in terms of their structure, i.e., the spatial distribution of their constituent atoms, in so far as this may prove possible. This might be called the chemical physics of biology.

—Francis Crick

(a) Possible DNA loops caused by the binding of the *lac* repressor

(b) Proposed model of the *lac* repressor binding to O_1 and O_3 based on crystallography studies

Figure 12.6 The *lac* Operator Sites. The *lac* operon has three operator sites: O_1, O_2, and O_3 (a). O_1 is the same operator shown in figure 12.4. As shown in (a) and (b), the *lac* repressor (violet) binds O_1 and one of the other operator sites (red) to block transcription, forming a DNA loop. The DNA loop contains the -35 and -10 binding sites (green) recognized by RNA polymerase. Thus these sites are inaccessible and transcription is blocked. The DNA loop also contains the CAP binding site and CAP (blue) is shown bound to the DNA (b). When the *lac* repressor is bound to the operator, CAP is unable to activate transcription.

cell, β-galactosidase converts lactose to allolactose, the inducer of the operon (figure 12.2). This occurs because, as noted previously, there is always a low level of permease and β-galactosidase synthesis. Allolactose binds to the *lac* repressor and causes the repressor to change to an inactive shape that is unable to bind any operator sites. The inactivated repressor leaves the DNA and transcription occurs (figure 12.7*b*).

Close examination of figures 12.4 and 12.6 clearly shows that the regulation of the *lac* operon is not as simple as has just

been described. That is because the *lac* operon is regulated by a second regulatory protein called CAP. CAP functions in a global regulatory network that allows *E. coli* to use glucose preferentially over all other carbon and energy sources by a mechanism called catabolite repression. The use of two different regulatory proteins to control the synthesis of an operon illustrates a point that is important in the discussion of the regulation of gene expression—there are often layers of regulation of any operon. As described in section 12.5, the use of two

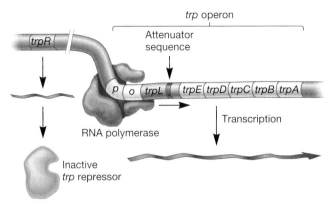

(a) Low trytophan levels, transcription of the entire *trp* operon occurs

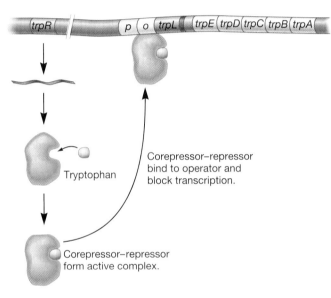

(b) High tryptophan levels, repression occurs

Figure 12.8 Regulation of the *trp* Operon by Tryptophan and the *trp* Repressor. The *trp* repressor is inactive when first synthesized and therefore is unable to bind the operator. It is activated by the binding of tryptophan, which serves as the corepressor. **(a)** When tryptophan levels are low, the repressor is inactive and transcription occurs. The enzymes encoded by the operon catalyze the reactions needed for tryptophan biosynthesis. **(b)** When tryptophan levels are sufficiently high, it binds the repressor. The repressor-corepressor complex binds the operator and transcription of the operon is inhibited.

Figure 12.7 Regulation of the *lac* Operon by the *lac* Repressor. **(a)** The *lac* repressor is active and can bind the operator as long as the inducer of the operon, allolactose, is not present. Binding of the repressor to the operator inhibits transcription of the operon by RNA polymerase. **(b)** When lactose is available, some of it is converted to allolactose by β-galactosidase. When sufficient amounts of allolactose are present, it binds and inactivates the *lac* repressor. The repressor leaves the operator and RNA polymerase is free to initiate transcription.

The Tryptophan Operon: Negative Transcriptional Control of Repressible Genes

The tryptophan (*trp*) operon of *E. coli* consists of five structural genes that encode enzymes needed for synthesis of the amino acid tryptophan (**figure 12.8**). It is regulated by the *trp* repressor, which is encoded by the *trpR* gene. Because the enzymes encoded by the *trp* operon function in a biosynthetic pathway, it is wasteful to make the enzymes needed for tryptophan synthesis when

regulatory proteins generates a continuum of expression levels. The highest levels of transcription occur when lactose is available and glucose is not; the lowest levels occur when lactose is not available and glucose is.

changes in gene expression; in many cases numerous genes and operons may be regulated by the same response regulator. Thus two-component systems often function in global regulatory networks. The effectiveness of two-component systems is illustrated by their abundance: most procaryotic cells use a variety of two-component signal transduction systems to respond to an array of environmental stresses. For example, the morphologically complex genus *Streptomyces* has over 50 such systems!

Two-component signal transduction systems involve a simple phosphorelay where the sensor kinase transfers its phosphoryl group directly to the response-regulator protein. However, there are instances when more proteins participate in the transfer of phosphoryl groups. These longer pathways are called **phosphorelay systems.** An important and well-studied phosphorelay system functions during sporulation in *Bacillus subtilis* and is described in section 12.5. It should be noted that some phosphorelay systems control protein activity rather than gene transcription. An example of this type of system is chemotaxis in *E. coli*, which is described in chapter 8.

1. Many genes and operons are regulated at the level of transcription initiation. Why do you think this is the case?
2. What are induction and repression? How do bacteria use them to respond to changing nutrient supplies?
3. Define negative control and positive control of transcription initiation. Describe how regulatory proteins function in these regulatory mechanisms. Define repressor protein, activator protein, operator, activator-binding site, inducer, corepressor, structural gene, and operon.
4. Using figure 12.5 as a guide, trace the "decision-making" pathway of an *E. coli* cell that is growing in a medium containing arabinose but lacking tryptophan.
5. Describe a two-component signal transduction system. How does it differ from a phosphorelay system? How are the *lac* repressor, the *trp* repressor, and the AraC protein similar to the response regulators of two-component and phosphorelay systems? How are they different?

12.3 REGULATION OF TRANSCRIPTION ELONGATION

Organisms can also regulate transcription by controlling the termination of transcription. In this type of regulation, transcription is initiated but prematurely stopped depending on the environmental conditions and the needs of the organism. The first demonstration of this level of regulation, called attenuation, occurred in the 1970s in studies of the *trp* operon. More recently, riboswitches have been discovered. These regulatory sequences in the leader of an mRNA both sense and respond to environmental conditions by either prematurely terminating transcription or blocking translation. Both attenuation and riboswitches are described in this section.

Attenuation

As noted earlier, the tryptophan (*trp*) operon of *E. coli* is under the control of a repressor protein, and excess tryptophan inhibits transcription of operon genes by acting as a corepressor and acti-

vating the repressor protein. Although the operon is regulated mainly by repression, the continuation of transcription also is controlled. That is, there are two decision points involved in transcriptional control, the initiation of transcription and the continuation of transcription past the leader region.

This additional level of control serves to adjust levels of transcription in a more subtle fashion. When the repressor is not active, RNA polymerase begins transcription of the leader region but it often does not progress to the first structural gene in the operon. Instead, transcription is terminated within the leader region; this is called **attenuation.** The ability to attenuate transcription is based on the nucleotide sequences in the leader region and on the fact that in procaryotes, transcription is coupled with translation (*see figure 11.39*). The leader of the *trp* operon mRNA is unusual in that it is translated. The product, which has never been isolated, is called the leader peptide. In addition to encoding the leader peptide, the leader contains **attenuator** sequences (**figure 12.11**). When transcribed, these sequences form stem-loop secondary structures in the newly formed mRNA. We define these sequences numerically (regions 1, 2, 3, and 4). When regions 1 and 2 pair with one another (1:2; figure 12.11*a*), they form a secondary structure called the pause loop, which causes RNA polymerase to slow down. The pause loop forms just prior to the formation of a second structure called the terminator loop, which is made when regions 3 and 4 base pair (3:4; figure 12.11*a*). A poly(U) sequence follows the 3:4 terminator loop, just as it does in other rho-independent transcriptional terminators (*see figure 11.31*). However, in this case, the terminator is in the leader rather than at the end of the gene. Another stem-loop structure can be formed in the leader region by the pairing of regions 2 and 3 (2:3, figure 12.11*b*). The formation of this antiterminator loop prevents the generation of both the 1:2 pause and 3:4 terminator loops.

How do these various loops control transcription termination? Three scenarios describe the process. In the first, translation is not coupled to transcription because protein synthesis is not occuring. In other words, no ribosome is associated with the mRNA. In this scenario, the pause and terminator loops form, stopping transcription before RNA polymerase reaches the *trpE* gene (figure 12.11*a*).

In the next two scenarios, translation and transcription are coupled; that is, a ribosome associates with the leader mRNA as the rest of the mRNA is being synthesized. The interaction between RNA polymerase and the nearest ribosome determines which stem-loop structures are formed. As a ribosome translates the mRNA, it will follow the RNA polymerase. Among the first several nucleotides of region 1 are two tryptophan (trp) codons; this is unusual because normally there is only one trp per 100 amino acids in *E. coli* proteins. If tryptophan levels are low, the ribosome will stall when it encounters the two trp codons. It stalls because the paucity of charged tRNA^{trp} molecules delays the filling of the A site of the ribosome (figure 12.11*b*). Meanwhile the RNA polymerase continues to transcribe mRNA, moving away from the stalled ribosome. The presence of the ribosome on region 1 will prevent it from base pairing with region 2. As RNA polymerase continues, region 3 is transcribed, enabling the formation of the 2:3 antiterminator loop. This prevents the formation of the

(a) No translation occuring

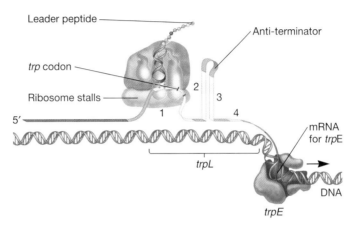

**(b) Translation occuring, low tryptophan levels,
2:3 forms ⟶ transcription continues**

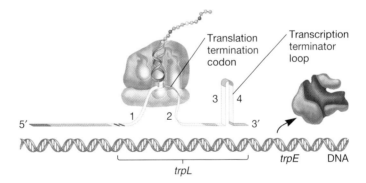

**(c) Translation occuring, high tryptophan levels,
3:4 forms ⟶ transcription is terminated**

Figure 12.11   Attenuation of the *trp* Operon.   (a) When protein synthesis has slowed, transcription and translation are not tightly coupled. Under these conditions, the most stable form of the mRNA occurs when region 1 hydrogen bonds to region 2 (RNA polymerase pause loop) and region 3 hydrogen bonds to region 4 (transcription terminator or attenuator loop). The formation of the transcription terminator causes transcription to stop just beyond *trpL* (*trp* leader). **(b)** When protein synthesis is occurring, transcription and translation are coupled, and the behavior of the ribosome on *trpL* influences transcription. If tryptophan levels are low, the ribosome pauses at the *trp* codons in *trpL* because of insufficient amounts of charged tRNA^{trp}. This blocks region 1 of the mRNA, so that region 2 can hydrogen bond only with region 3. Because region 3 is already hydrogen bonded to region 2, the 3:4 terminator loop cannot form. Transcription proceeds and the *trp* biosynthetic enzymes are made. **(c)** If tryptophan levels are high, translation of *trpL* progresses to the stop codon, blocking region 2. Regions 3 and 4 can hydrogen bond and transcription terminates.

3:4 terminator loop. Because the terminator loop is not formed, RNA polymerase is not ejected from the DNA and transcription continues into the *trp* biosynthetic genes. If, on the other hand, there is plenty of tryptophan in the cell, there will be an abundance of charged tRNA^trp and the ribosome will translate these two trp codons in the leader peptide sequence without hesitation. Thus the ribosome remains close to the RNA polymerase. As RNA polymerase and the ribosome continue through the leader region, regions 1 and 2 are transcribed and readily form a pause loop. Then regions 3 and 4 are transcribed, the terminator loop forms, and RNA polymerase is ejected from the DNA template. Finally, the presence of a UGA stop codon between regions 1 and 2 will cause early termination of translation (figure 12.11*c*). Although the leader peptide will be synthesized, it appears to be rapidly degraded. The genetic code (section 11.7)

Attenuation's usefulness is apparent. If the bacterium is deficient in an amino acid other than tryptophan, protein synthesis will slow and tryptophanyl-tRNA will accumulate. Transcription of the tryptophan operon will be inhibited by attenuation. When the bacterium begins to synthesize protein rapidly, tryptophan may be scarce and the concentration of tryptophanyl-tRNA may be low. This would reduce attenuation activity and stimulate operon transcription, resulting in larger quantities of the tryptophan biosynthetic enzymes. Acting together, repression and attenuation can coordinate the rate of synthesis of amino acid biosynthetic enzymes with the availability of amino acid end products and with the overall rate of protein synthesis. When tryptophan is present at high concentrations, any RNA polymerases not blocked by the activated repressor protein probably will not get past the attenuator sequence. Repression decreases transcription about seventyfold and attenuation slows it another eight- to tenfold; when both mechanisms operate together, transcription can be slowed about 600-fold.

Attenuation is important in regulating at least five other operons that include amino acid biosynthetic enzymes. In all cases, the leader peptide sequences resemble the tryptophan system in

organization. For example, the leader peptide sequence of the histidine operon codes for seven histidines in a row and is followed by an attenuator that is a terminator sequence.

Riboswitches

Regulation by riboswitches, or sensory RNAs, is a specialized form of transcription attenuation that does not involve ribosome behavior. In this case, the leader region of the mRNA can fold in different ways. If folded one way, transcription continues; if folded another, transcription is terminated. This leader region is called a **riboswitch** because it turns transcription on or off. What makes riboswitches unique and exciting is that they alter their folding pattern in direct response to the binding of an effector molecule—a capability previously thought to be associated only with proteins.

One of the first discoveries of this type of regulation was the riboflavin (*rib*) biosynthetic operon of *B. subtilis*. The synthesis of riboflavin biosynthetic enzymes is repressed by flavin mononucleotide (FMN), which is derived from riboflavin. When transcription of the *rib* operon begins, sequences in the leader region of the mRNA fold into a structure called the RFN-element. This element binds FMN, and in doing so alters the folding of the leader region, creating a terminator that stops transcription (**figure 12.12**).

It now appears that controlling transcription attenuation with sensory RNAs is an important method used by gram-positive bacteria to regulate amino acid-related genes. As in the case of the *rib* operon, the leader regions of these mRNAs contain a regulatory element. In this case, the region is called the T box. T box sequences give rise to competing terminator and antiterminator loops. The development of either a terminator or an antiterminator is determined by the binding of uncharged tRNA corresponding to the relevant amino acid (i.e., tRNA that is not carrying its cognate amino acid). For instance, expression of a tyrosyl-tRNA synthetase gene (i.e., a gene that encodes the enzyme that links tyrosine to a tRNA molecule) is governed by the presence of tRNA^Tyr. When the level

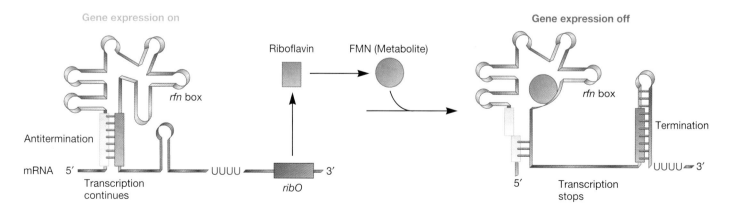

Figure 12.12 Riboswitch Control of the Riboflavin (*rib*) Operon of *Bacillus subtilis*. The *rib* operon produces enzymes needed for the synthesis of riboflavin, a component of flavin mononucleotide (FMN). Binding of FMN to the rfn (*rifampin*) box in the leader of *rib* mRNA causes a change in mRNA folding, which results in the formation of a transcription terminator and cessation of transcription.

Table 12.1	Regulation of Gene Expression by Riboswitches		
System	**Microbe (s)**	**Target genes encode:**	**Effector & Regulatory Response**
T box	Many gram-positive bacteria	Amino acid biosynthetic enzymes	Uncharged tRNA; anticodon base pairs to 5′ end of mRNA, preventing formation of transcriptional terminator
Vitamin B$_{12}$ element	E. coli	Cobalamine biosynthetic enzymes	Adenosylcobalamine (AdoCbl) binds to btuB mRNA and blocks translation
THI box	Rhizobium etli E. coli B. subtilis	Thiamin (Vitamin B$_1$) biosynthetic and transport proteins	Thiamin pyrophosphate (TPP) causes either premature transcriptional termination (R. etli, B. subtilis) or blocks ribosome binding (E. coli)
RFN-element	B. subtilis	Riboflavin biosynthetic enzymes	Flavin mononucleotide (FMN) cases premature transcriptional termination
S box	Low G + C gram-positive bacteria	Methionine biosynthetic enzymes	S-adenosylmethionine (SAM) causes premature transcriptional termination

of charged tRNA^{Tyr} falls, the anticodon of an uncharged tRNA binds directly to the "specifier sequence" codon in the leader of the mRNA. At the same time, the antiterminator loop is stabilized by base pairing between sequences in the loop and the acceptor end of the tRNA, which normally binds the amino acid. This prevents formation of the terminator structure, and transcription of the tyrosyl-tRNA synthetase gene continues. Genomic analysis now suggests that the T box mechanism may be involved in regulating over 300 genes and/or operons; some other genes that bear sensory RNA in their leader regions are listed in **table 12.1**. Other riboswitches have been shown to function at the level of translation. They are described in the next section.

12.4 REGULATION AT THE LEVEL OF TRANSLATION

It appears that in general, the riboswitches found in gram-positive bacteria function by transcriptional termination, while the riboswitches discovered in gram-negative bacteria regulate the translation of mRNA. Translation is usually regulated by blocking its initiation. As noted previously, some riboswitches work at this level. In addition, some small RNA molecules can control translation initiation. Both are described in this section.

Regulation of Translation by Riboswitches

Similar to the riboswitches described earlier, riboswitches that function at the translational level contain effector-binding elements at the 5′ end of the mRNA. Binding of the effector molecule alters the folding pattern of the leader region of the mRNA, which often results in occlusion of the Shine-Dalgarno sequence and other elements of the ribosome-binding site. This inhibits ribosome binding and initiation of translation (**figure 12.13**). An example of this type of regulation is observed for the thiamine biosynthetic operons of numerous bacteria and some archaea. The leader regions of thiamine operons contain a structure called the THI-element, which can bind thiamin pyrophosphate. Bind-

ing of thiamin pyrophosphate to the THI-element causes a conformational change in the leader region that sequesters the Shine-Dalgarno sequence and blocks translation initiation.

Regulation of Translation by Small RNA Molecules

A large number of RNA molecules have been discovered that do not function as mRNAs, tRNAs, or rRNAs. Microbiologists often refer to them as **small RNAs (sRNAs)** or as noncoding RNAs (ncRNAs). In E. coli, there are more than 40 sRNAs, ranging in size from around 40 to 400 nucleotides. It is thought that eucaryotes may have hundreds to thousands of sRNAs with lengths from 21 to over 10,000 nucleotides. Although some sRNAs have been implicated in the regulation of DNA replication and transcription, many function at the level of translation.

In E. coli, most sRNAs regulate translation by base pairing to the leader region of a target mRNA. Thus they are complementary to the mRNA and are called **antisense RNAs.** It seems intuitive that by binding to the leader, antisense RNAs would block ribosome binding and inhibit translation. Indeed, many antisense RNAs work in this manner. However, some antisense RNAs actually promote translation upon binding to the mRNA. Whether inhibitory or activating, most E. coli antisense RNAs work with a protein called Hfq to regulate their target RNAs. The Hfq protein is an RNA chaperone—that is, it interacts with RNA to promote changes in its structure. In addition, the Hfq protein may promote RNA-RNA interactions.

The regulation of synthesis of OmpF and OmpC porin proteins provides an example of translation control by an antisense RNA. In addition to regulation by the OmpR protein described previously (figure 12.10), expression of the ompF gene is regulated by an antisense RNA called MicF RNA, which is the product of the micF gene (mic for mRNA-interfering complementary RNA). The MicF RNA is complementary to ompF at the translation initiation site (**figure 12.14**). It base pairs with ompF mRNA and represses translation. MicF RNA is produced under conditions such as high

Table 13.1	Examples of Mutagens
Mutagen	**Effect(s) on DNA Structure**
Chemical	
5-Bromouracil	Base analog
2-Aminopurine	Base analog
Ethyl methanesulfonate	Alkylating agent
Hydroxylamine	Hydroxylates cytosine
Nitrogen mustard	Alkylating agent
Nitrous oxide	Deaminates bases
Proflavin	Intercalating agent
Acridine orange	Intercalating agent
Physical	
UV light	Promotes pyrimidine dimer formation
X rays	Causes base deletions, single-strand nicks, cross-linking, and chromosomal breaks

The enol tautomer forms hydrogen bonds like cytosine and directs the incorporation of guanine rather than adenine (**figure 13.3**). The mechanism of action of other base analogs is similar to that of 5-bromouracil.

DNA-modifying agents change a base's structure and therefore alter its base pairing characteristics. Some mutagens in this category are fairly selective; they preferentially react with some bases and produce a specific kind of DNA damage. An example of this type of mutagen is methyl-nitrosoguanidine, an alkylating agent that adds methyl groups to guanine, causing it to mispair with thymine (**figure 13.4**). A subsequent round of replication could then result in a GC-AT transition. Hydroxylamine is another example of a DNA-modifying agent. It hydroxylates the C-4 nitrogen of cytosine, causing it to base pair like thymine. There are many other DNA modifying agents that can cause mispairing.

Intercalating agents distort DNA to induce single nucleotide pair insertions and deletions. These mutagens are planar and insert themselves (intercalate) between the stacked bases of the helix. This results in a mutation, possibly through the formation of a loop in DNA. Intercalating agents include acridines such as proflavin and acridine orange.

Many mutagens, and indeed many carcinogens, directly damage bases so severely that hydrogen bonding between base pairs is impaired or prevented and the damaged DNA can no longer act as a template for replication. For instance, UV radiation generates cyclobutane type dimers, usually thymine dimers, between adjacent pyrimidines (**figure 13.5**). Other examples are ionizing radiation and carcinogens such as the fungal toxin aflatoxin B1 and other benzo(a)pyrene derivatives.

Retention of proper base pairing is essential in the prevention of mutations. Cells have developed extensive repair mechanisms. Often the damage can be repaired before a mutation is permanently es-

tablished. If a complete DNA replication cycle takes place before the initial lesion is repaired, the mutation frequently becomes stable and inheritable. Repair mechanisms are discussed in section 13.3.

Effects of Mutations

The effects of a mutation can be described at the protein level and in terms of traits or other easily observed phenotypes. In all cases, the impact is readily noticed only if it produces a change in phenotype. In general, the more prevalent form of a gene and its associated phenotype is called the **wild type.** A mutation from wild type to a mutant form is called a **forward mutation** (**table 13.2**). A forward mutation can be reversed by a second mutation that restores the wild-type phenotype. When the second mutation is at the same site as the original mutation, it is called a **reversion mutation.** A true reversion converts the mutant nucleotide sequence back to the wild-type sequence. If the second mutation is at a different site than the original mutation, it is called a **suppressor mutation.** Suppressor mutations may be within the same gene (intragenic suppressor mutation) or in a different gene (extragenic suppressor mutation). Because point mutations are the most common types of mutations, their effects will be the focus here.

Mutations in Protein-Coding Genes

Point mutations in protein-coding genes can affect protein structure in a variety of ways (table 13.2). Point mutations are named according to if and how they change the encoded protein. The most common types of point mutations are silent mutations, missense mutations, nonsense mutations, and frameshift mutations. These are described in more detail below.

Silent mutations change the nucleotide sequence of a codon, but do not change the amino acid encoded by that codon. This is possible because of the degeneracy of the genetic code. Therefore, when there is more than one codon for a given amino acid, a single base substitution may result in the formation of a new codon for the same amino acid. For example, if the codon CGU were changed to CGC, it would still code for arginine even though a mutation had occurred. The mutation can only be detected at the level of the DNA or mRNA. When there is no change in the protein, there is no change in the phenotype of the organism. The genetic code (section 11.7)

Missense mutations involve a single base substitution that changes a codon for one amino acid into a codon for another. For example, the codon GAG, which specifies glutamic acid, could be changed to GUG, which codes for valine. The effects of missense mutations vary. They alter protein structure, but the effect of this change may range from complete loss of activity to no change at all. This is because the effect of missense mutations on protein function depends on the type and location of the amino acid substitution. For instance, replacement of a nonpolar amino acid in the protein's interior with a polar amino acid can drastically alter the protein's three-dimensional structure and therefore its function. Similarly the replacement of a critical amino acid at the active site of an enzyme often destroys its activity. However, the replacement of one polar amino

(a) Base pairing of 5-BU with adenine or guanine

(b) How 5-BU causes a mutation in a base pair during DNA replication

Figure 13.3   Mutagenesis by the Base Analog 5-Bromouracil.
(a) Base pairing of the normal keto form of 5-BU is shown in the top illustration. The enol form of 5-BU (bottom illustration) base pairs with guanine rather than with adenine as might be expected for a thymine analog. **(b)** If the keto form of 5-BU is incorporated in place of thymine, its occasional tautomerization to the enol form (BU$_e$) will produce an AT to GC transition mutation.

Figure 13.4   Methyl-Nitrosoguanidine Mutagenesis.
Mutagensis by methyl-nitrosoguanidine due to the methylation of guanine.

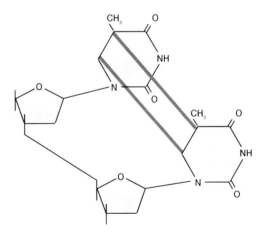

Figure 13.5   Thymine Dimer.   Thymine dimers are formed by ultraviolet radiation.

acid with another at the protein surface may have little or no effect. Missense mutations actually play a very important role in providing new variability to drive evolution because they often are not lethal and therefore remain in the gene pool.   Proteins (appendix 1)

Nonsense mutations cause the early termination of translation and therefore result in a shortened polypeptide. They are called nonsense mutations because they convert a sense codon to a nonsense or stop codon. Depending on the relative location of the mutation, the phenotype may be more or less severely af-

fected. Most proteins retain some function if they are shortened by only one or two amino acids; complete loss of normal function will almost certainly result if the mutation occurs closer to the beginning or middle of the gene.

Table 13.5	The Properties of Selected Composite Transposons			
Transposon	**Length (bp)**	**Terminal Repeat Length**	**Terminal Module**	**Genetic Markers**
Tn3	4,957	38		Ampicillin resistance
Tn501	8,200	38		Mercury resistance
Tn951	16,500	Unknown		Lactose utilization
Tn5	5,700		IS50	Kanamycin resistance
Tn9	2,500		IS1	Chloramphenicol resistance
Tn10	9,300		IS10	Tetracycline resistance
Tn903	3,100		IS903	Kanamycin resistance
Tn1681	2,061		IS1	Heat-stable enterotoxin
Tn2901	11,000		IS1	Arginine biosynthesis

The process of transposition in procaryotes can occur by two basic mechanisms. **Simple transposition,** also called **cut-and-paste transposition,** involves transposase-catalyzed excision of the transposon, followed by cleavage of a new target site and ligation of the transposon into this site (**figure 13.20**). Target sites are specific sequences about five to nine base pairs long. When a transposon inserts at a target site, the target sequence is duplicated so that short, direct-sequence repeats flank the transposon's terminal inverted repeats.

The second transposition mechanism is **replicative transposition.** In this mechanism, the original transposon remains at the parental site on the chromosome and a replicate is inserted at the target DNA site (**figure 13.21**).The transposition of the Tn3 transposon is a well-studied example of replicative transposition. In the first stage, DNA containing Tn3 fuses with the target DNA to form a cointegrate molecule (figure 13.21, step 1). This process requires the Tn3 transposase enzyme coded for by the *tnpA* gene (**figure 13.22**). Note that the cointegrate has two copies of the Tn3 transposon. In the second stage the cointegrate is resolved to yield two DNA molecules, each with a copy of the transposon (figure 13.21, step 3). Resolution involves a crossover and is catalyzed by a resolvase enzyme coded for by the *tnpR* gene (figure 13.22).

Transposable elements produce a variety of important effects. They can insert within a gene to cause a mutation or stimulate DNA rearrangement, leading to deletions of genetic material. Because some transposons carry stop codons or termination sequences, when transposed into genes they may block translation or transcription, respectively. Likewise, other transposons carry promoters and can activate genes near the point of insertion. Thus transposons can turn genes on or off. Transposons also are located in plasmids and participate in such processes as plasmid fusion, insertion of plasmids into chromosomes, and plasmid evolution.

The role of transposons in plasmid evolution is of particular note. Plasmids can contain several different transposon-target sites. Therefore, transposons frequently move between plasmids. Of concern is the fact that many transposons contain antibiotic-resistance genes. Thus as they move from one plasmid to another, resistance genes are introduced into the target plasmid, creating a resistance (R) plasmid. Multiple drug-resistance plasmids can arise from the accumulation of transposons in a plasmid (figure 13.22). Many R plasmids are able to move from one cell to another during conjugation, which spreads the resistance genes throughout a population. Finally, because transposons also move between plasmids and chromosomes, drug resistance genes can exchange between these two molecules, resulting in the further spread of antibiotic resistance.

Some transposons bear transfer genes and can move between bacteria through the process of conjugation, as discussed in section 13.7. A well-studied example of a **conjugative transposon** is Tn916 from *Enterococcus faecalis.* Although Tn916 cannot replicate autonomously, it can transfer itself from *E. faecalis* to a variety of recipients and integrate into their chromosomes. Because it carries a gene for tetracycline resistance, this conjugative transposon also spreads drug resistance.

13.6 BACTERIAL PLASMIDS

Conjugation, the transfer of DNA between bacteria involving direct contact, depends on the presence of an "extra" piece of DNA known as a plasmid. Plasmids play many important roles in the lives of procaryotes. They also have proved invaluable to microbiologists and molecular geneticists in constructing and transferring new genetic combinations and in cloning genes as described in chapter 14.

Recall from chapter 3 that **plasmids** are small double-stranded DNA molecules that can exist independently of host chromosomes. They have their own replication origins and autonomously replicate and are stably inherited. Some plasmids are **episomes,** plasmids that can exist either with or without being integrated into host chromosomes. Although there are a variety of plasmid types, our concern here is with **conjugative plasmids.** These plasmids can transfer copies of themselves to other bacteria during the process of conjugation, which is discussed in section 13.7.

Figure 13.20   Simple Transposition.   TE, transposable element; IR, inverted repeat.

Figure 13.21   Replicative Transposition.

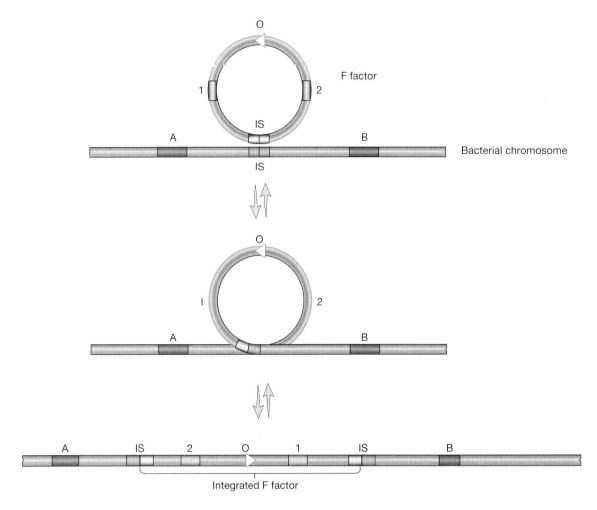

Figure 13.25 F Plasmid Integration. The reversible integration of an F plasmid or factor into a host bacterial chromosome. The process begins with association between plasmid and bacterial insertion sequences. The O arrowhead (white) indicates the site at which oriented transfer of chromosome to the recipient cell begins. A, B, 1, and 2 represent genetic markers.

F$^+$ × F$^-$ Mating

In 1952 William Hayes demonstrated that the gene transfer observed by Lederberg and Tatum was polar. That is, there were definite donor (F$^+$, or fertile) and recipient (F$^-$, or nonfertile) strains, and gene transfer was nonreciprocal. He also found that in F$^+$ × F$^-$ mating the progeny were only rarely changed with regard to auxotrophy (that is, chromosomal genes were not often transferred), but F$^-$ strains frequently became F$^+$.

These results are readily explained in terms of the F factor previously described (figure 13.23). The F$^+$ strain contains an extrachromosomal F factor carrying the genes for sex pilus formation and plasmid transfer. The **sex pilus** is used to establish contact between the F$^+$ and F$^-$ cells (**figure 13.28a**). Once contact is made, the pilus retracts, bringing the cells into close physical contact. The F$^+$ cell then prepares for DNA transfer by assembling a type IV secretion apparatus, using many of the same genes used for sex pilus biogenesis; the sex pilus is embedded in the secretion structure (**figure 13.29**). The F factor then replicates by a rolling-circle mechanism (*see figure 11.12*). Replication is initiated by a complex of proteins called the relaxosome, which

nicks one strand of the F factor at a site called *oriT* (for *ori*gin of *t*ransfer). Relaxase, an enzyme associated with the relaxosome, remains attached to the 5' end of the nicked strand. As F factor is replicated, the displaced strand and the attached relaxase enzyme move through the type IV secretion system to the recipient cell. Because the pilus is embedded in the secretion apparatus, it has been suggested that the DNA moves through a lumen in the pilus. However, studies of a related conjugation system, that of the plant pathogen *Agrobacterium tumefaciens,* provide strong evidence that the DNA does not move through the sex pilus. However, it should be noted that although the F factor system and the *Agrobacterium* system are related, there is one important difference between the two. The F factor system is used to transfer DNA from one bacterium to another, whereas the *Agrobacterium* system moves DNA from the bacterium into its plant host. Whatever the route of transfer, as the plasmid is transferred, the entering strand is copied to produce double-stranded DNA. The recombination frequency is low because chromosomal genes are rarely transferred with the independent F factor. Microorganism associations with vascular plants: *Agrobacterium* (section 29.5)

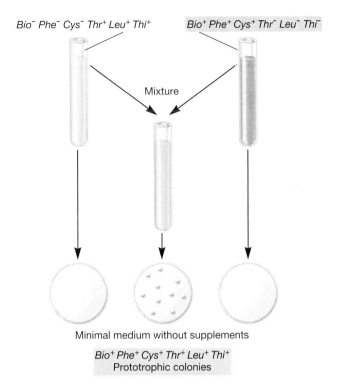

$Bio^- Phe^- Cys^- Thr^+ Leu^+ Thi^+$ $Bio^+ Phe^+ Cys^+ Thr^- Leu^- Thi^-$

Mixture

Minimal medium without supplements

$Bio^+ Phe^+ Cys^+ Thr^+ Leu^+ Thi^+$
Prototrophic colonies

Figure 13.26   Evidence for Bacterial Conjugation.
Lederberg and Tatum's demonstration of genetic recombination using triple auxotrophs. See text for details.

Pressure or suction

$Met^- Thr^+ Leu^+ Thi^+$ $Met^+ Thr^- Leu^- Thi^-$

Fritted glass filter

Figure 13.27   The U-tube Experiment.   The U-tube experiment used to show that genetic recombination by conjugation requires direct physical contact between bacteria. See text for details.

Hfr Conjugation

Not long after the discovery of $F^+ \times F^-$ mating, a second type of F factor-mediated conjugation was discovered. In this type of conjugation, the donor transfers chromosomal genes with great efficiency, but does not change the recipient bacteria into F^+ cells. Because of the *high frequency* of recombinants produced by this mating, it is referred to as **Hfr conjugation** and the donor is called an **Hfr strain.** Although initially the mechanism of Hfr

conjugation was not known, eventually it was determined that Hfr strains contain the F factor integrated into their chromosome, rather than free in the cytoplasm (figure 13.28*b*). When integrated, the F plasmid's *tra* operon is still functional; the plasmid can direct the synthesis of pili, carry out rolling-circle replication, and transfer genetic material to an F^- recipient cell. However, rather than transferring itself, the F factor directs the transfer of host chromosome. DNA transfer begins when the integrated F factor is nicked at its site of transfer origin. As it is replicated, the chromosome moves to the recipient (figure 13.28c). Because only part of the F factor is transferred, the F^- recipient does not become F^+ unless the whole chromosome is transferred. Transfer of the entire chromosome with the integrated F factor requires about 100 minutes in *E. coli,* and the connection between the cells usually breaks before this process is finished. Thus a complete F factor usually is not transferred, and the recipient remains F^-.

As mentioned earlier, when an Hfr strain participates in conjugation, bacterial genes are frequently transferred to the recipient. Gene transfer can be in either a clockwise or counterclockwise direction around the circular chromosome, depending on the orientation of the integrated F factor. After the replicated donor chromosome enters the recipient cell, it may be degraded or incorporated into the F^- genome by recombination.

F′ Conjugation

Because the F plasmid is an episome, it can leave the bacterial chromosome and resume status as an autonomous F factor. Sometimes during this process the plasmid makes an error in excision and picks up a portion of the chromosome. Because it is now genotypically distinct from the original F factor, it is called an **F′ plasmid (figure 13.30*a*).** It is not unusual to observe the inclusion of one or more chromosomal genes in excised F plasmids. A cell containing an F′ plasmid retains all of its genes, although some of them are on the plasmid. It mates only with an F^- recipient. F′ × F^- conjugation is similar to $F^+ \times F^-$ mating. Once again, the plasmid is transferred as it is copied by rolling-circle replication. However, bacterial genes on the chromosome usually are not transferred (figure 13.30*b*). Bacterial genes acquired during excision of the F′ plasmid are transferred with it and need not be incorporated into the recipient chromosome to be expressed. The recipient becomes F′ and is a partially diploid merozygote because the same bacterial genes present on the F′ plasmid are also found on the recipient's chromosome. In this way specific bacterial genes may spread rapidly throughout a bacterial population.

F′ conjugation is very important to the microbial geneticist. A partial diploid's behavior shows whether the allele carried by an F′ plasmid is dominant or recessive to the chromosomal gene. The formation of F′ plasmids also is useful in mapping the chromosome because if two genes are picked up by an F factor they must be neighbors.

Other Examples of Bacterial Conjugation

Although most research on plasmids and conjugation has been done using *E. coli* and other gram-negative bacteria, self-transmissible

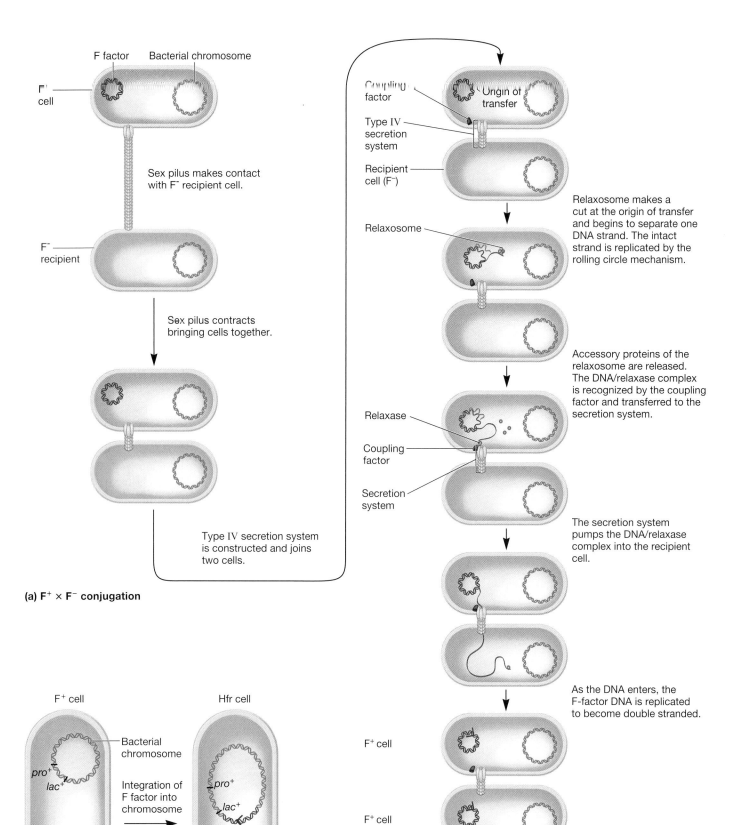

F factor   Bacterial chromosome

F⁺
cell

Sex pilus makes contact
with F⁻ recipient cell.

F⁻
recipient

Sex pilus contracts
bringing cells together.

Type IV secretion system
is constructed and joins
two cells.

(a) F⁺ × F⁻ conjugation

Coupling
factor   Origin of
                    transfer
Type IV
secretion
system

Recipient
cell (F⁻)

Relaxosome makes a
cut at the origin of transfer
and begins to separate one
DNA strand. The intact
strand is replicated by the
rolling circle mechanism.

Relaxosome

Accessory proteins of the
relaxosome are released.
The DNA/relaxase complex
is recognized by the coupling
factor and transferred to the
secretion system.

Relaxase

Coupling
factor

Secretion
system

The secretion system
pumps the DNA/relaxase
complex into the recipient
cell.

As the DNA enters, the
F-factor DNA is replicated
to become double stranded.

F⁺ cell

F⁺ cell

(b) Insertion of F factor into chromosome

F⁺ cell   Hfr cell

Bacterial
chromosome

pro⁺
lac⁺   Integration of
                         F factor into
                         chromosome

                                        pro⁺

                                        lac⁺

Origin of
transfer   Origin of
                                        transfer
F factor

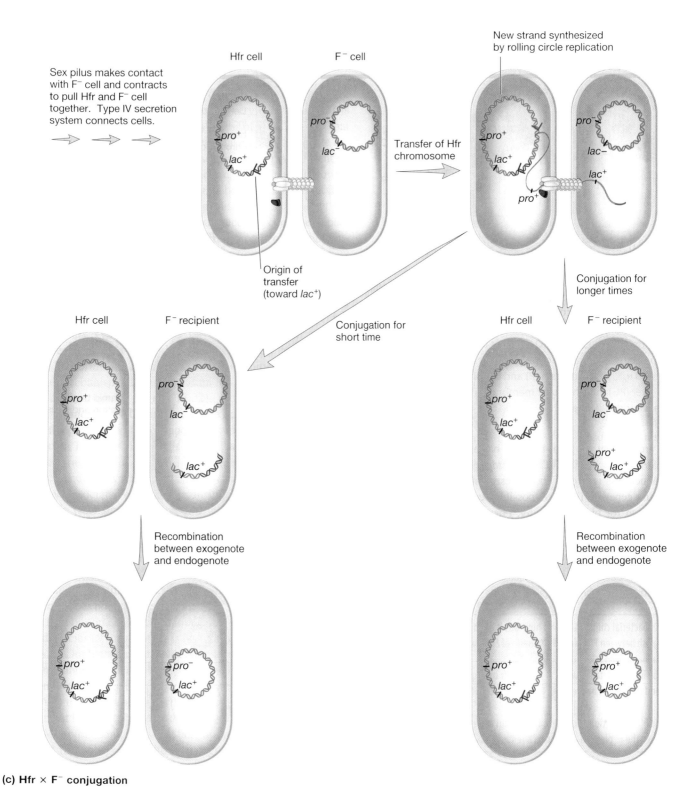

(c) Hfr × F⁻ conjugation

Figure 13.28   F Factor-Mediated Conjugation.   The F factor encodes proteins for building the sex pilus and proteins needed to construct the type IV secretion system that will transfer DNA from the donor to the F⁻ recipient. One protein, the coupling factor, is thought to guide the DNA to the secretion system. **(a)** During F⁺ × F⁻ conjugation, only the F factor is transferred because the plasmid is extrachromosomal. The recipient cell becomes F⁺. **(b)** Integration of the F factor into the chromosome creates an Hfr cell. **(c)** During Hfr × F⁻ conjugation, some plasmid genes and some chromosomal genes are transferred to the recipient. Note that only a portion of the F factor moves into the recipient. Because the entire plasmid is not transferred, the recipient remains F⁻. In addition, the incoming DNA must recombine into the recipient's chromosome if it is to be stably maintained.

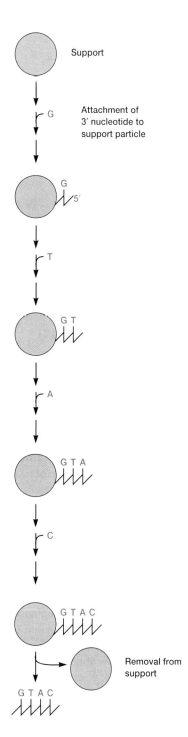

Figure 14.7 The Synthesis of a DNA Oligonucleotide.
During each cycle, the DNA synthesizer adds an activated nucleotide
(A, T, G, or C) to the growing end of the chain. At the end of the
process, the oligonucleotide is removed from its support.

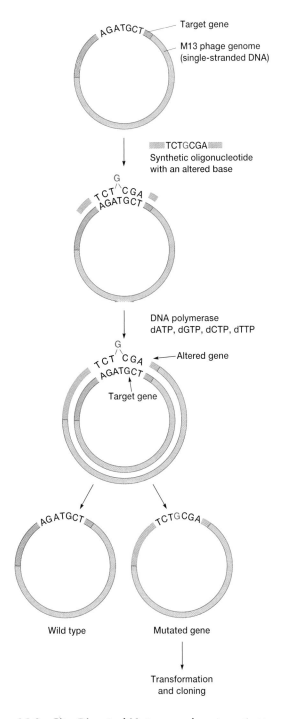

Figure 14.8 Site-Directed Mutagenesis. A synthetic
oligonucleotide is used to add a specific mutation to a gene. See
text for details.

quantitative). Thermocyclers specifically designed for real-time
PCR are used that record the amount of PCR product generated as
it happens; thus the term real-time PCR. Gene expression studies
often rely on real time-PCR, because mRNA transcripts can be
copied and amplified by reverse transcriptase (RT). Therefore the

procedure monitors the level of gene transcription of the gene targeted by the primers. This is sometimes called RT-RT-PCR.

The PCR technique is an essential tool in many areas of molecular biology, medicine, and biotechnology. As shown in **figure 14.10,** when PCR is used to obtain DNA for cloning, a

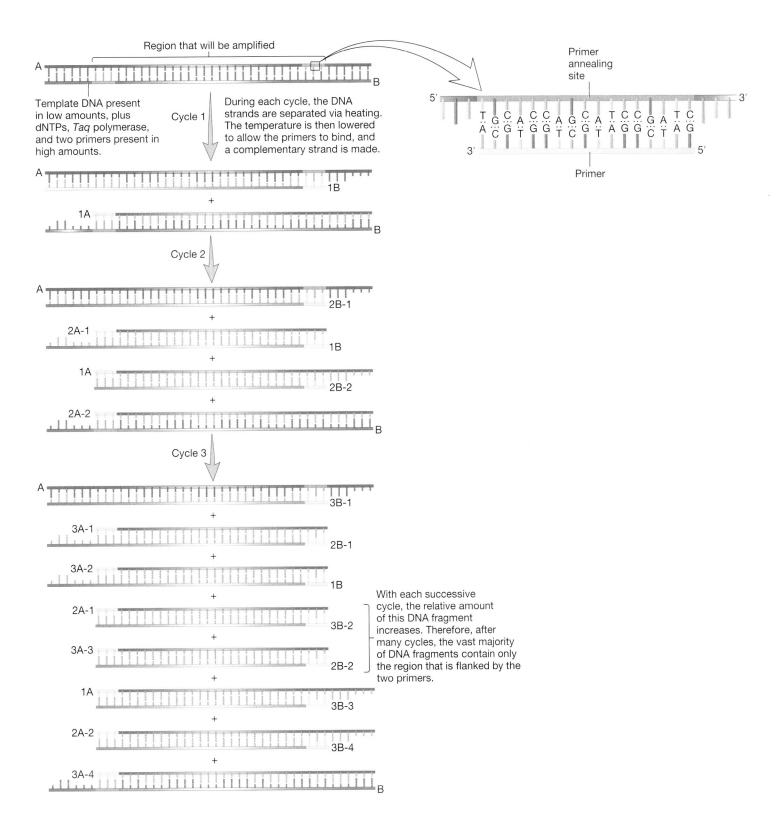

Figure 14.9 **The Technique of Polymerase Chain Reaction (PCR).** During each cycle, oligonucleotides that are complementary to the ends of the targeted DNA sequence bind to the DNA and act as primers for the synthesis of this DNA region. The primers used in actual PCR experiments are usually 15 to 20 nucleotides in length. The region between the two primers is typically hundreds of nucleotides in length, not just several nucleotides as shown here. The net result of PCR is the synthesis of many copies of DNA in the region that is flanked by the two primers.

Figure 14.10 Cloning Cellular DNA Fragments. The preparation of a recombinant clone from isolated DNA fragments or DNA generated by PCR.

number of steps traditionally employed are no longer required. PCR is also used to generate DNA for nucleotide sequencing. Because PCR amplifies DNA even if it is present in very small, initial quantities, it is used in a number of diagnostic tests, including those for AIDS, Lyme disease, chlamydia, tuberculosis, hepatitis, the human papilloma virus, and other infectious agents and diseases. The tests are rapid, sensitive, and specific. PCR is particularly valuable in the detection of genetic diseases such as sickle cell anemia, phenylketonuria, and muscular dystrophy. The technique is also employed in forensic science where it is used in criminal cases as a part of DNA fingerprinting technology. It is possible to exclude or incriminate suspects using extremely small samples of biological material discovered at the crime scene.

14.4 GEL ELECTROPHORESIS

Agarose or polyacrylamide gels usually are used to separate DNA fragments electrophoretically. In **gel electrophoresis,** charged molecules are placed in an electrical field and allowed to migrate toward the positive and negative poles. The molecules separate because they move at different rates due to their differences in charge and size. Because DNA is negatively charged, it is loaded into wells at the negative pole of the gel and migrates toward the positive (**figure 14.11**). Each fragment's migration rate is inversely proportional to the log of its molecular weight. That is to

say, the smaller a fragment is, the faster it moves through the gel. Migration rate is also a function of gel density. In practice, this means that higher concentrations of gel material (agarose or acrylamide) provide better resolution of small fragments and vice versa. DNA that has not been digested with restriction enzymes is usually supercoiled. For this and other reasons, DNA is usually cut with restriction endonucleases prior to electrophoresis. Small DNA molecules usually yield only a few bands because there are few restriction enzyme recognition sites. If the DNA fragment is large, or an entire chromosome is digested, many such sites are present and the DNA is cut in numerous places. When such DNA is electrophoresed, it produces a smear representing many thousands of DNA fragments of similar sizes that cannot be individually resolved. The region of the gel containing the desired DNA fragment must then be located using the Southern blot technique (figure 14.6). DNA from this region can be isolated from the gel material and electrophoresed again on a gel of another agarose or acrylamide concentration so that individual bands can be detected.

14.5 CLONING VECTORS AND CREATING RECOMBINANT DNA

Recombinant DNA technology depends on the propagation of many copies of the nucleotide sequence of choice. To accomplish this, genes or other genetic elements are inserted (i.e., cloned) into DNA vectors that replicate in a host organism. There are four major types of vectors: plasmids, bacteriophages and other viruses, cosmids, and artificial chromosomes (**table 14.3**). Each type has its own advantages, so the selection of the proper cloning vector is critical to the success of any cloning experiment. All engineered vectors share three features: an origin of replication; a region of DNA that bears unique restriction sites, called a multicloning site or polylinker; and a selectable marker. These elements are described in the discussion of plasmids, the most frequently used cloning vectors.

Plasmids

Plasmids make excellent cloning vectors because they replicate autonomously and are easy to purify. They can be introduced into microbes by conjugation and/or transformation. Many different plasmids are used in biotechnology, all derived from naturally occurring plasmids that have been genetically engineered (**figure 14.12**). Bacterial plasmids (section 3.6); Bacterial conjugation (section 13.7); Transformation (section 13.8)

Origin of Replication

The **origin of replication (*ori*)** allows the plasmid to replicate in the microbial host independently of the chromosome. pUC19, an *E. coli* plasmid, is said to have a high copy number because it replicates about 100 times in the course of one generation. That is to say, an *E. coli* cell with one chromosome can have as many as 100 copies of the plasmid. High copy number is often important because it facilitates plasmid purification and it can dramat-

(a)

(b)

Figure 14.11   Gel Electrophoresis of DNA.   (a) After cleavage into fragments, DNA is loaded into wells on one end of an agarose gel. When an electrical current is passed through the gel (from the negative pole to the positive pole), the DNA, being negatively charged, migrates toward the positive pole. The larger fragments, measured in numbers of base pairs, migrate more slowly and remain nearer the wells than the smaller (shorter) fragments. **(b)** An actual developed and stained gel reveals a separation pattern of the fragments of DNA. The size of a given DNA band can be determined by comparing it to a known set of molecular weight markers (lane 5) called a ladder.

ically increase the amount of cloned gene product produced by the cell. Some plasmids have two origins of replication, each recognized by different host organisms. These plasmids are called **shuttle vectors** because they can move or "shuttle" from one host to another. YEp24 is a shuttle vector that can replicate in yeast (*Saccharomyces cerevisiae*) and in *E. coli* because it has the 2μ circle yeast replication element and *E. coli* origin of replication, *ori* (figure 14.12).

Selectable Marker

Following the uptake of vector by host cells, one must be able to discriminate between those cells that successfully obtained vector from those that did not. Furthermore, one must be able to continue to select for the presence of plasmid, otherwise the host cell may stop replicating it. This is achieved by the presence of a gene that encodes a protein that is needed for the cell to survive under certain, selective conditions. Such a gene is called a **selectable marker.** In the case of pUC19, the selectable marker encodes the ampicillin resistance factor (*amp*^R, sometimes called *bla,* for *β-la*ctamase). The shuttle vector YEp24 bears both the *amp*^R gene for selection in *E. coli,* and *URA3,* which encodes a protein essential for uracil biosynthesis in yeast. Therefore when in *S. cerevisiae,* this plasmid must be maintained in uracil auxotrophs.

Multicloning Site (MCS) or Polylinker

A region of restriction enzyme cleavage sites found only once in the plasmid is essential for the insertion of foreign DNA. Cleavage at a unique restriction site generates a linear plasmid. Cleavage of the gene to be cloned with the same restriction enzyme results in compatible sticky ends, so that it may be inserted (ligated) into the **multicloning site (MCS).** Alternatively, two different, unique sites within the MCS may be cleaved and the DNA sequence between the two sites replaced with cloned DNA (**figure 14.13**). In either case, the plasmid and the DNA to be inserted are incubated in the presence of the enzyme **DNA ligase** so that when compatible sticky ends hydrogen bond, phosphodiester covalent bonds can be generated between the cloned DNA fragment and the vector. This requires the input of energy, thus ATP is added to this in vitro ligation reaction *(see figure 11.19).*

pUC19 has a number of unique restriction sites in its MCS (figure 14.12); this provides a number of cleavage options, making it easier to obtain the same, or compatible, sticky ends in both vector and the DNA to be inserted. In pUC19 the MCS is located within the 5′ end of the *lacZ* gene, which encodes β-galactosidase (β-Gal). This enzyme cuts the dissacharide lactose into galactose and glucose. When DNA has been cloned into the MCS, the *lacZ* gene is no longer intact, so a functional enzyme is not produced.

Table 14.3	Recombinant DNA Cloning Vectors		
Vector	**Insert Size (kb, 1 kb = 1,000 bp)**	**Example**	**Features**
Plasmid	<20 kb	pBR322, pUC19	Replicates independently of microbial chromosome so many copies may be maintained in a single cell
Bacteriophage	9–25 kb	λ1059, λ gt11, M13mp18, EMBL3	Packaged into lambda phage particles; single-stranded DNA viruses like M13 have been modified (e.g., M13mp18) to generate either double- or single-stranded DNA in the host
Cosmids	30–47 kb	pJC720, pSupercos	Can be packaged into lambda phage particles for efficient introduction into bacteria, then replicates as a plasmid
PACs (P1 artificial chromosomes)	75–100 kb	pPAC	Based on the bacteriophage P1 packaging mechanism
BACs (bacterial artificial chromosomes)	75–300 kb	pBAC108L	Modified F plasmid that can carry large DNA inserts; very stable within the cell
YACs (yeast artificial chromosomes)	100–1,000 kb	pYAC	Can carry largest DNA inserts, replicates in *Saccharomyces cerevisiae*

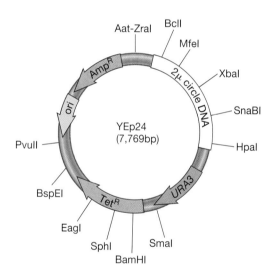

Figure 14.12 The Cloning Vectors pUC19 and YEp24. Restriction sites that are present only once in each vector are shown. pUC19 replicates only in *E. coli*, while YEp24 replicates in both *E. coli* and *S. cerevisiae*.

This can be detected by the color of colonies: cells turn blue when β-Gal splits the alternative substrate, X-Gal (5-bromo-4-chloro-3-indolyl-β-D-galactopyranoside), which is included in the medium (figure 14.13). This is important because the ligation of foreign DNA into a vector is never 100% efficient. Thus when the ligation mixture is introduced into host cells, one must be able to distinguish cells that carry just plasmid from those that carry plasmid into which DNA was successfully inserted. In the case of pUC19, all *E. coli* cells that take up plasmid (with or without insert) are selected by their resistance to ampicillin (Amp^R). Among these, colonies with plasmid lacking DNA insert will be blue (due to the presence of functional *lacZ* gene), while those that have pUC19 into which DNA was successfully cloned will be white. There are a number of other clever ways in which cells with vector versus those with vector with insert can be differentiated; the selection of blue versus white colonies is a common approach.

Phage Vectors

Phage vectors are engineered phage genomes that have been genetically modified to include useful restriction enzyme recognition sites for the insertion of foreign DNA. Once DNA has been inserted, the recombinant phage genome is packaged into viral capsids and used to infect host cells. The resulting phage lysate consists of thousands of phage particles that carry cloned DNA as well as the genes needed for host lysis. Two commonly used vectors are derived from the bacteriophages T7 and lambda (λ), both of which have double-stranded DNA genomes. Although these

Figure 14.13 Recombinant Plasmid Construction and Cloning. The construction and cloning of a recombinant plasmid vector using an antibiotic resistance gene to select for the presence of the plasmid. The interruption of the *lacZ* gene by cloned DNA is used to select for vectors with insert. The scale of the sticky ends of the fragments and plasmid has been enlarged to illustrate complementary base pairing. **(a)** The electron micrograph shows a plasmid that has been cut by a restriction enzyme and a donor DNA fragment. **(b)** The micrograph shows a recombinant plasmid. **(c)** After transformation, *E. coli* cells are plated on medium containing ampicillin and X-Gal so that only ampicillin-resistant transformants grow; X-Gal enables the visualization of colonies that were transformed with recombinant vector (vector + insert, white colonies).

phages infect *E. coli,* phage vectors have been engineered for a number of different bacterial host species. Viruses of *Bacteria* and *Archaea* (chapter 17)

Cosmids

Cosmids were developed when it became clear that cloning vectors were needed that could tolerate larger fragments of cloned DNA (table 14.3). Unlike phages and plasmids, cosmids do not exist in nature. Instead, these engineered vectors have been constructed to contain features from both. Cosmids have a selectable marker and MCS from plasmids, and a *cos* site from λ phage. In λ phage, the *cos* site is where multiple copies of phage genome are linked prior to packaging. Cleavage at the *cos* sites yields single genomes that are the right size for packaging. Cosmids take advantage of the fact that the only requirement for λ phage heads to package DNA is two *cos* sites on a linear DNA molecule or a single *cos* site on a circular one. As long as the cosmid with its cloned DNA is the appropriate size (about 37 to 52 kb), it will be packaged. The phage is then used to introduce the recombinant DNA into *E. coli,* where it replicates as a plasmid. Temperate bacteriophages and lysogeny (section 17.5)

Artificial Chromosomes

Artificial chromosomes are special cloning vectors used when particularly large fragments of DNA must be cloned, as when constructing a genomic library or when sequencing an organism's entire genome. In fact, **bacterial artificial chromosomes (BACs)** were crucial to the timely completion of the human genome project. Like natural chromosomes, artificial chromosomes replicate only once per cell cycle. **Yeast artificial chromosomes (YACs)** were developed first and consist of a yeast telomere at each end (*TEL*), a centromere sequence (*CEN*), a yeast origin of replication (*ARS, autonomously, replicating sequence*), a selectable marker such as URA3, and an MCS to facilitate the insertion of foreign DNA (**figure 14.14**). YACs are used when extraordinarily large DNA pieces (up to 1,000 kb; table 14.3) are to be cloned. BACs were developed, in part, because YACs tend to be unstable and may recombine with host chromosomes, thereby rearranging the cloned DNA. Although BACs accept smaller DNA inserts than do YACs (up to 300 kb), they are generally more stable. BACs are based on the F fertility factor of *E. coli.* The example shown in figure 14.14 is typical in that it includes genes that ensure a replication complex will be formed (*rep*E), as well as proper partitioning of one newly replicated BAC to each daughter cell (*sopA, sopB,* and *sopC*). It also includes features common to many plasmids such as an MCS within the *lacZ* gene for blue/white colony screening, and a selectable marker, in this case for resistance to the antibiotic chloramphenicol (Cm^R).

1. How are oligonucleotides synthesized? What is site-directed mutagenesis? How might site-directed mutagenesis be used to find the nucleotides that encode the active site of an enzyme?
2. Briefly describe the polymerase chain reaction. Explain the differences between reactions in which the products are collected after a defined number of cycles and real-time PCR. Suggest an application of each approach.
3. What is electrophoresis? How is it used in Southern blotting?
4. How are plasmids, cosmids, and artificial chromosomes different? What are some of the different purposes served by each?
5. Explain selection for antibiotic resistance followed by blue versus white screening of colonies containing recombinant plasmids. Why must both antibiotic selection and color screening be used? What would you conclude if, after transforming a ligation mixture into *E. coli,* only blue colonies were obtained?

14.6 CONSTRUCTION OF GENOMIC LIBRARIES

There are several ways in which the DNA to be cloned can be obtained. It can be synthesized by PCR or it can be located on the chromosome by Southern blotting. However, PCR amplification of a gene requires foreknowledge of its nucleotide sequence (or at least sequences flanking the gene), and a suitable probe must be obtained for Southern blotting. In both cases, once the DNA fragment is purified, it is cloned using a procedure like that described for recombinant plasmids and shown in figure 14.13. However, what if researchers wanted to clone a gene but they had no idea what its DNA sequence might be? A genomic library must then be constructed and screened.

The goal of **genomic library** construction is to have an organism's genome cloned as small fragments into separate vectors. Ideally the entire genome is represented; that is to say, the sum of the different fragments equals the whole genome. In this way specific groups of genes can be analyzed and isolated. The construction of a genomic library begins with cleaving the genome into small pieces by a restriction endonuclease (**figure 14.15**). These genomic

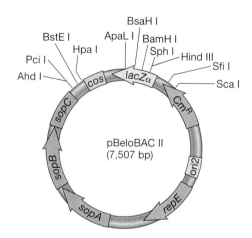

(a) Bacterial artificial chromosome (BAC)

| TEL | TRP1 | ARS | CEN | MCS | URA3 | TEL |

(b) Yeast artificial chromosome (YAC)

Figure 14.14 Artificial Chromosomes Can Be Used as Cloning Vectors. **(a)** The bacterial artificial chromosome pBelaBACIII and **(b)** a yeast artificial chromosome. See text for details.

fragments are then either cloned into vectors and introduced into a microbe or packaged into phage particles that are used to infect the host (**figure 14.16**). In either case, many thousands of different clones—each with a different genomic DNA insert—are created. To select the desired clone from the library, it is necessary to know something about the function of the target gene or genetic element. If the genomic library has been inserted into a microbe that expresses the foreign gene, it may be possible to assay each clone for a specific protein or phenotype. For example, if one is studying a newly isolated soil bacterium and wants to find the gene that encodes an enzyme needed for the biosynthesis of the amino acid alanine, the library could be expressed in an *E. coli* or *Bacillus subtilis* alanine auxotroph (figure 14.15). Recall that an alanine auxotroph requires the addition of this amino acid to the medium. Following introduction of the genomic library vector into host cells, those that now grow without alanine would be good candidates for the genomic library fragment that possesses the alanine biosynthetic gene or gene cluster. Success with this approach depends on the assumption that the function of the cloned gene product is similar in both organisms. If this is not the case, the host must be the same species from which the library was prepared. In this example, a soil bacterial mutant lacking the gene in question (e.g., an alanine auxotroph) is used as the genomic library host. The genetic complementation of a deficiency in the host cell is sometimes called **phenotypic rescue.** Detection and isolation of mutants (section 13.2; *see also figure 13.7*)

Alternatively, the library may be cloned into a phage vector (figure 14.16). The resulting plaques then contain phage particles whose genomes include the cloned DNA fragments. The plaques are screened by a technique based on hybridization of an oligonucleotide probe to the target DNA, much like in Southern blotting. However, in this case, DNA is transferred directly from the petri plate to the filter, which is then incubated with labeled probe. If the nucleic acid sequence of the DNA to be cloned is known, this approach can be used to screen transformed colonies as well.

If a genomic library is prepared from a eucaryote in an effort to isolate a structural gene, a cDNA library is usually constructed. In this way, introns are not present in the genomic library. Instead only the protein-coding regions of the genome are cloned. cDNA is prepared (figure 14.4) and cloned into a suitable vector. After the library is introduced into the host microbe, it may be screened by phenotypic rescue or by hybridization with an oligonucleotide as described earlier. In some cases, neither phenotypic rescue nor hybridization with a probe is possible. In such cases the researcher must develop a novel way that suits the particular set of circumstances to screen the genomic library.

14.7 INSERTING RECOMBINANT DNA INTO HOST CELLS

In cloning procedures, the selection of a host organism is as important as the choice of cloning vector. *E. coli* is the most frequent procaryotic host and *S. cerevisiae* is the most popular among eucaryotes. Host microbes that have been engineered to lack restriction enzymes and the recombination enzyme RecA make better hosts because it is less likely that the newly acquired DNA

will be degraded and/or recombined with the host chromosome. There are several ways to introduce recombinant DNA into a host microbe. Transformation and electroporation are two commonly employed techniques. Often the host microbe does not have the capacity to be transformed naturally. This is the case with *E. coli* and most gram-negative bacteria as well as many gram-positive bacteria. In these cases, the host cells may be rendered competent by treatment with divalent cations and artificially transformed by heat shocking the cells. DNA transformation (section 13.8)

Electroporation is a technique that has gained popularity because of its simplicity and wide application to a number of host organisms, including plant and animal cells. In this procedure, cells are mixed with the recombinant DNA and exposed to a brief pulse of high-voltage electricity. The plasma membrane becomes temporarily permeable and DNA molecules are taken up by some of the cells. The cells are then grown on media that select for the presence of the cloning vector as previously described.

With the exception of yeast, inserting DNA into eucaryotic cells is often more difficult and less efficient. The most direct approach is **microinjection,** wherein genetic material is micropipetted into the host cell. The DNA is then taken up by the nucleus and stably incorporated into the host genome. When this is done in a fertilized mammalian egg, the egg is then transplanted into the uterus of the host animal where it will develop into a transgenic animal. Transgenic mice have become an essential tool for biomedical research.

One of the most effective techniques to insert DNA into eucaryotic cells is to shoot microprojectiles coated with DNA into plant and animal cells. The **gene gun,** first developed at Cornell University, operates somewhat like a shotgun. A blast of compressed gas shoots a spray of DNA-coated metallic microprojectiles into the cells. The device has been used to transform corn and produce fertile corn plants bearing foreign genes. Other guns use electrical discharges to propel the DNA-coated projectiles. These guns are sometimes called biolistic devices, a name derived from biological and ballistic. They have been used to transform microorganisms (yeast, the mold *Aspergillus,* and the protist *Chlamydomonas*), mammalian cells, and a variety of plant cells (corn, cotton, tobacco, onion, and poplar).

Plant cells can also be transformed with vectors derived from the bacterium *Agrobacterium* (p. 378). Viruses increasingly are used to insert desired genes into eucaryotic cells. For example, genes may be placed in a retrovirus, which then infects the target cell and integrates a DNA copy of its RNA genome into the host chromosome. Adenoviruses also can transfer genes to animal cells. Recombinant baculoviruses will infect insect cells and promote the production of many proteins. Eucaryotic viruses and other acellular infectious agents (chapter 18)

14.8 EXPRESSING FOREIGN GENES IN HOST CELLS

When a gene from one organism is cloned into another, it is said to be a **heterologous gene.** Heterologous genes are not always expressed in the host cell without further modification of the recombinant vector. To be transcribed, the recombinant gene must

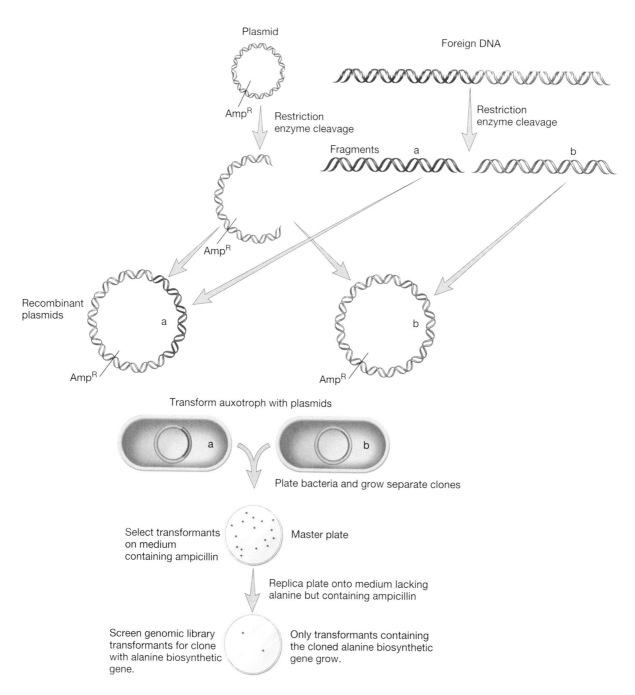

Plasmid

Foreign DNA

Amp^R

Restriction enzyme cleavage

Restriction enzyme cleavage

Amp^R

Fragments   a

b

Recombinant plasmids

a

b

Amp^R

Amp^R

Transform auxotroph with plasmids

a

b

Plate bacteria and grow separate clones

Select transformants on medium containing ampicillin

Master plate

Replica plate onto medium lacking alanine but containing ampicillin

Screen genomic library transformants for clone with alanine biosynthetic gene.

Only transformants containing the cloned alanine biosynthetic gene grow.

Figure 14.15 Construction of a Genomic Library and Screening by Phenotypic Rescue. A genomic library is made by cloning fragments of an organism's entire genome into a vector. For simplicity, only two possible recombinant vectors are shown. In reality, a large mixture of clones is generated. This mixture of clones is then introduced into a suitable host. Phenotypic rescue is one way to screen the clones for the gene of interest. It involves using a host with a genetic defect that can be complemented or "rescued" by the expression of a specific gene that has been cloned.

have a promoter that is recognized by the host RNA polymerase. Translation of its mRNA depends on the presence of leader sequences and mRNA modifications that allow proper ribosome binding. These are quite different in eucaryotes and procaryotes. For instance, if the host is a procaryote and the gene has been cloned from a eucaryote, a procaryotic leader must be provided and introns removed.

The problems of expressing recombinant genes in host cells are largely overcome with the help of special cloning vectors called **expression vectors.** These vectors contain the necessary transcription and translation start signals in addition to convenient polylinker sites. Some expression vectors contain regulatory regions of the *lac* operon so that the expression of the cloned genes can be controlled in the same manner as the operon.

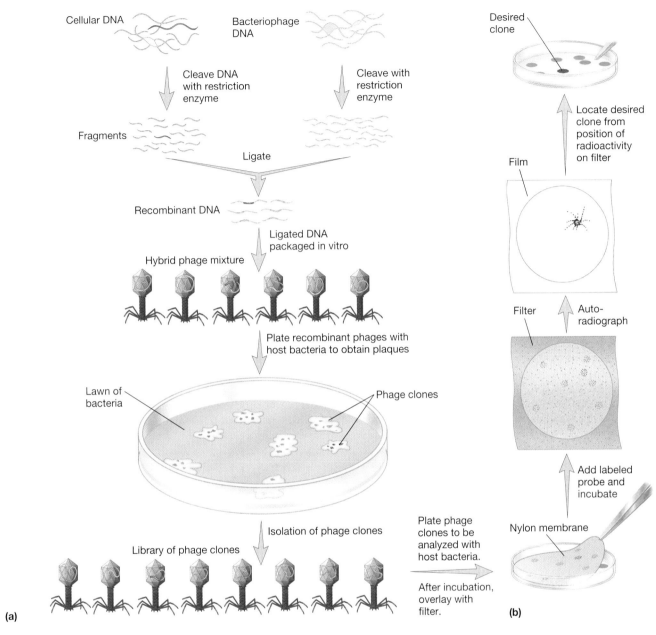

Figure 14.16 **The Use of Lambda Phage as a Vector.** **(a)** The preparation of a genomic library. Each plaque on the bacterial lawn contains a recombinant clone carrying a different DNA fragment. **(b)** Detection and cloning of the desired recombinant phage.

Somatostatin, the 14-residue hypothalamic peptide hormone that helps regulate human growth, provides an example of useful cloning and protein production. The gene for somatostatin was initially synthesized by chemical methods. Besides the 42 bases coding for somatostatin, the polynucleotide contained a codon for methionine at the 5′ end (which corresponds to the N-terminal end of the peptide) and two stop codons at the opposite end. To aid insertion into the plasmid vector, the 5′ ends of the synthetic gene were extended to form sticky ends complementary to those formed by the *Eco*RI and *Bam*HI restriction enzymes. A plasmid cloning vector was cut with both *Eco*RI and *Bam*HI to remove a part of the plasmid DNA. The synthetic gene was then

inserted into the vector by taking advantage of its sticky ends (**figure 14.17**). Finally, a fragment containing the initial part of the *lac* operon (including the promoter, operator, ribosome binding site, and much of the β-galactosidase gene) was inserted upstream to, or at the 5′ end of, the somatostatin gene. The plasmid now contained the somatostatin gene fused in the proper orientation to the remaining portion of the β-galactosidase gene.

After introduction of this recombinant plasmid into *E. coli,* the somatostatin gene was transcribed with the β-galactosidase gene fragment to generate mRNA. Translation formed proteins consisting of the hormone peptide attached to the β-galactosidase fragment by a methionine residue. Cyanogen bromide cleaves

Techniques & Applications

14.1 Visualizing Proteins with Green Fluorescence

What does a jellyfish that lives in the cold waters of the northern Pacific have to do with biotechnology? It turns out, a lot. The jellyfish, *Aequorea victoria,* produces a protein called **green fluorescent protein (GFP)** that scientists have adopted to visualize gene expression and protein localization in living cells. GFP is encoded by a single gene that, when translated, undergoes self-catalyzed modification to generate a strong, green fluorescence. This means that it is easily cloned and expressed in any organism. And GFP isn't just green anymore; site-directed mutagenesis of the GFP gene has generated a variety of proteins that glow throughout the blue-green-yellow spectrum.

One widely used GFP application is to tag proteins so their cellular placement, or localization, can be seen with a light microscope. Formerly, electron microscopy (EM) was the only method by which the location of proteins could be visualized. Sample preparation for EM involves harsh treatment with solvents and cellular dehydration—procedures that can damage cell structures and lead to artifacts. In contrast, when a protein is tagged with GFP, the timing of protein production and its localization can be followed in living cells. To accomplish this, a structural gene is genetically fused to the GFP gene to create a chimeric protein—a protein that consists of two parts: the structural protein being studied and GFP. Of course, care must be taken to ensure that the protein fusion still functions like the original protein. One way to do this is to test for phenotypic rescue of a mutant lacking the structural gene of interest.

GFP protein fusion technology has been used to examine some of the most basic questions in biology. One example is cell division, or cytokinesis. Genetic evidence clearly shows that the tubulin-like protein FtsZ is key for new septum formation in most procaryotic cells. Similarly, the isolation and analysis of cell division mutants has iden-

tified other proteins, such as FtsA, MinC, MinD, and MinE, that are also needed for cytokinesis. Such studies have led to the development of models predicting the events required for cytokinesis. The procaryotic cell cycle (section 6.1)

A special type of cell division that occurs during the formation of an endospore has been intensively studied in *Bacillus subtilis.* In this case, septum formation does not generate two equally sized daughter cells. Instead, asymmetric division gives rise to two different progeny: the endospore and the mother cell. Asymmetric cell division means that rather than FtsZ assembling in the center of the cell, it is polymerized at the pole that gives rise to the endospore. Originally, it was hypothesized that the cell accomplishes this by preventing FtsZ ring assembly in the middle of the cell while simultaneously activating FtsZ polymerization sites near the poles (see **Box figure**).

The fusion of the *B. subtilis ftsZ* structural gene to the gene encoding GFP generated a surprising result: the deployment of FtsZ to the cell pole is a much more dynamic process. Time lapse photomicroscopy shows that when the cell switches from vegetative growth and binary fission to endospore formation, FtsZ forms a spiral-like filament that moves from the cell center to the poles. Careful analysis reveals that many cells appear to have FtsZ spirals that are more abundant in one half of the cell. This suggests that the accumulation of a critical level of FtsZ at one pole before another may determine endospore placement. Surely GFP fusions will help to resolve this and other questions regarding protein localization. Global regulatory systems: Sporulation in *Bacillus subtilis* (section 12.5)

Ben-Yehuda, S., and Losick, R. 2002. Asymmetric cell division in B. subtilis *involves a spiral-like intermediate of the cytokinetic protein FtsZ. Cell. 109: 257–66.*

---

peptide bonds at methionine residues. Treatment of the fusion proteins with cyanogen bromide broke the peptide chains at the methionine and released the hormone (**figure 14.18**). Once free, the peptides were able to fold properly to become active. Because production of the fusion protein was under the control of the *lac* operon, it could be easily regulated. Many proteins have been produced since the synthesis of somatostatin. Examples include human growth hormone, interferons, and proteins used in vaccine production (**table 14.4**). In addition, the fusion of one protein to another has become a useful research tool (**Techniques & Applications 14.1**).

1. What is a genomic library? Describe two ways in which a genomic library might be screened for the clone of interest.

2. What is a transgenic animal? Describe how electroporation and gene guns are used to insert foreign genes into eucaryotic cells. What other approaches may be used? How are bacteria transformed?

3. How can one prevent recombinant DNA from undergoing recombination in a bacterial host cell?

4. List several reasons why a cloned gene might not be expressed in a host cell. What is an expression vector?

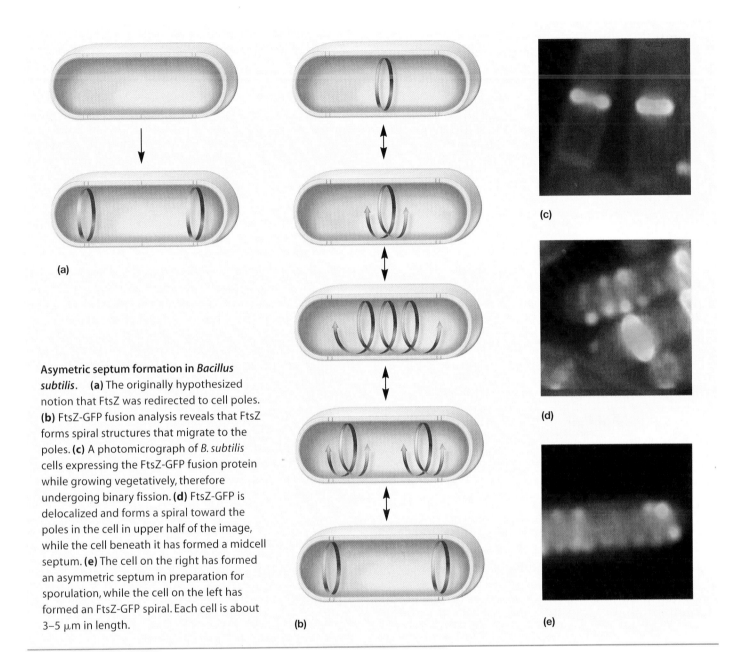

Asymmetric septum formation in *Bacillus subtilis*. **(a)** The originally hypothesized notion that FtsZ was redirected to cell poles. **(b)** FtsZ-GFP fusion analysis reveals that FtsZ forms spiral structures that migrate to the poles. **(c)** A photomicrograph of *B. subtilis* cells expressing the FtsZ-GFP fusion protein while growing vegetatively, therefore undergoing binary fission. **(d)** FtsZ-GFP is delocalized and forms a spiral toward the poles in the cell in upper half of the image, while the cell beneath it has formed a midcell septum. **(e)** The cell on the right has formed an asymmetric septum in preparation for sporulation, while the cell on the left has formed an FtsZ-GFP spiral. Each cell is about 3–5 μm in length.

(a)
(b)
(c)
(d)
(e)

14.9 APPLICATIONS OF GENETIC ENGINEERING

Genetic engineering and biotechnology will continue to contribute in the future to medicine, industry, and agriculture, as well as to basic research. In this section some practical applications are briefly discussed.

Medical Applications

The production of medically useful proteins such as somatostatin, insulin, human growth hormone, and some interferons (signaling molecules of the immune system) is of great practical importance (table 14.4). This is particularly true of substances that previously only could be obtained from human tissues. For example, in the past, human growth hormone for treatment of pituitary dwarfism was extracted from pituitaries obtained during autopsies and was available only in limited amounts. Interleukin-2 (a protein that helps regulate the immune response) and blood-clotting factor VIII have been cloned, as well as a number of other important peptides and proteins. It also is possible to use genetically engineered plants to produce cloned gene products such as oral vaccines. Genetically engineered mice produce human monoclonal antibodies. Synthetic vaccines—for instance, vaccines for malaria and rabies—are also

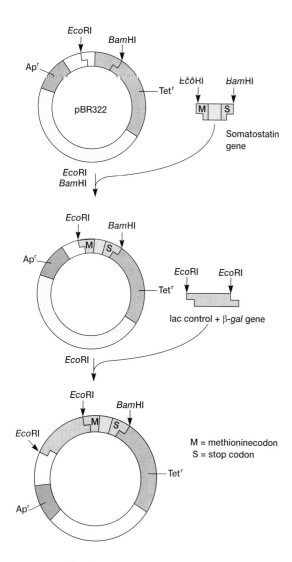

Figure 14.17 Cloning the Somatostatin Gene. An overview of the procedure used to synthesize a recombinant-plasmid containing the somatostatin gene. Ap^r, Tet^r, ampicillin- and tetracycline-resistance genes, respectively.

being developed with recombinant techniques. A recombinant hepatitis B vaccine is commercially available. Control of epidemics: Vaccines and immunization (section 36.8); Techniques & Applications 32.2: Monoclonal antibody technology

Livestock have become important in medical biotechnology through the use of an approach sometimes called molecular pharming. For instance, pig embryos injected with human hemoglobin genes develop into transgenic pigs that synthesize human hemoglobin. Current plans are to purify the hemoglobin and use it as a blood substitute. A pig could yield 20 units of blood substitute a year. Somewhat similar techniques have produced transgenic goats whose milk contains up to 3 grams of human tissue plasminogen activator (TPA) per liter. TPA dissolves blood clots and is used to treat cardiac patients.

However, genetic engineering has much more to offer medicine than the production of recombinant products. Probes are

now being used to screen individuals and the blood supply for infectious agents such as AIDS and West Nile virus. Individuals can be screened for mutant genes; this is common in testing future parents as well as in prenatal testing. A growing area of medical intervention is **gene therapy,** which seeks to replace malfunctioning genes with wild-type genes. Here the normal gene is most commonly delivered in a virus that has been engineered to enter host cells without causing harm. Gene therapy has received a lot of attention in the popular press for several reasons. Gene therapy may cure debilitating genetic diseases that are currently not curable, offering hope to patients and their loved ones. But some clinical gene therapy trials have failed and a few have resulted in death or serious injury to patient volunteers. Also, the potential for gene therapy to be used in germline cells has generated much controversy. Germline gene therapy changes the genetic make-up of gametes or fertilized ova. Thus the genetic change would be passed on to subsequent generations—germline gene therapy introduces heritable changes. One can envision two applications of germline therapy: (1) to correct a genetic defect of an embryo that if left unchanged would result in a debilitating or lethal disease (e.g., Tay Sachs disease, a neurodegenerative condition with a 100% mortality rate by age 5); or (2) to produce a "designer baby" that bears some predetermined phenotypic trait. Fears of the latter have prompted many governments to prohibit germline gene therapy in humans. In contrast, somatic cell gene therapy is nonheritable, and treats only cells in the affected organ. For instance, gene therapy trials with cystic fibrosis (CF) patients have sought to deliver a functional gene to lung tissue. CF patients produce copious amounts of respiratory secretions and are plagued by recurrent *Pseudomonas aeruginosa* infections; the average lifespan of a CF patient is about 30 years. Unfortunately gene therapy has so far provided only localized and short-term relief for these patients.

The challenges to gene therapy are many; however, the potential benefits are so great that medical biotechnologists persevere. One of the biggest obstacles is the method of delivering the wild-type gene. One approach is to remove cells from the patient, genetically alter them, and grow them in vitro. Once a sufficient number of genetically engineered cells are available, these cells are returned to the patient. These cells can be either adult differentiated cells or stem cells. Following genetic manipulation, cells can be introduced directly into the diseased organ or infused into the patient's blood, from which they must migrate to the affected site. Currently, there are at least 15 clinical trials testing **adult stem cell therapy**—somatic gene therapy using adult stem cells. More controversial is the use of **embryonic stem cells.** These cells are collected from 5-day-old embryos obtained from fertility clinics. Unlike adult stem cells, embryonic stem cells are pluripotent—they can differentiate into any cell type (e.g., cardiac, muscle, blood). Scientists believe that these cells are key to understanding embryonic development and may perhaps yield important new therapies.

Finally, it is important to understand the difference between therapeutic cloning and reproductive cloning. **Therapeutic cloning** is what has been described—the use of genetic engineer-

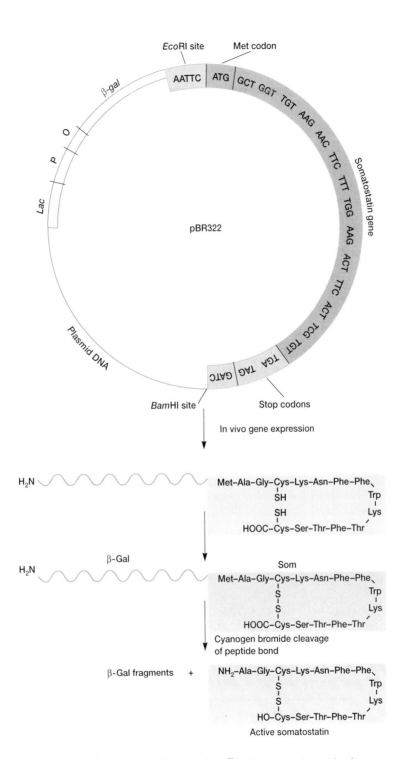

Figure 14.18   The Synthesis of Somatostatin by Recombinant *E. coli.* Cyanogen bromide cleavage at the methionine residue releases active hormone from the β-galactosidase fragment. The gene and associated sequences are shaded in color. Stop codons, the special methionine codon, and restriction enzyme sites are enclosed in boxes.

ing to alter a cell's genotype and return (or introduce) the modified cells to a patient. It is hoped that therapeutic cloning may one day provide treatments (if not cures) for conditions like Parkinson's disease, Alzheimer's disease, and spinal cord injuries. This differs from **reproductive cloning** in which an embryo is created

by replacing the nucleus of an unfertilized egg with a nucleus from a somatic cell. This embryo is implanted into a female's uterus with the intention of creating a new life. Human reproductive cloning is legally forbidden throughout most of the world. The U.S. government does not fund therapeutic cloning that uses

Table 14.4	Some Human Peptides and Proteins Synthesized by Genetic Engineering	
Peptide or Protein	**Potential Use**
α_1-antitrypsin	Treatment of emphysema
α-, β-, and γ-interferons	As antiviral, antitumor, and anti-inflammatory agents
Blood-clotting factor VIII	Treatment of hemophilia
Calcitonin	Treatment of osteomalacia
Epidermal growth factor	Treatment of wounds
Erythropoetin	Treatment of anemia
Growth hormone	Growth promotion
Insulin	Treatment of diabetes
Interleukins-1, 2, and 3	Treatment of immune disorders and tumors
Macrophage colony stimulating factor	Cancer treatment
Relaxin	Aid to childbirth
Serum albumin	Plasma supplement
Somatostatin	Treatment of acromegaly
Streptokinase	Anticoagulant
Tissue plasminogen activator	Anticoagulant
Tumor necrosis factor	Cancer treatment

embryonic stem cell technology, although some states do support such research.

Agricultural Applications

Cloned genes can be inserted into plant as well as animal cells. A popular way to insert genes into plants is with a recombinant **Ti plasmid** (tumor-inducing plasmid) obtained from the bacterium *Agrobacterium tumefaciens* (**Techniques & Applications 14.2**). It also is possible to donate genes by forming plant cell protoplasts, making them permeable to DNA, and then adding the desired recombinant DNA. The gene gun is also used in the production of transgenic plants. Microorganism associations with vascular plants: Agrobacterium (section 29.5)

Genetic engineering has made plants resistant to environmental stresses. For example, the genes for detoxification of glyphosate herbicides were isolated from *Salmonella,* cloned, and introduced into tobacco cells using the Ti plasmid. Plants regenerated from the recombinant cells were resistant to the herbicide. Herbicide-resistant varieties of cotton and fertile, transgenic corn also have been developed. This is of considerable importance because many crops suffer stress when treated with herbicides. Resistant plants are not stressed by the chemicals being used to control weeds.

U.S. farmers grow substantial amounts of genetically modified (GM) crops. About a third of the corn, half of the soybeans,

Techniques & Applications

14.2   Plant Tumors and Nature's Genetic Engineer

A plasmid from the plant pathogenic bacterium *Agrobacterium tumefaciens* is responsible for much success in the genetic engineering of plants. Infection of normal cells by the bacterium transforms them into tumor cells, and crown gall disease develops in dicotyledonous plants such as grapes and ornamental plants. Normally, the gall or tumor is located near the junction of the plant's root and stem. The tumor forms because of the insertion of genes into the plant cell genome, and only strains of *A. tumefaciens* possessing a large conjugative plasmid called the Ti plasmid are pathogenic (*see section 29.5* and *figure 29.19*). The Ti plasmid carries genes for virulence and the synthesis of substances involved in the regulation of plant growth. The genes that induce tumor formation reside between two 23 base pair direct-repeat sequences. This region is known as T-DNA and is very similar to a transposon. T-DNA contains genes for the synthesis of plant growth hormones (an auxin and a cytokinin) and an amino acid derivative called opine that serves as a nutrient source for the invading bacteria. In diseased plant cells, T-DNA is inserted into the chromosomes at various sites and is stably maintained in the cell nucleus.

When the molecular nature of crown gall disease was recognized, it became clear that the Ti plasmid and its T-DNA had great potential as a vector for the insertion of recombinant DNA into plant chromosomes. In one early experiment the yeast alcohol dehydrogenase gene was added to the T-DNA region of the Ti plasmid. Subsequent infec-

tion of cultured plant cells resulted in the transfer of the yeast gene. Since then, many modifications of the Ti plasmid have been made to improve its characteristics as a vector. Usually one or more antibiotic resistance genes are added, and the nonessential T-DNA, including the tumor inducing genes, is deleted. Those genes required for the actual infection of the plant cell by the plasmid are retained. T-DNA also has been inserted into *E. coli* plasmids to produce cloning vectors that can move between bacteria and plants (see **Box figure**). The gene or genes of interest are spliced into the T-DNA region between the direct repeats. Then the plasmid is returned to *A. tumefaciens,* plant culture cells are infected with the bacterium, and transformants are selected by screening for antibiotic resistance (or another trait encoded by T-DNA). Finally, whole plants are regenerated from the transformed cells. In this way several plants have been made herbicide resistant. Microbes as products: Biopesticides (section 41.8)

Unfortunately the *A. tumefaciens* Ti plasmid cannot be transferred naturally to monocotyledonous plants such as corn, wheat, and other grains. It has been used only to modify plants such as potato, tomato, celery, lettuce, and alfalfa. However, genetic modification of the Ti plasmid so that it can be productivity transferred to monocotyledonous plants is an active area of research. The creation of new procedures for inserting DNA into plant cells may well lead to the use of recombinant DNA techniques with many important crop plants.

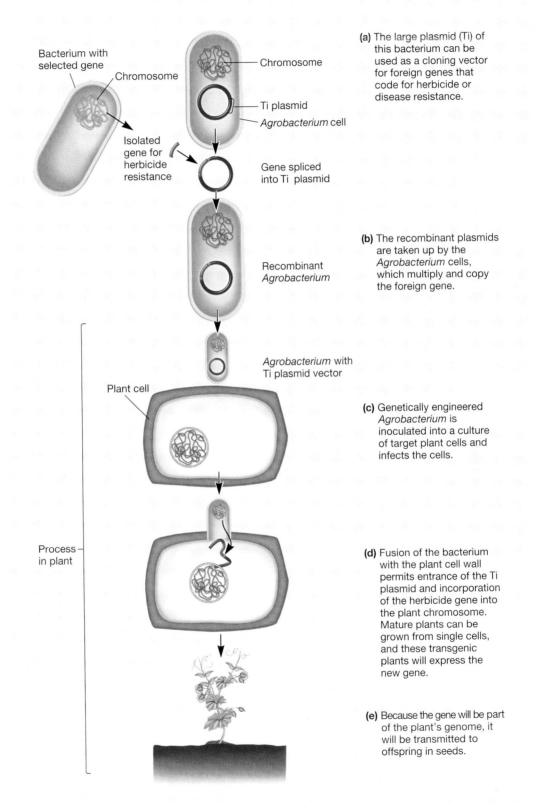

(a) The large plasmid (Ti) of this bacterium can be used as a cloning vector for foreign genes that code for herbicide or disease resistance.

Bacterium with selected gene

Chromosome

Chromosome

Ti plasmid

Agrobacterium cell

Isolated gene for herbicide resistance

Gene spliced into Ti plasmid

(b) The recombinant plasmids are taken up by the *Agrobacterium* cells, which multiply and copy the foreign gene.

Recombinant *Agrobacterium*

Agrobacterium with Ti plasmid vector

Plant cell

(c) Genetically engineered *Agrobacterium* is inoculated into a culture of target plant cells and infects the cells.

Process in plant

(d) Fusion of the bacterium with the plant cell wall permits entrance of the Ti plasmid and incorporation of the herbicide gene into the plant chromosome. Mature plants can be grown from single cells, and these transgenic plants will express the new gene.

(e) Because the gene will be part of the plant's genome, it will be transmitted to offspring in seeds.

Bioengineering of plants.   Most techniques employ a genetically modified strain of a natural tumor-producing bacterium called *Agrobacterium tumefaciens*.

and a significant fraction of cotton crops are genetically modified. Cotton and corn are resistant to herbicides and insects. Soybeans have herbicide resistance and lowered saturated fat content. Other examples of genetically engineered commercial crops are canola, potato, squash, and tomato. Golden rice is a GM crop that may have an important impact in the developing world. Compared to unmodified rice, this strain stores about twice as much iron because it has been genetically modified to overexpress some of its own genes. In addition, it serves as a good source of beta-carotene (a vitamin A precursor) thanks to the introduction of genes from the petunia and bacteria.

Many new agricultural applications are being explored. Much effort is being devoted to defending plants against pests without the use of chemical pesticides. A strain of *Pseudomonas fluorescens* carrying the gene for the *Bacillus thuringiensis* toxin has been developed. This toxin destroys many insect pests such as the cabbage looper and the European corn borer. A variety of corn with the *B. thuringiensis* toxin gene has been created. Unfortunately, some insects seem able to develop resistance to the toxin. There is considerable interest in insect-killing viruses and particularly in the baculoviruses. A scorpion toxin gene has been inserted into the autographa californica multicapsid nuclear polyhedrosis virus (AcMNPV). The engineered AcMNPV kills cabbage looper more rapidly than the normal virus and reduces crop damage significantly. Finally, virus-resistant strains of soybeans, potatoes, squash, rice, and other plants are under development. Microbes as products: Biopesticides (section 41.8)

---

1. List several important present or future applications of genetic engineering in medicine, industry, and agriculture.
2. What is the Ti plasmid and why is it so important?

14.10 SOCIAL IMPACT OF RECOMBINANT DNA TECHNOLOGY

Despite the positive social impact of recombinant DNA technology, the potential to alter an organism genetically raises serious scientific and philosophical questions. These issues are the subjects of vigorous debate, as briefly reviewed here.

In contrast to the use of biotechnology in basic and applied science, the use of gene therapy in human beings raises pressing ethical and moral questions. These problems are not extreme as long as adult stem cells are used. However, as witnessed in American political discourse and legislation since 2001, the use of embryonic stem cells is problematic. Is it morally acceptable to sacrifice an embryo to obtain these cells? Proponents point out that adult stem cells are rare and are not truly pluripotent. They also note that these donated embryos are unwanted and will eventually be destroyed by the fertility clinics where they are frozen. Those opposed to embryonic stem cell research believe just as strongly that human life should not be destroyed, even at the very earliest stages. In the summer of 2001, President G. W. Bush banned the use of U.S. federal funds for the establishment of any additional embryonic stem cell cultures or lines. However, in 2004, California passed a ballot initiative releasing $3 billion for embryonic stem cell study in that state, further underscoring the controversial nature of this research. Other countries, most notably the United Kingdom, have actively pursued embryonic stem cell research.

Another area of considerable controversy involves the use of recombinant organisms in agriculture. Many ecologists are worried that the release of recombinant plants without careful prior risk assessment may severely disrupt the ecosystem. Another commonly cited concern is allergic reactions in humans consuming genetically modified (GM) foods, although no such responses have been reported. Others are anxious that the recombinant DNA from GM plants could be transferred to nearby wild plants. This could occur by the movement of pollen by wind and insects to neighboring fields. This would be problematic for a number of reasons including the possibility that GM crops bearing herbicide resistance genes may pollinate weeds, generating "super-weeds" that would require the application of more (rather than less) herbicide. In addition, farmers whose crops are certified organic fear genetic contamination; organic crops are tested for the presence of genetic modification, which if found excludes them from the organic market. Clearly, the rate of development of GM foods has outpaced the consideration of genetic pollution. In Europe, GM crops are not commonly produced by farmers or purchased by consumers. Some large U.S. food producers have responded to public concern and quit using GM crops.

As with any technology, the potential for abuse exists. A case in point is the use of genetic engineering in biological warfare and terrorism. Although international agreements limit research in this area to defense against other biological weapons, knowledge obtained in such research can easily be used in offensive biological warfare. Effective vaccines constructed using recombinant DNA technology can protect the attacker's troops and civilian population. It is relatively easy and inexpensive to prepare bacteria capable of producing massive quantities of toxins or to develop particularly virulent strains of viral and bacterial pathogens, so even small countries and terrorist organizations might acquire biological weapons. Since September 11, 2001 governments worldwide have established tighter regulations to control the availability of pathogens that might be used in a bioterrorism attack. Bioterrorism preparedness (section 36.9)

Recombinant DNA technology has greatly enhanced our knowledge of genes and how they function, and it promises to improve our lives in many ways. Yet, as this brief discussion shows, problems and concerns remain to be resolved. Past scientific advances have sometimes led to unanticipated and unfortunate consequences, such as environmental pollution and nuclear weapons. With prudence and forethought we may be able to avoid past mistakes in the use of recombinant DNA technology.

---

1. Describe four major areas of concern about the application of genetic engineering. In each case give both the arguments for and against the use of genetic engineering.

Summary

14.1 Historical Perspectives

a. Genetic engineering became possible after the discovery of restriction enzymes and reverse transcriptase, and the development of essential methods in nucleic acid chemistry such as the Southern blotting technique (**table 14.1**).

14.2 Synthetic DNA

a. Oligonucleotides of any desired sequence can be synthesized by a DNA synthesizer machine. This made site-directed mutagenesis possible and is important for the synthesis of primers and DNA probes used in PCR and Southern blotting, respectively.

14.3 The Polymerase Chain Reaction

a. The polymerase chain reaction allows small amounts of specific DNA sequences to be increased in concentration thousands of times (**figure 14.9**).

14.4 Gel Electrophoresis

a. Gel electrophoresis is used to separate molecules according to charge and size.

b. DNA fragments are separated on agarose and acrylamide gels. Because DNA is acidic it migrates from the negative to the positive end of a gel (**figure 14.11**).

14.5 Cloning Vectors and Creating Recombinant DNA

a. There are four types of cloning vectors: plasmids, phages and viruses, cosmids, and artificial chromosomes. Cloning vectors generally have at least three components: an origin of replication, a selectable marker, and a multicloning site or polylinker (**table 14.3; figures 14.12** and **14.14**).

b. The most common approach to cloning is to digest both vector and DNA to be inserted with the same restriction enzyme or enzymes so that compatible sticky ends are generated. The vector and DNA to be cloned are then incubated in vitro in the presence of DNA ligase, which catalyzes the covalent insertion of the DNA fragment into the vector (**figure 14.5**).

c. Once the recombinant plasmid has been introduced into host cells, cells carrying vector must be selected. This is often accomplished by allowing the growth of only antibiotic-resistant cells because the vector bears the antibiotic-resistance gene. Cells that took up vector with inserted DNA can be distinguished from those that contain only vector in several ways. Often a blue versus white colony phenotype is used; this is based on the presence or absence of a functional *lacZ* gene, respectively (**figure 14.13**).

14.6 Construction of Genomic Libraries

a. It is sometimes necessary to find a gene on a chromosome without the knowledge of the gene's DNA sequence. A genomic library is constructed by cleaving an organism's genome into many fragments, each of which is cloned into a vector to make a unique recombinant plasmid.

b. Genomic libraries can be screened for the gene of interest by either phenotypic rescue (genetic complementation) or DNA hybridization with an oligonucleotide probe. However, there are instances when a novel approach to screening the library for a specific gene must be devised (**figures 14.15** and **14.16**).

14.7 Inserting Recombinant DNA into Host Cells

a. Bacteria and the yeast *S. cerevisiae* are the most common host species.

b. DNA can be introduced into microbes by transformation or electroporation.

c. A variety of techniques are used to introduce DNA into eucaryotic host cells including electroporation, microinjection, and the use of a gene gun. The Ti plasmid of the plant bacterial pathogen *Agrobacterium tumefaciens* has been engineered to transfer foreign DNA into plant genomes.

14.8 Expressing Foreign Genes in Host Cells

a. The recombinant vector often must be modified by the addition of promoters, leaders, and other elements. Eucaryotic gene introns also must be removed. An expression vector has the necessary features to express any recombinant gene it carries.

b. Many useful products, such as the hormone somatostatin, have been synthesized using recombinant DNA technology (**figures 14.17** and **14.18**).

14.9 Applications of Genetic Engineering

a. Recombinant DNA technology will provide many benefits in medicine, industry, and agriculture.

14.10 Social Impact of Recombinant DNA Technology

a. Despite the great promise of genetic engineering, it also brings with it potential challenges in areas of safety, human experimentation, potential ecological disruption, and biological warfare.

Key Terms

adult stem cell therapy 376
autoradiography 359
bacterial artificial chromosome (BAC) 370
biotechnology 357
cloning 357
complementary DNA (cDNA) 358
cosmid 370
DNA ligase 367
electroporation 371
embryonic stem cells 376

expression vector 372
gel electrophoresis 366
gene gun 371
gene therapy 376
genetic engineering 357
genomic library 370
green fluorescent protein (GFP) 374
heterologous gene 371
microinjection 371
multicloning site (MCS) 367
oligonucleotides 361

origin of replication (*ori*) 366
phenotypic rescue 371
polymerase chain reaction (PCR) 362
primers 362
probe 358
real-time PCR (RT PCR) 363
recombinant DNA technology 357
reproductive cloning 377
restriction enzymes 357
reverse transcriptase (RT) 358
selectable marker 367

shuttle vector 367
site-directed mutagenesis 362
Southern blot procedure 358
sticky ends 358
therapeutic cloning 376
thermocycler 362
Ti plasmid 378
vectors 358
yeast artificial chromosome (YAC) 370

Critical Thinking Questions

1. Could the Southern blotting technique be applied to RNA? How might this be done?

2. Initial attempts to perform PCR were carried out using the DNA polymerase from *E. coli*. What was the major difficulty?

3. What advantage might there be in creating a genomic library first rather than directly isolating the desired DNA fragment?

4. You have cloned a structural gene required for riboflavin synthesis in *E. coli*. You find that an *E. coli* riboflavin auxotroph carrying the cloned gene on a vector makes less riboflavin than does the wild-type strain. Why might this be the case?

5. Suppose that you inserted a simple plasmid (one containing an antibiotic resistance gene and a separate restriction site) carrying a human interferon gene into *E. coli*, but none of the transformed bacteria produced interferon. Give as many plausible reasons as possible for this result.

6. What do you consider to be the greatest potential benefit of genetic engineering? Discuss possible ethical problems with this potential application?

Learn More

Ben-Ari, E. 2002. *Bacillus thuringiensis* as a paradigm for transgenic organisms. *ASM News* 68(12):597–602.

Cockerill, F. R., and Smith, T. F. 2002. Rapid-cycle real-time PCR: A revolution for clinical microbiology. *ASM News* 68(2):77–83.

Drlica, K. 2004. *Understanding DNA and gene cloning: A guide for the curious,* 4th ed. New York: John Wiley & Sons.

Falkow, S. 2001. I'll have the chopped liver please, or how I learned to love the clone. *ASM News* 67(11):555–59.

Glick, B. R., and Pasternak, J. J. 2003. *Molecular biotechnology: Principles and applications of recombination DNA,* 3d ed. Washington, D.C.: ASM Press.

Murray, N. E. 2000. DNA restriction and modification. In *Encyclopedia of microbiology,* 2d ed., vol. 2, J. Lederberg, editor-in-chief, 91–105. San Diego: Academic Press.

Rieger, M. A.; Lamond, M.; Preston, C.; Powles, S. B.; and Roush, R. T. 2002. Pollen-mediated movement of herbicide resistance between commercial canola fields. *Science* 296: 2386–88.

Wolfenbarger, L. L., and Phifer, P. R. 2000. The ecological risks and benefits of genetically engineered plants. *Science* 290:2088–93.

Please visit the Prescott website at www.mhhe.com/prescott7 for additional references.

15

Microbial Genomics

Each dot in the microarray pictured here consists of an oligonucleotide fragment of a single gene bound to a glass slide. Gene expression of two types of cells (for example, one wild-type and one mutant) can be compared by labeling the cDNA from each cell with either a red or green fluorescent label and allowing the cDNAs to bind to homologous sequences attached to the microarray. The color of each spot reveals the relative level of expression of each gene.

PREVIEW

- Genomics is the study of the molecular organization of genomes, their information content, and the gene products they encode. It may be divided into a number of subdisciplines including structural genomics, functional genomics, and comparative genomics.

- Individual pieces of DNA can be sequenced using the Sanger method. The easiest way to analyze entire genomes is by whole-genome shotgun sequencing in which randomly produced fragments are sequenced individually and then aligned to give the complete genome.

- Newly sequenced genomes must be annotated to determine the location of genes and genetic elements such as promoters and ribosome-binding sites.

- Bioinformatics combines biology, mathematics, statistics, and computer science in the analysis of genomes and proteomes.

- Microarray technology enables the study of gene expression at the transcriptional level. In contrast, proteomics uses the entire collection of proteins produced at any given time by an organism to evaluate gene expression at the level of translation.

- Many microbial genomes have been sequenced and compared. The results tell us much about genome structure, microbial physiology, microbial phylogeny, and how pathogens cause disease. They will undoubtedly help in preparing new vaccines and drugs for the treatment of infectious disease.

Chapter 13 provides a brief introduction to microbial recombination, and chapter 14 describes the development of recombinant DNA technology. In this chapter, we carry these themes further with the discussion of the ongoing revolution in genomics. We begin with a general overview of the topic, followed by an introduction to DNA sequencing techniques.

Next, the whole-genome shotgun sequencing method is briefly described. Genome function and the analysis of the transcripts and proteins produced by microbes are then explored. We focus on annotation, DNA microarrays, and the use of two-dimensional gel electrophoresis to study the cellular pool of proteins. The chapter concludes with a discussion of examples in which genomic analysis has extended our knowledge of microbial physiology, pathogenicity, evolution, and ecology.

15.1 INTRODUCTION

Genomics is the study of the molecular organization of genomes, their information content, and the gene products they encode. It is a broad discipline, which may be divided into at least three general areas. **Structural genomics** is the study of the physical nature of genomes. Its primary goal is to determine and analyze the DNA sequence of the genome. **Functional genomics** is concerned with the way in which the genome functions. That is, it examines the transcripts produced by the genome and the array of proteins they encode. The third area of study is **comparative genomics,** in which genomes from different organisms are compared to look for significant differences and similarities. This helps identify important, conserved regions of the genome in an effort to discern patterns in function and regulation. These data also provide much information about microbial evolution, particularly with respect to phenomena such as horizontal gene transfer.

Genomics is an exciting and growing field that has changed the ways in which key questions in microbial physiology, genetics, ecology, and evolution are pursued. Prior to the advent of genomics,

A prerequisite to understanding the complete biology of an organism is the determination of its entire genome sequence.

—J. Craig Venter, et al.

analysis of gene expression was limited to the identification of a small subset of transcripts and proteins. As we will see, genomic technologies enable scientists to study the cell in a holistic way by capturing a snapshot of changes in the entire pool of mRNA transcripts or proteins. The cell can thus be viewed as a network of interconnected circuits, not as a series of individual pathways. Further, genomics provides a window into entire microbial communities—microbial ecologists no longer need to confine their studies to the tiny fraction of microorganisms that have been cultivated. Finally, our understanding of the evolution of all organisms can be illuminated by the insights we gain in procaryotic evolution from genomic approaches. In these ways, and more, genomics has truly revolutionized biology.

15.2 DETERMINING DNA SEQUENCES

Techniques for sequencing DNA were developed in 1977 by two groups: Alan Maxam and Walter Gilbert, and Frederick Sanger. Sanger's method is most commonly used and is discussed here. This method involves the synthesis of a new strand of DNA using the DNA to be sequenced as a template. The reaction begins when single strands of template DNA are mixed with primer (a short piece of DNA complementary to the 5′ end of the region to be sequenced), DNA polymerase, the four deoxynucleoside triphosphates (dNTPs), and dideoxynucleoside triphosphates (ddNTPs). ddNTPs differ from dNTPs in that the 3′ carbon lacks a hydroxyl group (**figure 15.1**). In such a reaction mixture, DNA synthesis will proceed until a ddNTP, rather than a dNTP, is added to the growing chain. Without a 3′-OH group to attack the 5′-PO_4 of the next dNTP to be incorporated, synthesis stops (*see figure 11.14*). Indeed, Sanger's technique is frequently referred to as the **chain-termination DNA sequencing method.** In order to obtain sequence information, four separate synthesis reactions must be prepared, one for each ddNTP (**figure 15.2**). When each DNA synthesis reaction is stopped, a collection of DNA fragments of varying lengths has been generated. The reaction prepared with ddATP produces fragments ending with an A, those with ddTTP produce fragments with T termini, and so forth. If the DNA is to be manually sequenced, radioactive dNTPs are used and each reaction is electrophoresed in a separate lane on a polyacrylamide gel. The molecular weight of each fragment is determined by its

length, so shorter fragments migrate faster than larger fragments (*see figure 14.11*). Because synthesis proceeds with the addition of a nucleotide to the 3′-OH of the primer (i.e., in the 5′ to 3′ direction) the ddNTP at the end of the shortest fragment is assigned as the 5′ end of the DNA sequence, while the largest fragment is the 3′ end. In this way, the DNA sequence is read directly from the gel from the smallest to the largest fragment. DNA replication (section 11.4); Gel electrophoresis (section 14.4)

DNA is more often prepared for automated sequencing. Here the four reaction mixtures can be combined and loaded into a single lane of a gel because each ddNTP is labeled with a different colored fluorescent dye (figure 15.2b). These fragments are then electrophoresed on a polyacrylamide gel and a laser beam determines the order in which they exit the gel. A chromatogram is generated in which the amplitude of each spike represents the fluorescent intensity of each particular fragment (figure 15.2c). The corresponding DNA sequence is listed above the chromatogram.

Fully automated capillary electrophoresis DNA analyzers are required for large projects. These sequencing machines are very fast and can run for 24 hours without operator attention. As many as 96 samples can be sequenced simultaneously, making it possible to sequence as many as 1 million bases per day, per sequencer. This level of automation, involving many sequencers running at the same time, is needed for the completion of whole-genome sequences.

15.3 WHOLE-GENOME SHOTGUN SEQUENCING

Although methods for sequencing relatively short regions of DNA have been in use for some time, efficient methods for sequencing whole genomes were not available until 1995, when J. Craig Venter, Hamilton Smith, and their collaborators developed **whole-genome shotgun sequencing** and the computer software needed to assemble sequence data into a complete genome. They used their new method to sequence the genomes of the bacteria *Haemophilus influenzae* and *Mycoplasma genitalium*. This was a significant accomplishment because prior to this only a few viral genomes had been fully sequenced. These are much smaller than the genome of *H. influenzae,* which contains about 1,743 genes and 1,830,137 base pairs, or about 1.8 Mb (*million base pairs*). Venter and Smith's contribution to biology has ushered in what has been called the genomic era. Within 10 years the number of complete genomes published grew from 2 to 249 with over 500 ongoing genome sequencing projects. **Table 15.1** lists just a few microbial genomes.

The process of whole-genome shotgun sequencing is fairly complex when considered in detail, and there are many procedures to ensure the accuracy of the results, but the following summary gives a general idea of the procedure. For simplicity, this approach may be broken into four stages: library construction, random sequencing, fragment alignment and gap closure, and editing.

1. *Library construction.* The DNA molecules are randomly broken into fairly small fragments using ultrasonic waves; the fragments are then purified (**figure 15.3**). These fragments are next attached to plasmid or cosmid vectors and plasmids or cosmids with a single insert are isolated. Special *Escherichia*

Figure 15.1 Dideoxyadenosine Triphosphate (ddATP).
Note the lack of a hydroxyl group on the 3′ carbon, which prevents further chain elongation by DNA polymerase.

① Isolated unknown DNA fragment

Original DNA to be sequenced

② DNA is denatured to produce single template strand.

③ Labeled specific primer molecule hybridizes to the DNA strand

④ DNA polymerase and regular nucleotide mixture (dATP, dCTP, dGTP, and dTTP) are added; ddG, ddA, ddC, and ddT are placed in separate reaction tubes with the regular nucleotides. The dd nucleotides are labeled with some type of tracer, which allows them to be visualized.

Incubate

⑤ Newly replicated strands are terminated at the point of addition of a dd nucleotide.

(a)

⑥ Schematic view of how all possible positions on the fragment are occupied by a labeled nucleotide

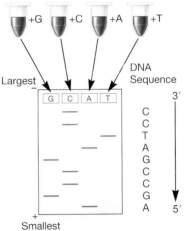

⑦ Running the reaction tubes in four separate gel lanes separates them by size and nucleotide type. Reading from bottom to top, one base at a time, provides the correct DNA sequence.

(b)

Figure 15.2 The Sanger Method for DNA Sequencing.

(a) Steps 1–6 are used for both manual and automated sequencing. Step (7) shows preparation of a gel for manual sequencing in which radiolabeled ddNTPs are used. (b) Part of an automated DNA sequencing run. Here the ddNTPs are labeled with fluorescent dyes. (c) Data generated during an automated DNA sequencing run. Bases 493 to 580 are shown.

(c)

Table 15.1 | **Examples of Complete Published Microbial Genomes**

Genome	Domain[a]	Number of Strains Sequenced	Size (Mb)	% G + C
Agrobacterium tumefaciens	B	2	4.92	60
Aquifex aeolicus	B	1	1.55	43
Archaeoglobus fulgidus	A	1	2.18	48
Bacillus anthracis	B	4	5.09–5.23	36
Bacillus subtilis	B	1	4.21	43
Borrelia burgdorferi	B	1	1.44	28
Campylobacter jejuni	B	1	1.64	31
Caulobacter crescentus	B	1	4.02	62–67
Chlamydia pneumoniae	B	2	1.23	40
Chlamydia trachomatis	B	2	1.05–1.07	41
Chlorobium tepidum	B	1	2.15	57
Clostridium perfringens	B	1	3.03	29
Corynebacterium glutamicum	B	1	3.31	55–58
Deinococcus radiodurans	B	1	3.06	67
Escherichia coli	B	6	4–5.45	50
Geobacter sulfurreducens PCA	B	1	3.81	61
Haemophilus influenzae Rd	B	1	1.83	39
Halobacterium sp. NRC-1	A	1	2.01	68
Helicobacter pylori	B	2	1.64–1.67	39
Listeria monocytogenes	B	2	2.9	37–39
Methanobacterium thermoautotrophicum	A	1	1.75	49
Methanocaldococcus jannaschii	A	1	1.66	31
Mycobacterium leprae	B	1	3.27	58
Mycobacterium tuberculosis	B	2	4.40	65
Mycoplasma genitalium	B	1	0.58	31
Mycoplasma pneumoniae	B	1	0.82	40
Nanobacterium equitans	A	1	0.49	32
Neisseria meningitidis	B	3	2.18–2.27	51
Prochlorococcus marinus	B	3	1.66–2.41	31–51
Pseudomonas aeruginosa	B	1	6.26	67
Pyrococcus abyssi	A	1	1.77	44
Pyrococcus horiksohii	A	1	1.74	42
Rhodopseudomonas palustris	B	1	5.46	65
Rickettsia prowazekii	B	1	1.11	29
Saccharomyces cerevisiae	E	1	12.14	38
Salmonella enterica serovar Typhimurium	B	1	4.86	50–53
Staphylococcus aureus	B	7	2.80–2.90	33
Streptococcus mutans	B	1	2.03	37
Streptococcus pneumoniae	B	2	2.16	40
Streptococcus pyogenes	B	6	1.84–1.90	39
Streptomyces coelicolor	B	1	8.67	72
Sulfolobus tokodaii	A	1	2.69	33
Synechocystis sp.	B	1	3.57	47
Thermoplasma acidophilum	A	1	1.56	46
Thermotoga maritima	B	1	1.86	46
Treponema pallidum	B	1	1.14	52
Vibrio cholerae	B	1	4.03	48
Yersinia pestis	B	3	4.60–4.65	48
Yersinia pseudotuberculosis	B	1	4.74	48

[a]The following abbreviations are used: A, Archaea; B, Bacteria; E, Eucarya.

Microbial chromosome

Sonication

DNA fragments

Agarose gel electrophoresis of fragments and DNA size markers

Fragment purification from gel

DNA fragments

Clonal library preparation

Sequence the clonal inserts, particularly the end sequences.

Assembly of a Contig

Overlap

B

Overlap

Clone A

Clone C

Overlap

Construct sequence contigs and align using overlaps; fill in gaps.

ACCAATATTCAATACCAATATTCAAT
TGTTACTCGGTCATGTTACTCGGTCA
ACCAATATTCAATACCAATATTCAAT
TGTTACTCGGTCATGTTACTCGGTCA
ACCAATATTCAATACCAATATTCAAT
TGTTACTCGGTCATGTTACTCGGTCA
ACCAATATTCAATACCAATATTCAAT
TGTTACTCGGTCATGTTACTCGGTCA
ACCAATATTCAATACCAATATTCAAT
TGTTACTCGGTCATGTTACTCGGTCA

coli strains lacking restriction enzymes are transformed with the plasmids to produce a library of the plasmid clones. Cloning vectors and creating recombinant DNA (section 14.5)

2. *Random sequencing.* After the clones are prepared and the DNA purified, thousands of DNA fragments are sequenced with automated sequencers, employing special dye-labeled primers. Thousands of templates are used, normally with primers that recognize the plasmid DNA sequences adjacent to the DNA insert. The nature of the process is such that almost all stretches of the genome are sequenced several times, and this increases the accuracy of the final results.

3. *Fragment alignment and gap closure.* Using special computer programs, the sequenced DNA fragment data are clustered and assembled into longer stretches of sequence by comparing nucleotide sequence overlaps between fragments. Two fragments are joined together to form a larger stretch of DNA if the sequences at their ends overlap and match. This overlap comparison process results in a set of larger, contiguous nucleotide sequences called contigs.

Finally, the contigs are aligned in the proper order to form the complete genome sequence. If gaps exist between two contigs, sometimes fragments with ends in two adjacent contigs are available. These fragments are analyzed and the gaps filled in with their sequences. When this approach is not possible, a variety of other techniques are used to align contigs and fill in gaps. For example, λ phage libraries containing large DNA fragments can be constructed. The large fragments in these libraries overlap the previously sequenced contigs. These fragments are then combined with oligonucleotide probes that match the ends of the contigs to be aligned. If the probes bind to a λ library fragment, it can be used to prepare a stretch of DNA that represents the gap region. Overlaps in the sequence of this new fragment with two contigs allow them to be placed side-by-side and fill in the gap between them. Construction of genomic libraries (section 14.6)

4. *Editing.* The sequence is then carefully proofread in order to resolve any ambiguities in the sequence. Also the sequence must be checked for unwanted frameshift mutations and corrected if necessary.

Using this approach, it took less than 4 months to sequence the *M. genitalium* genome (about 500,000 base pairs in size). The shotgun technique also has been used successfully by Celera Genomics in the Human Genome Project and to sequence the *Drosophila* genome. Researchers at the Wellcome Trust Sanger Centre (UK) have also sequenced (and continue to sequence) the genomes of many important microbial pathogens, while the U.S. Department of Energy Joint Genome Institute supports the sequencing of many bacteria of environmental relevance. Once an organism's genome has been sequenced, the level of inquiry and

Figure 15.3   Whole-Genome Shotgun Sequencing.

the pace of research are greatly enhanced. The following sections describe only some of the ways a genome sequence can be used to learn more about an organism.

1. What is the goal of each of the three general areas of genomics?
2. Why is the Sanger technique of DNA sequencing also called the chain-termination method?
3. How would one recognize a gap in the genome sequence following whole-genome shotgun sequencing?

15.4 BIOINFORMATICS

The analysis of entire genomes generates not only a tremendous amount of nucleotide sequence data but, as we shall see, a rapidly growing volume of information regarding genome content, structure, and arrangement, as well as data detailing protein structure and function. The only feasible way to organize and analyze these data is through the use of computers. This has led to the development of the field of **bioinformatics,** which combines biology, mathematics, computer science, and statistics. Determining the location and nature of genes or presumed genes on a newly sequenced genome is a complex process called **annotation.** Once genes have been identified, bioinformaticists can perform computer, or *in silico,* **analysis** to further examine the genome.

Genome Annotation

Obviously, obtaining genome sequence without any understanding of the location and nature of individual genes would be a pointless exercise. The process of genome annotation seeks to identify every potential (putative) protein-coding gene as well as each rRNA- and tRNA-coding gene. A protein-coding gene is usually first recognized as an **open reading frame (ORF);** to find all ORFs, both strands of DNA must be analyzed. A procaryotic ORF is generally defined as a sequence of at least 100 codons with the following three features: (1) it is not interrupted by a stop codon; (2) there is an apparent ribosomal binding site at the 5′ end; and (3) terminator

sequences at the 3′ end. Only if these genetic elements are present is an ORF considered a putative gene (**figure 15.4**). ORFs that are presumed to encode proteins (as opposed to tRNA or rRNA) are called **coding sequences (CDS).** Bioinformaticists have developed algorithms to compare the sequence of predicted CDS with those in large databases containing nucleotide and amino acid sequences of known proteins. If an ORF matches one in the database, it is assumed to encode the same protein or type of protein. Although such comparisons are not without errors, they can provide tentative functional assignments for about 40 to 80% of the CDS in a given genome. The remaining fall into two classes: (1) genes that have matches in the database but no function has yet been assigned. These are said to encode conserved hypothetical proteins. (2) Genes whose translated products are unique to that organism. Such genes are said to encode proteins of unknown function. However, as more genomes are published, future comparisons may reveal a match in another organism. It is helpful to organize identified genes according to product function and/or location in the cell, such as ribosomal and transfer RNAs, lipid metabolism, energy metabolism, cell wall-associated, etc. (**table 15.2**). Clearly bioinformatics is a dynamic field and its continued development is crucial for further progress in structural, functional, and comparative genomics.

15.5 FUNCTIONAL GENOMICS

Functional genomics seeks to explain how genes and genomes operate. The base-by-base comparison of two or more gene sequences is called **alignment.** Alignment of genes on the same genome (i.e., from a single organism) may show that the nucleotide sequences are so alike that they most probably arose through gene duplication; such genes are called **paralogs.** Alignments of genes found in two or more different organisms may reveal that they are so strikingly similar that they are predicted to have the same function; these genes are called **orthologs.** Because the DNA code is degenerate, such alignments are generally performed after the gene's nucleotide sequence has been translated to amino acids.

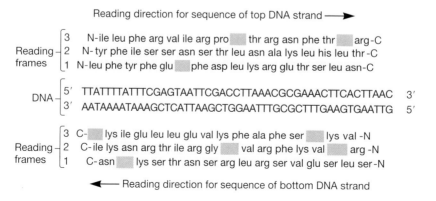

Figure 15.4 Finding Potential Protein Coding Genes. Annotation of genomic sequence requires that both strands of DNA be translated from the 5′ to 3′ direction in each of three possible reading frames. Stop codons are shown in green. See text for details.

Table 15.2	Estimated Number of Genes Involved in Various Cell Functions[a]								
Gene Function	*Escherichia coli K12*	*Bacillus subtilis*	*Mycoplasma genitalium*	*Treponema pallidum*	*Rickettsia prowazekii*	*Chlamydia trachomatis*	*Mycobacterium tuberculosis*	*Methanocaldococcus jannaschii*	*Pyrococcus abyssi*
Approximate total number of genes[b]	4,289	4,100	484	1,040	834	894	4,425	1,728	1,765
Cellular processes[c]	190	374	6	77	40	46	132	26	64
Cell envelope components	172	185	29	53	74	45	152	25	106
Transport and binding proteins	315	400	33	59	49	58	168	56	140
DNA metabolism	102	122	30	51	63	48	68	53	63
Transcription	41	114	13	25	26	18	40	21	37
Protein synthesis	122	161	90	99	104	133	110	118	108
Regulatory functions	176	293	5	22	26	12	165	19	66
Energy metabolism[d]	368	439	33	54	89	56	234	158	180
Central intermediary metabolism[e]	73	96	7	6	19	12	293	19	79
Amino acid biosynthesis	114	143	0	7	13	18	91	64	76
Fatty acid and phospholipid metabolism	67	84	8	11	22	27	158	9	18
Purines, pyrimidines, nucleosides, and nucleotides	77	81	19	21	19	14	57	37	51
Biosynthesis of cofactors and prosthetic groups	100	113	5	15	24	27	109	50	53

[a]Data adapted from TIGR (The Institute for Genomic Research) databases.

[b]The number of genes with known or hypothetical functions.

[c]Genes involved in cell division, chemotaxis and motility, detoxification, transformation, toxin production and resistance, pathogenesis, adaptations to atypical conditions, etc.

[d]Genes involved in amino acid and sugar catabolism, polysaccharide degradation and biosynthesis, electron transport and oxidative phosphorylation, fermentation, glycolysis/gluconeogenesis, pentose phosphate pathway, Entner-Doudoroff, pyruvate dehydrogenase, TCA cycle, photosynthesis, chemoautotrophy, etc.

[e]Amino sugars, phosphorus compounds, polyamine biosynthesis, sulfur metabolism, nitrogen fixation, nitrogen metabolism, etc.

Bioinformaticists also analyze the translated amino acid sequence of presumed genes to gain an understanding of potential protein structure and function. Often a short pattern of amino acids, called a **motif,** will represent a functional unit within a protein, such as the active site of an enzyme. For instance, **figure 15.5** shows the C-terminal domain of the cell division protein MinD from a number of microbes. Because these amino acids are found in such a wide range of organisms, they are said to be phylogenetically well conserved. In this case, the conserved region is predicted to form a coil needed for proper localization of the protein to the membrane.

Comparing sequences with other microbes can also provide information about the physical structure of the genome, such as the presence of transposable elements, operons, and repeat elements. The fraction of the genome that has been acquired from another organism by horizontal or lateral gene transfer can be inferred by this type of comparative genomics. In addition, specific aspects of an organism's physiology can be deduced by the presence or absence of specific genes. Genomic analysis can also provide information about the phylogenetic relationships among microbes (**figure 15.6**). Examples of some of the insights derived from such analyses are discussed in section 15.6.

Evaluation of RNA-Level Gene Expression: Microarray Analysis

Once the identity and function of the genes that comprise a genome are established, the key question remains, "Which genes are expressed at any given time?" Prior to the genomic era, researchers could identify only a limited number of genes whose expression was altered under specific circumstances. However, the development of **DNA micorarrays (gene chips)** now allows scientists to look at the expression level of a vast collection of genes at once. DNA microarrays are solid supports, usually of glass or silicon, upon which DNA is attached in an organized grid fashion. Each spot of DNA, called a **probe,** represents a single gene or ORF. The location of each gene on the grid is carefully recorded so that when analyzed, the genetic identity of each spot is known.

Figure 15.5 **Analysis of Conserved Regions of Phylogenetically Well-Conserved Proteins.** C-terminal amino acid residues of MinD from 20 organisms representing all three domains of life are aligned to show strong similarities. Amino acid residues identical to *E. coli* are boxed in yellow, and conservative substitutions (e.g., one hydrophobic residue for another) are boxed orange. The number of the last residue shown relative to the entire amino acid sequence is shown at the extreme right of each line.

Of the several ways in which microarrays can be constructed, two techniques are most commonly employed: **Spotted arrays** are prepared by the robotic application of probe to the chip. The probe may be a PCR product, cDNA, or a short DNA fragment within the gene or ORF, called an **oligonucleotide.** When eucaryotic genomes are to be analyzed, oligonucleotide probes are called **expressed sequence tags (ESTs)** because each is derived from cDNA, which itself is the product of an expressed gene (*see figure 14.4*). The genes to be represented by DNA on spotted arrays are carefully selected, so this technology is most commonly used for custom-made microarrays.

Commercially prepared microarrays are usually prepared by a technique known as photolithography (**figure 15.7**). Here, a mask is laid over the chip, and light is used to control the synthesis of the oligonucleotide directly on the chip surface. Each hole in the mask will eventually hold many copies of a different oligonucleotide, and each mask can control the synthesis of several hundred thousand squares. Thus a single microarray can contain hundreds of thousands of different probes, and each probe is present in millions of copies. Commercial microarrays are available with probes for every expressed gene or ORF on the genomes of a number of microbes (**figure 15.8**). These include *E. coli* (about 4,200 ORFs), pathogens such as *Helicobacter pylori* (1,590 ORFs) and *Mycobacterium tuberculosis* (4,000 ORFs), as well as the yeast *Saccharomyces cerevisiae* (approximately 6,600 ORFs).

The analysis of gene expression using microarray technology, like many other molecular genetic techniques, is based on hybridization between the probe DNA and the nucleic acids to be

Figure 15.6 **Phylogenetic Relationships of Some Procaryotes with Sequenced Genomes.** These procaryotes are discussed in the text. *Methanocaldococcus jannaschii* is in the domain *Archaea*, the rest are members of the domain *Bacteria*. Genomes from a broad diversity of procaryotes have been sequenced and compared. *Source: The Ribosomal Database Project.*

analyzed, often called the targets, which may be either mRNA or single-stranded cDNA (**figure 15.9**). The target nucleotides are labeled with fluorescent dyes and incubated with the chip under conditions that ensure proper binding of target mRNA (or cDNA) to its complementary probe. Unbound target is washed off and the chip is then scanned with laser beams. Fluorescence at each spot or probe indicates that mRNA hybridized. Analysis of the color and intensity of each probe shows which genes were expressed.

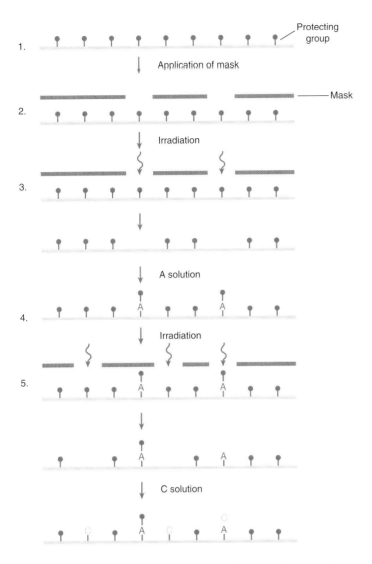

Figure 15.7 Construction of a DNA Chip with Attached Oligonucleotide Sequences. Only two cycles of synthesis are shown. The steps are as follows: (1) The glass support is coated with light-sensitive protecting groups that prevent random nucleoside attachment. (2) The surface is covered with a mask that has holes corresponding to the sites for attachment of the desired nucleosides. (3) Laser light passing through the mask holes removes the exposed protecting groups. (4) The chip is bathed in a solution containing the first nucleoside to be attached. The nucleoside will chemically couple to the light-activated sites. Each nucleoside has a light-removable protecting group to prevent addition of another nucleoside until the appropriate time. (5) Steps 2 through 4 are repeated using a new mask each time until all sequences on the chip have been completed.

DNA microarray analysis can be used to determine which genes are expressed during cellular differentiation or as the result of mutation. One common application of microarray technology in microbiology is to determine the genes whose expression is changed (either up or down regulated) in response to environmental changes (figure 15.9). In this sort of analysis, two different-col-

ored fluorochromes are used. For instance, the pathogen *Helicobacter pylori* dwells in the stomach where it causes ulcers; its genome has been sequenced. One might want to know which *H. pylori* genes are repressed or induced upon exposure to acidic conditions. To determine this, total cellular mRNA from bacteria grown in neutral conditions is prepared and tagged with a green fluorochrome to serve as a control or reference, while mRNA from cells exposed to acidic pH is labeled red. The green (reference) and red (experimental) mRNA samples are then mixed and hybridized to the same microarray. After the unattached mRNA is washed off, the chip is scanned and the image is computer analyzed. A yellow spot or probe indicates that roughly equal numbers of green and red mRNA molecules (targets) were bound, so there was no change in the level of gene expression for that gene. If a target is red, more mRNA from bacteria grown under the experimental acidic conditions was present when the two mRNAs were mixed, thus this gene was induced. Conversely, a green target indicates that the gene was repressed upon exposure to acid stress. Careful image analysis is used to determine the relative intensity of each spot, so that the magnitude of induction or repression of each gene whose expression is altered can be approximated.

Such analysis leads to the detection of genes whose expression falls into specific patterns. For any given microbial genome, only about half to two-thirds of the ORFs are known genes, so such patterns may be used to help assign tentative functions to some unknown ORFs. This is because genes that are expressed under the same conditions may be co-regulated, suggesting that they share a common function or are involved in a common process. However, it is important to keep in mind that microarray results represent only mRNAs present at the time of preparation. Therefore, if a gene is transiently expressed, it may be missed. In addition, if a gene product is regulated posttranslationally, for instance by phosphorylation, its mRNA will be present but the protein may not be active.

15.6 COMPARATIVE GENOMICS

As we have seen, one approach to learning more about the genome of any given microbe is to compare it with that of others. This is the domain of comparative genomics. The publication of over 200 published microbial genomes has truly changed our understanding of microbial biology. One very striking insight is the fact that microbial genomes are not as static as once thought. In fact, microbial genomes are amazingly fluid with a substantial portion of the genome transferred *between* cells, not from parent to offspring. As discussed in chapter 13, this is called **lateral** or **horizontal gene transfer (HGT)**. Broadly defined, HGT is the exchange of genetic material between organisms that need not be of similar evolutionary lineages. Genome analysis has revealed that HGT is frequently mediated by phages, and that lysogeny may be the rule, rather than the exception. In fact, some microbes carry multiple prophages. It has become clear that HGT is a major evolutionary force in short-term microbial evolution and long-term speciation. For instance, it is thought that *E. coli* may have acquired the lactose (*lac*) operon from another microbe and thus

Gene chip probe Array

Hybridized probe feature

Single-stranded, fluorescently labeled DNA target

Oligonucleotide probe

—24μm

Each probe contains millions of copies of a specific oligonucleotide probe.

1.28 cm—

Over 200,000 different probes complimentary to genetic information of interest

Image of hybridized probe array

Figure 15.8   A DNA Microarray.   The DNA chip manufactured by Affymetrix, Inc. contains probes designed to represent thousands or tens of thousands of genes.

became capable of colonizing the mammalian colon, where milk sugar is a common carbon source. On a larger scale, the methane-producing archaeon *Methanosarcina mazei* appears to have acquired about one-third of its genes from other procaryotes. It appears that microbes use HGT to quickly develop the capacity to colonize new ecological niches.

Comparative genomics can provide useful insights in trying to discern the origin and prevalence of particular phenotypic traits. An example is provided by the genome sequence of *Picrophilis torridus*. This archaeon grows optimally at a pH of 0.7 and a temperature of 65°C. It thrives in hot, acidic, sulfataric fields, a habitat it shares with other extremophilic archaea, like *Thermoplasma,* and the more distantly related *Sulfolobus* (**figure 15.10a**). Understanding the mechanisms by which these microbes survive under these circumstances is not purely academic, as proteins engineered to withstand harsh treatment have industrial importance. In most cases, microbes living at low pH are able to maintain a neutral internal pH, but this is not the case for *P. torridus,* whose intracellular pH is around 4.6. This suggests it has evolved a unique strategy to prevent irreversible macromolecular damage. By comparing the genes and their products that are found only in these acidophiles

and not in neutrophilic and alkalophilic microbes, researchers can focus on potential adaptive strategies (figure 15.10b).

Another area in which comparative genomics has recently been applied is vaccine development. The promise of new vaccines as a direct result of genome sequencing is covered extensively in the popular press. However, finding good targets for the development of new vaccines is complicated. The molecules (antigens) that form the basis of an effective vaccine must meet many requirements, including: (1) the antigen must be expressed by the pathogen during infection; (2) the antigen must be either secreted or found on the surface of the pathogen; (3) it must be found in all strains of the pathogen; (4) it must elicit a host immune response; and (5) the antigen must be essential for the survival of the pathogen, at least while it is in the host. Considering the thousands of ORFs that are revealed in each annotated genome, finding genes whose products meet all of these criteria might seem next to impossible. However, using comparative genomics, researchers can examine the genomes of multiple strains of a given pathogen to create a "short list" of candidates. This reduces the number of potential antigens to a manageable number. These antigens are then tested in a variety of assays to discover the best molecules for vaccine development. This

Figure 15.9   A Microarray System for Monitoring Gene Expression.   Cloned genes from an organism are amplified by PCR, and after purification, samples are applied by a robotic printer to generate a spotted microarray. To monitor enzyme expression, mRNA from test and reference cultures are converted to cDNA by reverse transcriptase and labeled with two different fluorescent dyes. The labeled mixture is hybridized to the microarray and scanned using two lasers with different excitation wavelengths. The fluorescence responses are measured as normalized ratios that show whether the test gene response is higher or lower than that of the reference.

process was used for the development of an experimental Group B *Streptococcus* (GBS) vaccine. GBS causes serious infections in infants so a vaccine is highly desired. Eight genomes of GBS clinical isolates were compared and 312 surface antigens were tested for their ability to protect mice from infection. Four were found to be effective, and when combined into a single vaccine, they protected animals against infection by all known clinical strains. This demonstrates two important concepts: (1) genome analysis is a powerful mechanism for gaining information, but results must be experimentally verified; and (2) genomics will meet at least some of the expectations described by the popular press.

15.7   PROTEOMICS

Genome function can be studied at the translation level as well as the transcription level. The entire collection of proteins that an organism produces is called its **proteome.** Thus **proteomics** is the study of the proteome or the array of proteins an organism can produce. It is an essential discipline because proteomics provides information about genome function that mRNA studies cannot. There is not always a direct correlation between mRNA and the

pool of cellular proteins because of differing levels of mRNA and protein stability and posttranslational regulation. Measurement of mRNA levels can show the dynamics of gene expression and tell what might occur in the cell, whereas proteomics discovers what is actually happening. Much of the research in this area is referred to as **functional proteomics.** It is focused on determining the function of different cellular proteins, how they interact with one another, and the ways in which they are regulated.

Although new techniques in proteomics are currently being developed, we will focus briefly only on the most common approach, **two-dimensional gel electrophoresis.** In this procedure, a mixture of proteins is separated using two different electrophoretic procedures (dimensions). This permits the visualization of thousands of cellular proteins, which would not otherwise be separated in a single electrophoretic dimension. As shown in **figure 15.11,** the first dimension makes use of isoelectric focusing, in which proteins move electrophoretically through a pH gradient (e.g., pH 3 to 10). First the protein mixture is applied to an acrylamide gel in a tube with an immobilized pH gradient and electrophoresed. Each protein moves along the pH gradient until the protein's net charge is zero and the protein stops moving. The pH at this point is equal to the protein's isoelectric point. Thus the first dimension separates the proteins

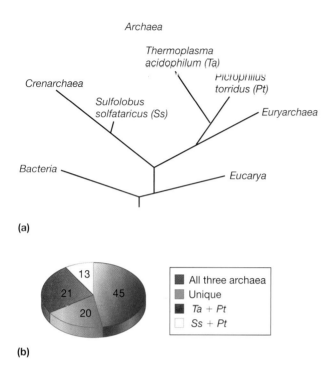

(a)

(b)

Figure 15.10   Comparative Genomics of *Picrophilus torridus.*   (a) Simplified phylogenetic tree with the positions of the three acidiphilic, thermophilic archaea *P. torridus, T. acidophilum*, and *S. sulfataricus.* **(b)** ORFs found in *P. torridus, T. acidophilum*, and *S. sulfataricus*, ORFs shared between *P. torridus* and *T. acidophilum* (*Ta + Pt*), or *S. sulfataricus* (*Sa + Pt*), and those that are unique to *P. torridus* (Unique).

based on their content of ionizable amino acids. The second dimension is SDS *p*oly*a*crylamide *g*el *e*lectrophoresis (SDS-PAGE). SDS (*s*odium *d*odecyl *s*ulfate) is an anionic detergent that denatures proteins and coats the polypeptides with a negative charge. After the isoelectric gel has been completed, it is soaked in SDS buffer and then placed at the edge of an SDS-PAGE gel. A voltage is then applied. Under these circumstances, polypeptides are separated according to their molecular weight—that is, the smallest polypeptide will travel fastest and farthest. Two-dimensional gel electrophesis can resolve thousands of proteins; each protein is visualized as a spot of varying intensity depending on its cellular abundance. Radiolabeled proteins are usually used, enabling greater sensitivity so that newly synthesized proteins can be distinguished and their rates of synthesis determined. Computer analysis is used to compare two-dimensional gels from microbes grown under different conditions or to compare wild-type and mutant strains. Websites have been developed for the deposition of such images, allowing researchers access to valuable and ever-growing databases.   Gel electrophoresis (section 14.4)

Two-dimensional gel electrophoresis is even more powerful when coupled with mass spectrometry (MS). The unknown protein spot is cut from the gel and cleaved into fragments by treatment with proteolytic enzymes. Then the fragments are analyzed by a mass spectrometer and the mass of the fragments is plotted. This mass fingerprint can be used to estimate the probable amino acid composition of each fragment and tentatively identify the

protein. Sometimes proteins or collections of fragments are run through two mass spectrometers in sequence, a process known as tandem MS (**figure 15.12**). The first spectrometer separates proteins and fragments, which are further fragmented. The second spectrometer then determines the amino acid sequence of each fragment produced in the first stage. The sequence of a whole protein often can be determined by analysis of such fragment sequence data. Alternatively, if the genome of the organism has been sequenced, only the sequence of the N-terminal amino acids need to be obtained. Computer analysis is then used to compare this partial amino acid sequence with the predicted translated sequences of all the annotated ORFs on the organism's genome. In this way, both the protein and the gene that encodes it can be identified. Further investigation of the protein may rely on one of the large databases of protein sequences that enable comparative analysis. Comparing amino acid sequences can provide information regarding the protein structure, function, and evolution.

Proteomics has been used to study the physiology of many microbes, including *E. coli.* Some areas of research have been the effect of phosphate limitation, proteome changes under anoxic conditions, heat-shock protein production, and the response to the toxicant 2,4-dinitrophenol. One particularly useful approach in studying genome function is to inactivate a specific gene and then look for changes in protein expression. Because changes in the whole proteome are followed, gene inactivation can tell much about gene function and the large-scale effects of gene activity. Gene-protein databases for a number of microbes have been established. These provide information about the conditions under which each protein is expressed and where it is located in the cell.

A second branch of proteomics is called **structural proteomics.** Here the focus is on determining the three-dimensional structures of many proteins and using these to predict the structures of other proteins and protein complexes. The assumption is that proteins fold into a limited number of shapes, and that proteins can be grouped into families of similar structures. When a number of protein structures are determined for a given family, the patterns of protein structure organization or protein-folding rules will be known. Then computational biologists (i.e., bioinformaticists) use this information and the amino acid sequence of a newly discovered protein to predict its final shape, a process known as **protein modeling.**

1. What is genome annotation? Why do you think it requires knowledge of mathematics and statistics as well as biology and computer science?
2. Describe how microarrays are constructed and used to analyze gene expression. How might the following scientists use DNA microarrays in their research? (a) An environmental microbiologist who is interested in how the soil microbe *Rhodopseudomonas palustris* degrades the toxic compound 3-chlorobenzene. (b) A medical microbiologist who wants to learn about how the pathogen *Salmonella* survives within a host cell.
3. Why does two-dimensional gel electrophoresis allow the visualization of many more cellular proteins than seen in electrophoresis in a single dimension?
4. What is the role of mass spectrometry in proteomics?
5. What is the difference between functional and structural proteomics? How do you think structural proteomics might be used in vaccine development?

Load a mixture of proteins onto an isoelectric focusing tube gel.

pH 4.0

Proteins migrate until they reach the pH where their net charge is 0. At this point, a single band could contain two or more different proteins.

pH 10.0

Lay the tube gel onto an SDS-gel and separate proteins according to their molecular mass.

SDS-gel

pH 4.0   pH 10.0

200 kDa

10 kDa

(a) The technique of 2-dimensional gel electrophoresis

Figure 15.11   Two-Dimensional Gel Electrophoresis.
(a) The technique involves two electrophoresis steps. First, a mixture of proteins is separated on an isoelectric focusing gel that has the shape of a tube. Proteins migrate to the point where their net charge is zero. This tube gel is placed into a long well on top of an SDS-polyacrylamide gel. This second gel separates the proteins according to their mass. In this diagram, only a few spots are seen, but an actual experiment would involve a mixture of hundreds or thousands of different proteins. **(b)** An autoradiograph of a 2-D gel. Each spot represents a unique protein.

pH 4.0   pH 10.0

200 kDa

10 kDa

(b) An autoradiograph of a 2-dimensional gel. Each protein is a discrete spot.

15.8   INSIGHTS FROM MICROBIAL GENOMES

The development of whole-genome shotgun sequencing and other genome sequencing techniques has led to the characterization of many microbial genomes in a relatively short time. These genomes represent great phylogenetic diversity and comparisons among them will contribute significantly to understanding evolutionary processes, deducing which genes are responsible for various cellular processes, dissecting the complexities of genetic regulation, and genome organization. In addition, genomics has become an important tool in developing new antimicrobial agents and may lead to new approaches to detoxify hazardous wastes and provide energy.

Examination of the genomes sequenced to date has provided valuable information, generated many questions, and stimulated new areas of research. In this section we learn how scientists are using experimental approaches to answer some unresolved questions that arise from genome analysis as well as how the genomic

era has brought new insights into microbial physiology, ecology, and evolution.

Identification of Genes with Unknown Functions

A common result of genomic studies is the identification of genes for which no function can be assigned. For instance, genome analysis of the important pathogen *Neisseria meningitidis*, which is one causative agent of meningitis, is fairly typical. While just over half of the ORFs can be assigned biological roles based on similarity to proteins of known function, 16% of the ORFs match genes of unknown function in other organisms and about a quarter have no database matches at all. A number of approaches have been developed in an attempt to solve the identity of unknown genes. Perhaps the most comprehensive has been tested in Baker's yeast, otherwise known as *Saccharomyces cerevisiae*.

When the genome of *S. cerevisiae* was fully sequenced in 1996, only about one-third of the ORFs had a known function. At

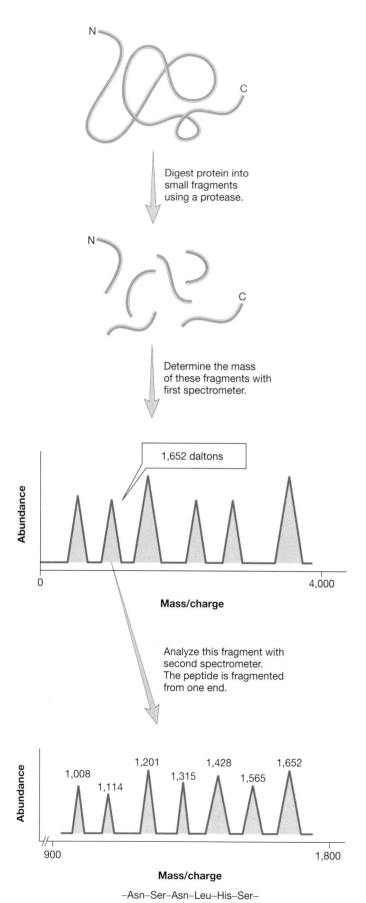

Digest protein into small fragments using a protease.

Determine the mass of these fragments with first spectrometer.

1,652 daltons

Abundance

0

Mass/charge

4,000

Analyze this fragment with second spectrometer. The peptide is fragmented from one end.

Abundance

1,008

1,114

1,201

1,315

1,428

1,565

1,652

900

1,800

Mass/charge

–Asn–Ser–Asn–Leu–His–Ser–

that time, an international consortium of scientists began a project to elucidate the role of ORFs to which no function could be assigned. They decided to create a collection of *S. cerevisiae* strains, each with a deletion in a specific ORF of unknown function. These mutants could then be used in studies designed to reveal their mutant phenotypes. The phenotype of each mutant would then be used to assign a tentative function for the gene.

The yeast deletions were created by a PCR-based gene strategy. In this approach, PCR was used to create DNA cassettes that could be introduced into yeast cells, and through recombination, replace a specific gene in the genome (**figure 15.13**). Each DNA cassette contained three major elements: (1) a gene encoding resistance to the antifungal agent geneticin (*KanMX*) that replaced the targeted ORF. Transformed cells were incubated in the presence of geneticin so only cells in which the targeted ORF had been replaced by the *KanMX* gene were able to grow. (2) Sequences flanking the start and stop codons of the target ORF were cloned on either side of the *KanMX* gene. Because these sequences matched those at the 5′ and 3′ ends of the target ORF, they were used to direct homologous recombination between the cassette and the chromosome, resulting in replacement of the ORF with the *KanMX* gene. (3) A unique DNA sequence called a "molecular bar code" was inserted in each cassette so once the cassette had integrated into the chromosome, individual strains in a mixed population could be identified, as described next. These cassettes were used to generate over 2,000 deletion mutant strains of *S. cerevisiae*. PCR (section 14.3); Creating genetic variability: Recombination at the molecular level (section 13.4)

The phenotypes of these "yeast knock-out" (YKO) mutants were determined in parallel competition assays. Here, large mixtures of different deletion mutants were cultured together under specific environmental conditions (e.g., high temperature or pH). The culture was periodically sampled, and PCR was used to amplify the molecular bar code of each deletant. Thus changes in mutant populations were followed over time. If the deleted gene was important for survival and growth under the test conditions, the deletant mutant missing that gene grew more slowly and the abundance of this mutant eventually declined. To assess the relative number of each mutant strain, a mixture of PCR-amplified molecular bar codes was hybridized to their complementary sequences on a microarray. The intensity of each spot on the microarray reflected the population size of each strain in the mixture. Mutations in genes whose products promote growth in the conditions being studied gave a less intense signal than did those whose products are not required.

Using this approach, many genes have been identified, including those that are involved in growth in rich and minimal media, at different pH values, and in the presence of certain drugs. Indeed, by the end of 2003, 80% of the annotated yeast genes had known functions (although some of these were revealed by analysis of newly identified orthologous genes) and all the genes on the genome are expected to be "known" sometime in 2007. Clearly, this approach to dissecting a genome provides important infor-

Figure 15.12 **The Use of Tandem Mass Spectrometry to Determine the Amino Acid Sequence of a Peptide.**

5' end of ORF

3' end of ORF

KanMX

Unique & specific
nucleotide sequence

Figure 15.13   DNA Cassette Used in the Construction of Yeast-Deletion Mutants.   The *KanMX* gene encodes resistance to the antifungal agent geneticin so yeast mutants can be selected and maintained. The blue sequences flanking the *KanMX* gene are the same as the sequences on either side of the ORF to be replaced. The molecular bar code is an additional genetic element inserted in each cassette so individual mutants can be identified by PCR amplification of genomic DNA.

mation, and similar projects are under way for *E. coli* and the endospore-forming bacterium *Bacillus subtilis*. Updates of these projects are posted on the Internet.

Genomic Analysis of Pathogenic Microbes

Analysis of the genomes of plant and animal pathogens provides considerable information about the evolution of virulence, host-pathogen interactions, potential treatment methods, and vaccine development. The genomes of numerous pathogens have been sequenced and compared to the genomes of phylogenetically related pathogens as well as close relatives that are not pathogenic.

The genome of *Mycoplasma genitalium* was one of the first to be sequenced. This microbe infects cells in the human genital and respiratory tracts. It has a genome of only 580 kilobases (0.58 Mb), one of the smallest genomes of any organism (**figure 15.14**). Thus the sequence data are of great interest because they help establish the minimal set of genes needed for free-living existence. There appear to be approximately 517 genes (480 protein-encoding genes and 37 genes for RNA species). About 90 proteins are involved in translation, and only around 29 proteins for DNA replication. Interestingly, 140 genes, or 29% of those in the genome, code for membrane proteins, and up to 4.5% of the genes seem to be involved in evasion of host immune responses. Only 5 genes have regulatory functions. Even in this small genome, about one-fifth of the genes do not match any known protein sequence. Comparison with the *M. pneumoniae* and *Ureaplasma urealyticum* genomes and studies of gene inactivation by transposon insertion suggest that about 108 to 121 *M. genitalium* genes may not be essential for survival. Thus for this microbe, the minimum gene set required for laboratory growth conditions seems to be approximately 300 genes; about 100 of these have unknown functions.

Haemophilus influenzae has a much larger genome, 1.8 megabases and 1,743 genes (**figure 15.15**). More than one-third of the genes have unknown functions. The bacterium lacks three Krebs cycle genes and thus a functional cycle. It devotes many genes (64 genes) to regulatory functions. *Haemophilus influenzae* is a species capable of natural transformation. The process must be very important to this bacterium because it contains 1,465 copies of the recognition sequence used in DNA uptake during transformation.   DNA transformation (section 13.8)

One of the most alarming trends in the treatment of infectious disease is the rise of antibiotic-resistant bacteria. This is particularly true among the staphylococci. These gram-positive microbes cause an estimated 1 million serious infections each year. The two predominant opportunistic pathogens are *Staphylococcus epidermidis* and *S. aureus*. *S. epidermidis* is commonly found on the skin and has emerged as a serious pathogen only in recent years. It has been found to infect implanted medical devices like artificial heart valves. In contrast, *S. aureus* is more aggressive and can cause conditions that range from minor skin infections to life-threatening abscesses, heart infections, and toxic shock syndrome. During the 1960s, methicillin and other semi-synthetic penicillin antibiotics were frequently prescribed to treat staphylococcal infections, giving rise to the development of meticillin-resistant *S. aureus* (MRSA) and *S. epidermidis* (MRSE). By 2005, 60% of *S. aureus* clinical isolates were resistant to methicillin; some strains were resistant to up to 20 different antibiotics! The only drug effective against such multiply resistant strains was vancomycin, but *S. aureus* strains resistant to vancomycin have now been isolated.   Antimicrobial chemotherapy (chapter 34)

Strains of *S. aureus* and *S. epidermidis* that were initially isolated between 1960 and 1998 vary in their resistance to a number of antibiotics and their virulence levels. The genomes of a number of such strains have been used in comparative genomic analysis to track the evolution of antibiotic resistance and virulence. The origin of about 1,700 genes that are shared by all strains of both species is difficult to ascertain. However, most genes that are strain-specific appear to have been introduced by prophages, transposons, insertion sequences, and plasmid-mediated gene transfer. For instance, this analysis reveals that an *Enterococcus faecalis* transposon introduced vancomycin resistance into *S. aureus*. Both *S. aureus* and *S. epidermidis* have acquired the genes that encode a capsule made of glutamate polymers from the gram-positive pathogen *Bacillus anthracis*, the causative agent of anthrax. In all three species, the polyglutamate capsule is a major virulence factor.   Direct contact diseases: Staphylococcal diseases (section 38.3)

The genome of another pathogen, *Bacillus anthracis*, underwent intense scrutiny following a series of letter-based bioterrorism attacks in the fall of 2001. Although the genes that encode the components of the anthrax toxin are encoded on plasmids, the *B. anthracis* chromosome has a number of virulence-enhancing genes

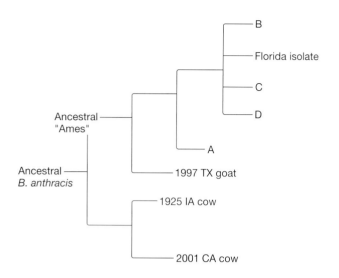

Figure 15.16   Proposed Phylogenetic Tree of the B. anthracis Ames Isolate Used in U.S. Bioterror Attacks in the Autumn of 2001.   Isolates A–D are laboratory strains; other isolates are indicated by state and year of initial cultivation. Isolates B and C are identical to the isolate recovered from the first victim of the attacks (Florida isolate). Isolate D differs only by the insertion of a single adenine in a key region used for comparison. Laboratory strain A has two additional bases and has lost one of the toxin-encoding plasmids. This whole-genome analysis demonstrated the presence of four genetic elements on the genome that vary, despite previous analysis suggesting that these strains were identical or nearly identical. This approach is a powerful new technique for tracking infectious disease outbreaks.

with unknown functions play a major role in cell division. Perhaps *Chlamydia* employs a mechanism of cell division different from that of other procaryotes. Finally, the genome contains at least 20 genes that have been obtained from eucaryotic host cells (most bacteria have no more than 3 or 4 such genes). Some of these genes are plantlike; originally *Chlamydia* may have infected a plantlike host and then moved to animals.   Phylum *Chlamydiae* (section 21.5); The procaryotic cell cycle: Cytokinesis (section 6.1)

One of the most difficult human pathogens to study is the causative agent of syphilis, *Treponema pallidum*. This is because it is not possible to grow *T. pallidum* outside the human body. We know little about its metabolism or the way it avoids host defenses, and no vaccine for syphilis has yet been developed. Naturally, the sequencing of the *T. pallidum* genome generated considerable excitement and hope. It turns out that *T. pallidum* is metabolically crippled. It can use carbohydrates as an energy source, but lacks the TCA cycle and oxidative phosphorylation enzymes (**figure 15.17**). *T. pallidum* also lacks many biosynthetic pathways (e.g., for enzyme cofactors, fatty acids, nucleotides, and some electron transport proteins) and must rely on molecules supplied by its host. In fact, about 5% of its genes code for transport proteins. Given the lack of several critical pathways, it is not surprising that the pathogen has not been cultured successfully. The genes for surface proteins are of particular interest. *T. pallidum* has a family of surface protein genes characterized by many repetitive sequences. Some have speculated that

these genes might undergo recombination in order to generate new surface proteins, enabling the organism to avoid attack by the immune system. It may be possible to develop a vaccine for syphilis using some of the newly discovered surface proteins. We also may be able to identify strains of *T. pallidum* using these surface proteins, which would be of great importance in syphilis epidemiology. The genome results should ultimately help us understand how *T. pallidum* causes syphilis. About 40% of the genes have unknown functions. Possibly some of them are responsible for avoiding host defenses and for the production of toxins and other virulence factors.   Direct contact diseases: Sexually transmitted diseases (section 38.3)

For centuries, tuberculosis has been one of the major scourges of humankind. About one-third of the human population is infected with the causative agent *Mycobacterium tuberculosis*. After establishing residence in immune system cells in the lung, it often remains in a dormant state until the host's immune system is compromised. The disease kills about 3 million people annually and is the direct cause of death for many AIDS patients. Predictably, *M. tuberculosis* is becoming ever more drug resistant. Genome studies could be of great importance in the fight to control the renewed spread of tuberculosis. The annotated *M. tuberculosis* genome was published in 1998; at that time it was predicted to have 3,974 genes. In 2002, its genome was reexamined and, based largely on inspection of small ORFs, an additional 82 genes were identified. The number of published genomes during the intervening four years is reflected by the change in the number of genes of unknown function: in 1998 there were 606 genes for which no function or ortholog could be assigned; by 2002, there were only 272 such genes. Significantly, the majority of these genes remain hypothetical because although orthologs were found, no functional assignment has yet been made. Undoubtedly, as more genomes are sequenced and proteomes examined, the number of genes of unknown function for all organisms whose genomes have been sequenced will decline.

What has the genome sequence of this ancient pathogen revealed? More than 250 genes are devoted to lipid metabolism (*E. coli* has only about 50 such genes), and *M. tuberculosis* may obtain much of its energy by degrading host lipids. There are a surprisingly large number of regulatory elements in the genome. This may mean that the infectious process is much more complex than previously thought. Two families of novel glycine-rich proteins with unknown functions are present and represent about 10% of the genome. They may be a source of antigenic variation involved in defense against the host immune system. One major medical problem has been the lack of a highly effective vaccine. A large number of proteins that are either secreted by the bacterium or on the bacterial surface have been identified from the genome sequence. It is hoped that some of these proteins can be used to develop better vaccines. This is particularly important in view of the spread of multiply drug resistant *M. tuberculosis*.   Mycobacterium tuberculosis (sections 24.4 and 38.1)

The *M. tuberculosis* genome has been compared to the genomes of two relatives—*M. leprae,* which causes leprosy, and *M. bovis,* the causative agent of tuberculosis in a wide range of animals, including cows and humans. The *M. bovis* and *M. leprae* genomes differ from that of *M. tuberculosis* in some important ways. The genomes of *M. bovis* and *M. tuberculosis* are most similar—about 99.5% identical at the sequence level. However, the *M.*

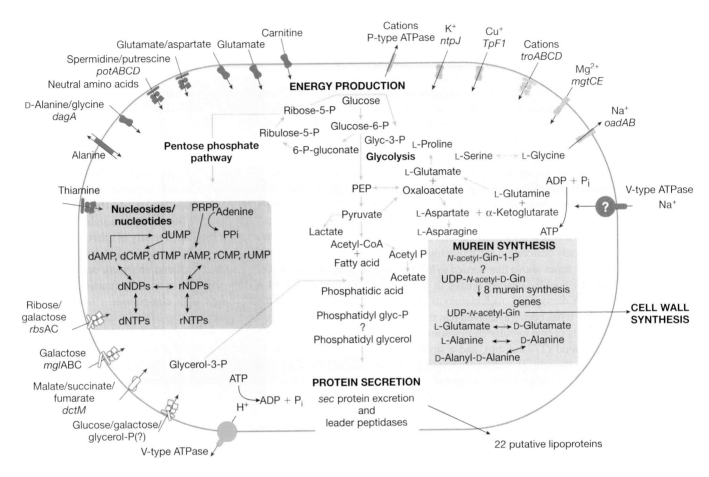

Figure 15.17   Metabolic Pathways and Transport Systems of *Treponema pallidum*.   This depicts *T. pallidum* metabolism as deduced from genome annotation. Note the limited biosynthetic capabilities and extensive array of transporters. Although glycolysis is present, the TCA cycle and respiratory electron transport are lacking. Question marks indicate where uncertainties exist or expected activities have not been found.

bovis genome is missing 11 separate regions, making its genome slightly smaller (4.3 Mb vs. 4.4. Mb). The sequence dissimilarities involve the inactivation of some genes, leading to major differences in the way the two bacteria respond to environmental conditions. This may account for the host range differences between these two closely related pathogens. The divergence between *M. tuberculosis* and *M. leprae* is even more striking. The *M. leprae* genome is a third smaller than that of *M. tuberculosis*. About half the genome is devoid of functional genes. Instead, there are over 1,000 degraded, nonfunctional genes called pseudogenes. In total, *M. leprae* seems to have lost as many as 2,000 genes during its career as an intracellular parasite. It even lacks some of the enzymes required for energy production and DNA replication. This might explain why the bacterium has such a long doubling time, about two weeks in mice. One hope from genomics studies is that critical surface proteins can be discovered and used to develop a sensitive test for early detection of leprosy. This would allow immediate treatment of the disease before nerve damage occurs. Direct contact diseases: Leprosy (section 38.3)

Another microbe that has reduced the size of its genome is *Rickettsia prowazekii*, a member of the α-proteobacteria, and an obligate intracellular parasite of lice and humans. It is the causative

agent of typhus, a disease that killed millions during the First and Second World Wars. It lacks genes for glycolysis and many genes for the biosynthesis of amino acids and nucleosides. Like *M. leprae*, its genome contains a number of pseudogenes. Gene inactivation and deletion have also been observed in plant pathogens. For instance, the plant pathogen *Phytoplasma asteris* is transmitted by insects and has an intracellular life cycle in both its insect vectors and plant hosts. It lacks many genes related to amino acid, nucleotide, and fatty acid biosynthesis. It also lacks many genes that function in energy metabolism. Presumably, microbes that obtain nutrients and ATP from their hosts lack the selective pressure needed to maintain the corresponding functional genes. Arthropod-borne diseases (section 38.2)

Genomic Analysis of Extremophiles

Microbes that live in harsh environments are called extremophiles. Genomic analysis of such microbes has been pursued with the goal of understanding how organisms can tolerate extreme environmental conditions. The genomes of over a dozen thermophiles and hyperthermophiles, including members of both the *Bacteria* and *Archaea*, have been analyzed. Originally, it was postulated that

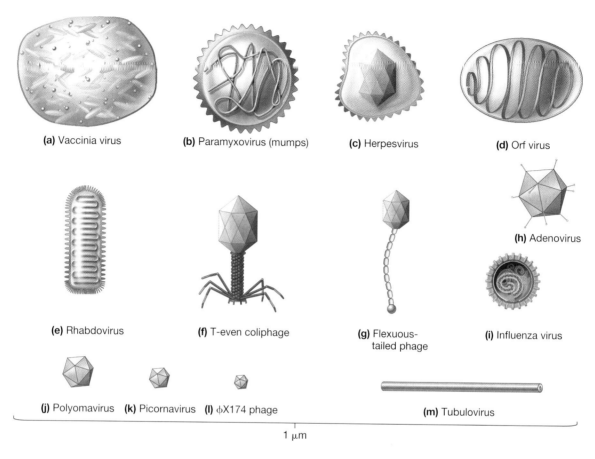

(a) Vaccinia virus   (b) Paramyxovirus (mumps)   (c) Herpesvirus   (d) Orf virus

(e) Rhabdovirus   (f) T-even coliphage   (g) Flexuous-tailed phage   (h) Adenovirus   (i) Influenza virus

(j) Polyomavirus   (k) Picornavirus   (l) φX174 phage   (m) Tubulovirus

1 μm

Figure 16.2   The Size and Morphology of Selected Viruses.   The viruses are drawn to scale. A 1 μm line is provided at the bottom of the figure.

for the coat protein. Suppose, however, that the TMV capsid was composed of six different protomers all about 150 amino acids in length. If this were the case, about 2,900 of the 6,000 nucleotides in the TMV genome would be required just for capsid construction, and much less genetic material would be available for other purposes.

The various morphological types of viruses primarily result from the combination of a particular type of capsid symmetry with the presence or absence of an envelope, which is a lipid layer external to the nucleocapsid. There are three types of capsid symmetry: helical, icosahedral, and complex. Those virions having an envelope are called **enveloped viruses;** whereas those lacking an envelope are called **naked viruses** (figure 16.1).

Helical Capsids

Helical capsids are shaped like hollow tubes with protein walls. The tobacco mosaic virus provides a well-studied example of helical capsid structure (figure 16.3). In this virus, the self-assembly of protomers in a helical or spiral arrangement produces a long, rigid tube, 15 to 18 nm in diameter by 300 nm long. The capsid encloses an RNA genome, which is wound in a spiral and lies within a groove formed by the protein subunits. Not all heli-

cal capsids are as rigid as the TMV capsid. The influenza virus genome is enclosed in thin, flexible helical capsids that are folded within an envelope (**figure 16.4**).

The size of a helical capsid is influenced by both its protomers and the nucleic acid enclosed within the capsid. The diameter of the capsid is a function of the size, shape, and interactions of the protomers. The nucleic acid appears to determine helical capsid length because the capsid does not extend much beyond the end of the DNA or RNA.

Icosahedral Capsids

The icosahedron is a regular polyhedron with 20 equilateral triangular faces and 12 vertices (figure 16.2h, j–l). It is one of nature's favorite shapes. The **icosahedral capsid** is the most efficient way to enclose a space. A few genes, sometimes only one, can code for proteins that self-assemble to form the capsid. In this way a small number of genes can specify a large three-dimensional structure.

When icosahedral viruses are negatively stained and viewed in the transmission electron microscope, a complex structure is revealed (**figure 16.5**). The capsids are constructed from ring- or knob-shaped units called **capsomers,** each usually made of five

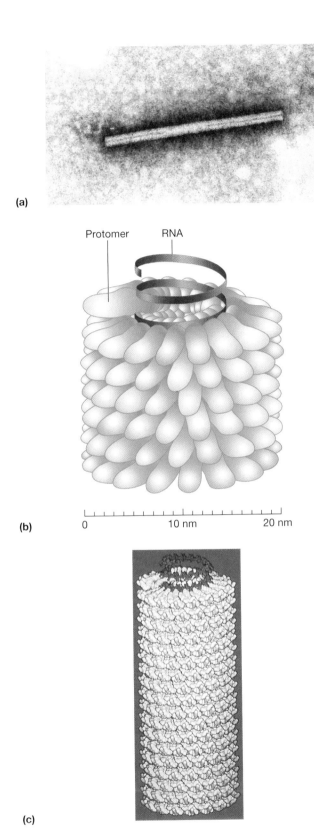

(a)

(b)

Protomer   RNA

|_I_I_I_I_I_I_I_I_I_I_I_|
0   10 nm   20 nm

(c)

Figure 16.3   Tobacco Mosaic Virus Structure.   **(a)** An electron micrograph of the negatively stained helical capsid (×400,000). **(b)** Illustration of TMV structure. Note that the nucleocapsid is composed of a helical array of protomers with the RNA spiraling on the inside. **(c)** A model of TMV.

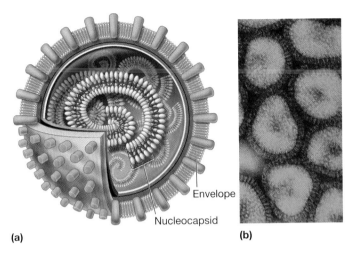

Envelope

Nucleocapsid

(a)   (b)

Figure 16.4   Influenza Virus.   Influenza virus is an enveloped virus with a helical nucleocapsid. **(a)** Schematic view. Influenza viruses have segmented genomes consisting of 7 to 8 different RNA molecules. Each is coated by capsid proteins. **(b)** Because there are 7 to 8 flexible nucleocapsids enclosed by an envelope, the virions are pleomorphic. Electron micrograph (×350,000).

or six protomers. **Pentamers (pentons)** have five subunits; **hexamers (hexons)** possess six. Pentamers are usually at the vertices of the icosahedron, whereas hexamers generally form its edges and triangular faces (**figure 16.6**). The icosahedron in figure 16.6 is constructed of 42 capsomers; larger icosahedra are made if more hexamers are used to form the edges and faces (e.g., adenoviruses have a capsid with 252 capsomers as shown in figure 16.5c,d). In some RNA viruses, both the pentamers and hexamers of a capsid are constructed with only one type of subunit. In other viruses, pentamers are composed of different proteins than are the hexamers.

The self-assembly of capsids is a remarkable process that is not fully understood. Enzymatic activity is not required to link protomers together. However, noncapsid proteins may be involved. They usually provide a scaffolding upon which the protomers are assembled.

Although most icosahedral capsids appear to contain both pentamers and hexamers, simian virus 40 (SV-40), a small, double-stranded DNA virus, has only pentamers (**figure 16.7a**). The virus is constructed of 72 cylindrical pentamers with hollow centers. Five flexible arms extend from the edge of each pentamer toward neighboring pentamers (figure 16.7b,c). The arms of adjacent pentamers twist around each other and act as ropes that tie the pentamers together.

Viruses with Capsids of Complex Symmetry

Although most viruses have either icosahedral or helical capsids, many viruses do not fit into either category. The poxviruses and large bacteriophages are two important examples.

The poxviruses are the largest of the animal viruses (about 400 × 240 × 200 nm in size) and can even be seen with a phase-contrast microscope or in stained preparations. They possess an

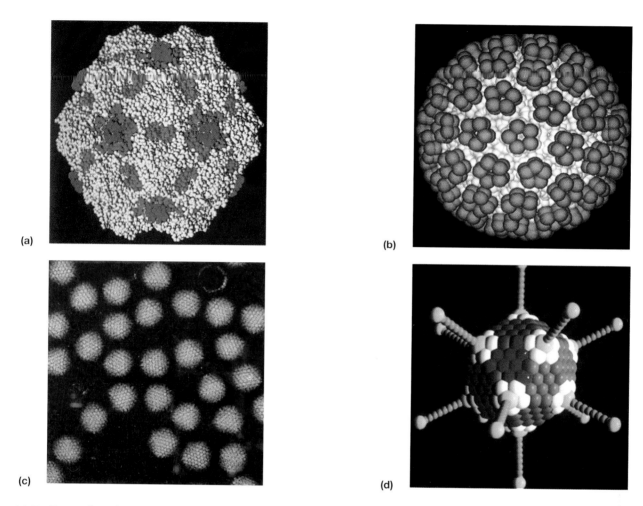

Figure 16.5 Examples of Icosahedral Capsids. (a) Canine parvovirus model, 12 capsomers. **(b)** Computer-simulated image of the polyomavirus (72 capsomers) that causes a rare demyelinating disease of the central nervous system. **(c)** Adenovirus, 252 capsomers (×171,000). **(d)** Computer-simulated model of adenovirus.

exceptionally complex internal structure with an ovoid- to brick-shaped exterior. **Figure 16.8** shows the morphology of vaccinia virus, a poxvirus. The double-stranded DNA is associated with proteins and contained in the nucleoid, a central structure shaped like a biconcave disk and surrounded by a membrane. Two elliptical or lateral bodies lie between the nucleoid and its outer envelope, a membrane and a thick layer covered by an array of tubules or fibers.

Some large bacteriophages are even more elaborate than the poxviruses. The T2, T4, and T6 phages (T-even phages) that infect *Escherichia coli* are said to have **binal symmetry** because they have a head that resembles an icosahedron and a tail that is helical. The icosahedral head is elongated by one or two rows of hexamers in the middle and contains the DNA genome (**figure 16.9**). The tail is composed of a collar joining it to the head, a central hollow tube, a sheath surrounding the tube, and a complex baseplate. The sheath is made of 144 copies of the gp18 protein arranged in 24 rings, each containing six copies. In T-even phages, the baseplate is hexagonal and has a pin and a jointed tail fiber at each corner.

There is considerable variation in structure among the large bacteriophages, even those infecting a single host. In contrast with the T-even phages, many other coliphages (phages that infect *E. coli*) have true icosahedral heads. T1, T5, and lambda phages have sheathless tails that lack a baseplate and terminate in rudimentary tail fibers. Coliphages T3 and T7 have short, noncontractile tails without tail fibers. Bacteriophages are discussed in more detail in chapter 17.

Viral Envelopes and Enzymes

Many animal viruses, some plant viruses, and at least one bacterial virus are bounded by an outer membranous layer called an **envelope** (**figure 16.10**). Animal virus envelopes usually arise from host cell nuclear or plasma membranes; their lipids and carbohydrates are normal host constituents. In contrast, envelope proteins are coded for by virus genes and may even project from the envelope surface as **spikes,** which are also called **peplomers.** In many cases, these spikes are involved in virus attachment to the host cell surface. Because they differ among viruses, they also

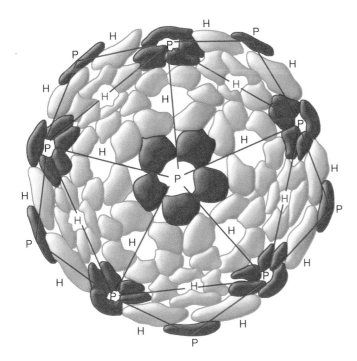

Figure 16.6   The Structure of an Icosahedral Capsid Formed from a Single Type of Protomer. The protomers associate to form either pentons (P), shown in red, or hexons (H), shown in gold. The blue lines define the triangular faces of the icosahedron. Notice that pentons are located at the vertices and that the hexons form the edges and faces of the icosahedron. This capsid contains 42 capsomers.

can be used to identify some viruses. The envelope is a flexible, membranous structure, so enveloped viruses frequently have a somewhat variable shape and are called pleomorphic. However, the envelopes of viruses like the bullet-shaped rabies virus are firmly attached to the underlying nucleocapsid and endow the virion with a constant, characteristic shape (figure 16.10b). In some viruses the envelope is disrupted by solvents like ether to such an extent that lipid-mediated activities are blocked or envelope proteins are denatured and rendered inactive. The virus is then said to be "ether sensitive."

Influenza virus (figure 16.10a) is a well-studied example of an enveloped virus. Spikes project about 10 nm from the surface at 7 to 8 nm intervals. Some spikes possess the enzyme neuraminidase, which functions in the release of mature virions from the host cell. Other spikes have hemagglutinin proteins, so named because they can bind the virions to red blood cell membranes and cause the red blood cells to clump together (agglutinate). This is called hemagglutination (*see figure 35.11*). Hemagglutinins participate in virion attachment to host cells. Proteins, like the spike proteins that are exposed on the outer envelope surface, are generally glycoproteins—that is, the proteins have carbohydrate attached to them. A nonglycosylated protein, the M or matrix protein, is found on the inner surface of the envelope and helps stabilize it.

(a)

(b)

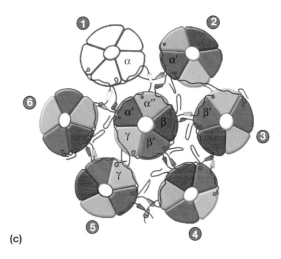
(c)

Figure 16.7   An Icosahedral Capsid Constructed of Pentamers. (a) The simian virus 40 capsid. The 12 pentamers at the icosahedron vertices are in white. The nonvertex pentamers are shown with each polypeptide chain in a different color. (b) A pentamer with extended arms. (c) A schematic diagram of the surface structure depicted in part a. The body of each pentamer is represented by a five-petaled flower design. Each arm is shown as a line or a line and cylinder (α-helix) with the same color as the rest of its protomer. The outer protomers are numbered clockwise beginning with the one at the vertex.

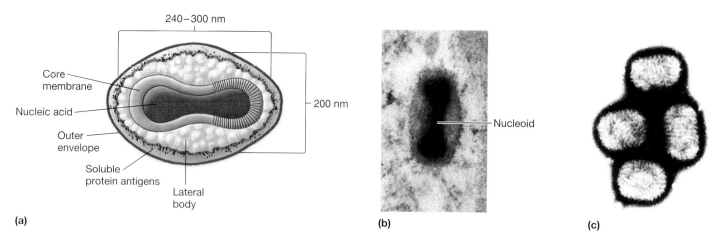

Figure 16.8 Vaccinia Virus Morphology.   (a) Diagram of vaccinia structure. **(b)** Micrograph of the virion clearly showing the nucleoid (×200,000). **(c)** Vaccinia surface structure. An electron micrograph of four virions showing the thick array of surface fibers (×150,000).

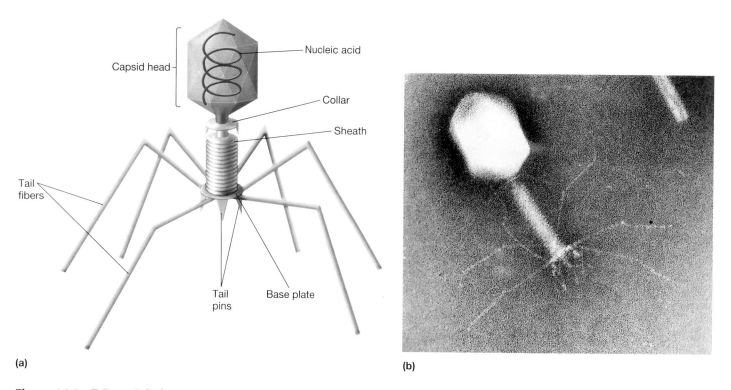

Figure 16.9   T-Even Coliphages.   (a) The structure of the T4 bacteriophage. **(b)** The micrograph shows the phage before injection of its DNA.

It was originally thought that all virions lacked enzymes. However, as just illustrated in the discussion of influenza virus, this is not the case. In some instances, enzymes are associated with the envelope or capsid (e.g., influenza neuraminidase), but most viral enzymes are located within the capsid. Many of these are involved in nucleic acid replication. For example, the influenza virus uses RNA as its genetic material and carries an enzyme that synthesizes RNA using an RNA template. Such enzymes are called RNA-dependent RNA polymerases. Thus although viruses lack true metabolism and cannot reproduce independently of living cells, they may carry one or more enzymes essential to the completion of their life cycles.

Viral Genomes

Viruses are exceptionally flexible with respect to the nature of their genomes. They employ all four possible nucleic acid types: single-stranded DNA, double-stranded DNA, single-stranded

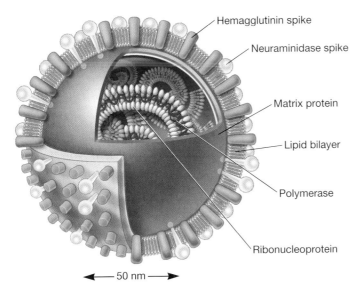

Hemagglutinin spike

Neuraminidase spike

Matrix protein

Lipid bilayer

Polymerase

Ribonucleoprotein

← 50 nm →

(a) Influenza virus

(b) Rabies virus

Envelope spikes

Envelope

Core

(c) HIV

(e) Semliki Forest virus

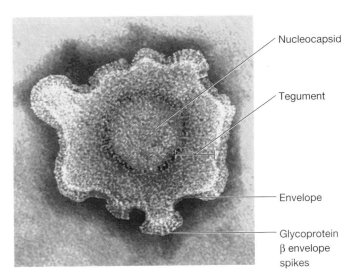

Nucleocapsid

Tegument

Envelope

Glycoprotein β envelope spikes

(d) Herpesvirus

Figure 16.10   Examples of Enveloped Viruses.   (a) Diagram of the influenza virion. **(b)** Negatively stained rabies virus. **(c)** Human immunodeficiency viruses. **(d)** Herpesviruses. **(e)** Computer image of the Semliki Forest virus, a virus that occasionally causes encephalitis in humans. Images (b), (c), and (d) are artificially colorized.

Table 16.1	Types of Viral Nucleic Acids	
Nucleic Acid Type	**Nucleic Acid Structure**	**Virus Examples**
DNA		
Single Stranded	Linear, single-stranded DNA	Parvoviruses
	Circular, single-stranded DNA	φX174, M13, fd phages
Double Stranded	Linear, double-stranded DNA	Herpesviruses (herpes simplex viruses, cytomegalovirus, Epstein-Barr virus), adenoviruses, T coliphages, lambda phage, and other bacteriophages
	Linear, double-stranded DNA with single chain breaks	T5 coliphage
	Double-stranded DNA with cross-linked ends	Vaccinia, smallpox viruses
	Closed, circular, double-stranded DNA	Polyomaviruses (SV-40), papillomaviruses, PM2 phage, cauliflower mosaic virus
RNA		
Single-Stranded	Linear, single-stranded, positive-strand RNA	Picornaviruses (polio, rhinoviruses), togaviruses, RNA bacteriophages, TMV, and most plant viruses
	Linear, single-stranded, negative-strand RNA	Rhabdoviruses (rabies), paramyxoviruses (mumps, measles)
	Linear, single-stranded, segmented, positive-strand RNA	Brome mosaic virus (individual segments in separate virions)
	Linear, single-stranded, diploid (two identical single strands), positive-strand RNA	Retroviruses (Rous sarcoma virus, human immunodeficiency virus)
	Linear, single-stranded, segmented, negative-strand RNA	Paramyxoviruses, orthomyxoviruses (influenza)
Double-Stranded	Linear, double-stranded, segmented RNA	Reoviruses, wound-tumor virus of plants, cytoplasmic polyhedrosis virus of insects, phage φ6, many mycoviruses

Modified from S. E. Luria, et al., *General Virology*, 3d edition, 1983. John Wiley & Sons, Inc., New York, NY.

RNA, and double-stranded RNA. All four types are found in animal viruses. Most plant viruses have single-stranded RNA genomes, and most bacterial viruses contain double-stranded DNA. **Table 16.1** summarizes many variations seen in viral nucleic acids. The size of viral genetic material also varies greatly. The smallest genomes (those of the MS2 and Qβ viruses) are around 4,000 nucleotides, just large enough to code for three or four proteins. MS2, Qβ, and some other viruses even save space by using overlapping genes. At the other extreme, T-even bacteriophages, herpesvirus, and vaccinia virus have genomes of 1.0 to 2.0×10^5 nucleotides and may be able to direct the synthesis of over 100 proteins. In the following paragraphs the nature of each nucleic acid type is briefly summarized. *Gene structure (section 11.5)*

Most DNA viruses use double-stranded DNA (dsDNA) as their genetic material. However, some have single-stranded DNA (ssDNA) genomes. In both cases, the genomes can be either linear or circular (**figure 16.11**). Some DNA genomes can switch from one form to the other. For instance, the *E. coli* phage lambda has a genome that is linear in the capsid, but is converted into a

circular form once the genome enters the host cell. Another important characteristic of DNA viruses is that their genomes often contain unusual nitrogenous bases. For example, the T-even phages of *E. coli* have 5-hydroxymethylcytosine (*see figure 17.9*) instead of cytosine, and the hydroxymethyl group is often modified by attachment of a glucose moiety.

RNA viruses also can be either double-stranded (dsRNA) or single-stranded (ssRNA). Although relatively few RNA viruses have dsRNA genomes, dsRNA viruses are known to infect animals, plants, fungi, and at least one bacterial species. More common are the viruses with ssRNA genomes. Some ssRNA genomes have a base sequence that is identical to that of viral mRNA, in which case the genomic RNA strand is called the **plus strand** or **positive strand.** In fact, plus strand RNAs can direct protein synthesis immediately after entering the cell. However, other viral RNA genomes are complementary rather than identical to viral mRNA, and are called **minus** or **negative strands.** Polio, tobacco mosaic, brome mosaic, and Rous sarcoma viruses are all positive strand RNA viruses; rabies, mumps, measles, and influenza

Figure 16.11 **Circular Phage DNA.** The closed circular DNA of the phage PM2 (×93,000). Note both the relaxed and highly twisted or supercoiled forms.

viruses are examples of negative strand RNA viruses. Many RNA viruses have **segmented genomes**—that is, the genome consists of more than one RNA strand or segment. In many cases, each segment codes for one protein. Usually all segments are enclosed in the same capsid even though some virus genomes may be composed of as many as 10 to 12 segments. However, it is not necessary that all segments be located in the same virion for successful reproduction. The genome of brome mosaic virus, a virus that infects certain grass species, is composed of four segments distributed among three different virus particles. All three of the largest segments are required for infectivity. Despite this complex and seemingly inefficient arrangement, the different brome mosaic virions manage to successfully infect the same host.

Plus strand viral RNA often resembles mRNA in more than the equivalence of its nucleotide sequence. Just as eucaryotic mRNA usually has a 5′ cap of 7-methylguanosine, many plant and animal viral RNA genomes are capped. In addition, most plus strand RNA animal viruses also have a poly-A sequence at the 3′ end of their genome, and thus closely resemble eucaryotic mRNA with respect to the structure of both ends. Strangely enough, a number of single-stranded plant viral RNAs have 3′ ends that resemble eucaryotic transfer RNA. Indeed, the genome of tobacco mosaic virus actually accepts amino acids. Transcription (section 11.6); Translation (section 11.8)

1. Define the following terms: nucleocapsid, capsid, icosahedral capsid, helical capsid, complex virus, binal symmetry, protomer, capsomer, pentamer or penton, and hexamer or hexon. How do pentamers and hexamers associate to form a complete icosahedron; what determines helical capsid length and diameter?
2. What is an envelope? What are spikes (peplomers)? Why are some enveloped viruses pleomorphic? Give two functions spikes might serve in the virus life cycle, and the proteins that the influenza virus uses in these processes.

3. All four nucleic acid forms can serve as virus genomes. Describe each, the types of virion possessing it, and any distinctive physical characteristics the nucleic acid can have. What are the following: plus strand, minus strand, and segmented genome?
4. What advantage would an RNA virus gain by having its genome resemble eucaryotic mRNA?

16.4 VIRUS REPRODUCTION[1]

The differences in virus structure and viral genomes have important implications for the mechanism a virus uses to reproduce within its host cell. Indeed, even among viruses with similar structures and genomes, each can exhibit unique life cycles. However, despite these differences a general pattern of virus reproduction can be discerned. Because viruses need a host cell in which to reproduce, the first step in the life cycle of a virus is attachment to a host (**figure 16.12**). This is followed by entry of either the nucleocapsid or the viral nucleic acid into the host. If the nucleocapsid enters, uncoating of the genome usually occurs before further steps can occur. Once free in the cytoplasm, genes encoded by the viral genome are expressed. That is, the viral genes are transcribed and translated. This allows the virus to take control of the host cell's biosynthetic machinery so that new virions can be made. The viral genome is then replicated and viral proteins are synthesized. New virions are constructed by self-assembly of coat proteins with the nucleic acids, and finally the mature virions are released from the host. As discussed in chapters 17 and 18, the details of virus reproduction can vary dramatically. For instance, some viruses are released by lysing their hosts, whereas others bud from the host without lysis.

16.5 THE CULTIVATION OF VIRUSES

Because they are unable to reproduce independent of living cells, viruses cannot be cultured in the same way as procaryotic and eucaryotic microorganisms. For many years researchers have cultivated animal viruses by inoculating suitable host animals or embryonated eggs—fertilized chicken eggs incubated about 6 to 8 days after laying (**figure 16.13**). To prepare the egg for cultivation of viruses, the shell surface is first disinfected with iodine and penetrated with a small sterile drill. After inoculation, the drill hole is sealed with gelatin and the egg incubated. Some viruses reproduce only in certain parts of the embryo; consequently they must be injected into the proper region. For example, the myxoma virus grows well on the chorioallantoic membrane, whereas the mumps

[1]Virologists usually refer to the production of new virus particles within a host cell as virus replication. Indeed, many virologists state that viruses do not reproduce, they replicate. However, to avoid confusion about the meaning of the term replication, we will use the term reproduction when discussing the production of new virions, and use the term replication when discussing the synthesis of new copies of viral genomes.

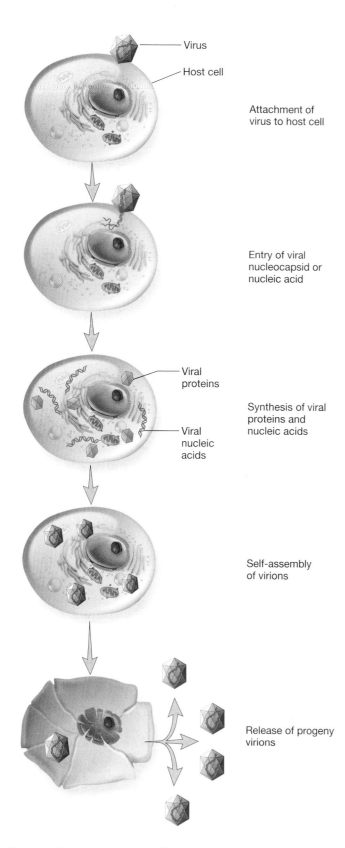

Figure 16.12 Generalized Illustration of Virus Reproduction. There is great variation in the details of virus reproduction for individual virus species.

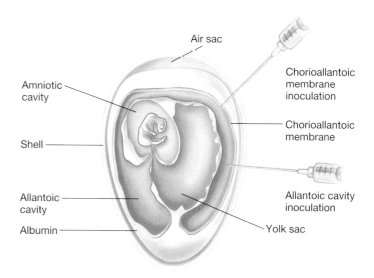

Figure 16.13 Cultivation of Viruses in an Embryonated Egg. Two sites that are often used to grow animal viruses are the chorioallantoic membrane and the allantoic cavity. The diagram shows a 9-day chicken embryo.

virus grows best in the allantoic cavity. The infection may produce a local tissue lesion known as a pock, whose appearance often is characteristic of the virus.

More recently animal viruses have been grown in tissue (cell) culture on monolayers of animal cells. This technique is made possible by the development of growth media for animal cells and by the use of antimicrobial agents that prevent bacterial and fungal contamination. Viruses are added to a layer of animal cells in a specially prepared petri dish and allowed time to attach to the cells. The cells are then covered with a thin layer of agar to limit virion spread so that only adjacent cells are infected by newly produced virions. As a result, localized areas of cellular destruction and lysis called **plaques** often are formed (**figure 16.14**) and may be detected if stained with dyes, such as neutral red or trypan blue, that can distinguish living from dead cells. Viral growth does not always result in the lysis of cells to form a plaque. Animal viruses, in particular, can cause microscopic or macroscopic degenerative changes or abnormalities in host cells and in tissues. These are called **cytopathic effects** (**figure 16.15**). Cytopathic effects may be lethal, but plaque formation from cell lysis does not always occur.

Bacterial and archaeal viruses are cultivated in either broth or agar cultures of young, actively growing cells. In some infected cultures, so many host cells are destroyed that turbid cultures clear rapidly because of cell lysis. Agar cultures are prepared by mixing viruses with cool, liquid agar and a suitable culture of host cells. The mixture is quickly poured into a petri dish containing a bottom layer of sterile agar. After hardening, cells in the layer of top agar grow and reproduce, forming a continuous, opaque layer or "lawn." Wherever a virion comes to rest in the top agar, the virus infects an adjacent cell and reproduces. Eventually, lysis of

Figure 16.14 Virus Plaques. **(a)** Poliovirus plaques in a monkey kidney cell culture. **(b)** Plaques formed by bacteriophages growing on a lawn of bacterial cells.

Figure 16.15 Cytopathic Effects of Viruses. **(a)** A monolayer of normal fibroblast cells from fetal tonsils. **(b)** Cytopathic effects caused by infection of fetal tonsil fibroblasts with adenovirus. **(c)** Cytopathic effects caused by infection of fetal tonsil fibroblasts with herpes simplex virus.

the cells generates a plaque or clearing in the lawn (figure 16.14*b* and **figure 16.16**). As shown in figure 16.16, plaque appearance often is characteristic of the virus being cultivated.

Plant viruses are cultivated in a variety of ways. Plant tissue cultures, cultures of separated cells, or cultures of protoplasts (cells lacking cell walls) may be used. Viruses also can be grown in whole plants. Leaves are mechanically inoculated when rubbed with a mixture of viruses and an abrasive. When the cell walls are broken by the abrasive, the viruses directly contact the plasma membrane and infect the exposed host cells. (In nature, the role of the abrasive is frequently filled by insects that suck or crush plant leaves and thus transmit viruses.) A localized **necrotic lesion** often develops due to the rapid death of cells in the infected area (**figure 16.17**). Even when lesions do not occur, the infected plant may show symptoms such as changes in pigmentation or leaf shape. Some plant viruses can be transmitted only if a diseased part is grafted onto a healthy plant.

16.6 Virus Purification and Assays

Virologists must be able to purify viruses and accurately determine their concentrations in order to study virus structure, reproduction, and other aspects of their biology. These methods are so important that the growth of virology as a modern discipline depended on their development.

Virus Purification

Purification makes use of several virus properties. Virions are very large relative to proteins, are often more stable than normal cell components, and have surface proteins. Because of these characteristics, many techniques useful for the isolation of proteins and organelles can be employed to isolate viruses. Four of the most widely used approaches are (1) differential and density gradient centrifugation, (2) precipitation of viruses,

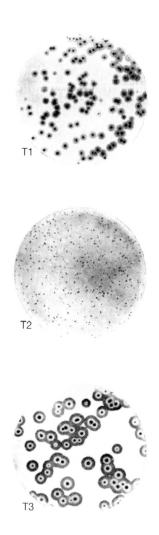

T1

T2

T3

Figure 16.16 Phage Plaques. Plaques produced on a lawn of *E. coli* by some of the T coliphages (T1, T2, and T3 phages). Note the large differences in plaque appearance. The photographs are about 1/3 full size.

(a)

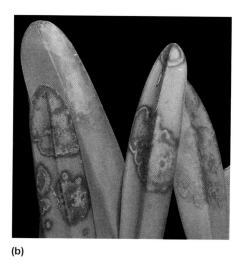

(b)

Figure 16.17 Necrotic Lesions on Plant Leaves. (a) Tobacco mosaic virus on *Nicotiana glutinosa*. (b) Tobacco mosaic virus infection of an orchid showing leaf color changes.

(3) denaturation of contaminants, and (4) enzymatic digestion of host cell constituents.

Differential and density gradient centrifugation often are used in the initial purification steps to separate virus particles from host cells. The process begins with host cells in later stages of infection because they contain mature virions. Infected cells are first disrupted in a buffer to produce an aqueous suspension or homogenate consisting of cell components and viruses. Viruses can then be isolated by **differential centrifugation,** the centrifugation of a suspension at various speeds to separate particles of different sizes (**figure 16.18**). Usually the homogenate is first centrifuged at high speed to sediment viruses and other large cellular particles. The supernatant, which contains the homogenate's soluble molecules, is discarded. The pellet is next resuspended and centrifuged at a low speed to remove substances heavier than viruses. Finally, higher speed centrifugation sediments the viruses. This process may be repeated to purify the virus particles further.

Additional purification of a virus preparation can be achieved by **gradient centrifugation** (**figure 16.19**). A sucrose solution is poured into a centrifuge tube so that its concentration smoothly and linearly increases from the top to the bottom of the tube. The virus preparation, often produced by differential centrifugation, is layered on top of the gradient and centrifuged. As shown in figure 16.19a, the particles settle under centrifugal force until they come to rest at the level where the virus and sucrose densities are equal (isopycnic gradient centrifugation). Viruses can be separated from other particles based on very small differences in density. Gradients also can separate viruses based on differences in their sedimentation rate (rate zonal gradient centrifugation). When this is done, particles are separated on the basis of both size and density; usually the largest virus will move most rapidly down the gradient. Figure 16.19b shows that viruses differ from one another and cell components with respect to either density (grams per milliliter) or sedimentation coefficient(s). Thus these

Figure 16.18   The Use of Differential Centrifugation to Purify a Virus.   At the beginning the centrifuge tube contains homogenate and icosahedral viruses (in green). First, the viruses and heavier cell organelles are removed from smaller molecules. After resuspension, the mixture is centrifuged just fast enough to sediment cell organelles while leaving the smaller virus particles in suspension; the purified viruses are then collected. This process can be repeated several times to further purify the virions.

(a)

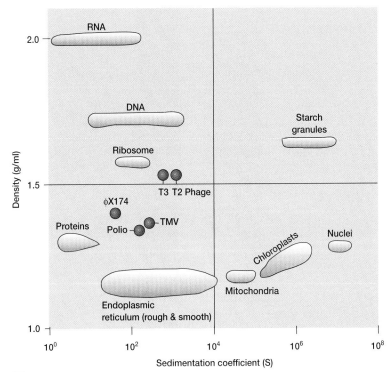

(b)

Figure 16.19   Gradient Centrifugation.
(a) A linear sucrose gradient is prepared, *1,* and the particle mixture is layered on top, *2* and *3.* Centrifugation, *4,* separates the particles on the basis of their density and sedimentation coefficient, (the arrows in the centrifuge tubes indicate the direction of centrifugal force). *5.* In isopycnic gradient centrifugation, the bottom of the gradient is denser than any particle, and each particle comes to rest at a point in the gradient equal to its density. Rate zonal centrifugation separates particles based on their sedimentation coefficient, a function of both size and density, because the bottom of the gradient is less dense than the densest particles and centrifugation is carried out for a shorter time so that particles do not come to rest. The largest, most dense particles travel fastest. **(b)** The densities and sedimentation coefficients of representative viruses (shown in color) and other biological substances.

two types of gradient centrifugation are very effective in virus purification.

Although centrifugation procedures remove much cellular material, some cell components can remain in the virus preparation. Viruses can be separated from any remaining cellular contaminants by precipitation, denaturing, or enzymatic degradation of the contaminants. Many precipitation procedures use ammonium sulfate to precipitate the cellular contaminants. Initially, ammonium sulfate is added to a concentration just below that needed to precipitate the virus particles. Thus many cell components precipitate while the virus remains in solution. After any precipitated contaminants are removed, more ammonium sulfate is added and the precipitated viruses are collected by centrifugation. Viruses sensitive to ammonium sulfate often are purified by precipitation with polyethylene glycol.

Because viruses frequently are less sensitive to denaturing conditions than many cell components, exposure of a virus preparation to heat or a change in pH can be used in the final steps of virus purification. Furthermore, some viruses also tolerate treatment with organic solvents like butanol and chloroform. Thus solvent treatment can be used to both denature protein contaminants and extract any lipids in the preparation. The solvent is thoroughly mixed with the virus preparation, then allowed to stand and separate into organic and aqueous layers. The unaltered virus remains suspended in the aqueous phase while lipids dissolve in the organic phase. Substances denatured by organic solvents collect at the interface between the aqueous and organic phases.

One of the last steps used to purify viruses is enzymatic degradation of contaminants. This often is used to remove any remaining cellular proteins and nucleic acids in the virus preparation. Although viruses are composed of a protein coat surrounding a nucleic acid, they usually are more resistant to attack by nucleases and proteases than are free nucleic acids and proteins. For example, ribonuclease and trypsin often degrade cellular ribonucleic acids and proteins while leaving virions unaltered.

Virus Assays

The quantity of viruses in a sample can be determined either directly by counting particle numbers or indirectly by measurement of an observable effect of the virus. The values obtained by the two approaches often do not correlate closely; however, both are of value.

Virions can be counted directly with the electron microscope. In one procedure the virus-containing sample is mixed with a known concentration of small latex beads and sprayed on a coated specimen grid. The beads and virions are counted; the virus concentration is calculated from these counts and from the bead concentration (**figure 16.20**). This technique often works well with concentrated preparations of viruses of known morphology. Viruses can be concentrated by centrifugation before counting if the preparation is too dilute. However, if the beads and viruses are not evenly distributed (as sometimes happens), the final count will be inaccurate.

An indirect method of counting virus particles is the **hemagglutination assay.** Many viruses can bind to the surface of red blood cells (*see figure 35.11*). If the ratio of viruses to cells is

Figure 16.20 Tobacco Mosaic Virus. A tobacco mosaic virus preparation viewed in the transmission electron microscope. Latex beads 264 nm in diameter (white spheres) have been added.

large enough, virus particles join the red blood cells together—that is, they agglutinate, forming a network that settles out of suspension. In practice, red blood cells are mixed with a series of virus dilutions and each mixture is examined. The hemagglutination titer is the highest dilution of virus (or the reciprocal of the dilution) that still causes hemagglutination. This assay is an accurate, rapid method for determining the relative quantity of viruses such as the influenza virus. If the actual number of viruses needed to cause hemagglutination is determined by another technique, the assay can be used to ascertain the number of virions present in a sample.

A variety of indirect assays determine virus numbers in terms of infectivity, and many of these are based on the same techniques used for virus cultivation. For example, in the **plaque assay** several dilutions of viruses are plated out with appropriate host cells. When the number of viruses plated are much lower than the number of host cells available for infection and when the viruses are distributed evenly, each plaque in a layer of host cells is assumed to have arisen from the reproduction of a single virion. Therefore a count of the plaques produced at a particular dilution will give the number of infectious virions, called **plaque-forming units (PFU),** and the concentration of infectious units in the original sample can be easily calculated. For instance, suppose that 0.10 ml of a 10^{-6} dilution of the virus preparation yields 75 plaques. The original concentration of plaque-forming units is

$$PFU/ml = (75\ PFU/0.10\ ml)(10^6) = 7.5 \times 10^8.$$

Viruses producing different plaque morphology types on the same plate may be counted separately. Although the number of PFU does not equal the number of virions, their ratios are proportional: a preparation with twice as many viruses will have twice the plaque-forming units.

The same approach employed in the plaque assay may be used with embryos and plants. Chicken embryos can be inoculated with a diluted preparation or plant leaves rubbed with a mixture of diluted virus and abrasive. The number of pocks on embryonic membranes or necrotic lesions on leaves is multiplied by the dilution factor and divided by the inoculum volume to obtain the concentration of infectious units.

When biological effects are not readily quantified in these ways, the amount of virus required to cause disease or death can be determined by the endpoint method. Organisms or cell cultures are inoculated with serial dilutions of a virus suspension. The results are used to find the endpoint dilution at which 50% of the host cells or organisms are killed (**figure 16.21**). The **lethal dose (LD_{50})** is the dilution that contains a dose large enough to destroy 50% of the host cells or organisms. In a similar sense, the **infectious dose (ID_{50})** is the dose that, when given to a number of hosts, causes an infection of 50% of the hosts under the conditions employed.

1. Discuss the ways that viruses can be cultivated. Define the terms pock, plaque, cytopathic effect, and necrotic lesion.
2. Give the four major approaches by which viruses may be purified, and describe how each works. Distinguish between differential and density gradient centrifugation in terms of how they are carried out.
3. How can one find the virus concentration, both directly and indirectly, by particle counts and measurement of infectious unit concentration? Define plaque-forming units, lethal dose, and infectious dose.

16.7 PRINCIPLES OF VIRUS TAXONOMY

The classification of viruses is in a much less satisfactory state than that of cellular microorganisms. In part, this is due to a lack of knowledge of their origin and evolutionary history (**Microbial Tidbits 16.2**). In 1971, the International Committee for Taxonomy of Viruses (ICTV) developed a uniform classification system. Since

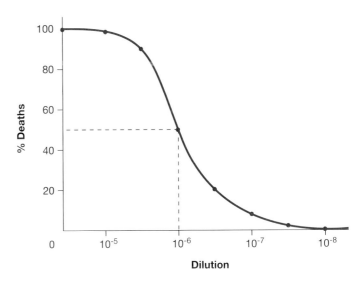

Figure 16.21   A Hypothetical Dose-Response Curve.   The LD_{50} is indicated by the dashed line.

then the number of viruses and taxonomic categories has continued to expand. In its eighth report, the ICTV described almost 2000 virus species and placed them in 3 orders, 73 families, 9 subfamilies, and 287 genera (**table 16.2**). The committee places greatest weight on specific properties to define families: nucleic acid type, nucleic acid strandedness, the sense (positive or negative) of ssRNA genomes, presence or absence of an envelope, symmetry of the capsid, and dimensions of the virion and capsid. Virus order

Microbial Tidbits

16.2   The Origin of Viruses

The origin and subsequent evolution of viruses are shrouded in mystery, in part because of the lack of a fossil record. However, recent advances in the understanding of virus structure and reproduction have made possible more informed speculation on virus origins. At present there are two major hypotheses entertained by virologists. It has been proposed that at least some of the more complex enveloped viruses, such as the poxviruses and herpesviruses, arose from small cells, probably procaryotic, that parasitized larger, more complex cells. These parasitic cells became ever simpler and more dependent on their hosts, much like multicellular parasites have done, in a process known as retrograde evolution. There are several problems with this hypothesis. Viruses are radically different from procaryotes, and it is difficult to envision the mechanisms by which such a transformation might have occurred or the selective pressures leading to it. In addition, one would expect to find some forms intermediate between procaryotes and at least the more complex enveloped viruses, but such forms have not been detected.

The second hypothesis is that viruses represent cellular nucleic acids that have become partially independent of the cell. Possibly a few mutations could convert nucleic acids, which are only synthe-

sized at specific times, into infectious nucleic acids whose replication could not be controlled. This conjecture is supported by the observation that the nucleic acids of retroviruses (*see section 18.2*) and a number of other virions contain sequences quite similar to those of normal cells, plasmids, and transposons (*see chapter 13*). The small, infectious RNAs called viroids (*see section 18.9*) have base sequences complementary to transposons (*see section 13.5*), the regions around the boundary of mRNA introns, and portions of host DNA. This has led to speculation that they have arisen from introns or transposons. It has been proposed that cellular proteins spontaneously assembled into icosahedra around infectious nucleic acids to produce primitive virions.

It is possible that viruses have arisen by way of both mechanisms. Because viruses differ so greatly from one another, it seems likely that they have originated independently many times during the course of evolution. Many viruses have evolved from other viruses just as cellular organisms have arisen from specific predecessors. The question of virus origins is complex and quite speculative; future progress in understanding virus structure and reproduction may clarify this question.

names end in *virales;* virus family names in *viridae;* subfamily names, in *virinae;* and genus (and species) names, in *virus.* An example of this nomenclature scheme is shown in **figure 16.22.**

Although the ICTV committee reports are the official authority on viral taxonomy, many virologists find it useful to use an alternative classification scheme devised by Nobel laureate David Baltimore. The Baltimore system complements the ICTV system but focuses on the genome of the virus and the process used to synthesize viral mRNA. Baltimore's original system recognized six groups of viruses. Since then the system has been expanded to include seven groups. This was done in part by considering genome replication as well as mRNA synthesis in the classification scheme (**table 16.3**). As discussed in chapters 17 and 18, such a system helps virologists (and microbiology students) simplify the vast array of viral life cycles into a relatively small number of basic types.

Table 16.2	**Some Common Virus Groups and Their Characteristics**							
ICTV Taxon (Baltimore System Group)[a]	**Genome Size (kbp or kb)**	**Nucleic Acid**	**Strandedness**	**Capsid Symmetry[b]**	**Number of Capsomers**	**Presence of Envelope**	**Size of Capsid (nm)[c]**	**Host Range[d]**
Picornaviridae (IV)	7–8	RNA	Single	I	32	−	22–30	A
Togaviridae (IV)	10–12		Single	I	32	+	40–70(e)	A
Retroviridae (VI)	7–12		Single	I?		I	100(e)	A
Orthomyxoviridae (V)	10–15		Single	H		+	9(h), 80–120(e)	A
Paramyxoviridae (V)	15		Single	H		+	18(h), 125–250(e)	A
Coronaviridae (IV)	27–31		Single	H		+	14–16(h), 80–160(e)	A
Rhabdoviridae (V)	11–15		Single	H		+	18(h), 70–80 × 130–240 (bullet shaped)	A
Bromoviridae (IV)	8–9		Single	I,B		−	26–35; 18–26 × 30–85	P
Tobamovirus (IV)	7		Single	H		−	18 × 300	P
Leviviridae [Qβ] (IV)	3–4		Single	I	32	−	26–27	B
Reoviridae (III)	19–32	RNA	Double	I	92	−	70–80	A,P
Cystoviridae (III)	13		Double	I		+	100(e)	B
Parvoviridae (II)	4–6	DNA	Single	I	12	−	20–25	A
Geminiviridae (II)	3–6		Single	I		−	18 × 30 (paired particles)	P
Microviridae (II)	4–6		Single	I		−	25–35	B
Inoviridae (II)	7–9		Single	H		−	6 × 900–1,900	B
Polyomaviridae (I)	5	DNA	Double	I	72	−	40	A
Papillomaviridae (I)	7–8		Double	I	72	−	55	A
Adenoviridae (I)	28–45		Double	I	252	−	60–90	A
Iridoviridae (I)	140–383		Double	I		−	130–180	A
Herpesviridae (I)	125–240		Double	I	162	+	100, 180–200(e)	A
Poxviridae (I)	130–375		Double	C		+	200–260 × 250–290(e)	A
Baculoviridae (I)	80–180		Double	H		+	40 × 300(e)	A
Hepadnaviridae (VII)	3		Double	C	42	+	28 (core), 42(e)	A
Caulimoviridae (I)	8		Double	I,B		−	50; 30 × 60–900	P
Corticoviridae (I)	9		Double	I		−	60	B
Myoviridae (I)	39–169		Double	Bi		−	80 × 110, 110[e]	B, Arch
Lipothrixviridae (I)	16		Double	H		+	38 × 410	Arch

[a]ICTV = International Committee on Virus Taxonomy. The ICTV and Baltimore Clarification Systems are discussed in section 16.7.

[b]Types of symmetry: I, icosahedral; H, helical; C, complex; Bi, binal; B, bacilliform.

[c]Diameter of helical capsid (h); diameter of enveloped virion (e).

[d]Host range: A, animal; P, plant; B, bacterium; Arch, archaeon.

[e]The first number is the head diameter; the second number, the tail length.

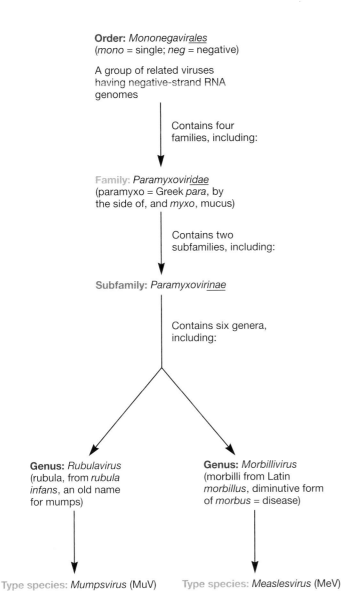

Order: *Mononegavirales*
(*mono* = single; *neg* = negative)

A group of related viruses having negative-strand RNA genomes

Contains four families, including:

Family: *Paramyxoviridae*
(paramyxo = Greek *para*, by the side of, and *myxo*, mucus)

Contains two subfamilies, including:

Subfamily: *Paramyxovirinae*

Contains six genera, including:

Genus: *Rubulavirus*
(rubula, from *rubula infans*, an old name for mumps)

Genus: *Morbillivirus*
(morbilli from Latin *morbillus*, diminutive form of *morbus* = disease)

Type species: *Mumpsvirus* (MuV)

Type species: *Measlesvirus* (MeV)

Figure 16.22   The Naming of Viruses. Because of the difficulty in establishing evolutionary relationships, most virus families have not been placed into an order. Virus names are derived from various aspects of their biology and history, including the features of their structure, diseases they cause, and locations where they were first identified or recognized.

Table 16.3	The Baltimore System of Virus Classification	
Group	**Description**	
I	Double-stranded DNA genome *genome replication: dsDNA → dsDNA* *mRNA synthesis: dsDNA → mRNA*	
II	Single-stranded DNA genome *genome replication: ssDNA → dsDNA → ssDNA* *mRNA synthesis: ssDNA → dsDNA → mRNA*	
III	Double-stranded RNA genome *replication: dsRNA → ssRNA → dsRNA* *mRNA synthesis: dsRNA → mRNA*	
IV	Plus-strand RNA genome *replication: +RNA → −RNA → +RNA* *mRNA synthesis: +RNA = mRNA*	
V	Negative-strand RNA genome *replication: −RNA → +RNA → −RNA* *mRNA synthesis: −RNA → mRNA*	
VI	Single-stranded RNA genome *replication: ssRNA → dsDNA → ssRNA* *mRNA synthesis: ssRNA → dsDNA → mRNA*	
VII	Double-stranded gapped DNA genome *replication: gapped dsDNA → dsDNA → +RNA →* *−DNA → gapped dsDNA* *mRNA synthesis: gapped dsDNA → dsDNA → mRNA*	

As is the case with the taxonomy of cellular life forms, the taxonomy of viruses is rapidly changing as more and more viral genomes are sequenced. They have been useful in establishing evolutionary relationships among viruses, and have led to the creation of new virus families and genera.

1. List some characteristics used in classifying viruses. Which seem to be the most important?
2. What are the endings for the names of virus families, subfamilies, and genera or species?

Summary

16.1 Early Development of Virology

a. Europeans were first protected from a viral disease when Edward Jenner developed a smallpox vaccine in 1798.

b. Chamberland's invention of a porcelain filter that could remove bacteria from virus samples enabled microbiologists to show that viruses were different from bacteria.

c. In the late 1930s Stanley, Bawden, and Pirie crystallized the tobacco mosaic virus and demonstrated that it was composed only of protein and nucleic acid.

16.2 General Properties of Viruses

a. A virion is composed of either DNA or RNA enclosed in a coat of protein (and sometimes other substances as well). It cannot reproduce independently of living cells.

16.3 The Structure of Viruses

a. All virions have a nucleocapsid composed of a nucleic acid, usually either DNA or RNA, held within a protein capsid made of one or more types of protein subunits called protomers (**figure 16.1**).

b. There are four types of viral morphology: naked icosahedral, naked helical, enveloped icosahedral and helical, and complex.

c. Helical capsids resemble long hollow protein tubes and may be either rigid or quite flexible. The nucleic acid is coiled in a spiral on the inside of the cylinder (**figure 16.3***b*).

d. Icosahedral capsids are usually constructed from two types of capsomers: pentamers (pentons) at the vertices and hexamers (hexons) on the edges and faces of the icosahedron (**figure 16.6**).

e. Complex viruses (e.g., poxviruses and large phages) have complicated morphology not characterized by icosahedral and helical symmetry. Large phages often have binal symmetry: their heads are icosahedral and their tails, helical (**figure 16.9**)

f. Viruses can have a membranous envelope surrounding their nucleocapsid. The envelope lipids usually come from the host cell; in contrast, many envelope proteins are viral and may project from the envelope surface as spikes or peplomers.

g. Viral nucleic acids can be either single stranded or double stranded, DNA or RNA. Most DNA viruses have double-stranded DNA genomes that may be linear or closed circles (**table 16.1**).

h. RNA viruses usually have ssRNA that may be either plus (positive) or minus (negative) when compared with mRNA (positive). Many RNA genomes are segmented.

i. Although viruses lack true metabolism, some contain a few enzymes necessary for their reproduction.

16.4 Virus Reproduction

a. Virus reproduction can be divided into five steps: (1) attachment to host; (2) entry into host; (3) synthesis of viral nucleic acid and proteins; (4) self-assembly of virions and; (5) release from host (**figure 16.12**).

16.5 The Cultivation of Viruses

a. Viruses are cultivated using tissue cultures, embryonated eggs, bacterial cultures, and other living hosts (**figure 16.13**).

b. Sites of animal viral infection may be characterized by cytopathic effects such as pocks and plaques. Phages produce plaques in bacterial lawns. Plant viruses can cause localized necrotic lesions in plant tissues (**figures 16.14** to **16.17**).

16.6 Virus Purification and Assays

a. Viruses can be purified by techniques such as differential and gradient centrifugation, precipitation, and denaturation or digestion of contaminants (**figures 16.18** and **16.19**).

b. Virus particles can be counted directly with the transmission electron microscope or indirectly by the hemagglutination assay (**figure 16.20**).

c. Infectivity assays can be used to estimate virus numbers in terms of plaque-forming units, lethal dose (LD_{50}), or infectious dose (ID_{50}).

16.7 Principles of Virus Taxonomy

a. Currently viruses are classified with a taxonomic system placing primary emphasis on the type and strandedness of viral nucleic acids, and on the presence or absence of an envelope (**table 16.2**).

b. The Baltimore system of virus classification is used by many virologists to organize viruses based on their genome type and the mechanisms they use to synthesize mRNA and replicate their genomes (**table 16.3**).

Key Terms

bacteriophage 409
binal symmetry 412
capsid 409
capsomers 410
cytopathic effects 418
differential centrifugation 420
envelope 412
enveloped virus 410

gradient centrifugation 420
helical capsid 410
hemagglutination assay 422
hexamers (hexons) 411
icosahedral capsid 410
infectious dose (ID_{50}) 423
lethal dose (LD_{50}) 423
minus strand or negative strand 416

naked virus 410
necrotic lesion 419
nucleocapsid 409
pentamers (pentons) 411
peplomer or spike 412
phage 409
plaque 418
plaque assay 422

plaque-forming units (PFU) 422
plus strand or positive strand 416
protomers 409
segmented genome 417
virion 409
virologist 407
virology 407
virus 409

Critical Thinking Questions

1. Many classification schemes are used to identify bacteria. These start with Gram staining, progress to morphology/ arrangement characteristics, and include a battery of metabolic tests. Build an analogous scheme that could be used to identify viruses. You might start by considering the host, or you might start with viruses found in a particular environment, such as a marine filtrate.

2. Consider the different perspectives on the origin of viruses in Microbial Tidbits 16.2. Discuss whether you think viruses evolved before the first procaryote, or whether they have coevolved, and are perhaps still coevolving with their hosts.

Learn More

Diamond, J. 1999. *Guns, germs, and steel.* New York: W.W. Norton.

Flint, S. J.; Enquist, L. W.; Racaniello, V. R.; and Skalka, A. M. 2004. *Principles of virology,* 2d ed. Washington, D.C.: ASM Press.

Foster, K. R.; Jenkins, M. F.; and Toogood, A. C. 1998. The Philadelphia yellow fever epidemic of 1793. *Scientific American* (August):88–93.

Hull, R. 2002. *Matthews' plant virology,* 4th ed. San Diego: Academic Press.

Knipe, D. M., and Howley, P. M., editors-in-chief. 2001. *Fields virology,* 4th ed. New York: Lippincott Williams & Wilkins.

Mayo, M. A.; Maniloff, J.; Desselberger, U.; Ball, L. A.; and Fauquet, C. M., editors. 2005. *Virus taxonomy: VIIIth report of the International Committee on Taxonomy of Viruses.* San Diego: Elservier Academic Press.

Nelson, D. 2004. Phage taxonomy: We agree to disagree. *J. Bacteriol.* 186(21):7029–31.

Oldstone, M. B. 1998. *Viruses, plagues, and history.* New York: Oxford University Press.

Zaitlin, M. 1999. Tobacco mosaic virus and its contributions to virology. *ASM News* 65(10):675–80.

17

The Viruses:
Viruses of *Bacteria* and *Archaea*

A scanning electron micrograph of T-even bacteriophages infecting *E. coli*. The phages are colored blue.

PREVIEW

- Viruses that infect procaryotic cells, both bacterial and archaeal, have been identified. These viruses are diverse in their morphology and reproductive strategies.

- Some procaryotic viruses are virulent viruses. Shortly after infecting their host, they begin reproduction. At the completion of their life cycle, they lyse their host. This type of life cycle is called a lytic cycle.

- Some procaryotic viruses begin reproduction upon entering their host but do not kill the host by lysis. Some are extruded from the cell. Although many cellular processes are slowed, the cell remains viable and can release many virus progeny over time.

- Many procaryotic viruses are temperate viruses. The genomes of many temperate viruses can integrate into the host cell's chromosome, where they are replicated as the host chromosome is replicated. This relationship is called lysogeny and it can be maintained for long periods.

- Analysis of viral genomes is providing insight into the evolution of viruses and their hosts.

Chapter 16 introduces many of the facts and concepts underlying the field of virology, including information about the nature of viruses, their structure and taxonomy, and how they are cultivated and studied. Clearly the viruses are a complex, diverse, and fascinating group, the study of which has done much to advance disciplines such as genetics and molecular biology.

In this chapter we are concerned with viruses that infect procaryotic cells. Because the discovery of archaeal viruses is relatively recent and little is known about their biology, our focus will be on the **bacteriophages**—viruses that infect bacteria. Because of their crucial role in the history of genetics and molecular biology, it is tempting to think of bacteriophages as useful only for laboratory research in these disciplines. However, bacteriophages are significant members of terrestrial and aquatic ecosystems. Indeed, they may be the most abundant form of life on the planet and are major agents of microbial evolution. It has been estimated that numerous bacteriophage species infect each species of bacteria. *Escherichia coli*, for example, is subject to infection by more than 20 phage species. Bacteriophages are also important in industry and medicine. For example, many phages destroy the gram-positive lactic acid bacteria that are critical to the production of fermented milk products such as yogurt and cheese. Bacteriophages also can carry a variety of virulence factors that convert their bacterial hosts into potent pathogens. This is the case for major human pathogens such as *Streptococcus pyogenes*, *Staphylococcus aureus*, *Corynebacterium diphtheriae*, *Vibrio cholerae*, *E. coli* O157:H7, and *Salmonella enterica*. On the positive side, Russian physicians have used bacteriophages for years to treat bacterial diseases. Research indicates that phages may be effective in treating bacterial infections, including those caused by antibiotic-resistant bacteria.

You might wonder how such naive outsiders get to know about the existence of bacterial viruses. Quite by accident, I assure you. Let me illustrate by reference to an imaginary theoretical physicist, who knew little about biology in general, and nothing about bacterial viruses in particular. . . . Suppose now that our imaginary physicist, the student of Niels Bohr, is shown an experiment in which a virus particle enters a bacterial cell and 20 minutes later the bacterial cell is lysed and 100 virus particles are liberated. He will say: "How come, one particle has become 100 particles of the same kind in 20 minutes? That is very interesting. Let us find out how it happens!. . . Is this multiplying a trick of organic chemistry which the organic chemists have not yet discovered? Let us find out."

—Max Delbrück

17.1 CLASSIFICATION OF BACTERIAL AND ARCHAEAL VIRUSES

Some of the families of bacterial and archaeal viruses are shown in **figure 17.1.** These families have been designated by the International Committee for the Taxonomy of Viruses (ICTV), the agency responsible for standardizing the classification of all viruses, including those that infect *Bacteria* and *Archaea*. Of the almost 2,000 virus species classified and catalogued by the ICTV, most are viruses of eucaryotes and bacteria. In fact, only about 40 archaeal viruses have been identified. Of those, only about eight have been assigned to virus taxa. Yet the discovery of archaeal viruses has had a significant impact on our understanding of viruses and on the ICTV classification scheme: archaeal viruses have led to the recognition of four new virus families, and at least three more families are awaiting ICTV approval (**Microbial Diversity and Ecology 17.1**). This is due primarily to the unusual morphologies observed among the known archaeal viruses, including viruses that are spindle-shaped and droplet-shaped. Furthermore, many are en-

veloped viruses. Among the archaeal viruses, only *Methanobacterium* virus ψM1 and *Halobacterium* virus H are placed with bacterial viruses (families *Siphoviridae* and *Myoviridae,* respectively). This is because these two viruses have capsid and tail structures similar to the phages in these existing bacteriophage families. The remaining six assigned archaeal viruses define the new archaeal virus families *Fuselloviridae, Guttaviridae, Lipothrixviridae,* and *Rudiviridae*. All of the other families illustrated in figure 17.1 contain bacteriophages.

Capsid structure is a critical phenotypic trait used to classify viruses of procaryotes. However, recent analyses of the genomes of bacteriophages have revealed problems with using phage morphology in taxonomy. This is especially true if one is trying to draw evolutionary relationships among the phages. As discussed in section 17.6, genetic modules appear to have been swapped across different species of viruses and even different families. This kind of lateral gene transfer makes the classification of bacteriophages difficult, and some virologists contend that current approaches to classifying bacteriophages need to be reconsidered and replaced by molecular approaches.

Bacteriophages exhibit a wide degree of diversity in terms of genome structure. Although most possess double-stranded DNA (dsDNA) genomes, phages with single-stranded DNA (ssDNA), single-stranded RNA (ssRNA), and double-stranded RNA (dsRNA) genomes have been identified and studied. Thus far, there is less diversity among the genomes of archaeal viruses. All are dsDNA viruses, either circular or linear. Because the nature of the genome correlates with the mechanisms used for synthesis of viral mRNA and with the replication of viral genomes, the discussion of bacteriophage reproduction that follows focuses primarily on these processes.

17.2 VIRULENT DOUBLE-STRANDED DNA PHAGES

After DNA bacteriophages have reproduced within the host cell, many of them are released when the cell is destroyed by lysis. A phage life cycle that culminates with the host cell bursting and releasing virions is called a **lytic cycle,** and viruses that reproduce solely in this way are called **virulent viruses.** The events taking place during the lytic cycle are reviewed in this section, with the primary focus on the T-even phages of *E. coli,* which are some of the most complex viruses known *(see figure 16.9).* T-even phages are double-stranded DNA bacteriophages with complex contractile tails. They are placed in the family *Myoviridae.*

The One-Step Growth Experiment

The development of the one-step growth experiment in 1939 by Max Delbrück and Emory Ellis marks the beginning of modern bacteriophage research. In a **one-step growth experiment,** the reproduction of a large phage population is synchronized so that the molecular events occurring during reproduction can be followed. A culture of susceptible bacteria such as *E. coli* is mixed with bacteriophage particles, and the phages are allowed a short interval to attach to their host cells. The culture is then greatly diluted so that any virions released upon host cell lysis do not immediately infect new

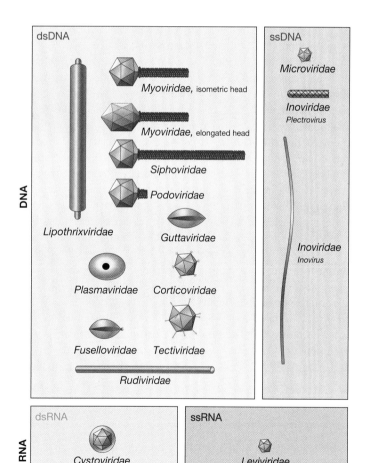

Figure 17.1 Families and Genera of Procaryotic Viruses.
The *Myoviridae* are the only family with contractile tails. *Plasmaviridae* are pleomorphic. *Tectiviridae* have distinctive double capsids, whereas the *Corticoviridae* have complex capsids containing lipid.

Microbial Diversity & Ecology

17.1   Host Independent Growth of an Archaeal Virus

The fact that viruses cannot replicate without first infecting a host cell has resulted in their classification as "acellular entities" or "forms"—they are not cells. So it was quite a surprise when an archaeal virus that develops long tails only when outside its host was discovered (**see Box figure**). This archaeal virus was found in acidic hot springs (pH 1.5, 85–93°C) in Italy where it infects the hyperthermophilic archaeon *Acidianus convivator*. When the virus infects its host, lemon-shaped virions are assembled. Following host lysis, two "tails" begin to form at either end of the virion. These projections continue to assemble until they reach a length at least that of the viral capsid. Curiously, tails are only produced if virions are incubated at high temperature, leading to the hypothesis that they are part of a survival strategy. The virus is thus called ATV for *Acidianus* two-tailed virus.

To find out more about the structure of the tails, the ATV genome was sequenced. ATV is a double-stranded DNA virus that encodes only nine structural proteins. The tail protein is an 800 amino acid protein that bears homology to eucaryotic intermediate filament proteins. Both intermediate filaments and purified ATV tail protein assemble into filamentous structures without additional energy or cofactors. The cytoplasmic matrix, microfilaments, intermediate filaments, and microtubules (section 4.3)

It is suspected that the development of tails only at high temperatures may be a survival strategy for the virus when host cell density is low. So far, ATV is the only virus that infects procaryotes living in acidic hot springs and induces lysis rather than lysogeny. It would thus seem all other such viruses have evolved lysogeny as a means to survive these harsh conditions. Why this virus has evolved lysis and tail development is unknown, but it suggests that viruses may be more complicated than simple "entities."

Häring, M.; Vestergaard, G.; Rachel, R.; Chen, L.; Garret, R. A.; and Prangishvili, D. 2005. Independent virus development outside a host. Nature 436:1101–02.

***Acidianus* Two-Tailed Virus, ATV.   (a)** Virions collected from an acidic hot spring bear two long projections, or tails, at each end. **(b, c)** The hyperthermophilic archaeon *A. convivator* extrudes lemon-shaped virions. **(d)** These subsequently develop tail-like structures independent of the host and only when at high temperature. Scale bars: *a–c*, 0.5 μm; *d*, 0.1 μm.

cells. This strategy works because phages lack a means of seeking out host cells and must contact them during random movement through the solution. Thus phages are less likely to contact host cells in a dilute mixture. The number of infective phage particles released from bacteria is subsequently determined at various intervals by a plaque assay. Virus purification and assays (section 16.6)

A plot of the bacteriophages released from host cells versus time shows several distinct phases (**figure 17.2**). During the **latent period,** which immediately follows phage addition, there is no release of virions. This is followed by the **rise period** (**burst**) when the host cells rapidly lyse and release infective phages. Finally, a plateau is reached and no more viruses are liberated. The total

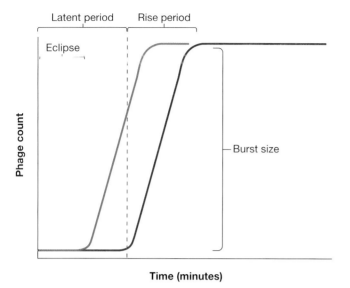

Figure 17.2 The One-Step Growth Curve. In the initial part of the latent period, the eclipse period, the host cells do not contain any complete, infective virions. During the remainder of the latent period, an increasing number of infective virions are present, but none are released. The latent period ends with host cell lysis and rapid release of virions during the rise period or burst. In this figure the blue line represents the total number of complete virions. The red line is the number of free viruses (the unadsorbed virions plus those released from host cells). When *E. coli* is infected with T2 phage at 37°C, the growth plateau is reached in about 30 minutes and the burst size is approximately 100 or more virions per cell. The eclipse period is 11–12 minutes, and the latent period is around 21–22 minutes.

number of phages released can be used to calculate the **burst size,** the number of viruses produced per infected cell.

The latent period is the shortest time required for virus reproduction and release. During the first part of this phase, host bacteria do not contain any complete, infective virions. This can be shown by lysing them with chloroform. This initial segment of the latent period is called the **eclipse period** because the virions detectable before infection are now concealed or eclipsed. The number of completed, infective phages within the host increases after the end of the eclipse period, and the host cell is prepared for lysis.

Between the beginning of a one-step growth experiment and the final burst, a carefully orchestrated series of events occurs **(figure 17.3)**. The major events—adsorption to the host cell, viral penetration of the host cell, synthesis of viral nucleic acids and proteins, assembly of phage particles and release of virions—are described next.

Adsorption and Penetration

Like all viruses, bacteriophages do not randomly attach to the surface of a host cell; rather, they fasten to specific surface structures called **receptors.** The nature of these receptors varies with the

phage; cell wall lipopolysaccharides and proteins, teichoic acids, flagella, and pili can serve as receptors. The T-even phages of *E. coli* use cell wall lipopolysaccharides or proteins as receptors. Variation in receptor properties is at least partly responsible for phage host preferences.

T-even phage adsorption involves several tail structures. Phage attachment begins when a tail fiber contacts the appropriate receptor **(figure 17.4a)**. As more tail fibers make contact, the baseplate settles down on the surface (figure 17.4b). Binding is probably due to electrostatic interactions and is influenced by pH and the presence of ions such as Mg^{2+} and Ca^{2+}. After the baseplate is seated firmly on the cell surface, conformational changes occur in the baseplate and sheath, and the tail sheath reorganizes so that it shortens from a cylinder 24 rings long to one of 12 rings (figure 17.4c, d). That is, the sheath becomes shorter and wider, and the central tube or core is pushed through the bacterial wall. The baseplate contains the protein gp5, which has lysozyme activity. This aids in the penetration of the tube through the peptidoglycan layer. Finally, the linear DNA is extruded from the head, through the tail tube, and into the host cell (figure 17.4e, f). The tube may interact with the plasma membrane to form a pore through which DNA passes.

The penetration mechanisms of other bacteriophages can be different from that of the T-even phages, but most have not been studied in as much detail. An exception is the PRD1 phage of the family *Tectiviridae,* which has a membrane under its icosahedral capsid (figure 17.1). It infects pseudomonads and members of the *Enterobacteriaceae.* The PRD1 phage attaches to a surface receptor by a spike structure at one of its capsid vertices **(figure 17.5)**. This causes a conformational change in the capsid proteins. The attached spike-penton complex of the capsid dissociates from the virion, and the underlying membrane forms a tubular structure that penetrates the bacterial envelope. Virus DNA is then injected through the membrane tube into the bacterium. Penetration of the membrane tube is made possible partly by two enzymes, both of which break the same glycosidic bond that is attacked by lysozyme.

Synthesis of Phage Nucleic Acids and Proteins

With the exception of the *Hepadnaviridae* (*see tables 16.1 and 16.3*), all double-stranded DNA viruses follow a similar route for synthesis of viral nucleic acids and proteins **(figure 17.6)**. The DNA genome serves as the template for mRNA synthesis, and the mRNA molecules made are translated to yield viral proteins. Sometime after the onset of mRNA synthesis, DNA replication ensues and more viral genomes are made. The details of nucleic acid and protein synthesis vary from virus to virus, but all are designed to manipulate the host cell to the advantage of the virus. T4 bacteriophage will serve as our example.

Within 2 minutes after injection of T4 DNA into a host *E. coli* cell, the *E. coli* RNA polymerase starts synthesizing T4 mRNA (figure 17.3). This mRNA is called **early mRNA** because it is made before viral DNA is made. Early mRNA directs the synthesis of protein factors and enzymes required to take over the host

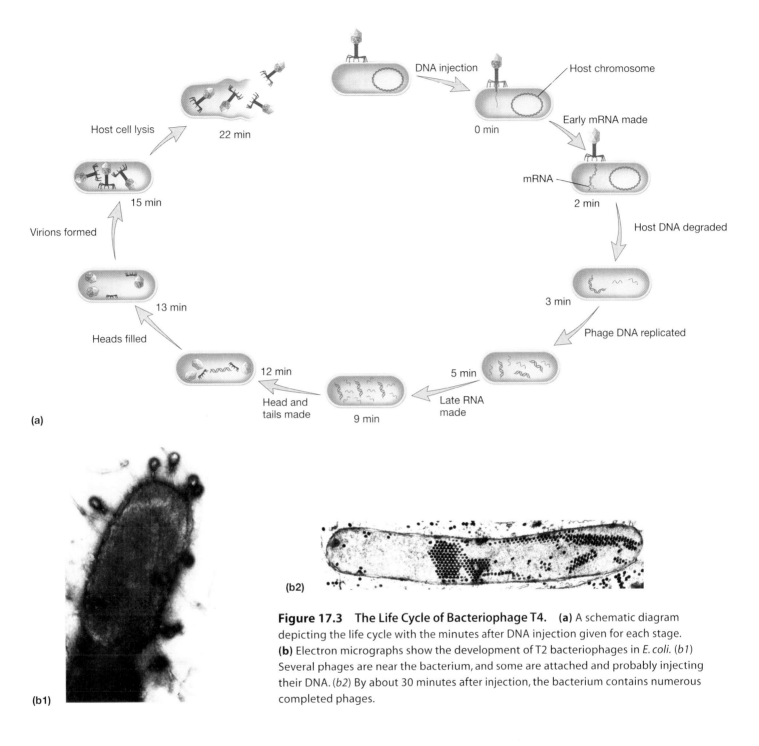

(a)

(b2)

(b1)

Figure 17.3   The Life Cycle of Bacteriophage T4.   (a) A schematic diagram depicting the life cycle with the minutes after DNA injection given for each stage. **(b)** Electron micrographs show the development of T2 bacteriophages in *E. coli.* (*b1*) Several phages are near the bacterium, and some are attached and probably injecting their DNA. (*b2*) By about 30 minutes after injection, the bacterium contains numerous completed phages.

cell and force it to manufacture additional viral constituents. Some early virus-specific enzymes degrade host DNA to nucleotides, thereby simultaneously halting host gene expression and providing raw material for viral DNA synthesis. Within 5 minutes, viral DNA synthesis commences. DNA replication is initiated from several origins of replication and it proceeds bidirectionally from each. Viral DNA replication is followed by the synthesis of **late mRNAs,** which are important in later stages of the infection. Thus expression of viral genes is temporally ordered. How does T4 accomplish this?

T4 controls the expression of its genes by regulating the activity of the *E. coli* RNA polymerase. Initially T4 genes are transcribed by the regular host RNA polymerase and the sigma factor σ^{70} (*see table 12.3*). After a short interval, a virus enzyme catalyzes the transfer of the chemical group ADP-ribose from NAD to an α-subunit of RNA polymerase (*see figure 11.27*). This modification of the host enzyme helps inhibit the transcription of host genes and promotes virus gene expression. Later the second α-subunit receives an ADP-ribosyl group. This turns off some of the early T4 genes, but not before the product of one early gene,

(a) Landing   **(b)** Attachment   **(c)** Tail contraction   **(d)** Penetration and unplugging   **(e)** DNA injection

Cell wall

(f)

Figure 17.4   T4 Phage Adsorption and DNA Injection.
(a–e) Adsorption and DNA injection is mediated by the phage's tail fibers and base plate as shown here. **(f)** An electron micrograph of an *E. coli* cell being infected by a T-even phage. These phages have injected their nucleic acid through the cell wall and now have empty capsids.

motA, stimulates transcription of somewhat later genes. One of these later genes encodes the sigma factor gp55. This sigma factor helps RNA polymerase bind to late promoters and transcribe late genes, which become active around 10 to 12 minutes after infection.   Global regulatory systems (section 12.5)

The tight regulation of expression of T4 genes is aided by the organization of the T4 genome. As can be seen in **figure 17.7,** genes with related functions—such as the genes for phage head or tail fiber construction—are usually clustered together. Early and late genes also are clustered separately on the genome; they are even transcribed in different directions—early genes in the counterclockwise direction and late genes, clockwise.

Also apparent in figure 17.7 is that a considerable portion of the T4 genome encodes products needed for its replication, including all the protein subunits of its replisome and enzymes needed to prepare for synthesis of DNA **(figure 17.8)**. Some of these enzymes synthesize an important component of T4 DNA, **hydroxymethylcytosine (HMC) (figure 17.9)**. HMC is a modified nucleotide that replaces cytosine in T4 DNA. Once HMC is synthesized, replication ensues by a mechanism similar to that seen in bacteria. After T4 DNA has been synthesized, it is glucosylated by the addition of glucose to the HMC residues. Glucosylated HMC residues protect T4 DNA from attack by *E. coli* endonucleases called **restriction enzymes,** which would otherwise cleave the viral DNA at specific points and destroy it. This

bacterial defense mechanism is called **restriction.** Other chemical groups also can be used to modify phage DNA and protect it against restriction enzymes. For example, methyl groups are added to the amino groups of adenine and cytosine in lambda phage DNA for the same reason.

The T4 genome is linear dsDNA and shows what is called terminal redundancy—that is, a base sequence is repeated at each end of the molecule. These two characteristics contribute to the formation of long DNA molecules called **concatemers,** which are composed of several genome units linked together in the same orientation **(figure 17.10)**. Why does this occur? As discussed in chapter 11, the ends of linear DNA molecules cannot be replicated without special machinery such as the enzyme telomerase observed in eucaryotes. T4 does not have telomerase activity. Therefore, each progeny DNA molecule has single-stranded 3′ ends. These ends participate in homologous recombination with double-stranded regions of other progeny DNA molecules, generating the concatemers. During assembly, concatemers are cleaved such that the genome packaged in the capsid is slightly longer than the T4 gene set. Thus each progeny virus has a genome unit that begins with a different gene. However, if each genome of the progeny viruses was circularized, the sequence of genes in each virion would be the same (figure 17.10). Therefore the T4 genome is said to be circularly permuted, and the genetic map of T4 is drawn as circular molecule.

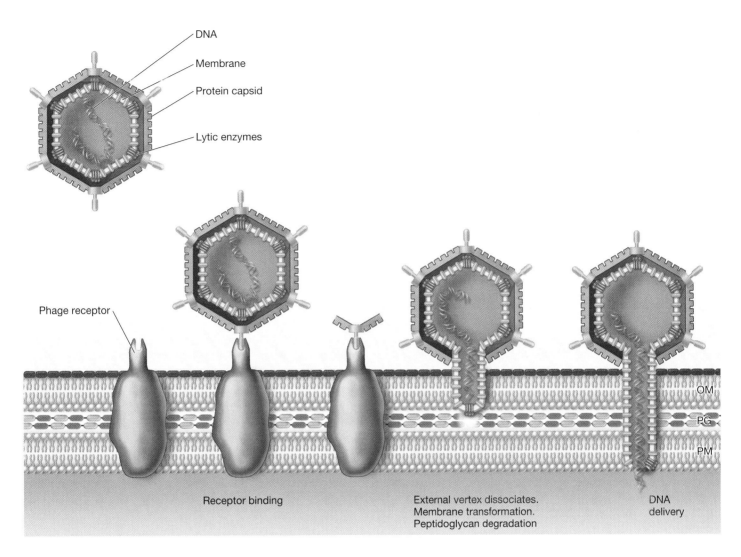

DNA

Membrane

Protein capsid

Lytic enzymes

Phage receptor

OM

PG

PM

Receptor binding

External vertex dissociates.
Membrane transformation.
Peptidoglycan degradation

DNA
delivery

Figure 17.5   Attachment and Penetration of Host Cell by Phage PRD1.   The gram-negative cell wall layers are indicated by OM (outer membrane), PG (peptidoglycan), and PM (plasma membrane). See text for details.

+ mRNA

Protein

± DNA

± DNA

Figure 17.6   Replication Strategy Used by Double-Stranded DNA Viruses.   Because the genome of double-stranded DNA viruses is similar to the host, the replication process closely resembles that of the host cell and can involve the use of host polymerases, viral polymerases, or both. The DNA serves as the template for DNA replication and mRNA synthesis. Translation of the mRNA by the host cell's translation machinery yields viral proteins, which are assembled with the viral DNA to make mature virions. These are eventually released from the host.

Assembly of Phage Particles

The assembly of T4 phage is an exceptionally complex self-assembly process that involves special virus proteins and some host cell factors. Late mRNA directs the synthesis of three kinds of proteins: (1) phage structural proteins, (2) proteins that help with phage assembly without becoming part of the virion structure, and (3) proteins involved in cell lysis and phage release. Late mRNA transcription begins about 9 minutes after T4 DNA injection into *E. coli*. All the phage proteins required for assembly are synthesized simultaneously and then used in four fairly independent subassembly lines **(figure 17.11)**. The baseplate is constructed of 16 gene products, which are assigned numbers rather than names **(figure 17.12)**. After the baseplate is finished, the tail tube is built on it and the sheath is assembled around the tube. The phage prohead (procapsid) is constructed of 10 proteins. The prohead is assembled with the aid of **scaffolding proteins** that are degraded or removed after construction is completed. A special portal protein is located at the base of the prohead where it connects to the tail. The portal protein partici-pates in DNA packaging, yielding the mature head (figure 17.11). After completion of the head, it spontaneously combines with the tail assembly. Figure 17.12 shows the mature virion and the proteins from which it is constructed.

DNA packaging within the T4 prohead is accomplished by a complex of proteins sometimes called the "packasome." The pack-asome consists of the portal protein just mentioned. It also contains a set of proteins called the terminase complex, which generates double-stranded ends at the ends of the concatemers created when the viral genome was replicated. These double-stranded ends are needed for packaging the T4 genome. Once generated, the termi-nase proteins and phage DNA join with the portal protein at the base of the prohead. The completed packasome then moves DNA into the prohead, a process powered by ATP hydrolysis. The con-catemer is cut when the phage head is filled with DNA—a DNA molecule roughly 3% longer than the length of one set of T4 genes.

Release of Phage Particles

Many phages lyse their host cells at the end of the intracellular phase. The lysis of *E. coli* by T4 takes places after about 150 virus particles have accumulated in the host cell **(figure 17.13)**. Two proteins are involved. One directs the synthesis of an enzyme that attacks peptidoglycan in the host's cell wall. It is sometimes called T4 lysozyme. Another T4 protein called holin creates holes in the *E. coli* plasma membrane, enabling T4 lysozyme to move from the cytoplasm to the peptidoglycan.

Figure 17.8   A Model of the T4 Replisome.   T4 encodes most of the proteins needed to replicate its dsDNA genome, including components of the T4 replisome. This figure illustrates the viral replisome at the replication fork. Compare this model with the bacterial replisome shown in figure 11.16.

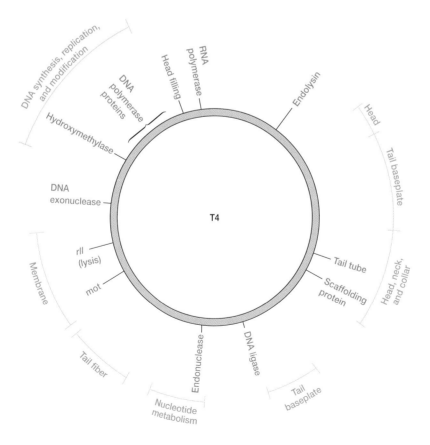

Figure 17.7   A Map of the T4 Genome.   Some of its genes and their functions are shown. Genes with related functions tend to be clustered together.

Figure 17.9 **5-Hydroxymethylcytosine (HMC).** In T4 DNA, the HMC often has glucose attached to its hydroxyl.

1. How is a one-step growth experiment carried out? Summarize what occurs in each phase. Define latent period, eclipse period, rise period (burst), and burst size.
2. Define the following terms: adsorption, penetration, phage assembly, phage release, lytic cycle, receptor site, early mRNA, late mRNA, hydroxymethylcytosine, restriction, restriction enzymes, concatemers, and scaffolding proteins. Use these terms to write a paragraph describing the life cycle of T4 bacteriophage.
3. Explain why the T4 phage genome is circularly permuted.

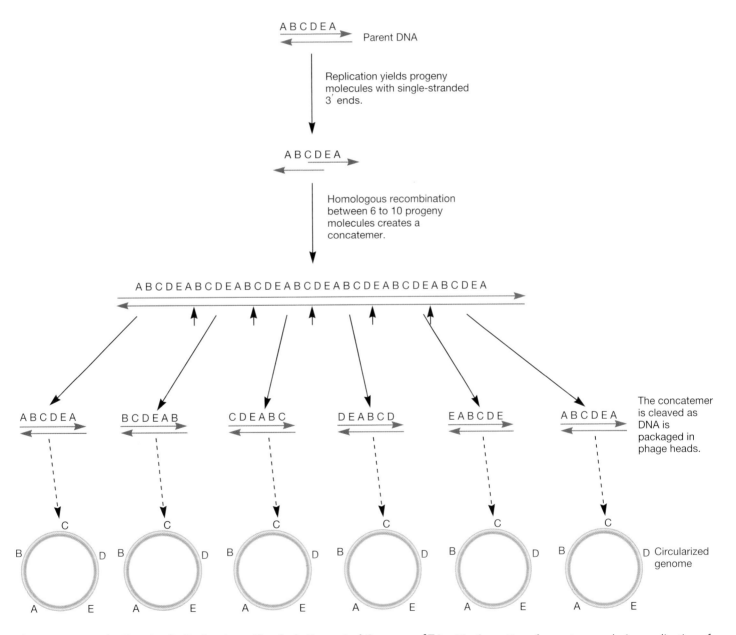

Figure 17.10 **The Terminally Redundant, Circularly Permuted Genome of T4.** The formation of concatemers during replication of the T4 genome is an important step in phage reproduction. During assembly of the virions, the phage head is filled with DNA cleaved from the concatemer. Because slightly more than one set of T4 genes is packaged in each head, each virion contains a different DNA fragment (note that the ends of the fragments are different). However, if each genome was circularized, the sequence of genes would be the same.

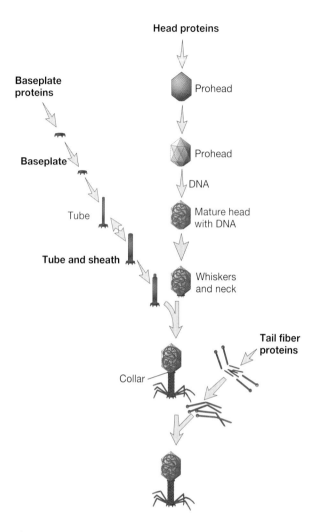

Figure 17.11 The Assembly of T4 Bacteriophage. Note the subassembly lines for the baseplate, tail tube and sheath, tail fibers, and head.

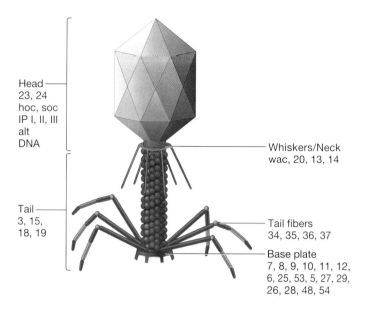

Figure 17.12 The Mature T4 Virion. T4 is composed of numerous proteins, most of which are designated with numbers rather than names. The proteins comprising each component of the phage are indicated. This image was reconstructed from electron micrographs. The phage is shown at 3 nm resolution.

Figure 17.13 Release of T4 Bacteriophages by Lysis of the Host Cell. The host cell has been lysed (upper right portion of the cell) and virions have been released into the surroundings. Progeny virions also can be seen in the cytoplasm. In addition, empty capsids of the infecting phages coat the outside of the cell (X36,500).

17.3 SINGLE-STRANDED DNA PHAGES

Although many bacterial viruses are double-stranded DNA viruses, several single-stranded DNA (ssDNA) phages have been identified and their replication studied. Two ssDNA phages of *E. coli*—φX174, an icosahedral phage belonging to the *Microviridae,* and fd phage, a filamentous phage belonging to the *Inoviridae*—are the focus of this section.

The life cycle of φX174 begins with its attachment to the cell wall of its host. The circular ssDNA genome is injected into the cell, while the protein capsid remains outside the cell. The φX174 ssDNA genome has the same base sequence as viral mRNA and is therefore said to be plus-strand DNA. In order for either transcription or genome replication to occur, the phage DNA must be converted to a double-stranded form called the **replicative form (RF) (figure 17.14)**. This is accomplished by the bacterial DNA polymerase. The replicative form directs the synthesis of more RF copies and plus-strand DNA, both by rolling-circle replication. After assembly of virions, the phage is released by host ly-

sis through a different method than that used by T4 phage. Rather than producing a lysozyme-like enzyme, as does T4, φX174 produces an enzyme (enzyme E) that blocks peptidoglycan synthesis. Enzyme E inhibits the activity of the bacterial protein MraY, which catalyzes the transfer of murein precursors to lipid carriers (*see figure 10.12*). Blocking cell wall synthesis weakens the host cell wall, causing the cell to lyse and release the progeny virions. DNA replication (section 11.4)

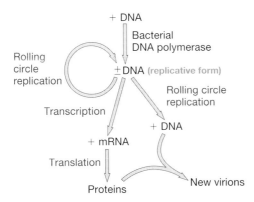

Figure 17.14   The Reproduction of ϕX174, a Plus-Strand DNA Phage.   See text for details.

Although the fd phage also has a circular, positive-strand DNA genome, it behaves quite differently from ϕX174 in many respects. It is shaped like a long fiber about 6 nm in diameter by 900 to 1,900 nm in length (figure 17.1). Its ssDNA lies in the center of the filament and is surrounded by a tube made of a coat protein organized in a helical arrangement. The virus infects F⁺, Hfr, and F′ *E. coli* cells by attaching to the tip of the pilus; the DNA enters the host along or possibly through the F factor-encoded sex pilus with the aid of a special adsorption protein. As with ϕX174, a replicative form is first synthesized and then transcribed. A phage-coded protein then aids in replication of the phage DNA by rolling-circle replication. Unlike either T4 or ϕX174, fd and other filamentous fd phages do not kill their host cell. Instead, they establish a relationship in which new virions are continually released by a secretory process. Filamentous phage coat proteins are first inserted into the membrane. The coat then assembles around the viral DNA as it is secreted through the host plasma membrane **(figure 17.15)**. Although the host cell is not lysed, it grows and divides at a slightly reduced rate.

17.4   RNA PHAGES

Although most bacteriophages are DNA viruses, numerous RNA phages have been identified and studied. Many of these are ssRNA viruses. Bacteriophages MS2 and Qβ, family *Leviviridae,* are small, tailless, icosahedral, plus-strand RNA viruses (figure 17.1). They attach to the side of the F pilus of their *E. coli* host. Retraction of the pilus brings the virions close to the outer membrane of the cell, from which they gain entry. Like many bacteriophages, the capsids of these viruses remain outside the cell and only the RNA genome enters. Because their genomes are plus stranded, the incoming RNA can act as messenger RNA and direct the synthesis of phage proteins. One of the first enzymes synthesized is a viral **RNA replicase,** an RNA-dependent RNA polymerase **(figure 17.16)**. The replicase then copies the plus strand to produce a double-stranded intermediate (±RNA), which is called the replicative form and is analogous to the ±DNA seen in the reproduction of ssDNA phages. The same replicase then uses this

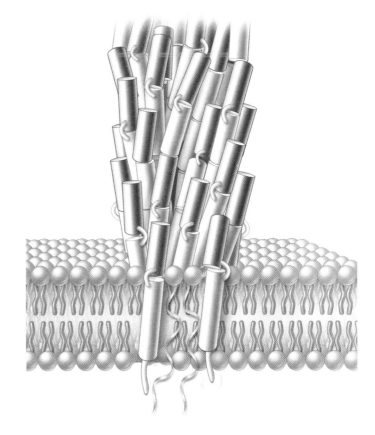

Figure 17.15   Release of Pf1 Phage.   The Pf1 phage is a filamentous bacteriophage that is released from *Pseudomonas aeruginosa* without lysis. In this illustration the blue cylinders are hydrophobic α-helices that span the plasma membrane, and the red cylinders are amphipathic helices that lie on the membrane surface before virus assembly. In each protomer the two helices are connected by a short, flexible peptide loop (yellow). It is thought that the blue helix binds with circular, single-stranded viral DNA (green) as it is extruded through the membrane. The red helix simultaneously attaches to the growing virus coat that projects from the membrane surface. Eventually the blue helix leaves the membrane and also becomes part of the capsid.

replicative form to synthesize thousands of copies of +RNA. Some of these plus strands are used to make more ±RNA in order to accelerate +RNA synthesis. Other +RNA strands act as mRNA and direct the synthesis of phage proteins. Finally, +RNA strands are incorporated into maturing virus particles. The mature virions are released by lysis. Release of virions has been studied for Qβ. It produces a protein that interferes with the bacterial protein MurA. MurA participates in the synthesis of UDP-NAM, a precursor for peptidoglycan synthesis. In the case of ϕX174, this interference weakens the cell wall and the host cell lyses. Synthesis of sugars and polysaccharides: Synthesis of peptidoglycan (section 10.4)

MS2 and Qβ have only three or four genes and are genetically the simplest phages known. In MS2, one protein is involved in phage adsorption to the host cell (and possibly also in virion construction or maturation). The other three genes code for a coat protein, RNA replicase, and a protein needed for cell lysis.

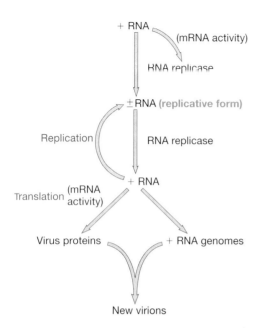

Figure 17.16 The Reproduction of Plus-Strand RNA Bacteriophages.

Several dsRNA phages have been discovered. The bacteriophage φ6 of *Pseudomonas syringae* pathovar phaseolicola (previously called *Pseudomonas phaseolicola*), a plant pathogen, is the best studied. It is an unusual bacteriophage for several reasons. One is that it is an enveloped virus. Within the envelope is a nucleocapsid containing a segmented genome consisting of three dsRNA segments and an RNA-dependent RNA polymerase. The life cycle of the virus also has unusual features. Like MS2 and Qβ, φ6 attaches to the side of a pilus. However, φ6 uses an envelope protein to facilitate adsorption. Retraction of the pilus brings the phage into contact with the outer membrane. The viral envelope then fuses with the cell's outer membrane, a process mediated by another envelope protein. Fusion of the two membranes delivers the nucleocapsid into the periplasmic space. Here, a protein associated with the nucleocapsid digests the peptidoglycan, allowing the nucleocapsid to cross that layer of the cell wall of its gram-negative host. Finally, the intact nucleocapsid enters the host cell by a process that resembles endocytosis. Because bacterial cells do not have the proteins and other factors needed for endocytosis, it is thought that viral proteins mediate this mechanism of entry. Once inside the host, the viral RNA polymerase acts as a **transcriptase,** catalyzing synthesis of viral mRNA from each dsRNA segment. The enzyme also acts as a replicase, synthesizing plus-strand RNA from each segment. These are enclosed within newly formed capsid proteins, where they serve as templates for the synthesis of the complementary negative strand, regenerating the dsRNA genome. Once the nucleocapsid is completed, a nonstructural viral protein called P12 functions in surrounding the nucleocapsid with a plasma membrane-derived envelope. Interestingly, the enveloped nucleocapsid is located within the cytoplasm. Finally, additional viral proteins are added to the envelope and the host cell is lysed, releasing the mature virions.

1. How does the reproduction of the ssDNA phages φX174 and fd differ from each other and from T4? How is their reproduction similar to the ssRNA phages? How does it differ?
2. What role does RNA replicase play in the reproduction of ssRNA phages? In dsRNA phages?
3. What is peculiar about the structure and life cycle of phage φ6?

17.5 TEMPERATE BACTERIOPHAGES AND LYSOGENY

Thus far in our discussion of bacteriophages, we have focused primarily on virulent phages—those that only have one reproductive option: to begin replication immediately upon entering their host, followed by release from the host by lysis. However, many DNA phages are **temperate phages** that have two reproductive options: they can reproduce lytically as do the virulent phages or they can remain within the host cell without destroying it. Many temperate phages accomplish this by integrating their genome into the host cell's chromosome.

The relationship between a temperate phage and its host is called **lysogeny.** The form of the virus that remains within its host is called a **prophage,** and the infected bacteria are called **lysogens** or **lysogenic bacteria.** Lysogenic bacteria reproduce and in most other ways appear to be perfectly normal. However, they have two distinctive characteristics. The first is that they cannot be reinfected by the same virus—that is, they have immunity to superinfection. The second is that under appropriate conditions they lyse and release phage particles. This occurs when conditions within the cell cause the prophage to initiate synthesis of phage proteins and to assemble new virions, a process called **induction.** Induction leads to destruction of infected cells and release of new phages—that is, induction initiates the lytic cycle. Why and how these phenomena occur are discussed shortly.

Another important outcome of lysogeny is **lysogenic conversion.** This occurs when a temperate phage induces a change in the phenotype of its host. Lysogenic conversions often involve alterations in surface characteristics of the host. Many other lysogenic conversions give the host pathogenic properties. An example of the former is seen when *Salmonella* is infected by epsilon phage. The phage changes the activities of several enzymes involved in construction of the carbohydrate component of the bacterium's lipopolysaccharide, thereby altering the antigenic properties of the bacterium. Interestingly, these changes also eliminate the receptor for epsilon phage, and the bacterium becomes immune to infection by another epsilon phage. An example of altered pathogenic properties is observed when *Corynebacterium diphtheriae,* the cause of diphtheria, is infected with phage β. The phage β genome encodes diphtheria toxin, which is responsible for the symptoms of the disease. Thus only those strains of *C. diphtheriae* that are infected by the phage cause disease. Airborne diseases: Diphtheria (section 38.1)

Clearly, the infection of a bacterium by a temperate phage has significant impact on the host but why would viruses evolve this alternate life cycle? Two advantages of lysogeny have been recognized. The first is that lysogeny allows a virus to remain viable within a dormant host. Bacteria often become dormant due to nutrient deprivation and while in this state, they do not synthesize nucleic acids or proteins. In such situations, a prophage would survive but most virulent bacteriophages would not be replicated, as they require active cellular biosynthetic machinery. The second advantage arises when there are many more phages in an environment than there are host cells, a situation virologists refer to as a high multiplicity of infection (MOI). In these conditions, lysogeny allows for the survival of host cells so that the virus can continue to reproduce.

It should be apparent from this discussion that once a temperate phage infects its host, it must "decide" which reproductive cycle to follow. How does it make this choice? How temperate phages make this decision is best illustrated by bacteriophage lambda. Lambda is a double-stranded DNA phage that infects the K12 strain of *E. coli*. It has an icosahedral head 55 nm in diameter and a noncontractile tail with a thin tail fiber at its end **(figure 17.17)**. Its DNA genome is a linear molecule with cohesive ends—single-stranded stretches, 12 nucleotides long, that are complementary to each other and can base pair.

Like most bacteriophages, lambda attaches to its host and then injects its genome into the cytoplasm, leaving the capsid outside. Once inside the cell, the linear genome is circularized when the two cohesive ends base pair with each other; the breaks in the strands are sealed by the host cell's DNA ligase **(figure 17.18)**. The lambda genome has been carefully mapped, and over 40 genes have been located **(figure 17.19)**. Most genes are clustered according to their function, with separate groups involved in head synthesis, tail synthesis, lysogeny, DNA replication, and cell lysis. This organization is important because once the genome is circularized, a cascade of regulatory events occurs that determine if the phage pursues a lytic cycle or establishes lysogeny. Regulation of appropriate genes is facilitated by clustering and coordinated transcription from the same promoter. Below, we first provide an overview of lysogeny and the lytic cyle of lambda. This is followed by a more detailed examination of the decision-making process.

The cascade of events leading to either lysogeny or the lytic cycle involves a number of regulatory proteins that function as repressors or activators or both. Two regulatory proteins are of particular importance: the **lambda repressor** (product of the *cI* gene) and the **Cro protein** (product of the *cro* gene). The lambda repressor promotes lysogeny, and the Cro protein promotes the lytic cycle. In essence, the decision to pursue lysogeny or to pursue a lytic cycle is the result of a race between the production of these two proteins. If lambda repressor prevails, the production of Cro protein is inhibited and lysogeny occurs; if the Cro protein prevails, the production of lambda repressor is inhibited and the lytic cycle occurs. This is because the lambda repressor prevents transcription of viral genes, while Cro does just the opposite: it ensures viral gene expression.

The lambda repressor is 236 amino acids long and folds into a dumbbell shape with globular domains at each end **(figure 17.20)**. One domain binds DNA while the other binds another lambda repressor molecule to form a dimer. The dimer is the most active form of the repressor. Lambda repressor binds two operator sites, O_L and O_R, thereby blocking transcription of most viral genes **(figure 17.21 and table 17.1)**. When bound at O_L, it represses transcription from the promoter P_L (promoter leftward). Likewise, when bound at O_R it represses transcription in the rightward direction from P_R (promoter rightward). However, it also activates transcription in the leftward direction from the *cI* promoter P_{RM} (RM stands for *re*pressor *m*aintenance). Recall that *cI* encodes the lambda repressor. Thus lambda repressor controls its own synthesis.

As noted earlier, if the lambda repressor wins the race with the Cro protein, lysogeny is established and the lambda genome is integrated into the host chromosome. Integration is catalyzed by the enzyme **integrase,** the product of the *int* gene, and takes place at a site in the host chromosome called the attachment site (*att*) (figure 17.19; *see also figure 13.40*). A homologous site is found on the phage genome, so the phage and bacterial *att* sites can base pair with each other. The bacterial site is located between the galactose (*gal*) and biotin (*bio*) operons, and as a result of integration, the circular lambda genome becomes a linear stretch of DNA located between these two host operons **(figure 17.22)**. The prophage can remain integrated indefinitely, being replicated as the bacterial genome is replicated.

Figure 17.17   Bacteriophage Lambda.

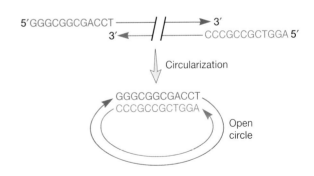

Figure 17.18   Lambda Phage DNA.   A diagram of lambda phage DNA showing its 12 base, single-stranded cohesive ends (printed in purple) and the circularization their complementary base sequences make possible.

Techniques & Applications

18.2   Constructing a Virus

The poliovirus has been constructed completely from scratch beginning with the base sequence of its RNA genome. Because DNA is easier to synthesize than RNA, a DNA copy of the poliovirus RNA genome attached to the promoter for T7 phage RNA polymerase was first produced using DNA synthesizer machines. The synthetic DNA was then incubated with the T7 phage RNA polymerase and the appropriate nucleotides. The polymerase used the DNA template to form complete RNA copies of the poliovirus genome. The RNA genomes were then incubated with a cell-free cytoplasmic extract of HeLa cells, which supplied the constituents necessary for protein synthesis. The RNA directed the synthesis of poliovirus proteins; the proteins and RNA then spontaneously assembled into complete, infectious poliovirus virions. These particles could infect a special mouse strain and cause a disease that resembled human poliomyelitis.

Some scientists believe that it will be possible to make almost any virus once the genome sequence is known. This could be a particular threat if bioterrorists managed to create smallpox. Others believe that it will not be that easy to create a virulent pathogen like smallpox. The smallpox genome is much larger than that of poliovirus and would be harder to synthesize. Unlike the case with poliovirus, smallpox DNA is not infectious by itself and requires the presence of virion proteins such as polymerases. Thus one would have to find a way to supply these enzymes, perhaps by using helper viruses. A potentially greater threat is the synthesis of special hybrid strains that are very infectious and also quite lethal. On a more positive note, it may be possible to construct substantially weakened strains of viruses such as poliovirus for the production of more effective vaccines.   Bioterrorism preparedness (section 36.9)

---

new +RNA genomes. Notice that during this process genetic information is transferred from RNA to DNA rather than from DNA to RNA as in cellular information flow.   DNA replication (section 11.4)

Another interesting aspect of the life cycles of RNA viruses is related to the synthesis of their proteins. A key feature of translation in eucaryotic cells is that their mRNA molecules encode a single protein. Yet RNA viruses must generate multiple proteins. Several strategies for achieving this have been described. As already noted for the picornaviruses, synthesis of a polyprotein from a genomic-sized RNA occurs. The polyprotein is later cleaved to give rise to all the proteins needed by the virus to complete its life cycle. Those RNA viruses that have segmented genomes can use each segment to encode a different protein. Some RNA viruses have an RNA-dependent RNA polymerase that is able to initiate mRNA synthesis at internal sites along the template RNA. This generates **subgenomic mRNA,** which is mRNA that is smaller than the RNA genome. RNA splicing is used by some RNA viruses to generate different transcripts and therefore different proteins. Finally, translation itself may lead to synthesis of multiple proteins through **ribosomal frameshifting,** which causes translation of a different reading frame yielding a different protein with each frame read. Many RNA viruses use several of these strategies to produce a full array of viral proteins.

Assembly of Virus Capsids

Some **late genes** direct the synthesis of capsid proteins, and these spontaneously self-assemble to form the capsid just as in bacteriophage morphogenesis. Recently the self-assembly process has been dramatically demonstrated (Techniques & Applications 18.2). It appears that during icosahedral virus assembly empty **procapsids** are first formed; then the nucleic acid is inserted in some unknown way. The site of morphogenesis varies with the virus (**table 18.2**). Large paracrystalline clusters of either complete virions or procapsids are often seen at the site of virus maturation (**figure 18.10**). The assembly of enveloped virus capsids is generally similar to that of naked virions, except for poxviruses. These are assembled in the cytoplasm by a lengthy, complex process that begins with the enclosure of a portion of the cytoplasmic matrix through construction of a new membrane. Then newly synthesized DNA condenses, passes through the membrane, and moves to the center of the immature virus. Nucleoid and elliptical body construction takes place within the membrane.

Virion Release

Mechanisms of virion release differ between naked and enveloped viruses. Naked virions appear to be released most often by host cell lysis. In contrast, the formation of envelopes and the release of enveloped viruses are usually concurrent processes, and the host cell may continue virion release for some time. All viral envelopes are derived from host cell membranes by a multistep process. First, virus-encoded proteins are incorporated into the plasma membrane. Then the nucleocapsid is simultaneously released and the envelope formed by membrane budding (**figures 18.11** and **18.12**). In several virus families, a matrix (M) protein attaches to the plasma membrane and aids in budding. Most envelopes arise from the plasma membrane. However, in herpesviruses, budding and envelope formation usually involve the nuclear membrane (table 18.2). The endoplasmic reticulum, Golgi apparatus, and other internal membranes also can be used to form envelopes.

Interestingly, it has been discovered that actin filaments can aid in virion release. Many viruses alter the actin microfilaments of the host cell cytoskeleton. For example, vaccinia virus appears to form long actin tails and use them to move intracellularly at up to 2.8 μm per minute. The actin filaments also propel vaccinia through the plasma membrane. In this way the virion escapes without destroying the host cell and infects adjacent cells. This is

Virus	Nucleic Acid Replication	Capsid Assembly	Membrane Used in Budding
Table 18.2 Intracellular Sites of Animal Virus Reproduction			
DNA Viruses			
Adenoviruses	Nucleus	Nucleus	
Hepadnaviruses	Cytoplasm	Cytoplasm	Endoplasmic reticulum
Herpesviruses	Nucleus	At nuclear membrane	Nucleus
Papillomaviruses	Nucleus	Nucleus	
Parvoviruses	Nucleus	Nucleus	
Polyomaviruses	Nucleus	Nucleus	
Poxviruses	Cytoplasm	Cytoplasm	
RNA Viruses			
Coronaviruses	Cytoplasm	Cytoplasm	Golgi apparatus and endoplasmic reticulum
Orthomyxoviruses	Nucleus	Cytoplasm	Plasma membrane
Paramyxoviruses	Cytoplasm	Cytoplasm	Plasma membrane
Picornaviruses	Cytoplasm	Cytoplasm	
Reoviruses	Cytoplasm	Cytoplasm	
Retroviruses	Cytoplasm and nucleus	At plasma membrane	Plasma membrane
Rhabdoviruses	Cytoplasm	Cytoplasm	Plasma membrane, intracytoplasmic membranes
Togaviruses	Cytoplasm	Cytoplasm	Plasma membrane, intracytoplasmic membranes

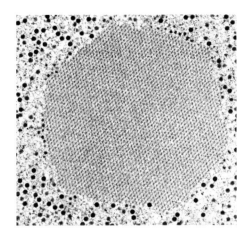

Figure 18.10   Paracrystalline Clusters.   A crystalline array of adenoviruses within the cell nucleus ($\times$35,000).

similar to the mechanism of pathogenesis used by the bacterium *Listeria monocytogenes.*   The cytoplasmic matrix, microfilaments, intermediate filaments, and microtubules (section 4.3); Disease 4.1: Getting around

1. Compare and contrast each stage of vertebrate virus reproduction with those seen for bacteriophages.
2. What probably plays the most important role in determining the tissue and host specificity of vertebrate viruses? Give some specific examples.
3. In general, DNA viruses can be much more dependent on their host cells than can RNA viruses. Why is this so? What strategies are used by RNA viruses to complete their life cycles?

4. Outline the retrovirus life cycle. What is proviral DNA? How is it synthesized? How do you think the ability to form proviral DNA is related to the asymptomatic stage of HIV infection?
5. Compare the retrovirus life cycle with that of a lysogenic bacteriophage such as lambda. What advantage might a virus have in incorporating its genome into that of the host cell?

18.3   CYTOCIDAL INFECTIONS AND CELL DAMAGE

An infection that results in cell death is a **cytocidal infection.** Vertebrate viruses can harm their host cells in many ways; often this leads to cell death. Microscopic or macroscopic degenerative changes or abnormalities in host cells and tissues are referred to as **cytopathic effects (CPEs).** Seven possible mechanisms of host cell damage are briefly described here. However, it should be emphasized that more than one of these mechanisms may be involved in a cytopathic effect.

1. Many viruses can inhibit host DNA, RNA, and protein synthesis. Cytocidal viruses (e.g., picornaviruses, herpesviruses, and adenoviruses) are particularly active in this regard. The mechanisms of inhibition are not yet clear.
2. Cell endosomes may be damaged, resulting in the release of hydrolytic enzymes and cell destruction.
3. Virus infection can drastically alter plasma membranes through the insertion of virus-specific proteins so that the infected cells are attacked by the immune system. When infected by viruses such as herpesviruses and measles virus, as many as 50 to 100 cells may fuse into one abnormal, giant,

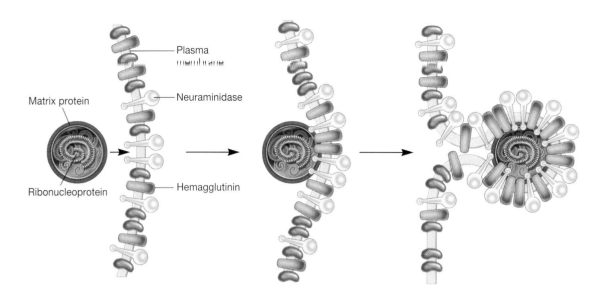

Figure 18.11 Release of Influenza Virus by Plasma Membrane Budding. First, viral envelope proteins (hemagglutinin and neuraminidase) are inserted into the host plasma membrane. Then the nucleocapsid approaches the inner surface of the membrane and binds to it. At the same time viral proteins collect at the site and host membrane proteins are excluded. Finally, the plasma membrane buds to simultaneously form the viral envelope and release the mature virion.

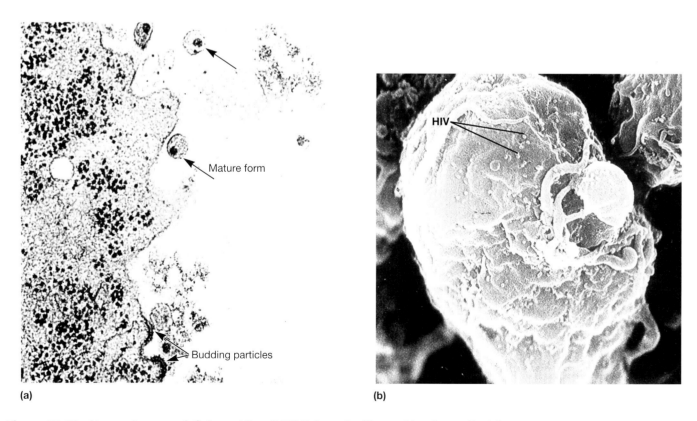

Figure 18.12 Human Immunodeficiency Virus (HIV) Release by Plasma Membrane Budding. **(a)** A transmission electron micrograph of HIV particles beginning to bud, as well as some mature particles. **(b)** A scanning electron micrograph view of HIV particles budding from a lymphocyte.

multinucleated cell called a **syncytium.** HIV appears to destroy CD4$^+$ T-helper cells at least partly through its effects on the plasma membrane.

4. High concentrations of proteins from several viruses (e.g., mumps virus and influenza virus) can have a direct toxic effect on cells and organisms.

5. Intracellular structures called **inclusion bodies** are formed during many virus infections. These may result from the clustering of subunits or virions within the nucleus or cytoplasm (e.g., the Negri bodies in rabies infections); they also may contain cell components such as ribosomes (arenavirus infections) or chromatin (herpesviruses). Regardless of their composition, these inclusion bodies can directly disrupt cell structure.

6. Chromosomal disruptions result from infections by herpesviruses and others.

7. Finally, the host cell may not be directly destroyed but transformed into a malignant cell. This is discussed in section 18.5.

18.4 PERSISTENT, LATENT, AND SLOW VIRUS INFECTIONS

Many virus infections (e.g., influenza) are **acute infections**—that is, they have a fairly rapid onset and last for a relatively short time **(figure 18.13)**. However, some viruses can establish **persistent infections** lasting many years. There are several kinds of persistent infections. In **chronic virus infections,** the virus is almost always detectable and clinical symptoms may be either mild or absent for long periods. Examples are hepatitis B virus and HIV. In **latent virus infections** the virus stops reproducing and remains dormant for a period before becoming active again. During latency, no symptoms or viruses are detectable, although antibodies to the virus may be present at low levels. Examples are herpes simplex virus, varicella-zoster virus, cytomegalovirus, and Epstein-Barr virus, which causes mononucleosis. Herpes simplex type 1 virus often infects children and then becomes dormant within the nervous system ganglia; years later it can be activated to cause cold sores. The varicella-zoster virus causes chicken pox in children and then, after years of inactivity, may produce the skin disease shingles (initial adult infections result in chicken pox). These and other examples of persistent infections are discussed in more detail in chapter 37.

The causes of persistence and latency are probably multiple, although the precise mechanisms are still unclear. The virus genome may be integrated into the host genome thereby becoming a provirus. Viruses may become less antigenic and thus less susceptible to attack by the immune system. Often virus infections are latent in a site not subject to immune attack, such as the central nervous system. Viruses may also mutate to less virulent and slower reproducing forms. Sometimes a deletion mutation produces a *defective interfering (DI) particle* that cannot reproduce but slows normal virus reproduction, thereby reducing host damage and establishing a chronic infection. Mutations and their chemical basis (section 13.1)

A small group of viruses causes extremely slowly developing infections, often called **slow virus diseases** or slow infections, in which symptoms may take years to emerge. Measles virus occasionally produces a slow infection. A child may have a normal case of measles, then 5 to 12 years later develop a degenerative brain disease called subacute sclerosing panencephalitis (SSPE). Lentiviruses such as HIV also cause slow diseases.

1. What is a cytocidal infection? What is a cytopathic effect? Outline the ways in which viruses can damage host cells during cytocidal infections.
2. Define the following: acute infection, persistent infection, chronic infection, latent virus infection, and slow virus disease.
3. Why might an infection be chronic or latent?

18.5 VIRUSES AND CANCER

Cancer [Latin *cancer,* crab] is one of the most serious medical problems in developed nations. It is the focus of an immense amount of research. A **tumor** [Latin *tumere,* to swell] is a growth or lump of tissue resulting from **neoplasia,** abnormal new cell growth and reproduction due to loss of regulation. Tumor cells have aberrant shapes and altered plasma membranes that may contain distinctive tumor antigens. Their unregulated proliferation and loss of differentiation result in invasive growth that forms unorganized cell masses. This reversion to a more primitive or less differentiated state is called **anaplasia.**

There are two major types of tumors with respect to overall form or growth pattern. If the tumor cells remain in place to form a compact mass, the tumor is benign. In contrast, cells from malignant or cancerous tumors can actively spread throughout the body in a process known as **metastasis,** often by floating in the blood and establishing secondary tumors. Some cancers are not solid, but cell suspensions. For example, leukemias are composed of undifferentiated malignant white blood cells that circulate throughout the body. Indeed, dozens of kinds of cancers arise from a variety of cell types and afflict all kinds of organisms.

As one might expect from the wide diversity of cancers, there are many causes of cancer, only a few of which are directly related to viruses. Possibly as many as 30 to 60% of cancers may be related to diet and cigarette smoke. Many chemicals in our surroundings are carcinogenic and may cause cancer by inducing gene mutations or interfering with normal cell differentiation. However, it is important to note that many cancers are not linked to environmental risk factors.

Carcinogenesis is a complex, multistep process. It can be initiated by a chemical, usually a mutagen, but a cancer does not appear to develop until at least one more triggering event (possibly exposure to another chemical carcinogen or a virus) takes place. Cancer-causing genes, or **oncogenes,** are directly involved. Some oncogenes are contributed to a cell by viruses, as is discussed later. Others arise from genes within the cell called **proto-oncogenes.** Proto-oncogenes are cellular genes required for normal growth. If they are mutated or over-expressed, they may become oncogenes. That is, their products contribute to the malignant

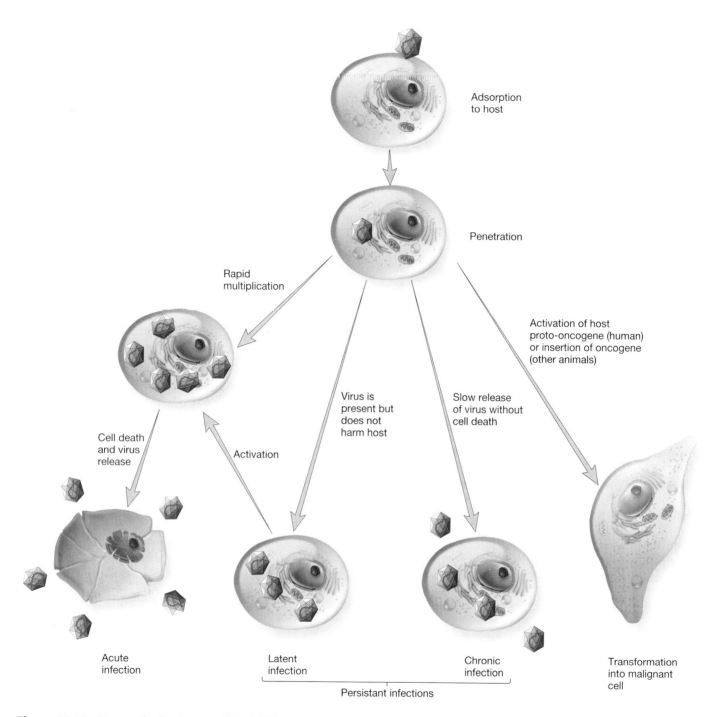

Adsorption to host

Penetration

Rapid multiplication

Activation of host proto-oncogene (human) or insertion of oncogene (other animals)

Virus is present but does not harm host

Slow release of virus without cell death

Cell death and virus release

Activation

Acute infection

Latent infection

Chronic infection

Transformation into malignant cell

Persistant infections

Figure 18.13 **Types of Infections and Their Effects on Host Cells.**

transformation of the cell. Many oncogenes are involved in the regulation of cell growth and signal transduction; for example, some code for growth factors that regulate cell reproduction. It may be that various cancers arise through different combinations of causes. Not surprisingly the chances of developing cancer rise with age because an older person will have had a longer time to accumulate the mutations needed for oncogenic transformation.

Immune surveillance and destruction of cancer cells also may be less effective in older people.

Although viruses are known to cause many animal cancers, it is very difficult to prove that this is the case with human cancers because indirect methods of study must be used and Koch's postulates can't be applied completely. One tries to find virus particles and components within tumor cells, using techniques such as electron microscopy, immunologic tests, DNA-based assays, and enzyme assays. Attempts are also made to isolate suspected cancer viruses by cultivation in tissue culture or other animals. Sometimes a good correlation between the presence of a virus and cancer can be detected.

At present, viruses have been implicated in the genesis of at least eight human cancers. With the exception of a few retroviruses, these viruses have dsDNA genomes.

1. The Epstein-Barr virus (EBV) is one of the best-studied human cancer viruses. EBV is a herpesvirus and the cause of two cancers. Burkitt's lymphoma is a malignant tumor of the jaw and abdomen found in children of central and western Africa. EBV also causes nasopharyngeal carcinoma. Both EBV particles and genomes have been found within tumor cells; Burkitt's lymphoma patients also have high blood levels of antibodies to EBV. Interestingly there is some evidence that a person also must have had malaria to develop Burkitt's lymphoma. Environmental factors must play a role, because EBV does not cause much cancer in the United States despite its prevalence. This may be due to a low incidence of malaria in the United States.
2. Hepatitis B virus appears to be associated with one form of liver cancer (hepatocellular carcinoma) and can be integrated into the human genome.
3. Hepatitis C virus causes cirrhosis of the liver, which can lead to liver cancer.
4. Human herpesvirus 8 and HIV are associated with the development of Kaposi's sarcoma.
5. Some strains of human papillomaviruses have been linked to cervical cancer.
6. At least two retroviruses, the human T-cell lymphotropic virus I (HTLV-1) and HTLV-2, are associated with adult T-cell leukemia and hairy-cell leukemia, respectively. Other retrovirus-associated cancers may well be discovered in the future.

Viruses known to cause cancer are called **oncoviruses.** All known human dsDNA oncoviruses trigger cancerous transformation of cells by a similar mechanism. They encode proteins that bind to and thereby inactivate cellular proteins known as **tumor suppressors.** Tumor-suppressor proteins regulate cell cycling or monitor and/or repair DNA damage. Two tumor suppressors known to be targets of human oncovirus proteins are called Rb and p53. Rb has multiple functions in the nucleus, all of which are critical to normal cell cycling. When Rb molecules are rendered inactive by an oncoviral protein, cells undergo uncontrolled reproduction. Thus they are said to be hyperproliferative. The protein p53 is often referred to as "the guardian of the genome." This is because p53 normally initiates either cell cycle arrest or programmed cell death in response to DNA damage. However, when p53 is inactivated by an oncoviral protein, it cannot do so and genetic damage persists. From the point of view of the virus, hyperproliferation and the lack of programmed cell death are beneficial. However, for the cell, it can

be catastrophic. Cells can rapidly accumulate the additional mutations needed for oncogenic transformation. *The nucleus and cell division (section 4.8)*

Retroviruses exert their oncogenic powers in a different manner. Some carry oncogenes captured from host cells many, many generations ago. Thus they transform the host cell by bringing the oncogene into the cell. For example, Rous sarcoma virus carries a mutated, oncogenic *src* gene that codes for an overactive tyrosine kinase. This enzyme is localized to the plasma membrane and phosphorylates the amino acid tyrosine in several cellular proteins. These Src-targeted proteins are essential in maintaining the cell's ability to respond to normal anti-growth signals from other cells and the extracellular matrix. When phosphorylated, they become active and override these signals, in effect signaling unregulated growth. The human retroviruses HTLV-1 and HTLV-2 transform a group of immune system cells called T cells by producing a regulatory protein that sometimes activates genes involved in cell division as well as stimulating virus reproduction. The second transformation mechanism used by retroviruses involves the integration of a viral genome into the host chromosome such that strong, viral regulatory elements are near a cellular proto-oncogene. This results in such a high level of expression of the cellular protein that the gene is now an oncogene. For example, some chicken retroviruses induce lymphomas when they are integrated next to the *c-myc* cellular proto-oncogene, which codes for a protein that is involved in the induction of either DNA or RNA synthesis.

1. What are the major characteristics of cancer?
2. How might viruses cause cancer? Are there other ways in which a malignancy might develop?
3. Define the following terms: tumor, neoplasia, anaplasia, metastasis, and oncogene.

18.6 PLANT VIRUSES

Although it has long been recognized that viruses can infect plants and cause a variety of diseases, plant viruses generally have not been as well studied as bacteriophages and animal viruses. This is partly because they are often difficult to cultivate and purify. Some viruses, such as tobacco mosaic virus (TMV), can be grown in isolated protoplasts of plant cells just as phages and some animal viruses are cultivated in cell suspensions. However, many cannot grow in protoplast cultures and must be inoculated into whole plants or tissue preparations. This makes studying the life cycle difficult. Many plant viruses require insect vectors for transmission; some of these can be grown in monolayers of cell cultures derived from aphids, leafhoppers, or other insects.

The essentials of capsid morphology are outlined in chapter 16 and apply to plant viruses because they do not differ significantly in construction from their animal virus and phage relatives **(figure 18.14)**. Many have either rigid or flexible helical capsids

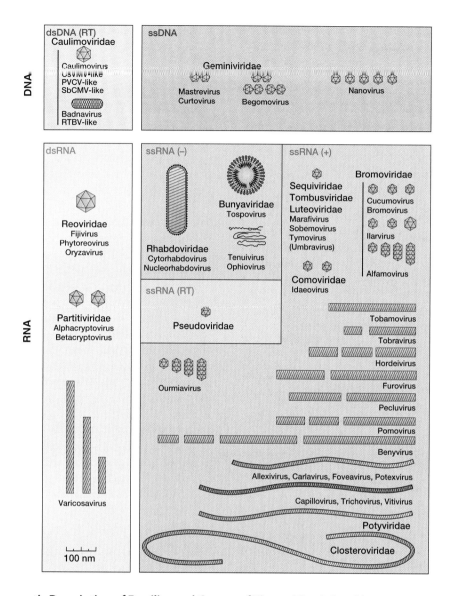

Figure 18.14 **A Diagrammatic Description of Families and Genera of Viruses That Infect Plants.** RT stands for reverse transcriptase.

(tobacco mosaic virus). Others are icosahedral or have modified the icosahedral pattern with the addition of extra capsomers (turnip yellow mosaic virus, **figure 18.15**). Most capsids are composed of one type of protein; no specialized attachment proteins have yet been detected. Almost all plant viruses are RNA viruses, either single stranded or double stranded. Caulimoviruses and geminiviruses with their DNA genomes are exceptions to this rule.

Like all viruses, plant viruses must penetrate a host cell before they can reproduce. However, penetration of plant cells is hampered by the fact that they are protected by complex outer layers, including cell walls. Entry of a plant virus into its host requires the presence of mechanical damage to the cell wall, and this is usually caused by insects or other animals that feed on plants. Particularly important are sucking insects such as aphids and leafhoppers. These insects not only create an entryway for

the virus, they are often responsible for transmitting plant viruses from plant to plant. As the insect feeds on the plant, it can pick up a virus particle on its mouthparts. In some cases, the virus is stored in the insect's foregut as the insect moves to another plant. However, some plant viruses can reproduce within insect tissues—these viruses use both plants and insects as hosts. Whether passively transmitted or actively reproducing in the insect host, when the insect feeds on another plant, virus particles can be transferred to the new plant.

As noted earlier, most plant viruses are RNA viruses and, of these, plus-strand RNA viruses are most common. TMV is the best studied plus-strand RNA virus and is the focus of this discussion. Recall that the plus-strand genomes of RNA viruses can serve as mRNA. However, following entry into its host, the TMV RNA genome is not immediately translated. Rather the RNA is processed by a mechanism that is not understood. The resulting mRNAs encode several proteins, including the coat protein and an RNA-dependent RNA polymerase. Thus TMV can replicate its own genome. This is not the case with all positive-strand RNA plant viruses. Most plants contain an enzyme with RNA-dependent RNA

Figure 18.15 Turnip Yellow Mosaic Virus (TYMV). An RNA plant virus with icosahedral symmetry.

polymerase activity. Thus it is possible that some plus-strand RNA plant viruses make use of the host enzyme.

After the coat protein and RNA genome of TMV have been synthesized, they spontaneously assemble into complete TMV virions in a highly organized process **(figure 18.16)**. The protomers come together to form disks composed of two layers of protomers arranged in a helical spiral. Association of coat protein with TMV RNA begins at a specific assembly initiation site close to the 3' end of the genome. The helical capsid grows by the addition of protomers, probably as disks, to the end of the rod. As the rod lengthens, the RNA passes through a channel in its center and forms a loop at the growing end. In this way the RNA can easily fit as a spiral into the interior of the helical capsid.

Reproduction within the host depends on the virus's ability to spread throughout the plant. Viruses can move long distances through the plant vasculature; usually they travel in the phloem. The spread of plant viruses in nonvascular tissue is hindered by the presence of tough cell walls. Nevertheless, a virus such as TMV spreads slowly, about 1 mm/day or less, moving from cell to cell through the plasmodesmata. These are slender cytoplasmic strands extending through holes in adjacent cell walls that join plant cells by narrow bridges. Viral "movement proteins" are required for transfer from cell to cell. The TMV movement protein accumulates in the plasmodesmata, but the way in which it promotes virus movement is not well understood.

Several cytological changes can take place in TMV-infected cells. Plant virus infections often produce microscopically visible intracellular inclusions, usually composed of virion aggregates.

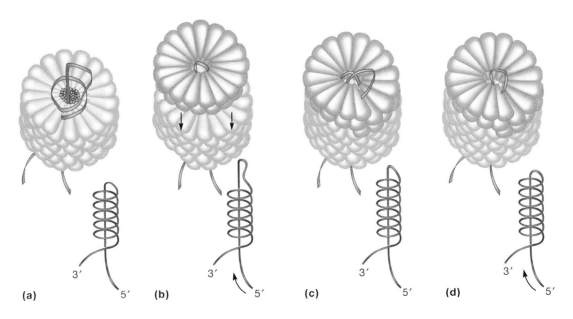

(a)   3'   5'   (b)   3'   5'   (c)   3'   5'   (d)   3'   5'

Figure 18.16 TMV Assembly. The elongation phase of tobacco mosaic virus nucleocapsid construction. The lengthening of the helical capsid through the addition of a protein disk to its end is shown in a sequence of four illustrations; line drawings depicting RNA behavior are included. The RNA genome inserts itself through the hole of an approaching disk and then binds to the groove in the disk as it locks into place at the end of the cylinder.

(a)

(b)

Figure 18.17 Intracellular TMV. (a) A crystalline mass of tobacco mosaic virions from a 10-day-old lesion in a *Chenopodium amaranticolor* leaf. **(b)** Freeze-fracture view of a crystalline mass of tobacco mosaic virions in an infected leaf cell. In both views the particles can be seen longitudinally and in cross section.

Hexagonal crystals of almost pure TMV virions sometimes develop in TMV-infected cells **(figure 18.17)**. The host cell chloroplasts become abnormal and often degenerate, while new chloroplast synthesis is inhibited.

---

1. Why have plant viruses not been as well studied as animal and bacterial viruses?
2. Describe in molecular terms the way in which TMV is reproduced.
3. How are plant viruses transmitted between hosts?
4. Compare plant and vertebrate viruses in terms of their entry into host cells. Why do they differ so greatly in this regard?

18.7 VIRUSES OF FUNGI AND PROTISTS

Most mycoviruses, viruses that infect fungi, have been isolated from higher fungi such as *Penicillium* and *Aspergillus*. They contain dsRNA and have isometric capsids (that is, their capsids are roughly spherical polyhedra), which are approximately 25 to 50 nm in diameter. Many appear to be latent viruses. Some mycoviruses induce disease symptoms in hosts such as the mushroom *Agaricus bisporus,* but cytopathic effects and toxic virus products have not yet been observed. The *Fungi* (chapter 26)

Much less is known about the viruses of lower fungi, and only a few have been examined in any detail. Both dsRNA and ssRNA genomes have been found; capsids usually are isometric or hexagonal and vary in size from 40 to over 200 nm. Unlike the viruses of higher fungi, virus reproduction in lower fungi is accompanied by host cell destruction and lysis.

Viruses that infect photosynthetic protists belong to the family *Phycodnaviridae.* Four genera are recognized by the ICTV. All have linear dsDNA genomes. Those algal viruses that have been studied have polyhedral capsids. One virus of *Uronema gigas* resembles many bacteriophages in having a tail. Another interesting group of viruses infects *Chlorella* strains that are endosymbionts of the ciliated protist *Paramecium busaria*. These viruses have very large genomes and encode proteins not usually encoded by viral genomes. They, along with the Mimivirus described below, have fostered considerable discussion about what features distinguish cellular organisms from acellular entities.

The viruses of only three genera of protozoa have been studied. It is known that *Giardia intestinalis* and *Leishmania* spp. are infected by members of the *Totiviridae,* a family of naked icosahedral viruses with dsRNA genomes. Finally, a giant dsDNA virus, named Mimivirus, has been discovered in the amoeba *Acanthamoeba polyphaga.* The virus is 400 nm in diameter and has a genome about 800 kilobase pairs in size; thus its genome is larger than that of some bacteria. Mimivirus is distantly related to the *Poxviridae* and *Phycodnaviridae.* The protists (chapter 25)

18.8 INSECT VIRUSES

Members of many virus families are known to infect insects. Some use insects as agents for spreading to populations of susceptible animals and plants. These include *Flaviviridae* and *Togaviridae,* whose members cause yellow fever, West Nile disease, and several types of viral encephalitis. Other virus families use insects as primary hosts. Of these, probably the most important are the *Baculoviridae, Reoviridae, Iridoviridae,* and *Polydnaviridae.*

The *Iridoviridae* are icosahedral viruses with lipid in their capsids and a linear dsDNA genome. They are responsible for the iridescent virus diseases of the crane fly and some beetles. The group's name comes from the observation that larvae of infected insects can have an iridescent coloration due to the presence of crystallized virions in their fat bodies.

Many insect virus infections are accompanied by the formation of inclusion bodies within the infected cells. Granulosis viruses form granular protein inclusions, usually in the cytoplasm. Nuclear polyhedrosis and cytoplasmic polyhedrosis virus infections produce polyhedral inclusion bodies in the nucleus or the cytoplasm of affected cells. Although all three types of viruses generate inclusion bodies, they belong to two distinctly different families. The cytoplasmic polyhedrosis viruses are reoviruses; they are icosahedral with double shells and have dsRNA genomes. Nuclear polyhedrosis viruses and granulosis viruses are baculoviruses—rod-shaped, enveloped viruses of helical symmetry and with dsDNA.

The inclusion bodies, both polyhedral and granular, are protein in nature and enclose one or more virions **(figure 18.18)**. Insect larvae are infected when they feed on leaves contaminated with inclusion bodies. Polyhedral bodies protect the virions against heat, low pH, and many chemicals; the viruses can remain viable in the soil for years. However, when exposed to the alkaline contents of insect guts, the inclusion bodies dissolve to liberate the virions, which then infect midgut cells. Some viruses remain in the midgut while others spread throughout the insect. Just as with bacterial and vertebrate viruses, insect viruses can persist in a latent state within the host for generations while producing no disease symptoms. A reappearance of the disease may be induced by chemicals, thermal shock, or even a change in the insect's diet.

Much of the current interest in insect viruses arises from their promise as biological control agents for insect pests. Many people hope that some of these viruses may partially replace the use of toxic chemical pesticides. Baculoviruses have received the most attention for at least three reasons. First, they attack only invertebrates and have considerable host specificity; this means that they should be safe for nontarget organisms. Second, because

they are encased in protective inclusion bodies, these viruses have a good shelf life and better viability when dispersed in the environment. Finally, they are well suited for commercial production because they often reach extremely high concentrations in larval tissue (as high as 10^{10} viruses per larva). The use of nuclear polyhedrosis viruses for the control of the cotton bollworm, Douglas fir tussock moth, gypsy moth, alfalfa looper, and European pine sawfly has either been approved by the U.S. Environmental Protection Agency or is being considered. The granulosis virus of the codling moth also is useful. Usually inclusion bodies are sprayed on foliage consumed by the target insects. More sensitive viruses are administered by releasing infected insects to spread the disease. As in the case with other pesticides, it is possible that resistance to these agents may develop in the future.

1. Describe the major characteristics of the viruses that infect higher fungi, lower fungi, and protists. In what ways do they seem to differ from one another?
2. Summarize the nature of granulosis, nuclear polyhedrosis, and cytoplasmic polyhedrosis viruses and the way in which they are transmitted by inclusion bodies.
3. What are baculoviruses and why are they so promising as biological control agents for insect pests?

18.9 VIROIDS AND VIRUSOIDS

Although some viruses are exceedingly small and simple, even simpler infectious agents exist. Viroids are infectious agents that consist only of RNA. Virusoids, formerly called satellite RNAs, are similar to viroids in that they also consist only of RNA. Finally, prions are infectious agents that consist only of protein. Prions are discussed in section 18.10.

Viroids cause over 20 different plant diseases, including potato spindle-tuber disease, exocortis disease of citrus trees, and chrysanthemum stunt disease. Viroids are covalently closed, circular, ssRNAs, about 250 to 370 nucleotides long **(figures 18.19 and 18.20)**. The circular RNA normally exists as a rodlike shape due to intrastrand base pairing, which forms double-stranded regions with single-stranded loops (figure 18.20). Some viroids are found in the nucleolus of infected host cells, where between 200 and 10,000 copies may be present. Others are located within chloroplasts. Interestingly, the RNA of viroids does not encode any gene products, so they cannot replicate themselves. Rather, it is thought that the viroid is replicated by one of the host cell's DNA-dependent RNA polymerases. The host polymerase evidently uses the viroid RNA as a template for RNA synthesis, rather than host DNA. The host polymerase synthesizes a complementary RNA molecule, a negative-strand RNA. This then serves as the template for the same host polymerase, and new viroid RNAs are synthesized. Both steps may occur by a rolling-circle-like mechanism.

A plant may be infected with a viroid without showing symptoms—that is, it may have a latent infection. However, the same viroid in another host species may cause severe disease.

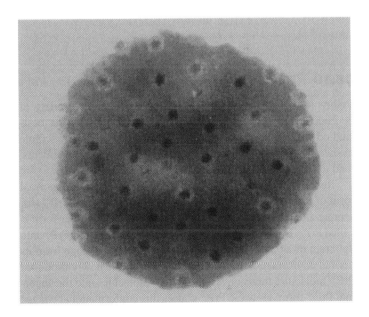

Figure 18.18 Inclusion Bodies. A section of a cytoplasmic polyhedron from a gypsy moth (*Lymantria dispar*). The occluded virus particles with dense cores are clearly visible (×50,000).

18.7 Viruses of Fungi and Protists

a. Mycoviruses from higher fungi have isometric capsids and dsRNA, whereas the viruses of lower fungi may have either dsRNA or dsDNA genomes.

b. Only a few viruses of protists have been isolated and studied. Some are of special note because they have extremely large dsDNA genomes that include genes not usually found in viral genomes.

18.8 Insect Viruses

a. Members of several virus families infect insects, and many of these viruses produce inclusion bodies that aid in their transmission.

b. Baculoviruses and other viruses are finding uses as biological control agents for insect pests.

18.9 Viroids and Virusoids

a. Infectious agents simpler than viruses exist. For example, several plant diseases are caused by short strands of infectious RNA called viroids (**figure 18.20**).

b. Virusoids are infectious RNAs that encode one or more gene products. They require a helper virus for replication.

18.10 Prions

a. Prions are small proteinaceous agents associated with at least six degenerative nervous system disorders: scrapie, bovine spongiform encephalopathy, kuru, fatal familial insomnia, the Gerstmann-Strässler-Scheinker syndrome, and Creutzfeldt-Jakob disease. The precise nature of prions is not yet clear.

b. Most evidence supports the hypothesis that prion proteins exist in two forms: the infectious, abnormally folded form and a normal cellular form. The interaction between the abnormal form and the cellular form converts the cellular form into the abnormal form.

Key Terms

acute infection 461	late genes 458	proto-oncogene 461	slow virus diseases 461
anaplasia 461	latent virus infection 461	proviral DNA 457	subgenomic mRNA 458
cancer 461	metastasis 461	replicase 455	syncytium 461
chronic virus infection 461	neoplasia 461	replicative form (RF) 455	transcriptase 455
cytocidal infection 459	oncogene 461	retrovirus 457	tropism 448
cytopathic effect (CPEs) 459	oncovirus 463	reverse transcriptase (RT) 457	tumor 461
defective interfering (DI) particle 461	persistent infection 461	ribonuclease H 457	tumor suppressor 463
early genes 454	prion 468	ribosomal frameshifting 458	viroid 467
inclusion bodies 461	procapsids 458	RNA silencing 468	virusoid 468

Critical Thinking Questions

1. Consider each of the following viral reproduction steps: adsorption, penetration, replication, and transcription. Suggest a strategy by which one could pharmacologically inhibit or discourage entry and propagation of viruses in animal cells. Can you explain host range using some of the same rationale?

2. Would it be advantageous for a virus to damage host cells? If it is not, why isn't damage to the host avoided? Is it possible that a virus might become less pathogenic when it has been associated with the host population for a longer time?

3. How does one prove that a virus is causing cancer? Try to think of approaches other than those discussed in the chapter. Give a major reason why it is so difficult to prove that a specific virus causes human cancer. Is it accurate to say that viruses cause cancer? Explain why or why not.

4. From what you know about cancer, is it likely that a single type of treatment can be used to cure it? What approaches might be effective in preventing cancer?

5. Propose some experiments that might be useful in determining what prions are and how they cause disease.

Learn More

Chien, P.; Weissman, J. S.; and DePace, A. H. 2004. Emerging principles of conformation-based prion inheritance. *Annu. Rev. Biochem.* 73:617–56.

Flint, S. J.; Enquist, L. W.; Racaniella, V. R.; and Skalka, A. M. 2004. *Principles of virology,* 2d ed. Washington, D.C.: ASM Press.

Gibbs, W. W. 2003. Untangling the roots of cancer. *Sci. Am.* 289(1):56–65.

Hull, R. 2002. *Matthew's Plant Virology,* 4th ed. San Diego: Academic Press.

Raoult, D.; Audic. S.; Robert, C; Abergel, C.; Renesto, P.; Ogata, H.; LaScola, B.; Suzan, M.; and Claverie, J.-M. 2004. The 1.2 megabase genome sequence of mimivirus. *Science* 306:1344–50.

Smith, A. E., and Helenius, A. 2004. How viruses enter animal cells. *Science* 304:237–41.

Wang. M. B.; Bian, X. Y.; Wu, L. M.; Liu, L. X.; Smith, N. A.; Isenegger, D.; Wu, R. M.; Masuta, C.; Vance, V. B.; Watson, J. M.; Rezaian, A.; Dennis, E. S.; and Waterhouse, P. M. 2004. On the role of RNA silencing in the pathogenicity and evolution of virioids and virus satellites. *Proc. Natl. Acad. Sci.* 101(9):3275–80.

19

Microbial Evolution, Taxonomy, and Diversity

The stromatolites shown here are layered rocks formed by incorporation of minerals into microbial mats. Fossilized stromatolites indicate that microorganisms existed early in Earth's history.

PREVIEW

- The evolutionary relationships among the three domains of life—*Bacteria, Archaea,* and *Eucarya*—are represented in the universal phylogenetic tree. The root, or origin, is placed early in the bacterial line of descent. This suggests that the *Archaea* and *Eucarya* share a common ancestry that is independent of the *Bacteria*. This long evolutionary history has generated a spectacular degree of microbial diversity.

- The RNA world hypothesis suggests that the first self-replicating entity was RNA and that this molecule formed the basis of the first primitive cell. Although it is unclear how the first eucaryotic nucleus arose, there is abundant evidence demonstrating that mitochondria and chloroplasts arose from endosymbiotic proteobacteria and cyanobacteria, respectively. Hydrogenosomes, found in some anaerobic protists, appear to share the same common ancestor as mitochondria.

- In order to make sense of the diversity of organisms, it is necessary to group similar organisms together and organize these groups in a nonoverlapping hierarchical arrangement. Taxonomy is the science of biological classification.

- A polyphasic approach is used to classify procaryotes. This combines information that is based on the analysis of microbial phenotypic, genotypic, and phylogenetic features. The results of these analyses are often summarized in treelike diagrams called dendrograms.

- Morphological, physiological, metabolic, ecological, genetic, and molecular characteristics are all useful in taxonomy because they reflect the organization and activity of the genome. Nucleic acid sequences are probably the best indicators of microbial phylogeny and relatedness because nucleic acids are either the genetic material itself or the products of gene transcription. Small subunit rRNAs and the genes that encode them display a number of features that make them useful in determining microbial phylogenies.

- Bacterial taxonomy is rapidly changing due to the acquisition of new data, particularly the use of molecular techniques such as the comparison of ribosomal RNA structure and chromosome sequences. This is leading to new phylogenetic classifications.

- The current edition of *Bergey's Manual of Systematic Bacteriology* is phylogenetically organized and distributes the *Bacteria* and *Archaea* among 25 phyla.

One of the most fascinating and attractive aspects of the microbial world is its extraordinary diversity. It seems that almost every possible shape, size, physiology, and lifestyle are represented. In this section of the text we focus on microbial diversity. Chapter 19 introduces the general principles of microbial evolution and taxonomy. This is followed by a five-chapter (20–24) survey of the most important procaryotic groups. Our survey of microbial diversity ends with an introduction to the major types of eucaryotic microorganisms: protists and fungi.

19.1 Microbial Evolution

Biological diversity is usually thought of in terms of plants and animals; yet, the assortment of microbial life forms is huge and largely uncharted. Consider the metabolic diversity of microorganisms—this alone suggests that the number of habitats occupied by microbes vastly exceeds that of all larger organisms. How has microbial life been able to radiate to such a bewildering level of diversity? To answer this question, one must consider microbial evolution. The field of microbial evolution, like any other scientific endeavor, is based on the formulation of hypotheses, the gathering of data, the analysis of the data, and the reformation of hypotheses based on newly acquired evidence. That is to say, the study of microbial evolution is based on the scientific method. To be sure, it is sometimes more difficult to amass evidence when considering events that occurred millions, and often billions, of

What's in a name? That which we call a rose by any other name would smell as sweet. . . .

—W. Shakespeare

tricarboxylic acid cycle (**figure 20.3***a*). This pathway is also present in the green sulfur bacteria. Methanogenic archaea and probably most extreme thermophiles incorporate CO_2 by the reductive acetyl-CoA pathway (figure 20.3*b*). A similar pathway also is present in acetogenic bacteria and autotrophic sulfate-reducing bacteria.

Archaeal Taxonomy

It should be clear by now that the *Archaea* are quite distinct from other living organisms. Within the domain, however, there is great diversity. As shown in **table 20.1,** the *Archaea* can be di-

vided into five major groups based on physiological and morphological differences.

On the basis of phylogenetic evidence, *Bergey's Manual* divides the *Archaea* into the phyla *Euryarchaeota* [Greek *eurus*, wide, and Greek *archaios*, ancient or primitive] and *Crenarchaeota* [Greek *crene*, spring or fount, and *archaios*]. The euryarchaeotes are given this name because they occupy many different ecological niches and have a variety of metabolic patterns. The phylum *Euryarchaeota* is very diverse with eight classes (*Methanobacteria, Methanococci, Halobacteria, Thermoplasmata, Thermococci, Archaeglobi, Methanopyri,* and the recently added *Methanomicro-*

Figure 20.3 Mechanisms of Autotrophic CO_2 Fixation. **(a)** The reductive tricarboxylic acid cycle. The cycle is reversed with ATP and reducing equivalents [H] to form acetyl-CoA from CO_2. The acetyl-CoA may be carboxylated to yield pyruvate, which can then be converted to glucose and other compounds. This sequence appears to function in *Thermoproteus neutrophilus.* **(b)** The synthesis of acetyl-CoA and pyruvate from CO_2 in *Methanobacterium thermoautotrophicum.* One carbon comes from the reduction of CO_2 to a methyl group, and the second is produced by reducing CO_2 to carbon monoxide through the action of the enzyme CO dehydrogenase (E_1). The two carbons are then combined to form an acetyl group. Corrin-E_2 represents the cobamide-containing enzyme involved in methyl transfers. Special methanogen coenzymes and enzymes are described in figures 20.10 and 20.11.

(a)

(b)

Table 20.1	**Characteristics of the Major Archaeal Physiological Groups**	
Group	**General Characteristics**	**Representative Genera**
Methanogenic archaea	Strict anaerobes. Methane is the major metabolic end product. S^0 may be reduced to H_2S without yielding energy production. Cells possess coenzyme M, factors 420 and 430, and methanopterin.	*Methanobacterium* *Methanococcus* *Methanomicrobium* *Methanosarcina*
Archaeal sulfate reducers	Irregular gram-negative coccoid cells. H_2S formed from thiosulfate and sulfate. Autotrophic growth with thiosulfate and H_2. Can grow heterotrophically. Traces of methane also formed. Extremely thermophilic and strictly anaerobic. Possess factor 420 and methanopterin but not coenzyme M or factor 430.	*Archaeoglobus*
Extremely halophilic archaea	Rods, cocci, or irregular shaped cells, that may include pyramids or cubes. Stain gram negative or gram positive, but like all archaea lack peptidoglycan. Primarily chemoorganoheterotrophs. Most species require sodium chloride ≥ 1.5 M, but some survive in as little as 0.5 M. Most produce characteristic bright-red colonies; some are unpigmented. Neutrophilic to alkalophilic. Generally mesophilic; however, at least one species is known to grow at 55°C. Possess either bacteriorhodopsin or halorhodopsin and can use light energy to produce ATP.	*Halobacterium* *Halococcus* *Natronobacterium*
Cell wall-less archaea	Pleomorphic cells lacking a cell wall. Thermoacidophilic and chemoorganotrophic. Facultatively anaerobic. Plasma membrane contains a mannose-rich glycoprotein and a lipoglycan.	*Thermoplasma*
Extremely thermophilic S^0-metabolizers	Gram-negative rods, filaments, or cocci. Obligately thermophilic (optimum growth temperature between 70–110°C). Usually strict anaerobes but may be aerobic or facultative. Acidophilic or neutrophilic. Autotrophic or heterotrophic. Most are sulfur metabolizers. S^0 reduced to H_2S anaerobically; H_2S or S^0 oxidized to H_2SO_4 aerobically.	*Desulfurococcus* *Pyrodictium* *Pyrococcus* *Sulfolobus* *Thermococcus* *Thermoproteus*

bia), nine orders, and 16 families. The methanogens, extreme halophiles, sulfate reducers, and many extreme thermophiles with sulfur-dependent metabolism are located in the *Euryarchaeota*. Methanogens are the dominant physiological group.

The crenarchaeotes (**figure 20.4**) are thought to resemble the ancestor of the *Archaea,* and almost all the well-characterized species are thermophiles or hyperthermophiles. The phylum *Crenarchaeota* has only one class, *Thermoprotei,* which is divided into four orders and six families. The order *Thermoproteales* contains gram-negative-staining anaerobic to facultative, hyperthermophilic rods. They often grow chemolithoautotrophically by reducing sulfur to hydrogen sulfide. Members of the order *Sulfolobales* are coccus-shaped thermoacidophiles. The order *Desulfurococcales* contains gram-negative-staining coccoid or disk-shaped hyperthermophiles. They grow either chemolithotrophically by hydrogen oxidation or organotrophically by fermentation or respiration with sulfur as the electron acceptor. The order *Caldisphaerales* was recently added. It has only one genus, *Caldisphaera,* whose members are thermoacidophilic, aerobic, heterotrophic cocci. The taxonomy of both phyla will undoubtedly undergo further revisions as more organisms are discovered. This is particularly the case with the crenarchaeotes because of the discovery of mesophilic forms in the ocean; these crenarchaeotes may constitute a significant fraction of the oceanic picoplankton.

1. What are the *Archaea?* Briefly describe the major ways in which they differ from *Bacteria* and eucaryotes.
2. How do archaeal cell walls differ from those of the *Bacteria?* What is pseudomurein?
3. In what ways do archaeal membrane lipids differ from those of *Bacteria* and eucaryotes? How do these differences contribute to the survival of thermophilic and hyperthermophilic archaea?
4. List the differences between *Archaea* and other organisms with respect to DNA replication, transcription, and translation.
5. Briefly describe the way in which archaea degrade and synthesize glucose. In what two unusual ways do they incorporate CO_2?
6. How are the phyla *Euryarchaeota* and *Crenarchaeota* distinguished?

20.2 PHYLUM *CRENARCHAEOTA*

As mentioned previously, most of the crenarchaeotes that have been cultured are extremely thermophilic, and many are acidophiles and sulfur dependent. The sulfur may be used either as an electron acceptor in anaerobic respiration or as an electron donor by lithotrophs. Many are strict anaerobes. They grow in geothermally heated water or soils that contain elemental sulfur. These environments are scattered all over the world. Examples are the

(a)

(b)

(c)

Figure 20.8 **The Phylum *Euryarchaeota*.** A phylogenetic tree developed from the nucleotide sequences of both small and large subunit rRNA sequences.

Figure 20.9 **Selected Methanogens.** **(a)** *Methanobrevibacter smithii.* **(b)** *Methanogenium marisnigri;* electron micrograph (×45,000). **(c)** *Methanosarcina mazei;* SEM. Bar = 5 μm.

is a nickel tetrapyrrole serving as a cofactor for the enzyme methyl-CoM methylreductase. The pathway for methane synthesis is thought to function as shown in **figure 20.11.** It appears that ATP synthesis is linked with methanogenesis by electron transport, proton pumping, and a chemiosmotic mechanism. Some methanogens can live autotrophically by forming acetyl-CoA from two molecules of CO_2 and then converting the acetyl-CoA to pyruvate and other products (figure 20.3*b*). Electron transport and oxidative phosphorylation: The electron transport chain (section 9.5)

Methanogens thrive in anoxic environments rich in organic matter: the rumen and intestinal system of animals, freshwater and marine sediments, swamps and marshes, hot springs, anoxic sludge digesters, and even within anaerobic protozoa. Methanogens often are of ecological significance. The rate of methane production can be so great that bubbles of methane sometimes rise to the surface of a lake or pond. Rumen methanogens are so active that a cow can belch 200 to 400 liters of methane a day. Microbial interactions: The rumen ecosystem (section 30.1)

Methanogenic archaea are potentially of great practical importance since methane is a clean-burning fuel and an excellent energy source. For many years sewage treatment plants have been using the methane they produce as a source of energy for heat and electricity. Anaerobic digester microbes degrade particulate wastes such as sewage sludge to H_2, CO_2, and acetate. CO_2-reducing methanogens form CH_4 from CO_2 and H_2, while aceticlastic methanogens cleave acetate to CO_2 and CH_4 (about 2/3 of the methane produced by an anaerobic digester comes from acetate). A kilogram of organic matter can yield up to 600 liters of methane. It is quite likely that future research will greatly increase the efficiency of methane production and make methanogenesis an important source of pollution-free energy. Wastewater treatment (section 41.2)

Methanogenesis also can be an ecological problem. Methane absorbs infrared radiation and thus is a greenhouse gas. There is ev-

idence that atmospheric methane concentrations have been rising over the last 200 years. Methane production may significantly promote future global warming. Recently it has been discovered that methanogens can oxidize Fe^0 and use it to produce methane and energy. This means that methanogens growing around buried or submerged iron pipes and other objects may contribute significantly to iron corrosion. Soil microorganisms and the atmosphere (section 29.6)

Table 20.2 | **Selected Characteristics of Representative Genera of Methanogens**

Genus	Morphology	% G + C	Wall Composition	Gram Reaction	Motility	Methanogenic Substrates Used
Order *Methanobacteriales*						
Methanobacterium	Long rods or filaments	32–61	Pseudomurein	+ to variable	−	$H_2 + CO_2$, formate
Methanothermus	Straight to slightly curved rods	33	Pseudomurein with an outer protein S-layer	+	+	$H_2 + CO_2$
Order *Methanococcales*						
Methanococcus	Irregular cocci	29–34	Protein	−	+	$H_2 + CO_2$, formate
Order *Methanomicrobiales*						
Methanomicrobium	Short curved rods	45–49	Protein	−	+	$H_2 + CO_2$, formate
Methanogenium	Irregular cocci	52–61	Protein or glycoprotein	−	−	$H_2 + CO_2$, formate
Methanospirillum	Curved rods or spirilla	47–52	Protein		+	$H_2 + CO_2$, formate
Order *Methanosarcinales*						
Methanosarcina	Irregular cocci, packets	36–43	Heteropolysaccharide or protein	+ to variable	−	$H_2 + CO_2$, methanol, methylamines, acetate

Microbial Diversity & Ecology

20.2 Methanotrophic Archaea

The marine environment may contain as much as 10,000 billion tons of methane hydrate buried in the ocean floor, around twice the amount of all known fossil fuel reserves. Although some methane rises toward the surface, it often is used before it escapes from the sediments in which it is buried. This is fortunate because methane is a much more powerful greenhouse gas than carbon dioxide. If the atmosphere were flooded with methane, the Earth could become too hot to support life as we know it. The reason for this disappearance of methane in sediments has been unclear until a recent discovery.

By using fluorescent probes for specific DNA sequences, an assemblage of archaea and bacteria has been discovered in anoxic, methane-rich sediments. These clusters of procaryotes contain a core of about 100 archaea from the order *Methanosarcinales* surrounded by a layer of sulfate-reducing bacteria related to the *Desulfosarcina* (see **Box figure**). These two groups appear to cooperate metabolically in such a way that methane is anaerobically oxidized and sulfate reduced; perhaps the bacteria use waste products of methane oxidation to derive energy from sulfate reduction. Isotope studies show that the archaea feed on methane and the bacteria get much of their carbon from the archaea. These methanotrophs may be crucial contributors to the Earth's carbon cycle because it is thought that they oxidize as much as 300 million tons of methane annually.

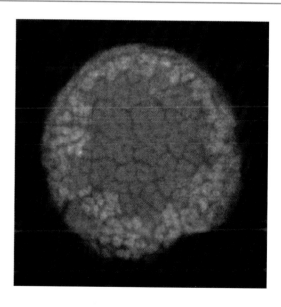

Methane-Consuming Archaea. A cluster of methanotrophic archaea, stained red by a specific fluorescent probe, surrounded by a layer of bacteria labeled by a green fluorescent probe.

(a) Methanofuran (MFR)

(b) Tetrahydromethanopterin (H₄MPT)

(c) Coenzyme M

(d) Coenzyme F₄₂₀

(e) Coenzyme F₄₃₀

Figure 20.10 Methanogen Coenzymes.
(a) Coenzyme MFR, **(b)** H₄MPT, and **(c)** coenzyme M are used to carry one-carbon units during methanogenesis. MFR and a simpler form of H₄MPT called methantopterin (MPT; not shown) also participate in the synthesis of acetyl-CoA. The portions of the coenzymes that carry the one-carbon units are shown in blue. H₄MPT carries carbon units on nitrogens 5 and 10, like the more common enzyme tetrahydrofolate. **(d)** Coenzyme F₄₂₀ participates in redox reactions. The part of the molecule that is reversibly oxidized and reduced is highlighted. **(e)** Coenzyme F₄₃₀ participates in reactions catalyzed by the enzyme methyl-CoM methylreductase.

The Halobacteria

The **extreme halophiles** or **halobacteria,** order *Halobacteriales,* are another major archaeal group, currently with 17 genera in one family, the *Halobacteriaceae* (**figure 20.12**). Most are aerobic chemoorganotrophs with respiratory metabolism. Extreme halophiles demonstrate a wide variety of nutritional capabilities. The first halophiles were isolated from salted fish in the 1880s and required complex nutrients such as yeast extract. More recent isolates grow best in defined media, using carbohydrates or simple compounds such as glycerol, acetate, or pyruvate as their carbon source. Halophiles can be motile or nonmotile and are found in a variety of cell shapes. These include cubes and pyramids in addition to rods and cocci.

The most obvious distinguishing trait of this family is its absolute dependence on a high concentration of NaCl. These procary-

otes require at least 1.5 M NaCl (about 8%, wt/vol), and usually have a growth optimum at about 3 to 4 M NaCl (17 to 23%). They will grow at salt concentrations approaching saturation (about 36%). The cell walls of most halobacteria are so dependent on the presence of NaCl that they disintegrate when the NaCl concentration drops below 1.5 M. Thus halobacteria only grow in high-salinity habitats such as marine salterns and salt lakes such as the Dead Sea between Israel and Jordan, and the Great Salt Lake in Utah. Halophiles are used in the production of many salted food products, including soy sauce. Halobacteria often have red-to-yellow pigmentation from carotenoids that are probably used as protection against strong sunlight. They can reach such high population levels that salt lakes, salterns, and salted fish actually turn red.

Probably the best-studied member of the family is *Halobacterium salinarium.* This archaeon is unusual because it produces a

Figure 20.11 Methane Synthesis. Pathway for CH_4 synthesis from CO_2 in *M. thermoautotrophicum*. Cofactor abbreviations: methanopterin (MPT), methanofuran (MFR), and 2-mercaptoethanesulfonic acid or coenzyme M (CoM). The nature of the carbon-containing intermediates leading from CO_2 to CH_4 are indicated in parentheses.

(a) *Halobacterium salinarium*

(b) *"Haloquadratum walsbyi"*

Figure 20.12 Examples of Halobacteria. **(a)** *Halobacterium*. A young culture that has formed long rods; SEM. Bar = 1 μm. **(b)** *"Haloquadratum walsbyi"*; SEM. Bar = 1 μm.

protein called **bacteriorhodopsin** that can trap light energy without the presence of chlorophyll. Structurally similar to the rhodopsin found in the mammalian eye, bacteriorhodopsin functions as a light-driven proton pump. Like all members of the rhodopsin family, bacteriorhodopsin has two distinct features: (1) a chromophore that is a derivative of retinal (an aldehyde of vitamin A), which is covalently attached to the protein by a Schiff base with the amino group of lysine (**figure 20.13**); and (2) seven membrane-spanning domains connected by loops on either side with the retinal resting within the membrane. Bacteriorhodpsin molecules form aggregates in a modified region of the cell membrane called **purple membrane.** When retinal absorbs light, the double bond between carbons 13 and 14 changes from a *trans* to a *cis* configuration and the Schiff base loses a proton. Protons move across

the plasma membrane to the periplasmic space during these alterations, and the Schiff base changes are directly involved in this movement (figure 20.13). Bacteriorhodopsin undergoes several conformational changes during the photocycle. These conformational changes also are involved in proton transport. The light-driven proton pumping generates a pH gradient that can be used to power the synthesis of ATP by a chemiosmotic mechanism. Electron transport and oxidative phosphorylation (section 9.5); Phototrophy: Rhodopsin-based phototrophy (section 9.12)

Halobacterium has three additional rhodopsins, each with a different function. Halorhodopsin uses light energy to transport chloride ions into the cell and maintain a 4 to 5 M intracellular KCl concentration. Two more are called sensory rhodopsin I (SRI) and SRII. **Sensory rhodopsins** act as photoreceptors, in this case,

Figure 20.13 The Photocycle of Bacteriorhodopsin. In this hypothetical mechanism the retinal component of bacteriorhodopsin is buried in the membrane and retinal interacts with two amino acids, A_1 and A_2 (aspartates 96 and 85), that can reversibly accept and donate protons. A_2 is connected to the cell exterior, and A_1 is closer to the cell interior. Light absorption by retinal in **step 1** triggers an isomerization from 13-*trans*-retinal to 13-*cis*-retinal. The retinal then donates a proton to A_2 in **steps 2 and 3,** while A_1 is picking up another proton from the interior and A_2 is releasing a proton to the outside. In **steps 4 and 5,** retinal obtains a proton from A_1 and isomerizes back to the 13-*trans* form. The cycle is then ready to begin again after **step 6.**

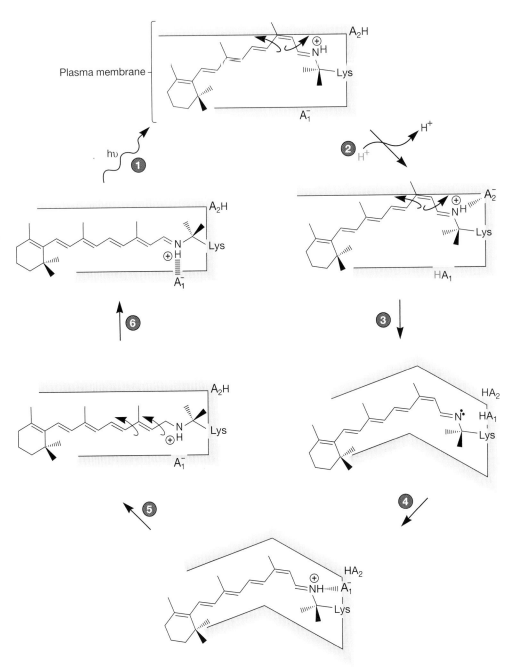

one for red light and one for blue. They control flagellar activity to position the organism optimally in the water column. *Halobacterium* moves to a location of high light intensity, but one in which ultraviolet light is not sufficiently intense to be lethal.

Surprisingly, it now appears rhodopsin is widely distributed among procaryotes. DNA sequence analysis of uncultivated marine bacterioplankton reveals the presence of rhodopsin genes among both α- and γ-proteobacteria. This newly discovered rhodopsin is called proteorhodopsin. Cyanobacteria also have rhodopsin proteins, which like SRI and SRII of the halobacteria, sense the spectral quality of light. Thus the cyanobacterial molecules are also considered sensory rhodopsins. These procaryotic

rhodopsins conserve the seven transmembrane helices through the cell membrane and the lysine residue that forms the Schiff base linkage with retinal. Microorganisms in marine environments: The photic zone of the open ocean (section 28.3)

The Thermoplasms

Procaryotes in the class *Thermoplasmata* are thermoacidophiles that lack cell walls. At present, three genera, *Thermoplasma, Picrophilus,* and *Ferroplasma* are known. They are sufficiently different from one another to be placed in separate families, *Thermoplasmataceae, Picrophilaceae,* and *Ferroplasmataceae.*

Figure 20.14 *Thermoplasma.* Transmission electron micrograph. Bar = 0.5 μm.

Thermoplasma grows in refuse piles of coal mines. These piles contain large amounts of iron pyrite (FeS), which is oxidized to sulfuric acid by chemolithotrophic bacteria. As a result the piles become very hot and acidic. This is an ideal habitat for *Thermoplasma* because it grows best at 55 to 59°C and pH 1 to 2. Although it lacks a cell wall, its plasma membrane is strengthened by large quantities of diglycerol tetraethers, lipid-containing polysaccharides, and glycoproteins. The organism's DNA is stabilized by association with archaeal histones that condense the DNA into structures resembling eucaryotic nucleosomes. At 59°C, *Thermoplasma* takes the form of an irregular filament, whereas at lower temperatures it is spherical (**figure 20.14**). The cells may be flagellated and motile.

Picrophilus is even more unusual than *Thermoplasma.* It originally was isolated from moderately hot solfataric fields in Japan. Although it lacks a regular cell wall, *Picrophilus* has an S-layer outside its plasma membrane. The cells grow as irregularly shaped cocci, around 1 to 1.5 μm in diameter, and have large cytoplasmic cavities that are not membrane bounded. *Picrophilus* is aerobic and grows between 47 and 65°C with an optimum of 60°C. It is most remarkable in its pH requirements: it grows only below pH 3.5 and has a growth optimum at pH 0.7. Growth even occurs at about pH 0!

Extremely Thermophilic S⁰-Metabolizers

This physiological group contains the class *Thermococci*, with one order, *Thermococcales.* The *Thermococcales* are strictly anaerobic and can reduce sulfur to sulfide. They are motile by flagella and have optimum growth temperatures around 88 to 100°C. The order contains one family and three genera, *Thermococcus, Paleococcus,* and *Pyrococcus.*

Sulfate-Reducing *Euryarchaeota*

Euryarchaeal sulfate reducers are found in the class *Archaeoglobi* and the order *Archaeoglobales.* This order has only one family and three genera. *Archaeoglobus* contains gram-negative-staining, irregular coccoid cells with cell walls consisting of glycoprotein subunits. It can extract electrons from a variety of electron donors (e.g., H_2, lactate, glucose) and reduce sulfate, sulfite, or thiosulfate to sulfide. Elemental sulfur is not used as an acceptor. *Archaeoglobus* is extremely thermophilic (the optimum is about 83°C) and has been isolated from marine hydrothermal vents. The organism is not only unusual in being able to reduce sulfate, unlike other archaea, but it also possesses the methanogen coenzymes F_{420} and methanopterin.

1. What are thermoacidophiles and where do they grow? In what ways do they use sulfur in their metabolism? Briefly describe *Sulfolobus* and *Thermoproteus.*
2. Generally characterize methanogenic archaea and distinguish them from other groups.
3. Briefly describe how methanogens produce methane and the roles of their unique cofactors in this process.
4. Where does one find methanogens? Discuss their ecological and practical importance.
5. Where are the extreme halophiles found and what is unusual about their cell walls and growth requirements?
6. What is the purple membrane and what pigment does it contain?
7. How is *Thermoplasma* able to live in acidic, very hot coal refuse piles when it lacks a cell wall? How is its DNA stabilized? What is so remarkable about *Picrophilus?*
8. Characterize *Archaeoglobus.* In what way is it similar to the methanogens and how does it differ from other extreme thermophiles?

Summary

20.1 Introduction to the *Archaea*

a. The *Archaea* are highly diverse with respect to morphology, reproduction, physiology, and ecology. Although best known for their growth in anoxic, hypersaline, and high-temperature habitats they also inhabit marine arctic, temperate, and tropical waters.

b. Archaeal cell walls do not contain peptidoglycan and differ from bacterial walls in structure. They may be composed of pseudomurein, polysaccharides, or glycoproteins and other proteins.

c. Archaeal membrane lipids differ from those of other organisms in having branched chain hydrocarbons connected to glycerol by ether links. Bacterial and eucaryotic lipids have glycerol connected to fatty acids by ester bonds.

d. Their tRNA, ribosomes, elongation factors, RNA polymerases, and other components distinguish *Archaea* from *Bacteria* and eucaryotes.

e. Although much of archaeal metabolism appears similar to that of other organisms, the *Archaea* differ with respect to glucose catabolism, pathways for CO_2 fixation, and the ability of some to synthesize methane (**figure 20.3**).

f. *Archaea* may be divided into five groups: methanogenic archaea, sulfate reducers, extreme halophiles, cell wall-less archaea, and extremely thermophilic S⁰-metabolizers (**table 20.1**).

g. The second edition of *Bergey's Manual* divides the *Archaea* into two phyla, the *Crenarchaeota* and *Euryarchaeota,* each with several orders (**figures 20.4 and 20.8**).

20.2 Phylum *Crenarchaeota*

a. The extremely thermophilic S⁰-metabolizers in the phylum *Crenarchaeota* depend on sulfur for growth and are frequently acidophiles. The sulfur may be used as an electron acceptor in anaerobic respiration or as an electron donor by lithotrophs. They are almost always strict anaerobes and grow in geothermally heated soil and water that is rich in sulfur.

20.3 Phylum *Euryarchaeota*

a. The phylum *Euryarchaeota* contains five major groups: methanogens, halobacteria, the thermoplasms, extremely thermophilic S⁰-metabolizers, and sulfate-reducing archaea.

b. Methanogenic archaea are strict anaerobes that can obtain energy through the synthesis of methane. They have several unusual cofactors that are involved in methanogenesis (**figures 20.10** and **20.11**).

c. Extreme halophiles or halobacteria are aerobic chemoheterotrophs that require at least 1.5 M NaCl for growth. They are found in habitats such as salterns, salt lakes, and salted fish.

d. *Halobacterium salinarum* can carry out phototrophy without chlorophyll or bacteriochlorophyll by using bacteriorhodopsin, which employs retinal to pump protons across the plasma membrane (**figure 20.13**).

e. The thermophilic archaeon *Thermoplasma* grows in hot, acidic coal refuse piles and survives despite the lack of a cell wall. Another thermoplasm, *Picrophilus,* can grow at pH 0.

f. The class *Thermococci* contains extremely thermophilic organisms that can reduce sulfur to sulfide.

g. Sulfate-reducing archaea are placed in the class *Archaeoglobi*. The extreme thermophile *Archaeoglobus* differs from other archaea in using a variety of electron donors to reduce sulfate. It also contains the methanogen cofactors F_{420} and methanopterin.

Key Terms

Archaea 503
bacteriorhodopsin 515
extreme halophiles 514

halobacteria 514
Korarchaeota 511
methanogens 510

pseudomurein 504
purple membrane 515

sensory rhodopsin 515
thermoacidophiles 508

Critical Thinking Questions

1. Do you think the *Archaea* should be separate from the *Bacteria* although both groups are procaryotic? Give your reasoning and evidence.

2. Explain why the fixation of CO_2 by *Thermoproteus* and possibly by *Sulfolobus* using a reductive reversal of the TCA cycle is *not* photosynthesis.

3. Often when the temperature increases, many procaryotes change their shapes from elongated rods into spheres. Suggest one reason for this change.

4. Why would ether linkages be more stable in membranes than ester lipids? How would the presence of tetraether linkages stabilize a thermophile's membrane?

5. Suppose you wished to isolate procaryotes from a hot spring in Yellowstone National Park. How would you go about it?

Learn More

Bell, S., and Jackson, S. P. 2001. Mechanism and regulation of transcription in archaea. *Curr. Opin. Microbiol.* 4:208–13.

Burggraf, S.; Huber, H.; and Stetter, K. O. 1997. Reclassification of the crenarchaeal orders and families in accordance with 16S rRNA sequence data. *Int. J. Syst. Bacteriol.* 47(3): 657–60.

Fuhrman, J. A., and Davis, A. A. 1997. Widespread archaea and novel bacteria from the deep sea as shown by 16S rRNA gene sequences. *Mar. Ecol. Prog. Ser.* 150: 275–85.

Gaasterland, T. 1999. Archaeal genomics. *Curr. Opin. Microbiol.* 2: 542–47.

Garrity, G. M., editor-in-chief. 2001. *Bergey's manual of systematic bacteriology,* 2d ed., vol. 1, D. R. Boone and R. W. Castenholz, editors. New York: Springer-Verlag.

Grabowski, B., and Kelman, Z. 2003. Archaeal DNA replication: Eukaryal proteins in a bacterial context. *Annu. Rev. Microbiol.* 57: 487–516.

Kashefi, K., and Lovley, D. 2004. Extending the upper temperature limit for life. *Science* 301: 934.

Kelman, L. M., and Kelman, A. 2003. *Archaea:* An archetype for replication initiation studies? *Mol. Microbiol.* 48: 605–15.

Oren, A. 1999. Bioenergetic aspects of halophilism. *Microbiol. Mol. Biol. Rev.* 63(2): 334–48.

Orphan, V. J.; House, C. H.; Hinrichs, K.-U.; McKeegan, K. D.; and DeLong, E. F. 2001. Methane-consuming archaea revealed by directly coupled isotopic and phylogenetic analysis. *Science* 293: 484–87.

Pereto, J.; Lopez-Garcia, P.; and Moreira, D. 2004. Ancestral lipid biosynthesis and early membrane evolution. *Trends Biochem. Sci.* 29: 469–77.

Sowers, K. R., and Schreier, H. J. 1999. Gene transfer systems for the *Archaea*. *Trends Microbiol.* 7(5): 212–19.

Walsby, A. E. 2005. *Archaea* with square cells. *Trends Microbiol.* 13: 193–95.

Waters, E., et al. 2003. The genome of *Nanobacterium equitans:* Insights into early *Archaeal* evolution and derived parasitism. *Proc. Natl. Acad. Sci. USA.* 100: 12984–88.

21

Bacteria:
The Deinococci and Nonproteobacteria Gram Negatives

Spirochetes are distinguished by their structure and mechanism of motility. *Treponema pallidum* shown here causes syphilis.

PREVIEW

- Some bacterial groups, such as those represented by the hyperthermophiles *Aquifex* and *Thermotoga,* are deeply branching and very old; other bacterial taxa have arisen more recently.

- All photosynthetic bacteria, including the cyanobacteria, were once considered a phenetically unified group. However, phylogenetic analysis reveals that the purple photosynthetic bacteria are proteobacteria, separating from them green sulfur and green nonsulfur bacteria. The cyanobacteria are separated from other photosynthetic bacteria because they resemble eucaryotic phototrophs in having both photosystems I and II and carrying out oxygenic photosynthesis. Their rRNA sequences also indicate that they are different from other photosynthetic bacteria.

- Bacteria, such as the chlamydiae, that are obligate intracellular parasites have relinquished some of their metabolic independence through loss of metabolic pathway genes. They use their host's energy supply and/or cell constituents.

- Gliding motility is widely distributed among bacteria and is very useful to organisms that digest insoluble nutrients or move over the surfaces of moist, solid substrata.

Chapter 20 surveys the *Archaea,* which are located in volume 1 of the second edition of *Bergey's Manual.* Volumes 1 and 5 of *Bergey's Manual* also describe a wide variety of other procaryotic groups that are members of the second domain: *Bacteria.* Chapter 21 is devoted to 10 of these bacterial phyla. Their phylogenetic locations are depicted in **figure 21.1.** We follow the general organization and perspective of the second edition of *Bergey's Manual* in most cases.

In describing each bacterial group, we include aspects such as distinguishing characteristics, morphology, reproduction, physiology, metabolism, and ecology. The taxonomy of each major group is summarized, and representative species are discussed. Students of microbiology should appreciate bacteria as living organisms rather than simply as agents of disease of little interest or importance in other contexts.

21.1 *AQUIFICAE* AND *THERMOTOGAE*

Thermophilic microbes are found in both the bacterial and archaeal domains, but to date, all hyperthermophilic procaryotes (those with optimum growth temperatures above 85°C) belong to the *Archaea.* The phyla *Aquificiae* and *Thermotoga* are two examples of bacterial thermophiles.

The phylum *Aquificae* is thought to represent the deepest or oldest branch of *Bacteria* (*see figure 19.3*). It contains one class, one order, and eight genera. Two of the best-studied genera are *Aquifex* and *Hydrogenobacter. Aquifex pyrophilus* is a gram-negative, microaerophilic rod. It is thermophilic with a temperature optimum of 85°C and a maximum of 95°C. *Aquifex* is a chemolithoautotroph that captures energy by oxidizing hydrogen, thiosulfate, and sulfur with oxygen as the terminal electron acceptor. Because *Aquifex* and *Hydrogenobacter* are both thermophilic chemolithoautotrophs, it has been suggested that the original bacterial ancestor was probably thermophilic and chemolithoautotrophic. Chemolithotrophy (section 9.11)

There are wide areas of the bacteriological landscape in which we have so far detected only some of the highest peaks, while the rest of the beautiful mountain range is still hidden in the clouds and the morning fogs of ignorance. The goal is still lying on the ground, but we have to bend down to grasp it.

—*Preface to* The Prokaryotes

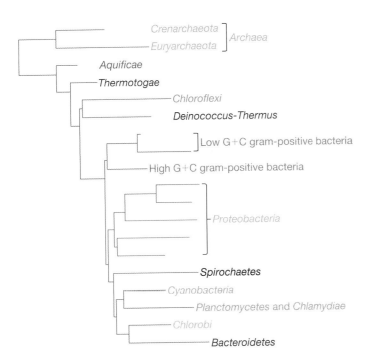

Figure 21.1 **Phylogenetic Relationships Among Procaryotes.** The *Deinococcus-Thermus* group and other nonproteobacterial gram-negatives are highlighted.

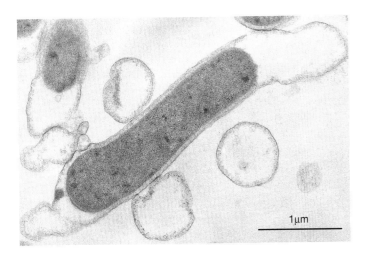

Figure 21.2 *Thermotoga maritima.* Note the loose sheath extending from each end of the cell.

The second oldest or deepest branch is the phylum *Thermotogae*, which has one class, one order, and six genera. The members of the genus *Thermotoga* [Greek *therme,* heat; Latin *toga,* outer garment], like *Aquifex,* are thermophiles with a growth optimum of 80°C and a maximum of 90°C. They are gram-negative rods with an outer sheathlike envelope (like a toga) that can extend or balloon out from the ends of the cell (**figure 21.2**). They grow in active geothermal areas found in marine hydrothermal systems and terrestrial solfataric springs. In contrast to *Aquifex, Thermotoga* is a

chemoheterotroph with a functional glycolytic pathway that can grow anaerobically on carbohydrates and protein digests.

The genome of *Aquifex* is about a third the size of the *Escherichia coli* genome and, as expected, contains the genes required for chemolithoautotrophy. The *Thermotoga* genome is somewhat larger and has genes for sugar degradation. About 24% of its coding sequences are similar to archaeal genes; this proportion is greater than that of other bacteria, including *Aquifex* (16% similarity) and may be due to lateral (horizontal) gene transfer. Comparative genomics (section 15.6)

21.2 DEINOCOCCUS-THERMUS

The phylum *Deinococcus-Thermus* contains the class *Deinococci* and the orders *Deinococcales* and *Thermales.* There are only three genera in the phylum; the genus *Deinococcus* is best studied. Deinococci are spherical or rod-shaped with distinctively different 16S rRNA. They often are associated in pairs or tetrads (**figure 21.3a**) and are aerobic, mesophilic, and catalase positive; usually they can produce acid from only a few sugars. Although they stain gram positive, their cell wall is layered with an outer membrane like the gram-negative bacteria (figure 21.3b). They also differ from gram-positive cocci in having L-ornithine in their peptidoglycan, lacking teichoic acid, and having a plasma membrane with large amounts of palmitoleic acid rather than phosphatidylglycerol phospholipids. Almost all strains are extraordinarily resistant to both desiccation and radiation; they can survive as much as 3 to 5 million rad of radiation (an exposure of 100 rad can be lethal to humans).

Much remains to be discovered about the biology of these bacteria. Deinococci can be isolated from ground meat, feces, air, freshwater, and other sources, but their natural habitat is not yet known. Their great resistance to radiation results from their ability to repair a severely damaged genome, which consists of a circular chromosome, a megaplasmid, and a small plasmid. When exposed to high levels of γ radiation, the genome is broken into many fragments. Within 12 to 24 hours, the genome is pieced back together, ensuring viability. It is unclear how this is accomplished. Genome studies have shown that *D. radiodurans* has a very efficient DNA repair system; however, novel DNA repair genes have not yet been reported (*see figure 15.18*). It appears that the ability to accumulate high levels of Mn(II) may help protect the microbe from high levels of radiation-induced toxic oxygen species. Currently, the mechanism by which Mn(II) confers protection is unclear. Insights from microbial genomes: Genomic analysis of extremophiles (section 15.8)

21.3 PHOTOSYNTHETIC BACTERIA

There are three groups of gram-negative photosynthetic bacteria: the purple bacteria, the green bacteria, and the cyanobacteria (**table 21.1**). The cyanobacteria differ most fundamentally from the green and purple photosynthetic bacteria in being able to carry out **oxygenic photosynthesis.** They have photosystems I

(a)

RS   PG   OM

(b)

Figure 21.3   The Deinococci.   (a) A *Deinococcus radiodurans* microcolony showing cocci arranged in tetrads (average cell diameter 2.5 μm). **(b)** The cell wall of *D. radiodurans* with a regular surface protein array (RS), a peptidoglycan layer (PG), and an outer membrane (OM). Bar = 100 nm.

and II, use water as an electron donor, and generate oxygen during photosynthesis. In contrast, purple and green bacteria have only one photosystem and use **anoxygenic photosynthesis.** Because they are unable to use water as an electron source, they employ reduced molecules such as hydrogen sulfide, sulfur, hydrogen, and organic matter as their electron source for the reduction of NAD(P)$^+$ to NAD(P)H. Consequently, many purple and green bacteria form sulfur granules. Purple sulfur bacteria accumulate granules within their cells, whereas green sulfur bacteria deposit the sulfur granules outside their cells. The purple nonsulfur bacteria normally use organic molecules as an electron source. There also are differences in photosynthetic pigments, the organization of photosynthetic membranes, nutritional requirements, and oxygen relationships. Phototrophy (section 9.12)

The differences in photosynthetic pigments and oxygen requirements among the photosynthetic bacteria have significant ecological consequences. As shown in **figure 21.4,** the chlorophylls, bacteriochlorophylls, and their associated accessory pigments have distinct absorption spectra. Oxygenic cyanobacteria and photosynthetic protists dominate the aerated upper layers of freshwater and marine microbial communities, where they absorb large amounts of red and blue light. Below these microbes, the anoxygenic purple and green photosynthetic bacteria inhabit the deeper anoxic zones that are rich in hydrogen sulfide and other reduced compounds that can be used as electron donors. Their bacteriochlorophyll and accessory

pigments enable them to use light in the far-red spectrum that is not used by other photosynthetic organisms **(table 21.2)**. In addition, the bacteriochlorophyll absorption peaks at about 350 to 550 nm, enabling them to grow at greater depths because shorter wavelength light can penetrate water farther. As a result, when the water is sufficiently clear, a layer of green and purple bacteria develops in the anoxic, hydrogen sulfide-rich zone.

Bergey's Manual places photosynthetic bacteria into seven major groups distributed between five bacterial phyla. The phylum *Chloroflexi* contains the green nonsulfur bacteria, and the phylum *Chlorobi,* the green sulfur bacteria. The cyanobacteria are placed in their own phylum, *Cyanobacteria.* Purple bacteria are divided between three groups. Purple sulfur bacteria are placed in the γ-proteobacteria, families *Chromatiaceae* and *Ectothiorhodospiraceae.* The purple nonsulfur bacteria are distributed between the α-proteobacteria (five different families) and one family of the β-proteobacteria. Finally, the gram-positive heliobacteria in the phylum *Firmicutes* are also photosynthetic. There appears to have been considerable horizontal transfer of photosynthetic genes among the five phyla. At least 50 genes related to photosynthesis are common to all five. In this chapter, we describe the cyanobacteria and green bacteria; purple bacteria are discussed in chapter 22, while heliobacteria are featured in chapter 23. Class *Gammaproteobacteria:* The purple sulfur bacteria (section 22.3); Class *Clostridia* (section 23.4)

| Characteristic | Anoxygenic Photosynthetic Bacteria | | | | Oxygenic Photosynthetic Bacteria |
	Green Sulfur[a]	Green Nonsulfur[b]	Purple Sulfur	Purple Nonsulfur	Cyanobacteria
Major photosynthetic pigments	Bacteriochlorophylls *a* plus *c*, *d*, or *e* (the major pigment)	Bacteriochlorophylls *a* and *c*	Bacteriochlorophyll *a* or *b*	Bacteriochlorophyll *a* or *b*	Chlorophyll *a* plus phycobiliproteins *Prochlorococcus* has divinyl derivatives of chlorophyll *a* and *b*
Morphology of photosynthetic membranes	Photosynthetic system partly in chlorosomes that are independent of the plasma membrane	Chlorosomes present when grown anaerobically	Photosynthetic system contained in spherical or lamellar membrane complexes that are continuous with the plasma membrane	Photosynthetic system contained in spherical or lamellar membrane complexes that are continuous with the plasma membrane	Thylakoid membranes lined with phycobilisomes
Photosynthetic electron donors	H_2, H_2S, S	Photoheterotrophic donors—a variety of sugars, amino acids, and organic acids; photoautotrophic donors—H_2S, H_2	H_2, H_2S, S	Usually organic molecules: sometimes reduced sulfur compounds or H_2	H_2O
Sulfur deposition	Outside of the cell		Inside the cell[c]	Outside of the cell in a few cases	
Nature of photosynthesis	Anoxygenic	Anoxygenic	Anoxygenic	Anoxygenic	Oxygenic (sometimes facultatively anoxygenic)
General metabolic type	Obligately anaerobic photolithoautotrophs	Usually photoheterotrophic; sometimes photoautotrophic or chemoheterotrophic (when aerobic and in the dark)	Obligately anaerobic photolithoautotrophs	Usually anaerobic photoorgano-heterotrophs; some facultative photolithoautotrophs (in the dark, chemo-organoheterotrophs)	Aerobic photo-lithoautotrophs
Motility	Nonmotile; some have gas vesicles	Gliding	Motile with polar flagella; some are peritrichously flagellated	Motile with polar flagella or nonmotile; some have gas vesicles	Nonmotile, swimming motility without flagella or gliding motility; some have gas vesicles
Percent G + C	48–58	53–55	45–70	61–72	35–71
Phylum or class	*Chlorobi*	*Chloroflexi*	α-, β-, and γ-proteobacteria	α-proteobacteria β-proteobacteria (*Rhodocyclus*)	*Cyanobacteria*

[a]Characteristics of *Chlorobi*.

[b]Characteristics of *Chloroflexus*.

[c]With the exception of *Ectothiorhodospira*.

Chlorophylls (chl)
& Bacteriochlorophylls (Bchl)

Phycobiliproteins

Carotenoids

Figure 21.4 Photosynthetic Pigments. Absorption spectra of five photosynthetic bacteria showing the differences in absorption maxima and the contributions of various accessory pigments.

Table 21.2	Procaryotic Bacteriochlorophyll and Chlorophyll Absorption Maxima	
	Long Wavelength Maxima (nm)	
Pigment	**In Ether or Acetone**	**Approximate Range of Values in Cells**
Chlorophyll *a*	665	680–685
Bacteriochlorophyll *a*	775	850–910 (purple bacteria)[a]
Bacteriochlorophyll *b*	790	1,020–1,035
Bacteriochlorophyll *c*	660	745–760
Bacteriochlorophyll *d*	650	725–745
Bacteriochlorophyll *e*	647	715–725

[a]The spectrum of bacteriochlorophyll *a* in green bacteria has a different maximum, 805–810 nm.

Phylum *Chlorobi*

The phylum *Chlorobi* has only one class (*Chlorobia*), order (*Chlorobiales*), and family (*Chlorobiaceae*). The **green sulfur bacteria** are a small group of obligately anaerobic photolitho-autotrophs that use hydrogen sulfide, elemental sulfur, and hydrogen as electron sources. The elemental sulfur produced by sulfide oxidation is deposited outside the cell. Their photosynthetic pigments are located in ellipsoidal vesicles called **chlorosomes** or chlorobium vesicles, which are attached to the plasma membrane but are not continuous with it. Chlorosomes are the most efficient light-harvesting complexes found in nature. The chlorosome membrane is not a normal lipid bilayer. Instead bacteriochlorophyll molecules are grouped into lateral arrays held to-

gether by carotenoids and lipids. Chlorosomes contain accessory bacteriochlorophyll pigments, but the reaction center bacteriochlorophyll is located in the plasma membrane where it obtains energy from chlorosome pigments. These bacteria flourish in the anoxic, sulfide-rich zones of lakes. Although they lack flagella and are nonmotile, some species have gas vesicles (**figure 21.5***a*) to adjust their depth for optimal light and hydrogen sulfide. Those forms without vesicles are found in sulfide-rich muds at the bottom of lakes and ponds.

The green sulfur bacteria are very diverse morphologically. They may be rods, cocci, or vibrios; some grow singly, and others form chains and clusters. They are either grass-green or chocolate-brown in color. Representative genera are *Chlorobium*, *Prosthecochloris*, and *Pelodictyon*.

Phylum *Chloroflexi*

The phylum *Chloroflexi* has both photosynthetic and nonphotosynthetic members. *Chloroflexus* is the major representative of the photosynthetic **green nonsulfur bacteria.** However, the term "green nonsulfur" is a misnomer because not all members of this group are green, and some use sulfur. *Chloroflexus* is a filamentous, gliding, thermophilic bacterium that often is isolated from neutral to alkaline hot springs where it grows in the form of orange-reddish mats, usually in association with cyanobacteria. It resembles the green sulfur bacteria with small chlorosomes and accessory bacteriochlorophyll *c*. However, like the purple bacteria, its light-harvesting complexes contain bacteriochlorophyll *a* and are in the plasma membrane. Finally, its metabolism is more similar to that of the purple nonsulfur bacteria. *Chloroflexus* can carry out anoxygenic photosynthesis with organic compounds as carbon sources or grow aerobically as a chemoheterotroph. It doesn't appear closely related to any other bacterial group based on 16S rRNA studies and is a deep and ancient branch of the bacterial tree (*see figure 19.3*). Genomic analysis of *Chloroflexus aurantiacus* should help elucidate the origin of photosynthesis.

Figure 21.5 Typical Green Sulfur Bacteria. **(a)** An electron micrograph of *Pelodictyon clathratiforme* (×105,000). Note the chlorosomes (dark gray areas) and gas vesicles (light gray areas with pointed ends). **(b)** *Chlorobium limicola* with extracellular sulfur granules.

(b) *Chlorobium limicola*

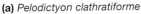
(a) *Pelodictyon clathratiforme*

The nonphotosynthetic, gliding, rod-shaped or filamentous bacterium *Herpetosiphon* also is included in this phylum. *Herpetosiphon* is an aerobic chemoorganotroph with respiratory metabolism that uses oxygen as the electron acceptor. It can be isolated from freshwater and soil habitats.

1. Give the major distinguishing characteristics of *Aquifex, Thermotoga,* and the deinococci. What is thought to contribute to the dessication and radiation resistance of the deinococci?
2. How do oxygenic and anoxygenic photosynthesis differ from each other and why? What is the ecological significance of these differences?
3. In general terms give the major characteristics of the following groups: purple sulfur bacteria, purple nonsulfur bacteria, and green sulfur bacteria. How do purple and green sulfur bacteria differ?
4. What are chlorosomes or chlorobium vesicles?
5. Compare the green nonsulfur (*Chloroflexi*) and green sulfur bacteria (*Chlorobi*).

Phylum *Cyanobacteria*

The **cyanobacteria** are the largest and most diverse group of photosynthetic bacteria. There is little agreement about the number of cyanobacterial species. Older classifications had as many as 2,000 or more species. In one recent system this has been reduced to 62 species and 24 genera. *Bergey's Manual of Systematic Bacteriology* describes 56 genera. Cyanobacterial diversity is reflected in the G + C content of the group, which ranges from 35 to 71%. Although cyanobacteria are gram-negative bacteria, their photosynthetic system closely resembles that of the eucaryotes because they have chlorophyll *a* and photosystems I and II, thereby performing oxygenic photosynthesis. Indeed, the cyanobacteria were once known as "blue-green al-

Figure 21.6 Cyanobacterial Thylakoids and Phycobilisomes. *Synechococcus lividus* with an extensive thylakoid system. The phycobilisomes lining these thylakoids are clearly visible as granules at location t (×85,000).

gae." Like the red algae, cyanobacteria use phycobiliproteins as accessory pigments. Photosynthetic pigments and electron transport chain components are located in thylakoid membranes lined with particles called **phycobilisomes (figure 21.6)**. These contain phycobilin pigments, particularly phycocyanin and phycoerythrin, that transfer energy to photosystem II. Carbon dioxide is assimilated through the Calvin cycle; the enzymes needed for this process are localized to internal structures called **carboxysomes.** The reserve carbohydrate is glycogen. Sometimes

Figure 21.7 Cyanobacterial Cell Structure. (**a**) Schematic diagram of a vegetative cell. The insert shows an enlarged view of the envelope with its outer membrane and peptidoglycan. (**b**) Thin section of *Synechocystis* during division. Many structures are visible. (a) Illustration copyright © Hartwell T. Crim, 1998.

they store extra nitrogen as polymers of arginine or aspartic acid in **cyanophycin** granules. Phosphate is stored in polyphosphate granules (**figure 21.7**). Because cyanobacteria lack the enzyme α-ketoglutarate dehydrogenase, they do not have a fully functional citric acid cycle. The pentose phosphate pathway plays a central role in their carbohydrate metabolism. Although many cyanobacteria are obligate photolithoautotrophs, a few can grow slowly in the dark as chemoheterotrophs by oxidizing glucose and a few other sugars. Under anoxic conditions *Oscillatoria limnetica* oxidizes hydrogen sulfide instead of water and carries out anoxygenic photosynthesis much like the green photosynthetic bacteria. Obviously, cyanobacteria are capable of considerable metabolic flexibility.

Cyanobacteria also vary greatly in shape and appearance. They range in diameter from about 1 to 10 μm and may be unicellular, exist as colonies of many shapes, or form filaments called trichomes (**figure 21.8c**). A **trichome** is a row of bacterial cells that are in close contact with one another over a large area. Although many appear blue-green because of phycocyanin, isolates from the open ocean are red or brown in color because of the pigment phycoerythrin. Cyanobacteria modulate the relative amounts of these pigments in a process known as **chromatic adaptation.** When orange light is perceived, phycocyanin production is stimulated, while blue and blue-green light promote the production of phycoerythrin. It is thought that sensory rhodopsins may play a role in

signaling the spectral quality of light (*see p. 515*). Many cyanobacterial species use gas vacuoles to position themselves in optimum illumination in the water column—a form of **phototaxis. Gliding motility** is used by other cyanobacteria (**Microbial Diversity & Ecology 21.1**). Although cyanobacteria lack flagella, about one-third of the marine *Synechococcus* strains swim at rates up to 25 μm/sec by an unknown mechanism. Swimming motility is not used for phototaxis; instead, it appears to be used in chemotaxis toward simple nitrogenous compounds such as urea.

Cyanobacteria show great diversity with respect to reproduction and employ a variety of mechanisms: binary fission, budding, fragmentation, and multiple fission. In the last process a cell enlarges and then divides several times to produce many smaller progeny, which are released upon the rupture of the parental cell. Fragmentation of filamentous cyanobacteria can generate small, motile filaments called **hormogonia.** Some species develop **akinetes,** specialized, dormant, thick-walled resting cells that are resistant to desiccation (**figure 21.9a**). Often these germinate to form new filaments.

Many filamentous cyanobacteria fix atmospheric nitrogen by means of special cells called **heterocysts** (figure 21.9). Around 5 to 10% of the cells develop into heterocysts when these cyanobacteria are deprived of both nitrate and ammonia, their preferred nitrogen sources. When individual cyanobacterial cells in

(a) *Chroococcus turgidus*

(b) *Nostoc*

(c) *Oscillatoria*

(d) *Anabaena spiroides* and *Microcystis aeruginosa*

Figure 21.8   Oxygenic Photosynthetic Bacteria.   Representative cyanobacteria. **(a)** *Chroococcus turgidus,* two colonies of four cells each (×600). **(b)** *Nostoc* with heterocysts (×550). **(c)** *Oscillatoria* trichomes seen with Nomarski interference-contrast optics (×250). **(d)** The cyanobacteria *Anabaena spiroides* and *Microcystis aeruginosa.* The spiral *A. spiroides* is covered with a thick gelatinous sheath (×1,000).

a filamentous chain differentiate into heterocysts, the heterocysts develop a very thick cell wall. Within these specialized cells, photosynthetic membranes are reorganized and the proteins that make up photosystem II and phycobilisomes are degraded. Photosystem I remains functional to produce ATP, but no oxygen is generated. This inability to generate O_2 is critical because the enzyme nitrogenase is extremely oxygen sensitive. The thick heterocyst wall slows or prevents O_2 diffusion into the cell, and any O_2 present is consumed during respiration. Heterocyst structure and physiology ensure that it remains anaerobic; it is dedicated to nitrogen fixation and does not replicate. It obtains nutrients from adjacent vegetative cells and contributes fixed nitrogen in the form of amino acids. Nitrogen fixation also is carried out by some cyanobacteria that lack heterocysts. Some fix nitrogen under dark, anoxic conditions in microbial mats. Planktonic forms such as *Trichodesmium* fix nitrogen and contribute significantly to the marine nitrogen budget. Synthesis of amino acids: Nitrogen fixation (section 10.5); Biogeochemical cycling: Nitrogen cycle (section 27.2)

The classification of cyanobacteria is still unsettled. At present all taxonomic schemes must be considered tentative and while many genera have been assigned, species names have not yet been designated in most cases. *Bergey's Manual* divides the cyanobacteria into five subsections with 56 genera (**table 21.3**). These are distinguished using cell or filament morphology and reproductive patterns. Some other properties important in cyanobacterial characterization are ultrastructure, genetic characteristics, physiology and biochemistry, and habitat/ecology (preferred habitat and growth habit). Subsection I contains unicellular rods or cocci. Most are nonmotile and all reproduce by binary fission or budding. Organisms in subsection II are also unicellular, though several individual cells may be held together in an aggregate by an outer wall. Members of this group reproduce by multiple fission to form spherical, very small, reproductive cells called **baeocytes,** which escape when the outer wall ruptures. Some baeocytes disperse through gliding motility. The other three subsections contain filamentous cyanobacteria. Filaments are often surrounded by a sheath or slime layer. Cyanobacteria in subsection III form unbranched trichomes composed only of vegetative cells, whereas the other two subsections produce heterocysts in the absence of an adequate nitrogen source and also may

Microbial Diversity & Ecology

21.1   The Mechanism of Gliding Motility

Gliding motility always occurs on a solid surface, but it varies greatly in rate (from about 2 μm per minute to over 600 μm per minute) and in the nature of the motion. Although first observed over 100 years ago, the mechanism by which bacteria glide remains a mystery. Bacteria such as *Myxococcus* and *Flexibacter* glide along in a direction parallel to the longitudinal axis of their cells. Others (*Saprospira*) travel with a screwlike motion or even move in a direction perpendicular to the long axis of the cells in their trichome (*Simonsiella*). *Beggiatoa,* cyanobacteria, and some other bacteria rotate around their longitudinal axis while gliding, but this is not always seen. Many will flex or twitch as well as glide. Such diversity in gliding movement may indicate that more than one mechanism for motility exists. This conclusion is supported by the observation that some gliders (e.g., *Cytophaga, Flexibacter,* and *Flavobacterium*) move attached latex beads over their surface, whereas others such as *Myxococcus* either do not move beads or move them very slowly. (That is, not all gliding bacteria have rapidly moving cell-surface components.) Although slime is required for gliding, it does not appear to propel bacteria directly; rather, it probably attaches them to the substratum and lubricates the surface for more efficient movement.

A variety of mechanisms for gliding motility have been proposed. Cytoplasmic fibrils or filaments are associated with the envelope of many gliding bacteria. In *Oscillatoria* they seem to be contractile and may produce waves in the outer membrane, resulting in movement. Gliding motility may be best understood in the bacterium *Myxococcus xanthus.* This rod-shaped microbe glides in a pattern of reversals, so it has two motility "engines"—one at each pole. When gliding forward, type IV pili located at the front pole pull cells forward. This type of gliding motility is possible only when the cells are in a group and able to contact one another. Thus it is called social, or S, motility. The second motility engine consists of nozzlelike structures at the rear pole that eject slime to push the cells along the surface. Because slime production can occur in solitary cells, gliding driven by this mechanism is called adventurous, or A, motility. While the pili are always localized to the front of the cell, the nozzle structures are very dynamic and switch from one pole to another during gliding reversals. The presence of two different gliding mechanisms in a single bacterium suggests that further investigation may reveal additional diversity among gliding microorganisms. Class *Deltaproteobacteria:* Order *Myxococcales* (section 22.4)

(a) *Cylindrospermum*

(b) *Anabaena*

(c) *Anabaena* heterocyst

Figure 21.9   Examples of Heterocysts and Akinetes.
(a) *Cylindrospermum* with terminal heterocysts (H) and subterminal akinetes (A) (×500). **(b)** *Anabaena,* with heterocysts. **(c)** An electron micrograph of an *Anabaena* heterocyst. Note the cell wall (W), additional outer walls (E), membrane system (M), and a pore channel to the adjacent cell (P).

Table 21.3	Characteristics of the Cyanobacterial Subsections					
Subsection	General Shape	Reproduction and Growth	Heterocysts	% G + C	Other Properties	Representative Genera
I	Unicellular rods or cocci; nonfilamentous aggregates	Binary fission, budding	−	31–71	Nonmotile or swim without flagella	*Chroococcus* *Gloeothece* *Gleocapsa* *Prochlorococcus* *Prochloron* *Synechococcus*
II	Unicellular rods or cocci; may be held together in aggregates	Multiple fission to form baeocytes	−	40–46	Only some baeocytes are motile	*Pleurocapsa* *Dermocarpella* *Chroococcidiopsis*
III	Filamentous, unbranched trichome with only vegetative cells	Binary fission in a single plane, fragmentation	−	34–67	Usually motile	*Lyngbya* *Oscillatoria* *Prochlorothrix* *Spirulina* *Pseudanabaena*
IV	Filamentous, unbranched trichome may contain specialized cells	Binary fission in a single plane, fragmentation to form hormogonia	+	38–47	Often motile, may produce akinetes	*Anabaena* *Cylindrospermum* *Aphanizomenon* *Nostoc* *Scytonema* *Calothrix*
V	Filamentous trichomes either with branches or composed of more than one row of cells	Binary fission in more than one plane, hormogonia formed	+	42–44	May produce akinetes; greatest morphological complexity and differentiation in cyanobacteria	*Fischerella* *Stigonema* *Geitleria*

form akinetes. Heterocystous cyanobacteria are subdivided into those that form unbranched filaments (subsection IV) and cyanobacteria that divide in a second plane to produce branches or aggregates (subsection V).

The cyanobacteria that are collectively referred to as prochlorphytes (genera *Prochloron, Prochlorococcus,* and *Prochlorothrix*) are distinguished by the presence of both chlorophyll *a* and *b* and their lack of phycobilins. This pigment arrangement imparts a grass-green color to these microbes. As the only procaryotes to possess chlorophyll *b,* the ancestors of unicellular prochlorophytes are considered by some to be the best candidates as the endosymbionts that gave rise to chloroplasts. Considerable controversy has surrounded their phylogeny, but small subunit rRNA sequence analysis places them with the cyanobacteria. Microbial evolution: The endosymbiotic origin of mitochondria and chloroplasts (section 19.1)

The three recognized prochlorophyte genera are quite different from one another. *Prochloron* was first discovered as an extracellular symbiont growing either on the surface or within the cloacal cavity of marine colonial ascidian invertebrates (**figure 21.10**). These bacteria are single-celled, spherical, and from 8 to 30 µm in diameter. Their mol% of G + C is 31 to 41. *Prochlorothrix* is free living, has cylindrical cells that form filaments, and grows in freshwater. Its DNA has a higher G + C content (53 mol%).

Prochlorococcus marinus, which is less than 1 µm in diameter, flourishes about 100 meters below the ocean surface. It differs from other prochlorophytes in having divinyl chlorophyll *a* and *b* and α-carotene instead of chlorophyll *a* and β-carotene. During the summer, it reaches concentrations of 5×10^5 cells per milliliter. It is one of the most numerous of the marine plankton and may be the most abundant oxygenic photosynthetic organism on Earth.

Another indication of the vast diversity among the cyanobacteria is the wide range of habitats they occupy. Thermophilic species may grow at temperatures of up to 75°C in neutral to alkaline hot springs. Because these photoautotrophs are so hardy, they are primary colonizers of soils and surfaces that are devoid of plant growth. Some unicellular forms even grow in the fissures of desert rocks. In nutrient-rich warm ponds and lakes, surface cyanobacteria such as *Anacystis* and *Anabaena* can reproduce

(a)   (b)

Figure 21.10   Prochloron.   (a) A scanning electron micrograph of *Prochloron* cells on the surface of a *Didemnum candidum* colony. **(b)** *Prochloron didemni* thin section; transmission electron micrograph (×23,500).

Figure 21.11   Bloom of Cyanobacteria and Algae in a Eutrophic Pond.

rapidly to form blooms (**figure 21.11**). The release of large amounts of organic matter upon the death of the bloom microorganisms stimulates the growth of chemoheterotrophic bacteria. These microbes subsequently deplete available oxygen. This kills fish and other organisms. Some species can produce toxins that kill livestock and other animals that drink the water. Other cyanobacteria (e.g., *Oscillatoria*) are so pollution resistant and

characteristic of freshwaters with high organic matter content that they are used as water pollution indicators. Disease 25.1: Harmful algal blooms; Microorganisms in marine and freshwater environments (chapter 28)

Cyanobacteria are particularly successful in establishing symbiotic relationships with other organisms. For example, they are the photosynthetic partner in most lichen associations. Cyanobacteria are symbionts with protozoa and fungi, and nitrogen-fixing species form associations with a variety of plants (liverworts, mosses, gymnosperms, and angiosperms). Microbial interactions (section 30.1)

1. Summarize the major characteristics of the cyanobacteria that distinguish them from other photosynthetic organisms.
2. Define or describe the following: phycobilisomes, hormogonia, akinetes, heterocysts, and baeocytes.
3. What is a trichome and how does it differ from a simple chain of cells?
4. Briefly discuss the ways in which cyanobacteria reproduce.
5. Describe how a vegetative cell, a heterocyst, and an akinete are different. How are heterocysts modified to carry out nitrogen fixation? Why do you think heterocysts are considered terminally differentiated cells?
6. Give the features of the five major cyanobacterial groups.
7. Compare the prochlorophytes with other cyanobacteria and chloroplasts. Why do you think the phylogeny of the prochlorophytes has been so difficult to establish?
8. List some important positive and negative impacts cyanobacteria have on humans and the environment.

(a)   **(b)**

Gemmata obscuriglobus   *"Candidatus* Brocadia *anammoxidans"*

(c)

Figure 21.12   Planctomycete Cellular Compartmentalization.   **(a)** An electron micrograph of *Gemmata obscuriglobus* showing the nuclear body envelope (E), the intracytoplasmic membrane (ICM), and the paryphoplasm (P). **(b)** An electron micrograph of the anaerobic ammonia-oxidizing planctomycete *"Candidatus Brocadia anammoxidans"* (quotation marks indicate unsettled nomenclature). The anamoxosome is labeled AM. **(c)** Schematic drawings corresponding to (a) and (b): cell wall (CW), cytoplasmic membrane (CM).

21.4   Phylum *Planctomycetes*

The phylum *Planctomycetes* contains one class, one order, and four genera. The planctomycetes are morphologically unique bacteria, having compartmentalized cells **(figure 21.12)**. Although each species is unique, all follow a basic cellular organization that includes a cytoplasmic membrane closely surrounded by the cell wall, which lacks peptidoglycan. The largest internal compartment, the intracytoplasmic membrane (ICM), is sepa-

rated from the cytoplasmic membrane by a peripheral, ribosome-free region called the paryphoplasm. The nucleoid of the planctomycete *Gemmata obscuriglobus* is located in the nuclear body, which is enclosed in a double membrane. The species *"Candidatus Brocadia anammoxidans"* does not localize its DNA within a nuclear body; however, it has another compartment: the anammoxosome. This is the site of anaerobic ammonia oxidation, a unique and recently discovered form of chemolithotrophy in which ammonium ion (NH_4^+) serves as the electron donor and nitrite (NO_2^-) as the terminal electron acceptor; it is reduced to nitrogen gas (N_2). Note that the nomenclature of this microbe is still in flux so in some cases genus and species names are enclosed in quotation marks.   Biogeochemical cycling: Nitrogen cycle (section 27.2)

The genus *Planctomyces* attaches to surfaces through a stalk and holdfast; the other genera in the order lack stalks. Most of these bacteria have life cycles in which sessile cells bud to produce motile swarmer cells. The swarmer cells are flagellated and swim for a while before settling down to attach and begin reproduction.

21.5 PHYLUM *CHLAMYDIAE*

The gram-negative *Chlamydiae* are obligate intracellular parasites. That is, they must grow and reproduce within host cells. Although their ability to cause disease is widely recognized, many species grow within protists and animal cells without adverse affects. It is thought that these hosts represent a natural reservoir for the *Chlamydiae*.

The phylum *Chlamydiae* has one class, one order, four families, and only six genera. The genus *Chlamydia* is by far the most important and best studied; it will be the focus of our attention. **Chlamydiae** are nonmotile, coccoid bacteria, ranging in size from 0.2 to 1.5 μm. They reproduce only within cytoplasmic vesicles of host cells by a unique developmental cycle involving the formation of two cell types: elementary bodies and reticulate bodies. Although their envelope resembles that of other gram-negative bacteria, the cell wall differs in lacking muramic acid and a peptidoglycan layer. Elementary bodies achieve osmotic stability by cross-linking their outer membrane proteins, and possibly periplasmic proteins, with disulfide bonds. Chlamydiae are extremely limited metabolically, relying on their host cells for key metabolites. This is reflected in the size of their genome. It is relatively small at 1.0 to 1.3 Mb; the G + C content is 41 to 44%.

Chlamydial reproduction begins with the attachment of an **elementary body (EB)** to the cell surface (**figure 21.13**). Elementary bodies are 0.2 to 0.6 μm in diameter, contain electron-dense nuclear material and a rigid cell wall, and are infectious. The host cell phagocytoses the EB, which are held in inclusion bodies where the EB reorganizes itself to form a **reticulate body (RB)** or **initial body.** The RB is specialized for reproduction rather than infection. Reticulate bodies are 0.6 to 1.5 μm in diameter and have less dense nuclear material and more ribosomes than EBs; their walls are also more flexible. About 8 to 10 hours after infection, the reticulate body undergoes binary fission and RB reproduction continues until the host cell dies. A chlamydia-filled inclusion can become large enough to be seen

(a)

(b)

Figure 21.13   The Chlamydial Life Cycle.   **(a)** An electron micrograph of an inclusion body containing a mixture of small, black elementary bodies (EB), larger reticulate bodies (RB), and an intermediate body (IB), a chlamydial cell intermediate in morphology between EB and RB. The RBs appear gray and granulated due to a high concentration of ribosomes. **(b)** A schematic representation of the infectious cycle of chlamydiae.

in a light microscope and even fill the host cytoplasm. After 20 to 25 hours, RBs begin to differentiate into infectious EBs and continue this process until the host cell lyses and releases the chlamydiae 40 to 72 hours after infection.

Chlamydial metabolism is very different from that of other gram-negative bacteria. It had been thought that chlamydiae cannot catabolize carbohydrates or other substances or synthesize ATP. *Chlamydia psittaci,* one of the best-studied species, lacks both flavoprotein and cytochrome electron transport chain carriers, but has a membrane translocase that acquires host ATP in exchange for ADP. Thus chlamydiae seem to be energy parasites that are completely dependent on their hosts for ATP. However, this might not be the complete story. The *C. trachomatis* genome sequence indicates that the bacterium may be able to synthesize at least some ATP. Although there are two genes for ATP/ADP translocases, there also are genes for substrate-level phosphorylation, electron transport, and oxidative phosphorylation. When supplied with precursors from the host, RBs can synthesize DNA, RNA, glycogen, lipids, and proteins. Presumably the RBs have porins and active membrane transport proteins, but little is known about these. They also can synthesize at least some amino acids and coenzymes. The EBs have minimal metabolic activity and cannot take in ATP or synthesize proteins. They are designed exclusively for transmission and infection. Insights from microbial genomes: Genomic analysis of pathogenic microbes (section 15.8)

Three chlamydial species are important pathogens of humans and other warm-blooded animals. *C. trachomatis* infects humans and mice. In humans it causes trachoma, nongonococcal urethritis, and other diseases. *C. psittaci* causes psittacosis in humans. However, unlike *C. trachomatis,* it also infects many other animals (e.g., parrots, turkeys, sheep, cattle, and cats) and invades the intestinal, respiratory, and genital tracts; the placenta and fetus; the eye; and the synovial fluid of joints. *Chlamydiophila pneumoniae* is a common cause of human pneumonia. Human diseases caused by bacteria (chapter 38)

21.6 PHYLUM *SPIROCHAETES*

The phylum *Spirochaetes* (Greek *spira,* a coil, and *chaete,* hair) contains gram-negative, chemoheterotrophic bacteria distinguished by their structure and mechanism of motility. They are slender, long bacteria (0.1 to 3.0 μm by 5 to 250 μm) with a flexible, helical shape (**figure 21.14**). Many species are so slim that they are only clearly visible in a light microscope by means of phase-contrast or dark-field optics. Spirochetes differ greatly from other bacteria with respect to motility and can move through very viscous solutions though they lack external rotating flagella. When in contact with a solid surface, they exhibit creeping or crawling movements. Their unique pattern of motility is due to an unusual morphological structure called the axial filament. The light microscope (section 2.2)

The distinctive features of spirochete morphology are evident in electron micrographs (**figure 21.15**). The central protoplasmic cylinder contains cytoplasm and the nucleoid, and is bounded by a plasma membrane and a gram-negative cell wall. Two to more than a hundred flagella, called **axial fibrils, periplasmic flagella,** or endoflagella, extend from both ends of the cylinder and often overlap one another in the center third of the cell (figure 21.16c,d). The whole complex of periplasmic flagella, the **axial filament,** lies inside a flexible outer sheath. The outer sheath contains lipid, protein, and carbohydrate and varies in structure between different genera. Its precise function is unknown, but the sheath is essential because spirochetes die if it is damaged or removed. The outer sheath of *Treponema pallidum* has few proteins exposed on

(a) *Cristispira*

(c) *Leptospira interrogans*

(b) *Treponema pallidum*

Figure 21.14 The Spirochetes. Representative examples. **(a)** *Cristispira* sp. from a clam; phase contrast (× 2,200). **(b)** *Treponema pallidum* (×1,000). **(c)** *Leptospira interrogans* (×2,200).

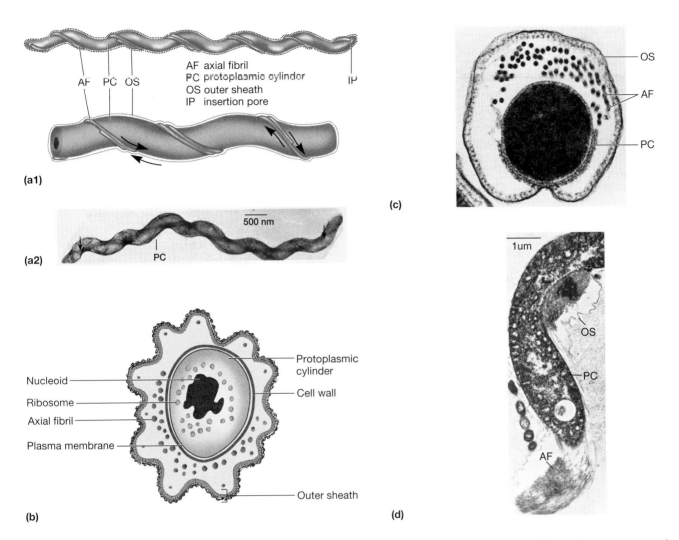

Figure 21.15   Spirochete Morphology.   (a1) A surface view of spirochete structure as interpreted from electron micrographs. **(a2)** A longitudinal view of *Treponema zuelzerae* with axial fibrils extending most of the cell length. **(b)** A cross section of a typical spirochete showing morphological details. **(c)** Electron micrograph of a cross section of *Clevelandina* from the termite *Reticulitermes flavipes* showing the outer sheath, protoplasmic cylinder, and axial fibrils (×70,000). **(d)** Longitudinal section of *Cristispira* showing the outer sheath (OS), the protoplasmic cylinder (PC), and the axial fibrils (AF).

its surface. This allows the syphilis spirochete to avoid attack by host antibodies.

The way in which periplasmic flagella propel the cell has not been fully established. Mutants with straight rather than curved flagella are nonmotile. The periplasmic flagella rotate like the external flagella of other bacteria. This causes the corkscrew-shaped outer sheath to rotate and move the cell through the surrounding liquid (**figure 21.16**). Flagellar rotation may also flex or bend the cell and account for the crawling movement seen on solid surfaces.

Spirochetes can be anaerobic, facultatively anaerobic, or aerobic. Carbohydrates, amino acids, long-chain fatty acids, and long-chain fatty alcohols may serve as carbon and energy sources.

Figure 21.16   Spirochete Motility.   A hypothetical mechanism for spirochete motility. See text for details.

The group is exceptionally diverse ecologically and grows in habitats ranging from mud to the human mouth. Members of the genus *Spirochaeta* are free-living and often grow in anoxic and sulfide-rich freshwater and marine environments. Some species of the genus *Leptospira* grow in oxic water and moist soil. Some spirochetes form symbiotic associations with other organisms and are found in a variety of locations: the hindguts of termites and wood-eating roaches, the digestive tracts of mollusks (*Cristispira*) and mammals, and the oral cavities of animals (*Treponema denticola, T. oralis*). Spirochetes from termite hindguts and freshwater sediments have nitrogenase and can fix nitrogen. There is evidence that they contribute significantly to the nitrogen nutrition of termites. Spirochetes coat the surfaces of many protozoa from termite and wood-eating roach hindguts (**figure 21.17**). For example, the flagellate *Myxotricha paradoxa* is covered with slender spirochetes (0.15 by 10 μm in length) that are firmly attached and help move the protozoan. Some members of the genera *Treponema, Borrelia,* and *Leptospira* are important pathogens; for example, *Treponema pallidum* causes syphilis, and *Borrelia burgdorferi* is responsible for Lyme disease. The study of *T. pallidum* and its role in syphilis has been hindered by the inability to culture the spirochete outside its human host. The *T. pallidum* genome sequence shows that this spirochete is metabolically crippled and quite dependent on its host. The *B. burgdorferi* genome consists of a linear chromosome of 910,725 base pairs and at least 17 linear and circular plasmids, which constitute another 533,000 base pairs. The plasmids have some genes that are normally found on chromosomes, and plasmid proteins seem to be involved in bacterial virulence.   Insights from microbial genomes: Genomic analysis of pathogenic microbes (section 15.8); Arthropod-borne diseases: Lyme disease (section 38.2); Direct contact diseases: Sexually transmitted diseases (section 38.3)

Bergey's Manual divides the phylum *Spirochaetes* into one class, one order (*Spirochaetales*), and three families (*Spirochaetaceae, Serpulinaceae,* and *Leptospiraceae*). At present, there are 13 genera in the phylum. **Table 21.4** summarizes some of the more distinctive properties of selected genera.

1. Describe the *Planctomycetes* and their more distinctive properties.
2. Give the major characteristics of the phylum *Chlamydiae*. What are elementary and reticulate bodies? Briefly describe the steps in the chlamydial life cycle. Why do you think *Chlamydiae* differentiate into specialized cell types for infection and reproduction?
3. How does chlamydial metabolism differ from that of other bacteria?
4. Define the following terms: protoplasmic cylinder, axial fibrils or periplasmic flagella, axial filament, and outer sheath. Draw and label a diagram of spirochete morphology, locating these structures. Why do you think this form of motility might be especially well suited for movement through viscous fluids?

21.7 PHYLUM *BACTEROIDETES*

The phylum *Bacteroidetes* is very diverse and seems most closely related to the phylum *Chlorobi*. The phylum has three classes (*Bacteroides, Flavobacteria,* and *Sphingobacteria*), 12 families, and 63 genera.

(a)

(b)

Figure 21.17   Spirochete-Protozoan Associations.   The surface spirochetes serve as organs of motility for protozoa. **(a)** The spirochete-*Personympha* association with the spirochetes projecting from the protist's surface. **(b)** Electron micrograph of small spirochetes (S) attached to the membrane of the flagellate protozoan *Barbulanympha*.

Genus	Dimensions (μm) and Flagella	G + C Content (mol%)	Oxygen Relationship	Carbon + Energy Source	Habitats
Spirochaeta	0.2–0.75 × 5–250; 2–40 periplasmic flagella (almost always 2)	51–65	Facultatively anaerobic or anaerobic	Carbohydrates	Aquatic and free-living
Cristispira	0.5–3.0 × 30–180; ≥ 100 periplasmic flagella	N.A.[a]	Facultatively anaerobic?	N.A.	Mollusk digestive tract
Treponema	0.1–0.4 × 5–20: 2–16 periplasmic flagella	25–53	Anaerobic or microaerophilic	Carbohydrates or amino acids	Mouth, intestinal tract, and genital areas of animals; some are pathogenic (syphilis, yaws)
Borrelia	0.2–0.5 × 3–20; 14–60 periplasmic flagella	27–32	Anaerobic or microaerophilic	Carbohydrates	Mammals and arthropods; pathogens (relapsing fever, Lyme disease)
Leptospira	0.1 × 6–24; 2 periplasmic flagella	35–53	Aerobic	Fatty acids and alcohols	Free-living or pathogens of mammals, usually located in the kidney (leptospirosis)
Leptonema	0.1 × 6–20; 2 periplasmic flagella	51–53	Aerobic	Fatty acids	Mammals
Brachyspira	0.2 × 1.7–6.0; 8 periplasmic flagella	25–27	Anaerobic	N.A.	Mammalian intestinal tract
Serpulina	0.3–0.4 × 7–9; 16–18 periplasmic flagella	25–26	Anaerobic	Carbohydrates and amino acids	Mammalian intestinal tract

Table 21.4 Characteristics of Spirochete Genera

[a]N.A., information not available

The class *Bacteroides* contains anaerobic, gram-negative, nonsporing, motile or nonmotile rods of various shapes. These bacteria are chemoheterotrophic and usually produce a mixture of organic acids as fermentation end products. They do not reduce sulfate or other sulfur compounds. The genera are identified using properties such as general shape, motility and flagellation pattern, and fermentation end products. These bacteria grow in habitats such as the oral cavity and intestinal tract of vertebrates and the rumen of ruminants. Microbial interactions: The rumen ecosystem (section 30.1); Normal microbiota of the human body: Large intestine (section 30.3)

Although difficulty culturing these anaerobes has hindered our understanding of them; they are clearly widespread and important. Often they benefit their host. *Bacteroides ruminicola* is a major component of the rumen flora; it ferments starch, pectin, and other carbohydrates. About 30% of the bacteria isolated from human feces are members of the genus *Bacteroides,* and these organisms may provide extra nutrition by degrading cellulose, pectins, and other complex carbohydrates (*see figure 30.18*). The family also is involved in human disease. Members of the genus *Bacteroides* are associated with diseases of major organ systems, ranging from the central nervous system to the skeletal system.

B. fragilis is a particularly common anaerobic pathogen found in abdominal, pelvic, pulmonary, and blood infections.

Another important group in the *Bacteroidetes* is the class *Sphingobacteria.* Besides the similarity in their 16S rRNA sequences, sphingobacteria often have sphingolipids in their cell walls. Some genera in this class are *Sphingobacterium, Saprospira, Flexibacter, Cytophaga, Sporocytophaga,* and *Crenothrix.*

The genera *Cytophaga, Sporocytophaga,* and *Flexibacter* differ from each other in morphology, life cycle, and physiology. Bacteria of the genus *Cytophaga* are slender rods, often with pointed ends (**figure 21.18a**). *Sporocytophaga* is similar to *Cytophaga* but forms spherical resting cells called microcysts (figure 21.18b,c). *Flexibacter* produces long, flexible threadlike cells when young (figure 21.18d) and is unable to use complex polysaccharides. Often colonies of these bacteria are yellow to orange because of carotenoid or flexirubin pigments. Some of the flexirubins are chlorinated, which is unusual for biological molecules.

Members of the genera *Cytophaga* and *Sporocytophaga* are aerobes that actively degrade complex polysaccharides. Soil cytophagas digest cellulose; both soil and marine forms attack

(a) *Cytophaga*

(b) *Sporocytophaga myxococcoides*

(c) *S. myxococcoides* microcysts

(d) *Flexibacter elegans*

Figure 21.18 Nonphotosynthetic, Nonfruiting, Gliding Bacteria. Representative members of the order *Cytophagales.* **(a)** *Cytophaga* sp. (×1,150). **(b)** *Sporocytophaga myxococcoides,* vegetative cells on agar (×1,170). **(c)** *Sporocytophaga myxococcoides,* mature microcysts (×1,750). **(d)** Long thread cells of *Flexibacter elegans* (×1,100).

chitin, pectin, and keratin. Some marine species even degrade agar, a component of seaweed. Cytophagas play a major role in the mineralization of organic matter and can cause great damage to exposed fishing gear and wooden structures. They also are a major component of the bacterial population in sewage treatment plants and presumably contribute significantly to the waste treatment process. Wastewater treatment (section 41.2)

Although most cytophagas are free-living, some can be isolated from vertebrate hosts and are pathogenic. *Cytophaga columnaris* and others cause diseases such as columnaris disease, cold water disease, and fin rot in freshwater and marine fish.

The gliding motility so characteristic of these organisms is quite different from flagellar motility. Gliding motility is present in a wide diversity of taxa: fruiting and nonfruiting aerobic chemoheterotrophs, cyanobacteria, green nonsulfur bacteria, and at least two gram-positive genera (*Heliobacterium* and *Desulfonema*). Gliding bacteria lack flagella and are stationary while suspended in liquid medium. When in contact with a surface, they glide along, leaving a slime trail. Movement can be very rapid; some cytophagas travel 150 μm in a minute, whereas filamentous gliding bacteria may reach speeds of more than 600 μm/minute. Young organisms are the most motile, and motility often is lost with age. Low nutrient levels usually stimulate glid-

ing. The gliding mechanism is not well understood (Microbial Diversity & Ecology 21.1).

Gliding motility gives a bacterium many advantages. Many aerobic chemoheterotrophic gliding bacteria actively digest insoluble macromolecular substrates such as cellulose and chitin, and gliding motility is ideal for searching these out. Because many of the digestive enzymes are cell wall associated, the bacteria must be in contact with insoluble nutrient sources; gliding motility makes this possible. Gliding movement is well adapted to drier habitats and to movement within solid masses such as soil, sediments, and rotting wood that are permeated by small channels. Finally, gliding bacteria, like flagellated bacteria, can position themselves at optimal conditions of light intensity, oxygen, hydrogen sulfide, temperature, and other factors that influence growth and survival.

---

1. Give the major properties of the class *Bacteroides.*
2. How do these bacteria benefit and harm their hosts?
3. List three advantages of gliding motility.
4. Briefly describe the following genera: *Cytophaga, Sporocytophaga,* and *Flexibacter.*
5. Why are the cytophagas ecologically important?

Summary

21.1 *Aquificae* and *Thermotogae*

a. *Aquifex* and *Thermotoga* are hyperthermophilic gram-negative rods that represent the two deepest or oldest phylogenetic branches of the *Bacteria*.

21.2 *Deinococcus-Thermus*

a. Members of the order *Deinococcales* are aerobic, gram-positive cocci and rods that are distinctive in their unusually great resistance to desiccation and radiation.

21.3 Photosynthetic Bacteria

a. Cyanobacteria carry out oxygenic photosynthesis; purple and green bacteria use anoxygenic photosynthesis.

b. The four most important groups of purple and green photosynthetic bacteria are the purple sulfur bacteria, the purple nonsulfur bacteria, the green sulfur bacteria, and the green nonsulfur bacteria (**table 21.1**).

c. The bacteriochlorophyll pigments of purple and green bacteria enable them to live in deeper, anoxic zones of aquatic habitats.

d. The phylum *Chlorobi* includes the green sulfur bacteria—obligately anaerobic photolithoautotrophs that use hydrogen sulfide, elemental sulfur, and hydrogen as electron sources.

e. Green nonsulfur bacteria such as *Chloroflexus* are placed in the phylum *Chloroflexi*. *Chloroflexus* is a filamentous, gliding thermophilic bacterium that is metabolically similar to the purple nonsulfur bacteria.

f. Cyanobacteria carry out oxygenic photosynthesis by means of a photosynthetic apparatus similar to that of the eucaryotes. Phycobilisomes contain the light-harvesting pigments phycocyanin and phycoerythrin (**figure 21.7**).

g. Cyanobacteria reproduce by binary fission, budding, multiple fission, and fragmentation by filaments to form hormogonia. Some produce a dormant akinete.

h. Many nitrogen-fixing cyanobacteria form heterocysts, specialized cells in which nitrogen fixation occurs (**figure 21.9b,c**).

i. *Bergey's Manual* divides the cyanobacteria into five subsections and includes the prochlorophytes in the phylum *Cyanobacteria* (**table 21.3**).

21.4 Phylum *Planctomycetes*

a. The phyla *Planctomycetes* and *Chlamydiae* lack peptidoglycan in their walls. The *Planctomycetes* have unusual cellular compartmentalization (**figure 21.12**).

21.5 Phylum *Chlamydiae*

a. Chlamydiae are nonmotile, coccoid, gram-negative bacteria that reproduce within the cytoplasmic vacuoles of host cells by a life cycle involving elementary bodies (EBs) and reticulate bodies (RBs) (**figure 21.13**). They are energy parasites.

21.6 Phylum *Spirochaetes*

a. The spirochetes are slender, long, helical, gram-negative bacteria that are motile because of the axial filament underlying an outer sheath or outer membrane (**figure 21.15**).

21.7 Phylum *Bacteroidetes*

a. Members of the class *Bacteroides* are obligately anaerobic, chemoheterotrophic, nonsporing, motile or nonmotile rods of various shapes. Some are important rumen and intestinal symbionts, others can cause disease.

b. Gliding motility is present in a diversity of bacteria, including the sphingobacteria.

c. Cytophagas degrade proteins and complex polysaccharides and are active in the mineralization of organic matter.

Key Terms

akinetes 525
anoxygenic photosynthesis 521
axial fibrils 532
axial filament 532
baeocytes 526
carboxysome 524

chlamydiae 531
chlorosomes 523
chromatic adaptation 525
cyanobacteria 524
cyanophycin 525
elementary body (EB) 531

gliding motility 525
green nonsulfur bacteria 523
green sulfur bacteria 523
heterocysts 525
hormogonia 525
initial body 531

oxygenic photosynthesis 520
periplasmic flagella 532
phototaxis 525
phycobilisomes 524
reticulate body (RB) 531
trichome 525

Critical Thinking Questions

1. The cyanobacterium *Anabaena* grows well in liquid medium that contains nitrate as the sole nitrogen source. Suppose you transfer some of these filaments to the same medium except it lacks nitrate and other nitrogen sources. Describe the morphological and physiologial changes you observe.

2. Many types of movement are employed by bacteria in these phyla. Review them and propose mechanisms by which energy (ATP or proton gradients) might drive the locomotion.

3. Propose two experimental approaches you might use to examine the mechanism by which *Cytophaga* glides.

Learn More

Daly, M. J.; Gaidamakova, E. K.; Matrosova, V. Y.; Valienko, A.; Zhai, M.; Venkateswaran, A.; Hess, M.; Omelchenko, M. V.; Kostandarithes, H. M.; Makarova, K. S.; Wackett, L. P.; Fredrickson, J. K.; and Ghosal, D. 2004. Accumulation of Mn(II) in *Deinococcus radiodurans* facilitates gamma-radiation resistance. *Science* 306:1025–28.

Everett, K. D. 2000. *Chlamydia* and *Chlamydiales:* More than meets the eye. *Vet. Microbiol.* 75:109–26.

Golden, J. W., and Yoon, H-S. 2003. Heterocyst development in *Anabaena. Curr. Opin. Microbiol.* 6:557–63.

Honda, D.; Yokota, A.; and Sugiyama, J. 1999. Detection of seven major evolutionary lineages in cyanobacteria based on the 16S rRNA gene sequence analysis with new sequences of five marine *Synechococcus* strains. *Mol. Evol.* 48: 723–39.

Lilburn, T. G.; Kim, K. S.; Ostrom, N. E.; Byzek, K. R.; Leadbetter, J. R.; and Breznak, J. A. 2001. Nitrogen fixation by symbiotic and free-living spirochetes. *Science* 292:2495–98.

Lindsay, M. R.; Webb, R. I.; Strous, M.; Jetten, M. S.; Butler, M. K.; Forde, R. J.; and Fuerst, J. A. 2001. Cell compartmentalization in planctomycetes: Novel types of structural organization for the bacterial cell. *Arch. Microbiol.* 175:413–29.

Litvaitis, M. K. 2002. A molecular test of cyanobacterial phylogeny: Inferences from constraint analysis. *Hydrobiologia* 468:135–45.

McBride, M. J. 2001. Bacterial gliding motility: Multiple mechanisms for cell movement over surfaces. *Annu. Rev. Microbiol.* 55:49–75.

Olson, I.; Paster, B. J.; and Dewhirst, F. E. 2000. Taxonomy of spirochetes. *Anaerobe* 6:39–57.

Partensky, F.; Hess, W. R.; and Vaulot, D. 1999. *Prochlorococcus,* a marine photosynthetic prokaryote of global significance. *Microbiol. Mol. Biol. Rev.* 63(1): 106–27.

Raymond, J.; Zhaxybayeva, O.; Gogarten, J. P.; Gerdes, S. Y.; and Blankenship, R. E. 2002. Whole-genome analysis of photosynthetic prokaryotes. *Science* 298:1616–20.

Strous, M.; Fuerst, J. A.; Kramer, E. H. M.; Logemann, S.; Muyzer, G.; van de Pas-Schoonen, K.; Webb, R.; Kuenen, J. G.; and Jetten, M. S. M. 1999. Missing lithotroph identified as new planctomycete. *Nature* 400:446–49.

Ting, C. S.; Rocap, G.; King, J.; and Chisholm, S. W. 2002. Cyanobacterial photosynthesis in the oceans: The origins and significance of divergent light-harvesting strategies. *Trends Microbiol.* 10(3):134–42.

**Please visit the Prescott website at www.mhhe.com/prescott7
for additional references.**

22

Bacteria:
The Proteobacteria

Salmonella enterica serovar Typhi, stained here with the fluorescent dye acridine orange, is a significant human pathogen. It causes typhoid fever.

PREVIEW

- The phylum *Proteobacteria* is the largest, phylogenetically coherent bacterial group with over 2,000 species assigned to more than 500 genera.

- Many of these gram-negative bacteria are of considerable importance, either as disease agents or because of their contributions to ecosystems. In addition, bacteria such as *Escherichia coli,* are major experimental organisms studied in many laboratories.

- These bacteria are very diverse in their metabolism and life-styles, which range from obligate intracellular parasitism to a free-living existence in soil and aquatic habitats.

- Chemolithotrophic bacteria obtain energy and electrons by oxidizing inorganic compounds rather than the organic nutrients employed by many bacteria. They often have substantial ecological impact because of their ability to oxidize many forms of inorganic nitrogen and sulfur.

- Some *Proteobacteria* produce specialized structures such as prosthecae, stalks, buds, sheaths, or complex fruiting bodies.

- Many bacteria that specialize in predatory or parasitic modes of existence, such as *Bdellovibrio* and the rickettsias, have relinquished some of their metabolic independence through the loss of metabolic pathways. They depend on the prey's or host's energy supply and/or cell constituents.

In chapters 20 and 21 we describe many of the groups found in volumes 1 and 5 of the second edition of *Bergey's Manual.* We now introduce the bacteria that are covered in volume 2 of the second edition of *Bergey's Manual.* We provide an overview of the major biological features of each group and a few selected representative bacteria of particular interest.

Volume 2 of the second edition of *Bergey's Manual* is devoted entirely to the **proteobacteria.** This is the largest and most diverse group of bacteria; currently there are over 500 genera. Although 16S rRNA studies show that they are phylogenetically related, proteobacteria vary markedly in many respects. The morphology of these gram-negative bacteria ranges from simple rods and cocci to genera with prosthecae, buds, and even fruiting bodies. Physiologically they are just as diverse. Photoautotrophs, chemolithotrophs, and chemoheterotrophs are all well represented. There is no obvious overall pattern in metabolism, morphology, or reproductive strategy that characterizes proteobacteria.

Comparison of 16S rRNA sequences has revealed five lineages of descent within the phylum *Proteobacteria*: *Alphaproteobacteria, Betaproteobacteria, Gammaproteobacteria, Deltaproteobacteria,* and *Epsilonproteobacteria* (**figure 22.1**). However, new sequence data suggest that the separation between the *Betaproteobacteria* and *Gammaproteobacteria* is less distinct than once thought. If further analysis supports this, the *Betaproteobacteria* may be considered a subgroup of the *Gammaproteobacteria.* Members of the purple photosynthetic bacteria are found among the α-, β-, and γ-proteobacteria. This has led to the proposal that the proteobacteria arose from a single photosynthetic ancestor, presumably similar to the purple bacteria. That is to say, the phylum is considered monophyletic, although this has recently been questioned. Subsequently photosynthesis would have been lost by various lines, and new metabolic capacities were acquired as these bacteria adapted to different ecological niches.

Microbes is a vigitable, an'ivry man is like a conservatory full iv millyons iv these potted plants.

—*Finley Peter Dunne*

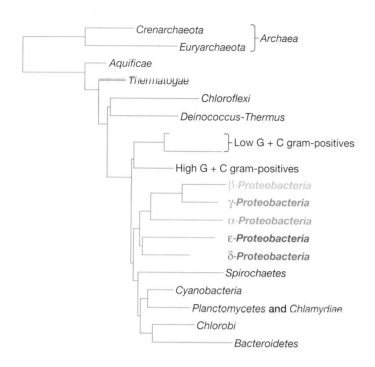

Figure 22.1 Phylogenetic Relationships Among the Procaryotes. The *Proteobacteria* are highlighted.

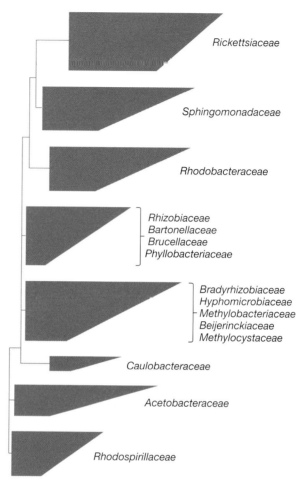

Figure 22.2 Phylogenetic Relationships Among Major Families Within the α-Proteobacteria. The relationships are based on 16S rRNA sequence data.

22.1 CLASS *ALPHAPROTEOBACTERIA*

The **α-proteobacteria** include most of the oligotrophic proteobacteria (those capable of growing at low nutrient levels). Some have unusual metabolic modes such as methylotrophy (*Methylobacterium*), chemolithotrophy (*Nitrobacter*), and the ability to fix nitrogen (*Rhizobium*). Members of genera such as *Rickettsia* and *Brucella* are important pathogens; in fact, *Rickettsia* is an obligate intracellular parasite. Many genera are characterized by distinctive morphology such as prosthecae.

The class *Alphaproteobacteria* has seven orders and 20 families. **Figure 22.2** illustrates the phylogenetic relationships among major groups within the α-proteobacteria, and **table 22.1** summarizes the general characteristics of many of the bacteria discussed in the following sections.

The Purple Nonsulfur Bacteria

All the purple bacteria use anoxygenic photosynthesis, possess bacteriochlorophylls *a* or *b*, and have their photosynthetic apparatus in membrane systems that are continuous with the plasma membrane. Most are motile by polar flagella. All purple nonsulfur bacteria are α-proteobacteria, with the exception of *Rhodocyclus* (β-proteobacteria).   Photosynthetic bacteria (section 21.3)

The **purple nonsulfur bacteria** are exceptionally flexible in their choice of an energy source. Normally they grow anaerobically as photoorganoheterotrophs; they trap light energy and employ organic molecules as both electron and carbon sources (*see table 21.1*). Although they are called nonsulfur bacteria, some

species can oxidize very low, nontoxic levels of sulfide to sulfate, but they do not oxidize elemental sulfur to sulfate. In the absence of light, most purple nonsulfur bacteria can grow aerobically as chemoorganoheterotrophs, but some species carry out fermentations anaerobically. Oxygen inhibits bacteriochlorophyll and carotenoid synthesis so that cultures growing aerobically in the dark are colorless.

Purple nonsulfur bacteria vary considerably in morphology (**figure 22.3**). They may be spirals (*Rhodospirillum*), rods (*Rhodopseudomonas*), half circles or circles (*Rhodocyclus*), or they may even form prosthecae and buds (*Rhodomicrobium*). Because of their metabolism, they are most prevalent in the mud and water of lakes and ponds with abundant organic matter and low sulfide levels. There also are marine species.

Rhodospirillum and *Azospirillum* (both in the family *Rhodospirillaceae*) are among several bacterial genera capable of forming cysts. These resting cells differ from the well-characterized endospores made by gram-positive bacteria such as *Bacillus* and *Clostridium*. Like spores, cysts are very resistant to desicca-

Table 22.1	Characteristics of Selected α-Proteobacteria				
Genus	Dimensions (μm) and Morphology	G + C Content (mol%)	Genome size (Mb)	Oxygen Requirement	Other Distinctive Characteristics
Agrobacterium	0.6–1.0 × 1.5–3.0; motile, nonsporing rods with peritrichous flagella	57–63	2.5	Aerobic	Chemoorganotroph that can invade plants and cause tumors
Caulobacter	0.4–0.6 × 1–2; rod- or vibrioid-shaped with a flagellum and prostheca and holdfast	62–65	4.0	Aerobic	Heterotrophic and oligotrophic; asymmetric cell division
Hyphomicrobium	0.3–1.2 × 1–3; rod-shaped or oval with polar prosthecae	59–65	Nd*	Aerobic	Reproduces by budding; methylotrophic
Nitrobacter	0.5–0.9 × 1.0–2.0; rod- or pear-shaped, sometimes motile by flagella	59–62	3.4	Aerobic	Chemolithotroph, oxidizes nitrite to nitrate
Rhizobium	0.5–1.0 × 1.2–3.0; motile rods with flagella	57–66	5.1	Aerobic	Invades leguminous plants to produce nitrogen-fixing root nodules
Rhodospirillum	0.7–1.5 wide; spiral cells with polar flagella	62–64	4.4	Anaerobic, microaerobic, aerobic	Photoheterotroph under anoxic conditions
Rickettsia	0.3–0.5 × 0.8–2.0; short nonmotile rods	29–33	1.1–1.3	Aerobic	Obligate intracellular parasite

*Nd: Not determined; genome not yet sequenced

tion but are less tolerant of other environmental stresses like heat and UV light. Cysts are made in response to nutrient limitation. They have a thick outer coat and store an abundance of poly-β-hydroxybutyrate (PHB). Cyst-forming bacteria are not limited to α-proteobacteria; for instance *Azotobacter,* a γ-proteobacterium, also forms cysts. The cytoplasmic matrix: Inclusion bodies (section 3.3); The bacterial endospore (section 3.11)

Rickettsia and *Coxiella*

The genus *Rickettsia* is placed in the order *Rickettsiales* and family *Rickettsiaceae* of the α-proteobacteria, whereas *Coxiella* is in the order *Legionellales* and family *Coxiellaceae* of the γ-proteobacteria. However, here we discuss *Rickettsia* and *Coxiella* together because of their similarity in life-style, despite their apparent phylogenetic distance.

These bacteria are rod-shaped, coccoid, or pleomorphic with typical gram-negative walls and no flagella. Although their size

varies, they tend to be very small. For example, *Rickettsia* is 0.3 to 0.5 μm in diameter and 0.8 to 2.0 μm long; *Coxiella* is 0.2 to 0.4 μm by 0.4 to 1.0 μm. All species are parasitic or mutualistic. The parasitic forms grow in vertebrate erythrocytes, macrophages, and vascular endothelial cells. Often they also live in blood-sucking arthropods such as fleas, ticks, mites, or lice, which serve as vectors or as primary hosts. Microbial interactions (section 30.1)

Because these genera include important human pathogens, their reproduction and metabolism have been intensively studied. Rickettsias enter the host cell by inducing phagocytosis. Members of the genus *Rickettsia* immediately escape the phagosome and reproduce by binary fission in the cytoplasm (**figure 22.4**). In contrast, *Coxiella* remains within the phagosome after it has fused with a lysosome and actually reproduces within the phagolysosome. Eventually the host cell bursts, releasing new organisms. Besides incurring damage from cell lysis, the host is harmed by the toxic effects of rickettsial cell walls (wall toxicity appears related to the mechanism of penetration into host cells). Phagocytosis (section 31.3)

(a) *Rhodospirillum rubrum*

(b) *R. rubrum*

(c) *Rhodopseudomonas acidophila*

(d) *Rhodocyclus purpureus*

(e) *Rhodomicrobium vannielii*

Figure 22.3 Typical Purple Nonsulfur Bacteria. **(a)** *Rhodospirillum rubrum*; phase contrast (×410). **(b)** *R. rubrum* grown anaerobically in the light. Note vesicular invaginations of the cytoplasmic membranes; transmission electron micrograph (×51,000). **(c)** *Rhodopseudomonas acidophila*; phase contrast. **(d)** *Rhodocyclus purpureus*; phase contrast. Bar = 10 μm. **(e)** *Rhodomicrobium vannielii* with vegetative cells and buds; phase contrast.

Rickettsias are very different from most other bacteria in physiology and metabolism. They lack glycolytic pathways and do not use glucose as an energy source, but rather oxidize glutamate and tricarboxylic acid cycle intermediates such as succinate. The rickettsial plasma membrane has carrier-mediated transport systems, and host cell nutrients and coenzymes are absorbed and directly used. For example, rickettsias take up both NAD^+ and uridine diphosphate glucose. Their membrane also has an adenylate exchange carrier that exchanges ADP for external ATP. Thus host ATP may provide much of the energy needed for growth. This metabolic dependence explains why many of these organisms must be cultivated in the yolk sacs of chick embryos or in tissue culture cells. Genome sequencing shows that *R. prowazekii* is similar in many ways to mitochondria. Possibly mitochondria arose from an endosymbiotic association with an ancestor of *Rickettsia*. Microbial evolution: Endosymbiotic origin of mitochondria and chloroplasts (section 19.1)

These orders contain many important pathogens. *Rickettsia prowazekii* and *R. typhi* are associated with typhus fever, and *R. rickettsii*, with Rocky Mountain spotted fever. *Coxiella burnetii* causes Q fever in humans. These diseases are discussed in some detail in chapter 38. Rickettsias are also important pathogens of domestic animals such as dogs, horses, sheep, and cattle.

The *Caulobacteraceae* and *Hyphomicrobiaceae*

A number of the proteobacteria are not simple rods or cocci but have some sort of appendage. These bacteria have interesting life

Figure 22.4 *Rickettsia* and *Coxiella.*
Rickettsial morphology and reproduction. **(a)** A human fibroblast filled with *Rickettsia prowazekii* (×1,200). **(b)** A chicken embryo fibroblast late in infection with free cytoplasmic *R. prowazekii* (×13,600). **(c)** *Coxiella burnetti* growing within fibroblast vacuoles (×9,000). **(d)** *R. prowazekii* leaving a disrupted phagosome (arrow) and entering the cytoplasmic matrix (× 46,000).

(a)   (b)

(c)   (d)

cycles that feature a prostheca or reproduction by budding. A **prostheca** (pl., prosthecae), also called a **stalk,** is an extension of the cell, including the plasma membrane and cell wall, that is narrower than the mature cell. **Budding** is distinctly different from the **binary fission** normally used by bacteria. The bud first appears as a small protrusion at a single point and enlarges to form a mature cell. Most or all of the bud's cell envelope is newly synthesized. In contrast, portions of the parental cell envelope are shared with the progeny cells during binary fission. Finally, the parental cell retains its identity during budding, and the new cell is often smaller than its parent. In binary fission the parental cell disappears as it forms progeny of equal size. The families *Caulobacteraceae* and *Hyphomicrobiaceae* of the α-proteobacteria contain two of the best studied prosthecate genera: *Caulobacter* and *Hyphomicrobium.*   The procaryotic cell cycle (section 6.1)

The genus *Hyphomicrobium* contains chemoheterotrophic, aerobic, budding bacteria that frequently attach to solid objects in freshwater, marine, and terrestrial environments. (They even grow in laboratory water baths.) The vegetative cell measures about 0.5 to 1.0 by 1 to 3 μm (**figure 22.5**). At the beginning of the reproductive cycle, the mature cell produces a prostheca (also called a hypha), 0.2 to 0.3 μm in diameter, that grows to several μm in length (**figure 22.6**). The nucleoid divides, and a copy moves into the hypha while a bud forms at its end. As the bud matures, it produces one to three flagella, and a septum divides the

1μm

Figure 22.5   Prosthecate, Budding Bacteria. *Hyphomicrobium facilis* with hypha and young bud.

bud from the hypha. The bud is finally released as an oval- to pear-shaped swarmer cell, which swims off, then settles down and begins budding. The mother cell may bud several times at the tip of its hypha.

Hyphomicrobium also has distinctive physiology and nutrition. Sugars and most amino acids do not support abundant

Figure 22.6 The Life Cycle of *Hyphomicrobium.*

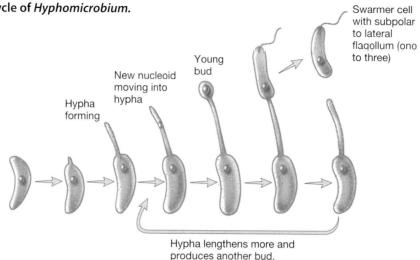

Swarmer cell
with subpolar
to lateral
flagellum (one
to three)

Young
bud

New nucleoid
moving into
hypha

Hypha
forming

Hypha lengthens more and
produces another bud.

growth; instead *Hyphomicrobium* grows on ethanol and acetate and flourishes with one-carbon compounds such as methanol, formate, and formaldehyde. That is, it is a facultative **methylotroph** and can derive both energy and carbon from reduced one-carbon compounds. It is so efficient at acquiring one-carbon molecules that it can grow in a medium without an added carbon source (presumably the medium absorbs sufficient atmospheric carbon compounds). *Hyphomicrobium* may comprise up to 25% of the total bacterial population in oligotrophic or nutrient-poor freshwater habitats.

Bacteria in the genus *Caulobacter* alternate between polarly flagellated rods and cells that possess a prostheca and **holdfast,** by which they attach to solid substrata (**figure 22.7**). Incredibly, the material secreted at the end of the *Caulobacter cresentus* holdfast is the strongest biological adhesion molecule known—sort of a bacterial superglue. Caulobacters are usually isolated from freshwater and marine habitats with low nutrient levels, but they also are present in the soil. They often adhere to bacteria, photosynthetic protists, and other microorganisms and may absorb nutrients released by their hosts. The prostheca differs from that of *Hyphomicrobium* in that it lacks cytoplasmic components and is composed almost totally of the plasma membrane and cell wall. It grows longer in nutrient-poor media and can reach more than 10 times the length of the cell body. The prostheca may improve the efficiency of nutrient uptake from dilute habitats by increasing surface area; it also gives the cell extra buoyancy.

The life cycle of *Caulobacter* is unusual (**figure 22.8**). When ready to reproduce, the cell elongates and a single polar flagellum forms at the end opposite the prostheca. The cell then undergoes asymmetric transverse binary fission to produce a flagellated swarmer cell that swims away. The swarmer, which cannot reproduce, comes to rest, ejects its flagellum, and forms a new prostheca on the formerly flagellated end. The new stalked

cell then starts the cycle anew. This process takes about two hours to complete. The species *C. cresentus* has become an important model organism in the study of microbial development and the bacterial cell cycle.

Family *Rhizobiaceae*

The order *Rhizobiales* of the α-proteobacteria contains 11 families with a great variety of phenotypes. This includes the family *Hyphomicrobiaceae,* which has already been discussed. An important family in this order is *Rhizobiaceae,* which includes the aerobic genera *Rhizobium* and *Agrobacterium.*

Members of the genus *Rhizobium* are 0.5 to 0.9 by 1.2 to 3.0 μm motile rods that become pleomorphic under adverse conditions (**figure 22.9**). Cells often contain poly-β-hydroxybutyrate inclusions. They grow symbiotically within root nodule cells of legumes as nitrogen-fixing bacteroids (figure 22.9*b, also see figure 29.13*). In fact, the *Leguminosae* is the most successful plant family on Earth, with over 18,000 species. Their proliferation reflects their capacity to establish symbiotic relationships with bacteria that form nodules on their roots. Within the nodules the microbes reduce or fix atmospheric nitrogen into ammonium, making it directly available to the plant host. The process by which bacteria perform this fascinating and important symbiosis is discussed in chapter 29.   Synthesis of amino acids: Nitrogen fixation (section 10.5); Microorganism associations with vascular plants: The Rhizobia (section 29.5)

The genus *Agrobacterium* is placed in the family *Rhizobiaceae* but differs from *Rhizobium* in not stimulating root nodule formation or fixing nitrogen. Instead agrobacteria invade the crown, roots, and stems of many plants and transform plant cells into autonomously proliferating tumor cells. Most of the genes that encode distinguishing characteristics are carried on plas-

(a)

(d)

(b)

(c)

Figure 22.7 *Caulobacter* **Morphology and Reproduction.**
(a) Rosettes of cells adhering to each other by their prosthecae; phase contrast (×600). **(b)** A cell dividing to produce a swarmer (×6,030). Note prostheca and flagellum. **(c)** A stalked cell and a flagellated swarmer cell (×6,030). **(d)** A rosette of cells as seen in the electron microscope.

mids (*see figure 29.19*). The best-studied species is *A. tumefaciens,* which enters many broad-leaved plants through wounds and causes crown gall disease (**figure 22.10**). The ability to produce tumors depends on the presence of a large Ti (for *tumor-inducing*) plasmid. Tumor production by *Agrobacterium* is discussed in greater detail in Techniques & Applications 14.2 and section 29.5. Plasmids (section 3.5)

Nitrifying Bacteria

The taxonomy of the aerobic chemolithotrophic bacteria, those bacteria that derive energy and electrons from reduced inorganic compounds, is quite complex. Normally these bacteria employ CO_2 as their carbon source and thus are chemolithoautotrophs, but some can function as chemolithoheterotrophs and use reduced organic carbon sources. In *Bergey's Manual,* the chemolithotrophic bacteria are distributed between the α-, β-, and γ-proteobacteria. The nitrifying bacteria are found in all three classes. Chemolithotrophy (section 9.11)

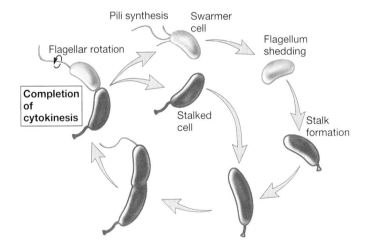

Figure 22.8 Caulobacter Life Cycle. Stalked cells attached to a substrate undergo asymmetric binary fission producing a stalked and a flagellated cell, called a swarmer cell. The swarmer cell swims freely and makes pili until it settles, ejects its flagella, and forms a stalk. Only stalked cells can divide.

(a)

(b)

Figure 22.9 Rhizobium. (a) *Rhizobium leguminosarum* with two polar flagella (×14,000). **(b)** Scanning electron micrograph of bacteroids in alfalfa root (×640).

The **nitrifying bacteria** are a very diverse collection of bacteria. *Bergey's Manual* places nitrifying genera in three classes and several families: *Nitrobacter* in the *Bradyrhizobiaceae*, α-proteobacteria; *Nitrosomonas* and *Nitrosospira* in the *Nitrosomonadaceae*, γ-proteobacteria; *Nitrococcus* in the *Ectothiorhodospiraceae*, γ-proteobacteria; and *Nitrosococcus* in the *Chromatiaceae*, γ-proteobacteria. All are aerobic, gram-negative organisms with the ability to capture energy from the oxidation of either ammonia or nitrite. However, they differ considerably in other properties (**table 22.2**). Nitrifiers may be rod-shaped, ellipsoidal, spherical, spirillar or lobate, and they may possess either polar or peritrichous flagella (**figure 22.11**). Often they have extensive membrane complexes in their cytoplasm. Identification is based on properties such as their preference for nitrite or ammonia, their general shape, and the nature of any cytomembranes present.

Nitrifying bacteria make important contributions to the nitrogen cycle. In soil, sewage disposal systems, and freshwater and marine habitats, the β-proteobacteria *Nitrosomonas* and *Nitrospira* and the γ-proteobacterium *Nitrosococcus* oxidize ammonia to nitrite. In the same niches, members of the γ-proteobacterial genus *Nitrococcus* then oxidize nitrite to nitrate. The whole process of converting ammonia to nitrite to nitrate is called **nitrification** and it occurs rapidly in oxic soil treated with fertilizers containing am-

Figure 22.10 Agrobacterium. Crown gall tumor of a tomato plant caused by *Agrobacterium tumefaciens*.

monium salts. Nitrate is readily used by plants, but it is also rapidly lost through leaching of water-soluble nitrates and by denitrification to nitrogen gas, so the benefits gained from nitrification can be fleeting. Biogeochemical cycling: Nitrogen cycle (section 27.2)

1. Describe the general properties of the α-proteobacteria.
2. Discuss the characteristics and physiology of the purple nonsulfur bacteria. Where would one expect to find them growing?
3. Briefly describe the characteristics and life cycle of the genus *Rickettsia*.
4. In what way does the physiology and metabolism of the rickettsias differ from that of other bacteria?
5. Name some important rickettsial diseases.
6. Define the following terms: prostheca, stalk, budding, swarmer cell, methylotroph, and holdfast.
7. Briefly describe the morphology and life cycles of *Hyphomicrobium* and *Caulobacter*.
8. What is unusual about the physiology of *Hyphomicrobium?* How does this influence its ecological distribution?
9. How do *Agrobacterium* and *Rhizobium* differ in life-style? What effect does *Agrobacterium* have on plant hosts?
10. What are chemolithotrophic bacteria?
11. Give the major characteristics of the nitrifying bacteria and discuss their ecological importance. How does the metabolism of *Nitrobacter* differ from that of *Nitrosomonas?*

22.2 CLASS *BETAPROTEOBACTERIA*

The **β-proteobacteria** overlap the α-proteobacteria metabolically but tend to use substances that diffuse from organic decomposition in the anoxic zone of habitats. Some of these bacteria use hydrogen,

Table 22.2	Selected Characteristics of Representative Nitrifying Bacteria					
Species	Cell Morphology and Size (μm)	Reproduction	Motility	Cytomembranes	G + C Content (mol%)	Habitat
Ammonia-Oxidizing Bacteria						
Nitrosomonas europaea (β-proteobacteria)	Rod; 0.8–1.1 × 1.0–1.7	Binary fission	−	Peripheral, lamellar	50.6–51.4	Soil, sewage, freshwater, marine
Nitrosococcus oceani (γ-proteobacteria)	Coccoid; 1.8–2.2 in diameter	Binary fission	+; 1 or more subpolar flagella	Centrally located parallel bundle, lamellar	50.5	Obligately marine
Nitrosospira briensis (β-proteobacteria)	Spiral; 0.3–0.4 in diameter	Binary fission	+ or −; 1 to 6 peritrichous flagella	Rare	53.8–54.1	Soil
Nitrite-Oxidizing Bacteria						
Nitrobacter winogradskyi (α-proteobacteria)	Rod, often pear-shaped; 0.5–0.9 × 1.0–2.0	Budding	+ or −; 1 polar flagellum	Polar cap of flattened vesicles in peripheral region of the cell	61.7	Soil, freshwater, marine
Nitrococcus mobilis (γ-proteobacteria)	Coccoid; 1.5–1.8 in diameter	Binary fission	+; 1 or 2 subpolar flagella	Tubular cytomembranes randomly arranged in cytoplasm	61.3 (1 strain)	Marine

From Brenner, D. J.; Krieg, N. R.; and Staley, J. T. Eds. 2005. Bergey's Manual to Systemic Bacteriology 2nd ed. Vol. 2: *The Proteobacteria.* Garrity, G. M. Ed-in-Chief. New York: Springer.

ammonia, methane, volatile fatty acids, and similar substances. As with the α-proteobacteria, there is considerable metabolic diversity; the β-proteobacteria may be chemoheterotrophs, photolithotrophs, methylotrophs, and chemolithotrophs.

The class *Betaproteobacteria* has seven orders and 12 families. **Figure 22.12** shows the phylogenetic relationships among major groups within the β-proteobacteria, and **table 22.3** summarizes the general characteristics of many of the bacteria discussed in this section. Here we discuss two genera with important human pathogens: *Neisseria* and *Bordetella*.

Order *Neisseriales*

The order *Neisseriales* has one family, *Neisseriaceae,* with 15 genera. The best-known and most intensely studied genus is *Neisseria*. Members of this genus are nonmotile, aerobic, gram-negative cocci that most often occur in pairs with adjacent sides flattened. They may have capsules and fimbriae. The genus is chemoorganotrophic, and produces the enzymes oxidase and catalase (thus they are said to be oxidase positive and catalase positive). They are inhabitants of the mucous membranes of

mammals, and some are human pathogens. *Neisseria gonorrhoeae* is the causative agent of gonorrhea; *Neisseria meningitidis* is responsible for some cases of bacterial meningitis. Direct contact diseases: Sexually transmitted diseases (section 38.3)

Order *Burkholderiales*

The order contains four families, three of them with well-known genera. The genus *Burkholderia* is placed in the family *Burkholderiaceae.* This genus was established when *Pseudomonas* was divided into at least seven genera based on rRNA data: *Acidovorax, Aminobacter, Burkholderia, Comamonas, Deleya, Hydrogenophaga,* and *Methylobacterium.* Members of the genus *Burkholderia* are gram-negative, aerobic, nonfermentative, nonsporing, mesophilic straight rods. With the exception of one species, all are motile with a single polar flagellum or a tuft of polar flagella. Catalase is produced and they often are oxidase positive. Most species use PHB as their carbon reserve. One of the most important species is *B. cepacia,* which can degrade over 100 different organic molecules and is very active in recycling organic materials in nature. Originally described as the plant pathogen that causes onion rot, it

(a)

(b)

(c)

Figure 22.11   Representative Nitrifying Bacteria.
(a) *Nitrobacter winogradskyi*; phase contrast (×2,500). (b) *N. winogradskyi*. Note the polar cap of cytomembranes (×213,000). (c) *N. europaea* with extensive cytoplasmic membranes (×81,700).

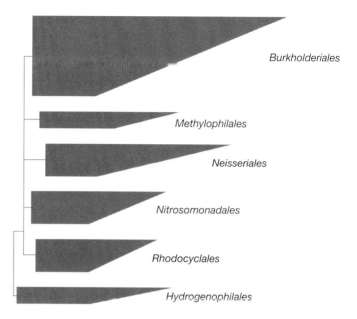

Figure 22.12   Phylogenetic Relationships Among Major Groups Within the β-Proteobacteria.   The relationships are based on 16S rRNA sequence data.

has emerged in the last 20 years as a major nosocomial pathogen. It is a particular problem for cystic fibrosis patients. Two other species, *B. mallai* and *B. pseudomallai* are human pathogens that could be misused as bioterrorism agents. Bioterrorism preparedness (section 36.9)

Surprisingly, two genera within the *Burkholderiaceae* family are capable of forming nitrogen-fixing symbioses with legumes much like the rhizobia that belong to the α-proteobacteria. Genome analysis of the nitrogen-fixing β-proteobacteria *Burkholderia* and *Ralstonia* reveals the presence of nodulation (*nod*) genes that are very similar to those of the rhizobia. This suggests a common genetic origin. It is thought the β-proteobacteria gained the capacity to form symbiotic, nitrogen-fixing nodules with legumes through lateral gene transfer.

The family *Alcaligenaceae* contains the genus *Bordetella*. This genus is composed of gram-negative, aerobic coccobacilli, about 0.2 to 0.5 by 0.5 to 2.0 μm in size. *Bordetella* is a chemoorganotroph with respiratory metabolism that requires organic sulfur and nitrogen (amino acids) for growth. It is a mammalian parasite that multiplies in respiratory epithelial cells. *Bordetella pertussis* is a nonmotile, encapsulated species that causes whooping cough. Airborne diseases: Diphtheria (section 38.1)

Some genera in the order have a **sheath**—a hollow, tubelike structure surrounding a chain of cells. Sheaths often are close fitting, but they are never in intimate contact with the cells they enclose and may contain ferric or manganic oxides. They have at least two functions. Sheaths help bacteria attach to solid surfaces and acquire nutrients from slowly running water as it flows past, even if it is nutrient-poor. Sheaths also protect against predators such as protozoa and the δ-proteobacterium *Bdellovibrio* (p. 563).

Table 22.3 | **Characteristics of Selected β-Proteobacteria**

Genus	Dimensions (μm) and Morphology	G + C Content (mol%)	Genome Size (Mb)	Oxygen Requirement	Other Distinctive Characteristics
Bordetella	0.2–0.5 × 0.5–2.0; nonmotile coccobacillus	66–70	3.7–5.3	Aerobic	Requires organic sulfur and nitrogen; mammalian parasite
Burkholderia	0.5–1.0 × 1.5–4; straight rods with single flagella or a tuft at the pole	59–69.5	4.1–7.2	Aerobic, some capable of anaerobic respiration with NO_3^-	Poly-β-hydroxybutyrate as reserve; can be pathogenic
Leptothrix	0.6–1.5 × 2.5–15; straight rods in chains with sheath, free cells flagellated	68–71	Nd*	Aerobic	Sheaths encrusted with iron and manganese oxides
Neisseria	0.6–1.9; cocci in pairs with flattened adjacent sides	48–56	2.2–2.3	Aerobic	Inhabitant of mucous membranes of mammals
Nitrosomonas	Size varies with strain; spherical to ellipsoidal cells with intracytoplasmic membranes	45–54	2.8	Aerobic	Chemolithotroph that oxidizes ammonia to nitrite
Sphaerotilus	1.2–2.5 × 2–10; single chains of cells with sheaths, may have holdfasts	70	Nd	Aerobic	Sheaths not encrusted with iron and manganese oxides
Thiobacillus	0.3–0.5 × 0.9–4; rods, often with polar flagella	52–68	Nd	Aerobic	All chemolithotrophic, oxidizes reduced sulfur compounds to sulfate, some also chemoorganotrophic

*Nd: Not determined; genome not yet sequenced

(a)   **(b)**

Figure 22.13   Sheathed Bacteria, *Sphaerotilus natans.*
(a) Sheathed chains of cells and empty sheaths. **(b)** Chains with holdfasts (indicated by the letter a) and individual cells containing poly-β-hydroxybutyrate granules.

Two well-studied sheathed genera are *Sphaerotilus* and *Leptothrix*. *Sphaerotilus* forms long sheathed chains of rods, 0.7 to 2.4 by 3 to 10 μm, attached to submerged plants, rocks, and other solid objects, often by a holdfast (**figure 22.13**). Single swarmer cells with a bundle of subpolar flagella escape the filament and form a new chain after attaching to a solid object at another site. *Sphaerotilus* prefers slowly running freshwater polluted with sewage or industrial waste. It grows so well in activated sewage sludge that it sometimes forms tangled masses of filaments and interferes with the proper settling of sludge. *Leptothrix* characteristically deposits large amounts of iron and manganese oxides in its sheath (**figure 22.14**). This seems to protect it and allow *Leptothrix* to grow in the presence of high concentrations of soluble iron compounds.

Order *Nitrosomonadales*

A number of chemolithotrophs are found in the order *Nitrosomonadales*. Two genera of nitrifying bacteria (*Nitrosomonas* and *Nitrosospira*) are members of the family *Nitrosomonadaceae*

(a)

(b)

Figure 22.14 Sheathed Bacteria, *Leptothrix* Morphology. (a) *L. lopholea* trichomes radiating from a collection of holdfasts. (b) *L. cholodnii* sheaths encrusted with MnO₂.

(a)

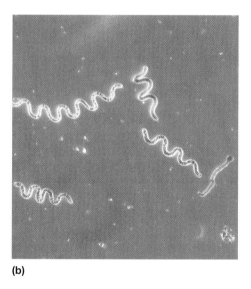

(b)

Figure 22.15 The Genus *Spirillum*. (a) *Spirillum volutans* with bipolar flagella visible (×450). (b) *Spirillum volutans*; phase contrast (×550).

but were discussed earlier with other genera of nitrifying bacteria (pp. 545–546). The stalked chemolithotroph *Gallionella* is in this order. The family *Spirillaceae* has one genus, *Spirillum* (**figure 22.15**).

Order *Hydrogenophilales*

This small order contains *Thiobacillus*, one of the best-studied chemolithotrophs and most prominent of the colorless sulfur bacteria. Like the nitrifying bacteria, **colorless sulfur bacteria** are a highly diverse group. Many are unicellular rod-shaped or spiral sulfur-oxidizing bacteria that are nonmotile or motile by

flagella (**table 22.4**). *Bergey's Manual* disperses these bacteria between two classes; for example, *Thiobacillus* and *Macromonas* are β-proteobacteria, whereas *Thiomicrospira*, *Thiobacterium*, *Thiospira*, *Thiothrix*, *Beggiatoa*, and others are γ-proteobacteria. Only some of these bacteria have been isolated and studied in pure culture. Most is known about the genera *Thiobacillus* and *Thiomicrospira*. *Thiobacillus* is a gram-negative rod, and *Thiomicrospira* is a long spiral cell (**figure 22.16**); both have polar flagella. They differ from many of the nitrifying bacteria in that they lack extensive internal membrane systems.

Table 22.4 | **Colorless Sulfur-Oxidizing Genera**

Genus	Motility; Cell Shape	G + C Content Flagella	Location of (mol%)	Sulfur Deposit[a]	Nutritional Type
Thiobacillus	Rods	+; polar	62–67	Extracellular	Obligate or facultative chemolithotroph
Thiomicrospira	Spirals, comma, or rod shaped	− or +; polar	39.6–49.9	Extracellular	Obligate chemolithotroph
Thiobacterium	Rods embedded in gelatinous masses	−	N.A.[b]	Intracellular	Probably chemoorganoheterotroph
Thiospira	Spiral rods, usually with pointed ends	+; polar (single or in tufts)	N.A.	Intracellular	Unknown
Macromonas	Rods, cylindrical or bean shaped	+; polar tuft	67	Intracellular	Probably chemoorganoheterotroph

[a]When hydrogen sulfide is oxidized to elemental sulfur.

[b]N.A., data not available.

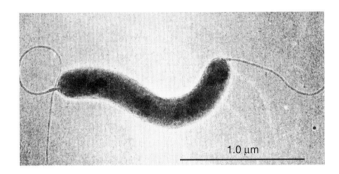

Figure 22.16 **Colorless Sulfur Bacteria.** *Thiomicrospira pelophila,* a γ-proteobacterium, with polar flagella.

The metabolism of *Thiobacillus* has been intensely studied. It grows aerobically by oxidizing a variety of inorganic sulfur compounds (elemental sulfur, hydrogen sulfide, thiosulfate) to sulfate. ATP is produced by a combination of oxidative phosphorylation and substrate-level phosphorylation by means of adenosine 5′-phosphosulfate. Although *Thiobacillus* normally uses CO_2 as its major carbon source, *T. novellus* and a few other strains can grow heterotrophically. Some species are very flexible metabolically. For example, *Thiobacillus ferrooxidans* also uses ferrous iron as an electron donor and produces ferric iron as well as sulfuric acid. *T. denitrificans* even grows anaerobically by reducing nitrate to nitrogen gas. Interestingly, some other sulfur-oxidizing bacteria such as *Thiobacterium* and *Macromonas* probably do not derive energy from sulfur oxidation. They may use the process to detoxify metabolically produced hydrogen peroxide.

Sulfur-oxidizing bacteria have a wide distribution and great practical importance. *Thiobacillus* grows in soil and aquatic habitats, both freshwater and marine. In marine habitats *Thiomicrospira* is more important than *Thiobacillus*. Because of their great acid tolerance (*T. thiooxidans* grows at pH 0.5 and cannot

grow above pH 6), these bacteria prosper in habitats they have acidified by sulfuric acid production, even though most other organisms cannot. The production of large quantities of sulfuric acid and ferric iron by *T. ferrooxidans* corrodes concrete and pipe structures. Thiobacilli often cause extensive acid and metal pollution when they release metals from mine wastes. However, sulfur-oxidizing bacteria also are beneficial. They may increase soil fertility when they release elemental sulfur by oxidizing it to sulfate. Thiobacilli are used in processing low-grade metal ores because of their ability to leach metals from ore. Biogeochemical cycling: Sulfur cycle (section 27.2)

1. Describe the general properties of the β-proteobacteria.
2. Briefly describe the following genera and their practical importance: *Neisseria, Burkholderia,* and *Bordetella.*
3. What is a sheath and of what advantage is it?
4. How does *Sphaerotilus* maintain its position in running water? How does it reproduce and disperse its progeny?
5. Characterize the colorless sulfur bacteria and discuss their placement in the second edition of *Bergey's Manual.*
6. How do colorless sulfur bacteria obtain energy by oxidizing sulfur compounds? What is adenosine 5′-phosphosulfate?
7. List several positive and negative impacts sulfur-oxidizing bacteria have on the environment and human activities.

22.3 CLASS *GAMMAPROTEOBACTERIA*

The **γ-proteobacteria** constitute the largest subgroup of proteobacteria with an extraordinary variety of physiological types. Many important genera are chemoorganotrophic and facultatively anaerobic. Other genera contain aerobic chemoorganotrophs, photolithotrophs, chemolithotrophs, or methylotrophs. According to some DNA-rRNA hybridization studies, the γ-proteobacteria are composed of several deeply branching groups. One consists of the purple sulfur bacteria; a second includes the intracellular parasites *Legionella* and *Coxiella*. The two largest groups contain a wide variety of nonphotosynthetic genera. Ribosomal RNA superfamily I is represented by the families *Vibrionaceae, Enterobacteriaceae*, and *Pasteurellaceae*. These bacteria use the Embden-Meyerhof and pentose phosphate pathways to catabolize carbohydrates. Most are facultative anaerobes. Ribosomal RNA superfamily II contains mostly aerobes that often use the Entner-Doudoroff and pentose phosphate pathways to catabolize many different kinds of organic molecules. The genera *Pseudomonas, Azotobacter, Moraxella, Xanthomonas,* and *Acinetobacter* belong to this superfamily.

The exceptional diversity of these bacteria is evident from the fact that *Bergey's Manual* divides the class *Gammaproteobacte-*

ria into 14 orders and 28 families. **Figure 22.17** illustrates the phylogenetic relationships among major groups and selected γ-proteobacteria, and **table 22.5** outlines the general characteristics of some of the bacteria discussed in this section.

The Purple Sulfur Bacteria

As mentioned previously, the purple photosynthetic bacteria are distributed between three subgroups of the proteobacteria. Despite the diversity of these organisms, they share some general characteristics, which are summarized in table 21.1 (*see p. 522*). Most of the purple nonsulfur bacteria are α-proteobacteria and were discussed earlier in this chapter (pp. 540–541). Because the purple sulfur bacteria are γ-proteobacteria, they are described here.

Bergey's Manual divides the purple sulfur bacteria into two families: the *Chromatiaceae* and *Ectothiorhodospiraceae* in the

(a)

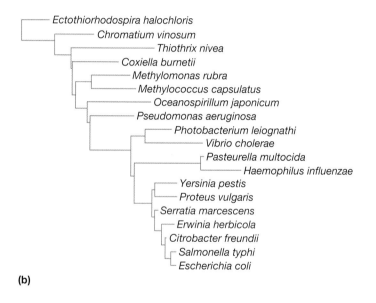

(b)

Figure 22.17 Phylogenetic Relationships Among γ-Proteobacteria. **(a)** The major phylogenetic groups based on 16S rRNA sequence comparisons. Representative genera are given in parentheses. Each tetrahedron in the tree represents a group of related organisms; its horizontal edges show the shortest and longest branches in the group. Multiple branching at the same level indicates that the relative branching order of the groups cannot be determined from the data. **(b)** The relationships of a few species based on 16S rRNA sequence data. *Source: The Ribosomal Database Project.*

	Dimensions (μm)	G + C Content	Oxygen	
Genus	**and Morphology**	**(mol %)**	**Requirement**	**Other Distinctive Characteristics**
Azotobacter	1.5–2.0; ovoid cells, pleomorphic, peritrichous flagella or nonmotile	63.2–67.5	Aerobic	Can form cysts; fix nitrogen nonsymbiotically
Beggiatoa	1–200 × 2–10; colorless cells form filaments, either single or in colonies	35–39	Aerobic or microaerophilic	Gliding motility; can form sulfur inclusions with hydrogen sulfide present
Chromatium	1–6 × 1.5–16; rod-shaped or ovoid, straight or slightly curved, polar flagella	48–50	Anaerobic	Photolithoautotroph that can use sulfide; sulfur stored within the cell
Ectothiorhodospira	0.7–1.5 in diameter; vibrioid- or rod-shaped, polar flagella	61.4–68.4	Anaerobic, some aerobic or microaerophilic	Internal lamellar stacks of membranes; deposits sulfur granules outside cells
Escherichia	1.1–1.5 × 2–6; straight rods, peritrichous flagella or nonmotile	48–59	Facultatively anaerobic	Mixed acid fermenter; formic acid converted to H_2 and CO_2, lactose fermented, citrate not used
Haemophilus	<1.0 in width, variable lengths; coccobacilli or rods, nonmotile	37–44	Facultative or aerobic	Fermentative; requires growth factors present in blood; parasites on mucous membranes
Leucothrix	Long filaments of short cylindrical cells, usually holdfast is present	46–51	Aerobic	Dispersal by gonidia, filaments don't glide; rosettes formed; heterotrophic
Methylococcus	0.8–1.5 × 1.0–1.5; cocci with capsules, nonmotile	59–65	Aerobic	Can form a cyst; uses methane, methanol, and formaldehyde as sole carbon and energy sources
Photobacterium	0.8–1.3 × 1.8–2.4; straight, plump rods with polar flagella	39–44	Facultatively anaerobic	Two species can emit blue-green light; Na^+ needed for growth
Pseudomonas	0.5–1.0 × 1.5–5.0; straight or slightly curved rods, polar flagella	58–69	Aerobic or facultatively anaerobic	Respiratory metabolism with oxygen or nitrate as acceptor; some use H_2 or CO as energy source
Vibrio	0.5–0.8 × 1.4–2.6; straight or curved rods with sheathed polar flagella	38–51	Facultatively anaerobic	Fermentative or respiratory metabolism; sodium ions stimulate or are needed for growth; oxidase positive

Table 22.5   Characteristics of Selected γ-Proteobacteria

order *Chromatiales.* The family *Ectothiorhodospiraceae* contains eight genera. *Ectothiorhodospira* has red, spiral-shaped, polarly flagellated cells that deposit sulfur globules externally (**figure 22.18**). Internal photosynthetic membranes are organized as lamellar stacks. The typical purple sulfur bacteria are located in the family *Chromatiaceae,* which is much larger and contains 26 genera.

The **purple sulfur bacteria** are strict anaerobes and usually photolithoautotrophs. They oxidize hydrogen sulfide to sulfur and deposit it internally as sulfur granules (usually within invaginated pockets of the plasma membrane); often they eventually oxidize the sulfur to sulfate. Hydrogen also may serve as an electron donor. *Thiospirillum, Thiocapsa,* and *Chromatium* are typical purple sulfur bacteria (**figure 22.19**). They are found in anoxic, sulfide-rich zones of lakes, bogs, and lagoons where large blooms can occur under certain conditions (**figure 22.20**).

Figure 22.18   Purple Bacteria.   *Ectothiorhodospira mobilis;* light micrograph.

(a) *Chromatium vinosum*

(b) *C. vinosum*

Figure 22.19 Typical Purple Sulfur Bacteria. **(a)** *Chromatium vinosum* with intracellular sulfur granules. **(b)** Electron micrograph of *C. vinosum*. Note the intracytoplasmic vesicular membrane system. The large white areas are the former sites of sulfur globules.

(a)

(b)

Figure 22.20 Purple Photosynthetic Sulfur Bacteria. **(a)** Purple photosynthetic sulfur bacteria growing in a bog. **(b)** A sewage lagoon with a bloom of purple photosynthetic bacteria.

Order *Thiotrichales*

The order *Thiotrichales* contains three families, the largest of which is the family *Thiotrichaceae*. This family has several genera that oxidize sulfur compounds (see the colorless sulfur bacteria [p. 550] and chapter 9 for sulfur oxidation and chemolithotrophy). Morphologically both rods and filamentous forms are present.

Two of the best-studied gliding genera in this family are *Beggiatoa* and *Leucothrix* (**figures 22.21** and **22.22**). *Beggiatoa* is microaerophilic and grows in sulfide-rich habitats such as sulfur springs, freshwater with decaying plant material, rice paddies, salt marshes, and marine sediments. Its filaments contain short, disklike cells and lack a sheath. *Beggiatoa* is very versatile metabolically. It oxidizes hydrogen sulfide to form large sulfur grains located in pockets formed by invaginations of the plasma membrane. *Beggiatoa* can subsequently oxidize the sulfur to sulfate.

The electrons are used by the electron transport chain in energy production. Many strains also can grow heterotrophically with acetate as a carbon source, and some incorporate CO_2 autotrophically.

Leucothrix (figure 22.22) is an aerobic chemoorganotroph that forms filaments or trichomes up to 400 μm long. It is usually marine and is attached to solid substrates by a holdfast. *Leucothrix* has a complex life cycle in which it is dispersed by the formation of gonidia. Rosette formation often is seen in culture (figure 22.22*d*). *Thiothrix* is a related genus that forms sheathed filaments and releases gonidia from the open end of the sheath (**figure 22.23**). In contrast with *Leucothrix*, *Thiothrix* is a chemolithotroph that oxidizes hydrogen sulfide and deposits sulfur granules internally. It also requires an organic compound for growth (i.e., it is a mixotroph). *Thiothrix* grows in sulfide-rich flowing water and activated sludge sewage systems.

Figure 22.21 *Beggiatoa alba.* A light micrograph showing part of a colony (×400). Note the dark sulfur granules within many of the filaments.

Order *Methylococcales*

The single family in ths order is *Methylococcaceae.* It contains rods, vibrios, and cocci that use methane, methanol, and other reduced one-carbon compounds as their sole carbon and energy sources under aerobic or microaerobic (low oxygen) conditions. That is, they are methylotrophs distinguishing them from bacteria that use methane exclusively as their carbon and energy source, which are called methanotrophs. The family contains seven genera, two of which are *Methylococcus* (spherical, nonmotile cells) and *Methylomonas* (straight, curved, or branched rods with a single, polar flagellum). When oxidizing methane, the bacteria contain complex arrays of intracellular membranes. Almost all are capable of forming cysts. Methanogenesis from substrates such as H_2 and CO_2 is widespread in anoxic soil and water, and methylotrophic bacteria grow above anoxic habitats all over the world.

Methane-oxidizing bacteria use methane as a source of both energy and carbon. Methane is first oxidized to methanol by the enzyme methane monooxygenase. The methanol is then oxidized to formaldehyde by methanol dehydrogenase, and the electrons

Figure 22.22 Morphology and Reproduction of *Leucothrix mucor.* **(a)** Life cycle of *L. mucor.* **(b)** Separation of gonidia from the tip of mature filament; phase contrast (×1,400). **(c)** Gonidia aggregating to form rosettes; phase contrast (×950). **(d)** Young developing rosettes (×1,500). **(e)** A knot formed by a *Leucothrix* filament.

(a)

(b)

(c)

(d)

(e)

Figure 22.23 Thiothrix. A *Thiothrix* colony viewed with phase-contrast microscopy (×1,000).

Pseudomonas cells

(a)

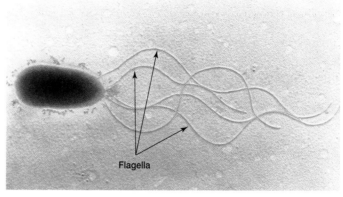

Flagella

(b)

Figure 22.24 The Genus *Pseudomonas*. (a) A phase-contrast micrograph of *Pseudomonas* cells containing PHB (poly-β-hydroxybutyrate) granules. **(b)** A transmission electron micrograph of *Pseudomonas putida* with five polar flagella, each flagellum about 5–7 μm in length.

from this oxidation are donated to an electron transport chain for ATP synthesis. Formaldehyde can be assimilated into cell material by the activity of either of two pathways, one involving the formation of the amino acid serine and the other proceeding through the synthesis of sugars such as fructose 6-phosphate and ribulose 5-phosphate.

Order *Pseudomonadales*

Pseudomonas is the most important genus in the order *Pseudomonadales*, the family *Pseudomonaceae*. These bacteria are straight or slightly curved rods, 0.5 to 1.0 μm by 1.5 to 5.0 μm in length and are motile by one or several polar flagella (**figure 22.24; *see figure 3.32a*).** These chemoheterotrophs usually carry out aerobic respiration. Sometimes nitrate is used as the terminal electron acceptor in anaerobic respiration. All pseudomonads have a functional tricarboxylic acid cycle and can oxidize substrates completely to CO_2. Most hexoses are degraded by the Entner-Doudoroff pathway rather than the Embden-Meyerhof pathway. ◀ The breakdown of glucose to pyruvate (section 9.3); The tricarboxylic acid cycle (section 9.4); also see appendix II.

The genus *Pseudomonas* is an exceptionally heterogeneous taxon currently composed of about 60 species. Many can be placed in one of seven rRNA homology groups. The three best characterized groups are subdivided according to properties such as the presence of poly-β-hydroxybutyrate (PHB), the production of a fluorescent pigment, pathogenicity, the presence of arginine dihydrolase, and glucose utilization. For example, the fluorescent subgroup does not accumulate PHB and produces a diffusible, water-soluble, yellow-green pigment that fluoresces under UV radiation (**figure 22.25**). *Pseudomonas aeruginosa, P. fluorescens, P. putida,* and *P. syringae* are members of this group.

The pseudomonads have a great practical impact in several ways, including these:

1. Many can degrade an exceptionally wide variety of organic molecules. Thus they are very important in the **mineraliza-**

tion process (the microbial breakdown of organic materials to inorganic substances) in nature and in sewage treatment. The fluorescent pseudomonads can use approximately 80 different substances as their carbon and energy sources. ◀ Microorganisms in the soil environment (section 29.3)

2. Several species (e.g., *P. aeruginosa*) are important experimental subjects. Many advances in microbial physiology and biochemistry have come from their study. For example, the study of *P. aeruginosa* has significantly advanced our understanding of how bacteria form biofilms and the role of extracellular signaling in bacterial communities and pathogenesis. The genome of *P. aeruginosa* has an unusually large number of genes for catabolism, nutrient transport, the efflux of organic molecules, and metabolic regulation. This may explain its ability to grow in many environments and resist antibiotics. ◀ Microbial growth in natural environments: Biofilms (section 6.6); Global regulatory systems: Quorum sensing (section 12.5)

3. Some pseudomonads are major animal and plant pathogens. *P. aeruginosa* infects people with low resistance such as cystic fibrosis patients. It also invades burns, and causes urinary tract infections. *P. syringae* is an important plant pathogen.
4. Pseudomonads such as *P. fluorescens* are involved in the spoilage of refrigerated milk, meat, eggs, and seafood because they grow at 4°C and degrade lipids and proteins.

The genus *Azotobacter* also is in the family *Pseudomonadaceae*. The genus contains large, ovoid bacteria, 1.5 to 2.0 μm in diameter, that may be motile by peritrichous flagella. The cells are often pleomorphic, ranging from rods to coccoid shapes, and form cysts as the culture ages (**figure 22.26**). The genus is aerobic, catalase positive, and fixes nitrogen nonsymbiotically. *Azotobacter* is widespread in soil and water.

Order *Vibrionales*

Three closely related orders of the γ-proteobacteria contain a number of important bacterial genera. Each order has only one family of facultatively anaerobic gram-negative rods. **Table 22.6** summarizes the distinguishing properties of the families *Enterobacteriaceae, Vibrionaceae,* and *Pasteurellaceae* from the orders *Enterbacteriales, Vibrionales,* and *Pasteurellales,* respectively.

The order *Vibrionales* contains only one family, the *Vibrionaceae*. Members of the family *Vibrionaceae* are gram-negative,

Figure 22.25 *Pseudomonas* **Fluorescence.** *Pseudomonas aeruginosa* colonies fluorescing under ultraviolet light.

Figure 22.26 *Azotobacter.* (a) Electron micrograph of vegetative *A. chroococcum.* (b) *Azotobacter* cyst.

straight or curved rods with polar flagella (**figure 22.27**). Most are oxidase positive, and all use D-glucose as their sole or primary carbon and energy source (table 22.6). The majority are aquatic microorganisms, widespread in freshwater and the sea. There are eight genera in the family: *Vibrio, Photobacterium, Salinivibrio, Listonella, Allomonas, Enterovibrio, Catencoccus,* and *Grimontia*.

Several vibrios are important pathogens. *Vibrio cholerae* causes cholera, and *V. parahaemolyticus* can cause gastroenteritis in humans following consumption of contaminated seafood. *V. anguillarum* and others are responsible for fish diseases. Food-borne and waterborne diseases: Cholera (section 38.4)

The *Vibrio cholerae* genome contains about 3,800 open reading frames distributed between two circular chromosomes, chromosome 1 (2.96 million base pairs) and chromosome 2 (1.07 million bp). The larger chromosome primarily has genes for essential cell functions such as DNA replication, transcription, and protein synthesis. It also has most of the virulence genes. For example, the cholera toxin gene is located in an integrated CTX phage on chromosome 1. Chromosome 2 also has essential genes such as transport genes and ribosomal protein genes. Copies of some genes are present on both chromosomes. Perhaps *V. cholerae* achieves faster genome duplication and cell division by distributing its genes between two chromosomes.

Several members of the family are unusual in being bioluminescent. *Vibrio fischeri, V. harveyi,* and at least two species of *Photobacterium* are among the few marine bacteria capable of **bioluminescence** and emit a blue-green light because of the activity of the enzyme luciferase (**Microbial Diversity & Ecology 22.1**). The peak emission of light is usually between 472 and 505 nm, but one strain of *V. fischeri* emits yellow light with a major peak at 545 nm. Although many of these bacteria are free-living, *V. fischeri, V. harveyi, P. phosphoreum,* and *P. leiognathi* live symbiotically in the luminous organs of fish (**figure 22.28**) and squid (*see figure 6.30*).

Order *Enterobacteriales*

The family *Enterobacteriaceae* is the largest of the families listed in table 22.6. It contains gram-negative, peritrichously flagellated or nonmotile, facultatively anaerobic, straight rods with simple

0.2 μm

(a)   (b)

Table 22.6	Characteristics of Families of Facultatively Anaerobic Gram-Negative Rods		
Characteristics	*Enterobacteriaceae*	*Vibrionaceae*	*Pasteurellaceae*
Cell dimensions	0.3–1.0 × 1.0–6.0 μm	0.3–1.3 × 1.0–3.5 μm	0.2–0.4 × 0.4–2.0 μm
Morphology	Straight rods; peritrichous flagella or nonmotile	Straight or curved rods; polar flagella; lateral flagella may be produced on solid media	Coccoid to rod-shaped cells, sometimes pleomorphic; nonmotile
Physiology	Oxidase negative	Oxidase positive; all can use D-glucose as sole or principal carbon source	Oxidase positive; heme and/or NAD often required for growth; organic nitrogen source required
G + C content	38–60%	38–51%	38–47%
Symbiotic relationships	Some parasitic on mammals and birds; some species plant pathogens	Most not pathogens; several inhabit light organs of marine organisms	Parasites of mammals and birds
Representative genera	*Escherichia, Shigella, Salmonella, Citrobacter, Klebsiella, Enterobacter, Erwinia, Serratia, Proteus, Yersinia*	*Vibrio, Photobacterium*	*Pasteurella, Haemophilus*

From G. M. Garrity editor-in-chief. *Bergey's Manual of Systematic Bacteriology,* vol. 2. Copyright © 2005 New York: Springer. Reprinted by permission.

Figure 22.27 The *Vibrionaceae.* Electron micrograph of *Vibrio alginolyticus* grown on agar, showing a sheathed polar flagellum and unsheathed lateral flagella (×18,000).

nutritional requirements. The order *Enterobacteriales* has only one family, *Enterobacteriaceae,* with 44 genera. The relationship between *Enterobacteriales* and the orders *Vibrionales* and *Pasteurellales* can be seen by inspecting figure 22.17.

The metabolic properties of the *Enterobacteriaceae* are very useful in characterizing its constituent genera. Members of the family, often called **enterobacteria** or **enteric bacteria** [Greek *enterikos,* pertaining to the intestine], all degrade sugars by means of the Embden-Meyerhof pathway and cleave pyruvic acid

to yield formic acid in formic acid fermentations. Those enteric bacteria that produce large amounts of gas during sugar fermentation, such as *Escherichia* spp., have the formic hydrogenlyase complex that degrades formic acid to H_2 and CO_2. The family can be divided into two groups based on their fermentation products. The majority (e.g., the genera *Escherichia, Proteus, Salmonella,* and *Shigella*) carry out mixed acid fermentation and produce mainly lactate, acetate, succinate, formate (or H_2 and CO_2), and ethanol. In contrast, *Enterobacter, Serratia, Erwinia,* and *Klebsiella* are butanediol fermenters. The major products of butanediol fermentation are butanediol, ethanol, and carbon dioxide. The two types of formic acid fermentation are distinguished by the methyl red and Voges-Proskauer tests.   Fermentations (section 9.7)

Because the enteric bacteria are so similar in morphology, biochemical tests are normally used to identify them after a preliminary examination of their morphology, motility, and growth responses (**figure 22.29** provides a simple example). Some more commonly used tests are those for the type of formic acid fermentation, lactose and citrate utilization, indole production from tryptophan, urea hydrolysis, and hydrogen sulfide production. For example, lactose fermentation occurs in *Escherichia* and *Enterobacter* but not in *Shigella, Salmonella,* or *Proteus.* **Table 22.7** summarizes a few of the biochemical properties useful in distinguishing between genera of enteric bacteria. The mixed acid fermenters are located on the left in this table and the butanediol fermenters on the right. The usefulness of biochemical tests in identifying enteric bacteria is shown by the popularity of commercial identification systems, such as the Enterotube and API 20-E systems, that are based on these tests.   Identification of microorganisms from specimens (section 35.2)

Microbial Diversity & Ecology

22.1   Bacterial Bioluminescence

Several species in the genera *Vibrio* and *Photobacterium* can emit light of a blue-green color. The enzyme luciferase catalyzes the reaction and uses reduced flavin mononucleotide, molecular oxygen, and a long-chain aldehyde as substrates.

$$FMNH_2 + O_2 + RCHO \xrightarrow{\text{luciferase}} FMH + H_2O + RCOOH + light$$

The evidence suggests that an enzyme-bound, excited flavin intermediate is the direct source of luminescence. Because the electrons used in light generation are probably diverted from the electron transport chain and ATP synthesis, the bacteria expend considerable energy on luminescence. Luminescence activity is regulated and can be turned off or on under the proper conditions.

There is much speculation about the role of bacterial luminescence and its value to bacteria, particularly because it is such an energetically expensive process. Luminescent bacteria occupying the luminous organs of fish do not emit light when they grow as free-living organisms in the seawater. Free-living luminescent bacteria can reproduce and infect young fish. Once settled in a fish's luminous organ, the quorum-sensing molecule autoinducer produced by the bacteria stimulates the emission of light. Other luminescent bacteria growing on potential food items such as small crustacea may use light to attract fish to the food source. After ingestion, they could establish a symbiotic relationship in the host's gut.

The mechanism by which autoinducer regulates light production in these marine bacteria is an important model for understanding quorum sensing in many gram-negative bacteria, including a number of pathogens.   Global regulatory systems: Quorum sensing (section 12.5)

Figure 22.28   Bioluminescence.   (a) A photograph of the Atlantic flashlight fish *Kryptophanaron alfredi*. The light area under the eye is the fish's luminous organ, which can be covered by a lid of tissue. **(b)** The masses of photobacteria in the SEM view are separated by thin epithelial cells. **(c)** Ultrathin section of the luminous organ of a fish, *Equulites novaehollandiae*, with the bioluminescent bacterium *Photobacterium leiognathi*, PL.

(a)

(b)

(c)

Members of the *Enterobacteriaceae* are so common, widespread, and important that they are probably more often seen in most laboratories than any other bacteria. *Escherichia coli* is undoubtedly the best-studied bacterium and the experimental organism of choice for many microbiologists. It is an inhabitant of the colon of humans and other warm-blooded animals, and it is quite useful in the analysis of water for fecal contamination. Some strains cause gastroenteritis or urinary tract infections. Several genera contain very important human pathogens responsible for a variety of diseases: *Salmonella* (**figure 22.30**), typhoid fever and gastroenteritis; *Shigella,* bacillary dysentery; *Klebsiella,* pneumonia; *Yersinia,* plague. Members of the genus *Erwinia* are major pathogens of crop plants and cause blights, wilts, and several other plant diseases. These and other members of the family are discussed in more detail in chapter 38.   Water purification and sanitary analysis (section 41.1)

Order *Pasteurellales*

The family *Pasteurellaceae* in the order *Pasteurellales* differs from the *Vibrionales* and the *Enterobacteriales* in several ways (table 22.6). Most notably, they are small (0.2 to 0.4 μm in diameter) and nonmotile, normally oxidase positive, have complex nutritional requirements of various kinds, and are parasitic in

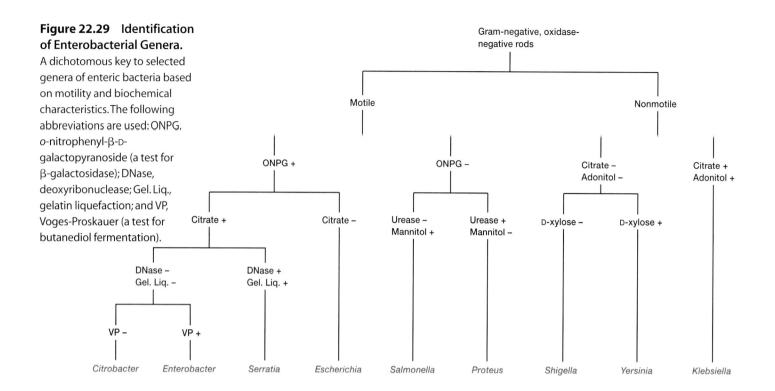

Figure 22.29 Identification of Enterobacterial Genera.
A dichotomous key to selected genera of enteric bacteria based on motility and biochemical characteristics. The following abbreviations are used: ONPG, o-nitrophenyl-β-D-galactopyranoside (a test for β-galactosidase); DNase, deoxyribonuclease; Gel. Liq., gelatin liquefaction; and VP, Voges-Proskauer (a test for butanediol fermentation).

Table 22.7	Some Characteristics of Selected Genera in the *Enterobacteriaceae*				
Characteristics	***Escherichia***	***Shigella***	***Salmonella***	***Citrobacter***	***Proteus***
Methyl red	+	+	+	+	+
Voges-Proskauer	−	−	−	−	d
Indole production	(+)	d	−	d	d
Citrate use	−	−	(+)	+	d
H₂S production	−	−	(+)	d	(+)
Urease	−	−	−	(+)	+
β-galactosidase	(+)	d	d	+	−
Gas from glucose	+	−	(+)	+	+
Acid from lactose	+	−	(−)	d	−
Phenylalanine deaminase	−	−	−	−	+
Lysine decarboxylase	(+)	−	(+)	−	−
Ornithine decarboxylase	(+)	d	(+)	(+)	d
Motility	d	−	(+)	+	+
Gelatin liquification (22°C)	−	−	−	−	+
% G + C	48–59	49–53	50–53	50–52	38–41
Genome size (Mb)	4.6–5.5	4.6	4.5–4.9	Nd[d]	Nd
Other characteristics	1.1–1.5 × 2.0–6.0 μm; peritrichous when motile	No gas from sugars	0.7–1.5 × 2–5 μm; peritrichous flagella	1.0 × 2.0–6.0 μm; peritrichous	0.4–0.8 × 1.0–3.0 μm; peritrichous

[a] (+) usually present

[b] (−) usually absent

[c] d, strains or species vary in possession of characteristic

[d] Nd: Not determined; genome not yet sequenced

vertebrates. The family contains seven genera: *Pasteurella, Haemophilus, Actinobacillus, Lonepinella, Mannheimia, Phocoenobacter,* and *Gallibacterium.*

As might be expected, members of this family are best known for the diseases they cause in humans and many animals. *Pasteurella multocida* and *P. haemolytica* are important animal pathogens. *P. multocida* is responsible for fowl cholera, which kills many chickens, turkeys, ducks, and geese each year. *P. haemolytica* is at least partly responsible for pneumonia in cattle, sheep, and goats (e.g., "shipping fever" in cattle). *H. influenzae* type b is a major human pathogen that causes a variety of diseases, including meningitis in children.   Airborne diseases: Meningitis (section 38.1)

1. Describe the general properties of the γ-proteobacteria.
2. What are the major characteristics of the purple sulfur bacteria? Contrast the families *Chromatiaceae* and *Ectothiorhodospiraceae.*
3. Describe the genera *Beggiatoa, Leucothrix,* and *Thiothrix.*

4. In what habitats would one expect to see the *Methylococcaceae* growing and why?
5. What is a methylotroph? How do methane-oxidizing bacteria use methane as both an energy source and a carbon source?
6. Give the major distinctive properties of the genera *Pseudomonas* and *Azotobacter.* Briefly discuss the taxonomic changes that have occurred in the genus *Pseudomonas.*
7. Why are the pseudomonads such important bacteria? What is mineralization?
8. List the major distinguishing traits of the families *Vibrionaceae, Enterobacteriaceae,* and *Pasteurellaceae.*
9. What major human disease is associated with the *Vibrionaceae,* and what species causes it?
10. Briefly describe bioluminescence and the way it is produced.
11. Into what two groups can the enteric bacteria be placed based on their fermentation patterns?
12. Give two reasons why the enterobacteria are so important.

Yersinia	*Klebsiella*	*Enterobacter*	*Erwinia*	*Serratia*
+	(+)ᵃ	(−)ᵇ	+	dᶜ
− (37°C)	(+)	+	(+)	+
d	d	−	(−)	(−)
(−)	(+)	+	(+)	+
−	−	−	(+)	−
d	(+)	(−)	−	−
+	(+)	+	+	+
(−)	(+)	(+)	(−)	d
(−)	(+)	(+)	d	d
−	−	(−)	(−)	−
(−)	(+)	d	−	d
d	−	(+)	−	d
− (37°C)	−	+	+	+
(−)	−	d	d	(+)
46–50	53–58	52–60	50–54	52–60
4.6	Nd	Nd	5.1	5.1
0.5–0.8 × 1.0–3.0 μm; peritrichous when motile	0.3–1.0 × 0.6–6.0 μm; capsulated	0.6–1.0 × 1.2–3.0 μm; peritrichous	0.5–1.0 × 1.0–3.0μm; peritrichous; plant pathogens and saprophytes	0.5–0.8 × 0.9–2.0 μm; peritrichous; colonies often pigmented

(a) *Salmonella enteritidis*   **(b)** *S. typhi*

Figure 22.30   The *Enterobacteriaceae*. *Salmonella* treated with fluorescent stains. **(a)** *Salmonella enterica* serovar Enteritidis with peritrichous flagella (×500). *S. enteritidis* is associated with gastroenteritis. **(b)** *S. enterica* serovar Typhi with acridine orange stain (×2,000). *S. typhi* causes typhoid fever.

22.4   Class *Deltaproteobacteria*

Although the **δ-proteobacteria** are not a large assemblage of genera, they show considerable morphological and physiological diversity. These bacteria can be divided into two general groups, all of them chemoorganotrophs. Some genera are predators such as the bdellovibrios and myxobacteria. Others are anaerobes that generate sulfide from sulfate and sulfur while oxidizing organic nutrients. The class has eight orders and 20 families. **Figure 22.31** illustrates the phylogenetic relationships among major groups within the δ-proteobacteria, and **table 22.8** summarizes the general properties of some representative genera.

Orders *Desulfovibrionales, Desulfobacterales,* and *Desulfuromonadales*

These sulfate- or sulfur-reducing bacteria are a diverse group united by their anaerobic nature and the ability to use elemental sulfur or sulfate and other oxidized sulfur compounds as electron acceptors during anaerobic respiration (**figure 22.32**). An electron transport chain reduces sulfur and sulfate to hydrogen sulfide and generates the proton motive force that drives the synthesis of ATP. The best-studied sulfate-reducing genus is *Desulfovibrio; Desulfuromonas* uses only elemental sulfur as an acceptor. Anaerobic respiration (section 9.6)

These bacteria are very important in the cycling of sulfur within the ecosystem. Because significant amounts of sulfate are present in almost all aquatic and terrestrial habitats, sulfate-reducing bacteria are widespread and active in locations made anoxic by microbial digestion of organic materials. *Desulfovibrio* and other sulfate-reducing bacteria thrive in habitats such as muds and sediments of polluted lakes and streams, sewage lagoons and digesters, and waterlogged soils. *Desulfuromonas* is

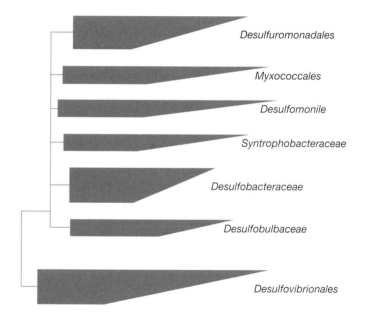

Figure 22.31   Phylogenetic Relationships Among Major Groups Within the δ-Proteobacteria. The relationships are based on 16S rRNA sequence.

most prevalent in anoxic marine and estuarine sediments. It also can be isolated from methane digesters and anoxic hydrogen sulfide-rich muds of freshwater habitats. It uses elemental sulfur, but not sulfate, as its electron acceptor. Often sulfate and sulfur reduction are apparent from the smell of hydrogen sulfide and the blackening of water and sediment by iron sulfide. Hydrogen sulfide production in waterlogged soils can kill animals, plants, and microorganisms. Sulfate-reducing bacteria negatively impact industry because of their primary role in the anaerobic corrosion of

Table 22.8 | Characteristics of Selected δ- and ε-Proteobacteria

Class Genus	Dimensions (μm) and Morphology	G + C Content (mol%)	Oxygen Requirement	Other Distinctive Characteristics
δ-Proteobacteria				
Bdellovibrio	0.2–0.5 × 0.5–1.4; comma-shaped rods with a sheathed polar flagellum	49.5–51	Aerobic	Preys on other gram-negative bacteria where it grows in the periplasm, alternates between predatory and intracellular reproductive phases
Desulfovibrio	0.5–1.5 × 2.5–10; curved or sometimes straight rods, motile by polar flagella	46.1–61.2	Anaerobic	Oxidizes organic compounds to acetate and reduces sulfate or sulfur to H_2S
Desulfuromonas	0.4–0.9 × 1.0–4.0; straight or slightly curved or ovoid rods, lateral or subpolar flagella	54–62	Anaerobic	Reduces sulfur to H_2S, oxidizes acetate to CO_2, forms pink or peach-colored colonies
Myxococcus	0.4–0.7 × 2–8; slender rods with tapering ends, gliding motility	68–71	Aerobic	Forms fruiting bodies with microcysts not enclosed in a sporangium
Stigmatella	0.7–0.8 × 4–8; straight rods with tapered ends, gliding motility	67–68	Aerobic	Stalked fruiting bodies with sporangioles containing myxospores (0.9–1.2 × 2–4 μm)
ε-Proteobacteria				
Campylobacter	0.2–0.8 × 0.5–5; spirally curved cells with a single polar flagellum at one or both ends	29–47	Microaerophilic	Carbohydrates not fermented or oxidized; oxidase positive and urease negative; found in intestinal tract, reproductive organs, and oral cavity of animals
Helicobacter	0.2–1.2 × 1.5–10; helical, curved, or straight cells with rounded ends; multiple, sheathed flagella	24–48	Microaerophilic	Catalase and oxidase positive; urea rapidly hydrolyzed; found in the gastric mucosa of humans and other animals

iron in pipelines, heating systems, and other structures. Biogeochemical cycling: Sulfur cycle (section 27.2)

Order *Bdellovibrionales*

The order has only the family *Bdellovibrionaceae* and four genera. The genus *Bdellovibrio* [Greek *bdella,* leech] contains aerobic gram-negative, curved rods with polar flagella (**figure 22.33**). The flagellum is unusually thick due to the presence of a sheath that is continuous with the cell wall. *Bdellovibrio* has a distinctive life-style: it preys on other gram-negative bacteria and alternates between a nongrowing predatory phase and an intracellular reproductive phase.

The life cycle of *Bdellovibrio* is complex although it requires only 1 to 3 hours for completion (**figure 22.34**). The free bacterium swims along very rapidly (about 100 cell lengths per second) until it collides violently with its prey. It attaches to the bacterial surface, begins to rotate as fast as 100 revolutions per second, and bores a hole through the host cell wall in 5 to 20 minutes with the aid of several hydrolytic enzymes that it releases. Its flagellum is lost during penetration of the cell.

After entry, *Bdellovibrio* takes control of the host cell and grows in the periplasmic space (between the cell wall and plasma membrane) while the host cell loses its shape and rounds up. The predator inhibits host DNA, RNA, and protein synthesis within minutes and disrupts the host's plasma membrane so that cytoplasmic constituents leak out of the cell. The growing bacterium uses host amino acids as its carbon, nitrogen, and energy source. It employs fatty acids and nucleotides directly in biosynthesis, thus saving carbon and energy. The bacterium rapidly grows into a long filament under the cell wall and then divides into many smaller, flagellated progeny, which escape upon host cell lysis. Such multiple fission is rare in procaryotes.

The *Bdellovibrio* life cycle resembles that of bacteriophages in many ways. Not surprisingly, if a *Bdellovibrio* culture is plated out on agar with host bacteria, plaques will form in the bacterial lawn. This technique is used to isolate pure strains and count the number of viable organisms just as with phages.

(a) *Desulfovibrio saprovorans*

(b) *Desulfovibrio gigas*

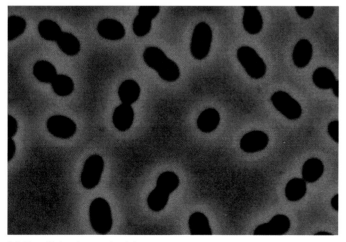

(c) *Desulfobacter postgatei*

Figure 22.32   The Dissimilatory Sulfate- or Sulfur-Reducing Bacteria.   Representative examples. **(a)** Phase-contrast micrograph of *Desulfovibrio saprovorans* with PHB inclusions ($\times$2,000). **(b)** *Desulfovibrio gigas;* phase contrast ($\times$2,000). **(c)** *Desulfobacter postgatei;* phase contrast ($\times$2,000).

0.2 µm

Figure 22.33   *Bdellovibrio* Morphology.   Negatively stained *Bdellovibrio bacteriovorus* with its sheathed polar flagellum.

Order *Myxococcales*

The **myxobacteria** are gram-negative, aerobic soil bacteria characterized by gliding motility, a complex life cycle with the production of fruiting bodies, and the formation of dormant myxospores. In addition, their G + C content is around 67 to 71%, significantly higher than that of most gliding bacteria. Myxobacterial cells are rods, about 0.4 to 0.7 by 2 to 8 µm long, and may be either slender with tapered ends or stout with rounded, blunt ends (**figure 22.35**). The order *Myxococcales* is divided into six families based on the shape of vegetative cells, myxospores, and sporangia.   Microbial diversity & ecology 21.1: The mechanism of gliding motility

Most myxobacteria are micropredators or scavengers. They secrete an array of digestive enzymes that lyse bacteria and yeasts. Many myxobacteria also secrete antibiotics, which may kill their prey. The digestion products, primarily small peptides, are absorbed. Most myxobacteria use amino acids as their major source of carbon, nitrogen, and energy. All are chemoheterotrophs with respiratory metabolism.

The myxobacterial life cycle is quite distinctive and in many ways resembles that of the cellular slime molds (**figure 22.36**). In the presence of a food supply, myxobacteria migrate along a solid surface, feeding and leaving slime trails. During this stage the cells often form a swarm and move in a coordinated fashion. Some species congregate to produce a sheet of cells that moves rhythmically to generate waves or ripples. When their nutrient supply is exhausted, the myxobacteria aggregate and differentiate into a **fruiting body.**   Protist classification: *Eumycetozoa* (section 25.6)

The life cycle of the species *Myxococcus xanthus* has been well studied. Development in this microbe is induced by nutrient limitation and involves the exchange of at least five different extracellular signaling molecules that allow the cells to communicate with one another. Two of these signals have been characterized. Both the A factor, a mixture of peptides and amino acids, and the protein C factor are released and help trigger the process. Fruiting body development also requires gliding motil-

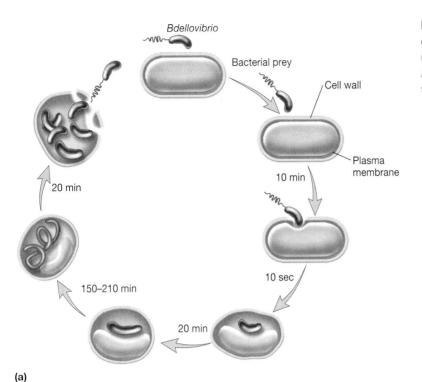

(a)

Figure 22.34 The Life Cycle of *Bdellovibrio*. (a) A general diagram showing the complete life cycle. **(b)** *Bdellovibrio bacteriovorus* penetrating the cell wall of *E. coli* (×55,000). **(c)** A *Bdellovibrio* encapsulated between the cell wall and plasma membrane of *E. coli* (×60,800).

(b)

(c)

ity. Two types of gliding have been characterized in *M. xanthus:* adventurous (A) motility is propelled by the extrusion of a gel-like material from the rear pole; social (S) motility is governed by the production of retractable pili from the front end of the cell. When the pili retract, the cell creeps forward. This type of motility was originally called social motility because it is only observed in cells that are close together. It is now known that cell-to-cell contact is required for S motility because cells share outer membrane lipoproteins involved in pili secretion.

A variety of new proteins are synthesized during fruiting body formation. Fruiting bodies range in height from 50 to 500 μm and often are colored red, yellow, or brown by carotenoid pigments. They vary in complexity from simple globular objects

made of about 100,000 cells (*Myxococcus*) to the elaborate, branching, treelike structures formed by *Stigmatella* and *Chondromyces* (**figure 22.37**). Some cells develop into dormant **myxospores** that frequently are enclosed in walled structures called sporangioles or sporangia. Each species forms a characteristic fruiting body.

Myxospores are not only dormant but desiccation-resistant, and they may survive up to 10 years under adverse conditions. They enable myxobacteria to survive long periods of dryness and nutrient deprivation. The use of fruiting bodies provides further protection for the myxospores and assists in their dispersal. (The myxospores often are suspended above the soil surface.) Because myxospores are kept together within the fruiting body, a colony

(a)   (b)   (c)

Figure 22.35 Gliding, Fruiting Bacteria (Myxobacteria). Myxobacterial cells and myxospores. **(a)** *Stigmatella aurantiaca* ($\times$1,200). **(b)** *Chondromyces crocatus* ($\times$950). **(c)** Myxospores of *Myxococcus xanthus* ($\times$1,100). All photographs taken with a phase-contrast microscope.

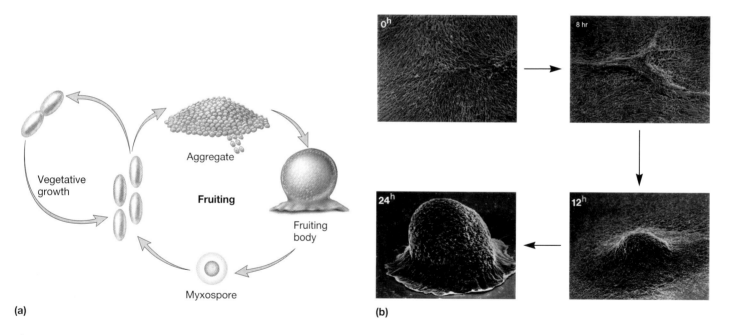

(a)   (b)

Figure 22.36 Life cycle of *Myxococcus xanthus*. **(a)** When nutrients are plentiful, *M. xanthus* grows vegetatively. However, when nutrients are depleted, a complex exchange of extracellular signaling molecules triggers the cells to aggregate and form fruiting bodies. Most of the cells within a fruiting body will become resting myxospores that will not germinate until nutrients are available. **(b)** Scanning electron micrographs taken during aggregate (0–12 hours) and fruiting body (24 hours) formation.

of myxobacteria automatically develops when the myxospores are released and germinate. This communal organization may be advantageous because myxobacteria obtain nutrients by secreting hydrolytic enzymes and absorbing soluble digestive products. A mass of myxobacteria can produce enzyme concentrations sufficient to digest their prey more easily than can an individual cell. Extracellular enzymes diffuse away from their source, and an in-

dividual cell will have more difficulty overcoming diffusional losses than a swarm of cells.

Myxobacteria are found in soils worldwide. They are most commonly isolated from neutral soils or decaying plant material such as leaves and tree bark, and from animal dung. Although they grow in habitats as diverse as tropical rain forests and the arctic tundra, they are most abundant in warm areas.

Myxospores   Wall of sporangiole

Sporangiole

Myxospores

Slime envelope

Myxospores

Stalk

Sporangiole

Fruiting body of
Myxococcus

Fruiting body of
Stigmatella

Fruiting body of
Polyangium

(a)

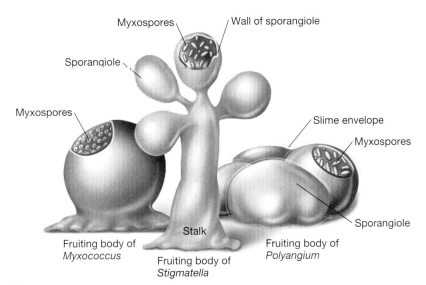

Figure 22.37   Myxobacterial Fruiting Bodies.
(a) An illustration of typical fruiting body structure.
(b) *Myxococcus fulvus.* Fruiting bodies are about
130–400 μm high. **(c)** *Myxococcus stipitatus.* The stalk is
as tall as 200 μm. **(d)** *Chondromyces crocatus* viewed
with the SEM. The stalk may reach 700 μm or more in
height.

(b) *Myxococcus fulvus*

(c) *Myxococcus stipitatus*

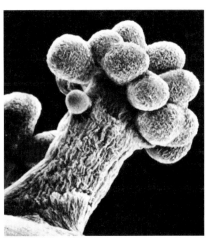

(d) *Chondromyces crocatus*

---

1. Briefly characterize the δ-proteobacteria.
2. Describe the metabolic specialization of the dissimilatory sulfate- or sulfur-reducing bacteria. Why are they important?
3. Characterize the genus *Bdellovibrio* and outline its life cycle in detail.
4. Give the major distinguishing characteristics of the myxobacteria. How do they obtain most of their nutrients?
5. Briefly describe the myxobacterial life cycle. What are fruiting bodies and myxospores?

22.5   CLASS *EPSILONPROTEOBACTERIA*

The ε-**proteobacteria** are the smallest of the five proteobacterial classes. They all are slender gram-negative rods, which can be straight, curved, or helical. The ε-proteobacteria have one order, *Campylobacterales,* and three families: *Campylobacteraceae, Helicobacteraceae,* and the recently added *Nautiliaceae.* Two patho-

genic genera, *Campylobacter* and *Helicobacter,* are microaerophilic, motile, helical or vibrioid, gram-negative rods. Table 22.8 summarizes some of the characteristics of these two genera.

The genus *Campylobacter* contains both nonpathogens and species pathogenic for humans and other animals. *C. fetus* causes reproductive disease and abortions in cattle and sheep. It is associated with a variety of conditions in humans ranging from **septicemia** (pathogens or their toxins in the blood) to **enteritis** (inflammation of the intestinal tract). *C. jejuni* causes abortion in sheep and enteritis diarrhea in humans.   Food-borne and waterborne diseases: *Campylobacter jejuni* gastroenteritis (section 38.4)

There are at least 23 species of *Helicobacter,* all isolated from the stomachs and upper intestines of humans, dogs, cats, and other mammals. In developing countries 70 to 90% of the population is infected; developed countries range from 25 to 50%. Most infections are probably acquired during childhood, but the precise mode of transmission is unclear. The major human pathogen is *Helicobacter pylori,* which causes gastritis

(a)

(b)

Figure 22.38   ε-Proteobacteria Dominate Filamentous Microbial Mats in a Wyoming Sulfidic Cave Spring.   (a) A channel formed by the spring appears white due to the high density of filamentous ε-proteobacteria. The stake in the center of photo (arrow) is about 25 cm high. **(b)** Filamentous ε-proteobacteria within the springs.

and peptic ulcer disease. *H. pylori* produces large quantities of urease, and urea hydrolysis appears to be associated with its virulence.   Direct-contact diseases: Peptic ulcer disease and gastritis (section 38.3)

The genomes of *C. jejuni* and *H. pylori* (both about 1.6 million base pairs in size) have been sequenced. They are now being studied and compared in order to understand the life styles and pathogenicity of these bacteria.

The ε-proteobacteria are now recognized to be more metabolically and ecologically diverse than previously thought. For instance, filamentous microbial mats in anoxic, sulfide-rich cave springs are dominated by members of the ε-proteobacteria (**figure 22.38**). The inclusion of a new family, the *Nautiliaceae*, in the

second edition of *Bergey's Manual* is a result of the recent isolation of two genera of moderately thermophilic (optimum growth temperature about 55°C) chemolithoautotrophs from deep-sea hydrothermal vents. Members of the genera *Nautilia* and *Caminibacter* are strict anaerobes that oxidize H_2 and use sulfur as an electron acceptor. Species are found as either freely living or as symbionts of vent macrofauna.   Microbial interactions: Sulfide-based mutualisms (section 30.1)

---

1. Briefly describe the properties of the ε-proteobacteria.
2. Give the general characteristics of *Campylobacter* and *Helicobacter*. What is their public health significance?

Summary

The proteobacteria are the largest and most diverse group of bacteria. On the basis of rRNA sequence data, they are divided into five classes: the α-, β-, γ-, δ-, and ε-proteobacteria.

22.1 Class *Alphaproteobacteria*

a. The purple nonsulfur bacteria can grow anaerobically as photoorganoheterotrophs and often aerobically as chemoorganoheterotrophs (**figure 22.3**). They are found in aquatic habitats with abundant organic matter and low sulfide levels.

b. Rickettsias are obligately intracellular parasites responsible for many diseases (**figure 22.4**). They have numerous transport proteins in their plasma membranes and make extensive use of host cell nutrients, coenzymes, and ATP.

c. Many proteobacteria have prosthecae, stalks, or reproduction by budding. Most of these bacteria are placed among the α-proteobacteria.

d. Two examples of budding and/or appendaged bacteria are *Hyphomicrobium* (budding bacteria that produce swarmer cells) and *Caulobacter* (bacteria with prosthecae and holdfasts) (**figures 22.5–22.8**).

e. *Rhizobium* carries out nitrogen fixation, whereas *Agrobacterium* causes the development of plant tumors. Both are in the family *Rhizobiaceae* of the α-proteobacteria (**figures 22.9** and **22.10**).

f. Chemolithotrophic bacteria derive energy and electrons from reduced inorganic compounds. Nitrifying bacteria are aerobes that oxidize either ammonia or nitrite to nitrate and are responsible for nitrification (**table 22.2**).

22.2 Class *Betaproteobacteria*

a. The genus *Neisseria* contains nonmotile, aerobic, gram-negative cocci that usually occur in pairs. They colonize mucous membranes and cause several human diseases

b. *Sphaerotilus, Leptothrix,* and several other genera have sheaths, hollow tube-like structures that surround chains of cells without being in intimate contact with the cells (**figure 22.14**).

c. The colorless sulfur bacteria such as *Thiobacillus* oxidize elemental sulfur, hydrogen sulfide, and thiosulfate to sulfate while generating energy chemolithotrophically.

22.3 Class *Gammaproteobacteria*

a. The γ-proteobacteria are the largest subgroup of proteobacteria with great variety in physiological types (**table 22.5** and **figure 22.17**).

b. The purple sulfur bacteria are anaerobes and usually photolithoautotrophs. They oxidize hydrogen sulfide to sulfur and deposit the granules internally (**figure 22.19**).

c. Bacteria like *Beggiatoa* and *Leucothrix* grow in long filaments or trichomes (**figures 22.21** and **22.22**). Both genera have gliding motility. *Beggiatoa* is primarily a chemolithotroph and *Leucothrix,* a chemoorganotroph.

d. The *Methylococcaceae* are methylotrophs; they use methane, methanol, and other reduced one-carbon compounds as their sole carbon and energy sources.

e. The genus *Pseudomonas* contains straight or slightly curved, gram-negative, aerobic rods that are motile by one or several polar flagella and do not have prosthecae or sheaths (**figure 22.24**).

f. The pseudomonads participate in natural mineralization processes, are major experimental subjects, cause many diseases, and often spoil refrigerated food.

g. The most important facultatively anaerobic, gram-negative rods are found in three families: *Vibrionaceae, Enterobacteriaceae,* and *Pasteurellaceae* (**table 22.6**).

h. The *Enterobacteriaceae,* often called enterobacteria or enteric bacteria, are gram-negative, peritrichously flagellated or nonmotile, facultatively anaerobic, straight rods with simple nutritional requirements.

i. The enteric bacteria are usually identified by a variety of physiological tests and are very important experimental organisms and pathogens of plants and animals (**table 22.7** and **figure 22.29**).

22.4 Class *Deltaproteobacteria*

a. The δ-proteobacteria contain chemorganotrophic gram-negative bacteria that are anaerobic and can use elemental sulfur and oxidized sulfur compounds as electron acceptors in anaerobic respiration (**table 22.8**). They are very important in sulfur cycling in the ecosystem. Other δ-proteobacteria are predatory aerobes.

b. *Bdellovibrio* is an aerobic curved rod with sheathed polar flagellum that preys on other gram-negative bacteria and grows within their periplasmic space (**figures 22.33** and **22.34**).

c. Myxobacteria are gram-negative, aerobic soil bacteria with gliding motility and a complex life cycle that leads to the production of dormant myxospores held within fruiting bodies (**figures 22.35–22.37**).

22.5 Class *Epsilonproteobacteria*

a. The ε-proteobacteria are the smallest of the proteobacterial classes and contain two important pathogenic genera: *Campylobacter* and *Helicobacter.* These are microaerophilic, motile, helical or vibrioid, gram-negative rods.

b. Recently a new family, the *Nautiliaceae,* has been added. Many of these bacteria are chemolithoautotrophs from deep-sea hydrothermal vents ecosystems.

Key Terms

α-proteobacteria 540
β-proteobacteria 546
binary fission 543
bioluminescence 557
budding 543
colorless sulfur bacteria 550
δ-proteobacteria 562

enteric bacteria (enterobacteria) 558
enteritis 567
ε-proteobacteria 567
fruiting body 564
γ-proteobacteria 551
holdfast 544
methylotroph 544

mineralization 556
myxobacteria 564
myxospores 565
nitrification 546
nitrifying bacteria 546
prostheca 543
proteobacteria 539

purple nonsulfur bacteria 540
purple sulfur bacteria 553
septicemia 567
sheath 548
stalk 543

Critical Thinking Questions

1. *Helicobacter pylori* produces large quantities of urease. Urease catalyzes the reaction:

$$H_2N-\overset{\displaystyle O}{\overset{\displaystyle \|}{C}}-NH_2 \rightarrow CO_2 + 2NH_3$$

    Suggest why this allows *H. pylori* to inhabit the acidic habitat of the gastric mucosa.

2. Methylotrophs oxidize methane to methanol, then to formaldehyde, and finally into acetate. Suggest mechanisms by which the bacterium protects itself from the toxic effects of the intermediates, methanol and formaldehyde.

3. *Bdellovibrio* is an intracellular predator. Once it invades the periplasm, it manages to inhibit many aspects of host metabolism. Suggest a mechanism by which this inhibition could occur so rapidly.

4. Why might the ability to form dormant cysts be of great advantage to *Agrobacterium* but not as much to *Rhizobium*?

5. Why are gliding and budding and/or appendaged bacteria distributed among so many different sections in *Bergey's Manual*?

Learn More

Balows, A.; Trüper, H. G.; Dworkin, M.; Harder, W.; and Schleifer, K.-H. 1992. *The prokaryotes*, 2d ed. New York: Springer-Verlag.

Engel, A., S.; Lee, N.; Porter, M. L.; Stern, L. A.; Bennett, P. C.; and Wagner, M. 2003. Filamentous *"Epsilonproteobacteria"* dominate microbial mats from sulfidic cave springs. *Appl. Environ. Microbiol.* 69:5503–11.

Eremeeva, M. E., and Dasch, G.A. 2000. Rickettsiae. In *Encyclopedia of microbiology*, 2d ed., vol. 4, J. Lederberg, editor-in-chief, 140–80. San Diego: Academic Press.

Garrity, G. M., editor-in-chief. 2005. *Bergey's Manual of Systematic Bacteriology*, 2d ed., vol. 2, D. J. Brenner, N. R. Krieg, and J. T. Staley, editors. New York: Springer-Verlag.

Kaplan, H. B. 2003. Multicellular development and gliding motility in *Myxococcus xanthus*. *Curr. Opin. Microbiol.* 6:572–77.

McGrath, P. T.; Vollier, P.; and McAdams, H. H. 2004. Setting the pace: Mechanisms tying *Caulobacter* cell-cycle progression to macroscopic cellular events. *Curr. Opin. Microbiol.* 7:192–97.

Miroshnichenko, M. L.; Haridon, S. L.; Schumann, P.; Spring, S.; Bonch-Osmolovskaya, E. A.; Jeanthon, C.; and Stackenbrandt, E. 2004. *Caminibacter profundus* sp. nov., a novel thermophile of *Nautiliales* ord. nov. within the class *"Epsilonproteobacteria,"* isolated from a deep-sea hydrothermal vent. *Int. J. Sys. Evol. Microbiol.* 54:41–5.

Miroshnichenko, M. L.; Kostrikina, N. A.; Haridon, S. L.; Jeathon, C.; Hippe, H.; Stackenbrandt, E.; and Bonch-Osmolovskaya, E. A. 2002. *Nautilia lithotrophica* gen. nov., sp. nov. a thermophilic sulfur-reducing ε-proteobacterium isolated from a deep-sea hydrothermal vent. *Int. J. Sys. Evol. Microbiol.* 52: 1299–1304.

Moulin, L. O.; Muinve, A.; Dreyfus, B.; and Boivin-Masson, C. 2001. Nodulation of legumes by members of the β-subclass of Proteobacteria. *Nature* 411: 948–50.

Parkhill, J., et al. 2000. The genome sequence of the food-borne pathogen *Campylobacter jejuni* reveals hypervariable sequences. *Nature* 403:655–68.

Shapiro, L.; McAdams, H. H.; and Losick, R. 2002. Generating and exploiting polarity in bacteria. *Science* 298:1942–46.

Zhu, J.; Oger, P. M.; Schrammeijer, B.; Hooykaas, P. J.; Farrand, S. K.; and Winans, S. C. 2000. The basis of crown gall tumorigenesis. *J. Bacteriol.* 182: 3885–95.

Please visit the Prescott website at www.mhhe.com/prescott7 for additional references.

23

Bacteria:
The Low G + C Gram Positives

Lactobacilli are indispensable to the food and dairy industry. They are not considered pathogens.

PREVIEW

• *Bergey's Manual* groups the gram-positive bacteria phylogenetically into two major groups: the low G + C gram-positive bacteria and the high G + C gram-positive bacteria. This classification is based primarily on nucleic acid sequences rather than phenotypic similarity.

• The low G + C gram positives contain (1) clostridia and relatives, (2) the mycoplasmas, and (3) the bacilli and lactobacilli. Endospore formers, cocci, and rods are found among the clostridia and bacilli groups. Thus common possession of a complex structure such as an endospore does not necessarily indicate close relatedness between the genera.

• Peptidoglycan structure varies among different groups in ways that are often useful in their identification.

• The majority of gram-positive bacteria are harmless, free-living saprophytes, but most major groups include pathogens of humans, other animals, and plants. Some gram-positive bacteria are very important in the food and dairy industries.

This chapter surveys many of the bacteria found in volume 3 of the second edition of *Bergey's Manual of Systematic Bacteriology*. This edition of *Bergey's Manual* divides the gram-positive bacteria phylogenetically between volumes 3 and 4. Here we focus on the mycoplasmas, *Clostridium* and its relatives, and the bacilli and lactobacilli.

23.1 GENERAL INTRODUCTION

Gram-positive bacteria were historically grouped on the basis of their general shape (e.g., rods, cocci, or irregular) and their ability to form endospores. However, analysis of the phylogenetic relationships within the gram-positive bacteria by comparison of 16S rRNA sequences shows that they are divided into a low G + C group and high G + C, or actinobacterial, group (**figure 23.1**). The most recent edition of *Bergey's Manual of Systematic Bacteriology* places the low G + C gram-positive bacteria in volume 3. This volume describes over 1,300 species placed in 255 genera. The bacterial cell wall: Gram-positive cell walls (section 3.6)

The low G + C gram-positive bacteria are placed in the phylum *Firmicutes* and divided into three classes: *Clostridia, Mollicutes*, and *Bacilli*. The phylum *Firmicutes* is large and complex; it has 10 orders and 34 families. The mycoplasmas, class *Mollicutes*, are also considered low G + C gram positives despite their lack of a cell wall. Ribosomal RNA data indicate that the mycoplasmas are closely related to the lactobacilli. **Figure 23.2** shows the phylogenetic relationships among some of the bacteria in this chapter.

23.2 CLASS MOLLICUTES (THE MYCOPLASMAS)

The class *Mollicutes* has five orders and six families. The best-studied genera are found in the orders *Mycoplasmatales (Mycoplasma, Ureaplasma), Entomoplasmatales (Entomoplasma,*

We noted, after having grown the bacterium through a series of such cultures, each fresh culture being inoculated with a droplet from the previous culture, that the last culture of the series was able to multiply and act in the body of animals in such a way that the animals developed anthrax with all the symptoms typical of this affection.

Such is the proof, which we consider flawless, that anthrax is caused by this bacterium.

—Louis Pasteur

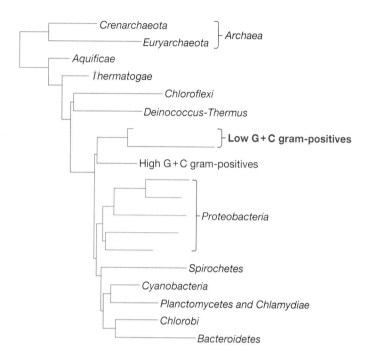

Figure 23.1 Phylogenetic Relationships Among the Procaryotes. The low G + C gram-positive bacteria are highlighted.

Mesoplasma, Spiroplasma), Acholeplasmatales (Acholeplasma), and *Anaeroplasmatales (Anaeroplasma, Asteroleplasma).* **Table 23.1** summarizes some of the major characteristics of these genera.

Members of the class *Mollicutes* are commonly called **mycoplasmas.** Although they evolved from ancestors with gram-positive cell walls, they now lack cell walls and cannot synthesize peptidoglycan precursors. Thus they are penicillin resistant but susceptible to lysis by osmotic shock and detergent treatment. Because they are bounded only by a plasma membrane, these procaryotes are pleomorphic and vary in shape from spherical or pear-shaped organisms, about 0.3 to 0.8 μm in diameter, to branched or helical filaments (**figure 23.3**). Some mycoplasmas (e.g., *M. genitalium*) have a specialized terminal structure that projects from the cell and gives them a flask or pear shape. This structure aids in attachment to eucaryotic cells. They are among the smallest bacteria capable of self-reproduction. Although most are nonmotile, some can glide along liquid-covered surfaces. Most species differ from the vast majority of bacteria in requiring sterols for growth, which are incorporated into the plasma membrane. Here sterols may facilitate osmotic stability. Most are facultative anaerobes, but a few are obligate anaerobes. When growing on agar, most species form colonies with a "fried-egg" appearance because they grow into the agar surface at the center while spreading outward on the surface at the colony edges (**figure 23.4**). Their genomes are among the smallest found in procaryotes, ranging from 0.7 to 1.7 Mb (table 23.1); the G + C content ranges from 23 to 41%. The complete genomes of the human pathogens *Mycoplasma genitalium, M. pneumoniae,* and *Ureaplasma urealyticum* have been sequenced. These genomes are

characteristically small with less than 1,000 genes; it seems that not many genes are required to sustain a free-living existence. Mycoplasmas can be saprophytes, commensals, or parasites, and many are pathogens of plants, animals, or insects. Insights from microbial genomes: Genomic analysis of pathogenic microbes (section 15.8)

Metabolically, the mycoplasmas are incapable of synthesizing a number of macromolecules. In addition to requiring sterols, they also need fatty acids, vitamins, amino acids, purines, and pyrimidines. Some produce ATP by the Embden-Meyerhof pathway and lactic acid fermentation. Others catabolize arginine or urea to generate ATP. The pentose phosphate pathway seems to be functional in at least some mycoplasmas; none appear to have the complete tricarboxylic acid cycle.

Mycoplasmas are remarkably widespread and can be isolated from animals, plants, the soil, and even compost piles. Although their complex growth requirements can make their growth in pure (axenic) cultures difficult, about 10% of the mammalian cell cultures in use are probably contaminated with mycoplasmas. This seriously interferes with tissue culture experiments. In animals, mycoplasmas colonize mucous membranes and joints and often are associated with diseases of the respiratory and urogenital tracts. Mycoplasmas cause several major diseases in livestock, for example, contagious bovine pleuropneumonia in cattle (*M. mycoides*), chronic respiratory disease in chickens (*M. gallisepticum*), and pneumonia in swine (*M. hyopneumoniae*). *M. pneumoniae* causes primary atypical pneumonia in humans. *Ureaplasma urealyticum* is commonly found in the human urogenital tract. It is now known to be associated with premature delivery of newborns, as well as neonatal meningitis and pneumonia. Spiroplasmas have been isolated from insects, ticks, and a variety of plants. They cause disease in citrus plants, cabbage, broccoli, corn, honey bees, and other hosts. Arthropods may often act as vectors and carry the spiroplasmas between plants. It is likely that more pathogenic mollicutes will be discovered as techniques for their detection, isolation, and study improve.

1. What morphological feature distinguishes the mycoplasmas? In what class are they found? Why have they been placed with the low G + C gram-positive bacteria?
2. What might mycoplasmas use sterols for?
3. What do you think the relationship between *Mycoplasma* genome size and growth requirements might be?
4. Where are mycoplasmas found in animals? List several animal and human diseases caused by them. What kinds of organisms do spiroplasmas usually infect?

23.3 PEPTIDOGLYCAN AND ENDOSPORE STRUCTURE

The gram-positive bacteria have traditionally been classified largely on the basis of observable characteristics such as cell shape, the clustering and arrangement of cells, the presence or absence of endospores, oxygen relationships, fermentation patterns, and peptidoglycan chemistry. Because of the importance of pep-

(a)

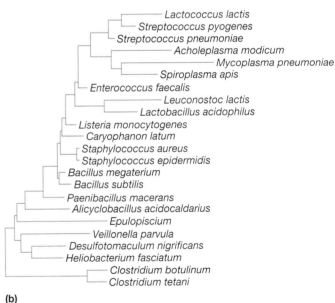

(b)

Figure 23.2   Phylogenetic Relationships in the Phylum *Firmicutes* (Low G + C Gram Positives).   (a) The major phylogenetic groups with representative genera in parentheses. Each tetrahedron in the tree represents a group of related organisms; its horizontal edges show the shortest and longest branches in the group. Multiple branching at the same level indicates that the relative branching order of the groups cannot be determined from the data. The quotation marks around some names indicate that they are not formally approved taxonomic names. **(b)** The relationships of a few species based on 16S rRNA sequence date.   *Source: The Ribosomal Database Project.*

tidoglycan and endospores in these bacteria, we briefly discuss these two important components.

Peptidoglycan structure varies considerably among different gram-positive groups. Many gram-positive bacteria (and most gram-negatives) have a peptidoglycan structure in which *meso*-di-aminopimelic acid in position 3 is directly linked through its free amino group with the free carboxyl of the terminal D-alanine of an adjacent peptide chain (**figure 23.5*a***). These include *Bacillus, Clostridium, Corynebacterium, Mycobacterium,* and *Nocardia.* In other gram-positive bacteria, lysine is substituted for diamino-pimelic acid in position 3, and the peptide subunits of the glycan

chains are cross-linked by interpeptide bridges containing mono-carboxylic L-amino acids or glycine, or both (figure 23.5*b*). Many genera including *Staphylococcus, Streptococcus, Micrococcus, Lactobacillus,* and *Leuconostoc* have this type of peptidoglycan. The high G + C gram-positive genus *Streptomyces* and several other actinobacterial genera have replaced *meso*-diaminopimelic acid with L,L-diaminopimelic acid in position 3 and have one glycine residue as the interpeptide bridge. The plant pathogenic corynebacteria (also a high G + C gram-positive) provide another example of peptidoglycan variation. In some of these bacteria, the interpeptide bridge connects positions 2 and 4 of the peptide

Table 23.1	**Properties of Some Members of the Class *Mollicutes***					
Genus	**No. of Recognized Species**	**G + C Content (mol%)**	**Genome Size (Mb)**	**Sterol Requirement**	**Habitat**	**Other Distinctive Features**
Acholeplasma	13	26–36	1.50–1.65	No	Vertebrates, some plants and insects	Optimum growth at 30–37°C
Anaeroplasma	4	29–34	1.50–1.60	Yes	Bovine or ovine rumen	Oxygen-sensitive anaerobes
Asteroleplasma	1	40	1.50	No	Bovine or ovine rumen	Oxygen-sensitive anaerobes
Entomoplasma	5	27–29	0.79–1.14	Yes	Insects, plants	Optimum growth, 30°C
Mesoplasma	12	27–30	0.87–1.10	No	Insects, plants	Optimum growth, 30°C; sustained growth in serum-free medium only with 0.04% Tween 80
Mycoplasma	104	23–40	0.60–1.35	Yes	Humans, animals	Optimum growth usually at 37°C
Spiroplasma	22	25–30	0.94–2.20	Yes	Insects, plants	Helical filaments; optimum growth at 30–37°C
Ureaplasma	6	27–30	0.75–1.20	Yes	Humans, animals	Urea hydrolysis

Adapted from J. G. Tully, et al., "Revised Taxonomy of the Class *Mollicutes*" in *International Journal of Systematic Bacteriology*, 43(2):378–85. Copyright © 1993 American Society for Microbiology, Washington, D.C. Reprinted by permission.

(a)   **(b)**

Figure 23.3   The *Mycoplasmas*.   Electron micrographs of *Mycoplasma pneumoniae* showing its pleomorphic nature. **(a)** A transmission electron micrograph of several cells (× 47,880). The central cell appears flask or pear-shaped because of its terminal structure. **(b)** A scanning electron micrograph (× 26,000).

subunits rather than 3 and 4 (figure 23.5c). Because the interpeptide bridge connects the carboxyl groups of glutamic acid and alanine, a diamino acid such as ornithine is used in the bridge. Many other variations in peptidoglycan structure are found, including other interbridge structures and large differences in the frequency of cross-linking between glycan chains. Bacilli and most gram-negative bacteria have fewer cross-links between chains than do gram-positive bacteria such as *Staphylococcus aureus* in which almost every muramic acid is cross-linked to another. These structural variants are often characteristic of particular groups and are therefore taxonomically useful. Synthesis of sugars and polysaccharides: Synthesis of peptidoglycan (section 10.4)

Bacterial endospores are assembled within the differentiating mother cell, which lyses to release the free spore. Mature spores

Figure 23.4 *Mycoplasma* **Colonies.** Note the "fried-egg" appearance, colonies stained before photographing (× 100).

have a complex structure with an outer coat, cortex, and inner spore membrane surrounding the protoplast (**figure 23.6**). They contain dipicolinic acid, are very heat resistant, and can remain dormant and viable for very long periods (**Microbial Tidbits 23.1**). Although endospore-forming bacteria are distributed widely, they are primarily soil inhabitants. Soil conditions are often extremely variable, and endospores are an obvious advantage in surviving periods of dryness or nutrient deprivation. In one well-documented experiment, spores remained viable for about 70 years. It has also been reported that viable spores have been recovered from Dominican bees that were encased in 25- to 40-million year-old amber. If this result is confirmed, spores from an ancestor of *Bacillus sphaericus* have survived for more than 25 million years! Similar

Figure 23.5 Representative Examples of Peptidoglycan Structure. **(a)** The peptidoglycan with a direct cross-linkage between positions 3 and 4 of the peptide subunits, which is present in most gram-negative and many gram-positive bacteria. **(b)** Peptidoglycan with lysine in position 3 and an interpeptide bridge. The bracket contains six typical bridges: (1) *Staphylococcus aureus*, (2) *S. epidermidis*, (3) *Micrococcus roseus* and *Streptococcus thermophilus*, (4) *Lactobacillus viridescens*, (5) *Streptococcus salvarius*, and (6) *Leuconostoc cremoris*. The arrows indicate the polarity of peptide bonds running in the C to N direction. **(c)** An example of the cross-bridge extending between positions 2 and 4 from *Corynebacterium poinsettiae*. The interbridge contains a D-diamino acid like ornithine, and L-homoserine (L-Hsr) is in position 3. The abbreviations and structures of amino acids in the figure are found in appendix I.

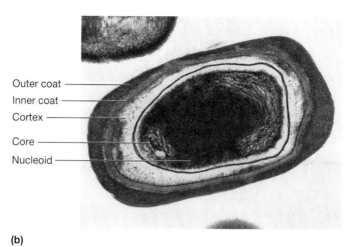

Outer coat
Inner coat
Cortex
Core
Nucleoid

(a)　　　　　　　　　　　　　　　　　　　(b)

Figure 23.6 Bacterial Endospores. **(a)** A colorized cross section of a *Bacillus subtilis* cell undergoing sporulation. The oval in the center is an endospore that is almost mature; when it reaches maturity, the mother cell will lyse to release it. **(b)** A cross section of a mature *B. subtilis* spore showing the cortex and spore coat layers that surround the core. The endospore in (a) is 1.3 μm; the spore in (b) is 1.2 μm.

Microbial Tidbits

23.1　Spores in Space

During the nineteenth-century argument over the question of the evolution of life, the panspermia hypothesis became popular. According to this hypothesis, life did not evolve from inorganic matter on Earth but arrived as viable bacterial spores that had escaped from another planet. More recently, the British astronomer Fred Hoyle has revived the hypothesis based on his study of the absorption of radiation by interstellar dust. Hoyle maintains that dust grains were initially viable bacterial cells that have been degraded, and that the beginning of life on Earth was due to the arrival of bacterial spores that had survived their trip through space.

Even more recently Peter Weber and J. Mayo Greenberg from the University of Leiden in the Netherlands have studied the effect of very high vacuum, low temperature, and UV radiation on the survival of *Bacillus subtilis* spores. Their data suggest that spores within an interstellar molecular cloud might be able to survive between 4.5 to 45 million years. Molecular clouds move through space at speeds sufficient to transport spores between solar systems in this length of time. Although these results do not prove the panspermia hypothesis, they are consistent with the possibility that bacteria might be able to travel between planets capable of supporting life.

reports have been made subsequently; all these studies will have to be reconfirmed. Usually endospores are observed either in the light microscope after spore staining or by phase-contrast microscopy of unstained cells. They also can be detected by heating a culture at 70 to 80°C for 10 minutes followed by incubation in the proper growth medium. Because only endospores and some thermophiles would survive such heating, bacterial growth tentatively confirms their presence. The bacterial endospore (section 3.11); Preparation and staining of specimens (section 2.3)

There are two classes of low G + C endospore-forming bacteria: *Clostridia* (the clostridia and relatives), and *Bacilli* (the bacilli and lactobacilli) (figure 23.2). Each class includes both rods and cocci and are discussed here.

23.4　CLASS *CLOSTRIDIA*

The class *Clostridia* has a very wide variety of gram-positive bacteria distributed into three orders and 11 families. The characteristics of some of the more important genera are summarized in **table 23.2**. Phylogenetic relationships are shown in figure 23.2.

By far the largest genus is *Clostridium*. It includes obligately anaerobic, fermentative, gram-positive bacteria that form endospores. The genus contains well over 100 species in several distinct phylogenetic clusters. The genus *Clostridium* may be subdivided into several genera in the future.

Members of the genus *Clostridium* have great practical impact. Because they are anaerobic and form heat-resistant en-

Table 23.2	Characteristics of Clostridia and Relatives			
Genus	**Dimensions (μm) and Morphology**	**G + C Content (mol%)**	**Oxygen Relationship**	**Other Distinctive Characteristics**
Clostridium	0.3–2.0 × 1.5–20; rod-shaped, often pleomorphic, nonmotile or peritrichous	22–55	Anaerobic	Does not carry out dissimilatory sulfate reduction; usually chemoorganotrophic, fermentative, and catalase negative; forms oval or spherical endospores
Desulfotomaculum	0.3–1.5 × 3–9; straight or curved rods, peritrichous or polar flagella	37–50	Anaerobic	Reduces sulfate to H_2S, forms subterminal to terminal endospores; stains gram negative but has gram-positive wall, catalase negative
Heliobacterium	1.0 × 4–10; rods that are frequently bent, gliding motility	52–55	Anaerobic	Photoheterotrophic with bacteriochlorophyll *g;* stains gram negative but has gram-positive wall, some form endospores
Veillonella	0.3–0.5; cocci in pairs, short chains, and masses; nonmotile	36–43	Anaerobic	Stains gram negative; pyruvate and lactate fermented, but not carbohydrates; acetate, propionate, CO_2, and H_2 produced from lactate; parasitic in mouths, intestines, and respiratory tracts of animals

dospores, they are responsible for many cases of food spoilage, even in canned foods. *C. botulinum* is the causative agent of botulism. Clostridia often can ferment amino acids to produce ATP by oxidizing one amino acid and using another as an electron acceptor in a process called the Stickland reaction (*see figure 9.19*). This reaction generates ammonia, hydrogen sulfide, fatty acids, and amines during the anaerobic decomposition of proteins. These products are responsible for many unpleasant odors arising during putrefaction.   Food-borne and waterborne diseases: Botulism (section 38.4)

Several clostridia produce toxins and are major disease agents. *C. tetani* (**figure 23.7**) is the causative agent of tetanus, and *C. perfringens,* of gas gangrene and food poisoning. *C. perfringens* genome sequence analysis reveals that the microbe possesses the genes for fermentation with gas production but lacks genes encoding enzymes for the TCA cycle or a respiratory chain. Nonetheless, *C. perfringens* has an extraordinary doubling time of only 8 to 10 minutes when in the human host. Clostridia also are industrially valuable; for example, *C. acetobutylicum* is used to manufacture butanol in some countries.

Desulfotomaculum is another anaerobic, endospore-forming genus. Unlike *Clostridium,* it reduces sulfate and sulfite to hydrogen sulfide during anaerobic respiration (**figure 23.8**). Although it stains gram negative, electron microscopic studies have shown that *Desulfotomaculum* has a gram-positive type cell wall. This concurs with phylogenetic studies that place it with the low G + C gram positives.

The heliobacteria are an excellent example of the diversity in this class. The genera *Heliobacterium* and *Heliophilum* are a group of unusual anaerobic, photosynthetic bacteria character-

C. tetani

Figure 23.7 *Clostridium tetani* **with Spores That Are Round and Terminal.**

ized by the presence of bacteriochlorophyll *g*. They have a photosystem I type reaction center like the green sulfur bacteria, but have no intracytoplasmic photosynthetic membranes; pigments are contained in the plasma membrane. Like *Desulfotomaculum,* they have a gram-positive type cell wall with lower than normal peptidoglycan content, and they stain gram negative. Some heliobacteria form endospores.   Phototrophy (section 9.12)

The phylogenetic placement of the genus *Veillonella* bears mentioning. Although these bacteria stain gram-negative, *Bergey's Manual* places them in the family *Acidominococcaceae,* in the order *Clostridiales.* Members of the genus *Veillonella* are anaerobic,

Figure 23.8 *Desulfotomaculum.* *Desulfotomaculum acetoxidans* with spores; phase contrast (×2,000).

chemoheterotrophic cocci ranging in diameter from about 0.3 to 2.5 μm. Usually they are diplococci (often with their adjacent sides flattened), but they may exist as single cells, clusters, or chains. All have complex nutritional requirements and ferment substances such as carbohydrates, lactate and other organic acids, and amino acids to produce gas (CO_2 and often H_2) plus a mixture of volatile fatty acids. They are parasites of homeothermic (warm-blooded) animals.

Like many groups of anaerobic bacteria, members of this genus have not been thoroughly studied. Some species are part of the normal biota of the mouth, the gastrointestinal tract, and the urogenital tract of humans and other animals. For example, *Veillonella* is plentiful on the tongue surface and dental plaque of humans and it can be isolated from the vagina. *Veillonella* is unusual in growing well on organic acids such as lactate, pyruvate, and malate while being unable to ferment glucose and other carbohydrates. It is well adapted to the oral environment because it can use the lactic acid produced from carbohydrates by the streptococci and other oral bacteria. Veillonella are found in infections of the head, lungs, and the female genital tract, but their precise role in such infections is unclear.

1. Describe, in a diagram, the chemical composition and structure of the peptidoglycan found in gram-negative bacteria and many gram-positive genera.
2. How do the cell walls of bacilli and most gram-negative bacteria differ from those of gram-positive bacteria such as *S. aureus* with respect to cross-linking frequency?
3. What is a bacterial endospore? Give its most important properties and two ways to demonstrate its presence. Why do you think reports of viable spores from ancient source have been met with skepticism?
4. Give the general characteristics of *Clostridium, Desulfotomaculum,* the heliobacteria, and *Veillonella.* Briefly discuss why each is interesting or of practical importance.
5. What do you think is the evolutionary significance of the discovery of a photosynthetic genus within the low G + C gram-positive bacteria?

23.5 CLASS *BACILLI*

The second edition of *Bergey's Manual* gathers a large variety of gram-positive bacteria into one class, *Bacilli,* and two orders, *Bacillales* and *Lactobacillales.* These orders contain 17 families and over 70 gram-positive genera representing cocci, endospore-forming rods and cocci, and nonsporing rods. The biology of some members of the order *Bacillales* will be described first; then important representatives of the order *Lactobacillales* will be considered. The phylogenetic relationships between some of these organisms are pictured in figure 23.1, and the characteristics of selected genera are summarized in **table 23.3.**

Order Bacillales

The genus *Bacillus,* family *Bacillaceae,* is the largest in the order (**figure 23.9**). The genus contains gram-positive, endospore-forming, chemoheterotrophic rods that are usually motile with peritrichous flagella. It is aerobic, or sometimes facultative, and catalase positive. Many species once included in this genus have been placed in other families and genera based on rRNA sequence data. For example, the genus *Alicyclobacillus* contains acidophilic, sporing, gram-positive or gram-variable rods that have ω-alicyclic fatty acids with 6- or 7-carbon rings in their membranes. Members are aerobic or facultative and have a G + C content of about 51 to 60%. Another genus, *Paenibacillus* [Latin *paene,* almost, and bacillus], contains gram-positive rods that are facultative, motile by peritrichous flagella, have ellipsoidal endospores and swollen sporangia, produce acid and sometimes gas from glucose and various sugars, and have a G + C content of 40 to 54%. Some examples of organisms that were formerly in the genus *Bacillus* are *Paenibacillus alvei, P. macerans,* and *P. polymyxa.*

Bacillus subtilis, the type species for the genus, is the most well-studied gram-positive bacterium. It is a useful model organism for the study of gene regulation, cell division, quorum sensing, and cellular differentiation. Its 4.2-Mb genome was one of the first genomes to be completely sequenced. Genome sequencing reveals a number of interesting elements. For instance, several families of genes have been expanded by gene duplication; the largest such family encodes ABC transporters, which are the most frequent type of protein in *B. subtilis.* There are 18 genes that encode sigma factors. Recall that the use of alternative sigma subunits of RNA polymerase is one way in which bacteria regulate gene expression. In this case, many of the sigma factors govern sporulation and other responses to stressful conditions. The genome contains genes for the catabolism of many diverse carbon sources and antibiotic synthesis. There are at least 10 integrated prophages or remnants of prophages. *Protein secretion in procaryotes (section 3.8); Global regulations systems: Sporulation in Bacillus subtilis (section 12.5); Comparative genomics (section 15.6)*

Many species of *Bacillus* are of considerable importance. Some produce the antibiotics bacitracin, gramicidin, and polymyxin. *B. cereus* (figure 23.9*b*) causes some forms of food

Table 23.3	Characteristics of Members of the Class *Bacilli*				
Genus	Dimensions (μm) and Morphology	G + C Content (mol%)	Genome Size (Mb)	Oxygen Relationship	Other Distinctive Characteristics
Bacillus	0.5–2.5 × 1.2–10; straight rods, peritrichous	32–69	4.2–5.4	Aerobic or facultative	Forms endospores; catalase positive; chemoorganotrophic
Caryophanon	1.5–3.0 × 10–20; multicellular rods with rounded ends, peritrichous	41–46	Nd*	Aerobic	Acetate only major carbon source; catalase positive; trichome cells have greater width than length, trichomes can be in short chains
Enterococcus	0.6–2.0 × 0.6–2.5; spherical or ovoid cells in pairs or short chains, nonsporing, sometimes motile	34–42	3.2	Facultative	Ferments carbohydrates to lactate with no gas; complex nutritional requirements; catalase negative; occurs widely, particularly in fecal material
Lactobacillus	0.5–1.2 × 1.0–10; usually long, regular rods, nonsporing, rarely motile	32–53	1.9–3.3	Facultative or microaerophilic	Fermentative, at least half the end-product is lactate; requires rich, complex media; catalase and cytochrome negative
Lactococcus	0.5–1.2 × 0.5–1.5; spherical or ovoid cells in pairs or short chains, nonsporing, nonmotile	38–40	2.4	Facultative	Chemoorganotrophic with fermentative metabolism; lactate without gas produced; catalase negative; complex nutritional requirements; in dairy and plant products
Leuconostoc	0.5–0.7 × 0.7–1.2; cells spherical or ovoid, in pairs or chains; nonmotile and nonsporing	38–44	Nd	Facultative	Requires fermentable carbohydrate and nutritionally rich medium for growth; fermentation produces lactate, ethanol, and gas; catalase and cytochrome negative
Staphylococcus	0.9–1.3; spherical cells occurring singly and in irregular clusters, nonmotile and nonsporing	30–39	2.5–2.8	Facultative	Chemoorganotrophic with both respiratory and fermentative metabolism, usually catalase positive, associated with skin and mucous membranes of vertebrates
Streptococcus	0.5–2.0; spherical or ovoid cells in pairs or chains, nonmotile and nonsporing	34–46	1.8–2.2	Facultative	Fermentative, producing mainly lactate and no gas; catalase negative; commonly attack red blood cells (α- or β-hemolysis); complex nutritional requirements; commensals or parasites on animals
Thermoactinomyces	0.4–1.0 in diameter; branched, septate mycelium resembles those of actinomycetes	52.0–54.8	Nd	Aerobic	Usually thermophilic; true endospores form singly on hyphae; numerous in decaying hay, vegetable matter, and compost

*Nd: Not determined; genome not yet sequenced.

Figure 23.9 Bacillus. **(a)** *B. anthracis,* spores elliptical and central (×1,600). **(b)** *B. cereus* stained with SYTOX Green nucleic acid stain and viewed by epifluorescence and differential interference contrast microscopy. The cells that glow green are dead.

(a) *Bacillus anthracis*

(b) *B. cereus*

Figure 23.10 The Parasporal Body. **(a)** An electron micrograph of a *B. sphaericus* sporulating cell containing a parasporal body just beneath the endospore. **(b)** The crystalline parasporal body at a higher magnification. The crystal is surrounded by a two-layered envelope (arrows).

0.4 μm

(a)

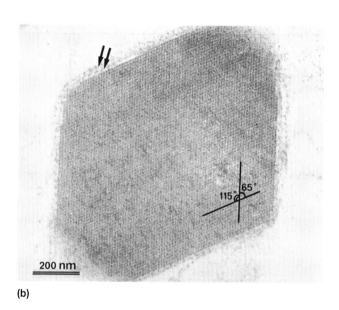

115° 65°

200 nm

(b)

poisoning and can infect humans. *B. anthracis* is the causative agent of the disease anthrax, which can affect both farm animals and humans. Several species are used as insecticides. For example, *B. thuringiensis* and *B. sphaericus* form a solid protein crystal, the **parasporal body,** next to their endospores during spore formation (**figure 23.10**). The *B. thuringiensis* parasporal body contains protein toxins that kill over 100 species of moths by dissolving in the alkaline gut of caterpillars and destroying the epithelium. The solubilized toxin proteins are cleaved by midgut proteases to smaller toxic polypeptides that attack the gut epithelial cells. The alkaline gut contents escape into the blood, causing paralysis and death. One of these toxins has been shown to form pores in the plasma membrane of the target insect's cells. These channels allow monovalent cations such as potassium to pass through. *B. thuringiensis* toxin genes have been engineered to make a variety of pest-resistant, genetically modified plants. The *B. sphaericus* paras-

poral body contains proteins toxic for mosquito larvae and may be useful in controlling the mosquitos that carry the malaria parasite *Plasmodium.* Zoonotic diseases: Anthrax (section 38.6); Microbes as products: Biopesticides (section 41.8)

The genus *Thermoactinomyces* has historically been classified as an actinomycete because it forms filaments that differentiate from soil-associated substrate hyphae into upwardly growing, aerial hyphae. However, phylogenetic analysis places it with the low G + C microbes in the order *Bacillales,* family *Thermoactinomycetaceae.* Its G + C content (52–55 mol%) is considerably lower than that of the *Actinobacteria.* This genus is thermophilic and grows between 45 and 60°C; it forms single spores on both its aerial and substrate mycelia (**figure 23.11**) and its 16S rRNA sequence suggests a relationship with the genus *Bacillus. Thermoactinomyces* is commonly found in damp haystacks, compost piles, and other high-temperature habitats. Unlike actinomycete exospores, *Thermoactinomyces* (figure 23.11*b*) forms true endospores that are

(a)

(b)

Figure 23.11 *Thermoactinomyces.* **(a)** *Thermoactinomyces vulgaris* aerial mycelium with developing endospores at tips of hyphae. **(b)** Thin section of a *T. sacchari* endospore. E, exosporium; OC, outer spore coat; IC, inner spore coat; CO, cortex; IM, inner forespore membrane; C, core.

heat resistant; they can survive at 90°C for 30 minutes. They are formed within hyphae and appear to have typical endospore structure, including the presence of calcium and dipicolinic acid. *Thermoactinomyces vulgaris* (figure 23.11*a*) from haystacks, grain storage silos, and compost piles is a causative agent of farmer's lung, an allergic disease of the respiratory system in agricultural workers. Recently spores from *Thermoactinomyces vulgaris* were recovered from the mud of a Minnesota lake and found to be viable after about 7,500 years of dormancy.

One of the more unusual genera in this order is *Caryophanon*. This gram-positive bacterium is strictly aerobic, catalase positive, and motile by peritrichous flagella. Its normal habitat is cow dung. *Caryophanon* morphology is distinctive. Individual cells are disk-shaped (1.5 to 2.0 μm wide by 0.5 to 1.0 μm long) and joined to form filaments called trichomes that are about 10 to 20 μm long (**figure 23.12**).

The family *Staphylococcaceae* contains four genera, the most important of which is the genus *Staphylococcus*. Members of this genus are facultatively anaerobic, nonmotile, gram-positive cocci that usually form irregular clusters (**figure 23.13**). They are catalase positive, oxidase negative, ferment glucose, and have teichoic acid in their cell walls. Staphylococci are normally associated with the skin, skin glands, and mucous membranes of warm-blooded animals.

Staphylococci are responsible for many human diseases. *S. epidermidis* is a common skin resident that is sometimes responsible for endocarditis and infections of patients with lowered resistance (e.g., wound infections, surgical infections, urinary tract infections, body piercing). *S. aureus* is the most important human staphylococcal pathogen and causes boils, abscesses, wound infections, pneumonia, toxic shock syndrome, and other diseases. Strains of **meticillin-resistant *Staphylococcus aureus* (MRSA;** formerly methicillin) and vancomycin-resistant *S. aureus* are

Figure 23.12 *Caryophanon* **Morphology.** *Caryophanon latum* in a trichome chain. Note the disk-shaped cells stacked side by side; phase contrast (×3,450).

among the most threatening antibiotic-resistant pathogens known. Vancomycin is considered the "drug of last resort" and infections caused by vancomycin-resistant *S. aureus* generally cannot be treated by antibiotic therapy. How did *S. aureus* so quickly evolve the capacity to defeat a diverse array of antibiotics? Comparative genome analysis of MRSA strains with antibiotic-sensitive *S. aureus* strains shows that this microbe is very adept at

(a)

(b)

Figure 23.13 *Staphylococcus.* **(a)** *Staphylococcus aureus,* Gram-stained smear (X1,500). **(b)** *Staphylococcus aureus* cocci arranged like clusters of grapes; color-enhanced scanning electron micrograph. Each cell is about 1μm in diameter.

acquiring genetic elements from other bacteria. In fact, large, mobile genetic elements appear to encode both antibiotic-resistance factors and proteins that increase virulence. Insights from microbial genomes: Genomic analysis of pathogenic microbes (section 15.8)

One of the virulence factors produced by *S. aureus* is the enzyme **coagulase,** which causes blood plasma to clot. Growth and hemolysis patterns on blood agar are also useful in identifying these staphylococci (figure 23.18). The structure of staphylococcal α-hemolysin has been determined. The toxin lyses a cell by forming solvent-filled channels in its plasma membrane. Water-soluble toxin monomers bind to the cell surface and associate with each other to form pores. The hydrophilic channels then al-

low free passage of water, ions, and small molecules. *S. aureus* usually grows on the nasal membranes and skin; it also is found in the gastrointestinal and urinary tracts of warm-blooded animals. Direct contact diseases: Staphylococcal diseases (section 38.3)

S. aureus is a major cause of food poisoning. For instance several years ago in Texas, 1,364 elementary school children were sickened by tainted chicken. A food service worker responsible for deboning the chicken was the most likely source of contamination. After the chicken was deboned, it was cooled to room temperature before further processing and refrigeration. This provided sufficient time for the bacterium to grow and produce toxins. This case is not uncommon. Poultry accounts for about one-quarter of all bacterial food poisoning cases in which the source of poisoning is known, and it must be cooked and handled carefully.

Listeria, family *Listeriaceae,* is another medically important genus in this order. The genus contains short rods that are aerobic or facultative, catalase positive, and motile by peritrichous flagella. It is widely distributed in nature, particularly in decaying matter. *Listeria monocytogenes* is a pathogen of humans and other animals and causes listeriosis, an important food infection. Food-borne and waterborne diseases: Listeriosis (section 38.4)

Order *Lactobacillales*

Many members of the order *Lactobacillales* produce lactic acid as their major or sole fermentation product and are sometimes collectively called **lactic acid bacteria (LAB)**. *Streptococcus, Enterococcus, Lactococcus, Lactobacillus,* and *Leuconostoc* are all members of this group. Lactic acid bacteria are nonsporing and usually nonmotile. They normally depend on sugar fermentation for energy. They lack cytochromes and obtain energy by substrate-level phosphorylation rather than by electron transport and oxidative phosphorylation. Nutritionally, they are fastidious and many vitamins, amino acids, purines, and pyrimidines must be supplied because of their limited biosynthetic capabilities. Lactic acid bacteria usually are categorized as facultative anaerobes, but some classify them as aerotolerant anaerobes. The influence of environmental factors on growth: Oxygen concentration (section 6.5); Fermentations (section 9.7)

The largest genus in this order is *Lactobacillus* with around 100 species. *Lactobacillus* contains nonsporing rods and sometimes coccobacilli that lack catalase and cytochromes, are usually facultative anaerobic or microaerophilic, produce lactic acid as their main or sole fermentation product, and have complex nutritional requirements (**figure 23.14**). Lactobacilli carry out either a homolactic fermentation using the Embden-Meyerhof pathway or a heterolactic fermentation with the pentose phosphate pathway. They grow optimally under slightly acidic conditions, when the pH is between 4.5 to 6.4. The genus is found on plant surfaces and in dairy products, meat, water, sewage, beer, fruits, and many other materials. Lactobacilli also are part of the normal flora of the human body in the mouth, intestinal tract, and vagina. They usually are not pathogenic.

(a) *Lactobacillus acidophilus*   **(b)** *L. lactis*   **(c)** *L. bulgaricus*

Figure 23.14   *Lactobacillus.*   (a) *L. acidophilus* (×1,000). **(b)** *L. lactis.* Gram stain (×1,000). **(c)** *L. bulgaricus;* phase contrast (×600).

Lactobacillus is indispensable to the food and dairy industry. Lactobacilli are used in the production of fermented vegetable foods (sauerkraut, pickles, silage), beverages (beer, wine, juices), sour dough bread, Swiss and other hard cheeses, yogurt, and sausage. Yogurt is probably the most popular fermented milk product in the United States. In commercial production, nonfat or low-fat milk is pasteurized, cooled to 43°C or lower, inoculated with *Streptococcus thermophilus* and *Lactobacillus bulgaricus*. *S. thermophilus* grows more rapidly at first and renders the milk anoxic and weakly acidic. *L. bulgaricus* then acidifies the milk even more. Acting together, the two species ferment almost all the lactose to lactic acid and flavor the yogurt with diacetyl (*S. thermophilus*) and acetaldehyde (*L. bulgaricus*). Fruits or fruit flavors are pasteurized separately and then combined with the yogurt. Microbiology of fermented foods: Fermented milks (section 40.6)

At least one species, *L. plantarum,* is sold commercially as a probiotic agent that may provide some health benefits for the consumer. On the other hand some lactobacilli also create problems. They sometimes are responsible for spoilage of beer, milk, and meat because their metabolic end products contribute undesirable flavors and odors.

Leuconostoc, family *Leuconostocaceae,* contains facultative gram-positive cocci, which may be elongated or elliptical and arranged in pairs or chains (**figure 23.15**). Leuconostocs lack catalase and cytochromes and carry out **heterolactic fermentation** by converting glucose to D-lactate and ethanol or acetic acid by means of the phosphoketolase pathway (**figure 23.16**). They can be isolated from plants, silage, and milk. The genus is used in wine production, in the fermentation of vegetables such as cabbage (sauerkraut; *see figure 40.19*) and cucumbers (pickles), and in the manufacture of buttermilk, butter, and cheese. *L. mesenteroides* synthesizes dextrans from sucrose and is important in in-

Figure 23.15   *Leuconostoc.*   *Leuconostoc mesenteroides;* phase-contrast micrograph.

dustrial dextran production. *Leuconostoc* species are involved in food spoilage and tolerate high sugar concentrations so well that they grow in syrup and are a major problem in sugar refineries.

Enterococcaceae (Enterococcus) and *Streptococcaceae (Streptococcus, Lactococcus)* are important families of chemoheterotrophic, mesophilic, nonsporing, gram-positive cocci. In practice, they are often distinguished primarily based on phenotypic properties such as oxygen relationships, cell arrangement, the presence of catalase and cytochromes, and peptidoglycan structure. The most important of these genera is *Streptococcus,* which is facultatively anaerobic and catalase negative. The

Figure 23.16   Heterolactic Fermentation and the Phosphoketolase Pathway.   The phosphoketolase pathway converts glucose to lactate, ethanol, and CO_2.

streptococci and their close relatives, the enterococci and lactococci, occur in pairs or chains when grown in liquid media (**figure 23.17**), do not form endospores, and usually are nonmotile. They all ferment sugars with lactic acid, but no gas, as the major product—that is, they carry out homolactic fermentation. A few species are anaerobic rather than facultative.

The genus *Streptococcus* is large and complex. These bacteria have been clustered into three groups: pyogenic streptococci, oral streptococci, and other streptococci. Many bacteria originally placed within the genus have been moved to two other gen-

era, *Enterococcus* and *Lactococcus*. Some major characteristics of these three closely related genera are summarized in **table 23.4. Table 23.5** lists a few properties of selected genera.

Many characteristics are used to identify these cocci. One of their most important taxonomic characteristics is the ability to lyse erythrocytes when growing on blood agar, an agar medium containing 5% sheep or horse blood (**figure 23.18**). In α-**hemolysis,** a 1 to 3 mm greenish zone of incomplete hemolysis forms around the colony; β-**hemolysis** is characterized by a zone of clearing or complete lysis without a marked color change. In addition, other hemolytic patterns are sometimes seen. Serological studies are also very important in identification because these genera often have distinctive cell wall antigens. Polysaccharide and teichoic acid antigens found in the cell wall or between the wall and the plasma membrane are used to identify these cocci, particularly pathogenic β-hemolytic streptococci, by the **Lancefield grouping system.** Biochemical and physiological tests are essential in identification (e.g., growth temperature preferences, carbohydrate fermentation patterns, acetoin production, reduction of litmus milk, sodium chloride and bile salt tolerance, and the ability to hydrolyze arginine, esculin, hippurate, and starch). Sensitivity to bacitracin, sulfa drugs, and optochin (ethylhydrocuprein) also are used to identify particular species. Some of these techniques are being replaced by molecular genetic approaches such as multilocus sequence typing (MSLT). Clinical microbiology and immunology (chapter 35): Techniques for determining microbial taxonomy and phylogeny: Molecular characteristics (section 19.4)

Members of the three genera have considerable practical importance. Pyogenic streptococci usually are pathogens and associated with pus formation (pyogenic means pus producing). Most species produce β-hemolysis on blood agar and form chains of cells. The major human pathogen in this group is *S. pyogenes,* which causes streptococcal sore throat, acute glomerulonephritis, and rheumatic fever. The normal habitat of oral streptococci is the oral cavity and upper respiratory tract of humans and other animals. In other respects oral streptococci are not necessarily similar. *S. pneumoniae* is α-hemolytic and grows as pairs of cocci (figures 23.17c and 23.18b). It is associated with lobar pneumonia and otitis media (inflammation of the middle ear). *S. mutans* is associated with the formation of dental caries. The enterococci such as *E. faecalis* are normal residents of the intestinal tracts of humans and most other animals. *E. faecalis* is an opportunistic pathogen that can cause urinary tract infections and endocarditis. Unlike the streptococci the enterococci will grow in 6.5% sodium chloride. They are major agents in the horizontal transfer of antibiotic-resistance genes. The lactococci ferment sugars to lactic acid and can grow at 10°C but not at 45°C. *L. lactis* is widely used in the production of buttermilk and cheese because it can curdle milk and add flavor through the synthesis of diacetyl and other products. Airborne diseases: Streptococcal diseases (section 38.1); Dental infections (section 38.7); Microbiology of fermented foods (section 40.6)

(a) *Streptococcus pyogenes*

(b) *S. agalactiae*

(c) *S. pneumoniae*

Figure 23.17   *Streptococcus.*   **(a)** *Streptococcus pyogenes* (×900). **(b)** *Streptococcus agalactiae,* the cause of Group B streptococcal infections. Note the long chains of cells; color-enhanced scanning electron micrograph (×4,800). **(c)** *Streptococcus pneumoniae* (×900).

Table 23.4   Classification of the Streptococci, Entercocci, and Lactococci

Characteristics	*Streptococcus*	*Enterococcus*	*Lactococcus*
Predominant arrangement (most common first)	Chains, pairs	Pairs, chains	Pairs, short chains
Capsule/slime layer	+	−	−
Habitat	Mouth, respiratory tract	Gastrointestinal tract	Dairy products
Growth at 45°C	Variable	+	−
Growth at 10°C	Variable	Usually +	+
Growth at 6.5% NaCl broth	Variable	+	−
Growth at pH 9.6	Variable	+	−
Hemolysis	Usually β (pyogenic) or α (oral)	α, β, −	Usually −
Serological group (Lancefield)	Variable (A–O)	Usually D	Usually N
Mol% G + C	34–46	34–42	38–40
Representative species	Pyogenic streptococci	*E. faecalis*	*L. lactis*
	S. agalactiae	*E. faecium*	*L. raffinolactis*
	S. pyogenes	*E. avium*	*L. plantarum*
	S. equi	*E. durans*	
	S. dysgalactiae	*E. gallinarum*	
	Oral streptococci		
	S. gordonii		
	S. salvarius		
	S. sanguis		
	S. oralis		
	S. pneumoniae		
	S. mitis		
	S. mutans		
	Other streptococci		
	S. bovis		
	S. thermophilus		

Table 23.5	Properties of Selected Streptococci and Relatives					
	Pyogenic Streptococci	**Oral Streptococci**			**Enterococci**	**Lactococci**
Characteristics	**S. pyogenes**	**S. pneumoniae**	**S. sanguis**	**S. mutans**	**E. faecalis**	**L. lactis**
Growth at 10°C	−ᵃ	−	−	−	+	+
Growth at 45°C	−	−	d	d	+	−
Growth at 6.5% NaCl	−	−	−	−	+	−
Growth at pH 9.6	−	−	−	−	+	−
Growth with 40% bile	−	−	d	d	+	+
α-hemolysis	−	+	+	−	−	d
β-hemolysis	+	−	−	−	+	−
Arginine hydrolysis	+	+	+	−	+	d
Hippurate hydrolysis	−	−	−	−	+	d
Mol% G + C of DNA	35–39	30–39	40–46	36–38	34–38	39
Genome size (Mb)	1.8	2.2	Nd*	2.0	3.2	2.3

Modified from *Bergey's Manual of Systematic Bacteriology*, Vol. 2, edited by P. H. A. Sneath, et al. Copyright © 1986 Williams and Wilkins, Baltimore, MD. Reprinted by permission.

ᵃSymbols: +, 90% or more of strains positive; −, 10% or less of strains positive; d, 11–89% of strains are positive.

*Nd: Not determined; genome not yet sequenced

(a)

(b)

(c)

Figure 23.18 Streptococcal and Staphylococcal Hemolytic Patterns. (a) *Streptococcus pyogenes* on blood agar, illustrating β-hemolysis. (b) *Streptococcus pneumoniae* on blood agar, illustrating α-hemolysis. (c) *Staphylococcus epidermidis* on blood agar with no hemolysis.

1. List the major properties of the genus *Bacillus*. What practical impacts does it have on society? Define parasporal body.
2. Briefly describe the genus *Thermoactinomyces,* with particular emphasis on its unique features. What disease does it cause?
3. What is distinctive about the morphology of *Caryophanon?* Can you think of a selective advantage that may be conferred by the morphology of *Thermoactinomyces* or *Caryophanon?*
4. Describe the genus *Staphylococcus*. How does the pathogen *S. aureus* differ from the common skin resident *S. epidermidis,* and where is it normally found?

5. List the major properties of the genus *Lactobacillus*. Why is it important in the food and dairy industries?
6. Describe the major distinguishing characteristics of the following taxa: *Streptococcus, Enterococcus, Lactococcus,* and *Leuconostoc.*
7. Of what practical importance is *Leuconostoc?* What are lactic acid bacteria?
8. What are α-hemolysis, β-hemolysis, and the Lancefield grouping system?
9. What is the difference between pyogenic and oral streptococci?

Summary

23.1 General Introduction

a. Traditionally, gram-positive bacteria were classified according to characteristics such as general shape, peptidoglycan structure, the possession of endospores, and their response to acid-fast staining.

b. Based on 16S rRNA analysis, gram-positive bacteria are now divided into low G + C and high G + C groups; this system is used in the current edition of *Bergey's Manual*.

c. The second edition of *Bergey's Manual* places the low G + C gram positives in the phylum *Firmicutes*, which contains three classes: *Clostridia, Mollicutes,* and *Bacilli* (**tables 23.1–23.3** and **figure 23.2**).

23.2 Class *Mollicutes* (The Mycoplasmas)

a. Mycoplasmas stain gram-negative because they lack cell walls and cannot synthesize peptidoglycan precursors. Many species require sterols for growth. They are one of the smallest bacteria capable of self-reproduction and usually grow on agar to give colonies a "fried-egg" appearance (**figure 23.4**).

23.3 Peptidoglycan and Endospore Structure

a. Peptidoglycan structure often differs between groups in taxonomically useful ways. Most variations are in amino acid 3 of the peptide subunit or in the interpeptide bridge (**figure 23.5**).

b. Endospores resist desiccation and heat; they are used by bacteria to survive harsh conditions, especially in the soil (**figure 23.6**).

23.4 Class *Clostridia*

a. Members of the genus *Clostridium* are anaerobic gram-positive rods that form endospores and don't carry out dissimilatory sulfate reduction (**figure 23.7**). They are responsible for botulism, tetanus, food spoilage, and putrefaction.

b. *Desulfotomaculum* is an anaerobic, endospore-forming genus that reduces sulfate to sulfide during anaerobic respiration (**figure 23.8**).

c. The heliobacteria are anaerobic, photosynthetic bacteria with bacteriochlorophyll *g*. Some form endospores.

d. The family *Veillonellaceae* contains anaerobic cocci that stain gram-negative. Some are parasites of vertebrates.

23.5 Class *Bacilli*

a. The class *Bacilli* is divided into two orders: *Bacillales* and *Lactobacillales*.

b. The genus *Bacillus* contains aerobic and facultative, catalase-positive, endospore-forming, chemoheterotrophic, gram-positive rods that are usually motile and have peritrichous flagella (**figure 23.9**). Species of *Bacillus* synthesize antibiotics and insecticides, and cause food poisoning and anthrax.

c. *Thermoactinomyces* is a gram-positive thermophile that forms a mycelium and true endospores (**figure 23.11**). It causes allergic reactions and leads to farmer's lung.

d. Members of the genus *Staphylococcus* are facultatively anaerobic, nonmotile, gram-positive cocci that form irregular clusters (**figure 23.13**). They grow on the skin and mucous membranes of warm-blood animals, and some are important human pathogens.

e. Several important genera such as *Lactobacillus, Listeria,* and *Caryophanon* contain regular, nonsporing, gram-positive rods. *Lactobacillus* carries out lactic acid fermentation and is extensively used in the food and dairy industries.

f. *Leuconostoc* carries out heterolactic fermentation using the phosphoketolase pathway (**figure 23.16**) and is involved in the production of fermented vegetable products, buttermilk, butter, and cheese.

g. The genera *Streptococcus, Enterococcus,* and *Lactococcus* contain gram-positive cocci arranged in pairs and chains that are usually facultative and carry out homolactic fermentation (**tables 23.4** and **23.5**). Some important species are the pyogenic coccus *S. pyogenes*, the oral streptococci *S. pneumoniae* and *S. mutans*, the enterococcus *E. faecalis* and lactococcus *L. lactis* (**figure 23.17**).

Key Terms

α-hemolysis 584
β-hemolysis 584
coagulase 582

heterolactic fermentation 583
lactic acid bacteria (LAB) 582
Lancefield grouping system 584

meticillin-resistant *Staphylococcus aureus* (MRSA) 581

mycoplasmas 572
parasporal body 580

Critical Thinking Questions

1. Many low G + C bacteria are parasitic. The dependence on a host might be a consequence of the low G + C content. Elaborate on this concept.

2. How might one go about determining whether the genome of *M. genitalium* is the smallest one compatible with a free-living existence?

3. Account for the ease with which anaerobic clostridia can be isolated from soil and other generally aerobic niches.

Learn More

Amesz, J. 1995. The heliobacteria, a new group of photosynthetic bacteria. *J. Photochem. Photobiol. B* 30:89–96.

Cunningham, M. W. 2000. Pathogenesis of group A streptococcal infections: *Clin. Microbiol. Rev.* 13(3):470–511.

Glass, J. I.; Lefkowitz, E. J.; Glass, J. S.; Heiner, C. R.; Chen, E. Y.; and Cassell, G. H. 2000. The complete sequence of the mucosal pathogen *Ureaplasma urealyticum. Nature* 470: 757–62.

Himmelreich, R.; Hilbert, H.; Plagens, H.; Pirkl, E.; Li, B.-C.; and Hermann, R. 1996. Complete sequence analysis of the genome of the bacterium *Mycoplasma pneumoniae. Nucleic Acids Res.* 24(22):4420–49.

Holden, M. T.; Feil, E. J.; et al., 2004. Complete genomes of two clinical *Staphylococcus aureus* strains: Evidence for the rapid evolution of virulence and drug resistance. *Proc. Natl. Acad. Sci. USA.* 101: 9786–91.

Holt, J. G., editor-in-chief. 1994. *Bergey's Manual of Determinative Bacteriology.* 9th ed. Baltimore, Md: Williams & Wilkins.

Johnson, E. A. 2000. Clostridia. In *Encyclopedia of microbiology,* 2d ed., vol. 1, J. Lederberg, editor-in-chief, 834–39. San Diego: Academic Press.

Karlin, S.; Theriot, J.; and Mrazek, J. 2004. Comparative analysis of gene expression among low G + C positive genomes. *Proc. Natl. Acad. Sci. USA.* 101: 6182–87.

Kleerevezem, M., et al. 2003. Complete genome sequence of *Lactobacillus plantarum* WCFS1. *Proc. Natl. Acad. Sci. USA.* 100: 1990–95.

Kunst, F., et al. 1997. The complete genome sequence of the gram-positive bacterium *Bacillus subtilis. Nature* 390:249–56.

Lambert, B., and Peferoen, M. 1992. Insecticidal promise of *Bacillus thuringiensis:* Facts and mysteries about a successful biopesticide. *BioScience* 42(2):112–22.

Nicholson, W. L.; Munakata, N.; Horneck, G.; Melosh, H. J.; and Setlow, P. 2000. Resistance of *Bacillus* endospores to extreme terrestrial and extraterrestrial environments. *Micro. Mol. Biol. Rev.* 64(3):548–72.

Vreeland, R. H.; Rosenweig, W. D.; and Powers, D. W. 2000. Isolation of a 250 million-year-old halotolerant bacterium from a primary salt crystal. *Nature* 407: 897–900.

Please visit the Prescott website at www.mhhe.com/prescott7 for additional references.

24

Bacteria:
The High G + C Gram Positives

Frankia forms nonmotile spores and is symbiotic with a number of higher plants such as alder trees.

PREVIEW

- *Bergey's Manual* classifies the actinomycetes and other high G + C gram-positive bacteria using 16S rRNA data. They are placed in the phylum *Actinobacteria,* which is a large, complex grouping with one class and six orders.

- The morphology and arrangement of spores, cell wall chemistry, and the types of sugars present in cell extracts are particularly important in actinomycete taxonomy and are used to divide these bacteria into different groups.

- Actinomycetes have considerable practical impact because they play a major role in the mineralization of organic matter in the soil and are the primary source of most naturally synthesized antibiotics. The genera *Corynebacterium, Mycobacterium,* and *Nocardia* include important human pathogens.

Chapter 24, the last of the survey chapters on bacteria, describes the high G + C gram-positive bacteria (**figure 24.1**). They are found in volume 4 of the second edition of *Bergey's Manual*. Many of these bacteria are called actinomycetes. Actinomycetes are gram-positive, aerobic bacteria, but are distinctive because they have filamentous hyphae that differentiate to produce asexual spores. Many closely resemble fungi in overall morphology. Presumably this resemblance results partly from adaptation to the same habitats. In this chapter, we first summarize the general characteristics of the actinomycetes. Representatives are described next, with emphasis on morphology, taxonomy, reproduction, and general importance. The **actinomycetes** [s., actinomycete] are a diverse group, but they share many properties.

24.1 GENERAL PROPERTIES OF THE ACTINOMYCETES

The actinomycetes are a fascinating group of microorganisms. They are the source of most of the antibiotics used in medicine today. They also produce metabolites that are used as anticancer drugs, antihelminthics (for instance ivermectin, which is given to dogs to prevent heart worm), and drugs that suppress the immune system in patients who have received organ transplants. This practical aspect of the actinomycetes is linked very closely to their mode of growth. Like the myxobacteria, the prosthecate bacteria, and several other microbes described in previous chapters, the actinomycetes undergo a complex life cycle. The life cycle of many actinomycetes includes the development of filamentous cells, called **hyphae,** and spores. When growing on a solid substratum such as soil or agar, the actinomycetes develop a branching network of hyphae. The hyphae grow both on the surface of the substratum and into it to form a dense mat of hyphae termed a **substrate mycelium.** Septae usually divide the hyphae into long cells (20 μm and longer) containing several nucleoids. In many actinomycetes, substrate hyphae differentiate into upwardly growing hyphae to form an **aerial mycelium** that extends above the substratum. It is at this time that medically useful compounds are formed. Because the physiology of the actinomycete has switched from actively growing vegetative cells into this special cell type, these compounds are often called **secondary metabolites.**

The aerial hyphae form thin-walled spores upon septation (**figure 24.2**). These spores are considered exospores because they do not develop within a mother cell like the endospores of

Actinomycetes are very important from a medical point of view. . . . They may be a nuisance, as when they decompose rubber products, grow in aviation fuel, produce odorous substances that pollute water supplies, or grow in sewage-treatment plants where they form thick clogging foams. . . . In contrast, actinomycetes are the producers of most of the antibiotics.

—H. A. Lechevalier and M. P. Lechevalier

Bacillus and *Clostridium*. If the spores are located in a sporangium, they may be called **sporangiospores.** The spores can vary greatly in shape (**figure 24.3**). Like spore formation in other bacteria, actinomycete sporulation is usually in response to nutrient deprivation. However, most actinomycete spores are not par-

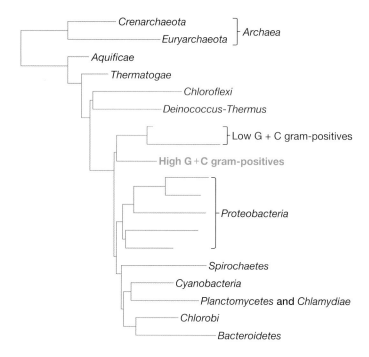

Figure 24.1 Phylogenetic Relationships Among Procaryotes. The high G + C gram-positive bacteria are highlighted.

ticularly heat resistant but withstand desiccation well, so they have considerable adaptive value. Most actinomycetes are not motile, and spores are dispersed by wind or adhering to animals; in this way they may find a new habitat that will provide needed nutrients. In the few motile genera, motility is confined to flagellated spores.

Actinomycete cell wall composition varies greatly among different groups and is of considerable taxonomic importance. Four major cell wall types can be distinguished according to three features of peptidoglycan composition and structure: the amino acid in tetrapeptide side chain position 3, the presence of glycine in interpeptide bridges, and peptidoglycan sugar content (**table 24.1;** *see also figure 23.5*). Whole cell extracts of actinomycetes with wall types II, III, and IV also contain characteristic sugars that are useful in identification (**table 24.2**). Some other taxonomically valuable properties are the morphology and color of the mycelium and sporangia, the surface features and arrangement of spores, the percent G + C in DNA, the phospholipid composition of cell membranes, and spore heat resistance. Of course 16S rRNA sequences has proven valuable. Several actinomycete genomes have been sequenced including *Mycobacterium tuberculosis, M. leprae, Streptomyces coelicolor,* and *S. avermitilis.*   The bacterial cell wall (section 3.6); Peptidoglycan and endospore structure (section 23.3)

Actinomycetes also have great ecological significance. They are primarily soil inhabitants and are very widely distributed. They can degrade an enormous number and variety of organic compounds and are extremely important in the mineralization of organic matter. Although most actinomycetes are free-living microorganisms, a few are pathogens of humans, other animals, and some plants.

Figure 24.2 An Actinomycete Colony.   The cross section of an actinomycete colony with living (blue-green) and dead (white) hyphae. The substrate mycelium and aerial mycelium with chains of spores are shown.

Figure 24.3 Examples of Actinomycete Spores as Seen in the Scanning Electron Microscope.
(a) Spores of *Pilimelia columellifera* on mouse hair (×520). (b) *Micromonospora echinospora*. (c) A chain of hairy streptomycete spores. (d) *Microbispora rosea*, paired spores on hyphae. (e) Aerial spore chain of *Kitasatosporia setae*.

Table 24.1	Actinomycete Cell Wall Types			
Cell Wall Type	Diaminopimelic Acid Isomer	Glycine in Interpeptide Bridge	Characteristic Sugars	Representative Genera
I	L, L	+	NA[a]	*Nocardioides, Streptomyces*
II	*meso*	+	NA	*Micromonospora, Pilimelia, Actinoplanes*
III	*meso*	−	NA	*Actinomadura, Frankia*
IV	*meso*	−	Arabinose, galactose	*Saccharomonospora, Nocardia*

[a]NA, either not applicable or no diagnostic sugars.

It has been clear for some time that some of the phenotypic traits traditionally used to determine actinomycete taxonomy do not always fit with 16S rRNA sequence data. Phylogenetic analyses based on 16S rRNA sequences are used to classify the high G + C gram-positive bacteria (i.e., those gram-positive bacteria with a DNA base composition above approximately 50 mol% G + C). All are placed in the phylum *Actinobacteria* and classified as shown in **figure 24.4.** The phylum is large and very complex; it contains one class (*Actinobacteria*), five subclasses, six orders, 14 suborders, and 44 families. In this system

Table 24.2 | Actinomycete Whole Cell Sugar Patterns

Sugar Pattern Types[a]	Characteristic Sugars	Representative Genera
A	Arabinose, galactose	*Nocardia, Rhodococcus, Saccharomonospora*
B	Madurose[b]	*Actinomadura, Streptosporangium, Dermatophilus*
C	None	*Thermomonospora, Actinosynnema, Geodermatophilus*
D	Arabinose, xylose	*Micromonospora, Actinoplanes*

[a]Characteristic sugar patterns are present only in wall types II-IV, those actinomycetes with *meso*-diaminopimelic acid.

[b] Madurose is 3-*O*-methyl-D-galactose.

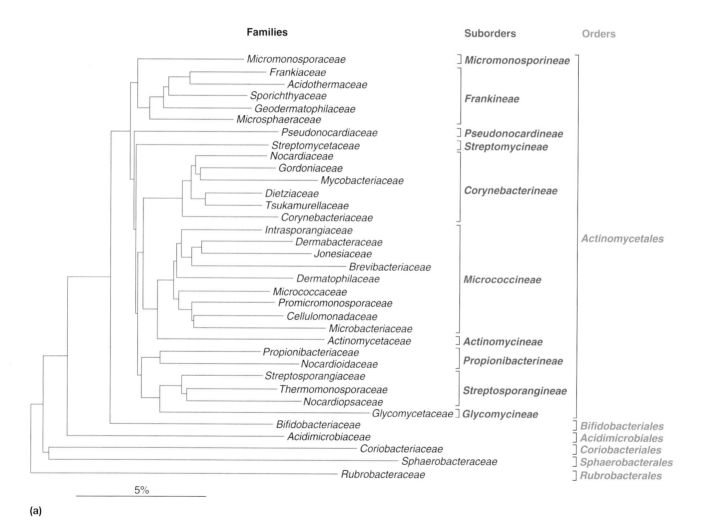

Figure 24.4 Classification of the Phylum *Actinobacteria.* **(a)** The phylogenetic relationships between orders, suborders, and families based on 16S rRNA data are shown. The bar represents 5 nucleotide substitutions per 100 nucleotides. *Source: (a) E. Stackebrandt, F. A. Rainey, and N. L. Ward-Rainey. Proposal for a new hierarchic classification system,* Actinobacteria, *classis nov.* Int. J. Syst. Bacteriol. *47(2):479–491, 1997. Figure 3, p. 482.*

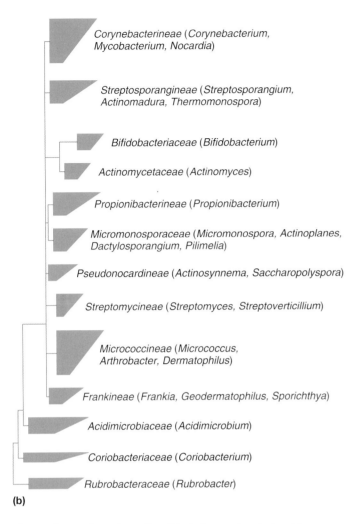

(b)

Figure 24.4 *(continued).* **(b)** Phylogenetic relationships among major actinobacterial groups based on 16S rRNA data. Representative genera are given in parentheses. Each tetrahedron in the tree represents a group of related organisms; its horizontal edges show the shortest and longest branches in the group. Multiple branching at the same level indicates that the relative branching order of the groups cannot be determined from the data.

the **actinobacteria** are composed of the actinomycetes and their high G + C relatives. **Figure 24.5** shows the phylogenetic relationships among selected representatives of the high G + C gram positives, and **table 24.3** summarizes the characteristics of some of the genera discussed in this chapter. Techniques for determining microbial taxonomy and phylogeny: Molecular characteristics (section 19.4)

Most of the genera discussed in the following survey are in the subclass *Actinobacteridae* and order *Actinomycetales* that is divided into 10 suborders. The survey focuses on several of these suborders. The order *Bifidobacteriales* also is briefly described.

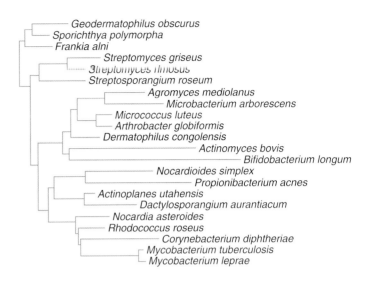

Figure 24.5 Phylogenetic Relationships Among Selected High G + C Gram-Positive Bacteria. The relationships among a few species based on 16S rRNA sequence data are shown. *Source: The Ribosomal Database Project.*

24.2 SUBORDER *ACTINOMYCINEAE*

There is one family with five genera in the suborder *Actinomycineae*. These include *Actinomyces, Actinobaculum, Arcanobacterium, Mobiluncus* and *Varibaculum*. Most are irregularly shaped, nonsporing, gram-positive rods with aerobic or faculative metabolism. The rods may be straight or slightly curved and usually have swellings, club shapes, or other deviations from normal rod-shape morphology.

Members of the genus *Actinomyces* are either straight or slightly curved rods that vary considerably in shape or slender filaments with true branching (**figure 24.6**). The rods and filaments may have swollen or clubbed ends. They are either facultative or strict anaerobes that require CO_2 for optimal growth. The cell walls contain lysine but not diaminopimelic acid or glycine. *Actinomyces* species are normal inhabitants of mucosal surfaces of humans and other warm-blooded animals; the oral cavity is their preferred habitat. *A. bovis* causes lumpy jaw in cattle. *Actinomyces* is responsible for actinomycoses, ocular infections, and periodontal disease in humans. The most important human pathogen is *A. israelii*.

24.3 SUBORDER *MICROCOCCINEAE*

The suborder *Micrococcineae* has 14 families and a wide variety of genera. Two of the best-known genera are *Micrococcus* and *Arthrobacter*.

The genus *Micrococcus* contains aerobic, catalase-positive cocci that occur mainly in pairs, tetrads, or irregular clusters and are usually nonmotile (**figure 24.7**). Unlike many other actinomycetes,

Table 24.3	Characteristics of Actinobacteria			
Genus	**Dimensions (μm) and Morphology**	**G + C Content (mol%)**	**Oxygen Relationship**	**Other Distinctive Characteristics**
Actinoplanes	Nonfragmenting, branching mycelium with little aerial growth; sporangia formed; motile spores with polar flagella	72–73	Aerobic	Hyphae often in palisade arrangement; highly colored; type II cell walls; found in soil and decaying plant material
Arthrobacter	0.8–1.2 × 1.0–8.0; young cells are irregular rods, older cells are small cocci; usually nonmotile	59–70	Aerobic	Have rod-coccus growth cycle; metabolism respiratory; catalase positive; mainly in soil
Bifidobacterium	0.5–1.3 × 1.5–8; rods of varied shape, usually curved; nonmotile	55–67	Anaerobic	Cells can be clubbed or branched, pairs often in V arrangement; ferment carbohydrates to acetate and lactate, but no CO_2; catalase negative
Corynebacterium	0.3–0.8 × 1.5–8.0; straight or slightly curved rods with tapered or clubbed ends; nonmotile	51–63	Facultatively anaerobic	Cells often arranged in a V formation or in palisades of parallel cells; catalase positive and fermentative; metachromatic granules
Frankia	0.5–2.0 in diameter; vegetative hyphae with limited to extensive branching and no aerial mycelium; multilocular sporangia formed	66–71	Aerobic to microaerophilic	Sporangiospores nonmotile; usually fixes nitrogen; type III cell walls; most strains are symbiotic with angiosperm plants and induce nodules
Micrococcus	0.5–2.0 diameter; cocci in pairs, tetrads, or irregular clusters; usually nonmotile	64–75	Aerobic	Colonies usually yellow or red; catalase positive with respiratory metabolism; primarily on mammalian skin and in soil
Mycobacterium	0.2–0.6 × 1.0–10; straight or slightly curved rods, sometimes branched; acid-fast; nonmotile and nonsporing	62–70	Aerobic	Catalase positive; can form filaments that are readily fragmented; walls have high lipid content; in soil and water; some parasitic
Nocardia	0.5–1.2 in diameter; rudimentary to extensive vegetative hyphae that can fragment into rod-shaped and coccoid forms	64–72	Aerobic	Aerial hyphae formed; catalase positive; type IV cell wall; widely distributed in soil
Propionibacterium	0.5–0.8 × 1–5; pleomorphic nonmotile rods, may be forked or branched; nonsporing	53–67	Anaerobic to aerotolerant	Fermentation produces propionate and acetate, and often gas; catalase positive
Streptomyces	0.5–2.0 in diameter; vegetative mycelium extensively branched; aerial mycelium forms chains of three to many spores	69–78	Aerobic	Form discrete lichenoid or leathery colonies that often are pigmented; use many different organic compounds as nutrients; soil organisms

Micrococcus does not undergo morphological differentiation. Micrococci colonies often are yellow, orange, or red. They are widespread in soil, water, and on mammalian skin, which may be their normal habitat.

The genus *Arthrobacter* contains aerobic, catalase-positive rods with respiratory metabolism and lysine in its peptidoglycan. Its most distinctive feature is a rod-coccus growth cycle (**figure 24.8**). When

Arthrobacter grows in exponential phase, the bacteria are irregular, branched rods that may reproduce by a process called **snapping division.** As they enter stationary phase, the cells change to a coccoid form. Upon transfer to fresh medium, the coccoid cells differentiate to form actively growing rods. Although arthrobacters often are isolated from fish, sewage, and plant surfaces, their most important habitat is the soil, where they constitute a significant

(a)

(b)

Figure 24.6 Representatives of the Genus *Actinomyces.* **(a)** *A. naeslundii;* Gram stain (×1,000). **(b)** *Actinomyces;* scanning electron micrograph (×18,000). Note filamentous nature of the colony.

Figure 24.7 Micrococcus. *Micrococcus luteus,* methylene blue stain (×1,000).

(a)　　　　　　　　　(b)

(c)　　　　　　　　　(d)

Figure 24.8 The Rod-Coccus Growth Cycle. The rod-coccus cycle of *Arthrobacter globiformis* when grown at 25°C. **(a)** Rods are outgrowing from cocci 6 hours after inoculation. **(b)** Rods after 12 hours of incubation. **(c)** Bacteria after 24 hours. **(d)** Cells after reaching stationary phase (3 day incubation). The cells used for inoculation resembled these stationary-phase cocci.

component of the microbial flora. They are well adapted to this niche because they are very resistant to desiccation and nutrient deprivation. This genus is unusually flexible nutritionally and can even degrade some herbicides and pesticides.

The mechanism of snapping division has been studied in *Arthrobacter.* These bacteria have a two-layered cell wall, and only the inner layer grows inward to generate a transverse wall dividing the new cells. The completed transverse wall or septum next thickens and puts tension on the outer wall layer, which still holds the two cells together. Eventually, increasing tension ruptures the outer layer at its weakest point, and a snapping movement tears the outer layer apart around most of its circumference. The new cells now rest at an angle to each other and are held together by the remaining portion of the outer layer that acts as a hinge.

A third genus in this suborder is *Dermatophilus. Dermatophilus* (type IIIB) also forms packets of motile spores with tufts of flagella, but it is a facultative anaerobe and a parasite of mammals responsible for the skin infection streptothricosis.

24.4 SUBORDER *CORYNEBACTERINEAE*

This suborder contains seven families with several well-known genera. Three of the most important genera are *Corynebacterium, Mycobacterium,* and *Nocardia.*

The family *Corynebacteriaceae* has one genus, *Corynebacterium,* which contains aerobic and facultative, catalase-positive, straight to slightly curved rods, often with tapered ends. Club-shaped forms are also seen. The bacteria often remain partially attached after snapping division, resulting in angular arrangements of the cells, or a palisade arrangement in which rows of cells are lined up side by side (**figure 24.9**). Corynebacteria form metachromatic granules, and their walls have *meso*-diaminopimelic acid. Although some species are harmless saprophytes, many corynebacteria are plant or animal pathogens. For example, *C. diphtheriae* is the causative agent of diphtheria in humans. Airborne diseases: Diphtheria (section 38.1)

The family *Mycobacteriaceae* contains the genus *Mycobacterium,* which is composed of slightly curved or straight rods that sometimes branch or form filaments (**figure 24.10**). Mycobacterial filaments differ from those of actinomycetes in readily fragmenting into rods and coccoid bodies. They are aerobic and catalase positive. Mycobacteria grow very slowly and must be incubated for 2 to 40 days after inoculation of a solidified complex medium to form a visible colony. Their cell walls have a very high lipid content and contain waxes with 60 to 90 carbon **mycolic acids.** These are complex fatty acids with a hydroxyl group on the β-carbon and an aliphatic chain attached to the α-carbon (**figure 24.11**). The presence of mycolic acids and other lipids outside the peptidoglycan layer makes mycobacteria **acid-fast** (basic fuchsin dye cannot be removed from the cell by acid alcohol treatment). Extraction of wall lipid with alkaline ethanol destroys acid-fastness. Preparation and staining of specimens: Differential staining (section 2.3)

Figure 24.9 *Corynebacterium diphtheriae.* Note the irregular shapes of individual cells, the angular associations of pairs of cells, and palisade arrangements (×1,000). These gram-positive rods do not form endospores.

Figure 24.10 The Mycobacteria. *Mycobacterium leprae.* Acid-fast stain (×400). Note the masses of red mycobacteria within blue-green host cells.

Although some mycobacteria are free-living saprophytes, they are best known as animal pathogens. *M. bovis* causes tuberculosis in cattle, other ruminants, and primates. Because this bacterium can produce tuberculosis in humans, dairy cattle are tested for the disease yearly; milk pasteurization kills the pathogen. Prior to widespread milk pasteurization, contaminated milk was a problematic source of transmission. Currently, *M. tuberculosis* is the chief source of tuberculosis in humans. The other major mycobacterial human disease is leprosy, caused by *M. leprae.* Airborne diseases: *Mycobacterium avium-M. intercellulare* and *M. tuberculosis* pulmonary infections (section 38.1); Insights from microbial genomes: Genomic analysis of pathogenic microbes (section 15.8)

The family *Nocardiaceae* is composed of two genera, *Nocardia* and *Rhodococcus.* These bacteria develop a substrate mycelium that readily breaks into rods and coccoid elements (**figure 24.12**). They also form an aerial mycelium that rises above the substratum and may produce conidia. Almost all are strict aerobes. Most species have peptidoglycan with *meso*-diaminopimelic acid and no peptide interbridge. The wall usually contains a carbohydrate composed of arabinose and galactose; mycolic acids are present in *Nocardia* and *Rhodococcus.* Because these and related genera resemble members of the genus *Nocardia* (named after Edmond Nocard [1850–1903], French bacteriologist and veterinary pathologist), they are collectively called **nocardioforms.**

Nocardia is distributed worldwide in soil and aquatic habitats. Nocardiae are involved in the degradation of hydrocarbons and waxes and can contribute to the biodeterioration of rubber joints in water and sewage pipes. Although most are free-living saprophytes, some species, particularly *N. asteroides,* are opportunistic pathogens that cause nocardiosis in humans and other animals. People with low resistance due to other health problems, such as individuals with HIV-AIDS, are most at risk. The lungs are most often infected, but the central nervous system and other organs may be invaded.

(a)

(b)

(c)

Figure 24.11 Mycolic Acid Structure. **(a)** The generic structure of mycolic acids, a family that includes over 500 different types. **(b)** A mycolic acid with two cyclopropane rings and **(c)** an unsaturated mycolic acid.

Nocardia

Figure 24.12 *Nocardia.* *Nocardia asteroides,* substrate mycelium and aerial mycelia with conidia illustration and light micrograph (×1,250).

Rhodococcus is widely distributed in soils and aquatic habitats. It is of considerable interest because members of the genus can degrade an enormous variety of molecules such as petroleum hydrocarbons, detergents, benzene, polychlorinated biphenyls (PCBs), and various pesticides. It may be possible to use rhodococci to remove sulfur from fuels, thus reducing air pollution by sulfur oxide emissions.

1. Define actinomycete, substrate mycelium, aerial mycelium, and exospore. Explain how these structures confer a survival advantage.
2. Describe how cell wall structure and sugar content are used to classify the actinomycetes. Include a brief description of the four major wall types.
3. Why are the actinomycetes of such practical interest?
4. Describe the phylum *Actinobacteria* and its relationship to the actinomycetes.
5. Describe the major characteristics of the following genera: *Actinomyces, Micrococcus, Arthrobacter,* and *Corynebacterium.* Include comments on their normal habitat and importance.
6. What is snapping division? the rod-coccus growth cycle?
7. Give the distinctive properties of the genus *Mycobacterium.*
8. List two human mycobacterial diseases and their causative agents. Which pathogen causes tuberculosis in cattle?
9. What is a nocardioform, and how can the group be distinguished from other actinomycetes?
10. Where is *Nocardia* found, and what problems may it cause? Consider both environmental and public health concerns.

24.5 SUBORDER *MICROMONOSPORINEAE*

The suborder *Micromonosporineae* contains only one family, *Micromonosporaceae.* Genera include *Micromonospora, Dactylosporangium, Pilimelia,* and *Actinoplanes.* Often the family is collectively referred to as the actinoplanetes [Greek *actinos,* a ray or beam, and *planes,* a wanderer]. They have an extensive substrate mycelium and are type IID cells. Often the hyphae are highly colored and diffusible pigments may be produced. Normally an aerial mycelium is absent or rudimentary. Spores are

Figure 24.13   Family *Micromonosporaceae.* Actinoplanete morphology. **(a)** *Actinoplanes* structure. **(b)** A scanning electron micrograph of mature *Actinoplanes* sporangia. **(c)** *Dactylosporangium* structure. **(d)** A *Dactylosporangium* colony covered with sporangia.

usually formed within a sporangium raised above the surface of the substratum at the end of a special hypha called a sporangiophore. The spores can be either motile or nonmotile. These bacteria vary in the arrangement and development of their spores. Some genera (*Actinoplanes, Pilimelia*) have spherical, cylindrical, or irregular sporangia with a few to several thousand spores per sporangium (figure 24.3*a* and **figure 24.13**). The sporangium develops above the substratum at the tip of a sporangiophore; the spores are arranged in coiled or parallel chains (**figure 24.14**). *Dactylosporangium* forms club-shaped, fingerlike, or pyriform sporangia with one to six spores (figure 24.13*c,d*). *Micromonospora* bears single spores, which often occur in branched clusters of sporophores (figure 24.3*b*).

Actinoplanetes grow in almost all soil habitats, ranging from forest litter to beach sand. They also flourish in freshwater, particularly in streams and rivers (probably because of abundant oxygen and plant debris). Some have been isolated from the ocean. The soil-dwelling species may have an important role in the decomposition of plant and animal material. *Pilimelia* grows in association with keratin. *Micromonospora* actively degrades chitin and cellulose, and produces antibiotics such as gentamicin.

24.6   SUBORDER *PROPIONIBACTERINEAE*

This suborder contains two families and 14 genera. The genus *Propionibacterium* contains pleomorphic, nonmotile, nonsporing rods that are often club-shaped with one end tapered and the other end rounded. Cells also may be coccoid or branched. They can be single, in short chains, or in clumps. The genus is facultatively anaerobic or aerotolerant; lactate and sugars are fermented to produce large quantities of propionic and acetic acids, and often carbon dioxide. *Propionibacterium* is usually catalase positive. The genus is found growing on the skin and in the digestive tract of animals, and in dairy products such as cheese. *Propionibacterium* contributes substantially to the production of Swiss cheese. *P. acnes* is involved with the development of body odor and acne vulgaris.   Microbiology of feremented foods (section 40.6)

24.7   SUBORDER *STREPTOMYCINEAE*

The suborder *Streptomycineae* has only one family, *Streptomycetaceae,* and three genera, the most important of which is *Streptomyces.* These bacteria have aerial hyphae that divide in a single

Figure 24.14   Sporangium Development in an Actinoplanete.   The developing sporangium is shown in purple with more mature stages to the right.

plane to form chains of 3 to 50 or more nonmotile spores with surface texture ranging from smooth to spiny and warty (figure 24.3*e;* **figures 24.15** and **24.16**). All have a type I cell wall and a G + C content of 69 to 78%. Filaments grow by tip extension rather than by fragmentation. Members of this family and similar bacteria are often called **streptomycetes** [Greek *streptos,* bent or twisted, and *myces,* fungus].

Streptomyces is a large genus; there are around 150 species. Members of the genus are strict aerobes, have cell wall type I, and form chains of nonmotile spores (figure 24.15*c* and 24.16). The three to many spores in each chain are often pigmented and can be smooth, hairy, or spiny in texture. *Streptomyces* species are determined by means of a mixture of morphological and physiological characteristics, including the following: the color of the aerial and substrate mycelia, spore arrangement, surface features of individual spores, carbohydrate use, antibiotic production, melanin synthesis, nitrate reduction, and the hydrolysis of urea and hippuric acid.

Streptomycetes are very important, both ecologically and medically. The natural habitat of most streptomycetes is the soil, where they may constitute from 1 to 20% of the culturable population. In fact, the odor of moist earth is largely the result of streptomycete production of volatile substances such as **geosmin.** Streptomycetes play a major role in mineralization. They are flexible nutritionally and can aerobically degrade resistant substances such as pectin, lignin, chitin, keratin, latex, agar, and aromatic compounds. Streptomycetes are best known for their synthesis of a vast array of antibiotics.

Stanley Waksman's discovery that *S. griseus* (**figure 24.17***a*) produces streptomycin was an enormously important contribution to science and public health. Streptomycin was the first drug to effectively combat tuberculosis, and in 1952 Waksman earned the Nobel Prize. In addition, this discovery set off a massive search resulting in the isolation of new *Streptomyces* species that produce other compounds of medicinal importance. In fact, since that time, the streptomycetes have been found to produce over 10,000 bioactive compounds. Hundreds of these natural products are now used in medicine and industry; about two-thirds of the antimicrobial agents used in human and veterinary medicine are derived from the streptomycetes. Examples include amphotericin B, chloramphenicol, erythromycin, neomycin, nystatin, and tetracycline. Some *Streptomyces* species produce more than one antibiotic. Antibiotic-producing bacteria have genes that encode proteins that make them resistant to such compounds. Antimicrobial chemotherapy (chapter 34)

The genome of *Streptomyces coelicolor,* which produces four antibiotics and serves as a model species for research, has been sequenced. At 8.67 Mbp, it is one of the largest procaryotic genomes. Its large number of genes (7,825) no doubt reflects the number of proteins required to undergo a complex life cycle. Many genes are devoted to regulation, with an astonishing 65 predicted RNA polymerase sigma subunits and over 50 two-component regulatory systems. The ability to exploit a variety of soil nutrients is also demonstrated by the presence of a large number of ABC transporters, the Sec protein translocation system, and secreted degradative enzymes. Finally, genes were discovered that are thought to encode an additional 18 secondary metabolites. Protein secretion in procaryotes (section 3.8); Regulation of transcription initiation: Two-component regulatory systems (section 12.2); Global regulatory systems (section 12.5)

(a) (b) (c)

Figure 24.19 *Frankia.* **(a)** An interference contrast micrograph showing hyphae, multilocular sporangia, and spores. **(b)** A colorized scanning electron micrograph of a sporangium surrounded by hyphae. **(c)** A nodule of the alder *Alnus rubra* showing cells filled with vesicles of *Frankia.* Scanning electron microscopy.

Another genus in this suborder, *Sporichthya,* is one of the strangest of the actinomycetes. It lacks a substrate mycelium. The hyphae remain attached to the substratum by holdfasts and grow upward to form aerial mycelia that release motile, flagellate spores in the presence of water.

24.10 ORDER *BIFIDOBACTERIALES*

The order *Bifidobacteriales* has one family, *Bifidobacteriaceae,* and 10 genera (five of which have unknown affiliation). *Falcivibrio* and *Gardnerella* are found in the human genital/urinary tract; *Gardnerella* is thought to be a major cause of bacterial vaginitis. *Bifidobacterium* probably is the best-studied genus. Bifidobacteria are nonmotile, nonsporing, gram-positive rods of varied shapes that are slightly curved and clubbed; often they are branched (**figure 24.20**). The rods can be single or in clusters and V-shaped pairs. *Bifidobacterium* is anaerobic and actively ferments carbohydrates to produce acetic and lactic acids, but no carbon dioxide. It is found in the mouth and intestinal tract of warm-blooded vertebrates, in sewage, and in insects. *B. bifidum* is a pioneer colonizer of the human intestinal tract, particularly when babies are breast fed. A few *Bifidobacterium* infections have been reported in humans, but the genus does not appear to be a major cause of disease. Direct contact diseases: Sexually transmitted diseases (section 38.3)

Figure 24.20 *Bifidobacterium.* *Bifidobacterium bifidum;* phase-contrast photomicrograph ($\times$1,500).

3. Describe the major properties of the genus *Streptomyces.*
4. Describe three ways in which *Streptomyces* is of ecological importance. Why do you think *Streptomyces* spp. produce antibiotics?
5. Briefly describe the genera of the suborder *Streptosporangineae.* What is madurose? Why is *Actinomadura* important?
6. Describe *Frankia* and discuss its importance.
7. Characterize the genus *Bifidobacterium.* Where is it found and why is it significant?

1. Give the distinguishing properties of the actinoplanetes.
2. Describe the genus *Propionibacterium* and comment on its practical importance.

Summary

24.1 General Properties of the Actinomycetes

a. Actinomycetes are aerobic, gram-positive bacteria that form branching, usually nonfragmenting, hyphae and asexual spores (**figure 24.2**).

b. The asexual spores borne on aerial mycelia are called spores if they are at the tip of hyphae or sporangiospores if they are within sporangia.

c. Actinomycetes have several distinctively different types of cell walls and often also vary in terms of the sugars present in cell extracts. Properties such as color and morphology are also taxonomically useful (**tables 24.1, 24.2, and 24.3**).

d. The second edition of *Bergey's Manual* classifies the high G + C bacteria phylogenetically using 16S rRNA data. The phylum *Actinobacteria* contains the actinomycetes and their high G + C relatives (**figure 24.4**).

24.2 Suborder *Actinomycineae*

a. The suborder *Actinomycineae* contains the genus *Actinomyces,* members of which are irregularly shaped, nonsporing rods that can cause disease in cattle and humans (**figure 24.6**).

24.3 Suborder *Micrococcineae*

a. The suborder *Micrococcineae* includes the genera *Micrococcus, Arthrobacter,* and *Dermatophilus. Arthrobacter* has an unusual rod-coccus growth cycle and carries out snapping division (**figures 24.7** and **24.8**).

24.4 Suborder *Corynebacterineae*

a. The genera *Corynebacterium, Mycobacterium,* and *Nocardia* are placed in the suborder *Corynebacterineae.* Mycobacteria form either rods or filaments that readily fragment. Their cell walls have a high lipid content and mycolic acids; the presence of these lipids makes them acid-fast (**figures 24.9–24.11**). The genera *Corynebacterium* and *Mycobacterium* contain several very important human pathogens.

b. Nocardioform actinomycetes have hyphae that readily fragment into rods and coccoid elements, and often form aerial mycelia with spores (**figure 24.12**).

24.5 Suborder *Micromonosporineae*

a. The suborder *Micromonosporineae* has genera that include *Micromonospora* and *Actinoplanes.* These actinomycetes have an extensive substrate mycelium and form special aerial sporangia (**figure 24.14**). They are present in soil, freshwater, and the ocean. The soil forms are probably important in decomposition.

24.6 Suborder *Propionibacterineae*

a. The genus *Propionibacterium* is in the suborder *Propionibacterineae.* Members of this genus are common skin and intestinal inhabitants and are important in cheese manufacture and the development of acne vulgaris.

24.7 Suborder *Streptomycineae*

a. The suborder *Streptomycineae* includes the genus *Streptomyces.* Members of this genus have type I cell walls and aerial hyphae bearing chains of 3 to 50 or more nonmotile spores (**figures 24.15** and **24.16**).

b. Streptomycetes are important in the degradation of more resistant organic material in the soil and produce many useful antibiotics. A few cause diseases in plants and animals.

24.8 Suborder *Streptosporangineae*

a. Many genera in suborder *Streptosporangineae* have the sugar derivative madurose and type III cell walls.

24.9 Suborder *Frankineae*

a. The genera *Frankia* and *Geodermatophilus* are placed in the suborder *Frankineae.* They produce clusters of spores at hyphal tips and have type III cell walls. *Frankia* grows in symbiotic association with nonleguminous plants and fixes nitrogen (**figure 24.19**).

24.10 Order *Bifidobacteriales*

a. The genus *Bifidobacterium* is placed in the order *Bifidobacteriales.* This irregular, anaerobic rod is one of the first colonizers of the intestinal tract in nursing babies.

Key Terms

acid-fast 596
actinobacteria 593
actinomycete 589
actinomycetoma 600

aerial mycelium 589
geosmin 599
hypha(e) 589
madurose 601

mycolic acids 596
nocardioforms 596
secondary metabolite 589
snapping division 594

sporangiospores 590
streptomycetes 599
substrate mycelium 589

Critical Thinking Questions

1. Even though these are "high G + C" organisms, there are regions of the genome that must be more AT-rich. Suggest a few such regions and explain why they must be more AT-rich.

2. Choose two different species in the phylum *Actinobacteria* and investigate their physiology and ecology. Compare and contrast these two organisms. Can you determine why having a high G + C genomic content might confer an evolutionary advantage?

3. *Streptomyces coelicolor* is studied as a model system for cellular differentiation. Some of the genes involved in sporulation contain a rare codon not used in vegetative genes. Suggest how *Streptomyces* might use the rare codon to regulate sporulation.

4. Suppose that you discovered a nodulated plant that could fix atmospheric nitrogen. How might you show that a bacterial symbiont was involved and that *Frankia* rather than *Rhizobium* was responsible?

Learn More

Anderson, A. S., and Wellington, E. M. H. 2001. The taxonomy of *Streptomyces* and related genera. *Int. J. Syst. Evol. Microbiol.* 51: 797–814.

Beaman, B. L.; Saubolle, M. A.; and Wallace R. J. 1995. *Nocardia, Rhodococcus, Streptomyces, Oerskovia,* and other aerobic actinomycetes of medical importance. In *Manual of Clinical Microbiology,* 6th ed., P. R. Murray, editor-in-chief, 379–99. Washington, D.C.: American Society for Microbiology.

Benson, D. R., and Silvester, W. B. 1993. Biology of *Frankia* strains, actinomycete symbionts of actinorhizal plants. *Microbiol. Rev.* 57(2): 293–319.

Bentley, S. D.; Chater, K. F.; Cerdeño-Tárraga, A.-M.; Challis, G. L.; Tomson, N. R.; James, K. D.; Harris, D. E.; Quail, M. A.; Kieser, H.; Harper, D.; Bateman, A.; Brown, S.; Chandra, G.; Chen, C. W.; Collins, M.; Cronin, A.; Fraser, A.; Goble, A.; Hidalgo, J.; Hornsby, T.; Howarth, S.; Huang, C.-H.; Kieser, T.; Larke, L.; Murphy, L.; Oliver, K.; O'Neil, S.; Rabbinowitsch, E.; Rajandream, M.-A.; Rutherford, K.; Rutter, S.; Segger, K.; Saunders, D.; Sharp, S.; Squares, R.; Squares, S.; Taylor, K.; Warren, T.; Wietzorrek, A.; Woodward, J.; Barrell, B. G.; Parkhill, J.; and Hopwood, D. A. 2002. Complete genome sequence of the model actinomycete *Streptomyces coelicolor* A3(2). *Nature* 417: 141–47.

Clawson, M. L.; Bourret, A.; and Benson, D. R. 2004. Assessing the phylogeny of *Frankia*-actinorhizal plant nitrogen-fixing root nodule symbioses with *Frankia* 16S rRNA and glutamine synthetase gene sequences. *Mol. Phylogen. Evol.* 31: 131–38.

Dyson, P. 2000. *Streptomyces* genetics. In *Encyclopedia of microbiology,* 2d ed., vol. 4, J. Lederberg, editor-in-chief, 451–66. San Diego: Academic Press.

Parenti, F., and Coronelli, C. 1979. Members of the genus *Actinoplanes* and their antibiotics. *Annu. Rev. Microbiol.* 33:389–411.

Stackebrandt, E.; Rainey, F. A.; and Ward-Rainey, N. L. 1997. Proposal for a new hierarchic classification system. *Actinobacteria* classis nov. *Int. J. Syst. Bacteriol.* 47(2):479–91.

Please visit the Prescott website at www.mhhe.com/prescott7 for additional references.

25

The Protists

This is a scanning electron micrograph (×2,160) of the protozan *Naegleria fowleri*. Three *N. fowleri* from an axenic culture, attacking and beginning to devour or engulf a fourth, presumably dead amoeba, with their amoebastomes (suckerlike structures that function in phagocytosis). This amoeba is the major cause of the disease in humans called primary amebic meningoencephalitis.

PREVIEW

- Protists are a polyphyletic collection of organisms. Most are unicellular and lack the level of tissue organization seen in higher eucaryotes.

- Protists are ubiquitous—they are found wherever there is adequate moisture. They are prominent members of marine planktonic communities where they contribute to both the amount of fixed CO_2 as well as the recycling of nutrients. Aquatic and terrestrial protists are also important in nutrient regeneration.

- Protists can be photoautotrophic, chemoorganotrophic, or mixotrophic. They can assimilate organic nutrients by saprotrophy or holotrophy, feeding on bacteria and other particulate forms of carbon. Many protists live in association with other organisms as mutualists, commensals, or parasites.

- Protists usually reproduce asexually by binary fission; some undergo multiple fission or budding. Many also have sexual cycles that involve meiosis and the fusion of gametes or gametic nuclei, resulting in a diploid zygote. The zygote is often a thick-walled, resistant, resting cell called a cyst.

- All protists have one or more nuclei; some have a macronucleus and a micronucleus. Energy metabolism occurs in mitochondria, hydrogenosomes, and chloroplasts, although some protists lack these organelles.

- The systematics and taxonomic classification of the protists are areas of active research and debate.

Chapter 25 presents the major biological features of the protists. We introduce protists of medical and ecological significance and follow the higher-level classification scheme of eucaryotes as proposed by the International Society of Protistologists. The kingdom *Protista*, as defined by Whittaker's five-kingdom scheme, is an artificial grouping of over 64,000 different single-celled life forms that lack common evolutionary heritage; that is to say, they are polyphyletic (**figure 25.1**). In fact, the protists are unified only by what they lack: absent is the level of tissue organization found in fungi, plants, and animals. The term **protozoa** [s., protozoan; Greek *protos*, first, and *zoon*, animal] has traditionally referred to chemoorganotrophic protists, and **protozoology** generally refers to the study of protozoa. The term **algae** can be used to describe photosynthetic protists. "Algae" was originally used to refer to all "simple aquatic plants," but this term has no phylogenetic utility. The study of photosynthetic protists (algae) is often referred to as **phycology** and is the realm of both botanists and protistologists. The study of all protists, regardless of their metabolic type, is called **protistology.**

For many years the protozoa were classified into four major groups based on their means of locomotion: flagellates (*Mastigophora*), ciliates (*Infusoria* or *Ciliophora*), amoebae (*Sarcodina*), and stationary forms (*Sporozoa*). Although nonprotistologists still use these terms, these divisions have no bearing on evolutionary relationships and should be avoided. It is now agreed that the old classification system is best abandoned, but for many years there was little agreement on what should take its place. Recent morphological, biochemical, and phylogenetic analyses have resulted in the development of a higher-level classification system for the eucaryotes. This scheme, as proposed by the International Society of Protistologists, is introduced in chapter 19 (*see table 19.9*) and is followed here. However, first we introduce some major ecological, morphological, and physiological elements that describe most protists.

And a pleasant sight they are indeed. Their shapes range from teardrops to bells, barrels, cups, cornucopias, stars, snowflakes, and radiating suns, to the common amoebas, which have no real shape at all. Some live in baskets that look as if they were fashioned of exquisitely carved ivory filigree. Others use colored bits of silica to make themselves bright mosaic domes. Some even form graceful transparent containers shaped like vases or wine glasses of fine crystal in which they make their homes.

—Helena Curtis

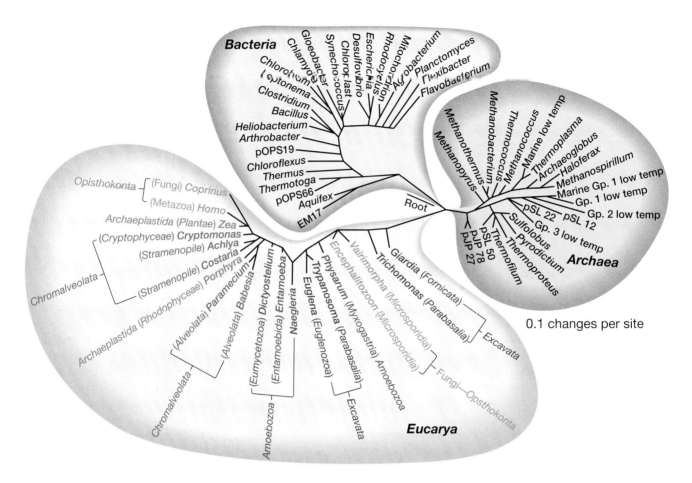

Figure 25.1 Universal Phylogenetic Tree, Highlighting Protists. Recent molecular phylogenetic evidence in combination with morphological and biochemical data has resulted in the establishment of super groups among the *Eucarya* (each super group is highlighted with a different color). This new, higher-order classification scheme is not entirely congruent with the placement of the protists on the Universal Tree of Life; the latter is based entirely on analysis of SSU rRNA. Nonetheless, it demonstrates that the protists are a highly polyphyletic group. Because the protists are not a monophyletic group, the term "protist" cannot be used to represent evolutionary histories. Instead the term "protist" as used in this chapter denotes a group of eucaryotic organisms that share some morphological, reproductive, ecological, and biochemical characteristics.

25.1 DISTRIBUTION

Protists grow in a wide variety of moist habitats. Moisture is absolutely necessary for their existence because they are susceptible to desiccation. Most protists are free living and inhabit freshwater or marine environments. Many terrestrial chemoorganotrophic forms can be found in decaying organic matter and in soil, where they are important in recycling the essential elements nitrogen and phosphorus. Others are **planktonic**—floating free in lakes and oceans. Planktonic microbes (both procaryotic and eucaryotic) are responsible for a majority of the nutrient cycling that occurs in these ecosystems.

Every major group of protists includes species that live in association with other organisms. For instance, some photosynthetic protists associate with fungi to form lichens while others live with corals where they provide fixed carbon to the coral animal. Thousands of others are parasites and cause important dis-

eases in humans and domesticated animals (**table 25.1**). Finally, protists are useful in biochemical and molecular biological research. Not only do they display an amazing array of unique adaptations, but many biochemical pathways used by protists are found in other eucaryotes, making them useful model organisms.

25.2 NUTRITION

Photosynthetic protists are exclusively aerobic. Like cyanobacteria and plants, they possess both photosystems I and II and perform oxygenic photosynthesis. Most photosynthetic forms are photoautotrophic, obtaining energy from light and fixing CO_2 to meet their carbon requirements. However, some are photoheterotrophic, using organic carbon rather than CO_2. Chemoheterotrophic protists can be holozoic or saprozoic. In **holozoic nutrition,** solid nutrients such as bacteria are acquired by phago-

Super Group (subrank)	Genus	Host	Preferred Site of Infection	Disease
Amoebozoa	Entamoeba	Mammals	Intestine	Amebiasis
	Iodamoeba	Swine	Intestine	Enteritis
Chromalveolata	Babesia	Cattle	Blood cells	Babesiosis
(Apicomplexa)	Therileria	Cattle, sheep, goats	Blood cells	Therileriasis
	Sarcocystis	Mammals, birds	Muscles	Sarcosporidiosis
	Taxoplasma	Cats	Intestine	Toxoplasmosis
	Isospora	Dogs	Intestine	Coccidiosis
	Eimeria	Cattle, cats, chickens, swine	Intestine	Coccidiosis
	Plasmodium	Many animals	Red blood cells, liver	Malaria
	Leucocytozoon	Birds	Spleen, lungs, blood	Leucocytozoonosis
	Cryptosporidium	Mammals	Intestine	Cryptosporidiosis
(Ciliophora)	Balantidium	Swine	Large intestine	Balantidiasis
Excavata (Kinetoplasta)	Leishmania	Dogs, cats, horses, sheep, cattle	Spleen, bone marrow, mucous membranes	Leishmaniasis
	Trypanosoma	Most animals	Blood	Chagas disease, Sleeping sickness
(Parabasalia)	Trichomonas	Horses, cattle	Genital tract	Trichmoniasis (abortion)
	Histomonas	Birds	Intestine	Blackhead disease
(Fornicata)	Giardia	Mammals	Intestine	Giardiasis

Table 25.1 Pathogenic Protists That Cause Major Diseases of Domestic Animals

cytosis and the subsequent formation of phagocytic vacuoles (**figure 25.2**). In **saprozoic nutrition,** soluble nutrients such as amino acids and sugars cross the plasma membrane by endocytosis, diffusion, or carrier-mediated transport (facilitated diffusion or active transport). The mechanisms by which soluble nutrients are assimilated are sometimes collectively called **osmotrophy.**

It is difficult to classify the nutritional strategies of some protists. Certain protists can derive energy through the oxidation of inorganic substrates but assimilate organic carbon compounds. Some simultaneously use both organic and inorganic forms of carbon; this is sometimes referred to as **mixotrophy.** This metabolic flexibility is clearly evident in those forms like the dinoflagellates that can perform photosynthesis and holozoic feeding concurrently.

25.3 MORPHOLOGY

Before we describe specific morphological features, it is helpful to define the terms amoeba and ciliate. Older, now discredited classification schemes placed all amoeboid and ciliated protists in the "Sarcodina" and "Infusoria," respectively. While these terms may remain useful in describing protists that demonstrate amoeboid and ciliated motility, they do not describe monophyletic groups. Thus in this chapter, "amoebae" and "ciliates" are not taxonomic terms—rather, they are adjectives, much like we might describe flagellated archaea or vibroid bacteria.

Because protists are eucaryotic cells, in many respects their morphology and physiology are the same as the cells of multicellular plants and animals. However, because all of life's various functions must be performed within a single cell, complexity arises at the level of specialized organelles rather than at the tissue level. Protists must remain relatively small because without multicellularity to facilitate the exchange of nutrients and metabolites, they need a high ratio of cell surface to intracellular volume. This is because the distance from the cell membrane to the center of the cell cannot exceed the distance a molecule can diffuse. Even the largest algae (the seaweeds) are long, thin, and often flattened—these are shapes that maximize the surface-to-volume ratio.

The protist cell membrane, called the **plasmalemma,** is identical to that of multicellular organisms. In some protists the cytoplasm immediately under the plasmalemma is divided into an outer gelatinous region called the **ectoplasm,** and an inner fluid region, the **endoplasm.** The ectoplasm imparts rigidity to the cell body. The plasmalemma and structures immediately beneath it are called the **pellicle.** One or more vacuoles are usually present in the cytoplasm of protozoa. These are differentiated into contractile, secretory, and food vacuoles. **Contractile vacuoles** function as osmoregulatory organelles in those protists that live in hypotonic environments, such as freshwater lakes. Osmotic balance is maintained by continuous water expulsion. Most marine and parasitic species are isotonic to their environment and lack

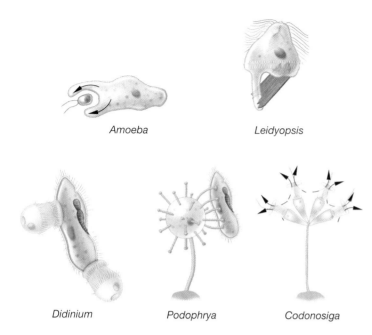

Amoeba   *Leidyopsis*

Didinium   *Podophrya*   *Codonosiga*

Figure 25.2 Holozoic Feeding Methods Among the Protists. *Amoeba* uses pseudopodia to surround bacteria. *Leidyopsis* lives in the guts of termites, where it uses pseudopodia to entrap wood particles. The ciliate *Didinium* eats only *Paramecium* (a larger ciliate) through a temporary cytostome. *Podophyra* uses tentacles to first attach to its prey and then siphon prey cytoplasm into its body where it is digested in food vacuoles. The sessile protist *Codonosiga* has a collar of microvilli; flagella help sweep food particles suspended in water into the collar. Despite this variety, all of these feeding methods are considered forms of pseudopodia.

such vacuoles. **Phagocytic vacuoles** are conspicuous in holozoic and parasitic species and are the sites of food digestion. In some organisms, they may occur anywhere on the cell surface, while others have a specialized structure for phagocytosis called the **cytostome** (cell mouth). When digestion commences, the phagocytic vacuole is acidic, but as it proceeds, the vacuolar pH increases and the membrane forms small blebs. These pinch off and carry nutrients throughout the cytoplasm. The undigested contents of the original phagocytic vacuole are expelled from the cell either at a random site on the cell membrane (as is the case with amoebae) or at a designated position called the **cytoproct.** Organelles of the biosynthetic-secretory and endocytic pathways (section 4.4)

Several energy-conserving organelles are observed in protists. Most aerobic chemoorganotrophic forms have mitochondria with cristae that are characterized by their physical appearance: they can be discoid, tubular, or lamellar; cristae morphology is of taxonomic value. The majority of anaerobic protists (such as *Trichonympha,* which lives in the gut of termites) lack mitochondria and cytochromes, and have an incomplete tricarboxylic acid cycle. However, some have small, membrane-bound organelles termed **hydrogenosomes.** Within these organelles, pyruvate derived from glycolysis is oxidized and decarboxylated to form CO_2, H_2, and the high-energy molecule acetyl-CoA. The conversion of each

acetyl-CoA to acetate yields an ATP (*see figure 19.4*). Because hydrogenosomes lack a complete tricarboxylic acid cycle, acetate must be excreted. Fermentation end products may be consumed by symbiotic procaryotes living within the protist. In some cases, methanogenic archaea consume the CO_2 and H_2 and generate methane. Recent evidence suggests that hydrogenosomes and mitochondria evolved from the same endosymbiotically derived organelle. Photosynthetic protists have chloroplasts featuring thylakoid membranes. A dense proteinaceous area, the **pyrenoid,** which is associated with the synthesis and storage of starch, may be present in the chloroplasts. Microbial evolution (section 19.1)

Many protists feature cilia or flagella at some point in their life cycle. Their formation is associated with a basal bodylike organelle called the **kinetosome,** which is similar in structure to the centriole. In addition to aiding in motility, these organelles may be used to generate water currents for feeding and respiration. Because there is no precise morphological difference between flagella and cilia, some scientists prefer to call them both **undulipodia** (meaning "wave foot"). Cilia and flagella (section 4.10)

---

1. How do the terms protozoa and algae differ from the term protist?
2. In what habitats can protists be found?
3. What roles do protists play in the trophic structure of their communities and in the organisms with which they associate?
4. What is a hydrogenosome? How does it differ from a mitochondrion?
5. Trace the path of a food item from the phagocytic vacuole to the cytoproct.

25.4 Encystment and Excystment

Many protists are capable of **encystment.** During encystment, the organism de-differentiates (becomes simpler in morphology) and develops into a resting stage called a **cyst.** The cyst is a dormant form marked by the presence of a cell wall and by very low metabolic activity. Cyst formation is particularly common among aquatic, free-living protists and parasitic forms. Cysts serve three major functions: (1) they protect against adverse changes in the environment, such as nutrient deficiency, desiccation, adverse pH, and low partial pressure of O_2; (2) they are sites for nuclear reorganization and cell division (reproductive cysts); and (3) they serve as a means of transfer between hosts in parasitic species. Although the exact stimulus for **excystment** (escape from the cysts) is unknown, it is generally triggered by a return to favorable environmental conditions. For example, cysts of parasitic species excyst after ingestion by the host and form the vegetative form called the **trophozoite.**

25.5 Reproduction

Most protists reproduce asexually and some also carry out sexual reproduction. The most common method of asexual reproduction is **binary fission.** During this process the nucleus first undergoes mitosis, then the cytoplasm divides by cytokinesis to form two identical individuals (**figure 25.3**). Multiple fission is also com-

(a) *Arcella*

(b) *Euglypha*

(c) *Trypanosoma*   **(d)** *Euglena*

Figure 25.3   Binary Fission in Protists.   (a) The two nuclei of the testate (shelled) amoeba *Arcella* divide as some of its cytoplasm is extruded and a new test for the daughter cell is secreted. **(b)** In another testate amoeba, *Euglypha*, secretion of new platelets is begun before cytoplasm begins to move out of the aperture. The nucleus divides while the platelets are used to construct the test for the daughter cell. For many protists, all organelles must be replicated before the cell divides. This is the case with **(c)** *Trypanosoma* and **(d)** *Euglena*.

mon, as is budding. Some filamentous, photosynthetic protists undergo fragmentation so that each piece of the broken filament grows independently.

Most protists also undergo sexual reproduction at some point in their life cycle. Protist cells that produce gametes are termed **gamonts.** The fusion of haploid gametes is called **syngamy.** Among protists, syngamy can involve the fusion of two morphologically similar gametes (**isogamy**) or the fusion of morphologically different types (**anisogamy**). Meiosis may occur either before the formation and union of gametes, as in most animals, or just after fertilization, as is also the case with lower plants. Furthermore, the exchange of nuclear material may occur in the familiar fashion—between two different individuals (**conjugation**)—or by the development of a genetically distinct nucleus within a single individual (**autogamy**).

With this level of reproductive complexity, perhaps it is not surprising that the nuclei among protists show considerable diversity. Most commonly, a **vesicular nucleus** is present. This is characterized by a nucleus that is 1 to 10 μm in diameter, spherical, with a distinct nucleolus and uncondensed chromosomes. **Ovular nuclei** are up to 10 times this size and possess many peripheral nucleoli. Still others have **chromosomal nuclei,** in which the chromosomes remain condensed throughout interphase

with a single nucleolus associated with one chromosome. Finally, the *Ciliophora* have two types of nuclei: a large **macronucleus** with distinct nucleoli and condensed chromatin and a smaller, diploid **micronucleus** with dispersed chromatin but lacking nucleoli. Macronuclei are engaged in trophic activities and regeneration processes while micronuclei are involved only in genetic recombination during sexual reproduction and the regeneration of the macronucleus.

1. What functions do cysts serve for a typical protist? What causes excystment to occur?
2. What is a gamont? What is the difference between anisogamy and syngamy?
3. How does conjugation differ from autogamy?
4. Describe vesicular, ovular, and chromosomal nuclei. Which is most like the nuclei found in higher eucaryotes?

25.6   PROTIST CLASSIFICATION

Ever since Antony van Leeuwenhoek described the first protozoan "animalcule" in 1674, the taxonomic classification of these microbes has remained in flux. During the 20th century, classification schemes based on functional morphology rather than evolutionary relationships were used. The application of molecular techniques is providing new insights into protist systematics and, in most cases, modern morphological and biochemical analyses are in agreement with molecular phylogenetic data. The 2005 International Society of Protistologists' publication of a higher-level classification scheme of all eucaryotes with an emphasis on the protists provides a long-needed consensus (**table 25.2**). Note that this scheme does not use formal hierarchical rank designations such as class and order, reflecting the fact that protist taxonomy remains an area of active research.

Super Group *Excavata*

The *Excavata* includes some of the most primitive, or deeply branching, eucaryotes (figure 25.1). Most possess a cytostome characterized by a suspension-feeding groove. This apparatus features a posteriorly directed flagellum to generate a current that enables the capture of small particles from a feeding current. Those that lack this morphological feature are presumed to have had it at one time during their evolution—that is to say, it is thought to have been secondarily lost.

Fornicata

Ever curious, Anton van Leeuwenhoek described *Giardia intestinalis* (**figure 25.4**) from his own diarrheic feces. Over 300 years later, this species continues to be a public health concern. Today it most often infects campers and other individuals who unwittingly consume contaminated water. Members of the *Fornicata* bear flagella and lack mitochondria. Unlike *Giardia,* most are harmless symbionts. The few free-living forms are most often found in waters that are heavily polluted with organic nutrients (a

Table 25.2 Classification of the Protists as Proposed by the International Society of Protistologists[a]

Super Group	Unifying Features	First Rank	General Description
Opisthokonta	During at least one stage of life cycle, cells have single posterior cilium without mastigonemes (hair-like projections on the flagella); possess kinetosomes or centrioles; when unicellular, mitochondria have flat cristae	Mesomycetozoa	Most have mitochondria with flat cristae; one or more life cycle stages feature spherical cells; one posterior flagellum or amoeboid; some have parasitic nonflagellated stages and endospores; some have trophic stages that feature a cell wall. Includes *Aphelidium, Dermocystidium, Ichthyophonus,* and *Nuclearia.*
		Choanomonada	Radially symmetric; phagotrophic with collar of microvilli encircling a single flagellum; mitochondria have flat cristae; solitary or colonial. Includes *Monosiga, Salpingoeca,* and *Stenphanoeca.*
Archaeplastids	Photosynthetic plastid with chlorophyll *a* from ancestral cyanobacterial endosymbiont, plastid later lost in some; uses starch as a storage product; usually have cell wall made of cellulose	Glaucophyta	Plastid is a cyanelle, which unlike chloroplasts has peptidoglycan between the two membranes; stacked thylakoids with chlorophyll *a* and phycobiliproteins and other pigments; includes flagellated and nonflagellated species or life cycle stages. Includes *Cyanophora, Glaucocystis,* and *Gloeochaeta.*
		Rhodophycae	Lack flagellated stages, kinetosome, and centrioles, unstacked thylakoids; chloroplast without external endoplasmic reticulum; also called red algae although traditional subgroups are no longer considered valid. Includes *Ceramium, Porphyra,* and *Sphaerococcus.*
		Chloroplastida	Pyrenoid often within plastid, which features chlorophylls *a* and *b*; cellulose-containing cell wall typical. Includes the *Charophyta* (green algae), *Chara, Nitella,* and *Volvox.*
Amoebozoa	Amoeboid motility with lobopodia; naked or testate; mitochondria with tubular cristae; usually uninucleate, sometimes multinucleate; cysts common	Tubulinea	Naked or testate with tubular pseudopodia; lacks centrosomes; motility based on actinomyosin cytoskeleton; cytoplasmic microtubules rare; lacks flagellated stages. Includes *Amoeba, Hydramoeba, Flabellula,* and *Rhizamoeba.*
		Flabellinea	Flattened amoebae lacking tubular pseudopodia; motility based on actinomyosin cytoskeleton; lacks centrosome and flagellated stages. Includes *Dermamoeba, Podostoma, Sappinia,* and *Thecamoeba.*
		Stereomyxida	Branched or reticulate plasmodial organisms; trilaminate centrosome. Includes *Corallomyxe* and *Stereomyxa.*
		Acanthamoebidae	Thin glycocalyx; prominent subpseudopodia that narrow to a blunt or fine tip (acanthopodia); single nucleus; cysts usually have a double wall; motility based on actinomyosin cytoskeleton; centriole-like body observed. Includes *Acanthamoeba, Balamuthia,* and *Protacanthamoeba.*
		Entamoebida	No flagella or centrioles; lacks mitochondria, hydrogenosomes, and peroxisomes. Includes *Entamoeba.*
		Mastigamoebidae	Amoeboid with several pseudopodia; single flagellum projecting forward; single kinetosome and nucleus, although some multinucleate; lacks mitochondria; inhabits low-oxygen to anoxic, nutrient-rich environments. Includes *Mastigella* and *Mastigamoeba.*
		Pelomyxa	Multiple cilia; anaerobic; lack mitochondria, hydrogenosomes, and peroxisomes; polymorphic life cycle with multinucleate stages; some are symbionts. Includes *Pelomyxa paulstris.*
		Eumycetozoa	Amoeboid organisms that produce fruiting bodylike structures; both cellular and acellular slime molds. Includes *Dictyostelium, Physarum, Hemitichia,* and *Stemonitis.*
Rhizaria	Possess thin pseudopodia (filopodia)	Cercozoa	Biciliated and/or amoeboid; most have mitochondria with tubular cristae; many encyst; kinetosomes connected to nucleus with cytoskeleton. Includes *Cercomonas, Katabia, Medusetta,* and *Sagosphaera.*

Taxon	Description
Haplosporidia	Plasmodial endoparasites of marine and freshwater animals; distinct mechanism of spore formation; mitochondria with tubular cristae. Includes *Haplosporidium*, *Minchinia*, and *Urosporidum*.
Foraminifera	Filopodia with granular cytoplasm that forms a complex network of reticulopodia. Simplest forms are open tubes or hollow spheres; in others, shells, called tests, are divided into chambers added during growth. Tests made of organic compounds or inorganic particles cemented together, or crystalline calcite. Includes *Allogromia*, *Carpenteria*, *Globigerinella*, *Lana*, and *Textularia*.
Gromia	Organic tests; branched filipodia; nongranular cytoplasm; flagellated dispersal cells or gametes. Includes *Gromia*.
Radiolaria	Many species exhibit radial symmetry, from which the name is derived. All have a porous capsular cell wall through which axopodia project. Skeletons, when present, made of amorphous silica (opal) or strontium sulfate; morphology can be simple to ornate; mitochondria with tubular cristae; axopodia supported by internal microtubules. Includes *Acanthometra*, *Lophospyris*, and *Stauroceu*.
Chromalveolata — Plastid from secondary endosymbiosis with an ancestral archaeplastid; plastid then lost in some; reacquired in others	
Cryptophyceae	Auto-, mixo-, and heterotrophic forms with ejectosomes or trichocysts (dartlike structures used for defense); mitochondria with flat cristae; biflagellated; tubular channels and/or longitudinal groove lined with ejectosomes; chlorophyll a and c_2 if chloroplasts present. Includes *Campylomonas*, *Cryptomonas*, *Goniomonas*, and *Rhodomonas*.
Haptophyta	Scales cover cell; solitary or colonial; motile cells biflagellate and usually have a haptonema (thin appendage between flagella used for prey capture and at attachment to substrates); outer nuclear membrane continuous with chloroplast membrane; auto-, mixo-, and heterotrophic forms. Includes the coccoliths and *Diacronema*, *Isochrysis*, and *Phaeocystis*.
Stramenopiles	Motile cells usually with two flagella, heterokont flagellation typical. Includes diatoms and labyrinthulids.
Alveolata	Mixo- or heterotrophic. Includes dinoflagellates, *Apixcomplexa* (e.g., *Plasmodium*), and Ciliphora (e.g., *Paramecium* and *Stentor*).
Excavata — Suspension feeding groove (cytostome) present or presumed to have been lost; feed by a flagella-generated current	
Fornicata	Lack typical mitochondria; uninucleate; usually have a feeding groove. Includes *Giardia*.
Malawimonas	Has mitochondria, two kinetosomes, and a single ventral flagellar vane. Includes *Malawinas*.
Parabasalia	Have parabasal structure; striated parabasal fibers connect Golgi to flagellar apparatus; up to thousands of flagella; hydrogenosomes present. Includes *Calonympha*, *Holomastigotes*, *Spirotrichosoma*, and *Trichomonas*.
Preaxostyla	Unicellular; flagellated; no mitochondria; heterotrophic. Includes *Dinenympha*, *Polymastix*, and *Streblomastix*.
Jakobida	Two flagella placed at top of wide ventral feeding groove. Includes *Jakoba* and *Histiona*.
Heterolobosea	Heterotrophic amoebae with eruptive pseudopodia; if flagellated, has 2 or 4 parallel flagella. Includes *Acrasis*, *Gruberella*, and *Rosculus*.
Euglenozoa	One or two flagella inserted into apical or subapical pocket; usually with tubular feeding apparatus; two kinetosomes; mitochondria have discoid cristae. Includes *Dinema*, *Euglena*, *Leishmania*, *Trypanoplasma* and *Trypanosoma*.

[a]Adapted from: Adl, S. M.; Simpson, A. G. B.; Farmer, M. A.; Anderson, R. A.; Anderson, O. R.; Barta, J. R.; Browser, S. S.; et al. 2005. The new higher level classification of Eukaryotes with emphasis on the taxonomy of protists. *J. Eukaryot. Microbiol.* 52,399–451.

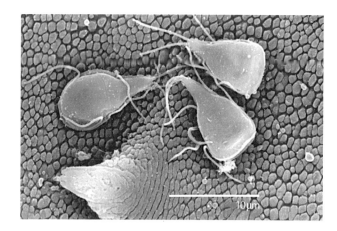

Figure 25.4 **Scanning Electron Micrograph of *Giardia* sp.** This pathogenic protist is shown attached to the parasitic flatworm *Echinostoma caproni.*

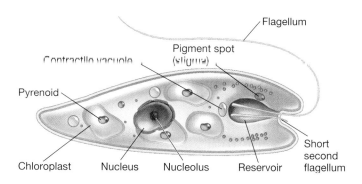

Figure 25.5 *Euglena.* **A Diagram Illustrating the Principle Structures Found in this Euglenoid.** Notice that a short second flagellum does not emerge from the anterior invagination. In some euglenoids both flagella are emergent.

condition known as eutrophication). Only asexual reproduction by binary fission has been observed. Other pathogenic species include *Hexamida salmonis,* a troublesome fish parasite found in hatcheries and fish farms, and *H. meleagridis,* a turkey pathogen that is responsible for the annual loss of millions of dollars in poultry revenue.   Food and waterborne diseases: *Giardiasis* (section 39.5)

Parabasalia

Members of the *Parabasalia* are flagellated; most are endosymbionts of animals. Without a distinct cytostome, they use phagocytosis to engulf food items. Here we consider two subgroups: the *Trichonymphida* and the *Trichomonadida. Trichonymphida* are obligate mutualists in the digestive tracts of wood-eating insects such as termites and wood roaches, where they secrete the enzyme cellulase needed for the digestion of wood. In fact, these protists use pseudopodia to entrap wood particles (figure 25.2). One species, *Trichonympha campanula,* can account for up to one-third of the biomass of an individual termite. This species is particularly large for a protist (several hundred micrometers) and can bear several thousand flagella. Although asexual reproduction is the norm, a hormone called ecdysone produced by the host when molting triggers sexual reproduction.

Trichomonadida, or simply the trichomonads, are symbionts of the digestive, reproductive, and respiratory tracts of many vertebrates, including humans. Four species infect humans: *Dientamoeba fragilis, Pentatrichomonas hominis, Trichomonas tenax,* and *T. vaginalis. D. fragilis* has recently been recognized as a cause of diarrhea, while *T. vaginalis* has long been known to be pathogenic. Found in the genitourinary tract of both men and women, most *T. vaginalis* strains are either not pathogenic or only mildly so. However, the sexual transmission of pathogenic strains results in painful inflammation associated with a whitish-green discharge. Microscopic examination of this discharge reveals very high populations of the protist. *Tritrichomonas foetus* is a cattle parasite and an important cause of spontaneous abortion in these animals. Trichomonads do not require oxygen and possess

hydrogenosomes rather than mitochondria. They undergo asexual reproduction only.   Direct contact diseases: Trichomoniasis (section 39.4)

Euglenozoa

These protists are commonly found in freshwater, although a few species are marine. About one-third of euglenids are photoautotrophic; the remaining are free-living chemoorganotrophs (principally saprotrophic) although a few parasitic species have been described. The representative genus is the photoautotroph *Euglena.* A typical *Euglena* cell (**figure 25.5**) is elongated and bounded by a plasmalemma. The pellicle consists of proteinaceous strips and microtubules; it is elastic enough to enable turning and flexing of the cell, yet rigid enough to prevent excessive alterations in shape. Photosynthetic protists can be characterized by the pigments and carbohydrates they possess. *Euglena* contains chlorophylls *a* and *b* together with carotenoids. The large nucleus contains a prominent nucleolus. The primary storage product is paramylon (a polysaccharide composed of β-1,3 linked glucose molecules), which is unique to euglenoids. A red eye spot or **stigma** helps the organism orient to light and is located near an anterior reservoir. A large contractile vacuole near the reservoir continuously collects water from the cell and empties it into the reservoir, thus regulating the osmotic pressure within the organism. Two flagella arise from the base of the reservoir, although only one emerges from the canal and actively beats to move the cell. Reproduction in euglenoids is by longitudinal mitotic cell division (figure 25.3*d*).

Several protists of medical relevance belong to the *Euglenozoa.* These include the trypanosomes, which exist only as parasites of plants and animals. Trypanosome diseases have global significance. Members of the genus *Leishmania* cause a group of conditions, collectively termed **leishmaniasis,** that include systemic and skin/mucous membrane afflictions affecting some 12 million people. Chagas' disease is caused by *Trypanosoma cruzi,* which is transmitted by "kissing bugs" (*Triatominae*), so called because they bite the face of sleeping victims (*see figure 39.10*). Two to three million citizens of South and Central America show the central and peripheral nervous system dysfunction that is characteristic of this disease; of these, about 45,000 die of the disease each year. *Try-*

panosoma gambiense and *T. rhodesiense* (often considered a sub-species of *T. brucei*) cause African sleeping sickness (**figure 25.6**). Ingestion of these parasites by the blood-sucking tsetse fly triggers a complex cycle of development and reproduction, first in the fly's gut and then in its salivary glands. From there it is easily transferred to a vertebrate host, where it often causes a fatal infection. It is estimated that about 65,000 people die annually of sleeping sickness. The presence of this dangerous parasite prevents the use of about 11 million square kilometers of African grazing land. Arthropod-borne diseases: Leishmaniasis and trypanosomiasis (section 39.3)

African trypanosomes have a thick glycoprotein layer coating the cell wall surface. The chemical composition of the glycoprotein layer is switched cyclically, expressing only one of 1,000 to 2,000 variable antigens at any given time. This process, known as antigenic variation, enables the parasite's escape from host immune surveillance. It is therefore not surprising that there are no vaccines for either Chagas' disease or African sleeping sickness and the few drugs available for treatment are not particularly effective. However, the annotated genome sequences of *T. cruzi* and *T. brucei* were reported in 2005. The release of these genomes allows scientists to compare the parasitic features and the mechanisms by which these protists so successfully evade the host's immune system. This may help identify new drug and vaccine targets. Comparative genomics (section 15.6)

---

1. Why is the classification of chemoorganotrophic protists according to their mode of locomotion no longer considered valid?
2. What are some features that distinguish the *Parabasalia* from the *Euglenozoa*?
3. What is the function of the stigma in *Euglena*? How does this protist maintain osmotic balance?
4. What *Euglenozoa* genera cause disease? What adaptations make these protists successful pathogens?

Super Group *Amoebozoa*

It is clear that the amoeboid form arose independently numerous times from various flagellated ancestors. Thus we see that some ameboid forms are placed in the super group *Amoebozoa* while others are placed in the *Rhizaria*. One of the morphological hallmarks of amoeboid motility is the use of **pseudopodia** (meaning "false feet") for both locomotion and feeding (**figure 25.7**). Pseudopodia can be rounded (**lobopodia**), long and narrow (**filopodia**), or form a netlike mesh (**reticulopodia**). Amoebae that lack a cell wall or other supporting structures and are surrounded

(a)

(b)

Figure 25.6 The Euglenozoan *Trypanosoma* and Its Insect Host. (a) *Trypanosoma* among red blood cells. Note the dark-staining nuclei, anterior flagella, and undulating changeable shape (×500). (b) The tsetse fly, shown here sucking blood from a human arm, is an important vector of the *Trypanosoma* species that causes African sleeping sickness.

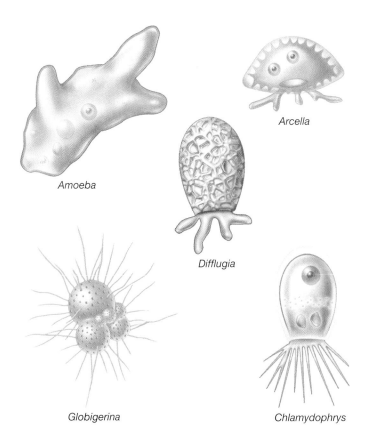

Figure 25.7 Pseudopodia. *Amoeba, Difflugia,* and *Arcella* have lobopodia; *Chlamydophrys* bears filopodia; the foraminiferan *Globigerina* has reticulopodia.

only by a plasma membrane are called **naked amoebae.** In contrast, the plasma membrane of the **testate amoebae** is covered by material that is either made by the protist itself or collected by the organism from the environment. Binary fission is the usual means of asexual division, although some amoebae form cysts that undergo multiple fission (figure 25.3a,b).

Tubulinea

These protists inhabit almost any environment where they will remain moist; this includes glacial meltwater, marine plankton, tidepools, lakes, and streams. Free-living forms are known to dwell in ventilation ducts and cooling towers where they feed on microbial biofilms. Others are endosymbionts, commensals, or parasites of invertebrates, fishes, and mammals. Some harbor intracellular symbionts, including algae, bacteria, and viruses, but the nature of these relationships is not well understood. *Amoeba proteus* (**figure 25.8**), a favorite among introductory biology laboratory instructors, is included in this group.

Entamoebida

Amoebic dysentery is the third leading cause of parasitic death worldwide and is caused by *Entamoeba histolytica*. Individuals acquire this pathogen by eating feces-contaminated food or by drinking water contaminated by *E. histolytica* cysts. These pass unharmed through the stomach and undergo multiple fission when introduced to the alkaline conditions in the intestines. There they not only graze on bacteria, but produce a suite of digestive enzymes that degrade gut epithelial cells. *E. histolytica* can penetrate into the bloodstream and migrate to the liver, lungs, and/or skin. Cysts in feces remain viable for weeks, but are killed by heat greater than 40°C. Food and waterborne diseases: Amebiasis (section 39.5)

Eumycetozoa

Since their first description in the 1880s, the *Eumycetozoa* or "slime molds" have been classified as plants, animals, and fungi. As we examine their morphology and behavior, the source of this confusion should become apparent. Analysis of certain proteins (for example, elongation factor EF-1α, β-tubulin, and actin) as well as physiological, behavioral, biochemical, and developmental data point to a monophyletic group (figure 25.1). The *Eumycetozoa* includes the *Myxogastria* and *Dictyostelia*. The **acellular slime mold,** or *Myxogastria,* life cycle includes a distinctive stage when the organisms exist as streaming masses of colorful protoplasm that creep along in amoeboid fashion over moist, rotting logs, leaves, and other organic matter, which they degrade (**figure 25.9**). Their name derives from the lack of individual cell membranes from which a large, multinucleate mass called a **plasmodium** is formed; there can be as many as 10,000 synchronously dividing nuclei within a single plasmodium (figure 25.9b). Feeding is by endocytosis. When starved or dried, the plasmodium develops ornate fruiting bodies. As these mature, they form stalks with cellulose walls that are resistant to environmental stressors (figure 25.9c,d,e). When conditions improve, spores germinate and release haploid amoeboflagellates. These fuse and as the resulting zygotes feed, nuclear division and synchronous mitotic divisions give rise to the multinucleate plasmodium.

The **cellular slime molds (*Dictyostelia*)** are strictly amoeboid and feed endocytically on bacteria and yeasts. Their complex life cycle involves true multicellularity, despite their primitive evolutionary status (**figure 25.10a**). The species *Dictyostelium discoideum* is an attractive model organism. The vegetative cells move as a mass, sometimes called a pseudoplasmodium, because individual cells retain their cell membranes. When starved, cells release cyclic AMP and a specific glycoprotein, which serve as molecular signals. Other cells sense these compounds and respond by forming an aggregate around the signal-producing cells (figure 25.10b). In this way large, motile, multicellular slugs develop and serve as precursors to fruiting body formation (figure 25.10c). Fruiting body morphogenesis commences when the slug stops and cells pile on top of each other. Cells at the bottom of this vertically oriented structure form a stalk by secreting cellulose, while cells at the tip differentiate into spores (figure 25.10d,e). Germinated spores become vegetative amoebae to start this asexual cycle anew.

Sexual reproduction in *D. discoideum* involves the formation of special spores call macrocysts. These arise by a form of conjugation that has some unusual features. First, a group of amoebae become enclosed within a wall of cellulose. Following conjugation a single, large amoeba forms and cannibalizes the remaining amoebae. The now giant amoeba matures into a macrocyst. Macrocysts can remain dormant within their cellulose walls for extended periods of time. Vegetative growth resumes after the diploid nucleus undergoes meiosis to generate haploid amoebae.

The 33.8-Mb genome of *D. discoideum* has been sequenced and annotated. Analysis of a number of proteins supports the notion that these soil-dwelling microbes are more primitive than the fungi. For instance, *D. discoideum* has 14 different histidine kinase receptor proteins; these proteins are generally thought to be distinctly procaryotic. Also of note is the presence of 40 genes that appear to be involved in cellulose biosynthesis or degradation. These genes could be involved in producing the cellulose the microbe needs during morphological differentiation and/or for degrading cellulose-containing microbes.

Figure 25.8 *Amoeba proteus.* The lobopodia are seen as long projections; the dark-staining nucleus is in the central region of the cell.

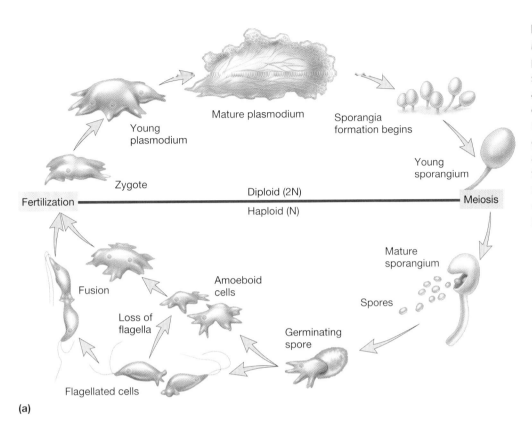

Mature plasmodium

Young plasmodium

Sporangia formation begins

Zygote

Young sporangium

Fertilization

Diploid (2N)

Haploid (N)

Meiosis

Mature sporangium

Spores

Fusion

Amoeboid cells

Loss of flagella

Germinating spore

Flagellated cells

(a)

Figure 25.9 Acellular Slime Molds. (a) The life cycle of a plasmodial slime mold includes sexual reproduction; when conditions are favorable for growth the adult diploid forms sporangia. Following meiosis, the haploid spores germinate, releasing haploid amoeboid or flagellated cells that fuse. (b) Plasmodium of the slime mold *Physarum* sp. (×175). Sporangia of (c) *Physarum polycephalum,* (d) *Hemitrichia,* and (e) *Stemonitis.*

(b) *Physarum* sp.

(c) *Physarum polycephalum*

(d) *Hemitrichia*

(e) *Stemonitis*

Figure 25.10 Development of *Dictyostelium discoideum,* a Cellular Slime Mold. **(a)** Life Cycle. **(b)** Aggregating *D. discoideum* become polar and begin to move in an oriented direction in response to the molecular signal cAMP. **(c)** Slug begins to right itself and **(d)** forms a spore-forming body called a sorocarp. **(e)** Electron micrograph of a sorocarp showing individual spores (×1,800).

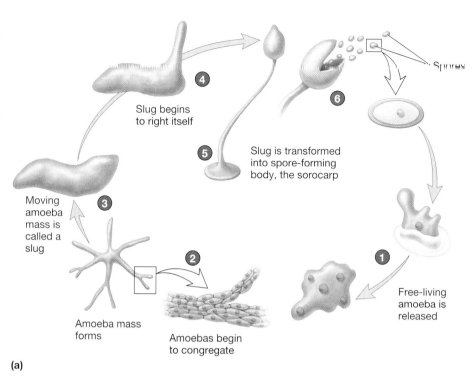

Slug begins to right itself

Moving amoeba mass is called a slug

Slug is transformed into spore-forming body, the sorocarp

Spores

Free-living amoeba is released

Amoeba mass forms

Amoebas begin to congregate

(a)

(b)

(c)

(d)

(e)

Super Group *Rhizaria*

These protists are amoeboid in morphology and thus were historically grouped with the members of what we now call the *Amoebozoa*. However, molecular phylogenetic analysis makes it clear that the *Amoebozoa* and *Rhizaria* are not monophyletic. Morphololgically, the *Rhizaria* can be distinguished by their fine pseudopodia (filopodia), which can be simple, branched, or connected. Filopodia supported by microtubules are known as an **axopodia.** Axopodia protrude from a central region of the cell called the axoplast and are primarily used in feeding (**figure 25.11**).

Radiolaria

Most *Radiolaria* have an internal skeleton made of siliceous material; however, members of the subgroup *Acantharia* have endoskeletons consisting of strontium sulfate. A few genera have an exoskeleton of siliceous spines or scales, while some lack a skeleton completely. Skeletal morphology is highly variable and often includes radiating spines that help the organisms float, as does the storage of oils and other low-density fluids (**figure 25.12**). The skeletal material of ancient radiolarians that settled on the ocean floor millions of years ago remains preserved and is useful to scientists. The strontium sulfate skeletons of acantharians are used to measure the relative amounts of natural versus anthropogenic radioactivity in marine sediments. Siliceous skeletons of radiolaria and diatoms (see *Stramenopiles*, p. 621) form deposits called siliceous ooze, which serve as ancient or paleoenvironmental indicators.

The *Radiolaria* feed by endocytosis using mucus-coated filopodia to entrap prey including bacteria, other protists, and even small invertebrates. Large prey items are partially digested extracellularly before becoming encased in a food vacuole. Many surface-dwelling radiolarians have algal symbionts thought to enhance their net carbon assimilation. Asexual reproduction is found in some forms; the acantharians reproduce only sexually by

(a)

(b) Radiolarian shells

Figure 25.12 Radiolaria. **(a)** The radiolarian *Acanthometra elasticum* demonstrates the internal skeleton. **(b)** Radiolarian shells made of silica.

Figure 25.11 Axopodia. The rhizarian *Actinosphaerium* has needlelike axopodia.

consecutive mitotic and meiotic divisions that release hundreds of biciliated isogametic cells. In others, asexual reproduction (binary or multiple fission, or budding) is most common, but sexual reproduction can be triggered by either nutrient limitation or a heavy feeding. In this case, two haploid nuclei fuse to form a diploid zygote encased in a cyst from which it is released when survival conditions improve.

Foraminifera

The *Foraminifera* (also simply called forams) range in size from roughly 20 μm to several centimeters. Their filopodia are arranged in a branching network called reticulopodia. They have characteristic tests arranged in multiple chambers that are sequentially added as the protist grows (**figure 25.13**).

Reticulopodia bear vesicles at their tips that secrete a sticky substance used to trap prey. Many species harbor endosymbiotic algae that can migrate out of reticulopodia (without being eaten) to expose themselves to more sunlight. It has been shown experimentally that individual forams with algal symbionts grow to larger sizes than those of the same species without symbionts, lending credibility to the notion that these algae contribute significantly to foram nutrition.

Foraminiferan life cycles can be complex. While some smaller species reproduce only asexually by budding and/or multiple fission, larger forms frequently alternate between sexual and asexual phases. During the sexual phase, flagellated gametes pair, fuse, and generate asexual individuals (agamonts). Meiotic division of the agamonts gives rise to haploid gamonts. There are several mechanisms by which gamonts return to the diploid condition. For instance, a variety of forams release flagellated gametes that become fertilized in the open water. In others, two or more gamonts attach to one another, enabling gametes to fuse within the chambers of the paired tests. When the shells separate, newly formed agamonts are released. True autogamy has been observed in at least one genus (*Rotaliella*): each gamont produces gametes that pair and fuse within a single test and the zygote is then released as an agamont.

Foraminifera are found in marine and estuarine habitats. Some forms are planktonic, but most are benthic. Foraminiferin tests accumulate on the sea floor where they constitute a fossil record dating back to the Early Cambrian (543 million years ago), which is helpful in oil exploration. Their remains, or ooze, can be up to hundreds of meters deep in some tropical regions. In Indonesia, the ancient tests are collected and used as paving material. Foram tests make up most of modern-day chalk, limestone, and marble, and are familiar to most as the White Cliffs of Dover in England (**figure 25.14**). They also formed the stones used to build the great pyramids.

---

1. Describe filopodia, lobopodia, and reticulopodia form and function.
2. List at least one mechanism used by *Entamoeba histolytica* to successfully infect a human host.
3. Why do you think the slime molds have been so hard to classify?
4. What is a plasmodium? How does it differ between the acellular and cellular slime molds?

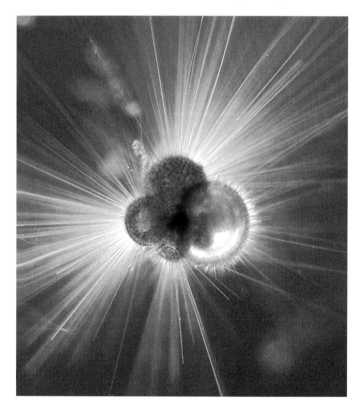

Figure 25.13 A Foraminiferan. The reticulopodia are seen projecting through pores in the calcareous test, or shell, of this protist.

Figure 25.14 White Cliffs of Dover. The limestone that forms these cliffs is composed almost entirely of fossil shells of protists, including foraminifera.

5. Describe the life cycle of *Dictyostelium discoideum*. Why is this organism a good model for the study of cellular differentiation, coordinated cell movement, and chemotaxis?
6. What adaptations do the planktonic radiolaria have to help them float?
7. Compare the means by which radiolaria use axopodia with the way foraminifera use reticulopodia.
8. Describe the forms of sexual reproduction in the *Foraminifera*.

Super Group *Chromalveolata*

The *Chromalveolata* are diverse and include autotrophic, mixotrophic, and heterotrophic protists. They are united in plastid origin, which appears to have been acquired by endosymbiosis with an ancestral archaeplastid (which itself acquired plastids from endosymbiotic cyanobacteria). Here we introduce three subgroups: the *Alveolata, Stramenopiles,* and *Haptophyta,* some members of which have been previously considered to be orders or super groups.

Alveolata

The *Alveolata* is a large group that includes the *Apicomplexa, Dinoflagellata* (dinoflagellates), and the *Ciliophora*. We begin our discussion with the *Apicomplexa*. All **apicomplexans** are either intra- or intercellular parasites of animals and are distinguished by a unique arrangement of fibrils, microtubules, vacuoles, and other organelles, collectively called the apical complex, which is located at one end of the cell (**figure 25.15a**). This unique combination of organelles is designed to penetrate host cells. Motility (flagellated or amoeboid) is confined to the gametes and zygotes of a few species.

Apicomplexans have complex life cycles in which certain stages sometimes occur in one host and other stages occur in a different host. The life cycle has both asexual (clonal) and sexual phases and is characterized by an alternation of haploid and diploid generations. The clonal and sexual stages are haploid, except for the zygote. The motile, infective stage is called the **sporozoite.** When this haploid form infects a host, it differentiates into a gamont; male and female gamonts pair and undergo multiple fission, which produces many gametes. Released gametes pair, fuse, and form zygotes. Each zygote secretes a protective covering and is then considered a spore. Within the spore, the nucleus undergoes meiosis (restoring the haploid condition) followed by mitosis to generate eight sporozoites ready to infect a new host (figure 25.15b).

A number of apicomplexans are important infectious agents. The most significant is *Plasmodium,* which causes malaria in some 500 million people annually, with a yearly death toll of 1 to 3 million (*see figure 39.5*). *Eimera* is the causative agent of cecal coccidiosis in chickens, a condition that costs hundreds of millions of dollars in lost animals each year in the United States. Toxoplasmosis, caused by members of the genus *Toxoplasma,* is transmitted either by consumption of undercooked meat or by fecal contamination from a cat's litterbox. Cryptosporidia are responsible for cryptosporidiosis, an infection that begins in the intestines but can disseminate to other parts of the body. Cryptosporidiosis has become problematic for AIDS patients and other immunocom-

(a)

(b)

Figure 25.15   The Apicomplexan Cell.   (a) The vegetative cell, or merozoite, illustrating the apical complex, which consists of the polar ring, conoid, rhoptries, subpellicular microtubules, and micropore. **(b)** The infective oocyte of *Eimeria*. The oocyst is the resistant stage and has undergone multiple fission after zygote formation (sporogony).

promised individuals.   Arthropod borne diseases: Malaria (section 39.3); Direct contact diseases: Toxoplasmosis (section 39.4); Food and waterborne diseases: Cryptosporidiosis (section 39.5)

Recently the genomes of the apicomplexans *Theilaria parva* and *Theilaria annulata* were sequenced and annotated. These tick-borne parasites cause diseases marked by rapid proliferation of white blood cells (lymphoproliferation). *T. parva* infects cattle and causes a fatal disease called East Coast Fever. This disease kills over a million cattle each year in sub-Saharan Africa, costing over $200 million and targeting farmers who can least afford such

(a) *Stentor*

(b) *Stylonychia*

10 µm

200 µm

(c)   Paramecium

Contractile vacuole
(partially full)

Cilia

Pellicle

Food vacuoles

Oral groove

Macronucleus

Micronucleus

Gullet

Anal pore

Cytoplasm

Contractile vacuole
(full)

Figure 25.17   The *Ciliophora*.   (a) *Stentor,* a large, vase-shaped, freshwater protozoan. **(b)** Two *Stylonychia* conjugating. **(c)** Structure of *Paramecium,* adjacent to an electron micrograph.

called a **frustule.** Diatom frustules are composed of two halves or thecae that overlap like a petri dish (**figure 25.19***a*). The larger half is the **epitheca,** and the smaller half is the **hypotheca.** Diatom frustules are composed of crystallized silica [$Si(OH)_4$] with very fine markings (figure 25.19*b*). They have distinctive, and often exceptionally beautiful, patterns that are unique for each species. Frustule morphology is very useful in diatom identification, and diatom frustules have a number of practical applications (**Techniques & Applications 25.2**). The fine detail and precise morphology of these frustules has made them attractive for nanotechnology.   Microbes as products: Nanotechnology (section 41.8)

Although the majority of diatoms are strictly photoautotrophic, some are facultative chemoorganotrophs, absorbing carbon-containing molecules through the holes in their walls. The vegetative cells of diatoms are diploid and can be unicellular, colonial, or filamentous. They lack flagella and have a single, large nucleus and smaller plastids. Reproduction consists of the organism dividing asexually, with each half then constructing a new theca within the old one. Because the epitheca and hypotheca are of different sizes, each time the hypotheca is used as a template to construct a new theca, the diatom gets smaller. However, when a cell has diminished to about 30% of its original size, sexual reproduction is usually triggered. The diploid vegetative cells

undergo meiosis to form gametes, which then fuse to produce a zygote. The zygote develops into an auxospore, which increases in size again and forms a new wall. The mature auxospore eventually divides mitotically to produce vegetative cells with frustules of the original size.

Diatoms are found in freshwater lakes, ponds, streams, and throughout the world's oceans. Marine planktonic diatoms produce 40 to 50% of the organic carbon in the ocean; they are therefore very important in global carbon cycling. In fact, marine diatoms are thought to contribute as much fixed carbon as all rain forests combined.   Biogeochemical cycling: Carbon cycle (section 27.2)

The complete genome of the marine planktonic diatom *Thalassiosira pseudonana* was recently sequenced. Its genome consists of 24 chromosomes (34 Mb), a plastid genome (0.139 Mb), and a mitochondrial genome (0.044 Mb). It contains novel genes for silicic acid transport and silica-based cell wall synthesis. Also discovered were genes for scavenging iron, multiple nitrate and ammonium transporters, and the enzymes for urea metabolism. The annotation of this genome will help scientists discover how this microbe has adapted so successfully to its nutrient-limited environment.

A group of protists once considered true fungi and traditionally called **öomycetes,** meaning "egg fungi," were recently assigned the name *Peronosporomycetes*. They differ from true fungi in a num-

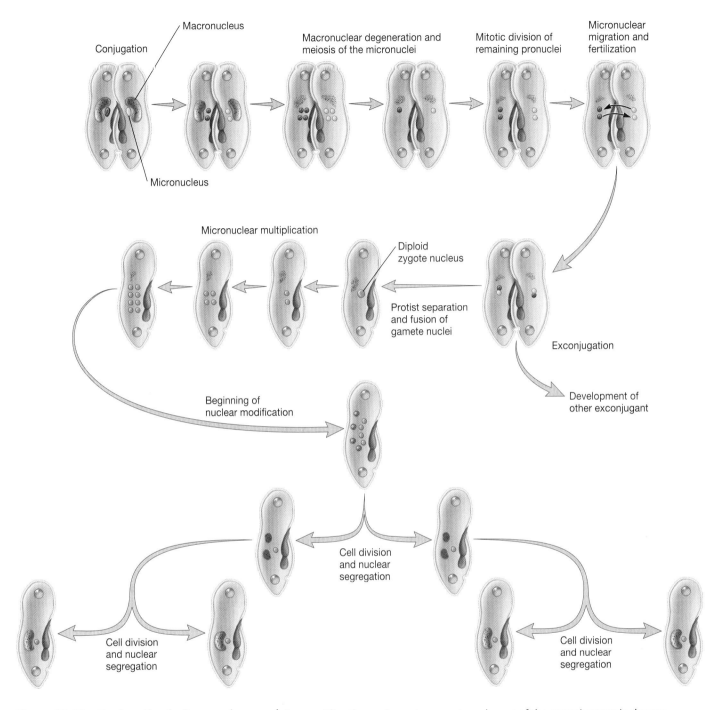

Figure 25.18   Conjugation in *Paramecium caudatum*.   After the conjugants separate, only one of the exconjugants is shown; however, a total of eight new protists result from each conjugation.

ber of features including their cell wall composition (cellulose and β-glucan instead of chitin) and the fact that they are diploid throughout their life cycle. When undergoing sexual reproduction, they form a relatively large egg cell (öogonium) that is fertilized by either a sperm cell or a smaller gametic cell (called an antheridium) to produce a zygote. When the zygote germinates, the asexual zoospores display heterokont flagellation. The *Fungi:* Reproduction (section 26.5)

Peronosporomycetes such as *Saprolegnia* and *Achlya* are saprophytes that grow as cottony masses on dead algae and animals, mainly in freshwater environments. Some öomycetes are parasitic on the gills of fish. *Peronospora hyoscyami* is responsible for "blue mold" on tobacco plants and grape downy mildew is caused by *Plasmopara viticola*. Certainly the most famous öomycete is *Phytophthora infestans*, which attacked the European potato crop in the mid-1840s, spawning the Irish famine.

Figure 25.19 Diatoms.
(a) The silaceous epitheca and hypotheca of the diatom *Cyclotella meneghiniana* fit together like a petri dish.
(b) A variety of diatoms show the intricate structure of the silica cell wall.

(a) *Cyclotella meneghiniana*   (b)

Techniques & Applications

25.2   Practical Importance of Diatoms

Diatoms have both direct and indirect economic significance for humans. Because diatoms make up most of the phytoplankton of the cooler parts of the ocean, they are the most important ultimate source of food for fish and other marine animals in these regions. It is not unusual for 1 liter of seawater to contain almost a million diatoms. They are also extremely important for the biochemical cycling of silica and as contributors to global fixed carbon.

When diatoms die, their frustules sink to the bottom. Because the siliceous part of the frustule is not affected by the death of the cell, diatom frustules tend to accumulate at the bottom of aquatic environments. These form deposits of material called diatomaceous earth. This material is used as an active ingredient in many commercial preparations, including detergents, fine abrasive polishes, paint removers, decolorizing and deodorizing oils, and fertilizers. Diatomaceous earth also is used extensively as a filtering agent, as a component in insulating (firebrick) and soundproofing products, and as an additive to paint to increase the night visibility of signs and license plates.

The use of diatoms as indicators of water quality and of pollution tolerance is becoming increasingly important. Specific tolerances for given species to various environmental parameters (concentrations of salts, pH, nutrients, nitrogen, temperature) have been compiled.

Diatomaceous earth can also be used to control insects. Insects have their soft body parts exposed but covered by a waxy film to prevent dehydration. When they contact the diatoms in diatomaceous earth the silica frustules break the waxy film on the insects, causing them to dehydrate and die. Insects cannot build up resistance to diatomaceous earth, and it can be fed to poultry, livestock, and pets with no ill effects.

The original classification of *P. infestans* as a fungus was misleading and for decades farmers attempted to control its growth with fungicide, to which it is (of course) resistant. This protist continues to take its toll; potato blight costs some $5 billion annually worldwide.

Labyrinthulids also have a complex taxonomic history: like the *Peronosporomycetes* they were formerly considered fungi. However, molecular phylogenetic evidence combined with the observation that they form heterokont flagellated zoospores places them among the *Stramenopiles*. The more familiar, nonflagellated stage of the life cycle features spindle-shaped cells that form complex colonies that glide rapidly along an ectoplasmic net made by the organism. This net is actually an external network of calcium-dependent contractile fibers made up of actinlike proteins that facilitate the movement of cells. Their feeding mechanism is like that of fungi: osmotrophy aided by the production of extracellular degradative enzymes. In marine habitats, the genus *Labyrinthula* grows on plants and algae and is thought to play a role in the "wasting disease" of eelgrass, an important intertidal plant. The *Fungi: Nutrition and metabolism (section 26.4)*

Haptophyta

One interesting subgroup of the haptophyta is the *Coccolithales*. These photosynthetic protists bear ornate calcite scales called coccoliths (**figure 25.20**). Together with the *Foraminifera*, the **coccolithophores** precipitate calcium carbonate ($CaCO_3$) in the open ocean, thereby influencing Earth's carbon budget. Cells are usually biflagellated and possess a unique organelle called a haptonema, which is somewhat similar to a flagellum but differs in microtubule arrangement. One species, *Emiliania huxleyi*, has been studied extensively. Like all coccolithophores, it is planktonic. High concentrations or blooms of *E. huxleyi* can significantly alter nutrient flux by emitting sulfur (as dimethyl sulfide) to the atmosphere and sequestering calcium carbonate in the sediments. Other coccolithophore species are known to cause toxic blooms. Marine and freshwater environments: Nutrient cycling (section 28.1)

Figure 25.20 The Haptophyte *Emiliana huxleyi*. Note the ornamental scales made of calcite.

Super Group *Archaeplastida*

The *Archaeplastida* includes all organisms with a photosynthetic plastid that arose through an ancient endosymbiosis with a cyanobacterium. It thus includes all higher plants as well as many protist species. Microbial evolution: Endosymbiotic origin of mitochondria and chloroplasts (section 19.1)

Chloroplastida

The *Chloroplastida* are often referred to as green algae [Greek *chloros*, green]. These phototrophs grow in fresh and salt water, in soil, on other organisms, and within other organisms. They have chlorophylls *a* and *b* along with specific carotenoids, and they store carbohydrates such as starch. Many have cell walls made of cellulose. They exhibit a wide diversity of body forms, ranging from unicellular to colonial, filamentous, membranous or sheetlike, and tubular types (**figure 25.21**). Some species have a holdfast structure that anchors them to the substratum. Both asexual and sexual reproduction are observed.

Chlamydomonas is a member of the subgroup *Chlorophyta* (**figure 25.22**). Individuals have two flagella of equal length at the anterior end by which they move rapidly in water. Each cell has a single haploid nucleus, a large chloroplast, a conspicuous pyrenoid, and a stigma (eyespot) that aids the cell in phototactic responses. Two small contractile vacuoles at the base of the flagella function as osmoregulatory organelles. *Chlamydomonas* reproduces asexually by producing zoospores through cell division. Sexual reproduction occurs when some products of cell division act as gametes and fuse to form a four-flagellated, diploid zygote that ultimately loses its flagella and enters a resting phase. Meiosis occurs at the end of this resting phase and produces four haploid cells that give rise to adults.

(a) *Chlorella*

(d) *Acetabularia*

(b) *Volvox*

(e) *Micrasterias*

(c) *Spirogyra*

Figure 25.21 *Chlorophyta* (Green Algae); Light Micrographs.
(a) *Chlorella*, a unicellular nonmotile Chlorophyte (×160). **(b)** *Volvox*, which demonstrates colonial growth (×450). **(c)** *Spirogyra* (×100). Four filaments are shown. Note the ribbonlike, spiral chloroplasts within each filament. **(d)** *Acetabularia*, the mermaid's wine goblet. **(e)** *Micrasterias* (×150).

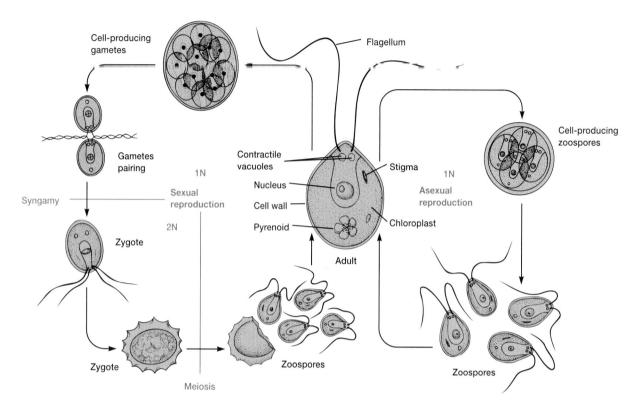

Figure 25.22 *Chlamydomonas:* **The Structure and Life Cycle of this Motile Green Alga.** During asexual reproduction, all structures are haploid; during reproduction, only the zygote is diploid.

Another *Chlorophyta, Chlorella* (figure 25.21a), is widespread both in fresh- and saltwater and also in soil. It only reproduces asexually and lacks flagella, eyespots, and contractile vacuoles; the nucleus is very small. Motile, colonial organisms such as *Volvox* represent another line of evolutionary specialization. A *Volvox* colony (figure 25.21b) is a hollow sphere made up of a single layer of 500 to 60,000 individual cells, each resembling a *Chlamydomonas* cell. The flagella of all the cells beat in a coordinated way to rotate the colony in a clockwise direction as it moves through the water. Only a few cells are reproductive and these are located at the posterior end of the colony. Some divide asexually and produce new colonies; others produce gametes. After fertilization, the zygote divides to form a daughter colony. In both cases the daughter colonies stay within the parental colony until it ruptures.

The chlorophyte *Prototheca moriformis* causes the disease protothecosis in humans and animals. *Prototheca* cells are fairly common in the soil and it is from this site that most infections occur. Severe systemic infections, such as massive invasion of the bloodstream, have been reported in animals. The subcutaneous type of infection is more common in humans. It starts as a small lesion and spreads slowly through the lymph glands, covering large areas of the body.

1. Explain the unique structural features of the diatoms. How does their morphology play a role in the alternation between asexual and sexual reproduction?
2. Why do you think the öomycetes and the *Labyrinthulids* were formerly considered fungi?
3. What is the ecological importance of the coccolithophores?
4. Compare the morphology and swimming behavior of *Chlamydomonas* and *Volvox*.

Summary

25.1 Distribution

a. Protists are found wherever other eucaryotic organisms exist.
b. They are important components of many terrestrial, aquatic, and marine ecosystems where they contribute to nutrient cycling. Many are parasitic in humans and animals (**table 25.1**) and some have become very useful in the study of molecular biology.

25.2 Nutrition

a. Photosynthetic protists are aerobic and perform oxygenic photosynthesis. They can be either photoautotrophic or photoheterotrophic.
b. Chemoorganotrophic protists (protozoa) can be either holozoic or saprozoic. Holozoic forms use phagocytosis to entrap and consume food particles (**figure 25.2**).

25.3 Morphology

a. Because protists are eucaryotic cells, in many respects their morphology and physiology resemble those of multicellular plants and animals. However, because all of their functions must be performed within the individual protist, many morphological and physiological features are unique.

b. The protistan cell membrane is called the plasmalemma and the cytoplasm can be divided into the ectoplasm and endoplasm. The cytoplasm contains contractile, secretory, and food (phagocytic) vacuoles (**figures 25.5** and **25.15**).

c. Energy metabolism occurs within mitochondria, hydrogenosomes, or chloroplasts. Some primitive chemoorganotrophic protists are fermentative, lacking mitochondria and hydrogenosomes.

25.4 Encystment and Excystment

a. Some protists can secrete a resistant covering and go into a resting stage (encystment) called a cyst. Cysts protect the organism against adverse environments, function as a site for nuclear reorganization, and serve as a means of transmission in parasitic species.

25.5 Reproduction

a. Most protists reproduce asexually by binary or multiple fission or budding (**figure 25.3**).

b. Some also use sexual reproduction via a variety of sexual reproductive strategies including conjugation, syngamy, and autogamy (**figure 25.18**).

25.6 Protist Classification

a. Protist phylogeny is the subject of active research and debate. The classification scheme presented here is that of the International Society of Protistologists (**table 25.2**).

b. The super group *Excavata* includes the *Fornicata*, the *Parabasalia*, and the *Euglenozoa*. Most have a cytoproct and use a flagellum for suspension feeding.

c. The human pathogen *Giardia* is a member of the *Fornicata*. It is considered one of the most primitive eucaryotes (**figure 25.4**).

d. Most *Parabasalia* are flagellated endosymbionts of animals. They include the obligate mutualists of wood-eating insects such as *Trichonympha* and the human pathogen *Trichomonas*.

e. Many of the members of the *Euglenozoa* are photoautotrophic (**figure 25.5**). The remainder are chemoorganotrophs, of which most are saprotrophic. Important human pathogens include members of the genus *Trypanosoma*. Trypanosomes cause a number of important human diseases including leishmaniasis, Chagas' disease, and African sleeping sickness (**figure 25.6**).

f. Amoeboid forms use pseudopodia, which can be lobopodia, filopodia, or reticulopodia (**figure 25.7**). Amoebae that bear external plates are called testate; those without plates are called naked or atestate amoebae.

g. Members of the *Amoebozoa* subclasses *Tubulinea* and *Entamoebida* include a number of endosymbiotic and parasitic protists, including *Entamoeba histolytica*, a major cause of parasitic death worldwide.

h. The *Amoebozoa* subclass *Eumycetozoa* includes the acellular and cellular slime molds, which were previously classified as plants, animals, or fungi. The acellular slime molds form a large mass of protoplasm, called a plasmodium, in which individual cells lack a cell membrane (**figure 25.9**). The cellular slime molds produce a pseudoplasmodium and each cell within has a cell wall. *Dictyostelium discoideum* is a cellular slime mold that is used as a model organism in the study of chemotaxis, cellular development, and behavior (**figure 25.10**).

i. The super group *Rhizaria* are amoeboid forms that include the *Radiolaria*, which have filopodia, and the *Foraminifera*, which bear netlike reticulopodia and tests that can be ornate. Most foraminifera are benthic and their tests accumulate on the ocean floor where they are useful in oil exploration (**figures 25.11** and **25.13**).

j. The super group *Chromalveolata* is diverse. It includes the *Alveolata*, which consists of the apicomplexans, dinoflagellates, the stramenopiles and the ciliophora.

k. Apicomplexans are parasitic with complex life cycles. The motile, infective stage is called the sporozoite (**figure 25.15**). The most important apicomplexan is *Plasmodium*, which causes malaria.

l. The dinoflagellates are a large group of nutritionally complex protists. Most are marine and planktonic. They are known for their phosphorescence and for causing toxic blooms. Symbiotic forms live in association with reef-building corals (**figure 25.16**).

m. The *Ciliophora* are chemorganotrophic protists that use cilia for locomotion and feeding (**figure 25.17**). In addition to asexual reproduction, conjugation is used in sexual reproduction (**figure 25.18**).

n. The *Stramenopila*, is extremely diverse and includes diatoms, golden and brown algae, the öomycetes, and labyrinthulids. Diatoms are found in fresh- and saltwater and are important components of marine plankton (**figure 25.19**). The öomycetes and labyrinthulids were once thought to be fungi.

o. The haptophytes include the coccolithophores, planktonic photosynthetic protists that contribute to the global carbon budget by precipitating calcium carbonate for their ornate scales (**figure 25.20**).

p. The *Archaeplastida* include the *Chlorophyta*, also known as green algae. All are photosynthetic with chlorophylls *a* and *b* along with specific carotenoids. They exhibit a wide range of morphologies (**figure 25.21**).

Key terms

acellular slime mold (Myxogastria) 614
algae 605
anisogamy 609
apicomplexan 619
autogamy 609
axopodia 617
binary fission 608
cellular slime mold (*Dictyostelia*) 614
chromosomal nuclei 609
ciliates (*Ciliophora*) 620
coccolithophore 624
conjugant 620
conjugation 609
contractile vacuole 607
cyst 608

cytoproct 608
cytostome 608
diatoms (*Bacillariophyta*) 621
dinoflagellates (*Dinoflagellata*) 620
ectoplasm 607
encystment 608
endoplasm 607
epitheca 622
excystment 608
filopodia 613
frustule 622
gamonts 609
heterokont flagella 621
holozoic nutrition 606
hydrogenosome 608
hypotheca 622

isogamy 609
kinetosome 608
labyrinthulid 624
leishmaniasis 612
lobopodia 613
macronucleus 609
micronucleus 609
mixotrophy 607
naked amoebae 614
öomycete 622
osmotrophy 607
ovular nuclei 609
pellicle 607
phagocytic vacuole 608
phycology 605
planktonic 606

plasmalemma 607
plasmodium 614
protistology 605
protozoa 605
protozoology 605
pseudopodia 613
pyrenoid 608
reticulopodia 613
saprozoic nutrition 607
sporozoite 619
stigma 612
syngamy 609
testate amoebae 614
trophozoite 608
vesicular nucleus 609
zooxanthellae 620

Critical Thinking Questions

1. Why do you think our knowledge of the biology of protists has lagged so far behind that of procaryotes, fungi, and higher eucaryotes?

2. Encystment is usually triggered by changes in the environment. How do you think protists perceive these changes? How might this be similar or different from endospore formation in gram-positive bacteria?

3. Suggest why, in some protists, the cytoplasmic material (ectoplasm) just under the plasma membrane is so rigid?

4. Which of the protists discussed in this chapter do you think are the most evolutionarily advanced or derived? Explain your reasoning.

5. Vaccine development for diseases caused by protists (e.g., malaria, Chagas disease) has been much less successful than that for bacterial diseases. Discuss one biological reason and one geopolitical reason for this fact.

Learn more

Adl, S. M.; Simpson, A. G. B.; Farmer, M. A.; Anderson, R. A.; Anderson, O. R.; Barta, J. R.; Bowser, S. S.; *et al.* 2005. The new higher level classification of Eukaryotes with emphasis on the taxonomy of protists. *J. Eukaryot. Microbiol.* 52: 399–451.

Armbrust, E. V.; Berges, J. A.; Bowler, C.; *et al.* 2004. The genome of the diatom *Thalassiosira pseudonana:* Ecology, evolution, and metabolism. *Science* 306: 79–86.

Baldauf, S. L.; Roger, A. J.; Senk-Siefert, I.; and Doolittle, W. F. 2000. A kingdom-level phylogeny of eukaryotes based on combined protein data. *Science* 290: 972–76.

Biron, D.; Libros, P.; Sagi, D.; Mirelman, D.; and Moses, E. 2001. "Midwives" assist dividing amoebae. *Nature* 410: 973–77.

Falkowski, P. G.; Katz, M. E.; Knoll, A. H.; Quigg, A.; Raven, J. A.; Schofield, O.; and Taylor, F. J. R. 2004. The evolution of modern eukaryotic phytoplankton. *Science* 305: 354–60.

Gardner, M. J.; Bishop, R.; Shah, T.; *et al.* 2005. Genome sequence of *Theileria parva,* a bovine pathogen that transforms lymphocytes. *Science* 309: 134–37.

Grell, K. B. 1973. *Protozoology.* New York: Springer-Verlag.

Lee, J. J.; Leedale, G. F.; and Bradbury, P., editors. 2000. *An illustrated guide to the protozoa,* 2d ed. Lawrence, Kansas: Society of Protozoologists.

Lipscomb, D. L.; Farris, J. S.; Kallersjo, M.; and Tehler, A. 1998. Support, ribosomal sequences and the phylogeny of the eukaryotes. *Cladistics* 14: 303–38.

Roberts, L. S., and Javovy, J. J. 2005. *Foundations in parasitology,* 7th ed. Dubuque, Iowa. McGraw-Hill Higher Education.

Sogin, M. L., and Silberman, J. D. 1998. Evolution of the protists and protistan parasites from the perspective of molecular systematics. *Int. J. Parasitol.* 28: 11–20.

Please visit the Prescott website at www.mhhe.com/prescott7 for additional references.

26

The *Fungi* (*Eumycota*)

This is a scanning electron micrograph of the microscopic, unicellular yeast, *Saccharomyces cerevisiae* (×21,000). *S. cerevisiae* is the most thoroughly investigated eucaryotic microorganism. This has led to a better understanding of the biology of the eucaryotic cell. Today it serves as a widely used biotechnological production organism as well as a eucaryotic model system.

PREVIEW

- Fungi are widely distributed and are found wherever moisture is present. They are of great importance to humans in both beneficial and harmful ways.

- Fungi exist primarily as filamentous hyphae. A mass of hyphae is called a mycelium.

- Like some bacteria and protists, fungi digest insoluble organic matter by secreting exoenzymes, then absorbing the solubilized nutrients.

- Two reproductive structures occur in fungi: (1) sporangia form asexual spores, and (2) gametangia form sexual gametes.

- Like the study of protists, fungal systematics is an area of active research. Eight fungal subdivisions are presented here, including the *Chytridiomycetes, Zygomycota, Ascomycota, Basidiomycota, Urediniomycetes, Ustilaginomycetes, Glomeromycota,* and *Microsporidia.*

- The *Chytridiomycetes* are a group of terrestrial and aquatic fungi that reproduce by motile zoospores with single, posterior, whiplash flagella.

- The *Zygomycota* are characterized by resting structures called zygospores—cells in which zygotes are formed.

- The *Ascomycota* form zygotes within a characteristic saclike structure, the ascus. The ascus contains two or more ascospores.

- Yeasts are unicellular fungi—most are ascomycetes.

- Basidiomycetes possess dikaryotic hyphae, one of each mating type. The hyphae divide uniquely, forming basidiocarps within which club-shaped basidia can be found. The basidia bear two or more basidiospores.

- The *Ustilaginomycetes* and *Urediniomycetes* include important plant pathogens, whereas the *Glomeromycota* form important associations with vascular plants and enhance plant nutrient uptake. Some members of the *Microsporidia* are considered emerging pathogens of humans.

In this chapter we introduce the *Fungi*. Like protists, fungi have a long and confused taxonomic history. Their relatively simple morphology, wide diversity, and lack of a fossil record limit the value of traditional taxonomic approaches. More recently, the application of molecular techniques including sequence comparisons of small subunit rRNA and conserved proteins has offered new insights into fungal evolution. For example, the division Deuteromycetes (also known as *fungi imperfecti*) is no longer recognized. Here we present eight fungal groups: the *Chytridiomycetes, Zygomycota, Ascomycota, Basidiomycota, Urediniomycetes, Ustilaginomycetes, Glomeromycota,* and *Microsporidia* (**figure 26.1**). The *Urediniomycetes* and the *Ustilaginomycetes* are commonly considered *Basidiomycota*. However, recent evidence suggests that they may be taxonomically distinct.

Microbiologists use the term **fungus** [pl., fungi; Latin *fungus*, mushroom] to describe eucaryotic organisms that are spore-bearing, have absorptive nutrition, lack chlorophyll, and reproduce sexually and asexually. Scientists who study fungi are **mycologists** [Greek *mykes*, mushroom, and *logos*, science], and the scientific discipline devoted to fungi is called **mycology.** The study of fungal toxins and their effects is called **mycotoxicology,** and the diseases caused by fungi in animals are known as **mycoses** (s., mycosis). According to the universal phylogenetic tree, fungi are members of the domain *Eucarya* (figure 26.1*a*). Morphological, biochemical and molecular phylogenetic analyses demonstrate

Yeasts, molds, mushrooms, mildews, and the other fungi pervade our world. They work great good and terrible evil. Upon them, indeed, hangs the balance of life; for without their presence in the cycle of decay and regeneration, neither man nor any other living thing could survive.

—Lucy Kavaler

(a)

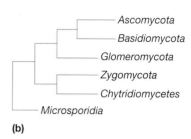

(b)

Figure 26.1 The *Fungi*. **(a)** The Universal Tree of Life, an unrooted tree, divides fungi so that with the exception of the *Microsporidia,* all are closely related to the *Metazoa (Homo). Coprinus* is shown as the representative fungal group. The *Microsporidia* have a confused taxonomic history but are now considered by most to be fungi. **(b)** A rooted tree showing most of the fungal groups discussed in this chapter. Unlike the unrooted three (a), here the *Microsporidia* are shown to be closely related to other fungal groups. Many consider *Urediniomycetes* and the *Ustilaginomycetes* to be *Basidiomycota.*

that the *Fungi* constitute a monophyletic group. They are sometimes referred to as the true fungi or ***Eumycota*** [Greek *eu,* true, and *mykes,* fungus].

26.1 DISTRIBUTION

Fungi are primarily terrestrial organisms, although a few are freshwater or marine. They have a global distribution from polar to tropical regions. Many are pathogenic and infect plants and animals. Fungi also form beneficial relationships with other organisms. For example, the vast majority of vascular plant roots form associations (called mycorrhizae) with fungi. Fungi also are found in the upper portions of many plants. These endophytic fungi affect plant reproduction and palatability to herbivores. Lichens are associations of fungi and photosynthetic protists or cyanobacteria. Microorganism associations with vascular plants: Mycorrhizae (section 29.5); Microbial interactions (section 30.1)

26.2 IMPORTANCE

About 90,000 fungal species have been described; however, some estimates suggest that 1.5 million species may exist. Fungi are important to humans in both beneficial and harmful ways. With bacteria and a few other groups of chemoorganotrophic organisms, fungi act as decomposers, a role of enormous significance. They degrade complex organic materials in the environment to simple organic compounds and inorganic molecules. In this way carbon, nitrogen, phosphorus, and other critical constituents of dead organisms are released and made available for living organisms. Microorganisms in the soil environment (section 29.3)

On the other hand, fungi are a major cause of disease. Plants are particularly vulnerable to fungal diseases because fungi can invade leaves through their stomates (**figure 26.2**). Over 5,000 species attack economically valuable crops, garden plants, and many wild plants. Fungi also cause many diseases of animals (**table 26.1**) and humans. In fact, about 20 new human fungal pathogens are documented each year.

Fungi, especially the yeasts, are essential to many industrial processes involving fermentation. Examples include the making of bread, wine, and beer. Fungi also play a major role in the preparation of some cheeses, soy sauce, and sufu; in the commercial production of many organic acids (citric, gallic) and certain drugs (ergometrine, cortisone); and in the manufacture of many antibiotics (penicillin, griseofulvin) and the immunosuppressive drug cyclosporin. These topics are discussed more fully in chapters 34 and 40.

Finally, fungi are important research tools in the study of fundamental biological processes. Cytologists, geneticists, biochemists, biophysicists, and microbiologists regularly use fungi in their research. The yeast *Saccharomyces cerevisiae* is the best understood eucaryotic cell. It has been a valuable model organism in the study of cell biology, genetics, and cancer.

Figure 26.2 Fungal Pathogens of Plants. Scanning electron micrograph of a young mycelial aggregate forming over a leaf stoma (×1,000).

26.3 STRUCTURE

The body or vegetative structure of a fungus is called a **thallus** [pl., thalli]. It varies in complexity and size, ranging from the single-cell microscopic yeasts to multicellular molds, macroscopic puffballs, and mushrooms (**figure 26.3**). The fungal cell usually is encased in a cell wall of **chitin.** Chitin is a strong but flexible nitrogen-containing polysaccharide consisting of *N*-acetylglucosamine residues.

A **yeast** is a unicellular fungus that has a single nucleus and reproduces either asexually by budding and transverse division or sexually through spore formation. Each bud that separates can grow into a new yeast, and some group together to form colonies. Generally yeast cells are larger than bacteria, vary considerably in size, and are commonly spherical to egg shaped. They lack flagella but possess most of the other eucaryotic organelles (**figure 26.4**).

The thallus of a **mold** consists of long, branched, threadlike filaments of cells called **hyphae** [s., hypha; Greek *hyphe,* web] that form a **mycelium** (pl., mycelia), a tangled mass or tissuelike aggregation of hyphae (**figure 26.5**). In some fungi, protoplasm streams through hyphae, uninterrupted by cross walls. These hyphae are

Table 26.1	Some Mycotoxicoses[a] Produced by Fungal Mycotoxins in Domestic Animals			
Disease	**Fungus**	**Mycotoxin**	**Contaminated Foodstuff**	**Animals Affected**
Aflatoxicosis	*Aspergillus flavus*	Aflatoxins	Rice, corn, sorghum, cereals, peanuts, soybeans	Poultry, swine, cattle, sheep, dogs
Ergotism	*Claviceps purpurea*	Ergot alkaloids	Seedheads of many grasses, grains	Cattle, horses, swine, poultry
Mushroom poisoning	*Amanita verna*	Amanitins	Eaten from pastures	Cattle
Poultry hemorrhagic syndrome	*Aspergillus flavus* and others	Aflatoxins	Toxic grain and meal	Chickens
Slobbers	*Rhizoctonia*	Alkaloid slaframine	Red clover	Sheep, cattle
Tall fescue toxicosis	*Acremonium coenophialum* (an endophytic fungus)	Ergot alkaloids	Endophyte-infected tall fescue plants	Cattle, horses

[a]A mycotoxicosis [pl., mycotoxicoses] is a poisoning caused by a fungal toxin.

(a) *Penicillium*

(b) *Lycoperdon*

(c) A mushroom

Figure 26.3 Fungal Thalli. (a) The multicellular common mold, *Penicillium,* growing on an apple. **(b)** A large group of puffballs, *Lycoperdon,* growing on a log. **(c)** A mushroom is made of densely packed hyphae that form the mycelium or visible structure (thallus).

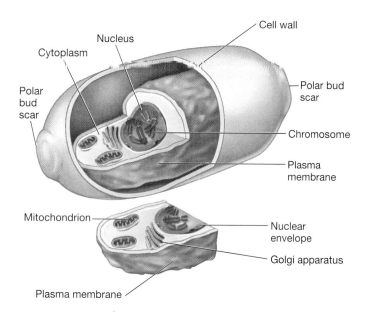

Figure 26.4   A Yeast.   Diagrammatic drawing of a yeast cell showing typical morphology. For clarity, the plasma membrane has been drawn separated from the cell wall. In a living cell the plasma membrane adheres tightly to the cell wall.

Figure 26.5   Mold Mycelia.   The hyphae that compose the fungal mycelium can form a macroscopic mass, as shown by this basidiomycete growing in and on soil.

called **coenocytic** or aseptate (**figure 26.6***a*). The hyphae of other fungi (figure 26.6*b*) have cross walls called **septa** (s., septum) with either a single pore (figure 26.6*c*) or multiple pores (figure 26.6*d*) that enable cytoplasmic streaming. These hyphae are termed **septate.**

Hyphae are composed of an outer cell wall and an inner lumen, which contains the cytosol and organelles (**figure 26.7**). A plasma membrane surrounds the cytoplasm and lies next to the cell wall. The filamentous nature of hyphae results in a large surface area relative to the volume of cytoplasm. This makes adequate nutrient absorption possible.

Many fungi, especially those that cause diseases in humans and animals, are dimorphic (**table 26.2**)—that is, they have two forms. Dimorphic fungi can change from the yeast (Y) form in the animal to the mold or mycelial form (M) in the external environment in response to changes in various environmental factors (nutrients, CO_2 tension, oxidation-reduction potentials, temperature). This shift is called the **YM shift.** In plant-associated fungi the opposite type of dimorphism exists: the mycelial form occurs in the plant and the yeast form in the external environment.

1. How can a fungus be defined?
2. With what organisms do fungi associate? What does the global distribution of fungi imply about the diversity of this group? Explain your answer.
3. What does the term coenocytic mean? Consider the fact that the genomes of coenocytic fungi are not separated by septa. How might this affect clonal growth?
4. What are some forms represented by different fungal thalli?
5. What organelles would you expect to find in the cytoplasm of a typical fungus?
6. What are the major differences between a yeast and a mold?

26.4   Nutrition and Metabolism

Fungi grow best in dark, moist habitats where there is little danger of desiccation, but they are found wherever organic material is available. Most fungi are **saprophytes,** securing their nutrients from dead organic material. Like many bacteria and protists, fungi release hydrolytic exoenzymes that digest external substrates. They then absorb the soluble products—a process sometimes called **osmotrophy.** They are chemoorganoheterotrophs and use organic compounds as a source of carbon, electrons, and energy. Glycogen is the primary storage polysaccharide in fungi. Most fungi use carbohydrates (preferably glucose or maltose) and nitrogenous compounds to synthesize their own amino acids and proteins.

Fungi usually are aerobic. Some yeasts, however, are facultatively anaerobic and can obtain energy by fermentation. Many fungal fermentations are of industrial importance, such as the production of ethyl alcohol in the manufacture of beer and wine. Obligately anaerobic fungi are found in the rumen of cattle.

26.5   REPRODUCTION

Reproduction in fungi can be either asexual or sexual. Asexual reproduction is accomplished in several ways:

1. A parent cell can undergo mitosis and divide into two daughter cells by a central constriction and formation of a new cell wall (**figure 26.8***a*).
2. Mitosis in vegetative cells may be concurrent with budding to produce a daughter cell. This is very common in the yeasts.
3. The most common method of asexual reproduction is spore production. Asexual spore formation occurs in an individual

Figure 26.6 Hyphae. Drawings of **(a)** coenocytic hyphae (aseptate) and **(b)** hyphae divided into cells by septa. **(c)** Electron micrograph ($\times$40,000) of a section of *Drechslera sorokiniana* showing wall differentiation and a single pore. **(d)** Drawing of a multiperforate septal structure.

Table 26.2	Some Medically Important Dimorphic Fungi
Fungus	**Disease**[a]
Blastomyces dermatitidis	Blastomycosis
Candida albicans	Candidiasis (Thrush)
Coccidioides immitis	Coccidioidomycosis
Histoplasma capsulatum	Histoplasmosis
Sporothrix schenckii	Sporotrichosis
Paracoccidioides brasiliensis	Paracoccidioidomycosis

[a]See chapter 40 for a discussion of each of these diseases.

Figure 26.7 Hyphal Morphology. Diagrammatic representation of a hyphal tip showing typical organelles and other structures.

fungus through mitosis and subsequent cell division. There are several types of asexual spores, each with its own name:

a. A hypha can fragment (by the separation of hyphae through splitting of the cell wall or septum) to form cells that behave as spores. These cells are called **arthroconidia** or **arthrospores** (figure 26.8*b*).

b. If the cells are surrounded by a thick wall before separation, they are called **chlamydospores** (figure 26.8*c*).

c. If the spores develop within a sac (**sporangium;** pl., sporangia) at a hyphal tip, they are called **sporangiospores** (figure 26.8*d*).

d. If the spores are not enclosed in a sac but produced at the tips or sides of the hypha, they are termed **conidiospores** (figures 26.8*e* and 26.13).

e. Spores produced from a vegetative mother cell by budding (figure 26.8*f*) are called **blastospores.**

Sexual reproduction in fungi involves the fusion of compatible nuclei. Homothallic fungal species are self-fertilizing and produce sexually compatible gametes on the same mycelium.

Heterothallic species require outcrossing between different but sexually compatible mycelia. It has long been held that sexual reproduction must occur between mycelia of opposite **mating types (MAT).** However, one instance of same-sex mating was discovered following an outbreak of the pathogenic yeast *Crytococcus gatti* in Canada. Depending on the species, sexual fusion may occur between haploid gametes, gamete-producing bodies called **gametangia,** or hyphae. Sometimes both the cytoplasm and haploid nuclei fuse immediately to produce the diploid zygote. Usually, however, there is a delay between cytoplasmic and nuclear fusion. This produces a **dikaryotic stage** in which cells contain two separate haploid nuclei (N + N), one from each parent (**figure 26.9**). After a period of dikaryotic existence, the two nuclei fuse and undergo meiosis to yield spores. For example, in the zygomycetes the zygote develops into a **zygospore** (**figure 26.10**); in the ascomycetes, an **ascospore** (figure 26.14); and in the basidomycetes; a **basidiospore** (figure 26.15).

Fungal spores are important for several reasons. The spores enable fungi to survive environmental stresses such as desiccation, nutrient limitation, and extreme temperatures, although they are not as stress resistant as bacterial endospores. Because they are often small and light, spores can remain suspended in air for long periods. Thus they frequently aid in fungal dissemination, a significant factor that helps explain the wide distribution of many fungi. Fungal spores often spread by adhering to the bodies of insects and other animals. The bright colors and fluffy textures of many molds often are due to their aerial hyphae and spores. The

Figure 26.8 **Diagrammatic Representation of Asexual Reproduction in the Fungi and Some Representative Spores.**
(a) Transverse fission. **(b)** Hyphal fragmentation resulting in arthroconidia (arthrospores) and **(c)** chlamydospores. **(d)** Sporangiospores in a sporangium. **(e)** Conidiospores arranged in chains at the end of a conidiophore. **(f)** Blastospores are formed from buds off of the parent cell.

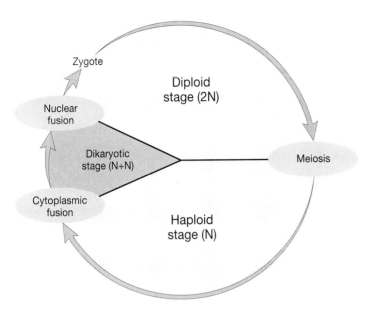

Figure 26.9 Reproduction in Fungi. A drawing of the generalized life cycle for fungi showing the alternation of haploid and diploid stages. Some fungal species do not pass through the dikaryotic stage indicated in this drawing. The asexual (haploid) stage is used to produce spores that aid in the dissemination of the species. The sexual (diploid) stage involves the formation of spores that survive adverse environmental conditions (e.g., cold, dryness, heat).

size, shape, color, and number of spores are useful in the identification of fungal species.

---

1. What are saprophytes? How do fungi usually obtain their nutrients?
2. Fungi were originally classified as plants. Why do you think early 20th-century biologists made this mistake?
3. How is asexual reproduction accomplished in the fungi? sexual reproduction?
4. Describe each of the following types of asexual fungal spores: sporangiospore, conidiospore, and blastospore.
5. How are fungi dispersed in the environment?

26.6 CHARACTERISTICS OF THE FUNGAL DIVISIONS

The long evolutionary history of fungi is rich in examples of **convergent** and **divergent evolution.** That is to say, many structurally and functionally similar structures evolved independently (converged) while other structures became dissimilar (diverged) over time. Without an extensive fossil record, the use of morphology in phylogenetic analysis is limited. However, sequence analysis of 18S rRNA and certain protein-coding genes has shown that the *Fungi* comprise a monophyletic group with eight subdivisions. Four of these—the *Chytridiomycetes, Zygomycota, Ascomycota,* and *Basidiomytota*—have been recognized as separate groups for some time. The other four—the *Urediniomycetes, Ustilaginomycetes, Glomeromycota,* and *Microsporidia*—have

Figure 26.10 The *Zygomycota.* Diagrammatic representation of the life cycle of *Rhizopus stolonifer.* Both the sexual and asexual phases are illustrated.

only recently been proposed as separate groups. **Table 26.3** surveys all eight groups; we focus most of our discussion on the better-known groups.

Chytridiomycota

The simplest of the fungi belong to the *Chytridiomycota,* or **chytrids.** They are unique among fungi in the production of a zoospore with a single, posterior, whiplash flagellum. This is considered a primitive feature that was lost in more evolved fungi. Free-living members of this taxon are saprotrophic—living on plant or animal matter in freshwater, mud, or soil. Parasitic forms infect aquatic plants and animals including insects. A few are found in the anoxic rumen of herbivores. Based on zoospore morphology, the subdivisions within the chytridiomycetes include the *Blastocladiales, Monoblepharidales, Neocallimastigaceae, Spizellomycetales,* and the *Chytridiales.*

Chytridiomycota display a variety of life cycles involving both asexual and sexual reproduction. Members of this group are microscopic in size and may consist of a single cell, a small multinucleate mass, or a true mycelium with hyphae capable of penetrating porous substrates. Sexual reproduction results in the release of sporangiospores from sporangia (figure 26.8d) at the surface. Many are capable of degrading cellulose and even keratin, which enables the degradation of crustacean exoskeletons. The genus *Allomyces* is used to study morphogenesis.

Zygomycota

The *Zygomycota* contains fungi called **zygomycetes.** Most live on decaying plant and animal matter in the soil; a few are parasites of

Table 26.3	Abbreviated Classification of the Fungi as proposed by International Society of Protistologists[a]	
Subclass	**Characteristics**	**Examples**
Chytridiomycetes	Flagellated cells in at least one stage of life cycle; may have one or more flagella. Cell walls with chitin and β-1,3-1,6-glucan; glycogen is used as a storage carbohydrate. Sexual reproduction often results in a zygote that becomes a resting spore or sporangium; saprophytic or parasitic. Currently six major subdivisions.	*Allomyces* *Blastocladiella* *Coelomomyces* *Physoderma* *Synchytrium*
Zygomycota	Thalli usually filamentous and nonseptate, without cilia; sexual reproduction gives rise to thick-walled zygospores that are often ornamented. Includes seven subdivisions: *Basidiobolus, Dimargaritales, Endogonales, Entomophthorales, Harpellales, Kickxellales, Mucorales,* and *Zoopagales.* Human pathogens found among the *Mucorales* and *Entomophthorales.*	*Amoebophilus* *Mucor* *Phycomyces* *Rhizopus* *Thamnidium*
Ascomycota	Sexual reproduction involves meiosis of a diploid nucleus in an ascus, giving rise to haploid ascospores; most also undergo asexual reproduction with the formation of conidiospores with specialized aerial hyphae called conidiophores. Many produce asci within complex fruiting bodies called ascocarps. Includes saprophytic, parasitic forms; many form mutualisms with phototrophic microbes to form lichens. Four monophyletic subdivisions including: *Saccharomycetes, Pezizomycotina, Taphrinomycotina,* and *Neolecta.*	*Ascobolus* *Aspergillis* *Candida* *Crinula* *Neurospora* *Penicillium* *Pneumocystis* *Saccharomyces*
Basidiomycota	Includes many common mushrooms and shelf fungi. Sexual reproduction involves formation of a basidium (small, club-shaped structure that typically forms spores at the ends of tiny projections) within which haploid basidiospores are formed. Usually 4 spores per basidium, but can range from 1 to 8. Sexual reproduction involves fusion with opposite mating type resulting in a dikaryotic mycelium with parental nuclei paired but not initially fused. No subdivisions recognized.	*Agaricus* *Boletes* *Dacrymyces* *Lycoperdon* *Polyporus* *Russula* *Tremella*
Urediniomycetes	Mycelial or yeast forms. Sexual reproduction involves fusion of parental nuclei in probasidium followed by meiosis in a separate compartment. Many are plant pathogens called rusts, animal pathogens, nonpathogenic endophytes, and rhizosphere species. No subdivisions recognized.	*Caeoma* *Melampsora* *Uromyces*
Ustilaginomycetes	Plant parasites that cause rusts and smuts. Mycelial in parasitic phase; meiospores formed on septate or aseptate basidia; cell wall principally composed of glucose. No subdivisions recognized.	*Malassezia* *Tilletia* *Ustilago*
Glomeromycota	Filamentous, most are endomycorrhizal, arbuscular; lack cilium; form asexual spores outside of host plant; lack centrioles, conidia, and aerial spores. No subdivisions recognized.	*Acaulospora* *Entrophospora* *Glomus*
Microsporidia	Obligate intracellular parasites usually of animals. Lack mitochondria, peroxisomes, kinetosomes, cilia, and centrioles; spores have an inner chitin wall and outer wall of protein; produce a tube for host penetration. Subdivisions currently uncertain.	*Amblyospora* *Encephalitozoon* *Enterocytozoon* *Nosema*

[a]Adapted from: Adl, S. M.; Simpson, A. G. B.; Farmer, M. A.; Anderson, R. A.; Anderson, O. R.; Barta, J. R.; Bowser, S. S.; et al. 2005. The new higher level classification of Eukaryotes with emphasis on the taxonomy of protists. *J. Eukaryot. Microbiol.* 52:399–451.

plants, insects, other animals, and humans. The hyphae of zygomycetes are coenocytic, with many haploid nuclei. Asexual spores, usually wind dispersed, develop in sporangia at the tips of aerial hyphae. Sexual reproduction produces tough, thick-walled zygotes called zygospores that can remain dormant when the environment is too harsh for growth of the fungus.

The bread mold, *Rhizopus stolonifer,* is a very common member of this division. This fungus grows on the surface of moist, carbohydrate-rich foods, such as breads, fruits, and vegetables. On breads, for example, *Rhizopus*'s hyphae rapidly cover the surface. Hyphae called rhizoids extend into the bread, and absorb nutrients (figure 26.10). Other hyphae (stolons) become erect, then arch back into the substratum forming new rhizoids. Still others remain erect and produce at their tips asexual sporangia filled with the black spores, giving the mold its characteristic color. Each spore, when liberated, can germinate to start a new mycelium.

Rhizopus usually reproduces asexually, but if food becomes scarce or environmental conditions unfavorable, it begins sexual reproduction. Sexual reproduction requires compatible strains of opposite mating types (figure 26.10). These have traditionally been labeled + and − strains because they are not morphologically distinguishable as male and female. When the two mating strains are close, hormones are produced that cause their hyphae to form projections called **progametangia** [Greek *pro,* before], and then mature gametangia. After fusion of the gametangia, the nuclei of the two gametes fuse, forming a zygote. The zygote develops a thick, rough, black coat and becomes a dormant zygospore. Meiosis often occurs at the time of germination; the zygospore then splits open and produces a hypha that bears an asexual sporangium and the cycle begins anew.

The genus *Rhizopus* is also important because it is involved in the rice disease known as seedling blight. If one considers that rice feeds more people on Earth than any other crop, the implications of this disease are obvious. It was thought that *Rhizopus* secreted a toxin that kills rice seedlings, so scientists set about isolating the toxin and the genes that produce it. Much to everyone's surprise, a β-proteobacterium, *Burkholderia,* found growing within the fungus produces the toxin. Although the evolutionary history of the *Rhizopus-Burkholderia* association remains to be clarified, there is at least one interesting twist to this story: the same toxin has been shown to stop cell division in some human cancer cells and is now being investigated as an antitumor agent.

The zygomycetes also contribute to human welfare. For example, one species of *Rhizopus* is used in Indonesia to produce a food called tempeh from boiled, skinless soybeans. Another zygomycete (*Mucor* spp.) is used with soybeans in Asia to make a curd called sufu. Others are employed in the commercial preparation of some anesthetics, birth control agents, industrial alcohols, meat tenderizers, and the yellow coloring used in margarine and butter substitutes. Microbiology of food (chapter 40)

Ascomycota

The *Ascomycota* contain fungi called **ascomycetes,** commonly known as sac fungi. Ascomycetes are ecologically important in freshwater, marine, and terrestrial habitats because they degrade many chemically stable organic compounds including lignin, cellulose, and collagen. Many species are quite familiar and economically important (**figure 26.11**). For example, most of the red, brown, and blue-green molds that cause food spoilage are ascomycetes. The powdery mildews that attack plant leaves and the fungi that cause chestnut blight and Dutch elm disease are ascomycetes. Many yeasts as well as edible morels and truffles are ascomycetes. The

pink bread mold *Neurospora crassa,* also an ascomycete, is an extremely important research tool in genetics and biochemistry.

Many ascomycetes are parasites on higher plants. *Claviceps purpurea* parasitizes rye and other grasses, causing the plant disease **ergot. Ergotism,** the toxic condition in humans and animals who eat grain infected with the fungus, is often accompanied by gangrene, psychotic delusions, nervous spasms, abortion, and convulsions. During the Middle Ages ergotism, then known as St. Anthony's fire, killed thousands of people. For example, over 40,000 deaths from ergot poisoning were recorded in France in the year 943. It has been suggested that the widespread accusations of witchcraft in Salem Village and other New England communities in the late 1690s may have resulted from outbreaks of ergotism. The pharmacological activities are due to an active ingredient, lysergic acid diethylamide (LSD). In controlled dosages other active compounds can be used to induce labor, lower blood pressure, and ease migraine headaches.

The ascomycetes are named for their characteristic reproductive structure, the saclike **ascus** [pl., asci; Greek *askos,* sac]. Many ascomycetes are yeast. The term yeast is used to refer to unicellular fungi that reproduce asexually by either budding or binary fission (**figure 26.12***a*); the life cycle of the yeast *Saccharomyces cerevisiae* is well understood. This ascomycete alternates between haploid and diploid states (figure 26.12*b*). As long as nutrients remain plentiful, haploid and diploid cells undergo mitosis to produce haploid and diploid daughter cells, respectively. Each daughter cell leaves a scar on the mother cell as it separates, and daughter cells bud only from unscarred regions of the cell wall. When a mother cell has no more unscarred cell wall remaining, it can no longer reproduce and will senesce (die). When nutrients are limited, diploid *S. cerevisiae* cells undergo meiosis to produce four haploid cells that remain bound within a common cell wall, the acsus. Upon the addition of nutrients, if two haploid cells of opposite mating types (*a* and α) come into contact, they will fuse to create a diploid. Typically only cells of opposite mating types can fuse; this process is tightly regulated by the action of pheromones.

Filamentous ascomycetes form septate hyphae. Asexual reproduction is common in these ascomycetes and takes place by means of conidiospores (**figure 26.13**). Sexual reproduction also

(a) *Morchella esculenta*

(b) *Sarcoscypha coccinea*

(c) *Tuber brumale*

Figure 26.11 The *Ascomycota*. **(a)** The common morel, *Morchella esculenta,* is one of the choicest edible fungi. It fruits in the spring. **(b)** Scarlet cups, *Sarcoscypha coccinea,* with open ascocarps (apothecia). **(c)** The black truffle, *Tuber brumale,* is highly prized for its flavor by gourmet cooks. Truffles are mycorrhizal associations on oak trees.

(a) *Saccharomyces cerevisiae:* budding division

Figure 26.12 The Life Cycle of the Yeast *Saccharomyces cerevisiae*. **(a)** Budding division results in asymmetric septation and the formation of a smaller daughter cell. **(b)** When nutrients are abundant, haploid and diploid cells undergo mitosis and grow vegetatively. When nutrients are limited, diploid *S. cerevisiae* cells undergo meiosis to produce four haploid cells that remain bound within a common cell wall, the ascus. Upon the addition of nutrients, two haploid cells of opposite mating types (*a* and α) fuse to create a diploid cell.

(b) *S.cereviseae* life cycle

Figure 26.13 Asexual Reproduction in *Ascomyota*. Characteristic conidiospores of *Aspergillus* as viewed with the electron microscope (×1,200).

involves ascus formation, with each ascus usually bearing eight haploid ascospores, although some species can produce over 1,000 (**figure 26.14a**). In the more complex ascomycetes, ascus formation is preceded by the development of special **ascogenous hyphae** into which pairs of nuclei migrate (figure 26.14b). One

nucleus of each pair originates from a "male" mycelium (**antheridium**) or cell and the other from a "female" organ or cell (**ascogonium**) that has fused with it. As the ascogenous hyphae grow, the paired nuclei divide so that there is one pair of nuclei in each cell. After the ascogenous hyphae have matured, nuclear fusion occurs at the hyphal tips in the ascus mother cells. The diploid zygote nucleus then undergoes meiosis, and the resulting four haploid nuclei divide mitotically again to produce a row of eight nuclei in each developing ascus. These nuclei are walled off from one another. Thousands of asci may be packed together in a cup- or flask-shaped fruiting body called an **ascocarp** (figure 26.14b). When the ascospores mature, they often are released from the asci with great force. If the mature ascocarp is jarred, it may appear to belch puffs of "smoke" consisting of thousands of ascospores. Upon reaching a suitable environment, the ascospores germinate and start the cycle anew.

The genomes of three *Aspergillus* species have been sequenced, annotated, and compared. This ascomycete genus is important for a number of reasons. *A. fumigatus* is ubiquitous, commonly found in homes and in the environment. It is known to trigger allergic responses and is implicated in the increased incidence in severe asthma and sinusitis. It is also pathogenic, infecting immunocompromised individuals with a mortality rate of nearly 50%. Its 29.4-Mb genome consists of eight chromosomes and about 10,000 genes. *A. nidulans* is a model organism that is

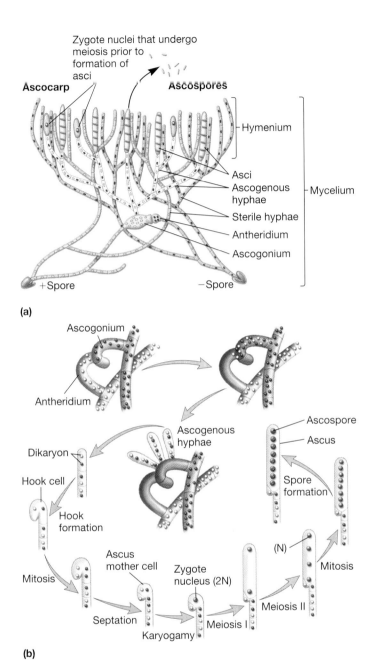

+Spore   −Spore

(a)

Ascogonium
Antheridium
Ascogenous hyphae
Dikaryon
Hook cell
Hook formation
Mitosis
Ascus mother cell
Septation
Karyogamy
Zygote nucleus (2N)
Meiosis I
Meiosis II
(N)
Mitosis
Spore formation
Ascus
Ascospore

(b)

Figure 26.14   The Typical Life Cycle of a Filamentous Ascomycete. Sexual reproduction involves the formation of asci and ascospores. Within the ascus, karyogamy is followed by meiosis to produce the ascospores. **(a)** Sexual reproduction and ascocarp morphology of a cup fungus. **(b)** The details of sexual reproduction in ascogenous hyphae. The nuclei of the two mating types are represented by unfilled and filled circles.

used to study questions of eucaryotic cell and developmental biology. *A. oryzae* is used in the production of traditional fermented foods and beverages in Japan including saki and soy sauce. Because it secretes many industrially useful proteins and can be genetically manipulated, it has become an important organism in biotechnology. Its metabolic and genetic versatility is reflected in the size of its genome—at 37 Mb, it is considerably larger than

either *A. fumigatus* or *A. nidulans*. Much of this additional sequence is involved in the production of secreted hydrolytic enzymes, nutrient transport systems, and secondary metabolites.

Comparative analysis of the three genomes reveals over 5,000 nonprotein-coding regions that are conserved in all three species. A variety of regulatory elements are present, including a riboswitch and other forms of translational control. These genome sequences will be useful to scientists seeking to understand the interaction between *Aspergillus* and the immune system, its role in food and industrial microbiology, and eucaryotic evolution. Regulation at the level of translation: Riboswitches (section 12.4); Comparative genomics (section 15.6); Microbiology of fermented foods (section 40.6)

Basidiomycota

The *Basidiomycota* includes the **basidiomycetes,** commonly known as club fungi. Examples include jelly fungi, rusts, shelf fungi, stinkhorns, puffballs, toadstools, mushrooms, and bird's nest fungi. Basidiomycetes are named for their characteristic structure or cell, the **basidium,** which is involved in sexual reproduction (**figure 26.15**). A basidium [Greek *basidion,* small base] is produced at the tip of hyphae and normally is club shaped. Two or more basidiospores are produced by the basidium, and basidia may be held within fruiting bodies called **basidiocarps.**

The basidiomycetes affect humans in many ways. Most are saprophytes that decompose plant debris, especially cellulose and lignin. For example, the common fungus *Polyporus squamosus* forms large, shelflike structures that project from the lower portion of dead trees, which they help decompose. The fruiting body can reach 2 feet in diameter and has many pores (hence the name *Polyporus*), each lined with basidia that produce basidiospores. Thus a single fruiting body can produce millions of spores. Many mushrooms are used as food throughout the world. The cultivation of the mushroom *Agaricus campestris* is a multimillion-dollar business (*see figure 40.20*). Of course not all mushrooms are edible; as suggested by its name, ingestion of *Russula emetica* induces vomiting.

Many mushrooms produce specific alkaloids that act as either poisons or hallucinogens. One such example is the "death angel" mushroom, *Amanita phalloides.* Two toxins isolated from this species are phalloidin and α-amanitin. Phalloidin primarily attacks liver cells where it binds to plasma membranes, causing them to rupture and leak their contents. Alpha-amanitin attacks the cells lining the stomach and small intestine and is responsible for the severe gastrointestinal symptoms associated with mushroom poisoning.

The basidiomycete *Cryptococcus neoformans* is an important human and animal pathogen. It produces the disease called cryptococcosis, a systemic infection primarily involving the lungs and central nervous system. The production of an elaborate capsule is an important virulence factor for the microbe. Analysis of the *C. neoformans* 19-Mb genome has uncovered over 30 new genes involved in capsule biosynthesis. This may help researchers develop new antifungal agents. Airborne diseases (section 39.2)

Urediniomycetes and *Ustilaginomycetes*

Often considered *Basidiomycota*, both the *Urediniomycetes* and the *Ustilaginomycetes* include important plant pathogens causing

26.6 Characteristics of the Fungal Divisions

a. *Chytridiomycota* produce motile spores. Most are saprophytic; some reside in the rumen of herbivores.

b. *Zygomycota* are coenocytic. Most are saprophytic. One example is the common bread mold, *Rhizopus stolonifer.* Sexual reproduction occurs through a form of conjugation involving + and − strains (**figure 26.10**).

c. *Ascomycota* are known as the sac fungi because they form a sac-shaped reproductive structure called an ascus (**figure 26.11**). In asexual reproduction they produce characteristic conidia (**figure 26.12**). Sexual reproduction involves strains of different mating types (**figure 26.13**).

d. *Basidiomycota* are the club fungi. They are named after their basidium that produces basidiospores (**figure 26.15**).

e. Urediniomycetes and Ustilaginomycetes are often considered *Basidiomycota.* Genera from both groups include important plant pathogens (**figure 26.16**).

f. *Glomeromycota* include the mycorrhizal fungi that grow in association with plant roots. They serve to increase plant nutrient uptake.

g. *Microsporidia* have a unique morphology and are still sometimes considered protists. They include virulent human pathogens; some infect other vertebrates and insects (**figure 26.17**).

Key Terms

antheridium 638	basidiocarp 639	convergent evolution 635	mold 631	septa 632
appressorium 640	basidiomycetes 639	dikaryotic stage 634	mycelium 631	septate 632
arthroconidia 633	basidiospore 634	divergent evolution 635	mycologist 629	sporangiospore 633
arthrospore 633	basidium 639	ergot 637	mycology 629	sporangium 633
ascocarp 638	blastospore 633	ergotism 637	mycosis 629	thallus 631
ascogenous hypha 638	chitin 631	*Eumycota* 630	mycorrhizal fungi 640	yeast 631
ascogonium 638	chlamydospore 633	fungus 629	mycotoxicology 629	YM shift 632
ascomycetes 637	chytrids 635	gametangium 634	osmotrophy 632	zygomycetes 635
ascospore 634	coenocytic 631	hypha 631	progametangium 636	zygospore 634
ascus 637	conidiospore 633	mating type (MAT) 634	saprophyte 632	

Critical Thinking Questions

1. What are some logical targets to exploit in treating animals or plants suffering from fungal infections? Are they different from the targets you would use when treating infections caused by bacteria? By viruses? Explain.

2. Fungi tend to reproduce asexually when nutrients are plentiful and conditions are favorable for growth, but reproduce sexually when environmental or nutrient conditions are not favorable. Why is this an evolutionarily important and successful strategy?

3. Because asexual spores are such a rapid way of reproducing for some fungi, what adaptive "use" is there for an additional sexual phase?

4. Some fungi can be viewed as coenocytic organisms that exhibit little differentiation. When differentiation does occur, such as in the formation of reproductive structures, it is preceded by septum formation. Why does this occur?

5. Both bacteria and fungi are major environmental decomposers. Obviously competition exists in a given environment, but fungi usually have an advantage. What characteristics specific to the fungi provide this advantage?

Learn More

Adl, S. M.; Simpson, A. G. B.; Farmer, M. A.; Anderson, R. A.; Anderson, O. R.; Barta, J. R.; Bowser, S. S.; *et al.* 2005. The new higher level classification of Eukaryotes with emphasis on the taxonomy of protists. *J. Eukaryot. Microbiol.* 52: 399–451.

Alexopoulos, C. J.; Mims, C. W.; Blackwell, M. 1996. *Introductory mycology,* 4th ed. New York: John Wiley and Sons.

Barr, D. J. S. 1992. Evolution and kingdoms of organisms from the perspective of a mycologist *Mycologia* 84:1–11.

Carlile, M. J., and Gooday, G. W. 2001. The *Fungi,* 2d ed. New York: Academic Press.

Feofilova, E. P. 2001. The Kingdom Fungi: Heterogeneity of physiological and biochemical properties and relationships with plants, animals, and prokaryotes (Review). *Appl. Biochem. Microbiol.* 37:124–37.

Griffin, D. H. 1996. *Fungal physiology,* 2d ed. New York: Wiley-Liss.

Loftus, B. J.; Fung, E.; Roncaglia, P.; *et al.* 2005. The genome of the basidiomycetous yeast and human pathogen *Cryptococcus neoformans. Science* 307: 1321–24.

Mueller, G. M.; Bills, G. F.; and Foster, M. S., editors. 2004. *Biodiversity of Fungi: Inventory and monitoring methods.* New York: Elsevier Academic Press.

Ostergaard, S.; Olsson, L.; and Nielson, J. 2000. Metabolic engineering of *Saccharomyces cerevisiae. Microbiol. Mol. Biol. Rev.* 64(1):34–50.

Partida-Martinez, L. P., and Herweck, C. 2005. Pathogenic fungus harbours endosymbiotic bacteria for toxin production. *Nature* 437: 884–88.

Scholte, E.-J.; Ng'habi, K.; Kihonda, J.; Takken, W.; Paaijmans, K.; Abdulla, S.; Killeen, G. F.; and Knols, B. G. J. 2005. An entomopathogenic fungus for control of adult African malaria mosquitoes. *Science* 308: 1641–42.

27

Biogeochemical Cycling and Introductory Microbial Ecology

Microorganisms living in environments where most known organisms cannot survive are important for understanding microbial diversity. These strands of iron-oxidizing *Ferroplasma* were discovered growing at pH 0 in an abandoned mine near Redding, California. This hardy microorganism has only a plasma membrane to protect itself from the rigors of this harsh environment.

PREVIEW

- Life on Earth would not be possible without microbes. Despite their importance, less than 1% of all microbial species have been cultured, identified, and studied.

- Microbial ecology is the study of community dynamics and the interaction of microbes (both procaryotic and eucaryotic) with each other and with plants, animals, and the environment in which they live.

- Energy, electrons, and nutrients must be available in a suitable physical environment for microorganisms to function. Microbes interact with their environment to obtain energy (from light or chemical sources), electrons, and nutrients, which leads to a process called biogeochemical cycling. Microorganisms change the physical state and mobility of many nutrients.

- Biogeochemical cycling refers to the biological and chemical processes that elements such as carbon, nitrogen, sulfur, iron, and magnesium undergo during microbial metabolism. Life on Earth would not be possible without these microbial activities.

- Microorganisms are an important part of ecosystems—self-regulating biological communities and their physical environments. The microbial loop describes the many functions microbes play in the cycling of nutrients.

- Environments that seem extreme from a human point of view support a wide diversity of bacterial and archaeal species that offer insights into the adaptive capabilities of cells and the dynamics of community structure.

- Microbial ecologists employ a variety of diverse analytical techniques to understand the critical role of microbes in specific ecosystems and in maintaining life on Earth.

In previous chapters microorganisms are usually considered as isolated entities. However, microorganisms exist in communities. Recent estimates suggest that most microbial communities have between 10^{10} to 10^{17} individuals representing at least 10^7 different taxa. How can such huge populations exist and, moreover, survive together in a productive fashion? To answer this question, one must know which microbes are present and how they interact—that is, one must study **microbial ecology.** In this chapter we begin our consideration of this multidisciplinary field by discussing the process of elemental cycling—the microbe-mediated exchange of nutrients that makes life on Earth possible. This is sometimes called **environmental microbiology (Microbial Diversity & Ecology 27.1**). We then present an overview of some of the physical and biological features that define microbial habitats. Finally, an overview of some of the more important tools and techniques used to study microbial ecology is presented. This chapter thus provides the foundation for a more detailed review of microbial communities in marine and freshwater environments (chapter 28) and terrestrial ecosystems (chapter 29).

27.1 FOUNDATIONS IN MICROBIAL DIVERSITY AND ECOLOGY

Microorganisms function as **populations** or assemblages of similar organisms, and as **communities,** or mixtures of different microbial populations. These microorganisms have evolved while interacting with each other, with higher organisms, and with the inorganic world. They largely play beneficial and vital roles; disease-causing organisms are only a minor component of the microbial world. Microorganisms, as they interact with other organisms and their environment, also contribute to the functioning of **ecosystems**—self-regulating biological communities and their physical environment. Knowledge of these interactions is

Everything is everywhere, the environment selects.

—*M. W. Beijerinck*

new magnetically sensitive microorganisms. Magnetotactic bacteria may be described as **magneto-aerotactic bacteria** because they are thought to use magnetic fields to migrate to the position in a bog or swamp where the oxygen level is best suited for their functioning. Microorganisms have also been described that use ferrous ion as an electron donor in anoxygenic photosynthesis. Thus with production of ferric ion in lighted anoxic zones by iron-oxidizing bacteria, the stage is set for subsequent chemotrophic-based iron reduction, such as that done by *Geobacter* and *Shewanella*, creating a strictly anoxic oxidation/reduction cycle for iron.

Manganese Cycle

The importance of microorganisms in manganese cycling is becoming much better appreciated. The manganese cycle involves the transformation of manganous ion (Mn^{2+}) to MnO_2 (equivalent to manganic ion [Mn^{4+}]), which occurs in hydrothermal vents and bogs (**figure 27.8**). *Leptothrix, Arthrobacter,* and *Pedomicrobium,* are important in Mn^{2+} oxidation. *Shewanella, Geobacter,* and other chemoorganotrophs can carry out the complementary manganese reduction process.

---

1. Trace the fate of a single phosphorus atom from a rock to a stream and back to the earth.
2. Describe the difference between assimilatory and dissimilatory sulfate reduction.
3. Why do you think some bacteria can reduce Fe_3^+ only in acidic conditions?
4. What are some important microbial genera that contribute to manganese cycling?

Microorganisms and Metal Toxicity

In addition to metals such as iron and manganese, which are largely nontoxic to microorganisms and animals, there are a series of metals that have varied toxic effects on microorganisms and homeothermic animals. Microorganisms play important roles in modifying the toxicity of these metals (**table 27.3**).

The metals can be considered in broad categories. The noble metals have distinct effects on microorganisms, including growth inhibition. The second group includes metals or metalloids that microorganisms can methylate to form more mobile products called organometals. Organometals contain carbon-metal bonds. Some organometals can cross the blood-brain barrier and affect the central nervous system and organ function of vertebrates.

The mercury cycle is of particular interest and illustrates many characteristics of those metals that can be methylated. Mercury compounds were widely used in industrial processes over the centuries. Indeed Lewis Carroll alluded to this problem when he wrote of the Mad Hatter in *Alice in Wonderland*. At that time mercury was used in the shaping of felt hats. Microorganisms methylated some of the mercury, thus rendering it more toxic to the hatmakers.

A devastating situation developed in southwestern Japan when large-scale mercury poisoning occurred in the Minamata Bay region because of industrial mercury released into the marine environment. Inorganic mercury that accumulated in bottom muds of the bay was methylated by anaerobic bacteria of the genus *Desulfovibrio* (**figure 27.9**). Such methylated mercury forms are volatile and lipid soluble, and the mercury concentrations increased in the food chain by the process of biomagnification (the progressive accumulation of refractile compounds by successive trophic levels). The mercury was ultimately ingested by the human population, the "top consumers," through their primary food source—fish—leading to severe neurological disorders.

The third group of metals occurs in ionic forms directly toxic to microorganisms and more complex organisms. In higher organisms, plasma proteins react with the ionic forms of

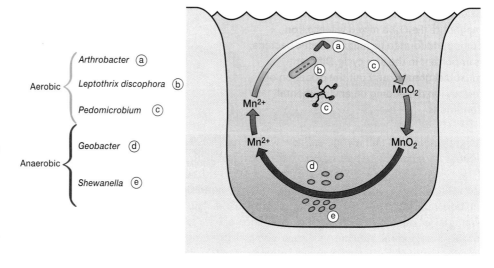

Figure 27.8 The Manganese Cycle, Illustrated in a Stratified Lake.
Microorganisms make many important contributions to the manganese cycle. After diffusing from anoxic (pink) to oxic (blue) zones, manganous ion (Mn^{2+}) is oxidized chemically and by many morphologically distinct microorganisms in the oxic water column to manganic oxide—$MnO_2^{(IV)}$, valence equivalent to 4+. When the $MnO_2^{(IV)}$ diffuses into the anoxic zone, bacteria such as *Geobacter* and *Shewanella* carry out the complementary reduction process. Similar processes occur across oxic/anoxic transitions in soils, muds, and other environments.

Table 27.3 | **Examples of Microorganism-Metal Interactions and Relations to Effects on Microorganisms and Homeothermic Animals**

Metal Group	Metal		Interactions and Transformations	
			Microorganisms	**Homeothermic Animals**
Noble metals	Ag Au Pt	Silver Gold Platinum	Microorganisms can reduce ionic forms to the elemental state. Low levels of ionized metals released to the environment have antimicrobial activity.	Many of these metals can be reduced to elemental forms and do not tend to cross the blood-brain barrier. Silver reduction can lead to inert deposits in the skin.
Metals that form stable carbon metal bonds	As Hg Se	Arsenic Mercury Selenium	Microorganisms can transform inorganic and organic forms to methylated forms, some of which tend to bioaccumulate in higher trophic levels.	Methylated forms of some metals can cross the blood-brain barrier, resulting in neurological effects or death.
Other metals	Cu Zn Co	Copper Zinc Cobalt	In the ionized form, at higher concentrations, these metals can directly inhibit microorganisms. They are often required at lower concentrations as trace elements.	At higher levels, clearance from higher organisms occurs by reaction with plasma proteins and other mechanisms. Many of these metals serve as trace elements at lower concentrations.

these metals and aid in their excretion, unless excessive long-term contact and ingestion occur. Relatively high doses of these metals are required to cause lethal effects. At lower concentrations many of these metals serve as required trace elements. *The use of chemical agents in control: Heavy metals (section 7.5)*

---

1. What are examples of the three groups of metals in terms of their toxicity to microorganisms and homeothermic animals?
2. How can microbial activity render some metals more or less toxic to warm-blooded animals?
3. Why do metals such as mercury have such major effects on higher organisms?

27.3 THE PHYSICAL ENVIRONMENT

Microorganisms, as they interact with each other and with other organisms in biogeochemical cycling, also are influenced by their immediate physical environment, whether this is soil, water, the deep marine environment, or a plant or animal host. It is important to consider the specific environments where microorganisms interact with each other, other organisms, and the physical environment.

The Microenvironment and Niche

The specific physical location of a microorganism is its **microenvironment.** In this physical microenvironment, the flux of required electron donors and acceptors, and nutrients to the actual location of the microorganism can be limited. At the same time, waste products may not be able to diffuse away from the microorganism at rates sufficient to avoid growth inhibition by high waste product concentrations. These fluxes and gradients create a unique **niche,** which includes the microorganism, its physical

habitat, the time of resource use, and the resources available for microbial growth and function (**figure 27.10**).

This physically structured environment also can limit the predatory activities of protozoa. If the microenvironment has pores with diameters of 3 to 6 μm, bacteria in the pores will be protected from predation, while allowing diffusion of nutrients and waste products. If the pores are larger, perhaps greater than 6 μm in diameter, protozoa may be able to feed on the bacteria. It is important to emphasize that microorganisms can create their own microenvironments and niches. For example, microorganisms in the interior of a colony have markedly different microenvironments and niches than those of the same microbial populations located on the surface or edge of the colony. Microorganisms also can associate with clays and form inert microhabitats called "clay hutches" for protection. Microbial growth in soils is discussed more fully in chapter 29.

Biofilms and Microbial Mats

One way microorganisms create their own microenvironments and niches is by forming **biofilms.** These are organized microbial systems consisting of layers of microbial cells associated with surfaces (**figure 27.11a**). Biofilms are an important factor in almost all areas of microbiology. *Microbial growth in natural environments: Biofilms (section 6.6); Global regulatory systems: Quorum sensing (section 12.5)*

Simple biofilms develop when microorganisms attach and form a monolayer of cells. Depending on the particular microbial growth environment (e.g., light, nutrients present, and diffusion rates), biofilms can become more complex with layers of organisms of different types (figure 27.11b). A typical example would involve photosynthetic organisms on the surface, facultative chemoorganotrophs in the middle, and possibly sulfate-reducing microorganisms on the bottom.

Figure 27.9 The Mercury Cycle. Interactions between the atmosphere, oxic water, and anoxic sediment are critical. Microorganisms in anoxic sediments, primarily *Desulfovibrio*, can transform mercury to methylated forms that can be transported to water and the atmosphere. These methylated forms also undergo biomagnification. The production of volatile elemental mercury (Hg^0) releases this metal to waters and the atmosphere. Sulfide, if present in the anoxic sediment, can react with ionic mercury to produce less soluble HgS.

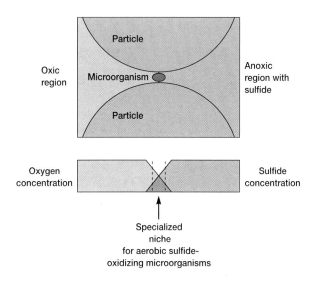

Figure 27.10 The Creation of a Niche from a Microenvironment. As shown in this illustration, two nearby particles create a physical microenvironment for possible use by microorganisms. Chemical gradients, as with oxygen from the oxic region, and sulfide from the anoxic region, create a unique niche. This niche is the physical environment and the resources available for use by specialized aerobic sulfide-oxidizing bacteria.

More complex biofilms can develop to form a four-dimensional structure (X, Y, Z, and time) with cell aggregates, interstitial pores, and conduit channels (figure 27.11*c*). This developmental process involves the growth of attached microorganisms, resulting in accumulation of additional cells on the surface, together with the continuous trapping and immobilization of free-floating microorganisms that move over the expanding biofilm. This structure allows nutrients to reach the biomass, and the channels are shaped by protozoa that graze on bacteria.

These more complex biofilms, in which microorganisms create unique environments, can be observed by the use of confocal scanning laser microscopy (CSLM) as discussed in chapter 2 (*see figure 2.25*). The diversity of nonliving and living surfaces that can be exploited by biofilm-forming microorganisms include surfaces in catheters and dialysis units, which have intimate contact with human body fluids. Control of such microorganisms and their establishment in these sensitive medical devices is an important part of modern hospital care.

Biofilms also can protect pathogens from disinfectants, create a focus for later occurrence of disease, or release microorganisms and microbial products that may affect the immune system of a susceptible host. Biofilms are critical in ocular diseases because *Chlamydia, Staphylococcus,* and other pathogens survive in ocular devices such as contact lenses and in cleaning solutions.

Figure 27.11   The Growth of Biofilms.   Biofilms, or microbial growths on surfaces such as in freshwater and marine environments, can develop and become extremely complex, depending on the energy sources that are available. **(a)** Initial colonization by a single type of bacterium. **(b)** Development of a more complex biofilm with layered microorganisms of different types. **(c)** A mature biofilm with cell aggregates, interstitial pores, and conduits.

Depending on environmental conditions, biofilms can become so large that they are visible and have macroscopic dimensions. Bands of microorganisms of different colors are evident as shown in **figure 27.12.** These thick biofilms, called **microbial mats,** are found in many freshwater and marine environments. These mats are complex layered microbial communities that can form at the surface of rocks or sediments in hypersaline and freshwater lakes, lagoons, hot springs, and beach areas. They consist of filamentous microbes, including cyanobacteria. A major characteristic of mats is the extreme gradients that are present. Visible light only penetrates approximately 1 mm into these communities, and below this photosynthetic zone, anoxic conditions occur and sulfate-reducing bacteria play a major role. The sulfide that these organisms produce diffuses to the anoxic lighted region, allowing sulfur-dependent photosynthetic microorganisms to grow. Some believe that microbial mats could have allowed the formation of terrestrial ecosystems prior to the development of vascular plants, and fossil microbial mats, called stromatolites, have been dated at over 3.5 billion years old *(see figure 19.2).* Molecular techniques and stable isotope measurements (see sec-

tion 27.4) are being used to better understand these unique microbial communities.

1. What are the similarities and differences between a microenvironment and a niche?
2. Why might pores in soils, waters, and animals be important for survival of bacteria if protozoa are present?
3. Why might conditions vary for a bacterium on the edge of a colony in comparison with the center of the colony?
4. What are biofilms? What types of surfaces on living organisms can provide a site for biofilm formation?
5. Why are biofilms important in human health?
6. What are microbial mats, and where are they found?

Microorganisms and Ecosystems

Microorganisms, as they interact with each other and other organisms, and influence nutrient cycling in their specific microenvironments and niches, also contribute to the functioning of ecosystems.

Figure 27.12 Microbial Mats. Microorganisms, through their metabolic activities, can create environmental gradients resulting in layered ecosystems. A vertical section of a hot spring (55°C) microbial mat, showing the various layers of microorganisms.

Ecosystems have been defined as "communities of organisms and their physical and chemical environments that function as self-regulating units." These self-regulating biological units respond to environmental changes by modifying their structure and function.

Microorganisms in ecosystems can have two complementary roles: (1) the synthesis of new organic matter from CO_2 and other inorganic compounds during **primary production** and (2) decomposition of accumulated organic matter. The general relationship between **primary producers** that synthesize organic matter and chemoorganotrophic **decomposers** was once thought to be quite simple. It was held that different organisms performed these nonoverlapping processes. We now understand that microbial communities are more complicated. Because aquatic systems are easier to investigate than terrestrial microbial communities, much of our current understanding of the relative contributions of microbes to ecosystem function comes from studies of these environments. As shown in **figure 27.13a,** the traditional food chain is only part of the picture. Larger plants and animals contribute to a common pool of **dissolved organic matter (DOM)** that is consumed by a variety of procaryotic and eucaryotic microbes. These microbes then return some of this DOM to larger animals in the form of particulate material—that is to say, protists eat bacteria and primary consumers eat protists (figure 27.13b,c). In addition, the metabolism and death of these microbes recycles some organic matter back to the general pool of DOM. This complex web of interactions

is called the **microbial loop.** In terrestrial systems, the roles of microbes appear to be similar, with the exception of primary production, which is performed chiefly by vascular plants instead of microbes. The microbial loop is discussed in further detail in chapter 28 (*see sections 28.1 and 28.3; figure 28.15*).

Microorganisms thus carry out many important functions as they interact in ecosystems, including:

1. Contributing to the formation of organic matter through photosynthetic and chemosynthetic processes.
2. Decomposing organic matter, often with the release of inorganic compounds (e.g., CO_2, NH_4^+, CH_4, H_2) in mineralization processes.
3. Serving as a nutrient-rich food source for other chemoheterotrophic microorganisms, including protozoa and animals.
4. Modifying substrates and nutrients used in symbiotic growth processes and interactions, thereby contributing to biogeochemical cycling.
5. Changing the amounts of materials in soluble and gaseous forms. This occurs either directly by metabolic processes or indirectly by modifying the environment (e.g., altering the pH).
6. Producing inhibitory compounds that decrease microbial activity or limit the survival and functioning of plants and animals.
7. Contributing to the functioning of plants and animals through positive and negative symbiotic interactions, as discussed in chapter 30.

Microorganism Movement between Ecosystems

Microorganisms are moved constantly between ecosystems. This often happens naturally in many ways: (1) soil is transported around the Earth by windstorms and falls on land areas and waters far from its origin; (2) rivers transport eroded materials, sewage plant effluents, and urban wastes to the ocean; and (3) insects and animals release urine, feces, and other wastes to environments as they migrate around the Earth. When plants and animals die after moving to a new environment, they decompose and their specially adapted and coevolved microorganisms (and their nucleic acids) are released. The fecal-oral route of disease transmission, often involving foods and waters, and the acquisition of diseases in hospitals (nosocomial infections) are important examples of pathogen movement between ecosystems. Each time a person coughs or sneezes, microorganisms also are being transported to new ecosystems.

Humans also both deliberately and unintentionally move microorganisms between different ecosystems. This occurs when microbes are added to environments to speed up microbially mediated degradation processes or when a plant-associated inoculum, such as *Rhizobium,* is added to soil to increase the formation of nitrogen-fixing nodules on legumes. One of the most important accidental modes of microbial movement is the use of modern transport vehicles such as automobiles, trains, ships, and airplanes. These often rapidly move microorganisms long distances. Biodegradation and bioremediation in natural communi-

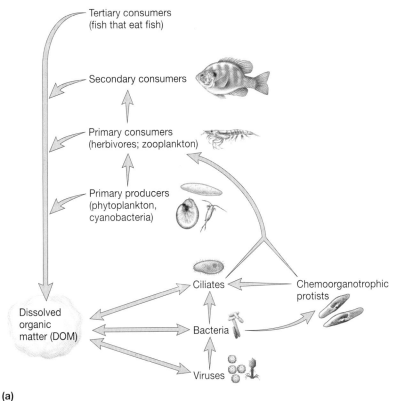

Tertiary consumers
(fish that eat fish)

Secondary consumers

Primary consumers
(herbivores; zooplankton)

Primary producers
(phytoplankton,
cyanobacteria)

Ciliates

Chemoorganotrophic
protists

Dissolved
organic
matter (DOM)

Bacteria

Viruses

(a)

Figure 27.13   The Microbial Loop.   (a) Microorganisms play vital roles in ecosystems as primary producers, decomposers, and primary consumers. The microbial loop describes the exchange of dissolved organic matter (DOM) between organisms so that it is recycled many times and not immediately lost to the system. **(b)** Protists consume bacteria; in this case a naked amoeba is consuming the cyanobacterium *Synechococcus* which fluoresce red. **(c)** Protists consume protists; here the ciliate *Didinium* (gold) is preying upon another ciliate, *Paramecium*.

(b)

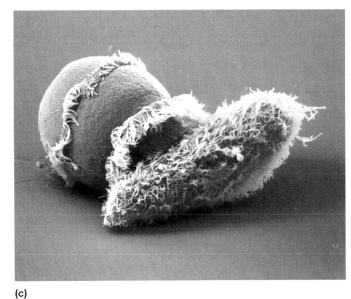

(c)

ties (section 41.6); Microorganism associations with vascular plants: The rhizobia (section 29.5)

The fate of microorganisms placed in environments where they normally do not live, or of microorganisms returned to their original environments, is of theoretical and practical importance. Pathogens that are normally associated with an animal host largely have lost their ability to compete effectively for nutrients with microorganisms indigenous to other environments. Upon moving to a new environment, the population of viable and culturable pathogens gradually decreases. However, more sensitive viability assessment procedures, particularly molecular techniques, indicate that **vibiable but nonculturable**

(**VBNC**) **microorganisms** may play critical roles in disease occurrence (p. 660). The growth curve: Senescence and death (section 6.2)

---

1. Define the following terms: ecosystem, primary production, decomposer.
2. List important functions of higher consumers in natural environments.
3. What are the important functions of microorganisms in ecosystems?
4. How can microorganisms move between different ecosystems?

Extreme Environments

Microorganisms are found in a wide range of environments that differ in pH, temperature, atmospheric pressure, salinity, water availability, and ionizing radiation. In some environments, these conditions can be at either end of a continuum (e.g., very alkaline or acidic; extremely hot or cold). Such environments are called **extreme environments** (**figure 27.14**), and some of the characteristics of such ecosystems are summarized in **table 27.4.** The microorganisms that survive in such environments are described as **extremophiles.** Although extreme environments are usually considered to have decreased microbial diversity, as judged by the microorganisms that can be cultured, with the increased use of molecular detection techniques, it appears that many contain a surprising diversity of microorganisms. Work to establish relationships between the microorganisms that can be observed and detected by molecular techniques and culturable microorganisms is ongoing. The influence of environmental factors on growth (section 6.5)

Many microbial genera have specific requirements for survival in extreme environments. For example, a high sodium ion concentration is required to maintain membrane integrity in many halophilic procaryotes, including members of the genus *Halobacterium*. Halobacteria require a sodium ion concentration of at least 1.5 M, and about 3 to 4 M for optimum growth. Phylum *Euryarchaeota:* The halobacteria (section 20.3)

Some bacteria found in deep-sea environments provide another example of extremophiles. These bacteria can be described as **baro-** or **piezotolerant bacteria** (growth from approximately 1 to 500 atm), **moderately barophilic bacteria** (growth optimum 5,000 meters, and still able to grow at 1 atm), and **extreme barophilic bacteria,** which require approximately 400 atm or higher for growth. Microorganisms in marine environments: Benthic marine environments (section 28.3)

Intriguing changes in basic physiological processes occur in microorganisms functioning under extreme acidic or alkaline conditions. These acidophilic and alkalophilic microorganisms have markedly different problems in maintaining a more neutral internal pH and chemiosmotic processes. Obligately acidophilic microorganisms can grow at a pH of 3.0 or lower, and major pH differences can exist between the interior and exterior of the cell. These acidophiles include members of the genera *Thiobacillus, Sulfolobus,* and *Thermoplasma*. The higher relative internal pH is maintained by a net outward translocation of protons. This may occur as the result of unique membrane lipids, hydrogen ion re-

(a)

(b)

(c)

Figure 27.14 Microorganisms Growing in Extreme Environments. Many microorganisms are especially suited to survive in extreme environments. **(a)** Salterns turned red by halophilic algae and halobacteria. **(b)** A hot spring colored green and blue by cyanobacterial growth. **(c)** A source of acid drainage from a mine into a stream. The soil and water have turned red due to the presence of precipitated iron oxides caused by the activity of bacteria such as *Thiobacillus*.

Table 27.4	Characteristics of Extreme Environments in Which Microorganisms Grow	
Stress	**Environmental Conditions**	**Microorganisms Observed**
High temperature	121°C	*Geogemma barossii*
	110–113°C, deep marine trenches	*Pyrolobus fumarii* *Methanopyrus kandleri* *Pyrodictium abyssi*
	67–102°C, marine basins	*Pyrococcus abyssi*
	85°C, hot springs	*Thermus* *Sulfolobus*
	75°C, sulfur hot springs	*Thermothrix thiopara*
Low temperature	−12°C, antarctic ice	*Psychromonas ingrahamii*
Osmotic stress	13–15% NaCl	*Chlamydomonas*
	25% NaCl	*Halobacterium* *Halococcus*
pH	pH 10.0 or above	*Bacillus*
	pH 3.0 or lower	*Saccharomyces* *Thiobacillus*
	pH 0.5	*Picrophilus oshimae*
	pH 0.0	*Ferroplasma acidarmanus*
Low water availability	a_w = 0.6–0.65	*Torulopsis* *Candida*
Temperature and low pH	85°C, pH 1.0	*Cyanidium* *Sulfolobus acidocaldarum*
Pressure	500–1,035 atm	*Colwellia hadaliensis*
Radiation	1.5 million rads	*Deinococcus radiodurans*

Figure 27.15   Massive Growth of the Extreme Acidophile *Ferroplasma* in a California Mine.   Slime streamers of *Ferroplasma acidarmanus*, an archaeon, which have developed within pyritic sediments at and near pH 0. This unique procaryote has a plasma membrane and no cell wall.

nal proton concentrations may be maintained by means of coordinated hydrogen and sodium ion fluxes.

Observations of microbial growth at 121°C in thermal vent areas, or of **hyperthermophiles** (**Techniques & Applications 27.2**), indicate that this area will continue to be a fertile field for investigation. For some successful microorganisms, an extreme environment may not be "extreme" but required and even, perhaps, ideal. Phylum *Euryarchaeota* (section 20.3)

---

1. What factors can create extreme environments?
2. Define extremophile and discuss an example of adaptation to extremes alkalinity or acidity.
3. Why are molecular techniques changing our view of extreme environments?
4. What is unique about *Ferroplasma acidarmanus?*

---

moval during reduction of oxygen to water, or the pH-dependent characteristics of membrane-bound enzymes. An archaeal iron-oxidizing acidophile, *Ferroplasma acidarmanus,* capable of growth at pH 0, has been isolated from a sulfide ore body in California. This unique procaryote, capable of massive surface growth in flowing waters in the subsurface, possesses a single peripheral cytoplasmic membrane and no cell wall (**figure 27.15**).

The extreme alkalophilic microorganisms grow at pH values of 10.0 and higher and must maintain a net inward flux of protons. These obligate alkalophiles cannot grow below a pH of 8.5 and are often members of the genus *Bacillus; Micrococcus* and *Exiguobacterium* representatives have also been reported. Some cyanobacteria also have similar characteristics. Increased inter-

27.4   MICROBIAL ECOLOGY AND ITS METHODS: AN OVERVIEW

Microbial ecologists study natural microbial communities that may exist in soils, waters, or in association with other organisms, including humans. Regardless of the habitat they study, these scientists seek to answer several fundamental questions. First they are interested in the microbial population: How many microbes are present? What genera and species are represented in the ecosystem? Once these questions have been addressed, community dynamics can be explored: How do these microorganisms interact with one another, with higher eucaryotes, and with the abiotic features found in the environment? The multidisciplinary nature of microbial ecology has generated an enormous assortment of methods including microscopic, cultural, physical,

(a)

(b)

Figure 28.5 *Thioploca* the "Spaghetti Bacterium." *Thioploca* ("sulfur braid") is an unusual microorganism that links separated resources of sulfide from the mud and nitrate from the overlying water. **(a)** Bundles that join the oxic surface and the lower anoxic mud. **(b)** An individual *Thioploca,* showing the elemental sulfur globules and tapered ends. Bar = 40 μm.

usually 100 to 300 μm in diameter (750 μm cells occasionally occur with a biovolume of 200,000,000 μm^3). Sulfide and nitrate are used as the electron donor and acceptor, respectively. In this case nitrate, from the overlying seawater, penetrates the anoxic sulfide-containing muds only during storms. When this short-term mixing occurs, *Thiomargarita* takes up and stores the nitrate in a huge internal vacuole, which may occupy 98% of the organism's volume. The vacuolar nitrate can approach a concentration of 800 mM. The elemental sulfur granules appear near the cell edge in a thin layer of cytoplasm. Between storms, the organism lives using the stored nitrate as an electron acceptor. These unique bacteria are important in sulfur and nitrogen cycling in these environments. Microbial Diversity & Ecology 3.1: Monstrous microbes

A critical adaptation of microorganisms in aquatic systems is the ability to link and use resources that are in separate locations, or that are available at the same location only for short intervals such as during storms. One of the most interesting bacteria linking widely separated resources is *Thioploca* spp., which lives in bundles surrounded by a common sheath (**figure 28.5**). These microbes are found in upwelling areas along the coast of Chile, where oxygen-poor but nitrate-rich waters are in contact with sulfide-rich bottom muds (much like the environment off the coast of western Africa). The individual cells are 15 to 40 μm in diameter and many centimeters long, making them, like *T. namibiensis,* one of the largest bacteria known. They form filamentous sheathed structures, and the individual cells can glide 5 to 15 cm deep into the sulfide-rich sediments. These unique microorganisms are found in expanses off the coast of Chile.

In addition to living a planktonic existence, many microorganisms take advantage of surfaces. These include sessile microorganisms of the genera *Sphaerotilus* and *Leucothrix* and the prosthecate and budding bacteria of the genera *Caulobacter* and *Hyphomicrobium.* There are also a wide range of aerobic gliding bacteria such as the genera *Flexithrix* and *Flexibacter,* which move over surfaces where organic matter is adsorbed. These organisms are characterized by their exploitation of surfaces and nutrient gradients. They are obligate aerobes, although sometimes they can carry out denitrification, as occurs in the genus *Hyphomicrobium.* In addition, bacteria may be primary colonizers of submerged surfaces, allowing subsequent development of a complex biofilm. Biofilms (sections 6.6 and 27.3); Class *Alphaproteobacteria: The Caulobacter* and *Hyphomicrobium* (section 22.1); Class *Gammaproteobacteria:* Order *Thiotrichales* (section 22.3); Microbial Diversity & Ecology 21.1: The mechanism of gliding motility

Microscopic fungi, which usually are thought to be terrestrial organisms living in soils and on fruits and other foods, also grow in freshwater and marine environments. Zoosporic organisms adapted to life in the water include the chytrids, which have motile asexual reproductive spores with a single whiplash flagellum. Chytrids are important because of their role in decomposing dead organic matter. In addition, many chytrids attack algae (**figure 28.6**). Characteristics of fungal divisions: *Chytridiomycota* (sections 26.6)

Another important group includes filamentous fungi that can sporulate under water. These hyphomycetes include the **Ingoldian fungi,** named after C. T. Ingold. In 1942 Ingold discovered fungi that produce unique tetraradiate forms (**figure 28.7**). The ecology

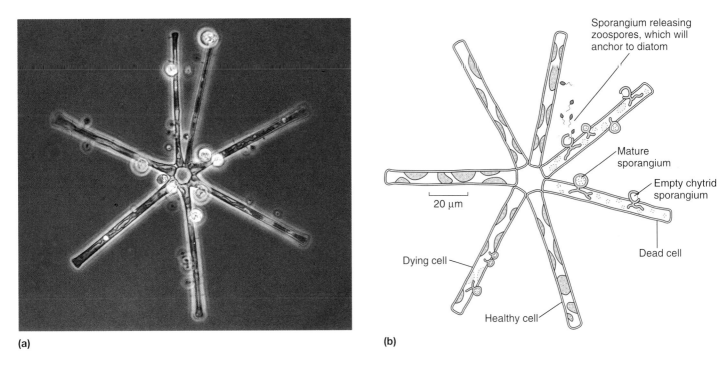

(a)

(b)

Figure 28.6 Chytrids and Aquatic Environments.   Chytrids play important roles in aquatic environments. **(a)** The infection of the diatom by the chytrid *Rhizophydium* is shown in the photograph, and **(b)** in the illustration showing details of the parasitic process.

of these aquatic fungi is very interesting. The tetraradiate conidium forms on a vegetative mycelium, which grows inside decomposing leaves. When the vegetative hyphae differentiate into an aerial mycelium, conidia are released into the water. Released conidia are then transported and often are present in surface foam. When they contact leaves, the conidia attach and establish new centers of growth. These uniquely adapted fungi contribute significantly to the processing of organic matter, especially leaves. Often aquatic insects will feed only on leaves that contain fungi.

---

1. What are ultramicrobacteria? What is the evidence that they have adapted genetically to oligotrophic environments as opposed to simply responding to nutrient limitation?
2. Compare the environments in which *Thiomargarita namibiensis* and *Thioploca* spp. occur. How are these two microbes similar?
3. Describe the life cycle of Ingoldian fungi.

28.3   MICROORGANISMS IN MARINE ENVIRONMENTS

As terrestrial organisms, we must remind ourselves that 97% of the Earth's water is in marine environments. Although much of this is in the deep sea, from a microbiological perspective, the surface waters have been most intensely studied. This is where the photosynthesis that drives all the marine ecosystems occurs. Only recently have scientists had the capacity to probe the deep-sea sediments and the subsurface (the benthos), and investiga-

tions of this kind are revealing a number of surprises. We begin our discussion of marine ecosystems with estuaries, and then discuss the microbial communities that inhabit the open ocean and finally the dark, cold, high-pressure benthos.

Coastal Marine Systems: Estuaries and Salt Marshes

An estuary is a semi-enclosed coastal region where a river meets the sea. Estuaries are defined by tidal mixing between freshwater and salt water. They feature a characteristic salinity profile called a **salt wedge** (**figure 28.8**). Salt wedges are formed because saltwater is denser than freshwater, so seawater sinks below overlying freshwater. As the contribution from the incoming river increases and that of the ocean decreases, the relative amount of seawater declines with the estuary's increased distance from the sea. The distance the salt wedge intrudes up the estuary is not static. Most estuaries undergo large-scale tidal flushing; this forces organisms to adapt to changing salt concentrations on a daily basis. Microbes that live under such conditions combat the resulting osmotic stress by adjusting their intracellular osmolarity to limit the difference with that of the surrounding water. Most protists and fungi produce osmotically active carbohydrates for this purpose, whereas procaryotic microbes regulate internal concentrations of potassium or special amino acids such as ecoine and betaine. Thus most microbes that inhabit estuaries are **halotolerant,** which is distinct from halophilic. Halotolerant microbes can withstand significant changes in salinity; halophilic microorganisms have an absolute requirement for high salt concentrations.

Figure 28.7 Ingoldian Fungi. These aquatic fungi are capable of sporulation under water. They play important roles in the processing of complex organic matter, such as leaves, which fall into streams and lakes. These microbes include types with tetraradiate conidia. **(a)** Fungal hyphae grow inside the decomposing leaf and give rise to tetraradiate conidia. These aerial structures project from the leaf surface into the water column. **(b)** The new tetraradiate conidium then can be released and attach to a leaf surface, repeating the process and accelerating leaf decomposition and the release of nutrients.

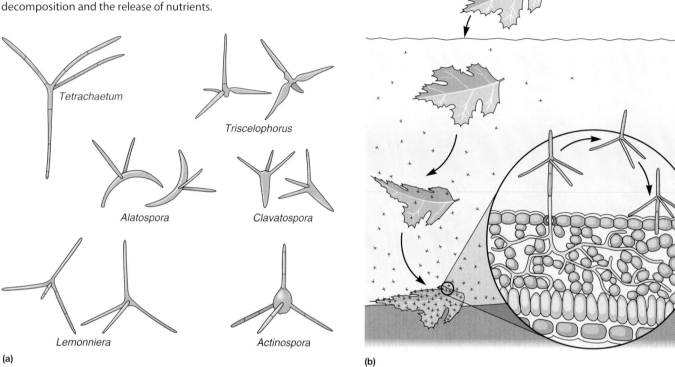

(a)

Tetrachaetum

Triscelophorus

Alatospora

Clavatospora

Lemonniera

Actinospora

(b)

Figure 28.8 A Salt Wedge. An estuary contains both freshwater and saltwater. Because seawater is denser than freshwater, the water masses do not mix. Rather, the seawater remains below the freshwater with the relative amount of seawater decreasing in the upper reaches of the estuary.

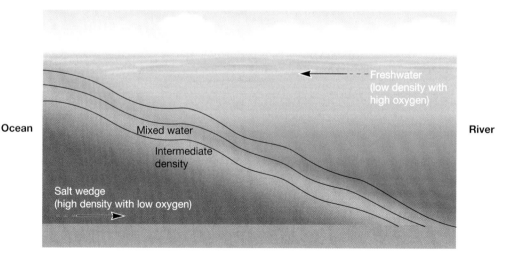

Ocean

Freshwater (low density with high oxygen)

Mixed water

Intermediate density

Salt wedge (high density with low oxygen)

River

Estuaries are unique in many respects. Their calm, nutrient-rich waters serve as nurseries for juvenile forms of many commercially important fish and invertebrates. However, despite their importance to the commercial fishing industry, estuaries are among the most polluted marine environments. They are the receptacles of waste that is dumped in rivers and pollutants that have been discharged during industrial processes. Recall that

from the 17th century through most of the 20th century, industries dumped their wastes without fear of punitive consequences. The cleanup of rivers and estuaries contaminated with industrial wastes such as polychlorinated biphenyls (PCBs) continues to this day. Often industrial pollution includes organic materials that can be used as nutrients. This further magnifies the problem because chemoorganotrophic microbes consume available oxygen, forming anoxic dead zones. Such anoxic regions, which are devoid of almost all macroscopic life, are now present in the Gulf of Mexico, Chesapeake Bay, and fjords entering the North Sea.

Organic pollution can also create the opposite problem: too much growth. However, in this case, a single microbial species, either algal or cyanobacterial, grows at the expense of all other organisms in the community. This phenomenon, called a bloom, often results from the introduction of nutrients combined with mixing sediments. If the microbes produce a toxic product or are in themselves toxic to other organisms such as shellfish or fish, the term **harmful algal bloom (HAB)** is used. Some HABs are called red tides because the microbial density is so great that the water becomes red or pink (the color of the algae). The number of HABs has dramatically increased in the last decade or so. HABs are sometimes responsible for killing large numbers of fish or marine mammals. For instance, off the coast of California, an HAB was responsible for sudden, large-scale deaths among sea lions. In this case, the bloom species was a diatom of the genus *Pseudonitzschia*. Anchovies consumed the diatoms and the potent neurotoxin, domoic acid, accumulated in the fish. The mammals were poisoned after they ingested large quantities of the fish, an important component of the sea lion diet. Disease 25.1: Harmful algal blooms (HABs)

HABs are often caused by dinoflagellates. Some bloom-causing dinoflagellates produce a potent neurotoxin, called a **brevetoxin.** Dinoflagellates of the genus *Alexandrium* produce a brevetoxin responsible for most of the *paralytic shellfish poisoning* (PSP), which affects humans as well as other animals in coastal, temperate North America. The brevetoxin produced by the dinoflagellate *Karinia brevis* killed a number of endangered manatees and bottlenose dolphins in 2002 in a bloom in Florida. Another dinoflagellate, *Pfiesteria piscicida,* has become a problem in the Chesapeake Bay and regions south. This protist produces lethal lesions in fish (**figure 28.9**) and has had a devastating effect on the local fishery industry. Exposure to this microbe also causes neurological damage to humans, including short-term memory loss. Our understanding of this microbe has been limited by a number of factors, including a debate about its life cycle and problems in isolating its toxin. Protist classification: *Alveolata* and *Stramenopiles* (section 25.6)

Salt marshes generally differ from estuaries in that they lack freshwater input from a single source. Smaller streams enter salt marshes, which are flatter and have a wider expanse of sediment and plant life exposed at low tide. For this reason, salt marshes are sometimes called salt meadows. The microbial communities within salt marsh sediments are very dynamic. These ecosystems can be modeled in **Winogradsky columns (figure 28.10)**, named after the pioneering microbial ecologist Sergei Winogradsky (1856–1953). A Winogradsky column is easily constructed using a glass cylinder into which either marine or freshwater sediment is placed and then overlaid with saltwater or freshwater, respectively. Winogradsky columns prepared with saltwater and salt marsh sediments contain a considerable amount of sulfur because sulfate is found in seawater. Anaerobic microbes use the sulfate as a terminal electron acceptor and produce hydrogen sulfide. When the top of the column is sealed, much of the cylinder will eventually become anoxic. The addition of shredded newspaper introduces a source of cellulose that is degraded to fermentation products by the genus *Clostridium*. With these fermentation products available as electron donors and using sulfate as an acceptor, *Desulfovibrio* produces hydrogen sulfide. The hydrogen sulfide diffuses upward toward the oxygenated zone, creating a stable hydrogen sulfide gradient. In this gradient the photoautotrophs *Chlorobium* and *Chromatium* develop as visible olive green and purple zones. These microorganisms use hydrogen sulfide as an electron source, and CO_2, from sodium carbonate, as a carbon source. Above this region the purple nonsulfur bacteria of the genera *Rhodospirillum* and *Rhodopseudomonas* can grow. These photoorganotrophs use organic matter as an electron donor under anoxic conditions and function in a zone where the sulfide level is lower. Both O_2 and hydrogen sulfide may be present higher in the column, allowing specially adapted microorganisms to function. These include the chemolithotrophs *Beggiatoa* and *Thiothrix,* which use reduced sulfur compounds as electron donors and O_2 as an acceptor. In the upper portion of the column, diatoms and cyanobacteria may be visible. Phototrophy (section 9.12); Photosynthetic bacteria (section 21.3); Class *Alphaproteobacteria:* Purple nonsulfur bacteria (section 22.1); Class *Gammaproteobacteria* (section 22.3); Class *Clostridia* (section 23.4)

---

1. Define salt wedge and explain its influence on estuarine microbial communities.
2. Consider the fact that during droughts, rivers that normally flow into an estuary have a significantly diminished flow. Examine figure 28.8 and explain how this could result in salt intrusion into a public water supply. What do you think the consequences would be on estuarine microbial communities?
3. Name two groups of protists known to cause HABs in marine ecosystems.
4. Describe the ecosystem that develops within a Winogradsky column. How is this similar to the microbial community you would expect to find in a salt marsh? How might it be different?

Figure 28.9 *Pfiesteria* **Lesions.** Lesions on menhaden resulting from parasitism by the dinoflagellate *Pfiesteria piscicida*.

Figure 28.10 The Winogradsky Column. A microcosm in which microorganisms and nutrients interact over a vertical gradient. Fermentation products and sulfide migrate up from the reduced lower zone, and oxygen penetrates from the surface. This creates conditions similar to those in a lake or salt marsh with nutrient-rich sediments. Light is provided to simulate the penetration of sunlight into the anoxic lower region, which allows photosynthetic microorganisms to develop.

Important reactions and microorganisms

Component

Water layer — Diatoms and cyanobacteria

O_2-dominated mud (light brown) — Algae and aerobic sulfide oxidizing microorganisms
Beggiatoa
Thiobacillus
Thiothrix

Rust-colored zone — Photoheterotrophs
Rhodospirillum
Rhodopseudomonas

Mud

Red zone — *Chromatium*

Green zone — *Chlorobium*

Mud plus sulfate, carbonate, and newspaper (as a cellulose source)

H_2S diffusion

Anoxic H_2S-dominated zone (black)

Cellulose ⟶ fermentation products (*Clostridium*)

Fermentation products plus sulfate ⟶ sulfide (*Desulfovibrio*)

Figure 28.11 Mean Annual Surface Chlorophyll Levels. Global chlorophyll as measured by ocean satellites. The dark-blue regions correspond to subtropical gyres (massive spirals of water) where nutrient levels are especially low; note scale below image. Image courtesy of NASA-Goddard Space Flight Center and the Orbital Sciences Corporation.

0.01 0.02 0.03 0.04 0.06 0.1 0.2 0.3 0.5 1.0 2.0 3.0 5.0 10 20 30 40 50 60
Chlorophyll *a* Concentration (mg / m³)

The Photic Zone of the Open Ocean

Sometimes called "the invisible rain forest," the upper 200 to 300 meters of the open ocean is home to a diverse collection of photosynthetic microbes. Open ocean regions are also called **pelagic.** The use of satellite imagery to measure chlorophyll (**figure 28.11**) shows that although chlorophyll levels are lower in the sea than on

land, the sheer volume of the oceans accounts for the fact that about half the world's photosynthesis is performed by marine microbes. The open ocean is an oligotrophic environment—that is to say, nutrient levels are very low. Recall that the influx of new nutrients is limited, so primary productivity (and thus the whole ecosystem) depends on rapid recycling of nutrients. Unlike terrestrial ecosys-

Figure 28.12   The Biological Carbon Pump.   The vast majority of the carbon fixed by microbial autotrophs in the open ocean remains in the photic zone. However, a small fraction is "pumped" to the seafloor, which in turn returns much of it to the surface in regions where deep and surface waters mix (upwelling regions). The flux of CO_2 in and out of the world's oceans is in equilibrium, but this equilibrium may be upset by the increase in atmospheric CO_2, the hallmark of global warming. Note depth scale on right.

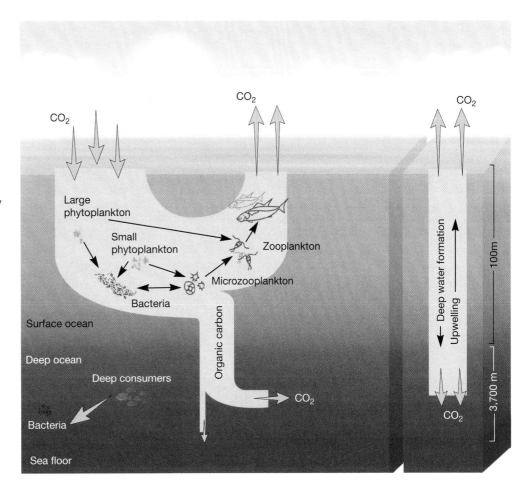

tems, at least 90% of the nutrients required by the microbial community are recycled within the microbial loop (figure 28.3).

Global warming has focused intense scrutiny on determining how much CO_2 phytoplankton can "draw down" out of the atmosphere and sequester in the benthos. As shown in **figure 28.12,** there is a constant exchange of CO_2 at the ocean surface, as well as limited export of carbon to the seafloor. Organic matter escapes the photic zone and falls through the depths as **marine snow (figure 28.13)**. This material gets its name from its appearance, sometimes seen in video images of mid- and deep-ocean water, where drifting flocculate particles look much like snow. Marine snow consists primarily of fecal pellets, diatom frustules, and other materials that are not easily degraded. During its fall to the bottom, marine snow is colonized by a community of microbes so that mineralization continues. By the time it reaches the seafloor, less than 1% of photosynthetically derived materials remains unaltered. Once there, only a tiny fraction will be buried in the sediments; most will return to the surface in upwelling regions (figure 28.12). An urgent question is whether or not the oceans can draw down more CO_2 as atmospheric concentrations continue to increase. One approach to reversing (or at least slowing) global warming is to fertilize specific regions of the oceans. These regions, known as high-nutrient, low-chlorophyll (HNLC)

Figure 28.13   Marine Snow.   The export of organic matter out of the photic zone to deeper water occurs through the sinking of marine snow. Shown here is material collected in a sediment trap at 5,367 meters on the Sohm Abyssial Plain in the Sargasso Sea. It includes cylindrical fecal pellets, planktonic tests (round white objects), transparent snail-like pteropod shells, radiolarians, and diatoms.

areas, are limited by iron. A number of experiments have been performed wherein large transects of the southern Pacific were fertilized with iron to trigger diatom blooms. As might be expected, the notion of fertilizing vast regions of the ocean to rid the atmosphere of CO_2 is controversial for a number of reasons. First, it is not apparent that this will be effective—while all studies show a temporary increase in primary production, only some studies show a measurable loss of carbon from the photic zone; others report no increase in CO_2 draw-down. Second, many scientists are skeptical about the long-term effects of altering such a large and important ecosystem.

The cycling of other elements, in addition to carbon, occurs within the photic zone. In general, the open ocean is limited by nitrogen, not iron. Two recent discoveries have led marine microbiologists and oceanographers to reexamine the traditionally accepted nitrogen cycle. First, the filamentous cyanobacterium *Trichodesmium* is fixing far more N_2 than previously thought. Second, unicellular cyanobacteria collected at sea are also fixing N_2. These microbes were known to fix nitrogen under laboratory conditions but researchers assumed they did not do so in the open ocean. Thus there appears to be at least two recently identified sources of "new" organic nitrogen. The other recent discovery that has revolutionized our understanding of nitrogen cycling is the presence of bacteria that perform the anammox reaction below the photic zone where oxygen concentrations reach a minimum. In nitrogen-limited areas, it is generally understood that there is a net loss of ammonium, nitrate, and nitrite. A large fraction of this loss has been attributed to denitrification (the anaerobic reduction of nitrate to N_2) occurring at the oxygen-minimum zones. It now appears that consortia of bacteria capable of the anammox reaction—the anaerobic oxidation of NH_4^+ to N_2—are responsible for much of the loss of nitrogen that could otherwise support life. This consortia includes members of the interesting phylum *Planctomycetes*. Once again, the urgency to understand the global oceanic nitrogen budget reflects concern over increasing levels of atmospheric CO_2, global warming, and the capacity of the world's oceans to remove more CO_2 while maintaining overall ecosystem equilibrium.

Photosynthetic bacteria: Phylum *Cyanobacteria* (section 21.3); Phylum *Planctomycetes* (section 21.4); Biogeochemical cycling: Nitrogen cycle (section 27.2)

The pelagic cyanobacterium *Trichodesmium* has long been of interest not only because it can fix nitrogen, but because it forms extensive blooms in the open ocean. These blooms, which look similar to floating straw, can cover up to 300,000 square kilometers. Considering the fact that nutrient levels, especially nitrogen and phosphorus, are extremely low in the open ocean, it has been clear for a long time that the cyanobacterium must have evolved clever mechanisms to survive. As a nitrogen-fixing cyanobacterium, neither carbon nor nitrogen would be limiting. It appears that *Trichodesmium* must be able to outcompete other phytoplankton for phosphorus. Recent experiments that combine genomics with real-time PCR show how *Trichodesmium* has mastered phosphorus limitation. Annotation of the *Trichodesmium* genome reveals genes predicted to encode proteins associated with high-affinity transport and assimiliation of phosphonate, compounds with a C—P bond. Phosphonate is an im-

portant part of DOM, but most photosynthetic planktonic microbes are unable to use it. Researchers at the Woods Hole Oceanographic Institution in Massachusetts used real-time PCR to show that the genes thought to be involved in phosphonate assimilation (in cultured and Sargasso Sea field populations) were expressed only under phosphate-limiting conditions. Thus *Trichodesmium* may be so successful in the oligotrophic oceans because it uses a source of phosphorus not available to most other phytoplankton.   Biogeochemical cycling: Phosphorus cycle (section 27.2)

It is clear that when researchers consider microbial activities in the open ocean, they are assessing global processes. So perhaps it is not surprising that the most abundant group of monophyletic organisms on Earth is marine. Members of the α-proteobacterial clade called **SAR11** have been detected by rRNA gene cloning from almost all open-ocean samples taken worldwide. In addition, it has been found at depths of 3,000 meters as well as in coastal waters and even in some freshwater lakes. Using a technique called *fluorescence in situ hybridization (FISH),* in which specific DNA fragments are labeled with fluorescent dye, SAR11 has been found to constitute 25 to 50% of the total procaryotic community in the surface waters in both nearshore and open-ocean samples. Indeed SAR11 bacteria are estimated to constitute about 25% of all microbial life on the planet.

But what exactly are SAR11 microbes and what are they doing? The answer to that question took about a decade to emerge because SAR11 (named after the *Sar*gasso Sea, where it was first detected) could not be coaxed into growing in the laboratory. In 2002 Stephen Giovannoni, the scientist who first discovered SAR11, and his graduate students isolated several members of the SAR11 clade in pure culture, and named the microbe *Pelagibacter ubique*. It grows only in seawater cultures with very low nutrients, conditions much like the oligotrophic ocean. *P. ubique* is vibrioid and only 0.4 to 0.9 μm in length, making it one of the smallest known, free-living microbes. Unlike most microbes in culture, which grow to densities exceeding 10^8 cells per milliliter, SAR11 isolates stop replicating after reaching about 10^6 cells per milliliter. Curiously, this is the density at which they are generally found in nature, suggesting that natural factors in seawater somehow control population growth. By measuring how much and how fast SAR11 microbial isolates assimilate radiolabeled amino acids, glucose, and complex biomolecules, scientists have begun to get a picture of their role in the microbial loop. These microbes contribute as much as 50% to the bacterial biomass production and DOM flux in some marine environments, and it appears that they may selectively degrade the kinds of complex biomolecules that comprise marine snow.   Microbial ecology and its methods: An overview (section 27.4)

The genome of *P. ubique* was recently sequenced and annotated. At 1.31 Mb, it is the smallest genome of any independently replicating cell sequenced to date. Unlike the small genomes of parasitic bacteria that have lost genes for energy capture and other essentials that can be obtained from the host, SAR11 accomplishes this feat by eliminating "genomic waste." There are no pseudogenes, phage genes, or recent gene duplications. In addition, the number of nucleotides between coding regions is also very limited.

Nonetheless, *P. ubique* has the genes necessary for the Entner-Douderoff pathway, the TCA cycle, and a complete electron transport chain. It has adapted to life in the oligotrophic ocean by encoding a number of high-affinity nutrient transport systems and maintaining its small size, thereby optimizing its surface area to volume ratio. In addition, it uses both respiration and a proteorhodopsin proton pump to capture energy. Insights from microbial genomes (section 15.8); Phylum *Euryarchaeota: The Halobacteria* (section 20.3)

The genome sequence of *Silicobacter pomeroyi*, another member of the bacterioplankton community, demonstrates that we have much to learn about the strategies bacteria use to cope in the oligotrophic ocean. Like *P. ubique*, *S. pomeroyi* is an α-proteobacterium but is part of the marine *Roseobacter* clade, which constitutes 10 to 20% of coastal and oceanic bacterioplankton. Originally thought to be a chemoorganotroph, the microbe depends on chemolithoheterotrophy, supplementing its heterotrophic existence with the use of inorganic compounds such as sulfide and carbon monoxide (CO) as sources of energy and electrons. Sulfide is found on marine snow and the *S. pomeroyi* genome encodes two putative quorum-sensing systems that would be useful to the microbe when at high cell densities, as might occur on heavily colonized pieces of marine snow. Two operons encode CO dehydrogenases; these enzymes catalyze the oxidation of CO to CO_2. CO is abundant in seawater due to photooxidation of DOM. However, because *S. pomeroyi* lacks the genes for carbon fixation, CO does not appear to be a carbon source. Rather CO oxidation is used to donate electrons for energy production. Once again, this has implications for the global carbon budget as *Silicobacter*-like consumption of CO may remove as much as 10 to 60 *trillion* grams of C annually, thereby buffering the partial pressure of CO_2 in the ocean.

One of the most interesting discoveries in recent years is the widespread presence of large numbers of marine archaea, once thought to inhabit only extreme environments. In fact, the *Archaea* are ubiquitous in the marine environment—they have been found in polar and tropical regions, and in estuarine, planktonic, and deep-sea communities. Archaea are also found in freshwater environments. 16S rRNA analysis and direct visualization of archaea using epifluorescence has revealed that at least 20% of oceanic picoplankton are *Crenarchaeota*. Many of these populations are closely related, suggesting that a shared collection of adaptations has been fine-tuned to meet the needs of the different niches within specific picoplankton communities. The distribution of archaea relative to that of bacteria differs widely with the particular ecosystem. **Figure 28.14** shows typical results for the open ocean. Bacteria are most numerous in the upper 150 to 200 meters (i.e., the photic zone), but archaea increase in relative abundance with depth until they approximate, or sometimes even exceed, bacteria.

As mentioned in our discussion of the microbial loop, viruses are important members of marine and freshwater microbial communities. In fact, **virioplankton** are the most numerous members of marine ecosystems. It was not until the 1990s that the abundance of marine viruses was recognized; their study has become one of the most active and exciting areas of marine microbiology.

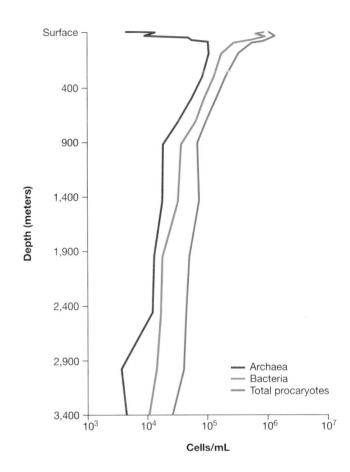

Figure 28.14   Archaea Are Plentiful in Ocean Depths. The distribution of archaea and bacteria, together with an estimate of total procaryotes, over a depth of 3,400 meters is shown at a Pacific Ocean location. These results indicate that archaea make up a significant part of the observable picoplankton below the surface zone.

Quantifying viruses is tricky: the traditional method of plaque formation requires knowledge of virus and host, as well as the ability to grow the host in the laboratory. Recall that only about 1% of all microbes have been cultured; this often prevents the measurement of virus diversity by examining actual virus infection. Instead, virus particles may be visualized directly. This requires the concentration of many tens of liters of water for direct examination by transmission electron microscopy. Because this does not prove that any given virus can actually infect a host cell, viruses enumerated in this way are called **virus/like particles (VLPs).** Using this approach, the average VLP density in seawater is between 10^6 to 10^7 per milliliter (although it may be closer to 10^8 per milliliter); their numbers decline to roughly 10^6 below about 250 meters. Marine viruses are so abundant that virus particles are now recognized as the most abundant life forms on Earth. As might be expected by their numbers, these viruses are very diverse, including single- and double-stranded RNA and DNA viruses that infect archaea, bacteria, and protists. Measurement of viral lysis of bacteria and archaea in the field is difficult

Figure 28.15   The Role of Viruses in the Microbial Loop.   Viral lysis of autotrophic and heterotrophic microbes accelerates the rate at which these microbes are converted to particulate and dissolved organic matter (P-D-OM). This is thought to increase net community respiration and decrease the efficiency of nutrient transfer to higher trophic levels.

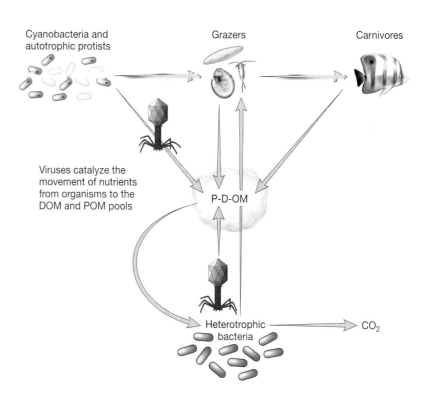

and yields variable estimates, but appears to account for between 10 and 50% of the total mortality of the procaryotic members of the marine microbial community. Computer modeling and model experiments indicate that viruses contribute to nutrient cycling by accelerating the rate at which their microbial hosts are converted to POM and DOM, thereby "feeding" other microorganisms without first making them available for protists and other bacteriovores. This "short-circuits" the microbial loop (**figure 28.15**).

Genome sequencing of DNA cloned directly from marine ecosystems suggests that phages are important vectors for horizontal gene transfer. In fact, it has been calculated that in the oceans, phage-mediated gene transfer occurs at an astounding rate of 20 billion times per second. Recently, the importance of phage-mediated lateral gene transfer was demonstrated when it was discovered that cyanophages thought to infect the cyanobacteria *Synechococcus* and *Prochlorococcus* carry the structural gene for a photosynthetic reaction center protein. These phages are presumably shuttling this essential gene between the two genera of cyanobacteria, and may thereby play a critical role in the evolution of these important primary producers.   Transmission electron microscopy (section 2.4); Viruses of *Bacteria* and *Archaea* (chapter 17); Transduction (section 13.9); Environmental genomics (section 15.9); Microbial ecology and its methods: An overview (section 27.4)

Benthic Marine Environments

The majority of the Earth's crust is under the sea, which means that of all the world's microbial ecosystems, we know the least about the largest. However, the combination of deep-ocean drilling projects and the exploration of geologically active sites,

such as submarine volcanoes and hydrothermal vents, has revealed that the study of ocean sediments, or **benthos,** can be rewarding and surprising. Marine sediments range from the very shallow to the deepest trenches, from dimly illuminated to completely dark, and from the newest sediment on Earth to material that is millions of years old. The temperature of such sediments depends on the proximity of geologically active areas. The exciting discovery of hydrothermal vent communities with large and diverse invertebrates, some of which depend on endosymbiotic chemolithotrophic bacteria, has been intensely investigated since their discovery in the late 1970s. These microbes are discussed in chapter 30. Because the vast majority of Earth's crust lies at great depth far from geothermally active regions, most benthic marine microbes live under high pressure, without light, and at temperatures between 1°C to 4°C.

Deep-ocean sediments were once thought to be devoid of all life and were therefore not considered worth the considerable effort it takes to study them. In fact, it took an international consortium of scientists to organize the Ocean Drilling Project in 1985, which has now been expanded to the Integrated Ocean Drilling Program through 2013. Researchers aboard the research vessel *JOIDES Resolution* drill cores of sediments from water depths up to 8,200 meters (at its deepest, the ocean is about 11,000 m). Microbiologists now know that far from being sterile, it appears the total subsurface (intraterrestrial) microbial biomass equals that of all terrestrial and marine plants. This is possible because benthic marine microbes inhabit not just the surface of the sea floor, but within sediments to a depth of at least 0.6 km. To survive at these depths, microorganisms must be able to tolerate atmospheric pressures up to 1,100 atm (pressure increases about 1 atm/10 meters depth). Such microbes are said to be **barophilic** (Greek, *baro*, weight, and *philein*, to love). Some are obligate barophiles and must be cultured in special hyperbaric incubation chambers. In fact, scientists have yet to find the subterranean depth limit of microbial growth. However, it appears that it will not be governed by pressure; rather, it will be determined by temperature. It seems unlikely that the maximum temperature at which life can be sustained has been identified.

One outcome of deep-ocean sediment exploration has been the discovery of methane hydrates. These pools of trapped methane are produced by methanogenic archaea that convert acetate to methane. This accumulates in lattice-like cages of crystalline water 500 meters or more below the sediment surface in many regions of the world's oceans. The formation of methane hydrates requires both cold temperatures and high pressure. This discovery is very significant because there may be up to 10^{13} metric tons of methane hydrate worldwide—80,000 times the world's current known natural gas reserve.

It is also exciting that recent deep-sea sediment drilling has turned our understanding of bacterial energetics literally upside down. As discussed in chapter 9, it is generally understood that anaerobic respiration occurs such that there is preferential use of available terminal electron acceptors. That which yields the most energy (ΔG) from the oxidation of NADH or an inorganic reduced compound (e.g., H_2, H_2S) will be used before those electron acceptors that produce a smaller ΔG (*see table 8.1 and figure 8.8*). Thus following oxygen depletion, nitrate will be reduced; then manganese, iron, sulfate, and finally carbon dioxide. When sediment cores up to 420 meters deep were collected off the coast of Peru, this predictable profile of electron acceptors and their microbial-derived reduced products were observed within the upper strata of the sediments (**figure 28.16**). However, when researchers measured these signature chemical compounds at great depth, the profile was upside down. This suggests the presence of unknown sources of these electron acceptors at subsurface depths of more than 420 meters. In addition, contrary to our long-held notion of thermodynamic limits, methane formation (methanogenesis) and iron and manganese reduction seem to be co-occurring in sediments with high sulfate concentrations. Although the identity of the microbes that make up this community awaits further study, it is clear that with densities of 10^8 cells per gram of sediment at the seafloor surface and 10^4 cells per gram in deep subsurface sediments, these communities are important. Free energy and reactions (section 8.4); Oxidation-reduction reactions, electron carriers, and electron transport systems (section 8.6); Anaerobic respiration (section 9.6)

---

1. What is marine snow? Why is it important in CO_2 draw-down?
2. Name two sources of organic nitrogen in the open ocean that have only recently been recognized.
3. Draw a diagram of the anammox reaction and explain its importance to the global carbon and nitrogen budgets.
4. Why do you think that despite its great abundance, SAR11 was not discovered until the late 20th century?
5. Why do you think marine viruses are difficult to study?

Figure 28.16   Microbial Activity in Deep-Ocean Sediments. At the surface of the seafloor, reduction of oxidized substrates that serve as electron acceptors in anaerobic respiration follows a predictable stratification based on thermodynamic considerations. This sequence is just the opposite in very deep subsurface sediments, suggesting a source of electron acceptors from deep within Earth's crust. Meters below seafloor, mbsf.

6. Describe the role of marine viruses in the microbial loop.
7. What are methane hydrates?
8. Explain what is meant by "upside-down microbial energetics" as described for deep subsurface ocean sediments.

28.4 MICROORGANISMS IN FRESHWATER ENVIRONMENTS

While the vast majority of water on Earth is in marine environments, freshwater is crucial to our terrestrial existence. Here we discuss freshwater (aquatic) environments and describe them as habitats for a diverse collection of microbes.

Microorganisms in Glaciers and Permanently Frozen Lakes

We begin our discussion of freshwater microbes with those that reside in ice that has remained frozen for thousands of years. Although this may seem like an extreme environment, it is important to note that a majority of the Earth's surface never exceeds a temperature of 5°C. This includes polar regions, the deep ocean, as well as high-altitude terrestrial locations throughout the world. Surprisingly, microbes within glaciers are not dormant. Rather, evidence that active microbial communities exist in these environments has been growing over the last decade. In fact, this is an exciting time in glacial microbiology; determining the diversity in these systems and assessing the role of these microorganisms in biogeochemical cycling can involve novel and creative techniques. The results may be of great consequence because glaciers have traditionally been regarded as areas that do not contribute to the global carbon budget. In addition, since the discovery of ice on Mars and on Jupiter's moon Europa, astrobiologists have become very interested in ice-dwelling psychrophilic microbes.

The influence of environmental factors on growth: Temperature (section 6.5)

Among the frozen landscapes of interest to microbiologists are permanently frozen lakes, such as Antarctica's McMurdo Dry Valley Lakes where the ice is 3 to 6 meters deep. Life in these ecosystems depends on the photosynthetic activity of microbial psychrophiles. In contrast, lakes that lie below glaciers are blocked from solar radiation. These communities are driven by chemosynthesis. One of the most intriguing and well-studied Antarctic habitats is Lake Vostok, one of 68 lakes located 3 to 4 kilometers below the East Antarctic Ice Sheet (**figure 28.17**). Geothermal heating, pressure, and the insulation of the overlying ice keep these lakes in a liquid state. It is thought that Lake Vostok was formed approximately 420,000 years ago and that its water is about a million years old. This stable environment supports a number of microbes including gram-negative proteobacteria and gram-positive actinomycetes. The overlying ice also harbors similar microbes, although at lower population densities. To find out if these microbial populations are active, radiolabeled substrates, including ^{14}C-acetate and ^{14}C-glucose, were added to samples incubated at an Antarctic laboratory. Indeed, the respiration of these compounds, measured as ^{14}C-CO_2, demonstrates that these communities are dynamic and of great interest.

Microorganisms in Streams and Rivers

As freshwater glaciers melt, their waters enter streams and rivers. This marks a departure from an environment that is stable on a ge-

Figure 28.17 The East Antarctic Ice Sheet and Lake Vostok.
(a) Lake Vostok lies beneath several kilometers of the ice sheet.
(b) The deep drilling station from which ice and lake water samples were obtained is indicated by the yellow dot. The yellow outline identifies a smooth plateau of snow and ice that floats on top of Lake Vostok. The surrounding rough ice is formed when the ice sheet moves over bedrock rather than liquid water.
(c) Microbes have been found in both the overlying ice and the lake water, as seen in this epifluorescent image.

ologic time scale to one that is extremely changeable. The continuous flow of water in streams and all but the largest rivers prevents the development of significant planktonic communities. Instead, most of the microbial biomass is attached to surfaces. Depending on the size of the stream or river, the source of nutrients may vary. The source may be in-stream, called **autochthonous** production based on photosynthetic microorganisms (**figure 28.18a**). Nutrients also may come from outside the stream, including runoff sediment from riparian areas (the edge of a river), or leaves and other organic matter falling directly into the water (figure 28.18b). Such nutrients are called **allochthonous.** Chemoorganotrophic microorganisms metabolize the available organic material, recycling nutrients within the ecosystem. Autotrophic microorganisms grow using the minerals released from the organic matter. This leads to the production of O_2 during the daylight hours; respiration occurs at night farther down the river, resulting in diurnal oxygen shifts. Eventually the O_2 level approaches saturation, completing a self-purification process. Thus when the amount of organic matter added to streams and rivers does not exceed the system's oxidative capacity, productive and aesthetically pleasing streams and rivers are maintained.

(a)

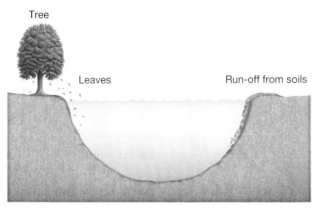

Tree

Leaves

Run-off from soils

(b)

Figure 28.18 Organic Matter Sources for Lakes and Rivers.
Organic matter used by microorganisms in lakes and rivers can be synthesized in the water (autochthonous), or can be added to the water from outside sources (allochthonous). **(a)** Autochthonous sources of organic matter, primarily photosynthesis, and **(b)** allochthonous sources of organic matter.

The capacity of streams and rivers to process added organic matter is limited. If too much organic matter is added, the system is said to be **eutrophic** and the rate of respiration exceeds that of photosynthesis. Thus the water may become anoxic. This is especially the case with urban and agricultural areas located adjacent to streams and rivers. The release of inadequately treated municipal wastes and other materials from a specific location along a river or stream represents **point source pollution.** Such point source additions of organic matter can produce distinct and predictable changes in the microbial community and available oxygen, creating an oxygen sag curve (**figure 28.19**). Runoff from agriculturally active fields and feedlots is an example of **nonpoint source pollution.** This can cause disequilibrium in the microbial community leading to algal or cyanobacterial blooms.

Along with the stresses of added nutrients, removal of silica from rivers by the construction of dams and trapping of sediments causes major ecological disturbances. For example, construction of the dam at the "iron gates" on the Danube (600 miles above the Black Sea) has led to a decrease in silica to 1/60th of the previous concentration. This decreased silica availability inhibits the growth of diatoms because the ratio of silica to nitrate has been altered (silica is required for diatom frustule formation). With this shift in resources, Black Sea diatoms are not able to grow and immobilize nutrients. The result has been a 600-fold increase in nitrate levels and a massive development of toxic algae. This example shows that the delicate balance of river ecosystems can be altered in unexpected ways by dams changing microbiological processes.

Although there are over 36,000 dams worldwide, many nations have recognized the devastating consequences of unrestricted or improper dam construction and efforts are under way to remove

Figure 28.19 The Dissolved Oxygen Sag Curve. Microorganisms and their activities can create gradients over distance and time when nutrients are added to rivers. An excellent example is the dissolved oxygen sag curve, caused when organic wastes are added to a clean river system. During the later stages of self-purification, the phototrophic community will again become dominant, resulting in diurnal changes in river oxygen levels.

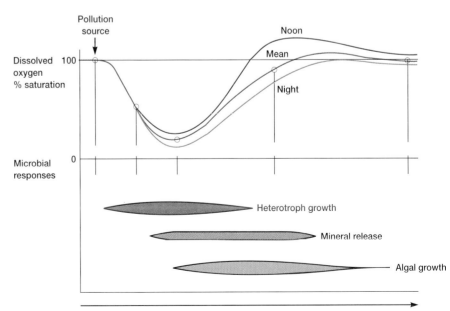

River flow (time/distance downstream from pollution source)

some of the worst dams. These structures must be breached to restore normal water flow and enable important commercial fish renewed access to spawning grounds. Clearly this will be an area
requiring microbiological expertise for many years to come.

---

1. Describe the Lake Vostok ecosystem. Why is this a chemosynthesis based,
   rather than a photosynthetically based, microbial community?
2. What is an oxygen sag curve? What changes in a river cause these effects?
3. What are point and nonpoint souce pollution? Can you think of examples in
   your community?
4. Why might dams influence microorganisms and microbial processes in
   rivers?

Microorganisms in Lakes

Lakes offer a completely different set of physical and biological
features for microbial communities, although not all lake environments are the same. Lakes vary in nutrient status. Some are oligotrophic (**figure 28.20***a*), others are eutrophic (figure 28.20*b*).
Nutrient-poor lakes remain oxic throughout the year, and seasonal
temperature shifts usually do not result in distinct oxygen stratification. In contrast, eutrophic lakes usually have bottom sediments
that are rich in organic matter. In thermally stratified lakes the **epilimnion** (warmer, upper layer) is oxic, while the **hypolimnion**
(colder, bottom layer) often is anoxic (particularly if the lake is nutrient-rich). The epilimnion and hypolimnion are separated by a
thermocline.

Because there is little mixing between the epilimnion and the
hypolimnion, the bottom waters may become deprived of oxygen. This is a permanent situation in tropical eutrophic lakes and
occurs in the summer in temperate eutrophic lakes. If nutrient levels are high, this bottom, or benthic, zone becomes dominated by
anaerobic microbial activity. In very warm eutrophic lakes,
anaerobic microbes release gases such as H_2S into the water. In
addition, human activities (e.g., septic and agricultural runoff)
may add high levels of nitrogen and phosphorus to the lake waters. This can result in a bloom of algae, plants, and/or bacteria in
the epilimnion. During autumn cooling, temperate lakes lose their
thermocline because surface waters increase in density and
storms mix the two layers. This sometimes happens within a 24-
hour period; if bottom water has become filled with anaerobic by-
products, the sudden upwelling causes fish kills.

In oligotrophic lakes, phosphorus is often the limiting nutrient. If added to oligotrophic freshwater, cyanobacteria capable of
nitrogen fixation may bloom. Several genera, notably *Anabaena,
Nostoc,* and *Cylindrospermum,* can fix nitrogen under oxic conditions. The genus *Oscillatoria,* using hydrogen sulfide as an
electron donor for photosynthesis, can fix nitrogen under anoxic
conditions. If both nitrogen and phosphorus are present,
cyanobacteria compete with algae. Cyanobacteria function more
efficiently in alkaline waters (8.5 to 9.5) and higher temperatures
(30 to 35°C). Photosynthetic protists, in comparison, generally
grow best at neutral pH and have lower optimum temperatures.
By using CO_2 at rapid rates, cyanobacteria also increase the pH,
making the environment less suitable for protists. Photosynthetic
bacteria: Phylum *Cyanobacteria* (section 21.3)

Cyanobacteria have additional competitive advantages. Many
produce hydroxamates, which bind iron, making this important

**Figure 28.20 Oligotrophic and Eutrophic
Lakes.** Lakes can have different levels of
nutrients, ranging from low nutrient to extremely
high nutrient systems. The comparison of **(a)** an
oligotrophic (nutrient-poor) lake, which is oxygen
saturated and has a low microbial population,
with **(b)** a eutrophic (nutrient-rich) lake. The
eutrophic lake has a bottom sediment layer and
can have an anoxic hypolimnion. As microbial
biomass increases with nutrient levels, light
penetration is diminished. Thus the bottom of
eutrophic lakes may be become dark, anoxic, and
even poisonous from H_2S production.

(a)

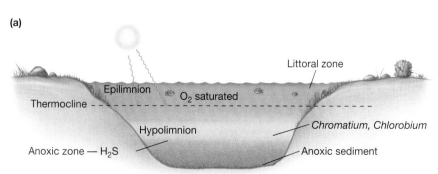

(b)

trace nutrient less available for protists. Some cyanobacteria also resist predation because they produce toxins. In addition, some synthesize odor-producing compounds that affect the quality of drinking water. However, both cyanobacteria and protists can contribute to massive blooms in strongly eutrophied lakes. Lake management can improve the situation by removing or sealing bottom sediments or adding coagulating agents to speed sedimentation.

1. What terms can be used to describe the different parts of a lake?
2. What are some important effects of eutrophication on lakes?
3. Why are cyanobacteria so important in waters that have been polluted by phosphorus additions?

Summary

28.1 Marine and Freshwater Environments

a. Most water on the Earth is marine (97%). The majority of this is cold (2 to 5°C) and at high pressure. Fresh water is a minor but important part of Earth's biosphere.

b. Oxygen solubility and diffusion rates in surface waters are limited; waters are low oxygen diffusion rate environments, in comparison with soils. Carbon dioxide, nitrogen, hydrogen, and methane are also important gases for microbial activity in waters.

c. The carbonate equilibrium system keeps the oceans buffered at pH 7.6 to 8.3 (**figure 28.2**).

d. The penetration of light into the surface water determines the depth of the photic zone. Warming the surface waters can lead to the development of a thermocline.

e. The nutrient composition of the ocean influences the C:N:P ratio of the phytoplankton, which is called the Redfield ratio. This ratio is important for predicting nutrient cycling in oceans. Atmospheric additions of minerals, including iron and nitrogen, affect this ratio and global-level oceanic processes.

f. The microbial loop describes the transfer of nutrients between trophic levels while taking into account the multiple contributions of microbes to recycling nutrients. Nutrients are recycled so efficiently, the majority remain in the photic zone (**figure 28.3**).

28.2 Microbial Adaptations to Marine and Freshwater Environments

a. The marine microbial community is dominated, in terms of numbers and biomass, by ultramicrobacteria.

b. Many unusual microbial groups are found in waters, especially when oxidants and reductants can be linked. These include *Thioploca* and *Thiomargarita*, both of which are found in coastal areas where nutrient mixing occurs. *Thiomargarita* is the world's largest known bacterium (**figure 28.4**).

c. Aquatic fungi are important members of the aquatic microbial community. These include the chytrids, with a motile zoosporic stage, and the Ingoldian fungi, which often have tetraradiate structures. Both of these are uniquely adapted to an aquatic existence, and the chytrids may contribute to disease in amphibians (**figures 28.6 and 28.7**).

28.3 Microorganisms in Marine Environments

a. Tidal mixing in estuaries, as characterized by a salt wedge, is osmotically stressful to microbes in this habitat. Thus they have evolved mechanisms to cope with rapid changes in salinity (**figure 28.8**).

b. Coastal regions like estuaries and salt marshes can be the sites of harmful algal blooms such as those caused by the diatom *Pseudonitzchia* and the dinoflagellates *Alexanderium* and *Pfiesteria* (**figure 28.9**).

c. Autotrophic microbes in the photic zone within the open ocean account for about one-half of all the carbon fixation on Earth.

d. The carbon and nitrogen budgets of the open-ocean photic zone are intensely studied because of their implications for controlling global warming (**figure 28.12**).

e. Two members of the α-proteobacteria—SAR11 and *Silicobacter pomeroyi*—demonstrate unique adaptations to life in the oligotrophic open ocean. SAR11 is the most numerous organism on Earth.

f. Archaea are important components of the microbial community. Viruses are present at high concentrations in many waters, and occur at 10-fold higher levels than the bacteria. In marine systems they may play a major role in controlling cyanobacterial development and nutrient turnover (**figures 28.14 and 28.15**).

g. Sediments deep beneath the ocean's surface are home to about one-half of the world's procaryotic biomass.

h. Methane hydrates, the result of psychrophilic archaeal methanogenesis under extreme atmospheric pressure, may contain more natural gas than is currently found in known reserves.

i. Deep sediment drilling reveals that microbes within this habitat may employ unique energetic strategies (**figure 28.16**).

28.4 Microorganisms in Freshwater Environments

a. Glaciers and permanently frozen lakes are sites of active microbial communities. The East Antarctic Ice Sheet and Lake Vostok, which lies beneath it, are productive study sites (**figure 28.17**).

b. Nutrient sources for streams and rivers may be autochthonous or allochthonous. Often allochthonous inputs include urban, industrial, and agricultural runoff (**figure 28.18**).

c. Lakes can be oligotrophic or eutrophic. Eutrophication can cause increased growth of chemoorganotrophic microbes and the system may become anoxic (**figure 28.20**).

Key Terms

allochthonous 682
anoxic 668
autochthonous 682
barophilic 681

benthos 680
brevetoxin 675
carbonate equilibrium system 668
dissolved organic matter (DOM) 671

epilimnion 684
eutrophic 683
*f*luorescence *in situ* *h*ybridization (FISH) 678

halotolerant 673
harmful algal bloom (HAB) 675
hypolimnion 684
hypoxic 668

Critical Thinking Questions

1. In what habitats might you find microbes like *Thioploca* and *Thiomargarita*? What culture conditions would you use to purify such microbes from their natural habitat?

2. How might it be possible to cleanse an aging eutrophic lake? Consider chemical, biological, and physical approaches as you formulate your plan.

3. *Clostridium botulinum*, the causative agent of botulism, sometimes causes fish kills in lakes where people swim. Currently there are few, if any, monitoring procedures for this potential source of disease transmission to humans. Do you think a monitoring program is needed and, if so, how would you implement such a program?

4. Do you think fertilization of the ocean with iron to increase CO_2 drawdown is a good approach to controlling global warming? Why or why not? Why do you think similar experiments yield different results?

Learn More

Arrigo, K. R. 2005. Marine micro-organisms and global nutrient cycles. *Nature* 437:349–55.

Capone, D. G. 2001. Marine nitrogen fixation: What's all the fuss? *Curr. Opin. Microbiol.* 4:341–48.

DeLong, E. F., and Karl, D. M. 2005. Genomic perspectives in microbial oceanography. *Nature* 437:336–42.

Dyhrman, S. T.; Chapell, P. D.; Haley, S. T.; Moffett, J. W.; Orchard, E. D.; Waterbury, J. B.; and Webb, E. A. 2006. Phosphonate utilization by the globally important marine diazotrophic *Trichodesmium*. *Nature* 439:68–71.

Edwards, K. J.; Bach, W.; and McCollom, T. M. 2005. Geomicrobiology in oceanography: Microbe-mineral interactions at and below the seafloor. *Trends Microbiol.* 13:449–56.

Giovannoni, S. J., and Stingl, U. 2005. Molecular diversity and ecology of microbial plankton. *Nature* 437:343–48.

Karl, D. M.; Bird, D. F.; Björkman, K.; Houlihan, T.; Shackelford, R.; and Tupas, L. 1999. Microorganisms in the accreted ice of Lake Vostok, Antarctica. *Science* 286:2144–47.

Kuypers, M. M.; Lavik, G.; Woebken, D.; Schmid, M.; Fuchs, B. M.; Amann, R.; Jørgensen, B. B.; and Jetten, M. S. M. 2005. Massive nitrogen loss from the Benguela upwelling system through anaerobic ammonium oxidation. *Proc. Natl. Acad. Sci. USA* 102:6478–83.

Moran, M. A.; Buchan, A.; González, J. M.; Heidelberg, J. F.; Whitman, W. B.; Kiene, R. P., et al. 2004. Genome sequence of *Silicobacter pomeryi* reveals adaptations to the marine environment. *Nature* 432:910–13.

Rappé, M. S.; Connon, S. A.; Vergin, K. L.; and Giovinonni, S. J. 2002. Cultivation of the ubiquitous SAR11 marine bacterioplankton clade. *Nature* 418:630–33.

Schippers, A.; Neritin, L. N.; Kallmeyer, J.; Ferdelman, T. G.; Cragg, B. A.; Parkes, R. J., and Jørgensen, B. B. 2005. Prokaryotic cells of the deep sub-seafloor biosphere identified as living bacteria. *Nature* 433:861–64.

Smatacek, V., and Nicol, S. 2005. Polar ocean ecosystems in a changing world. *Nature* 437:362–68.

Suttle, C. A. 2005. Viruses in the sea. *Nature* 437:356–61.

Please visit the Prescott website at www.mhhe.com/prescott7 for additional references.

29

Microorganisms in Terrestrial Environments

Terrestrial plants and filamentous fungi have developed long-term relationships that benefit both partners. The tips of pine tree roots are usually surrounded by dense fungal sheaths that are part of a hyphal network which extends out into the soil. The plant supplies organic matter to maintain the fungus, and the fungus, in turn, provides the plant with nutrients and water.

PREVIEW

- Soil is a complex environment offering a variety of microhabitats. This is one reason why microbial diversity in soils is much greater than that found in aquatic environments.

- In terrestrial ecosystems, primary production is performed by plants, but the nutrient recycling that occurs though a microbial loop is also essential. Each climate and soil type has a community of microbes specifically adapted to that particular microhabitat.

- Many microbes inhabit the pores between soil particles; others live in association with plants. The plant root surface (rhizoplane) and the region close to plant roots (the rhizosphere) are important sites for microbial growth.

- Mycorrhizal fungi associate with most plants. In this relationship, the fungi provide their plant partner with essential nutrients like nitrogen and phosphorus, while the plant supplies organic carbon to the fungi.

- Rhizobia include α- and β-proteobacteria that form nodules within the roots of leguminous plants, where they fix nitrogen. This process has been best studied in the genus *Rhizobium* and its relatives; it involves a complex plant-microbe communication system and the differentiation of the bacterium into a form that can fix nitrogen. A variety of other bacteria, including the actinomycete *Frankia*, also fix nitrogen while interacting with plants.

- The plant pathogen *Agrobacterium* also relies on an intercellular communication system with host plants, in which it causes tumors called galls. These arise following the insertion of a fragment of bacterial DNA, called T DNA, into the plant cell's chromosome.

- Subsurface microbiology is a relatively new and exciting field that explores vast microbial communities living deep beneath the topsoil. Recent studies show that the biomass within this microbial world equals at least one-third of that living above ground.

hapter 28 introduces aquatic and marine environments; we now turn our attention to land. The microbiology of soil is important for a variety of reasons. These include the contribution terrestrial microbes make to global biogeochemical cycles and the essential role soil microbes play in agriculture and in maintaining environmental quality. These are just some of the reasons that the microbial ecology of soil is a dynamic and growing field. In the past, culture-based investigations limited scientists' understanding to an estimated 1% of the soil microbes in any given community. As we shall see, the ability to study these complex communities without relying on the direct isolation and growth of individual species has had a profound impact on the appreciation of the soil as a complex and vital environment.

We begin by describing the soil habitat and how microbes contribute to the development of soils. This is followed by a discussion of specific soil microbial communities and the interaction of microbes with vascular plants. We pay particular attention to two very important relationships: that between fungi and plants and between nitrogen-fixing bacteria and leguminous plants. Finally, the new and exciting field of deep subsurface microbiology is introduced.

29.1 SOILS AS AN ENVIRONMENT FOR MICROORGANISMS

A soil scientist would describe soil as weathered rock combined with organic matter and nutrients. An agronomist would point out that soil supports plant life. However, a microbial ecologist knows

They [the leaves] that waved so loftily, how contentedly, they return to dust again and are laid low, resigned to lie and decay at the foot of the tree and afford nourishment to new generations of their kind, as well as to flutter on high!

—Henry D. Thoreau

that the formation of organic matter and the growth of plants depend on the microbial community within the soil. Historically, the complexity of soil as a habitat has been a challenge to understanding soil microbial ecology. Soil is very dynamic and is formed in a wide variety of environments. These environments range from Arctic tundra regions, where approximately 11% of the Earth's soil carbon pool is stored, to Antarctic dry valleys, where there are no vascular plants. In addition, deeper subsurface zones, where plant roots and their products cannot penetrate, also have surprisingly large microbial communities. Microbial activities in these environments can lead to the formation of minerals such as dolomite; microbial activity also occurs in deep continental oil reservoirs, in stones, and even in rocky outcrops. These microbes are dependent on energy sources from photosynthetic protists and nutrients in rainfall and dust.

Most soils are dominated by inorganic geological materials, which are modified by the biotic community, including microorganisms and plants, to form soils. The spaces between soil particles

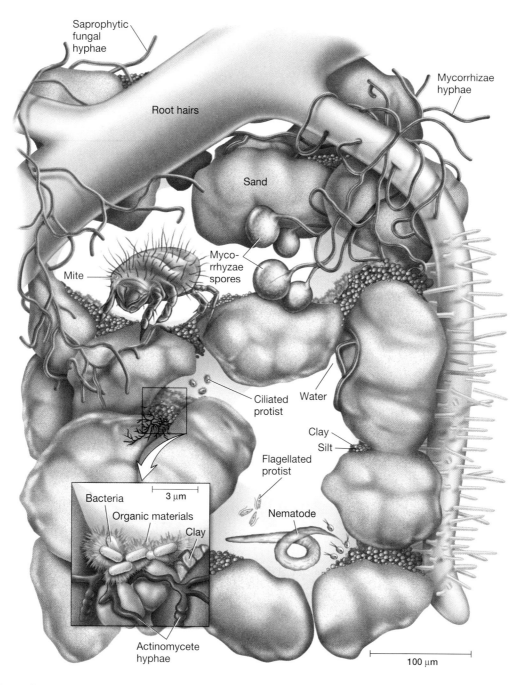

Figure 29.1 The Soil Habitat. A typical soil habitat contains a mixture of clay, silt, and sand along with soil organic matter. Roots, animals (e.g., nematodes and mites), as well as chemoorganotrophic bacteria consume oxygen, which is rapidly replaced by diffusion within the soil pores where the microbes live. Note that two types of fungi are present: mycorrhizal fungi, which derive their organic carbon from their symbiotic partners—plant roots; and saprophytic fungi, which contribute to the degradation of organic material.

are critical for the movement of water and gases (**figure 29.1**). Total pore space, and thus gas diffusion, is determined by the texture of the soil. For instance, sandy soils have larger pore spaces than do clay soils, so sandy soils tend to drain quickly. Pores are also critical because they provide the optimum environment for microbial growth. Here the microbes are within thin water films on the particle surfaces where oxygen is present at high levels and can be easily replenished by diffusion. Oxygen diffusion through air in the soil occurs about 10,000 times faster than it does through water (**figure 29.2**). The oxygen concentrations and flux rates in pores and channels is high, whereas within water-filled zones the oxygen flux rate is much lower. As an example, particles as small as about 2.0 mm can be oxic on the outside and anoxic on the inside.

Depending on the physical characteristics of the soil, rainfall or irrigation may rapidly change a soil from being well aerated to an environment with isolated pockets of water, which are "miniaquatic" habitats. If this process of flooding continues, a waterlogged soil can be created that is more like a lake sediment. If oxygen consumption exceeds that of oxygen diffusion, waterlogged soils can become anoxic.

Shifts in water content and gas fluxes also affect the concentrations of CO_2, CO, and other gases present in the soil atmosphere, as noted in **table 29.1.** These changes are accentuated in

Table 29.1	Concentrations of Oxygen and Carbon Dioxide in the Atmosphere of a Tropical Soil under Wet and Dry Conditions			
	Oxygen Content (%)		Carbon Dioxide Content (%)	
Soil Depth (cm)	Wet	Dry	Wet	Dry
10	13.7	20.7	6.5	0.5
25	12.7	19.8	8.5	1.2
45	12.2	18.8	9.7	2.1
90	7.6	17.3	10.0	3.7
120	7.8	16.4	9.6	5.1

From E. W. Russell, *Soil Conditions and Plant Growth*, 10th edition. Copyright © 1973 Longman Group Limited, Essex, United Kingdom. Reprinted by permission.

Note: Normal air contains approximately 21% oxygen and 0.035% carbon dioxide.

the smaller pores where many bacteria are found. The roots of plants growing in aerated soils also consume oxygen and release CO_2, influencing the concentrations of these gases in the root environment.

1. What is the importance of soil pores?
2. Contrast differences in oxygen flux rates and concentrations in a soil with those in a miniaquatic environment.
3. How do the concentrations of oxygen and carbon dioxide differ between the atmosphere and the soil interior?

29.2 SOILS, PLANTS, AND NUTRIENTS

Soils can be divided into two general categories: A **mineral soil** contains less than 20% organic carbon whereas an **organic soil** possesses at least this amount. By this definition, the vast majority of Earth's soils are mineral. The importance of organic matter within soils cannot be underestimated. **Soil organic matter (SOM)** helps to retain nutrients, maintain soil structure, and hold water for plant use. SOM is subject to gains and losses, depending on changes in environmental conditions and agricultural management practices. Plowing and other disturbances expose SOM to more oxygen, leading to extensive microbiological degradation of organic matter. Irrigation causes periodic wetting and drying, which can also lead to increased degradation of SOM, especially at higher temperatures.

Microbial degradation of plant material results in the evolution of CO_2 and the incorporation of the plant carbon into additional microbial biomass. However, a small fraction of the decomposed plant material remains in the soil as SOM. When considering this material, it is convenient to divide the SOM into humic and nonhumic fractions (**table 29.2**). **Nonhumic SOM** has not undergone significant biochemical degradation. It can represent up to about 20% of the soil organic matter. **Humic SOM,** or **humus,** is dark brown to black. It results when the products of microbial metabolism have undergone chemical transformation within the soil. Although there is no precise

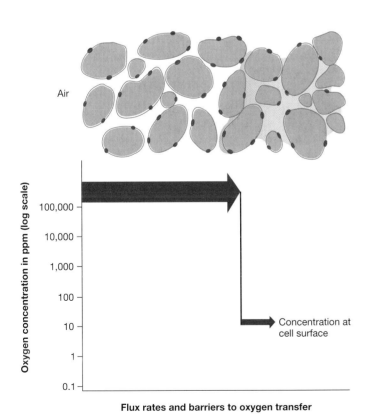

Figure 29.2 Oxygen Concentrations and Fluxes in a Soil.
Microorganisms in thin water films on the surface of soil particles have ample access to oxygen. In comparison, microbes in isolated water volumes have limited oxygen fluxes, creating miniaquatic environments.

Table 29.2	Fractions of Soil Organic Matter	
SOM Fraction	**Definition**	**Physical Appearance**
Humic substances	High-molecular-weight organic material produced by secondary synthesis reactions	Dark brown to black
Nonhumic substances	Unaltered remains of plants, animals, and microbes, from which macromolecules have not yet been extracted	Light brown
Humic acid	Organic matter extracted from soils by various reagents (often dilute alkali treatment) that is then precipitated by acidification	Dark brown to black
Fulvic acid	Soluble organic matter that remains after humic acid extraction	Yellow
Humin	SOM that cannot be extracted from soil with dilute alkali	

chemical composition of humus, it can be described as a complex blend of phenolic compounds, polysaccharides, and proteins. The recalcitrant nature of this material to degradation is evident by ^{14}C dating: the average age of most SOM ranges from 150 to 1,500 years.

The degradation of plant material and the development of SOM can be thought of as a three-step process. First, easily degraded compounds such as soluble carbohydrates and proteins are broken down. About half the carbon is respired as CO_2 and the remainder is rapidly incorporated into new biomass. During the second stage, complex carbohydrates, such as the plant structural polysaccharide **cellulose,** are degraded. Fungi and members of the bacterial genera *Streptomyces, Pseudomonas,* and *Bacillus* produce extracellular cellulase enzymes that break down cellulose into two to three glucose units called cellobiose and cellotriose, respectively. These smaller compounds are readily degraded and assimilated as glucose monomers. Finally, very resistant material, in particular lignin, is attacked. **Lignin** is an important structural component of woody plants. While its exact structure differs among species, the common building block is the phenylpropene unit. This consists of a hydroxylated six-carbon aromatic benzene ring and a three-carbon linear side chain (**figure 29.3**). A single lignin molecule can consist of up to 600 cross-linked phenylpropene units. It is therefore not surprising that lignin degradation is much slower than that of cellulose. Basidiomycete fungi and actinomycetes (e.g., *Streptomyces* spp.) are capable of extracellular lignin degradation. These microbes produce extracellular phenoloxidase enzymes needed for aerobic lignin degradation. Lignin decomposition is also limited by the physical nature of the material. For example, healthy woody plants are saturated with sap, which limits oxygen diffusion. In addition, high ethylene and CO_2 levels and the presence of phenolic and terpenoid compounds retard the growth of lignin-degrading fungi. It follows that no more than 10% of the carbon found in lignin is recycled into new microbial biomass. Lignin can be degraded anaerobically but this process is very slow, so lignin tends to accumulate in wet, poorly oxygenated soils, like peat bogs.

SOM represents only a small part of the total soil volume, but exerts a disproportionate influence on the biological, chemical, and physical dynamics of soil. Therefore, steps are frequently taken to maximize the SOM content of soil. The practice of re-

Figure 29.3 Example Phenylpropene Units. These molecules are polymerized to form lignin.

duced tillage has been employed to retain the amount of crop residue that contributes to SOM formation and water retention in the soil, which in turn helps prevent erosion.

Nitrogen is another important element of the soil ecosystem. Many soils are nitrogen limited; this is why each year tons of nitrogen fertilizer are added to agricultural soils. Nitrogen in soil is often considered in relation to the soil carbon content as the organic **carbon to nitrogen ratio (C/N ratio).** A C/N ratio of 20 or more (i.e., much more carbon than nitrogen) results in loss of soluble nitrogen from the system. Because the addition of nitrogen under such conditions does not stimulate growth, it is said that the soil has reached its **nitrogen saturation point.** Conversely, ratios below 20 enable microbes to convert ammonium and nitrate to biomass (e.g., proteins and nucleic acids).

Throughout the world, soils are increasingly being impacted by mineral nitrogen releases resulting from human activities. This nitrogen has two major sources: (1) agricultural fertilizers containing chemically synthesized nitrogen, and (2) fossil fuel combustion. Fossil fuel-based releases occur especially in eastern North America, Europe, eastern China, and Japan. The major types of nitrogen fertilizers used in agriculture are liquid ammonia and ammonium nitrate. Ammonium ion usually is added because it will be attracted to the negatively charged clays in a soil and be retained on the clay surfaces until used as a nutrient by the plants. However, the nitrifier populations in a soil can oxidize the ammonium ion to nitrite and nitrate, and these anions can be leached from the plant environment and enter surface waters and groundwaters. Biogeochemical cycling: Nitrogen cycle (section 27.2)

An inevitable consequence of the application of nitrogen fertilizers has been higher nitrate levels in waters, which can contribute to infant respiratory problems, and possibly to the production of nitrites and the formation of nitrosamine carcinogens. In addition, plants grown in high nitrate-concentration soils may accumulate nitrate to a level that is harmful for animals. Cereal grains, many weeds, and grass hay contain high nitrate levels when grown in such soils. Nitrogen fertilizers also can affect microbial community structure and function. The use of nitrogen fertilizers has led to decreases in filamentous fungal development in a wide variety of soils. The loss of fungi affects the soil structure because fungi are important conduits of nitrogen and phosphorus

to most plants. With a weakened and decreased fungal community, many plants become more susceptible to stresses such as drought and toxic metals.

Phosphorus in fertilizers also is critical. The binding of this anionic fertilizer component to soils is dependent on the cation exchange capacity (CEC) and soil pH. As the soil phosphorus sorption capacity is reached, the excess, together with phosphorus that moves to lakes, streams, and estuaries with soil erosion can stimulate the growth of freshwater organisms, particularly cyanobacteria, in the process of eutrophication. The cyanobacteria are then able to fix more nitrogen, because in most freshwater systems, phosphorus is the critical limiting element. As noted in **Microbial Tidbits 29.1,** excessive fertilizer can have unexpected global consequences. Microorganisms in marine environment: Coastal marine systems (section 28.3)

1. List three reasons why soil organic matter (SOM) is important.
2. What is the difference between humic and nonhumic SOM? Which is more abundant in most soils? Why?
3. Describe the three phases of plant degradation and SOM formation.
4. What are possible effects of nitrogen-containing fertilizers on microbial communities?
5. Most nitrogen fertilizer is added as ammonium ion. Why is this preferred over nitrate?
6. Why is nitrate of concern when it reaches rivers, lakes, and groundwaters?
7. Why might enrichment of fresh waters with phosphorus be even more critical than nitrogen enrichment?

Microbial Tidbits

29.1   An Unintended Global-Scale Nitrogen Experiment

Technological advances can have many unexpected consequences. An excellent example is the discovery of the Haber-Bosch Process at the beginning of the twentieth century that made possible the synthesis of ammonium nitrogen from inert dinitrogen gas. This led to the low-cost availability of mineral nitrogen for crop fertilization, which until then was largely dependent on animal-derived nitrogen sources (such as guano, primarily from Chile, and animal manure) and crop rotation schemes, often involving planting nitrogen-fixing legumes. With the availability of ample ammonium ion as fertilizer from the Haber-Bosch Process, there was no further need to continue such "inefficient" agricultural systems; a green cash crop could be generated each year without the tiresome use of manure and crop rotations. This, however, had unexpected consequences; without the addition of manures, organic matter was lost from soils in these more intensive cropping systems, and there were changes in plant and microbial communities, as well as increased water pollution with nitrates.

Nitrogen production by the Haber-Bosch Process, together with burning fossil fuels, adds mineral nitrogen to the atmosphere with other unexpected consequences. Studies of Northern Hemisphere

forests indicate that nitrogen is released from soils and drainage waters primarily as nitrate. It has been assumed that this reflected global-level natural processes as the phenomenon was so widespread. However, studies of forests in South America, which are far from sources of mineral nitrogen, suggest a very different view. Steven Perakis and Lars Hedin of Cornell University have shown that remote South American forests release only small amounts of nitrate, and instead that most nitrogen is released in organic forms. What appears to have happened? By having large-scale releases of mineral nitrogen in the Northern Hemisphere for almost a century, human activities have dramatically changed the nitrogen cycle of soils and waters in most areas of the populated world. It appears that the entire Earth has been turned into a giant nitrogen-amendment experiment. Agricultural and industrial pollution has strongly impacted the Northern Hemisphere, and is now affecting the Southern Hemisphere as well. Most researchers and policymakers have not been aware of this global-level nitrogen impact. The problem is that we cannot return to the beginning and modify such a global-scale experiment.

29.3 MICROORGANISMS IN THE SOIL ENVIRONMENT

If we look at a soil in greater detail (figure 29.1), we find that bacteria, archaea fungi, and protists use different functional strategies to take advantage of this complex physical matrix. Most soil procaryotes are located on the surfaces of soil particles and require water and nutrients that must be located in their immediate vicinity. Procaryotes are found most frequently on surfaces within smaller soil pores (2 to 6 μm in diameter). Here they are probably less liable to be eaten by protozoa, unlike those located on the exposed outer surface of a sand grain or organic matter particle.

Terrestrial filamentous fungi, in comparison, bridge open areas between soil particles or aggregates, and are exposed to high levels of oxygen. These fungi will tend to darken and form oxygen-impermeable structures called sclerotia and hyphal cords. This is particularly important for basidiomycetes, which form such structures as an oxygen-sealing mechanism. Within these structures, the filamentous fungi move nutrients and water over great distances, including across air spaces, a unique part of their functional strategy. These oxidatively polymerized, oxygen-impermeable hyphal boundaries do not usually occur in fungi growing in aquatic environments. Characteristics of the fungal divisions: *Basidiomycota* (section 26.6)

The microbial populations in soils can be very high. In a surface soil the bacterial population can approach 10^9 to 10^{10} cells per gram dry weight of soil as measured microscopically. Fungi can be present at up to several hundred meters of hyphae per gram of soil. We tend to think of soil fungi as small structures like the mushrooms sprouting from our lawns. However, the vast majority of

fungal biomass is below ground. For instance, an individual clone of the fungus *Armillaria bulbosa,* which lives associated with tree roots in hardwood forests, was discovered that covers about 30 acres in the Upper Peninsula of Michigan. It is estimated to weigh a minimum of 100 tons (an adult blue whale weighs about 150 tons) and be at least 1,500 years old. Thus some fungal mycelia are among the largest and most ancient living organisms on Earth.

Soil microbial communities appear to be more diverse than those found in most freshwater and marine environments. Recall that less than 1% of all soil microbes have been cultured (**figure 29.4**), so molecular techniques have been key to gaining an understanding of these complex ecosystems. Small subunit rRNA analysis and the re-association rate of denatured (single-stranded) DNA extracted directly from environmental samples have been used to measure community diversity. DNA reassociation measures the rate at which denatured DNA returns to the double-stranded state; this depends on the number of homologous chromosomes in the sample. Procaryotic community genome size is then calculated based on the assumption that the average procaryotic genome is about the same size as that of *E. coli* (4.1 Mb). Community genome size thus reflects the genomic diversity within the sample. Such analyses show that diversity is highest in pristine organic soils, with as many as 11,000 different genomes per cubic cm. The lowest diversity is found in extreme environments such as hypersaline ecosystems (**table 29.3**). Microbial ecology and its methods: Examination of microbial community structure (section 27.4)

Molecular techniques have been key to identifying microbes not previously thought to inhabit soils. The *Crenarchaeota* have been discovered in soil and ocean sediments by extracting microbial DNA and amplifying it with the polymerase chain reaction (*see figure 27.17*). Examination of soils from different areas of the world continues to yield surprises. Microorganisms are present and prolific in subsurface environments, including oil reservoirs. Hyperthermophilic archaea have been found in such harsh sub-

Figure 29.4 Cultured Soil Microbes. Although only about 1% of microorganisms have been cultured, those that grow in the laboratory show morphological diversity.

Table 29.3	Procaryotic Diversity Determined by Direct Cell Counts Using Fluorescence Microscopy (Abundance) and the Re-association Rate of DNA Isolated from Each Community Listed (Genome Equivalents)	
DNA Source	**Abundance (cells/cm³)**	**Genome Equivalents[1]**
Forest soil	4.8×10^9	6,000
Pasture soil	1.8×10^{10}	3,500–8,800
Arable soil	2.1×10^{10}	140–350
Marine fish farm	7.7×10^9	50
Hypersaline pond (22% salinity)	6.0×10^9	7

From V. Torsvik, L. Ovraes, and T. F. Thingstad. 2002. *Science* 296:1064–66.

[1] Genome equivalents are based on the assumption that the average procaryotic genome is about the size of the *E. coli* genome (4.1 Mb).

surface environments and are probably indigenous to these poorly understood regions of our world.   *Crenarchaeota (section 20.2)*

The coryneforms, the nocardioforms, and the true filamentous bacteria (the streptomycetes) (**table 29.4**) are an important part of the soil microbial community. These gram-positive bacteria play a major role in the degradation of hydrocarbons, older plant materials, and soil humus. In addition, some members of these groups actively degrade pesticides. The filamentous actinomycetes, primarily of the genus *Streptomyces,* produce an odor-causing compound called **geosmin,** which gives soils their characteristic earthy odor. Polyprosthecate bacteria such as the genera *Verrucomicrobium, Pedomicrobium,* and *Prosthecobacter* are present in soils at high levels. With their small size, and the difficulties involved in culturing them, they have been largely overlooked in assessments of soil microbial diversity.   *High G + C gram-positive bacteria (chapter 24)*

As discussed in section 29.6, the microbial community in soil makes important contributions to the carbon, nitrogen, sulfur, iron, and manganese biogeochemical cycles. Because the soil is primarily an oxidized environment, the inorganic forms of these elements tend to be in the oxidized state. If there are localized water-saturated, lower oxygen-flux environments, the biogeochemical cycles shift toward reduced species.   *Biogeochemical cycling (section 27.2)*

Soil insects and other animals such as nematodes and earthworms also contribute to organic matter transformations in soils. These organisms carry out decomposition, often leading to the release of minerals, and physically "reducing" the size of organic particles such as plant litter. This increases the surface area and makes organic materials more available for use by bacteria and fungi. Earthworms also mix substrates with their internal gut microflora and enzymes; this contributes substantially to decomposition and has major effects on soil structure and the soil microbial community.

Nutrients are regenerated in soils through a microbial loop that differs from that which operates in the photic zone of the open ocean. A major distinction is that plants rather than microbes account for most primary production in nearly all terrestrial systems. But much like the microbial loop in marine waters, microbes in soils rapidly recycle the organic material derived from plants and animals, including the many nematodes and insects. In turn, microbes themselves are preyed upon by soil protists, whose numbers can reach 100,000 per gram of soil. This makes microbial organic matter available to other trophic levels. Another difference between marine and terrestrial microbial loops reflects the physical and biological properties of soil. Degradative enzymes released by plants, insects, and other animals do not rapidly diffuse away; instead, they represent a significant contribution to the biological activity in soil ecosystems. In fact, these free enzymes contribute to many hydrolytic degradation reactions, such as proteolysis; catalase and peroxidase activities also have been detected.

1. What are the differences in preferred soil habitats between bacteria and filamentous fungi?
2. What types of archaea have been detected in soils?
3. How can earthworms, nematodes, and insects influence microbial communities?
4. How does the microbial loop function in soils?

29.4 MICROORGANISMS AND THE FORMATION OF DIFFERENT SOILS

Soils are formed when geologic materials are exposed. This may occur after a dramatic event such as a volcanic eruption or from a simple disturbance. Soil formation is the result of the combined action of weathering and colonization of geologic material by microbes. The microbial community that initially colonizes a newly disturbed environment will bring changes to the local environment by degrading and recycling organic material. For example, if extreme soil erosion has removed topsoil, phosphorus may be present, but nitrogen and carbon must be imported by physical or biological processes. Under these circumstances, autotrophic, N_2-fixing cyanobacteria are active in pioneer-stage nutrient accumulation. These primary microflora will eventually be replaced by a new community of microorganisms that further alter the ecosystem. This sets in motion successive waves of microbial communities until a community that is sufficiently diverse and physiologically well-suited is established and a stable climax ecosystem is developed. The major types of plant-soil systems are shown in **figure 29.5** and are discussed here.

Tropical and Temperate Region Soils

In warm, moist tropical soils, organic matter is decomposed very quickly and the mobile inorganic nutrients can be leached out of the surface soil environment, causing a rapid loss of fertility. To limit nutrient loss, many tropical plant root systems penetrate the rapidly decomposing litter layer. As soon as organic material and minerals are released during decomposition, the roots take them

Table 29.4	Easily Cultured Gram-Positive Irregular Branching and Filamentous Bacteria Common in Soils	
Bacterial Group	**Representative Genera**	**Comments and Characteristics**
Coryneforms	*Arthrobacter*	Rod-coccus cycle
	Cellulomonas	Important in degradation of cellulose
	Corynebacterium	Club-shaped cells
Mycobacteria	*Mycobacterium*	Acid-fast
Nocardioforms	*Nocardia*	Rudimentary branching
Streptomycetes	*Streptomyces*	Aerobic filamentous bacteria
Bacilli	*Thermoactinomyces*	Higher temperature growth

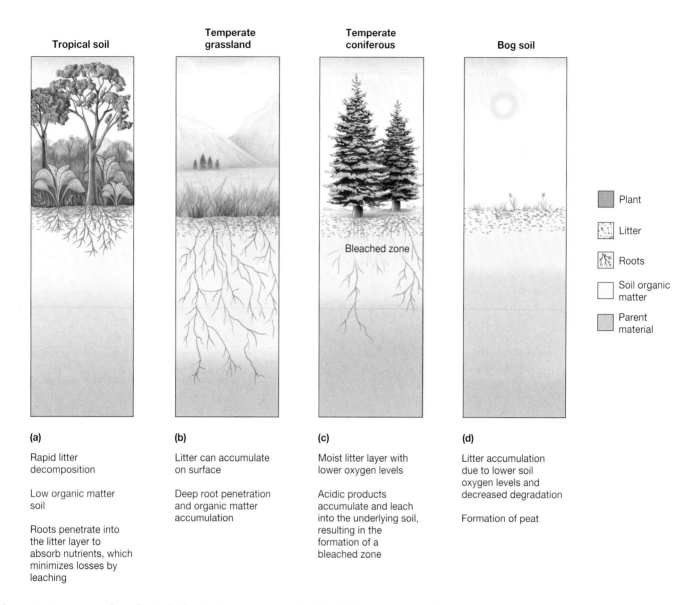

(a)

Rapid litter
decomposition

Low organic matter
soil

Roots penetrate into
the litter layer to
absorb nutrients, which
minimizes losses by
leaching

(b)

Litter can accumulate
on surface

Deep root penetration
and organic matter
accumulation

(c)

Moist litter layer with
lower oxygen levels

Acidic products
accumulate and leach
into the underlying soil,
resulting in the
formation of a
bleached zone

(d)

Litter accumulation
due to lower soil
oxygen levels and
decreased degradation

Formation of peat

Figure 29.5   Examples of Tropical and Temperate Region Plant-Soil Biomes.   Climate, parent material, plants, topography, and microorganisms interact over time to form different plant-soil systems. In these figures the characteristics of **(a)** tropical, **(b)** temperate grassland, **(c)** temperate coniferous forest, and **(d)** bog soils are illustrated.

up to avoid losses in leaching. Thus it is possible to recycle nutrients before they are lost due to water movement through the soil (figure 29.5*a*). With deforestation, there is no leaf litter so nutrients are not recycled, leading to their loss from the soil and decreased soil fertility.

Tropical plant-soil communities are often used in **slash-and-burn agriculture.** The vegetation on a site is chopped down and burned to release the trapped nutrients. For a few years, until the minerals are washed from the soils, crops can be grown. When the minerals are lost from these low organic matter soils, the farmer must move to a new area and start over by again cutting and burning the native plant community. This cycle of slash-and-burn agriculture is stable if there is sufficient time for the plant community to regenerate before it is again cut and burned. If the cy-

cle is too short, rapid and almost irreversible degradation of the soil occurs.

In many temperate region soils, in contrast, the decomposition rates are less than that of primary production, leading to litter accumulation. Deep root penetration in temperate grasslands results in the formation of fertile soils, which provide a valuable resource for the growth of crops in intensive agriculture (figure 29.5*b*).

The soils in many cooler coniferous forest environments suffer from an excessive accumulation of organic matter as plant litter (figure 29.5*c*). In winter, when moisture is available, the soils are cool, and this limits decomposition. In summer, when the soils are warm, water is not as available for decomposition. Organic acids are produced in the cool, moist litter layer, and they leach into the underlying soil. These acids solubilize soil components such as aluminum and iron, and a bleached zone may form. Litter continues to accumulate, and fire becomes the major means by which nutrient cycling is maintained. Controlled burns have become part of the environmental management in this type of plant-soil system.

Bog soils provide a unique set of conditions for microbial communities (figure 29.5d). In these soils, decomposition is slowed by the waterlogged, predominantly anoxic conditions, which lead to peat accumulation. When such areas are drained, they become more oxic and SOM is degraded, resulting in soil subsidence. Under oxic conditions the lignin-cellulose complexes of the accumulated organic matter are more susceptible to decomposition by filamentous fungi.

Cold Moist Soils

Soils in cold environments, whether in Arctic, Antarctic, or alpine regions, are of extreme interest because of their wide distribution and impacts on global-level processes. The colder mean soil temperatures at these sites decrease the rates of both decomposition and plant growth. In these cases SOM accumulates, and plant growth can become limited due to the immobilization of nutrients. Often, below the plant growth zone, these soils are permanently frozen. These permafrost soils hold about 11% of the Earth's soil carbon and 95% of its organically bound nutrients. These soils are very sensitive to physical disturbance and pollution, and the widespread exploration of such areas for oil and minerals can have long-term effects on their structure and function.

In water-saturated bog areas, oxygen limitation means that bacteria are more important than fungi in decomposition processes, and there is decreased degradation of lignified materials. As in other soils, the nutrient cycling processes of nitrification, denitrification, nitrogen fixation, and methane synthesis and utilization, although occurring at slower rates, can have major impacts on global gaseous cycles.

Desert Soils

Soils of hot and cold arid and semiarid deserts are dependent on periodic and infrequent rainfall. When it rains, water can puddle in low areas and be retained on the soil surface by microbial communities called **desert crusts.** These consist of cyanobacteria and associated microbes, including *Anabaena, Microcoleus, Nostoc,* and *Scytonema.* The depth of the photosynthetic layer is perhaps 1 mm, and the cyanobacterial filaments and slime link the sand particles (**figure 29.6**), which change the surface soil albedo (the amount of sunlight reflected), water infiltration rate, and susceptibility to erosion. These crusts are quite fragile, and vehicle damage can be evident for decades. After a rainfall, nitrogen fixation begins within approximately 25 to 30 hours, and when the rain evaporates or drains, the crust dries up and nitrogen is released for use by other microorganisms and the plant community. Photosynthetic bacteria: *Phylum Cyanobacteria* (section 21.3)

Figure 29.6   A Desert Crust as Observed with the Scanning Electron Microscope.   The crust has been disturbed to show the extracellular sheaths and filaments of the cyanobacterium *Microcoleus vaginatus.* The sand grains are linked by these filamentous growths, creating a unique ecological structure.

Geologically Heated Hyperthermal Soils

Geologically heated soils are found in such areas as Iceland, the Kamchatka peninsula in eastern Russia, Yellowstone National Park, and at many mining waste sites. These soils are populated by bacteria and archaea, many of which are chemolithoautotrophs. A wide variety of chemoorganotrophic genera also are found in these environments; these include the aerobes *Thermomicrobium, Thermoleophilum,* and also the anaerobes *Thermosipho* and *Thermotoga.* An important microorganism found in heated mining wastes is *Thermoplasma.* Such geothermal soils have been of great interest as a source of new microbes to use in biotechnology, and the search for new, unique microorganisms in these areas is intensifying all over the world. Phylum *Euryarchaeota: Thermoplasms* (section 20.3)

---

1. Characterize each major soil type discussed in this section in terms of the balance between primary production and organic matter decomposition.
2. What is slash-and-burn agriculture? Describe the roles of microorganisms in this process.
3. What is unique about bogs in terms of organic matter degradation?
4. Describe desert crusts. What types of microorganisms function in these unique environments?
5. What unique microbial genera are found in geothermally heated soils?

Figure 29.10   Endomycorrhizae.   Endomycorrhizae, or arbuscular mycorrhizae, form characteristic structures within roots. These can be observed with a microscope after the roots are stained. The arbuscules of *Gigaspora margarita* can be seen inside the root cortex cells of cotton.

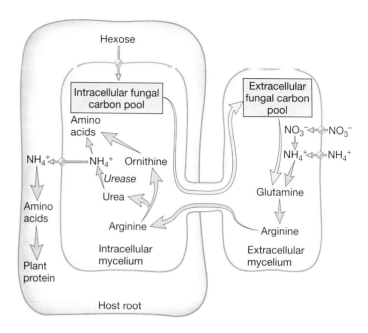

Figure 29.11   Nitrogen Exchange between Arbuscular Mycorrhizal Fungi and Host Plant.   Nitrate and ammonium are taken up by the fungal mycelium that is outside the host plant cell (extracellular mycelium) and converted to arginine. This amino acid is transferred to the mycelium within the host plant cell and broken down so that only the ammonium enters the plant.

were performed in which AM-colonized plants (and the appropriate negative controls lacking mycorrhizae) were treated with $^{15}NO_3^-$ and $^{15}NH_4^+$ (**figure 29.11**). These forms of nitrogen are incorporated into fungal tissue through the glutamine synthetase-glutamate synthase pathway (GS-GOGAT). Thus ^{15}N-containing glutamine (which is converted to arginine) is recovered. However, prior to transferring the nitrogen to host cells, intracellular fungal hyphae degrade the amino acids, transferring only the $^{15}NH_4^+$. Thus while the fungus provides its host with much needed ammonium, it retains the carbon skeleton it needs for nitrogen uptake.
Synthesis of amino acids: Nitrogen assimilation (section 10.5)

In addition to these two most abundant mycorrhizal types, the orchidaceous mycorrhizae are of particular interest (figure 29.8*b*). Orchids are unusual plants in that many never produce chlorophyll, while others only do so after they have matured past the seedling stage. Therefore all orchids have an absolute dependence on their endomycorrhizal partners for at least part of their lives. Indeed, orchid seeds will not germinate unless first colonized by a basidiomycete orchid mycorrhiza. Because the host orchid cannot produce photosynthetically derived fixed carbon (or produces very little), unlike AM and ECM fungi, orchid mycorrhizal fungi are saprophytic. They must degrade organic matter to obtain carbon, which the orchids then also consume. In this case, the orchid functions as a parasite.

Depending on the environment of the plant, mycorrhizae can increase a plant's competitiveness. In wet environments they increase the availability of nutrients, especially phosphorus. In arid environments, where nutrients do not limit plant functioning to the same degree, the mycorrhizae aid in water uptake, allowing increased transpiration rates in comparison with nonmycorrhizal plants. These benefits have distinct energy costs for the plant in the form of photosynthate required to support the plant's "mycorrhizal habit." Based on the ubiquity of mycorrhizae, most plants

are apparently willing to trade photosynthate—produced with the increased water acquisition—for water.

Bacteria are also associated with the mycorrhizal fungi. As the external hyphal network radiates out into the soil, a **mycorrhizosphere** is formed due to the flow of carbon from the plant into the mycorrhizal hyphal network and then into the surrounding soil. In addition, "mycorrhization helper bacteria" can play a role in the development of mycorrhizal relationships with ectomycorrhizal fungi. Bacterial symbionts also are found in the cytoplasm of AM fungi, as shown in **figure 29.12.** Such bacteria-like organisms (BLOs) appear to be related to *Burkholderia cepacia.* It has been suggested that these "trapped" bacteria contribute to the nitrogen metabolism of the plant-fungal complex by assisting with the synthesis of essential amino acids.

---

1. Describe the two-way relationship between mycorrhizal fungi and the plant host.
2. List three major differences between arbuscular mycorrhizae and ectomycorrhizae.
3. What is the function of the rhizomorph and Hartig's net?
4. Describe the uptake and transfer of ammonium by arbuscular mycorrhizae to the plant host. Why do you think only the ammonium is transferred?
5. What makes orchid mycorrhizae different from other mycorrhizal associations?
6. Propose two potential functions for mycorrhization helper bacteria.

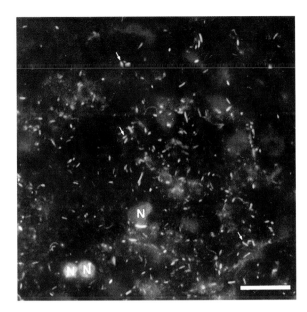

Figure 29.12 Mycorrhization Helper Bacteria. Stained bacterial endosymbionts in unfixed spores of the AM fungus *Gigaspora margarita.* Living bacteria fluoresce bright yellow-green; lipids and fungal nuclei (N) appear as diffuse masses. Bar = 7 μm.

Nitrogen fixation

The enzymatic conversion of gaseous nitrogen (N_2) to ammonia (NH_3) often occurs as part of a symbiotic relationship between bacteria and plants. These symbioses produce more than 100 million metric tons of fixed nitrogen annually and are a vital part of the global nitrogen cycle. In addition, symbiotic nitrogen fixation accounts for more than half of the nitrogen used in agriculture. The provision of fixed nitrogen enables the growth of host plants in soils that would otherwise be nitrogen limiting, and simultaneously reduces loss of nitrogen by denitrification and leaching. Indeed, with over 18,000 species, legumes are the most successful plants on Earth, probably due to their symbiotic relationships with nitrogen-fixing bacteria. For these reasons nitrogen fixation, particularly that by members of the genus *Rhizobium* and related α-proteobacteria in association with their leguminous host plants have been the subject of intense investigation.

The Rhizobia

Several microbial genera are able to form nitrogen-fixing nodules with legumes. These include the α-proteobacteria *Allorhizobium, Azorhizobium, Bradyrhizobium, Mesorhizobium, Sinorhizobium,* and *Rhizobium.* Collectively these genera are often called the **rhizobia.** Recently the phylogenetic diversity of the rhizobia has been extended by the discovery that the β-proteobacteria *Burkholderia caribensis* and *Ralstonia taiwanensis* also form nitrogen-fixing nodules on legumes. Here, the general process of nodulation is presented followed by molecular details that have been revealed largely though studies of the genus *Rhizobium.* Class *Alphaproteobacteria* (section 22.1); Class *Betaproteobacteria* (section 22.2)

Rhizobia live freely in the soil, but when they approach the plant root, they are assumed to be an alien invader. The plant responds with an **oxidative burst,** producing a mixture that can contain superoxide radicals, hydrogen peroxide, and N_2O. This redox-based oxidative burst, involving glutathione and homoglutathione, is critical for determining the fate of the infection process and influences whether further steps in the early infection process will occur. The rhizobia, if they are to be effective colonizers, must use antioxidant defenses to survive and continue the infection process. Only rhizobia and related genera with sufficient antioxidant abilities are able to proceed to the next steps in the infection process.

The plant roots also release **flavonoid inducer molecules** that stimulate rhizobial colonization of the root surfaces (**figure 29.13a**). In response to this molecular message, rhizobia produce their own signaling compounds called **Nod factors.** The precise structure of individual Nod factors depends on the bacterial species, but all consist of four to five units of β-1,4 linked N-acetyl-D-glucosamine bearing an acyl chain at the nonreducing terminal residue, and a sulfate attached to the reducing end (figure 29.13c). Upon receipt of the Nod factor message, gene expression in the outer (epidermal) cells of the roots is altered so that the root hairs become deformed. In some cases the root hairs will curl to resemble a shepherd's crook, entrapping bacteria (figure 29.13c,d). In these regions, the plant cell wall is locally modified, the plant plasma membrane invaginates, and new plant material is laid down. These modifications lead to the development of a bacteria-filled, tubelike structure called the **infection thread** (29.13e,f). The infection thread grows toward the base of the root hair cell to a region called the primordium. Division of these root cells ultimately gives rise to the nodule. When bacteria are released from the infection thread into the primodium, they remain surrounded by a plant cell membrane called the peribacteroid membrane (figure 29.13h). It is here that each bacterial cell differentiates into the nitrogen-fixing form called a **bacteroid.** Bacteroids are terminally differentiated—they can neither divide nor revert back to the nondifferentiated state. Further growth and differentiation leads to the development of a structure called a **symbiosome.** Recall that the nitrogenase enzyme is very sensitive to oxygen. To help protect the nitrogenase, a protein called **leghemoglobin,** which binds to oxygen and helps maintain microaerobic conditions within the mature nodule, is produced (figure 29.13i, j). This protein is similar in structure to myo- and hemoglobins found in animals; however, it has a higher affinity for oxygen. Interestingly, the protein moiety is encoded by plant genes whereas the heme group is the product of bacterial genes. Synthesis of amino acids: Nitrogen fixation (section 10.5)

The symbiosomes within mature **root nodules** are the site of nitrogen fixation. Within these nodules, the differentiated bacteriods reduce atmospheric N_2 to ammonium, and in return they receive carbon and energy in the form of dicarboxylic acids from their host legume. It had long been thought the principal form of nitrogen that was transferred to the host plant was ammonium, but more recent evidence shows that a more complex cycling of amino acids occurs. Apparently, the plant provides amino acids to the bacteroids so that

Figure 29.13 Root Nodule Formation by *Rhizobium*. Root nodule formation on legumes by *Rhizobium* is a complex process that produces the nitrogen-fixing symbiosis. **(a)** The plant root releases flavonoids that stimulate the production of various Nod metabolites by *Rhizobium*. There are many different Nod factors that control infection specificity. **(b)** Attachment of *Rhizobium* to root hairs involves specific bacterial proteins called rhicadhesins and host plant lectins that affect the pattern of attachment and *nod* gene expression. **(c)** Structure of a typical Nod factor that promotes root hair curling and plant cortical cell division. The bioactive portion (nonreducing *N*-fatty acyl glucosamine) is highlighted. These Nod factors enter root hairs and migrate to their nuclei. **(d)** A plant root hair covered with *Rhizobium* and undergoing curling. **(e)** Initiation of bacterial penetration into the root hair cell and infection thread growth coordinated by the plant nucleus "N." **(f)** A branched infection thread shown in an electron micrograph. **(g)** Cell-to-cell spread of *Rhizobium* through transcellular infection threads followed by release of rhizobia and infection of host cells. **(h)** Formation of bacteroids surrounded by plant-derived peribacteroid membranes and differentiation of bacteroids into nitrogen-fixing symbiosomes. The bacteria change morphologically and enlarge around 7 to 10 times in volume. The symbiosome contains the nitrogen-fixing bacteroid, a peribacteroid space, and the peribacteroid membrane. **(i)** Light micrograph of two nodules that develop by cell division (×5). This section is oriented to show the nodules in longitudinal axis and the root in cross section. **(j)** *Sinorhizobium meliloti* nitrogen-fixing nodules on roots of white sweet clover *(Melilotus alba)*.

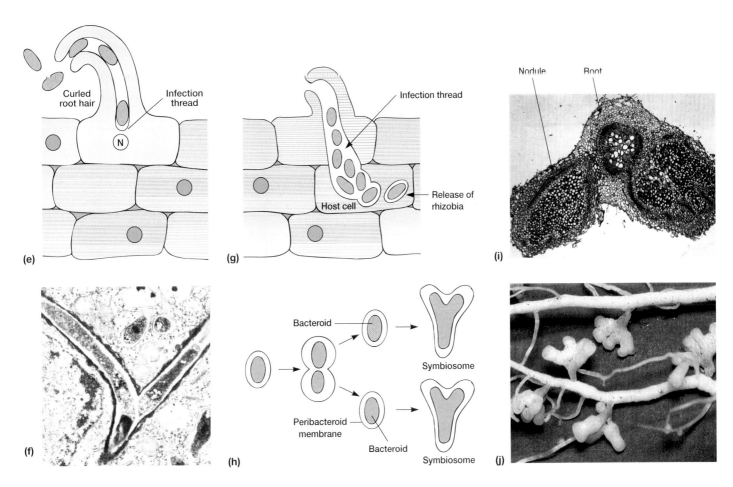

Figure 29.13   *(Continued).*

they do not need to assimilate ammonium. In return, the bacteroids shuttle amino acids (which bear the newly fixed nitrogen) back to the plant. This creates an interdependent relationship, providing selective pressure for the evolution of mutualism.

The genes essential for nodulation (*nod, nol,* and *noe* genes) and nitrogen fixation (*nif* and *fix* genes) show homology among the α- and β-proteobacterial rhizobia. However, the arrangement of these genes on the bacterial genome is not conserved. In some cases, these genes are clustered; in others they are not. For instance, in *Rhizobium meliloti,* which forms nodules on alfalfa, clusters of symbiosis genes are encoded on huge megaplasmids that are over a million base pairs long. In *Bradyrhizobium japonicum* (which forms a symbiosis with soybean plants), the *nod* genes are located on the chromosome. Evidence for transfer of a symbiosis gene island has been demonstrated for *Mesorhizobium loti,* which infects the model legume *Lotus japonicus* (bird's-foot trefoil). This genomic diversity has led to taxonomic confusion in the reclassification of the rhizobia that were previously identified only by phenotypic features.

Interestingly, plants use the same initial response to the establishment of productive nitrogen-fixing symbionts as they use in establishing arbuscular mycorrhizae associations. For example, the plant genetic locus called DMI (*d*oesn't *m*ake *i*nfections— recall that genes are frequently named for their mutant phenotype)

is induced upon the initial colonization of rhizobia or arbuscular mycorrhizal fungi. Thus some plant mutants unable to form nodules are also unable to interact with arbuscular mycorrhizae.

The molecular mechanisms by which both the legume host and the rhizobial symbionts establish productive nitrogen-fixing bacteriods within nodules continues to be an intense area of research. A major goal of biotechnology is to introduce nitrogen-fixation genes into plants that do not normally form such associations. It has been possible to produce modified lateral roots on nonlegumes such as rice, wheat, and oilseed rape; the roots are invaded by nitrogen-fixing bacteria. It appears that infection begins with bacterial attachment to the root tips. Although these modified root structures have not yet been found to fix useful amounts of nitrogen, they do enhance rice production and intense work is expected to continue in this area.

1. List several bacteria that are considered rhizobia.
2. Describe the communication system between a rhizobia bacterium and its legume host.
3. What is the function of the infection thread? Why do you think it is important that the bacteria do not enter plant cell cytoplasm until they reach the primordium?

4. What does the term "terminally differentiated" mean? Can you think of other cells that are also terminally differentiated?

5. How is leghemoglobin made and what is its function?

Actinorhizae

Another example of symbiotic nitrogen fixation occurs between the actinomycete *Frankia* and eight nonleguminous host plant families (**table 29.7**). These bacterial associations with plant roots are called **actinorhizae** or actinorhizal relationships (**figure 29.14**). *Frankia* fixes nitrogen and is important, particularly in trees and shrubs. As examples, these associations occur in areas where Douglas fir forests have been clear-cut, and in bog and heath environments where bayberries and alders are dominant. The nodules of some plants (*Alnus, Ceanothus*) are as large as baseballs. The nodules of *Casuarina* (Australian pine) approach soccer ball size.

Members of the genus *Frankia* are slow-growing and were impossible to culture apart from the plant until 1978. Since then, this actinomycete has been grown on specialized media supple-

mented with metabolic intermediates such as pyruvate. Major advances in understanding the physiology, genetics, and molecular biology of these microorganisms are now taking place.

As in all plant-microbe associations, the actinorhizal relationship costs the plant energy. However, the plant benefits and is better able to compete in nature. This association provides a unique opportunity for microbial management to improve plant growth processes.

Stem-Nodulating Rhizobia

Other associations of nitrogen-fixing microorganisms with plants also occur. A particularly interesting association is caused by **stem-nodulating rhizobia,** found primarily in tropical legumes (**figure 29.15**). These nodules form at the base of adventitious roots branching out of the stem just above the soil surface and, because they contain oxygen-producing photosynthetic tissues, they have unique mechanisms to protect the oxygen-sensitive nitrogen fixation enzymes. One microorganism that forms such

Table 29.7	Nonleguminous Nodule-Bearing Plants with *Frankia* Symbioses[a]		
Family	**Genus**	**Frankia Isolated?**	**Isolated Strains Infective?**
Casuarinaceae	Allocasuarina	+	+
	Casuarina	+	+
	Ceuthostoma	−	−
	Gymnostoma	+	+
Coriariaceae	Coriaria	+	+
Datiscaceae	Datisca	+	−
Betulaceae	Alnus	+	+
Myricaceae	Comptonia	+	+
	Myrica	+	+
Elaeagnaceae	Elaeagnus	+	+
	Hippophae	+	+
	Shepherdia	+	+
Rhamnaceae	Adolphia	−	−
	Ceanothus	+	−
	Colletia	+	+
	Discaria	+	+
	Kentrothamnus	−	−
	Retanilla	+	+
	Talguenea	+	+
	Trevoa	+	+
Rosaceae	Cercocarpus	+	−
	Chaemabatia	−	−
	Cowania	+	−
	Dryas	−	−
	Purshia	+	−

Source: Data from Dr. D. Baker, MDS Panlabs and Dr. J. Dawson, University of Illinois. Personal communications.

[a]*Frankia* isolation from nodules and ability of these isolated strains to initiate nodulation are also noted.

Figure 29.14 Actinorhizae. *Frankia*-induced actinorhizal nodules in *Ceanothus* (Buckbrush).

Figure 29.15 Stem-Nodulating Rhizobia. Nitrogen-fixing microorganisms also can form nodules on stems of some tropical legumes. Nodules formed on the stem of a tropical legume by a stem-nodulating *Rhizobium*.

Figure 29.16 Fungal Endophytes. Fungi can invade the upper parts of some plants. A fungal endophyte growing inside the leaf sheath of a grass, tall fescue, is shown.

Figure 29.17 Parasitic Castration of Plants by Endophytic Fungi. Stroma of the fungus *Atkinsonella hypoxylon* infecting *Danthonia compressa* and causing abortion of the terminal spikelets.

root and stem nodules is *Azorhizobium caulinodans,* which forms nodules on the tropical legume *Sesbania rostrata.* It has been shown that some of these stem-nodulating rhizobia are photosynthetic. Thus they can obtain their energy not only from the plant's organic compounds, but also from the light.

Fungal and Bacterial Endophytes

Specialized fungi and bacteria can live within some plants as endophytes and may be beneficial. Specialized clavicipitaceous fungi form systemic fungal infections in which the endophyte grows between the plant's cortex cells (**figure 29.16**). Plants infected with these endophytes may be less susceptible to attack by various chewing insects due to the production of alkaloids, a form of "chemical defense." *Rhizobium leguminosarum* bv. *trifolii,* which forms a nitrogen-fixing symbiosis with clover, can

also form a natural endophytic association with rice roots. This interaction, first observed in the Nile Delta, is supported by the rotation of clover with the rice. The association promotes rice root and shoot growth, resulting in increased rice grain yield at maturity. The rice/*Rhizobium*/clover association provides approximately one-quarter of the nitrogen needed by the rice crop.

Not all endophytic relationships are mutualistic; some are parasitic. Parasitic fungal endophytes can actually reduce the genetic variability of the plant by sterilizing their host (**figure 29.17**). This "parasitic castration of plants" by systemic fungi, which promotes increased fungal spread in a less variable plant community, is suggested to be of major importance in the co-evolution of plants and fungi.

Endophytic bacteria have been discovered in sugar cane, cotton, pears, and potatoes. Some are plant pathogens that can survive for extended periods in a quiescent state. The majority have no known positive or deleterious effect on plant growth or development. The use of these bacteria as microbial delivery systems in agriculture is a current topic in agricultural biotechnology. It also has been possible to establish *Azorhizobium,* a root and stem nodulating bacterium of *Sesbana rostrata,* in the lateral roots of wheat plants, leading to a possible increased plant dry weight and nitrogen content.

1. What is the major contribution of *Frankia* to plant functioning? Which types of plants are infected?
2. What are stem-nodulating rhizobia?
3. Describe some possible effects of endophytic fungi on plants.

Agrobacterium

Clearly, some microbe-plant interactions are beneficial for both partners. However, others involve microbial pathogens that harm or even kill their host. *Agrobacterium tumefaciens* is an α-proteobacterium that has been studied intensely for several decades. Initially, research focused on how this microbe causes **crown gall disease,** which results in the formation of tumorlike growths in a wide variety of plants (**figure 29.18**). Within the last 15 years or so, however, *A. tumefaciens* has become one of biotechnology's most important tools. The molecular genetics by which this pathogen infects its host is the basis for plant genetic engineering. Techniques & Applications 14.2: Plant tumors and nature's genetic engineer: Class *Alphaproteobacteria* (section 22.1)

The genes for plant infection and virulence are encoded on an *A. tumefaciens* plasmid, called the **Ti (*t*umor-*i*nducing) plasmid.** These genes include 21 *vir* genes (*vir* stands for *vir*ulence), which are found in six separate operons. Two of these genes, *virD1* and *virD2,* encode proteins that cleave a separate region of the Ti plas-

mid, called the **T DNA.** After excision, this T DNA fragment is integrated into the host plant's genome. Once incorporated into a plant cell's genome, T DNA directs the overproduction of phytohormones that cause unregulated growth and reproduction of plant cells, thereby generating a tumor or gall in the plant.

The *vir* genes are not expressed when *A. tumefaciens* is living saprotrophically in the soil. Instead, they are induced by the presence of plant phenolics and monosaccharides present in an acidic (pH 5.2–5.7) and cool (below 30°C) environment (**figure 29.19*a***). The microbe usually infects its host through a wound. Upon reception of the plant signal, a two-component signal transduction system is activated: VirA is a sensor kinase that, in the presence of a phenolic signal, phosphorylates the response regulator VirG. Activated VirG then induces transcription of the other *vir* genes. This enables the bacterial cell to become adequately positioned relative to the plant cell, at which point the *virB* operon expresses the apparatus that will transfer the T DNA. This transfer is similar to bacterial conjugation and involves a type IV secretion system. After VirD1 and VirD2 excise the T DNA from the Ti plasmid, the T DNA, with the VirD2 protein attached to the 5′ end, is delivered to the plant cell cytoplasm. The protein VirE2 is also transferred and, together with VirD2, the T DNA is shepherded to the plant cell nucleus where it is integrated into the host's genome. Here the T DNA has two specific functions. First, it directs the host cell to overproduce phytohormones that cause tumor formation. Second, it stimulates the plant to produce special amino acid and sugar derivatives called opines (figure 29.19*b*). Opines are not metabolized by the plant but *A. tumefaciens* is attracted to opines; chemotaxis of bacteria from the surrounding soil population will further advance the infection because the bacterium can use opines as sources of carbon, energy, nitrogen, and phosphorus. Regulation of transcription initiation: Two-component signal transduction systems (section 12.2), Bacterial conjugation (section 13.7)

Other Plant Pathogens

In addition to *Agrobacterium,* a variety of other bacteria cause an array of spots, blights, wilts, rots, cankers, and galls, as shown in **table 29.8.** The soft rots caused by the enterobacteria *Erwinia chrysanthemi* and *E. carotovora* have significant economic impact. These bacteria digest plant tissue by producing extracellular enzymes that degrade pectin, cellulose, and proteins. These exoenzymes are secreted by type II and type I secretion systems; mutants lacking these secretion systems are no longer pathogenic. Similarly, proteobacteria belonging to the genera *Ralstonia, Pseudomonas, Pantoea,* and *Xanthomonas* rely on type III secretion systems to deliver virulence proteins. Although these microbes cause a diverse collection of plant diseases, all colonize the spaces between plant cells to kill their hosts. Another group of important plant pathogens includes the wall-less phytoplasms that infect vegetable and fruit crops such as sweet potatoes, corn, and citrus. Protein secretion in procaryotes (section 3.8)

As discussed in chapters 25 and 26, respectively, protists and fungi can be devastating plant pathogens. Examples are the fungus *Puccinia graminis,* which causes wheat rust, and the öomycete

Figure 29.18 *Agrobacterium. Agrobacterium*-caused tumor on a *Kalanchoe* sp. plant.

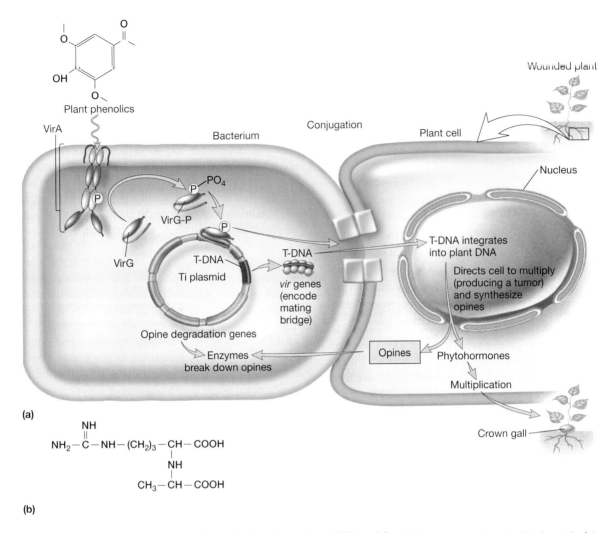

(a)

(b)

Figure 29.19 Functions of Genes Carried on the *Agrobacterium* Ti Plasmid. **(a)** Genes carried on the Ti plasmid of *Agrobacterium* control tumor formation by a two-component regulatory system that stimulates formation of the mating bridge and excision of the T-DNA. The T-DNA is moved by transfer genes, which lead to integration of the T-DNA into the plant nucleus. T-DNA encodes plant hormones that cause the plant cells to divide, producing the tumor. The tumor cells produce opines (shown in **b**) that can serve as a carbon source for the infecting *Agrobacterium*. Ultimately a crown gall is formed on the stem of the wounded plant above the soil surface.

Phytophthora infestans, which was responsible for the Irish potato famine.

A wide range of viruses and virusoids infect plants, as described in sections 18.6 and 18.9 respectively, and are of worldwide importance in terms of plant disease and economic losses. These include tobacco mosaic virus (TMV), the first virus to be characterized. A virus of particular interest in terms of plant-pathogen interactions is one member of the hypoviruses that infects the fungus *Cryphonectria parasitica,* the cause of chestnut blight. Based on pioneering studies carried out in Italy and France, workers in Connecticut and West Virginia noted that if they infected the fungus with the hypovirus, the rate and occurrence of blight was decreased. They are hoping to treat trees with the less lethal virus strains and eventually transform the indigenous lethal strains of *Cryphonectria* into more benign fungi.

Tripartite and Tetrapartite Associations

An additional set of interactions occurs when the same plant develops relationships with two or three different types of microorganisms. These more complex interactions are important to a variety of plant types in both temperate and tropical agricultural systems. First described in 1896, these symbiotic associations involve the interaction of the plant-associated microorganisms with each other and the host plant. Several **tripartite associations** are known to occur: the plant plus (1) endomycorrhizae plus rhizobia, including *Rhizobium* and *Bradyrhizobium;* (2) endomycorrhizae and actinorhizae; and (3) ectomycorrhizae and actinorhizae. Nodulated and mycorrhizal plants are better suited for coping with nutrient-deficient environments. **Tetrapartite associations** also occur. These consist of endomycorrhizae, ectomycorrhizae, *Frankia,* and the host plant. These complex associations, in spite

Klein, D. A. 2000. The rhizosphere. In *Encyclopedia of microbiology,* 2d ed., vol. 4, J. Lederberg, editor-in-chief, 117–26. San Diego: Academic Press.

Klironomos, J. N. 2002. Feedback with soil biota contributes to plant rarity and invasiveness in communities. *Nature* 417;67–70.

Kuhn, D. M., and Ghannoum, M. A. 2003. Indoor mold, toxigenic fungi, and *Stachybotrys chartarum:* Infectious disease perspective. *Clin. Microbial. Rev.* 16(1):144–72.

Lu, Y., and Conrad, R. 2005. In situ stable isotope probing of methanogenic *Archaea* in the rice rhizosphere. *Science* 309: 1088–90.

Pounds, J. A.; Bustamante, M. R.; Coloma, L. A.; Consuegra, J. A.; Fogden, M. P. L.; Foster, P. N.; La Marca, E., et al. 2006. Widespread amphibian extinctions from epidemic disease driven by global warming. *Nature* 439:161–67.

Tursvik, V.; Øvreås, L.; and Thingstad, T. F. 2002. Prokaryotic diversity—Magnitude, dynamics, and controlling factors. *Science* 296:1064–66.

Vandenkoornhuyse, P.; Baldauf, S. L.; Leyval, C.; Straczek, J.; and Young, J. P. W. 2002. Extensive fungal diversity in plant roots. *Science* 295:2051–61.

Please visit the Prescott website at www.mhhe.com/prescott7 for additional references.

30

Microbial Interactions

A "garden" of tube worms (*Riftia pachyptila*) at the Galápagos Rift hydrothermal vent site (depth 2,550 m). Each worm grows to more than a meter in length thanks to endosymbiotic chemolithoautotrophic bacteria, which provide carbohydrate to these gutless worms. The bacteria use the Calvin cycle to fix CO_2 and H_2S as an electron donor.

PREVIEW

- The term symbiosis, or "together-life," can be used to describe many of the interactions between microorganisms, and also microbial interactions with higher organisms, including plants and animals. These interactions may be positive or negative.

- Symbiotic interactions include mutualism, cooperation, commensalism, parasitism, predation, amensalism, and competition. These interactions are important in natural processes and in the occurrence of disease. The interactions can vary depending on the environment and changes in the interacting organisms.

- Microorganisms, as they interact, can form complex physical assemblages that include biofilms. These form on living and inert surfaces and have major impacts on microbial survival and the occurrence of disease.

- Microorganisms also interact by the use of chemical signal molecules, which allow the microbial population to respond to increased population density. Such responses include quorum sensing, which controls a wide variety of microbial activities.

- Gnotobiotic refers to a microbiologically monitored environment or animal in which the identities of all microorganisms present are known or to an environment or animal that is germfree.

- Most microorganisms associated with the human body are bacteria; they normally colonize specific sites. There are both positive and negative aspects of these normal microbial associations. Sometimes they compete with pathogens, other times they are capable of producing opportunistic infections.

Our discussion of microbial ecology has so far considered microbial communities in complex ecosystems. However, the ecology of microorganisms also involves the physiology and behavior of microbes as they interact with one another and with higher organisms. In this chapter, we begin by defining types of microbial interactions and present a number of illustrative examples. We conclude this chapter with a consideration of perhaps the most intimate, yet still relatively unexplored, microbial habitat: the human body.

Our discussion of interactions between microbes and between microbes and eucaryotes focuses on the nature of symbioses. Although **symbiosis** is often used in a nonscientific sense to mean a mutually beneficial relationship, here we use the term in its original broadest sense, as an association of two or more different species of organisms, as suggested by H. A. deBary in 1879.

30.1 MICROBIAL INTERACTIONS

Microorganisms can associate physically with other organisms in a variety of ways. One organism can be located on the surface of another, as an **ectosymbiont.** In this case, the ectosymbiont usually is a smaller organism located on the surface of a larger organism. In contrast, one organism can be located within another organism as an **endosymbiont.** While the simplest microbial interactions involve two members, a symbiont and its host, a number of interesting organisms host more than one symbiont. The term **consortium** can be used to describe this physical relationship. For example, *Thiothrix* species, a sulfur-using bacterium, is attached to the surface of a mayfly larva and itself

. . . every organic being is related, in the most essential yet often hidden manner, to that of all other organic beings, with which it comes into competition for food or residence, or from which it has to escape, or on which it preys.

Charles Darwin

contains a parasitic bacterium. Fungi associated with plant roots (mycorrhizal fungi) often contain endosymbiotic bacteria, as well as having bacteria living on their surfaces. Microorganism associations with vascular plants: Mycorrhizae (section 29.5)

These physical associations can be intermittent and cyclic or permanent. Examples of intermittent and cyclic associations of microorganisms with plants and marine animals are shown in **table 30.1.** Important human diseases, including listeriosis, malaria, leptospirosis, legionellosis, and vaginosis also involve such intermittent and cyclic symbioses. These diseases are discussed in chapters 38 and 39. Interesting permanent relationships also occur between bacteria and animals, as shown in **table 30.2.** Hosts include squid, leeches, aphids, nematodes, and mollusks. In each of these cases, an important characteristic of the host animal is conferred by the permanent bacterial symbiont.

Although it is possible to observe microorganisms in these varied physical associations with other organisms, the fact that there is some type of physical contact provides no information on the nature of the interactions that might be occurring. These interactions include mutualism, cooperation, commensalism, predation, parasitism, amensalism, and competition (**figure 30.1**). These interactions are now discussed.

1. Define the term symbiosis.
2. List several important diseases that involve cyclic and intermittent symbioses. Compare these with permanent relationships.

Mutualism

Mutualism [Latin *mutuus,* borrowed or reciprocal] defines the relationship in which some reciprocal benefit accrues to both partners. This is an obligatory relationship in which the **mutualist** and the host are dependent on each other. When separated, in many cases, the individual organisms will not survive. Several examples of mutualism are presented next.

Microorganism-Insect Mutualisms

Mutualistic associations are common in the insects. This is related to the foods used by insects, which often include plant sap or animal fluids lacking in essential vitamins and amino acids. The required vitamins and amino acids are provided by bacterial symbionts in exchange for a secure habitat and ample nutrients (**Microbial Diversity & Ecology 30.1**). The aphid is an excellent example of this mutualistic relationship. This insect harbors the γ-proteobacterium *Buchnera aphidicola* in its cytoplasm, and a mature insect contains literally millions of these bacteria in its body. The *Buchnera* provides its host with 10 essential amino acids, and if the insect is treated with antibiotics, it dies. Likewise, *B. aphidicola* is an obligate mutualistic symbiont. The inability of either partner to grow without the other indicates that the two organisms demonstrate coevolution, or have evolved together. It is estimated that the *B. aphidicola*-aphid endosymbiosis was established about 150 million years ago. The genomes of two different *B. aphidicola* strains have been sequenced and annotated to reveal extreme genomic stability. These strains diverged 50 to 70 million years ago, and since that time there have

Table 30.1	Intermittent and Cyclical Symbioses of Microorganisms with Plants and Marine Animals	
Symbiosis	**Host**	**Cyclical Symbiont**
Plant-bacterial	*Gunnera* (tropical angiosperm)	*Nostoc* (cyanobacterium)
	Azolla (rice paddy fern)	*Anabaena* (cyanobacterium)
	Phaseolus (bean)	*Rhizobium* (N_2 fixer)
	Ardisia (angiosperm)	*Protobacterium*
Marine animals	Coral coelenterates	*Symbiodinium* (dinoflagellate)
	Luminous fish	*Vibrio, Photobacterium*

Adapted from L. Margulis and M. J. Chapman, 1998. Endosymbioses: Cyclical and permanent in evolution. *Trends in Microbiology* 6(9):342–46, tables 1, 2, and 3.

Table 30.2	Examples of Permanent Bacterial-Animal Symbioses and the Characteristics Contributed by the Bacterium to the Symbiosis	
Animal Host	**Symbiont**	**Symbiont Contribution**
Sepiolid squid (*Euprymna scolopes*)	Luminous bacterium (*Vibrio fischeri*)	Luminescence
Medicinal leech (*Hirudo medicinalis*)	Enteric bacterium (*Aeromonas veronii*)	Blood digestion
Aphid (*Schizaphis graminum*)	Bacterium (*Buchnera aphidicola*)	Amino acid synthesis
Nematode worm (*Heterorhabditis* spp.)	Luminous bacterium (*Photorhabdus luminescens*)	Predation and antibiotic synthesis
Shipworm mollusk (*Lyrodus pedicellatus*)	Gill cell bacterium	Cellulose digestion and nitrogen fixation

Source: From E. G. Ruby, 1999. Ecology of a benign "infection": Colonization of the squid luminous organ by *Vibrio fischeri*. In *Microbial ecology and infectious disease*, E. Rosenberg, editor, American Society for Microbiology, Washington, D.C., 217–31, table 1.

Interaction type	Interaction example

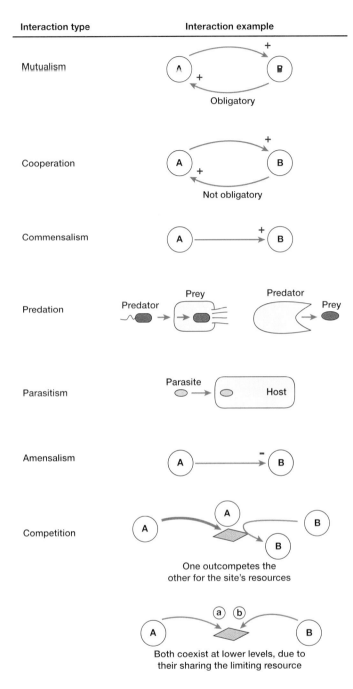

Figure 30.1   Microbial Interactions.   Basic characteristics of symbiotic interactions that can occur between different organisms.

been no gene duplications, translocations, inversions, or genes acquired by horizontal transfer. The genomes are small, only 0.64 Mb each with 93% of their genes common to both strains. Furthermore, only two genes have no orthologs in their close relative *E. coli*. This tremendous degree of stability implies that although the initial acquisition of the endosymbiont by ancestral aphids enabled their use of an otherwise deficient food source (sap), the bacteria have not continued to expand the ecological niche of their insect host through the acquisition of new traits

that might be advantageous.   Bioinformatics: Genome annotation (section 15.4); Microbial evolution (section 19.1)

The protozoan-termite relationship is another classic example of mutualism in which the flagellated protozoa live in the gut of termites and wood roaches (**figure 30.2a**). These flagellates exist on a diet of carbohydrates, acquired as cellulose ingested by their host (figure 28.2b). The protozoa engulf wood particles, digest the cellulose, and metabolize it to acetate and other products. Termites oxidize the acetate released by their flagellates. Because the host is almost always incapable of synthesizing cellulases (enzymes that catalyse the hydrolysis of cellulose), it is dependent on the mutualistic protozoa for its existence. This mutualistic relationship can be readily tested in the laboratory if wood roaches are placed in a bell jar containing wood chips and a high concentration of O_2. Because O_2 is toxic to the flagellates, they die. The wood roaches are unaffected by the high O_2 concentration and continue to ingest wood, but they soon die of starvation due to a lack of cellulases.

Zooxanthellae

Many marine invertebrates (sponges, jellyfish, sea anemones, corals, ciliates) harbor endosymbiotic dinoflagellates called **zooxanthellae** within their tissue (**figure 30.3a**). Because the degree of host dependency on the mutualistic protist is somewhat variable, only one well-known example is presented.   Protist classification: *Alveolata* (section 25.6)

The hermatypic (reef-building) corals (figure 30.3b) satisfy most of their energy requirements using their zooxanthellae, which are found at densities between 5×10^5 and 5×10^6 cells per square centimeter of coral animal. In exchange for up to 95% of their photosynthate (fixed carbon), zooxanthellae receive nitrogenous compounds, phosphates, CO_2, and protection from UV light from their hosts. This efficient form of nutrient cycling and tight coupling of trophic levels accounts for the stunning success of reef-building corals in developing vibrant ecosystems. However, during the past several decades, the number of coral bleaching events has increased dramatically. **Coral bleaching** is defined as a loss of either the photosynthetic pigments from the corals or expulsion of the zooxanthallae. It has been determined that damage to photosystem II of the zooxanthellae generates reactive oxygen species (ROS); it is these ROS that appear to be the direct cause of damage (recall that photosystem II uses water as the electron source resulting in the evolution of oxygen). Coral bleaching appears to be caused by a variety of stressors, but it has been experimentally determined, as well as observed in field sites, that temperature increases as small as 2°C above the average summer maxima can cause coral bleaching. Sadly, evidence suggests that many corals will be unable to evolve quickly enough to keep pace with the observed and predicted increases in ocean temperatures if global warming continues unchecked.   Phototrophy: Light reaction in oxygenic photosynthesis (section 9.12)

Sulfide-Based Mutualisms

Tube worm-bacterial relationships exist several thousand meters below the surface of the ocean, where the Earth's crustal plates are spreading apart (**figure 30.4**). Vent fluids are anoxic, contain

Microbial Diversity & Ecology

30.1 *Wolbachia pipientis:* The World's Most Infectious Microbe?

Most people have never heard of the bacterium *Wolbachia pipientis,* but this rickettsia infects more organisms than does any other microbe. It is known to infect a variety of crustaceans, spiders, mites, millipedes, and parasitic worms and may infect more than a million insect species worldwide. *Wolbachia* inhabits the cytoplasm of these animals where it apparently does no harm. To what does *Wolbachia* owe its extraordinary success? Quite simply, this endosymbiont is a master at manipulating its hosts' reproductive biology. In some cases it can even change the sex of the infected organism.

Wolbachia is transferred from one generation of host to the next through the eggs of infected female hosts. So in order to survive, this microbe must ensure the fertilization and viability of infected eggs while decreasing the likelihood that uninfected eggs survive. In the 1970s, scientists noticed that if a male wasp infected with *Wolbachia* mated with an uninfected female, few if any of these uninfected offspring survived. However, if infected females of the same wasp species mated with either infected or uninfected males, all of the eggs were viable—and infected with *Wolbachia.* Although scientists had no clear understanding of the cellular or molecular mechanisms involved, they suggested that "cytoplasmic incompatibility" might be responsible. They proposed that the cytoplasm of infected sperm was toxic to uninfected eggs, but that eggs carrying *Wolbachia* produced an antidote to the hypothetical poison.

It wasn't until the 1990s that the mechanism of cytoplasmic incompatibility became clear. Researchers at the University of California, Santa Cruz used an assortment of dyes so they could visualize fertilization of eggs of the wasp *Nasonia vitripennis.* They discovered cytoplasmic incompatibility involved timing, not toxins. Specifically, when infected sperm fertilize uninfected eggs, the sperm chromosomes try to align with those of the egg while the egg's chromosomes are still confined to the pronucleus. These eggs ultimately divide as if never fertilized and develop into males. However, chromosomes behave normally when a male and infected female mate. This yields a normal sex distribution and all progeny are infected with *Wolbachia.*

Another important breakthrough in the 1990s was an understanding of how prevalent *Wolbachia* infection is in the insect world. John Werren, a microbiologist at the University of Rochester in New York, and his colleagues used the polymerase chain reaction (PCR) to survey insects from Panama, England, and the United States, and found that roughly 20% of all species they sampled harbored *Wolbachia.* Other researchers found infection rates to be as high as 75% in smaller geographic locations. These high rates of global infection result because once a few individuals harbor *Wolbachia,* it spreads quickly through a population. In studying these newly identified infected species, scientists discovered that in addition to cytoplasmic incompatibility, the endosymbiont has developed other means to ensure its endurance (**Box figure a**). In some insect species, *Wolbachia* simply kills all the male offspring and induces parthenogenesis of infected females—that is, the mothers sim-

(a)

(a) *Wolbachia pipientis* Within the Egg Cytoplasm of the Ant *Gnamptogenys menadensis.* In this insect, *Wolbachia* is maternally transmitted, so the bacterium has evolved mechanisms to manipulate the sex distribution of the offspring so that the host produces mostly females. The *Wolbachia* cell is indicated by (B) and host mitochondria are labeled m.

ply clone themselves. This limits genetic diversity, but allows 100% transmission of *Wolbachia* to the next generation. In other species, the microbe allows the birth of males but then modifies their hormones so that the males become feminized and produce eggs.

Wolbachia's ability to manipulate the reproduction of its hosts has brought it to the attention of evolutionary biologists. These scientists are interested in the bacterium's potential role in speciation. It was noticed that two North American wasp species (*N. giraulti* and *N. longicornis*), each carrying a different strain of *Wolbachia,* appeared to be morphologically, behaviorally, and most importantly genetically similar. Predictably, when the two species mate with each other, there are no viable offspring. But much to the surprise of the scientific community, when wasps are treated with an antibiotic to cure them of their *Wolbachia* infections, the two wasp species mate and produce viable, fertile offspring. Could it be that while there are no genetic barriers to reproduction, the presence of two different *Wolbachia* endosymbionts forms a reproductive wall that could ultimately drive the evolution of new species?

Finally, *Wolbachia* may be driving more than speciation. They are required for the embryogenesis of filarial nemotodes, which cause diseases like elephantitis and river blindness. *Onchocerca volvulus* is the filarial nematode that causes river blindness. It is

(continued)

transmitted by blackflies in Africa, Latin America, and Yemen with a worldwide incidence of about 18 million people. When a fly bites a person, the nematode establishes its home in small nodules beneath the skin, where it can survive for as long as 14 years. During that time, it releases millions of larvae, many of which migrate to the eye. Eventually the host mounts an inflammatory response that results in progressive vision loss (**Box figure b**). It is now recognized that this inflammatory response is principally directed at the *Wolbachia* infecting the nematodes, not the worms themselves. German researchers discovered that in patients treated with a single course of antibiotic to kill the endosymbiont, nematode reproduction stopped. While inflammatory damage cannot be reversed, disease progression is halted. This may prove a more effective and cost-efficient means of treatment than the current anti-parasitic method that must be repeated every six months.

Our understanding of the distribution of *Wolbachia* and the mechanistically clever ways it has evolved to assure its survival will continue to grow. The microbe may ultimately become a valuable tool in investigating the complexities of speciation as well as the key to curing devastating diseases that strike those in developing countries.

Source: A. St. Andre, N. M. Blackwell, L. R. Hall, A. Hoerauf, N, W. Brattig, L. Volkman, M. J. Taylor, L. Ford, A. G. Hise, J. H. Lass, E. Diaconu, and E. Pearlman. 2002. The role of endosymbiotic Wolbachia *bacteria in the pathogenesis of river blindness.* Science *295: 1892–95.*

C. Zimmer. 2001. Wolbachia: *A tale of sex and survival.* Science *292:1093–95.*

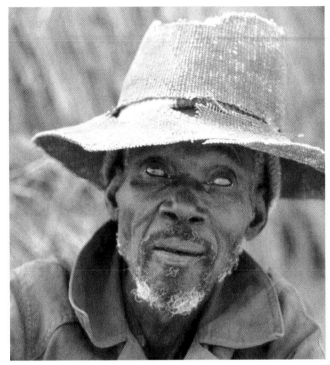

(b)

(b) River blindness. This is the second-leading cause of blindness worldwide. Evidence suggests that it is not the nematode but its endosymbiont, *Wolbachia pipientis*, that causes the severe inflammatory response that leaves many, like the man shown here, blind.

(a)

(b)

Figure 30.2 Mutualism. Light micrographs of **(a)** a worker termite of the genus *Reticulitermes* eating wood (×10), and **(b)** *Trichonympha*, a multiflagellated protozoan from the termite's gut (×135). Notice the many flagella that occur over most of its length. The ability of *Trichonympha* to break down cellulose allows termites to use wood as a food source.

(a)

(b)

Figure 30.3   Zooxanthellae.   (a) Zooxanthellae (green) within the tip of a hydra tentacle (×150). **(b)** The green color of this rose coral (*Manilina*) is due to the abundant zooxanthellae within its tissues.

Figure 30.4   Hydrothermal Vents and Related Geological Activity.   The chemical reactions between seawater and rocks that occur over a range of temperatures on the seafloor supplies the carbon and energy that support a diverse collection of microbial communities in specific niches within the vent system.

Figure 30.5   The Tube Worm–Bacterial Relationship.   (a) A community of tube worms (*Riftia pachyptila*) at the Galápagos Rift hydrothermal vent site (depth 2,550 m). Each worm is more than a meter in length and has a 20 cm gill plume. **(b, c)** Schematic illustration of the anatomical and physiological organization of the tube worm. The animal is anchored inside its protective tube by the vestimentum. At its anterior end is a respiratory gill plume. Inside the trunk of the worm is a trophosome consisting primarily of endosymbiotic bacteria, associated cells, and blood vessels. At the posterior end of the animal is the opisthosome, which anchors the worm in its tube. **(d)** Oxygen, carbon dioxide, and hydrogen sulfide are absorbed through the gill plume and transported to the blood cells of the trophosome. Hydrogen sulfide is bound to the worm's hemoglobin ($HSHbO_2$) and carried to the endosymbiont bacteria. The bacteria oxidize the hydrogen sulfide and use some of the released energy to fix CO_2 in the Calvin cycle. Some fraction of the reduced carbon compounds synthesized by the endosymbiont is translocated to the animal's tissues.

high concentrations of hydrogen sulfide, and can reach a temperature of 350°C. However, because of increased atmospheric pressure, the water does not boil. The seawater surrounding these vents has sulfide concentrations around 250 μM and temperatures 10 to 20°C above the ambient seawater temperature of about 2°C.

Giant (>1 m in length), red, gutless tube worms (*Riftia* spp.; **figure 30.5a**) near these hydrothermal vents provide an example of a unique form of mutualism and animal nutrition in which chemolithotrophic bacterial endosymbionts are maintained within specialized cells of the tube worm host (figure 30.5 b,c,d). The *Riftia* tube worms live at the interface between the hot, anoxic fluids of the vents and the cold, oxygen-containing seawater. Here, reduced sulfides from the vents react rapidly and spontaneously with oxygen in the seawater. In order to provide both reduced sulfur and oxygen to their bacterial endosymbionts, *Riftia's* blood contains a unique kind of hemoglobin, which ac-

counts for the bright-red plume extending out of their tubes. Hydrogen sulfide (H_2S) and O_2 are removed from the seawater by the worm's hemoglobin, and delivered to a special organ called the trophosome. The trophosome is packed with chemolithotrophic bacterial endosymbionts that fix CO_2 using the Calvin cycle (*see figure 10.4*) with electrons provided by H_2S. The CO_2 is carried to the endosymbionts in three ways: (1) freely in the bloodstream, (2) bound to hemoglobin, and (3) as organic acids such as malate and succinate. When these acids are decarboxylated, they release CO_2. This process is similar to carbon fixation by plants and cyanobacteria, but it occurs in the deepest, darkest reaches of the ocean. This mutualism is enormously successful: not only do the worms grow to an astounding size in densely packed communities, but the bacteria (which have not yet been cultured in the laboratory) reach densities of up to 10^{11} cells per gram of worm tissue.

Methane-Based Mutualisms

Other unique food chains involve methane-fixing microorganisms. By converting methane to carbohydrate, these bacteria perform the first step in providing organic matter for consumers. Methanotrophs are bacteria capable of using methane as a sole carbon source. They occur as intracellular symbionts of methane-vent mussels. In these mussels the thick, fleshy gills are filled with bacteria. In the Barbados Trench, methanotrophic carnivorous sponges have been discovered in a mud volcano at a depth of 4,943 m. Abundant methanotrophic symbionts were confirmed by the presence of enzymes related to methane oxidation in sponge tissues. These sponges are not satisfied with just bacteria; they also trap swimming prey to give variety to their diet.

Methanotrophic microorganisms are important in ecosystems outside of methane vents. For example, a methanotrophic endosymbiont was recently discovered to reduce the flux of methane from peat bogs; wetlands are the largest natural source of this greenhouse gas. *Sphagnum* moss, the principal plant in peat bogs (and a favorite among florists), can grow when submerged in water. Methanotrophic α-proteobacteria living within the outer cortex cells of *Sphagnum* stems oxidize methane as it diffuses through the water column:

$$CH_4 + 2O_2 \rightarrow CO_2 + H_2O$$

The resulting CO_2 is then readily fixed by the plant, which uses the Calvin cycle:

$$2CO_2 + 2H_2O \rightarrow 2CH_2O + 2O_2$$

This enables extremely efficient carbon recycling within this ecosystem:

$$CH_4 + 2CO_2 \rightarrow 2CH_2O$$

The Rumen Ecosystem

Ruminants are the most successful and diverse group of mammals on Earth today. Examples include cattle, deer, elk, bison, water buffalo, camels, sheep, goats, giraffes, and caribou. These animals spend vast amounts of time chewing their cud—a small ball of partially digested grasses that the animal has consumed but not yet completely digested. It is thought that the ruminants evolved an "eat now, digest later" strategy because their grazing can often be interrupted by predator attacks.

These herbivorous animals have stomachs that are divided into four chambers (**figure 30.6**). The upper part of the ruminant stomach is expanded to form a large pouch called the **rumen** and a smaller, honeycomb-like region, the reticulum. The lower portion is divided into an antechamber, the omasum, followed by the "true" stomach, the abomasum. The rumen is a highly muscular, anaerobic fermentation chamber where huge amounts of grasses eaten by the animal are digested by a diverse microbial community that includes bacteria, archaea, fungi, and protists. This microbial community is large—about 10^{12} organisms per milliliter of digestive fluid. When the animal eats plant material, it is

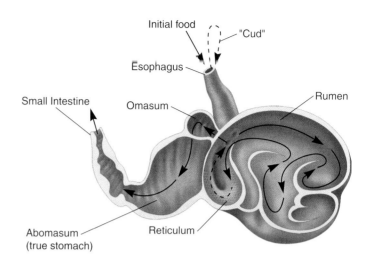

Figure 30.6 Ruminant Stomach. The stomach compartments of a cow. The microorganisms are active mainly in the rumen. Arrows indicate direction of food movement.

mixed with saliva and swallowed without chewing to enter the rumen. Here the material is churned and thoroughly mixed. Eventually, microbial attack and mixing coats the grass with microbes, reducing it to a pulpy, partially digested, mass. At this point the mass moves into the reticulum where it is regurgitated as cud, chewed, and re-swallowed by the animal. As this process proceeds, the grass becomes progressively more liquefied and flows out of the rumen into the omasum and then the abomasum. Here the nutrient-enriched grass material meets the animal's digestive enzymes and soluble organic and fatty acids are absorbed into the animal's bloodstream.

The microbial community in the rumen is extremely dynamic. The rumen is slightly warmer than the rest of the animal and with a redox potential of about −30 mV, all resident microorganisms must carry out anaerobic metabolism. Not only does the animal have a mutualistic relationship with the microbial community, but within the microbial community there are very specific interactions. One population of bacteria produce extracellular cellulases that cleave the β(1 → 4) linkages between the successive D-glucose molecules that form plant cellulose. The D-glucose is then fermented to organic acids such as acetate, butyrate, and propionate. These organic acids, as well as fatty acids are the true energy source for the animal. In some ruminants, the processing of organic matter stops at this stage (**Microbial Diversity & Ecology 30.2**). In others, such as cows, acetate, CO_2, and H_2 are used by methanogenic archaea to generate methane (CH_4), a greenhouse gas. In fact, a single cow can produce as much as 200 to 400 liters of CH_4 per day. The animal releases this CH_4 by a process called eructation (Latin *eructare*, to belch). Although the methanogens consume acetate that could be used by their animal hosts, they provide most of the vitamins needed by the ruminant. In fact, rumen microbes are so effective in fortifying the grass consumed by the animal, unlike humans, most ruminants have no required dietary amino acids. Fermentations (section 9.7)

Microbial Diversity & Ecology

30.2 Coevolution of Animals and Their Gut Microbial Communities

Organisms with digestive tracts had to make an interesting evolutionary choice: will the microbial community produce methane or not? The use of plant materials as a major food source does not always lead to methane production in the digestive tract. For example, kangaroos do not produce methane, whereas sheep and cattle do. The kangaroo has a distinct advantage in terms of nutrition. When complex plant materials are degraded only to organic acids such as acetate, the animal's digestive system can directly absorb the acids. In sheep and cattle, the microbial community is more complex and converts acetate-level substrates to methane and carbon dioxide, leading to nutrient loss from the original plant material. This loss can be substantial, with 10 to 15% of the organic matter in the feed lost to the atmosphere as methane.

An examination of over 250 reptiles, birds, and mammals showed that their maintenance of methanogenic microorganisms and methane production is under phylogenetic and not dietary control (see **Box figure**). Although low levels of methanogens can be detected in vertebrates that do not produce much of this important greenhouse gas, the lack of methane production seems to result from absence of methanogen receptor sites in the digestive system. As shown in this figure, the ability to maintain methanogens often has been lost. A similar situation occurs with arthropods: methane is produced by only a few organisms, including tropical millipedes, cockroaches, termites, and scarab beetles.

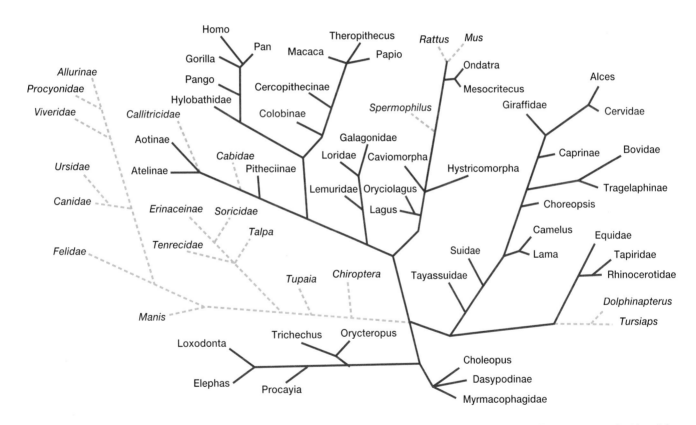

Coevolution of animals and their gut microbial communities: the methane choice. Methane-producing vertebrates are noted with solid red lines and Roman letters; nonmethane producers are noted with blue dotted lines and italics.

1. What is the critical characteristic of a mutualistic relationship?
2. How might one test to see if an insect-microbe relationship is mutualistic?
3. What is the role of *Riftia* hemoglobin to the success of the tube worm-endosymbiont mutualistic relationship?
4. How is the *Riftia* endosymbiont similar to cyanobacteria? How is it different?

5. Describe how the Sphagnum–methanotroph mutualism results in efficient carbon cycling.
6. What structural features of the rumen make it suitable for an herbivorous diet? Why do ruminants chew their cud?
7. Why is it important that the rumen is a reducing environment?

Cooperation

Cooperation and commensalism are two positive but not obligatory types of symbioses found widely in the microbial world. These involve syntrophic relationships. **Syntrophism** [Greek *syn*, together, and *trophe*, nourishment] is an association in which the growth of one organism either depends on or is improved by growth factors, nutrients, or substrates provided by another organism growing nearby. Sometimes both organisms benefit.

Cooperation benefits both organisms (figure 30.1). A cooperative relationship is not obligatory and, for most microbial ecologists, this nonobligatory aspect differentiates cooperation from mutualism. Unfortunately, it is often difficult to distinguish obligatory from nonobligatory because that which is obligatory in one habitat may not be in another (e.g., the laboratory). Nonetheless, the most useful distinction between cooperation and mutualism is the observation that cooperating organisms can be separated from one another and remain viable, although they may not function as well.

Two examples of a cooperative relationship include the association between *Desulfovibrio* and *Chromatium* (**figure 30.7a**), in which the carbon and sulfur cycles are linked, and the interaction of a nitrogen-fixing microorganism with a cellulolytic organism such as *Cellulomonas* (figure 30.7b). In the second example, the cellulose-degrading microorganism liberates glucose from the cellulose, which can be used by nitrogen-fixing microbes. An excellent example of a cooperative biodegradative association is shown in **figure 30.8.** In this case degradation of the toxin 3-chlorobenzoate depends on the functioning of microorganisms with complementary capabilities. If any one of the three microorganisms is not present and active, the degradation of the substrate will not occur. This example points out how the sum of the microbes in a community can be considered greater than the contribution made by any single microorganism.

In other cooperative relations, sulfide-dependent autotrophic filamentous microorganisms fix carbon dioxide and synthesize organic matter that serves as a carbon and energy source for a heterotrophic organism. Some of the most interesting include the polychaete worms *Alvinella pompejana* (**figure 30.9**), the Pompeii worm, and also *Paralvinella palmiformis,* the Palm worm. Both have filamentous bacteria on their dorsal surfaces. These filamentous bacteria can tolerate high levels of metals such as arsenic, cadmium, and copper. When growing on the surface of the animal, they may provide protection from these toxic metals, as well as thermal protection; in addition, they appear to be used as a food source. A deep-sea crustacean has been discovered that uses sulfur-oxidizing autotrophic bacteria as its food source. This shrimp, *Rimicaris exoculata* (**figure 30.10**) has filamentous sulfur-oxidizing bacteria growing on its surface (figure 30.10b). When these are dislodged the shrimp ingests them. This nominally "blind" shrimp use a reflective organ to respond to the glow emitted by geothermally active black smoker chimneys. The organ is sensitive to a light wavelength that is not detectable by humans.

Another interesting example of bacterial epigrowth is shown by nematodes, including *Eubostrichus parasitiferus,* that live at the interface between oxic and anoxic sulfide-containing marine sediments (**figure 30.11a,b**). These animals are covered by sulfide-oxidizing

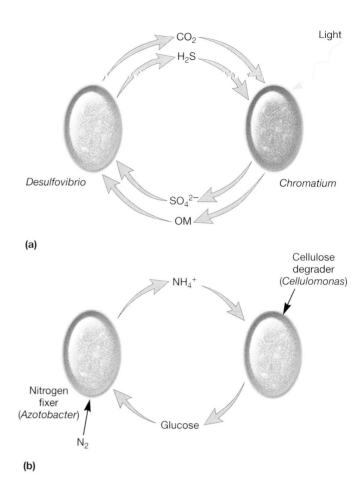

(a)

(b)

Figure 30.7 Examples of Cooperative Symbiotic Processes. **(a)** The organic matter (OM) and sulfate required by *Desulfovibrio* are produced by the *Chromatium* in its photosynthesis-driven reduction of CO_2 to organic matter and oxidation of sulfide to sulfate. **(b)** *Azotobacter* uses glucose provided by a cellulose-degrading microorganism such as *Cellulomonas,* which uses the nitrogen fixed by *Azotobacter.*

bacteria that are present in intricate patterns (figure 30.11b). The bacteria not only decrease levels of toxic sulfide, which often surround the nematodes, but they also serve as a food supply.

In 1990, hydrothermal vents were discovered in a freshwater environment at the bottom of Lake Baikal, the oldest (25 million years old) and deepest lake in the world. This lake is located in the far east of Russia (**figure 30.12a,b**) and has the largest volume of any freshwater lake (not the largest area—which is Lake Superior). Microbial mats featuring long, white strands are in the center of the vent field where the highest temperatures are found. At the edge of the vent field, where the water temperature is lower, the microbial mats end, and sponges, gastropods, and other organisms, which use the sulfur-oxidizing bacteria as a food source, are present (figure 30.12b). Similar although less developed areas have been found in Yellowstone Lake, Wyoming.

A form of cooperation also occurs when a population of similar microorganisms monitors its own density—the process of

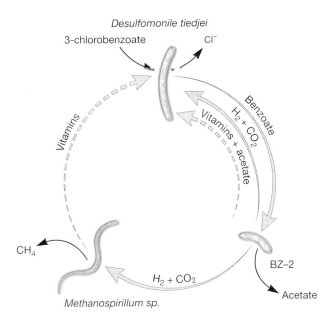

Figure 30.8 **Associations in a Defined Three-Membered Cooperative and Commensalistic Community That Can Degrade 3-Chlorobenzoate.** If any member is missing, degradation will not take place. The solid arrows demonstrate nutrient flows, and the dashed lines represent hypothesized flows.

Figure 30.9 **A Marine Worm-Bacterial Cooperative Relationship.** *Alvinella pompejana*, a 10 cm long worm, forms a cooperative relationship with bacteria that grow as long threads on the worm's surface. The bacteria and *Alvinella* are found near the black smoker-heated water fonts.

quorum sensing, which was discussed in sections 6.6 and 12.5. The microorganisms produce specific autoinducer compounds, and as the population increases and the concentration of these compounds reaches critical levels, specific genes are expressed.

(a)

(b)

Figure 30.10 **A Marine Crustacean-Bacterial Cooperative Relationship.** **(a)** A picture of the marine shrimp *Rimicaris exoculata* clustered around a hydrothermal vent area, showing the massive development of these crustaceans in the area where chemolithotrophic bacteria grow using sulfide as an electron and energy source. The bacteria, which grow on the vent openings and also on the surface of the crustaceans, fix carbon, and serve as the nutrient for the shrimp. **(b)** An electron micrograph of a thin section across the leg of the marine crustacean *Rimicaris exoculata*, showing the chemolithotrophic bacteria that cover the surface of the shrimp. The filamentous nature of these bacteria, upon which this commensalistic relationship is based, is evident in this thin section.

These responses are important for microorganisms that form associations with each other, plants, and animals. Intercellular communication is critical for the establishment of biofilms and for colonization of hosts by pathogens.

1. How does cooperation differ from mutualism? What might be some of the evolutionary implications of both types of symbioses?
2. What is syntrophism? Is physical contact required for this relationship?
3. Why are *Alvinella, Rimicaris,* and *Eubostrichus* good examples of cooperative microorganism-animal interactions?
4. Where is Lake Baikal located, and why is it unique in terms of its microbial communities?

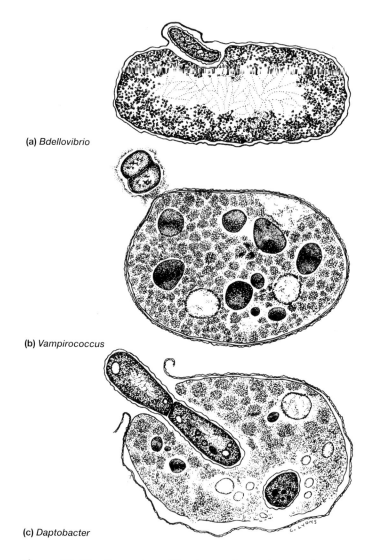

(a) *Bdellovibrio*

(b) *Vampirococcus*

(c) *Daptobacter*

Figure 30.13 Examples of Predatory Bacteria Found in Nature. **(a)** *Bdellovibrio*, a periplasmic predator that penetrates the cell wall and grows outside the plasma membrane, **(b)** *Vampirococcus* with its unique epibiotic mode of attacking a prey bacterium, and **(c)** *Daptobacter* showing its cytoplasmic location as it attacks a susceptible bacterium.

that with the rise in antibiotic-resistant pathogens, such forms of treatments may become viable alternatives. Class *Deltaproteobacteria* (section 22.4); Microbiology of fermented foods: Probiotics (section 40.6)

Although *Vampirococcus* and *Daptobacter* also kill their prey, they gain entry in a less-dramatic fashion. *Vampirococcus* attaches itself as an epibiont to the outer membrane of its prey. It then secretes degradative enzymes that result in the release of the prey's cytoplasmic contents. In contrast, *Daptobacter* penetrates the prey cell and consumes the cytoplasmic contents directly.

A surprising finding is that predation has many beneficial effects, especially when one considers interactive populations of predators and prey, as summarized in **table 30.3.** Simple ingestion and assimilation of a prey bacterium can lead to increased rates of

nutrient cycling, critical for the functioning of the microbial loop *(see figure 27.13)*. In this process, organic matter produced through photosynthetic and chemotrophic activity is mineralized before it reaches higher consumers, allowing the minerals to be made available to the primary producers. Ingestion and short-term retention of bacteria also are critical for ciliate functioning in the rumen, where methanogenic bacteria contribute to the health of the ciliates by decreasing toxic hydrogen levels by using H_2 to produce methane, which then is passed from the rumen. The physical environmnet: Microorganisms and ecosystems (section 27.3)

Predation also can provide a protective, high-nutrient environment for particular prey. Ciliates ingest the gram-positive bacterium *Legionella* and protect this important pathogen from chlorine, which often is used in an attempt to control *Legionella* in cooling towers and air-conditioning units. The ciliate serves as a reservoir host. *Legionella pneumophila* also has been found to have a greater potential to invade macrophages and epithelial cells after predation, indicating that ingestion not only provides protection but also may enhance pathogenicity. A similar phenomenon of survival in protozoa has been observed for *Mycobacterium avium*, a pathogen of worldwide concern. These protective aspects of predation have major implications for survival and control of disease-causing microorganisms in the biofilms present in water supplies and air-conditioning systems. Airborne diseases (section 38.1)

Fungi often show interesting predatory skills. Some fungi can trap protozoa by the use of sticky hyphae or knobs, sticky networks of hyphae, or constricting or nonconstricting rings. A classic example is *Arthrobotrys,* which traps nematodes by use of constricting rings. After the nematode is trapped, hyphae grow into the immobilized prey and the cytoplasm is used as a nutrient. Other fungi have conidia that, after ingestion by an unsuspecting predator, grow and attack the susceptible host from inside the intestinal tract. In this situation the fungus penetrates the host cells in a complex interactive process.

Clearly predation in the microbial world is not straightforward. It often has a fatal and final outcome for an individual prey organism but it can have a wide range of beneficial effects on prey populations. Predation is clearly critical in the functioning of natural environments.

Parasitism

Parasitism is one of the most complex microbial interactions; the line between parasitism and predation is difficult to define (figure 30.1). This is a relationship between two organisms in which one benefits from the other, and the host is usually harmed. This can involve nutrient acquisition and/or physical maintenance in or on the host. In parasitism there is always some co-existence between host and parasite. Successful parasites have evolved to co-exist in equilibrium with their hosts. This is because a host that dies immediately after parasite invasion may prevent the microbe from reproducing to sufficient numbers to ensure colonization of a new host. But what happens if the host-parasite equilibrium is upset? If the balance favors the host (perhaps by a strong immune defense or

Table 30.3	The Many Faces of Predation

Predation Result	Example
Digestion	The microbial loop. Soluble organic matter from primary producers is normally used by bacteria, which become a particulate food source for higher consumers. Flagellates and ciliates prey on these bacteria and digest them, making the nutrients they contain available again in mineral form for use in primary production. In this way a large portion of the carbon fixed by the photosynthetic microbes is mineralized and recycled and does not reach the higher trophic levels of the ecosystem (*see figure 27.13*).
	Predation also can reduce the density-dependent stress factors in prey populations, allowing more rapid growth and turnover of the prey than would occur if the predator were not active.
Retention	Bacteria retained within the predator serve a useful purpose, as in the transformation of toxic hydrogen produced by ciliates in the rumen to methane. Also, trapping of chloroplasts (kleptochloroplasty) by protozoa provides the predator with photosynthate.
Protection and increased fitness	The intracellular survival of *Legionella* ingested by ciliates protects it from stresses such as heating and chlorination. Ingestion also results in increased pathogenicity when the prey is again released to the external environment, and this may be required for infection of humans. The predator serves as a reservoir host.
	Nanoplankton may be ingested by zooplankton and grow in the zooplankton digestive system. They are then released to the environment in a more fit state. Dissemination to new locations also occurs.

antimicrobial therapy), the parasite loses its habitat and may be unable to survive. On the other hand, if the equilibrium is shifted to favor the parasite, the host becomes ill, and depending on the specific host-parasite relationship, may die. One good example is the disease typhus. This disease is caused by the rickettsia *Rickettsia typhi,* which is harbored in fleas that live on rats. It is transmitted to humans who are bitten by such fleas, so in order to contract typhus, one must be in close proximity to rats. Humans often live in association with rats, and in such communities there is always a small number of people with typhus—that is to say, typhus is endemic. However, during times of war or when people are forced to become refugees, lack of sanitation and overcrowding result in an increased number of rat-human interactions. Typhus can then reach epidemic proportions. During the Crimean War (1853–1856), about 213,000 men were killed or wounded in combat while over 850,000 were sickened or killed by typhus.

On the other hand, a controlled parasite-host relationship can be maintained for long periods of time. For example, **lichens (figure 30.14)** are the association between specific ascomycetes (a fungus) and certain genera of either green algae or cyanobacteria. In a lichen, the fungal partner is termed the **mycobiont** and the algal or cyanobacterial partner, the **phycobiont.** In the past the lichen symbiosis was considered to be a mutualistic interaction. It recently has been found that a lichen forms only when the two potential partners are nutritionally deprived. In nutrient-limited environments, the relationship between the fungus and its photosynthetic partner has coevolved to the point where lichen morphology and metabolic relationships are extremely stable. In fact, lichens are assigned generic and species names. The characteristic morphology of a given lichen is a property of the association and is not exhibited by either symbiont individually. Characteristics of fungal divisions: *Ascomycota* (section 26.6); Photosynthetic bacteria: Phylum *Cyanobacteria* (section 21.3)

Because the phycobiont is a photoautotroph—dependent only on light, carbon dioxide, and mineral nutrients—the fungus can

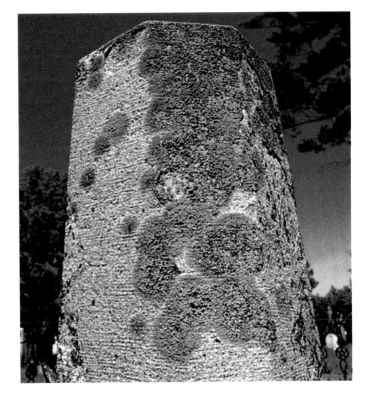

Figure 30.14   Lichens.   Crustose (encrusting) lichens growing on a granite post.

get its organic carbon directly from the alga or cyanobacterium. The fungus often obtains nutrients from its partner by projections of fungal hyphae called haustoria, which penetrate the phycobiont cell wall. It also uses the O_2 produced during phycobiont photophosphorylation in carrying out respiration. In turn the fungus protects the phycobiont from high light intensities, provides water and minerals to it, and creates a firm substratum within which the phycobiont can grow protected from environmental stress. The invasive nature of the fungal partner is why lichens are considered parasitic relationships.

An important aspect of many symbiotic relationships, including parasitism, is that over time, the symbiont, once it has established a relationship with the host, will tend to discard excess, unused genomic information, a process called **genomic reduction.** This is clearly the case with the aphid endosymbiont *Buchnera aphidicola* (p. 718) and it has also occurred with the parasite *Mycobacterium leprae,* and with the microsporidium *Encephalitozoon cuniculi.* The latter organism, which parasitizes a wide range of animals, including humans, now can only survive inside the host cell. Insights from microbial genomes (section 15.8)

1. Define predation and parasitism. How are these similar and different?
2. How can a predator confer positive benefits on its prey? Think of the responses of individual organisms versus populations as you consider this question.
3. What are examples of parasites that are important in microbiology?
4. What is a lichen? Discuss the benefits the phycobiont and mycobiont provide each other.

Amensalism

Amensalism (from the Latin for *not* at the same table) describes the adverse effect that one organism has on another organism (figure 30.1). This is a unidirectional process based on the release of a specific compound by one organism which has a negative effect on another organism. A classic example of amensalism is the production of antibiotics that can inhibit or kill a susceptible microorganism (**figure 30.15a**). Community complexity is demonstrated by the capacity of attine ants (ants belonging to a New World tribe) to take advantage of an amensalistic relationship between an actinomycete and the parasitic fungi *Escovopsis.* This amensalistic relationship enables the ant to maintain a mutualism with another fungal species, *Leucocoprini.* Amazingly, these ants cultivate a garden of *Leucocoprini* for their own nourishment (figure 30.15b). To prevent the parasitic fungus *Escovopsis* from decimating their fungal garden, the ants also promote the growth of an actinomycete of the genus *Pseudonocardia,* which produces an antimicrobial compound that inhibits the growth of *Escovopsis.* This unique amensalistic process appears to have evolved 50 to 65 million years ago in South America. Thus this relationship has been subject to millions of years of coevolution, such that particular groups of ants cultivate specific strains of fungi that are then subject to different groups of *Escovopsis* parasites. In addition, the ants have developed intricate crypts within their exoskeletons for the growth of the antibiotic-producing *Pseudonocardia.* As shown in figure 30.15c, these crypts have been modified throughout the ants' evolutionary history. The most primitive "paleo-attine" ants carry the bacterium on their forelegs, "lower" and "higher" attines have evolved special plates on their ventral surfaces, while the entire surface of the most recent attines, leaf-cutter ants of the genus

Acromyrmex, are covered with the bacterium. Related ants that do not cultivate fungal gardens (e.g., *Atta* sp.) do not host *Pseudonocardia.* This unique multipartner relationship has enabled scientists to explore the behavioral, physiological, and structural aspects of the organisms involved.

Other important amensalistic relationships involve microbial production of specific organic compounds that disrupt cell wall or plasma membrane integrity of target microorganisms. These include the bacteriocins. The bacteriocin nisin has been used as an additive for controlling the growth of undesired pathogens in dairy products for over 40 years. Antibacterial peptides also can be released by the host in the intestine and other sites. These molecules, called cecropins in insects and defensins in mammals, are effector molecules that play significant roles in innate immunity. In vertebrates these molecules are released by phagocytes and intestinal cells, and have powerful antimicrobial activity. Human sweat is also antimicrobial. Sweat glands produce an antimicrobial peptide called dermicidin. The skin also produces similar compounds including an antimicrobial peptide called cathelicidin. Finally, metabolic products, such as organic acids formed in fermentation, can produce amensalistic effects. These compounds inhibit growth by changing the environmental pH, for example, during natural milk spoilage. Chemical mediators in nonspecific (innate) resistance: Antimicrobial peptides (section 31.6)

Competition

Competition arises when different organisms within a population or community try to acquire the same resource, whether this is a physical location or a particular limiting nutrient (figure 30.1). If one of the two competing organisms can dominate the environment, whether by occupying the physical habitat or by consuming a limiting nutrient, it will outgrow the other organism. This phenomenon was studied by E. F. Gause, who in 1934 described it as the **competitive exclusion principle.** He found that if two competing ciliates overlapped too much in terms of their resource use, one of the two protozoan populations was excluded. In chemostats, competition for a limiting nutrient may occur among microorganisms with transport systems of differing affinity. This can lead to the exclusion of the slower-growing population under a particular set of conditions. If the dilution rate is changed, the previously slower-growing population may become dominant. Often two microbial populations that appear to be similar nevertheless coexist. In this case, they share the limiting resource (space, a limiting nutrient) and coexist while surviving at lower population levels. The continuous culture of microorganisms: The chemostat (section 6.4)

1. What is the origin of the term amensalism?
2. The production of antimicrobial agents and their effects on target species has been called a "microbial arms race." Explain what you think this phrase means and its implications for the attine ant system.
3. What are bacteriocins?
4. What is the competitive exclusion principle?

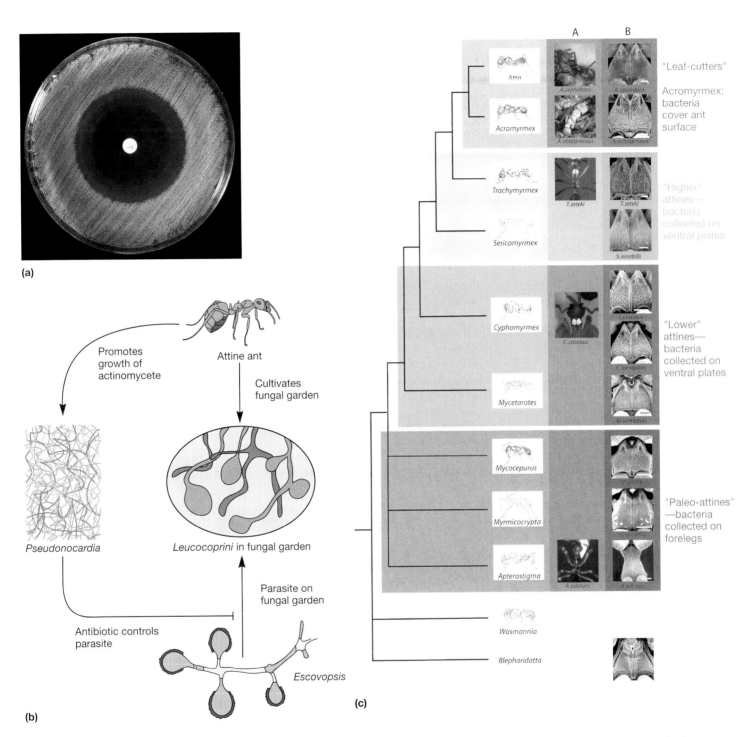

Figure 30.15 Amensalism: An Adverse Microbe-Microbe Interaction. (a) Antibiotic production and inhibition of growth of a susceptible bacterium on an agar medium. (b) A schematic diagram describing the use of antibiotic-producing streptomycetes by ants to control fungal parasites in their fungal garden. (c) Coevolution of attine ants and the antibiotic-producing *Pseudonocardia* has resulted in specialized localization of the bacterium on the ant. This rooted tree illustrates the phylogeny of fungus-growing ants; column A shows the placement of the bacteria on the ants' body, column B presents scanning electron micrographs of the areas colonized by the microbe.

30.2 HUMAN-MICROBE INTERACTIONS

As we have seen, many microorganisms live much of their lives in a special ecological relationship; an important part of their environment is a member of another species. Here we discuss microorganisms normally associated with the human body, the **normal microbial flora** or **microbiota.** If we consider the human body as a diverse environment in and on which specific niches are formed, the normal flora may be discussed as the microbial ecology of a human. The application of ecological principles can assist in our understanding of host-microbe relationships. Interactions between host and microbe are dynamic, permitting niche fulfillment that maximizes benefit to the microbe and, in some cases, the host. Tolerating a normal flora likewise suggests that the host derives benefit. Acquisition of a normal microbial flora represents a selective process, where a niche may be defined by cellular receptors, surface properties, or secreted products. It should be noted that microbial niche variations are also related to age, gender, diet, nutrition, and developmental stage of the host. In general, the adult human microbial flora is relatively constant and can thus be mapped.

The survival of a host, such as a human, depends upon an elaborate network of defenses that keeps harmful microorganisms and other foreign material from entering the body. Should they gain access, additional host defenses are summoned to prevent them from establishing another type of relationship, one of parasitism or **pathogenicity.** Pathogenicity is the ability to produce pathologic changes or disease. A **pathogen** [Greek *patho,* and *gennan,* to produce] is any disease-producing microorganism. Here we introduce the normal human microbiota, which function not as pathogens but as symbionts that are part of the host's first line of defense against harmful infectious agents.

Gnotobiotic Animals

To determine the role of the normal microorganisms associated with a host and evaluate the consequences of colonization, it is possible to deliver an animal by cesarean section and raise that animal in the absence of microorganisms—that is, germfree. These microorganism-free animals provide suitable experimental models for investigating the interactions of animals and their microorganisms. Comparing animals possessing normal microorganisms (conventional animals) with germfree animals permits the elucidation of many complex relationships between microorganisms, hosts, and specific environments. Germfree experiments also extend and challenge the microbiologist's "pure culture concept" to in vivo research.

The term **gnotobiotic** [Greek *gnotos,* known, and *biota,* the flora and fauna of a region] has been defined in two ways. Some think of a gnotobiotic environment or animal as one in which all the microbiota are known; they distinguish it from one that is truly germfree. We shall use the term in a more inclusive sense. Gnotobiotic refers to a microbiologically monitored environment or animal that is germfree (**axenic** [*a,* neg, and Greek *Xenos,* a stranger]) or in which the identities of all microbiota are known.

Development of a lifelong symbiotic relationship with microbes begins during birth. The infant's exposure to the vaginal mucosa, skin, hair, food, and other nonsterile objects quickly results in the acquisition of a predominantly commensal normal flora. The human fetus in utero (as is the case in most mammals) is usually free from microorganisms. As an infant begins to acquire a normal microbiota, the microbial population stabilizes during the first week or two of life. Colonization of the newborn varies with respect to its environment. The newborn likely acquires external flora from those who provide its care. Likewise, internal flora are acquired through its diet. Bifidobacteria represent more than 90% of the total intestinal bacteria in breast-fed infants, with *Enterobacteriaceae* and enterococci in smaller proportions. This suggests that human milk may act as a selective medium for nonpathogenic bacteria, as bottle-fed babies appear to have a much smaller proportion of intestinal bifidobacteria. Switching to cow's milk or solid food (mostly polysaccharide) appears to result in the loss of bifidobacteria predominance, as *Enterobacteriaceae,* enterococci, bacteroides, lactobaccili, and clostridia increase in number. Bacterial chemotaxis and trophism may explain the high frequency of bacterial-tissue associations. Additionally, the host may partly direct microbe-tissue associations as seen in the selective destruction of gram-positive bacteria by the antimicrobial peptide angiogenin-4 secreted by special intestinal cells called **Paneth cells.**

Louis Pasteur first suggested that animals could not live in the absence of microorganisms. Attempts between 1899 and 1908 to grow germfree chickens had limited success because the birds died within a month. Thus it was believed that intestinal bacteria were essential for the adequate nutrition and health of the chickens. It was not until 1912 that germfree chickens were shown to be as healthy as normal birds when they were fed an adequate diet. Since then, gnotobiotic animals and systems have become commonplace in research laboratories (**figure 30.16**).

What have we learned from germfree animals? Germfree animals are usually more susceptible to pathogens. With the normal commensal microbiota absent, foreign and pathogenic microorganisms establish themselves very easily. The number of microorganisms necessary to infect a germfree animal and produce a diseased state is much smaller. Conversely, germfree animals are almost completely resistant to the intestinal protozoan *Entamoeba histolytica* that causes amebic dysentery. This resistance results from the absence of the bacteria that *E. histolytica* uses as a food source. Germfree animals also do not show any dental caries or plaque formation. However, if they are inoculated with cariogenic (caries or cavity-causing) streptococci of the *Streptococcus mutans–Streptococcus gordonii* group and fed a high-sucrose diet, they develop caries. Protist classification: Super group *Amoebozoa* (section 25.6); Food- and water-borne diseases (sections 37.4, 38.4, and 39.5)

---

1. Define gnotobiotic.
2. Compare a germfree mouse to a normal one with regard to overall susceptibility to pathogens. What benefits does an animal gain from its microbiota?

(a)

(b)

Figure 30.16 Raising Gnotobiotic Animals. (a) Schematic of a gnotobiotic isolator. The microbiological culture media monitor the sterile environment. If growth occurs on any of the cultures, gnotobiotic conditions do not exist. **(b)** Gnotobiotic isolators for rearing colonies of small mammals.

30.3 NORMAL MICROBIOTA OF THE HUMAN BODY

In a healthy human the internal tissues (e.g., brain, blood, cerebrospinal fluid, muscles) are normally free of microorganisms. Conversely, the surface tissues (e.g., skin and mucous membranes) are constantly in contact with environmental microorganisms and become readily colonized by various microbial species. The mixture of microorganisms regularly found at any anatomical site is referred to as the normal microbiota, the indigenous microbial population, the microflora, or the normal flora. For consistency, the term normal microbiota is used in this chapter. An overview of the microbiota native to different regions of the body is presented next (**figure 30.17**).

Because bacteria make up most of the normal microbiota, they are emphasized over the fungi (mainly yeasts) and protists.

There are many reasons to acquire knowledge of the normal human microbiota. Three specific examples include:

1. An understanding of the different microorganisms at particular locations provides greater insight into the possible infections that might result from injury to these body sites.
2. A knowledge of the normal microbiota helps the physician-investigator understand the causes and consequences of colonization and growth by microorganisms normally absent at a specific body site.
3. An increased awareness of the role that these normal microbiota play in stimulating the host immune response can be gained. This awareness is important because the immune system provides protection against potential pathogens.

As noted previously, three of the most important types of symbiotic relationships are commensalism, mutualism, and parasitism. Within each category the association may be either ectosymbiotic or endosymbiotic. In the following subsections, examples are presented of both ecto- and endosymbiotic relationships. Both commensalistic and mutualistic relationships are also considered. Parasitism and pathogenicity are presented in chapter 33.

Skin

The adult human is covered with approximately 2 square meters of skin. It has been estimated that this surface area supports about 10^{12} bacteria. Recall that commensalism is a symbiotic relationship in which one species benefits and the other is unharmed. Commensal microorganisms living on or in the skin can be either resident (normal) or transient microbiota. Resident organisms normally grow on or in the skin. Their presence becomes fixed in well-defined distribution patterns. Those that are temporarily present are transient microorganisms. Transients usually do not become firmly entrenched and are unable to multiply.

It should be emphasized that the skin is a mechanically strong barrier to microbial invasion. Few microorganisms can penetrate the skin because its outer layer consists of thick, closely packed cells called keratinocytes. In addition to direct resistance to penetration, continuous shedding of the outer epithelial cells removes many of those microorganisms adhering to the skin surface.

The anatomy and physiology of the skin vary from one part of the body to another, and the normal resident microbiota reflect these variations. The skin surface or epidermis is not a favorable environment for microbial colonization. In addition to a slightly acidic pH, a high concentration of sodium chloride, and a lack of moisture in many areas, certain inhibitory substances (bactericidal and/or bacteriostatic) on the skin help control microbial colonization. For example, the sweat glands release lysozyme (muramidase), an enzyme that lyses *Staphylococcus epidermidis* and other gram-positive bacteria by hydrolyzing the $\beta(1 \rightarrow 4)$ glycosidic bond connecting *N*-acetylmuramic acid and *N*-acetylglucosamine microorganism or foreign material encountered by the vertebrate host. It includes general mechanisms inherited as part of the innate structure and function of each animal (such as skin, mucus, and constitutively produced antimicrobial mediators like lysozyme), and acts as a first line of defense. The nonspecific immune response defends against foreign particles equally and lacks immunological memory—that is, nonspecific responses occur to the same extent each time a microorganism or foreign body is encountered.

In contrast, the **specific immune responses,** also known as **acquired, adaptive,** or **specific immunity,** resist a particular foreign agent. Moreover, the effectiveness of specific immune responses increases on repeated exposure to foreign agents such as viruses, bacteria, or toxins; that is to say, specific responses have "memory." Substances that are recognized as foreign and provoke immune responses are called **antigens** (contraction of the words *anti*body and *gen*erators). The antigens cause specific cells to replicate and manufacture a variety of proteins that function to protect the host. One such cell, the B cell, produces and secretes glycoproteins called antibodies. **Antibodies** bind to specific antigens and inactivate them or contribute to their elimination. Other immune cells become activated to destroy cells harboring intracellular pathogens. The nonspecific and specific responses usually work together to eliminate pathogenic microorganisms and other foreign agents (**figure 31.1**).

The distinction between the innate and adaptive systems is, to a degree, artificial. Although innate systems predominate imme-

diately upon initial exposure to foreign substances, multiple bridges occur between innate and adaptive immune system components (figure 31.1). Importantly, there are a variety of cells that function in both innate and adaptive immunity. These cells are known as the **white blood cells,** or leukocytes (as they were first identified as part of a white "buffy coat" between red blood cells and the plasma of centrifuged blood). Blood cell development occurs in the bone marrow of mammals during the process of **hematopoesis.** White blood cells (WBCs) are divided between the innate and adaptive immune systems according to specific cellular behaviors. Cells that typically mature prior to leaving the bone marrow and provide the same physiological response regardless of antigen are assigned to the innate immune system (e.g., macrophages and dendritic cells). Cells that are not completely functional after leaving the bone marrow but differentiate in response to various types of antigens are assigned to the adaptive immune system (e.g., B and T cells). The WBCs form the basis for immune responses to invading microbes and foreign substances. Many of these cells reside in specific tissues and/or organs. Some tissues and organs provide supportive functions in nurturing the cells so that they can mature and/or respond correctly to antigens.

The bridges that interconnect the innate and specific immune responses are numerous and present a difficulty for any author trying to describe immunity. We begin our discussion with an introduction to the various cells, tissues, and organs of the innate defenses. In this discussion, reference will be made to processes and chemical mediators that are described in detail either later in this chapter or in chapter 32 (specific defenses). For your convenience, cross-references to the detailed discussion are provided. Thus the discussion of innate immunity proceeds from the general to the ever-more specific and detailed.

1. Define each of the following terms: immune system, immunity, immunology, antigen, antibody.
2. Compare and contrast the specific and nonspecific immune responses.
3. Describe how the activity of white blood cells is often used to assign their role in immunity.

31.2 CELLS, TISSUES, AND ORGANS OF THE IMMUNE SYSTEM

As indicated, the immune system is an organization of molecules, cells, tissues, and organs, each with a specialized role in defending against viruses, microorganisms, cancer cells, and nonself proteins (e.g., organ transplants). Immune system cells and tissue are now considered.

Cells of the Immune System

The cells responsible for both nonspecific and specific immunity are the **leukocytes** [Greek *leukos,* white, and *kytos,* cell], or white blood cells. All of the leukocytes originate from pluripotent stem cells in the fetal liver and in the bone marrow of the animal host (**figure 31.2**). Pluripotency indicates that these stem cells have

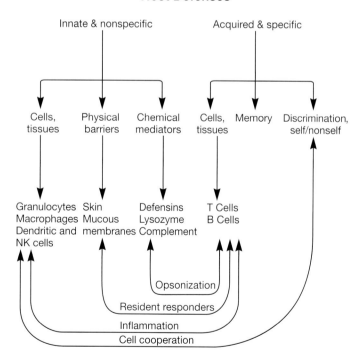

Figure 31.1 Some of the Major Components that Make Up the Mammalian Immune System. Double-headed arrows indicate potential bridging events that unite innate and acquired forms of immunity.

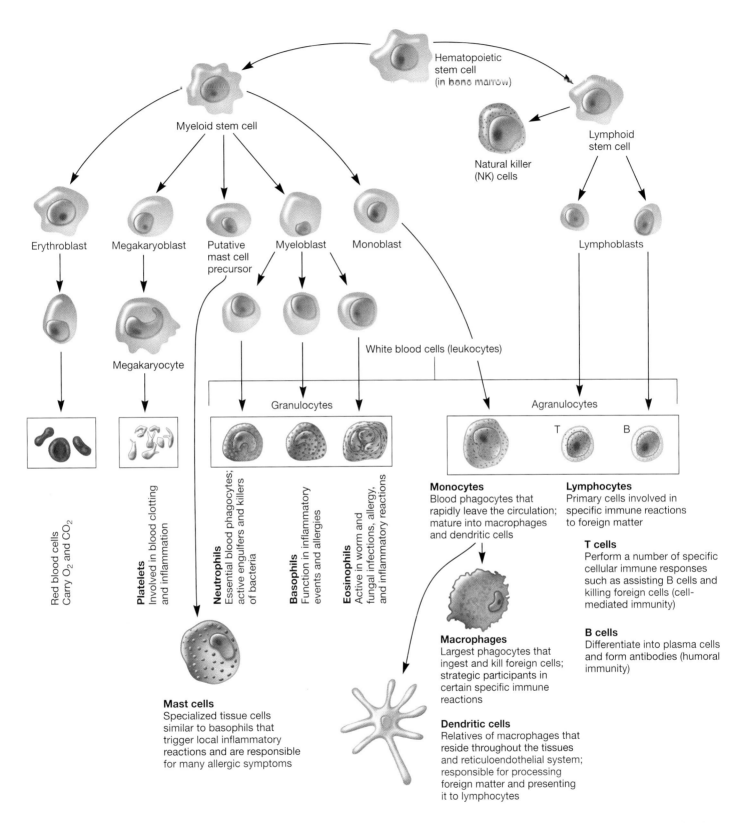

Figure 31.2 **The Different Types of Human Blood Cells.** Pluripotent stem cells in the bone marrow divide to form two blood cell lineages: (1) the lymphoid stem cell gives rise to B cells that become antibody-secreting plasma cells, T cells that become activated T cells, and natural killer cells. (2) The common myeloid progenitor cell gives rise to the granulocytes (neutrophils, eosinophils, basophils), monocytes that give rise to macrophages and dendritic cells, an unknown precursor that gives rise to mast cells, megakaryocytes that produce platelets, and the erythroblast that produces erythrocytes (red blood cells).

Table 31.1	Normal Adult Blood Count	
Cell Type	**Cells/mm³**	**Percent WBC**
Red blood cells	5,000,000	
Platelets	250,000	
White blood cells	7,200	100
Neutrophils	4,320	60
Lymphocytes	2,160	30
Monocytes	430	6
Eosinophils	215	3
Basophils	70	1

not yet committed to differentiating into one specific cell type. When they migrate to other body sites, some differentiate into hematopoietic stem cells that are destined to become leukocytes. When stimulated to undergo further development, some leukocytes become residents within tissues, where they respond to local trauma. These cells may sound the alarm that signals invasion by foreign organisms. Other leukocytes circulate in body fluids and are recruited to the sites of infection after the alarm has been raised. The average adult has approximately 7,400 leukocytes per mm³ of blood (**table 31.1**). This average value shifts substantially during infectious and allergic responses of the host. Thus the complete blood count (CBC) and the phenotype differential (DIFF), which determines the relative number of each type of blood cell, are used clinically to assist in the diagnosis of disease and infection. In defending the host against pathogenic microorganisms, leukocytes cooperate with each other first to recognize the pathogen as an invader and then to destroy it. These different leukocytes are now briefly examined.

Monocytes and Macrophages

Monocytes and macrophages are highly phagocytic and make up the **monocyte-macrophage system** (**figure 31.3**). Although the specifics of phagocytosis will be discussed shortly, recall that it involves the engulfment of large particles and microorganisms that are then enclosed in a phagocytic vacuole or phagosome. **Monocytes** [Greek *monos,* single, and *cyte,* cell] are mononuclear leukocytes with an ovoid- or kidney-shaped nucleus and granules in the cytoplasm that stain gray-blue with basic dyes (figure 31.2). They are produced in the bone marrow and enter the blood, circulate for about eight hours, enlarge, migrate to the tissues, and mature into macrophages or dendritic cells.

Because **macrophages** [Greek *macros,* large, and *phagein,* to eat] are derived from monocytes, they are also classified as mononuclear phagocytic leukocytes. However, they are larger than monocytes, contain more organelles that are critical for phagocytosis, and have a plasma membrane covered with microvilli (**figure 31.4**). Macrophages have surface molecules that function as receptors to nonspecifically recognize common components of pathogens. These receptors include mannose and fucose receptors and a special class of molecules called toll-like receptors (p. 753), which bind lipopolysaccharide (LPS), peptido-

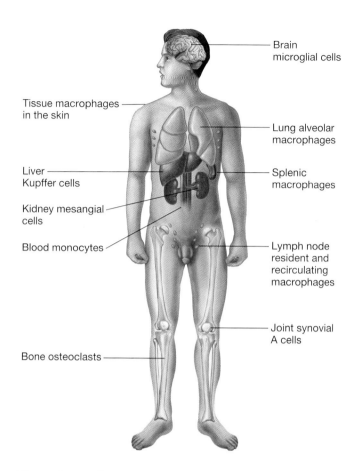

Figure 31.3   The Monocyte-Macrophage System.   This system consists of tissue (such as found within the liver, spleen, and lymph nodes) containing "fixed" or immobile phagocytes that have specific names depending on their location.

glycan, fungal cell wall components called zymosan, viral nucleic acids, and foreign DNA. In addition, macrophages possess specialized scavenger receptors including a surface protein called CD14, which also binds LPS. Macrophages also have receptors for antibodies and serum glycoproteins known as complement (p. 763). Both antibody and complement proteins can coat microorganisms or foreign material and enhance their phagocytosis. This enhancement is termed opsonization and is discussed in further detail in section 31.6. Macrophages spread throughout the animal body and take up residence in specific tissues where they are given special names (figure 31.3). Since macrophages are highly phagocytic, their function in nonspecific resistance is discussed in more detail in the context of phagocytosis.   Antigens: Cluster of differentiation molecules (section 32.2); Antibodies (section 32.7)

Granulocytes

Granulocytes have irregular-shaped nuclei with two to five lobes, and as such, are often called **polymorphonuclear leukocytes.** Their cytoplasmic matrix has granules that contain reactive substances that kill microorganisms and enhance inflammation (figure 31.2). Three types of granulocytes exist: basophils, eosinophils, and neutrophils. Because of the irregular-shaped nuclei, neutrophils are also called **polymorphonuclear neutrophils** or **PMNs.**

Figure 31.4   Phagocytosis by a Macrophage. One type of nonspecific host resistance involves white blood cells called macrophages and the process of phagocytosis. This scanning electron micrograph ($\times$3,000) shows a macrophage devouring a colony of bacteria. Phagocytosis is one of many nonspecific defenses humans and other animals use to combat microbial pathogens.

Basophils [Greek *basis,* base, and *philein,* to love] have an irregular-shaped nucleus with two lobes, and granules that stain bluish-black with basic dyes (figure 31.2). Basophils are non-phagocytic cells that function by releasing specific compounds from their cytoplasmic granules in response to certain types of stimulation. These molecules include histamine, prostaglandins, serotonin, and leukotrienes. Because these physiological mediators influence the tone and diameter of blood vessels, they are termed vasoactive mediators. Basophils (and mast cells) possess high-affinity receptors for one type of antibody, known as immunoglob-ulin E (IgE), that is associated with allergic responses. When these cells become coated with IgE antibodies, binding of antigen to the IgE can trigger the secretion of vasoactive mediators. As discussed in chapter 32, these inflammatory mediators play a major role in certain allergic responses such as eczema, hay fever, and asthma. Antibodies (section 32.7); Immune disorders: Hypersensitivities (section 32.11)

Eosinophils [Greek *eos,* dawn, and *philien*] have a two-lobed nucleus connected by a slender thread of chromatin, and granules that stain red with acidic dyes (figure 31.2). Unlike basophils, eosinophils migrate from the bloodstream into tissue spaces, especially mucous membranes. Their role is important in the defense against protozoan and helminth parasites, mainly by releasing cationic peptides (p. 762) and reactive oxygen intermediates (p.755) into the extracellular fluid. These molecules damage the parasite plasma membrane, destroying it. Eosinophil numbers are often increased during allergic reactions, especially type 1 hypersensitivities. Eosinophils also play a role in allergic reactions, as they have granules containing histaminase and aryl sulphatase, down-regulators of the inflammatory mediators histamine and leukotrienes, respectively.

Neutrophils [Latin *neuter,* neither, and *philien*] stain readily at a neutral pH, have a nucleus with three to five lobes connected by slender threads of chromatin, and contain inconspicuous organelles known as the primary and secondary granules. Lytic enzymes and bactericidal substances are contained within larger primary and smaller secondary granules. Primary granules contain peroxidase, lysozyme, defensins, and various hydrolytic enzymes, whereas secondary granules have collagenase, lactoferrin, catheli-cidins, and lysozyme. Both of these granules help accomplish intracellular digestion of foreign material after it is phagocytosed. Neutrophils also use oxygen-dependent and oxygen-independent pathways that generate additional antimicrobial substances to kill ingested microorganisms. Like macrophages, neutrophils have receptors for antibodies and complement proteins and are highly phagocytic. However, unlike macrophages, neutrophils do not reside in healthy tissue but circulate in blood so as to rapidly migrate to the site of tissue damage and infection, where they become the principal phagocytic and microbicidal cells. Neutrophils and their antimicrobial compounds are described in more detail in the contexts of the inflammatory response (section 31.4) and phagocytosis (section 31.3).

---

1. Describe the structure and function of each of the following blood cells: monocyte, macrophage, basophil, eosinophil, and neutrophil.
2. What is the significance of the respective blood cell percentages in blood?

Mast Cells
Mast cells are bone marrow-derived cells that differentiate in the blood and connective tissue. Although they contain granules with histamine and other pharmacologically active substances similar to those in basophils, they arise from a different cellular lineage (figure 31.2). Mast cells, along with basophils, play an important role in the development of allergies and hypersensitivities.

Dendritic Cells
Dendritic cells constitute only 0.2% of white blood cells in the blood but play an important role in nonspecific resistance (**figure 31.5**). They are present in the skin and mucous membranes of the nose, lungs, and intestines where they readily contact invading pathogens, phagocytose and process antigens, and display foreign antigens on their surface. This is a process known as **antigen presentation,** which is discussed in more detail in section 32.2.

Dendritic cells recognize specific *pathogen-associated molecular patterns* (PAMPs) on microorganisms (p. 753). These molecular patterns enable dendritic cells to distinguish between potentially harmful microbes and other host molecules. After the pathogen is recognized, the dendritic cell's *pattern recognition receptors* (PRRs) bind the pathogen and phagocytose it. The dendritic cells then migrate to lymphoid tissues where, as activated cells, they present antigen to T cells. Antigen presentation triggers the activation of T cells, which are critical for the initiation and regulation of an effective specific immune response. Thus not only do dendritic cells destroy invading pathogens as part of the innate response, but they also help trigger specific immune responses. T cell biology (section 32.5)

Figure 31.5 The Dendritic Cell. The dendritic cell was named for its cellular extensions, which resemble the dendrites of nerve cells. Dendritic cells reside in most tissue sites where they survey their local environments for pathogens and altered host cells.

Lymphocytes

Lymphocytes [Latin *lympha*, water, and *cyte*, cell] are the major cells of the specific immune system. Lymphocytes can be divided into three populations: T cells, B cells, and null cells (which include special cells called natural killer, or NK, cells). Lymphocytes leave the bone marrow in a kind of cellular stasis. In other words, they are not actively replicating like other somatic cells. B and T lymphocytes differentiate from their respective lymphoid stem cell precursor cells and are then blocked from exiting the G_0 phase of the cell cycle. This is important because it ensures that their gene products are made only when they are needed. In general, lymphocytes require a specific antigen to bind to a surface receptor (the B-cell receptor on B cells or the T-cell receptor on T cells) so that cellular activation can occur. Activation stimulates the cell to enter mitosis. Once activated, the lymphocytes continue to replicate as they circulate throughout the host, leaving several clones of activated lymphocytes to populate various lymphoid tissues. In addition to activated cells, some of the replicated lymphocytes are inhibited from further replication, waiting to be activated by the same antigen sometime later in the life of the host. These cells are called **memory cells.** The nucleus and cell division: Mitosis and meiosis (section 4.8)

After **B lymphocytes** or **B cells** reach maturity within the bone marrow, they circulate in the blood and disperse into various lymphoid organs, where they become activated. The activated B cell becomes more ovoid. Its nuclear chromatin condenses and numerous folds of endoplasmic reticulum become more visible. A mature, activated B cell is called a **plasma cell.** The phenotypic changes reflect functional changes occur-

ring within the plasma cell as it begins to secrete large quantities of glycoproteins called antibodies. Some of these antibodies can directly neutralize toxins and viruses, and are important in stimulating an efficient phagocytic response (**figure 31.6**). Actions of antibodies (section 32.8)

Lymphocytes destined to become **T lymphocytes** or **T cells** leave the bone marrow to mature in the thymus gland; they can remain in the thymus, circulate in the blood, or reside in lymphoid organs such as the lymph nodes and spleen, like B cells do. Also like B cells, T cells require a specific antigen to bind to their receptor to signal the continuation of replication. Unlike B cells, however, T cells do not secrete antibodies. Activated T cells produce and secrete proteins called **cytokines** (figure 31.6 and p. 766). Cytokines can have various effects on other cells including other T cells, B cells, granulocytes, and other somatic cells. In some cases the cytokines stimulate cells to mature and differentiate, produce new effector products, and even cause some cells to die. Because B and T cells must be activated by specific antigens, they are included in the adaptive or specific immune system. B cells and T cells are discussed further in chapter 32.

Natural killer (NK) cells are a small population of large, nonphagocytic granular lymphocytes that play an important role in innate immunity (figure 31.2). The major NK cell function is to destroy malignant cells and cells infected with microorganisms. They recognize their targets in one of two ways. They can bind to antibodies that coat infected or malignant cells; thus the antibody bridges the two cell types. This process is called **antibody-dependent cell-mediated cytotoxicity (ADCC)** (**figure 31.7**) and can result in the death of the target cell. The second way that NK cells recognize infected cells and cancer cells relies on the presence of specialized proteins on the surface of all nucleated host cells, known as the class I *m*ajor *h*istocompatibility (MHC) antigen. If a host cell loses this MHC protein, as when some viruses or cancers overtake the cell, the NK cell kills it by releasing pore-forming proteins and cytotoxic enzymes called granzymes. Together the pore-forming proteins and the granzymes cause the target cell to lyse (**figure 31.8**). Recognition of foreignness (section 32.4)

---

1. How might mast cells, lymphocytes, and dendritic cells work together in innate immunity?
2. Discuss the role of NK cells in protecting the host.
3. What is the purpose of antibody-dependent cell-mediated cytotoxicity?

Organs and Tissues of the Immune System

Based on function, the organs and tissues of the immune system can be divided into primary or secondary lymphoid organs and tissues (**figure 31.9**). The primary organs and tissues are where immature lymphocytes mature and differentiate into antigen-sensitive B and T cells. The thymus is the primary lymphoid organ for T cells, and the bone marrow is the primary lymphoid tissue for B cells. The secondary organs and tissues serve as areas where lymphocytes may encounter and bind antigen, whereupon they proliferate and differentiate into fully active, antigen-specific effector

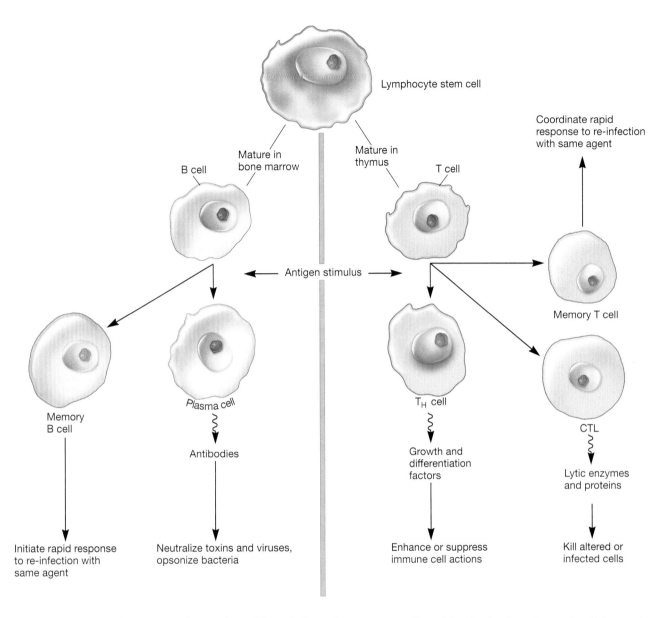

Figure 31.6   The Development and Function of B and T Lymphocytes.   B cells and T cells arise from the same cell lineage but diverge into two different functional types. Immature B cells and T cells are indistinguishable by histological staining. However, they express different proteins on their surfaces that can be detected by immunohistochemistry. Additionally, the final secreted products of mature B and T cells can be used to identify the cell type.

cells. The spleen is a secondary lymphoid organ and the lymph nodes and mucus-associated tissues (GALT, gut-associated lymphoid tissue and SALT, skin-associated lymphoid tissues) are the secondary lymphoid tissues. The thymus, bone marrow, lymph nodes, and spleen are now discussed in more detail. GALT and SALT are described in section 31.5 as part of the host's physical and mechanical barriers.

Primary Lymphoid Organs and Tissues

Immature undifferentiated lymphocytes are generated in the bone marrow. They mature and become committed to a specific antigen within the primary lymphoid organ/tissues. The two most important primary sites in mammals are the thymus and bone marrow.

The **thymus** is a highly organized lymphoid organ located above the heart. Precursor cells from the bone marrow migrate into the outer cortex of the thymus where they proliferate. As they mature and acquire T-cell surface markers, approximately 90% die. This is due to a selection process in which T cells that recognize host (self) antigens are destroyed. The remaining 10% move into the medulla of the thymus, become mature T cells, and subsequently enter the bloodstream (figure 31.9*a*).

In mammals, the bone marrow (figure 31.9*b*) is the site of B-cell maturation. Like thymic selection during T-cell maturation,

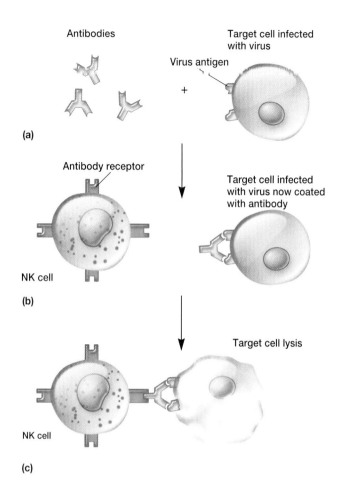

Figure 31.7 Antibody-Dependent Cell-Mediated Cytotoxicity. **(a)** In this mechanism, IgG antibodies bind to a target cell infected with a virus. **(b)** NK cells have specific antibody receptors on their surface. **(c)** When the NK cells encounter virus, infected cells coated with antibody, they kill the target cell.

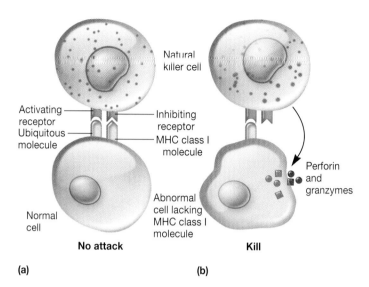

Figure 31.8 The System Used by Natural Killer Cells to Recognize Normal Cells and Abnormal Cells That Lack the Major Histocompatability Complex Class I Surface Molecule. **(a)** The killer-activating receptor recognizes a normal ubiquitous molecule on the plasma membrane of a normal cell. Since the killer-inhibitory receptor recognizes the MHC class I molecule, there is no attack. **(b)** In the absence of the inhibitory signal, the receptor issues an order to the NK cell to attack and kill the abnormal cell. The cytotoxic granules of the NK cell contain perforin and granzymes. With no inhibitory signal, the granules release their contents, killing the abnormal cell.

a selection process within the bone marrow eliminates B cells bearing self-reactive antigen receptors. In birds, undifferentiated lymphocytes move from the bone marrow to the **bursa of Fabricius** where B cells mature; this is where B cells were first identified and how they came to be known as "B" (for bursa) cells.

Secondary Lymphoid Organs and Tissues

The spleen is the most highly organized secondary lymphoid organ. The **spleen** is a large organ located in the abdominal cavity (figure 31.9). It specializes in filtering the blood and trapping blood-borne microorganisms and antigens. Once trapped by splenic macrophages or dendritic cells, the pathogen is phagocytosed, killed, and digested. The resulting peptide (protein fragment consisting of less than about 50 amino acid residues) antigens are delivered to the macrophage or dendritic cell surface where they are presented to B and T cells. This is the most common means by which lymphocytes become activated to carry out their immune functions.

Lymph nodes lie at the junctions of lymphatic vessels where they filter out harmful microorganisms and antigens from the lymph; pathogens and antigens are trapped by phagocytic macrophages and dendritic cells (figure 31.9c). They then phagocytose the foreign material and present antigen to lymphocytes. It is within the lymph nodes that B cells differentiate into memory cells and antibody-secreting plasma cells. This involves specialized T cells, called T-helper cells, which are also found here. Dendritic cells and macrophages (antigen-presenting cells) present antigens to the T-helper cells, which subsequently secrete cytokines that promote B-cell immune responses. B cell biology (section 32.6)

Lymphoid tissues are found throughout the body and act as regional centers of antigen sampling and processing (figure 31.9). Lymphoid tissues are found as highly organized or loosely associated cellular complexes. Some lymphoid cells are closely associated with specific tissues such as skin (*skin*-associated *l*ymphoid *t*issue, or SALT) and mucous membranes (*m*ucus-associated *l*ymphoid *t*issue, or MALT). SALT and MALT are good examples of highly organized lymphoid tissues, typically seen histologically as macrophages surrounded by specific areas of B and T lymphocytes, and sometimes dendritic cells. Loosely associated lymphoid tissue is best represented by the *b*ronchial-

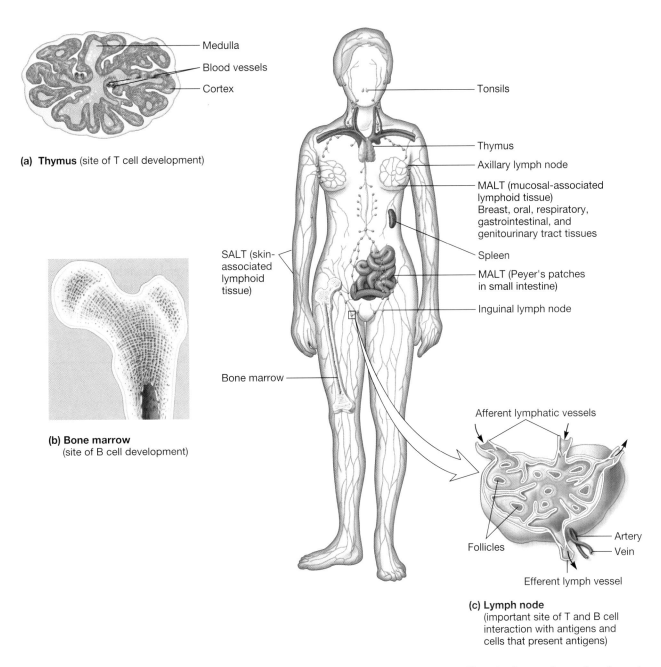

(a) **Thymus** (site of T cell development)

- Medulla
- Blood vessels
- Cortex

(b) **Bone marrow**
(site of B cell development)

SALT (skin-associated lymphoid tissue)

Bone marrow

- Tonsils
- Thymus
- Axillary lymph node
- MALT (mucosal-associated lymphoid tissue) Breast, oral, respiratory, gastrointestinal, and genitourinary tract tissues
- Spleen
- MALT (Peyer's patches in small intestine)
- Inguinal lymph node

Afferent lymphatic vessels

Follicles

Artery
Vein

Efferent lymph vessel

(c) **Lymph node**
(important site of T and B cell interaction with antigens and cells that present antigens)

Figure 31.9   Anatomy of the Lymphoid System.   Lymph is distributed through a system of lymphatic vessels, passing through many lymph nodes and lymphoid tissues. For example, **(a)** the thymus is involved in T-cell development and atrophies with age; **(b)** the bone marrow is the site of B-cell development; and **(c)** lymph enters a lymph node through the afferent lymph vessels, percolates through and around the follicles in the node, and leaves through the efferent lymphatic vessels. The lymphoid follicles are the site of cellular interactions and extensive immunologic activity.

*a*ssociated *l*ymphoid *t*issue, or BALT, characterized by the lack of cellular partitioning. The primary role of these lymphoid tissues is to efficiently organize leukocytes to increase interaction between the innate and the acquired arms of the immune response. In other words, the lymphoid tissues serve as the interface between the innate and acquired immunity of a host. Thus as we shall see, a microbe attempting to invade a potential host is greeted by nonspecific, physical, chemical, and granulocyte barriers that are designed to kill the invader, digest the carcass into small antigens, and assist the lymphocytes in formulating long-term protection against the next invasion. We now examine the phagocytic processes in more detail and then consider how the host integrates many of the innate immune activities into a substantial barrier to microbial invasion, known as the inflammatory response.

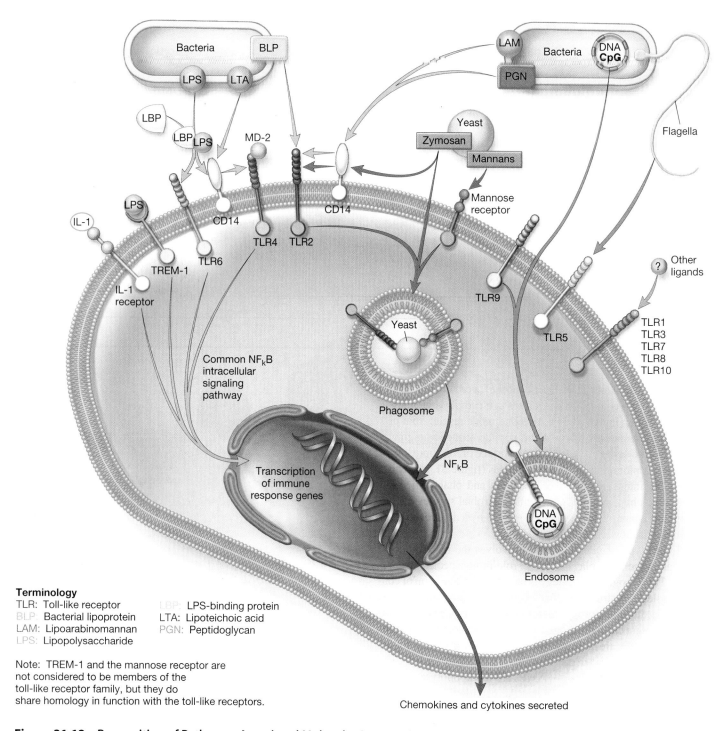

Terminology
TLR: Toll-like receptor
BLP: Bacterial lipoprotein
LAM: Lipoarabinomannan
LPS: Lipopolysaccharide

LBP: LPS-binding protein
LTA: Lipoteichoic acid
PGN: Peptidoglycan

Note: TREM-1 and the mannose receptor are
not considered to be members of the
toll-like receptor family, but they do
share homology in function with the toll-like receptors.

Chemokines and cytokines secreted

Figure 31.12   Recognition of Pathogen-Associated Molecular Patterns (PAMPs) by Toll-like Receptors (TLRs).   PAMP binding of
TLR results in a signaling process that upregulates gene expression. A common NF_kB signal transduction pathway is used.

host cell nucleus to initiate appropriate gene expression and host response. There are at least 10 distinct proteins in this family of mammalian receptors. For example, TLR-4 signals the presence of
bacterial lipopolysaccharide (LPS) and heat-shock proteins (table
31.2). TLR-9 signals the dinucleotide CpG motif present on DNA
released by dying bacteria. TLR-2 signals the presence of bacterial lipoproteins and peptidoglycans. Binding of TLRs triggers an

evolutionarily ancient signaling pathway that activates transcription factor NF_kB by degrading its inhibitor, I_kB. This induces expression of a variety of genes, including genes for cytokines,
chemokines, and costimulatory molecules that play essential roles
in calling forth and directing the adaptive immune response later in
an infection. Thus binding of specific microbial components to
phagocyte receptors is an important first step in phagocytosis. Once

bound, the microbe and/or its components can be internalized as part of a **phagosome** that is then united with a lysosome to facilitate microbial killing and digestion.

Intracellular Digestion

Once ingested by phagocytosis, microorganisms in membrane-enclosed vesicles are delivered to a lysosome by fusion of the phagocytic vesicle, called a phagosome, with the lysosome membrane, forming a new vacuole called a **phagolysosome** (figure 31.10 c,d). This is when the killing begins because lysosomes deliver a variety of hydrolases such as lysozyme, phospholipase A$_2$, ribonuclease, deoxyribonuclease, and proteases. The activity of these degradative enzymes is enhanced by the acidic vacuolar pH. Collectively, these enzymes participate in the destruction of the entrapped microorganisms. In addition to these oxygen-independent lysosomal hydrolases, macrophage and neutrophil lysosomes contain oxygen-dependent enzymes that produce toxic **reactive oxygen intermediates (ROIs)** such as the superoxide radical ($O_2^-\cdot$), hydrogen peroxide (H_2O_2), singlet oxygen ($^1O_2$), and hydroxyl radical (OH^-). The NADPH required for this process is supplied by a large increase in pentose phosphate pathway activity (*see figure 9.6*). Neutrophils also contain the heme-protein myeloperoxidase, which catalyzes the production of hypochlorous acid. Some reactions that form ROIs are shown in **table 31.3**. These reactions result from the **respiratory burst** that accompanies the increased oxygen consumption and ATP generation needed for phagocytosis. These reactions occur within the lysosome as soon as the phagosome is formed; lysosome fusion is not necessary for the respiratory burst. ROIs are effective in killing invading microorganisms. The influence of environmental factors on growth: Oxygen concentration (section 6.5)

Macrophages, neutrophils, and mast cells have also been shown to form **reactive nitrogen intermediates (RNIs).** These molecules include nitric oxide (NO) and its oxidized forms, nitrite (NO_2^-) and nitrate (NO_3^-). The RNIs are very potent cytotoxic agents and may be either released from cells or generated within cell vacuoles. Nitric oxide is probably the most effective

RNI. Macrophages produce it from the amino acid arginine. Nitric oxide can block cellular respiration by complexing with the iron in electron transport proteins. Macrophages use RNIs in the destruction of a variety of infectious agents including the herpes simplex virus, the protozoa *Toxoplasma gondii* and *Leishmania major,* the opportunistic fungus *Cryptococcus neoformans,* and the metazoan pathogen *Schistosoma mansoni.* RNIs are also used to kill tumor cells.

Neutrophil granules contain a variety of other microbicidal substances such as several cationic peptides, the bactericidal permeability-increasing protein (BPI), and the family of broad-spectrum antimicrobial peptides including defensins. There are four human defensins produced by neutrophils called (HNPs): HNP-1, 2, 3, and 4. These defensins are synthesized by myeloid (mononuclear granulocyte) precursor cells during their sojourn in the bone marrow, and are then stored in the cytoplasmic granules of mature neutrophils. This compartmentalization strategically locates defensins (and other antimicrobial products) for extracellular secretion or delivery to phagocytic vacuoles. Susceptible microbial targets include a variety of gram-positive and gram-negative bacteria, yeasts and molds, and some viruses. Defensins act against bacteria and fungi by permeabilizing cell membranes. They form voltage-dependent membrane channels that allow ionic efflux. Antiviral activity involves direct neutralization of enveloped viruses, so they can no longer bind host cell receptors; nonenveloped viruses are not affected by defensins. Defensins are described in more detail in section 31.6.

Exocytosis

Once the microbial invaders have been killed and digested into small antigenic fragments, the phagocyte may do one of two things. Neutrophils tend to expel the microbial fragments by the process of exocytosis. This is essentially a reverse of the phagocytic process whereby the phagolysosome unites with the cell membrane resulting in the extracellular release of the microbial fragments. Other phagocytic cells, such as macrophages and dendritic cells, continue to process the microbial fragments by passing them from the phagolysosome to the endoplasmic reticulum. Here the peptide components of the fragments are united with glycoproteins destined for the cell membrane. The glycoproteins bind the peptides within their extracellular domain so that they are presented outward from the cell once the glycoprotein is secured in the cell membrane. This so-called antigen presentation is critical because it is the event that permits wandering lymphocytes to evaluate killed microbes (as antigens) and be activated. Thus antigen presentation links a nonspecific immune response to a specific immune response. Recognition of foreignness (section 32.4)

Oxygen Intermediate	Reaction
Table 31.3	**Formation of Reactive Oxygen Intermediates**
Superoxide ($O_2^-\cdot$)	NADPH oxidase $NADPH + 2O_2 \longrightarrow 2O_2^-\cdot + H^+ + NADP^+$
Hydrogen peroxide (H_2O_2)	Superoxide dismutase $O_2^-\cdot + 2H^+ \longrightarrow H_2O_2 + O_2$
Hypochlorous acid (HOCl)	Myeloperoxidase $H_2O_2 + Cl^- \longrightarrow HOCl + OH^+$
Singlet oxygen ($^1O_2$)	Peroxidase $ClO^- + H_2O_2 \longrightarrow {}^1O_2 + Cl^- + H_2O$
Hydroxyl radical (OH^-)	Peroxidase $O_2^-\cdot + H_2O_2 \longrightarrow OH^- + OH^- + O_2$

1. What is the role of opsonin-independent phagocytosis?
2. Once a phagolysosome forms, how is the entrapped microorganism destroyed?
3. What is the purpose of the respiratory burst that occurs within macrophages and other phagocytic cells? Describe the nature and function of reactive oxygen and nitrogen intermediates.

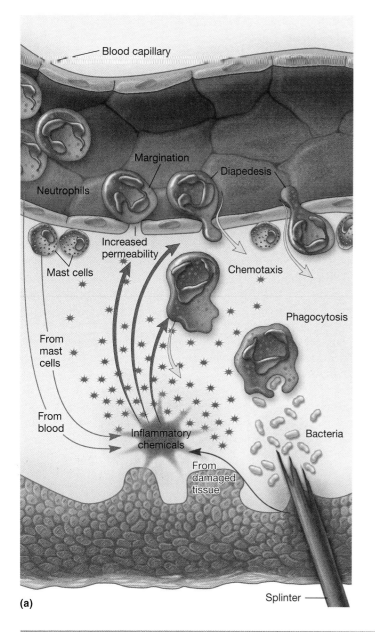

(a)

31.4 INFLAMMATION

So far we have discussed the cellular, tissue, and organ components of innate immunity, along with an explanation of how pathogens are recognized by the innate immune system, destroyed, and their presence is communicated to the acquired immune system. But how do the innate immune cells perceive an impending invasion by pathogens so as to be recruited for host defense? One part of the answer lies in the process known as inflammation. **Inflammation** [Latin, *inflammatio,* to set on fire] is an important nonspecific defense reaction to tissue injury, such as that caused by a pathogen or wound. Acute inflammation is the immediate response of the body to injury or cell death. The gross features were described over 2,000 years ago and are still known as the cardinal signs of inflammation. These signs include redness (*rubor*), warmth (*calor*), pain (*dolor*), swelling (*tumor*), and altered function (*functio laesa*).

The acute inflammatory response begins when injured tissue cells release chemical signals (chemokines) that activate the inner lining (endothelium) of nearby capillaries (**figure 31.13**). Within the capillaries, **selectins** (a family of cell adhesion molecules) are displayed on the activated endothelial cells. These adhesion molecules attract and attach wandering neutrophils to the endothelial cells. This slows the neutrophils and causes them to roll along the endothelium where they encounter the inflammatory chemicals that act as activating signals (figure 31.13*b*). These signals activate **integrins** (adhesion receptors) on the neutrophils. The neutrophil integrins then attach tightly to the selectins. This causes the neutrophils to stick to the endothelium and stop rolling (margination). The neutrophils now undergo dramatic shape changes, squeeze through the endothelial wall (diapedesis) into the interstitial tissue fluid, migrate to the site of injury (extravasion), and attack the pathogen or other cause of the tissue damage. Neutrophils and other leukocytes are attracted to the infection site by chemotactic factors, which are also called chemotaxins. They include substances released by bacteria, endothelial cells, mast cells, and tissue breakdown products. Depending on the severity and nature of tissue damage, other types

(b)

Figure 31.13 Physiological Events of the Acute Inflammatory Response. **(a)** At the site of injury (splinter), chemical messengers are released from the damaged tissue, mast cells, and the blood plasma. These inflammatory chemicals stimulate neutrophil migration, diapedesis, chemotaxis, and phagocytosis. **(b)** Neutrophil integrins interact with endothelial selectins (1) to facilitate margination (2) and diapedesis (3).

of leukocytes (e.g., lymphocytes, monocytes, and macrophages) may follow the neutrophils.

The release of inflammatory mediators from injured tissue cells sets into motion a cascade of events that result in the development of the signs of inflammation. The mediators increase the acidity in the surrounding extracellular fluid, which activates the extracellular enzyme **kallikrein (figure 31.14)**. Kallikrein cleavage releases the peptide bradykinin from its long precursor chain. Bradykinin then binds to receptors on the capillary wall, opening the junctions between cells and allowing fluid and infection-fighting leukocytes to leave the capillary and enter the infected tissue. Simultaneously, bradykinin binds to mast cells in the connective tissue associated with most small blood vessels. This activates the mast cells by causing an influx of calcium ions, which leads to degranulation and release of preformed mediators such as histamine. If nerves in the infected area are damaged, they release substance P, which also binds to mast cells, boosting preformed-mediator release. Hista-

mine in turn makes the intercellular junctions in the capillary wall wider so that more fluid, leukocytes, kallikrein, and bradykinin move out, causing swelling or edema. Bradykinin then binds to nearby capillary cells and stimulates the production of prostaglandins (PGE_2 and $PGF_{2\alpha}$) to promote tissue swelling in the infected area. Prostaglandins also bind to free nerve endings, making them fire and start a pain impulse.

Activated mast cells also release a small molecule called arachidonic acid, the product of a reaction catalyzed by phospholipase A_2. Arachidonic acid is metabolized by the mast cell to form potent mediators including prostaglandins E_2 and $F_{2\alpha}$, thromboxane A_2, slow-reacting substance (SRS), and leukotrienes (LTC_4 and LTD_4). All of these mediators play specific roles in the inflammatory response. During acute inflammation, the offending pathogen is neutralized and eliminated by a series of important events:

1. The increase in blood flow and capillary dilation bring into the area more antimicrobial factors and leukocytes that destroy the pathogen. Dead host cells also release antimicrobial factors.
2. Blood leakage into tissue spaces increases the temperature and further stimulates the inflammatory response and may inhibit microbial growth.
3. A fibrin clot often forms and may limit the spread of the invaders so that they remain localized.
4. Phagocytes collect in the inflamed area and phagocytose the pathogen. In addition, chemicals stimulate the bone marrow to release neutrophils and increase the rate of granulocyte production.

Chronic Inflammation

In contrast to acute inflammation, which is a rapid and transient process, chronic inflammation is a slow process characterized by the formation of new connective tissue, and it usually causes permanent tissue damage. Regardless of the cause, chronic inflammation lasts two weeks or longer. Chronic inflammation can occur as a distinct process without much acute inflammation. The persistence of bacteria by a variety of mechanisms can stimulate chronic inflammation. For example, mycobacteria, which include species that cause tuberculosis and leprosy, have cell walls with a very high lipid and wax content, making them relatively resistant to phagocytosis and intracellular killing. These bacteria and a number of other pathogens can survive within the macrophage, as do some protozoan pathogens such as *Leishmania*. In addition, some bacteria produce toxins that stimulate tissue-damaging reactions even after bacterial death. Suborder *Corynebacterineae* (section 24.4); Protist classification: *Excavata* (section 25.6)

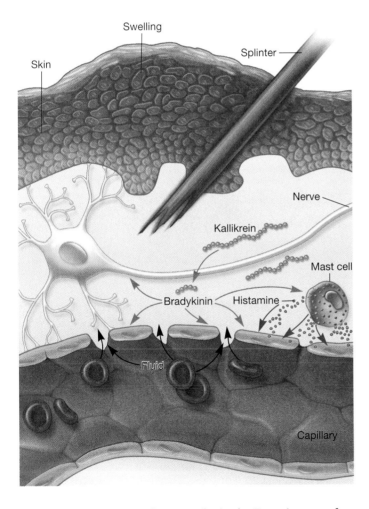

Figure 31.14   Tissue Injury Results in the Recruitment of Kallikrein, From Which Bradykinin Is Released.   Bradykinin acts on endothelial and nerve cells resulting in edema and pain, respectively. It also stimulates mast cells to release histamine. Histamine also acts on endothelial cells, further increasing fluid leakage into injured tissue sites.

Chronic inflammation is characterized by a dense infiltration of lymphocytes and macrophages. If the macrophages are unable to protect the host from tissue damage, the body attempts to wall off and isolate the site by forming a **granuloma** [Latin, *granulum*, a small particle; Greek, *oma*, to form]. Granulomas are formed when neutrophils and macrophages are unable to destroy the microorganism during inflammation. Infections caused by some bacteria (listeriosis, brucellosis), fungi (histoplasmosis, coccidioidomycosis), helminth parasites (schistosomiasis), protozoa (leishmaniasis), and large antibody antigen complexes (rheumatoid arthritis) result in granuloma formation and chronic inflammation. These infectious diseases are discussed in chapters 38 and 39.

---

1. What major events occur during an inflammatory reaction, and how do they contribute to pathogen destruction?
2. How does chronic inflammation differ from acute inflammation?

31.5 PHYSICAL BARRIERS IN NONSPECIFIC (INNATE) RESISTANCE

With few exceptions, a potential microbial pathogen invading a human host immediately confronts a vast array of nonspecific (innate) defense mechanisms (**figure 31.15**). Although the effectiveness of some individual mechanisms is not great, collectively their defense is formidable. Many direct factors (nutrition, physiology, fever, age, genetics) and equally as many indirect factors (personal hygiene, socioeconomic status, living conditions) influence all host-microbe relationships. At times they favor the establishment of the microorganism within the host; at other times they provide some measure of defense to the host. For example, when the host is either very young or very old, susceptibility to infection increases. Babies are at a particular risk after their maternal immunity has waned and before their own immune systems have matured. Very old persons experience a decline in the immune system itself and in the homeostatic functioning of many organs, which reduce host defenses. In addition to these direct and indirect factors, a vertebrate host has some specific physical and mechanical barriers.

Physical and Mechanical Barriers

Physical and mechanical barriers, along with the host's secretions (flushing mechanisms), are the first line of defense against microorganisms. Protection of the most important body surfaces by these mechanisms is discussed next.

Skin

The intact skin contributes greatly to nonspecific host resistance. It forms a very effective mechanical barrier to microbial invasion. Its outer layer consists of thick, closely packed cells called keratinocytes, which produce keratins. Keratins are scleroproteins (i.e., insoluble proteins) that make up the main components of hair, nails, and the outer skin cells. These outer skin cells shed continuously, removing any grime or microorganisms that manage to adhere to their surface. The skin is slightly acidic (around pH 5 to 6) due to skin oil, secretions from sweat glands, and organic acids produced by commensal staphylococci. It also contains a high concentration of sodium chloride and is subject to periodic drying.

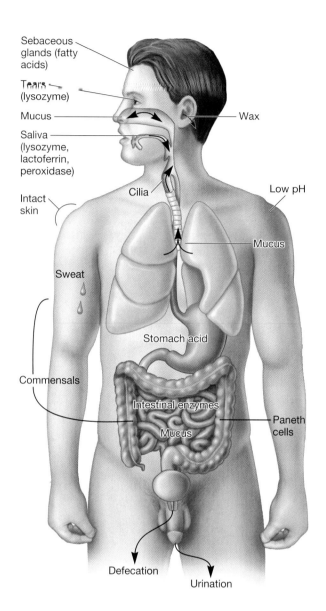

Figure 31.15 Host Defenses. Some nonspecific (innate) host defense mechanisms that help prevent entry of microorganisms into the host's tissues.

Despite the skin's defenses, at times some pathogenic microorganisms gain access to the tissue under the skin surface. Here they encounter a specialized set of cells called the ***skin-associated lymphoid tissue* (SALT)** (**figure 31.16**). The major function of SALT is to confine microbial invaders to the area immediately underlying the epidermis and to prevent them from gaining access to the bloodstream. One type of SALT cell is the **Langerhans cell,** a specialized myeloid cell that can phagocytose antigens. Once the Langerhans cell has internalized the antigen, it migrates from the epidermis to nearby lymph nodes where it differentiates into a mature dendritic cell. Recall that dendritic cells can present antigens and activate nearby lymphocytes to induce the acquired immune system. This dendritic cell-lymphocyte interaction illustrates another bridge between the innate and acquired immune systems.

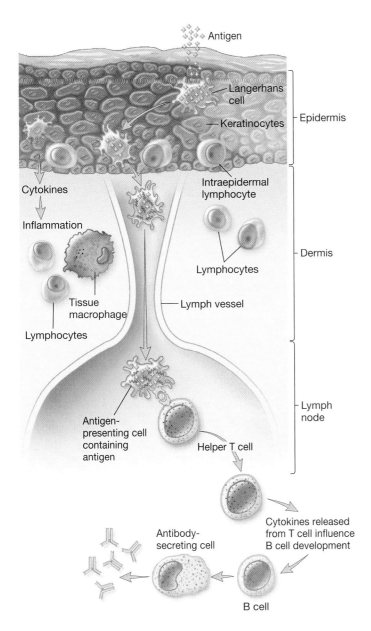

Figure 31.16 Skin-Associated Lymphoid Tissue (SALT).
Keratinocytes make up 90% of the epidermis. They are capable of secreting cytokines that cause an inflammatory response to invading pathogens. Langerhans cells internalize antigen and move to a lymph node where they differentiate into dendritic cells that present antigen to helper T cells. The intraepidermal lymphocytes may function as T cells that can activate B cells to induce an antibody response.

The epidermis also contains another type of SALT cell called the **intraepidermal lymphocyte** (figure 31.16). These cells are strategically located in the skin so that they can intercept any antigens that breach the first line of defense. Most of these specialized SALT cells are T cells. Unlike other T cells, they have limited receptor diversity and have likely evolved to recognize common skin pathogen patterns. A large number of tissue macrophages (figure 31.3) are also located in the dermal

layer of the skin and phagocytose most microorganisms they encounter (figure 31.4).

Mucous Membranes

The mucous membranes of the eye (conjunctiva) and the respiratory, digestive, and urogenital systems withstand microbial invasion because the intact stratified squamous epithelium and mucous secretions form a protective covering that resists penetration and traps many microorganisms. This mechanism contributes to nonspecific immunity. Furthermore, many mucosal surfaces are bathed in specific antimicrobial secretions. For example, cervical mucus, prostatic fluid, and tears are toxic to many bacteria. One antibacterial substance in these secretions is **lysozyme** (muramidase), an enzyme that lyses bacteria by hydrolyzing the $\beta(1 \rightarrow 4)$ bond connecting N-acetylmuramic acid and N-acetylglucosamine of the bacterial cell wall peptidoglycan—especially in gram-positive bacteria (**figure 31.17**). These mucous secretions also contain specific immune proteins that help prevent the attachment of microorganisms. They also contain significant amounts of the iron-binding protein, lactoferrin. **Lactoferrin** is released by activated macrophages and polymorphonuclear leukocytes (PMNs). It sequesters iron from the plasma, reducing the amount of iron available to invading microbial pathogens and limiting their ability to multiply. Finally, mucous membranes produce lactoperoxidase, an enzyme that catalyzes the production of superoxide radicals, a reactive oxygen intermediate that is toxic to many microorganisms (table 31.3). The bacterial cell wall: Peptidoglycan structure (section 3.6)

Like the skin, mucous membranes also have a specialized immune barrier called **m**ucus-**a**ssociated **l**ymphoid **t**issue **(MALT).** There are several types of MALT. The system most studied is the **g**ut-**a**ssociated **l**ymphoid **t**issue **(GALT).** GALT includes the tonsils, adenoids, diffuse lymphoid areas along the gut, and specialized regions in the intestine called Peyer's patches. Less well-organized MALT also occurs in the respiratory system and is called **b**ronchial **a**ssociated **l**ymphoid **t**issue **(BALT);** the diffuse MALT in the urogenital system does not have a specific name. MALT can operate by two basic mechanisms. First, when an antigen arrives at the mucosal surface, it contacts a type of cell called the **M cell** (**figure 31.18a**). The M cell does not have the brush border or microvilli found on adjacent columnar epithelial cells. Instead it has a large pocket containing B cells, T cells, and macrophages. When an antigen contacts the M cell, it is endocytosed and released into the pocket. Macrophages engulf the antigen or pathogen and try to destroy it. An M cell also can endocytose an antigen and transport it to a cluster of cells called an organized lymphoid follicle (figure 31.18b). The B cells within this follicle recognize the antigen and mature into antibody-producing plasma cells. The plasma cells leave the follicle and secrete a class of mucous membrane-associated antibody called secretory (s) IgA. sIgA is then transported into the lumen of the gut where it interacts with the antigen that caused its production. Similar to the SALT, GALT intra- and inter-epithelial lymphocytes are strategically distributed so the likelihood of antigen detection is increased should the intestinal membrane be breached. Antibodies (section 32.7)

(a)   (b)   (c)

Figure 31.17   Action of Lysozyme on the Cell Wall of Gram-Positive Bacteria.   **(a)** In the structure of the cell wall peptidoglycan backbone, the $\beta(1\rightarrow4)$ bonds connect alternating *N*-acetylglucosamine (NAG) and *N*-acetylmuramic acid (NAM) residues. The chains are linked through cross-bridges. Lysozyme splits the molecule as indicated by the arrow. **(b)** The $\beta(1\rightarrow4)$ bond fits into the active site of lysozyme (shaded area) facilitating **(c)** bond hydrolysis.

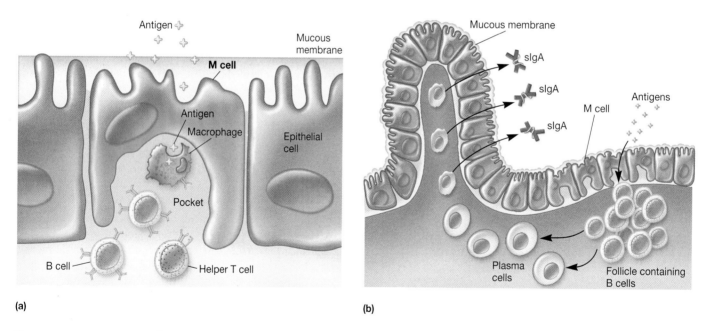

(a)   (b)

Figure 31.18   Function of M Cells in Mucosal-Associated Immunity.   **(a)** Structure of an M cell located between two epithelial cells in a mucous membrane. The M cell endocytoses the pathogen and releases it into the pocket containing helper T cells, B cells, and macrophages. It is within the pocket that the pathogen often is destroyed. **(b)** The antigen is transported by the M cell to the organized lymphoid follicle containing B cells. The activated B cells mature into plasma cells, which produce secretory IgA and release it into the lumen where it reacts with the antigen that caused its production.

1. Why is the skin such a good first line of defense against pathogenic microorganisms?
2. How do intact mucous membranes resist microbial invasion of the host?
3. Describe SALT function in the immune response.
4. How do M cells function in MALT?

Respiratory System

The mammalian respiratory system has formidable defense mechanisms. The average person inhales at least eight microorganisms a minute, or 10,000 each day. Once inhaled, a microorganism must first survive and penetrate the air-filtration system of the upper and lower respiratory tracts. Because the airflow in these tracts is very turbulent, microorganisms are deposited on the moist, sticky mucosal surfaces. Microbes larger than 10 μm usually are trapped by hairs and cilia lining the nasal cavity. The cilia in the nasal cavity beat toward the pharynx, so that mucus with its trapped microorganisms is moved toward the mouth and expelled (**figure 31.19**). Humidification of the air within the nasal cavity causes many hygroscopic (attracting moisture from the air) microorganisms to swell, and this aids phagocytosis. Microbes smaller than 10 μm often pass through the nasal cavity and are trapped by the **mucociliary blanket** that coats the mucosal surfaces of lower portions of the respiratory system. The trapped microbes are transported by ciliary action (**mucociliary escalator**) that moves them away from the lungs. Coughing and sneezing reflexes clear the respiratory system of microorganisms by expelling air forcefully from the lungs through the mouth and nose, respectively. Salivation also washes microorganisms from the mouth and nasopharyngeal areas into the stomach. Microorganisms that succeed in reaching the alveoli of the lungs encounter a population of fixed phagocytic cells called **alveolar macrophages** (figure 31.3). These cells can ingest and kill most bacteria by phagocytosis.

Gastrointestinal Tract

Most microorganisms that reach the stomach are killed by gastric juice (a mixture of hydrochloric acid, proteolytic enzymes, and mucus). The very acidic gastric juice (pH 2 to 3) is sufficient to destroy most organisms and their toxins, although exceptions exist (protozoan cysts, *Helicobacter pylori, Clostridium* and *Staphylococcus* toxins). However, organisms embedded in food particles are protected from gastric juice and reach the small intestine. Once in the small intestine, microorganisms often are damaged by various pancreatic enzymes, bile, enzymes in intestinal secretions, and the GALT system. **Peristalsis** [Greek *peri*, around, and *stalsis*, contraction] and the normal loss of columnar epithelial cells act in concert to purge intestinal microorganisms. In addition, the normal microbiota of the large intestine (*see figure 30.17*) is extremely important in preventing the establishment of pathogenic organisms. For example, many normal commensals in the intestinal tract produce metabolic products, such as fatty acids, that prevent unwanted microorganisms from becoming established. Other normal microbiota outcompete potential pathogens for attachment sites and nutrients. The mucous membranes of the intestinal tract contain cells called **Paneth cells.** These cells produce lysozyme (figure 31.17) and a set of peptides called **cryptins.** Cryptins are toxic for some bacteria, although their mode of action is not known.

Genitourinary Tract

Under normal circumstances the kidneys, ureters, and urinary bladder of mammals are sterile. Urine within the urinary bladder also is sterile. However, in both the male and female, a few

Figure 31.19 The Bronchial-Associated Lymphoid Tissue (BALT). (a) The respiratory tract is lined with a mucous membrane of ciliated epithelial cells. The circuitous passages of the nasal cavity prevent large particles from entering deeper into the respiratory tract. Mucus traps particles along the tract and the (b) cilia (×5,000) sweep them upward toward the throat to be expectorated.

Nasal cavity
Nostril
Oral cavity
Pharynx
Epiglottis
Larynx
Trachea
Bronchus
Bronchioles
Right lung
Left lung

(a)

Cilia
Microvilli

(b)

Degli-Esposti, M.; and Smyth, M. 2005. Close encounters of a different kind: Dendritic cells and NK cells take center stage. *Nature Rev. Immunol.* 5:112–24.

Djaldetti, M.; Salman, H.; Bergman, M.; Djaldetti, R.; and Bessler, H. 2002. Phagocytosis the mighty weapon. *Microsc. Res. Tech.* 57:421–31.

Goldberg, A. L. 2000. Probing the proteosome pathway. *Nature Biotechnol.* 18:494–96.

Iwasaki, A., and Medzhitov, R. 2004. Toll-like receptor control of the adaptive immune response. *Nature Immunol.* 5:987–95.

Marshall, S. H., and Arenas, G. 2003. Antimicrobial peptides: A natural alternative to chemical antibiotics and a potential for applied biotechnology. *Electronic J. Biotechnol.* 6: 271–84.

Medzhitov, R.; Preston-Hurlburt, P.; and Janeway, C. A. 1997. A human analogue of the *Drosophila* Toll protein signals activation of adaptive immunity. *Nature* 388: 394–97.

Roy, C., and Sansonetti, P. 2004. Host-microbe interactions: Bacteria, host-pathogen interactions: Interpreting the dialog. *Curr. Opin. Microbiol.* 7:1–3.

**Please visit the Prescott website at www.mhhe.com/prescott7
for additional references.**

32

Specific (Adaptive) Immunity

Nude (athymic) mice have a genetic defect (*nu* mutation) that affects thymus gland development. Thus T cells do not form. They, however, do have a B-cell component. This unique deficiency provides animals in which to study B/T-cell dichotomy and environmental influences on the maturation and differentiation of T cells, as well as many different immune disorders.

PREVIEW

- The major function of the specific immune response in vertebrates is to provide protection (immunity) against harmful microorganisms, toxins, and abnormal (cancer) cells through the recognition of foreign (nonself) antigens.

- Antigens are substances, such as proteins, nucleoproteins, polysaccharides, and some glycolipids, to which lymphocytes respond.

- Specific (adaptive) immunity has two branches: humoral and cell-mediated. Cell-mediated immunity involves specialized white blood cells called T cells that act against microbe-infected cells and foreign tissues. They also regulate the activation and proliferation of other immune system cells such as macrophages, B cells, and other T cells. Humoral immunity, or antibody-mediated immunity, involves the production of glycoprotein antibodies by plasma cells derived from B cells.

- In order to function, B and T cells must be exposed to a specific antigen and other required signals. This results in their proliferation and differentiation in a process known as clonal selection.

- Activated T cells produce cytokines that assist in the development and function of other immunocytes.

- Activated B cells produce antibody, also called immunoglobulin (Ig). Two distinguishing characteristics of Ig are their diversity and specificity. There are five human Ig classes based on physicochemical and biological properties: IgG, IgM, IgD, IgA, and IgE.

- The binding of antigen to antibody initiates the participation of other elements that determine the fate of the antigen. For example, the classical complement pathway can be activated, leading to cell lysis or phagocytosis. Other defensive mechanisms include toxin neutralization, adherance inhibition, and opsonization.

- Immune disorders range from mild conditions like hay fever to life-threatening diseases. They can be categorized as hypersensitivities, autoimmune diseases, transplantation (tissue) rejection, and immunodeficiencies.

The immune system is responsible for the ability of a host to resist foreign invaders. It includes an array of cells and molecules with specialized roles in defending the host against the continuous onslaught of microbes, toxins, and cancer cells. Chapter 31 discusses nonspecific host resistance and the innate mechanisms by which the host protects itself from invading microorganisms. Recall that the innate resistance system responds to a foreign substance in the same manner and to the same magnitude each time, and that its activation can assist in the formation of specific immune responses.

In chapter 32 we continue our discussion of the immune response by describing the system of specific (adaptive) responses used to protect the host. Specific immunity is acquired and requires sufficient time to fully develop. However, upon subsequent exposure to the same substance, activation of a specific immune response is significantly faster and stronger than that of the initial response. Thus the cooperation between the host's innate resistance mechanisms and its specific killing responses prevents pathogen invasion and maintains host integrity through the combination of immediate, nonspecific resistance and delayed, specific responses.

After a brief overview of how the specific immune responses work, we launch into chapter 32 by first defining the nature of molecules (antigens) that elicit specific immune reactions, including the methods by which specific immunity can be induced, and the role recognition of "self" plays in a host's detection of invaders. We then elaborate on the structure and function of T and B cells, along with their effector products, that we began in chapter 31. Finally, we end the chapter with a brief overview of why the host shouldn't attack itself (tolerance) and the disorders that occur when the host does.

The remarkable capacity of the immune system to respond to many thousands of different substances with exquisite specificity saves us all from certain death by infection

—*Martin C. Raff*

32.1 OVERVIEW OF SPECIFIC (ADAPTIVE) IMMUNITY

The specific (adaptive) immune system of vertebrates has three major functions: (1) to recognize anything that is foreign to the body ("nonself"); (2) to respond to this foreign material; and (3) to remember the foreign invader. The recognition response is highly specific. The immune system is able to distinguish one pathogen from another, to identify cancer cells, and to discriminate the body's own "self" proteins and cells as different from "nonself" proteins, cells, tissues, and organs. After recognition of an invader has occurred, the specific immune system responds by amplifying and activating specific lymphocytes to attack it. This is called an **effector response.** A successful effector response either eliminates the foreign material or renders it harmless to the host, thus preventing disease. If the same invader is encountered at a later time, the immune system is prepared to mount a more intense and rapid memory or **anamnestic response** that eliminates the invader once again and protects the host from disease. Four characteristics distinguish specific (adaptive) immunity from nonspecific (innate) resistance:

1. Discrimination between self and nonself. The (adaptive) specific immune system almost always responds selectively to nonself and produces specific responses against the stimulus.
2. Diversity. The system is able to generate an enormous diversity of molecules such as antibodies that recognize trillions of different foreign substances.
3. Specificity. Immunity is also selective in that it can be directed against one particular pathogen or foreign substance (among trillions); the immunity to this one pathogen or substance usually does not confer immunity to others.
4. Memory. When re-exposed to the same pathogen or substance, the body reacts so quickly that there is usually no noticeable pathogenesis. By contrast, the reaction time for inflammation and other nonspecific (innate) defenses is just as long for a later exposure to a given antigen as it was for the initial one.

The recognition of foreign substances by a mammalian host occurs because host cells express a unique protein on their surface, marking them as residents of that host, or as "self." Thus the introduction of materials lacking that unique self-marker results in their destruction by the host. There are, of course, exceptions to this process. For example, some materials are highly conserved throughout evolution—that is, their three-dimensional structures are identical, or very similar, regardless of the host that produces them, and as such, they are not distinguished as foreign.

Two branches or arms of specific (adaptive) immunity are recognized (**figure 32.1**): humoral (antibody-mediated) immunity and cellular (cell-mediated) immunity. **Humoral (antibody-mediated) immunity,** named for the fluids or "humors" of the body, is based on the action of soluble glycoproteins called antibodies that occur in body fluids and on the plasma membranes of B lymphocytes. Circulating antibodies bind to microorganisms,

toxins, and extracellular viruses, neutralizing them or "tagging or marking" them for destruction by mechanisms as described in section 31.2. **Cellular (cell-mediated) immunity** is based on the action of specific kinds of T lymphocytes that directly attack cells infected with viruses or parasites, transplanted cells or organs, and cancer cells. T cells can lyse these cells or release chemicals (cytokines) that enhance specific immunity and nonspecific (innate) defenses such as phagocytosis and inflammation. Because the activity of the acquired immune response is so potent, it is imperative that T and B cells consistently discriminate between self and nonself with great accuracy. How they accomplish this is discussed next.

32.2 ANTIGENS

The immune system distinguishes between "self" and "nonself" through an elaborate recognition process. During their development, B and T cells that would recognize components of their host (self-determinants) are induced to undergo apoptosis (programmed cell death). This is critical for the removal of effector lymphocytes that could react against their host. This ensures that lymphocytes can produce specific immunologic reactions only against foreign materials and organisms, leading to their removal. Self and nonself substances that elicit an immune response and react with the products of that response are often called **antigens.** Antigens include molecules such as proteins, nucleoproteins, polysaccharides, and some glycolipids. While the term "immunogen" (*immun*ity *gen*erator) is a more precise descriptor for a substance that elicits a specific immune response, "antigen" is used more frequently. Most antigens are large, complex molecules with a molecular weight generally greater than 10,000 Daltons (Da). The ability of a molecule to function as an antigen depends on its size, structural complexity, chemical nature, and degree of foreignness to the host. Cells, tissues, and organs of the immune system: Lymphocytes (section 31.2)

Each antigen can have several **antigenic determinant sites,** or **epitopes (figure 32.2).** Epitopes are the regions or sites in the antigen that bind to a specific antibody or T-cell receptor through an antigen-binding site. Antibodies are formed most readily in response to epitopes that project from the foreign molecule or to terminal residues of a specific polymer chain. Chemically, epitopes include sugars, organic acids and bases, amino acid side chains, hydrocarbons, and aromatic groups. The number of epitopes on the surface of an antigen is its **valence.** The valence determines the number of antibody molecules that can combine with the antigen at one time. If one determinant site is present, the antigen is monovalent. Most antigens, however, have more than one copy of the same epitope and are termed multivalent. Multivalent antigens generally elicit a stronger immune response than do monovalent antigens. As we will see, each antibody molecule has at least two antigen-binding sites so multivalent antigens can be "cross-linked" by antibodies, a phenomenon that can result in precipitation or agglutination of antigen. **Antibody affinity** relates to the strength with which an antibody binds to its antigen at a given antigen-binding site. Affinity tends to

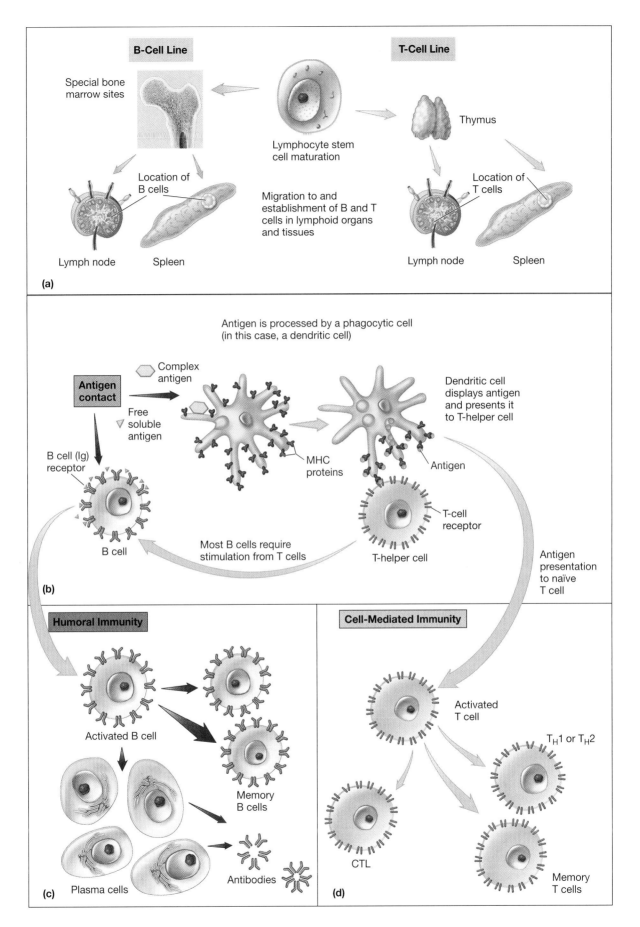

Figure 32.1   Acquired Immune System Development.   (a) Lymphocyte stem cells develop into B- and T-cell precursors that migrate to the bone marrow or thymus, respectively. Mature B and T cells seed secondary lymphoid tissues. **(b)** Lymphocyte receptor binding of antigen activates B and T cells to become effector cells. **(c)** B lymphocytes develop into memory cells and antibody-secreting plasma cells. **(d)** T cells develop into memory cells, helper T cells, and cytotoxic T cells.

(a)

Microbial cells,
viruses

Foreign human
or animal cells

Plant molecules

Epitopes

(b)

(c)

Figure 32.2   Antigen Characteristics Are Numerous and Diverse.   (a) Whole cells and viruses make good immunogens. **(b)** Complex molecules with several epitopes make good immunogens. **(c)** Poor immunogens include small molecules not attached to a carrier molecule (*1*), simple molecules (*2*), and large but repetitive molecules (*3*).

increase during the course of an immune response and is discussed in section 32.7. The **avidity** of an antibody relates to its overall ability to bind antigen at all antigen-binding sites.

Haptens

Many small organic molecules are not antigenic by themselves but become antigenic if they bond to a larger carrier molecule such as a protein. These small antigens are called **haptens** [Latin *haptein,* to grasp]. When lymphocytes are stimulated by the combined hapten-carrier molecule, they can react to either the hapten or the larger carrier molecule. This occurs because the hapten functions as one epitope of the carrier. When the carrier is processed (section 31.2) and presented to T cells, responses to both the hapten and the carrier protein can be elicited. As a result, both hapten-specific and carrier-specific antibodies can be made. One example of a hapten is penicillin. By itself penicillin is a small molecule

and is not antigenic. However, when it is combined with certain serum proteins of sensitive individuals, the resulting molecule becomes immunogenic, activates lymphocytes, and initiates a severe and sometimes fatal allergic immune reaction. In these instances the hapten is acting as an antigenic determinant on the carrier molecule.

Cluster of Differentiation Molecules (CDs)

Recall that antigens are molecules that elicit an immune response. It is important to recognize that the molecules of a given host can elicit an immune reaction when exposed to lymphocytes obtained from a different host; they are indeed, antigens. This is the basis for identifying and studying cell receptors. One practical example of this is seen in the identification of cell surface proteins that have specific roles in intercellular communication. For example, lymphocytes and other cells of the immune system bear particular membrane proteins called ***c*luster of *d*ifferentiation (CDs) molecules or antigens.** CDs are cell surface proteins and many are receptors. CDs can be measured in situ and from peripheral blood, biopsy samples, or other body fluids. They often are used in a classification system to differentiate between leukocyte subpopulations. To date, over 300 CDs have been characterized. **Table 32.1** summarizes some of the functions of several CDs.

CDs have both biological and diagnostic significance. The presence of various CDs on the cell's surface can be used to determine the cell's identity. For example, it has been established that the CD4 molecule is a cell surface receptor for human immunodeficiency virus (HIV; the virus that causes AIDS), and CD34 is the cell surface indicator of stem cells. As we will see, using the CD antigen system to name cells is more efficient than describing all of a cell's functions. We also use this approach in naming specific cell types as we discuss their relative functions in immunity.

1. Distinguish between "self" and "nonself" substances.
2. Define and give several examples of an antigen. What is an antigenic determinant site or epitope?
3. Define hapten and CD antigens.
4. Give some examples of the biological significance of cluster of differentiation molecules (CDs).

32.3 TYPES OF SPECIFIC (ADAPTIVE) IMMUNITY

Acquired immunity refers to the type of specific (adaptive) immunity a host develops after exposure to foreign substances, or after transfer of antibodies or lymphocytes from an immune donor. Acquired immunity can be obtained actively or passively by natural or artificial means (**figure 32.3**).

Naturally Acquired Immunity

Naturally acquired active immunity occurs when an individual's immune system contacts a foreign stimulus (antigen) such as a pathogen that causes an infection. The immune system responds by producing antibodies and activated lymphocytes that inactivate or

Table 32.1	Functions of Some Cluster of Differentiation (CD) Molecules
Molecule	**Function**
CD1 a,b,c	MHC class I-like receptor used for lipid antigen presentation
CD3 δ,ε,γ	T-cell antigen receptor
CD4	MHC class II co-receptor on T cells, monocytes, and macrophages; HIV-1 and HIV-2 (gp120) receptor
CD8	MHC class I co-receptor on cytotoxic T cells
CD11 a,b,c,d	α-subunits of integrin found on various myeloid and lymphoid cells; used for binding to cell adhesion molecules
CD19	B-cell antigen co-receptor
CD25	IL-2 α chain on activated T cells, B cells, and monocytes
CD34	Stem cell protein that binds to sialic acid residues
CD45	Tyrosine phosphatase common to all hematopoeitic cells
CD56	NK cell and neural cell adhesion molecule
CD97 α/β	Igα/Igβ receptor on B cells
CD106	Endothelial cell vascular cell adhesion molecule-1
CD209	Dendritic cell-specific c-type lectin

Acquired Immunity

Natural immunity
is acquired through the normal life experiences of
a human and is not induced through medical means.

Artificial immunity
is that produced purposefully through
medical procedures (also called immunization).

Active immunity
is the consequence of
a person developing his
own immune response
to a microbe.

Passive immunity
is the consequence of
one person receiving
preformed immunity
made by another person.

Active immunity
is the consequence of a
person developing his
own immune response
to a microbe.

Passive immunity
is the consequence
of one person receiving
preformed immunity
made by another person.

Figure 32.3   Immunity Can Be Acquired by Various Means.   Naturally acquired immunity can be either active or passive. Artificially acquired immunity can also be either active or passive.

destroy the pathogen. The immunity produced can be either life-long, as with measles or chickenpox, or last for only a few years, as with influenza. **Naturally acquired passive immunity** involves the transfer of antibodies from one host to another. For example, some of a pregnant woman's antibodies pass across the placenta to her fetus. If the female is immune to diseases such as polio or diphtheria, this placental transfer also gives her fetus and newborn temporary

immunity to these diseases. Certain other antibodies can pass from the female to her offspring in the first secretions (called colostrum) from the mammary glands. These maternal antibodies are essential for providing immunity to the newborn for the first few weeks or months of life, until its own immune system matures. Protection of the newborn by antibodies from colostrum is especially important in certain animal species (such as cattle and horses), which have less antibody transfer across the placenta than do primates. Naturally acquired passive immunity generally lasts only a short time (weeks to months, at most).

Artificially Acquired Immunity

Artificially acquired active immunity results when an animal is intentionally exposed to a foreign material and induced to form antibodies and activated lymphocytes. This foreign material is called a vaccine and the procedure is vaccination (immunization). A vaccine may consist of a preparation of killed microorganisms; living, weakened (attenuated) microorganisms; genetically engineered organisms; or inactivated bacterial toxins (toxoids) that are administered to induce immunity artificially. Vaccines and immunizations are discussed in detail in section 36.8.

Artificially acquired passive immunity results when antibodies or lymphocytes that have been produced outside the host are introduced into a host. Although this type of immunity is immediate, it is short lived, lasting only a few weeks to a few months. An example would be botulinum antitoxin produced in a horse and given to a human suffering from botulism food poisoning, or a bone marrow transplant given to a patient with genetic immunodeficiency.

---

1. What are the three related activities mediated by the specific (adaptive) immune systems?
2. What distinguishes specific immunity from nonspecific (innate) resistance?
3. What are the two arms of specific (adaptive) immunity?
4. How does naturally acquired immunity occur? Contrast active and passive immunity.

32.4 RECOGNITION OF FOREIGNNESS

Distinguishing between self and nonself is essential in maintaining host integrity. This distinction is highly specific and selective so as to eliminate invading pathogens but not destroy host tissue.

An important extension of this exquisite survival mechanism is seen in the modern use of tissue transplantation, where organs, tissues, and cells from unrelated persons are carefully matched to the self-recognition markers on the recipient's cells when used in disease treatment. While entire textbooks are devoted to this subject, we only discuss the aspects of foreignness recognition that assist us in understanding why and how lymphocytes respond. Without this ability to recognize foreign materials, lymphocytes have no reason to differentiate into effector cells.

Recall that each cell of a particular host needs to be identified as a member of that host so it can be distinguished from foreign invaders. To accomplish this, each cell must express gene products that mark it as a resident of that host. Furthermore, it is not enough to simply identify resident (self) cells—in addition, effective cooperation between cells must occur so that efficient information sharing and selective effector activities occur. Such a system has evolved in mammals and is encoded in the major histocompatibility gene complex.

The **major histocompatibility complex (MHC)** is a collection of genes on chromosome 6 in humans and chromosome 17 in mice. This term is derived from the Greek word for tissue [*histo*] and the ability to get along [compatibility]. The MHC is called the **human leukocyte antigen (HLA)** complex in humans and the H-2 complex in mice. Almost all human cells contain HLA molecules on their plasma membranes. HLA molecules can be divided into three classes: class I molecules are found on all types of nucleated body cells; class II molecules appear only on cells that can process nonself materials and present antigens to other cells (i.e., macrophages, dendritic cells, and B cells); and class III molecules include various secreted proteins that have immune functions (**figure 32.4**). Unlike class I and II MHC molecules, class III molecules are mostly secreted products whose presence is not required to discriminate between self and nonself. Furthermore, the class III MHC molecules are not membrane proteins, are not related to class I or II molecules, and have no role in antigen presentation. Thus they are not discussed further.

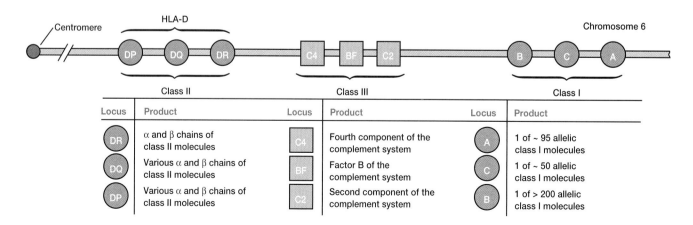

Figure 32.4   Major Histocompatibility Complex.   The MHC region of human chromosome 6 and the gene products associated with each locus.

Each individual has two sets of MHC genes—one from each parent, and both are expressed (i.e., they are codominant). Thus a person expresses many different HLA products (figure 32.4). The HLA proteins differ among individuals; the closer two people are related, the more similar are their HLA molecules. In addition, many forms of each HLA gene exist. This is because multiple alleles of each gene have arisen by high gene mutation rates, gene recombination, and other mechanisms (i.e., each gene locus is polymorphic). The differences in the HLA products expressed by individuals appear to account for some of the variation in infectious disease susceptibility. Class I molecules comprise HLA types A, B, and C and serve to identify almost all cells of the body as "self." They also stimulate antibody production when introduced from one host into another host with different class I molecules. This is the basis for MHC typing when a patient is being prepared for an organ transplant (**Techniques & Applications 32.1**).

Class II HLA molecules are produced only by certain white blood cells, such as activated macrophages, dendritic cells, mature B cells, some T cells, and certain cells of other tissues. Class II molecules are required for T-cell communication with macrophages, dendritic cells, and B cells. As discussed later in section 32.7, part of the T-cell receptor must recognize a peptide within a class II molecule on the antigen-presenting cell before the T cell can secrete cytokines necessary for the immune response.

Class I MHC molecules consist of a complex of two protein chains, one with a mass of 45,000 Daltons (Da), known as the alpha chain, and the other with a mass of 12,000 Da (β2-microglobulin) (**figure 32.5a**). The alpha chain can be divided into three functional domains, designated α_1, α_2, and α_3. The α_3 domain is attached to the plasma membrane by a short amino acid sequence that extends into the cell interior, while the rest of the protein protrudes to the outside. The β_2-microglobulin (β_2m) protein and α_3 segment of the alpha chain are noncovalently associated with one another and are close to the plasma membrane. The α_1 and α_2 domains lie to the outside and form the antigen-binding pocket. Only nonnucleated cells (red blood cells) lack class I MHC molecules.

Class II MHC molecules are also transmembrane proteins consisting of α and β chains of mass 34,000 Da and 28,000 Da, respectively (figure 32.5b). Both chains combine to form a three-dimensional protein pocket, the antigen-binding pocket, into

Techniques & Applications

32.1   Donor Selection for Tissue or Organ Transplants

The successful transplantation of tissues began in 1908 when Alexis Carrel reported survival of nine cats with functional kidneys transplanted among them. Since then a number of different tissues and organs have been successfully transplanted between humans. Tissues that are not rejected and do not induce an immunological response when transplanted are said to be "histocompatible." Thus when the genes encoding the cellular proteins responsible for recognition of self and nonself were identified, they were labeled the major histocompatibility complex. Over the years many of the major histocompatibility complex (MHC) molecules have been identified and diagnostic antibodies made against them.

Successful tissue transplantation requires that the ABO blood group and the MHC molecules of the donor and the recipient be as closely matched as possible. Other than identical twins, no two persons have the same MHC (called human leukocyte antigens [HLAs] in humans) proteins on their cells. Thus it is necessary to identify these antigens so that a donated tissue or organ is less likely to be recognized as foreign by the recipient's immune system.

The likelihood of tissue or organ transplant acceptance can be increased by using persons who have a great degree of genetic similarity. The greater the similarity, the more likely the persons will have similar MHC complexes. The reason for this is that 77 histocompatibility alleles are determined by just four human histocompatibility genes—A, B, C, and D (which includes DQ, DR, and DP). These genes are transmitted from parents to offspring according to Mendelian genetics. For example, two siblings (brother and sister) may have approximately a 25% chance of being B identical, a 50% chance of sharing one B allele, and only a 25% chance of having completely different B alleles. Therefore, when tissue or organ transplants are performed, a deliberate attempt is made to use tissues from siblings or other genetically related, histocompatible people.

Family members are the logical first choice in the search for close HLA matches; the HLA genes are usually inherited as a complete set from parents. After matching the blood types of donor and recipient, identification of HLA proteins is performed. Potential donors are first screened using a modification of the complement fixation assay. In this assay donor lymphocytes are incubated with recipient serum (containing antibodies) in the presence of complement proteins. If the donor lymphocytes are killed (lysed by complement), the recipient serum contains antibodies against donor HLA proteins on the lymphocytes. This rules out that potential donor. If the donor lymphocytes survive exposure to the recipient serum and complement, HLA typing ensues. Clinical immunology: Complement fixation (section 35.3)

HLA typing is typically accomplished using the microcytotoxicity test. In this test, potential donor and recipient leukocytes are placed separately into wells of a microtiter plate. Antibodies specific for various class I and class II MHC (HLA-A, B, C, or D) proteins are added to different cell-filled wells, along with complement. Cytotoxicity (cell damage and death) occurs if cells express the HLA protein recognized by the antibody. These dead cells are readily measured using a specific dye. Once enough HLA molecules are identified, the donor and recipient HLA profiles are compared to determine compatibility. Additional testing can determine the degree of antibody recognition as a potential measure of rejection severity.

(a) Class I MHC

(b) Class II MHC

Figure 32.5 The Membrane-Bound Class I and Class II Major Histocompatibility Complex Molecules. **(a)** The class I molecule is a heterodimer composed of the alpha protein, which is divided into three domains: α_1, α_2, and α_3, and the protein β_2 microglobulin. **(b)** The class II molecule is a heterodimer composed of two distinct proteins called alpha and beta. Each is divided into two domains, α_1, α_2, and β_1, β_2, respectively. **(c)** This space-filling model of a class I MHC protein illustrates that it holds shorter peptide antigens (blue) than does a class II MHC. **(d)** The difference is because the peptide binding site of the class I molecule is closed off, whereas the binding site of the class II molecules is open on both ends.

which a nonself peptide fragment can be captured for presentation to other cells of the immune system (immunocytes). Although MHC class I and class II molecules are structurally distinct, both fold into similar shapes. Each MHC molecule has a deep groove into which a short peptide derived from a foreign substance can bind (figure 32.5c,d). Because this peptide is not part of the MHC molecule, it can vary from one MHC molecule to the next. Foreign peptides (antigen fragments) in the MHC groove must be present to activate T cells, which in turn activate other immunocytes.

By binding and presenting foreign peptides, class I and class II molecules inform the immune system of the presence of nonself. These peptides arise in different places within cells as the result of a process known as **antigen processing.** Class I molecules bind to peptides that originate in the cytoplasm. Foreign peptides within the cytoplasm of mammalian cells come from replicating viruses or other intracellular pathogens, or are the result of cancerous transformation. These intracellular antigenic proteins are digested by a cytoplasmic structure called the proteasome (*see figure 4.9*), as part of the natural process by which a cell continually renews its protein

contents. The resulting short peptide fragments are pumped by specific transport proteins from the cytoplasm into the endoplasmic reticulum. Within the endoplasmic reticulum the class I MHC alpha chain is synthesized and associates with β_2-microglobulin. This dimer appears to bind antigen as soon as the foreign peptide enters the endoplasmic reticulum. The class I MHC molecule and antigenic peptide are then carried to, and anchored in, the plasma membrane. In this way, the host cell presents the antigen to a subset of T cells called CD8$^+$, or cytotoxic, T lymphocytes. CD8$^+$ T cells bear a receptor that is specific for class I MHC molecules that present antigen; as will be discussed in section 32.5, these T cells bind and ultimately kill infected host cells.

Class II MHC molecules bind to fragments that initially come from antigens outside the cell. This pathway functions with bacteria, viruses, and toxins that have been taken up by endocytosis. An **antigen-presenting cell (APC),** such as a macrophage, dendritic cell, or B cell, takes in the antigen or pathogen by receptor-mediated endocytosis or phagocytosis, and produces antigen fragments by digestion in the phagolysosome. Fragments then combine with preformed class II MHC molecules and are delivered to the cell surface. It is here that the peptide is recognized by $CD4^+$ T-helper cells. Unlike $CD8^+$ T cells, $CD4^+$ T cells do not directly kill target cells. Instead they respond in two distinct ways. One is to proliferate, thereby increasing the number of $CD4^+$ cells that can react to the antigen. Some of these cells will become memory T cells, which can respond to subsequent exposures to the same antigen. The second response is to secrete cytokines (e.g., interleukin-2) that either directly inhibit the pathogen that produced the antigen or recruit and stimulate other cells to join in the immune response.

Phagocytosis (section 31.3); Chemical mediators in nonspecific resistance: Cytokines (section 31.6)

32.5 T CELL BIOLOGY

In order for acquired immunity to develop, T cells and B cells must be activated. T cells are major players in the cell-mediated immune response (figure 32.1), and also have a major role in B-cell activation. They are immunologically specific, can carry a vast repertoire of immunologic memory, and can function in a variety of regulatory and effector ways. Because of their paramount importance, we discuss them first.

T-Cell Receptors

T cells have specific **T-cell receptors (TCRs)** for antigens on their plasma membrane surface. The receptor site is composed of two parts: an alpha polypeptide chain and a beta polypeptide chain (**figure 32.6a**). Each chain is stabilized by disulfide bonds. The re-

ceptor is anchored in the plasma membrane and parts of the α and β chains extend into the cytoplasm. The recognition sites of the T-cell receptor extend out from the membrane and have a terminal variable section complementary to antigen fragments. T-cells respond to antigen fragments presented in the MHC molecules.

Types of T Cells

T cells originate from stem cells in the bone marrow, but T cell precursors migrate to the thymus for further differentiation. This includes destruction of T cells that recognize self antigens (so-called self-reactive T cells). Like all lymphocytes, T cells that have survived the process of development are called mature cells, but they are also considered to be "naïve" cells because they have not yet been activated by a specific MHC-antigen peptide combination. This activation of T cells involves specific molecular signaling events inside the cell, which is discussed in more detail shortly. Once activation occurs, T cells proliferate to form memory cells, as well as activated or "effector" cells, which carry out specific functions to protect the host against the invading antigen. The two major types of T cells, the T-helper (T_H) cells and the cytotoxic T lymphocytes (CTLs) (figure 32.3), are discussed first, then some of the details of T-cell activation are examined (**table 32.2**).

T-Helper Cells

T-helper (T_H) cells, also known as **$CD4^+$ T cells,** are activated by antigen presented by class II MHC molecules on APCs. They can be further subdivided into T_H0 cells, T_H1 cells, and T_H2 cells.

Figure 32.6 The Role of the T-cell Receptor Protein in T-helper Cell Activation. **(a)** A schematic illustration of the proposed overall structure of the antigen receptor site on a T-cell plasma membrane. **(b)** An antigen-presenting cell begins the activation process by displaying an antigen fragment (e.g., peptide) on its surface as part of a complex with the histocompatibility molecules. A T-helper cell is activated after the variable region of its receptor (designated V_α and V_β) reacts with the antigen fragment in a class II MHC molecule on the presenting cell surface.

Table 32.2	Some Functional Examples of T Cell Subpopulations
Type	**Function**
CD4$^+$ T-helper (T$_H$0) cell	Precursor cell activated by specific antigen presented on class II MHC to differentiate into T$_H$1 or T$_H$2 cells
T$_H$1 cell	Produces IL-2, IFN-γ and TNF-α, which activate macrophages and CTLs to promote cellular immune responses
T$_H$2 cell	Produces IL-4, IL-5, IL-6, IL-10, and IL-13 to promote B cell maturation and humoral immune responses
CD8$^+$ T cell	Precursor cell activated by specific antigen presented on MHC-I to differentiate into cytotoxic lymphocyte
Cytotoxic T lymphocyte (CTL), also called cytotoxic T cell	Kills cells expressing foreign specific antigen on class I MHC by perforin and granzyme release or induction of apoptosis by CD95L (Fas ligand)

T$_H$0 cells are simply undifferentiated precursors of T$_H$1 cells and T$_H$2 cells, while T$_H$1 and T$_H$2 cells are distinguished by the different types of cytokines they produce (**figure 32.7**). Activated **T$_H$1 cells** promote cytotoxic T lymphocyte (CTL) activity, activate macrophages, and mediate inflammation by producing interleukin (IL)-2, interferon (IFN)-γ, and tumor necrosis factor (TNF)-α. These cytokines are also responsible for delayed-type (type IV) hypersensitivity reactions, in which host cells and tissues are damaged nonspecifically by activated T cells (pp. 807–809). **T$_H$2 cells** tend to stimulate antibody responses in general, and defend against helminth parasites by producing cytokines such as IL-4, IL-5, IL-6, IL-10, and IL-13. An overabundance of T$_H$2 type responses may also be involved in promoting allergic reactions. Allergic and hypersensitivity reactions are discussed in section 32.11. Chemical mediators in nonspecific (innate) resistance: Cytokines (section 31.6)

Cytotoxic T Lymphocytes
Cytotoxic T lymphocytes (CTLs) are **CD8$^+$ T cells** that function to destroy host cells that have been infected by an intracellular pathogen, such as a virus. Activation of CTLs can be thought of as a two step process. First, naïve CD8$^+$ cells must interact with an APC (i.e., a macrophage or dendritic cell) that has processed the antigen and presents it to the immature CTL on its class I MHC molecule (**figure 32.8**). Note that unlike CD4$^+$ cells, CD8$^+$ cells interact with APCs through their class I MHCs. This is important because CD8$^+$ cells then mature into CTLs that can respond to the same antigen as presented in the class I MHC of any host cells that have been infected by the intracellular pathogen (recall that such cells do not possess class II MHCs). All host cells that present the same antigen thus become target cells. Once activated, these CTLs kill target cells in at least two ways: the perforin pathway and the CD95 pathway.

In the **perforin pathway,** binding of the CTL to the target cell triggers movement of cytoplasmic granules toward the part of the plasma membrane that is in contact with the target cell. These CTL granules fuse with the plasma membrane, releasing molecules called perforin and granzymes into the intercellular space. Perforin, which has considerable homology to the C9 component of complement that forms the membrane attack complex, polymerizes in the target cell's membrane to form pores. These allow the granzymes to enter the target cell, where they induce pro-

grammed cell death (apoptosis). Chemical mediators of innate resistance: Complement (section 31.6)

In the **CD95** or **Fas-FasL pathway,** the activated CTL increases expression of a protein called Fas ligand (FasL; also known as CD95L) on its surface. FasL can interact with the transmembrane Fas protein receptor found on the target cell surface. This induces the target cell to undergo apoptosis. By inducing target cell apoptosis rather than cell lysis, both the perforin and the CD95 pathways stimulate membrane changes that are thought to trigger phagocytosis and slow destruction of the apoptotic cell by macrophages. Thus any infectious agent (e.g., a virus) that caused the cell to be initially attacked by the CTL is also destroyed. If the target cell were simply to be lysed, any infectious agent it harbored could potentially be released unharmed and infect surrounding cells.

T-Cell Activation

In order to respond to a foreign substance, lymphocytes must be activated by the antigen to which they ultimately respond. Lymphocyte activation is a complex process that is still not completely understood. However, in general, the binding of an antigen within the lymphocyte receptor initiates a signaling cascade involving other membrane-bound proteins and intracellular messengers. Lymphocyte proliferation, differentiation, and expression of specific cytokine genes, however, occurs only when a second signal is communicated along with the antigen. A general discussion of this process in T cells now follows.

All naïve T cells, whether CD4$^+$ or CD8$^+$ cells, require two signals to be activated by antigen. Signal 1 occurs when an antigen fragment, presented in a MHC molecule of an antigen-presenting cell (APC), fills the appropriate T-cell receptor. Signal 1 differs for T$_H$ and CTL cells. In the case of T$_H$ cells, an exogenous antigen is taken into an APC by phagocytosis or endocytosis, processed, and presented to the T$_H$ cell by class II MHC molecules. CD4 co-receptors on the T$_H$ cell interact with the antigen-bound MHC molecule to assist with signal 1 recognition. For CTLs to be activated, an endogenous (cytoplasmic) antigen is processed by the APC proteasome and presented on class I MHC molecules. In this case, the CD8 co-receptor on the CTL interacts with the antigen-bound MHC molecule on the APC, assisting with signal 1 recognition. Organelles of the biosynthetic-secretory and endocytic pathways (section 4.4)

Figure 32.7 T-Cell Responses.   A virus is phagocytosed by a macrophage and a small antigen fragment (peptide) presented to naïve T_H cells in association with class II MHC molecules. Once activated (activation signals 1 and 2), the T_H0 cell may differentiate into a T_H2 cell that secretes the cytokines IL-4, IL-5, and IL-6, followed by IL-10, and IL-13 (not shown) which causes B cell proliferation and subsequent secretion of specific antiviral antibodies. Alternatively they differentiate into T_H1 cells that secrete IL-2, IFN-γ, and TNF-α. IL-2 regulates the proliferation of cytotoxic T cells (CTLs). Once a CTL proliferates and differentiates into an activated effector cell, it attacks and causes lysis or programmed cell death (apoptosis) of a virus-infected cell by either the perforin or Fas pathways, respectively.

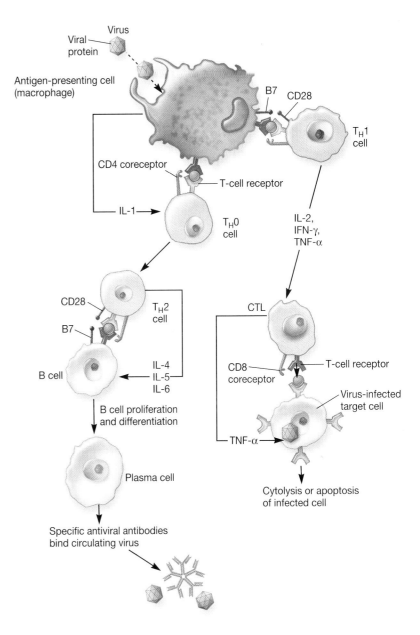

In addition to signal 1, both naïve cell types require a second, co-stimulatory signal (signal 2) to become activated. Without signal 2, a T cell receiving signal 1 only will often become **anergic,** or unresponsive to that antigen. There may be more than one factor that contributes to signal 2, but the most important seems to be the B7 (CD80) protein on the surface of an APC. One type of APC that is particularly good at stimulating naïve T cells is the dendritic cell. This type of phagocytic cell expresses high levels of B7 constitutively (at all times). Thus the combination of signals 1 and 2 provided by a mature dendritic cell presenting the antigen fragment stimulates molecular events inside the T cell, which causes it to proliferate and differentiate.

In T_H cells, signal 1 stimulates a signal transduction pathway that results in the production of key cytokines. First, signal 1 activates a tyrosine kinase located in the cytoplasm (**figure 32.9**). Tyrosine kinases add phosphate groups to the amino acid tyrosine in

proteins. In this case, phosphorylation of the enzyme phospholipase C_γ1 stimulates it to cleave the molecule phosphatidylinositol bisphosphate located in the T-helper cell plasma membrane. Two cleavage products are formed and each contributes to a separate pathway within the cell. One of the cleavage products, diacylglycerol, activates protein kinase C (PKC). PKC moves into the nucleus where it catalyzes the formation of a protein complex called AP-1 from separate components. The other cleavage product, inositol triphosphate, causes a calcium channel to open; calcium ions then rush into the cytoplasm, leading to further enzymatic activity and the activation of calmodulin, calcineurin, and nuclear factor of activated T_H1 cells (NF-AT). NF-AT then migrates into the nucleoplasm where it binds to the newly formed AP-1, forming a NF-AT/AP-1 complex. This functions as a transcription factor, causing interleukin-2 mRNA to be expressed. Interleukin-2 (IL-2) mRNA moves out of the nucleus to the ribosomes where the IL-2

T-Independent Antigen Triggering

Not all antibody responses require direct T cell help. There are a few specific antigens that trigger B cells into antibody production without T cell cooperation. These are called **T-independent antigens** and their stimulation of B cells is known as **T-independent antigen triggering.** Examples include bacterial lipopolysaccharides, certain tumor-promoting agents, antibodies against other antibodies, and antibodies to certain B cell differentiation antigens. The T-independent antigens tend to be polymeric—that is, they are composed of repeating polysaccharide or protein subunits. The resulting antibody generally has a relatively low affinity for antigen.

The mechanism for activation by T-independent antigens probably depends on their polymeric structure. Large molecules present a large array of identical epitopes to a B cell specific for that determinant. The repeating epitopes cross-link membrane-bound BCRs such that cell activation occurs and antibody is secreted. Because there is no T cell help, the B cell cannot alter its antibody production, and no memory B cells are formed. Thus T-independent B cell activation is less effective than T-dependent B cell activation: the antibodies produced have a low affinity for antigen and no immunologic memory is formed. **Tables 32.3** and **32.4** summarize and compare many of the important properties of lymphocytes that we have discussed.

Table 32.3	Comparison of Lymphocytes Involved in the Immune Response	
Property	**T Cells**	**B Cells**
Origin	Bone marrow	Bone marrow
Maturation and expression of antigen receptors	Thymus	Bone marrow; bursa of Fabricius in birds
Differentiation	Lymphoid tissue	Lymphoid tissue
Mobility	Great	Very little (some stages circulate)
Complement receptors	Absent	Present
Surface immunoglobulins	Absent	Present
Proliferation	Upon antigenic stimulation, proliferate and differentiate into effector and memory cells	Upon antigenic stimulation, proliferate and differentiate into plasma and memory cells
Immunity type	Cell mediated and humoral (B cell activation by T_H cells)	Humoral
Distribution	High in blood, lymph, and lymphoid tissue	High in spleen, lymph nodes, bone marrow, and other lymphoid tissue; low in blood
Secretory product	Cytokines	Antibodies
Subsets and functions	T-helper (T_H) cell: necessary for B-cell activation by T-dependent antigens and T-effector cells. There are three types of T-helper cells: T_H1, T_H2, and T_H0.	Plasma cell: a cell arising from a B cell that manufactures specific antibodies
	Cytotoxic T cell: differentiates into a CTL that lyses cells recognized as nonself and virus or parasite-infected cells	Memory Cell: a long-lived cell responsible for the anamnestic response
	Memory cells: a long-lived cell responsible for the anamnestic response	

Table 32.4	Antigen Recognition by T and B Cells	
Characteristic	**T Cells**	**B Cells**
Binds soluble antigen	No	Yes
Biochemistry of the antigens	Mostly proteins, but some glycolipids presented on MHC molecules	Proteins, glycolipids, polysaccharides
Antigen recognition	Antigens processed internally and presented as linear peptides bound to MHC molecules	Accessible areas of protein structure containing sequential amino acids and nonsequential amino acids
	Involve two cells: T-cell receptor, with processed antigen in MHC molecule on APC	Immunoglobulin receptor binds antigen in native conformation
		Involves two partners: antigen and membrane immunoglobulin

1. What are B cell receptors? How are they involved in B cell activation?
2. Briefly compare and contrast B cells and T cells with respect to their formation, structure, and roles in the immune response.
3. How does antigen-antibody binding occur? What is the basis for antibody specificity?
4. How does T-independent antigen triggering of B cells differ from T-dependent triggering?

32.7 ANTIBODIES

An **antibody** or **immunoglobulin (Ig)** is a glycoprotein that is made by activated B cells called plasma cells. Certain antibodies serve as the antigen receptor (BCR) on the B cell surface. After antigen capture by the membrane-bound immunoglobulin and B cell activation, B cells differentiate into plasma cells, which secrete a soluble form of antibody that circulates through the bloodstream to recognize and bind the antigen that induced its synthesis. Antibodies are present in the blood serum, tissue fluids, and mucosal surfaces of vertebrate animals. There are five classes of human antibodies. **Table 32.5** summarizes some of the more important physiochemical properties of the human immunoglobulin classes. The classes differ from each other in molecular size, structure, charge, amino acid composition, and carbohydrate content. Before we examine each of the immunoglobulin types, a thorough evaluation of immunoglobulin structure will help us appreciate the differences among immunoglobulin classes.

Table 32.5	Physicochemical Properties of Human Immunoglobulin Classes				
	Immunoglobulin Classes				
Property	**IgG**[a]	**IgM**	**IgA**[b]	**IgD**	**IgE**
Heavy chain	γ_1	μ	α_1	δ	ε
Mean serum concentration (mg/ml)	9	1.5	3.0	0.03	0.00005
Percent of total serum antibody	80–85	5–10	5–15	<1	0.002–0.05
Valency	2	5(10)	2(4)	2	2
Mass of heavy chain (kDa)	51	65	56	70	72
Mass of entire molecule (kDa)	146	970	160[c]	184	188
Placental transfer	+	−	−	−	−
Half-life in serum (days)[d]	23	5	6	3	2
Complement activation					
Classical pathway	++	+++	−	−	−
Alternative pathway	−	−	+	−	−
Induces mast cell degranulation	−	−	−	−	+
Major characteristics	Most abundant Ig in body fluids; neutralizes toxins; opsonizes bacteria; activates complement; transplacental antibody	First to appear after antigen stimulation; very effective agglutinator; expressed as membrane-bound antibody on B cells	Secretory antibody; protects external surfaces	Present on B-cell surface; B-cell recognition of antigen	Anaphylactic-mediating antibody; resistance to helminths
% carbohydrate	3	7–10	7	12	11

[a] Properties of IgG subclass 1.

[b] Properties of IgA subclass 1.

[c] sIgA = 360 − 400 kDa

[d] Time required for half of the antibodies to disappear.

Immunoglobulin Structure

Recall that antigen is captured by its respective antibody through the antigen-binding or more appropriately, antigen-combining, site. Each B cell has numerous surface-bound (transmembrane) antibodies, which have two combining sites—that is, they are bivalent. Some bivalent antibody molecules can combine to form multimeric antibodies that have up to 10 combining sites. All immunoglobulin molecules have a basic structure composed of four polypeptide chains: two identical heavy and two identical light chains connected to each other by disulfide bonds (**figure 32.12**). Each light chain polypeptide usually consists of about 220 amino acids and has a mass of approximately 25,000 Da. Each heavy chain consists of about 440 amino acids and has a mass of about 50,000 to 70,000 Da. The heavy chains are structurally distinct for each immunoglobulin class or subclass. Both light (L) and heavy (H) chains contain two different regions. **Constant (C) regions**

(C_L and C_H) have amino acid sequences that do not vary significantly between antibodies of the same class. The **variable (V) regions** (V_L and V_H) from different antibodies have different amino acid sequences. It is the variable regions (V_L and V_H) that, when folded together, form the antigen-binding sites.

The four chains are arranged in the form of a flexible "Y" with a hinge region. This hinge allows the antibody molecule to be more flexible, adjusting to the different spatial arrangements of epitopes or antigenic determinants of antigens. The stalk of the Y is termed the **crystallizable fragment (Fc)** and binds to a cell by interacting with the cell surface Fc receptor. The top of the Y consists of two **antigen-binding fragments (Fab)** that bind with compatible epitopes. The Fc fragments are composed only of constant regions, whereas the Fab fragments have both constant and variable regions. Both the heavy and light chains contain several homologous units of about 100 to 110 amino acids. Within each unit, called a **domain,** disulfide bonds form a loop of approximately 60 amino

(a)

(b)

(c)

Figure 32.12 Immunoglobulin (Antibody) Structure.
(a) An immunoglobulin molecule. The molecule consists of two identical light chains and two identical heavy chains held together by disulfide bonds. **(b)** A computer-generated model of antibody structure showing the arrangement of the four polypeptide chains. **(c)** Within the immunoglobulin unit structure, intrachain disulfide bonds create loops that form domains. All light chains contain a single variable domain (V_L) and a single constant domain (C_L). Heavy chains contain a variable domain (V_H) and either three or four constant domains (C_H1, C_H2, C_H3, and C_H4). The variable regions (V_H, V_L), when folded together in three-dimensions, form the antigen-binding sites.

acids (figure 32.12c). Interchain disulfide bonds also link heavy and light chains together.

The light chain may be either of two distinct forms called kappa (κ) and lambda (λ). These can be distinguished by the amino acid sequence of the constant (C) portion of the chain (**figure 32.13a**). In humans, the constant regions of all κ chains are identical. However, there are four similar λ protein sequences possible, reflecting four, slightly different λ subtypes. Regardless of immunoglobulin class, each antibody molecule produced by a sole B cell will contain either κ or λ light chains, but never both. Within the light chain variable (V) domain are hypervariable regions, or **complementarity-determining regions (CDRs),** that differ in amino acid sequence more frequently than the rest of the variable domain. These figure prominently in determining antigen specificity.

The amino-terminal domain of the heavy chain has a pattern of variability similar to that of the variable (V) region of the kappa chain (V_κ) and the variable region of the lambda chain (V_λ) domains, and is termed the V_H domain. The other domains of the heavy chains are termed constant (C) domains (figure 32.13b). The constant domains of the heavy chain form the constant (C_H) region. The amino acid sequence of this region determines the classes of heavy chains. In humans there are five classes of heavy chains designated by lowercase Greek letters: gamma (γ), alpha (α), mu (μ), delta (δ), and epsilon (ε), and sometimes written as G, A, M, D or E. The properties of these heavy chains determine, respectively, the five immunoglobulin (Ig) classes—IgG, IgA, IgM, IgD, and IgE (table 32.5). Each immunoglobulin class differs in its general properties, half-life, distribution in the body, and interaction with other components of the host's defensive systems. There are variants of immunoglobulins that can be classified as: (1) **Isotypes** are the variations in the heavy chain constant regions associated with the different classes that are normally present in all individuals (**figure 32.14a**). Therefore there are five isotypes corresponding to the five antibody classes. (2) **Allotypes** are the genetically controlled, allelic forms of immunoglobulin molecules (figure 32.14b) that are not present in all individuals. They arise by genetic recombination (pp. 796–798). (3) **Idiotypes** are individual, specific immunoglobulin molecules that differ in the hypervariable region of the Fab portion due to mutations that occur during B cell development (figure 32.14c). These variations of immunoglobulin structure reflect the diversity of antibodies generated by the immune response.

1. What is the variable region of an antibody? The hypervariable or complementarity-determining region? The constant region?
2. What is the function of the Fc region of an antibody? The Fab region?
3. Name the two types of antibody light chains.
4. What determines the class of heavy chain of an antibody?
5. Name the five immunoglobulin classes.
6. Distinguish among isotype, allotype, and idiotype.

Immunoglobulin Function

Each end of the immunoglobulin molecule has a unique role. The Fab region is concerned with binding to antigen, whereas the Fc region mediates binding to Fc receptors found on various cells of the immune system, or the first component of the classical complement system. The binding of an antibody to an antigen usually

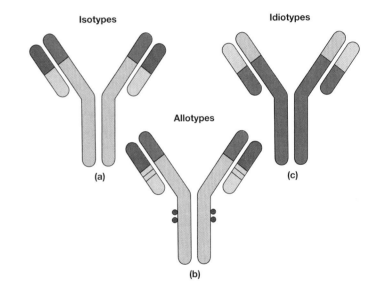

Figure 32.14   Variants of Immunoglobulins.   (a) Isotypes represent variants present in serum of a normal individual. **(b)** Allotypes represent alternative forms coded for by different alleles and so are not present in all individuals of a population. **(c)** Idiotypes are individually specific to each immunoglobulin molecule. Carbohydrate side chains are in red.

Figure 32.13   Constant and Variable Domains.   Location of constant (C) and variable (V) domains within **(a)** light chains and **(b)** heavy chains. The dark blue bands represent hypervariable regions or complementarity-determining regions within the variable domains.

does not cause destruction of the antigen or of the microorganism, cell, or agent to which it is attached. Rather the antibody serves to mark and identify the nonself agent as a target for immunological attack and to activate nonspecific immune responses that can destroy the target.

An antigen binds to an antibody at the antigen-binding site within the Fab region of the antibody. More specifically, a pocket is formed by the folding of the V_H and V_L regions (figure 32.12). At this site, specific amino acids contact the antigen's epitope or haptenic groups and form multiple noncovalent bonds between the antigen and amino acids of the binding site (**figure 32.15**). Because binding is due to weak, noncovalent bonds such as hydrogen bonds and electrostatic attractions, the antigen's shape must exactly match that of the antigen-binding site. If the shape of the epitope and binding site are not truly complementary, the antibody will not effectively bind the antigen. Although a lock-and-key mechanism normally may operate, in at least one case, the antigen-binding site does change shape when it complexes with the antigen (an induced-fit mechanism). Regardless of the precise mechanism, antibody specificity results from the nature of antibody-antigen binding.

Phagocytes have Fc receptors for immunoglobulin on their surface, so bacteria that are covered with antibodies are better targets for phagocytosis by neutrophils and macrophages. This is termed **opsonization.** Other cells may kill antibody-coated cells through a process called antibody-dependent cell-mediated cytotoxicity. Immune destruction also is promoted by antibody-induced activation of the classical complement system.　Chemical mediators of innate immunity: Complement (section 31.6)

Immunoglobulin Classes

Immunoglobulin γ, or **IgG,** is the major immunoglobulin in human serum, accounting for 80% of the immunoglobulin pool (**figure 32.16a**). IgG is present in blood plasma and tissue fluids. The IgG class acts against bacteria and viruses by opsonizing the invaders and neutralizing toxins. It is also one of the two immunoglobulin classes that activate complement by the classical pathway. IgG is the only immunoglobulin molecule able to cross the placenta and provide natural immunity in utero and to the neonate at birth.

There are four human IgG subclasses (IgG1, IgG2, IgG3, and IgG4) that vary chemically in their heavy chain composition and the number and arrangement of interchain disulfide bonds (figure 32.16b). About 65% of the total serum IgG is IgG1, and 23% is IgG2. Differences in biological function have been noted in these subclasses. For example, IgG2 antibodies are opsonic and develop in response to toxins. IgG1 and IgG3, upon recognition of their specific antigens, bind to Fc receptors expressed on neutrophils and macrophages. This increases phagocytosis by these cells. The IgG4 antibodies function as skin-sensitizing immunoglobulins.

Immunoglobulin μ, or **IgM,** accounts for about 10% of the immunoglobulin pool. It is usually a polymer of five monomeric units (pentamer), each composed of two heavy chains and two light chains (**figure 32.17**). The monomers are arranged in a pinwheel array with the Fc ends in the center, held together by disulfide bonds and a special **J** (joining) **chain.** IgM is the first immunoglobulin made during B-cell maturation and individual IgM monomers are expressed on B cells, serving as the antibody

(a)

(b)

Figure 32.15 Antigen-Antibody Binding. **(a)** An example of antigen-antibody binding is represented in this model of the monoclonal antibody mAb17-IA bound to the surface of the human rhinovirus. The heavy chains are red, light chains are blue, and the antigen capsid protein is yellow. The RNA interior of the virus would be toward the bottom of the diagram. The antibody is bound bivalently across icosahedral twofold axes of the virus. **(b)** Based on X-ray crystallography, the hapten molecule nestles in a pocket formed by the antibody combining site. In the illustration the hapten makes contact with only 10 to 12 amino acids in the hypervariable regions of the light and heavy chains. The numbers represent contact amino acids.

Figure 32.16   Immunoglobulin G.   (a) The basic structure of human IgG. **(b)** The structure of the four human IgG subclasses. Note the arrangement and numbers of disulfide bonds (shown as thin black lines). Carbohydrate side chains are shown in red.

component of the BCR. Pentameric IgM is secreted into serum during a primary antibody response (as is discussed shortly). IgM tends to remain in the bloodstream where it agglutinates (or clumps) bacteria, activates complement by the classical pathway, and enhances the ingestion of pathogens by phagocytic cells.

Although most IgM appears to be pentameric, around 5% or less of human serum IgM exists in a hexameric form. This molecule contains six monomeric units but seems to lack a J chain. Hexameric IgM activates complement up to 20-fold more effectively than does the pentameric form. It has been suggested that bacterial cell wall antigens such as gram-negative lipopolysaccharides may directly stimulate B cells to form hexameric IgM without a J chain. If this is the case, the immunoglobulins formed during primary immune responses are less homogeneous than previously thought.

Immunoglobulin α, or **IgA,** accounts for about 15% of the immunoglobulin pool. Some IgA is present in the serum as a monomer. However, IgA is most abundant in mucous secretions where it is a dimer held together by a J chain (**figure 32.18**). IgA has special features that are associated with secretory mucosal surfaces. IgA, when transported from the mucus-associated lymphoid tissue to mucosal surfaces, acquires a protein called the secretory component. **Secretory IgA (sIgA),** as the modified molecule is called, is the primary immunoglobulin of mucus-

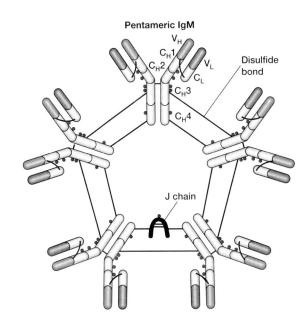

Figure 32.17   Immunoglobulin M.   The pentameric structure of human IgM. The disulfide bonds linking peptide chains are shown in black; carbohydrate side chains are in red. Note that 10 antigen-binding sites are present.

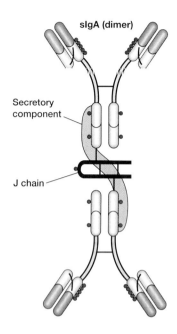

Figure 32.18 Immunoglobulin A. The dimeric structure of human secretory IgA. Notice the secretory component (tan) wound around the IgA dimer and attached to the constant domain of each IgA monomer. Carbohydrate side chains are shown in red.

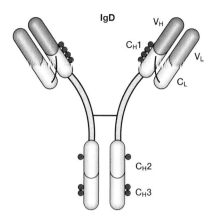

Figure 32.19 Immunoglobulin D. The structure of human IgD. The disulfide bonds linking protein chains are shown in black; carbohydrate side chains are in red.

Figure 32.20 Immunoglobulin E. The structure of human IgE.

associated lymphoid tissue. Secretory IgA is also found in saliva, tears, and breast milk. In these fluids and related body areas, sIgA plays a major role in protecting surface tissues against infectious microorganisms by the formation of an immune barrier. For example, in breast milk sIgA helps protect nursing newborns. In the intestine, sIgA attaches to viruses, bacteria, and protozoan parasites such as *Entamoeba histolytica*. This prevents pathogen adherence to mucosal surfaces and invasion of host tissues, a phenomenon known as immune exclusion. In addition, sIgA binds to antigens within the mucosal layer of the small intestine; subsequently the antigen-sIgA complexes are excreted through the adjacent epithelium into the gut lumen. This rids the body of locally formed immune complexes and decreases their access to the circulatory system. Secretory IgA also plays a role in the alternative complement pathway. Chemical mediators of innate immunity: Complement (section 31.6)

Immunoglobulin δ, or **IgD,** is an immunoglobulin found in trace amounts in blood serum. It has a monomeric structure (**figure 32.19**) similar to that of IgG. IgD antibodies do not fix complement and cannot cross the placenta, but they are abundant in combination with IgM on the surface of B cells and thus are part of the B cell receptor complex. Therefore their function is to signal the B cell to start antibody production upon antigen binding.

Immunoglobulin ε, or **IgE** (**figure 32.20**), makes up only a small percent of the total immunoglobulin pool. The classic skin-sensitizing and anaphylactic antibodies belong to this class. The Fc portion of IgE can bind to special Fc_ε receptors on mast cells, eosinophils, and basophils. When two IgE molecules on the sur-

face of these cells are cross-linked by binding to the same antigen, the cells degranulate. This degranulation releases histamine and other pharmacological mediators of inflammation. It also stimulates eosinophilia (an excessive number of eosinophils in the blood) and gut hypermotility (increased rate of movement of the intestinal contents), which aid in the elimination of helminthic parasites. Thus although IgE is present in small amounts, this class of antibodies has potent biological capabilities, as is discussed in section 32.11.

1. Explain the different functions of antibody when it is bound to B cells versus when it is soluble in serum.
2. Describe the major functions of each immunoglobulin class.
3. Why is the structure of IgG considered the model for all five immunoglobulin classes?
4. Which immunoglobulin can cross the placenta?
5. Which immunoglobulin is most prevalent in the immunoglobulin pool? The least prevalent?

Antibody Kinetics

The synthesis and secretion of antibody can also be evaluated with respect to time. Monomeric IgM serves as the B-cell receptor for antigen, and pentameric IgM is secreted after B-cell activation. Furthermore, under the influence of T-helper cells (responding to other stimuli), the IgM-secreting plasma cells may stop producing and secreting IgM in favor of another antibody class (IgG, IgA, IgE, for example). This is known as **class switching.** These events take time to unfold.

The Primary Antibody Response

When an individual is exposed to an antigen (for example, an infection or vaccine), there is an initial lag phase, or latent period, of several days to weeks before an antibody response is mounted. During this latent period no antibody can be detected in the blood (**figure 32.21**). Once B cells have differentiated into plasma cells, antibody is secreted and can be detected. This explains why antibody-based HIV tests, for example, are not accurate until weeks after exposure. The **antibody titer,** which is a measurement of serum antibody concentration (the reciprocal of the highest dilution of an antiserum that gives a positive reaction in the test being used), then rises logarithmically to a plateau during the second, or log, phase. In the plateau phase the antibody titer stabilizes. This is followed by a decline phase, during which antibodies are naturally metabolized or bound to the antigen and cleared from the circulation. During the primary antibody response, IgM appears first, then IgG, or another antibody class. The affinity of the antibodies for the antigen's determinants is low to moderate during this primary antibody response.

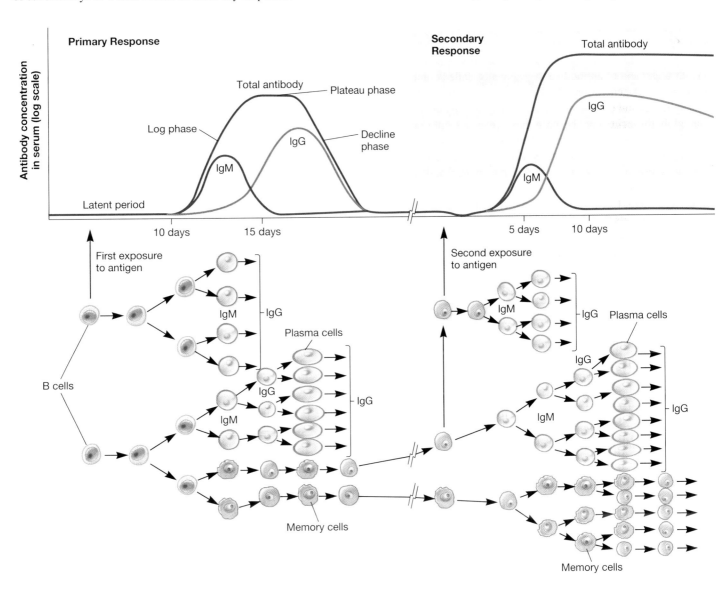

Figure 32.21   Antibody Production and Kinetics.   The four phases of a primary antibody response correlate to the clonal expansion of the activated B cell, differentiation into plasma cells, and secretion of the antibody protein. The secondary response is much more rapid and total antibody production is nearly 1,000 times greater than that of the primary response.

Critical Thinking Questions

1. Why do you think antibodies are proteins rather than polysaccharides or lipids? List all properties of proteins that make them suitable molecules from which to make antibodies.

2. How did the clonal selection theory inspire the development of monoclonal antibody techniques?

3. What is the difference in the kinetics of antibody formation in response to a first and second exposure to the same antigen?

4. Why do MHC, TCR, and BCR molecules require accessory proteins or co-receptors for a signal to be sent within the cell?

5. Why do you think two signals are required for B- and T-cell activation, but only one signal is required for activation of an APC?

6. Most immunizations require multiple exposures to the vaccine (i.e., boosters). Why is this the case?

Learn More

Ansel, K. M.; Harris, R. B. S.; and Cyster, J. G. 2002. CXCL13 is required for B1 cell homing, natural antibody production and body cavity immunity. *Immunity* 16:67–76.

Dempsey, P.; Allicson, M.; Akkaraju, S.; Goodnow C.; and Fearon, D. 1996. C3d of complement as a molecular adjuvant: Bridging innate and acquired immunity. *Science* 271:348–50.

Fruend, J., and McDermott, K. 1942. Sensitization to horse serum by means of adjuvants. *Proc. Soc. Exp. Biol. Med.* 49:548–53.

Horton, R.; Wilming, L.; and Rand, V. *et al.* 2004. Gene map of the extended human MHC. *Nature Rev. Genetics* 5:889–99.

Janeway, C. 2004. *Immunobiology*, 6th ed. New York: Garland Science.

Kondilis, H. D.; Kor, B.; Steckman, B.; and Kargel, M. 2005. Regulation of T-cell receptor β allelic exclusion at a level beyond accessibility. *Nature Immunol.* 8:189–97.

Linton, P., and Dorskind, K. 2004. Age-related changes in lymphocyte development and function. *Nature Immunol.* 5:133–39.

Liu, C. *et al.* 2005. The role of CCL2 in recruitment of T-precursors to fetal thymi. *Blood* 105: 31–39.

Modlin, R., and Sieling, P. 2005. Now presenting: γδ T cells. *Science* 309:252–53.

Sallusto, F.; Cella, M.; Danieli C.; and Lanzavecchia, A. 1995. Dendritic cells use macropinocytosis and the mannose receptor to concentrate macromolecules in the major histocompatibility complex class II compartment: down regulation by cytokines and bacterial products. *J. Exp. Med.* 182:389–400.

Silva-Santos, B.; Pennington, D.; and Heyday, A. 2005. Lymphotoxin-mediated regulation of γδ cell differentiation by αβ T cell progenitors. *Science* 307: 925–28.

33

Pathogenicity of Microorganisms

Three *Streptococcus pneumoniae*, each surrounded by a slippery mucoid capsule (shown as a layer of white spheres around the diplococcus bacteria). The polysaccharide capsule is vital to the pathogenicity of this bacterium because it prevents phagocytic cells from accomplishing phagocytosis.

PREVIEW

- If a microorganism (symbiont) either harms or lives at the expense of another organism, it is called a parasitic organism and the relationship is termed parasitism. In this relationship the infected organism is referred to as the host.

- Those organisms capable of causing disease are called pathogens. Disease is any change in the host from a healthy to an unhealthy, abnormal state in which part or all of the host's body is not capable of carrying on its normal functions.

- The steps for the infectious process involving viral diseases include the following: entry into a potential host, attachment to a susceptible cell, penetration of viral nucleic acid, replication of virus particles within the host cell, and ultimate release from the host cell. Newly replicated virus particles are available to infect other susceptible cells. Viral infection can result in cellular injury, stimulation of immune responses, or evasion of the virus from immune detection resulting in chronic infection.

- The steps of the infectious process involving bacterial diseases usually include the following: the bacterium is transmitted to a suitable host, attaches to and/or colonizes the host, grows and multiplies within or on the host, and interferes with or impairs the normal physiological activities of the host.

- During coevolution with human hosts, pathogenic bacteria have evolved complex signal transduction pathways to regulate the genes necessary for virulence.

- The genes that encode virulence factors are often located on large segments of DNA within the bacterial genome, called pathogenicity islands, that carry genes responsible for virulence.

- Two distinct categories of disease can be recognized based on the role bacteria play in the disease causing process: infections (invasion and growth) and intoxications.

- Toxins produced by pathogenic bacteria are either exotoxins or endotoxins.

- Viruses and bacteria are continuously evolving and producing unique mechanisms that enable them to escape the host's arsenal of defenses.

Chapter 30 introduces the concept of symbiosis and deals with several of its subordinate categories, including commensalism and mutualism. In this chapter the process of parasitism is presented along with one of its possible consequences—pathogenicity. The parasitic way of life is so successful, that it has evolved independently in nearly all groups of organisms. In recent years concerted efforts to understand organisms and their relationships with their hosts have developed within the disciplines of virology, bacteriology, mycology, parasitology (protozoology and helminthology), entomology, and zoology. This chapter examines the parasitic way of life in terms of health and disease and emphasizes viral and bacterial disease mechanisms. We conclude the chapter with some viral and bacterial mechanisms used to evade host defenses.

33.1 HOST-PARASITE RELATIONSHIPS

Relationships between two organisms can be very complex. A larger organism that supports the survival and growth of a smaller organism is called the host. The interaction of the two is symbiotic. **Symbiosis** refers to the "living together" of organisms and includes

Pathogenicity is not the rule. Indeed, it occurs so infrequently and involves such a relatively small number of species, considering the huge population of bacteria on earth, that it has a freakish aspect. Disease usually results from inconclusive negotiations for symbiosis, an overstepping of the line by one side or the other, a biological misinterpretation of borders.

—Lewis Thomas

Table 33.3	Microbial Products (Virulence Factors) Involved in Bacterial Pathogen Dissemination Throughout a Mammalian Host	
Product	**Organism Involved**	**Mechanism of Action**
Coagulase	*Staphylococcus aureus*	Coagulates (clots) the fibrinogen in plasma. The clot protects the pathogen from phagocytosis and isolates it from other host defenses.
Collagenase	*Clostridium* spp.	Breaks down collagen that forms the framework of connective tissues; allows the pathogen to spread.
Deoxyribonuclease (along with calcium and magnesium)	Group A streptococci, staphylococci, *Clostridium perfringens*	Lowers viscosity of exudates, giving the pathogen more mobility.
Elastase and alkaline protease	*Pseudomonas aeruginosa*	Cleaves laminin associated with basement membranes.
Hemolysins	Staphylococci, streptococci, *Escherichia coli*, *Clostridium perfringens*	Lyse erythrocytes; make iron available for microbial growth.
Hyaluronidase	Groups A, B, C, and G streptococci, staphylococci, clostridia	Hydrolyzes hyaluronic acid, a constituent of the extracellular matrix that cements cells together and renders the intercellular spaces amenable to passage by the pathogen.
Hydrogen peroxide (H_2O_2) and ammonia (NH_3)	*Mycoplasma* spp., *Ureaplasma* spp.	Are produced as metabolic wastes. These are toxic and damage epithelia in respiratory and urogenital systems.
Immunoglobulin A protease	*Streptococcus pneumoniae*	Cleaves immunoglobulin A into Fab and Fc fragments.
Lecithinase or phospholipase	*Clostridium* spp.	Destroys the lecithin (phosphatidylcholine) component of plasma membranes, allowing pathogen to spread.
Leukocidins	Staphylococci, pneumococci, streptococci	Pore-forming exotoxins that kill leukocytes; cause degranulation of lysosomes within leukocytes, which decreases host resistance.
Porins	*Salmonella enterica* serovar Typhimurium	Inhibit leukocyte phagocytosis by activating the adenylate cyclase system.
Protein A Protein G	*Staphylococcus aureus* *Streptococcus pyogenes*	Located on cell wall. Immunoglobulin G (IgG) binds to protein A by its Fc end, thereby preventing complement from interacting with bound IgG.
Pyrogenic exotoxin B (cysteine protease)	Group A streptococci, (*Streptococcus pyogenes*)	Degrades proteins.
Streptokinase (fibrinolysin, staphylokinase)	Group A, C, and G streptococci, staphylococci	A protein that binds to plasminogen and activates the production of plasmin, thus digesting fibrin clots; this allows the pathogen to move from the clotted area.

Pathogenicity Islands

The genes that encode major virulence factors in many bacteria (e.g., *Yersinia* spp., *Pseudomonas aeruginosa, Shigella flexneri, Salmonella,* enteropathogenic *E. coli*) are found on large segments of DNA, called **pathogenicity islands,** which carry genes responsible for virulence. Pathogenicity islands have been acquired during evolution by horizontal gene transfer. A pathogen may have more than one pathogenicity island. They have several common sequence characteristics. The 3′ and 5′ ends of the islands contain insertion-like elements, suggesting their promiscuity as mobile genetic elements. The G + C nucleotide content of pathogenicity islands differs significantly from the G + C content of the remaining bacterial genome. The pathogenicity island DNA also exhibits several open reading frames, suggesting other putative genes. Interestingly, pathogenicity islands are typically associated with genes that encode tRNA. An excellent example of virulence genes

carried in a pathogenicity island are those involved in protein secretion. So far, five pathways of protein secretion (types I to V) have been described in gram-negative bacteria. A set of approximately 25 genes encodes a pathogenicity mechanism termed the **type III secretion system (TTSS)** that enables gram-negative bacteria to secrete and inject virulence proteins into the cytoplasm of eucaryotic host cells.   Protein secretion in procaryotes (section 3.8)

Many gram-negative bacteria that live in close relationships with host organisms are able to modulate host activities by secreting proteins directly into the interior of the host cell using the TTSS. Perhaps the best studied TTSS is that of *Yersinia pestis* and *Y. enterocolitica,* which cause bubonic plague and gastroenteritis, respectively. Both bacteria use the same plasmid-encoded TTSS consisting of the Yop (*Yersinia outer protein*) secretion (Ysc) injectisome and secreted Yop products (**figure 33.4a**). The TTSS injectisome is composed of a basal body and a needle. The

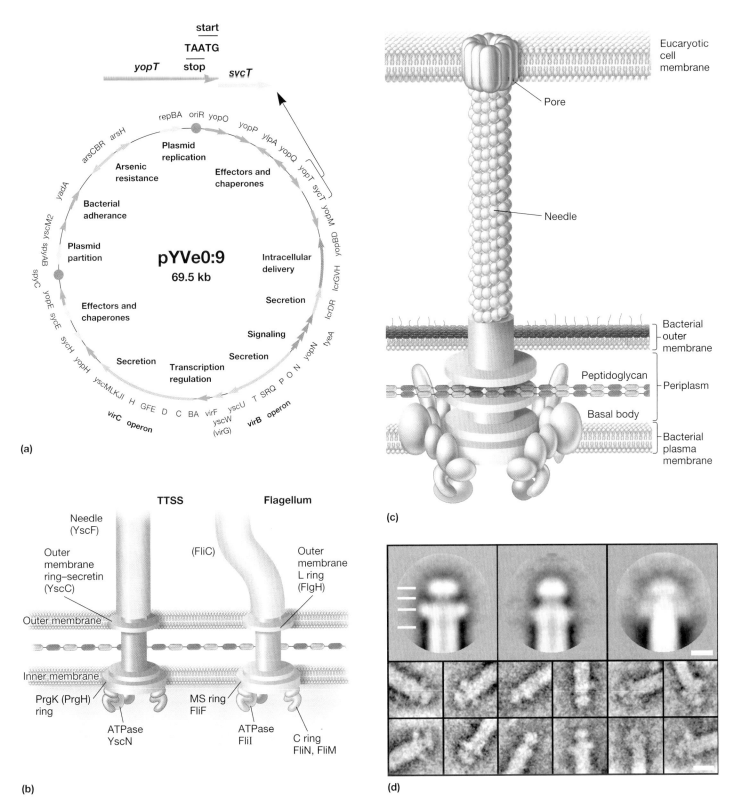

(a)

(b)

(c)

(d)

Figure 33.4 Type III Secretion System. (a) The type III secretion system (TTSS) and other virulence genes of *Yersinia* are encoded on the pYV plasmid. The TTSS genes encoding the *Yersinia* outer proteins (Yop) are homologous to many of the genes encoding flagellar proteins. (b) Both the TTSS injectisome and the flagella are anchored in the plasma membrane by similar basal body structures. (c) X-ray fiber diffraction resolves the injectisome as a helical structure. (d) Scanning tunneling electron microscopy reveals the injectisome tip, indicating how it may lock into the translocator pore on the target cell.

basal body is made from a number of proteins that are homologous (having similar amino acid sequences) to proteins that make up the basal body of bacterial flagella. This suggests that the injectisome is held in the bacterial envelope, similar to the ring system that holds a flagellum (figure 33.4*b*). The injectisome employs an ATPase that "energizes" the transport of other TTSS proteins, called "effectors," through "translocator pores" (formed by YopB and YopD proteins), into the host cell. Another protein, LcrV (also known as V antigen), is required for the correct assembly of the translocator pores. X-ray fiber diffraction of TTSS of *Shigella* demonstrates that the needle component of the TTSS has a helical arrangement (figure 33.4*c*). Scanning tunneling electron microscopy of the injectisome needle reveals a characteristic tip (head, neck, and base) through which a central channel is seen (figure 34.4*d*). LcrV localizes at the tip of the needle, which may explain its critical role in facilitating transport of TTSS-mediated proteins and their role in *Yersinia* virulence.

Unlike other bacterial secretory systems, the type III system is triggered specifically by contact with host cells, which helps avoid inappropriate activation of host defenses. Secretion of these virulence proteins into a host cell allows the pathogen to subvert the host cell's normal signal transduction system. Redirection of cellular signal transduction can disarm host immune responses or reorganize the cytoskeleton, thus establishing subcellular niches for bacterial colonization and facilitating "stealth and interdiction" of host defense communication lines.

Pathogenicity islands generally increase microbial virulence and are absent in nonpathogenic members of the same genus or species. One specific example is found in *E. coli*. The enteropathogenic *E. coli* strains possess large DNA fragments, 35 to 170 kilobases in size, that contain several virulence genes absent from commensal *E. coli* strains. Some of these genes code for proteins that alter actin microfilaments within a host intestinal cell. As a consequence, the host cell surface bulges and develops into a cuplike pedestal to which the bacterium tightly binds.

1. What seven steps are involved in the infection process and pathogenesis of bacterial diseases?
2. What are some ways in which bacterial pathogens are transmitted to their hosts? Define vector and fomite.
3. Describe several specific adhesins by which bacterial pathogens attach to host cells.
4. Once under the mucus and epithelial surfaces, what are some mechanisms that bacterial pathogens possess to promote their dissemination throughout the body of a host?
5. What are virulence factors? Pathogenicity islands?

33.4 TOXIGENICITY

Two distinct categories of disease can be recognized based on the role of the bacteria in the disease-causing process: infections and intoxications. An infectious disease results partly from the pathogen's growth and reproduction (or invasiveness) that often produce tissue alterations.

Intoxications are diseases that result from a specific toxin (e.g., botulinum toxin) produced by bacteria. Some toxins are only produced during host infection. Toxins can even induce disease in the absence of the organism that produced them. A **toxin** [Latin *toxicum*, poison] is a substance, such as a metabolic product of the organism, that alters the normal metabolism of host cells with deleterious effects on the host. The term **toxemia** refers to the condition caused by toxins that have entered the blood of the host. Some toxins are so potent that even if the bacteria that produced them are eliminated (for instance, by antibiotic chemotherapy), the disease conditions persist. Toxins produced by bacteria can be divided into two main categories: exotoxins and endotoxins. The primary characteristics of the two groups are compared in **table 33.4**.

Exotoxins

Exotoxins are soluble, heat-labile, proteins (a few are enzymes) that usually are released into the surroundings as the bacterial pathogen grows. In general, exotoxins are produced by gram-positive bacteria, although some gram-negative bacteria also make exotoxins. Often exotoxins may travel from the site of infection to other body tissues or target cells in which they exert their effects.

Exotoxins are usually synthesized by specific bacteria that often have plasmids or prophages bearing the toxin genes. They are associated with specific diseases and often are named for the disease they produce (e.g., the diphtheria toxin). Exotoxins are among the most lethal substances known; they are toxic in microgram-per-kilogram concentrations (e.g., botulinum toxin), but are typically heat-labile (inactivated at 60 to 80°C). Exotoxins are proteins that exert their biological activity by specific mechanisms. As proteins, the toxins are highly immunogenic and can stimulate the production of neutralizing antibodies called **antitoxins.** The toxin proteins can also be inactivated by formaldehyde, iodine, and other chemicals to form immunogenic **toxoids** (tetanus toxoid, for example). In fact, the tetanus vaccine is a solution of tetanus toxoid.

Exotoxins can be grouped into four types based on their structure and physiological activities. (1) One type is the AB toxin, which gets its name from the fact that the portion of the toxin (B) that binds to a host cell receptor is separate from the portion (A) that has the enzyme activity that causes the toxicity (**figure 33.5***a*). (2) A second type, which also may be an AB toxin, consists of those toxins that affect a specific host site (nervous tissue [neurotoxins], the intestines [enterotoxins], general tissues [cytotoxins]) by acting extracellularly or intracellularly on the host cells. (3) A third type does not have separable A and B portions and acts by disorganizing host cell membranes. Examples include the leukocidins, hemolysins, and phospholipases. (4) A fourth type is the superantigen that acts by stimulating T cells directly to release cytokines. Examples of these types are now discussed. The general properties of some AB exotoxins are presented in **table 33.5**.

AB Toxins

AB toxins are composed of an enzymatic subunit (A) that is responsible for the toxic effect once inside the host cell and a binding subunit (B) (figure 33.5). Isolated A subunits are enzymatically active but lack binding and cell entry capability, whereas isolated

Table 33.4	Characteristics of Exotoxins and Endotoxins	
Characteristic	Exotoxins	Endotoxins
Chemical composition	Protein, often with two components (A and B)	Lipopolysaccharide complex on outer membrane; lipid A portion is toxic
Disease examples	Botulism, diphtheria, tetanus	Gram-negative infections, meningococcemia
Effect on host	Highly variable between different toxins	Similar for all endotoxins
Fever	Usually do not produce fever	Produce fever by induction of interleukin-1 and TNF
Genetics	Frequently carried by extrachromosomal genes such as plasmids	Synthesized directly by chromosomal genes
Heat stability	Most are heat sensitive and inactivated at 60-80°C	Heat stable to 250°C
Immune response	Antitoxins provide host immunity; highly antigenic	Weakly immunogenic; immunogenicity associated with polysaccharide
Location	Usually excreted outside the living cell	Part of outer membrane of gram-negative bacteria
Production	Produced by both gram-positive and gram-negative bacteria	Found only in gram-negative bacteria; Released on bacterial death and some liberated during growth
Toxicity	Highly toxic and fatal in nanogram quantities	Less potent and less specific than exotoxin; causes septic shock
Toxoid production	Converted to antigenic, nontoxic toxoids; toxoids are used to immunize (e.g., tetanus toxoid)	Toxoids cannot be made

B subunits bind to target cells but are nontoxic and biologically inactive. The B subunit interacts with specific receptors on the target cell or tissue such as the gangliosides GM1 for cholera toxin, GT1 and/or GD1 for tetanus toxin, and SV2 for botulinum toxin. The B subunit therefore determines what cell type the toxin will affect.

Several mechanisms for the entry of A subunits or fragments into target cells have been proposed. In one mechanism the B subunit inserts into the plasma membrane and creates a pore through which the A subunit enters (figure 33.5a). In another mechanism entry is by receptor-mediated endocytosis (figure 33.5b).

The mechanism of action of an AB toxin can be quite complex, as shown by the example of diphtheria toxin (figure 33.5b). The diphtheria toxin is a protein of about 62,000 Daltons. It binds to cell surface receptors by the B subunit and is taken into the cell through the formation of a clathrin-coated vesicle. The toxin then enters the vesicle membrane and the two subunits are separated; the A subunit escapes into the cytosol. The A subunit is an enzyme that catalyzes the addition of an ADP-ribose group to the eucaryotic elongation factor EF2 that aids in translocation during protein synthesis. The substrate for this reaction is the coenzyme NAD^+.

$$NAD^+ + EF2 \rightarrow ADP\text{-ribosyl-}EF2 + nicotinamide$$

The modified EF2 protein cannot participate in the elongation cycle of protein synthesis, and the cell dies because it can no longer synthesize proteins. ADP-ribosylation is a common mechanism for the A subunit of a number of toxins; however, the specific host molecule to which the ADP-ribose group is attached differs. AB exotoxins vary widely in their relative contribution to the disease process with which they are associated.

A variation of this AB toxin is the cytolethal distending toxin (CDT) produced by *Campylobacter* spp. Discovered in 1987, CDT is a tripartite holotoxin complex encoded by three tandem genes; *cdtA, cdtB, and cdtC*. CDT binding and internalization appear to be encoded by the *cdtA* and *cdtC* genes, while the active

component of the holotoxin is encoded within the *cdtB* gene. The predicted amino acid sequence of CdtB is homologous to proteins having deoxyribonuclease (DNase) I activity. However, the mechanism of CDT in disease is unclear. *C. jejuni* has all three genes. In culture with epithelial cells, the CDT of *C. jejuni* induces a progressive epithelial cell distension resulting from an irreversible blockage of the cell cycle at the G2/M phase. This leads to oversized cells without cell division (distension) and cell death.

Specific Host Site Exotoxins

The second type of exotoxin is categorized on the basis of the site affected: **neurotoxins** (nerve tissue), **enterotoxins** (intestinal mucosa), and **cytotoxins** (general tissues). Some of the bacterial pathogens that produce these exotoxins are presented in table 33.5: neurotoxins (botulinum toxin and tetanus toxin), enterotoxins (cholera toxin, *E. coli* heat labile toxins), and cytotoxins (diphtheria toxin, Shiga toxin). Note that many AB toxins are also host site specific, thus these categories are not mutually exclusive.

Neurotoxins usually are ingested as preformed toxins that affect the nervous system and indirectly cause enteric (pertaining to the small intestine) symptoms. Examples include staphylococcal enterotoxin B, *Bacillus cereus* emetic toxin [Greek *emetos*, vomiting], and botulinum toxin.

True enterotoxins [Greek *enter*, intestine] have a direct effect on the intestinal mucosa and elicit profuse fluid secretion (diarrhea). The classic enterotoxin, cholera toxin (choleragen), has been studied extensively. It is an AB toxin. The B subunit is made of five parts arranged as a donut-shaped ring. The B subunit ring anchors itself to the epithelial cell's plasma membrane and then inserts the smaller A subunit into the cell. The A subunit ADP-ribosylates and thereby activates tissue adenylate cyclase to increase intestinal cyclic AMP (cAMP) concentrations. High concentrations of cAMP provoke the movement of massive quantities of water and electrolytes across the intestinal cells into

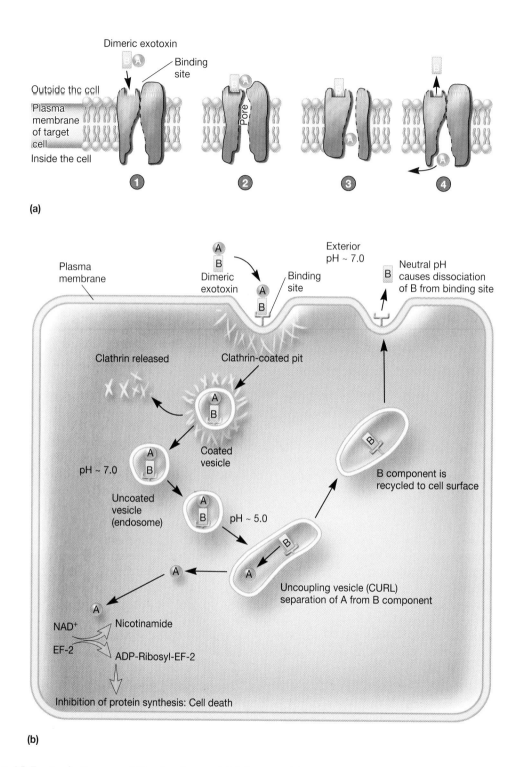

(a)

(b)

Figure 33.5   Two AB Exotoxin Transport Mechanisms   **(a)** Subunit B of the dimeric exotoxin (AB) binds to a specific membrane receptor of a target cell [1]. A conformational change [2] generates a pore [3] through which the A subunit crosses the membrane and enters the cytosol, followed by re-creation [4] of the binding site. **(b)** Receptor-mediated endocytosis of the diphtheria toxin involves the dimeric exotoxin binding to a receptor-ligand complex that is internalized in a clathrin-coated pit that pinches off to become a coated vesicle. The clathrin coat depolymerizes resulting in an uncoated endosome vesicle. The pH in the endosome decreases due to the H^+-ATPase activity. The low pH causes A and B components to separate. An endosome in which this separation occurs is sometimes called a CURL (*c*ompartment of *u*ncoupling of *r*eceptor and *l*igand). The B subunit is then recycled to the cell surface. The A subunit moves through the cytosol, catalyzes the ADP-ribosylation of EF-2 (elongation factor 2) and inhibits protein synthesis, leading to cell death.

Table 33.5 | Properties of Some AB Model Bacterial Exotoxins

Toxin	Organism	Gene Location	Subunit Structure	Target Cell Receptor	Enzymatic Activity	Biologic Effects
Anthrax toxins	*Bacillus anthracis*	Plasmid	Three separate proteins (EF, LF, PA)[a]	Capillary morphogenesis protein 2 (CMP-2) and tumor endothelium marker 8 (TEM8)	EF is a calmodulin-dependent adenylate cyclase; LF is a zinc-dependent protease that cleaves a host signal transduction molecule (MAPKK)	EF + PA: increase in target cell cAMP level, localized edema; LF + PA: altered cell signaling; death of target cells
Bordetella adenylate cyclase toxin	*Bordetella* spp.	Chromosomal	A-B[b]	CR3 intergrin (CD11–CD18)	Calmodulin-activated adenylate cyclase	Increase in target cell cAMP level; decrease ATP production; modified cell function or cell death
Botulinum toxin	*Clostridium botulinum*	Phage	A-B[c]	Synaptic vesicle 2 (SV2)	Zinc-dependent endoprotease cleavage of presynaptic protein (SNARE)	Decrease in peripheral, presynaptic acetylcholine release; flaccid paralysis
Cholera toxin	*Vibro cholera*	Phage	A-5B[d]	Ganglioside (GM$_1$)	ADP ribosylation of adenylate cyclase regulatory protein, G$_S$	Activation of adenylate cyclase, increase in cAMP level; secretory diarrhea
Diphtheria toxin	*Corynebacterium diphtheriae*	Phage	A-B[e]	Heparin-binding, EGF-like growth factor precursor	ADP ribosylation of elongation factor 2	Inhibition of protein synthesis; cell death
Heat-labile enterotoxins[f]	*E. coli*	Plasmid	——————————————— Similar or Identical to Cholera Toxin ———————————————			
Pertussis toxin	*Bordetella pertussis*	Chromosomal	A-5B[g]	Asparagine-linked oligosaccharide and lactosylceramide sequences	ADP ribosylation of signal-transducing G proteins	Block of signal transduction mediated by target G proteins
Pseudomonas exotoxin A	*P. aeruginosa*	Chromosomal	A-B	α$_2$-Macroglobulin/LDL receptor	—— Similar or Identical to Diphtheria Toxin ——	
Shiga toxin	*Shigella dysenteriae*	Chromosomal	A-5B[h]	Globotriaosylceramide (Gb$_3$)	RNA *N*-glycosidase	Inhibition of protein synthesis, cell death
Shiga-like toxin 1	*Shigella* spp., *E. coli*	Phage	——————————————— Similar or Identical to Shiga Toxin ———————————————			
Tetanus toxin	*C. tetani*	Plasmid	A-B[c]	Ganglioside (GT$_1$ and/or GD$_{1b}$)	Zinc-dependent endopeptidase cleavage of synaptobrevin	Decrease in neurotransmitter release from inhibitory neurons; spastic paralysis

Adapted from G. L. Mandell, et al., *Principles and Practice of Infectious Diseases*, 3d edition Copyright © 1990 Churchill-Livingstone, Inc., Medical Publishers, New York, NY. Reprinted by permission.

[a]The binding component (known as protective antigen [PA]) catalyzes/facilitates the entry of either edema factor (EF) or lethal factor (LF).

[b]Apparently synthesized as a single polypeptide with binding and catalytic (adenylate cyclase) domains.

[c]Holotoxin is apparently synthesized as a single polypeptide and cleaved proteolytically as diphtheria toxin; subunits are referred to as L: light chain, A equivalent; H: heavy chain, B equivalent.

[d]The A subunit is proteolytically cleaved into A1 and A2, with A1 possessing the ADP-ribosyl transferase activity; the binding component is made up of five identical B units.

[e]Holotoxin is synthesized as a single polypeptide and cleaved proteolytically into A and B components held together by disulfide bonds.

[f]The heat-labile enterotoxins of *E. coli* are now recognized to be a family of related molecules with identical mechanisms of action.

[g]The binding portion is made up of two dissimilar heterodimers labeled S2-S3 and S2-S4 that are held together by a bridging peptide, SS.

[h]Subunit composition and structure similar to cholera toxin.

the lumen of the gut. To maintain osmotic homeostasis, the cell then releases this water; this results in severe diarrhea (cholera victims can lose 20% of their water per day). The genes for this enterotoxigenicity are encoded on a filamentous phage within *Vibrio cholerae*.   Food-borne and waterborne diseases: Cholera (section 38.4)

Cytotoxins have a specific toxic action upon cells/tissues of special organs and are named according to the type of cell/tissue or organ for which they are specific. Examples include nephrotoxin (kidney), hepatotoxin (liver), and cardiotoxin (heart).

Membrane-Disrupting Exotoxins

The third type of exotoxin lyses host cells by disrupting the integrity of the plasma membrane. There are two subtypes of **membrane-disrupting exotoxins.** The first, is a protein that binds to the cholesterol portion of the host cell plasma membrane, inserts itself into the membrane, and forms a channel (pore) (**figure 33.6***a*). This causes the cytoplasmic contents to leak out. Also, because the osmolality of the cytoplasm is higher than the extracellular fluid, this causes a sudden influx of water into the cell,

(a)

(b)

Figure 33.6   Two Subtypes of Membrane-Disrupting Exotoxins.   (a) A channel-forming (pore-forming) type of exotoxin inserts itself into the normal host cell membrane and makes an open channel (pore). Formation of multiple pores causes cytoplasmic contents to leave the cell and water to move in, leading to cellular lysis and death of the host cell. **(b)** A phospholipid-hydrolyzing phospholipase exotoxin destroys membrane integrity. The exotoxin removes the charged polar head groups from the phospholipid part of the host cell membrane. This destabilizes the membrane and causes the host cell to lyse.

causing it to swell and rupture. Two specific examples of this type of membrane-disrupting exotoxin are now presented.

Some pathogens produce membrane-disrupting toxins that kill phagocytic leukocytes. These are termed **leukocidins** [*leukocyte* and Latin *caedere,* to kill]. Most leukocidins are produced by pneumococci, streptococci, and staphylococci. Since the pore-forming exotoxin produced by these bacteria destroys leukocytes, this in turn decreases host resistance. Other toxins, called **hemolysins** [*haima,* blood, and Greek *lysis,* dissolution], also can be secreted by pathogenic bacteria. Many hemolysins probably form pores in the plasma membrane of erythrocytes through which hemoglobin and/or ions are released (the erythrocytes lyse or, more specifically, hemolyze). **Streptolysin-O (SLO)** is a hemolysin, produced by *Streptococcus pyogenes,* that is inactivated by O_2 (hence the "O" in its name). SLO causes beta hemolysis of erythrocytes on agar plates incubated anaerobically. A complete zone of clearing around the bacterial colony growing on blood agar is called **beta hemolysis,** and a partial clearing of the blood (leaving a greenish halo of hemoglobin) is called **alpha hemolysis. Streptolysin-S (SLS)** is also produced by *S. pyogenes* but is insoluble and bound to the bacterial cell. It is O_2 stable (hence the "S" in its name) and causes beta hemolysis on aerobically incubated blood-agar plates. In addition to hemolysins, SLO and SLS are also leukocidins and kill leukocytes. It should also be noted that hemolysins attack the plasma membranes of many cells, not just erythrocytes and leukocytes.

The second subtype of membrane-disrupting toxins are the **phospholipase** enzymes. Phospholipases remove the charged head group (figure 33.6*b*) from the lipid portion of the phospholipids in the host-cell plasma membrane. This destabilizes the membrane so that the cell lyses and dies. One example of the pathogenesis caused by phospholipases is observed in the disease gas gangrene. In this disease, the *Clostridium perfringens* α-toxin almost completely destroys the local population of white blood cells (that are drawn in by inflammation to fight the infection) through phospholipase activity.

Superantigens

As discussed in chapter 32, superantigens are bacterial and viral proteins that can provoke as many as 30% of a person's T cells to release massive concentrations of cytokines. The best-studied superantigen is also a staphylococcal enterotoxin. Staphylococcal entertoxin B (SEB) exhibits biological activity as a superantigen at nanogram concentrations and is therefore classified as a select agent; it has the potential to be misused as a bioterror agent. SEB exerts its superantigen activity by bridging the unfilled class II MHC molecules of antigen-presenting cells to T-cell receptors. Because no processed antigen is involved, many T cells are activated at once. This activation of T cells results in normal cytokine release; however, the sum total of the combined cytokines overwhelms cells and tissues. Cytokines stimulate endothelial damage, circulatory shock, and multiorgan failure.   T-cell biology: Superantigens (section 32.5)

Roles of Exotoxins in Disease

Humans are exposed to bacterial exotoxins in three main ways: (1) ingestion of preformed exotoxin, (2) colonization of a mucosal surface followed by exotoxin production, and (3) colonization of a wound or abscess followed by local exotoxin production. Each of these is now briefly discussed.

Figure 33.7   Roles of Exotoxins in Disease Pathogenesis. Three ways **(a, b, c)** in which bacterial exotoxins can contribute to the progression of disease in a human.

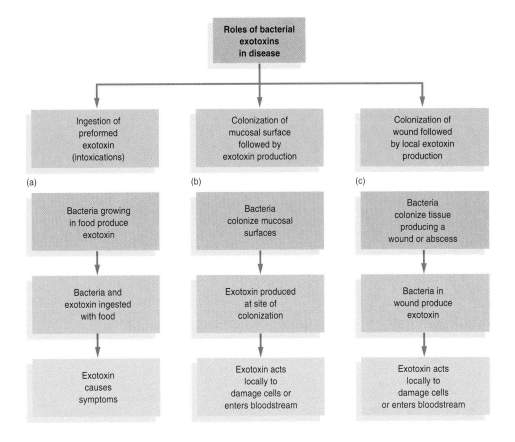

In the first example (**figure 33.7***a*), the exotoxin is produced by bacteria growing in food. When food is consumed, the preformed exotoxin is also consumed. The classical example is staphylococcal food poisoning caused solely by the ingestion of preformed enterotoxin. Since the bacteria (*Staphylococcus aureus*) cannot colonize the gut, they pass through the body without producing any more exotoxin; thus, this type of bacterial disease is self-limiting.

In the second example (figure 33.7*b*), bacteria colonize a mucosal surface but do not invade underlying tissue or enter the bloodstream. The toxin either causes disease locally or enters the bloodstream and is distributed systemically where it can cause disease at distant sites. The classical example here is the disease cholera caused by *Vibrio cholerae*. Once the bacteria enter the body, they adhere to the intestinal mucosa. They are not invasive but secrete the cholera toxin. As a result, cholera toxin stimulates hypersecretion of water and chloride ions and the patient loses massive quantities of water through the gastrointestinal tract.

The third example of exotoxins in disease pathogenesis occurs when bacteria grow in a wound or abscess (figure 33.7*c*). The exotoxin causes local tissue damage or kills phagocytes that enter the infected area. A disease of this type is gas gangrene in which the exotoxin (α-toxin) of *Clostridium perfringens* lyses red blood cells, induces edema, and causes tissue destruction in the wound.

---

1. What is the difference between an infectious disease and an intoxication? Define toxemia.
2. Describe some general characteristics of exotoxins.
3. How do exotoxins get into host cells?
4. Describe the biological effects of several bacterial exotoxins.

5. Discuss the mechanisms by which exotoxins can damage cells.
6. What are the four types of exotoxins?
7. What is the mode of action of a leukocidin? Of a hemolysin?
8. Name two specific hemolysins.
9. What are the three main roles exotoxins have in human disease pathogenesis?

Endotoxins

Gram-negative bacteria have lipopolysaccharide (LPS) in the outer membrane of their cell wall that, under certain circumstances, is toxic to specific hosts. This LPS is called an **endotoxin** because it is bound to the bacterium and is released when the microorganism lyses (**Techniques & Applications 33.1**). Some is also released during bacterial multiplication. The toxic component of the LPS is the lipid portion, called lipid A. Lipid A is not a single macromolecular structure but appears to be a complex array of lipid residues. The lipid A component exhibits all the properties associated with endotoxicity and gram-negative bacteremia. The bacterial cell wall: Gram-negative cell walls (section 3.6)

Besides the preceding characteristics, bacterial endotoxins are

1. Heat stable
2. Toxic (nanogram amounts)
3. Weakly immunogenic
4. Generally similar, despite source
5. Usually capable of producing general systematic effects: fever (are pyrogenic), shock, blood coagulation, weakness, diarrhea, inflammation, intestinal hemorrhage, and fibrinolysis (enzymatic breakdown of fibrin, the major protein component of blood clots)

Techniques & Applications

33.1 Detection and Removal of Endotoxins

Bacterial endotoxins plagued the pharmaceutical industry and medical device producers for years. For example, administration of drugs contaminated with endotoxins resulted in complications—even death—to patients. In addition, endotoxins can be problematic for individuals and firms working with cell cultures and genetic engineering. The result has been the development of sensitive tests and methods to identify and remove these endotoxins. The procedures must be very sensitive to trace amounts of endotoxins. Most firms have set a limit of 0.25 **endotoxin units (E.U.),** 0.025 ng/ml, or less as a release standard for their drugs, media, or products.

One of the most accurate tests for endotoxins is the in vitro *Limulus* amoebocyte lysate (LAL) assay. The assay is based on the observation that when an endotoxin contacts the clot protein from circulating amoebocytes of the horseshoe crab (*Limulus*), a gel-clot forms. The assay kits available today contain calcium, proclotting enzyme, and procoagulogen. The proclotting enzyme is activated by bacterial endotoxin (lipopolysaccharide) and calcium to form active clotting enzyme (see **Box figure**). Active clotting enzyme then catalyzes the cleavage of procoagulogen into polypeptide subunits (coagulogen). The subunits join by disulfide bonds to form a gel-clot. Spectrophotometry is then used to measure the protein precipitated by the lysate. The LAL test is sensitive at the nanogram level but must be standardized against Food and Drug Administration Bureau of Biologics endotoxin reference standards. Results are reported in endotoxin units per milliliter and reference made to the particular reference standards used.

Removal of endotoxins presents more of a problem than their detection. Those present on glassware or medical devices can be inactivated if the equipment is heated at 250°C for 30 minutes. Soluble endotoxins range in size from 20 kDa to large aggregates with diameters up to 0.1 μm. Thus they cannot be removed by conventional filtration systems. Manufacturers have developed special filtration systems and filtration cartridges that retain these endotoxins and help alleviate contamination problems.

The characteristics of endotoxins and exotoxins are contrasted in table 33.4. The main biological effect of lipid A is an indirect one, being mediated by host molecules and systems rather than by lipid A directly. For example, endotoxins can initially activate Hageman Factor (blood clotting factor XII), which in turn activates up to four humoral systems: coagulation, complement, fibrinolytic, and kininogen systems (*see figure 38.26*).

Gram-negative endotoxins also indirectly induce a fever in the host by causing macrophages to release **endogenous pyrogens** that reset the hypothalamic thermostat. One important endogenous pyrogen is the cytokine interleukin-1 (IL-1). Other cytokines released by macrophages, such as the tumor necrosis factor, also produce fever.

Evidence indicates that LPS affects macrophages, monocytes, and neutrophils by binding to the soluble pattern-recognition receptor (formerly LPS-binding protein) for transfer to the membrane-bound CD14 on these cells. LPS-bound CD14 then complexes with toll-like receptor (TLR) 4 to initiate a signaling process that upregulates the phagocyte response to LPS. Part of this response is the synthesis and release of cytokines IL-1, IL-6, IL-8, tumor necrosis factor α, and platelet-activating factor. These and other pro-inflammatory mediators signal target cells resulting in fever, complement activation, prostaglandin synthesis, and activation of the coagulation cascade. Phagocytosis: Toll-like receptors (section 31.3); Chemical mediators in nonspecific resistance: Cytokines (section 31.6)

1. Describe the chemical structure of the LPS endotoxin.
2. List some general characteristics of endotoxins.
3. How do gram-negative endotoxins induce fever in a mammalian host?

33.5 HOST DEFENSE AGAINST MICROBIAL INVASION

In host-pathogen relationships, the balance between host integrity and resource utilization results in the co-evolution of survival strategies (**figure 33.8**). Competition, harsh environments, and cellular biowarfare have demanded unique solutions from host and pathogen alike. The complex, albeit "sneaky," methods employed by microbes to gain access to host resources have invariably been met with equally complex countermeasures. The host-pathogen relationship is indeed defined by ecological principles. The host provides a myriad of niches for microbes that have adapted to various temperature optima, nutrient content, oxygen concentration, and tolerance for host, as well as other microorganisms (antagonism).

Chapters 31 and 32 detail the innate and adaptive responses, respectively, available to the human host in preventing or limiting infection. A variety of physical, chemical, and biological barriers establish a formidable defense against microbial invasion.

Figure 33.8   The Balance Between Microorganism Activity and Host Immunity. The interaction between the two determines the final host-parasite balance.

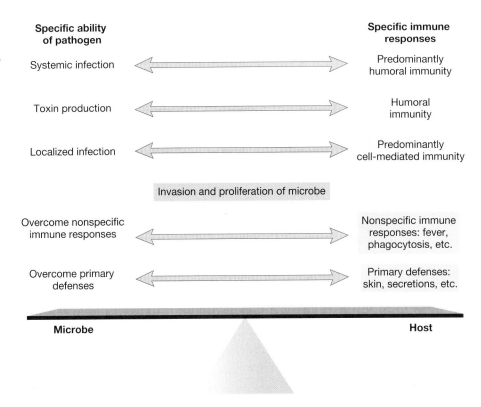

Specific ability of pathogen		Specific immune responses
Systemic infection		Predominantly humoral immunity
Toxin production		Humoral immunity
Localized infection		Predominantly cell-mediated immunity
Invasion and proliferation of microbe		
Overcome nonspecific immune responses		Nonspecific immune responses: fever, phagocytosis, etc.
Overcome primary defenses		Primary defenses: skin, secretions, etc.

Microbe   **Host**

Nonetheless, microorganisms often breech these sophisticated barriers, occasionally prevailing. However, more recent evidence suggests that host cells are equipped with sensitive inter- and intracellular surveillance systems that recognize unique pathogen-associated patterns. As in medical diagnostics, early immune detection usually leads to early clearance of disease. Coupled with the ability to produce highly specific receptors and antibodies, the host prevents many more microbial invasions than those that are noticed.

Primary Defenses

Recall that the host has evolved several strategies to prevent pathogen entry. This is intuitively the most logical mechanism to develop. In other words, there would be no need for secondary defenses if the primary systems were 100% effective. An impenetrable suit of armor might restrict microbial entry, but limits other essential functions. Thus a multilayered skin, speckled with glands producing antimicrobial substances and an army of formidable microbial allies, is a sound compromise. Mucous membranes with cilia, pH regulation, flushing mechanisms, and additional antimicrobial products are efficient transition sites where the host tissues interface with their environment.

Secondary Defenses

Because the defenses of the skin and mucous membranes can be overcome, a secondary system has evolved. Composed of soluble antimicrobial products and cells capable of sensing and responding to invading microbes, the secondary system is quite effective

in controlling infection. The host blood and interstitial fluids contain evolutionarily conserved, antimicrobial proteins that are very efficacious. These proteins include a variety of low-molecular-weight, "pore-forming" peptides and the ubiquitous lysozyme. Additionally, host cells have evolved strategies to exploit pathogen sensitivities to toxic oxygen radicals. Finally, sophisticated, soluble receptors police the host in search of microbial ligands. Binding of such ligands to their receptors initiates processes designed to amplify microbial detection and destruction. Examples of these processes include complement activation, inflammation, fever, and phagocytosis.

Factors Influencing Host Defenses

There are a variety of factors that influence the primary and secondary immune responses of the host. Age, stress, nutritional deficiencies, and genetic background all play substantial roles in host defense. The very young and the very old tend to be more susceptible to diseases. Some individuals are inherently (genetically) more resistant or more susceptible to particular diseases. Nutrition also influences these factors. In fact, historians have discovered a link between times of famine and times of disease. For example, nutritional deficiencies can decrease epithelial integrity, weaken antibody responses, and facilitate changes in the normal flora. Stress too can inhibit the immune response by stimulating the production of corticosteroids, which depress immune responses. Lastly, long-term exposure to environmental pollutants, drug abuse, or certain prescribed medicines may also inhibit normal host defenses against infection.

33.6 MICROBIAL MECHANISMS FOR ESCAPING HOST DEFENSES

So far, we have discussed some of the ways viral and bacterial pathogens cause disease in a host. During the course of microbe and human evolution, these same pathogens have evolved ways for evading host defenses. Many of these mechanisms are found throughout the microbial world and several are now discussed.

Evasion of Host Defenses by Viruses

As noted earlier in this chapter, the pathology arising from a viral infection is due to either (1) the host's immune response, which attacks virus-infected cells or produces hypersensitivity reactions, or (2) the direct consequence of viral multiplication within host cells. Viruses have evolved a variety of ways to suppress or evade the host's immune response. These mechanisms are now becoming recognized through genomics and the functional analysis of specific gene products.   Immune disorders: Hypersensitivities (section 32.11)

Some viruses may mutate and change antigenic sites (antigenic drift) on the virion proteins (e.g., the influenza virus) or may down-regulate the level of expression of viral cell surface proteins (e.g., the herpesvirus). Other viruses (HIV) may infect cells (T cells) of the immune system and diminish their function. HIV as well as the measles virus and cytomegalovirus cause the fusion of host cells. This allows these viruses to move from an infected cell to an uninfected cell without exposure to the antibody-containing fluids of the host. The herpesvirus may infect neurons that express little or no major histocompatibility complex molecules. The adenovirus produces proteins that inhibit major histocompatibility complex function. Finally, hepatitis B virus-infected cells produce large amounts of antigens not associated with the complete virus. These antigens bind the available neutralizing antibody so that there is insufficient free antibody to bind with the complete virion.   Airborne diseases: Influenza (section 37.1); Recognition of foreignness (section 32.4)

Evasion of Host Defenses by Bacteria

Bacteria also have evolved many mechanisms to evade host defenses. Because bacteria would not be well served either by the death of their host or their own death, their survival strategy is protection against host defenses rather than host destruction.

Evading the Complement System

To evade the activity of complement, some bacteria have capsules (see chapter opening figure) that prevent complement activation. Some gram-negative bacteria can lengthen the O chains in their lipopolysaccharide to prevent complement activation. Others such as *Neisseria gonorrhoeae* generate **serum resistance.** These bacteria have modified lipooligosaccharides on their surface that interfere with proper formation of the membrane attack complex (*see figure 31.23*) during the complement cascade. The virulent forms of *N. gonorrhoeae* that possess serum resistance are able to spread throughout the body of the host and cause systemic disease, whereas those *N. gonorrhoeae* that lack serum resistance remain localized in the genital tract.   Chemical mediators in nonspecific resistance: Complement (section 31.6)

Resisting Phagocytosis

As noted previously, before a phagocytic cell can engulf a bacterium, it must first directly contact the bacterium's surface. Some bacteria such as *Streptococcus pneumoniae, Neisseria meningitidis,* and *Haemophilus influenzae* can produce a slippery mucoid capsule that prevents the phagocyte from effectively contacting the bacterium. Other bacteria evade phagocytosis by producing specialized surface proteins such as the M protein on *S. pyogenes*. Like capsules, these proteins interfere with adherence between a phagocytic cell and the bacterium.

Bacterial pathogens use other mechanisms to resist phagocytosis. For example, *Staphylococcus* produces leukocidins that destroy phagocytes before phagocytosis can occur. *S. pyogenes* releases a protease that cleaves the C5a complement factor and thus inhibits complement's ability to attract phagocytes to the infected area.

Survival Inside Phagocytic Cells

Some bacteria have evolved the ability to survive inside neutrophils, monocytes, and macrophages. They are very pathogenic because they are impervious to a most important host protective mechanism. One method of evasion is to escape from the phagosome before it merges with the lysosome, as seen with *Listeria monocytogenes, Shigella,* and *Rickettsia*. These bacteria use actin-based motility to move within mammalian host cells and spread between them. Upon lysing the phagosome, they gain access to the cytoplasm. Each bacterium then recruits to its surface host cell actin and other cytoskeletal proteins and activates the assembly of an actin tail (**figure 33.9a**). The actin tails propel the bacteria through the cytoplasm of the infected cell to its surface where they push out against the plasma membrane and form protrusions (figure 33.9b). The protrusions are engulfed by adjacent cells, and the bacteria once again enter phagosomes and escape into the cytoplasm. In this way the infection spreads to adjacent cells. The lysosomes never have a chance to merge with the phagosomes. Another approach is to resist the toxic products released into the phagolysosome after fusion occurs. A good example of a bacterium that is resistant to the lysosomal enzymes is *Mycobacterium tuberculosis*, probably at least partly because of its waxy external layer. Still other bacteria prevent fusion of phagosomes with lysosomes (e.g., *Chlamydia*).   Phagocytosis (section 31.3)

Evading the Specific Immune Response

To evade the specific immune response, some bacteria (e.g., *S. pyogenes*) produce capsules that are not antigenic because they resemble host tissue components. *N. gonorrhoeae* can evade the specific immune response by two mechanisms: (1) it makes genetic variations in its pili (phase variation) so that specific antibodies are useless against the new pili and adherence to host tissue occurs, and (2) it produces IgA proteases that destroy secretory IgA and allow adherence. Finally, some bacteria produce proteins (such as staphylococcal protein A and protein G of *S. pyogenes*) that interfere with antibody-mediated opsonization by binding to the Fc portion of immunoglobulins.

1. What are some mechanisms viruses use to evade host defenses?
2. How do bacteria evade each of the following host defenses: the complement system, phagocytosis, and the specific immune response?

Figure 33.9   Formation of Actin Tails by Intracellular Bacterial Pathogens.   (a) Transmission electron micrograph of *Lysteria monocytogenes* in a host macrophage. The bacterium has polymerized host actin into a long tail that it uses for intracellular propulsion and to move from one host cell to another. **(b)** *Burkholderia pseudomallei* (stained red) also forms actin tails (stained dark green) as shown in this confocal micrograph. Note that the actin tails enable the bacterial cells to be propelled out of the host cell.

(a)

(b)

Summary

33.1 Host-Parasite Relationships

a. Parasitism is a type of symbiosis between two species in which the smaller organism is physiologically dependent on the larger one, termed the host. The parasitic organism usually harms its host in some way.

b. An infection is the colonization of the host by a parasitic organism. An infectious disease is the result of the interaction between the parasitic organism and its host, causing the host to change from a state of health to one of a diseased state. Any organism that produces such a disease is a pathogen (**figure 33.1**).

c. Pathogenicity refers to the quality or ability of an organism to produce pathological changes or disease. Virulence refers to the degree or intensity of pathogenicity of an organism and is measured experimentally by the LD_{50} or ID_{50} (**figure 33.2**).

33.2 Pathogenesis of Viral Diseases

a. The fundamental process of viral infection is the expression of the viral replicative cycle in a host cell. To produce disease a virus enters a host, comes into contact with susceptible cells, and reproduces.

b. Viruses spread to adjacent cells when they are released by either host cell lysis or budding.

c. Host cell damage caused by viruses stimulates a host immune response involving neutralizing antibodies for free virions and activated killer cells for intracellular viruses.

d. Recovery from infection results when the virus has either been cleared from the body of the host, establishes a persistent infection, or kills the host.

e. The viral infection cycle is complete when the virus is shed back into the environment, to be acquired by another host.

33.3 Overiew of Bacterial Pathogenesis

a. Pathogens or their products can be transmitted to a host by either direct or indirect means. Transmissibility is the initial requisite in the establishment of an infectious disease.

b. Special adherence factors allow pathogens to bind to specific receptor sites on host cells and colonize the host (**table 33.2** and **figure 33.3**).

c. Pathogens can enter host cells by both active and passive mechanisms. Once inside, they can produce specific products and/or enzymes that promote dissemination throughout the body of the host. These are termed virulence factors (**table 33.3**).

d. The pathogen generally is found in the area of the host's body that provides the most favorable conditions for its growth and multiplication.

e. During coevolution with human hosts, some pathogenic bacteria have evolved complex signal transduction pathways to regulate the genes necessary for virulence.

f. Many bacteria are pathogenic because they have large segments of DNA called pathogenicity islands that carry genes responsible for virulence.

33.4 Toxigenicity

a. Intoxications are diseases that result from the entrance of a specific toxin into a host. The toxin can induce the disease in the absence of the toxin-producing organism. Toxins produced by pathogens can be divided into two main categories: exotoxins and endotoxins (**table 33.4**).

b. Exotoxins are soluble, heat-labile, potent, toxic proteins produced by the pathogen. They have very specific effects and can be categorized as neurotoxins, cytotoxins, or enterotoxins. Most exotoxins conform to the AB model in which the A subunit or fragment is enzymatic and the B subunit or fragment, the binding portion (**table 33.5**). Several mechanisms exist by which the A component enters target cells (**figure 33.5**).

c. Exotoxins can be divided into four types: (1) the AB toxins, (2) specific host site toxins (neurotoxins, enterotoxins, cytotoxins), (3) toxins that disrupt plasma membranes of host cells (leukocidins, hemolysins, and phospholipases), and (4) superantigens.

d. Bacterial exotoxins cause disease in a human host in three main ways: (1) ingestion of preformed exotoxin, (2) colonization of a mucosal surface followed

Table 34.3 | **Inhibition Zone Diameter of Selected Chemotherapeutic Drugs**

Chemotherapeutic Drug	Disk Content	Zone Diameter (Nearest mm)		
		Resistant	Intermediate	Susceptible
Carbenicillin (with *Proteus* spp. and *E. coli*)	100 μg	≤17	18–22	≥23
Carbenicillin (with *Pseudomonas aeruginosa*)	100 μg	≤13	14–16	≥17
Ceftriaxone	30 μg	≤13	14–20	≥21
Chloramphenicol	30 μg	≤12	13–17	≥18
Erythromycin	15 μg	≤13	14–17	≥18
Penicillin G (with staphylococci)	10 U[a]	≤20	21–28	≥29
Penicillin G (with other microorganisms)	10 U	≤11	12–21	≥22
Streptomycin	10 μg	≤11	12–14	≥15
Sulfonamides	250 or 300 μg	≤12	13–16	≥17
Tetracycline	30 μg	≤14	15–18	≥19

[a]One milligram of penicillin G sodium = 1,600 units (U).

Figure 34.3   Interpretation of Kirby-Bauer Test Results.
The relationship between the minimal inhibitory concentrations of a
hypothetical drug and the size of the zone around a disk in which
microbial growth is inhibited. As the sensitivity of microorganisms to
the drug increases, the MIC value decreases and the inhibition zone
grows larger. Suppose that this drug varies from 7–28 μg/ml in the
body during treatment. Dashed line A shows that any pathogen with
a zone of inhibition less than 12 mm in diameter will have an MIC
value greater than 28 μg/ml and will be resistant to drug treatment.
A pathogen with a zone diameter greater than 17 mm will have an
MIC less than 7 μg/ml and will be sensitive to the drug (see line B).
Zone diameters between 12 and 17 mm indicate intermediate
sensitivity and usually signify resistance.

Figure 34.4   The Etest®.   An example of a bacterial culture plate
with Etest® strips arranged radially on it. The strips are arranged so
that the lowest antibiotic concentration in each is at the center. The
MIC concentration is read from the scale at the point it intersects the
zone of inhibition as shown by the arrow in this example. Etest® is a
registered trademark of AB BIODISK and patented in all major
markets.

Figure 34.5 Penicillins. The structures and characteristics of representative penicillins. All are derivatives of 6-aminopenicillanic acid; in each case the shaded portion of penicillin G is replaced by the side chain indicated. The β-lactam ring is also shaded (blue), and an arrow points to the bond that is hydrolyzed by penicillinase.

6-aminopenicillanic acid

Penicillin G

High activity against most gram-positive bacteria, low against gram negative; destroyed by acid and penicillinase

Penicillinases attack here on the β-lactam ring

Penicillin V

Same spectrum but more acid resistant than penicillin G

Ampicillin

Active against gram-positive and gram-negative bacteria; acid stable

Carbenicillin

Active against gram-negative bacteria like *Pseudomonas* and *Proteus*; acid stable; not well absorbed by small intestine

Methicillin

Penicillinase-resistant, but less active than penicillin G; acid-labile

Ticarcillin

Similar to carbenicillin, but more active against *Pseudomonas*

respect to the side chain attached to its amino group (**figure 34.5**). The most crucial feature of the molecule is the **β-lactam ring,** which is essential for bioactivity. Many penicillin-resistant bacteria produce **penicillinase** (also called **β-lactamase**), an enzyme that inactivates the antibiotic by hydrolyzing a bond in the β-lactam ring.

Although the complete mechanism of action of penicillins is still not completely known, their structures resemble the terminal D-alanyl-D-alanine found on the peptide side chain of the peptidoglycan subunit. It has been proposed that this structural similarity blocks the enzyme catalyzing the transpeptidation reaction that forms the peptidoglycan cross-links (*see figure 10.12*). Thus formation of a complete cell wall is blocked, leading to osmotic lysis. This mechanism is consistent with the observation that penicillins act only on growing bacteria that are synthesizing new peptidoglycan.

Evidence has indicated that the mechanism of penicillin action is even more complex than previously imagined. It has been discovered that penicillins bind to several periplasmic proteins (penicillin-binding proteins, or PBPs) and may also destroy bacteria by activating their own autolytic enzymes. However, there is also some evidence that penicillin kills bacteria even in the absence of autolysins or murein hydrolases. Lysis could occur after bacterial viability has already been lost. Penicillin may stimulate special proteins called bacterial holins to form holes or lesions in the plasma membrane, leading directly to membrane leakage and death. Murein hydrolases also could move through the holes, disrupt the peptidoglycan, and lyse the cell.

Penicillins differ from each other in several ways. The two naturally occurring penicillins, penicillin G and penicillin V, are narrow-spectrum drugs. Penicillin G is effective against gonococci, meningococci, and several gram-positive pathogens such as streptococci and staphylococci. However, it must be administered by injection (parenterally) because it is destroyed by stomach acid. Penicillin V (figure 34.5) is similar to penicillin G in spectrum of activity, but can be given orally because it is more resistant to acid. The semisynthetic penicillins, on the other hand, have a broader spectrum of activity. Ampicillin can be administered orally and is effective against gram-negative bacteria such as *Haemophilus, Salmonella,* and *Shigella.* Carbenicillin and ticarcillin are potent against *Pseudomonas* and *Proteus.*

An increasing number of bacteria have become resistant to natural penicillins and many of the semisynthetic analogs. Physicians frequently employ specific semisynthetic penicillins that are not destroyed by β-lactamases to combat antibiotic-resistant pathogens. These include methicillin (figure 34.5), nafcillin, and oxacillin. However, this practice has been confounded by the emergence of meticillin-resistant bacteria.

Although penicillins are the least toxic of the antibiotics, about 1 to 5% of the adults in the United States are allergic to them. Occasionally, a person will die of a violent allergic response; therefore, patients should be questioned about penicillin allergies before treatment is begun. ▶ Immune disorders: Hypersensitivities (section 32.11)

Cephalosporins

Cephalosporins are a family of antibiotics originally isolated in 1948 from the fungus *Cephalosporium.* They contain a **β-lactam** structure that is very similar to that of the penicillins (**figure 34.6**). As might be expected from their structural similarities to penicillins, cephalosporins also inhibit the transpeptidation reaction during peptidoglycan synthesis. They are broad-spectrum drugs frequently given to patients with penicillin allergies (although about 10% of patients allergic to penicillin are also allergic to cephalosporins).

Many cephalosporins are in use. Cephalosporins are broadly categorized into four generations (groups of drugs that are sequentially developed) based on their spectrum of activity. First-generation cephalosporins are more effective against gram-positive pathogens than gram-negatives. Second-generation drugs, developed after the first generation, have improved effects on gram-

Figure 34.6 Cephalosporin Antibiotics. These drugs are derivatives of 7-aminocephalosporanic acid and contain a β-lactam ring.

negative bacteria with some anaerobe coverage. Third-generation drugs are particularly effective against gram-negative pathogens, and some reach the central nervous system. This is of particular note because many antimicrobial agents do not cross the blood-brain barrier. Finally, fourth-generation cephalosporins are broad spectrum with excellent gram-positive and gram-negative coverage and, like their third-generation predecessors, inhibit the growth of the difficult opportunistic pathogen *Pseudomonas aeruginosa.*

Vancomycin and Teicoplanin

Vancomycin is a glycopeptide antibiotic produced by *Streptomyces oreintalis.* It is a cup-shaped molecule composed of a peptide linked to a disaccharide. Vancomycin's peptide portion blocks the transpeptidation reaction by binding specifically to the D-alanine-D-alanine terminal sequence on the pentapeptide portion of peptidoglycan. The antibiotic is bactericidal for *Staphylococcus* and some members of the genera *Clostridium, Bacillus, Streptococcus,* and *Enterococcus.* It is given both orally and intravenously and has been particularly important in the treatment of antibiotic-resistant staphylococcal and enterococcal infections. However, vancomycin-resistant strains of *Enterococcus* have become widespread and cases of resistant *Staphylococcus aureus* have appeared. This poses a serious public health threat—vancomycin has been considered the "drug of last resort" in cases of antibiotic-resistant *S. aureus.* Clearly newer drugs must be developed.

Teicoplanin is a glycopeptide antibiotic from the actinomycete *Actinoplanes teichomyceticus* that is similar in structure and mechanism of action to vancomycin, but has fewer side effects. It is active against staphylococci, enterococci, streptococci, clostridia, *Listeria,* and many other gram-positive pathogens.

Protein Synthesis Inhibitors

Many antibiotics inhibit protein synthesis by binding with the procaryotic ribosome. Because these drugs discriminate between procaryotic and eucaryotic ribosomes, their therapeutic index is fairly high, but not as high as that of cell wall inhibitors. Some drugs bind to the 30S (small) ribosomal subunit, while others attach to the 50S (large) subunit. Several different steps in protein synthesis can be affected: aminoacyl-tRNA binding, peptide bond formation, mRNA reading, and translocation. Translation (section 11.8)

Aminoglycosides

Although there is considerable variation in structure among several important **aminoglycoside antibiotics,** all contain a cyclohexane ring and amino sugars (**figure 34.7**). **Streptomycin,** kanamycin, neomycin, and tobramycin are synthesized by different species of the genus *Streptomyces,* whereas gentamicin comes from another actinomycete, *Micromonospora purpurea.* Aminoglycosides bind to the 30S (small) ribosomal subunit and interfere with protein synthesis by directly inhibiting the synthesis process and also by causing misreading of the mRNA.

These antibiotics are bactericidal and tend to be most effective against gram-negative pathogens. Streptomycin's usefulness has decreased greatly due to widespread drug resistance, but it is still

Figure 34.7   Representative Aminoglycoside Antibiotics.

Figure 34.8   Tetracyclines.   Three members of the tetracycline family. Tetracycline lacks both of the groups that are shaded. Chlortetracycline (aureomycin) differs from tetracycline in having a chlorine atom (blue); doxycycline consists of tetracycline with an extra hydroxyl (light blue).

effective in treating tuberculosis and plague. Gentamicin is used to treat *Proteus, Escherichia, Klebsiella,* and *Seratia* infections. Aminoglycosides can be quite toxic, however, and can cause deafness, renal damage, loss of balance, nausea, and allergic responses.

Tetracyclines

The **tetracyclines** are a family of antibiotics with a common four-ring structure to which a variety of side chains are attached (**figure 34.8**). Oxytetracycline and chlortetracycline are produced naturally by *Streptomyces* species while others are semisynthetic drugs. These antibiotics are similar to the aminoglycosides and combine with the 30S (small) subunit of the ribosome. This inhibits the binding of aminoacyl-tRNA molecules to the A site of the ribosome. Because their action is only bacteriostatic, the effectiveness of treatment depends on active host resistance to the pathogen.

Tetracyclines are broad-spectrum antibiotics that are active against gram-negative, as well as gram-positive, bacteria, rickettsias, chlamydiae, and mycoplasmas. High doses may result in nausea, diarrhea, yellowing of teeth in children, and damage to the liver and kidneys. Although their use has declined in recent years, they are still sometimes used to treat acne.

Macrolides

The **macrolide antibiotics** contain 12- to 22-carbon lactone rings linked to one or more sugars (**figure 34.9**). **Erythromycin** is usually bacteriostatic and binds to the 23S rRNA of the 50S (large) ribosomal subunit to inhibit peptide chain elongation during protein synthesis. Erythromycin is a relatively broad-spectrum antibiotic effective against gram-positive bacteria, mycoplasmas, and a few gram-negative bacteria. It is used with patients who are allergic to penicillins and in the treatment of whooping cough, diphtheria, diarrhea caused by *Campylobacter,* and pneumonia from *Legionella* or *Mycoplasma* infections. Clindamycin is effective against a variety of bacteria including staphylococci, and anaerobes such as *Bacteroides*. Azithromycin, which has surpassed erythromycin in use, is particularly effective against *Chlamydia trachomatis.*

Chloramphenicol

Chloramphenicol (**figure 34.10**) was first produced from cultures of *Streptomyces venezuelae* but it is now synthesized chemically. Like erythromycin, this antibiotic binds to 23S rRNA on the 50S ribosomal subunit to inhibit the peptidyl transferase re-

action. It has a very broad spectrum of activity but, unfortunately, is quite toxic. One may see allergic responses or neurotoxic reactions. The most common side effect is depression of bone marrow function, leading to aplastic anemia and a decreased number of white blood cells. Consequently, this antibiotic is used only in life-threatening situations when no other drug is adequate.

Metabolic Antagonists

Several valuable drugs act as **antimetabolites**—they antagonize, or block, the functioning of metabolic pathways by competitively inhibiting the use of metabolites by key enzymes. These drugs can act as structural analogs, molecules that are structurally similar to naturally occurring metabolic intermediates. These analogs compete with intermediates in metabolic processes because of their similarity, but are just different enough so that they prevent normal cellular metabolism. As such they are bacteriostatic but broad spectrum.

Sulfonamides or Sulfa Drugs

The first antimetabolites to be used successfully as chemotherapeutic agents were the sulfonamides, discovered by G. Domagk. **Sulfonamides,** or sulfa drugs, are structurally related to sulfanilamide, an analog of *p*-aminobenzoic acid, or PABA (**figures 34.11** and **34.12**). PABA is used in the synthesis of the cofactor folic acid (folate). When sulfanilamide or another sulfonamide enters a bacterial cell, it competes with PABA for the active site of an enzyme involved in folic acid synthesis, causing a decline in folate concentration. This decline is detrimental to the bacterium because folic acid is a precursor of purines and pyrimidines, the bases used in the construction of DNA, RNA, and other important cell constituents. The resulting inhibition of purine and pyrimidine synthesis leads to cessation of protein synthesis and DNA replication, thus the pathogen dies. Sulfonamides are selectively toxic for many pathogens because these bacteria manufacture their own fo-

Figure 34.9 Erythromycin, a Macrolide Antibiotic. The 14-member lactone ring is connected to two sugars.

Figure 34.10 Chloramphenicol.

Figure 34.11 Sulfanilamide. Sulfanilamide and its relationship to the structure of folic acid.

Sulfamethoxazole

Sulfisoxazole

Figure 34.12   Two Sulfonamide Drugs.   The blue shaded areas are side chains substituted for a hydrogen in sulfanilamide (figure 35.4).

late and cannot effectively take up this cofactor, whereas humans do not synthesize folate (we must obtain it in our diet). Sulfonamides thus have a high therapeutic index.

The increasing resistance of many bacteria to sulfa drugs limits their effectiveness. Furthermore, as many as 5% of the patients receiving sulfa drugs experience adverse side effects, chiefly allergic responses such as fever, hives, and rashes.

Trimethoprim

Trimethoprim is a synthetic antibiotic that also interferes with the production of folic acid. It does so by binding to dihydrofolate reductase (DHFR), the enzyme responsible for converting dihydrofolic acid to tetrahydrofolic acid, competing against the dihydrofolic acid substrate (**figure 34.13**). It is a broad-spectrum antibiotic often used to treat respiratory and middle ear infections, urinary tract infections, and traveler's diarrhea. It can be combined with sulfa drugs to increase efficacy of treatment by blocking two key steps in the folic acid pathway (**figure 34.14**). The inhibition of two successive steps in a single biochemical pathway means that less of each drug is needed in combination than when used alone. This is termed a synergistic drug interaction.

The most common side effects associated with trimethoprim are abdominal pain, abnormal taste, diarrhea, loss of appetite, nausea, swelling of the tongue, and vomiting. Taking trimethoprim with food may reduce some of these side effects. Some patients are allergic to trimethoprim, exhibiting rash and itching. Some patients develop photosensitivity reactions (i.e., rashes due to sun exposure).

Nucleic Acid Synthesis Inhibition

The antibacterial drugs that inhibit nucleic acid synthesis function by inhibiting DNA polymerase and DNA helicase or RNA

(a) Dihydrofolic acid (DFA)

(b) Dihydrofolate reductase

(c) Trimethoprim

Figure 34.13   Competitive Inhibition of Dihydrofolate Reductase (DHFR) by Trimethoprim.   (a) Dihydrofolic acid (DFA) is the natural substrate for the DHFR enzyme of the folic acid pathway. **(b)** DHFR structure and its interaction with DFA (red). Note the chemical structure and how it fits into the active site of the enzyme. **(c)** Trimethoprim mimics the structural orientation of the DFA and thus competes for the active site of the enzyme. The consequence of this is delayed or absent folic acid synthesis because the DFA cannot be converted to tetrahydrofolic acid when trimethoprim occupies the DHFR active site.

polymerase, thus blocking processes of replication or transcription, respectively. These drugs are not as selectively toxic as other antibiotics because procaryotes and eucaryotes do not differ greatly with respect to nucleic acid synthesis.

Quinolones

The **quinolones** are synthetic drugs that contain the 4-quinolone ring. The quinolones are important antimicrobial agents that inhibit nucleic acid synthesis. They are increasingly used to treat a wide variety of infections. The first quinolone, nalidixic acid

Figure 34.14 Synergistic Drug Interaction Between the Sulfonamides and Trimethoprim. Two successive steps in the biochemical pathway for folic acid synthesis are blocked by these drugs. Thus the efficacy of the drug combination is greater than that of either drug used alone.

(**figure 34.15**), was synthesized in 1962. Since that time, generations of fluoroquinolones have been produced. Three of these—ciprofloxacin, norfloxacin, and ofloxacin—are currently used in the United States, and more fluoroquinolones are being synthesized and tested. Ciprofloxacin (Cipro) gained notoriety during the 2001 bioterror attacks in the United States as one treatment for anthrax. Bioterrorism preparedness (section 36.9)

Quinolones act by inhibiting the bacterial DNA gyrase and topoisomerase II. DNA gyrase introduces negative twist in DNA and helps separate its strands (**figure 34.16**). Inhibition of DNA gyrase disrupts DNA replication and repair, bacterial chromosome separation during division, and other cell processes involving DNA. Fluoroquinolones also inhibit topoisomerase II, another enzyme that untangles DNA during replication. It is not surprising that quinolones are bactericidal. DNA replication (section 11.4)

The quinolones are broad-spectrum antibiotics. They are highly effective against enteric bacteria such as *E. coli* and *Klebsiella pneumoniae*. They can be used with *Haemophilus, Neiserria, P. aeruginosa,* and other gram-negative pathogens. The quinolones also are active against gram-positive bacteria such as *S. aureus, Streptococcus pyogenes,* and *Mycobacterium tuberculosis.* Currently, they are used in treating urinary tract infections, sexually transmitted diseases caused by *Neisseria* and *Chlamydia,* gastrointestinal infections, respiratory infections, skin infections, and osteomyelitis (bone infection). Quinolones are effective when administered orally but can sometimes cause diverse side effects, particularly gastrointestinal upset.

Figure 34.15 Quinolone Antimicrobial Agents.
Ciprofloxacin and norfloxacin are newer generation fluoroquinolones. The 4-quinolone ring in nalidixic acid has been numbered.

1. Explain five ways in which chemotherapeutic agents kill or damage bacterial pathogens.
2. Why do penicillins and cephalosporins have a higher therapeutic index than most other antibiotics?

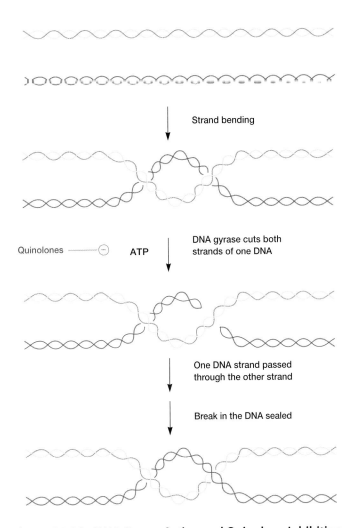

Strand bending

Quinolones —— ⊖   ATP

DNA gyrase cuts both
strands of one DNA

One DNA strand passed
through the other strand

Break in the DNA sealed

Figure 34.16   **DNA Gyrase Action and Quinolone Inhibition.**

3. Would there be any advantage to administering a bacteriostatic agent along with penicillins? Any disadvantage?
4. What are antimetabolites?
5. Why are some antibiotics toxic?

34.5 FACTORS INFLUENCING ANTIMICROBIAL DRUG EFFECTIVENESS

It is crucial to recognize that drug therapy is not a simple matter. Drugs may be administered in several different ways, and they do not always spread rapidly throughout the body or immediately kill all invading pathogens. A complex array of factors influence the effectiveness of drugs.

First, the drug must actually be able to reach the site of infection. Understanding the factors that control drug activity, stability, and metabolism in vivo are essential in drug formulation. For example, the mode of administration plays an important role. A drug such as penicillin G is not suitable for oral administration because it is relatively unstable in stomach acid. Some antibiotics—for

example, gentamicin and other aminoglycosides—are not well absorbed from the intestinal tract and must be injected intramuscularly or given intravenously. Other antibiotics (neomycin, bacitracin) are so toxic that they can only be applied topically to skin lesions. Non-oral routes of administration often are called **parenteral routes.** Even when an agent is administered properly, it may be excluded from the site of infection. For example, blood clots or necrotic tissue can protect bacteria from a drug, either because body fluids containing the agent may not easily reach the pathogens or because the agent is absorbed by materials surrounding it.

Second, the pathogen must be susceptible to the drug. Bacteria in biofilms or abscesses may be replicating very slowly and are therefore resistant to chemotherapy, because many agents affect pathogens only if they are actively growing and dividing. A pathogen, even though growing, may simply not be susceptible to a particular agent. For example, penicillins and cephalosporins, which inhibit cell wall synthesis (table 34.1), do not harm mycoplasmas, which lack cell walls. To control resistance, drug cocktails can be used to treat some infections. A notable example of this is the use of clavulonic acid (to inactivate penicillinase) combined with ampicillin (Augmentin) to treat penicillin-resistant bacteria.

Third, the chemotherapeutic agent must exceed the pathogen's MIC value if it is going to be effective. The concentration reached will depend on the amount of drug administered, the route of administration and speed of uptake, and the rate at which the drug is cleared or eliminated from the body. It makes sense that a drug will remain at high concentrations longer if it is absorbed over an extended period and excreted slowly.

Finally, chemotherapy has been rendered less effective and much more complex by the spread of drug-resistance genes.

1. Briefly discuss the factors that influence the effectiveness of antimicrobial drugs.
2. What is parenteral administration of a drug?

34.6 DRUG RESISTANCE

The spread of drug-resistant pathogens is one of the most serious threats to public health in the 21st century (**Disease 34.2**). This section describes the ways in which bacteria acquire drug resistance and how resistance spreads within a bacterial population.

Mechanisms of Drug Resistance

The long-awaited "superbug" arrived in the summer of 2002. *S. aureus,* a common but sometimes deadly bacterium, had acquired a new antibiotic-resistance gene. The new strain was isolated from foot ulcers on a diabetic patient in Detroit, Michigan. Meticillin-resistant (formerly methicillin-resistant) *S. aureus* (MRSA) had been well known as the bane of hospitals. This newer strain had developed resistance to vancomycin, one of the few antibiotics that was still able to control *S. aureus*. This new vancomycin-resistant

The sale of antimicrobial drugs is big business. In the United States millions of pounds of antibiotics valued at billions of dollars are produced annually. As much as 70% of these antibiotics are added to livestock feed.

Because of the massive quantities of antibiotics being prepared and used, an increasing number of diseases are resisting treatment due to the spread of drug resistance. A good example is *Neisseria gonorrhoeae*, the causative agent of gonorrhea. Gonorrhea was first treated successfully with sulfonamides in 1936, but by 1942 most strains were resistant and physicians turned to penicillin. Within 16 years a penicillin-resistant strain emerged in Asia. A penicillinase-producing gonococcus reached the United States in 1976 and is still spreading in this country. Thus penicillin is no longer used to treat gonorrhea.

In late 1968 an epidemic of dysentery caused by *Shigella* broke out in Guatemala and affected at least 112,000 persons; 12,500 deaths resulted. The strains responsible for this devastation carried an R plasmid conferring resistance to chloramphenicol, tetracycline, streptomycin, and sulfonamide. In 1972 a typhoid epidemic swept through Mexico producing 100,000 infections and 14,000 deaths. It was due to a *Salmonella* strain with the same multiple-drug-resistance pattern seen in the previous *Shigella* outbreak.

Haemophilus influenzae type b is responsible for many cases of childhood pneumonia and middle ear infections, as well as respiratory infections and meningitis. It is now becoming increasingly resistant to tetracyclines, ampicillin, and chloramphenicol. Similarly, the worldwide rate of penicillin-nonsusceptible (i.e., resistant) *Streptococcus pneumoniae* (PNSP) continues to increase. There is a direct correlation between the daily use of antibiotics (expressed as defined daily dose [DDD] per day) and the percent of PNSP isolates cultured (**Box figure**). This dramatic correlation is alarming. More alarming is the continued indiscriminant use of antibiotics in light of these data.

In 1946 almost all strains of *Staphylococcus* were penicillin sensitive. Today most hospital strains are resistant to penicillin G, and some are now also resistant to methicillin and/or gentamicin and only can be treated with vancomycin. Some strains of *Enterococcus* have become resistant to most antibiotics, including vancomycin. Recently a few cases of vancomycin-resistant *S. aureus* have been reported in the United States and Japan. At present these strains are only intermediately resistant to vancomycin. If full vancomycin resistance spreads, *S. aureus* may become untreatable.

It is clear from these and other examples (e.g., multiresistant *Mycobacterium tuberculosis*) that drug resistance is an extremely serious public health problem. Much of the difficulty arises from drug misuse. Drugs frequently have been overused in the past. It has been estimated that over 50% of the antibiotic prescriptions in hospitals are given without clear evidence of infection or adequate medical indication. Many physicians have administered antibacterial drugs to patients with colds, influenza, viral pneumonia, and other viral diseases. A recent study showed that over 50% of the patients diagnosed with colds and upper respiratory infections and 66% of those with chest colds (bronchitis) are given antibiotics, even though over 90% of these cases are caused by viruses. Frequently antibiotics are prescribed without culturing and identifying the pathogen or without determining bacterial sensitivity to the drug. Toxic, broad-spectrum antibiotics are sometimes given in place of narrow-spectrum drugs as a substitute for culture and sensitivity testing, with the consequent risk of dangerous side effects, opportunistic infections, and the selection of drug-resistant mutants. The situation is made worse by patients not completing their course of medication. When antibiotic treatment is ended too early, drug-resistant mutants may survive. Drugs are available without prescription to the public in many coun-

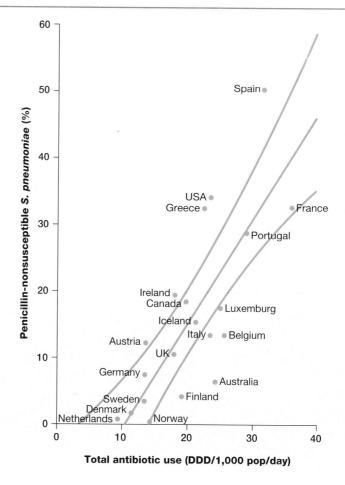

Total antibiotic use (DDD/1,000 pop/day)

tries; people may practice self-administration of antibiotics and further increase the prevalence of drug-resistant strains.

The use of antibiotics in animal feeds is undoubtedly another contributing factor to increasing drug resistance. The addition of low levels of antibiotics to livestock feeds raises the efficiency and rate of weight gain in cattle, pigs, and chickens (partially because of infection control in overcrowded animal populations). However, this also increases the number of drug-resistant bacteria in animal intestinal tracts. There is evidence for the spread of bacteria such as *Salmonella* from animals to human populations. In 1983, 18 people in four midwestern states were infected with a multiple-drug-resistant strain of *Salmonella newport*. Eleven were hospitalized for salmonellosis and one died. All 18 patients had recently been infected by eating hamburger from beef cattle fed subtherapeutic doses of chlortetracycline for growth promotion. Resistance to some antibiotics has been traced to the use of specific farmyard antibiotics. Avoparcin resembles vancomycin in structure, and virginiamycin resembles Synercid. There is good circumstantial evidence that extensive use of these two antibiotics in animal feed has led to an increase in vancomycin and Synercid resistance among enterococci. The use of the quinolone antibiotic enrofloxacin in swine herds appears to have promoted ciprofloxacin resistance in pathogenic strains of *Salmonella*. In 2005, the use of fluoroquinolones in U.S. poultry farming was banned in recognition of this public health threat.

The spread of antibiotic resistance can be due to quite subtle factors. For example, products such as soap and deodorants often now contain triclosan and other germicides. There is increasing evidence that the widespread use of triclosan actually favors an increase in antibiotic resistance (*see section 7.5*).

S. aureus (VRSA) strain also resisted most other antibiotics including ciprofloxacin, methicillin, and penicillin. Isolated from the same patient was another dread of hospitals—vancomycin-resistant enterococci (VRE). Genetic analyses revealed that the patient's own vancomycin-sensitive *S. aureus* had acquired the vancomycin-resistance gene, *vanA,* from VRE through conjugation. So was born a new threat to the health of the human race. Bacterial conjugation (section 13.7); Bacterial plasmids (sections 3.5 and 13.6)

Bacteria often become resistant in several different ways (**figure 34.17**). Unfortunately, a particular type of resistance mechanism is not confined to a single class of drugs (**figure 34.18**). Two bacteria may use different resistance mechanisms to withstand the same chemotherapeutic agent. Furthermore, resistant mutants arise spontaneously and are then selected for in the presence of the drug.

Pathogens often become resistant simply by preventing entrance of the drug. Many gram-negative bacteria are unaffected by penicillin G because it cannot penetrate the envelope's outer membrane. Genetic mutations that lead to changes in penicillin binding proteins also render a cell resistant. A decrease in permeability can lead to sulfonamide resistance. Mycobacteria resist many drugs because of the high content of mycolic acids (*see figure 24.11*) in a complex lipid layer outside their peptidoglycan. This layer is impermeable to most water soluble drugs. Suborder *Corynebacterineae:* Genus *Mycobacterium* (section 24.4)

A second resistance strategy is to pump the drug out of the cell after it has entered. Some pathogens have plasma membrane translocases, often called efflux pumps, that expel drugs. Because they are relatively nonspecific and can pump many different drugs, these transport proteins often are called multidrug-resistance pumps. Many are drug/proton antiporters—that is, protons enter the cell as the drug leaves. Such systems are present in *E. coli, P. aeruginosa,* and *S. aureus* to name a few.

Many bacterial pathogens resist attack by inactivating drugs through chemical modification. The best-known example is the hydrolysis of the β-lactam ring of penicillins by the enzyme penicillinase. Drugs also are inactivated by the addition of chemical groups. For example, chloramphenicol contains two hydroxyl groups (figure 34.10) that can be acetylated in a reaction catalyzed by the enzyme chloramphenicol acyltransferase with acetyl CoA as the donor. Aminoglycosides (figure 34.7) can be modified and inactivated in several ways. Acetyltransferases catalyze the acetylation of amino groups. Some aminoglycoside-modifying enzymes catalyze the addition to hydroxyl groups of either phosphates (phosphotransferases) or adenyl groups (adenyltransferases).

Because each chemotherapeutic agent acts on a specific target, resistance arises when the target enzyme or cellular structure is modified so that it is no longer susceptible to the drug.

Figure 34.17 Antibiotic Resistance Has Many Sources. Incomplete and indiscriminant use of antibiotics in people and animals leads to increased selective pressure on bacteria. Bacteria capable of resisting antibiotics survive and spread these traits by horizontal gene transfer.

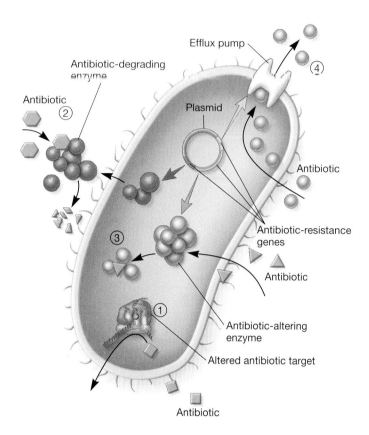

Figure 34.18 Antibiotic Resistance Mechanisms. Bacteria can resist the action of antibiotics by (1) preventing access to (or altering) the target of the antibiotic, (2) degrading the antibiotic, (3) altering the antibiotic, and/or (4) rapid extrusion of the antibiotic.

For example, the affinity of ribosomes for erythromycin and chloramphenicol can be decreased by a change in the 23S rRNA to which they bind. Enterococci become resistant to vancomycin by changing the terminal D-alanine-D-alanine in their peptidoglycan to a D-alanine-D-lactate. This drastically reduces antibiotic binding. Antimetabolite action may be resisted through alteration of susceptible enzymes. In sulfonamide-resistant bacteria the enzyme that uses *p*-aminobenzoic acid during folic acid synthesis (the dihydropteroic acid synthetase; figure 34.14) often has a much lower affinity for sulfonamides.

Finally, resistant bacteria may either use an alternate pathway to bypass the sequence inhibited by the agent or increase the production of the target metabolite. For example, some bacteria are resistant to sulfonamides simply because they use preformed folic acid from their surroundings rather than synthesize it themselves. Other strains increase their rate of folic acid production and thus counteract sulfonamide inhibition.

The Origin and Transmission of Drug Resistance

Genes for drug resistance may be present on bacterial chromosomes, plasmids, transposons, and integrons. Because they are often found on mobile genetic elements, they can freely exchange between bacteria. Spontaneous mutations in the bacterial chromosome, although they do not occur very often, can make bacteria drug resistant. Usually such mutations result in a change in the drug target; therefore the antibiotic cannot bind and inhibit growth (e.g., the protein target to which streptomycin binds on bacterial ribosomes). Many mutants are probably destroyed by natural host resistance mechanisms. However, when a patient is being treated extensively with antibiotics, some resistant mutants may survive and flourish because of their competitive advantage over nonresistant strains.

Frequently a bacterial pathogen is drug resistant because it has a plasmid bearing one or more resistance genes; such plasmids are called **R plasmids** (*r*esistance plasmids). Plasmid resistance genes often code for enzymes that destroy or modify drugs; for example, the hydrolysis of penicillin or the acetylation of chloramphenicol and aminoglycoside drugs. Plasmid-associated genes have been implicated in resistance to the aminoglycosides, choramphenicol, penicillins and cephalosporins, erythromycin, tetracyclines, sulfonamides, and others. Once a bacterial cell possesses an R plasmid, the plasmid (or its genes) may be transferred to other cells quite rapidly through normal gene exchange processes such as conjugation, transduction, and transformation (**figure 34.19**). Because a single plasmid may carry genes for resistance to several drugs, a pathogen population can become resistant to several antibiotics simultaneously, even though the infected patient is being treated with only one drug. Bacterial conjugation (section 13.7); Transduction (section 13.9); DNA transformation (section 13.8); Bacterial plasmids (sections 3.5 and 13.6)

Antibiotic resistance genes can be located on genetic elements other than plasmids. Many composite transposons contain genes for antibiotic resistance, and some bear more than one resistance gene. They are found in both gram-negative and gram-positive bacteria. Some examples and their resistance markers are Tn5 (kanamycin, bleomycin, streptomycin), Tn9 (chloramphenicol), Tn10 (tetracycline), Tn21 (streptomycin, spectinomycin, sulfonamide), Tn551 (erythromycin), and Tn4001 (gentamicin, tobramycin, kanamycin). Resistance genes on composite transposons can move rapidly between plasmids and through a bacterial population. Often several resistance genes are carried together as gene cassettes in association with a genetic element known as an integron. An **integron** is composed of an integrase gene and sequences for site-specific recombination. Thus integrons can capture genes and gene cassettes. **Gene cassettes** are genetic elements that may exist as circular nonreplicating DNA when moving from one site to another, but which normally are a linear part of a transposon, plasmid, or bacterial chromosome. Cassettes usually carry one or two genes and a recombination site. Several cassettes can be integrated sequentially in an integron. Thus integrons also are important in spreading resistance genes. Finally, conjugative transposons, like composite transposons, can carry resistance genes. Because they are capable of moving between bacteria by conjugation, they are also effective in spreading resistance. Transposable elements (section 13.5)

Extensive drug treatment favors the development and spread of antibiotic-resistant strains because the antibiotic destroys

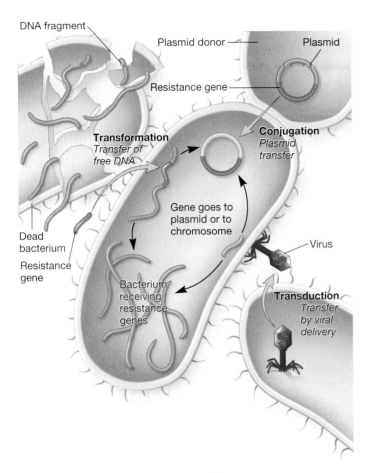

Figure 34.19 Horizontal Gene Exchange. Bacteria exchange genetic information, like antibiotic resistance genes, through conjugation, transformation, and/or transduction.

*Labels in figure:*
DNA fragment
Plasmid donor — Plasmid
Resistance gene
Transformation *Transfer of free DNA*
Conjugation *Plasmid transfer*
Gene goes to plasmid or to chromosome
Dead bacterium
Resistance gene
Bacterium receiving resistance genes
Virus
Transduction *Transfer by viral delivery*

susceptible bacteria that would usually compete with drug-resistant strains. The result may be the emergence of drug-resistant pathogens. Several strategies can be employed to discourage the emergence of drug resistance. The drug can be given in a high enough concentration to destroy susceptible bacteria and most spontaneous mutants that might arise during treatment. Sometimes two or even three different drugs can be administered simultaneously with the hope that each drug will prevent the emergence of resistance to the other. This approach is used in treating tuberculosis and malaria. Finally, chemotherapeutic drugs, particularly broad-spectrum drugs, should be used only when definitely necessary. If possible, the pathogen should be identified, drug sensitivity tests run, and the proper narrow-spectrum drug employed.

Despite efforts to control the emergence and spread of drug resistance, the situation continues to worsen. Of course, antibiotics should be used in ways that reduce the development of resistance. Another approach is to search for new antibiotics that microorganisms have never encountered. Pharmaceutical companies collect and analyze samples from around the world in a search for completely new antimicrobial agents. Structure-based or rational drug design is a third option. If the three-dimensional structure of

a susceptible target molecule such as an enzyme essential to microbial function is known, computer programs can be used to design drugs that precisely fit the target molecule. These drugs might be able to bind to the target and disrupt its function sufficiently to destroy the pathogen. Pharmaceutical companies are using this approach to attempt to develop drugs for the treatment of AIDS, cancer, septicemia caused by lipopolysaccharide (LPS), and the common cold. At least one company is developing "enhancers." These are cationic peptides that disrupt bacterial membranes by displacing their magnesium ions. Antibiotics then penetrate and rapidly exert their effects. Other pharmaceutical companies are developing efflux-pump inhibitors to administer with antibiotics and prevent their expulsion by the resistant pathogen.

There has been some progress in developing new antibiotics that are effective against drug-resistant pathogens. Two new drugs are fairly effective against vancomycin-resistant enterococci. Synercid is a mixture of the streptogramin antibiotics quinupristin and dalfopristin that inhibits protein synthesis. A second drug, linezolid (Zyvox), is the first drug in a new family of antibiotics, the oxazolidinones. It inhibits protein synthesis and is active against both vancomycin-resistant enterococci and meticillin-resistant *Staphylococcus aureus*.

Information that is coming from the sequencing and analysis of pathogen genomes is also useful in identifying new targets for antimicrobial drugs. For example, genomics studies are providing data for research on inhibitors of both aminoacyl-tRNA synthetases and the enzyme that removes the formyl group from the N-terminal methionine during bacterial protein synthesis. Bacteria must synthesize the fatty acids they require for growth rather than acquiring the acids from their environment. The drug susceptibility of enzymes in the fatty acid synthesis system is being analyzed by screening pathogens for potential targets. *Genomics (chapter 15)*

A most interesting response to the current crisis is the renewed interest in an idea first proposed early in the twentieth century by Felix d'Herelle, one of the discoverers of bacterial viruses or bacteriophages. d'Herelle proposed that bacteriophages could be used to treat bacterial diseases. Although most microbiologists did not pursue his proposal actively due to technical difficulties and the advent of antibiotics, Russian scientists developed the medical use of bacteriophages. Currently Russian physicians use bacteriophages to treat many bacterial infections. Bandages are saturated with phage solutions, phage mixtures are administered orally, and phage preparations are given intravenously to treat *Staphylococcus* infections. Three American companies are actively conducting research on phage therapy and preparing to carry out clinical trials. *Viruses of Bacteria and Archaea (chapter 17)*

1. Briefly describe the five major ways in which bacteria become resistant to drugs and give an example of each.
2. Define plasmid, R plasmid, integron, and gene cassette. How are R plasmids involved in the spread of drug resistance?
3. List several ways in which the development of antibiotic-resistant pathogens can be slowed or prevented.

34.7 ANTIFUNGAL DRUGS

Treatment of fungal infections generally has been less successful than that of bacterial infections largely because eucaryotic fungal cells are much more similar to human cells than are bacteria. Many drugs that inhibit or kill fungi are therefore quite toxic for humans. In addition, most fungi have a detoxification system that modifies many antifungal agents. As a result the added antibiotics are fungistatic only as long as repeated application maintains high levels of unmodified antibiotic. Despite their relatively low therapeutic index, a few drugs are useful in treating many major fungal diseases. Effective antifungal agents frequently either extract membrane sterols or prevent their synthesis. Similarly, because animal cells do not have cell walls, the enzyme chitin synthase is the target for fungal-active antibiotics such as polyoxin D and nikkomycin.

Fungal infections are often subdivided into infections of superficial tissues or superficial mycoses and systemic mycoses. Treatment for these two types of disease is very different. Several drugs are used to treat superficial mycoses. Three drugs containing imidazole—miconazole, ketoconazole (**figure 34.20**), and clotrimazole—are broad-spectrum agents available as creams and solutions for the treatment of dermatophyte infections such as athlete's foot, and oral and vaginal candidiasis. They are thought to disrupt fungal membrane permeability and inhibit sterol synthesis. Tolnaftate is used topically for the treatment of cutaneous infections, but is not as effective against infections of the skin and hair. **Nystatin** (figure 34.20), a polyene antibiotic from *Streptomyces,* is used to control *Candida* infections of the skin, vagina, or alimentary tract. It binds to sterols and damages the membrane, leading to fungal membrane leakage. **Griseofulvin** (figure 34.20), an antibiotic formed by *Penicillium,* is given orally to treat chronic dermatophyte infections. It is thought to disrupt the mitotic spindle and inhibit cell division; it also may inhibit protein and nucleic acid synthesis. Side effects of griseofulvin include headaches, gastrointestinal upset, and allergic reactions. Human diseases caused by fungi and protists (chapter 39)

Systemic infections are very difficult to control and can be fatal. Three drugs commonly used against systemic mycoses are **amphotericin B,** 5-flucytosine, and fluconazole (figure 34.20). Amphotericin B from *Streptomyces* spp. binds to the sterols in fungal membranes, disrupting membrane permeability and caus-

Figure 34.20 Antifungal Drugs. Six commonly used drugs are shown here.

ing leakage of cell constituents. It is quite toxic and used only for serious, life-threatening infections. The synthetic oral antimycotic agent 5-flucytosine (5-fluorocytosine) is effective against most systemic fungi, although drug resistance often develops rapidly. The drug is converted to 5-fluorouracil by the fungi, incorporated into RNA in place of uracil, and disrupts RNA function. Its side effects include skin rashes, diarrhea, nausea, aplastic anemia, and liver damage. Fluconazole is used in the treatment of candidiasis, cryptococcal meningitis, and coccidioidal meningitis. Because adverse effects to fluconazole are relatively uncommon, it is used prophylactically to prevent life-threatening fungal infections in AIDS patients and other individuals who are severely immunosuppressed.

---

1. Summarize the mechanism of action and the therapeutic use of the following antifungal drugs: miconazole, nystatin, griseofulvin, amphotericin B, and 5-flucytosine.

34.8 ANTIVIRAL DRUGS

For many years the possibility of treating
drugs appeared remote because viruses enter host cells and
use of host cell enzymes and constituents. A drug that would block virus reproduction also was thought to be toxic for the host. Inhibitors of virus-specific enzymes and life cycle processes have now been discovered, and several drugs are used therapeutically. Some important examples are shown in **figure 34.21.** Reproduction of vertebrate viruses (section 18.2)

Most antiviral drugs disrupt either critical stages in the virus life cycle or the synthesis of virus-specific nucleic acids. **Amantadine** and rimantadine can be used to prevent influenza A infections. When given early in the infection (in the first 48 hours), they reduce the incidence of influenza by 50 to 70% in an exposed population. Amantadine blocks the penetration and uncoating of influenza virus particles. **Adenine arabinoside** or **vidarabine**

Figure 34.21 Representative Antiviral Drugs.

disrupts the activity of DNA polymerase and several other enzymes involved in DNA and RNA synthesis and function. It is given intravenously or applied as an ointment to treat herpes infections. A third drug, **acyclovir**, is also used in the treatment of herpes infections. Upon phosphorylation, acyclovir resembles deoxy-GTP and inhibits the virus DNA polymerase. Unfortunately acyclovir-resistant strains of herpes are already developing. Effective acyclovir derivatives and relatives are now available. Valacyclovir is an orally administered prodrug form of acyclovir. Prodrugs are inactive until metabolized. Ganciclovir, penciclovir, and its oral form famciclovir are effective in treatment of herpesviruses. Another kind of drug, foscarnet, inhibits the virus DNA polymerase in a different way. Foscarnet is an organic analog of pyrophosphate (figure 34.21) that binds to the polymerase active site and blocks the cleavage of pyrophosphate from nucleoside triphosphate substrates. It is used in treating herpes and cytomegalovirus infections. Airborne diseases: Influenza (section 37.1)

Several broad-spectrum anti-DNA virus drugs have been developed. A good example is the drug HPMPC or cidofovir (figure 34.21). It is effective against papovaviruses, adenoviruses, herpesviruses, iridoviruses, and poxviruses. The drug acts on the viral DNA polymerase as a competitive inhibitor and alternative substrate of dCTP. It has been used primarily against cytomegalovirus but also against herpes simplex and human papillomavirus infections.

Research on anti-HIV drugs has been particularly active. Many of the first drugs to be developed were reverse transcriptase inhibitors such as **azidothymidine (AZT)** or **zidovudine,** lamivudine (3TC), didanosine (ddI), zalcitabine (ddC), and stavudine (d4T) (figure 34.21). These interfere with reverse transcriptase activity and therefore block HIV reproduction. More recently **HIV protease inhibitors** have also been developed. Three of the most used are saquinvir, indinavir, and ritonavir (figure 34.21). Protease inhibitors are effective because HIV, like many viruses, translates multiple proteins as a single polypeptide. This polypeptide must then be cleaved into individual proteins required for virus replication. Protease inhibitors mimic the peptide bond that is normally attacked by the protease. The most successful treatment regimen involves a cocktail of agents given at high dosages to prevent the development of drug resistance. For example, the combination of AZT, 3TC, and ritonavir is very effective in reducing HIV plasma concentrations almost to zero. However, the treatment does not eliminate latent proviral HIV DNA that still resides in memory T cells, and possibly elsewhere. Reproduction of vertebrate viruses: Genome replication, transcription, and protein synthesis in RNA viruses (section 18.2); Direct contact diseases: Acquired immune deficiency syndrome (AIDS) (section 37.3)

Probably the most publicized antiviral agent has been Tamiflu (generically, oseltamivir phosphate). Tamiflu is a neuraminidase inhibitor that has received much attention in light of 21st-century predictions of an influenza pandemic, including avian influenza ("bird flu"). While Tamiflu is not a cure for neurominidase-expressing viruses, two clinical trials showed that patients who took Tamiflu were relieved of flu symptoms 1.3 days faster than patients who did not take Tamiflu. Prophylactic use has resulted in viral resistance to Tamiflu. Tamiflu is not a substitute for yearly flu vaccination and frequent hand-washing.

34.9 ANTIPROTOZOAN DRUGS

The mechanism of drug action for most antiprotozoan drugs is not completely elucidated. Drugs such as chloroquine, atovaquone, mefloquine, iodoquinol, metronidazole, nitazoxanide, and pentamidine, for example, have potent antiprotozoan action but a clear mechanism of action for each class of protozoa is unknown. It may be that each drug has more than one activity and that the relative role of each mechanism to the overall antiprotozoan activity may be different for the various species of protozoa. However, most antiprotozoan drugs appear to act on protozoan nucleic acid or some metabolic event.

Chloroquine is used to treat malaria. Several mechanisms of action have been reported. It can raise the internal pH, clump the plasmodial pigment, and intercalate into plasmodial DNA. Chloroquine also inhibits heme polymerase, an enzyme that converts toxic heme into nontoxic hemazoin. Inhibition of this enzyme leads to a buildup of toxic heme. Mefloquine is also used to treat malaria and has been found to swell the *Plasmodium falciparum* food vacuoles, where it may act by forming toxic complexes that damage membranes and other plasmodial components.

Metronidazole is used to treat *Entamoeba* infections. Anaerobic organisms readily reduce it to the active metabolite within the cytoplasm. Aerobic organisms appear to reduce it using ferrodoxin (a protein of the electron transport system). Reduced metronidazole interacts with DNA altering its helical structure and causing DNA fragmentation; it prevents normal nucleic acid synthesis, resulting in cell death.

A number of antibiotics that inhibit bacterial protein synthesis are also used to treat protozoan infection. These include the aminoglycosides clindamycin, and paromomycin. Aminoglycosides can be considered polycationic molecules that have a high affinity for nucleic acids. Specifically, aminoglycosides possess high affinities for RNAs. Different aminoglycoside antibiotics bind to different sites on RNAs. RNA binding interferes with the normal expression and function of the RNA, resulting in cell death.

Interference of eucaryotic electron transport is one common activity of some antiprotozoan drugs. Atovaquone is used to treat *Pneumocystis jiroveci* (formerly called *P. carinii*) and *Toxoplasma gondii*. It is an analog of ubiquinone, an integral component of the eucaryotic electron transport system. As an analog of ubiquinone, atovaquone can act as a competitive inhibitor and thus suppress electron transport. The ultimate metabolic effects of electron transport blockade include inhibited or delayed synthesis of nucleic acids and ATP. Another drug that interferes with electron transport is nitazoxanide, which used to treat cryptosporidiosis. It appears to exert its effect through interference with the pyruvate:ferredoxin oxidoreductase. It has also been reported to form toxic free radicals once the nitro group is reduced intracellularly.

Pentamidine is used to treat *Pneumocystis* infection. Some reports indicate that it interferes with protozoan metabolism, although the drug only moderately inhibits glucose metabolism, protein synthesis, RNA synthesis, and intracellular amino acid transport in vitro. Pyrimethamine, used to treat *Toxoplasma* infection, and dapsone, used for *Pneumocystis* infection, appear to act in the same way as trimethoprim—interfering with folic acid synthesis by inhibition of dihydrofolate reductase.

As with other antimicrobial therapies, traditional drug development starts by identifying a unique target to which a drug can bind and thus prevent some vital function. A second consideration is often related to drug spectrum—how many different species have that target so that the proposed drug can be used broadly as a chemotherapeutic agent. This is also true for use of agents needed to remove protozoan parasites from their hosts. However, because protozoa are eucaryotes, the potential for drug action on host cells and tissues is greater than it is when targeting procary-

otes. Most of the drugs used to treat protozoan infection have significant side effects; nonetheless, the side effects are usually acceptable when weighed against the parasitic alternative.

1. Why do you think drugs that inhibit bacterial protein synthesis are also effective against some protists?
2. Why do you think malaria, like tuberculosis, is now treated with several drugs simultaneously?
3. What special considerations must be taken into account when treating infections caused by protozoan parasites?

Summary

Chemotherapeutic agents are compounds that destroy pathogenic microorganisms or inhibit their growth and are used in the treatment of disease. Most are antibiotics: microbial products or their derivatives that can kill susceptible microorganisms or inhibit their growth.

34.1 The Development of Chemotherapy

a. The modern era of chemotherapy began with Paul Ehrlich's work on drugs against African sleeping sickness and syphilis. Other early pioneers were Gerhard Domagk, Alexander Fleming, Howard Florey, Ernst Chain, Norman Heatley, and Selman Waksman.

34.2 General Characteristics of Antimicrobial Drugs

a. An effective chemotherapeutic agent must have selective toxicity. A drug with great selective toxicity has a high therapeutic index and usually disrupts a structure or process unique to the pathogen. It has fewer side effects.

b. Antibiotics can be classified in terms of the range of target microorganisms (narrow spectrum versus broad spectrum); their source (natural, semisynthetic, or synthetic); and their general effect (static versus cidal) (**table 34.1**).

34.3 Determining the Level of Antimicrobial Activity

a. Antibiotic effectiveness can be estimated through the determination of the minimal inhibitory concentration and the minimal lethal concentration with dilution susceptibility tests. Tests like the Kirby-Bauer test (a disk diffusion test) and the Etest are often used to estimate a pathogen's susceptibility to drugs quickly (**figures 34.2, 34.3,** and **34.4**).

34.4 Antibacterial Drugs

a. Members of the penicillin family contain a β-lactam ring and disrupt bacterial cell wall synthesis, resulting in cell lysis (**figure 34.5**). Some, like penicillin G, are usually administered by injection and are most effective against gram-positive bacteria. Others can be given orally (penicillin V), are broad spectrum (ampicillin, carbenicillin), or are penicillinase resistant (methicillin).

b. Cephalosporins are similar to penicillins, and are given to patients with penicillin allergies (**figure 34.6**).

c. Vancomycin is a glycopeptide antibiotic that inhibits the transpeptidation reaction during peptidoglycan synthesis. It is used against drug-resistant staphylococci, enterococci, and clostridia.

d. Aminoglycoside antibiotics like streptomycin and gentamicin bind to the small ribosomal subunit, inhibit protein synthesis, and are bactericidal (**figure 34.7**).

e. Tetracyclines are broad-spectrum antibiotics having a four-ring nucleus with attached groups (**figure 34.8**). They bind to the small ribosomal subunit and inhibit protein synthesis.

f. Erythromycin is a bacteriostatic macrolide antibiotic that binds to the large ribosomal subunit and inhibits protein synthesis (**figure 34.9**).

g. Chloramphenicol is a broad-spectrum, bacteriostatic antibiotic that inhibits protein synthesis (**figure 34.10**). It is quite toxic and used only for very serious infections.

h. Sulfonamides or sulfa drugs resemble *p*-aminobenzoic acid and competitively inhibit folic acid synthesis (**figure 34.12**).

i. Trimethoprim is a synthetic antibiotic that inhibits the dihydrofolate reductase, which is required by organisms in the manufacture of folic acid (**figure 34.13**).

j. Quinolones are a family of bactericidal synthetic drugs that inhibit DNA gyrase and thus inhibit such processes as DNA replication (**figure 34.15**).

34.5 Factors Influencing Antimicrobial Drug Effectiveness

a. A variety of factors can greatly influence the effectiveness of antimicrobial drugs during actual use.

34.6 Drug Resistance

a. Bacteria can become resistant to a drug by excluding it from the cell, pumping the drug out of the cell, enzymatically altering it, modifying the target enzyme or organelle to make it less drug sensitive, as examples. The genes for drug resistance may be found on the bacterial chromosome, a plasmid called an R plasmid, or other genetic elements such as transposons (**figures 34.18** and **34.19**).

b. Chemotherapeutic agent misuse fosters the increase and spread of drug resistance, and may lead to superinfections.

34.7 Antifungal Drugs

a. Because fungi are more similar to human cells than bacteria, antifungal drugs generally have lower therapeutic indexes than antibacterial agents and produce more side effects.

b. Superficial mycoses can be treated with miconazole, ketoconazole, clotrimazole, tolnaftate, nystatin, and griseofulvin (**figure 34.20**). Amphotericin B, 5-flucytosine, and fluconazole are used for systemic mycoses.

34.8 Antiviral Drugs

a. Antiviral drugs interfere with critical stages in the virus life cycle (amantadine, rimantadine, and ritonavir) or inhibit the synthesis of virus-specific nucleic acids (zidovudine, adenine arabinoside, acyclovir) (**figure 34.21**).

34.9 Antiprotozoan Drugs

a. The mechanisms by which most drugs used to treat protozoan infection are unknown.

b. Antiprotozoan drugs interfere with critical steps in nucleic acid synthesis, protein synthesis, and electron transport of folic acid synthesis.

Key Terms

acyclovir 856
adenine arabinoside or vidarabine 855
amantadine 855
aminoglycoside antibiotic 845
amphotericin B 854
antibiotic 835
antimetabolites 846
azidothymidine (AZT) 856
β-lactam 844
β-lactam ring 843
β-lactamase 843

broad-spectrum drugs 837
cephalosporin 844
chemotherapeutic agent 835
chloramphenicol 846
cidal 837
dilution susceptibility tests 840
erythromycin 846
gene cassette 852
griseofulvin 854
HIV protease inhibitors 856
integron 852

Kirby-Bauer method 840
macrolide antibiotic 846
minimal inhibitory concentration
 (MIC) 840
minimal lethal concentration
 (MLC) 840
narrow-spectrum drugs 837
nystatin 854
parenteral route 849
penicillinase 843
penicillins 841

quinolones 847
R plasmids 852
selective toxicity 837
static 837
streptomycin 845
sulfonamide 846
tetracycline 845
therapeutic index 837
trimethoprim 847
vancomycin 845
zidovudine 856

Critical Thinking Questions

1. What advantage might soil bacteria and fungi gain from the synthesis of antibiotics?

2. Why might it be desirable to prepare a variety of semisynthetic antibiotics?

3. Why is it so difficult to find or synthesize effective antiviral drugs?

4. How might the use of antibiotics as growth promoters in livestock contribute to antibiotic resistance among human pathogens?

5. A recent study found that 480 *Streptomyces* strains freshly isolated from the soil are resistant to at least six different antibiotics. In fact, some isolates are resistant to 20 different antibiotic drugs. Why do you think these bacteria, which are neither pathogenic nor exposed to human use of antibiotics, are resistant to so many drugs? What might be the implications for human bacterial pathogens?

6. You are a pediatrician treating a child with an upper respiratory infection that is clearly caused by a virus. The child's mother insists that you prescribe antibiotics—she's not leaving without them! How do you convince the child's mother that antibiotics will do more harm than good?

Learn More

DeClercq, E. 2005. Antivirals and antiviral strategies. *Nature Rev. Microbiol.* 2:704–20.

D'Costa, V. M.; McGrann, K. M.; Hughes, D. W.; and Wright, G. D. 2006. Sampling the antibiotic resistome. *Science* 311:374–77.

Fischetti, V.A. 2005. Bacteriophage lytic enzymes: Novel anti-infectives. *Trends Microbiol.* 13:491–96.

Furuya, E. Y., and Lowy, F. D. 2006. Antimicrobial-resistant bacteria in the community setting. *Nature Rev. Microbiol.* 4:36–45.

Harbarth, S., and Samore, M. H. 2005. Antimicrobial resistance determinants and future control. *Emerg. Infect. Dis.* 11:794–801.

Klugman, K. P., and Lonks, J. R. 2005. Hidden epidemic of macrolide-resistant pneumococci. *Emerg. Infect. Dis.* 11:802–7.

Payne, D., and Tomasz, A. 2004. The challenge of antibiotic-resistant bacterial pathogens: The medical need, the market and prospects for new antimicrobial agents. *Curr. Opin. Microbiol.* 7:435–38.

Schmid, M. 2005. Seeing is believing: The impact of structural genomics on antimicrobial drug discovery. *Nature Rev. Microbiol.* 2:739–46.

Walsh, C. 2003. Where will the new antibiotics come from? *Nature Rev. Microbiol.* 1:65–79.

Walsh, F. M., and Amyes, S. G. B. 2004. Microbiology and drug resistance mechanisms of fully resistant pathogens. *Curr. Opin. Microbiol.* 7:439–44.

White, D. G.; Zhao, S.; Singh, R.; and McDermott, P. F. 2004. Antimicrobial resistance among gram-negative foodborne bacterial pathogens associated with foods of animal origin. *Foodborne Pathol. Dis.* 1:137–52.

**Please visit the Prescott website at www.mhhe.com/prescott7
for additional references.**

35

Clinical Microbiology and Immunology

The major objective of the clinical microbiologist is to isolate and identify pathogens from clinical specimens rapidly. In this illustration, a clinical microbiologist is picking up suspect bacterial colonies for biochemical, immunologic, or molecular testing.

PREVIEW

- Clinical microbiologists perform many services related to the identification and control of pathogens and the detection of immune dysfunction.

- Success in clinical microbiology depends on (1) using the proper aseptic technique; (2) correctly obtaining the clinical specimen from the infected patient by swabs, needle aspiration, intubation, or catheters; (3) correctly handling the specimen; and (4) quickly transporting the specimen to the laboratory.

- Once the clinical specimen reaches the laboratory, it is cultured to identify the infecting pathogens. Identification techniques include microscopy; growth on enrichment, selective, differential, or characteristic media; specific biochemical tests; rapid test methods; immunologic techniques; bacteriophage typing; and molecular methods such as nucleic acid-based hybridization techniques, gas-liquid chromatography, and genomic fingerprinting.

- After the microorganism has been isolated, cultured, and/or identified, samples are used in susceptibility tests to find which method of control will be most effective.

- Clinical specimens can also be tested for the presence and/or concentration of either antigen or antibody. These immunological tests use principles of antigen-antibody binding. Various methods of reporting the binding events are used including colorimetric, enzyme-substrate reactions, radionucleotide detection, and precipitation in agar.

Pathogens, particularly bacteria and yeasts, coexist with harmless microorganisms on or in the host. These pathogens must be properly identified as the actual cause of infectious diseases. This is the purpose of clinical microbiology and immunology. The clinical microbiologist identifies agents and organisms based on morphological, biochemical, im-

munologic, and molecular procedures. Time is a significant factor in the identification process, especially in life-threatening situations. Advances in technology for rapid identification have greatly aided the clinical microbiologist. Molecular methods allow identification of microorganisms based on highly specific genomic and biochemical properties. Once isolated and identified, the microorganism can then be subjected to antimicrobial sensitivity tests. Even in the absence of a culture, immunologic tests can detect pathogens by measuring antigens or antibodies in the specimen. In the final analysis the patient's well-being and health can benefit significantly from information provided by the clinical laboratory—the subject of this chapter.

35.1 SPECIMENS

Infection is the invasion and multiplication in body tissues by bacteria, fungi, viruses, protozoa or helminths that often results in localized cellular injury due to competition for nutrients, toxin production, and/or intracellular replication. The major goal of the **clinical microbiologist** is to isolate and identify pathogenic microorganisms from clinical specimens rapidly. The purpose of the clinical laboratory is to provide the physician with information concerning the presence or absence of microorganisms that may be involved in the infectious disease process (**figure 35.1**). These individuals and facilities also determine the susceptibility of microorganisms to antimicrobial agents. Clinical microbiology makes use of information obtained from research on such diverse topics as microbial biochemistry and physiology, immunology, molecular biology, genomics, and the host-parasite relationships

The specimen is the beginning. All diagnostic information from the laboratory depends upon the knowledge by which specimens are chosen and the care with which they are collected and transported.

—Cynthia A. Needham

(a) The identification of the microorganism begins at the patient's beside. The nurse is giving instructions to the patient on how to obtain a sputum specimen.

(b) The specimen is sent to the laboratory to be processed. Notice that the specimen and worksheet are in different bags.

(c) Specimens such as sputum are plated on various types of media under a laminar airflow hood. This is to prevent specimen aerosols from coming in contact with the microbiologist.

(d) Sputum and other specimens are usually Gram stained to determine whether or not bacteria are present and to obtain preliminary results on the nature of any bacteria found.

(e) After incubation, the plates are examined for significant isolates. The Gram stain may be repeated for correlation.

(f) Suspect colonies are picked for biochemical, immunologic, or molecular testing.

(g) Colonies are prepared for identification by rapid test systems.

(h) In a short time, sometimes 4 hours, computer-generated information is obtained that consist of biochemical identification and antibiotic susceptibility results.

(I) All information about the specimen is now entered into a computer and the data are transmitted directly to the hospital ward.

Figure 35.1   Isolation and Identification of Microorganisms in a Clinical Laboratory.

involved in the infectious disease process. Importantly, tests developed to exploit the antigen-antibody binding capabilities, the focus of clinical immunology, can often detect microorganisms in specimens by identifying microbial antigens and/or quantifying the type and amount of responding antibody.

In clinical microbiology a clinical specimen represents a portion or quantity of human material that is tested, examined, or studied to determine the presence or absence of particular microorganisms. Safety for the patients, hospital, and laboratory staff is of utmost importance. The guidelines presented in **Tech-**

Techniques & Applications

35.1   Standard Microbiological Practices

The identification of potentially fatal, blood-borne infectious agents (HIV, hepatitis B virus, and others) spurred the codification of standard microbiological practices to limit exposure to such agents. These standard microbiological practices are *minimum* guidelines that should be supplemented with other precautions based on the potential exposure risks and biosafety level regulations for the lab. Briefly:

1. Eating, drinking, manipulation of contact lenses, and the use of cosmetics, gum, and tobacco products are strictly prohibited in the lab.
2. Hair longer than shoulder length should be tied back. Hands should be kept away from face at all times. Items (e.g., pencils) should not be placed in the mouth while in the lab. Protective clothing (lab coat, smock, etc.) is recommended while in the lab. Exposed wounds should be covered and protected.

3. Lab personnel should know how to use the emergency eyewash and/or shower stations.
4. Work space should be disinfected at the beginning and completion of lab time. Hands should be washed thoroughly after any exposure and before leaving the lab.
5. Precautions should be taken to prevent injuries caused by sharp objects (needles, scalpels, etc.). Sharp instruments should be discarded for disposal in specially marked containers.

Recommended guidelines for additional precautions should reflect the laboratory's biosafety level (BSL). The following table defines the BSL for the four categories of biological agents and suggested practices.

BSL	Agents	Practices
1	Not known to consistently cause disease in healthy adults (e.g., *Lactobacillus casei, Vibrio fischeri*)	Standard Microbiological Practices
2	Associated with human disease, potential hazard if percutaneous injury, ingestion, mucous membrane exposure occurs (e.g., *Salmonella typhi, E. coli* O157:H7, *Staphylococcus aureus*)	BSL-1 practice plus: • Limited access • Biohazard warning signs • "Sharps" precautions • Biosafety manual defining any needed waste decontamination or medical surveillance policies
3	Indigenous or exotic agents with potential for aerosol transmission; disease may have serious or lethal consequences (e.g., *Coxiella burnetti, Yersinia pestis,* Herpes viruses)	BSL-2 practice plus: • Controlled access • Decontamination of all waste • Decontamination of lab clothing before laundering • Baseline serum values determined in workers using BSL-3 agents
4	Dangerous/exotic agents that pose high risk of life-threatening disease, aerosol-transmitted lab infections; or related agents with unknown risk of transmission (e.g., *Variola major* (smallpox virus), *Ebola* virus, hemorrhagic fever viruses)	BSL-3 practices plus: • Clothing change before entering • Shower on exit • All material decontaminated on exit from facility

niques & Applications 35.1 were established by the Centers for Disease Control and Prevention (CDC) to address areas of personal protection and specimen handling. Other important concerns regarding specimens need emphasis:

1. The specimen selected should adequately represent the diseased area and also may include additional sites (e.g., urine and blood specimens) in order to isolate and identify potential agents of the particular disease process.
2. A quantity of specimen adequate to allow a variety of diagnostic testing should be obtained.
3. Attention must be given to specimen collection in order to avoid contamination from the many varieties of microorganisms indigenous to the skin and mucous membranes (*see figure 30.17*).

4. The specimen should be collected in appropriate containers and forwarded promptly to the clinical laboratory.
5. If possible, the specimen should be obtained before antimicrobial agents have been administered to the patient.

Collection

Overall, the results obtained in the clinical laboratory are only as good as the quality of the specimen collected for analysis. Specimens may be collected by several methods using aseptic technique. In this case, aseptic technique refers to specific procedures used to prevent unwanted microorganisms from contaminating the clinical specimen. Each method is designed to ensure that only the proper material is sent to the clinical laboratory.

The most common method used to collect specimens from the anterior nares or throat is the sterile **swab.** A sterile swab is a rayon-, calcium alginate, or dacron-tipped polystyrene applicator. Manufacturers of swabs have their own unique container design and instructions for proper use. For example, many commercially manufactured swabs contain a transport medium designed to preserve any microorganisms and to prevent multiplication of rapidly growing members of the population (**figure 35.2a**). However, with the exception of the nares or throat, the use of swabs for the collection of specimens is of little value and should be discouraged for two major reasons: swabs are associated with a greater risk of contamination with surface and subsurface microorganisms, and they have a limited volume capacity (<0.1 ml).

Needle aspiration is used to collect specimens aseptically (e.g., anaerobic bacteria) from cerebrospinal fluid, pus, and blood. Stringent antiseptic techniques are used to avoid skin contamination. To prevent blood from clotting and entrapping microorganisms, various anticoagulants (e.g., heparin, sodium citrate) are included within the specimen bottle or tube (figure 35.2b).

Intubation [Latin *in,* into, and *tuba,* tube] is the insertion of a tube into a body canal or hollow organ. For example, intubation can be used to collect specimens from the stomach. In this procedure a long sterile tube is attached to a syringe, and the tube is either swallowed by the patient or passed through a nostril (figure 35.2c) into the patient's stomach. Specimens are then withdrawn periodically into the sterile syringe. The most common intubation tube is the Levin tube.

A **catheter** is a tubular instrument used for withdrawing or introducing fluids from or into a body cavity. For example, urine specimens may be collected with catheters to detect urinary tract infections caused by bacteria and from newborns and neonates who cannot give a voluntary urinary specimen. Three types are commonly used for urine. The hard catheter is used when the urethra is very narrow or has strictures. The French catheter is a soft tube used to obtain a single specimen sample. If multiple samples are required over a prolonged period, a Foley catheter is used (figure 35.2d).

The most common method used for the collection of urine is the clean-catch method. After the patient has cleansed the urethral meatus (opening), a small container is used to collect the urine. The optimal time to use the clean-catch method is early morning because the urine contains more microorganisms as a result of being in the bladder overnight. In the clean-catch midstream method, the first urine voided is not collected because it becomes contaminated with those transient microorganisms normally found in the lower portion of the urethra. Only the midstream portion is collected since it most likely will contain those microorganisms found in the urinary bladder. If warranted for some patients, needle aspirations also are done directly into the urinary bladder.

Sputum is the most common specimen collected in suspected cases of lower respiratory tract infections. **Sputum** is the mucous secretion expectorated from the lungs, bronchi, and trachea through the mouth, in contrast to saliva, which is the secretion of the salivary glands that contains oral microflora. Sputum is collected in specially designed sputum cups (figure 35.2e).

Handling

Immediately after collection the specimen must be properly labeled and handled. The person collecting the specimen is responsible for ensuring that the name, hospital, registration number, location in the hospital, diagnosis, current antimicrobial therapy, name of attending physician, admission date, and type of specimen are correctly and legibly written or imprinted on the culture request form. This information must correspond to that written or imprinted on a label affixed to the specimen container. The type or source of the sample and the choice of tests to be performed also must be specified on the request form.

Transport

Speed in transporting the specimen to the clinical laboratory after it has been obtained from the patient is of prime importance. Some laboratories refuse to accept specimens if they have been in transit too long.

Microbiological specimens may be transported to the laboratory by various means (figure 35.1b). For example, certain specimens should be transported in a medium that preserves the microorganisms and helps maintain the ratio of one organism to another. This is especially important for specimens in which normal microorganisms may be mixed with microorganisms foreign to the body location.

Under some circumstances, transport media may require supplementation to support microbial survival or to inhibit normal flora microbes that may be in the specimen. For example, the use of 50,000 U of penicillin, 10 mg of streptomycin, or 0.2 mg of chloramphenicol can be added per mL of specimen to ensure recovery of fungi. Alternatively, polyvinyl alcohol-based preservatives can be used for fixation of ova and parasites in clinical specimens. Importantly, the Select Agents legislation (Federal Register 12/20/02 and 4/18/05) governs policy for the possession, use, and transport (outside of the clinical collection point) of potential biothreat agents. Thus cultivation, storage, and transport of clinical and/or environmental samples known to contain select agents (microbes of potential bioterrorism threat) are now regulated by the Select Agents Program. In most cases, specific packaging and approvals for transport (including postal) are required for specimens containing these organisms.

Special treatment is required for specimens when the microorganism is thought to be anaerobic. The material is aspirated with a needle and syringe. Most of the time it is practical to remove the needle, cap the syringe with its original seal, and bring the specimen directly to the clinical laboratory. Transport of these specimens should take no more than 10 minutes; otherwise, the specimen must be injected immediately into an anaerobic transport vial (**figure 35.3**). Vials should contain a transport medium with an indicator, such as resazurin, to show that the interior of the vial is anoxic at the time the specimen is introduced. Swabs for anaerobic culture usually are less satisfactory than aspirates or tissues, even if they are transported in an anaerobic vial, because of greater risk of contamination and poorer recovery of anaerobes.

- Tamper-evident seal
- Plastic case
- Long swab with rayon tip
- Transport medium
- Squeeze container to release medium

(a) Sterile swab

(b) Vacu-tainer blood collection tubes

(c) Nasotracheal intubation

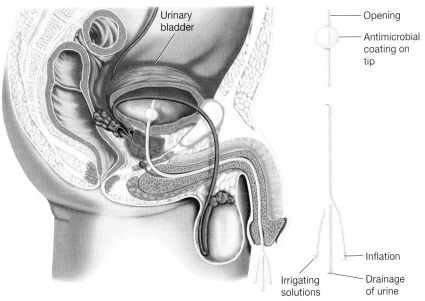

- Urinary bladder
- Opening
- Antimicrobial coating on tip
- Inflation
- Irrigating solutions
- Drainage of urine

(d) Catheter

(e) Sputum cup

Figure 35.2   Collection of Clinical Specimens.   **(a)** A drawing of a sterile swab containing a specific transport medium. **(b)** Sterile Vacu-tainer tubes for the collection of blood. **(c)** Nasotracheal intubation. **(d)** A drawing of a Foley catheter. Notice that three separate lumens are incorporated within the round shaft of the catheter for drainage of urine, inflation, and introducing irrigating solutions into the urinary bladder. After the Foley catheter has been introduced into the urinary bladder, the tip is inflated to prevent it from being expelled. **(e)** This specially designed sputum cup allows the patient to expectorate a clinical specimen directly into the cup. In the laboratory, the cup can be opened from the bottom to reduce the chance of contamination from extraneous pathogens.

Figure 35.3 Some Anaerobic Transport Systems. A vial and syringe. These systems contain a nonnutritive transport medium that retards diffusion of oxygen after specimen addition and helps maintain microorganism viability up to 72 hours. A built-in color indicator is clear and turns lavender in the presence of oxygen.

Many clinical laboratories insist that stool specimens for culture be transported in special buffered preservatives. Preparation of these transport media is described in various manuals.

Transport of urine specimens to the clinical laboratory must be done as soon as possible. No more than 1 hour should elapse between the time the specimen is obtained and the time it is examined. If this time schedule cannot be followed, the urine sample must be refrigerated immediately.

Cerebrospinal fluid (CSF) from patients suspected of having meningitis should be examined immediately by personnel in the clinical microbiology laboratory. CSF is obtained by lumbar puncture under conditions of strict asepsis, and the sample is transported to the laboratory within 15 minutes. Specimens for the isolation of viruses are iced before transport, and can be kept at 4°C for up to 72 hours; if the sample will be stored longer than 72 hours, it should be frozen at −72°C.

---

1. What is the function of the clinical microbiologist? the clinical microbiology laboratory?
2. What general guidelines should be followed in collecting and handling clinical specimens?
3. Define the following terms: specimen, swab, catheter, and sputum.
4. What are some transport problems associated with stool specimens? anaerobic cultures? urine specimens?

35.2 IDENTIFICATION OF MICROORGANISMS FROM SPECIMENS

The clinical microbiology laboratory can provide preliminary or definitive identification of microorganisms based on (1) microscopic examination of specimens, (2) study of the growth and biochemical characteristics of isolated microorganisms (pure cultures), (3) immunologic tests that detect antibodies or microbial antigens, (4) bacteriophage typing (restricted to research settings and the CDC), and (5) molecular methods.

Microscopy

Wet-mount, heat-fixed, or chemically fixed specimens can be examined with an ordinary bright-field microscope. Examination can be enhanced with either phase-contrast or dark-field microscopy. The latter is the procedure of choice for the detection of spirochetes in skin lesions associated with early syphilis or Lyme disease. The fluorescence microscope can be used to identify certain acid-fast microorganisms (*Mycobacterium tuberculosis*) after they are stained with fluorochromes such as auramine-rhodamine. Some morphological and genetic features used in classification and identification of microorganisms are presented in section 19.4 and in table 19.4. Direct microscopic examination of most specimens suspected of containing fungi can be made as well. Identification of hyphae in clinical specimens is a presumptive positive result for fungal infection. Definitive identification of most fungi is based on the morphology of reproductive structures (spores). The light microscope (section 2.2)

Many stains that can be used to examine specimens for specific microorganisms have been described. Two of the more widely used bacterial stains are the Gram stain and the acid-fast stain. Because these stains are based on the chemical composition of cell walls, they are not useful in identifying bacteria without cell walls (e.g., mycoplasmas). Lactophenol aniline (cotton) blue is typically used to stain fungi from cultures. Fungal infections (i.e., mold and yeast infections) often are diagnosed by direct microscopic examination of specimens using fluorescence. For example, the identification of molds often can be made if a portion of the specimen is mixed with a drop of 10% Calcofluor White stain on a glass slide. Concentrated wet mounts of blood, stool, or urine specimens can be examined microscopically for the presence of eggs, cysts, larvae, or vegetative cells of parasites. D'Antoni's iodine (1%) is often used to stain internal structures of parasites. Blood smears for apicomplexan (malaria) and flagellate (trypanosome) parasites are stained with Giemsa. Refer to standard references, such as the *Manual of Clinical Microbiology* published by the American Society for Microbiology, for details about other reagents and staining procedures.

The field of clinical microbiology changed dramatically in the 1980s when immunologists created hybrid cells (hybridomas) that secrete antibodies and live a very long time (*see Techniques & Applications 32.2*). Recall that each **hybridoma** cell and its progeny normally produce a **monoclonal antibody (mAb)** of a single specificity. Thus antibodies recognizing a single epitope are produced and used for diagnostics. One such method, known as immunofluorescence or immunohistochemistry, results from the chemical attachment of fluorescent molecules to mAbs; the mAb binds to a single epitope and the fluorescent molecule "reports" that binding. The technique is used to "stain" microorganisms, or clinical specimens thought to contain microorganisms. In the clinical microbiology laboratory, fluorescently labeled mAbs to viral or bacterial antigens have replaced polyclonal antisera for use in culture confirmation when accurate, rapid identification is required. With the use of sensitive techniques such as fluorescence microscopy, it is possible to perform antibody-based microbial identifications with improved accuracy, speed, and fewer organisms. Antibodies (section 32.7). Action of antibodies: Immune complex formation (section 32.8).

Immunofluorescence is a process in which fluorochromes (fluorescent dyes) are exposed to UV, violet, or blue light to make them fluoresce or emit visible light. Dyes such as rhodamine B or fluorescein isothiocyanate (FITC) can be coupled to antibody molecules without changing the antibody's capacity to bind to a specific antigen. Fluorochrome dyes also can be attached to antigens. There are two main kinds of fluorescent antibody assays: direct and indirect.

Direct immunofluorescence involves fixing the specimen (cell or microorganism) containing the antigen of interest onto a slide (**figure 35.4a**). Fluorescein-labeled antibodies are then added to the slide and incubated. The slide is washed to remove any unbound antibody and examined with the fluorescence microscope (*see figure 2.12*) for a yellow-green fluorescence. The pattern of fluorescence reveals the antigen's location. Direct immunofluorescence is used

Figure 35.4 Direct and Indirect Immunofluorescence. **(a)** In the direct fluorescent-antibody (DFA) technique, the specimen containing antigen is fixed to a slide. Fluorescently labeled antibodies that recognize the antigen are then added, and the specimen is examined with a fluorescence microscope for yellow-green fluorescence. **(b)** The indirect fluorescent-antibody technique (IFA) detects antigen on a slide as it reacts with an antibody directed against it. The antigen-antibody complex is located with a fluorescent antibody that recognizes antibodies. **(c)** Three infected nuclei in a cytomegalovirus (CMV) positive tissue culture. **(d)** Several infected cells in a herpes simplex virus positive tissue culture.

to identify antigens such as those found on the surface of group A streptococci and to diagnose enteropathogenic *Escherichia coli, Neisseria meningitidis, Salmonella* spp., *Shigella sonnei, Listeria monocytogenes, Haemophilus influenzae* type b, and the rabies virus. *The light microscope: The fluorescence microscope (section 2.2)*

Indirect immunofluorescence (figure 35.4*b*) is used to detect the presence of antibodies in serum following an individual's exposure to microorganisms. In this technique a known antigen is fixed onto a slide. The test antiserum is then added, and if the specific antibody is present, it reacts with antigen to form a complex. When fluorescein-labeled antibodies are added, they react with the fixed antibody. After incubation and washing, the slide is examined with the fluorescence microscope. The occurrence of fluorescence shows that antibody specific to the test antigen is present in the serum. Indirect immunofluorescence is used to identify the presence of *Treponema pallidum* antibodies in the diagnosis of syphilis (treponemal antibody absorption, FTA-ABS), as well as antibodies produced in response to other microorganisms.

Growth and Biochemical Characteristics

Typically microorganisms have been identified by their particular growth patterns and biochemical characteristics. These characteristics vary depending on whether the clinical microbiologist is dealing with viruses, fungi (yeasts, molds), parasites (protozoa, helminths), common gram-positive or gram-negative bacteria, rickettsias, chlamydiae, or mycoplasmas.

Viruses

Viruses are identified by isolation in conventional cell (tissue) culture, by immunodiagnosis (fluorescent antibody, enzyme immunoassay, radioimmunoassay, latex agglutination, and immunoperoxidase) tests, and by molecular detection methods such as nucleic acid probes and PCR amplification assays. Several types of systems are available for virus cultivation: cell cultures, embryonated hen's eggs, and experimental animals; these are discussed further in section 35.3.

Cell cultures are divided into three general classes:

1. Primary cultures consist of cells derived directly from tissues such as monkey kidney and mink lung cells that have undergone one or two passages (subcultures) since harvesting.
2. Semicontinuous cell cultures or low-passage cell lines are obtained from subcultures of a primary culture and usually consist of diploid fibroblasts that undergo a finite number of divisions.
3. Continuous or immortalized cell cultures, such as HEp-2 cells, are derived from transformed cells that are generally epithelial in origin. These cultures grow rapidly, are heteroploid (having a chromosome number that is not a simple multiple of the haploid number), and can be subcultured indefinitely.

Each type of cell culture favors the growth of a different array of viruses, just as bacterial culture media have differing selective and restrictive properties for growth of bacteria. Viral replication in cell cultures is detected in two ways: (1) by observing the presence or absence of cytopathic effects (CPEs), and (2) by hemadsorption. A **cytopathic effect** is an observable morphological change that occurs in cells because of viral replication. Examples include ballooning, binding together, clustering, or even death of the culture cells (*see figure 16.15*). During the incubation period of a cell culture, red blood cells can be added. Several viruses alter the plasma membrane of infected culture cells so that red blood cells adhere firmly to them. This phenomenon is called **hemadsorption.** *Virus reproduction (section 16.4)*

Embryonated hen's eggs can be used for virus isolation. There are three main routes of egg inoculation for virus isolation as different viruses grow best on different cell types. (1) the allantoic cavity, (2) the amniotic cavity, and (3) the chorioallantoic membrane (*see figure 16.13*). Egg tissues are inoculated with clinical specimens to determine the presence of virus; virus is revealed by the development of pocks on the chorioallantoic membrane, by the development of hemagglutinins in the allantoic and amniotic fluid, and by death of the embryo.

Embryonated chicken eggs and laboratory animals, especially suckling mice, may be used for virus isolation. Inoculated animals are observed for specific signs of disease or death. Several serological tests for viral identification make use of mAb-based immunofluorescence. These tests (figure 35.4*c,d*) detect viruses such as the cytomegalovirus and herpes simplex virus in tissue cultures.

---

1. Name two specimens for which microscopy would be used in the initial diagnosis of an infectious disease.
2. Name three general classes of cell cultures.
3. Explain two ways by which the presence of viral replication is detected in cell culture.
4. What are the three main routes of egg inoculation for virus isolation?
5. What are the advantages of using monoclonal antibody (mAb) immunofluorescence in the identification of viruses?

Fungi

Fungal cultures remain the standard for the recovery of fungi from patient specimens; however, the time needed to culture fungi varies anywhere from a few days to several weeks depending on the organism. For this reason, fungal cultures demonstrating no growth should be maintained for a minimum of 30 days before they are discarded as a negative result. Cultures should be evaluated for rate and appearance of growth on at least one selective and one nonselective agar medium, with careful examination of colonial morphology, color, and dimorphism. Fungal serology (e.g., complement fixation and immunodiffusion) is designed to detect serum antibody but is limited to a few fungi (e.g., *Blastomyces dermatitidis, Coccidioides immitis, Histoplasma capsulatum*). The cryptococcal latex antigen test is routinely used for the direct detection of *Cryptococcus neoformans* in serum and cerebrospinal fluid. In the clinical laboratory, nonautomated and automated methods for rapid identification (minutes to hours) are used to detect most yeasts. Any biochemical methods used to detect fungi should always be accompanied by morphological studies examining for pseudohyphae, yeast cell structure, chlamydospores, and so on.

Parasites
Culture of parasites from clinical specimens is not routine. Identification and characterization of ova, trophozoites, and cysts in the specimen result in the definitive diagnosis of parasitic infection. This is typically accomplished by direct microscopic evaluation of the clinical specimen. Typical histological staining of blood, negative staining of other body fluids, and immunofluorescence staining are routinely used in the identification of parasites. Some serological tests also are available.

Bacteria
Isolation and growth of bacteria are required before many diagnostic tests can be used to confirm the identification of the pathogen. The presence of bacterial growth usually can be recognized by the development of colonies on solid media or turbidity in liquid media. The time for visible growth to occur is an important variable in the clinical laboratory. For example, most pathogenic bacteria require only a few hours to produce visible growth, whereas it may take weeks for colonies of mycobacteria or mycoplasmas to become evident. The clinical microbiologist as well as the clinician should be aware of reasonable reporting times for various cultures.

The initial identity of a bacterial organism may be suggested by (1) the source of the culture specimen; (2) its microscopic appearance and gram reaction; (3) its pattern of growth on selective, differential, or metabolism-determining media (**table 35.1**); and (4) its hemolytic, metabolic, and fermentative properties on the various media (table 35.1; *see also table 19.5*). After the microscopic and growth characteristics of a pure culture of bacteria are examined, specific biochemical tests can be performed. Some of the most common biochemical tests used to identify bacterial isolates are listed in **table 35.2.** Classic dichotomous keys are coupled with the biochemical tests for the identification of bacteria from specimens. Generally, fewer than 20 tests are required to identify clinical bacterial isolates to the species level (**figure 35.5**). Microbial nutrition (chapter 5)

Rickettsias
Although rickettsias, chlamydiae, and mycoplasmas are bacteria, they differ from other bacterial pathogens in a variety of ways. Therefore the identification of these three groups is discussed separately. Rickettsias can be diagnosed by immunoassays or by isolation of the microorganism. Because isolation is both hazardous to the clinical microbiologist and expensive, immunological methods are preferred. Isolation of rickettsias and diagnosis of rickettsial diseases are generally confined to reference and specialized research laboratories.

Chlamydiae
Chlamydiae can be demonstrated in tissues and cell scrapings with Giemsa staining, which detects the characteristic intracellular inclusion bodies (*see figure 21.14*). Immunofluorescent stain-

ing of tissues and cells with monoclonal antibody reagents is a more sensitive and specific means of diagnosis. The most sensitive methods for demonstrating chlamydiae in clinical specimens involve nucleic acid sequencing and PCR-based methods. Techniques for determining microbial taxonomy and phylogeny: Molecular characteristics (section 19.4)

Mycoplasmas
The most routinely used techniques for identification of the mycoplasmas are immunological (hemagglutinin), complement-fixing antigen-antibody reactions using the patient's sera, and PCR (depending on the lab). These microorganisms are slow growing; therefore positive results from isolation procedures are rarely available before 30 days—a long delay with an approach that offers little advantage over standard techniques. DNA probes are also used for the detection of *Mycoplasma pneumoniae* in clinical specimens.

---

1. How can fungi and parasites be detected in a clinical specimen? Rickettsias? Chlamydiae? Mycoplasmas?
2. Why must the clinical microbiologist know what are reasonable reporting times for various microbial specimens?
3. How can a clinical microbiologist determine the initial identity of a bacterium?
4. Describe a dichotomous key that could be used to identify a bacterium.

Rapid Methods of Identification
Clinical microbiology has benefited greatly from technological advances in equipment, computer software and data bases, molecular biology, and immunochemistry (**Microbial Tidbits 35.2**). With respect to the detection of microorganisms in specimens, there has been a shift from the multistep methods previously discussed to unitary procedures and systems that incorporate standardization, speed, reproducibility, miniaturization, mechanization, and automation. These rapid identification methods can be divided into three categories: (1) manual biochemical "Kit" systems, (2) mechanized/automated systems, and (3) immunologic systems.

One example of a "kit approach" biochemical system for the identification of members of the family *Enterobacteriaceae* and other gram-negative bacteria is the API 20E system. It consists of a plastic strip with 20 microtubes containing dehydrated biochemical substrates that can detect certain biochemical characteristics (**figure 35.6**). The biochemical substrates in the 20 microtubes are inoculated with a pure culture of bacteria evenly suspended in sterile physiological saline. After 5 hours or overnight incubation, the 20 test results are converted to a seven- or nine-digit profile (**figure 35.7**). This profile number can be used with a computer or a book called the *API Profile Index* to identify the bacterium.

Clinical laboratory scientists (medical technologists) are the trained and certified workforce that is the front line in laboratory-based disease detection. They staff the sentinel laboratories that receive patient specimens. The production of faster and more specific detection technologies has allowed them to rapidly and

Table 35.1	Isolation of Pure Bacterial Cultures from Specimens

Selective Media

A selective medium is prepared by the addition of specific substances to a culture medium that will permit growth of one group of bacteria while inhibiting growth of some other groups. These are examples:

Salmonella-Shigella agar (SS) is used to isolate *Salmonella* and *Shigella* species. Its bile salt mixture inhibits many groups of coliforms. Both *Salmonella* and *Shigella* species produce colorless colonies because they are unable to ferment lactose. Lactose-fermenting bacteria will produce pink colonies.

Mannitol salt agar (MS) is used for the isolation of staphylococci. The selectivity is obtained by the high (7.5%) salt concentration that inhibits growth of many groups of bacteria. The mannitol in this medium helps in differentiating the pathogenic from the nonpathogenic staphylococci, as the former ferment mannitol to form acid while the latter do not. Thus this medium is also differential.

Bismuth sulfite agar (BS) is used for the isolation of *Salmonella enterica* serovar Typhi, especially from stool and food specimens. *S. enterica* serovar Typhi reduces the sulfite to sulfide, resulting in black colonies with a metallic sheen.

The addition of blood, serum, or extracts to tryptic soy agar or broth will support the growth of many fastidious bacteria. These media are used primarily to isolate bacteria from cerebrospinal fluid, pleural fluid, sputum, and wound abscesses.

Differential Media

The incorporation of certain chemicals into a medium may result in diagnostically useful growth or visible change in the medium after incubation. These are examples:

Eosin methylene blue agar (EMB) differentiates between lactose fermenters and nonlactose fermenters. EMB contains lactose, salts, and two dyes—eosin and methylene blue. *E. coli,* which is a lactose fermenter, will produce a dark colony or one that has a metallic sheen. *S. enterica* serovar Typhi, a nonlactose fermenter, will appear colorless.

MacConkey agar is used for the selection and recovery of *Enterobacteriaceae* and related gram-negative rods. The bile salts and crystal violet in this medium inhibit the growth of gram-positive bacteria and some fastidious gram-negative bacteria. Because lactose is the sole carbohydrate, lactose-fermenting bacteria produce colonies that are various shades of red, whereas nonlactose fermenters produce colorless colonies.

Hektoen enteric agar is used to increase the yield of *Salmonella* and *Shigella* species relative to other microbiota. The high bile salt concentration inhibits the growth of gram-positive bacteria and retards the growth of many coliform strains.

Blood agar: addition of citrated blood to tryptic soy agar makes possible variable hemolysis, which permits differentiation of some species of bacteria. Three hemolytic patterns can be observed on blood agar.

1. α-hemolysis—greenish to brownish halo around the colony (e.g., *Streptococcus gordonii, Streptococcus pneumoniae*).
2. β-hemolysis—complete lysis of blood cells resulting in a clearing effect around growth of the colony (e.g., *Staphylococcus aureus* and *Streptococcus pyogenes*).
3. Nonhemolytic—no change in medium (e.g., *Staphylococcus epidermidis* and *Staphylococcus saprophyticus*).

Media to Determine Biochemical Reactions

Some media are used to test bacteria for particular metabolic activities, products, or requirements. These are examples:

Urea broth is used to detect the enzyme urease. Some enteric bacteria are able to break down urea, using urease, into ammonia and CO_2.

Triple sugar iron (TSI) agar contains lactose, sucrose, and glucose plus ferrous ammonium sulfate and sodium thiosulfate. TSI is used for the identification of enteric organisms based on their ability to attack glucose, lactose, or sucrose and to liberate sulfides from ammonium sulfate or sodium thiosulfate.

Citrate agar contains sodium citrate, which serves as the sole source of carbon, and ammonium phosphate, the sole source of nitrogen. Citrate agar is used to differentiate enteric bacteria on the basis of citrate utilization.

Lysine iron agar (LIA) is used to differentiate bacteria that can either deaminate or decarboxylate the amino acid lysine. LIA contains lysine, which permits enzyme detection, and ferric ammonium citrate for the detection of H_2S production.

Sulfide, indole, motility (SIM) medium is used for three different tests. One can observe the production of sulfides, formation of indole (a metabolic product from tryptophan utilization), and motility. This medium is generally used for the differentiation of enteric organisms.

accurately identify agents of disease. However, the bioterror incidents of 2001 spawned a renewed demand for "better, faster, and smarter" microbial detection and identification technologies. While nucleic acid-based detection systems, like PCR, have garnered much attention as the basis of newer detection systems, antibody-based identification technologies are still considered more flexible and easier to modify. Traditional antibody-based detection technologies are being linked to sophisticated reporting systems that provide "med techs" with an ever-increasing cadre of cutting-edge technology. Examples of more recent microbial identification technologies include biosensors based on: (1) microfluidic antigen sensors, (2) real time (20-minute) PCR, (3) highly

Table 35.2	Some Common Biochemical Tests Used by Clinical Microbiologists in the Diagnosis of Bacteria from a Patient's Specimen	
Biochemical Test	**Description**	**Laboratory Application**
Carbohydrate fermentation	Acid and/or gas are produced during fermentative growth with sugars or sugar alcohols.	Fermentation of specific sugars used to differentiate enteric bacteria as well as other genera or species.
Casein hydrolysis	Detects the presence of caseinase, an enzyme able to hydrolyze milk protein casein. Bacteria that use casein appear as colonies surrounded by a clear zone.	Used to cultivate and differentiate aerobic actinomycetes based on casein utilization. For example, *Streptomyces* uses casein and *Nocardia* does not.
Catalase	Detects the presence of catalase, which converts hydrogen peroxide to water and O_2.	Used to differentiate *Streptococcus* ($-$) from *Staphylococcus* ($+$) and *Bacillus* ($+$) from *Clostridium* ($-$).
Citrate utilization	When citrate is used as the sole carbon source, this results in alkalinization of the medium.	Used in the identification of enteric bacteria. *Klebsiella* ($+$), *Enterobacter* ($+$), *Salmonella* (often $+$); *Escherichia* ($-$), *Edwardsiella* ($-$).
Coagulase	Detects the presence of coagulase. Coagulase causes plasma to clot.	This is an important test to differentiate *Staphylococcus aureus* ($+$) from *S. epidermidis* ($-$).
Decarboxylases (arginine, lysine, ornithine)	The decarboxylation of amino acids releases CO_2 and amine.	Used in the identification of enteric bacteria.
Esculin hydrolysis	Tests for the cleavage of a glycoside.	Used in the differentiation of *Staphylococcus aureus*, *Streptococcus mitis,* and others ($-$) from *S. bovis, S. mutans,* and enterococci ($+$).
β-galactosidase (ONPG) test	Demonstrates the presence of an enzyme that cleaves lactose to glucose and galactose.	Used to separate enterics (*Citrobacter* $+$, *Salmonella* $-$) and to identify pseudomonads.
Gelatin liquefaction	Detects whether or not a bacterium can produce proteases that hydrolyze gelatin and liquify solid gelatin medium.	Used in the identification of *Clostridium, Serratia, Pseudomonas,* and *Flavobacterium.*
Hydrogen sulfide (H_2S)	Detects the formation of hydrogen sulfide from the amino acid cysteine due to cysteine desulfurase.	Important in the identification of *Edwardsiella, Proteus,* and *Salmonella.*
IMViC (indole; methyl red; Voges-Proskauer; citrate)	The indole test detects the production of indole from the amino acid tryptophan. Methyl red is a pH indicator to determine whether the bacterium carries out mixed acid fermentation. VP (Voges-Proskauer) detects the production of acetoin. The citrate test determines whether or not the bacterium can use sodium citrate as a sole source of carbon.	Used to separate *Escherichia* (MR$+$, VP$-$, indole$+$) from *Enterobacter* (MR$-$, VP$+$, indole$-$) and *Klebsiella pneumoniae* (MR$-$, VP$+$, indole$-$); also used to characterize members of the genus *Bacillus.*
Lipid hydrolysis	Detects the presence of lipase, which breaks down lipids into simple fatty acids and glycerol.	Used in the separation of clostridia.
Nitrate reduction	Detects whether a bacterium can use nitrate as an electron acceptor.	Used in the identification of enteric bacteria which are usually $+$.
Oxidase	Detects the presence of cytochrome *c* oxidase that is able to reduce O_2 and artificial electron acceptors.	Important in distinguishing *Neisseria* and *Moraxella* spp. ($+$) from *Acinetobacter* ($-$), and enterics (all $-$) from pseudomonads ($+$).
Phenylalanine deaminase	Deamination of phenylalanine produces phenylpyruvic acid, which can be detected colorimetrically.	Used in the characterization of the genera *Proteus* and *Providencia.*
Starch hydrolysis	Detects the presence of the enzyme amylase, which hydrolyzes starch.	Used to identify typical starch hydrolyzers such as *Bacillus* spp.
Urease	Detects the enzyme that splits urea to NH_3 and CO_2.	Used to distinguish *Proteus, Providencia rettgeri,* and *Klebsiella pneumoniae* ($+$) from *Salmonella, Shigella* and *Escherichia* ($-$).

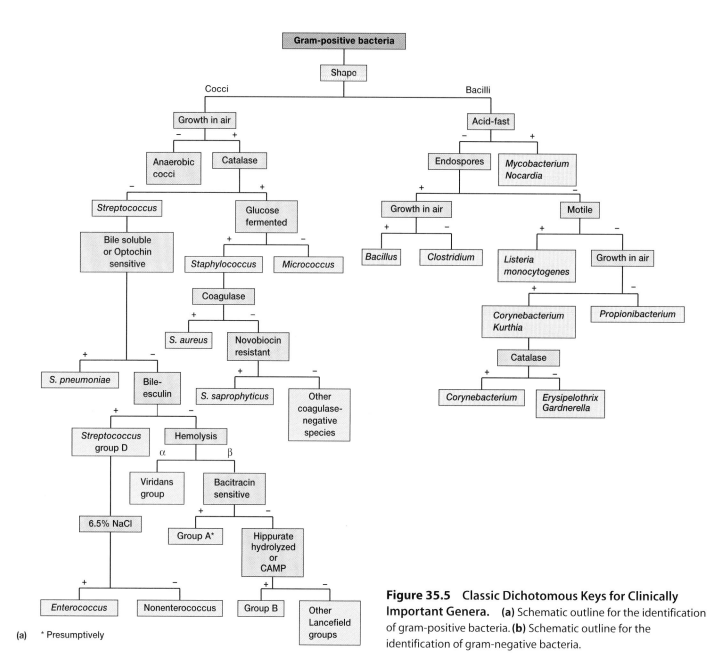

Figure 35.5 Classic Dichotomous Keys for Clinically Important Genera. **(a)** Schematic outline for the identification of gram-positive bacteria. **(b)** Schematic outline for the identification of gram-negative bacteria.

sensitive spectroscopy systems, and (4) liquid crystal amplification of microbial immune complexes. Some of these technologies are being used as part of military sentinel detection programs; others are awaiting approval by various licensing agencies before deployment into clinical laboratories. Additional technologies are expected as the demand for immediate, highly sensitive microbial detection increases globally. Thus the rapidly growing discipline of immunology has greatly aided the clinical microbiologist. Numerous technologies now exist that exploit the specificity and sensitivity of monoclonal antibodies to detect and identify microorganisms. These are briefly discussed here and expanded upon in section 35.3.

Recall that hybridomas have the antibody-producing capacity of their original plasma cells and are long-lived. The monoclonal anti-

bodies (mAbs) they produce have many applications. For example, they are routinely used in the typing of tissue; in the identification and epidemiological study of infectious microorganisms, tumors, and other surface antigens; in the classification of leukemias; in the identification of functional populations of different types of T cells; and in the identification and mapping of antigenic determinants (epitopes) on proteins. Importantly, mAbs can be conjugated with molecules that provide colorimetric, fluorometric, or enzymatic activity to report the binding of the mAb to specific microbial antigens. Numerous microbial detection kits are available to screen clinical specimens for the presence of specific microorganisms (**table 35.3**). mAbs have been produced against a wide variety of bacteria, viruses, fungi, and protozoans and have been made as cross-species or cross-genus reactivity to be used as an adjunct

Figure 35.5 *continued*

Microbial Tidbits

35.2 Biosensors: The Future Is Now

The 120-plus-year-old pathogen detection systems based on culture and biochemical phenotyping are being challenged. Fueled by the release of anthrax spores in the U.S. postal system, government agencies have been calling for newer technologies for the near-immediate detection and identification of microbes. In the past, detection technologies have traded speed for cost and complexity. The agar plate technique, refined by Robert Koch and his contemporaries in the 1880s, is a trusted and highly efficient method for the isolation of bacteria into pure cultures. Subsequent phenotyping biochemical methods, often using differential media in a manner similar to that used in the isolation step, then identifies common bacterial pathogens. Unfortunately, reliable results from this process often take several days. More rapid versions of the phenotyping systems can be very efficient, yet still require pure culture inoculations. The rapid immunological tests offer faster detection responses but may sacrifice sensitivity. Even DNA sequence comparisons, which are extremely accurate, may require significant time for DNA amplification and significant cost for reagents and sensitive readers. As usual, necessity has begat invention.

The more recent microbial detection systems, many of which are still untested in the clinical arena, sound like science fiction gizmos, yet promise a new age for near-immediate detection and identification of pathogens. These technologies are collectively referred to as "biosensors," and if the biosensor is integrated with a computer microchip for information management, it is then called a "biochip." Biosensors should ideally be capable of highly specific recognition so as to discriminate between nearest relatives, and "communicate" detection through some type of transducing system. Biosensors that detect specific DNA sequences, expressed proteins, and metabolic products have been developed that use optical (mostly fluorescence), electrochemical, or even mass displacement, to report detection. The high degree of recognition required to reduce false-positive results has demanded the uniquely specific, receptor-like capture that is associated with nucleic acid hybridization and antibody binding. Several microbial biosensors employ single-stranded DNA or RNA sequences, or antibody, for the detection component. The transducing or sensing component of biosensors may be markedly different, however. For example, microcantilever systems detect the increased mass of the receptor-bound ligand; the surface acoustic wave device detects change in specific gravity; the bulk quartz resonator monitors fluid density and viscosity; the quartz crystal microbalance measures frequency change in proportion to the mass of material deposited on the crystal; the micromirror sensor uses an optical fiber waveguide that changes reflectivity; and the liquid crystal-based system reports the reorientation of polarized light. Thus the specific capture of a ligand is reflected in the net change measured by each system and results in a signal that announces the initial capture event. Microchip control of the primary and subsequent secondary signals has resulted in automation of the detection process. The reliable detection of pathogens in complex specimens will be the real test as each of these technologies continues to compete for a place in the clinical laboratory.

Figure 35.6 A "Kit Approach" to Bacterial Identification. The API 20E manual biochemical system for microbial identification. **(a)** Positive and **(b)** negative results.

(a)

(b)

Figure 35.7 The API 20E Profile Number. The conversion of API 20E test results to the codes used in identification of unknown bacteria. The test results read top to bottom (and right to left in part *b*) correspond to the 7- and 9-digit codes when read in the right-to-left order. The tests required for obtaining a 7-digit code take an 18–24 hour incubation and will identify most members of the *Enterobacteriaceae*. The longer procedure that yields a 9-digit code is required to identify many gram-negative nonfermenting bacteria. The following tests are common to both procedures: ONPG (β-galactosidase); ADH (arginine dihydrolase); LDC (lysine decarboxylase); ODC (ornithine decarboxylase); CIT (citrate utilization); H_2S (hydrogen sulfide production); URE (urease); TDA (tryptophane deaminase); IND (indole production); VP (Voges-Proskauer test for acetoin); GEL (gelatin liquefaction); the fermentation of glucose (GLU), mannitol (MAN), inositol (INO), sorbitol (SOR), rhamnose (RHA), sucrose (SAC), melibiose (MEL), amygdalin (AMY), and arabinose (ARA); and OXI (oxidase test).

(a) Normal 7-digit code 5 144 572 = *E. coli.*

(b) 9-digit code 2 212 004 63 = *Pseudomonas aeruginosa*

Construction of a 9-digit profile

To the 7-digit profile illustrated in part *a*, 2 digits are added corresponding to the following characteristics:

$NO_2; N_2$	Reduction of nitrate to nitrite only
GAS:	Complete reduction of nitrate to N_2 gas or amines
MOT:	Observation of motility
MAC:	Growth on MacConkey medium
OF/O:	Oxidative utilization of glucose (OF-open)
OF/F:	Fermentative utilization of glucose (OF-closed)

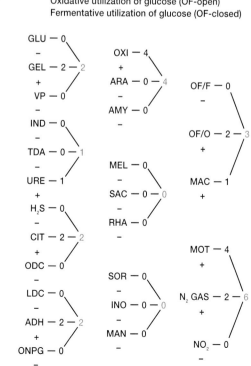

Table 35.3	Some Common Rapid Immunologic Test Kits for the Detection of Bacteria and Viruses in Clinical Specimens

Bactigen (Wampole Laboratories, Cranburg, N.J.)

The Bactigen kit is used for the detection of *Streptococcus pneumoniae, Haemophilus influenzae* type b, and *Neisseria meningitidis* groups A, B, C, and Y from cerebrospinal fluid, serum, and urine.

Culturette Group A Strep ID Kit (Marion Scientific, Kansas City, Mo.)

The Culturette kit is used for the detection of group A streptococci from throat swabs.

Directigen (Hynson, Wescott, and Dunning, Baltimore, Md.)

The Directigen Meningitis Test kit is used to detect *H. influenzae* type b, *S. pneumoniae,* and *N. meningitidis* groups A and C.

The Directigen Group A Strep Test kit is used for the direct detection of group A streptococci from throat swabs.

Gono Gen (Micro-Media Systems, San Jose, Calif.)

The Gono Gen kit detects *Neisseria gonorrhoeae.*

Ora Quick (OraAure Technologies, Bethleham, Pa.) Detects HIV in saliva in 10 minutes.

QuickVue *H. pylori* Test (Quidel, San Diego, Calif.)

A 7-minute test for detection of IgG antibodies against *Helicobacter pylori* in human serum or plasma.

Staphaurex (Wellcome Diagnostics, Research Triangle Park, N.C.)

Staphaurex screens and confirms *Staphylococcus aureus* in 30 seconds.

Directigen RSV (Becton Dickinson Microbiology Systems, Cockeysville, Md.)

By using a nasopharyngeal swab, the respiratory syncytial virus can be detected in 15 minutes.

SureCell Herpes (HSV) Test (Kodak, Rochester, N.Y.)

Detects the herpes (HSV) 1 and 2 viruses in minutes.

SUDS HIV-1 Test (Murex Corporation, Norcross, Ga.)

Detects antibodies to HIV-1 antigens in about 10 minutes.

method in the taxonomic identification of microorganisms. Those monoclonal antibodies that define species-specific antigens are extremely valuable in diagnostic reagents. Monoclonal antibodies that exhibit more restrictive specificity can be used to identify strains of biotypes within a species and in epidemiological studies involving the matching of microbial strains. Coupling sensitive visualization technologies such as fluorescence or scanning tunneling microscopy to mAb detection systems makes it possible to perform microbial identifications with improved accuracy, speed, and fewer organisms. Techniques & Applications 32.1: Donor selection for tissue or organ transplants

---

1. Describe in general how biochemical tests are used in the API 20E system to identify bacteria.
2. Why might cultures for some microorganisms be unavailable?

Bacteriophage Typing

Bacteriophages are viruses that attack members of a particular bacterial species, or strains within a species. **Bacteriophage (phage) typing** is based on the specificity of phage surface receptors for cell surface receptors. Only those bacteriophages that can attach to these surface receptors can infect bacteria and cause lysis. On a petri dish culture, lytic bacteriophages cause plaques on lawns of sensitive bacteria. These plaques represent infection by the virus (*see figure 16.14*). Viruses of *Bacteria* and *Archaea* (chapter 17)

In bacteriophage typing the clinical microbiologist inoculates the bacterium to be tested onto a petri plate. The plate is heavily and uniformly inoculated so that the bacteria will grow to form a solid sheet or lawn of cells. The plate is then marked off into squares (15 to 20 mm per side), and each square is inoculated with a drop of suspension from the different phages available for typing. After the plate is incubated for 24 hours, it is observed for plaques. The phage type is reported as a specific genus and species followed by the types that can infect the bacterium. For example, the series 10/16/24 indicates that this bacterium is sensitive to phages 10, 16, and 24, and belongs to a collection of strains, called a **phagovar,** that have this particular phage sensitivity. Bacteriophage typing remains a tool of the research and reference laboratory.

Molecular Methods and Analysis of Metabolic Products

The application of molecular technology enables the analysis of molecular characteristics of microorganisms in the clinical laboratory. Some of the most accurate approaches to microbial identification are through the analysis of proteins and nucleic acids. Examples include comparison of proteins; physical, kinetic, and regulatory properties of microbial enzymes; nucleic acid–base composition; nucleic acid hybridization; and nucleic acid sequencing (*see figure 19.12*). Three other molecular methods being widely used are nucleic acid probes, gas-liquid chromatography, and DNA fingerprinting. Techniques for determining microbial taxonomy and phylogeny (section 19.4)

Nucleic Acid–Based Detection Methods

Nucleic acid–based diagnostic methods for the detection and identification of microorganisms have become routine in many clinical microbiology laboratories. For example, DNA hybridization technology can identify a microorganism by probing its genetic composition. The use of cloned DNA as a probe is based upon the capacity of single-stranded DNA to bind (hybridize) with a complementary nucleic acid sequence present in test specimens to form a double-stranded DNA hybrid (**figure 35.8**). Thus DNA derived from one microorganism (the probe) is used to search for others containing the same sequence. Hybridization reactions may be applied to purified DNA preparations, to bacterial colonies, or to clinical specimens such as tissue, serum, sputum, and pus. DNA probes have been developed that bind to complementary strands of ribosomal RNA. These DNA:rRNA hybrids are more sensitive than conventional DNA probes, give

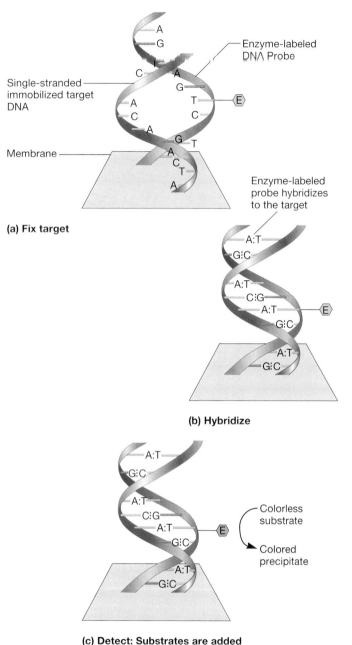

(a) Fix target

(b) Hybridize

(c) Detect: Substrates are added

Figure 35.8 Basic Steps in a DNA Probe Hybridization Assay. **(a)** Single-stranded target nucleic acid is bound to a membrane. A DNA probe with attached enzyme (E) also is employed. **(b)** The probe is added to the membrane. If the probe hybridizes to the target DNA, a double-stranded DNA hybrid is formed. **(c)** A colorless substrate is added. The enzyme attached to the probe converts the substrate to a colored precipitate. This detection system is semiquantitative, in that color intensity is proportional to the quantity of hybridized target nucleic acid present. **(d)** The pattern shows results for various strains isolated from 17 patients. Bands were developed by DNA hybridization using probes specific to genes from several *M. tuberculosis* strains. Lanes 1, 10, and 20 provide size markers for reference. Patients A and B are infected with the same common strain of the pathogen.

(d)

results in 2 hours or less, and require the presence of fewer microorganisms. DNA probe sensitivity can be increased by over one million-fold if the target DNA is first amplified using PCR. DNA:rRNA probes are available or are currently being developed for most clinically important microorganisms. The polymerase chain reaction (section 14.3)

The nucleotide sequence of small subunit ribosomal RNA (rRNA) can be used to identify bacterial genera (*see figure 19.10*). Usually, the rRNA encoding gene or gene fragment is amplified by PCR. After nucleotide sequencing, the rRNA gene is compared with those in the international database maintained by the National Center for Biotechnology (NCBI). This method of bacterial identification, called **ribotyping,** is based on the high level of 16s rRNA conservation among bacteria. Another approach, genomic fingerprinting, is also used in identifying pathogens. This does not involve nucleotide sequencing; rather, it compares the similarity of specific DNA fragments generated by restriction endonuclease digestion. BOX-, ERIC-, and REP PCR are also described in section 19.4 (*see figure 19.11*).

Gas-Liquid Chromatography

During chromatography a chemical mixture carried by a liquid or gas is separated into its individual components due to processes such as adsorption, ion-exchange, and partitioning between different solvent phases. In gas-liquid chromatography (GLC), specific microbial metabolites, cellular fatty acids, and products from the pyrolysis (a chemical change caused by heat) of whole bacterial cells are analyzed and identified. These compounds are easily removed from growth media by extraction with an organic solvent such as ether. The ether extract is then injected into the GLC system. Both volatile and nonvolatile acids can be identified. Based on the pattern of fatty acid production, common bacteria isolated from clinical specimens can be identified.

The reliability, precision, and accuracy of GLC have been improved significantly with continued advances in instrumentation; the introduction of instruments for high-performance liquid chromatography; and the use of mass spectrometry, nuclear magnetic resonance spectroscopy, and associated analytical techniques for the identification of components separated by the chromatographic process. These combined techniques can be used to discover specific chemical markers of various infectious disease agents by direct analysis of body fluids.

Plasmid Fingerprinting

As presented in section 3.5, a plasmid is an autonomously replicating extrachromosomal molecule of DNA. **Plasmid fingerprinting** identifies microbial isolates of the same or similar strains; related strains often contain the same number of plasmids with the same molecular weights and similar phenotypes. In contrast, microbial isolates that are phenotypically distinct have different plasmid fingerprints. Plasmid fingerprinting of many *E. coli, Salmonella, Campylobacter,* and *Pseudomonas* strains and species has demonstrated that this method often is more accurate than other phenotyping methods such as biotyping, antibiotic resistance patterns, phage typing, and serotyping.

The technique of plasmid fingerprinting involves five steps:

1. The bacterial strains are grown in broth or on agar plates.
2. The cells are harvested and lysed with a detergent.
3. The plasmid DNA is separated from the chromosomal DNA and then cut with specific restriction endonucleases.
4. The plasmid DNA is applied to agarose gels and electrophoretically separated.
5. The gel is stained with ethidium bromide, which binds to DNA, causing it to fluoresce under UV light. The plasmid DNA bands are then located.

Because the migration rate of plasmid DNA in agarose is inversely proportional to the molecular weight, plasmids of a different size appear as distinct bands in the stained gel. The molecular weight of each plasmid species can then be determined from a plot of the distance that each species has migrated versus the log of the molecular weights of plasmid markers of known size that have been electrophoresed simultaneously in the same gel (**figure 35.9**).

1. What is the basis for bacteriophage typing?
2. How can nucleic acid–based detection methods be used by the clinical microbiologist? Gas-liquid chromatography?
3. How can a suspect bacterium be plasmid fingerprinted?

Figure 35.9 Plasmid Fingerprinting. Agarose gel electrophoresis of plasmid DNA. A, B, C: plasmids that have not been digested by endonucleases. a, b, c: the same plasmids following restriction enzyme digestion.

35.3 CLINICAL IMMUNOLOGY

The culturing of certain viruses, bacteria, fungi, and protozoa from clinical specimens may not be possible because the methodology remains undeveloped (e.g., *Treponema pallidum;* hepatitis A, B, C; and Epstein-Barr virus), is unsafe (rickettsias and HIV), or is impractical for all but a few microbiology laboratories (e.g., mycobacteria, strict anaerobes, *Borrelia*). Cultures also may be negative because of prior antimicrobial therapy. (This is why it is so important to obtain a reliable sample prior to starting antimicrobial chemotherapy.) Under these circumstances, detection of antibodies or antigens may be quite valuable diagnostically.

Immunologic systems for the detection and identification of pathogens from clinical specimens are easy to use, give relatively rapid reaction endpoints, and are sensitive and specific (they give a low percentage of false positives and negatives). Some of the more popular immunologic rapid test kits for viruses and bacteria are presented in table 35.3.

Due to dramatic advances in clinical immunology, there has been a marked increase in the number, sensitivity, and specificity of serological tests. This increase reflects a better understanding of (1) immune cell surface antigens (CD antigens), (2) lymphocyte biology, (3) the production of monoclonal antibodies, and (4) the development of sensitive antibody-binding reporter systems. Furthermore, each individual's immunologic response to a microorganism is quite variable. As a result the interpretation of immunologic tests is sometimes difficult. For example, a single, elevated IgM titer does not distinguish between active and past infections. Rather an elevated IgG titer typically indicates an active infection, especially when subsidence of symptoms correlates with a four-fold (or greater) decrease in antibody titer. Furthermore, a lack of a measurable antibody titer may reflect an organism's lack of immunogenicity or an insufficient time for an antibody response to develop following the onset of the infectious disease. Some patients are also immunosuppressed due to other disease processes and/or treatment procedures (e.g., cancer and AIDS patients) and therefore do not respond. For these reasons, test selection and timing of specimen collection are essential to the proper interpretation of immunologic tests. In this section some of the more common antibody-based techniques that are employed in the diagnosis of microbial and immunological diseases are discussed.

Serotyping

Serum is the liquid portion of blood (devoid of clotting factors) that contains many different components, especially the immunoglobulins or antibodies. **Serotyping** refers to the use of serum (antibodies) to specifically detect and identify other molecules. Serotyping can be used to identify specific white blood cells or the proteins on cell surfaces (*see Techniques & Applications 32.1*). Serotyping can also be used to differentiate strains (serovars or serotypes) of microorganisms that differ in the antigenic composition of a structure or product (**Techniques & Applications 35.3**). The serological identification of a pathogenic strain has diagnostic value. Often the symptoms of infections depend on the nature of the cell products released by the pathogen. Therefore it is sometimes possible to identify a pathogen serologically by testing for cell wall antigens. For example, there are 90 different strains of *Streptococcus pneumoniae*, each unique in the nature of its capsular material. These differences can be detected by antibody-induced capsular swelling

(termed the **Quellung reaction**) when the appropriate antiserum for a specific capsular type is used (**figure 35.10**).

Agglutination

As noted in figure 32.25, when an immune complex is formed by cross-linking cells or particles with specific antibodies, it is called an agglutination reaction. Agglutination reactions usually form visible aggregates or clumps (**agglutinates**) that can be seen with the unaided eye. Direct agglutination reactions are very useful in the diagnosis of certain diseases. For example, the **Widal test** is a reaction involving the agglutination of typhoid bacilli when they are mixed with serum containing typhoid antibodies from an individual who has typhoid fever. Action of antibodies: Immune complex formation (section 32.8)

Techniques have also been developed that employ microscopic synthetic latex spheres coated with antigens. These coated microspheres are extremely useful in diagnostic agglutination reactions. For example, the modern pregnancy test detects the elevated level of human chorionic gonadotropin (hCG) hormone that occurs in a woman's urine and blood early in pregnancy. Latex agglutination tests are also used to detect antibodies that develop during certain mycotic, helminthic, and bacterial infections, and in drug testing.

Hemagglutination usually results from antibodies cross-linking red blood cells through attachment to surface antigens and is routinely used in blood typing. In addition, some viruses can accomplish **viral hemagglutination.** For example, if a person has a certain viral disease, such as measles, antibodies will be present in the serum to react with the measles viruses and neutralize them. Normally, hemagglutination occurs when measles viruses and red blood cells are mixed. However, red blood cells may be mixed first with a person's serum followed by the addition of virions. If no hemagglutination occurs, the serum antibodies have neutralized the measles viruses. This is considered a positive test result for the presence of virus-specific antibodies (**figure 35.11**).

Techniques & Applications

35.3   History and Importance of Serotyping

In the early 1930s Rebecca Lancefield (1895–1981) recognized the importance of serological tests. She developed a classification system for the streptococci based on the antigenic nature of cell wall carbohydrates. Her system is now known as the **Lancefield system** in which each different serotype is a Lancefield group and identified by a letter (A through T). This scheme is based on specific antibody agglutination reactions with cell wall carbohydrate antigens (C polysaccharides) extracted from the streptococci. Lancefield also showed that further subdividing of the group A streptococci into specific serological types was possible, based on the presence of type-specific M (protein) antigens.

Escherichia coli, Salmonella, and other bacteria are routinely serotyped with specific antigen-antibody reactions involving flagella (H) antigens, capsular (K) antigens, and lipopolysaccharide (O) antigens. Among *E. coli* there are over 167 different O antigens.

The current value of serotyping may be seen in the fact that *E. coli* O55, O111, and O127 serotypes are the ones most frequently associated with infantile diarrhea and *E. coli* O157:H7 is largely to blame for many life-threatening *E. coli* outbreaks. Thus the serotype of *E. coli* from stool samples is of diagnostic value and aids in identifying the source of the infection.

Figure 35.10   Serotyping.   *Streptococcus pneumoniae* has reacted with a specific pneumococcal antiserum leading to capsular swelling (the Quelling reaction). The capsules seen around the bacteria indicated potential virulence.

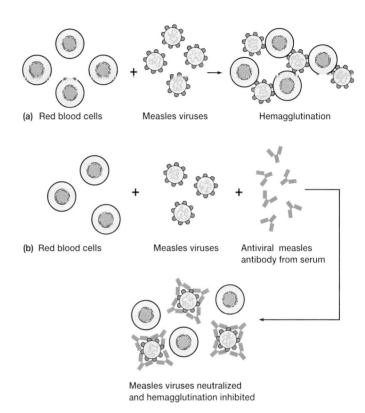

Figure 35.11   Viral Hemagglutination.   (a) Certain viruses can bind to red blood cells causing hemagglutination. **(b)** If serum containing specific antibodies to the virus is mixed with the red blood cells, the antibodies will neutralize the virus and inhibit hemagglutination (a positive test).

This hemagglutination inhibition test is widely used to diagnose influenza, measles, mumps, mononucleosis, and other viral infections.

Agglutination tests are also used to measure antibody titer. In the tube or well agglutination test, a specific amount of antigen is added to a series of tubes or shallow wells in a microtiter plate (**figure 35.12**). Serial dilutions of serum (1/20, 1/40, 1/80, 1/160, etc.) containing the antibody are then added to each tube or well. The greatest dilution of serum showing an agglutination reaction is determined, and the reciprocal of this dilution is the serum antibody titer.

---

1. What is serology?
2. When would you use the Widal test?
3. Why does hemagglutination occur and how can it be used in the clinical laboratory?

Complement Fixation

When complement binds to an antigen-antibody complex, it becomes "fixed" and "used up." Complement fixation tests are very sensitive and can be used to detect extremely small amounts of an antibody for a suspect microorganism in an individual's serum. A known antigen is mixed with test serum lacking complement (**figure 35.13a**). When immune complexes have had time to form, complement is added (figure 35.13b) to the mixture. If immune complexes are present, they will fix and consume complement. Afterward, sensitized indicator cells, usually sheep red blood cells previously coated with complement-fixing antibodies, are added to the mixture. Lysis of the indicator cells (figure 35.13c) results if immune complexes do not form in part *a* of the test because the antibodies are not present in the test serum. In the absence of antibodies, complement remains and lyses the indicator cells. On the other hand, if specific antibodies are present in the test serum and complement is consumed by the immune complexes, insufficient amounts of complement will be available to lyse the indicator cells. Absence of lysis shows that specific antibodies are present in the test serum. Complement fixation was once used in the diagnosis of syphilis (the Wassermann test) and is still used as a rapid, inexpensive screening method in the diagnosis of certain viral, fungal, rickettsial, chlamydial, and protozoan diseases.

Enzyme-Linked Immunosorbent Assay

The *enzyme-linked immunosorbent assay* (**ELISA**) has become one of the most widely used serological tests for antibody or antigen detection. This test involves the linking of various "label" enzymes to either antigens or antibodies. Two basic methods are used: the direct (also called the double antibody sandwich assay) and the indirect immunosorbent assay.

The double antibody sandwich assay is used for the detection of antigens (**figure 35.14a**). In this assay, specific antibody is placed in wells of a microtiter plate (or it may be attached to a membrane). The antibody is absorbed onto the walls, coating the plate. A test antigen (in serum, urine, etc.) is then added to each well. If the antigen reacts with the antibody, the antigen is retained

Figure 35.12   Agglutination Tests.   (a) Tube agglutination test for determining antibody titer. The titer in this example is 160 because there is no agglutination in the next tube in the dilution series (1/320). The blue in the dilution tubes indicates the presence of the patient's serum. **(b)** A microtiter plate illustrating hemagglutination. The antibody is placed in the wells (1–10). Positive controls (row 11) and negative controls (row 12) are included. Red blood cells are added to each well. If sufficient antibody is present to agglutinate the cells, they sink as a mat to the bottom of the well. If insufficient antibody is present, they form a pellet at the bottom. Can you read the different titers in rows A–H?

1/20   1/40   1/80   1/160   1/320   1/640   Control

Titer = 160

(a)

(b)

Agglutinated mat   Nonagglutinated pellet

Enlarged side view of wells

No immune complexes   Immune complexes

Free complement   Fixed complement

Indicator cell lysis (negative test)   No cell lysis (positive test)

(a)   **(b)**   **(c)**

Figure 35.13   Complement Fixation.   (a) Test serum is added to one test tube. A fixed amount of antigen is then added to both tubes. If antibody is present in the test serum, immune complexes form. **(b)** When complement is added, if complexes are present, they fix complement and consume it. **(c)** Indicator cells and a small amount of anti-erythrocyte antibody are added to the two tubes. If there is complement present, the indicator cells will lyse (a negative test): if the complement is consumed, no lysis will occur (a positive test).

when the well is washed to remove unbound antigen. A commercially prepared antibody-enzyme conjugate specific for the antigen is then added to each well. The final complex formed is an outer antibody-enzyme, middle antigen, and inner antibody—that is, it is a layered (Ab-Ag-Ab) sandwich. A substrate that the enzyme will convert to a colored product is then added, and any resulting product is quantitatively measured by optical density scanning of the plate. If the antigen has reacted with the absorbed

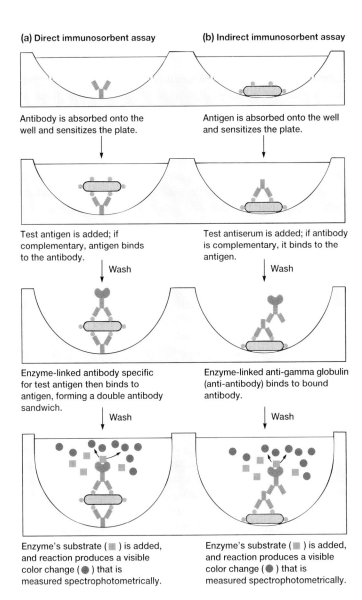

(a) Direct immunosorbent assay

Antibody is absorbed onto the well and sensitizes the plate.

Test antigen is added; if complementary, antigen binds to the antibody.

Wash

Enzyme-linked antibody specific for test antigen then binds to antigen, forming a double antibody sandwich.

Wash

Enzyme's substrate (▦) is added, and reaction produces a visible color change (●) that is measured spectrophotometrically.

(b) Indirect immunosorbent assay

Antigen is absorbed onto the well and sensitizes the plate.

Test antiserum is added; if antibody is complementary, it binds to the antigen.

Wash

Enzyme-linked anti-gamma globulin (anti-antibody) binds to bound antibody.

Wash

Enzyme's substrate (▦) is added, and reaction produces a visible color change (●) that is measured spectrophotometrically.

Figure 35.14   The ELISA or EIA Test.   (a) The direct or double antibody sandwich method for the detection of antigens. **(b)** The indirect immunosorbent assay for detecting antibodies. See text for details.

antibodies in the first step, the ELISA test is positive (i.e., it is colored). If the antigen is not recognized by the absorbed antibody, the ELISA test is negative because the unattached antigen has been washed away, and no antibody-enzyme is bound (it is colorless). This assay is currently being used for the detection of *Helicobacter pylori* infections, brucellosis, salmonellosis, and cholera. Many other antigens also can be detected by the sandwich method. For example, there are ELISA kits on the market that can test for many different food allergens.

The indirect immunosorbent assay detects antibodies rather than antigens. In this assay, antigen in appropriate buffer is incubated in the wells of a microtiter plate (figure 35.14*b*) and is absorbed onto the walls of the wells. Free antigen is washed away. Test antiserum is added, and if specific antibody is present, it binds

to the antigen. Unbound antibody is washed away. Alternatively the test sample can be incubated with a suspension of latex beads that have the desired antigen attached to their surface. After allowing time for antibody-antigen complex formation, the beads are trapped on a filter and unbound antibody is washed away. An anti-antibody that has been covalently coupled to an enzyme, such as horseradish peroxides, is added next. The antibody-enzyme complex (the conjugate) binds to the test antibody, and after unbound conjugate is washed away, the attached ligand is visualized by the addition of a chromogen. A **chromogen** is a colorless substrate acted on by the enzyme portion of the ligand to produce a colored product. The amount of test antibody is quantitated in the same way as an antigen is in the double antibody sandwich method. The indirect immunosorbent assay currently is being used to test for antibodies to human immunodeficiency virus (HIV) and rubella virus (German measles), and to detect certain drugs in serum. For example, antigen-coated latex beads are used in the SUDS HIV-1 test to detect HIV serum antibodies in about 10 minutes.

Immunoblotting (Western Blot)

Another immunologic technique used in the clinical microbiology laboratory is **immunoblotting.** Immunoblotting involves polyacrylamide gel electrophoresis of a protein specimen followed by transfer of the separated proteins to nitrocellulose sheets. Protein bands are then visualized by treating the nitrocellulose sheets with solutions of enzyme-tagged antibodies. This procedure demonstrates the presence of common and specific proteins among different strains of microorganisms (**figure 35.15**). Immunoblotting also can be used to show strain-specific immune responses to microorganisms, to serve as an important diagnostic indicator of a recent infection with a particular strain of microorganism, and to allow for prognostic implications with severe infectious diseases.

Immunoprecipitation

The **immunoprecipitation** technique detects soluble antigens that react with antibodies called precipitins. The precipitin reaction occurs when bivalent or multivalent antibodies and antigens are mixed in the proper proportions. The antibodies link the antigen to form a large antibody-antigen network or lattice that settles out of solution when it becomes sufficiently large (**figure 35.16***a*). Immunoprecipitation reactions occur only at the equivalence zone when there is an optimal ratio of antigen to antibody so that an insoluble lattice forms. If the precipitin reaction takes place in a test tube (figure 35.16*b*), a precipitation ring forms in the area in which the optimal ratio or equivalence zone develops. Action of antibodies: Immune complex formation (section 32.8)

Immunodiffusion

Immunodiffusion refers to a precipitation reaction that occurs between an antibody and antigen in an agar gel medium. Two techniques are routinely used: single radial immunodiffusion and double diffusion in agar.

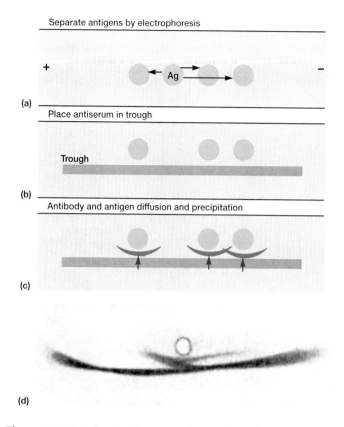

Separate antigens by electrophoresis

(a)

Place antiserum in trough

Trough

(b)

Antibody and antigen diffusion and precipitation

(c)

(d)

Figure 35.18   Classical Immunoelectrophoresis. (a) Antigens are separated in an agar gel by an electrical charge. **(b)** Antibody (antiserum) is then placed in a trough cut parallel to the direction of the antigen migration. **(c)** The antigens and antibodies diffuse through the agar and form precipitin arcs. **(d)** After staining, better visualization is possible.

the stream of cells flows past the laser beam, each fluorescent cell can be detected, counted, and even separated from the other cells in the suspension. The cytometer also can measure a cell's shape, size, and content of DNA and RNA. This technique has enabled the development of quantitative methods to assess antimicrobial susceptibility and drug cytotoxicity in a rapid, accurate, and highly reproducible way. The most outstanding contribution of this technique is the ability to detect the presence of heterogenous microbial populations with different responses to antimicrobial treatments.

Radioimmunoassay

The **radioimmunoassay (RIA)** technique has become an extremely important tool in biomedical research and clinical practice (e.g., in cardiology, blood banking, diagnosis of allergies, and endocrinology). Indeed, Rosalyn Yalow won the 1977 Nobel Prize in Physiology or Medicine for its development. RIA uses a purified antigen that is radioisotope-labeled and competes for antibody with unlabeled standard antigen or test antigen in experimental samples. The radioactivity associated with the antibody is

then detected by means of radioisotope analyzers and autoradiography (photographic emulsions that show areas of radioactivity). If there is much antigen in an experimental sample, it will compete with the radioisotope-labeled antigen for antigen-binding sites on the antibody, and little radioactivity will be bound. A large amount of bound radioactivity indicates that there is little antigen present in the experimental sample.

1. Explain how flow cytometry is both qualitative and quantitative (i.e., it can determine both the identity and the number of a specific cell type).
2. Describe the RIA technique.

35.4   SUSCEPTIBILITY TESTING

Many clinical microbiologists believe that determining the susceptibility of a microorganism to specific antibiotics is one of the most important tests performed in the clinical microbiology laboratory. Results can show the antibiotics to which a microorganism is most susceptible and the proper therapeutic dose needed to treat the infectious disease (*see figures 34.2b and 34.4*). Dilution susceptibility tests, disk diffusion tests (Kirby-Bauer method), the Etest, and drug concentration measurements in the blood are discussed in detail in section 34.3.

35.5   COMPUTERS IN CLINICAL MICROBIOLOGY

In the United States and other developed nations, computer systems in the clinical microbiology laboratory have replaced the handwritten mode of information acquisition and transmission. Computers improve the efficiency of the laboratory operation and increase the speed and clarity with which results can be reported. From a work-flow standpoint, the major functions involving the computer are test ordering, result entry, analysis of results, and report preparation.

Besides reporting laboratory tests, computers manage specimen logs, reports of overdue tests, quality control statistics, antimicrobial susceptibility probabilities, hospital epidemiological data, and many other items. The computer can be interfaced with various automated instruments for rapid and accurate calculation and transfer of clinical data. Direct entry into electronic notebooks is replacing written laboratory records. The personal digital assistant (PDA) has found a home in patient management, reporting histories, making diagnoses, and ordering prescriptions. The more stringent requirement for security and the increased applications will require the electronic laboratory notebook to be larger and more versatile than current PDAs, however.

1. Why is susceptibility testing so important in clinical microbiology?
2. What are some different ways in which computers can be used in the clinical microbiology laboratory? What are their major functions from the standpoint of work flow?

Summary

35.1 Specimens

a. The major focus of the clinical microbiologist is to isolate and identify microorganisms from clinical specimens accurately and rapidly. A clinical specimen represents a portion or quantity of biological material that is tested, examined, or studied to determine the presence or absence of specific microorganisms (**figure 35.1**).

b. Specimens may be collected by various methods that include swabs, needle aspiration, intubation, catheters, and clean-catch techniques. Each method is designed to ensure that only the proper material will be sent to the clinical laboratory (**figures 35.2** and **35.3**).

c. Immediately after collection the specimen must be properly handled and labeled. Speed in transporting the specimen to the clinical laboratory after it has been collected is of prime importance.

35.2 Identification of Microorganisms from Specimens

a. The clinical microbiology laboratory can provide preliminary or definitive identification of microorganisms based on (1) microscopic examination of specimens; (2) growth and biochemical characteristics of microorganisms isolated from cultures; and (3) immunologic techniques that detect antibodies or microbial antigens.

b. Viruses are identified by isolation in living cells or immunologic tests. Several types of living cells are available: cell culture, embryonated hen's eggs, and experimental animals. Rickettsial disease can be diagnosed immunologically or by isolation of the organism. Chlamydiae can be demonstrated in tissue and cell scrapings with Giemsa stain, which detects the characteristic intracellular inclusion bodies. The most routinely used techniques for identification of the mycoplasmas are immunologic. Identification of fungi often can be made if a portion of the specimen is mixed with a drop of 10% Calcofluor White stain. Wet mounts of stool specimens or urine can be examined microscopically for the presence of parasites.

c. Immunofluorescence is a process in which fluorochromes are irradiated with UV, violet, or blue light to make them fluoresce. These dyes can be coupled to an antibody. There are two main kinds of fluorescent antibody assays: direct and indirect (**figure 35.4**).

d. The initial identity of a bacterial organism may be suggested by (1) the source of the culture specimen; (2) its microscopic appearance; (3) its pattern of growth on selective, differential, enrichment, or characteristic media; and (4) its hemolytic, metabolic, and fermentative properties.

e. Rapid methods for microbial identification can be divided into three categories: (1) manual biochemical systems (**figure 35.6**), (2) mechanized/automated systems, and (3) immunologic systems.

f. Bacteriophage typing for bacterial identification is based on the fact that phage surface receptors bind to specific cell surface receptors. On a petri plate culture, bacteriophages cause plaques on lawns of bacteria with the proper receptors.

g. Various molecular methods and analyses of metabolic products also can be used to identify microorganisms. Examples include nucleic acid-based detection, gas-liquid chromatography, and plasmid fingerprinting.

35.3 Clinical Immunology

a. Serotyping refers to serological procedures used to differentiate strains (serovars or serotypes) of microorganisms that have differences in the antigenic composition of a structure or product (**figure 35.10**).

b. Agglutination reactions in vitro usually form aggregates or clumps (agglutinates) that are visible with the naked eye. Tests have been developed, such as the Widal test, latex microsphere agglutination reaction, hemagglutination, and viral hemagglutination, to detect antigen as well as to determine antibody titer (**figures 35.11** and **35.12**).

c. The complement fixation test can be used to detect a specific antibody for a suspect microorganism in an individual's serum (**figure 35.13**).

d. The enzyme-linked immunosorbent assay (ELISA) involves linking various enzymes to either antigens or antibodies. Two basic methods are involved: the double antibody sandwich method and the indirect immunosorbent assay (**figure 35.14**). The first method detects antigens and the latter, antibodies.

e. Immunoblotting involves polyacrylamide gel electrophoresis of a protein specimen followed by transfer of the separated proteins to nitrocellulose sheets and identification of specific bands by labeled antibodies (**figure 35.15**).

f. Immunoprecipitation reactions occur only when there is an optimal ratio of antigen and antibody to produce a lattice at the zone of equivalence, which is evidenced by a visible precipitate (**figure 35.16**).

g. Immunodiffusion refers to a precipitation reaction that occurs between antibody and antigen in an agar gel medium. Two techniques are routinely used: double diffusion in agar and single radial diffusion (**figure 35.17**).

h. In classical immunoelectrophoresis antigens are separated based on their electrical charge, then visualized by precipitation and staining (**figure 35.18**).

i. Flow cytometry and fluorescence allow single- or multiple-microorganism detection based on their cytometric parameters or by means of certain dyes called fluorochromes.

35.4 Susceptibility Testing

a. After the microorganism has been isolated, cultured, and/or identified, samples are used in susceptibility tests to find which method of control will be most effective. The results are provided to the physician as quickly as possible.

35.5 Computers in Clinical Microbiology

a. Computer systems in clinical microbiology are designed to replace handwritten information exchange and to speed data evaluation and report preparation.

Key Terms

agglutinates 876
bacteriophage (phage) typing 873
catheter 862
chromogen 879
clinical microbiologist 859
cytopathic effect 866
double diffusion agar assay
    (Öuchterlony technique) 880
enzyme-linked immunosorbent assay
    (ELISA) 877

flow cytometry 881
hemadsorption 866
hybridoma 864
immunoblotting 879
immunodiffusion 879
immunoelectrophoresis 881
immunofluorescence 865
immunoprecipitation 879

intubation 862
Lancefield system 876
monoclonal antibody (mAb) 864
needle aspiration 862
phagovar 873
plasmid fingerprinting 875
Quellung reaction 876
radioimmunoassay (RIA) 882

ribotyping 874
serotyping 876
single radial immunodiffusion (RID)
    assay 880
sputum 862
swab 862
viral hemagglutination 876
Widal test 876

Critical Thinking Questions

1. As more new ways of identifying the characteristics of microorganisms emerge, the number of distinguishable microbial strains also seems to increase. Why do you think this is the case?

2. Why are miniaturized identification systems used in clinical microbiology? Describe one such system and its advantage over classic dichotomous keys.

3. It has been speculated that as good as nucleic acid tests are in identifying viruses, antibody-based identification tests will be just as specific but easier and faster to develop and get to the marketplace. Do you agree or disagree? Defend your answer.

4. ELISA tests usually use a primary and secondary antibody. Why? What are the necessary controls one would need to perform to ensure that the antibody specificities are valid (that is, no false-positive or false-negative reactions)?

Learn More

de Las Rivas B.; Marcobal, A.; and Munoz, R. 2006. Development of a multilocus sequence typing method for analysis of *Lactobacillus plantarum* strains. *Microbiol.* 152:85–93.

Heikens, E.; Fleer, A.; Paauw, A.; Florijn, A.; and Fluit, A. C. 2005. Comparison of genotypic and phenotypic methods for species-level identification of clinical isolates of coagulase-negative staphylococci. *J. Clin. Microbiol.* 43:2286–90.

Jardi, R.; Rodriguez, F.; Buti, M.; Costa, X.; Cotrina, M.; Valdes, A.; Galimany, R.; Esteban, R.; and Guardia, J. 2001. Quantitative detection of hepatitis B virus DNA in serum by a new rapid real-time fluorescence PCR assay. *J. Virol. Hepat.* 8:465–71.

Tumpey, T. M.; Basler, C. F.; Aguilar, P. V.; Zeng, H.; Solórzano, A.; Swayne, D. E.; Cox, N. J.; Katz, J. M.; Taubenberger, J. K.; Palese, P.; and García-Sastre, A. 2005. Characterization of the reconstructed 1918 Spanish influenza pandemic virus. *Science* 310:77–80.

**Please visit the Prescott website at www.mhhe.com/prescott7
for additional references.**

36

The Epidemiology of Infectious Disease

This laboratory worker at the Centers for Disease Control and Prevention (CDC) is in the highest level of isolation (Level 4) to avoid contact with microorganisms and to prevent their escape into the environment.

PREVIEW

- The science of epidemiology deals with the occurrence and distribution of disease within a given population. Infectious disease epidemiology is concerned with organisms or agents responsible for the spread of infectious diseases in human and other animal populations.

- Statistics, an important working tool in this discipline, are used to determine morbidity and mortality rates.

- To trace the origin and manner of spread of an infectious disease outbreak, it is necessary to learn what pathogen is responsible.

- Epidemiologists identify the pathogen and investigate five links in the infectious disease cycle: (1) characteristics of the pathogen, (2) source and/or reservoir of the pathogen, (3) mode of transmission, (4) susceptibility of the host, and (5) exit mechanisms.

- Emerging and reemerging diseases and pathogens are a major global concern, as is the threat of bioterrorism.

- Bioterrorism will likely appear as a sudden increase in infectious disease—an epidemic. Newer laws now regulate use and transport of "select" agents that could be misused as agents of biocrimes or bioterrorism.

- Global travel requires global health considerations.

- The control of nosocomial (hospital acquired) infections has received increasing attention in recent years because of the number of individuals involved, increasing costs, and the length of hospital stays.

In this chapter we describe the epidemiological parameters that are studied in the infectious disease cycle. The practical goal of epidemiology is to establish effective recognition, control, prevention, and eradication measures within a given population. Because emerging and reemerging diseases and pathogens, as well as the threat of bioterrorism, are worldwide concerns, these topics are covered here. Global travel requires global health considerations that are continually monitored by epidemiologists. Nosocomial (hospital) acquired infections have increased in recent years, and a brief synopsis of their epidemiology also is presented.

The science of epidemiology originated and evolved in response to the great epidemic diseases such as cholera, typhoid fever, smallpox, and yellow fever (**Historical Highlights 36.1**). More recent epidemics of ebola, HIV/AIDS, cryptosporidiosis, enteropathogenic *Escherichia coli*, SARS, and avian influenza have underscored the importance of epidemiology in preventing global catastrophies caused by infectious diseases. Today its scope encompasses all diseases: infectious diseases, genetic abnormalities, metabolic dysfunction, malnutrition, neoplasms, psychiatric disorders, and aging. This chapter emphasizes only infectious disease epidemiology.

By definition, **epidemiology** [Greek *epi*, upon, and *demos*, people or population, and *logy*, study] is the science that evaluates the occurrence, determinants, distribution, and control of health and disease in a defined human population (**figure 36.1**). **Health** is the condition in which the organism (and all of its parts) performs its vital functions normally or properly. It is a state of physical and mental well-being and not merely the absence of disease. A **disease** [French *des*, from, and *aise*, ease] is an impairment of the normal state of an organism or any of its components that hinders the performance of vital functions. It is a response to environmental factors (e.g., malnutrition, industrial hazards, climate),

Epidemics of infectious disease are often compared with forest fires. Once fire has spread through an area, it does not return until new trees have grown up. Epidemics in humans develop when a large population of susceptible individuals is present. If most individuals are immune, then an epidemic will not occur.

—Andrew Cliff and Peter Haggett

Historical Highlights

36.1   John Snow—The First Epidemiologist

Much of what we know today about the epidemiology of cholera is based on the classic studies conducted by the British physician John Snow between 1849 and 1854. During this period a series of cholera outbreaks occurred in London, England, and Snow set out to find the source of the disease. Some years earlier when he was still a medical apprentice, Snow had been sent to help during an outbreak of cholera among coal miners. His observations convinced him that the disease was usually spread by unwashed hands and shared food, not by "bad" air or casual direct contact.

Thus when the outbreak of 1849 occurred, Snow believed that cholera was spread among the poor in the same way as among the coal miners. He suspected that water, and not unwashed hands and shared food, was the source of the cholera infection among the wealthier residents. Snow examined official death records and discovered that most of the victims in the Broad Street area had lived close to the Broad Street pump or had been in the habit of drinking from it. He concluded that cholera was spread by drinking water from the Broad Street pump, which was contaminated with raw sewage containing the disease agent. When the pump handle was removed, the number of cholera cases dropped dramatically.

In 1854 another cholera outbreak struck London. Part of the city's water supply came from two different suppliers: the Southwark and Vauxhall Company and the Lambeth Company. Snow interviewed cholera patients and found that most of them purchased their drinking water from the Southwark and Vauxhall Company. He also discovered that this company obtained its water from the Thames River below locations where Londoners had discharged their sewage. In contrast, the Lambeth Company took its water from the Thames before the river reached the city. The death rate from cholera was over eightfold lower in households supplied with Lambeth Company water. Water contaminated by sewage was transmitting the disease. Finally, Snow concluded that the cause of the disease must be able to multiply in water. Thus he nearly recognized that cholera was caused by a microorganism, though Robert Koch didn't discover the causative bacterium (*Vibrio cholerae*) until 1883.

To commemorate these achievements, the John Snow Pub now stands at the site of the old Broad Street pump. Those who complete the Epidemiologic Intelligence Program at the Centers for Disease Control and Prevention receive an emblem bearing a replica of a barrel of Whatney's Ale—the brew dispensed at the John Snow Pub.

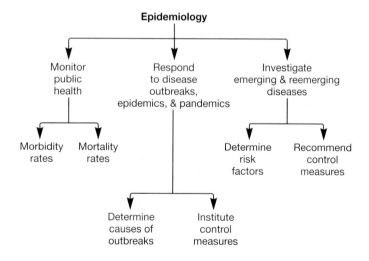

Figure 36.1 Epidemiology. Epidemiology is a multifaceted science that investigates diseases to discover their origin, evaluates diseases to assess their risk, and controls diseases to prevent future outbreaks.

specific infective agents (e.g., viruses, bacteria, fungi, protozoa, helminths), inherent defects of the body (e.g., various genetic or immunologic anomalies), or combinations of these.

Any individual who practices epidemiology is an **epidemiologist.** Epidemiologists are, in effect, disease detectives. Their major concerns are the discovery of the factors essential

to disease occurrence and the development of methods for disease prevention. In the United States, the Centers for Disease Control and Prevention (CDC, headquartered in Atlanta, GA) serves as the national agency for developing and applying disease prevention and control, environmental health, and health promotion and education activities. Its worldwide counterpart is the World Health Organization (WHO) located in Geneva, Switzerland.

36.1 EPIDEMIOLOGICAL TERMINOLOGY

When a disease occurs occasionally, and at irregular intervals in a human population, it is a **sporadic disease** (e.g., bacterial meningitis). When it maintains a steady, low-level frequency at a moderately regular interval, it is an **endemic** [Greek *endemos,* dwelling in the same people] **disease** (e.g., the common cold). **Hyperendemic diseases** gradually increase in occurrence frequency beyond the endemic level but not to the epidemic level (e.g., the common cold during winter months). An **outbreak** is the sudden, unexpected occurrence of a disease, usually focally or in a limited segment of a population (e.g., Legionnaires' disease). An **epidemic** [Greek *epidemios,* upon the people], on the other hand, is an outbreak affecting many people at once (i.e., there is a sudden increase in the occurrence of a disease above the expected level) (**figure 36.2**). Influenza is an example of a disease that may occur suddenly and unexpectedly in a family and often achieves epidemic status in a community. The first case in an epi-

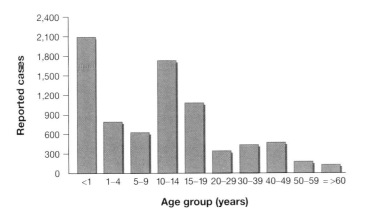

(a) Pertussis—Reported cases by age group, United States, 2000

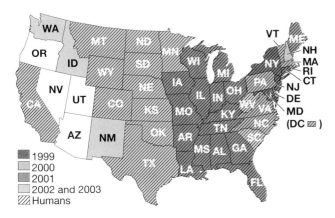

■ 1999
□ 2000
■ 2001
□ 2002 and 2003
▨ Humans

(b) West Nile virus cases were first reported in 1999 among a variety of animals and humans. Colors track its rapid progress across the United States in just 4 years.

Figure 36.2 Graphic Representation of Epidemiological Data. The Centers for Disease Control and Prevention collects and evaluates a number of parameters related to human health and disease. The data are then presented in various formats including **(a)** age and **(b)** geographic region.

demic is called the **index case.** Finally, a **pandemic** [Greek *pan*, all] is an increase in disease occurrence within a large population over a very wide region (usually the world). Usually, pandemic diseases spread among continents. The global H1N1 influenza outbreak of 1918 is a good example.

36.2 MEASURING FREQUENCY: THE EPIDEMIOLOGIST'S TOOLS

In order to determine if an outbreak, epidemic, or pandemic is occurring, epidemiologists must measure disease frequency at single time points and over time. The epidemiologist then uses statistics to analyze the data and determine risk factors and other factors associated with disease. **Statistics** is the branch of math-

ematics dealing with the collection, organization, and interpretation of numerical data. As a science particularly concerned with rates and the comparison of rates, epidemiology was the first medical field in which statistical methods were extensively used.

Measures of frequency usually are expressed as fractions. The numerator is the number of individuals experiencing the event—infection or other problem—and the denominator is the number of individuals in whom the event could have occurred, that is, the population at risk. The fraction is a proportion or ratio but is commonly called a rate because a time period is always specified. (A rate also can be expressed as a percentage.) In population statistics, rates usually are stated per 1,000 individuals, although other powers of 10 may be used for particular diseases (e.g., per 100 for very common diseases and per 10,000 or 100,000 for uncommon diseases).

A **morbidity rate** measures the number of individuals that become ill due to a specific disease within a susceptible population during a specific time interval. It is an incidence rate and reflects the number of new cases in a period. The rate is commonly determined when the number of new cases of illness in the general population is known from clinical reports. It is calculated as follows:

$$\text{Morbidity rate} = \frac{\substack{\text{number of new cases of a disease} \\ \text{during a specified period}}}{\text{number of individuals in the population}}$$

For example, if in one month there were 700 new cases of influenza per 100,000 individuals, then the morbidity rate would be expressed as 700 per 100,000 or 0.7%.

The **prevalence rate** refers to the total number of individuals infected in a population at any one time no matter when the disease began. The prevalence rate depends on both the incidence rate and the duration of the illness.

The **mortality rate** is the relationship of the number of deaths from a given disease to the total number of cases of the disease. The mortality rate is a simple statement of the proportion of all deaths that are assigned to a single cause. It is calculated as follows:

$$\text{Mortality rate} = \frac{\substack{\text{number of deaths due} \\ \text{to a given disease}}}{\substack{\text{size of the total population} \\ \text{with the same disease}}}$$

For example, if there were 15,000 deaths due to AIDS in a year, and the total number of people infected was 30,000, the mortality rate would be 15,000 per 30,000 or 1 per 2 or 50%.

The determination of morbidity, prevalence, and mortality rates aids public health personnel in directing health-care efforts to control the spread of infectious diseases. For example, a sudden increase in the morbidity rate of a particular disease may indicate a need for the implementation of preventive measures designed to reduce mortality.

1. What is epidemiology?
2. How would you define a disease? health?
3. What terms are used to describe the occurrence of a disease in a human population? in an animal population?
4. Define morbidity rate, prevalence rate, and mortality rate.

36.3 Recognition of an Infectious Disease in a Population

An infectious disease is a disease resulting from an infection by microbial agents such as viruses, bacteria, fungi, protozoa, and helminths. A **communicable disease** is an infectious disease that can be transmitted from person to person (not all infectious diseases are communicable; for example, rabies is an infectious disease acquired only through the bite of a rabid animal). The manifestations of an infectious or communicable disease can range from mild to severe to deadly depending on the agent and host. An epidemiologist studying an infectious disease is concerned with the causative agent, the source and/or reservoir of the disease agent (section 36.6), how it was transmitted, what host and environmental factors could have aided development of the disease within a defined population, and how best to control or eliminate the disease. These factors describe the natural history or cycle of an infectious disease.

Epidemiologists recognize an infectious disease in a population by using various surveillance methods. Surveillance is a dynamic activity that includes gathering information on the development and occurrence of a disease, collating and analyzing the data, summarizing the findings, and using the information to select control methods. Some combination of these surveillance methods is used most often for the:

1. Generation of morbidity data from case reports
2. Collection of mortality data from death certificates
3. Investigation of actual cases
4. Collection of data from reported epidemics
5. Field investigation of epidemics
6. Review of laboratory results: surveys of a population for antibodies against the agent and specific microbial serotypes, skin tests, cultures, stool analyses, etc.
7. Population surveys using valid statistical sampling to determine who has the disease
8. Use of animal and vector disease data
9. Collection of information on the use of specific biologics—antibiotics, antitoxins, vaccines, and other prophylactic measures
10. Use of demographic data on population characteristics such as human movements during a specific time of the year
11. Use of remote sensing and geographic information systems

As noted, surveillance may not always require the direct examination of cases. However, to accurately interpret surveillance data and study the course of a disease in individuals, epidemiologists and other medical professionals must be aware of the pattern of infectious diseases. Often infectious diseases have characteristic signs and symptoms. **Signs** are objective changes in the body, such as a fever or rash, that can be directly observed. **Symptoms** are subjective changes, such as pain and loss of appetite, that are personally experienced by the patient. The term symptom is often used in a broader scope to include the clinical signs. A **disease syndrome** is a set of signs and symptoms that are characteristic of the disease. Frequently additional laboratory tests are required for an accurate diagnosis because symptoms and readily observable signs are not sufficient for diagnosis.

The course of an infectious disease usually has a characteristic pattern and can be divided into several phases (**figure 36.3**). Knowledge of the pattern is essential in accurately diagnosing the disease.

1. The **incubation period** is the period between pathogen entry and the expression of signs and symptoms. The pathogen is spreading but has not reached a sufficient level to cause clinical manifestations. This period's length varies with disease.
2. The **prodromal stage** is the period in which there is an onset of signs and symptoms, but they are not yet specific enough to make a diagnosis. The patient often is contagious.
3. The illness period is the phase in which the disease is most severe and has characteristic signs and symptoms. The immune response has been triggered; B and T cells are becoming active.
4. In the period of decline, the signs and symptoms begin to disappear. The recovery stage often is referred to as **convalescence.**

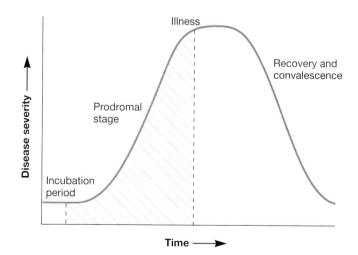

Figure 36.3 The Course of an Infectious Disease. Most infectious diseases occur in four stages. The duration of each stage is a characteristic feature of each disease. This is one of the major reasons the Centers for Disease Control and Prevention study the course of infectious diseases. The shaded area represents when a disease is typically communicable.

Remote Sensing and Geographic Information Systems: Charting Infectious Diseases

Remote sensing and geographic information systems are map-based tools that can be used to study the distribution, dynamics, and environmental correlates of microbial diseases. **Remote sensing (RS)** is the gathering of digital images of the Earth's surface from satellites and transforming the data into maps. A **geographic information system (GIS)** is a data management system that organizes and displays digital map data from RS and facilitates the analysis of relationships between mapped features. Statistical relationships often exist between mapped features and diseases in natural host or human populations. Examples include the location of the habitats of the malaria parasite and mosquito vectors in Mexico and Asia, Rift Valley fever in Kenya, Lyme disease in the United States, and African trypanosomiasis and schistosomiasis in both humans and livestock in the southeastern United States. RS and GIS may also permit the assessment of human risk from pathogens such as Sin Nombre virus (the virus that causes hantavirus pulmonary syndrome in North America). RS and GIS are most useful if disease dynamics and distributions are clearly related to mapped environmental variables. For example, if a microbial disease is associated with certain vegetation types or physical characteristics (e.g., elevation, precipitation), RS and GIS can identify regions where risk is relatively high.

Correlation with a Single Causative Agent

After an infectious disease has been recognized in a population, epidemiologists correlate the disease outbreak with a specific organism—its exact cause must be discovered (**Historical Highlights 36.2**). At this point the clinical or diagnostic microbiology laboratory enters the investigation. Its purpose is to isolate and identify the organism responsible for the disease.

36.4 RECOGNITION OF AN EPIDEMIC

As previously noted, an infectious disease epidemic is usually a short-term increase in the occurrence of the disease in a particular population. Two major types of epidemic are recognized: common source and propagated.

A **common-source epidemic** is characterized as having reached a peak level within a short period of time (1 to 2 weeks) followed by a moderately rapid decline in the number of infected patients (**figure 36.4a**). This type of epidemic usually results from a single common contaminated source such as food (food poisoning) or water (Legionnaires' disease).

A **propagated epidemic** is characterized by a relatively slow and prolonged rise and then a gradual decline in the number of individuals infected (figure 36.4b). This type of epidemic usually results from the introduction of a single infected individual into a susceptible population. The initial infection is then propagated to others in a gradual fashion until many individuals within the population are infected. An example is the increase in strep throat cases that coincides with new populations of sensitive children who arrive in classrooms. Only one infected child is necessary to initiate the epidemic.

To understand how epidemics are propagated, consider **figure 36.5.** At time 0, all individuals in this population are susceptible to a hypothetical pathogen. The introduction of an infected individual initiates the epidemic outbreak (lower curve), which spreads and reaches a peak (day 15). As individuals recover from the disease, they become immune and no longer transmit the pathogen (upper curve). The number of susceptible individuals therefore decreases. The decline in the number of susceptibles to the threshold density (the minimum number of individuals necessary to continue propagating the disease) coincides with the peak of the epidemic wave, and the incidence of new cases declines because the pathogen cannot propagate itself.

Historical Highlights

36.2   "Typhoid Mary"

In the early 1900s there were thousands of typhoid fever cases, and many died of the disease. Most of these cases arose when people drank water contaminated with sewage or ate food handled by or prepared by individuals who were shedding the typhoid fever bacterium (*Salmonella enterica* serovar Typhi). The most famous carrier of the typhoid bacterium was Mary Mallon.

Between 1896 and 1906 Mary Mallon worked as a cook in seven homes in New York City. Twenty-eight cases of typhoid fever occurred in these homes while she worked in them. As a result the New York City Health Department had Mary arrested and admitted to an isolation hospital on North Brother Island in New York's East River. Examination of Mary's stools showed that she was shedding

large numbers of typhoid bacteria though she exhibited no external symptoms of the disease. An article published in 1908 in the *Journal of the American Medical Association* referred to her as "Typhoid Mary," an epithet by which she is still known today. After being released when she pledged not to cook for others or serve food to them, Mary changed her name and began to work as a cook again. For five years she managed to avoid capture while continuing to spread typhoid fever. Eventually the authorities tracked her down. She was held in custody for 23 years until she died in 1938. As a lifetime carrier, Mary Mallon was positively linked with 10 outbreaks of typhoid fever, 53 cases, and 3 deaths.

Figure 36.4   Epidemic Curves.   (a) In a common-source epidemic, there is a rapid increase up to a peak in the number of individuals infected and then a rapid but more gradual decline. Cases usually are reported for a period that equals approximately one incubation period of the disease. **(b)** In a propagated epidemic the curve has a gradual rise and then a gradual decline. Cases usually are reported over a time interval equivalent to several incubation periods of the disease.

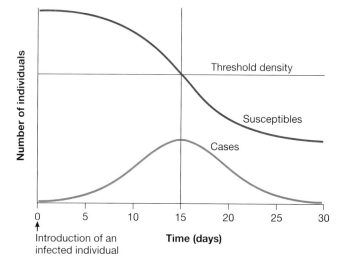

Figure 36.5   The Spread of an Imaginary Propagated Epidemic.   The lower curve represents the number of cases and the upper curve the number of susceptible individuals. Notice the coincidence of the peak of the epidemic wave with the threshold density of susceptible people.

Herd immunity is the resistance of a population to infection and pathogen spread because of the immunity of a large percentage of the population. The larger the proportion of those immune, the smaller the probability of effective contact between infective and susceptible individuals—that is, many contacts will be immune, and thus the population will exhibit a group resistance. A susceptible member of such an immune population enjoys an immunity that is not of his or her own making (not self-made) but instead arises because of membership in the group.

At times public health officials immunize large portions of the susceptible population in an attempt to maintain a high level of herd immunity. Any increase in the number of susceptible individuals may result in an endemic disease becoming epidemic. The proportion of immune to susceptible individuals must be constantly monitored because new susceptible individuals continually enter a population through migration and birth. In addition, pathogens can change through processes such as antigenic shift (see next paragraph) whereby immune individuals become susceptible again.

Pathogens cause endemic diseases because infected humans continually transfer them to others (e.g., sexually transmitted diseases) or because they continually reenter the human population from animal reservoirs (e.g., rabies). Other pathogens continue to evolve and may produce epidemics (e.g., AIDS, influenza virus [A strain], and *Legionella* bacteria). One way in which a pathogen changes is by **antigenic shift,** a major genetically determined change in the antigenic character of a pathogen (**figure 36.6**). An antigenic shift can be so extensive that the pathogen is no longer recognized by the host's immune system. For example, influenza viruses frequently change by recombination from one antigenic type to another. Antigenic shift also occurs through the hybridization of different influenza virus serovars; two serovars of a virus intermingle to form a new antigenic type. Hybridization may occur between an animal strain and a human strain of the virus. Even though resistance in the human population becomes so high that the virus can no longer spread (herd immunity), it can be transmitted to animals, where the hybridization takes place. Smaller antigenic changes also can take place by mutations in pathogen strains that help the pathogen avoid host immune responses. These smaller changes are called **antigenic drift.**

Whenever antigenic shift or drift occurs, the population of susceptible individuals increases because the immune system has not been exposed to the new mutant strain. If the percentage of susceptible people is above the threshold density (figure 36.5), the level of protection provided by herd immunity will decrease and the morbidity rate will increase. For example, the morbidity rates of influenza among school children may reach epidemic levels if the number of susceptible people rises above 30% for the whole population. As a result the goal of public health agencies is to make sure that at least 70% of the population is immunized against these diseases to provide the herd immunity necessary for protection of those who are not immunized.

---

1. How can epidemiologists recognize an infectious disease in a population? Define sign, symptom, and disease syndrome. What are the four phases seen during the course of an infection?
2. How can remote sensing and geographic information systems chart infectious diseases?
3. Differentiate between common-source and propagated epidemics.
4. Explain herd immunity. How does this protect the community?
5. What is the significance of antigenic shift and drift in epidemiology?

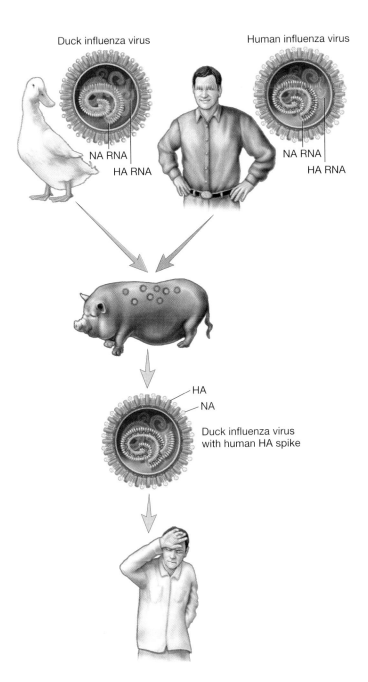

Figure 36.6   Antigenic Shift in Influenza Viruses.   Close habitation of humans, pigs, and ducks permits the multiple infection of the pigs with both human and duck influenza viruses. Exchange of genetic information then results in a new strain of influenza that is new to humans; humans have no immunity to the new strain.

36.5   THE INFECTIOUS DISEASE CYCLE: STORY OF A DISEASE

To continue to exist, a pathogen must reproduce and be disseminated among its hosts. Thus an important aspect of infectious disease epidemiology is a consideration of how reproduction and dissemination occur. The **infectious disease cycle** or **chain of in-**

fection represents these events in the form of an intriguing epidemiological mystery story (**figure 36.7**).

What Pathogen Caused the Disease?

The first link in the infectious disease cycle is the pathogen. After an infectious disease has been recognized in a population, epidemiologists must correlate the disease outbreak with a specific pathogen. The disease's exact cause must be discovered. This is where Koch's postulates, and modifications of them, are used to determine the etiology or cause of an infectious disease. At this point the clinical or diagnostic microbiology laboratory enters the investigation. Its purpose is to isolate and identify the pathogen that caused the disease and to determine the pathogen's susceptibility to antimicrobial agents or methods that may assist in its eradication.   The golden age of microbiology: Koch's postulates (section 1.4); Clinical microbiology and immunology (chapter 35)

Many pathogens can cause infectious diseases in humans and are discussed in detail in chapters 37 to 39. Often these pathogens are transmissible from one individual to another resulting in a communicable disease. Pathogens have the potential to produce disease (pathogenicity); this potential is a function of such factors as the number of pathogens, their virulence, and the nature and magnitude of host defenses.

What Was the Source and/or Reservoir of the Pathogen?

The source and/or reservoir of a pathogen is the second link in the infectious disease cycle. Identifying the source and/or reservoir is an important aspect of epidemiology. If the source or reservoir of the infection can be eliminated or controlled, the infectious disease cycle itself will be interrupted and transmission of the pathogen will be prevented (Historical Highlights 36.1 and 36.2).

A **source** is the location from which the pathogen is immediately transmitted to the host, either directly through the environment or indirectly through an intermediate agent. The source can be either animate (e.g., humans or animals) or inanimate (e.g., water, soil, or food). The **period of infectivity** is the time during which the source is infectious or is disseminating the pathogen.

The **reservoir** is the site or natural environmental location in which the pathogen normally resides. It is also the site from which a source acquires the pathogen and/or where direct infection of the host can occur. Thus a reservoir sometimes functions as a source. Reservoirs also can be animate or inanimate.

Much of the time, human hosts are the most important animate sources of the pathogen and are called carriers. A **carrier** is an infected individual who is a potential source of infection for others. Carriers play an important role in the epidemiology of disease. Four types of carriers are recognized:

1. An **active carrier** is an individual who has an overt clinical case of the disease.
2. A **convalescent carrier** is an individual who has recovered from the infectious disease but continues to harbor large numbers of the pathogen.

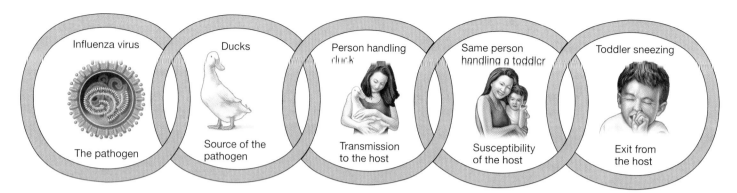

Figure 36.7 Chain of Infectious Disease.

3. A **healthy carrier** is an individual who harbors the pathogen but is not ill.
4. An **incubatory carrier** is an individual who is incubating the pathogen in large numbers but is not yet ill.

Convalescent, healthy, and incubatory carriers may harbor the pathogen for only a brief period (hours, days, or weeks) and then are called **casual, acute,** or **transient carriers.** If they harbor the pathogen for long periods (months, years, or life), they are called **chronic carriers.**

Animal diseases that can be transmitted to humans are termed **zoonoses** (Greek *zoon,* animal, and *nosos,* disease); thus animals also can serve as reservoirs. Humans contract the pathogen by several mechanisms: coming into direct contact with diseased animal flesh (e.g., tularemia); drinking contaminated cow's milk (e.g., tuberculosis and brucellosis); inhaling dust particles contaminated by animal excreta or products (e.g., Q fever, hantavirus pulmonary infection, anthrax); or eating insufficiently cooked infected flesh (e.g., anthrax, trichinosis). In addition, being bitten by arthropod **vectors** (organisms that spread disease from one host to another) such as mosquitoes, ticks, fleas, mites, or biting flies (e.g., equine encephalomyelitis, malaria, Lyme disease, Rocky Mountain spotted fever, plague, scrub typhus, and tularemia); or being bitten by a diseased animal (e.g., rabies) can lead to infection.

Table 36.1 lists some common zoonoses found in the Western Hemisphere. This table is noninclusive in scope; it merely abbreviates the enormous spectrum of zoonotic diseases that are relevant to human epidemiology. Domestic animals are the most common source of zoonoses because they live in greater proximity to humans than do wild animals. Diseases of wild animals that are transmitted to humans tend to occur sporadically because close contact is infrequent. Other major reservoirs of pathogens are water, soil, and food. These reservoirs are discussed in detail in chapters 40 and 41.

How Was the Pathogen Transmitted?

To maintain an active infectious disease in a human population, the pathogen must be transmitted from one host or source to another. Transmission is the third link in the infectious disease cycle and occurs by four main routes: airborne, contact, vehicle, and vector-borne (**figure 36.8**).

Airborne Transmission

Because air is not a suitable medium for the growth of pathogens, any pathogen that is airborne must have originated from a source such as humans, other animals, plants, soil, food, or water. In **airborne transmission** the pathogen is truly suspended in the air and travels over a meter or more from the source to the host. The pathogen can be contained within droplet nuclei or dust. **Droplet nuclei** can be small particles, 1 to 4 μm in diameter, that result from the evaporation of larger particles (10 μm or more in diameter) called droplets. Droplet nuclei can remain airborne for hours or days and travel long distances. Chicken pox and measles are examples of droplet-spread diseases.

When animals or humans are the source of the airborne pathogen, it usually is propelled from the respiratory tract into the air by an individual's coughing, sneezing, or vocalization. For example, enormous numbers of moisture droplets are aerosolized during a typical sneeze (**figure 36.9**). Each droplet is about 10 μm in diameter and initially moves about 100 m/second or more than 200 mi/hour!

Dust also is an important route of airborne transmission. At times a pathogen adheres to dust particles and contributes to the number of airborne pathogens when the dust is resuspended by some disturbance. A pathogen that can survive for relatively long periods in or on dust creates an epidemiological problem, particularly in hospitals, where dust can be the source of hospital acquired infections. **Table 36.2** summarizes some human airborne pathogens and the diseases they cause.

Contact Transmission

Contact transmission implies the coming together or touching of the source or reservoir of the pathogen and the host (**Historical Highlights 36.3**). Contact can be direct or indirect. Direct contact implies an actual physical interaction with the infectious source (figure 36.8). This route is frequently called person-to-person contact. Person-to-person transmission occurs primarily by touching, kissing, or sexual contact (sexually transmitted diseases); by contact with oral secretions or body lesions (e.g., herpes and boils);

Table 36.1	Infectious Organisms in Nonhuman Reservoirs That May Be Transmitted to Humans		
Disease	Etiologic Agent	Usual or Suspected Nonhuman Host	Usual Method of Human Infection
Anthrax	*Bacillus anthracis*	Cattle, horses, sheep, swine, goats, dogs, cats, wild animals, birds	Inhalation or ingestion of spores; direct contact
Babesiosis	*Babesia bovis, B. divergens, B. microti, B. equi*	*Ixodes* ticks of various species	Bite of infected tick
Brucellosis (undulant fever)	*Brucella melitensis, B. abortus, B. suis*	Cattle, goats, swine, sheep, horses, mules, dogs, cats, fowl, deer, rabbits	Milk; direct or indirect contact
Campylobacteriosis	*Campylobacter fetus, C. jejuni*	Cattle, sheep, poultry, swine, pets	Contaminated water and food
Cat-scratch disease	*Bartonella henselae*	Cats, dogs	Cat or dog scratch
Colorado tick fever	*Coltivirus*	Squirrels, chipmunks, mice, deer	Tick bite
Cowpox	Cowpox virus	Cattle, horses	Skin abrasions
Cryptosporidiosis	*Cryptosporidium* spp.	Farm animals, pets	Contaminated water
Encephalitis (California)	Arbovirus	Rats, squirrels, horses, deer, hares, cows	Mosquito
Encephalitis (St. Louis)	Arbovirus	Birds	Mosquito
Encephalomyelitis (Eastern equine)	Arbovirus	Birds, ducks, fowl, horses	Mosquito
Encephalomyelitis (Venezuelan equine)	Arbovirus	Rodents, horses	Mosquito
Encephalomyelitis (Western equine)	Arbovirus	Birds, snakes, squirrels, horses	Mosquito
Giardiasis	*Giardia intestinalis*	Rodents, deer, cattle, dogs, cats	Contaminated water
Glanders	*Burkholderia mallei*	Horses	Skin contact; inhalation
Hantavirus pulmonary syndrome	Pulmonary syndrome hantavirus	Deer mice	Contact with the saliva, urine, or feces of deer mice; aerosolized viruses
Herpes B viral encephalitis	*Herpesvirus simiae*	Monkeys	Monkey bite; contact with material from monkeys
Influenza	Influenza virus	Water fowl, pigs	Direct contact or inhalation
Leptospirosis	*Leptospira interrogans*	Dogs, rodents, wild animals	Direct contact with urine, infected tissue, and contaminated water
Listeriosis	*Listeria monocytogenes*	Sheep, cattle, goats, guinea pigs, chickens, horses, rodents, birds, crustaceans	Food-borne
Lyme disease	*Borrelia burgdorferi*	Ticks (*Ixodes scapularis* or related ticks)	Bite of infected tick
Lymphocytic choriomeningitis	Arenavirus	Mice, rats, dogs, monkeys, guinea pigs	Inhalation of contaminated dust; ingestion of contaminated food
Mediterranean fever (boutonneuse fever, African tick typhus)	*Rickettsia conorii*	Dogs	Tick bite
Melioidosis	*Burkholderia pseudomallei*	Rats, mice, rabbits, dogs, cats	Arthropod vectors, water, food
Orf (contagious ecthyma)	*Parapoxvirus*	Sheep, goats	Through skin abrasions
Pasteurellosis	*Pasteurella multocida*	Fowl, cattle, sheep, swine, goats, mice, rats, rabbits	Animal bite

(continued)

Table 36.2	**Some Airborne Pathogens and the Diseases They Cause in Humans**		
Microorganisms	**Disease**	**Microorganism**	**Disease**
Viruses		**Bacteria**	
Varicella	Chickenpox	*Actinomyces* spp.	Lung infections
Influenza	Flu (Influenza)	*Bordetella pertussis*	Whooping cough
Rubeola	Measles	*Chlamydia psittaci*	Psittacosis
Rubella	German measles	*Corynebacterium diphtheriae*	Diphtheria
Mumps	Mumps	*Mycoplasma pneumoniae*	Pneumonia
Polio	Poliomyelitis	*Mycobacterium tuberculosis*	Tuberculosis
Acute respiratory viruses	Viral pneumonia	*Neisseria meningitidis*	Meningitis
Pulmonary syndrome hantavirus	Hantavirus pulmonary syndrome	*Streptococcus* spp.	Pneumonia, sore throat
Variola	Smallpox	**Fungi**	
		Blastomyces spp.	Lung infections
		Coccidioides spp.	Coccidioidomycosis
		Histoplasma capsulatum	Histoplasmosis

Historical Highlights

36.3   The First Indications of Person-to-Person Spread of an Infectious Disease

In 1773 Charles White, an English surgeon and obstetrician, published his "Treatise on the Management of Pregnant and Lying-In Women." In it, he appealed for surgical cleanliness to combat childbed or puerperal fever. (Puerperal fever is an acute febrile condition that can follow childbirth and is caused by streptococcal infection of the uterus and/or adjacent regions.) In 1795 Alexander Gordon, a Scottish obstetrician, published his "Treatise on the Epidemic Puerperal Fever of Aberdeen," which demonstrated for the first time the contagiousness of the disease. In 1843 Oliver Wendell Holmes, a noted physician and anatomist in the United States, published a paper entitled "On the Contagiousness of Puerperal Fever" and also appealed for surgical cleanliness to combat this disease.

However, the first person to realize that a pathogen could be transmitted from one person to another was the Hungarian physician Ignaz Phillip Semmelweis. Between 1847 and 1849 Semmelweis observed that women who had their babies at the hospital with the help of medical students and physicians were four times as likely to contract puerperal fever as those who gave birth with the help of midwives. He concluded that the physicians and students were infecting women with material remaining on their hands after autopsies and other activities. Semmelweis thus began washing his hands with a calcium chloride solution before examining patients or delivering babies. This simple procedure led to a dramatic decrease in the number of cases of puerperal fever and saved the lives of many women. As a result Semmelweis is credited with being the pioneer of antisepsis in obstetrics. Unfortunately, in his own time, most of the medical establishment refused to acknowledge his contribution and adopt his procedures. After years of rejection Semmelweis had a nervous breakdown in 1865. He died a short time later of a wound infection. It is very probable that it was a streptococcal infection, arising from the same pathogen he had struggled against his whole professional life.

Vector-Borne Transmission

As noted earlier, living transmitters of a pathogen are called vectors. Most vectors are arthropods (e.g., insects, ticks, mites, fleas) or vertebrates (e.g., dogs, cats, skunks, bats). **Vector-borne transmission** can be either external or internal. In external (mechanical) transmission the pathogen is carried on the body surface of a vector. Carriage is passive, with no growth of the pathogen during transmission. An example would be flies carrying *Shigella* organisms on their feet from a fecal source to a plate of food that a person is eating.

In internal transmission the pathogen is carried within the vector. Here it can go into either a harborage or biologic transmission phase. In **harborage transmission** the pathogen does not undergo morphological or physiological changes within the vector. An example would be the transmission of *Yersinia pestis* (the etiologic agent of plague) by the rat flea from rat to human. **Biologic transmission** implies that the pathogen does go through a morphological or physiological change within the vector. An example would be the developmental sequence of the malarial parasite inside its mosquito vector.   Arthropod-borne diseases: Malaria (section 39.3)

Why Was the Host Susceptible to the Pathogen?

The fourth link in the infectious disease cycle is the host. The susceptibility of the host to a pathogen depends on both the pathogenicity of the organism and the nonspecific and specific defense mechanisms of the host. These susceptibility factors are the basis for chapters 31 and 32. In addition to host defense mechanisms, issues of nutrition, genetic predisposition, and stress are also germaine when considering host susceptibility to infection.

How Did the Pathogen Leave the Host?

The fifth and last link in the infectious disease cycle is release or exit of the pathogen from the host. From the point of view of the pathogen, successful escape from the host is just as important as its initial entry. Unless a successful escape occurs, the disease cycle will be interrupted and the pathogenic species will not be perpetuated. Escape can be active or passive, although often a combination of the two occurs. Active escape takes place when a pathogen actively moves to a portal of exit and leaves the host. Examples include the many parasitic helminths that migrate through the body of their host, eventually reaching the surface and exiting. Passive escape occurs when a pathogen or its progeny leaves the host in feces, urine, droplets, saliva, or desquamated cells. Microorganisms usually employ passive escape mechanisms.

---

1. What are some epidemiologically important characteristics of a pathogen? What is a communicable disease?
2. Define source, reservoir, period of infectivity, and carrier.
3. What types of infectious disease carriers does epidemiology recognize?
4. Describe the four main types of infectious disease transmission and give examples of each.
5. Define the terms droplet nuclei, vehicle, fomite, and vector.

36.6 VIRULENCE AND THE MODE OF TRANSMISSION

There is evidence that a pathogen's virulence may be strongly influenced by its mode of transmission and ability to live outside its host. When the pathogen uses a mode of transmission such as direct contact, it cannot afford to make the host so ill that it will not be transmitted effectively. This is the case with the common cold, which is caused by rhinoviruses and several other respiratory viruses. If the virus reproduced too rapidly and damaged its host extensively, the person would be bedridden and not contact others. The efficiency of transmission would drop because rhinoviruses shed from the cold sufferer could not contact new hosts and would be inactivated by exposure. Cold sufferers must be able to move about and directly contact others. Thus virulence is low and people are not incapacitated by the common cold.

On the other hand, if a pathogen uses a mode of transmission not dependent on host health and mobility, then the person's health will not be a critical matter. The pathogen might be quite successful—that is, transmitted to many new hosts even though it kills its host relatively quickly. Host death will mean the end of any resident pathogens, but the species as a whole can spread and flourish as long as the increased transmission rate outbalances the loss due to host death. This situation may arise in several ways.

When a pathogen is transmitted by a vector, it will benefit by extensive reproduction and spread within the host. If pathogen levels are very high in the host, a vector such as a biting insect has a better chance of picking up the pathogen and transferring it to a new host. Indeed, pathogens transmitted by biting arthropods often are very virulent (e.g., malaria, typhus, sleeping sickness). It is important that such pathogens do not harm their vectors, and the vector generally remains healthy, at least long enough for pathogen transmission.

Virulence also is often directly correlated with a pathogen's ability to survive in the external environment. If a pathogen cannot survive well outside its host and does not use a vector, it depends on host survival and will tend to be less virulent. When a pathogen can survive for long periods outside its host, it can afford to leave the host and simply wait for a new one to come along. This seems to promote increased virulence. Host health is not critical, but extensive multiplication within the host will increase the efficiency of transmission. Good examples are tuberculosis and diphtheria. *Mycobacterium tuberculosis* and *Corynebacterium diphtheriae* survive for a long time, at least weeks to months, outside human hosts.

Human cultural patterns and behavior almost certainly also affect pathogen virulence. Waterborne pathogens such as *Vibrio cholerae* (which causes cholera) are transmitted through drinking water systems. They can be virulent because immobile hosts still shed pathogens, which frequently reach the water. This is why the establishment of an uncontaminated drinking water supply is critical in limiting a cholera outbreak, particularly in refugee camps. The same appears to be true of *Shigella* and dysentery. Often one of the best ways to reduce virulence may be to reduce the frequency of transmission.

36.7 EMERGING AND REEMERGING INFECTIOUS DISEASES AND PATHOGENS

Only a few decades ago, a grateful public trusted that science had triumphed over infectious diseases by building a fortress of health protection. Antibiotics, vaccines, and aggressive public health campaigns had yielded a string of victories over old enemies like whooping cough, pneumonia, polio, and smallpox. In developed countries, people were lulled into believing that microbial threats were a thing of the past. Trends in the number of deaths caused by infectious diseases in the United States from 1900 through 1982 supported this conclusion (**figure 36.10**). However, this downward trend ended in 1982 and the death rate has since risen. The world has seen the global spread of AIDS, the resurgence of tuberculosis, and the appearance of new enemies like hantavirus pulmonary syndrome, hepatitis C and E, Ebola virus, Lyme disease, cryptosporidiosis, and *E. coli* O157:H7. In addition, during this same time period:

- A "bird flu" virus that had never before attacked humans began to kill people in southeast Asia.
- A new variant of a fatal prion disease of the brain, Creutzfeldt-Jakob disease was identified in the United

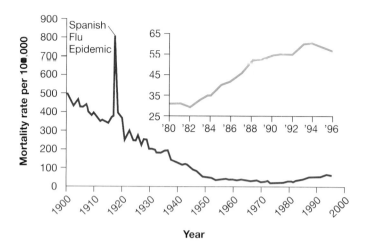

Figure 36.10 Infectious Disease Mortality in the United States Decreased Greatly during Most of the Twentieth Century. The insert is an enlargement of the right-hand portion of the graph and shows that the death rate from infectious diseases increased between 1980 and 1994.

Kingdom, transmitted by beef from animals with "mad cow disease."

• *Staphylococcus* bacteria with resistance to methicillin and vancomycin, long the antibiotics of first choice and last resort, respectively, were seen for the first time.

• Several major multistate foodborne outbreaks occurred in the United States, including those caused by parasites on raspberries, viruses on strawberries, and bacteria in produce, ground beef, cold cuts, and breakfast cereal.

• A new strain of tuberculosis that is resistant to many drugs, and occurs most often in people infected with HIV, arose in the city of New York and other large cities.

By the 1990s, the idea that infectious diseases no longer posed a serious threat to human health was obsolete. In the 21st century, it is clear that globally, humans will continually be faced with both new infectious diseases and the reemergence of older diseases once thought to be conquered (e.g., tuberculosis, dengue hemorrhagic fever, yellow fever). William McNeill, in *Plagues and Peoples* (1976), addresses this problem as follows: "Ingenuity, knowledge, and organization alter but cannot cancel humanity's vulnerability to invasion by parasitic forms of life. Infectious disease which antedated the emergence of humankind will last as long as humanity itself, and will surely remain, as it has been hitherto, one of the fundamental parameters and determinants of human history."

The Centers for Disease Control and Prevention has defined these diseases as "new, reemerging, or drug-resistant infections whose incidence in humans has increased within the past three decades or whose incidence threatens to increase in the near future." Some of the most recent examples of these diseases are

shown in **figure 36.11.** The increased importance of emerging and reemerging infectious diseases has stimulated the establishment of a field called **systematic epidemiology,** which focuses on the ecological and social factors that influence the development of these diseases.

After a century marked by dramatic advances in medical research and drug discovery, technology development, and sanitation, why are viruses, bacteria, fungi, and parasites posing such a problem and challenge? Many factors characteristic of the modern world undoubtedly favor the development and spread of these microorganisms and their diseases. Examples include:

1. Unprecedented worldwide population growth, population shifts (demographics), and urbanization
2. Increased international travel
3. Increased worldwide transport (commerce), migration, and relocation of animals and food products
4. Changes in food processing, handling, and agricultural practices
5. Changes in human behavior, technology, and industry
6. Human encroachment on wilderness habitats that are reservoirs for insects and animals that harbor infectious agents
7. Microbial evolution (e.g., selective pressure) and the development of resistance to antibiotics and other antimicrobial drugs (e.g., penicillin-resistant *Streptococcus pneumoniae,* meticillin-resistant *Staphylococcus aureus,* and vancomycin-resistant enterococci)
8. Changes in ecology and climate
9. Modern medicine (e.g., immunosuppression)
10. Inadequacy of public infrastructure and vaccination programs
11. Social unrest and civil wars
12. Bioterrorism (section 36.9)
13. Virulence-enhancing mechanisms of pathogens (e.g., the mobile genetic elements—bacteriophages, plasmids, transposons)

As population density increases in cities, the dynamics of microbial exposure and evolution increase in humans themselves. Urbanization often crowds humans and increases exposure to microorganisms. Crowding leads to unsanitary conditions and hinders the effective implementation of adequate medical care, enabling more widespread transmission and propagation of pathogens. In modern societies, crowded workplaces, community-living settings, day-care centers, large hospitals, and public transportation all facilitate microbial transmission. Furthermore, land development and the exploration and destruction of natural habitats have increased the likelihood of human exposure to new pathogens and may put selective pressures on pathogens to adapt to new hosts and changing environments. The introduction of pathogens to a new environment or host can alter transmission and exposure patterns, leading to sudden proliferation of disease. For example, the spread of Lyme disease in New England probably was due partly to ecological disruption that eliminated predators of deer. An increase in deer and the deer tick populations provided a favorable situation for pathogen spread to humans. Whenever there is alteration of the

Figure 36.11   Some Examples of Emerging and Reemerging Infectious Diseases.   Although diseases such as HIV are indicated in only one or two significant locations, they are very widespread and a threat in many regions.

environment and new environments are created, this may not only confer a survival advantage, but may also increase a pathogen's virulence and alter its drug susceptibility profile. When there are changes in climate or ecology, it should not be surprising to find changes in both beneficial and detrimental microorganisms. Global warming also affects microorganism selection and survival. Finally, mass migrations of refugees, workers, and displaced persons have led to a steady growth of urban centers at the expense of rural areas.

Microbiologists are all too familiar with the development of resistance to antibiotics used in human medicine. The distribution of nosocomial pathogens (**figure 36.12**) has changed throughout the antibiotic era. Hospital acquired infections were dominated early on by staphylococci, which initially responded to penicillin. During subsequent years, the emergence of meticillin-resistant *Staphylococcus aureus* (MRSA) increased dramatically; similar patterns are emerging for penicillin-resistant *Streptococcus pneumoniae*. Vancomycin-resistant *S. aureus* infections are now a major nosocomial threat. Glycopeptide-resistant *Enterococcus faecium* was first reported in the late 1980s; vancomycin-resistant enterococci (VRE) are now common in U.S. hospitals. Newly recognized nosocomial, gram-positive species include *Corynebacterium jeikeium* and *Rhodococcus equi*. The incidences of infections by the gram-negative pathogens *Pseudomonas aerug-*

inosa and *Acinetobacter* ssp. and the recently renamed gram-negative bacterial pathogens *Burkholderia cepacia* and *Stenotrophomonas maltophilia* have increased. With respect to the extended-spectrum β-lactam-resistant gram-negative bacilli, bacteria such as *Klebsiella pneumoniae, E. coli,* other *Klebsiella* spp., *Proteus* spp., *Morganella* spp., *Citrobacter* spp., *Salmonella* spp., and *Serratia marcescens* are resistant to penicillins; first-generation cephalosporins; and some third-generation cephalosporins such as cefotaxime (Claforan), ceftriaxone (Rocephin), ceftazidime, and aztreonam (Azactam).

Without a doubt, the key factors responsible for the rise in drug-resistant pathogens have been the excessive or inappropriate use of antimicrobial therapy and the sometimes indiscriminant use of broad-spectrum antibiotics. Although the CDC acknowledges that it is too late to solve the resistance problem simply by using antimicrobial agents more prudently, it is no less true that the problem of drug resistance will continue to worsen if this does not occur. Also needed (especially in underdeveloped countries) is a renewed emphasis on alternate prevention and control strategies that prevailed in the years before antimicrobial chemotherapy. These include improved sanitation and hygiene, isolation of infected persons, antisepsis, and vaccination. Drug resistance (section 34.6)

In this new millennium, the speed and volume of international travel are major factors contributing to the global emergence of

Figure 36.12 Nosocomial Infections. Relative frequency by body site. These data are from the National Nosocomial Infections Surveillance, which is conducted by the Centers for Disease Control and Prevention (CDC).

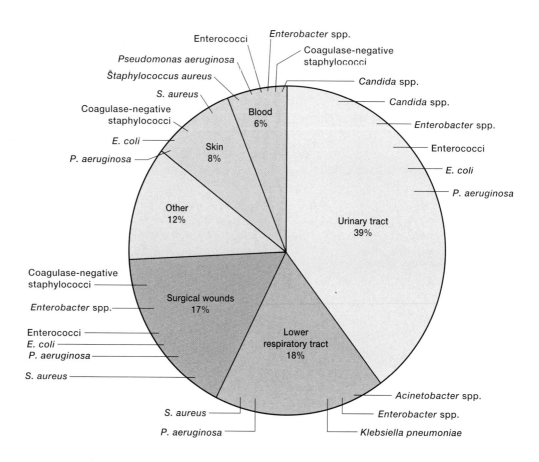

infectious diseases. The spread of a new disease often used to be limited by the travel time needed to reach a new host population. If the travel time was sufficiently long, as when a ship crossed the ocean, the infected travelers would either recover or die before reaching a new population. Because travel by air has obliterated time between exposure and disease outbreak, a traveler can spread virtually any human disease in a matter of hours. Vehicles of human transport, such as aircraft and ships, also transport the infectious agents and their vectors.

It is probably best to view emerging and reemerging pathogens and their diseases as the outcome of many different factors. Because the world is now so interconnected, we cannot isolate ourselves from other countries and continents. Changes in the disease status of one part of the world may well affect the health of the remainder. As Nobel laureate Joshua Lederberg has so eloquently stated, "The microbe that felled one child in a distant continent yesterday can reach your child today and seed a global pandemic tomorrow."

1. Describe how virulence and the mode of transmission may be related. What might cause the development of new human diseases?
2. How would you define emerging or reemerging infectious diseases?
3. What are some of the factors responsible for the emergence or reemergence of pathogens?
4. What are some key factors responsible for the rise in drug-resistant bacteria?

36.8 CONTROL OF EPIDEMICS

The development of an infectious disease is a complex process involving many factors, as is the design of specific epidemiological control measures. Epidemiologists must consider available resources and time constraints, adverse effects of potential control measures, and human activities that might influence the spread of the infection. Many times control activities reflect compromises among alternatives. To proceed intelligently, one must identify components of the infectious disease cycle that are primarily responsible for a particular epidemic. Control measures should be directed toward that part of the cycle that is most susceptible to control—the weakest link in the chain (figure 36.7).

There are three types of control measures. The first type is directed toward reducing or eliminating the source or reservoir of infection:

1. Quarantine and isolation of carriers
2. Destruction of an animal reservoir of infection
3. Treatment of water sewage to reduce contamination (*see chapter 41*)
4. Therapy that reduces or eliminates infectivity of the individual

The second type of control measure is designed to break the connection between the source of the infection and susceptible individuals. Examples include general sanitation measures:

1. Chlorination of water supplies
2. Pasteurization of milk and other beverages
3. Supervision and inspection of food and food handlers
4. Destruction of vectors

The third type of control measure reduces the number of susceptible individuals and raises the general level of herd immunity by immunization. Examples include:

1. Passive immunization to give a temporary immunity following exposure to a pathogen or when a disease threatens to take an epidemic form
2. Active immunization to protect the individual from the pathogen and the host population from the epidemic

Vaccines and Immunization

A **vaccine** [Latin *vacca,* cow] is a preparation of one or more microbial antigens used to induce protective immunity. It may consist of killed microorganisms, living, weakened microorganisms (**attenuated vaccine**); inactivated bacterial toxins (**toxoids**); purified cellular subunits; recombinant vectors (e.g., modified polio vaccine); or DNA. **Immunization** is the result achieved by the successful delivery of vaccines; it stimulates immunity. Vaccination attempts to induce antibodies and activated T cells to protect a host from future infection. Many epidemics have been stayed by mass prophylactic immunization. Vaccines have eradicated smallpox, pushed polio to the brink of extinction, and spared countless individuals from hepatitis A and B, influenza, measles, rotavirus disease, tetanus, typhus, and other dangerous diseases. **Vaccinomics,** the application of genomics and bioinformatics to vaccine development, is bringing a fresh approach to the Herculean problem of making vaccines against various microorganisms and parasites.

To promote a more efficient immune response, antigens in vaccines can be mixed with an **adjuvant** [Latin *adjuvans,* aiding], which enhances the rate and degree of immunization. Adjuvants can be any nontoxic material that prolongs antigen interaction with immune cells, assists in the antigen-presenting cell (APC) processing of antigens, or otherwise nonspecifically stimulates the immune response to the antigen. Several types of adjuvants can be used. Oil in water emulsions (Freund's incomplete adjuvant), aluminum hydroxide salts (alum), beeswax, and various combinations of bacteria (live or killed) are used in vaccine adjuvants. In most cases, the adjuvant materials trap the antigen, thereby promoting a sustained release as APCs digest and degrade the preparation. In other cases, the adjuvant activates APCs so that antigen recognition, processing, and presentation are more efficient.

The modern era of vaccines and immunization began in 1798 with Edward Jenner's use of cowpox as a vaccine against smallpox (**Historical Highlights 36.4**) and in 1881 with Louis Pasteur's anthrax vaccine. Vaccines for other diseases did not emerge until later in the 19th century, when largely through a process of trial and error, methods for inactivating and attenuating microorganisms were improved and vaccines were produced. Vaccines were eventually developed against most of the epidemic diseases that plagued western Europe and North America (diphtheria, measles, mumps, pertussis, German measles and polio, for example). Indeed, toward the end of the of the 20th century it seemed that the combination of vaccines and antibiotics would temper the problem of microbial infections. Such optimism was cut short by the emergence of new and previously unrecognized pathogens, and antibiotic resistance among existing pathogens. Nevertheless, vaccination is still one of the most cost-effective weapons for prevention of microbial disease.

Vaccination of most children should begin at about age 2 months (**table 36.3**). Before that age, they are protected by passive natural immunity from maternal antibodies. Further vaccination of teens and adults depends on their relative risk for exposure to infectious disease. Individuals living in close quarters (for instance, college students in residence halls, military personnel), the elderly, and individuals with reduced immunity (e.g., those with chronic and metabolic diseases), should receive vaccines for influenza, meningitis, and pneumonia. International travelers may be immunized against cholera, hepatitis A, plague, polio, typhoid, typhus, and yellow fever, depending on the country visited. Veterinarians, forest rangers, and others whose jobs involve contact with animals may be vaccinated against rabies, plague, and anthrax. Health-care workers are typically immunized against hepatitis B virus. The role of immunization as a protective therapy cannot be overstated; immunizations save lives.

Whole-Cell Vaccines

Many of the current vaccines used for humans that are effective against viral and bacterial diseases consist of whole microorganisms that are either inactivated (killed) or attenuated (live but avirulent) (**table 36.4**). These are termed **whole-cell vaccines.** The major characteristics of these vaccines are compared in **table 36.5**. Inactivated vaccines are effective, but they are less immunogenic so often require several boosters and normally do not adequately stimulate cell-mediated immunity or secretory IgA production. In contrast, attenuated vaccines usually are given in a single dose and stimulate both humoral and cell-mediated immunity.

Even though whole-cell vaccines are considered the "gold standard" of existing vaccines, they can be problematic in their own way. For example, whole-organism vaccines fail to shield against some diseases. Attenuated vaccines that work can also cause full-blown illness in an individual whose immune system is compromised (e.g., AIDS patients, cancer patients undergoing chemotherapy, the elderly). These same individuals may also contract the disease from healthy people who have been vaccinated recently. Moreover, attenuated viruses can at times mutate in ways that restore virulence, as has happened in some monkeys given an attenuated simian form of the AIDS virus.

Acellular or Subunit Vaccines

A few of the common risks associated with whole-cell vaccines can be avoided by using only specific, purified macromolecules

Since the time of the ancient Greeks, it has been recognized that people who have recovered from plague, smallpox, yellow fever, and various other infectious diseases rarely contract the diseases again. The first scientific attempts at artificial immunizations were made in the late eighteenth century by Edward Jenner (1749–1823), who was a country doctor from Berkley, Gloucestershire, England. Jenner investigated the basis for the widespread belief of the English peasants that anyone who had vaccinia (cowpox) never contracted smallpox. Smallpox was often fatal—10 to 40% of the victims died—and those who recovered had disfiguring pockmarks. Yet most English milkmaids, who were readily infected with cowpox, had clear skin because cowpox was a relatively mild infection that left no scars.

It was on May 14, 1796, that Jenner extracted the contents of a pustule from the arm of a cowpox-infected milkmaid, Sarah Nelmes, and injected it into the arm of eight-year-old James Phipps. As Jenner expected, immunization with the cowpox virus caused only mild symptoms in the boy. When he subsequently inoculated the boy with smallpox virus, the boy showed no symptoms of the disease.

Jenner then inoculated large numbers of his patients with cowpox pus, as did other physicians in England and on the European continent (see **Box figure**). By 1800 the practice known as variolation had begun in America, and by 1805 Napoleon Bonaparte had ordered all French soldiers to be vaccinated.

Further work on immunization was carried out by Louis Pasteur (1822–1895). Pasteur discovered that if cultures of chicken cholera bacteria were allowed to age for two or three months the bacteria

Nineteenth-Century Physicians Performing Vaccinations on Children

produced only a mild attack of cholera when inoculated into chickens. Somehow the old cultures had become less pathogenic (attenuated) for the chickens. He then found that fresh cultures of the bacteria failed to produce cholera in chickens that had been previously inoculated with old, attenuated cultures. To honor Jenner's work with cowpox, Pasteur gave the name vaccine to any preparation of a weakened pathogen that was used (as was Jenner's "vaccine virus") to immunize against infectious disease.

Table 36.3	Recommended Childhood and Adolescent Immunization Schedule, United States, 2006

Vaccine ▼ / Age ▶	Birth	1 Month	2 Months	4 Months	6 Months	12 Months	15 Months	18 Months	24 Months	4–6 Years	11–12 Years	13–14 Years	15 Years	16–18 Years
Hepatitis B	HepB	HepB		HepB		HepB					HepB Series			
Diptheria, Tetanus, Pertussis			DTaP	DTaP	DTaP		DTaP			DTaP	Tdap		Tdap	
Haemophilus influenzae type b			Hib	Hib	Hib	Hib								
Inactivated Poliovirus			IPV	IPV		IPV				IPV				
Measles, Mumps, Rubella						MMR				MMR		MMR		
Varicella						Varicella					Varicella			
Meningococcal										MPSV4	MCV4		MCV4 MPSV4	
Pneumococcal			PCV	PCV	PCV	PCV			PCV		PPV			
Influenza						Influenza (yearly)					Influenza (yearly)			
Hepatitis A										HepA Series				

Vaccines within broken line are for selected populations

This schedule indicates the recommended ages for routine administration of currently licensed childhood vaccines, as of December 1, 2005, for children through age 18 years. Any dose administered at the recommended age should be administered at any subsequent visit when indicated and feasible. ▨ Indicates age groups that warrant special effort to administer those vaccines not previously administered. Additional vaccines may be licensed and recommended during the year. Licensed combination vaccines may be used whenever any components of the combination are indicated and other components of the vaccine are not contraindicated and if approved by the Food and Drug Administration for that dose of the series. Providers should consult the respective Advisory Council on Immunization Practices (ACIP) statement for detailed recommendations. (www.cdc.gov/nip/acip). Clinically significant adverse events that follow immunization should be reported to the Vaccine Adverse Event Reporting System (VAERS).

▨ Range of recommended ages   ▨ Catch-up immunization   ▨ 11–12 year old assessment

Table 36.4	Examples of Vaccines to Prevent Viral and Bacterial Diseases in Humans		
Disease	**Vaccine**	**Booster**	**Recommendation**
Viral Diseases			
Brain infection	Inactivated Japanese encephalitis virus	None	Residents of and travelers to areas of endemic disease
Chickenpox	Attenuated Oka strain (Varivax)	None	Children 12–18 months: older children who have not had chickenpox
Hepatitis A	Inactivated virus (Havrix)	6–12 months	International travelers
Hepatitis B	HB viral antigen (Engerix-B, Recombivax HB)	None	High-risk medical personnel: children, birth to 18 months and 11–12 years of age
Influenza A	Inactivated virus or live attenuated	Yearly	All persons
Measles, Mumps, Rubella	Attenuated viruses (combination MMR vaccine)	None	Children 15–19 months old
Poliomyelitis	Attenuated (oral poliomyelitis vaccine, OPV) or inactivated vaccine	Adults as needed	Children 2–3 years old
Rabies	Inactivated virus	None	For individual in contact with wildlife, animal control personnel, veterinarians
Respiratory disease	Live attenuated adenovirus	None	Military personnel
Smallpox	Live attenuated vaccinia virus	None	Laboratory, health-care, and military personnel
Yellow fever	Attenuated virus	10 years	Military personnel and individuals traveling to endemic areas
Bacterial Diseases			
Anthrax	Extracellular components of unencapsulated *B. anthracis*	None	Agricultural workers, veterinary, and military personnel
Cholera	Fraction of *Vibrio cholerae*	6 months	Individuals in endemic areas, travelers
Diphtheria, Pertussis, Tetanus	Diphtheria toxoid, killed *Bordetella pertussis*, tetanus toxoid (DPT vaccine) or with acellular *pertussis* (DtaP); or tetanus toxoid, reduced diphtheria toxoid, and acellular pertussis vaccine, adsorbed (Tdap)	10 years	Children from 2–3 months old to 12 years, and adults; children 10–18 years, at least 5 years after DPT series, should receive Tdap
Haemophilus influenzae type b	Polysaccharide-protein conjugate (HbCV) or bacterial polysaccharide (HbPV)	None	Children under 5 years of age
Meningococcal infections	Bacterial polysaccharides of serotypes A/C/Y/W-135	None	Military; high-risk individuals; college students living in dormatories; elderly in nursing homes
Plague	Fraction of *Yersinia pestis*	Yearly	Individuals in contact with rodents in endemic areas
Pneumococcal pneumonia	Purified *S. pneumoniae* polysaccharide of 23 pneumococcal types	None	Adults over 50 with chronic disease
Q fever	Inactivated *Coxiella burnetii*	None	Workers in slaughter houses and meat-processing plants
Tuberculosis	Attenuated *Mycobacterium bovis* (BCG vaccine)	3–4 years	Individuals exposed to TB for prolonged periods of time; not licensed for use in the U.S.
Typhoid fever	Ty21a (live attenuated, polysaccharide)	None	Resident of and travelers to areas of endemic disease
Typhus fever	Killed *Rickettsia prowazekii*	Yearly	Scientists and medical personnel in areas where typhus is endemic

Table 36.5 | **A Comparison of Inactivated (Killed) and Attenuated (Live) Vaccines**

Major Characteristic	Inactivated Vaccine	Attenuated Vaccine
Booster shots	Multiple boosters required	Only a single booster typically required
Production	Virulent microorganism inactivated by chemicals or irradiation	Virulent microorganism grown under adverse conditions or passed through different hosts until avirulent
Reversion tendency	None	May revert to a virulent form
Stability	Very stable, even where refrigeration is unavailable	Less stable
Type of immunity induced	Humoral	Humoral and cell-mediated

Source: Adapted from Goldsby, T. J. Kindt, and B. A. Osborne, *Kuby Immunology.* 2003, New York: W. H. Freeman

Table 36.6 | **Purified Acellular or Subunit Vaccines Currently Available for Human Use**

Type of Subunit (Disease or Microorganism)	Form of Vaccine
Capsular polysaccharide	
Haemophilus influenzae type b	Polysaccharide-protein conjugate (HbCV) or bacterial polysaccharide (HbPV)
Neisseria meningitidis	Polysaccharides of serotypes A/C/Y/W-135
Streptococcus pneumoniae	23 distinct capsular polysaccharides
Surface antigen	
Hepatitis B	Recombinant surface antigen (HbsAg)
Toxoids	
Diphtheria	Inactivated exotoxin
Tetanus	Inactivated exotoxin

derived from pathogenic microorganisms. There are three general forms of **subunit vaccines:** (1) capsular polysaccharides, (2) recombinant surface antigens, and (3) inactivated exotoxins called toxoids. The purified microbial subunits or their secreted products can be prepared as nontoxic antigens to be used in the formulation of vaccines (**table 36.6**).

Recombinant-Vector and DNA Vaccines

Genes isolated from a pathogen that encode major antigens can be inserted into nonvirulent viruses or bacteria. Such recombinant microorganisms serve as vectors, replicating within the host and expressing the gene product of the pathogen-encoded anti-genic proteins. The antigens elicit humoral immunity (i.e., antibody production) when they escape from the vector, and they also elicit cellular immunity when they are broken down and properly displayed on the cell surface (just as occurs when host cells harbor an active pathogen). Several microorganisms, such as adenovirus and attenuated *Salmonella,* have been used in the production of these **recombinant-vector vaccines.**

On the other hand, **DNA vaccines** introduce DNA directly into the host cell (often via an air pressure or gene gun). When injected into muscle cells, the DNA is taken into the nucleus and the pathogen's DNA fragment is expressed, generating foreign proteins to which the host immune system responds. DNA vaccines are very stable; refrigeration is often unnecessary. At present, there are human trials underway with several different DNA vaccines against malaria, AIDS, influenza, hepatitis B, and herpesvirus. Vaccines against a number of cancers (such as lymphomas, prostate, colon) are also being tested.

The Role of the Public Health System: Epidemiological Guardian

The control of an infectious disease relies heavily on a well-defined network of clinical microbiologists, nurses, physicians, epidemiologists, and infection control personnel who supply epidemiological information to a network of local, state, national, and international organizations. These individuals and organizations comprise the public health system. For example, each state has a public health laboratory that is involved in infection surveillance and control. The communicable disease section of a state laboratory includes specialized laboratory services for the examination of specimens or cultures submitted by physicians, the local health department, hospital laboratories, sanitarians, epidemiologists, and others. These groups share their findings with other health agencies in the state, the Centers for Disease Control and Prevention, and the World Health Organization.

36.9 BIOTERRORISM PREPAREDNESS

Bioterrorism (Greek *bios*, life, and terrorism, the systematic use of terror to demoralize, intimidate, and subjugate) is defined as "the intentional or threatened use of viruses, bacteria, fungi, or toxins from living organisms to produce death or disease in humans, animals, and plants." The use of biological agents to effect personal or political outcome is not new (**Historical Highlights 36.5**), and the modern use of biological agents is a reality. The most notable intentional uses of biological agents for criminal or terror intent are (1) the use of *Salmonella enterica* serovar Typhimurium in 10 restaurant salad bars (Rajneeshee religious cult in The Dalles, OR, 1984); (2) the intentional release of *Shigella dysentariae* in a hospital laboratory break room (Texas, 1996); and (3) the use of weaponized *Bacillus anthracis* spores delivered through the U.S. postal system (perpetrator[s] still unknown, seven eastern U.S. states, 2001). The *Salmonella*-contaminated salads resulted in 751 documented cases and 45 hospitalizations due to salmonellosis. The *Shigella* release resulted in eight confirmed cases and four hospitalizations for shigellosis. The *Bacillus* spores infected 22 people (11 cases of inhalation anthrax and 11 cases of cutaneous anthrax) and were the cause of five deaths. The list of biological agents that could pose the greatest public health risk in the event of a bioterrorist attack is relatively short and includes viruses, bacteria, parasites, and toxins (**table 36.7**). Although short, the list includes agents that, if acquired and properly disseminated, could become a difficult public health challenge in terms of limiting the numbers of casualties and controlling panic. The agents are catagorized (A, B, or C) based on (1) ease of dissemination, (2) communicability, and (3) morbidity and mortality.

Biological agents are likely to be chosen as a means of localized attack (**biocrime**) or mass casualty (bioterrorism) for several reasons. They are mostly invisible, odorless, tasteless, and difficult to detect. Use of biological agents for terrorism also means that perpetrators may escape undetected as it may take hours to days before signs and symptoms of their use become evident. Additionally, the general public is not likely to be protected immunologically against agents that are thought to be used in bioterrorism. Furthermore, the use of biological agents in terrorism results in fear, panic and chaos.

There are several key indicators of a bioterrorism event. These include sudden increased numbers of sick individuals, especially with unusual (nonendemic) diseases (for that place and/or time of year). Also, sudden increased numbers of zoonoses, diseased animals, or vehicle-borne illnesses may indicate bioterrorism. Among weapons of mass destruction, biological weapons can be more destructive than chemical weapons, including nerve gas. In certain circumstances, biological weapons can be as devastating as a nuclear explosion—a few kilograms of anthrax could kill as many people as a Hiroshima-size nuclear bomb.

In 1998, the U.S. government launched the first national effort to create a biological weapons defense. The initiatives included

Historical Highlights

36.5   1346—The First Recorded Biological Warfare Attack

The Black Death *(see pp. 962–963)*, which swept through Europe, Asia, and North Africa in the mid-fourteenth century, was probably the greatest public health disaster in recorded history. Europe, for example, lost an estimated quarter to third of its population. This is not only of great historical interest but also relevant to current efforts to evaluate the threat of military or terrorist use of biological weapons.

Some believe that evidence for the origin of the Black Death in Europe is found in a memoir by the Genoese Gabriele de' Mussi. According to this fourteenth-century memoir, the Black Death reached Europe from the Crimea (a region of the Ukraine) in 1346 as a result of a biological warfare attack. The Mongol army hurled plague-infected cadavers into the besieged Crimean city of Caffa (now Feodosija, Ukraine), thereby transmitting the disease to the inhabitants; fleeing survivors then spread the plague from Caffa to the Mediterranean Basin. Such transmission was especially likely at Caffa where cadavers would have been badly mangled by being hurled, and the defenders probably often had cut or abraded hands from coping with the bombardment. Because many cadavers were involved, the opportunity for disease transmission was greatly increased. Disposal of victims' bodies in a major disease outbreak is always a problem, and the Mongol army used their hurling machines as a solution to limited mortuary facilities. It is possible that thousands of cadavers were disposed of this way; de' Mussi's description of "mountains of dead" might have been quite literally true. Indeed, Caffa could be the site of the most spectacular incident of biological warfare ever, with the Black Death as its disastrous consequence. It is a powerful reminder of the horrific consequences that can result when disease is successfully used as a weapon.

Table 36.7	Pathogens and Toxins Defined by the CDC as Select Agents	
Category	**Definition**	**Disease (Agent)**
A	Easily disseminated or transmitted from person to person; high mortality rates; potential for major public health impact; cause public panic and social disruption; require special action for public health preparedness	**Anthrax** (*Bacillus anthracis*) **Botulism** (*Clostridium botulinum* toxin) **Plague** (*Yersinia pestis*) **Smallpox** (*Variola major*) **Tularemia** (*Francisella tularensis*) **Viral hemorrhagic fever** (filoviruses and arenaviruses)

B	Moderately easy to disseminate, moderate morbidity and mortality rates; require specific enhancements of CDC's diagnostic capacity and enhanced disease surveillance	**Brucellosis** (*Brucella* species)
		Glanders (*Burkholderia mallei*)
		Melioidosis (*Burkholderia pseudomallei*)
		Psittacosis (*Chlamydia psittaci*)
		Q fever (*Coxiella burnetii*)
		Typhus fever (*Rickettsia prowazekii*)
		Viral encephalitis (alphaviruses)
		Toxemia
		    Ricin from castor beans
		    Staphylococcal enterotoxin B
		    Epsilon toxin *Clostridium perfringens*
		Other
		    Water safety threats (e.g., *Vibrio cholerae*, *Cryptosporidium parvum*) Food safety threats (e.g., *Salmonella* spp., *E. coli* O157:H7, *Shigella* spp.)

C	Emerging pathogens that could be engineered for mass dissemination; potential for high morbidity and mortality rates; major health impact potential	Nipah virus Hantaviruses Tickborne hemorrhagic fever viruses Tickborne encephalitis viruses Yellow fever virus Multidrug-resistant *Mycobacterium tuberculosis*

(1) the first-ever procurement of specialized vaccines and medicines for a national civilian protection stockpile; (2) invigoration of research and development in the science of biodefense; (3) investment of more time and money in genome sequencing, new vaccine research, and new therapeutic research; (4) development of improved detection and diagnostic systems; and (5) preparation of clinical microbiologists and the clinical microbiology laboratory as members of the "first responder" team to respond in a timely manner to acts of bioterrorism. In 2002, the U.S. government passed the Public Health Security and Bioterrorism Preparedness and Response Act, which identified "select" agents whose use is now tightly regulated. A final rule implementing the provisions of the act that govern the possession, use, and transport of biological agents that are considered likely to be used for biocrimes or bioterrorism was issued in 2005.

In 2003, the U.S. government established the Department of Homeland Security to coordinate the defense of the United States against terrorist attacks. As one of many duties, the Secretary of Homeland Security is responsible for developing and maintaining a National Incident Management System (NIMS) to monitor large-scale hazards. Bioterrorism and other public health inci-

dents are managed within this system. The Department of Health and Human Services has the initial responsibility for the national public health and will deploy assets as needed within the areas of its statutory responsibility (e.g., the Public Health Service Act and the Federal Food Drug and Cosmetic Act) while keeping the Secretary of Homeland Security apprised during an incident and the nature of the response. The Secretary of Health and Human Services directs the CDC to effect the necessary integration of public health activities.

The events of September and October 2001 in the United States have changed the world. Global efforts to prevent terrorism, especially using biological agents, are evolving from cautious planning to proactive preparedness. In the United States, the CDC has partnered with academic institutions across the country to educate, train, and drill public health employees, traditional first responders, and numerous environmental and health-care providers. Centers for Public Health Preparedness were established to bolster the overall response capability to bioterrorism. Another CDC-managed program that began in 1999, the Laboratory Response Network (LRN), serves to ensure an effective laboratory response to bioterrorism by helping to improve the nation's public health laboratory infrastructure, through its partnership with the FBI and the Association of Public Health Laboratories (APHL). The LRN maintains an integrated network that links state and local public health, federal, military, and international laboratories so that a rapid and coordinated response to bioterrorism or other public health emergencies (including veterinary, agriculture, military, and water- and food- related) can occur.

In the absence of overt terrorist threats and without the ability to rapidly detect bioterrorism agents, it is likely that an act of bioterrorism will be defined by sudden spikes in illnesses reported to the public health system. Thus important guidelines have been prepared for all sentinel (local hospital, contract, clinic, etc.) laboratories to assist in the management of clinical specimens containing select agents. These guidelines were developed by the American Society for Microbiology in coordination with the CDC and the APHL to offer standardized protocols to assist microbiologists in ruling out critical agents so specimens containing select agents can then be referred to the public health reference laboratories (of the LRN) for final confirmation. A summary of the rule-out tests for six bacterial agents is presented in **table 36.8.** The disease and microbiology associated with specific select agents are discussed in chapters 37, 38, and 39.

---

1. In what three general ways can epidemics be controlled? Give one or two specific examples of each type of control measure.
2. Name some of the microorganisms that can be used to commit biocrimes. From this list, which pose the greatest risk for causing large numbers of casualties?

3. Why are biological weapons more destructive than chemical weapons?
4. What is the Public Health Security and Bioterrorism Preparedness and Response Act designed to do?

36.10 GLOBAL TRAVEL AND HEALTH CONSIDERATIONS

From a global health perspective, developed countries such as Australia, the European countries, Israel, New Zealand, and the United States have highly effective public health systems. About 25% of the over 6 billion people on our planet Earth live in these countries. As a result, of the approximately 12 million deaths in these countries per year, only about 500,000 (about 4%) are due to infectious diseases. Less developed areas such as Africa, Central and South America, India, the parts of eastern Europe, and Asia have less developed public health systems and represent 75% of the human population. It is in these regions that infectious diseases are the major cause of death; for example, of approximately 38.5 million deaths per year, about 18 million (about 47%) are attributed to infectious microbial diseases. Despite the efforts of many governmental and nongovernmental agencies, it will take many years before all people have access to clean water, decent sanitation, and a basic health care infrastructure.

The high incidence of infectious diseases in less developed countries must be of great concern for people traveling to these destinations. Each year 1 billion passengers travel by air, and over

Table 36.8	Criteria for Presumptive Identification of Six Bacterial Select Agents		
Pathogen	**Gram Morphology**	**Colonial Morphology**	**Biochemical Results**
Bacillus anthracis	Gram positive endospore-forming rod	Grey, flat, "Medusa head" irregularity, nonhemolytic on sheep's blood agar	Catalase positive, oxidase positive, urea negative, Voges-Proskauer (VP) positive, phenylalanine (Phe) deaminase negative, NO_3^- to NO_2^- positive
Brucella suis	Gram negative rod (tiny)	Nonpigmented, convex-raised, pin-point after 48 hr, nonhemolytic	Catalase positive, oxidase variable, urea positive, VP negative
Burkholderia mallei	Gram negative straight or slightly curved cocobacilli, bundles	Grey, smooth, translucent after 48 hr, nonhemolytic	Catalase positive, oxidase variable, indole negative, Arginine dihydrolase positive, NO_3^- to NO_2^- positive
Clostridium botulinum	Gram positive endospore-forming rod	Creamy, irregular, rough, broad, nonhemolytic	Catalase negative, urea negative, gelatinase positive, indole negative, VP negative, Phe deaminase negative, NO_3^- to NO_2^- negative
Francisella tularensis	Gram negative rod (tiny)	Grey-white, shiny, convex, pin-point after 72 hr, nonhemolytic	Catalase positive (weak), oxidase negative, β-lactamase positive, urea negative
Yersinia pestis	Gram negative rod (bipolar staining)	Grey-white, "fried-egg" irregularity, nonhemolytic, grows faster and larger at 28°C	Catalase positive, oxidase negative, urea negative, indole negative, VP negative, Phe deaminase negative

50 million people from developed countries visit less developed countries. Furthermore, the time required to circumnavigate the globe has decreased from 365 days to fewer than 2 days.

Several kinds of precautions can be taken by individuals to prevent travel-related infectious diseases. Examples include:

1. Wash hands with soap and water frequently, especially before each meal.
2. Get or update vaccinations appropriate for specific destinations. Check the CDC travel advisory web site (www.cdc.gov/travel) for precautions regarding specific locations.
3. Avoid uncooked food, nonbottled water and beverages, and unpasteurized dairy products. Use bottled water for drinking, making ice cubes, and brushing teeth.
4. Use barrier protection if engaging in sexual activity.
5. Minimize skin exposure and use repellents to prevent arthropod-borne illnesses (e.g., malaria, dengue, yellow fever, Japanese encephalitis).
6. Avoid skin-perforating procedures (e.g., acupuncture, body piercing, tattooing, venipuncture, sharing of razors).
7. Do not pet or feed animals, especially dogs and monkeys.
8. Avoid swimming or wading in nonchlorinated fresh water.

Vaccinations are one of the most important strategies of prophylaxis in travel medicine. A medical consultation before travel is an excellent opportunity to update routine immunizations. Selection of immunizations should be based on requirements and risk of infection at the travel destination. According to International Health Regulations, many countries require proof of yellow fever vaccination on the International Certificate of Vaccination. Additionally, a few countries still require proof of vaccination against cholera, diphtheria, and meningococcal disease. The basic CDC immunization recommendations for those traveling abroad are presented in **table 36.9.** In the not-too-distant future, it is hoped that travelers will be offered a variety of oral vaccines against the microorganisms causing dengue fever and travelers' diarrhea (e.g., enterotoxigenic *E. coli*, *Campylobacter*, *Shigella*); vaccines against malaria and AIDS, the infections causing most deaths in travelers, are much farther in the future.

In addition, a traveler should:

1. Read carefully CDC information about the destination and follow recommendations
2. Begin the vaccination process early
3. Find a travel clinic for information and specialized immunizations
4. Plan ahead when traveling with children or if there are special needs
5. Learn about safe food and water (contaminated food and water are the major sources of stomach or intestinal illness while traveling), protection against insects, and other precautions

Table 36.9 Vaccine Recommendations for Travelers*

Category	Vaccine
Routine recommended vaccination**	Diphtheria/Tetanus/Pertussis (DPT)#†
	Hepatitis B (HBV)#
	Haemophilus influenzae type b (Hib)#
	Influenza
	Measles (MMR)#
	Poliomyelitis (IPV)#
	Varicella#
Selective vaccination based on exposure risk	Cholera
	Hepatitis A (HAV)
	Japanese Encephalitis
	Meningococcal (polysaccharide)
	Pneumococcal (polysaccharide)
	Rabies
	Tick-borne Encephalitis
	Typhoid Fever
	Yellow Fever
Mandatory vaccination for entry§	Meningococcal (polysaccharide)
	Yellow Fever

* Based on the CDC and WHO 2005 recommendations

** For travelers aged 2 years or older

Normally administered during childhood; should be updated if indicated for travel

† Adult booster of Tetanus/diphtheria (Td) vaccine should be every 10 years

§ Meningococcal vaccination to enter Saudi Arabia; Yellow Fever vaccination to enter various countries in the endemic zone of South America and Africa

1. Give some examples where population movements affect microbial disease transmission.
2. In addition to vaccinations, what are some additional precautions global travelers should take?

36.11 Nosocomial Infections

Nosocomial infections [Greek *nosos*, disease, and *komeion*, to take care of] result from pathogens that develop within a hospital or other type of clinical care facility and are acquired by patients while they are in the facility (figure 36.12). Besides harming patients, nosocomial infections can affect nurses, physicians, aides, visitors, salespeople, delivery personnel, custodians, and anyone who has contact with the hospital. Most nosocomial infections become clinically apparent while patients are still hospitalized; however, disease onset can occur after patients have been discharged. Infections that are incubating when patients are admitted to a hospital are not nosocomial; they are community acquired. However, because such infections can serve as a ready source or reservoir of pathogens for other patients or personnel, they are also considered in the total epidemiology of nosocomial infections.

The CDC estimates that about 10% of all hospital patients acquire some type of nosocomial infection. Because approximately 40 million people are admitted to hospitals annually, about 2 to 4 million people may develop an infection they did not have upon entering the hospital. Thus nosocomial infections represent a significant proportion of all infectious diseases acquired by humans.

Nosocomial diseases are usually caused by bacteria, most of which are noninvasive and part of the normal microbiota; viruses, protozoa, and fungi are rarely involved. Figure 36.12 summarizes the most common types of nosocomial infections and the most common nosocomial pathogens.

Source

The nosocomial pathogens that cause diseases come from either endogenous or exogenous sources. Endogenous sources are the patient's own microbiota; exogenous sources are microbiota other than the patient's. Endogenous pathogens are either brought into the hospital by the patient or are acquired when the patient becomes colonized after admission. In either case the pathogen colonizing the patient may subsequently cause a nosocomial disease (e.g., when the pathogen is transported to another part of the body or when the host's resistance drops). If it cannot be determined that the specific pathogen responsible for a nosocomial disease is exogenous or endogenous, then the term autogenous is used. An **autogenous infection** is one that is caused by an agent derived from the microbiota of the patient, despite whether it became part of the patient's microbiota following his or her admission to the hospital.

There are many potential exogenous sources in a hospital. Animate sources are the hospital staff, other patients, and visitors. Some examples of inanimate exogenous sources are food, computer keyboards, urinary catheters, intravenous and respiratory therapy equipment, and water systems (e.g., softeners, dialysis units, and hydrotherapy equipment).

Control, Prevention, and Surveillance

In the United States nosocomial infections prolong hospital stays by 4 to 13 days, result in over 4.5 billion dollars a year in direct hospital charges, and lead to over 20,000 direct and 60,000 indirect deaths annually. The enormity of this problem has led most hospitals to allocate substantial resources to the development of methods and programs for the surveillance, prevention, and control of nosocomial infections.

All personnel involved in the care of patients should be familiar with basic infection control measures such as isolation policies of the hospital; aseptic techniques; proper handling of equipment, supplies, food, and excreta; and surgical wound care and dressings. To adequately protect their patients, hospital personnel must practice proper aseptic technique and handwashing procedures, and must wear gloves when contacting mucous membranes and secretions. Patients should be monitored with respect to the frequency, distribution, symptomatology, and other characteristics common to nosocomial infections. A dynamic control and surveillance program can be invaluable in preventing many nosocomial infections, patient discomfort, extended stays, and further expense.

The Hospital Epidemiologist

Because of nosocomial infections, all hospitals desiring accreditation by the Joint Commission on Accreditation of Healthcare Organizations (JCAHO) must have a designated individual directly responsible for developing and implementing policies governing control of infections and communicable diseases. This individual is often a registered nurse known as a hospital epidemiologist, nurse epidemiologist, infection control nurse, infection control practitioner, or a clinical microbiologist/ technologist. In larger hospitals, a physician is the hospital epidemiologist and should be trained in infectious diseases. He or she oversees a staff that includes nurse epidemiologists, quality assurance specialists, fellows in infectious disease/hospital epidemiology, and epidemiology technicians. The hospital epidemiologist must meet with an infection control committee composed of various professionals who have expertise in the different aspects of infection control and hospital operation. The infection control committee periodically evaluates laboratory reports, patients' charts, and surveys done by the hospital epidemiologist to determine whether there has been any increase in the frequency of particular infectious diseases or potential pathogens.

Overall, the services provided by the hospital epidemiologist should include the following:

1. Research in infection control
2. Evaluation of disinfectants, rapid test systems, and other products
3. Efforts to encourage appropriate legislation related to infection control, particularly at the state level
4. Efforts to contain hospital operating costs, especially those related to fixed expenses such as the DRGs (diagnosis-related groups)
5. Surveillance and comparison of endemic and epidemic infection frequencies
6. Direct participation in a variety of hospital activities relating to infection control and maintenance of employee health

7. Education of hospital personnel in communicable disease control and disinfection and sterilization procedures

8. Establishment and maintenance of a system for identifying, reporting, investigating, and controlling infections and communicable diseases of patients and hospital personnel

9. Maintenance of a log of incidents related to infections and communicable diseases

10. Monitoring trends in the antimicrobial drug resistance of infectious agents

Computer software packages are available to aid the infection control practitioner. Such packages generate standard reports, cause-and-effect tabulations, and graphics for the daily epidemiological monitoring that must be done.

1. Describe a nosocomial infection.
2. What two general sources are responsible for nosocomial infections? Give some specific examples of each general source.
3. Why are nosocomial infections important?
4. What does a hospital epidemiologist do to control nosocomial infections?

Summary

36.1 Epidemiological Terminology

a. Epidemiology is the science that evaluates the determinants, occurrence, distribution, and control of health and disease in a defined population.

b. Specific epidemiological terminology is used to communicate disease incidence in a given population. Frequently used terms include sporadic disease, endemic disease, hyperendemic disease, epidemic, index case, outbreak, and pandemic.

36.2 Measuring Frequency: The Epidemiologist's Tools

a. Statistics is an important tool used in the study of modern epidemiology.

b. Epidemiological data can be obtained from such factors as morbidity, prevalence, and mortality rates.

36.3 Recognition of an Infectious Disease in a Population

a. An infectious disease is caused by microbial agents such as viruses, bacteria, fungi, protozoa, and helminths. A communicable disease can be transmitted from person to person.

b. The manifestations of an infectious disease can range from mild to severe to deadly, depending on the agent and host.

c. Surveillance is necessary for recognizing a specific infectious disease within a given population. This consists of gathering data on the occurrence of the disease, collating and analyzing the data, summarizing the findings, and applying the information to control measures.

d. It is important to recognize the signs and symptoms that compose a disease syndrome. This includes the characteristic course of the disease, such aspects as the incubation period and prodromal stage (**figure 36.3**).

e. Remote sensing and geographic information systems can be used to gather epidemiological data on the environment.

36.4 Recognition of an Epidemic

a. A common-source epidemic is characterized by a sharp rise to a peak and then a rapid, but not as pronounced, decline in the number of individuals infected (**figure 36.4**). A propagated epidemic is characterized by a relatively slow and prolonged rise and then a gradual decline in the number of individuals infected (**figure 36.5**).

b. Herd immunity is the resistance of a population to infection and pathogen spread because of the immunity of a large percentage of the individuals within the population.

36.5 The Infectious Disease Cycle: Story of a Disease

a. The infectious disease cycle or chain involves the characteristics of the pathogen, the source and/or reservoir of the pathogen, the transmission of the

pathogen, the susceptibility of the host, the exit mechanism of the pathogen from the body of the host, and its spread to a new reservoir or host (**figure 36.7**).

b. There are four major modes of transmission: airborne, contact, vehicle, and vector-borne (**figure 36.8**).

36.6 Virulence and the Mode of Transmission

a. The degree of virulence may be influenced by the pathogen's preferred mode of transmission. New human diseases may arise and spread because of ecosystem disruption, rapid transportation, human behavior, and other factors.

36.7 Emerging and Reemerging Infectious Diseases and Pathogens

a. It is now clear that globally, humans will continually be faced with both new infectious diseases and the reemergence of older diseases once thought to be conquered.

b. CDC has defined these diseases as "new, reemerging, or drug-resistant infections whose incidence in humans has increased within the past two decades or whose incidence threatens to increase in the near future" (**figure 36.11**).

c. Many factors characteristic of the modern world undoubtedly favor the development and spread of these microorganisms and their diseases.

36.8 Control of Epidemics

a. The public health system consists of individuals and organizations that function in the control of infectious diseases and epidemics.

b. Vaccination is one of the most cost-effective weapons for microbial disease prevention, and vaccines constitute one of the greatest achievements of modern medicine.

c. Artificially acquired immunity to pathogens can be accomplished by either active or passive immunization.

d. Many of the current vaccines in use for humans (**table 36.4**) consists of whole organisms that are either inactivated (killed) or attenuated (live but avirulent).

e. Some of the risks associated with whole-cell vaccines can be avoided by using only specific purified macromolecules derived from pathogenic microorganisms. Currently, there are three general forms of subunit or acellular vaccines: capsular polysaccharides, recombinant surface antigens, and inactivated exotoxins (toxoids) (**table 36.6**).

f. A number of microorganisms have been used for recombinant-vector vaccines. The attenuated microorganism serves as a vector, replicating within the host, and expressing the gene product of the pathogen-encoded antigenic proteins. The proteins can elicit humoral immunity when the proteins escape from the cells and cellular immunity when they are broken down and properly displayed on the cell surface.

g. Other genetic vaccines are termed DNA vaccines. DNA vaccines elicit protective immunity against a pathogen by activating both branches of the immune system: humoral and cellular.

h. Epidemiological control measures can be directed toward reducing or eliminating infection sources, breaking the connection between sources and susceptible individuals, or isolating the susceptible individuals and raising the general level of herd immunity by immunization.

36.9 Bioterrorism Preparedness

a. Today bioterrorism is a reality. Terrorist incidents and hoaxes involving toxic or infectious agents have been on the rise.

b. Among weapons of mass destruction, biological weapons are more destructive than chemical weapons. The list of biological agents that could pose the greatest public health risk in the event of a bioterrorist attack is short and includes viruses, bacteria, parasites, and toxins (**table 36.7**).

36.10 Global Travel and Health Considerations

a. Certain precautions and health considerations should be taken into consideration when traveling globally.

b. Vaccinations are one of the most important strategies of prophylaxis in travel medicine (**table 36.9**).

36.11 Nosocomial Infections

a. Nosocomial infections are infections acquired during hospitalization and are produced by a pathogen acquired during a patient's stay. These infections come from either endogenous or exogenous sources (**figure 36.12**).

b. Hospitals must designate an individual to be responsible for identifying and controlling nosocomial infections. This person is known as a hospital epidemiologist, nurse epidemiologist, infection control nurse, or infection control practitioner.

Key Terms

active carrier 891
acute carrier 892
adjuvant 901
airborne transmission 892
antigenic drift 890
antigenic shift 890
attenuated vaccine 901
autogenous infection 909
biocrime 905
biologic transmission 896
bioterrorism 905
carrier 891
casual carrier 892
chronic carrier 892
common-source epidemic 889
common vehicle transmission 894
communicable disease 888
contact transmission 892

convalescence 888
convalescent carrier 891
disease 885
disease syndrome 888
DNA vaccine 904
droplet nuclei 892
endemic disease 886
epidemic 886
epidemiologist 886
epidemiology 885
fomite 894
geographic information system (GIS) 889
harborage transmission 896
health 885
healthy carrier 892
herd immunity 890
hyperendemic disease 886

immunization 901
incubation period 888
incubatory carrier 892
index case 887
infectious disease cycle (chain of infection) 891
morbidity rate 887
mortality rate 887
nosocomial infection 908
outbreak 886
pandemic 887
period of infectivity 891
prevalence rate 887
prodromal stage 888
propagated epidemic 889
recombinant-vector vaccine 904
remote sensing (RS) 889

reservoir 891
signs 888
source 891
sporadic disease 886
statistics 887
subunit vaccine 904
symptoms 888
systematic epidemiology 898
toxoid 901
transient carrier 892
vaccine 901
vaccinomics 901
vector 892
vector-borne transmission 896
vehicle 894
whole-cell vaccines 901
zoonoses 892

Critical Thinking Questions

1. Why is international cooperation a necessity in the field of epidemiology? What specific problem can you envision if there were no such organizations?

2. What common sources of infectious disease are found in your community? How can the etiologic agents of such infectious diseases spread from their source or reservoir to members of your community?

3. How could you prove that an epidemic of a given infectious disease was occurring?

4. How can changes in herd immunity contribute to an outbreak of a disease on an island?

5. College dormitories are notorious for outbreaks of flu and other infectious diseases. This is particularly prevalent during final exam weeks. Using your knowledge of the immune response and epidemiology, suggest practices that could be adopted to minimize the risks at such a critical time of the term.

6. Why does an inactivated vaccine induce only a humoral response, whereas an attenuated vaccine induces both humoral and cell-mediated responses?

7. Why is a DNA vaccine delivered intramuscularly and not by intravenous or oral routes?

Learn More

Buckland, D. C. 2005. The process development challenge for a new vaccine. *Nature Medicine* 11:S16–S19.

Curtis, R. 2002. Bacterial infectious disease control by vaccine development. *J. Clin. Invest.* 110:1061–66.

Desrosiers, R. 2004. Prospects for an AIDS vaccine. *Nature Medicine* 10:221–23.

Diamond, J. M. 1997. *Guns, germs, and steel.* New York: W.W. Norton & Company.

Friedman, D. S.; Heisey-Grove, D.; Argyros, F.; Berl, E.; Nsubuga, J.; Stiles, T.; Fontana, J.; Board, R. S.; Monroe, S.; McGrath, M. E.; Sutherby, H.; Dicker, R. C.; DeMaria, A.; and Matyas, B. T. 2005. An outbreak of norovirus gastroenteritis associated with wedding cakes. *Epidemiology and Infection* 133: 1057–63.

Soares, C. 2005. Cooping up avian flu: Buying time to arm for a pandemic is possible—maybe. *Sci. Am.* April 25, 2005.

Wayt Gibbs, W., and Soares, C. 2005. Preparing for a pandemic. *Sci. Am.* Nov 24, 2005.

**Please visit the Prescott website at www.mhhe.com/prescott7
for additional references.**

37 Human Diseases Caused by Viruses and Prions

Numerous global epidemics are attributed to influenza viruses. The rapid mutation of their surface antigens results in sudden antigenic shifts that lead to widespread human infections. More frequent mutations leading to antigenic drifting can be anticipated such that yearly immunization with influenza vaccine may prevent infection. The H5N1 strain may be the agent of the next influenza pandemic.

PREVIEW

- Hundreds of viruses are known to cause illness in humans. There are currently very few treatment options to control viruses; thus viruses are the source of many deadly human diseases.

- Some viruses can be transmitted through the air and directly or indirectly involve the respiratory system. Most of these viruses are highly communicable and cause diseases such as chickenpox, influenza, measles, mumps, respiratory syndromes and viral pneumonia, rubella, and hantavirus pulmonary syndrome.

- The arthropod-borne diseases are transmitted by arthropod vectors from human to human or animal to human. Examples include the various encephalitides, Colorado tick fever, West Nile fever, and historically important yellow fever.

- Some viruses are so sensitive to environmental influences that they are unable to survive for significant periods outside their hosts. These viruses are transmitted from host to host by direct contact and cause diseases such as HIV-AIDS, cold sores, the common cold, cytomegalovirus inclusion disease, genital herpes, human herpesvirus 6 infections, human parvovirus B19 infections, certain leukemias, infectious mononucleosis, human papillomavirus, viral hepatitides and warts.

- Viruses can be transmitted by food and wate. They usually either grow in or pass through the intestinal system and are acquired through fecal-oral transmission. Examples of such diseases include viral gastroenteritis, hepatitis A and E, and poliomyelitis.

- Many highly contagious and often fatal viral infections are initially contracted from animals; they are then spread from human to human. Some of the agents responsible for these zoonotic diseases, such as Ebola hemorrhagic and Lassa fevers, are potential agents of bioterrorism.

- The transmissible spongiform encephalopathies are caused by prions that remain clinically silent during a prolonged period of months or years, after which progressive disease becomes apparent, usually ending months later in profound disability or death. Examples include new variant Creutzfeldt-Jakob disease, kuru, and Gerstmann-Straussler-Scheinker disease.

Chapters 16, 17, and 18 review the general biology of viruses and introduce basic virology. In chapter 37 we continue this coverage by discussing some of the most important viruses that are pathogenic to humans. The viral diseases are grouped according to their mode of acquisition and transmission; viral diseases that occur in the United States are emphasized. Diseases caused by viruses that are listed as Select Agents are identified within the chapter by two asterisks (**).

More than 400 different viruses can infect humans. Human diseases caused by viruses are particularly interesting, considering the small amount of genetic information introduced into a host cell. This apparent simplicity belies the severe pathological features and clinical consequences that result from many viral diseases. With few exceptions, only prophylactic or supportive treatment is available. Collectively these diseases are some of the most common and yet most puzzling of all infectious diseases. The resulting frustration is compounded when year after year familiar diseases

Only once in human history have we witnessed the total eradication of a dreaded disease, and that was smallpox more than two decades ago. Now humanity stands on the brink of a second: the global eradication of polio.

—From UNICEF's Polio Website

of unknown etiology or new diseases become linked to virus infections (**table 37.1**).

37.1 AIRBORNE DISEASES

Because air does not support virus growth, any virus that is airborne must have originated from a living source. When humans are the source of the airborne virus, it usually is propelled from the respiratory tract by coughing, sneezing, or vocalizing.

Chickenpox (Varicella) and Shingles (Herpes Zoster)

Chickenpox (varicella) is a highly contagious skin disease primarily of children 2 to 7 years of age. Humans are the reservoir and the source for this virus, which is acquired by droplet inhalation into the respiratory system. The virus is highly infectious with secondary infection rates in susceptible household contacts of 65% to 86%. In the pre-vaccine era, about 4 million cases of chickenpox occurred annually in the United States, resulting in approximately 11,000 hospitalizations and 100 deaths.

The causative agent is the enveloped, DNA varicella-zoster virus (VZV), a member of the family *Herpesviridae*. The virus produces at least six glycoproteins that play a role in viral attachment to specific receptors on respiratory epithelial cells. Their recognition by the human immune system results in humoral and cellular immunity. This virus has been shown to inhibit the expression of MHC molecules by infected cells; however, this inhibition only temporarily interferes with immune recognition of the virus, perhaps as a way of increasing its transmission. Following an incubation period of 10 to 23 days, small vesicles erupt on the face or upper trunk, fill with pus, rupture, and become covered by scabs (**figure 37.1**). Healing of the vesicles occurs in about 10 days. During this time intense itching often occurs.

Laboratory confirmation of varicella virus is by detection of varicella-zoster immunoglobulin M (IgM) antibody; detection of VZV, demonstration of VZV antigen by direct fluorescent antibody and by polymerase chain reaction in clinical specimens or a significant rise in serum IgG antibody level to VZV. However, laboratory testing for VZV is not normally required, as the diagnosis of chickenpox is typically made by clinical assessment. Laboratory confirmation is recommended, though, to confirm the diagnosis of severe or unusual cases of chickenpox. As the incidence of chickenpox continues to decline due to vaccination, fewer cases are seen clinically, resulting in the likelihood of misdiagnosis. Furthermore, in persons who have previously received varicella vaccination, the disease is usually mild or atypical, and can pose particular challenges for clinical diagnosis. Therefore, laboratory confirmation of varicella cases is becoming more important. Chickenpox can be prevented or the infection shortened with an attenuated varicella vaccine (Varivax; *see table 36.4*) or the drug acyclovir (Zovirax or Valtrex). It should be noted that Valtrex (valacyclovir) is an orally administered prodrug of Zovirax or acyclovir (*see figure 34.21*). Valtrex is the valyl ester of acyclovir and is rapidly hydrolyzed to acyclovir in the body.

Table 37.1	Some Examples of Human Viral Diseases Recognized Since 1967	
Year	**Virus**	**Disease**
1967	Marburg virus	Hemorrhagic fever
1973	Rotavirus	Major cause of infantile diarrhea worldwide
1974	Parvovirus B19	Aplastic crisis in chronic hemolytic anemia
1977	Ebola virus	Ebola hemorrhagic fever
1977	Hantavirus	Hemorrhagic fever with renal syndrome
1980	Human T-cell lymphotrophic virus 1 (HTLV-1)	Adult T-cell leukemia
1982	Human T-cell lymphotrophic virus 2 (HTLV-2)	Hairy-cell leukemia
1983	Human immunodeficiency virus (HIV)	Acquired immunodeficiency syndrome (AIDS)
1988	Human herpesvirus 6 (HHV-6)	Sixth disease (roseola subitum); may be associated with multiple sclerosis
1988	Hepatitis E	Enterically transmitted non-A, non-B hepatitis
1989	Hepatitis C	Parenterally transmitted non-A, non-B liver infection
1991	Guanarito virus	Venezuelan hemorrhagic fever
1992	Lymphocytic choriomeningitis virus	Central nervous system infection often leading to meningitis, encephalomyelitis, or other diseases
1993	Sin Nombre virus	Hantavirus respiratory syndrome
1994	Sabia virus	Brazilian hemorrhagic fever
1994	Ross River virus	Ross River viral disease (Australia)
1994	Human herpesvirus 8 (HHV-8)	Associated with Kaposi's sarcoma in AIDS patients
1996	O'nyoung-nyong virus	Epidemic O'nyong fever
1997	Deer tick virus	Enzootic tick-borne encephalitis
1997	Avian flu (H5N1) virus	Avian influenza illness
1997	Transfusion-transmitted virus (TTV)	Hepatitis
1999	Australian bat lyssavirus (ABL)	ABL infection
2000	Hepatitis G	Chronic liver inflammation
2002	Metapneumovirus	Respiratory tract infections
2003	SARS-Coronavirus	Severe acute respiratory syndrome (SARS)

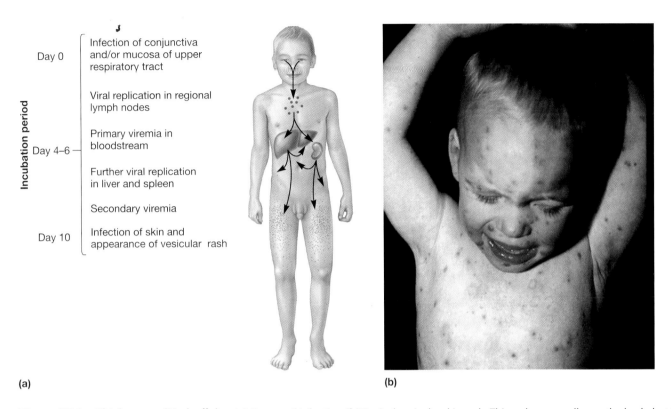

Incubation period

Day 0 — Infection of conjunctiva and/or mucosa of upper respiratory tract

Viral replication in regional lymph nodes

Day 4–6 — Primary viremia in bloodstream

Further viral replication in liver and spleen

Secondary viremia

Day 10 — Infection of skin and appearance of vesicular rash

(a)

(b)

Figure 37.1   Chickenpox (Varicella).   (a) Course of infection. **(b)** Typical vesicular skin rash. This rash occurs all over the body, but is heaviest on the trunk and diminishes in intensity toward the periphery.

Individuals who recover from chickenpox are subsequently immune to this disease; however, they are not free of the virus, as viral DNA resides in a dormant (latent) state within the nuclei of cranial nerves and sensory neurons in the dorsal root ganglia. This viral DNA is maintained in infected cells but virions cannot be detected (**figure 37.2***a*). When the infected person becomes immunocompromised by such factors as age, neoplastic diseases, organ transplants, AIDS, or psychological or physiological stress, the viruses may become activated (figure 37.2*b*). They migrate down sensory nerves, initiate viral replication, and produce painful vesicles because of sensory nerve damage (figure 37.2*c*). This syndrome is called **postherpetic neuralgia.** To manage the intense pain, corticosteroids or the drug gabapentin (Neurontin) can be prescribed. The reactivated form of chickenpox is called **shingles (herpes zoster).** Most cases occur in people over 50 years of age. Except for the pain of postherpetic neuralgia, shingles does not require specific therapy; however, in immunocompromised individuals, acyclovir, valacyclovir, vidarabine (Vira-A), or famciclovir (Famvir) are recommended. More than 500,000 cases of herpes zoster occur annually in the United States.

Influenza (Flu)

Influenza [Italian, to be influenced by the stars—*un influenza di freddo*], or the **flu,** is a respiratory system disease caused by negative-strand RNA viruses that belong to the family *Orthomyxoviridae*. There are four groups: influenza A, influenza B,

influenza C, and Thogoto viruses. They contain 7 to 8 segments of linear RNA, with a genome length between 12,000 to 15,000 nucleotides. The enveloped virion can be spherical (50–120 nm in diameter) or filamentous (200–300 nm long, 20 nm in diameter) (*see figure 16.4*). Influenza A infections are responsible for the majority of clinical influenza cases, with influenza B accounting for approximately 3% of flu in the United States. Influenza A infections usually peak in the winter and involve 10% or more of the population, with rates as high as 50 to 75% in school-age children. Influenza A viruses are widely distributed, infect a variety of mammal and bird hosts, and are further classified into subtypes (or strains) based on their membrane surface glycoproteins, hemagglutinin (HA), and neuraminidase (NA). HA and NA function in viral attachment and virulence. There are 16 HA and 9 NA antigenic forms known; they can recombine to produce various HA/NA subtypes of influenza. All subtype combinations infect birds. Influenza A viruses having H1, H2, and H3 HA antigens, along with N1 and N2 NA antigens, are predominant in nature, infecting humans since the early 1900s. H1N1 viruses appeared in 1918 and were replaced in 1957 by H2N2 subtypes as the predominant subtype. The H2N2 viruses were subsequently replaced by H3N2 as the principle subtypes in 1968. The H1N1 subtype reappeared in 1977 and co-circulates today with H2N1, H3N2, H5N2, H7N2, H7N3, H7N7, H9N2, H10N7, and H5N1 viruses.

The H5N1 subtype (also know as bird flu) appears to be the most likely candidate to initiate another influenza pandemic (global epidemic). The H5N1 subtype was responsible for six

(a) Primary infection—Chickenpox

(b) Recurrence—Shingles

(c)

Figure 37.2 Pathogenesis of the Varicella-Zoster Virus.
(a) After an initial infection with varicella (chickenpox), the viruses migrate up sensory peripheral nerves to their dorsal root ganglia, producing a latent infection. **(b)** When a person becomes immunocompromised or is under psychological or physiological stress, the viruses may be activated. **(c)** They migrate down sensory nerve axons, initiate viral replication, and produce painful vesicles. Since these vesicles usually appear around the trunk of the body, the name *zoster* (Greek for girdle) was used.

deaths in 1997, although it had been sporadically detected prior to that. Throughout 2003 to 2006, however, H5N1 was responsible for a substantial number of bird infections, resulting in the culling of millions of birds. The birds were destroyed to help prevent transmission of the virus to susceptible humans. Yet, by September 2006, H5N1 was responsible for 241 confirmed cases of influenza and 141 deaths in six countries.

One of the most important features of the influenza viruses is the frequency with which changes in antigenicity occur. If the variation is small, it is called **antigenic drift** (*see figure 36.6*). Antigenic drift results from the accumulation of mutations of HA and NA in a single strain of flu virus within a geographic region. This usually occurs every 2 to 3 years, causing local increases in the number of flu infections. **Antigenic shift** is a large antigenic change resulting from the reassortment of genomes when two different strains of flu viruses (from both animals and humans) infect

the same host cell and are incorporated into a single new capsid (*see figure 36.6*). Because there is a greater change with antigenic shift than antigenic drift, antigenic shifts can yield major epidemics and pandemics. Antigenic variation occurs almost yearly with influenza A virus, less frequently with the B virus, and has not been demonstrated in the C virus. Recognition of an epidemic (section 36.4)

The standard nomenclature system for influenza virus subtypes includes the following information: group (A, B, or C), host of origin, geographic location, strain number, and year of original isolation. Antigenic descriptions of the HA and NA are given in parentheses for type A. The host of origin is not indicated for human isolates—for example, A/Hong Kong/03/68 (H3N2). However, the host origin is given for others—for example, A/swine/Iowa/15/30 (H1N1).

Animal reservoirs are critical to the epidemiology of human influenza. For example, rural China is one region of the world where chickens, pigs, and humans live in close, crowded conditions. Influenza is widespread in chickens; although chickens can't usually transmit the virus to humans, they can transfer it to pigs. Pigs can transfer it to humans, and humans back to pigs. Recombination between human and avian strains thus occurs in pigs, leading to major antigenic shifts. This explains why influenza continues to be a major epidemic disease and frequently produces worldwide pandemics. Hippocrates described influenza in 412 B.C. The first well-documented global epidemic of influenza-like disease occurred in 1580. Since then, 31 possible influenza pandemics have been documented, with four occurring in the twentieth century. The worst pandemic on record occurred in 1918 and killed between 20 million and 50 million people. This disaster, traced to the Spanish influenza virus (*see figure 36.10*), was followed by pandemics of Asian flu (1957), Hong Kong flu (1968), and Russian flu (1977). (The names reflect popular impressions of where the episodes began, although all are now thought to have originated in China.)

The 2003–2004 U.S. influenza season began earlier than most and was moderately severe. Influenza A (H1N1, H1N2, and H3N2) and influenza B viruses co-circulated, with the predominant strain being influenza A (H3N2). In 2005, influenza activity in the United States peaked early in February and then declined, surprising many who predicted a severe flu season. However, widespread outbreaks of avian influenza A (H5N1) continued from late 2003 and were reported in Southeast Asia in 2004 through 2006, predominantly among poultry. In a number of Asian countries, though, these outbreaks were associated with severe human illnesses and deaths. Because of potentially severe consequences of pandemic H5N1 infection of humans, an international effort coordinated by the World Health Organization monitors virus surveillance, epidemiology, and control efforts. The National Institutes of Health have contracted for the manufacture of a vaccine to H5N1. It began clinical trials in 2005.

The virus (*see figure 18.9*) is acquired by inhalation or ingestion of virus-contaminated respiratory secretions. During an incubation period of 1 to 2 days, the virus adheres to the epithelium of the respiratory system (the neuraminidase present in envelope spikes may hydrolyze the mucus that covers the epithelium). The virus attaches to the epithelial cell by its hemagglutinin spike protein, causing part of the cell's plasma membrane to bulge inward, seal off, and form a vesicle (receptor-mediated endocytosis). This encloses the virus in an endosome (*see figure 18.11*). The hemagglutinin molecule in the virus envelope undergoes a dramatic conformational change when the endosomal pH decreases. The hydrophobic ends of the hemagglutinin spring outward and extend toward the endosomal membrane. After they contact the membrane, fusion occurs and the RNA nucleocapsid is released into the cytoplasmic matrix.

Influenza is characterized by chills, fever (usually > 102°F, 39°C), headache, malaise, cough, sore throat, and general muscular aches and pains. These symptoms arise from the death of respiratory epithelial cells, probably due to attacks by activated T cells. These symptoms are more debilitating than are symptoms of the common cold. Recovery usually occurs in 3 to 7 days, during which coldlike symptoms appear as the fever subsides. Influenza

alone usually is not fatal. However, death may result from pneumonia caused by secondary bacterial invaders such as *Staphylococcus aureus, Streptococcus pneumoniae,* and *Haemophilus influenzae.* A commercially available identification technique is Directigen FLU-A (an enzyme immunoassay [EIA] rapid test). This test can detect influenza A virus in clinical specimens in less than 15 minutes.

As with many other viral diseases, only the symptoms of influenza usually are treated. Amantadine (Symmetrel) (*see figure 34.21*), rimantadine (Flumadine), zanamivir (Relenza), and oseltamivir (Tamiflu) have been shown to reduce the duration and symptoms of type A influenza if administered during the first two days of illness. Unfortunately, 91% of the virus samples (representing the predominant influenza strain) tested by the CDC in December 2005 were resistant to rimantidine and amantidine, compared to 11% in the previous year. Amantadine and rimantadine are chemically related, antiviral drugs known as adamantanes. These usually have activity against influenza A viruses but not influenza B viruses. Amantadine and rimantadine are thought to interfere with influenza A virus M2 protein, a membrane ion channel protein. They also inhibit virus uncoating, which inhibits virus replication, resulting in decreased viral shedding. Zanamivir and oseltamivir are chemically related antiviral drugs known as neuraminidase inhibitors that have activity against both influenza A and B viruses. Neuraminidase inhibitors attack the virus directly by plugging the catalytic site of the enzyme neuraminidase. With the enzyme inactivated, viral particles can't travel from cell to cell. Importantly, aspirin (salicylic acid) should be avoided in children younger than 14 years to reduce the risk of Reye's syndrome (**Disease 37.1**).

The mainstay for prevention of influenza since the late 1940s has been inactivated virus vaccines, (*see table 36.4*), especially for the chronically ill, individuals over age 65, residents of nursing homes, and health-care workers in close contact with people at risk. Clinical disease in these patients is most likely to be severe. Because of influenza's high genetic variability, efforts are made each year to incorporate new virus subtypes into the vaccine. Even when no new subtypes are identified in a given year, annual immunization is still recommended because immunity using the inactivated virus vaccine typically lasts only 1 to 2 years.

Antiviral drugs (section 34.8)

Measles (Rubeola)

Measles [**rubeola:** Latin *rubeus,* red] is a highly contagious skin disease that is endemic throughout most of the world. In March 2000, a group of experts convened by the CDC concluded that fortunately, measles is no longer endemic in the United States. It seems that all cases of measles (less than 200 per year since 1998) in the United States were imported from other countries, usually Europe and Asia. We discuss measles because it remains of global importance. It is a negative-strand, enveloped RNA virus, in the genus *Morbillivirus* and the family *Paramyxoviridae.* The measles virus is monotypic, but small variations at the epitope level have been described.

Disease

37.1 Reye's and Guillain-Barré Syndromes

An occasional complication of influenza, chickenpox, and a few other viral diseases in children under 14 years of age is **Reye's syndrome.** After the initial infection has disappeared, the child suddenly begins to vomit persistently and experiences convulsions followed by delirium and a coma. Pathologically, the brain swells with injury to the neuronal mitochondria, fatty infiltration into the liver occurs, blood ammonia is elevated, and both serum glutamic oxaloacetic transaminase (SGOT) and serum glutamic pyruvic transaminase (SGPT) are elevated in the blood. Diagnosis is made by the measurement of the levels of these enzymes and ammonia.

The relationship between the initial viral infection and the brain and liver damage is unknown. Treatment is nonspecific and directed toward reducing intracranial pressure and correcting metabolic and electrolyte abnormalities. Some children who recover have residual neurological deficits—impaired mental capacity, seizures, and hemi-

plegia (paralysis on one side of the body). Mortality ranges from 10 to 40%. It is suspected that the use of aspirin or salicylate-containing products to lower the initial viral fever increases a child's chances of acquiring Reye's syndrome.

Another condition that involves the central nervous system and is associated with influenza infections is **Guillain-Barré syndrome** (sometimes called **French polio**). In this disorder the individual suffers a delayed reaction (usually within 8 weeks) either to the actual virus infection or to vaccines against influenza. The virus or viral antigen in the vaccine damages the Schwann cells that myelinate the peripheral nerves and thus causes demyelination. As a result a prominent feature of this syndrome is a symmetric weakness of the extremities and sensory loss. Fortunately recovery usually is complete because the remaining undamaged Schwann cells eventually proliferate and wrap around the demyelinated nerves.

(a)

(b)

Figure 37.3 Measles (Rubeola). **(a)** The rash of small, raised spots is typical of measles. The rash usually begins on the face and moves downward to the trunk. **(b)** Signs and symptoms of a measles infection.

The variations are based on genetic variability in the virus genes. Such variations, however, have no effect on protective function since a measles infection still provides a lifelong immunity against reinfection. The virus enters the body through the respiratory tract or the conjunctiva of the eyes. The receptor for the measles virus is the complement regulator CD46, also known as membrane cofactor protein.

The incubation period is usually 10 to 21 days, and the first symptoms begin about the tenth day with a nasal discharge, cough, fever, headache, and conjunctivitis. Within 3 to 5 days

skin eruptions occur as faintly pink maculopapular lesions that are at first discrete, but gradually become confluent (**figure 37.3**). The rash normally lasts about 5 to 10 days. Lesions of the oral cavity include the diagnostically useful bright-red **Koplik's spots** with a bluish-white speck in the center of each. Koplik's spots represent a viral exanthem (a skin eruption) occurring in the form of macules or papules as a result of the viral infection. Very infrequently a progressive degeneration of the central nervous system called **subacute sclerosing panencephalitis** occurs. No specific treatment is available for measles. The use of attenuated measles vaccine (Attenuvax) or in combination (MMR vaccine; *measles, mumps, rubella*) is recommended for all children (*see table 36.4*). Since public health immunization programs began in 1963, there has been near eradication of measles in the United States. In less well-developed countries, however, the morbidity and mortality in young children from measles infection remain high. It has been estimated that measles infects 50 million people and kills about 4 million a year worldwide. Serious outbreaks of measles are still reported in North America and Europe, especially among college students.

Mumps

Mumps is an acute, generalized disease that occurs primarily in school-age children. The number of mumps cases in the United States has decreased 99% since the widespread use of the MMR vaccine, with fewer than 300 cases reported annually. Occasional outbreaks occur in nonimmunized populations. The mumps virus is a member of the genus *Rubulavirus* in the family *Paramyxoviridae*. This virus is a pleomorphic, enveloped virus that contains a helical nucleocapsid containing negative-strand RNA. The virus is transmitted in saliva and respiratory droplets. The portal of entry is the respiratory tract. Mumps is about as contagious as influenza and rubella, but less so than measles or chickenpox. The virus replicates in the nasopharynx and lymph nodes of an infected person. Viral transmission is airborne or through direct contact with contaminated droplets or saliva. The most prominent manifestations of mumps are swelling and tenderness of the salivary (parotid) glands 16 to 18 days after infection of the host by the virus (**figure 37.4**). The swelling usually lasts for 1 to 2 weeks and is accompanied by a low-grade fever. Severe complications

Figure 37.4   Mumps.   Child with diffuse swelling of the salivary (parotid) glands due to the mumps virus.

of mumps are rare, however, meningitis and inflammation of the epididymis and testes (**orchitis**) can be important complications associated with this disease—especially in the postpubescent male. Therapy of mumps is limited to symptomatic and supportive measures. A live, attenuated mumps virus vaccine is available. It usually is given as part of the trivalent MMR vaccine (*see table 36.4*).

Respiratory Syndromes and Viral Pneumonia

Acute viral infections of the respiratory system are among the most common causes of human disease. The infectious agents are called the acute respiratory viruses and collectively produce a variety of clinical manifestations, including rhinitis (inflammation of the mucous membranes of the nose), tonsillitis, laryngitis, and bronchitis. The adenoviruses, coxsackievirus A, coxsackievirus B, echovirus, influenza viruses, parainfluenza viruses, poliovirus, respiratory syncytial virus, and reovirus are thought to be responsible. It should be emphasized that for most of these viruses there is a lack of specific correlation between the agent and the clinical manifestation—hence the term syndrome. Immunity is not complete, and reinfection is common. The best treatment is rest.   Taxonomy of eucaryotic viruses (section 18.1)

In those cases of pneumonia for which no cause can be identified, viral pneumonia may be assumed if mycoplasmal pneumonia has been ruled out. The clinical picture is nonspecific. Symptoms may be mild, or there may be severe illness and death.

Respiratory syncytial virus (RSV) often is described as the most dangerous cause of lower respiratory infections in young children. In the United States over 90,000 infants are hospitalized each year and at least 4,000 die. RSV is a member of the negative-strand RNA virus family *Paramyxoviridae*. The virion is variable in shape and size (average diameter of between 120 and 300 nm). It is enveloped with two virally specific glycoproteins as part of the structure. One of the two glycoproteins, G, is responsible for the binding of the virus to the host cell; the other, the fusion protein or F, permits fusion of the viral envelope with the host cell plasma membrane, leading to entry of the virus. The F protein also induces the fusion of the plasma membranes of infected cells. RSV thus gets its name from the resulting formation of a syncytium or multinucleated mass of fused cells. The multinucleated syncytia are responsible for inflammation, alveolar thickening, and the filling of alveolar spaces with fluid. The source of the RSV is hand contact and respiratory secretions of humans. The virus is unstable in the environment (surviving only a few hours on environmental surfaces), and is readily inactivated with soap and water and disinfectants.

Clinical manifestations consist of an acute onset of fever, cough, rhinitis, and nasal congestion. In infants and young children, this often progresses to severe bronchitis and viral pneumonia. Diagnosis is by either Directigen RSV or Test-Pack RSV rapid test kits. The virus is found worldwide and causes seasonal (November to March) outbreaks lasting several months. Treatment is with inhaled ribavirin (Virazole). A series of antibody (RSV-immune globulin) injections has been shown to reduce the severity of this

This virus is transmitted by contact with virus-contaminated blood, by the fecal-oral route, by in utero transmission from mother to fetus, sexually, or through organ transplantation. Diagnosis is made by enzyme-linked immunosorbent assay (ELISA), which detects serum antibody to a recombinant antigen of HCV, and nucleic acid detection by PCR. HCV is found worldwide. Prior to routine screening, HCV accounted for more than 90% of hepatitis cases developed after a blood transfusion. Worldwide, hepatitis C has reached epidemic proportions, with more than 1 million new cases reported annually. In the United States, nearly 4 million persons are infected and 25,000 new cases occur annually. Currently, HCV is responsible for about 8,000 deaths annually in the United States. Furthermore, HCV is the leading reason for liver transplantation in the United States. Treatment is with Ribovirin and pegylated (coupled to polyethylene glycol) recombinant interferon-alpha (Intron A, Roferon-A). This combination therapy can rid the virus in 50% of those infected with genotype 1 and in 80% of those infected with genotype 2 or 3.

In 1977 a cytopathic hepatitis agent termed the **Delta agent** was discovered. Later it was called the hepatitis D virus (HDV) and the disease **hepatitis D** was designated. HDV is a unique agent in that it is dependent on the hepatitis B virus to provide the envelope protein (HBsAg) for its RNA genome. Thus HDV only replicates in liver cells co-infected with HBV. Both must be actively replicating. Furthermore, the RNA of the HDV is smaller than the RNA of the smallest picornaviruses and its circular conformation differs from the linear structure typical of animal RNA viruses. Thus its similarity to the plant virusoids has led some to also call this agent a virusoid. HDV is spread only to persons who are already infected with HBV (superinfection) or to individuals who get HBV and the virusoid at once (coinfection). The primary laboratory tools for the diagnosis of an HDV infection are serological tests for anti-delta antibodies. Treatment of patients with chronic HDV remains difficult. Some positive results can be obtained with alpha interferon treatment for 3 months to 1 year. Liver transplantation is the only alternative to chemotherapy. Worldwide, there are approximately 300 million HBV carriers, and available data indicate that no fewer than 5% of these are infected with HDV. Thus because of the propensity of HDV to cause fulminant as well as chronic liver disease, continued incursion of HDV into areas of the world where persistent hepatitis B infection is endemic has serious implications. Prevention and control involves the widespread use of the hepatitis B vaccine.

Two other forms of hepatitis have been identified: **hepatitis F** (causing fulminant, posttransfusion hepatitis) and a syncytial giant-cell hepatitis (hepatitis G) with viruslike particles resembling the measles virus. **Hepatitis G** (HGV) is a member of the *Flaviviridae* family. It has been cloned and is widely distributed in humans. HGV can be transmitted through needles or sexually. The significance of HGV infections in liver disease is not yet clear. However, infection causes chronic liver inflammation with its associated sequelae. Further virologic, epidemiological, and molecular efforts to characterize these new agents and their diseases are currently being undertaken.

Warts

Warts, or verrucae [Latin *verruca*, wart], are horny projections on the skin caused by the human papillomaviruses. The papillomaviruses are placed in the family *Papillomaviridae* (formerly they were in the *Papovaviridae*). These viruses have naked icosahedral capsids with a double-stranded, supercoiled, circular DNA genome. At least eight distinct genotypes produce benign epithelial tumors that vary in respect to their location, clinical appearance, and histopathologic features. Warts occur principally in children and young adults and are limited to the skin and mucous membranes. The viruses are spread between people by direct contact; autoinoculation occurs through scratching. Four major kinds of warts are **plantar warts, verrucae vulgaris, flat** or **plane warts, and anogenital condylomata (venereal warts) (figure 37.20)**. Treatment includes physical destruction of the wart(s) by electrosurgery, cryosurgery with liquid nitrogen or solid CO_2, laser fulguration (drying), direct application of the drug podophyllum to the wart(s), or injection of IFN-α (Intron A, Alferon N). Anogenital condylomata (venereal warts) are sexually transmitted and caused by types 6, 11, and 42 human papillomavirus (HPV). Once the virus enters the body, the incubation period is 1 to 6 months. The warts (figure 37.20d) are soft, pink, cauliflowerlike growths that occur on the external genitalia, in the vagina, on the cervix, or in the rectum. They often are multiple and vary in size. In addition to being a common sexually transmitted disease, genital infection with HPV is of considerable importance because specific types of genital HPV play a major role in the pathogenesis of epithelial cancers of the male and female genital tracts. Over the last decade, many studies have convincingly demonstrated that specific types of HPV are the causal agents of at least 90% of cervical cancers. The most common types conferring a high risk for cervical cancer include HPV types 16, 18, 31, 33, 35, 45, 51, 52, and 56. There is also a possible link between papillomaviruses and nonmelanoma squamous and basal cell cancers. Thirty percent of humans with a rare syndrome of persistent warts (not common warts, but a particular type of warty growth) eventually develop skin cancer, and HPV viral DNA is found in the malignant cells. The epidemiology, molecular biology, and role of HPV in the development of such cancers is an area of active research. Recently a vaccine has been licensed for use in the U.S. that is highly effective in preventing cervical cancer caused by HPVs 16 and 18. These two viruses cause an estimated 80% of all such cancers.

---

1. Describe the AIDS virus and how it cripples the immune system. How is the virus transmitted? What types of pathological changes can result?
2. Why do people periodically get cold sores? Describe the causative agent.
3. Why do people get the common cold so frequently? How are cold viruses spread?
4. Give two major ways in which herpes simplex virus type 2 is spread. Why do herpes infections become active periodically?
5. What two types of leukemias are caused by viruses?
6. Describe the causative agent and some symptoms of mononucleosis and exanthem subitum.

(a)   (b)

(c)   (d)

Figure 37.20   Warts.   (a) Common warts on fingers. **(b)** Flat warts on the face. **(c)** Plantar warts on the feet. **(d)** Perianal condyloma acuminata.

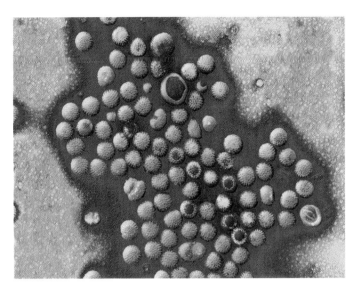

Figure 37.21   Viral Gastroenteritis.   Electron micrograph of rotaviruses (reoviruses) in a human gastroenteritis stool filtrate (×90,000). Note the spokelike appearance of the icosahedral capsids that surround double-stranded RNA within each virion.

7. What are the different causative viruses of hepatitis and how do they differ from one another? How can one avoid hepatitis? Do you know anyone who is a good candidate for infection with these viruses?
8. What kind of viruses cause the formation of warts? Describe the formation of venereal warts.

37.4 FOOD-BORNE AND WATERBORNE DISEASES

Food and water have been recognized as potential carriers (vehicles) of disease since the beginning of recorded history. Collectively more infectious diseases occur by these two routes than any other. A few of the many human viral diseases that are food- and waterborne are now discussed.   Food-borne diseases (section 40.4); Water purification and sanitary analysis (section 41.1)

Gastroenteritis (Viral)

Acute viral gastroenteritis (inflammation of the stomach or intestines) is caused by six major categories of viruses: rotaviruses (**figure 37.21**), adenoviruses, caliciviruses, astroviruses, Norwalk virus, and a group of Noroviruses (previously known as Norwalk-like viruses). The medical importance of these viruses is summarized in **table 37.5.**

The viruses responsible for gastroenteritis are transmitted by the fecal-oral route. Infection with rotaviruses and astroviruses is most common during the cooler months, while infection with adenovirus occurs year-round. Bacteria-caused diarrheal diseases usually occur in the warmer months of the year. The average incubation period for most of these diseases is 1 to 2 days. The clin-

ical manifestations typically range from asymptomatic to a relatively mild diarrhea with headache and fever; to a severe, watery, nonbloody diarrhea with abdominal cramps. Fatal dehydration is most common in young children. Vomiting is almost always present, especially in children. Viral gastroenteritis is usually self-limited. Treatment is designed to provide relief through the use of oral fluid replacement with isotonic liquids, analgesics, and antiperistaltic agents. Symptoms usually last for 1 to 5 days and recovery often results in protective immunity to subsequent infection.

Diarrheal diseases are the leading cause of childhood deaths (5 to 10 million deaths per year) in developing countries where malnutrition is common. Current estimates are that viral gastroenteritis produces 30 to 40% of the cases of infectious diarrhea in the United States, far outnumbering documented cases of bacterial and protozoan diarrhea (the cause of approximately 40% of presumed cases of diarrhea remains unknown). In the United States, rotavirus is responsible for the hospitalization of approximately 55,000 children each year and, worldwide, the death of over 600,000 children annually. Norovirus, on the other hand, is estimated to cause 23 million cases of acute gastroenteritis representing at least 50% of all food-borne outbreaks of gastroenteritis. Viral gastroenteritis is seen most frequently in infants 1 to 11 months of age, where the virus attacks the epithelial cells of the upper intestinal villi, causing malabsorption, impairment of sodium transport, and diarrhea.

Hepatitis A

Hepatitis A (infectious hepatitis) usually is transmitted by fecal-oral contamination of food, drink, or shellfish that live in contaminated water and contain the virus in their digestive system. The

Table 37.5	Medically Important Gastroenteritis Viruses	
Virus	**Epidemiological Characteristics**	**Clinical Characteristics**
Rotaviruses		
Group A	Endemic diarrhea in infants worldwide	Dehydrating diarrhea for 5–7 days; fever, abdominal cramps, nausea, and vomiting common
Group B	Large outbreaks in adults and children in China	Severe watery diarrhea for 3–5 days
Group C	Sporadic cases in children in Japan	Similar to group A
Norwalk virus and Norovirus	Epidemics of vomiting and diarrhea in older children and adults; occurs in families, communities, and nursing homes; often associated with shellfish, other food, or water and infected food handlers	Acute vomiting, fever, myalgia, and headache lasting 1–2 days, diarrhea
Caliciviruses other than the Norwalk group	Pediatric diarrhea; associated with shellfish and other foods in adults	Rotavirus-like illness in children; Norwalk-like in adults
Astroviruses	Pediatric diarrhea; reported in nursing homes	Watery diarrhea for 1–3 days

disease is caused by the hepatitis A virus (HAV) of the genus *Hepatovirus* in the family *Picornaviridae.* The hepatitis A virus is an icosahedral, linear, positive-strand RNA virus that lacks an envelope. Once in the digestive system, the viruses multiply within the intestinal epithelium. Usually only mild intestinal symptoms result. Occasionally viremia (the presence of viruses in the blood) occurs and the viruses may spread to the liver. The viruses reproduce in the liver, enter the bile, and are released into the small intestine. This explains why feces are so infectious. Symptoms last from 2 to 20 days and include anorexia, general malaise, nausea, diarrhea, fever, and chills. If the liver becomes infected, jaundice ensues. Laboratory diagnosis is by detection of anti-hepatitis A antibody. During epidemic years, about 30,000 cases were reported annually in the United States. The number of new cases has been dramatically reduced since the introduction of the hepatitis A vaccine in the 1990s. Fortunately the mortality rate is low (less than 1%), and infections in children are usually asymptomatic. Most cases resolve in 4 to 6 weeks and yield a strong immunity. Approximately 40 to 80% of the United States population have serum antibodies though few have been aware of the disease. Control of infection is by simple hygienic measures, the sanitary disposal of excreta, and the killed HAV vaccine (Havrix). This vaccine is recommended for travelers (*see table 36.9*) going to regions with high evidence rates of hepatitis A.

Hepatitis E

Hepatitis E is implicated in many epidemics in certain developing countries in Asia, Africa, and Central and South America. It is uncommon in the United States but is occasionally imported by infected travelers. The monopartite, positive-strand, RNA viral genome (7,900 nucleotides) is linear. The virion is spherical, nonenveloped, and 32 to 34 nm in diameter. Based on biologic and physicochemical properties, HEV has been provisionally classified in the *Caliciviridae* family; however, the organization of the HEV genome is substantially different from that of other

caliciviruses and, therefore, HEV may eventually be classified in a separate family.

Infection usually is associated with feces-contaminated drinking water. Presumably HEV enters the blood from the gastrointestinal tract, replicates in the liver, is released from hepatocytes into the bile, and is subsequently excreted in the feces. Like hepatitis A, an HEV infection usually runs a benign course and is self-limiting. The incubation period varies from 15 to 60 days, with an average of 40 days. The disease is most often recorded in patients that are 15 to 40 years of age. Children are typically asymptomatic or present mild signs and symptoms, similar to those of other types of viral hepatitis, including abdominal pain, anorexia, dark urine, fever, hepatomegaly, jaundice, malaise, nausea, and vomiting. Case fatality rates are low (1 to 3%) except for pregnant women (15 to 25%), who may die from fulminant hepatic failure. Diagnosis of HEV is by ELISA (IgM or IgG to recombinant HEV) or reverse transcriptase PCR. There are no specific measures for preventing HEV infections other than those aimed at improving the level of health and sanitation in affected areas.

Poliomyelitis

Poliomyelitis [Greek *polios,* gray, and *myelos,* marrow or spinal cord], **polio,** or **infantile paralysis** is caused by the poliovirus, a member of the family *Picornaviridae* (**Historical Highlights 37.3**). The poliovirus is a naked, positive-strand RNA virus with three different serotypes—P1, P2, and P3. The virus is very stable, especially at acidic pH, and can remain infectious for relatively long periods in food and water—its main routes of transmission. The average incubation period is 6 to 20 days. Once ingested, the virus multiplies in the mucosa of the throat and/or small intestine. From these sites the virus invades the tonsils and lymph nodes of the neck and terminal portion of the small intestine. Generally, there are either no symptoms or a brief illness characterized by fever, headache, sore throat, vomiting, and loss of appetite. The virus sometimes enters the bloodstream and causes a viremia. In most cases (more than 99%), the viremia is transient and clinical disease does not result. In

Like many other infectious diseases, polio is probably of ancient origin. Various Egyptian hieroglyphics dated approximately 2000 B.C. depict individuals with wasting, withered legs and arms (see **Box figure**). In 1840 the German orthopedist Jacob von Heine described the clinical features of poliomyelitis and identified the spinal cord as the problem area. Little further progress was made until 1890, when Oskar Medin, a Swedish pediatrician, portrayed the natural history of the disease as epidemic in form. He also recognized that a systemic phase, characterized by minor symptoms and fever, occurred early and was complicated by paralysis only occasionally. Major progress occurred in 1908, when Karl Landsteiner and William Popper successfully transmitted the disease to monkeys. In the 1930s much public interest in polio occurred because of the polio experienced by Franklin D. Roosevelt. This led to the founding of the March of Dimes campaign in 1938; the sole purpose of the March of Dimes was to collect money for research on polio. In 1949 John Enders, Thomas Weller, and Frederick Robbins discovered that the poliovirus could be propagated in vitro in cultures of human embryonic tissues of nonneural origin. This was the keystone that later led to the development of vaccines.

In 1952 David Bodian recognized that there were three distinct serotypes of the poliovirus. Jonas Salk successfully immunized humans with formalin-inactivated poliovirus in 1952, and this vaccine (IPV) was licensed in 1955. The live attenuated poliovirus vaccine (oral polio vaccine, OPV) developed by Albert Sabin and others had been employed in Europe since 1960 and was licensed for U.S. use in 1962. Both the Salk and Sabin vaccines led to a dramatic decline of paralytic poliomyelitis in most developed countries and, as such, have been rightfully hailed as two of the great accomplishments of medical science.

Ancient Egyptian with Polio. Note the withered leg.

the minority of cases (less than 1%), the viremia persists and the virus enters the central nervous system and causes paralytic polio. The virus has a high affinity for anterior horn motor nerve cells of the spinal cord. Once inside these cells, it multiplies and destroys the cells; this results in motor and muscle paralysis. Since the licensing of the formalin-inactivated Salk vaccine (1955) and the attenuated virus Sabin vaccine (1962), the incidence of polio has decreased markedly. No endogenous reservoir of polioviruses exists in the United States. An on-going global effort to eliminate polio has been very successful. However, sporadic cases were reported in 2005, mostly in areas where religious views and misinformation diminish vaccination efforts. Nonetheless, it is likely that polio will be the next human disease to be completely eradicated.

1. What two virus groups are associated with acute viral gastroenteritis? How do they cause the disease's symptoms?
2. Describe some symptoms of hepatitis A.
3. Why was hepatitis A called infectious hepatitis?
4. At what specific sites within the body can the poliomyelitis virus multiply? What is the usual outcome of an infection?

37.5   ZOONOTIC DISEASES

The diseases discussed here are caused by viruses that are normally zoonotic (animal-borne). The RNA virus families *Arenaviridae, Bunyaviridae, Flaviviridae, Filoviridae,* and *Picornoviridae* represent notable examples of human viral infections found in animal reservoirs before transmission to and between humans. Some of these viruses are exotic and rare; others are being irradicated by public health efforts. Some of the virus types are found in relatively small geographic areas; others are distributed across continents. Several of these viruses cause diseases with substantial morbidity and mortality. It is for these reasons that many are placed on the Select Agents list as potential bioweapons, as indicated by double asterisks.

**Ebola and Marburg Hemorrhagic Fevers

Viral hemorrhagic fever (VHF) is the term used to describe a severe, multisystem syndrome caused by several distinct viruses. Characteristically, the overall host vascular system is damaged, resulting in vascular leakage (hemorrhage) and dysfunction transmission of rabies to humans. It should be noted, however, that not all rabid animals exhibit signs of agitation and aggression (known as furious rabies). In fact, paralysis (dumb rabies) is the more common sign exhibited by rabid animals.

The virus multiplies in the salivary glands of an infected host. It is transmitted to humans or other animals by the bite of an infected animal whose saliva contains the virus; by aerosols of the virus that can be spread in caves where bats dwell; or by contamination of scratches, abrasions, open wounds, and mucous membranes with saliva from an infected animal. After inoculation, a region of the virions' glycoprotein envelope spike attaches to the plasma membrane of nearby skeletal muscle cells, the virus enters the cells, and multiplication of the virus occurs. When the concentration of the virus in the muscle is sufficient, the virus enters the nervous system through unmyelinated sensory and motor terminals; the binding site is the nicotinic acetylcholine receptor.

The virus spreads by retrograde axonal flow at 8 to 20 mm per day until it reaches the spinal cord, when the first specific symptoms of the disease—pain or paresthesia at the wound site—may occur. A rapidly progressive encephalitis develops as the virus quickly disseminates through the central nervous system. The virus then spreads throughout the body along the peripheral nerves, including those in the salivary glands, where it is shed in the saliva.

Within brain neurons the virus produces characteristic **Negri bodies,** masses of viruses or unassembled viral subunits that are visible in the light microscope. In the past, diagnosis of rabies consisted solely of examining nervous tissue for the presence of these bodies. Today diagnosis is based on *d*irect immuno*f*luorescent *a*ntibody (DFA) of brain tissue, virus isolation, detection of Negri bodies, and a rapid rabies enzyme-mediated immunodiagnosis test.

Symptoms of rabies usually begin 2 to 16 weeks after viral exposure and include anxiety, irritability, depression, fatigue, loss of appetite, fever, and a sensitivity to light and sound. The disease quickly progresses to a stage of paralysis. In about 50% of all cases, intense and painful spasms of the throat and chest muscles occur when the victim swallows liquids. The mere sight, thought, or smell of water can set off spasms. Consequently, rabies has been called hydrophobia (fear of water). Death results from destruction of the regions of the brain that regulate breathing. Safe and effective vaccines (human diploid-cell rabies vaccine HDCV [Imovax Rabies] or rabies vaccine adsorbed [RVA]) against rabies are available; however, to be effective they must be given soon after the person has been infected. Veterinarians and laboratory personnel, who have a high risk of exposure to rabies, usually are immunized every 2 years and tested for the presence of suitable antibody titer. About 30,000 people annually receive this treatment. In the United States fewer than 10 cases of rabies occur yearly in humans, although about 8,000 cases of animal rabies are reported each year from various sources (figure 37.22b). Prevention and control involves pre-exposure vaccination of dogs and cats, postexposure vaccination of humans, and pre-exposure vaccination of humans at special risk (persons spending a month or more in countries where rabies is common in dogs).

Some states and countries (Hawaii and Great Britain, for example) retain their rabies-free status by imposing quarantine periods on any entering dog or cat. If an asymptomatic, unvaccinated dog or cat bites a human, the animal is typically confined and observed by a veterinarian for at least 10 days. If the animal shows no signs of rabies in that time, it is determined to be uninfected. Animals demonstrating signs of rabies are killed and brain tissue submitted for rabies testing. Postexposure prophylaxis—rabies immune globulin for passive immunity and rabies vaccine for active immunity—is initiated to exploit the relatively long incubation period of the virus. This is usually recommended for anyone bitten by one of the common reservoir species (raccoons, skunks, foxes, and bats), unless it is proven that the animal was uninfected. Once symptoms of rabies develop in a human, death usually occurs.

---

1. Why are Ebola and Marburg hemorrhagic fever diseases so deadly?
2. What precautions can be taken to prevent Hantavirus and Lassa virus transmission to humans?
3. How does the rabies virus cause death in humans?

37.6 PRION DISEASES

Prion diseases, also called **transmissible spongiform encephalopathies** (TSEs), are fatal neurodegenerative disorders that have attracted enormous attention not only for their unique biological features but also for their impact on public health. Prions (protein infectious particles) are thought to consist of abnormally folded proteins (PrP^{sc}), which can induce normal forms of the protein (PrP^c) to fold abnormally. This group of diseases includes kuru, Creutzfeldt-Jakob disease (CJD), new variant Creutzfeldt-Jakob disease (vCJD), Gerstmann-Sträussler disease (GSD), and fatal familial insomnia (FFI; **table 37.6**). The first of these diseases to be studied in humans was kuru, discovered in the Fore tribe of New Guinea. Carlton Gadjusek and others showed that the disease was transmitted by ritual cannibalism (especially where brains and spinal cords were eaten). The primary symptom of the human disorders is dementia, usually accompanied by manifestations of motor dysfunction such as cerebral ataxia (inability to coordinate muscle activity) and myoclonus (shocklike contractions of muscle groups). FFI is also characterized by dysautonomia (abnormal functioning of the autonomic nervous system) and sleep disturbances. These symptoms appear insidiously in middle to late adult life and last from months (CJD, FFI, and kuru) to years (GSD) prior to death. Neuropathologically, these disorders produce a characteristic spongiform degeneration of the brain, as well as deposition of amyloid plaques. Prion diseases thus share important clinical, neuropathological, and cell biological features with another, more common cerebral amyloidosis, Alzheimer's disease. A familial (inherited) form of CJD has also been described, suggesting that certain genetic mutations cause the PrP^c protein to more easily assume the PrP^{sc} conformation. Prions (section 18.10)

New variant CJD is transmitted from cattle that have bovine spongiform encephalopathy (BSE, or mad cow disease) as described in section 40.4. There have been two confirmed cases of

Table 37.6	Prion Diseases of Humans		
Disease	**Incubation Period**	**Nature of Disease**	
Creutzfeldt-Jakob disease (CJD) (sporadic, iatrogenic, familial, new-variant)	Months to years	Spongiform encephalopathy (degenerative changes in the central nervous system)	
Kuru	Months to years	Spongiform encephalopathy	
Gerstmann-Sträussler-Scheinker disease (GSD)	Months to years	Genetic neurodegenerative disease	
Fatal familial insomnia (FFI)	Months to years	Genetic neurodegenerative disease with progressive, untreatable insomnia	

BSE in the United States (Washington state, 2004 and Texas, 2005) compared to approximately 40,000 BSE cases in the United Kingdom. Cattle experimentally infected by the oral route have tested positive for the BSE agent in the brain, spinal cord, retina, dorsal root ganglia, distal ileum, and bone marrow, suggesting that (1) the BSE agent survives passage along the GI tract, (2) the BSE agent is neurotropic, and (3) these tissues represent a source of infectious material that may be transmitted to humans and other animals. In fact, much evidence has accumulated to suggest that human vCJD can be acquired by individuals who eat meat products (especially if they contain brain and spinal cord) prepared from infected cattle. Data from the U.K. report 107 confirmed and 42 probable deaths attributed to vCJD; compared to 260 total worldwide deaths attributed to vCJD (as of September 2006). Estimates of the final total

vCJD cases (extrapolated from analysis of positive results from a tonsil and appendix tissue bank) expected in the United Kingdom by 2080 range from a few hundred to 140,000. Of additional concern is the report of four vCJD cases associated with blood transfusion in the United Kingdom. Iatrogenic CJD is induced by a physician or surgeon, a medical treatment, or diagnostic procedures. It has been transmitted by prion-contaminated human growth hormone, corneal grafts, and grafts of dura mater (tissue surrounding the brain). Donor screening and more thorough testing of grafts have decreased the frequency of prion transmission.

1. How are prions different from viruses; how are they similar?
2. In what way are spongiform encephalopathies commonly acquired?

Summary

37.1 Airborne Diseases

a. More than 400 different viruses can infect humans. These viruses can be grouped and discussed according to their mode of transmission and acquisition.

b. Most airborne viral diseases involve either directly or indirectly the respiratory system. Examples include chickenpox (varicella, **figure 37.1**), shingles (herpes zoster, **figure 37.2**), rubella (German measles, **figure 37.5**), influenza (flu), measles (rubeola, **figure 37.3**), mumps (**figure 37.4**), the acute respiratory viruses such as the respiratory syncytial virus, the eradicated smallpox (variola, **figure 37.6**), and viral pneumonia.

37.2 Arthropod-Borne Diseases

a. The arthropod-borne viral diseases are transmitted by arthropod vectors from human to human or animal to human (**table 37.2**). Examples include Rift Valley fever; St. Louis encephalitis; eastern, western, and Venezuelan equine encephalitis; West Nile fever; and yellow fever. All these diseases are characterized by fever, headache, nausea, vomiting, and characteristic encephalitis.

37.3 Direct Contact Diseases

a. Person-to-person contact is another way of acquiring or transmitting a viral disease. Examples of such diseases include AIDS (**figures 37.7–37.12**), cold sores (**figure 37.13**), the common cold, cytomegalovirus inclusion disease, genital herpes (**figure 37.16b,c**), human herpesvirus 6 infections, human par-

vovirus B19 infection, certain leukemias, infectious mononucleosis, and hepatitis (**table 37.4**): hepatitis B (serum hepatitis); hepatitis C; hepatitis D (delta hepatitis); hepatitis F; and hepatitis G.

37.4 Food-Borne and Waterborne Diseases

a. The viruses that are transmitted in food and water usually grow in the intestinal system and leave the body in the feces (**table 37.5**). Acquisition is generally by the oral route. Examples of diseases caused by these viruses include acute viral gastroenteritis (rotavirus and others), infectious hepatitis A, hepatitis E, and poliomyelitis.

37.5 Zoonotic Diseases

a. Diseases transmitted from animals are zoonotic. Several animal viruses can cause disease in humans. Examples of viral zoonoses include Ebola and Marburg fevers, hantavirus pulmonary syndrome, Lassa fever, and rabies.

37.6 Prion Diseases

a. A prion disease is a pathological process caused by a transmissible agent (a prion) that remains clinically silent for a prolonged period, after which the clinical disease becomes apparent. Examples include (**table 37.6**) Creutzfeldt-Jakob disease, kuru, Gerstmann-Sträussler-Scheinker disease, and fatal familial insomnia. These diseases are chronic infections of the central nervous system that result in progressive degenerative changes and eventual death.

tions. The microorganisms involved in dental infections are also described. Diseases caused by bacteria that are now listed as Select Agents (potential bioterror agents) are identified within the chapter by two asterisks (**).

Of all the known bacterial species, only a few are pathogenic to humans. Some human diseases have been only recently recognized **(table 38.1)**; others have been known since antiquity. In the following sections the more important disease-causing bacteria are discussed according to their mode of acquisition/transmission.

Table 38.1	Some Examples of Human Bacterial Diseases Recognized Since 1977	
Year	**Bacterium**	**Disease**
1977	*Legionella pneumophila*	Legionnaires' disease
1977	*Campylobacter jejuni*	Enteric disease (gastroenteritis)
1981	*Staphylococcus aureus*	Toxic shock syndrome
1982	*Escherichia coli* O157:H7	Hemorrhagic colitis; hemolytic uremic syndrome (HUS)
1982	*Borrelia burgdorferi*	Lyme disease
1982	*Helicobacter pylori*	Peptic ulcer disease
1984	Meticillin-resistant *Staphylococcus aureus*	Epidemic nosocomial infections
1986	*Ehrlichia chaffeensis*	Human ehrlichiosis
1988	*Chlamydia pneumoniae*	Atherosclerosis
1988	*Salmonella enteritidis* F14	Egg-borne salmonellosis
1989	*Enterococcus faecium;* vancomycin-resistant enterococci	Colitis and enteritis
1990	*Streptococcus pyogenes*	"Flesh-eating" and streptococcal toxic shock
1992	*Vibrio cholerae* O139	New strain associated with epidemic cholera in Asia
1992	*Bartonella henselae*	Cat-scratch disease; bacillary angiomatosis
1994	*Ehrlichia* spp.	Human granulocytic ehrlichiosis
1995	*Neisseria meningitidis*	Meningococcal supraglottitis
1996	Vancomycin-resistant *S. aureus*	Nosocomial infections
1997	*Kingella kingae*	Pediatric infections
2000	*Tropheryma whipplei*	Whipple's disease

38.1 Airborne Diseases

Most airborne diseases caused by bacteria involve the respiratory system. Other airborne bacteria can cause skin diseases. Some of the better known of these diseases are now discussed.

Chlamydial Pneumonia

Chlamydial pneumonia is caused by *Chlamydophila (Chlamydia) pneumoniae.* Clinically, infections are generally mild; pharyngitis, bronchitis, and sinusitis commonly accompany some lower respiratory tract involvement. Symptoms include fever, a productive cough (respiratory secretion brought up by coughing), sore throat, hoarseness, and pain on swallowing. Infections with *C. pneumoniae* are common but sporadic; about 50% of adults have antibody to the chlamydiae. Evidence suggests that *C. pneumoniae* is primarily a human pathogen directly transmitted from human to human by droplet (respiratory) secretions. Diagnosis of chlamydial pneumonia is based on symptoms and a microimmunofluorescence test. Tetracycline and erythromycin are routinely used for treatment. In seroepidemiological studies, *C. pneumoniae* infections have been linked with coronary artery disease as well as vascular disease at other sites. Following a demonstration of *C. pneumoniae*-like particles in atherloscerotic plaque tissue by electron microscopy, *C. pneumoniae* genes and antigens have been detected in artery plaque. Rarely, however, has the microorganism been recovered in cultures of atheromatous tissue (i.e., artery plaque). As a result of these findings, the possible etiologic role of *C. pneumoniae* in coronary artery disease and systemic atherosclerosis is under intense scrutiny. Phylum *Chlamydiae* (section 21.5).

Diphtheria

Diphtheria [Greek *diphthera*, membrane, and *ia*, condition] is an acute, contagious disease caused by the gram-positive bacterium *Corynebacterium diphtheriae* (*see figure 24.9*). *C. diphtheriae* is well-adapted to airborne transmission by way of nasopharyngeal secretions because it is very resistant to drying. Diphtheria mainly affects unvaccinated, poor people living in crowded conditions. Once within the respiratory system, bacteria that carry the prophage β containing the *tox* gene produce diphtheria toxin; *tox*+ phage infection of *C. diphtheriae* is required for toxin production. This toxin is an exotoxin that causes an inflammatory response and the formation of a grayish pseudomembrane on the pharynx and respiratory mucosa **(figure 38.1)**. The pseudomembrane consists of dead host cells and cells of *C. diphtheriae.* Diphtheria toxin is absorbed into the circulatory system and distributed throughout the body, where it may cause destruction of cardiac, kidney, and nervous tissues by inhibiting protein synthesis. The toxin is composed of two polypeptide subunits: A and B. The A subunit consists of the catalytic domain; the B subunit is composed of the receptor and transmembrane domains (*see figure 33.5*). The receptor domain binds to the heparin-binding epidermal growth factor receptor on the surface of various eucaryotic cells. Once bound, the toxin enters the cytoplasm by

Figure 38.1 Diphtheria Pathogenesis. **(a)** Diphtheria is a well-known, exotoxin-mediated infectious disease caused by *Corynebacterium diphtheriae*. The disease is an acute, contagious, febrile illness characterized by local oropharyngeal inflammation and pseudomembrane formation. If the exotoxin gets into the blood, it is disseminated and can damage the peripheral nerves, heart, and kidneys. **(b)** The clinical appearance includes gross inflammation of the pharynx and tonsils marked by grayish patches (a pseudomembrane) and swelling of the entire area.

endocytosis. The transmembrane domain of the toxin embeds itself into the target cell membrane causing the catalytic domain to be cleaved and translocated into the cytoplasm. The cleaved catalytic domain becomes an active enzyme, catalyzing the attachment of ADP-ribose (from NAD^+) to elongation factor-2 (EF-2). A single enzyme (i.e., catalytic domain) can exhaust the entire supply of cellular EF-2 within hours, resulting in protein synthesis inhibition and cell death. Suborder *Corynebacterineae* (section 24.4); Toxigenicity: AB toxins (section 33.4)

Typical symptoms of diphtheria include a thick mucopurulent (containing both mucus and pus) nasal discharge, pharyngitis, fever, cough, paralysis, and death. (*C. diphtheriae* can also infect the skin, usually at a wound or skin lesion, causing a slow-healing ulceration termed **cutaneous diphtheria.**) Diagnosis is made by observation of the pseudomembrane in the throat and by bacterial culture. Diphtheria antitoxin is given to neutralize any unabsorbed exotoxin in the patient's tissues; penicillin and erythromycin are used to treat the infection. Prevention is by active immunization with **DPT** (*d*iphtheria-*p*ertussis-*t*etnus) **vaccine;** and then boosted with DTap (*d*iphtheria *t*oxoid, *t*etanus toxoid, *a*cellular *B. pertussis* vaccine); or Tdap (*t*etanus toxoid, reduced *d*iphtheria toxoid, *a*cellular *p*ertussis vaccine, adsorbed), approved in 2005 (*see table 36.4*). Most cases involve people over 30 years of age who have a weakened immunity to the diphtheria toxin and live in tropical areas. Since 1980, fewer than six diphtheria cases have been reported annually in the United States, and most occur in nonimmunized individuals. Control of epidemics: Vaccines and immunization (section 36.8)

Legionnaire's Disease and Pontiac Fever

In 1976 the term **Legionnaires' disease,** or **legionellosis,** was coined to describe an outbreak of pneumonia that occurred at the Pennsylvania State American Legion Convention in Philadelphia. The bacterium responsible for the outbreak was *Legionella pneumophila,* a nutritionally fastidious, aerobic, gram-negative rod **(figure 38.2)**. It is now known that this bacterium is part of the natural microbial community of soil and freshwater ecosystems, and it has been found in large numbers in air-conditioning systems and shower stalls. Class *Gammaproteobacteria* (section 22.3)

An increasing body of evidence suggests that environmental protozoa are the most important factor for the survival and growth of *Legionella* in nature. A variety of free-living amoebae and ciliated protozoa that contain *Legionella* spp. have been isolated from water sites suspected as sources of *Legionella* infections. *Legionella* spp. multiply intracellularly within the amoebae, just as they do within human monocytes and macrophages. This might explain why there is no human-to-human spread of legionellosis.

Infection with *L. pneumophila* and other *Legionella* spp. results from the airborne spread of bacteria from an environmental reservoir to the human respiratory system. Males over 50 years of age most commonly contract the disease, especially if their immune system is compromised by heavy smoking, alcoholism, or chronic illness. The bacteria reside within the phagosomes of alveolar macrophages, where they multiply and produce localized tissue destruction through export of a cytotoxic exoprotease. Symptoms start 2 to 10 days after exposure and include a high

Figure 38.2 Legionnaires' Disease. *Legionella pneumophila,* the causative agent of Legionnaires' disease, with many lateral flagella; SEM (× 10,000).

fever, nonproductive cough (respiratory secretions are not brought up during coughing), headache, neurological manifestations, and severe bronchopneumonia. Diagnosis depends on isolation of the bacterium, documentation of a rise in antibody titer over time, or the presence of *Legionella* antigens in the urine as detected by a rapid test kit. Treatment begins with supportive measures and the administration of erythromycin or rifampin. Death occurs in 10 to 15% of cases.

Prevention of Legionnaires' disease depends on the identification and elimination of the environmental source of *L. pneumophila.* Chlorination, the heating of water, and the cleaning of water-containing devices can help control the multiplication and spread of *Legionella.* These control measures are effective because the pathogen does not appear to be spread from person to person. Since the initial outbreak of this disease in 1976, many outbreaks during summer months have been recognized in all parts of the United States. About 1,000 to 1,600 cases are diagnosed each year, and about 18,000 or more additional mild or subclinical cases are thought to occur. It is estimated that 23% of all nosocomial pneumonias are due to *L. pneumophila,* especially among immunocompromised patients.

L. pneumophila also causes a milder illness called **Pontiac fever.** This disease, which resembles an allergic disease more than an infection, is characterized by an abrupt onset of fever, headache, dizziness, and muscle pains. It is indistinguishable clinically from the various respiratory syndromes caused by viruses. Pneumonia does not occur. The disease resolves spontaneously within 2 to 5 days. No deaths from Pontiac fever have been reported.

Pontiac fever was first described from an outbreak in a county health department in Pontiac, Michigan. Ninety-five percent of the employees became ill and eventually showed elevated serum titers against *L. pneumophila.* These bacteria were later isolated from the lungs of guinea pigs exposed to the air of the building. The likely source was water from a defective air conditioner.

Meningitis

Meningitis [Greek *meninx,* membrane, and *–itis,* inflammation] is an inflammation of the brain or spinal cord meninges (membranes). Based on the specific cause, it can be divided into **bacterial (septic) meningitis** and **aseptic meningitis syndrome.** As shown in **table 38.2,** there are many causes of the aseptic meningitis syndrome, only some of which can be treated with antimicrobial agents. Thus accurate identification of the causative agent is essential for proper treatment of the disease. The immediate sources of the bacteria responsible for meningitis are respiratory secretions from carriers. The bacteria initially colonize the nasopharynx, after which they cross the mucosal barrier. They can enter the bloodstream and cross the blood-brain barrier to enter the cerebral spinal fluid (CSF), where they produce inflammation of the meninges.

The usual symptoms of meningitis include an initial respiratory illness or sore throat interrupted by one of the meningeal syndromes: vomiting, headache, lethargy, confusion, and stiffness in the neck and back. Bacterial meningitis can be diagnosed by a Gram stain and culture or rapid tests of the bacteria from CSF. Once bacterial meningitis is suspected, specific antibiotics (penicillin, chloramphenicol, cefotaxime, ceftriazone, ofloxacin) are administered immediately. In fact, antibiotics are often administered prophylactically to patient contacts.

Bacterial meningitis can be caused by various gram-positive and gram-negative bacteria. However, three organisms tend to be

Table 38.2	Causative Agents of Meningitis by Diagnostic Category	
Type of Meningitis	**Causative Agent**	
Bacterial (Septic) Meningitis		
	Streptococcus pneumoniae	
	Neisseria meningitidis	
	Haemophilus influenzae type b	
	Group B streptococci	
	Listeria monocytogenes	
	Mycobacterium tuberculosis	
	Nocardia asteroides	
	Staphylococcus aureus	
	Staphylococcus epidermidis	
Aseptic Meningitis Syndrome		
Agents Requiring Antimicrobials	Fungi	
	Amoebae	
	Treponema pallidum	
	Mycoplasmas	
	Leptospires	
Agents Requiring Other Treatments	Viruses	
	Cancers	
	Parasitic cysts	
	Chemicals	

associated with meningitis more frequently than others: *Streptococcus pneumoniae, Neisseria meningitidis,* and *Haemophilus influenzae* (serotype b). *S. pneumoniae* is discussed with other streptococcal diseases later in this chapter. *N. meningitidis,* often referred to as the meningococcus, is a normal inhabitant of the human nasopharynx (5 to 15% of humans carry the nonpathogenic serotypes). Most disease-causing *N. meningitidis* strains belong to serotypes A, B, C, Y and W-135. In general, serotype A strains are the cause of epidemic disease in developing countries, while serotype C and W-135 strains are responsible for meningitis outbreaks in the United States. Infection results from airborne transmission of the bacteria, typically through close contact with a primary carrier. The disease process is initiated by pili-mediated colonization of the nasopharynx by pathogenic bacteria. The bacteria cross the nasopharyngeal epithelium (typically through endocytosis) and invade the bloodstream (meningococcemia) where they proliferate. Symptoms caused by *N. meningitidis* are variable depending on the degree of bacterial dissemination. Infection of the CSF leads to meningitis; untreated meningitis is fatal.

Control of *N. meningitidis* infection is with vaccination and antibiotics. There are currently two vaccines available: the meningococcal polysaccharide (MPSV4) and the meningococcal conjugate vaccine (MCV4). Both vaccines are effective against serotypes A, C, Y and W-135. Vaccination is recommended for all college students living in residence halls. MCV4 is recommended for preteen children, teens, and adults less than 55 years of age. MPSV4 should be used for children 2 to 10 years of age and adults over 55 who are at risk.

Another agent of meningitis is *H. influenzae,* a small, gramnegative bacterium. Transmission is by inhalation of droplet nuclei shed by infectious individuals or carriers. *H. influenzae* serotype b can infect mucous membranes, resulting in sinusitis, pneumonia, and bronchitis. It can disseminate to the bloodstream and cause a bacteremia. *H. influenzae* serotype b can cross into the CSF, resulting in inflammation of the meninges (meningitis). *H. influenzae* disease (including pneumonia and meningitis) is primarily observed in children less than 5 years of age. It is estimated that *H. influenzae* serotype b causes at least 3 million cases of serious disease, and several hundreds of thousands of deaths each year.

A sharp reduction in the incidence of *H. influenzae* serotype b infections began in the mid-1980s due to administration of the *H. influenzae* type b conjugate vaccine, rifampin prophylaxis of disease contacts, and the availability of more efficacious therapeutic agents. From 1987 through 1999, the incidence of invasive infection among U.S. children under 5 years of age declined by 95%. Three to six percent of all *H. influenzae* infections are fatal. Furthermore, up to 20% of surviving patients have permanent hearing loss or other long-term sequelae. Currently, all children should be vaccinated with the *H. influenzae* type b conjugate vaccine at the age of 2 months.

Complicating diagnostic practices is the fact that a person may have meningitis symptoms but show no microbial agent in gram-stained specimens, and have negative cultures. In such a case the diagnosis often is called aseptic meningitis syndrome. Aseptic meningitis is typically more difficult to treat.

Mycobacterium avium-M. intracellulare and *M. tuberculosis* Pulmonary Diseases

M. avium-M. intracellulare Infections

An extremely large group of mycobacteria are normal inhabitants of soil, water, and house dust. Two of these are noteworthy pathogens in the United States—the two, *Mycobacterium avium* and *Mycobacterium intracellulare,* are so closely related that they are referred to as the *M. avium* complex (MAC). Globally, *M. tuberculosis* has remained more prevalent in developing countries, where as MAC has become the most common cause of mycobacterial infections in the United States. Suborder *Corynebacterineae* (section 24.4)

These mycobacteria are found worldwide and infect a variety of insects, birds, and animals. Both the respiratory and the gastrointestinal tracts have been proposed as entry portals for the *M. avium* complex; however, person-to-person transmission is not very efficient. The gastrointestinal tract is thought to be the most common site of colonization and dissemination in AIDS patients. MAC causes a pulmonary infection in humans similar to that caused by *M. tuberculosis.* Pulmonary MAC is more common in non-AIDS patients, particularly in elderly persons with preexisting pulmonary disease.

Shortly after the recognition of AIDS and its associated opportunistic infections, it became apparent that one of the more common AIDS-related infections was caused by MAC. In the United States, disseminated infection with MAC occurs in 15 to 40% of AIDS patients with CD4$^+$ cell counts of less than 100 per cubic millimeter. Disseminated infection with MAC produces disabling symptoms, including fever, malaise, weight loss, night sweats, and diarrhea. Carefully controlled epidemiological studies have shown that MAC shortens survival by 5 to 7 months among persons with AIDS. With more effective antiviral therapy for AIDS and with prolonged survival, the number of cases of disseminated MAC is likely to increase substantially, and its contribution to AIDS mortality will increase. Direct contact diseases: AIDS (section 37.3)

MAC can be isolated from sputum, blood, and aspirates of bone marrow. Acid-fast stains are of value in making a diagnosis. The most sensitive method for detection is the commercially available lysis-centrifugation blood culture system (Wampole Laboratories). Although no drugs are currently approved by the FDA for the therapy of MAC, every regimen should contain either azithromycin or clarithromycin and ethambutol as a second drug. One or more of the following can be added: closfazimine, rifabutin, rifampin, ciprofloxacin, and amikacin.

Mycobacterium tuberculosis Infections

Over a century ago Robert Koch identified *Mycobacterium tuberculosis* as the causative agent of **tuberculosis (TB)**. At the time, TB was rampant, causing one-seventh of all deaths in

**Q Fever

Q fever (Q for query because the cause of the fever was not known for some time) is an acute zoonotic disease caused by the γ-proteobacterium *Coxiella burnetii,* an intracellular, gram-negative bacterium. *C. burnetii* can survive outside host cells by forming a resistant, endospore-like body. It does not need an arthropod vector for transmission. This bacterium infects both wild animals and livestock. Cattle, sheep, and goats are the primary reservoirs. In animals, ticks (many species) transmit *C. burnetii,* whereas human transmission is primarily by inhalation of dust contaminated with bacteria from dried animal feces or urine, or consumption of unpasteurized milk. The disease can occur in epidemic form among slaughterhouse workers and sporadically among farmers and veterinarians. Each year, fewer than 30 cases of Q fever are reported in the United States. Class *Alphaproteobacteria: Rickettsia* and *Coxiella* (section 22.1)

In humans, Q fever is an acute illness characterized by the sudden onset of severe headache, malaise, confusion, sore throat, chills, sweats, nausea, chest pain, myalgia (muscle pain), and high fever. It is rarely fatal, but endocarditis—inflammation of the heart muscle—occurs in about 10% of the cases. Five to ten years may elapse between the initial infection and the appearance of the endocarditis. During this interval the bacteria reside in the liver and often cause hepatitis.

Diagnosis is most commonly made in national reference labs using an indirect immunofluorescence assay or PCR. Treatment is with doxycycline or quinolone antibiotics. Prevention and control consists of vaccinating researchers and others at high occupational risk and in areas of endemic Q fever; cow and sheep milk should be pasteurized before consumption.

Rocky Mountain Spotted Fever

Rocky Mountain spotted fever is caused by *Rickettsia rickettsii.* Although this disease was originally detected in the Rocky Mountain area, most cases now occur east of the Mississippi River. The disease is transmitted by ticks and usually occurs in people who are or have been in tick-infested areas. There are two principal vectors: *Dermacentor andersoni,* the wood tick, is distributed in the Rocky Mountain states and is active during the spring and early summer. *D. variabilis,* the dog tick, has assumed greater importance and is almost exclusively confined to the eastern half of the United States. Unlike other rickettsias, *R. rickettsii* can pass from generation to generation of ticks through their eggs in a process known as **transovarian passage.** No humans or mammals are needed as reservoirs for the continued propagation of this rickettsia in the environment.

When humans contact infected ticks, the rickettsias are either deposited on the skin (if the tick defecates after feeding) and then subsequently rubbed or scratched into the skin, or the rickettsias are deposited into the skin as the tick feeds. Once inside the skin, the rickettsias enter the endothelial cells of small blood vessels, where they multiply and produce a characteristic vasculitis.

The disease is characterized by the sudden onset of a headache, high fever, chills, and a skin rash (**figure 38.10**) that initially appears on the ankles and wrists and then spreads to the

Figure 38.10 Rocky Mountain Spotted Fever. Typical rash occurring on the arms and chest consists of generally distributed, sharply defined macules.

trunk of the body. If the disease is not treated, the rickettsias can destroy the blood vessels in the heart, lungs, or kidneys and cause death. Usually, however, severe pathological changes are avoided by antibiotic therapy (chloramphenicol, chlortetracycline), the development of immune resistance, and supportive therapy. Diagnosis is made through observation of symptoms and signs such as the characteristic rash and by serological tests. The best means of prevention remains the avoidance of tick-infested habitats and animals. There are roughly 400 to 800 reported cases of Rocky Mountain spotted fever annually in the United States.

---

1. How is epidemic typhus spread? Ehrlichiosis? Murine typhus? What are their symptoms?
2. Why are two antibiotics used to treat most rickettsial infections?
3. What is the causative agent of Lyme disease and how is it transmitted to humans? How does the illness begin? Describe the three stages of Lyme disease.
4. Why is plague sometimes called the Black Death? How is it transmitted? Distinguish between bubonic and pneumonic plague.
5. Describe the symptoms of Rocky Mountain spotted fever.
6. How does transovarian passage occur?

38.3 DIRECT CONTACT DISEASES

Most of the direct contact bacterial diseases involve the skin or underlying tissues. Others can become disseminated through specific regions of the body. Some of the better-known of these diseases are now discussed.

Gas Gangrene or Clostridial Myonecrosis

Clostridium perfringens, C. novyi, and *C. septicum* are gram-positive, endospore-forming rods termed the histotoxic clostridia. They can produce a necrotizing infection of skeletal muscle called

gas gangrene [Greek *gangraina,* an eating sore] or **clostridial myonecrosis** [*myo,* muscles, and *necrosis,* death]; however, *C. perfringens* is the most common cause. Class *Clostridia* (section 23.4)

Histotoxic clostridia occur in the soil worldwide and also are part of the normal endogenous microflora of the human large intestine. Contamination of injured tissue with spores from soil containing histotoxic clostridia or bowel flora is the usual means of transmission. Infections are commonly associated with wounds resulting from abortions, automobile accidents, military combat, or frostbite. If the spores germinate in anoxic tissue, the bacteria grow and secrete α-toxin, which breaks down muscle tissue. Growth often results in the accumulation of gas (mainly hydrogen as a result of carbohydrate fermentation), and of the toxic breakdown products of skeletal muscle tissue (**figure 38.11**). Toxigenicity: Exotoxins (section 33.4)

Clinical manifestations include severe pain, edema, drainage, and muscle necrosis. The pathology arises from progressive skeletal muscle necrosis due to the effects of α-toxin (a lecithinase). Lecithinase disrupts cell membranes leading to cell lysis. Other enzymes produced by the bacteria degrade collagen and tissue, facilitating spread of the disease.

Gas gangrene is a medical emergency. Laboratory diagnosis is through recovery of the appropriate species of clostridia accompanied by the characteristic disease symptoms. Treatment is extensive surgical debridement (removal of all dead tissue), the administration of antitoxin, and antimicrobial therapy with penicillin and tetracycline. Hyperbaric oxygen therapy (the use of high concentrations of oxygen at elevated pressures) also is considered effective. The oxygen saturates the infected tissue and thereby prevents the growth of the obligately anaerobic clostridia. Prevention and control include debridement of contaminated traumatic wounds plus antimicrobial prophylaxis and prompt treatment of all wound infections. Amputation of limbs often is necessary to prevent further spread of the disease.

Group B Streptococcal Disease

Streptococcus agalactiae, or **Group B streptococcus (GBS)**, is a gram-positive bacterium that causes illness in newborn babies, pregnant women, the elderly, and adults compromised by other severe illness. GBS is the most common cause of sepsis and meningitis in newborns, a frequent cause of newborn pneumonia, and a common cause of life-threatening, neonatal infections. GBS in neonates is more common than rubella, congenital syphilis, and respiratory syncytial virus, for example. Nearly 75% of infected newborns present symptoms in the first week of life; premature babies are more susceptible. Most GBS cases are apparent within hours after birth. This is called "early-onset disease." GBS can also develop in infants who are one week to several months old ("late-onset disease"), however this is very rare. Meningitis is more common with late-onset disease. Interestingly, about half of the infants with late-onset GBS disease are associated with a mother who is a GBS carrier; the source of infection for the other newborns with late-onset GBS disease is unknown. Infant mortality due to GBS disease is about 5%. Babies that survive GBS disease, particularly those with meningitis, may have permanent disabilities, such as hearing or vision loss or developmental disabilities.

GBS is transmitted directly from person to person. Many people are asymptomatic GBS carriers—they are colonized by GBS but do not become ill from it. Adults can carry GBS in the bowel, vagina, bladder, or throat. Twenty to 25% of pregnant women carry GBS in the rectum or vagina. Thus a fetus may be exposed to GBS before or during birth if the mother is a carrier. GBS carriers appear to be so temporarily—that is, they do not harbor the bacteria for life.

GBS disease is diagnosed when the gram-positive, beta-hemolytic, streptococcal bacterium is grown from cultures of otherwise sterile body fluids, such as blood or spinal fluid. Confirmation of GBS is by latex agglutination immunoassay. GBS cultures may take 24 to 72 hours for strong growth to appear. GBS infections in both newborns and adults are usually treated with penicillin or ampicillin, unless resistance or hypersensitivity is indicated.

GBS carriage can be detected during pregnancy by the presence of the bacterium in culture specimens of both the vagina and the rectum. An FDA-approved DNA test for GBS is also available if rapid diagnosis of disease is necessary or culture results

Figure 38.11   Gas Gangrene (Clostridial Myonecrosis).   Necrosis of muscle and other tissues results from the numerous toxins produced by *Clostridium perfringens.*

① Tissue where *S. aureus* is often found but does not normally cause disease

Diseases that may be caused by *S. aureus* are:

② Pimples and impetigo

③ Boils and carbuncles on any surface area

④ Wound infections and abscesses

⑤ Spread to lymph nodes and to blood (septicemia), resulting in widespread seeding

⑥ Osteomyelitis

⑦ Endocarditis

⑧ Meningitis

⑨ Enteritis and enterotoxin poisoning (food poisoning)

⑩ Nephritis

⑪ Respiratory infections:
Pharyngitis
Laryngitis
Bronchitis
Pneumonia

Figure 38.16   Staphylococcal Diseases.   The anatomical sites of the major staphylococcal infections of humans are indicated by the corresponding numbers.

is caused by strains of *S. aureus* that produce the **exfoliative toxin,** or **exfoliatin.** This protein is usually plasmid encoded, although in some strains the toxin gene is on the bacterial chromosome. In this disease the epidermis peels off to reveal a red area underneath—thus the name of the disease. SSSS is seen most commonly in infants and children, and neonatal nurseries occasionally suffer large outbreaks of the disease.

The definitive diagnosis of staphylococcal disease can be made only by isolation and identification of the staphylococcus involved. This requires culture, catalase, coagulase, and other biochemical tests. Commercial rapid test kits also are available. There is no specific prevention for staphylococcal disease. The mainstay of treatment is the administration of specific antibiotics: penicillin, cloxacillin, methicillin, vancomycin, oxacillin, cefotaxime, ceftriaxone, a cephalosporin, or rifampin and others. Because of the prevalence of drug-resistant strains (e.g., meticillin-resistant staph), all staphylococcal isolates should be tested for antimicrobial susceptibility (**Disease 38.3**). Cleanliness, good hygiene, and aseptic management of lesions are the best means of control.   Drug resistance (section 34.6)

Sexually Transmitted Diseases

Sexually transmitted diseases (STDs) represent a worldwide public health problem. The various viruses that cause STDs are presented in chapter 37, the responsible bacteria in this chapter, and the yeasts and protozoa in chapter 39. **Table 38.5** lists the various microorganisms that can be sexually transmitted and the diseases they cause.

The spread of most sexually transmitted diseases is currently out of control. The World Health Organization estimates that 300 million new cases of sexually transmitted diseases occur annually, with the predominant number of infections in 15 to 30-year-old individuals. In the United States, sexually transmitted diseases remain a major public health challenge. The 2004 STD Surveillance Report published by the CDC indicated that while substantial progress has been made in preventing, diagnosing, and treating certain STDs, 19 million new infections occur each year, nearly half of them among people aged 15 to 24. Sexually transmitted diseases also exact a tremendous economic toll—with direct medical costs estimated at $13 billion annually in the United States alone. Many cases of STDs go undiagnosed; others, like human papillomavirus and genital herpes, are not reported at all. Some STDs can also lead to infertility or cancer.

STDs were formerly called venereal diseases (from Venus, the Roman goddess of love), and may sometimes be referred to as sexually transmitted infections (STIs). They occur most frequently in the most sexually active age group—15 to 30 years of age—but anyone who has sexual contact with an infected individual is at increased risk. In general, the more sexual partners a person has, the more likely the person will acquire an STD (**Disease 38.4**).

As noted in previous chapters, some of the microorganisms that cause STDs can also be transmitted by nonsexual means. Examples include transmission by contaminated hypodermic nee-

Table 38.4	Various Enzymes and Toxins Produced by Staphylococci
Product	**Physiological Action**
β-lactamase	Breaks down penicillin
Catalase	Converts hydrogen peroxide into water and oxygen and reduces killing by phagocytosis
Coagulase	Reacts with prothrombin to form a complex that can cleave fibrinogen and cause the formation of a fibrin clot; fibrin may also be deposited on the surface of staphylococci, which may protect them from destruction by phagocytic cells; coagulase production is synonymous with invasive pathogenic potential
DNase	Destroys DNA
Enterotoxins	Are divided into heat-stable toxins of six known types (A, B, C1, C2, D, E); responsible for the gastrointestinal upset typical of food poisoning
Exfoliative toxins A and B (superantigens)	Causes loss of the surface layers of the skin in scalded-skin syndrome
Hemolysins	Alpha hemolysin destroys erythrocytes and causes skin destruction
	Beta hemolysin destroys erythrocytes and sphingomyelin around nerves
Hyaluronidase	Also known as spreading factor; breaks down hyaluronic acid located between cells, allowing for penetration and spread of bacteria
Panton-Valentine leukocidin	Inhibits phagocytosis by granulocytes and can destroy these cells by forming pores in their phagosomal membranes
Lipases	Break down lipids
Nuclease	Breaks down nucleic acids
Protein A	Is antiphagocytic by competing with neutrophils for the Fc portion of specific opsonins
Proteases	Break down proteins
Toxic shock syndrome toxin-1 (a superantigen)	Is associated with the fever, shock, and multisystem involvement of toxic shock syndrome

dles and syringes shared among intravenous drug users, and transmission from infected mothers to their infants. Some STDs can be cured quite easily, but others, especially those caused by viruses, are presently difficult or impossible to cure. Because treatments are often inadequate, prevention is essential. Preventive measures are based mainly on better education of the total population and when possible, control of the sources of infection and treatment of infected individuals with chemotherapeutic agents.

---

1. Describe how *H. pylori* can survive the acidic conditions of the stomach.
2. Compare and contrast the diseases caused by the TSST and SSSS proteins of *S. aureus*.
3. Describe several diseases caused by the staphylococci.
4. Name four ways in which a person may contract an STD.

Bacterial Vaginosis

Bacterial vaginosis is considered a sexually transmitted disease (table 38.5). It has a polymicrobial etiology that includes *Gardnerella vaginalis* (a gram-positive to gram-variable, pleomorphic, nonmotile rod), *Mobiluncus* spp., and various anaerobic bacteria. The finding that these microorganisms inhabit the vagina and rectum of 20 to 40% of healthy women indicates a potential source of autoinfection in addition to sexual transmission. Although it is a mild disease, it is a risk factor for obstetric infections, various adverse outcomes of pregnancy, and pelvic inflammatory disease. Vaginosis is characterized by a copious, frothy, fishy-smelling discharge with varying degrees of pain or itching. Diagnosis is based on this fishy odor and the microscopic observation of clue cells in the discharge. **Clue cells** are sloughed-off vaginal epithelial cells covered with bacteria, mostly *G. vaginalis*. Treatment for bacterial vaginosis is with metronidazole (Flagyl, MetroGel-Vaginal), a drug that kills anaerobic streptococci and the *Mobiluncus* spp. that appear to be needed for the continuation of the disease.

Chancroid

Chancroid [French *chancre*, a destructive sore, and Greek *eidos*, to form], also known as **genital ulcer disease,** is a sexually transmitted disease caused by the pleomorphic gram-negative bacillus *Haemophilus ducreyi* (table 38.5). The bacterium enters the skin through a break in the epithelium. After an incubation period of 4 to 7 days, a papular lesion develops within the epithelium, causing swelling and white blood cell infiltration. Within several days a pustule forms and ruptures, producing a painful, circumscribed ulcer with a ragged edge; hence the term genital ulcer disease. Most of the ulcers in males are on the penis and, in females, at the

Figure 38.17 Staphylococcal Skin Infections. **(a)** Superficial folliculitis in which raised, domed pustules form around hair follicles. **(b)** In deep folliculitis the microorganism invades the deep portion of the follicle and dermis. **(c)** A furuncle arises when a large abscess forms around a hair follicle. **(d)** A carbuncle consists of a multilocular abscess around several hair follicles. **(e)** Impetigo on the neck of 2-year-old male. **(f)** Scalded skin syndrome in a 1-week-old premature male infant. Reddened areas of skin peel off, leaving "scalded"-looking moist areas.

Disease

38.3   Antibiotic-Resistant Staphylococci

During the late 1950s and early 1960s, *Staphylococcus aureus* caused considerable morbidity and mortality as a nosocomial, or hospital-acquired, pathogen. Penicillinase-resistant, semisynthetic penicillins have been successful antimicrobial agents in the treatment of staphylococcal infections. Unfortunately *m*eticillin-resistant *S. aureus* (MRSA) strains have emerged as a major nosocomial problem. One way in which staphylococci become resistant is through acquisition of a chromosomal gene (*mecA*) that encodes an alternate target protein which is not inactivated by methicillin and its relatives, all belonging to the metacillin family of β-lactam drugs. The majority of the strains are also resistant to several of the most commonly used antimicrobial agents, including macrolides, aminoglycosides, and other beta-lactam antibiotics, including the latest generation of cephalosporins. Serious infections by meticillin-resistant strains have been most often successfully treated with an older, potentially toxic antibiotic, vancomycin. However, strains of *Enterococcus* and *Staphylococcus* recently have become resistant to vancomycin.

Meticillin-resistant *S. epidermidis* strains also have emerged as a nosocomial problem, especially in individuals with prosthetic heart valves or in people who have undergone other forms of cardiac surgery. Resistance to methicillin also may extend to the cephalosporin antibiotics. Difficulties in performing in vitro tests that adequately recognize cephalosporin resistance of these strains continue to exist. Most serious infections due to meticillin-resistant *S. epidermidis* have been successfully treated with combination therapy, including vancomycin plus rifampin or an aminoglycoside.

Table 38.5	Summary of the Major Sexually Transmitted Diseases (STDs)		
Microorganism	**Disease**	**Comments**	**Treatment**
Viruses			
Human immunodeficiency virus (HIV)	Acquired immune deficiency syndrome (AIDS)	Pandemic in many parts of the world	A cocktail of reverse transcriptase and protease inhibitors
Herpes simplex virus (HSV-2)	Genital herpes	Painful blisters; enters latent stage, with reactivation due to stress; also oral, pharyngeal, and rectal herpes; no cure; very prevalent in the U.S.	Acyclovir and similar drugs alleviate the symptoms
Human papillomavirus (HPV) various serotypes	Condyloma acuminata (genital warts)	Predisposes to cervical cancer; no cure; very common in the U.S. HPV vaccine recently developed	Removal by various mechanical and chemical means; interferon injection
Hepatitis B virus (HBV)	Hepatitis B (serum hepatitis)	Transmitted in body fluids; cirrhosis, primary hepatocarcinoma	No treatment; recombinant HBV vaccine for prevention
Cytomegalovirus (CMV)	Congenital cytomegalic inclusion disease	Avoid sexual contact with an infected person	Ganciclovir and cidofovir for high-risk patients
Molluscum contagiosum	Genital molluscum contagiosum	Localized wartlike skin lesions	None
Bacteria			
Calymmatobacterium granulomatis	Granuloma inguinale (donovanosis)	Rare in the U.S.; draining ulcers can persist for years	Tetracycline, erythromycin, newer quinolones
Chlamydia trachomatis	Nongonococcal urethritis (NGU); cervicitis, pelvic inflammatory disease (PID), lymphogranuloma venereum	Serovars D-K cause most of the STDs in the U.S.; lymphogranuloma venereum rare in the U.S.	Tetracyclines, erythromycin, doxycycline, ceftriaxone
Gardnerella vaginalis	Bacterial vaginosis	Clue cells present	Metronidazole
Haemophilus ducreyi	Chancroid ("soft chancre")	Open sores on the genitals can lead to scarring without treatment; on the rise in the U.S.	Erythromycin or ceftriaxone
Helicobacter cinaedi, H. fennelliae	Diarrhea and rectal inflammation in homosexual men	Common in immunocompromised individuals	Metronidazole, macrolides
Mycoplasma genitalium	Implicated in some cases of NGU	Only recently described as an STD	Tetracyclines or erythromycin
Mycoplasma hominis	Implicated in some cases of PID	Widespread, often asymptomatic but can cause PID in women	Tetracyclines or erythromycin
Neisseria gonorrhoeae	Gonorrhea, PID	Most commonly reported STD in the U.S.; usually symptomatic in men and asymptomatic in women; antibiotic-resistant strains	Third-generation cephalosporins and/or quinolones
Treponema pallidum subsp. *pallidum*	Syphilis, congenital syphilis	Manifests many clinical syndromes	Benzathine penicillin G
Ureaplasma urealyticum	Urethritis	Widespread, often asymptomatic but can cause PID in women and NGU in men; premature birth	Tetracyclines or erythromycin
Yeasts			
Candida albicans	Candidiasis (moniliasis)	Produces a thick white vaginal discharge and severe itching	Nystatin, terconazole
Protozoa			
Trichomonas vaginalis	Trichomoniasis	Produces a frothy vaginal discharge; very common in the U.S.	Oral metronidazole

Disease

38.4 A Brief History of Syphilis

Syphilis was first recognized in Europe near the end of the fifteenth century. During this time the disease reached epidemic proportions in the Mediterranean areas. According to one hypothesis, syphilis is of New World origin and Christopher Columbus (1451–1506) and his crew acquired it in the West Indies and introduced it into Spain after returning from their historic voyage. Another hypothesis is that syphilis had been endemic for centuries in Africa and may have been transported to Europe at the same time that vast migrations of the civilian population were occurring (1500). Others believe that the Vikings, who reached the New World well before Columbus, were the original carriers.

Syphilis was initially called the Italian disease, the French disease, and the great pox as distinguished from smallpox. In 1530 the Italian physician and poet Girolamo Fracastoro wrote *Syphilis sive Morbus Gallicus* (Syphilis or the French Disease). In this poem a Spanish shepherd named Syphilis is punished for being disrespectful to the gods by being cursed with the disease. Several years later Fracastoro published a series of papers in which he described the possible mode of transmission of the "seeds" of syphilis through sexual contact.

Its venereal transmission was not definitely shown until the eighteenth century. The term venereal is derived from the name Venus, the Roman goddess of love. Recognition of the different stages of syphilis was demonstrated in 1838 by Philippe Ricord, who reported his observations on more than 2,500 human inoculations. In 1905 Fritz Schaudinn and Erich Hoffmann discovered the causative bacterium, and in 1906 August von Wassermann introduced the diagnostic test that bears his name. In 1909 Paul Ehrlich introduced an arsenic derivative, arsphenamine or salvarsan, as therapy. During this period, an anonymous limerick aptly described the course of this disease:

There was a young man from Black Bay
Who thought syphilis just went away
He believed that a chancre
Was only a canker
That healed in a week and a day.

But now he has "acne vulgaris"—
(Or whatever they call it in Paris);
On his skin it has spread
From his feet to his head,
And his friends want to know where his hair is.

There's more to his terrible plight:
His pupils won't close in the light
His heart is cavorting,
His wife is aborting,
And he squints through his gun-barrel sight.

Arthralgia cuts into his slumber;
His aorta is in need of a plumber;
But now he has tabes,
And saber-shinned babies,
While of gummas he has quite a number.

He's been treated in every known way,
But his spirochetes grow day by day;
He's developed paresis,
Has long talks with Jesus,
And thinks he's the Queen of the May.

---

entrance of the vagina. The disease is frequently accompanied by very swollen lymph nodes in the groin. Genital ulcer disease occurs commonly in the tropics; however, in the past decade there have been major outbreaks in the United States. Worldwide, genital ulcer disease is an important cofactor in the transmission of the AIDS virus. Diagnosis is by isolating *H. ducreyi* from the ulcers; treatment is with azithromycin or ceftriaxone. No vaccine is available. Control is the same as for other STDs: avoid contact with infected tissues by the use of barrier protection or abstinence.

Genitourinary Mycoplasmal Diseases

The mycoplasmas *Ureaplasma urealyticum* and *Mycoplasma hominis* are common parasitic microorganisms of the genital tract and their transmission is related to sexual activity (table 38.5). Both mycoplasmas can opportunistically cause inflammation of the reproductive organs of males and females. Because mycoplasmas are not usually cultured by clinical microbiologists, management and treatment of these infections depend on recognition of clinical syndromes and provision for adequate therapy. Tetracyclines are active against most strains; resistant organisms can be treated with erythromycin. Class *Mollicutes* (section 23.2)

Gonorrhea

Gonorrhea [Greek *gono,* seed, and *rhein,* to flow] is an acute, infectious, sexually transmitted disease of the mucous membranes of the genitourinary tract, eye, rectum, and throat (table 38.5). It is caused by the gram-negative, oxidase-positive, diplococcus, *Neisseria gonorrhoeae*. These bacteria are also referred to as **gonococci** [pl. of gonococcus; Greek *gono,* seed, and *coccus,* berry] and have a worldwide distribution. Over 500,000 cases are reported annually in the United States; the actual incidence is significantly higher. Class *Betaproteobacteria:* Order *Neisseriales* (section 22.2)

Once inside the body the gonococci attach to the microvilli of mucosal cells by means of pili and protein II, which function as adhesins. This attachment prevents the bacteria from being washed away by normal cervical and vaginal discharges or by the flow of urine. They are then phagocytosed by the mucosal cells and may even be transported through the cells to the intercellular spaces and subepithelial tissue. Phagocytes, such as neutrophils, also may contain gonococci inside vesicles **(figure 38.18)**. Because the gonococci are intracellular at this time, the host's defenses have little effect on the bacteria. Following penetration of the bacteria, the host tissue responds locally by the infiltration of mast cells, more

Figure 38.18   Gonorrhea. Gram stain of male urethral exudate showing *Neisseria gonorrhoeae* (diplococci) inside a PMN; light micrograph (×500). Although the presence of gram-negative diplococci in exudates is a probable indication of gonorrhea, the bacterium should be isolated and identified.

PMNs, and anitbody-secreting plasma cells. These cells are later replaced by fibrous tissue that may lead to urethral closing, or stricture, in males. Cells, tissues, and organs of the immune system (section 31.2); Phagocytosis (section 31.3)

In males the incubation period is 2 to 8 days. The onset consists of a urethral discharge of yellow, creamy pus, and frequent, painful urination that is accompanied by a burning sensation. In females, the cervix is the principal site infected. The disease is more insidious in females and few individuals are aware of any symptoms. However, some symptoms may begin 7 to 21 days after infection. These are generally mild; some vaginal discharge may occur. The gonococci also can infect the Fallopian tubes and surrounding tissues, leading to **pelvic inflammatory disease (PID)**. This occurs in 10 to 20% of infected females. Gonococcal PID is a major cause of sterility and ectopic pregnancies because of scar formation in the Fallopian tubes. Gonococci disseminate most often during menstruation, a time in which there is an increased concentration of free iron available to the bacteria. In both genders, disseminated gonococcal infection with bacteremia may occur. This can lead to involvement of the joints (gonorrheal arthritis), heart (gonorrheal endocarditis), or pharynx (gonorrheal pharyngitis). Gonorrheal eye infections can occur in newborns as they pass through an infected birth canal. The resulting disease is called **ophthalmia neonatorum, or conjunctivitis of the newborn.** This was once a leading cause of blindness in many parts of the world. To prevent this disease, tetracycline, erythromycin, povidone-iodine, or silver nitrate in dilute solution is placed in the eyes of newborns. This type of treatment is required by law in the United States and many other nations.

Laboratory diagnosis of gonorrhea relies on the successful growth of *N. gonorrhoeae* in culture to determine oxidase reaction, Gram stain reaction, and colony and cell morphology. The performance of confirmation tests also is necessary. Because the gonococci are very sensitive to adverse environmental conditions and survive poorly outside the body, specimens should be plated directly; when this is not possible, special transport media are necessary. A DNA probe (Gen-Probe Pace) for *N. gonorrhoeae* has been developed and is used to supplement other diagnostic techniques.

The Centers for Disease Control and Prevention recommends as treatment five single doses of cefixime, cefriaxone, ciprofloxacin, ofloxacin, and levofloxacin to eradicate the infection. Penicillin-resistant strains of gonococci occur worldwide. Most of these strains carry a plasmid that directs the formation of penicillinase, a β-lactamase enzyme that inactivates penicillin G and ampicillin. Since 1980, strains of *N. gonorrhoeae* with chromosomally mediated penicillin resistance have developed. Instead of producing a penicillinase, these strains have altered penicillin-binding proteins. Since 1986, tetracycline-resistant *N. gonorrhoeae* also have developed. Therefore, neither penicillins nor tetracyclines are recommended for treating gonococcal infections. Recently, the CDC has recommended discontinuation of quinolone use to treat gonorrhea in men who have sex with other men, as quinolone-resistant strains are increasing in this group. Instead, they recommend ceftriaxone or spectinomycin as an alternative. Drug resistance (section 34.6)

The most effective method for control of this sexually transmitted disease is public education, diagnosing and treating the asymptomatic patient, barrier protection, and treating infected individuals quickly to prevent further spread of the disease. More than 60% of all cases occur in the 15- to 24-year-old age group. Repeated gonococcal infections are common. Protective immunity to reinfection does not arise because of antigenic variation in which a single strain changes its pilin gene by a recombinational event and alters the expression of the various protein II genes by slipped strand mispairing. This can thus be viewed as a programmed evasion technique employed by the bacterium rather than a mere reflection of strain variation.

Lymphogranuloma Venereum

Lymphogranuloma venereum (LGV) is a sexually transmitted disease (table 38.5) caused by *Chlamydia trachomatis* serotypes L1–L3. It has a worldwide distribution but is more common in tropical climates. LGV proceeds through three phases. (1) In the primary phase a small ulcer appears several days to several weeks after a person is exposed to the chlamydiae. The ulcer may appear on the penis in males or on the labia or vagina in females. The ulcer heals quickly and leaves no scar. (2) The secondary phase begins 2 to 6 weeks after exposure, when the chlamydiae infect lymphoid cells, causing the regional lymph nodes to become enlarged and tender; such nodes are called buboes **(figure 38.19)**. Systemic symptoms such as fever, chills, and anorexia are common. (3) If the disease is not treated, a late phase ensues. This results from fibrotic changes and abnormal lymphatic drainage that produces fistulas (abnormal passages leading from an abscess or a hollow organ to the body surface or from one hollow organ to another), and urethral or rectal strictures (a decrease in size). An untreatable fluid accumulation in the penis, scrotum, or vaginal area may result. Phylum Chlamydiae (section 21.5)

The disease is detected by staining infected cells with iodine to observe inclusions (chlamydia-filled vacuoles), culture of the

Critical Thinking Questions

1. Why is tetanus a concern only when one has a deep puncture-type wound and not a surface cut or abrasion?

2. Think about our modern, Western lifestyles. Can you name and describe bacterial diseases that result from this life of relative luxury? Refer to *Infections of Leisure,* second edition, edited by David Schlossberg (1999), published by the American Society for Microbiology Press.

3. You have been assigned the task of eradicating gonorrhea in your community. Explain how you would accomplish this.

4. You are a park employee. How would you prevent visitors from acquiring arthropod-borne diseases?

5. Visit the World Heath Organization website at www.who.org and identify an infectious disease that is a problem in a developing country but not in the United States. List the reasons why the disease is not controlled in the developing country but is in North America. What policies and initiatives would you implement if you were in charge of the reducing the mortality and morbidity rate of this disease?

Learn More

Davis, D. H., and Elzer, P. H. (2002). *Brucella* vaccines in wildlife. *Vet. Microbiol.* 90:533–44.

Fux, C. A.; Costerton, J. W.; Stewart, P. S.; and Stoodley, P. 2005. Survival strategies of infectious biofilms. *Trends Microbiol.* 13:34–40.

Golden, M. R., and Manhart, L. F. 2005. Innovative approaches to the prevention and control of bacterial sexually transmitted infections. *Infect. Dis. Clin. North Am.* 19:513–40.

Jenkinson, H. F., and Lamont, R. J. 2005. Oral microbial communities in sickness and in health. *Trends Microbiol.* 13:590–95.

Kirn, T. J.; Jude, B. A.; and Taylor, R. K. 2005. A colonization factor links *Vibrio cholerae* environmental survival and human infection. *Nature* 438:863–66.

Liu, Y. M.; Chi, C. Y.; Ho, M. W.; Chen, C. M.; Liao, W. C.; Ho, C. M.; Lin, P. C.; and Wang, J. H. 2005. Microbiology and factors affecting mortality in necrotizing fasciitis. *J. Microbiol. Immunol. Infect.* 38:430–5.

Marketon, M. M.; DePaolo, R. W.; DeBord, K. L.; Jabri, B.; and Schneewind, O. 2005. Plague bacteria target immune cells during infection. *Science.* 309:1739–76.

Moayeri, M., and Leppla, S. H. 2004. The roles of anthrax toxin in pathogenesis. *Curr. Opin. Microbiol.* 7:19–24.

Oyston, P. C. F.; Sjostedt, A.; and Titball, R. W. 2004. Tularaemia: Bioterrorism defence renews interest in *Francisella tularensis. Nature Rev. Microbiol.* 2:967–72.

Russell, D. G.; Purdy, G. E.; Owens, R. M.; Rohde, K. H.; and Yates, R. M. 2005. *Mycobacterim tuberculosis* and the four-minute phagosome. *ASM News* 10:459–63.

Soriani, M.; Santi, I.; Taddei, A.; Rappuoli, R.; Grandi, G.; and Telford, J. L. 2006. Group B streptococcus crosses human epithelial cells by a paracellular route. *J. Infect. Dis.* 193:241–50.

Smith, R. P. 2006. Current diagnosis and treatment of lyme disease. *Compr. Ther.* 31:284–90.

Taylor, Z.; Nolan, C. M.; Blumberg, H. M.; American Thoracic Society; Centers for Disease Control and Prevention; and Infectious Diseases Society of America. 2005. Controlling tuberculosis in the United States. Recommendations from the American Thoracic Society, CDC, and the Infectious Diseases Society of America. *MMWR Recomm. Rep.* 4:1–81.

Tomaso, H.; Bartling, C.; Al Dahouk, S.; Hagen, R. M.; Scholz, H. C.; Beyer, W.; and Neubauer, H. 2006. Growth characteristics of *Bacillus anthracis* compared to other *Bacillus* spp. on the selective nutrient media Anthrax Blood Agar® and Cereus Ident Agar®. *Syst. Appl. Microbiol.* 29:24–28.

Please visit the Prescott website at www.mhhe.com/prescott7 for additional references.

39

Human Diseases Caused by Fungi and Protists

Malaria parasites (yellow) bursting out of red blood cells. Malaria is one of the worst scourges of humanity. Indeed, malaria has played an important part in the rise and fall of nations (see the chapter opening quote), and has killed untold millions the world over. Despite the combined efforts of 102 countries to eradicate malaria, it remains the most important disease in the world today in terms of lives lost and economic burden.

PREVIEW

- Fungal diseases (mycoses) are usually divided into four groups according to the level of infected tissue and mode of entry into the host: (1) superficial, (2) cutaneous, (3) subcutaneous, and (4) systemic.

- About 20 different protists cause human diseases that afflict hundreds of millions of people throughout the world.

- While seemingly different, eucaryotic fungal and protist pathogens gain access to humans by transmission routes previously identified for bacteria and viruses: air, arthropods, direct contact, food, water, and the host itself.

- The systemic mycoses, which are typically transmitted through air, are the most serious of the fungal infections in the normal host because they can disseminate throughout the body. Examples include blastomycosis, coccidioidomycosis, cryptococcosis, and histoplasmosis.

- Fungal pathogens do not appear to be transmitted by arthropods; however, several notable protist pathogens are. These include *Leishmania, Plasmodium,* and *Trypanosoma.*

- A number of fungal and protist pathogens are transmitted by direct contact. Examples include the superficial mycoses, which occur mainly in the tropics and include black piedra, white piedra, and tinea versicolor; the cutaneous mycoses generally called ringworms, tineas, or dermatomycoses (occurring worldwide and representing the most common fungal diseases in humans); and the dermatophytes, which cause the subcutaneous mycoses (chromomycosis, maduromycosis, and sporotrichosis).

- Fungal and protist diseases can also be transmitted through food and water. Examples include amebiasis, cryptosporidiosis, and giardiasis.

- Opportunistic diseases typically arise from the endogenous microbial flora when the host can no longer control them. The opportunistic mycoses can create life-threatening situations in the compromised host. Examples of these diseases include aspergillosis, candidiasis, microsporidiosis and *Pneumocystis* pneumonia.

In this chapter we describe some of the fungi and protists that are pathogenic to humans and discuss the clinical manifestations, diagnosis, epidemiology, pathogenesis, and treatment of the diseases caused by them. The biology of these organisms is covered in chapters 25 and 26, respectively. This chapter follows the format of the previous two chapters in identifying their diseases by route of transmission.

39.1 PATHOGENIC FUNGI AND PROTISTS

Fungi are eucaryotic saprophytes that are ubiquitous in nature. Although hundreds of thousands of fungal species are found in the environment, only about 50 produce disease in humans. **Medical mycology** is the discipline that deals with the fungi that cause human disease. These fungal diseases, known as **mycoses** [s., mycosis; Greek *mykes,* fungus], are typically divided into five groups according to the route of infection: superficial, cutaneous, subcutaneous, systemic, and opportunistic mycoses (**tables 39.1** and **39.2**). Superficial, cutaneous, and subcutaneous mycoses are direct contact infections of the skin, hair, and nails. Systemic mycoses are fungal infections that have disseminated to visceral tissues. Except for *Cryptococcus neoformans,* which has only a yeast form, the fungi that cause the systemic or deep mycoses are dimorphic—they exhibit a parasitic yeast-like phase (Y) and a saprophytic mold or mycelial phase (M). Most systemic mycoses are acquired by the inhalation of spores from soil in which the mold-phase of the fungus resides. If a susceptible person inhales enough spores, an infection begins as a lung lesion, becomes chronic, and spreads through the bloodstream to

Historians believe that malaria has probably had a greater impact on world history than any other infectious disease, influencing the outcome of wars, various population movements, and the development and decline of various civilizations.

—*Lynne S. Garcia*

Table 39.1	Examples of Some Medically Important Fungi		
Group	**Pathogen**	**Location**	**Disease**
Superficial mycoses	*Piedraia hortae*	Scalp	Black piedra
	Trichosporon beigelii	Beard, mustache	White piedra
	Malassezia furfur	Trunk, neck, face, arms	Tinea versicolor
Cutaneous mycoses	*Trichophyton mentagrophytes, T. verrucosum, T. rubrum*	Beard hair	Tinea barbae
	Trichophyton, Microsporum canis	Scalp hair	Tinea capitis
	Trichophyton rubrum, T. mentagrophytes, Microsporum canis	Smooth or bare parts of the skin	Tinea corporis
	Epidermophyton floccosum, T. mentagrophytes, T. rubrum	Groin, buttocks	Tinea cruris (jock itch)
	T. rubrum, T. mentagrophytes, E. floccosum	Feet	Tinea pedis (athlete's foot)
	T. rubrum, T. mentagrophytes, E. floccosum	Nails	Tinea unguium (onychomycosis)
Subcutaneous mycoses	*Phialophora verrucosa, Fonsecaea pedrosoi*	Legs, feet	Chromoblastomycosis
	Madurella mycetomatis	Feet, other areas of body	Maduromycosis
	Sporothrix schenckii	Puncture wounds	Sporotrichosis
Systemic mycoses	*Blastomyces dermatitidis*	Lungs, skin	Blastomycosis
	Coccidioides immitis	Lungs, other parts of body	Coccidioidomycosis
	Cryptococcus neoformans	Lungs, skin, bones, viscera, central nervous system	Cryptococcosis
	Histoplasma capsulatum	Within phagocytes	Histoplasmosis
Opportunistic mycoses	*Aspergillus fumigatus, A. flavus*	Respiratory system	Aspergillosis
	Candida albicans	Skin or mucous membranes	Candidiasis
	Pneumocystis jiroveci	Lungs, sometimes brain	*Pneumocystis* pneumonia
	Encephalitozoon, Nosema, Vitta forma, Pleistophora, Enterocytozoon, Trachipleistophora, Microsporidium	Lungs, sometimes brain	Microsporidiosis

other organs (the target organ varies with the species). The *Fungi* (chapter 26)

Protozoa, single-celled eucaryotic chemoorganotrophs, have become adapted to practically every type of habitat on the face of the Earth, including the human body. Many protists are transmitted to humans by arthropod vectors or by food and water vehicles. However, some protozoan diseases are transmitted by direct contact. Although fewer than 20 genera of protists cause disease in humans (**tables 39.3** and **39.4**), their impact is formidable. For example, there are over 150 million cases of malaria in the world each year. In tropical Africa alone, malaria is responsible for the deaths of more than a million children under the age of 14 annually. It is estimated that there are at least 8 million cases of trypanosomiasis, 12 million cases of leishmaniasis, and over 500 million cases

of amebiasis yearly. There is also an increasing problem with *Cryptosporidium* and *Cyclospora* contamination of food and water supplies (**table 39.5**). More of our population is elderly, and a growing number of persons are immunosuppressed due to HIV infection, organ transplantation, or cancer chemotherapy. These populations are at increased risk for protozoan infections. The protists (chapter 25)

While seemingly different, fungi and protists share a number of phenotypic features that serve them in their ability to cause infection: microscopic size, eucaryotic physiology, cell walls or wall-like structures, alternative stages for survival outside of the host, degradative enzymes, and others. Some fungi and protists also share transmission routes. We now discuss diseases of fungi and protists based on how they are acquired by the human host.

39.2 AIRBORNE DISEASES

Blastomycosis is the systemic mycosis caused by *Blastomyces dermatitidis*, a fungus that grows as a budding yeast in humans but as a mold on culture media and in the environment. It is found pre- dominately in moist soil enriched with decomposing organic debris, as in the Mississippi and Ohio River basins. *B. dermatitidis* is endemic in parts of the south-central, southeastern and midwestern United States. Additionally, microfoci have been reported in Central and South America and in parts of Africa. The disease occurs in three clinical forms: cutaneous, pulmonary, and disseminated. The initial infection begins when blastospores are inhaled into the lungs. The fungus can then spread rapidly, especially to the skin, where cutaneous ulcers and abscess formation occur (**figure 39.1**). *B. dermatitidis* can be isolated from pus and biopsy sections. Diagnosis requires the demonstration of thick-walled, yeast-like cells, 8 to 15 μm in diameter. Complement-fixation, immunodiffusion, and skin tests (blastomycin) are also useful. Amphotericin B (Fungizone), itraconazole (Sporanox), or ketoconazole (Nizoral) are the drugs of choice for treatment. Surgery may be necessary for the drainage of large abscesses. Mortality is

Table 39.2	Examples of Some Human Fungal Diseases Recognized Since 1974	
Year	**Fungus**	**Disease**
	Molds	
1974	*Phialophora parasitica*	Phaeohyphomycosis
1992	*Penicillium marneffei*	Disseminated infection
	Yeasts	
1985	*Enterocytozoon bieneusi*	Diarrhea, microsporidiosis
1989	*Candida lusitaniae*	Fungemia
1989	*Malassezia furfur*	Fungemia
1990	*Rhodotorula rubra*	Fungemia
1991	*Candida ciferrii*	Fungemia
1993	*Hansenula anomala*	Fungemia
1993	*Trichosporon beigelii*	Fungemia
1993	*Encephalitozoon cuniculi*	Disseminated microsporidiosis
1996	*Trachipleistophora hominis*	Disseminated microsporidiosis

Table 39.4	Examples of Human Protozoan Diseases Recognized Since 1976	
Year	**Protozoan**	**Disease**
1976	*Cryptosporidium parvum*	Acute and chronic diarrhea, cryptosporidiosis
1986	*Cyclospora cayatanensis*	Persistent diarrhea
1991	*Babesia* spp.	Atypical babesiosis
1998	*Brachiola vesicularum*	Myositis

Table 39.3	Examples of Some Medically Important Protozoa	
Morphological Group	**Pathogen**	**Disease**
Amoebae	*Entamoeba histolytica*	Amebiasis, amebic dysentery
	Acanthamoeba spp., *Naegleria fowleri*	Amebic meningoencephalitis
Apicomplexa	*Cryptosporidium parvum*	Cryptosporidiosis
	Cyclospora cayetanensis	Cyclosporidiosis
	Isospora belli	Isosporiasis
	Plasmodium falciparum, P. malariae, P. ovale, P. vivax	Malaria
	Toxoplasma gondii	Toxoplasmosis
Ciliates	*Balantidium coli*	Balantidiasis
Blood and tissue flagellates	*Leishmania tropica*	Cutaneous leishmaniasis
	L. braziliensis	Mucocutaneous leishmaniasis
	L. donovani	Kala-azar (visceral leishmaniasis)
	Trypanosoma cruzi	American trypanosomiasis
	T. brucei gambiense, T. brucei rhodesiense	African sleeping sickness
Digestive and genital organ flagellates	*Giardia intestinalis*	Giardiasis
	Trichomonas vaginalis	Trichomoniasis

Table 39.5	Water-Based Protozoan Pathogens That Can Be Maintained in the Environment Independent of Humans	
Organism	**Reservoir**	**Comments**
Acanthamoeba	Sewage sludge disposal areas	Can cause granulomatous amebic encephalitis (GAE); keratitis, corneal ulcers
Cryptosporidium	Many species of domestic and wild animals	Causes acute enterocolitis; important with immunologically compromised individuals; cysts resistant to chemical disinfection; not antibiotic sensitive
Cyclospora cayetanensis	Waters—does not withstand drying; possibly other reservoirs	Causes long-lasting (43 days average) diarrheal illness; infection self-limiting in immunocompetent hosts; sensitive to prompt treatment with sufonamide and trimethoprim
Giardia intestinalis	Beavers, sheep, dogs, cats	Major cause of early spring diarrhea; important in cold mountain water
Naegleria fowleri	Warm water (hot tubs), swimming pools, lakes	Inhalation in nasal passages; central nervous system infection; causes primary amebic meningoencephalitis (PAM)

Figure 39.1 Systemic Mycosis. Blastomycosis of the forearm caused by *Blastomyces dermatitidis.*

Figure 39.2 Systemic Mycosis: Coccidioidomycosis. *Coccidioides immitis* mature spherules filled with endospores within a tissue section; light micrograph (×400).

about 5%. There are no preventive or control measures. Antifungal drugs (section 34.7)

Coccidioidomycosis, also known as valley fever, San Joaquin fever, or desert rheumatism because of the geographical distribution of the fungus, is caused by *Coccidioides immitis. C. immitis* exists in the semi-arid, highly alkaline soils of the southwestern United States and parts of Mexico and South America. It has been estimated that in the United States about 100,000 people are infected annually, resulting in 50 to 100 deaths. Endemic areas have been defined by massive skin testing with the antigen coccidioidin, where 10 to 15% positive cases are reported. In the soil and on culture media, this fungus grows as a mold that forms arthroconidia at the tips of hyphae (*see figure 26.8*). Because arthroconidia are so abundant in these endemic areas, immunocompromised individuals can acquire the disease by inhalation as they simply move through the area. Wind turbulence and even construction of outdoor structures have been associated with increased exposure and infection. In humans, the fungus

grows as a yeast-forming, thick-walled spherule filled with spores (**figure 39.2**). Most cases of coccidioidomycosis are asymptomatic or indistinguishable from ordinary upper respiratory infections. Almost all cases resolve in a few weeks, and a lasting immunity results. A few infections result in a progressive chronic pulmonary disease or disseminated infections of other tissues; the fungus can spread throughout the body, involving almost any organ or site.

Diagnosis is accomplished by identification of the large spherules (approximately 80 μm in diameter) in pus, sputum, and aspirates. Culturing clinical samples in the presence of penicillin

and streptomycin on Sabouraud agar (used to isolate fungi) also is diagnostic. Newer methods of rapid confirmation include the testing of supernatants of liquid media cultures for antigens, serology, and skin testing. Miconazole (Lotrimin), itraconazole, ketoconazole, and amphotericin B are the drugs of choice for treatment. Prevention involves reducing exposure to dust (soil) in endemic areas. Antifungal drugs (section 34.7); Identification of microorganisms from specimens (section 35.2)

Cryptococcosis is a systemic mycosis caused by *Cryptococcus neoformans*. This fungus always grows as a large, budding yeast. In the environment, *C. neoformans* is a saprophyte with a worldwide distribution. Aged, dried pigeon droppings are an apparent source of infection. Cryptococcosis is found in approximately 15% of AIDS patients. The fungus enters the body by the respiratory tract, causing a minor pulmonary infection that is usually transitory. Some pulmonary infections spread to the skin, bones, viscera, and the central nervous system. Once the nervous system is involved, cryptococcal meningitis usually results. Diagnosis is accomplished by detection of the thick-walled, spherical yeast cells in pus, sputum, or exudate smears using India ink to define the organism (**figure 39.3**). The fungus can be easily cultured on Sabouraud dextrose agar. Identification of the fungus in body fluids is made by immunologic procedures. Treatment includes amphotericin B or itraconazole. There are no preventive or control measures.

Histoplasmosis is caused by *Histoplasma capsulatum* var. *capsulatum,* a facultative, parasitic fungus that grows intracellularly. It appears as a small, budding yeast in humans and on culture media at 37°C. At 25°C it grows as a mold, producing small microconidia (1 to 5 μm in diameter) that are borne singly at the tips of short conidiophores (*see figure 26.8*). Large macroconidia

(8 to 16 μm in diameter) are also formed on conidiophores (**figure 39.4***a*). In humans, the yeastlike form grows within phagocytic cells (figure 39.4*b*). *H. capsulatum* var. *capsulatum* is found as the mycelial form in soils throughout the world and is localized in areas that have been contaminated with bird or bat excrement. The microconidia can become airborne when contaminated soil is disturbed. Infection ensues when the microconidia are inhaled. Histoplasmosis is not, however, transmitted from an infected person. Within the United States, histoplasmosis is endemic within the Mississippi, Kentucky, Tennessee, Ohio, and Rio Grande River basins. More than 80% of the people who reside in parts of these areas have antibodies against the fungus. It has been estimated that in endemic areas of the United States, about 500,000 individuals are infected annually: 50,000 to 200,000 become ill; 3,000 require hospitalization; and about 50 die. The total number of infected individuals may be over 40 million in the United States alone. Histoplasmosis is a common disease among poultry farmers, spelunkers (people who explore caves), and bat guano miners (bat guano is used as fertilizer).

Humans acquire histoplasmosis from airborne microconidia that are produced under favorable environmental conditions. Microconidia are most prevalent where bird droppings—especially from starlings, crows, blackbirds, cowbirds, sea gulls, turkeys, and chickens—have accumulated. It is noteworthy that the birds themselves are not infected because of their high body temperature; their droppings simply provide the nutrients for this fungus. Only bats and humans demonstrate the disease and harbor the fungus.

Because histoplasmosis is a disease of the monocyte-macrophage system, many organs of the body can be infected (*see figure 31.3*). More than 95% of "histo" cases have either no symptoms or mild symptoms such as coughing, fever, and joint pain. Lesions may appear in the lungs and show calcification; the disease may resemble tuberculosis. Most infections resolve on their own. Only rarely does the disease disseminate. Laboratory diagnosis is accomplished by complement-fixation tests and isolation of the fungus from tissue specimens. Most individuals with this disease exhibit a hypersensitive state that can be demonstrated by the histoplasmin skin test. Currently the most effective treatment is with amphotericin B, ketoconazole, or itraconazole. Prevention and control involve wearing protective clothing and masks before entering or working in infested habitats. Soil decontamination with 3 to 5% formalin is effective where economically and physically feasible. Antifungal drugs (section 34.7)

Figure 39.3 Systemic Mycosis: Cryptococcosis. India ink preparation showing *Cryptococcus neoformans*. Although these microorganisms are not budding, they can be differentiated from artifacts by their doubly refractile cell walls, distinctly outlined capsules surrounding all cells, and refractile inclusions in the cytoplasm; light micrograph (×150).

39.3 Arthropod-Borne Diseases

Malaria

The most important human parasite among the protozoa is *Plasmodium*, the causative agent of **malaria (Disease 39.1)**. It has been estimated that more than 300 million people are infected each year, and over 1 million die annually of malaria in Africa alone. About 1,000 cases are reported each year in the United States, divided between returning U.S. travelers and non-U.S.

Microconidium

Macroconidium

Mycelium

(a)

Figure 39.4 Morphology of *Histoplasma capsulatum* var. *capsulatum*. (a) Mycelia, microconidia, and chlamydospores as found in the soil. These are the infectious particles; light micrograph (×125). **(b)** Yeastlike cells in a macrophage. Budding *H. capsulatum* within a vacuole. Tubular structures, ts, are observed beneath the cell wall, cw; electron micrograph (×23,000).

Macrophage

Budding yeast

Vacuole

(b)

Disease

39.1 A Brief History of Malaria

No other single infectious disease has had the impact on humans that malaria has. The first references to its periodic fever and chills can be found in early Chaldean, Chinese, and Hindu writings. In the late 5th century B.C., Hippocrates described certain aspects of malaria. In the 4th century B.C., the Greeks noted an association between individuals exposed to swamp environments and the subsequent development of periodic fever and enlargement of the spleen (splenomegaly). In the 17th century the Italians named the disease *mal' aria* (bad air) because of its association with the ill-smelling vapors from the swamps near Rome. At about the same time, the bark of the quinaquina (cinchona) tree of South America was used to treat the intermittent fevers, although it was not until the mid-19th century that quinine was identified as the active alkaloid. The major epidemiological breakthrough came in 1880, when French army surgeon Charles Louis Alphonse Laveran observed gametocytes in fresh blood. Five years later the Italian histologist Camillo Golgi observed the multiplication of the asexual blood forms. In the late 1890s Patrick Manson postulated that malaria was transmitted by mosquitoes. Sir Ronald Ross, a British army surgeon in the Indian Medical Service, subsequently observed developing plasmodia in the intestine of mosquitoes, supporting Manson's theory. Using birds as experimental models, Ross definitively established the major features of the life cycle of *Plasmodium* and received the Nobel Prize in 1902.

Human malaria is known to have contributed to the fall of the ancient Greek and Roman empires. Troops in both the U.S. Civil War and the Spanish-American War were severely incapacitated by the disease. More than 25% of all hospital admissions during these wars were malaria patients. During World War II malaria epidemics severely threatened both the Japanese and Allied forces in the Pacific. The same can be said for the military conflicts in Korea and Vietnam.

In the 20th century efforts were directed toward understanding the biochemistry and physiology of malaria, controlling the mosquito vector, and developing antimalarial drugs. In the 1960s it was demonstrated that resistance to *P. falciparum* among West Africans was associated with the presence of hemoglobin-S (Hb-S) in their erythrocytes. Hb-S differs from normal hemoglobin-A by a single amino acid, valine, in each half of the Hb molecule. Consequently these erythrocytes—responsible for sickle cell disease—have a low binding capacity for oxygen. Because the malarial parasite has a very active aerobic metabolism, it cannot grow and reproduce within these erythrocytes.

In 1955 the World Health Organization began a worldwide malarial eradication program that finally collapsed by 1976. Among the major reasons for failure were the development of resistance to DDT by the mosquito vectors and the development of resistance to chloroquine by strains of *Plasmodium*. Scientists are exploring new approaches, such as the development of vaccines and more potent drugs. For example, in 1984 the gene encoding the sporozoite antigen was cloned, permitting the antigen to be mass-produced by genetic engineering techniques. In 2002, the complete DNA sequences of *P. falciparum* and *Anopheles gambiae* (the mosquito that most efficiently transmits this parasite to humans in Africa) were determined. Together with the human genome sequence, researchers now have in hand the genetic blueprints for the parasite, its vector, and its victim. This has made possible a holistic approach to understanding how the parasite interacts with the human host, leading to new antimalarial strategies including vaccine design. Overall, no greater achievement for molecular biology could be imagined than the control of malaria—a disease that has caused untold misery throughout the world since antiquity and remains one of the world's most serious infectious diseases.

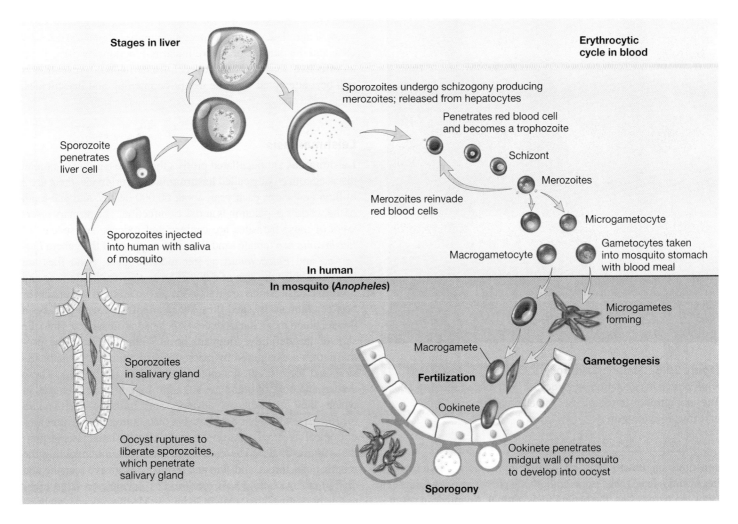

Figure 39.5 **Malaria.** Life cycle of *Plasmodium vivax*.

citizens. Human malaria is caused by four species of *Plasmodium: P. falciparum, P. malariae, P. vivax,* and *P. ovale*. The life cycle of *P. vivax* is shown in **figure 39.5.** The parasite first enters the bloodstream through the bite of an infected female *Anopheles* mosquito. As she feeds, the mosquito injects a small amount of saliva containing an anticoagulant along with small haploid sporozoites. The sporozoites in the bloodstream immediately enter hepatic cells of the liver. In the liver they undergo multiple asexual fission (schizogony) and produce merozoites. After being released from the liver cells, the merozoites attach to erythrocytes and penetrate these cells. Protists classification: *Alveolata* (section 25.6)

Once inside the erythrocyte, the *Plasmodium* begins to enlarge as a uninucleate cell termed a trophozoite. The trophozoite's nucleus then divides asexually to produce a schizont that has 6 to 24 nuclei. The schizont divides and produces mononucleated merozoites. Eventually the erythrocyte lyses, releasing the merozoites into the bloodstream to infect other erythrocytes. This erythrocytic stage is cyclic and repeats itself approximately every 48 to 72 hours or longer, depending on the species of *Plasmodium* involved. The sudden release of merozoites, toxins, and erythro-

cyte debris triggers an attack of the chills, fever, and sweats characteristic of malaria. Occasionally, merozoites differentiate into macrogametocytes and microgametocytes, which do not rupture the erythrocyte. When these are ingested by a mosquito, they develop into female and male gametes, respectively. In the mosquito's gut, the infected erythrocytes lyse and the gametes fuse to form a diploid zygote called the ookinete. The ookinete migrates to the mosquito's gut wall, penetrates, and forms an oocyst. In a process called sporogony, the oocyst undergoes meiosis and forms sporozoites, which migrate to the salivary glands of the mosquito. The cycle is now complete, and when the mosquito bites another human host, the cycle begins anew.

The pathological changes caused by malaria involve not only the erythrocytes but also the spleen and other visceral organs. Classic symptoms first develop with the synchronized release of merozoites and erythrocyte debris into the bloodstream, resulting in the malarial paroxysms—shaking chills, then burning fever followed by sweating. It may be that the fever and chills are caused partly by a malarial toxin that induces macrophages to release TNF-α and interleukin-1. Several of these paroxysms

39.4 DIRECT CONTACT DISEASES

Superficial Mycoses

The superficial mycoses are extremely rare in the United States, and most occur in the tropics. The fungi responsible are limited to the outer surface of hair and skin and hence are called superficial. Infections of the hair shaft are collectively called **piedras** (Spanish for stone because they are associated with the hard nodules formed by mycelia on the hair shaft). For example, **black piedra** is caused by *Piedraia hortae* and forms hard, black nodules on the hairs of the scalp (**figure 39.11**). **White piedra** is caused by the yeast *Trichosporon beigelii* and forms light-colored nodules on the beard and mustache. Some superficial mycoses are called **tineas** [Latin for grub, larva, worm], the specific type is designated by a modifying term. Tineas are superficial fungal infections involving the outer layers of the skin, nails, and hair. **Tinea versicolor** is caused by the yeast *Malassezia furfur* and forms brownish-red scales on the skin of the trunk, neck, face, and arms. Treatment involves removal of the skin scales with a cleansing agent and removal of the infected hairs. Good personal hygiene prevents these infections.

Cutaneous Mycoses

Cutaneous mycoses—also called **dermatomycoses, ringworms,** or tineas—occur worldwide and represent the most common fungal diseases in humans. Three genera of cutaneous fungi, or **dermatophytes,** are involved in these mycoses: *Epidermophyton, Microsporum,* and *Trichophyton.* Diagnosis is by microscopic examination of biopsied areas of the skin cleared with 10% potassium hydroxide and by culture on Sabouraud dextrose agar. Treatment is with topical ointments such as miconazole (Monistat-Derm), tolnaftate (Tinactin), or clotrimazole (Lotrimin) for 2 to 4 weeks. Griseofulvin (Grifulvin V) and itraconazole (Sporanox) are the only oral antifungal agents currently approved by the FDA for treating dermatophytoses.

Tinea barbae [Latin *barba,* the beard] is an infection of the beard hair caused by *Trichophyton mentagrophytes* or *T. verrucosum.* It is predominantly a disease of men who live in rural areas and acquire the fungus from infected animals. **Tinea capitis** [Latin *capita,* the head] is an infection of the scalp hair (**figure 39.12a**). It is characterized by loss of hair, inflammation, and scaling. Tinea capitis is primarily a childhood disease caused by *Trichophyton* or *Microsporum* species. Person-to-person transmission of the fungus occurs frequently when poor hygiene and overcrowded conditions exist. The fungus also occurs in domestic animals, who can transmit it to humans. A Wood's lamp (a UV light) can help with the diagnosis of tinea capitis because fungus-infected hair fluoresces when illuminated by UV radiation (figure 39.12b).

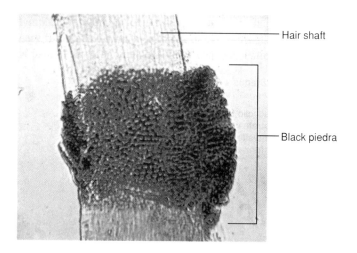

Hair shaft

Black piedra

Figure 39.11   Superficial Mycosis: Black Piedra.   Hair shaft infected with *Piedraia hortae;* light micrograph (×200).

(a)

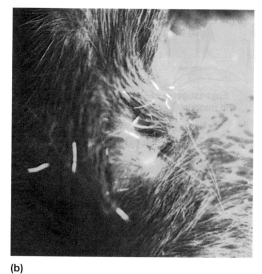
(b)

Figure 39.12   Cutaneous Mycosis: Tinea Capitis.   **(a)** Ringworm of the head caused by *Microsporum audouinii.* **(b)** Close-up using a Wood's light (a UV lamp).

Tinea corporis [Latin *corpus*, the body] is a dermatophytic infection that can occur on any part of the skin (**figure 39.13**). The disease is characterized by circular, red, well-demarcated, scaly, vesiculopustular lesions accompanied by itching. Tinea corporis is caused by *Trichophyton rubrum, T. mentagrophytes*, or *Microsporum canis*. Transmission of the disease agent is by direct contact with infected animals and humans or by indirect contact through fomites (inanimate objects).

Tinea cruris [Latin *crura*, the leg] is a dermatophytic infection of the groin (**figure 39.14**). The pathogenesis and clinical manifestations are similar to those of tinea corporis. The responsible fungi are *Epidermophyton floccosum, T. mentagrophytes*, or *T. rubrum*. Factors predisposing one to recurrent disease are moisture, occlusion, and skin trauma. Wet bathing suits, athletic supporters (**jock itch**), tight-fitting slacks, panty hose, and obesity are frequently contributing factors.

Tinea pedis [Latin *pes*, the foot], also known as **athlete's foot,** and **tinea manuum** [Latin *mannus*, the hand] are dermatophytic infections of the feet (**figure 39.15**) and hands, respectively. Clinical symptoms vary from a fine scale to a vesiculopustular eruption. Itching is frequently present. Warmth, humidity, trauma, and occlusion increase susceptibility to infection. Most infections are caused by *T. rubrum, T. mentagrophytes*, or *E. floccosum*. Tinea pedis and tinea manuum occur throughout the world, are most commonly found in adults, and increase in frequency with age.

Tinea unguium [Latin *unguis*, nail] is a dermatophytic infection of the nail bed (**figure 39.16**). In this disease the nail becomes discolored and then thickens. The nail plate rises and separates from the nail bed. *Trichophyton rubrum* or *T. mentagrophytes* are the causative fungi.

Subcutaneous Mycoses

The dermatophytes that cause subcutaneous mycoses are normal saprophytic inhabitants of soil and decaying vegetation. Because they are unable to penetrate the skin, they must be introduced into the subcutaneous tissue by a puncture wound. Most infections involve barefooted agricultural workers. Once in the subcutaneous tissue, the disease develops slowly—often over a period of years. During this time the fungi produce a nodule that eventually ulcerates and the organisms spread along lymphatic channels, producing more subcutaneous nodules. At times, such nodules drain to the skin surface. The administration of oral 5-fluorocytosine,

Figure 39.13   Cutaneous Mycosis: Tinea Corporis.
Ringworm of the body—in this case, the forearm—caused by *Trichophyton mentagrophytes*. Notice the circular patches (arrows).

Figure 39.14   Cutaneous Mycosis: Tinea Cruris.   Ringworm of the groin caused by *Epidermophyton floccosum*.

Figure 39.15   Cutaneous Mycosis: Tinea Pedis.   Ringworm of the foot caused by *Trichophyton rubrum, T. mentagrophytes*, or *Epidermophyton floccosum*.

Figure 39.16 Cutaneous Mycosis: Tinea Unguium.
Ringworm of the nails caused by *Trichophyton rubrum*.

Figure 39.17 Subcutaneous Mycosis. Chromoblastomycosis of the foot caused by *Fonsecaea pedrosoi*.

Figure 39.18 Subcutaneous Mycosis. Eumycotic mycetoma of the foot caused by *Madurella mycetomatis*.

Figure 39.19 Subcutaneous Mycosis. Sporotrichosis of the arm caused by *Sporothrix schenckii*.

iodides, or amphotericin B, and surgical excision, are the usual treatments. Diagnosis is accomplished by culture of the infected tissue.

One type of subcutaneous mycosis is **chromoblastomycosis.** The nodules are dark brown. This disease is caused by the black molds *Phialophora verrucosa* and *Fonsecaea pedrosoi*. These fungi exist worldwide, especially in tropical and subtropical regions. Most infections involve the legs and feet (**figure 39.17**). Another subcutaneous mycosis is **maduromycosis,** caused by *Madurella mycetomatis,* which is distributed worldwide and is especially prevalent in the tropics. Because the fungus destroys subcutaneous tissue and produces serious deformities, the resulting infection is often called a **eumycotic mycetoma,** or fungal tu-

mor (**figure 39.18**). One form of mycetoma, known as Madura foot, occurs through skin abrasions acquired while walking barefoot on contaminated soil.

Sporotrichosis is the subcutaneous mycosis caused by the dimorphic fungus *Sporothrix schenckii*. The disease occurs throughout the world and is the most common subcutaneous mycotic disease in the United States. The fungus can be found in the soil, on living plants, such as barberry shrubs and roses, or in plant debris, such as sphagnum moss, baled hay, and pine-bark mulch. Infection occurs by a puncture wound from a thorn or splinter contaminated with the fungus. The disease is an occupational hazard for florists, gardeners, and forestry workers. It is not spread from person to person. After an incubation period of 1 to 12 weeks, a small, red papule arises and begins to ulcerate (**figure 39.19**). New lesions appear along lymph channels and can remain localized or spread throughout the body, producing **extracutaneous sporotrichosis.**

Sporotrichosis is typically treated by ingestion of potassium iodide or itraconazole (Sporanox) until the lesions are healed, usually several weeks. Preventative measures include gloves and other protective clothing, as well as avoidance of contaminated landscaping materials, especially sphagnum moss.

Toxoplasmosis

Toxoplasmosis is a disease caused by the protist *Toxoplasma gondii*. This apicomplexan protist has been found in nearly all animals and most birds; cats are the definitive host and are required for completion of the sexual cycle (**figure 39.20**). Animals shed oocysts in the feces; the oocysts enter another host by way of the nose or mouth; and the parasites colonize the intestine. Toxoplasmosis also can be transmitted by the ingestion of raw or undercooked meat, congenital transfer, blood transfusion, or a tissue transplant. Originally toxoplasmosis gained public notice when it was discovered that in pregnant women the protist might also infect the fetus, causing serious congenital defects or death. Most cases of toxoplasmosis are asymptomatic. Adults usually complain of an "infectious mononucleosis-like" syndrome. In immunoincompetent or immunosuppressed individuals, it frequently results in fatal disseminated disease with a heavy cerebral involvement.

Acute toxoplasmosis is usually accompanied by lymph node swelling (lymphadenopathy) with reticular cell hyperplasia (enlargement). Pulmonary necrosis, myocarditis, and hepatitis caused by tissue necrosis are common. Retinitis (inflammation of the retina of the eye) is associated with necrosis due to the proliferation of the parasite within retinal cells. Toxoplasmosis is found worldwide, although most of those infected do not exhibit symptoms of the disease because the immune system usually prevents illness. However, *Toxoplasma* in the immunocompromised, such

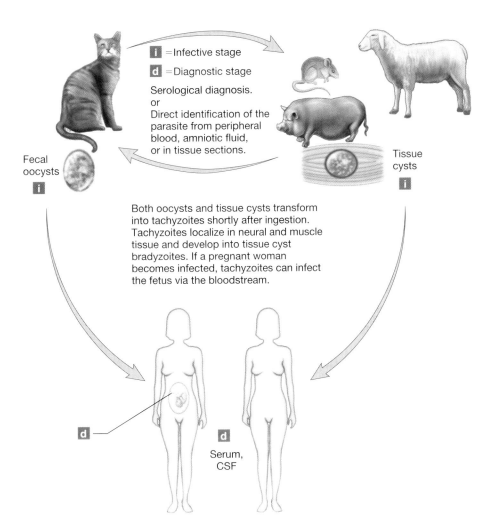

Figure 39.20   Life Cycle of *Toxoplasma gondii*. *Toxoplasma gondii* is a protozoan parasite of numerous mammals and birds. However, the cat, the definitive host, is required for completion of the sexual cycle. Oocysts are shed in the feces of infected animals where they may be ingested by another host. Ingested oocysts transform into tachyzoites, which migrate to various tissue sites via the bloodstream.

as AIDS or transplant patients, can produce a unique encephalitis with necrotizing lesions accompanied by inflammatory infiltrates. It continues to cause more than 3,000 congenital infections per year in the United States. Protist classification: *Alveolata* (section 25.6)

Laboratory diagnosis of toxoplasmosis is by serological tests. Epidemiologically, toxoplasmosis is ubiquitous in all higher animals. Treatment of toxoplasmosis is with a combination of pyrimethamine (Daraprim) and sulfadiazine. Prevention and control require minimizing exposure by the following: avoiding eating raw meat and eggs, washing hands after working in the soil, cleaning cat litterboxes daily, keeping household cats indoors if possible, and feeding them commercial food.

Trichomoniasis

Trichomoniasis is a sexually transmitted disease caused by the protozoan flagellate *Trichomonas vaginalis* (**figure 39.21**). It is one of the most common sexually transmitted diseases, with an estimated 7 million cases annually in the United States and 180 million cases annually worldwide. In response to the parasite, the body accumulates leukocytes at the site of the infection. In females this usually results in a profuse, purulent vaginal discharge that is yellowish to light cream in color and characterized by a disagreeable odor. The discharge is accompanied by itching. Males are generally asymptomatic because of the trichomonacidal action of prostatic secretions; however, at times a burning sensation occurs during urination. Diagnosis is made in females by microscopic examination of the discharge and identification of the protozoan. Infected males demonstrate protozoa in semen or urine. Treatment is by administration of metronidazole (Flagyl). Protist classification: *Parabasalia* (section 25.6)

---

1. Describe two piedras that infect humans?
2. Briefly describe the major tineas that occur in humans.
3. Describe the three types of subcutaneous mycoses that affect humans.
4. In what two ways does *Toxoplasma* affect human health?
5. How would you diagnose trichomoniasis in a female? In a male?

39.5 FOOD-BORNE AND WATERBORNE DISEASES

Amebiasis

It is now accepted that two species of *Entamoeba* infect humans: the nonpathogenic *E. dispar* and the pathogenic *E. histolytica*. *E. histolytica* is responsible for **amebiasis (amebic dysentery).** This very common parasite is endemic in warm climates where adequate sanitation and effective personal hygiene are lacking. Within the United States about 3,000 to 5,000 cases are reported annually. However, it is a major cause of parasitic death worldwide; about 500 million people are infected and as many as 100,000 die of amebiasis each year. Protists classification: *Entamoebida* (section 25.6)

Infection occurs by ingestion of mature cysts from fecally contaminated water, food, or hands, or from fecal exposure during sexual contact. After excystation in the lower region of the small intestine, the metacyst divides rapidly to produce eight small trophozoites (**figure 39.22**). These trophozoites move to the large intestine where they can invade the host tissue, live as commensals in the lumen of the intestine, or undergo encystation. In many hosts, the trophozoites remain in the intestinal lumen, resulting in an asymptomatic carrier state with cysts shed with the feces.

If the infective trophozoites invade the intestinal tissues, they multiply rapidly and spread laterally, while feeding on erythrocytes, bacteria, and yeasts. The invading trophozoites destroy the epithelial lining of the large intestine by producing a cysteine protease. This protease is a virulence factor of *E. histolytica* and may play a role in intestinal invasion by degrading the extracellular matrix and circumventing the host immune response through cleavage of secretory immunoglobulin A (sIgA), IgG, and complement factors. These cysteine proteases are encoded by at least seven genes, several of which are found in *E. histolytica* but not *E. dispar*. Lesions (ulcers) are characterized by minute points of entry into the mucosa and extensive enlargement of the lesion after penetration into the submucosa. *E. histolytica* also may invade

Figure 39.21 Trichomoniasis. *Trichomonas vaginalis,* showing the characteristic undulating membranes and flagella; scanning electron micrograph (×12,000).

(a)

(b)

(c)

Figure 39.22   Amebiasis Caused by *Entamoeba histolytica*. **(a)** Light micrographs of a trophozoite (×1,000) and **(b)** a cyst (×1,000). **(c)** Life cycle. Infection occurs by the ingestion of a mature cyst. Excystment occurs in the lower region of the small intestine and the metacyst rapidly divides to give rise to eight small trophozoites (only four are shown). These enter the large intestine, undergo binary fission, and may (1) invade the host tissues, (2) live in the lumen of the large intestine without invasion, or (3) undergo encystment and pass out of the host in the feces.

and produce lesions in other tissues, especially the liver, to cause hepatic amebiasis. However, all extraintestinal amebic lesions are secondary to those established in the large intestine. The symptoms of amebiasis are highly variable, ranging from an asymptomatic infection to fulminating dysentery (exhaustive diarrhea accompanied by blood and mucus), appendicitis, and abscesses in the liver, lungs, or brain.

Laboratory diagnosis of amebiasis can be difficult and is based on finding trophozoites in fresh, warm stools and cysts in ordinary stools. Serological testing for *E. histolytica* is available but often unreliable. The therapy for amebiasis is complex and depends on the location of the infection within the host and the host's condition. Asymptomatic carriers who are passing cysts should always be treated with iodoquinol or paromomycin because they represent the most important reservoir of the protozoan in the population. In symptomatic intestinal amebiasis, metronidazole (Flagyl)

or iodoquinol (Yodoxin) are the drugs of choice. Prevention and control of amebiasis is achieved by practicing good personal hygiene and avoiding water or food that might be contaminated with human feces. Viable cysts in water can be destroyed by hyperchlorination or iodination. Antiprotozoan drugs (section 34.9)

Amebic Meningoencephalitis

Free-living amoebae of the genera *Naegleria* and *Acanthamoeba* are facultative (opportunistic) parasites responsible for causing **primary amebic meningoencephalitis** in humans. They are among the most common protists found in freshwater and moist soil. In addition, several *Acanthamoeba* spp. are known to infect the eye, causing a chronically progressive, ulcerative *Acanthamoeba* **keratitis**—inflammation of the cornea—that may result in blindness. Wearers of soft contact lenses may be

Historical Highlights

40.1 An Army Travels on Its Stomach

The movement and maintenance of large numbers of military personnel have always been limited by food supplies. The need to maintain large numbers of troops under hostile and inclement conditions led the French government in 1795 to offer a prize of 12,000 francs to the individual who could preserve foods for use under field conditions. Eventually the prize was awarded to Nicholas Appert, a candy maker, for his development of a heating process in which meats and other products could be preserved under sealed conditions.

Appert's work was based on the assumptions that heating and boiling control "ferments" and that sealing the food in bottles be-

fore heating it avoids the effects of air on spoilage. Despite Leeuwenhoek's earlier work, Appert did not have the concept of microorganisms to assist him in explaining the effectiveness of his process. His containers were large glass bottles, sealed with laminated corks and fish glue. With extreme care and attention to detail, he was able to heat these bottles in boiling water to provide food that could be stored for several years. Appert's work was an important foundation for the later studies of Louis Pasteur.

Figure 40.6 Food Preparation for Canning. Microbial control is important in the processing and preservation of many foods. Worker pouring peas into a large, clean vat during the preparation of vegetable soup. After preparation the soup is transferred to cans. Each can is heated for a short period, sealed, processed at temperatures around 110–121°C in a canning retort to destroy spoilage microorganisms, and finally cooled.

processing of foods, especially with less acidic (pH values greater than 4.6) products such as green beans or meats. The use of physical methods in control (section 7.4)

Despite efforts to eliminate spoilage microorganisms during canning, sometimes canned foods become spoiled. This may be due to spoilage before canning, underprocessing during canning, and leakage of contaminated water through can seams during cooling. Spoiled food can be altered in such characteristics as color, texture, odor, and taste. Organic acids, sulfides, and gases (particularly CO_2 and H_2S) may be produced. If spoilage microorganisms produce gas, both ends of the can will bulge outward. Sometimes the swollen ends can be moved by thumb pressure (soft swells); in other cases the gas pressure is so great that the ends cannot be dented by hand (hard swells).

It should be noted that swelling is not always due to microbial spoilage. Acid in high-acid foods may react with the iron of the can to release hydrogen and generate a hydrogen swell. Hydrogen sulfide production by *Desulfotomaculum* can cause "sulfur stinkers."

Pasteurization involves heating food to a temperature that kills disease-causing microorganisms and substantially reduces the levels of spoilage organisms. In the processing of milk, beers, and fruit juices by conventional low-temperature holding (LTH) pasteurization, the liquid is maintained at 62.8°C for 30 minutes. Products can also be held at 71°C for 15 seconds, a high-temperature, short-time (HTST) process; milk can be treated at 141°C for 2 seconds for ultra-high-temperature (UHT) processing. Shorter-term processing results in improved flavor and extended product shelf life. Such heat treatment is based on a statistical probability that the number of remaining viable microorganisms will be below a certain level after a particular heating time at a specific temperature. This process is discussed in detail in section 7.4.

Water Availability

Dehydration, such as lyophilization to produce freeze-dried foods, is a common means of eliminating microbial growth. The modern process is simply an update of older procedures in which grains, meats, fish, and fruits were dried. The combination of free-water loss with an increase in solute concentration in the remaining water makes this type of preservation possible.

Chemical-Based Preservation

Various chemical agents can be used to preserve foods, and these substances are closely regulated by the U.S. Food and Drug Administration and are listed as being "generally recognized as safe" or **GRAS (table 40.5)**. They include simple organic acids, sulfite, ethylene oxide as a gas sterilant, sodium nitrite, and ethyl formate. These chemical agents may damage the microbial plasma membrane or denature various cell proteins. Other compounds

| Table 40.5 | Major Groups of Chemicals Used in Food Preservation | | |

Preservatives	Approximate Maximum Use Range	Organisms Affected	Foods
Propionic acid/propionates	0.32%	Molds	Bread, cakes, some cheeses, inhibitor of ropy bread dough
Sorbic acid/sorbates	0.2%	Molds	Hard cheeses, figs, syrups, salad dressings, jellies, cakes
Benzoic acid/benzoates	0.1%	Yeasts and molds	Margarine, pickle relishes, apple cider, soft drinks, tomato ketchup, salad dressings
Parabens[a]	0.1%	Yeasts and molds	Bakery products, soft drinks, pickles, salad dressings
SO_2/sulfites	200–300 ppm	Insects and microorganisms	Molasses, dried fruits, wine, lemon juice (not used in meats or other foods recognized as sources of thiamine)
Ethylene/propylene oxides	700 ppm	Yeasts, molds, vermin	Fumigant for spices, nuts
Sodium diacetate	0.32%	Molds	Bread
Dehydroacetic acid	65 ppm	Insects	Pesticide on strawberries, squash
Sodium nitrite	120 ppm	Clostridia	Meat-curing preparations
Caprylic acid	—	Molds	Cheese wraps
Ethyl formate	15–200 ppm	Yeasts and molds	Dried fruits, nuts

From James M. Jay. 2000. *Modern Food Microbiology*, 6th edition. Reprinted by permission of Aspen Publishing, Frederick, Md.

[a]Methyl-, propyl-, and heptyl-esters of *p*-hydroxybenzoic acid.

interfere with the functioning of nucleic acids, thus inhibiting cell reproduction.

The effectiveness of many of these chemical preservatives depends on the food pH. As an example, sodium propionate is most effective at lower pH values, where it is primarily undissociated and able to be taken up by lipids of microorganisms. Breads, with their low pH values, often contain sodium propionate as a preservative. Chemical preservatives are used with grain, dairy, vegetable, and fruit products. Sodium nitrite is an important chemical used to help preserve ham, sausage, bacon, and other cured meats by inhibiting the growth of *Clostridium botulinum* and the germination of its spores. This protects against botulism and reduces the rate of spoilage. Besides increasing meat safety, nitrite decomposes to nitric acid, which reacts with heme pigments to keep the meat red in color. Current concern about nitrite arises from the observation that it can react with amines to form carcinogenic nitrosamines.

Radiation

Radiation, both ionizing and nonionizing, has an interesting history in relation to food preservation. Ultraviolet radiation is used to control populations of microorganisms on the surfaces of laboratory and food-handling equipment, but it does not penetrate food. The major method used for radiation sterilization of food is gamma irradiation from a cobalt-60 source; however, cesium-137 is used in some facilities. Gamma radiation has excellent penetrating power, but must be used with moist foods because the radiation produces peroxides from water in the microbial cells,

resulting in oxidation of sensitive cellular constituents. This process of **radappertization,** named after Nicholas Appert, can extend the shelf life of seafoods, fruits, and vegetables. To sterilize meat products, commonly 4.5 to 5.6 megarads are used.

Electron beams can also be used to irradiate foods. The electrons are generated electrically, so they can be turned on only when needed. Also, this approach does not generate radioactive waste. On the other hand, electron beams do not penetrate food items as deeply as does gamma radiation. It is important to note that regardless of the radiation source (gamma rays or electron beams), the food itself does not become radioactive.

Microbial Product-Based Inhibition

There is increasing interest in the use of **bacteriocins** for the preservation of foods. Bacteriocins are bactericidal proteins active against closely related bacteria, which bind to specific sites on the cell, and often affect cell membrane integrity and function. The only currently approved product is nisin. Nisin, produced by some strains of *Lactococcus lactis*, is a small hydrophobic protein. It is nontoxic to humans and affects mainly gram-positive bacteria, especially *Enterococcus faecalis*. Nisin can be used particularly in low-acid foods to improve inactivation of *Clostridium botulinum* during the canning process or to inhibit germination of any surviving spores.

Bacteriocins have a wide variety of names, depending on the organisms that produce them. They can function by several mechanisms. They may dissipate the proton motive force (PMF) of

Figure 40.13 Examples of Bifid-Amended Dairy Products.
These are produced in many countries.

B-complex vitamins. It has also been suggested that bifidobacteria reduce or prevent the excretion of rotaviruses, a cause of diarrhea among children. *Bifidobacterium*-amended fermented milk products, including yogurt, are now available in various parts of the world (**figure 40.13**).

Yeast-Lactic Fermentation

Yeast-lactic fermentations include **kefir,** a product with an ethanol concentration of up to 2%. This unique fermented milk originated in the Caucasus Mountains and it is produced east into Mongolia. Kefir products tend to be foamy and frothy, due to active carbon dioxide production. This process is based on the use of kefir "grains" as an inoculum. These are coagulated lumps of casein that contain yeasts, lactic acid bacteria, and acetic acid bacteria. In this fermentation, the grains are used to inoculate the fresh milk and then recovered at the end of the fermentation. Originally, kefir was produced in leather sacks hung by the front door during the day, and passersby were expected to push and knead the sack to mix and stimulate the fermentation. Fresh milk could be added occasionally to maintain activity.

Mold-Lactic Fermentation

Mold-lactic fermentation results in a unique Finnish fermented milk called viili. The milk is placed in a cup and inoculated with a mixture of the fungus *Geotrichium candidum* and lactic acid bacteria. The cream rises to the surface, and after incubation at 18 to 20°C for 24 hours, lactic acid reaches a concentration of 0.9%. The fungus forms a velvety layer across the top of the final product, which also can be made with a bottom fruit layer.

Cheese Production

Cheese is one of the oldest human foods and is thought to have been developed approximately 8,000 years ago. About 2,000 distinct varieties of cheese are produced throughout the world, representing ap-

proximately 20 general types (**table 40.8** and **figure 40.14**). Often cheeses are classified based on texture or hardness as soft cheeses (cottage, cream, Brie), semisoft cheeses (Muenster, Limburger, blue), hard cheeses (cheddar, Colby, Swiss), or very hard cheeses (Parmesan). All cheese results from a lactic acid fermentation of milk, which results in coagulation of milk proteins and formation of a curd. Rennin, an enzyme from calf stomachs, but now produced by genetically engineered microorganisms, can also be used to promote curd formation. After the curd is formed, it is heated and pressed to remove the watery part of the milk (called the whey), salted, and then usually ripened (**figure 40.15**). The cheese curd can be packaged for ripening with or without additional microorganisms.

Lactococcus lactis is used as a starter culture for a number of cheeses including Gouda (figure 40.14*a*) and cheddar (figure 40.15). Starter culture density is often over 10^9 colony-forming units (CFUs) per gram of cheese before ripening. However, the high salt, low pH, and the temperatures that characterize the cheese microenvironment reduce these numbers rather quickly. This enables other bacteria, sometimes called **nonstarter lactic acid bacteria (NSLAB)** to grow; their numbers can reach 10^7 to 10^9 CFUs/g after several months of aging. Thus both starter and nonstarter LAB contribute to the final taste, texture, odor, and appearance of the cheese.

In some cases, molds are used to further enhance the cheese. Obvious examples are Roquefort and blue cheese. For these cheeses, *Penicillium roqueforti* spores are added to the curds just before the final cheese processing. Sometimes the surface of an already formed cheese is inoculated at the start of ripening; for example, Camembert cheese is inoculated with spores of *Penicillium camemberti*. The final hardness of the cheese is partially a function of the length of ripening. Soft cheeses are ripened for only about 1 to 5 months, whereas hard cheeses need 3 to 12 months, and very hard cheeses like Parmesan require 12 to 16 months ripening.

The ripening process also is critical for Swiss cheese. Gas production by *Propionibacterium* contributes to final flavor development and hole or eye formation in this cheese. Some cheeses are soaked in brine to stimulate the development of specific fungi and bacteria; Limburger is one such cheese.

Meat and Fish

Besides the fermentation of dairy products, a variety of meat products, especially sausage, can be fermented: country-cured hams, summer sausage, salami, cervelat, Lebanon bologna, fish sauces (processed by halophilic *Bacillus* species), izushi, and katsuobushi. *Pediococcus cerevisiae* and *Lactobacillus plantarum* are most often involved in sausage fermentations. Izushi is based on the fermentation of fresh fish, rice, and vegetables by *Lactobacillus* spp.; katsuobushi results from the fermentation of tuna by *Aspergillus glaucus*. Both meat fermentations originated in Japan.

1. What are the major types of milk fermentations?
2. Briefly describe how buttermilk, sour cream, and yogurt are made.
3. What is unique about the morphology of *Bifidobacterium?* Why is it used in milk?

Table 40.8	Major Types of Cheese and Microorganisms Used in Their Production	
	Contributing Microorganisms[a]	
Cheese (Country of Origin)	**Earlier Stages of Production**	**Later Stages of Production**
Soft, unripened		
Cottage	*Lactococcus lactis*	*Leuconostoc cremoris*
Cream	*L. cremoris, L. diacetylactis, Streptococcus thermophilus,* *L. delbrueckii* subspecies *bulgaricus*	
Mozzarella (Italy)	*S. thermophilus, L. bulgaricus*	
Soft, ripened		
Brie (France)	*Lactococcus lactis, L. cremoris*	*Penicillium camemberti,* *P. candidum,* *Brevibacterium linens*
Camembert (France)	*L. lactis, L. cremoris*	*Penicillium camemberti,* *B. linens*
Semisoft		
Blue, Roquefort (France)	*Lactococcus lactis, L. cremoris*	*P. roqueforti*
Brick, Muenster (United States)	*L. lactis, L. cremoris*	*B. linens*
Limburger (Belgium)	*L. lactis, L. cremoris*	*B. linens*
Hard, ripened		
Cheddar, Colby (Britain)	*Lactococcus lactis, L. cremoris*	*Lactobacillus casei, L. plantarum*
Swiss (Switzerland)	*L. lactis, L. helveticus, S. thermophilus*	*Propionibacterium shermanii,* *P. freudenreichii*
Very hard, ripened		
Parmesan (Italy)	*Lactococcus lactis, L. cremoris, S. thermophilus*	*Lactobacillus bulgaricus*

[a]*Lactococcus lactis* stands for *L. lactis* subsp. *lactis. Lactococcus cremoris* is *L. lactis* subsp. *cremoris,* and *Lactococcus diacetilactis* is *L. lactis* subsp. *diacetilactis.*

4. How and where are kefir and viili made?
5. What major steps are used to produce cheese? How is the cheese curd formed in this process? What is whey? How does Swiss cheese get its holes?
6. Which fungal genus is often used in cheese making? What cheeses are produced using this genus?
7. Give a microbial genus used in meat fermentations.

Production of Alcoholic Beverages

A variety of plants that contain adequate carbohydrates can be used to produce alcoholic beverages. When carbohydrates are available in readily fermentable form, the fermentation can be started immediately. For example, grapes are crushed to release the juice or **must,** which can be allowed to ferment without further delay. The must also can be treated by pasteurization or the use of sulfur dioxide, and then the desired microbial culture added.

In contrast, before cereals and other starchy materials can be used as substrates for the production of alcohol, their complex carbohydrates must be hydrolyzed. They are mixed with water and incubated in a process called **mashing.** The insoluble material is then removed to yield the **wort,** a clear liquid containing fermentable sugars and other simple molecules. Much of the art of beer and ale production involves the controlled hydrolysis of protein and carbohydrates to provide the desired body and flavor of the final product.

Wines and Champagnes

Wine production, or the science of **enology** [Greek *oinos,* wine, and *ology,* the science of], starts with the collection of grapes, continues with their crushing and the separation of the liquid (must) before fermentation, and concludes with a variety of storage and aging steps (**figure 40.16**). All grapes have white juices. To make a red wine from a red grape, the grape skins are allowed to remain in contact with the must before fermentation to release their skin-coloring components. Wines can be produced by using the natural grape skin microorganisms, but this natural mixture of bacteria and yeasts gives unpredictable fermentation results. To avoid such problems, fresh must is treated with a sulfur dioxide fumigant and a desired strain of the yeast *Saccharomyces cerevisiae* or *S. ellipsoideus* is added. After inoculation the juice is fermented for 3 to 5 days at temperatures between 20 and 28°C. Depending on the alcohol tolerance of the yeast strain (the alcohol eventually kills the yeast that produced it), the final product may contain 10 to 14% alcohol. Clearing and development of flavor occur during the aging process. The malolactic fermentation is an important part of wine production. Grape juice contains high levels of organic acids, including malic and tartaric acids. If the levels of these acids are not decreased during the fermentation process, the wine will be too acidic, and have poor stability and "mouth feel." This essential fermentation is carried out by the bacteria *Leuconostoc oenos, L. plantarum, L. hilgardii, L. brevis,* and *L. casei.* The activities of

Figure 40.14 Cheese. A vast array of cheeses are produced around the world using microorganisms. **(a)** Gouda (top left) and cheddar cheese (lower right). Note the typical indentations on the surface of Gouda caused by the cheesecloth and red wax covering. **(b)** Roquefort cheese crumbled for use in salad dressing. The dark areas are the result of extensive *Penicillium* growth. **(c)** Swiss cheese, a hard, ripened cheese, contains holes formed by carbon dioxide from a *Propionibacterium* fermentation. **(d)** Brie (left) and Limburger (right) cheeses are soft, ripened cheeses. Ripening results from the surface growth of microorganisms like *Penicillium camemberti* (Brie) and *Brevibacterium linens* (Limburger). **(e)** Cottage cheese and cream cheese (spread on crackers) are soft, unripened cheeses. They are sold immediately after production, and the curd is consumed without further modification by microorganisms.

these microbes transform malic acid (a four-carbon tricarboxylic acid) to lactic acid (a three-carbon monocarboxylic acid) and carbon dioxide. This results in deacidification (pH increase), improvement of flavor stability, and in some cases the possible accumulation of bacteriocins in the wines. Characteristics of the fungal division: *Ascomycota* (section 26.6); Fermentations (section 9.7)

A critical part of wine making involves the choice of whether to produce a dry (no remaining free sugar) or a sweeter (varying amounts of free sugar) wine. This can be controlled by regulating the initial must sugar concentration. With higher levels of sugar, alcohol will accumulate and inhibit the fermentation before the sugar can be completely used, thus producing a sweeter wine.

Figure 40.15 Cheddar Cheese Production. Cheddar, a village in England, has given its name to a cheese made in many parts of the world. "Cheddaring" is the process of turning and piling the curd to express whey and develop desired cheese texture.

During final fermentation in the aging process, flavoring compounds accumulate and influence the bouquet of the wine.

Microbial growth during the fermentation process produces sediments, which are removed during **racking.** Racking can be carried out at the time the fermented wine is transferred to bottles or casks for aging or even after the wine is placed in bottles.

Many processing variations can be used during wine production. The wine can be distilled to make a "burned wine" or brandy. *Acetobacter* and *Gluconobacter* can be allowed to oxidize the ethanol to acetic acid and form a **wine vinegar.** In the past an acetic acid generator was used to recirculate the wine over a bed of wood chips, where the desired microorganisms developed as a surface growth. Today the process is carried out in large aerobic submerged cultures under much more controlled conditions.

Natural champagnes are produced by continuing the fermentation in bottles to produce a naturally sparkling wine. Sediments that remain are collected in the necks of inverted champagne bottles after the bottles have been carefully turned. The necks of the bottles are then frozen and the corks removed to disgorge the accumulated sediments. The bottles are refilled with clear champagne from another disgorged bottle, and the product is ready for final packaging and labeling.

Beers and Ales

Beer and ale production uses cereal grains such as barley, wheat, and rice. The complex starches and proteins in these grains must be changed to a more readily usable mixture of simpler carbohydrates and amino acids. This process, shown in **figure 40.17,** involves germination of the barley grains and activation of their enzymes to pro-

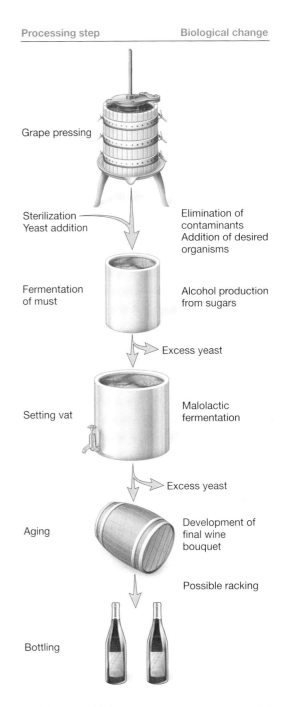

Figure 40.16 Wine Making. Once grapes are pressed, the sugars in the juice (the must) can be immediately fermented to produce wine. Must preparation, fermentation, and aging are critical steps.

duce a **malt.** The malt is then mixed with water and the desired grains, and the mixture is transferred to the mash tun or cask in order to hydrolyze the starch to usable carbohydrates. Once this process is completed, the **mash** is heated with hops (dried flowers of the female vine *Humulus lupulis*), which were originally added to the mash to inhibit spoilage microorganisms (**figure 40.18**). The hops also provide flavor and assist in clarification of the wort. In this heating step the hydrolytic enzymes are inactivated and the wort can be **pitched**—inoculated—with the desired yeast.

Processing step		Biological change

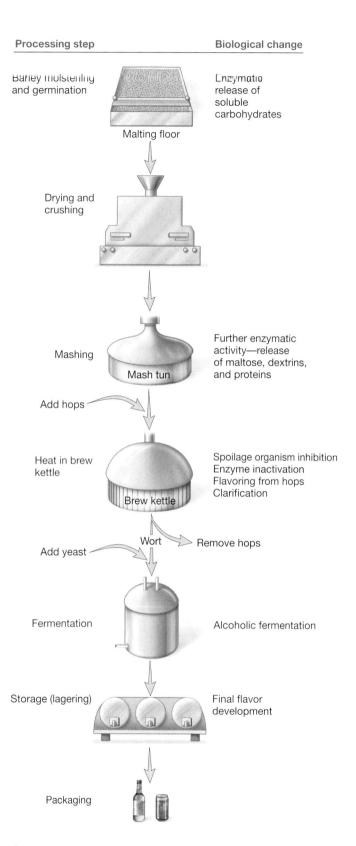

Barley moistening and germination — Enzymatic release of soluble carbohydrates

Malting floor

Drying and crushing

Mashing — Further enzymatic activity—release of maltose, dextrins, and proteins

Mash tun

Add hops

Heat in brew kettle — Spoilage organism inhibition / Enzyme inactivation / Flavoring from hops / Clarification

Brew kettle

Wort → Remove hops

Add yeast

Fermentation — Alcoholic fermentation

Storage (lagering) — Final flavor development

Packaging

Figure 40.17 Producing Beer. To make beer, the complex carbohydrates in the grain must first be transformed into a fermentable substrate. Beer production thus requires the important steps of malting, and the use of hops and boiling for clarification, flavor development, and inactivation of malting enzymes, to produce the wort.

Figure 40.18 Brew Kettles Used for Preparation of Wort. In large-scale processes, copper brew kettles can be used for wort preparation, as shown here at the Carlsberg Brewery, Copenhagen, Denmark.

Most beers are fermented with **bottom yeasts,** related to *Saccharomyces pastorianus,* which settle at the bottom of the fermentation vat. The beer flavor also is influenced by the production of small amounts of glycerol and acetic acid. Bottom yeasts require 7 to 12 days of fermentation to produce beer with a pH of 4.1 to 4.2. With a top yeast, such as *Saccharomyces cerevisiae,* the pH is lowered to 3.8 to produce ales. Freshly fermented (green) beers are aged or **lagered,** and when they are bottled, CO_2 is usually added. Beer can be pasteurized at 140°F or higher or sterilized by passage through membrane filters to minimize flavor changes.

Distilled Spirits

Distilled spirits are produced by an extension of beer production processes. The fermented liquid is boiled, and the volatile components are condensed to yield a product with a higher alcohol content than beer. Rye and bourbon are examples of whiskeys. Rye whiskey must contain at least 51% rye grain, and bourbon must contain at least 51% corn. Scotch whiskey is made primarily of barley. Usually a **sour mash** is used; the mash is inoculated with a homolactic (lactic acid is the major fermentation product) bacterium such as *Lactobacillus delbrueckii* subspecies *bulgaricus* (figure 40.11*b*), which can lower the mash pH to around 3.8 in 6 to 10 hours. This limits the development of undesirable organisms. Vodka and grain alcohols are also produced by distillation. Gin is vodka to which resinous flavoring agents—often juniper berries—have been added to provide a unique aroma and flavor.

Production of Breads

Bread is one of the most ancient of human foods, and is produced with the help of microorganisms. The use of yeasts to leaven bread is carefully depicted in paintings from ancient Egypt. A bakery at the Giza Pyramid area, from the year 2575 B.C., has been excavated. It is estimated that 30,000 people a day were provided with bread from this bakery. Samples of bread from

2100 B.C. are on display in the British Museum. In breadmaking, yeast growth is carried out under aerobic conditions. This results in increased CO_2 production and minimum alcohol accumulation. The fermentation of bread involves several steps: alpha- and beta-amylases present in the moistened dough release maltose and sucrose from starch. Then a baker's strain of the yeast *Saccharomyces cerevisiae*, which produces maltase, invertase, and zymase enzymes, is added. The CO_2 produced by the yeast results in the light texture of many breads, and traces of fermentation products contribute to the final flavor. Usually bakers add sufficient yeast to allow the bread to rise within 2 hours—the longer the rising time, the more additional growth by contaminating bacteria and fungi can occur, making the product less desirable.

By using more complex assemblages of microorganisms, bakers can produce special breads such as sour doughs. The yeast *Saccharomyces exiguus,* together with a *Lactobacillus* species, produces the characteristic acidic flavor and aroma of such breads.

Bread products can be spoiled by *Bacillus* species that produce ropiness. If the dough is baked after these organisms have grown, stringy and ropy bread will result, leading to decreased consumer acceptance.

Other Fermented Foods

Many other plant products can be fermented, as summarized in **table 40.9.** These include sufu, which is produced by the fermentation of tofu, a chemically coagulated soybean milk product. To carry out the fermentation, the tofu curd is cut into small chunks and dipped into a solution of salt and citric acid. After the cubes are heated to pasteurize their surfaces, the fungi *Actinimucor elegans* and some *Mucor* species are added. When a white mycelium develops, the cubes, now called pehtze, are aged in salted rice wine. This product has achieved the status of a delicacy in many parts of the Western world. Another popular product is tempeh, a soybean mash fermented by *Rhizopus.*

Sauerkraut or sour cabbage is produced from wilted, shredded cabbage, as shown in **figure 40.19.** Usually the mixed microbial community of the cabbage is used. A concentration of 2.2 to 2.8% sodium chloride restricts the growth of gram-negative bacteria while favoring the development of the lactic acid bacteria. The primary microorganisms contributing to this product are *Leuconostoc mesenteroides* and *Lactobacillus plantarum.* A predictable microbial succession occurs in sauerkraut's development. The activities of the lactic acid-producing cocci usually cease when the acid content reaches 0.7 to 1.0%. At this point *Lactobacillus plantarum* and *Lactobacillus brevis* continue to function. The final acidity is generally 1.6 to 1.8, with lactic acid comprising 1.0 to 1.3% of the total acid in a satisfactory product.

Pickles are produced by placing cucumbers and such components as dill seeds in casks filled with a brine. The sodium chloride concentration begins at 5% and rises to about 16% in 6 to 9 weeks. The salt not only inhibits the growth of undesirable bacteria but also extracts water and water-soluble constituents from the cucumbers. These soluble carbohydrates are converted to lactic acid. The fermentation, which can require 10 to 12 days, involves the development of the gram-positive bacteria *L. mesenteroides, Enterococcus faecalis, Pediococcus cerevisiae, L. brevis,* and *L. plantarum. L. plantarum* plays the dominant role

Table 40.9	Fermented Foods Produced from Fruits, Vegetables, Beans, and Related Substrates		
Foods	**Raw Ingredients**	**Fermenting Microorganisms**	**Area**
Coffee	Coffee beans	*Erwinia dissolvens, Saccharomyces* spp.	Brazil, Congo, Hawaii, India
Gari	Cassava	*Corynebacterium manihot, Geotrichum* spp.	West Africa
Kenkey	Corn	*Aspergillus* spp., *Penicillium* spp., lactobacilli, yeasts	Ghana, Nigeria
Kimchi	Cabbage and other vegetables	Lactic acid bacteria	Korea
Miso	Soybeans	*Aspergillus oryzae, Zygosaccharomyces rouxii*	Japan
Ogi	Corn	*Lactobacillus plantarum, Lactococcus lactis, Zygosaccharomyces rouxii*	Nigeria
Olives	Green olives	*Leuconostoc mesenteroides, Lactobacillus plantarum*	Worldwide
Ontjom	Peanut presscake	*Neurospora sitophila*	Indonesia
Peujeum	Cassava	Molds	Indonesia
Pickles	Cucumbers	*Pediococcus cerevisiae, L. plantarum, L. brevis*	Worldwide
Poi	Taro roots	Lactic acid bacteria	Hawaii
Sauerkraut	Cabbage	*L. mesenteroides, L. plantarum, L. brevis*	Worldwide
Soy sauce	Soybeans	*Aspergillus oryzae* or *A. soyae, Z. rouxii, Lactobacillus delbrueckii*	Japan
Sufu	Soybeans	*Actinimucor elegans, Mucor* spp.	China
Tao-si	Soybeans	*A. oryzae*	Philippines
Tempeh	Soybeans	*Rhizopus oligosporus, R. oryzae*	Indonesia, New Guinea, Surinam

Adapted from James M. Jay. 2000. *Modern Food Microbiology,* 6th edition. Reprinted by permission of Aspen Publishing, Frederick, Md.

Processing step   Biological change

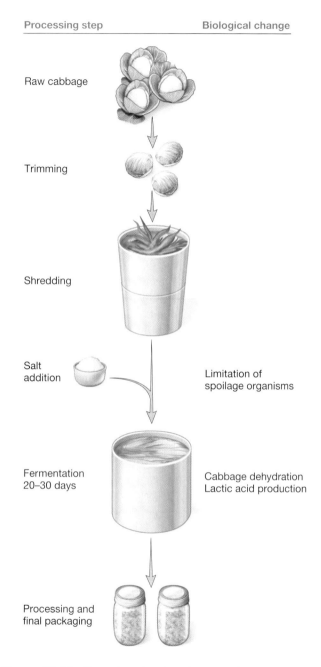

Raw cabbage

Trimming

Shredding

Salt
addition   Limitation of
                                   spoilage organisms

Fermentation   Cabbage dehydration
20–30 days   Lactic acid production

Processing and
final packaging

Figure 40.19   Sauerkraut.   Sauerkraut production employs a
lactic acid fermentation. The basic process involves fermentation of
shredded cabbage in the presence of 2.25–2.5% by weight of salt to
inhibit spoilage organisms.

in this fermentation process. Sometimes, to achieve more uniform pickle quality, natural microorganisms are first destroyed
and the cucumbers are fermented using pure cultures of *P. cerevisiae* and *L. plantarum.*

Grass, chopped corn, and other fresh animal feeds, if stored
under moist anoxic conditions, will undergo a lactic-type mixed
fermentation that produces pleasant-smelling **silage.** Trenches or
more traditional vertical steel or concrete silos are used to store
the silage. The accumulation of organic acids in silage can cause

rapid deterioration of these silos. Older wooden stave silos, if not
properly maintained, allow the outer portions of the silage to become oxic, resulting in spoilage of a large portion of the plant
material.

1. Describe and contrast the processes of wine and beer production. How
    are red wines produced when the juice of all grapes is white?
2. How do champagnes differ from wines?
3. Describe how distilled spirits like whiskey are produced.
4. How are bread, sauerkraut, and pickles produced? What microorganisms
    are most important in these fermentations?

40.7   MICROORGANISMS AS FOODS AND FOOD AMENDMENTS

Besides microorganisms' actions in fermentation as agents of
physical and biological change, they themselves can be used as a
food source. A variety of bacteria, yeasts, and other fungi have
been used as animal and human food sources. Mushrooms (*Agaricus bisporus*) are one of the most important fungi used directly
as a food source. Large caves provide optimal conditions for the
production of this delicacy (**figure 40.20**). Microorganisms can
be used directly as a food source or as a supplement to other foods
and are then called single-cell protein. One of the more popular
microbial food supplements is the cyanobacterium *Spirulina.* It is
used as a food source in Africa and is now being sold in United
States health food stores as a dried cake or powdered product.

An interesting application of probiotic microbes (primarily
Lactobacillus acidophilus) is to decrease *E. coli* occurrence in
beef cattle. The desired bacteria are sprayed on feed, and the cattle have been shown to have markedly lower (60% in some experiments) carriage of the toxic *E. coli* strain O157:H7. This can

Figure 40.20   Mushroom Farming.   Growing mushrooms
requires careful preparation of the growth medium and control of
environmental conditions. The mushroom bed is a carefully developed
compost, which can be steam sterilized to improve mushroom growth.

make it easier to produce beef that will meet current standards for microbiological quality at the time of slaughter. *Techniques & Applications 30.3: Probiotics for humans and animals*

In addition, there is a greater appreciation of the role of oligosaccharide polymers or **prebiotics,** which are not processed until they enter the large intestine. The combination of prebiotics and probiotic microorganisms is described as a **synbiotic** system. This synbiotic combination can result in an increase in the levels of butyric and propionic acids, as well as an increase in *Bifidobacterium* in the human intestine. The butryrate, especially, may play a role in the possible beneficial effects of probiotics on intestinal processes.

Probiotics are used successfully with poultry. For instance, the USDA has designated a probiotic *Bacillus* strain for use with chickens as GRAS (p. 1030). Feeding chickens a strain of *Bacillus subtilis* (Calsporin) leads to increased body weight and feed conversion. There is also a reduction in coliforms and *Campylobacter* in the processed carcasses. It has been suggested that

this probiotic decreases the need for antibiotics in poultry production and pathogen levels on farms. *Salmonella* can be controlled by spraying a patented blend of 29 bacteria, isolated from the chicken cecum, on day-old chickens. As they preen themselves, the chicks ingest the bacterial mixture, establishing a functional microbial community in the cecum and limiting *Salmonella* colonization of the gut in a process called **competitive exclusion.** In 1998, this product, called PREEMPT, was approved for use in the United States by the Food and Drug Administration.

1. What conditions are needed to have most efficient production of edible mushrooms? What is the scientific name of the most important fungus used for this purpose?
2. A cyanobacterium is widely used as a food supplement. What is this genus, and in what part of the world was it first used as a significant food source?
3. What are prebiotics, probiotics, and synbiotics?
4. What probiotic has recently been recognized as GRAS?

Summary

40.1 Microorganism Growth in Foods

a. Most foods, especially when raw, provide an excellent environment for microbial growth. This growth can lead to spoilage or preservation, depending on the microorganisms present and environmental conditions.

b. The course of microbial development in a food is influenced by the intrinsic characteristics of the food itself—pH, salt content, substrates present, water presence and availability—and extrinsic factors, including temperature, relative humidity, and atmospheric composition (**figure 40.1**).

c. Microorganisms can spoil meat, dairy products, fruits, vegetables, and canned goods in several ways. Spices, with their antimicrobial compounds, sometimes protect foods.

d. Modified atmosphere packaging (MAP) is used to control microbial growth in foods and to extend product shelf life. This process involves decreasing oxygen and increased carbon dioxide levels in the space between the food surface and the wrapping material.

40.2 Microbial Growth and Food Spoilage

a. Food spoilage is a major concern throughout the world. This can occur at any point in the food production process: growth, harvesting, transport, storage, or final preparation. Spoilage also can occur if foods are not stored properly.

b. Fungi, if they can grow in foods, especially cereals and grains, can produce important disease-causing chemicals, including the carcinogens aflatoxins (**figure 40.4**) and fumonisins (**figure 40.5**), and ergot alkaloids (mind-altering drugs).

c. Prior illness with hepatitis B can increase susceptibility to liver cancer from aflatoxins. Control of hepatitis B is suggested to be more critical than control of aflatoxins.

d. Algal toxins can be transmitted to humans by marine products. These can have severe amnesic, diarrhetic and neurotoxic effects.

40.3 Controlling Food Spoilage

a. Foods can be preserved in a variety of physical and chemical ways, including filtration, alteration of temperature (cooling, pasteurization, sterilization), drying, the addition of chemicals, radiation, and fermentation (**table 40.4**).

b. There is an increasing interest in using bacteriocins for food preservation. Nisin, a product of *Lactococcus lactis,* is the major substance approved for use in foods. Bacteriocins are especially important for control of *Listeria monocytogenes.*

40.4 Food-Borne Diseases

a. Foods can be contaminated by pathogens at any point in the food production, storage, or preparation processes. Pathogens such as *Salmonella, Campylobacter, Listeria,* and *E. coli* can be transmitted by the food to the susceptible consumer, where they grow and cause disease, or a food-borne infection (**table 40.6**). If the pathogen grows in the food before consumption and forms toxins that affect the food consumer without further microbial growth, the disease is a food-borne intoxication. Examples are intoxications caused by *Staphylococcus, Clostridium,* and *Bacillus.*

b. Noroviruses, *Campylobacter,* and *Salmonella* are thought to be the most important causes of food-borne illness. The cause of most food-borne illnesses is not known.

c. *E. coli* O157:H7 is an enterohemorrhagic bacterium that produces shigalike verocytotoxins, which especially affect the young. Proper food handling and thorough cooking are critical in control.

d. New variant Creutzfeldt-Jakob disease (vCJD) is of increasing worldwide concern as a food-borne infectious agent, which is related to the occurrence of "mad cow disease." The major means of vCJD transmission between animals is the use of mammalian tissue in ruminant animal feeds. There are difficulties in detecting such prohibited animal products in ruminant feeds.

e. Raw foods such as sprouts, seafood, and raspberries provide routes for disease transmission. Increases in the international shipment of fresh foods contribute to this problem. *Cyclospora* is a protozoan of concern, and the major source is contaminated waters (**figure 40.8**).

40.5 Detection of Food-Borne Pathogens

a. Detection of food-borne pathogens is a major part of food microbiology. The use of immunological and molecular techniques such as DNA and RNA hybridization, PCR, and pulsed field electrophoresis is making it possible to link disease occurrences to a common infection source (**figures 40.9** and **40.10**). PulseNet and FoodNet programs are being used to coordinate these control efforts.

40.6 Microbiology of Fermented Foods

a. Dairy products can be fermented to yield a wide variety of cultured milk products (**table 40.7**). These include mesophilic, therapeutic, thermophilic, lactoethanolic, and mold-lactic products.

b. Growth of lactic acid-forming bacteria (**figure 40.11**), often with the additional use of rennin, can coagulate milk solids. These solids can be processed to yield a wide variety of cheeses, including soft unripened, soft ripened, semisoft, hard, and very hard types (**table 40.8** and **figure 40.14**). Both bacteria and fungi are used in these cheese production processes.

c. Wines are produced from pressed grapes and can be dry or sweet, depending on the level of free sugar that remains at the end of the alcoholic fermentation (**figure 40.16**). Champagne is produced when the fermentation, resulting in CO_2 formation, is continued in the bottle.

d. Beer and ale are produced from cereals and grains. The starches in these substrates are hydrolyzed, in the processes of malting and mashing, to produce a fermentable wort. *Saccharomyces cerevisiae* is a major yeast used in the production of beer and ale (**figure 40.17**).

e. Many plant products can be fermented with bacteria, yeasts, and molds. Important products are breads, soy sauce, sufu, and tempeh (**table 40.9**). Sauer-kraut and pickles are produced in a fermentation process in which natural populations of lactobacilli play a major role (**figure 40.19**).

40.7 Microorganisms as Foods and Food Amendments

a. Microorganisms themselves can serve as an important food source. Mushrooms (*Agaricus bisporus*) are one of the most important fungi used as a food source. *Spirulina*, a cyanobacterium, also is a popular food source sold in specialty stores.

b. Many microorganisms, including some of those used to ferment milks, can be used as food amendments or microbial dietary adjuvants. Microorganisms such as *Lactobacillus* and *Bifidobacterium*, termed probiotics, can be used with oligopolysaccharides, termed prebiotics, to yield synbiotics. Several types of probiotic microorganisms are being used successfully in poultry production.

Key Terms

aflatoxin 1027
algal toxin 1028
bacteriocin 1031
bottom yeast 1044
competitive exclusion 1047
enology 1041
ergotism 1027
extrinsic factor 1024
food-borne infection 1032
food intoxication 1034
FoodNet 1036

fumonisin 1027
GRAS 1030
intrinsic factor 1024
kefir 1040
lactic acid bacteria (LAB) 1038
lager 1044
malt 1043
mash 1043
mashing 1041
modified atmosphere packaging (MAP) 1026

must 1041
nonstarter lactic acid bacteria (NSLAB) 1040
osmophilic microorganism 1024
pasteurization 1030
pitching 1043
prebiotic 1047
probiotic 1039
PulseNet 1036
putrefaction 1024

racking 1043
radappertization 1031
silage 1046
sour mash 1044
starter culture 1039
synbiotic 1047
wine vinegar 1043
wort 1041
xerophilic microorganism 1024

Critical Thinking Questions

1. Fresh lemon slices are often served with raw or steamed seafood (oysters, crab, shrimp). From a food microbiology perspective, provide an explanation for their being served. Are there other examples in either the cooking or the serving of foods that not only enhance flavor, but might have an antimicrobial strategy? Consider the example of marinades.

2. You are going through a salad line in a cafeteria at the end of the day. Which types of foods would you tend to avoid, and why?

3. Why were aflatoxins not discovered before the 1960s? Do you think this was the first time they had grown in a food product to cause disease?

4. What advantage might the shigalike toxin give *E. coli* O157:H7? Can we expect to see other "new" pathogens appearing, and what should we do, if anything?

5. Keep a record of what you eat for a day or two. Determine if the food, beverages, and snacks you ate could have been produced (at any level) without the aid of microorganisms. Indicate at what level(s) microorganisms were deliberately used. Be sure to consider ingredients such as citric acid, which is produced at the industrial level by several species of fungi.

6. Colonization of a susceptible human is critical to food-borne disease microorganisms. How might it be possible to modify foods to decrease these attachment processes?

Learn More

Barrett, J. R. 2000. Mycotoxins: of molds and maladies. *Environ. Health Perspect.* 108:A20–A23.

Beale, B. 2002. Probiotics: Their tiny worlds are under scrutiny. *The Scientist* 16(15):20–22.

Broadbent, J. R., and Steele, J. L. 2005. Cheese flavor and the genomics of lactic acid bacteria. *ASM News* 71:121–28.

Burgess, C.; O'Connell-Motherway, M.; Sybesma, W.; Hugenholtz, J.; and van Sinderen, D. 2004. Riboflavin production in *Lactococcus lactis:* Potential for in situ production of vitamin-enriched foods. *Appl. Environment. Microbiol.* 70:5769–77.

Farmworth. E. R. 2003. *Handbook of fermented functional foods.* Boca Raton. FL: CRC Press.

Fung, D. Y. C. 2000. Food spoilage and preservation. In *Encyclopedia of microbiology,* 2d ed., vol. 2, J. Leading, ed-in-chief, 412–20. San Diego: Academic Press.

Henry, S. H.; Bosch, F. X.; Troxell, T. C.; and Bolger, P. M. 2000. Reducing liver cancer—global control of aflatoxin. *Science* 286:2453–54.

Hui, Y. H.; Pierson, M. D.; and Gorham, J. R. 2001. *Foodborne disease handbook,* 2d ed., vol. 1., *Bacterial pathogens.* New York: Marcel Dekker.

Hui, Y. H.; Sattar, S. A.; Murrell, K. D.; Nip, W.-K.; and Stanfield, P. S. 2001. *Foodborne disease handbook,* 2d ed., vol. 2. *Viruses, parasites, pathogens and HACCP.* New York: Marcel Dekker.

Montville, T. J., and Matthews, K. 2004. *Food microbiology: An introduction.* Washington, D.C.: ASM Press.

Robinson, R. K.; Batt, C. A.; and Patel, P. D. 2000. *Encyclopedia of food microbiology.* San Diego; Academic Press.

Rodríguez-Lázaro, D.; Jofré, A.; Aymerich, T.; Hugas, M.; and Pla, M. 2004. Rapid quantitative analysis of *Listeria monocytogenes* in meat products by real-time PCR. *Appl. Environment. Microbiol.* 70:6299–301.

41

Applied and Industrial Microbiology

Biodegradation often can be facilitated by changing environmental conditions. Polychlorinated biphenyls (PCBs) are widespread industrial contaminants that accumulate in anoxic river muds. Although reductive dechlorination occurs under these conditions, oxygen is required to complete the degradation process. In this experiment, muds are being aerated to allow the final biodegradation steps to occur.

PREVIEW

- In this text, "applied microbiology" refers to the use of microorganisms in a variety of ecosystems to perform a specific process that helps meet a specific goal. Thus wastewater treatment and bioremediation are types of applied microbiology.

- The presence of clean freshwater is critical to all terrestrial organisms. Contaminated waters are a significant source of disease and worldwide death. Water purification systems must be designed and maintained to cleanse wastewater and ensure that human, agricultural, and industrial wastes do not foul our planet.

- Sewage treatment can be carried out using large vessels where mixing and aeration can be controlled (conventional treatment). Constructed wetlands, where aquatic plants and their associated microorganisms are used, now are finding widespread applications in the treatment of liquid wastes.

- Indicator organisms, which usually die off at slower rates than many disease-causing microorganisms, can be used to evaluate the microbiological quality of water.

- Groundwater is an important source of drinking water, especially in suburban and rural areas. In too many cases, this resource is being contaminated by disease-causing microorganisms and nutrients, especially from septic tanks.

- Microorganisms are used in industrial microbiology to create a wide variety of products and to assist in maintaining and improving the environment.

- Most work in industrial microbiology has traditionally been carried out using microorganisms isolated from nature or modified through mutations. Microorganisms with specific genetic characteristics are now more commonly genetically engineered to meet desired objectives.

- Protein evolution is used to generate variants with desired functions. Both in vivo and in vitro approaches are used.

- In controlled growth systems, different products are synthesized during growth and after growth is completed. Most antibiotics are produced after the completion of active growth.

- Antibiotics and other microbial products continue to contribute to animal and human welfare. Newer products include anticancer drugs. Combinatorial biology is making it possible to produce hybrid antibiotics with unique properties.

- The products of industrial microbiology also include bulk chemicals that are used as food supplements and acidifying agents. Other products are used as biosurfactants and emulsifiers in a wide variety of applications.

- The use of microbes to degrade toxic compounds in the environment is called bioremediation. Anaerobic degradation processes are important for the initial modification of many compounds, especially those with chlorine and other halogenated functions. Degradation can produce simpler or modified compounds that may not be less toxic than the original compound.

- Bacteria, fungi, and viruses are increasingly employed as biopesticides, thus reducing dependence on chemical pesticides.

In this final chapter, we use the term "applied microbiology," to refer to the use of microbes in their natural environment to perform processes useful to humankind. Such processes include wastewater treatment and bioremediation, both of which are discussed in this chapter. Like applied microbiology, industrial microbiology involves the use of microorganisms to achieve specific goals. Industrial microbiology, however, generally focuses on products such as pharmaceutical and medical compounds (e.g., antibiotics, hormones, transformed steroids), solvents, organic acids, chemical feedstocks, amino acids, and enzymes that have economic value. The microorganisms employed by industry have been isolated from nature, and in many cases, were modified using classic mutation-selection procedures. Genetic engineering has replaced this more traditional approach to developing microbial strains of industrial importance.

The microbe will have the last word.

—*Louis Pasteur*

In developed countries, the processes and products of applied and industrial microbiology are taken for granted, but this is not true globally. For instance, the provision of clean drinking water and the sanitary treatment of contaminated water is beyond reach for an alarmingly high number of people. According to the World Health Organization and UNICEF, over 1 billion people worldwide do not have access to safe, drinkable water and about 40% of the world's population lacks basic sanitation. We begin this chapter by presenting ways in which water can be purified so that it can be consumed without fear of disease transmission. We then describe several approaches to treating wastewater to keep our rivers, streams, lakes, and groundwater—which may be sources of drinking water—clean. The remainder of the chapter focuses on industrial microbiology. It is clear that many concepts presented throughout the text are integrated in these important subjects.

41.1 WATER PURIFICATION AND SANITARY ANALYSIS

Many important human pathogens are maintained in association with living organisms other than humans, including many wild animals and birds. Some of these bacterial and protozoan pathogens can survive in water and infect humans. As examples, *Vibrio vulnificus*, *V. parahaemolyticus*, and *Legionella* are of con-

tinuing concern. When waters are used for recreation or are a source of seafood that is consumed uncooked, the possibility for disease transmission exists. In many countries, such waters are the source of drinking water. Food-borne and waterborne viral diseases (section 37.4); Food-borne and waterborne bacterial diseases (section 38.4); Food-borne and waterborne diseases of fungi and protists (section 39.5)

Water purification is a critical link in controlling disease transmission in waters. As shown in **figure 41.1,** water purification can involve a variety of steps, depending on the type of impurities in the raw water source. Usually municipal water supplies are purified by a process that consists of at least three or four steps. If the raw water contains a great deal of suspended material, it often is first routed to a **sedimentation basin** and held so that sand and other very large particles can settle out. The partially clarified water is then mixed with chemicals such as alum and lime and moved to a **settling basin** where more material precipitates out. This procedure is called **coagulation** or flocculation and removes microorganisms, organic matter, toxic contaminants, and suspended fine particles. After these steps the water is further purified by passing it through **rapid sand filters,** which physically trap fine particles and flocs. This removes up to 99% of the remaining bacteria. After filtration the water is treated with a disinfectant. This step usually involves chlorination, but ozonation is becoming increasingly popular. When chlorination is employed, the chlorine dose must be large enough to leave residual free chlorine at a concentration of 0.2 to 2.0 mg/liter. A concern is the creation of **dis-**

Water purification steps

Untreated water

Color and precipitate removal

Softening (Ca, Mg removal)

Turbidity removal

Taste and odor removal

Disinfection

Aeration
Chemical oxidation
Ion exchange
Sedimentation

Chemical precipitation
• dosing
• mixing
• flocculation
• settling
Ion exchange

Chemical coagulation
• dosing
• mixing
• flocculation
• settling
Filtration

Aeration
Chemical oxidation
Adsorption

Irradiation
Ozonation
Chlorination

Drinking water

Water purification processes

Figure 41.1 Water Purification. Several alternatives can be used for drinking water treatment depending on the initial water quality. A major concern is disinfection: chlorination can lead to the formation of *dis*infection-*by*products (DBPs), including potentially carcinogenic *tri*halomethanes (THMs).

infection by-products (DBPs) such as **trihalomethanes (THMs)** that are formed when chlorine reacts with organic matter. Some of these compounds are carcinogens.

This purification process removes or inactivates disease-causing bacteria and indicator organisms (coliforms). Unfortunately, the use of coagulants, rapid filtration, and chemical disinfection often does not remove cysts of the protist *Giardia intestinalis*, *Cryptosporidium* oocysts, *Cyclospora*, and viruses. *Giardia*, a cause of human diarrhea, is now recognized as the most common identified waterborne pathogen in the United States. More consistent removal of *Giardia* cysts, which are about 7 to 10 by 8 to 12 μm in size, can be achieved with **slow sand filters.** This treatment involves the slow passage of water through a bed of sand in which a microbial layer covers the surface of each sand grain. Waterborne microorganisms are removed by adhesion to the gelatinous surface microbial layer (**Techniques & Applications 41.1**).   Food-borne and waterborne diseases: Giardiasis (section 39.5)

In the last few years, *Cryptosporidium* has become a significant problem. This protozoan parasite is smaller than *Giardia* and is even more difficult to remove from water. A major source of *Giardia* and *Cryptosporidium* contamination of soils, plant materials, and waters is the Canada goose, protected as a migratory bird. Each goose produces about 0.68 kg of manure per day, and geese are documented carriers of these important pathogens. The goose population also is rapidly increasing. For example, in the Atlantic flyway, there are at least a million geese, and these are estimated to be increasing at 17% per year. They gather on golf courses, pastures, croplands, and vulnerable watersheds used for public water supplies, which of course adds to the problem.

Viruses in drinking water also must be inactivated or removed. Coagulation and filtration reduce virus levels about 90 to 99%. Further inactivation of viruses by chemical oxidants, high pH, and photooxidation may yield a reduction as great as 99.9%. None of these processes, however, is considered sufficient protection. New standards for virus inactivation are being developed. Bacteriophages, which can be easily grown and assayed, now are being used to monitor disinfection. If sufficient reductions in bacteriophage infectivity occur with a given disinfection process, it is assumed that viruses capable of infecting humans will also be reduced to satisfactory levels.

1. What steps are usually taken to purify drinking water?
2. Why is chlorination, although beneficial in terms of bacterial pathogen control, of environmental concern?
3. Which important waterborne pathogens are not controlled reliably by chlorination?

Sanitary Analysis of Waters

Monitoring and detection of indicator and disease-causing microorganisms are a major part of sanitary microbiology. Bacteria from the intestinal tract generally do not survive in the aquatic environment, are under physiological stress, and gradually lose their ability to form colonies on differential and selective media. Their die-out rate depends on the water temperature, the effects of sunlight, the populations of other bacteria present, and the chemical composition of the water. Procedures have been developed to attempt to "resuscitate" these stressed coliforms using selective and differential media.

A wide range of viral, bacterial, and protozoan diseases result from the contamination of water with human and other animal fecal wastes. Although many of these pathogens can be detected directly, environmental microbiologists have generally used **indicator organisms** as an index of possible water contamination by human pathogens. Researchers are still searching for the "ideal" indicator organism to use in sanitary microbiology. These are among the suggested criteria for such an indicator:

1. The indicator bacterium should be suitable for the analysis of all types of water: tap, river, ground, impounded, recreational, estuary, sea, and waste.
2. The indicator bacterium should be present whenever enteric pathogens are present.
3. The indicator bacterium should survive longer than the hardiest enteric pathogen.

Techniques & Applications

41.1   Waterborne Diseases, Water Supplies, and Slow Sand Filtration

Slow sand filtration, in which drinking water is passed through a sand filter that develops a layer of microorganisms on its surface, has had a long and interesting history. After London's severe cholera epidemic of 1849, Parliament, in an act of 1852, required that the entire water supply of London be passed through slow sand filters before use.

The value of this process was shown in 1892, when a major cholera epidemic occurred in Hamburg, Germany, and 10,000 lives were lost. The neighboring town, Altona, which used slow sand filtration, did not have a cholera epidemic. Slow sand filters were installed in many cities in the early 1900s, but the process fell into disfavor with the advent of rapid sand filters, chlorination, and the use of coagulants such as alum. Slow sand filtration, a time-tested process, is regaining favor because of its filtration effectiveness and lower maintenance costs. Slow sand filtration is particularly effective for the removal of *Giardia* cysts. For this reason slow sand filtration is used in many mountain communities where *Giardia* is a problem.

4. The indicator bacterium should not reproduce in the contaminated water and produce an inflated value.
5. The assay procedure for the indicator should have great specificity; in other words, other bacteria should not give positive results. In addition, the procedure should have high sensitivity and detect low levels of the indicator.
6. The testing method should be easy to perform.
7. The indicator should be harmless to humans.
8. The level of the indicator bacterium in contaminated water should have some direct relationship to the degree of fecal pollution.

Coliforms, including *Escherichia coli,* are members of the family *Enterobacteriaceae.* These bacteria make up approximately 10% of the intestinal microorganisms of humans and other animals and have found widespread use as indicator organisms. They lose viability in freshwater at slower rates than most of the major intestinal bacterial pathogens. When such "foreign" enteric indicator bacteria are not detectable in a specific volume (100 ml) of water, the water is considered **potable** [Latin *potabilis,* fit to drink], or suitable for human consumption. Class *Gammaproteobacteria:* Order *Enterobacteriales* (section 22.3)

The coliform group includes *E. coli, Enterobacter aerogenes,* and *Klebsiella pneumoniae.* Coliforms are defined as facultatively anaerobic, gram-negative, nonsporing, rod-shaped bacteria that ferment lactose with gas formation within 48 hours at 35°C. The original test for coliforms that was used to meet this definition involved the presumptive, confirmed, and completed tests, as shown in **figure 41.2.** The presumptive step is carried out by means of tubes inoculated with three different sample volumes to give an estimate of the **most probable number (MPN)** of coliforms in the water. The complete process, including the confirmed and completed tests, requires at least 4 days of incubations and transfers.

Unfortunately the coliforms include a wide range of bacteria whose primary source may not be the intestinal tract. To address this difficulty, tests have been developed that allow waters to be tested for the presence of **fecal coliforms.** These are coliforms derived from the intestine of warm-blooded animals, which can grow at the more restrictive temperature of 44.5°C.

To test for coliforms and fecal coliforms, and more effectively recover stressed coliforms, a variety of simpler and more specific tests have been developed. These include the membrane filtration technique, the **presence-absence (P-A) test** for coliforms and the related Colilert **defined substrate test** for detecting both coliforms and *E. coli.*

The **membrane filtration technique** has become a common and often preferred method of evaluating the microbiological characteristics of water. The water sample is passed through a membrane filter. The filter with its trapped bacteria is transferred to the surface of a solid medium or to an absorptive pad containing the desired liquid medium. Use of the proper medium enables the rapid detection of total coliforms, fecal coliforms, or fecal enterococci by the presence of their characteristic colonies. Samples can be placed on a less selective resuscitation medium, or

incubated at a less stressful temperature, prior to growth under the final set of selective conditions. An example of a resuscitation step is the use of a 2 hour incubation on a pad soaked with lauryl sulfate broth, as is carried out in the LES Endo procedure. A resuscitation step often is needed with chlorinated samples, where the microorganisms are especially stressed. The advantages and disadvantages of the membrane filter technique are summarized in **table 41.1.** Membrane filters have been widely used with water that does not contain high levels of background organisms, sediment, or heavy metals.

More simplified tests for detecting coliforms and fecal coliforms are now available. The presence-absence test (P-A test) can be used for coliforms. This is a modification of the MPN procedure, in which a larger water sample (100 ml) is incubated in a single culture bottle with a triple-strength broth containing lactose broth, lauryl tryptose broth, and bromcresol purple indicator. The P-A test is based on the assumption that no coliforms should be present in 100 ml of drinking water. A positive test results in the production of acid (a yellow color) and constitutes a positive presumptive test requiring confirmation.

To test for both coliforms and *E. coli,* the related Colilert defined substrate test can be used. A water sample of 100 ml is added to a specialized medium containing *o*-nitrophenyl-β-D-galactopyranoside (ONPG) and 4-methylumbelliferyl-β-D-glucuronide (MUG) as the only nutrients. If coliforms are present, the medium will turn yellow within 24 hours at 35°C due to the hydrolysis of ONPG, which releases *o*-nitrophenol, as shown in **figure 41.3.** To check for *E. coli,* the medium is observed under long-wavelength UV light for fluorescence. When *E. coli* is present, MUG is modified to yield a fluorescent product. If the test is negative for the presence of coliforms, the water is considered acceptable for human consumption. The main change from previous standards is the requirement to have water free of coliforms and fecal coliforms. If coliforms are present, fecal coliforms or *E. coli* must be tested for.

Molecular techniques are now used routinely to detect coliforms in waters and other environments, including foods. 16S rRNA gene-targeted primers for coliforms enable the detection of one colony-forming unit (CFU) of *E. coli* per 100 ml of water, if a short enrichment step precedes the use of the PCR amplification. This allows the differentiation of nonpathogenic and enterotoxigenic strains, including the shigalike-toxin producing *E. coli* O157:H7. PCR (section 14.3)

In the United States a set of general guidelines for microbiological quality of drinking waters has been developed, including standards for coliforms, viruses, and *Giardia* (**table 41.2**). If unfiltered surface waters are being used, one coliform test must be run each day when the waters have higher turbidities.

Other indicator microorganisms include **fecal enterococci.** The fecal enterococci are increasingly being used as an indicator of fecal contamination in brackish and marine water. In salt water these bacteria die at a slower rate than the fecal coliforms, providing a more reliable indicator of possible recent pollution. Class *Bacilli:* Order *Lactobacillales* (section 23.5)

Inoculate 15 tubes: 5 with 10 ml of sample, 5 with 1.0 ml of sample, and 5 with 0.1 ml of sample.

Double-strength broth   Single-strength broth

10   10   10   10   10   1.0   1.0   1.0   1.0   1.0   0.1   0.1   0.1   0.1   0.1
(ml)   (ml)   (ml)

Lactose or lauryl tryptose broth

Presumptive

Negative persumptive.
The absence of gas in
broth tubes indicates
coliforms are absent.
Incubate an additional
24 hours to be sure.

← 24 ± 2 hours →
35°C

After 24 hours of
incubation, the tubes of
lactose broth are examined
for gas production.

Negative   **Positive**

Confirmed

No gas produced,
coliform group absent.

All positive presumptive
cultures used to inoculate
tubes of brilliant green lactose
bile broth. Incubation for 48 ± 3
hours at 35°C.

Positive test: gas production,
use positive confirmed
tubes to determine MPN.

Negative   **Positive**

Completed

Brilliant
green
lactose
bile broth
or lauryl
tryptose
broth

Nutrient
agar slant

Use coliform colonies
to inoculate nutrient agar
slant and a broth tube.

Plates of Levine's EMB or LES Endo
agar are streaked from positive
tubes and incubated at 35°C for
24 ± 2 hours.

After 24 hours of incubation make a
Gram-stained slide from the slant.
If the bacteria are gram-negative
nonsporing rods and produce gas from
lactose, the completed test is positive.

Figure 41.2   The Multiple-Tube Fermentation Test.   The multiple-tube fermentation technique has been used for many years for the
sanitary analysis of water. Lactose broth tubes are inoculated with different water volumes in the presumptive test. Tubes that are positive for gas
production are inoculated into brilliant green lactose bile broth in the confirmed test, and positive tubes are used to calculate the most probable
number (MPN) value. The completed test is used to establish that coliform bacteria are present.

Table 41.1	Advantages and Disadvantages of the Membrane Filter Technique for Evaluation of the Microbial Quality of Water

Advantages

Good reproducibility

Single-step results often possible

Filters can be transferred between different media

Large volumes can be processed to increase assay sensitivity

Time savings are considerable

Ability to complete filtrations on site

Lower total cost in comparison with MPN procedure

Disadvantages

High-turbidity waters limit volumes sampled

High populations of background bacteria cause overgrowth

Metals and phenols can adsorb to filters and inhibit growth

Source: Data from L. S. Clesceri, et al., *Standard Methods for the Examination of Water and Wastewater*, 20th edition, pages 9–56, 1998. American Public Health Association, Washington, D.C.

Figure 41.3 The Defined Substrate Test. This much simpler test is used to detect coliforms and fecal coliforms in single 100 ml water samples. The medium uses ONPG and MUG (see text) as defined substrates. **(a)** Uninoculated control. **(b)** Yellow color due to the presence of coliforms. **(c)** Fluorescent reaction due to the presence of fecal coliforms.

1. What is an indicator organism, and what properties should it have?
2. How is a coliform defined? How does this definition relate to presumptive, confirmed, and completed tests?
3. How does one differentiate between coliforms and fecal coliforms in the laboratory?

Table 41.2	Current Drinking Water Standards in the United States

Agent	Allowable Maximum Contaminant Level Goal (MCLG) or Maximum Contaminant Level (MCL)
Coliforms	MCLG = 0
	MCL = No more than 5% positive total coliform samples/month for water systems that collect > 40 samples/month. For water systems that collect < 40 routine samples/month, no more than 1 can be coliform positive. Every sample that has total coliforms must be analyzed for fecal coliforms. There cannot be any fecal coliforms or *E. coli*.
Cryptosporidium	MCLG = 0
Giardia intestinalis	MCLG = 0
Legionella	MCLG = 0
Viruses (enteric)	MCLG = 0

Source: Environmental Protection Agency, USA, July, 2002.

4. In what type of environment is it better to use fecal enterococci rather than fecal coliforms as an indicator organism? Why?
5. What are the advantages and disadvantages of membrane filters for microbiological examinations of water?
6. Why has the defined substrate test with ONPG and MUG been accepted as a test of drinking water quality?

41.2 WASTEWATER TREATMENT

Waters often contain high levels of organic matter from industrial and agricultural wastes (e.g., from food processing, petrochemical and chemical plants, and plywood plant resin wastes), and from human wastes. It is necessary to remove organic matter by the process of wastewater treatment. Depending on the effort given to this task, it may still produce waters containing nutrients and some microorganisms, which can be released to rivers and streams.

The process of wastewater treatment, when performed at a municipal level, must be monitored to ensure that waters released into the environment do not pose environmental and health risks. Our discussion of wastewater treatment must therefore begin with the means by which water quality is monitored. We then discuss large-scale wastewater treatment processes followed by home treatment systems.

Measuring Water Quality

Carbon removal during wastewater treatment can be measured several ways, including (1) as **total organic carbon (TOC)**, (2) as chemically oxidizable carbon by the **chemical oxygen de-**

mand **(COD)** test, or (3) as biologically usable carbon by the **biochemical oxygen demand (BOD)** test. The TOC includes all carbon, whether or not it is usable by microorganisms. This is determined by oxidizing the organic matter in a sample at high temperature in an oxygen stream and measuring the resultant CO_2 by infrared or potentiometric techniques. The COD gives a similar measurement, except that lignin often will not react with the oxidizing chemical, such as permanganate, that is used in this procedure. The BOD test, in comparison, measures only the portion of the total carbon that can be oxidized by microorganisms in a 5-day period under standard conditions.

The biochemical oxygen demand is an indirect measure of organic matter in aquatic environments. It is the amount of dissolved O_2 needed for microbial oxidation of biodegradable organic matter. When O_2 consumption is measured, the O_2 itself must be present in excess and not limit oxidation of the nutrients (**table 41.3**). To achieve this, the waste sample is diluted to assure that at least 2 mg/liter of O_2 are used while at least 1 mg/liter of O_2 remains in the test bottle. Ammonia released during organic matter oxidation can also exert an O_2 demand in the BOD test, so nitrification or the **nitrogen oxygen demand (NOD)** is often inhibited by 2-chloro-6-(trichloromethyl) pyridine (nitrapyrin). In the normal BOD test, which is run for 5 days at 20°C on untreated samples, nitrification is not a major concern. However, when treated effluents are analyzed, NOD can be a problem.

In terms of speed, the TOC is fastest, but less informative in terms of biological processes. The COD is slower and involves the use of wet chemicals with higher waste chemical disposal costs. The TOC, COD, and BOD provide different but complementary information on the carbon in a water sample. It is critical to note that these measurements, concerned with carbon removal, do not directly address concerns for removal of minerals such as nitrate, phosphate, and sulfate from waters. These minerals have global impacts on cyanobacterial and algal growth in lakes, rivers, and the oceans by contributing to the process of eutrophication. The removal of dissolved organic matter and possibly inorganic nutrients, plus inactivation and removal of pathogens, are important parts of wastewater treatment. Marine and freshwater environments: Nutrient cycling (section 28.1)

Table 41.3	The Biochemical Oxygen Demand (BOD) Test: A System with Excess and Limiting Components

Components in Excess at the End of the Incubation Period
Nitrogen
Phosphorus
Iron
Trace elements
Microorganisms
Oxygen
Component Limiting at the End of the Incubation Period
Organic matter

1. What are TOC, COD, and BOD and how are these similar and different?
2. What factors can lead to a nitrogen oxygen demand (NOD) in water?
3. What components should limit the reactions in a BOD test, and what components should not limit reaction rates? Why?
4. What minerals can contribute to eutrophication?

Wastewater Treatment Processes

The aerobic self-purification sequence that occurs when organic matter is added to lakes and rivers can be carried out under controlled conditions in which natural processes are intensified. This often involves the use of large basins (conventional sewage treatment) where mixing and gas exchange are carefully controlled.

An aerial photograph of a modern sewage treatment plant is shown in **figure 41.4a. Wastewater treatment** involves a number of steps that are spatially segregated (figure 41.4b). The first three steps are called primary, secondary, and tertiary treatment (**table 41.4**). At the end of the process, the water is usually chlorinated (itself an emerging environmental and human health problem) before it is released.

Primary treatment physically removes 20 to 30% of the BOD that is present in particulate form. In this treatment, particulate material is removed by screening, precipitation of small particulates, and settling in basins or tanks. The resulting solid material is usually called **sludge.**

Secondary treatment promotes the biological transformation of dissolved organic matter to microbial biomass and carbon dioxide. About 90 to 95% of the BOD and many bacterial pathogens are removed by this process. Several approaches can be used in secondary treatment to biologically remove dissolved organic matter. All of these techniques involve similar microbial activities. Under oxic conditions, dissolved organic matter will be transformed into additional microbial biomass plus carbon dioxide. When microbial growth is completed, under ideal conditions the microorganisms will aggregate and form a settleable stable floc structure. Minerals in the water also may be tied up in microbial biomass. When microorganisms grow, flocs can form. As shown in **figure 41.5a**, a healthy settleable floc is compact. In contrast, poorly formed flocs have a network of filamentous microbes that will retard settling (figure 41.5b).

When these processes occur with lower O_2 levels or with a microbial community that is too young or too old, unsatisfactory floc formation and settling can occur. The result is a **bulking sludge,** caused by the massive development of filamentous bacteria such as *Sphaerotilus* and *Thiothrix,* together with many poorly characterized filamentous organisms. These important filamentous bacteria form flocs that do not settle well, and thus produce effluent quality problems. Class *Betaproteobacteria: Order Burkholderiales* (section 22.2), Class *Gammaproteobacteria: Order Thiotrichales* (section 22.3)

An aerobic **activated sludge** system (**figure 41.6a**) involves a horizontal flow of materials with recycling of sludge—the active biomass that is formed when organic matter is oxidized and

(a)

1 = Primary clarifiers
2 = Activated sludge vessels
3 = Final clarifiers
4 = Chlorination vessels

(b)

Figure 41.4 An Aerial View of a Modern Conventional Sewage Treatment Plant. Sewage treatment plants allow natural processes of self-purification that occur in rivers and lakes to be carried out under more intense, managed conditions in large concrete vessels. **(a)** A plant in New Jersey. **(b)** A diagram of flows in the plant.

Table 41.4	Major Steps in Primary, Secondary, and Tertiary Treatment of Wastes
Treatment Step	**Processes**
Primary	Removal of insoluble particulate materials by settling, screening, addition of alum and other coagulation agents, and other physical procedures
Secondary	Biological removal of dissolved organic matter
	Trickling filters
	Activated sludge
	Lagoons
	Extended aeration systems
	Anaerobic digesters
Tertiary	Biological removal of inorganic nutrients
	Chemical removal of inorganic nutrients
	Virus removal/inactivation
	Trace chemical removal

degraded by microorganisms. Activated sludge systems can be designed with variations in mixing. In addition, the ratio of organic matter added to the active microbial biomass can be varied. A low-rate system (low nutrient input per unit of microbial biomass), with slower growing microorganisms, will produce an effluent with low residual levels of dissolved organic matter. A high-rate system (high nutrient input per unit of microbial biomass), with faster growing microorganisms, will remove more dissolved organic carbon per unit time, but produce a poorer quality effluent.

Aerobic secondary treatment also can be carried out with a **trickling filter** (figure 41.6b). The waste effluent is passed over rocks or other solid materials upon which microbial biofilms have developed, and the microbial community degrades the organic waste. A sewage treatment plant can be operated to produce less

sludge by employing the **extended aeration** process (figure 41.6c). Microorganisms grow on the dissolved organic matter, and the newly formed microbial biomass is eventually consumed to meet maintenance energy requirements. This requires extremely large aeration basins and long aeration times. In addition, with the biological self-utilization of the biomass, minerals originally present in the microorganisms are again released to the water.

All aerobic processes produce excess microbial biomass, or sewage sludge, which contains many recalcitrant organics. Often the sludges from aerobic sewage treatment, together with the materials settled out in primary treatment, are further treated by anaerobic digestion. Anaerobic digesters are large tanks designed to operate with continuous input of untreated sludge and removal of the final, stabilized sludge product. Methane is vented and often burned for heat and electricity production. This digestion process involves three steps: (1) the fermentation of the sludge components to form organic acids, including acetate; (2) production of the methanogenic substrates: acetate, CO_2, and hydrogen; and finally, (3) methanogenesis by the methane producers. These methanogenic processes, summarized in **table 41.5,** involve critical balances between electron acceptors and donors. To function most efficiently, the hydrogen concentration must be maintained at a low level. If hydrogen and organic acids accumulate, methane production can be inhibited, resulting in a stuck digester. Phylum *Euryarchaeota:* Methanogens (section 20.3)

Anaerobic digestion has many advantages. Most of the microbial biomass produced in aerobic growth is used for methane production in the anaerobic digester. Also, because the process of methanogenesis is energetically very inefficient, the microbes must consume about twice the nutrients to produce an equivalent biomass as that of aerobic systems. Consequently, less sludge is

(a)

(b)

Figure 41.5 Proper Floc Formation in Activated Sludge. Microorganisms play a critical role in the functioning of activated sludge systems. **(a)** The operation is dependent on the formation of settleable flocs. **(b)** If the plant does not run properly, poorly settling flocs can form due to such causes as low aeration, sulfide, and acidic organic substrates. These flocs do not settle properly because of their open or porous structure. As a consequence, the organic material is released with the treated water and lowers the quality of the final effluent.

Figure 41.6 Aerobic Secondary Sewage Treatment.
(a) Activated sludge with microbial biomass recycling. The biomass is maintained in a suspended state to maximize oxygen, nutrient, and waste transfer processes. **(b)** Trickling filter, where waste water flows over biofilms attached to rocks or other solid supports, resulting in transformation of dissolved organic matter to new biofilm biomass and carbon dioxide. Excess biomass and treated water flow to a final clarifier. **(c)** An extended aeration process, where aeration is continued beyond the point of microbial growth, allows the microbial biomass to self-consume due to microbial energy of maintenance requirements. (The extended length of the reactor allows this process of biomass self-consumption to occur.) Minerals originally incorporated in the microbial biomass are released to the water as the process occurs.

Table 41.5	Sequential Reactions in the Anaerobic Digester		
Process Step	**Substrates**	**Products**	**Major Microorganisms**
Fermentation	Organic polymers	Butyrate, propionate, lactate, succinate, ethanol, acetate,[a] H_2,[a] CO_2[a]	*Clostridium* *Bacteroides* *Peptostreptococcus* *Peptococcus* *Eubacterium* *Lactobacillus*
Acetogenic reactions	Butyrate, propionate, lactate, succinate, ethanol	Acetate, H_2, CO_2	*Syntrophomonas* *Syntrophobacter*
Methanogenic reactions	Acetate	$CH_4 + CO_2$	*Acetobacterium* *Methanosarcina*
	H_2 and HCO_3^-	CH_4	*Methanothrix* *Methanobrevibacter* *Methanomicrobium* *Methanogenium* *Methanobacterium* *Methanococcus* *Methanospirillum*

[a]Methanogenic substrates produced in the initial fermentation step.

produced and it can be easily dried. Dried sludge removed from well-operated anaerobic systems can even be sold as organic garden fertilizer. In contrast, sludge can be dangerous if the system is not properly managed. Then heavy metals and other environmental contaminants can be concentrated in the sludge. There may be longer-term environmental and public health effects from disposal of this material on land or in water.

Tertiary treatment further purifies wastewaters. It is particularly important to remove nitrogen and phosphorus compounds that can promote eutrophication. Organic pollutants can be removed with activated carbon filters. Phosphate usually is precipitated as calcium or iron phosphate (for example, by the addition of lime). Excess nitrogen may be removed by "stripping," volatilization as NH_3 at high pHs. Ammonia itself can be chlorinated to form dichloramine, which is then converted to molecular nitrogen. In some cases, biological processes can be used to remove nitrogen and phosphorus. A widely used process for nitrogen removal is denitrification. Here nitrate, produced under aerobic conditions, is used as an electron acceptor under conditions of low oxygen with organic matter added as an energy source. Nitrate reduction yields nitrogen gas and nitrous oxide (N_2O) as the major products. Currently there is a great deal of interest in anaerobic nitrogen removal processes. These include the anammox process where ammonium ion (used as the electron donor), is reacted with nitrite (the electron acceptor) produced by partial nitrification. The anammox process can convert up to 80% of the beginning ammonium ion to N_2 gas. To remove phosphorus, oxic and anoxic conditions are used alternately in a series of treatments, and phosphorus accumulates in specially adapted microbial biomass as polyphosphate. Tertiary treatment is expensive and is usually not employed except where necessary to prevent obvious ecological disruption. Biogeochemical cycling: Nitrogen cycle (section 27.2)

Wetlands are a vital natural resource and a critical part of our environment, and increasingly efforts are being made to protect these fragile aquatic communities from pollution. A major means of wastewater treatment is the use of **constructed wetlands,** where the basic components of natural wetlands (soils, aquatic plants, waters) are used as a functional waste treatment system. Constructed wetlands now are increasingly employed in the treatment of liquid wastes and for bioremediation, which is discussed in section 41.6. This system uses floating, emergent, or submerged plants, as shown in **figure 41.7.** The aquatic plants provide nutrients in the root zone, which can support microbial growth. Especially with emergent plants, the root zone can be maintained in an anoxic state in which sulfide, synthesized by *Desulfovibrio* using root zone organic matter as an energy source, can trap metals. Constructed wetlands also are being used to treat acid mine drainage (AMD) in many parts of the world. Higher-strength industrial wastes also can be treated.

1. Explain how primary, secondary, and tertiary treatments are accomplished.
2. What is bulking sludge? Name several important microbial groups that contribute to this problem.
3. What are the steps of organic matter processing that occur in anaerobic digestion? Why is acetogenesis such an important step?
4. After anaerobic digestion is completed, why is sludge disposal still of concern?
5. Why might different aquatic plant types be used in constructed wetlands?

Home Treatment Systems

Groundwater, or water in gravel beds and fractured rocks below the surface soil, is a widely used but often unappreciated water resource. In the United States groundwater supplies at least 100 million people with drinking water, and in rural and suburban ar-

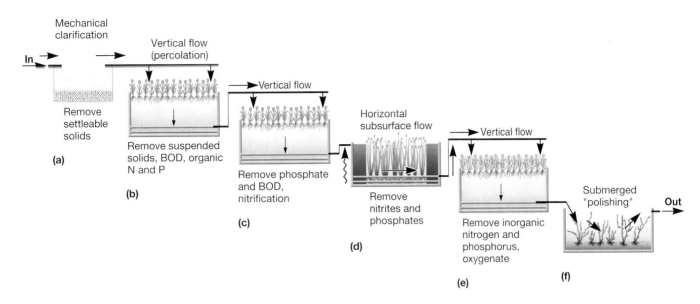

Figure 41.7   Constructed Wetland for Wastewater Treatment. Multistage constructed wetland systems can be used for organic matter and phosphate removal **(a)**. Free-floating macrophytes **(b,c,e)**, such as duckweed and water hyacinth, can be used for a variety of purposes. Emergent macrophytes **(d)**, such as bulrush, allow surface flow as well as vertical and horizontal subsurface flow. Submerged vegetation **(f)**, such as waterweed, allows final "polishing" of the water. These wetlands also can be designed for nitrification and metal removal from waters.

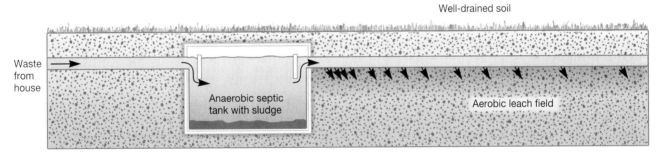

Figure 41.8   The Conventional Septic Tank Home Treatment System. This system combines an anaerobic waste liquefaction unit (the septic tank) with an aerobic leach field. Biological oxidation of the liquefied waste takes place in the leach field, unless the soil becomes flooded.

eas beyond municipal water distribution systems, 90 to 95% of all drinking water comes from this source.

The great dependence on this resource has not resulted in a corresponding understanding of microorganisms and microbiological processes that occur in the groundwater environment. Increasing attention is now being given to predicting the fate and effects of groundwater contamination on the chemical and microbiological quality of this resource. Pathogenic microorganisms and dissolved organic matter are removed from water during subsurface passage through adsorption and trapping by fine sandy materials, clays, and organic matter. Microorganisms associated with these materials—including predators such as protozoa—can use the trapped pathogens as food. This results in purified water with a lower microbial population.

This combination of adsorption-biological predation is used in home treatment systems (**figure 41.8**). Conventional **septic tank** systems include an anaerobic liquefaction and digestion step that occurs in the septic tank itself (the tank functions as a simple anaerobic digester). This is followed by organic matter adsorption and entrapment of microorganisms in an aerobic leach-field where biological oxidation occurs. A septic tank may not operate correctly for several reasons. If the retention time of the waste in the septic tank is too short, undigested solids move into the leach field, gradually plugging the system. If the leach field floods and becomes anoxic, biological oxidation does not occur, and effective treatment ceases. Other problems can occur, especially when a suitable soil is not present and the septic tank outflow from a conventional system drains too rapidly to the deeper subsurface.

Fractured rocks and coarse gravel materials provide little effective adsorption or filtration. This may result in the contamination of well water with pathogens and the transmission of disease. In addition, nitrogen and phosphorus from the waste can pollute the groundwater. This leads to nutrient enrichment of ponds, lakes, rivers, and estuaries as the subsurface water enters these environmentally sensitive water bodies.

Domestic and commercial on-site septic systems are now being designed with nitrogen and phosphorus removal steps. Nitrogen is usually removed by nitrification-denitrification processes with organic matter provided by sawdust or a similar material. Currently used systems function with essentially no maintenance for up to 8 years. For phosphorus removal, a reductive iron dissolution process can be used. With the need to control nitrogen and phosphorus releases from septic systems around the world, there will be increased emphasis on use of these and similar technologies in the future.

Subsurface zones also can become contaminated with pollutants from other sources. Land disposal of sewage sludges, illegal dumping of septic tank pumpage, improper toxic waste disposal, and runoff from agricultural operations all contribute to groundwater contamination with chemicals and microorganisms.

Many pollutants that reach the subsurface will persist and may affect the quality of groundwater for extended periods. Much research is being conducted to find ways to treat groundwater in place—**in situ treatment.** As will be further discussed, microorganisms and microbial processes are critical in many of these remediation efforts.

---

1. In rural areas, approximately what percentage of the water used for human consumption is groundwater?
2. What factors can limit microbial activity in subsurface environments? Consider the energetic and nutritional requirements of microorganisms in your answer.
3. How, in principle, are a conventional septic tank system and a leach-field system supposed to work? What alternatives are available to remove nitrogen and phosphorus from effluents? What factors can reduce the effectiveness of this system?

41.3 MICROORGANISMS USED IN INDUSTRIAL MICROBIOLOGY

The use of microorganisms in industrial microbiology follows a logical sequence. It is necessary first to identify or create a microorganism that carries out the desired process in the most efficient manner. This microorganism or its cloned genes are then used, either in a controlled environment such as a fermenter or in complex natural systems, such as in soils or waters, to achieve specific goals.

Thus the first task for an industrial microbiologist is to find a suitable microorganism, one that is genetically stable, easy to maintain and grow, and well suited for extraction or separation of desired products. A wide variety of alternative approaches are available, ranging from isolating microorganisms from the environment to using sophisticated molecular techniques to modify an existing microorganism. Here we present some of the commonly used approaches.

Finding Microorganisms in Nature

Until relatively recently, microbial cultures used in industrial microbiology were most often obtained from natural materials such as soil samples, waters, and spoiled bread and fruit. Cultures from all areas of the world continue to be examined to identify new strains with desirable characteristics. Interest in hunting for new microorganisms, or **bioprospecting,** continues today.

Less than 1% of the microbial species estimated to exist in most environments has been isolated or cultured (**table 41.6**). With increased interest in microbial diversity, microbial ecology, and especially in microorganisms from extreme environments (**Techniques & Applications 41.2**), microbiologists are exploring new ways to grow these previously uncultured microbes. For instance, single-cell gel microencapsulation allows nutrient diffusion and microbial communication, while minimizing overgrowth by more rapid-growing competitors.

Genetic Manipulation of Microorganisms

Genetic manipulations are used to produce microorganisms with new and desirable characteristics. The classical methods of genetic exchange coupled with recombinant DNA technology (*see chapters 13 and 14*) play a vital role in the development of cultures for industrial microbiology.

Mutagenesis

Once a promising microorganism is found, a variety of techniques can be used for its improvement, including chemical mutagens, ultraviolet light, and transposon mutagenesis. As an example, the first cultures of *Penicillium notatum*, which could be

Table 41.6	Estimates of the Percent "Cultured" Microorganisms in Various Environments
Environment	**Estimated Percent Cultured**
Seawater	0.001–0.100
Freshwater	0.25
Mesotrophic lake	0.1–1.0
Unpolluted estuarine waters	0.1–3.0
Activated sludge	1–15
Sediments	0.25
Soil	0.3

Source: D. A. Cowan. 2000. Microbial genomes—the untapped resource. *Tibtech* 18:14–16. Table 2, p. 15.

Techniques & Applications

41.2   The Potential of Thermophilic Archaea in Biotechnology

There is great interest in the characteristics of archaea isolated from the outflow mixing regions above deep hydrothermal vents that release water at 250 to 350°C. This is because these hardy organisms can grow at temperatures as high as 121°C. The problems in growing these microorganisms in a laboratory are formidable. For example, to grow some of them, it is necessary to use special culturing chambers and other specialized equipment to maintain water in the liquid state at these high temperatures.

Such microorganisms, termed hyperthermophiles, with optimum growth temperatures of 85°C or above, confront unique challenges in nutrient acquisition, metabolism, nucleic acid replication, and growth. Many are anaerobes that depend on elemental sulfur or ferric ion as electron acceptors. Enzyme stability is critical. Some DNA

polymerases are inherently stable at 140°C, whereas many other enzymes are stabilized in vivo with unique thermoprotectants. When these enzymes are separated from their protectant, they lose their unique thermostability.

These enzymes may have important applications in methane production, metal leaching and recovery, and in immobilized enzyme systems. In addition, the possibility of selective stereochemical modification of compounds normally not in solution at lower temperatures may provide new routes for directed chemical syntheses. This is an exciting and expanding area of the modern biological sciences to which environmental microbiologists can make significant contributions.

---

grown only under stationary conditions, yielded low concentrations of penicillin. In 1943 a strain of *Penicillium chrysogenum* was isolated—strain NRRL 1951—which was further improved through mutation (**figure 41.9**). Today most penicillin is produced with *Penicillium chrysogenum,* grown in aerobic stirred fermenters, which gives 55-fold higher penicillin yields than the original static cultures. Mutations and their chemical basis (section 13.1); Transposable elements (section 13.5)

Short lengths of chemically synthesized DNA sequences can be inserted into recipient microorganisms by the process of **site-directed mutagenesis.** This can create small genetic alterations leading to a change of one or several amino acids in the target protein. Such minor amino acid changes have been found to lead, in many cases, to unexpected changes in protein characteristics, and have resulted in new products such as more environmentally resistant enzymes and enzymes that can catalyze desired reactions. These approaches are part of the field of **protein engineering.** Synthetic DNA (section 14.2)

Protoplast Fusion

Most yeasts and molds are asexual or of a single mating type, which decreases the chance of random mutations that could lead to strain degeneration. **Protoplast fusion** can be used in genetic studies with these microorganisms. Protoplasts—cells lacking a cell wall—are prepared by growing the cells in an isotonic solution while treating them with enzymes, including cellulase and beta-galacturonidase. The protoplasts are then regenerated using osmotic stabilizers such as sucrose. After regeneration of the cell wall, the new protoplasm fusion product can be used in further studies.

A major advantage of the protoplast fusion technique is that protoplasts of different microbial species can be fused, even if they are not closely linked taxonomically. For example, protoplasts of *Penicillium roquefortii* have been fused with those of *P. chrysogenum.* Even yeast protoplasts and erythrocytes can be fused.

---

1. Why is the recovery of previously uncultured microorganisms from the environment an important goal?
2. What is protoplast fusion and what types of microorganisms are used in this process?
3. What is the goal of protein engineering?

Transfer of Genetic Information between Different Organisms

The transfer and expression of genes between different organisms can give rise to novel metabolic processes and products. This is part of the rapidly developing field of **combinatorial biology** (**table 41.7**). An important early example of this approach was the creation of the "superbug," patented by A. M. Chakarabarty in 1974, which had an increased capability of hydrocarbon degradation. Similarly, the genes for antibiotic production can be transferred to a microorganism that produces another antibiotic, or even to a non-antibiotic-producing microorganism. Other examples are the expression, in *E. coli,* of the enzyme creatininase from *Pseudomonas putida* and the production of pediocin, a bacteriocin, in a yeast used in wine fermentation for the purpose of controlling bacterial contaminants. When functional genes from one organism are transcribed and translated in another, it is called **heterologous gene expression.**

Heterologous gene expression can improve production efficiency and minimize the purification steps required before the product is ready for use. For example, recombinant baculoviruses can be replicated in insect larvae to achieve rapid large-scale production of a desired virus or protein. Transgenic plants may be used to manufacture large quantities of a variety of metabolic products. A gene encoding a foot-and-mouth disease virus antigen has been incorporated into *E. coli,* enabling the expression of this genetic information and synthesis of the gene product for use in vaccine production (**figure 41.10**).

Genetic information transfer allows the production of specific proteins and peptides without contamination by other products

Figure 41.9 Mutation Makes It Possible to Increase Fermentation Yields. A "genealogy" of the mutation processes used to increase penicillin yields with *Penicillium chrysogenum* using X-ray treatment (X), UV treatment (UV), and mustard gas (N). By using these mutational processes, the yield was increased from 120 International Units (IU) to 2,580 IU, a 20-fold increase. Unmarked transfers were used for mutant growth and isolation. Yields in international units/ml in brackets.

that might be synthesized in the original organism. This approach can decrease the time and cost of recovering and purifying a product. Another major advantage of engineered protein production is that only biologically active stereoisomers are produced. This specificity is required to avoid the possible harmful side effects of inactive stereoisomers.

1. What is combinatorial biology and what is the basic approach used in this technique? What types of major products have been created using combinatorial biology?
2. Why might one want to insert a gene in a foreign cell and how is this done?
3. Why is it important to produce specific isomers of products for use in animal and human health?

Modification of Gene Expression

In addition to inserting new genes in organisms, it also is possible to modify gene regulation by modifying regulatory molecules or the DNA sites to which they bind. These approaches make it possible to overproduce a wide variety of products, as shown in **table 41.8.**

The modification of gene expression also can be used to intentionally alter metabolic pathways by inactivation or deregulation of specific genes, which is the field of **pathway architecture.** Understanding pathway architecture makes it possible to design a pathway that will be most efficient by avoiding slower or energetically more costly routes. This approach has been used to improve penicillin production by *m*etabolic *path*way *e*ngineering **(MPE).**

An interesting development in modifying gene expression, which illustrates **metabolic control engineering,** is that of altering controls for the synthesis of lycopene, an important antioxidant thought to protect against some kinds of cancers and normally present at high levels in tomatoes. In this case, an engineered regulatory circuit was designed to control lycopene synthesis in response to the internal metabolic state of *E. coli.* Another recent development is the use of modified gene expression to produce variants of the antibiotic erythromycin. Blocking specific biochemical steps in pathways for the synthesis of an antibiotic precursor results in modified final products (**figure 41.11**). These altered products, which have slightly different structures, are tested for their possible antimicrobial effects. In addition this approach enables a better understanding of the structure-function relationships of antibiotics.

Protein Evolution

One of the newest approaches for creating novel metabolic capabilities in a given microorganism is protein evolution, which employs **forced evolution, adaptive mutations,** and **in vitro evolution (table 41.9).** Forced evolution and adaptive mutation involve the application of specific environmental stresses to "force" microorganisms to mutate and adapt, thus creating microorganisms with new biological capabilities. The mechanisms of these adaptive mutational processes include DNA rearrangements in which transposable elements and various types of recombination play critical roles.

In vitro evolution starts with purified nucleic acids rather than a whole organism. DNA templates (e.g., mutagenized versions of

Table 41.7 | Combinatorial Biology in Biotechnology: The Expression of Genes in Other Organisms to Improve Processes and Products

Property or Product Transferred	Microorganism Used	Combinatorial Process
Ethanol production	*Escherichia coli*	Integration of pyruvate decarboxylase and alcohol dehydrogenase II from *Zymomonas mobilis*.
1,3-Propanediol production	*E. coli*	Introduction of genes from the *Klebsiella pneumoniae dha* region into *E. coli* makes possible anaerobic 1,3-propanediol production.
Cephalosporin precursor synthesis	*Penicillium chrysogenum*	Production of 7-ADC and 7-ADCA[a] precursors by incorporation of the expandase gene of *Cephalosoporin acremonium* into *Penicillium* by transformation.
Lactic acid production	*Saccharomyces cerevisiae*	A muscle bovine lactate dehydrogenase gene (LDH-A) expressed in *S. cerevisiae*.
Xylitol production	*S. cerevisiae*	95% xylitol conversion from xylose was obtained by transforming the *XYLI* gene of *Pichia stipitis* encoding a xylose reductase into *S. cerevisiae*, making this organism an efficient organism for the production of xylitol, which serves as a sweetener in the food industry.
Creatininase[b]	*E. coli*	Expression of the creatininase gene from *Pseudomonas putida* R565. Gene inserted in a plasmid vector.
Pediocin[c]	*S. cerevisiae*	Expression of bacteriocin from *Pediococcus acidilactici* in *S. cerevisiae* to inhibit wine contaminants.
Acetone and butanol production	*Clostridium acetobutylicum*	Introduction of a shuttle vector into *C. acetobutylicum* results in acetone and butanol formation.

[a]7-ACA = 7-aminocephalosporanic acid; 7-ADCA = 7-aminodecacetoxycephalosporonic acid.

[b]T.-Y. Tang; C.-J. Wen; and W.-H. Liu. 2000. Expression of the creatininase gene from *Pseudomonas putida* RS65 in *Escherichia coli*. J. Ind. Microbiol. Biotechnol. 24:2–6.

[c]H. Schoeman; M. A. Vivier; M. DuToit; L. M. Y. Dicks; and I. S. Pretorius. 1999. The development of bactericidal yeast strains by expressing the *Pediococcus acidilactici* pediocin gene (pedA) in *Saccharomyces cerevisiae*. Yeast 15:647–656.

Adapted from S. Ostergaard; L. Olsson; and J. Nielson. 2000. Metabolic engineering of *Saccharomyces cerevisiae*. Microbiol. Mol. Biol. Rev. 64(1):34–50.

genes whose product is of interest) are transcribed in vitro by a phage RNA polymerase into RNA molecules that are selected based on their capacity to perform a specific function. The enzyme reverse transcriptase is then used to copy the selected RNA molecules into cDNA, which can then be amplified by PCR. After a number of such cycles, a gene that might be of industrial importance will "evolve."

Laboratory-based protein evolution was introduced in the 1970s. Since that time, recombinant DNA approaches have not only provided more efficient means of generating mutations, but **high-throughput screening (HTS)** now enables the rapid selection of a single desirable mutant or molecule from tens of thousands of newly constructed strains, molecules, or compounds. HTS employs a combination of robotics and computer analysis to screen samples, usually present in 96-well microtitre plates for a specific trait. Not only is HTS used to sift through whole cells for a particular phenotype, it is also essential for the identification of new natural and synthetic compounds that have a desired activity. The combination of molecular biological approaches and HTS has propelled protein evolution to a new level of efficiency not previously envisioned.

Preservation of Microorganisms

Once a microorganism or virus has been selected or created to serve a specific purpose, it must be preserved in its original form for further use and study. Periodic transfers of cultures have been used in the past, but this can lead to mutations and phenotypic changes in microorganisms. To avoid these problems, a variety of culture preservation techniques may be used to maintain desired culture characteristics (**table 41.10**). **Lyophilization,** or freeze-drying, and storage in liquid nitrogen are frequently employed with microorganisms. Although lyophilization and liquid nitrogen storage are complicated and require expensive equipment, they allow microbial cultures to be stored for years without loss of viability or an accumulation of undesirable mutations.

1. What types of recombinant DNA techniques are being used to modify gene expression in microorganisms?
2. Define metabolic control engineering, metabolic pathway engineering, forced evolution, and adaptive mutations.
3. What is high-throughput screening and why has it become so important?
4. What approaches can be used for the preservation of microorganisms?

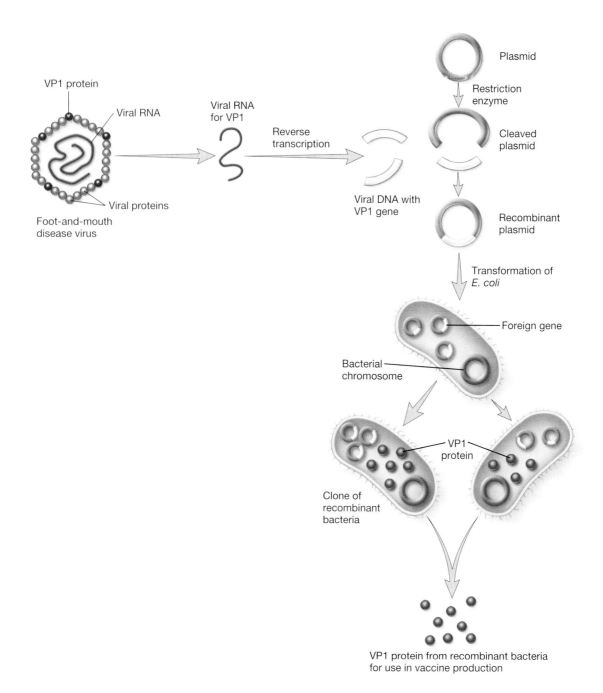

Figure 41.10 Recombinant Vaccine Production. Genes coding for desired products can be expressed in different organisms. By the use of recombinant DNA techniques, a foot-and-mouth disease vaccine is produced through cloning the vaccine genes into *E. coli.*

41.4 MICROORGANISM GROWTH IN CONTROLLED ENVIRONMENTS

For many industrial processes, microorganisms must be grown using specifically designed media under carefully controlled conditions, including temperature, aeration, and nutrient feeding. The development of appropriate culture media and the growth of microorganisms under industrial conditions are the subjects of this section.

Before proceeding, it is necessary to clarify terminology. The term **fermentation,** used in a physiological sense in earlier sections of the book, is employed in a much more general way in relation to industrial microbiology and biotechnology. As noted in **table 41.11,** the term can have several meanings. To industrial microbiologists, fermentation means the mass culture of microorganisms (or even plant and animal cells). Industrial fermentations requires the development of appropriate culture media and the transfer of small-scale technologies to a much larger scale.

Table 41.8	Examples of Recombinant DNA Systems Used to Modify Gene Expression	
Product	**Microorganism**	**Change**
Actinorhodin	*Streptomyces coelicolor*	Modification of gene transcription
Cellulase	*Clostridium* genes in *Bacillus*	Increased secretion through chromosomal DNA amplification
Recombinant protein albumin	*Saccharomyces cerevisiae*	Fusion to a high-production protein
Heterologous protein	*Saccharomyces cerevisiae*	Use of the inducible strong hybrid promoter UAS_{gal}/CYCl
Enhanced growth rate[a]	*Aspergillus nidulans*	Overproduction of glyceraldehyde-3-phosphate dehydrogenase
Amino acids[b]	*Corynebacterium*	Isolation of biosynthetic genes that lead to enhanced enzyme activities or removal of feedback regulation

[a,b]S. Ostergaard; L. Olsson; and J. Nielson. 2000. Metabolic engineering of *Saccharomyces cerevisiae*. *Microbiol. Mol. Biol. Rev.* 64(1):34–50. Table 1, p. 35.

Figure 41.11   Metabolic Engineering to Create Modified Antibiotics.   (a) Model for six elongation cycles (modules) in the normal synthesis of 6-deoxyerythonilide B (DEB), a precursor to the important antibiotic erythromycin. **(b)** Changes in structure that occur when the enoyl reductase enzyme of module 4 is blocked. **(c)** Changes in structure that occur when the keto reductase enzyme of module 5 is blocked. These changed structures (the highlighted areas) may lead to the synthesis of modified antibiotics with improved properties.

Table 41.9 Protein Evolution in Bacteria

Genetic Engineering Mechanisms	DNA Changes Mediated
Localized SOS mutagenesis	Base substitutions, frameshifts
Adapted frameshifting	-1 frameshifting
Tn5, Tn9, Tn10 precise excision	Reciprocal recombination of flanking 8/9 bp repeats; restores original sequence
In vivo deletion, inversion, fusion, and duplication formation	Generally reciprocal recombination of short sequence repeats; occasionally nonhomologous
Type II topoisomerase recombination	Deletions and fusions by nonhomologous recombination, sometimes at short repeats
Site-specific recombination (type I topoisomerases)	Insertions, excisions/deletions, inversions by concerted or successive cleavage-ligation reactions at short sequence repeats; tolerates mismatches
Transposable elements (many species)	Insertions, transpositions, replicon fusions, adjacent deletions/excisions, adjacent inversions by ligation of 3′ OH transposon ends of 5′ PO_4 groups from staggered cuts at nonhomologous target sites
DNA uptake (transformation competence)	Uptake of single strand independent of sequence, or of double-stranded DNA carrying species identifier sequence

Adapted from J. A. Shapiro. 1999. Natural genetic engineering, adaptive mutation, and bacterial evolution. In *Microbial Ecology of Infectious Disease*, E. Rosenberg, editor, 259–75. Washington, D.C.: American Society for Microbiology. Derived from Table 2, pp. 263–64.

Table 41.10 Methods Used to Preserve Cultures of Interest for Industrial Microbiology and Biotechnology

Method	Comments
Periodic transfer	Variables of periodic transfer to new media include transfer frequency, medium used, and holding temperature; this can lead to increased mutation rates and production of variants
Mineral oil slant	A stock culture is grown on a slant and covered with sterilized mineral oil; the slant can be stored at refrigerator temperature
Minimal medium, distilled water, or water agar	Washed cultures are stored under refrigeration; these cultures can be viable for 3 to 5 months or longer
Freezing in growth media	Not reliable; can result in damage to microbial structures; with some microorganisms, however, this can be a useful means of culture maintenance
Drying	Cultures are dried on sterile soil (soil stocks), on sterile filter paper disks, or in gelatin drops; these can be stored in a desiccator at refrigeration temperature, or frozen to improve viability
Freeze-drying (lyophilization)	Water is removed by sublimation, in the presence of a cryoprotective agent; sealing in an ampule can lead to long-term viability, with 30 years having been reported
Ultrafreezing	Liquid nitrogen at $-196°C$ is used, and cultures of fastidious microorganisms have been preserved for more than 15 years

Table 41.11 Fermentation: A Word with Many Meanings for the Microbiologist

1. Any process involving the mass culture of microorganisms, either aerobic or anaerobic
2. Any biological process that occurs in the absence of O_2
3. Food spoilage
4. The production of alcoholic beverages
5. Use of an organic substrate as the electron donor and acceptor
6. Use of an organic substrate as an electron donor, and of the same partially degraded organic substrate as an electron acceptor
7. Growth dependent on substrate-level phosphorylation

Medium Development

The medium used to grow a microorganism is critical because it can determine the level of microbial growth and product formation. In order to maximize competitiveness, lower-cost crude materials are used as sources of carbon, nitrogen, and phosphorus (**table 41.12**). Crude plant hydrolysates often are used as complex sources of carbon, nitrogen, and growth factors. By-products from the brewing industry frequently are employed because of their lower cost and greater availability. Other useful carbon sources include molasses and whey from cheese manufacture.   Culture media (section 5.7)

Table 41.12	Major Components of Growth Media Used in Industrial Processes
Source	**Raw Material**
Carbon and energy	Molasses
	Whey
	Grains
	Agricultural wastes (corncobs)
Nitrogen	Corn-steep liquor
	Soybean meal
	Stick liquor (slaughterhouse products)
	Ammonia and ammonium salts
	Nitrates
	Distiller's solubles
Vitamins	Crude preparations of plant and animal products
Iron, trace salts	Crude inorganic chemicals
Buffers	Chalk or crude carbonates
	Fertilizer-grade phosphates
Antifoam agents	Higher alcohols
	Silicones
	Natural esters
	Lard and vegetable oils

The levels and balance of minerals (especially iron) and growth factors can be critical in medium formulation. For example, biotin and thiamine, by influencing biosynthetic reactions, control product accumulation in many fermentations. The medium also may be designed so that carbon, nitrogen, phosphorus, iron, or a specific growth factor will become limiting after a given time during the fermentation. In such cases the limitation often causes a shift from growth to production of desired metabolites.

Growth of Microorganisms in an Industrial Setting

Once a medium is developed, the physical environment for optimum microbial growth in the mass culture system must be defined. This often involves precise control of agitation, temperature, pH, and oxygenation. Phosphate buffers can be used to control pH while also providing a source of phosphorus. Oxygen limitations can be critical in aerobic growth processes.

The O_2 concentration and flux rate must be sufficiently high to have O_2 in excess within the cells. This is especially true when a dense microbial culture is growing. When filamentous fungi and actinomycetes are cultured, aeration can be even further limited by filamentous growth (**figure 41.12**). Such filamentous growth results in a viscous, plastic medium, known as a **non-Newtonian broth,** which offers even more resistance to stirring and aeration.

It is essential to assure that these physical factors are not limiting microbial growth. This is most critical during **scaleup,** where a successful procedure developed in a small shake flask is modified for use in a large fermenter. The microenvironment of the small culture must be maintained despite increases in the culture volume. If a successful transition is made from a process originally developed in a 250 ml Erlenmeyer flask to a 100,000 liter reactor, then the process of scaleup has been carried out successfully.

Microorganisms can be grown in culture tubes, shake flasks, and stirred fermenters or other mass culture systems. Stirred fermenters can range in size from 3 or 4 liters to 100,000 liters or larger, depending on production requirements. A typical industrial stirred fermentation unit is illustrated in **figure 41.13b.** Not only must the medium be sterilized but aeration, pH adjustment,

(a)   (b)

Figure 41.12   Filamentous Growth During Fermentation.   Filamentous fungi and actinomycetes can change their growth form during the course of a fermentation. The development of pelleted growth by fungi has major effects on oxygen transfer and energy required to agitate the culture. **(a)** Initial culture. **(b)** After 18 hours growth.

Figure 41.13 Industrial Stirred Fermenters. **(a)** Large fermenters used by a pharmaceutical company for the microbial production of antibiotics. **(b)** Details of a fermenter unit. This unit can be run under oxic or anoxic conditions, and nutrient additions, sampling, and fermentation monitoring can be carried out under aseptic conditions. Biosensors and infrared monitoring can provide real-time information on the course of the fermentation. Specific substrates, metabolic intermediates, and final products can be detected.

(a)

(b)

sampling, and process monitoring must be carried out under rigorously controlled conditions. When required, foam control agents must be added, especially with high-protein media. Computers are used to monitor outputs from probes that determine microbial biomass, levels of critical metabolic products, pH, input and exhaust gas composition, and other parameters. Environmental conditions can be changed or held constant over time, depending on the goals for the particular process.

Frequently a critical component in the medium, often the carbon source, is added continuously—**continuous feed**—so that the microorganism will not have excess substrate available at any given time. This is particularly important with glucose and other carbohydrates. If excess glucose is present at the beginning of a fermentation, it can be catabolized to yield ethanol, which is lost as a volatile product and reduces the final yield. This can occur even under oxic conditions.

Besides the traditional stirred aerobic or anaerobic fermenter, other approaches can be used to grow microorganisms. These alternatives, illustrated in **figure 41.14,** include lift-tube fermenters (figure 41.14*a*), which eliminate the need for stir-

rers that can be fouled by filamentous fungi. Also available is solid-state fermentation (figure 41.14*b*), in which a particulate substrate is kept moist to maintain a thin surface water film where microbes can grow and oxygen is available. In various types of fixed- (figure 41.14*c*) and fluidized-bed reactors (figure 41.14*d*), the microorganisms are associated with inert surfaces as biofilms and medium flows past the fixed or suspended particles.

Dialysis culture units also can be used (figure 41.14*e*). These units allow toxic waste metabolites or end products to diffuse away from the microbial culture and permit new substrates to diffuse through the membrane toward the culture. Continuous culture techniques using chemostats (figure 41.14*f*) can markedly improve cell outputs and rates of substrate use because microorganisms can be maintained in a continuous logarithmic phase. However, continuous maintenance of an organism in an active growth phase is undesirable in many industrial processes.

Microbial products often are classified as primary and secondary metabolites. As shown in **figure 41.15, primary metabolites** consist of compounds related to the synthesis of microbial

Figure 41.14   Alternate Methods for Mass Culture.   In addition to stirred fermenters, other methods can be used to culture microorganisms in industrial processes. In many cases these alternate approaches will have lower operating costs and can provide specialized growth conditions needed for product synthesis.

(a) Lift-tube fermenter
Density difference of
gas bubbles entrained
in medium results in
fluid circulation

← Air in

(b) Solid-state fermentation
Growth of culture
without presence of
added free water

Flow in →

(c) Fixed-bed reactor
Microorganisms on surfaces
of support material;
flow can be up or down

Fixed
support
material

→ Flow out

(d) Fluidized-bed reactor
Microorganisms on surfaces
of particles suspended
in liquid or gas stream–
upward flow

→ Flow out

Suspended
support particles

Flow in →

(e) Dialysis culture unit
Waste products diffuse
away from the culture.
Substrate may diffuse
through membrane to
the culture

Membrane

Culture

Medium
or buffer

Medium in →

(f) Continuous culture unit (Chemostat)
Medium in and excess
medium and cells to waste

Medium and
cells out

cells during balanced growth. They include amino acids, nucleotides, and fermentation end products such as ethanol and organic acids. In addition, industrially useful enzymes, either associated with the microbial cells or exoenzymes, often are synthesized by microorganisms during growth.

Secondary metabolites usually accumulate during the period of nutrient limitation or waste product accumulation that follows the active growth phase. These compounds have only a limited relationship to the synthesis of cell materials and normal growth. Most antibiotics and the mycotoxins fall into this category.   The growth curve (section 6.2)

1. How is the cost of media reduced during industrial operations? Discuss the effect of changing balances in nutrients such as minerals, growth factors, and the sources of carbon, nitrogen, and phosphorus.
2. What are non-Newtonian broths and why are these important in fermentations?
3. Discuss scaleup and the objective of the scaleup process.
4. What parameters can be monitored in a modern, large-scale industrial fermentation?
5. Besides the aerated, stirred fermenter, what other alternatives are available for the mass culture of microorganisms in industrial processes? What is the principle by which a dialysis culture system functions?

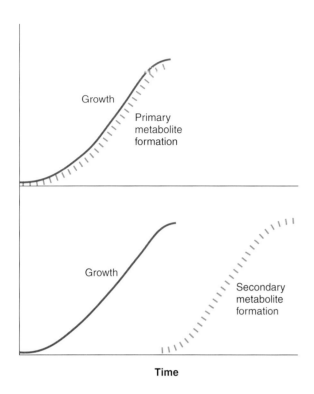

Figure 41.15   Primary and Secondary Metabolites.
Depending on the particular organism, the desired product may be formed during or after growth. Primary metabolites are formed during the active growth phase, whereas secondary metabolites are formed after growth is completed.

41.5   MAJOR PRODUCTS OF INDUSTRIAL MICROBIOLOGY

Industrial microbiology has provided products that have profoundly changed our lives and life spans. They include industrial and agricultural products, food additives, products for human and animal health, and biofuels (**table 41.13**). Antibiotics and other natural products used in medicine and health have made major contributions to the improved well-being of animal and human populations. Only major products in each category are discussed here.

Antibiotics

Many antibiotics are produced by microorganisms, predominantly by actinomycetes in the genus *Streptomyces* and by filamentous fungi (*see table 34.2*). We now discuss the synthesis of several of the most important antibiotics to illustrate the critical role of medium formulation and environmental control in the production of these important compounds.   Antimicrobial chemotherapy (chapter 34)

Penicillin and Semisynthetic Penicillins

Although penicillin is less widely used because of antibiotic resistance, this drug produced by *Penicillium chrysogenum,* is an ex-

Table 41.13	Major Microbial Products and Processes of Interest in Industrial Microbiology
Substances	**Microorganisms**
Industrial Products	
Ethanol (from glucose)	*Saccharomyces cerevisiae*
Ethanol (from lactose)	*Kluyveromyces fragilis*
Acetone and butanol	*Clostridium acetobutylicum*
2,3-butanediol	*Enterobacter, Serratia*
Enzymes	*Aspergillus, Bacillus, Mucor, Trichoderma*
Agricultural Products	
Gibberellins	*Gibberella fujikuroi*
Food Additives	
Amino acids (e.g., lysine)	*Corynebacterium glutamicum*
Organic acids (citric acid)	*Aspergillus niger*
Nucleotides	*Corynebacterium glutamicum*
Vitamins	*Ashbya, Eremothecium, Blakeslea*
Polysaccharides	*Xanthomonas*
Medical Products	
Antibiotics	*Penicillium, Streptomyces, Bacillus*
Alkaloids	*Claviceps purpurea*
Steroid transformations	*Rhizopus, Arthrobacter*
Insulin, human growth hormone, somatostatin, interferons	*Escherichia coli, Saccharomyces cerevisiae,* and others (recombinant DNA technology)
Biofuels	
Hydrogen	Photosynthetic microorganisms
Methane	*Methanobacterium*
Ethanol	*Zymomonas, Thermoanaerobacter*

cellent example of a fermentation for which careful adjustment of the medium composition is used to achieve maximum yields. Provision of the slowly hydrolyzed disaccharide lactose, in combination with limited nitrogen availability, stimulates a greater accumulation of penicillin after growth has stopped (**figure 41.16**). The same result can be achieved by using a slow continuous feed of glucose. If a particular penicillin is needed, the specific precursor is added to the medium. For example, phenylacetic acid is added to maximize production of penicillin G, which has a benzyl side chain (*see figure 34.5*). The fermentation pH is maintained around neutrality by the addition of sterile alkali, which assures maximum stability of the newly synthesized penicillin. Once the fermentation is completed, normally in 6 to 7 days, the broth is separated from the fungal mycelium and processed by absorption, precipitation, and crystallization to yield the final product. This basic product can then be modified by chemical procedures to yield a variety of semisynthetic penicillins.

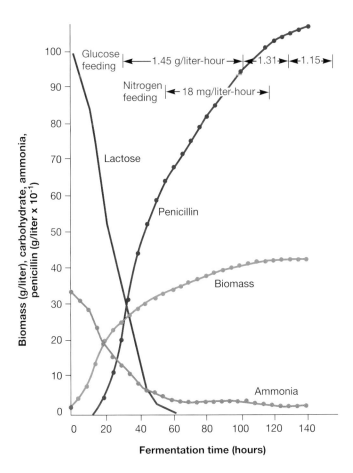

Figure 41.16   Penicillin Fermentation Involves Precise Control of Nutrients.   The synthesis of penicillin begins when nitrogen from ammonia becomes limiting. After most of the lactose (a slowly catabolized disaccharide) has been degraded, glucose (a rapidly used monosaccharide) is added along with a low level of nitrogen. This stimulates maximum transformation of the carbon sources to penicillin. The scale factor is presented using the convention recommended by the ASM. That is, a number on the axis should be multiplied by 0.10 to obtain the true value.

Streptomycin

Streptomycin is a secondary metabolite produced by *Streptomyces griseus*. In this fermentation a soybean-based medium is used with glucose as a carbon source. The nitrogen source is thus in a combined form (soybean meal), which limits growth. After growth the antibiotic levels in the culture begin to increase (**figure 41.17**) under conditions of controlled nitrogen limitation.

Amino Acids

Amino acids such as lysine and glutamic acid are used in the food industry as nutritional supplements in bread products and as flavor-enhancing compounds such as monosodium glutamate (MSG).

Amino acid production is typically carried out by means of **regulatory mutants,** which have a reduced ability to limit synthe-sis of a specific amino acid or key intermediate. Normal microorganisms avoid overproduction of biochemical intermediates by the careful regulation of cellular metabolism. Production of glutamic acid and several other amino acids is now carried out using mutants of *Corynebacterium glutamicum* that lack, or have only a limited ability to process, the TCA cycle intermediate α-ketoglutarate (*see appendix II*) to succinyl-CoA as shown in **figure 41.18.** A controlled low biotin level and the addition of fatty acid derivatives results in increased membrane permeability and excretion of high concentrations of glutamic acid. The impaired bacteria use the glyoxylate pathway to meet their needs for essential biochemical intermediates, especially during the growth phase. After growth becomes limited because of changed nutrient availability, an almost complete molar conversion (or 81.7% weight conversion) of isocitrate to glutamate occurs.   Synthesis of amino acids (section 10.5)

Lysine, an essential amino acid (i.e., required in the human diet) is used to supplement cereals and breads. It was originally produced in a two-step microbial process. This has been replaced by a single-step fermentation in which the bacterium *C. glutamicum,* blocked in the synthesis of homoserine, accumulates lysine. Over 44 g/liter can be produced in a 3-day fermentation.

Organic Acids

Organic acid production by microorganisms is important in industrial microbiology and illustrates the effects of trace metal levels and balances on organic acid synthesis and excretion. Citric, acetic, lactic, fumaric, and gluconic acids are major products (**table 41.14**). Until microbial processes were developed, the major source of citric acid was citrus fruit. Today most citric acid is produced by microorganisms; 70% is used in the food and beverage industry, 20% in pharmaceuticals, and the balance in other industrial applications.

The essence of citric acid fermentation involves limiting the amounts of trace metals such as manganese and iron to stop *Aspergillus niger* growth at a specific point in the fermentation. The medium often is treated with ion exchange resins to ensure low and controlled concentrations of available metals, and is carried out in aerobic stirred fermenters. Generally, high sugar concentrations (15 to 18%) are used, and copper has been found to counteract the inhibition of citric acid production by iron above 0.2 ppm. The success of this fermentation depends on the regulation and functioning of the glycolytic pathway and the tricarboxylic acid cycle. After the active growth phase, when the substrate level is high, citrate synthase activity increases and the activities of aconitase and isocitrate dehydrogenase decrease. This results in citric acid accumulation and excretion by the stressed microorganism.

In comparison, the production of gluconic acid involves a single microbial enzyme, glucose oxidase, found in *Aspergillus niger. A. niger* is grown under optimum conditions in a corn-steep liquor medium. Growth becomes limited by nitrogen, and the resting cells transform the remaining glucose to gluconic acid in a single-step reaction. Gluconic acid is used as a carrier for calcium and iron and as a component of detergents.

overlying PCB-contaminated sand-clay soils have been observed to create their own "clay hutches" by binding clays to their outer surfaces with exopolysaccharides. Thus the application of principles of microbial ecology can facilitate the successful management of microbial communities in nature.

---

1. What factors might limit the ability of microorganisms, after addition to a soil or water, to persist and carry out desired functions?
2. What types of microhabitats can be used with microorganisms when they are added to a complex natural environment?
3. Compare natural and inert microhabitats. Under what conditions might each be used?

41.8 MICROBES AS PRODUCTS

So far we have discussed the use of microbial products like antibiotics and organic acids, or the use of microbial communities, to meet defined goals. However, single microbial species can be marketed as valuable products. Perhaps the most common example is the inoculation of legume seeds with rhizobia to ensure efficient nodulation and nitrogen fixation, as discussed previously. Here we introduce several other microbes and microbial structures that are of industrial and/or agricultural relevance.

Nanotechnology

Diatoms have aroused the interest of nanotechnologists. These photosynthetic protists produce intricate silica shells that differ according to species (**figure 41.28**). Nanotechnologists are interested in diatoms because they create precise structures at the micrometer scale. Three-dimensional structures in nanotechnology are currently built plane by plane and meticulous care must be taken to etch each individual structure to its final, exact shape. Diatoms, on the other hand, build directly in three dimensions and do so while growing exponentially, making them attractive for nanotechnology. There have been a number of ideas and approaches to harness these microbial "factories," but one technique is especially fascinating. Diatoms shells are incubated at 900°C in an atmosphere of magnesium for several hours. Amazingly, this results in an atom-for-atom substitution of silicon with magnesium without loss of 3D structure. Thus silicon oxide, which is of little use in nanotechnology, is converted to highly useful magnesium oxide. Protist classification: Stramenopiles (section 25.6)

Magnetotactic bacteria are also of interest to nanotechnologists. Magnetosomes are formed by certain bacteria through the accumulation of iron into magnetite. The sizes and shapes of the magnetosomes differ among species, but like diatom shells, they

Figure 41.28 Marine Diatom Surface Features. (a) *Glyphodiscus stellatus*; scale bar is 20 μm. (b) *G. stellatus* close-up; scale bar is 5 μm. (c) *Roperia tesselata*; scale bar is 10 μm. (d) *R. tesselata* close-up; scale bar is 5 μm.

(a)

(b)

(c)

(d)

are perfectly formed despite the fact that they are only tens of nanometers in diameter. Although tiny, magnetic beads can be chemically synthesized, they are not as precisely formed and lack the membrane that surrounds magnetosomes. This membrane en ables the attachment of useful biological molecules like enzymes and antibodies. Potential applications for magnetosomes currently under investigation include their use as a contrast medium to improve magnetic resonance tomography (MRI) and as biological probes to detect cancer at early stages. The cytoplasmic matrix: Inclusion bodies (section 3.3)

Biosensors

A rapidly developing area of biotechnology, is that of **biosensor** production. In this field of bioelectronics, living microorganisms (or their enzymes or organelles) are linked with electrodes, and biological reactions are converted into electrical currents (**figure 41.29**). Biosensors have been developed to measure specific components in beer, to monitor pollutants, to detect flavor compounds in food, and to study environmental processes such as changes in biofilm concentration gradients. It is possible to measure the concentration of substances from many different environments (**table 41.17**). Applications include the detection of glucose, acetic acid, glutamic acid, ethanol, and biochemical oxygen demand. Biosensors have been developed using immunochemical-based detection systems. These new biosensors will detect pathogens, herbicides, toxins, proteins, and DNA. Since the bioterrorism attacks of 2001, the U.S. government has stepped up funding for research and development of biosensors capable of detecting minute levels of potential airborne pathogens. Many of these biosensors are based on the use of a streptavidin-biotin recognition system (**Techniques & Applications 41.5**). Microbial tidbits 35.2: Biosensors

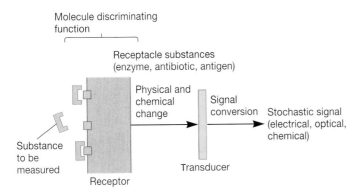

Figure 41.29   Biosensor Design.   Biosensors are finding increasing applications in medicine, industrial microbiology, and environmental monitoring. In a biosensor a biomolecule or whole microorganism carries out a biological reaction, and the reaction products are used to produce an electrical signal.

Table 41.17	Biosensors: Potential Biomedical, Industrial, and Environmental Applications

Clinical diagnosis and biomedical monitoring

Agricultural, horticultural, veterinary analysis

Detection of pollution, and microbial contamination of water

Fermentation analysis and control

Monitoring of industrial gases and liquids

Measurement of toxic gas in mining industries

Direct biological measurement of flavors, essences, and pheromones

---

1. What are biosensors and how do they detect substances?
2. What areas are biosensors being used in to assist in chemical and biological monitoring efforts?
3. Describe streptavidin-biotin systems and how they work. Why is this technique important?

Biopesticides

There has been a long-term interest in the use of bacteria, fungi, and viruses as **bioinsecticides** and **biopesticides** (**table 41.18**). These are defined as biological agents, such as bacteria, fungi, viruses, or their components, which can be used to kill a susceptible insect. In this section, we discuss some of the major uses of bacteria, fungi, and viruses to control populations of insects.

Bacteria

Bacterial agents include a variety of *Bacillus* species; however, *B. thuringiensis* is most widely used. This bacterium is only weakly toxic to insects as a vegetative cell, but during sporulation, it produces an intracellular protein toxin crystal, the parasporal body, that can act as a microbial insecticide for specific insect groups. Class *Bacilli:* Order *Bacillale:* (section 23.5)

The parasporal crystal, after exposure to alkaline conditions in the insect hindgut, fragments to release the protoxin. After this reacts with a protease enzyme, the active toxin is produced (**figure 41.30**). Six of the active toxin units integrate into the plasma membrane (figure 41.30*b,c*) to form a hexagonal-shaped pore through the midgut cell, as shown in figure 41.30*d*. This leads to the loss of osmotic balance and ATP, and finally to cell lysis.

B. thuringiensis can be grown in fermenters. The spores and crystals are released into the medium when cells lyse. The medium is then centrifuged and made up as a dust or wettable powder for application to plants. This insecticide, known as Bt, has been used on a worldwide basis for over 40 years. Unlike chemical insecticides, Bt does not accumulate in the soil or in nontarget animals. Rather it is readily lost from the environment by microbial and abiotic degradation.

Techniques & Applications

41.5   Streptavidin-Biotin Binding and Biotechnology

Egg white contains many proteins and glycoproteins with unique properties. One of the most interesting, which binds tenaciously to biotin, was isolated in 1963. This glycoprotein, called avidin due to its "avid" binding of biotin, was suggested to play an important role: making egg white antimicrobial by "tying up" the biotin needed by many microorganisms. Avidin, which functions best under alkaline conditions, has the highest known binding affinity between a protein and a ligand. Several years later, scientists at Merck & Co., Inc. discovered a similar protein produced by the actinomycete *Streptomyces avidini,* which binds biotin at a neutral pH and which does not contain carbohydrates. These characteristics make streptavidin an ideal binding agent for biotin, and it has been used

in an almost unlimited range of applications, as shown in the **Box figure.** The streptavidin protein is joined to a probe. When a sample is incubated with the biotinylated binder, the binder attaches to any available target molecules. The presence and location of target molecules can be determined by treating the sample with a streptavidin probe because the streptavidin binds to the biotin on the biotinylated binder, and the probe is then visualized. This detection system is employed in a wide variety of biotechnological applications, including use as a nonradioactive probe in hybridization studies and as a critical component in biosensors for a wide range of environmental monitoring and clinical applications.

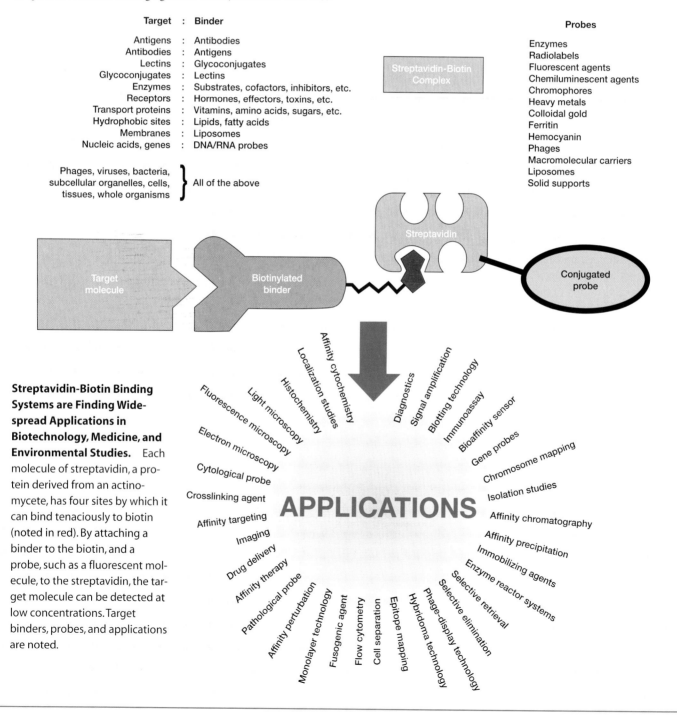

Streptavidin-Biotin Binding Systems are Finding Widespread Applications in Biotechnology, Medicine, and Environmental Studies. Each molecule of streptavidin, a protein derived from an actinomycete, has four sites by which it can bind tenaciously to biotin (noted in red). By attaching a binder to the biotin, and a probe, such as a fluorescent molecule, to the streptavidin, the target molecule can be detected at low concentrations. Target binders, probes, and applications are noted.

Microbial Group	Major Organisms and Applications
Bacteria	*Bacillus thuringiensis* and *Bacillus popilliae* are the two major bacteria of interest. *Bacillus thuringiensis* is used on a wide variety of vegetable and field crops, fruits, shade trees, and ornamentals. *B. popilliae* is used primarily against Japanese beetle larvae. Both bacteria are considered harmless to humans. *Pseudomonas fluorescens*, which contains the toxin-producing gene from *B. thuringiensis*, is used on maize to suppress black cutworms.
Viruses	Three major virus groups that do not appear to replicate in warm-blooded animals are used: nuclear polyhedrosis virus (NPV), granulosis virus (GV), and cytoplasmic polyhedrosis virus (CPV). These viruses are more protected in the environment.
Fungi	Over 500 different fungi are associated with insects. Infection and disease occur primarily through the insect cuticle. Four major genera have been used. *Beauveria bassiana* and *Metarhizium anisopliae* are used for control of the Colorado potato beetle and the froghopper in sugarcane plantations, respectively. *Verticillium lecanii* and *Entomophthora* spp., have been associated with control of aphids in greenhouse and field environments. *Coelomyces*, a chytrid, also is used for control of mosquitoes.

Figure 41.30 **The Mode of Action of the *Bacillus thuringiensis* Toxin.** (a) Release of the protoxin from the parasporal body and modification by proteases in the hindgut. (b) Insertion of the 68 kDa active toxin molecules into the membrane. (c) Aggregation and pore formation, showing a cross section of the pore. (d) Final hexagonal pore which causes an influx of water and cations as well as a loss of ATP, resulting in cell imbalance and lysis.

Unlike other genetically modified organisms (GMOs), transgenic plants expressing the *B. thuringiensis* toxin gene, *cry* (for *cry*stal), have generally been well accepted. In 1996, commercialized Bt-corn, Bt-potato, and Bt-cotton were introduced into the United States and soon farmers in other countries such as Australia, Canada, China, France, Indonesia, Mexico, Spain, and Ukraine followed. The widespread acceptance of these plants reflects the history of safe application of Bt as an insecticide without adverse environmental or health impacts. In addition, it is well understood that the Cry protein can only be activated in the target insect. Long-term studies have shown that Bt is nontoxic to mammals and is not an allergen in humans. One potential problem, the horizontal gene flow of the *cry* gene to weeds and other plants, has not been reliably demonstrated.

Viruses

Viruses that are pathogenic for specific insects include nuclear polyhedrosis viruses (NPVs), granulosis viruses (GVs), and cytoplasmic polyhedrosis viruses (CPVs). Currently over 125 types of NPVs are known, of which approximately 90% affect the *Lepidoptera*—butterflies and moths. Approximately 50 GVs are known, and they, too, primarily affect butterflies and moths. CPVs are the least host-specific viruses, affecting about 200 different types of insects. An important commercial viral pesticide is marketed under the trade name Elcar for control of the cotton bollworm *Heliothis zea.* Insect viruses (section 18.8)

Fungi

Fungi also can be used to control insect pests. Fungal bioinsecticides, as listed in table 41.18, are finding increasing use in agriculture. However, large-scale production of fungal insect pathogens is difficult and when introduced into natural ecosystems, they often do not persist. Nonetheless, the development of fungal biopesticides continues to progress.

1. What two important bacteria have been used as bioinsecticides?
2. Briefly describe how the *Bacillus thuringiensis* toxin kills insects.
3. What types of viruses are being used to attempt to control insects?
4. Which fungi presently are being used as biopesticides?

41.9 IMPACTS OF MICROBIAL BIOTECHNOLOGY

The use of microorganisms in industrial microbiology and biotechnology, as discussed in this chapter, does not take place in an ethical and ecological vacuum. Decisions to make a particular product, and also the methods used, can have long-term and often unexpected effects, as with the appearance of antibiotic-resistant pathogens around the world.

Microbiology is a critical part of the area of **industrial ecology,** concerned with tracking the flow of elements and compounds through the natural and social worlds, or the biosphere and the anthrosphere. Microbiology, especially as an applied discipline, should be considered within its supporting social world.

Microorganisms have been of immense benefit to humanity through their role in food production and processing, the use of their products to improve human and animal health, in agriculture, and for the maintenance and improvement of environmental quality. Other microorganisms, however, are important pathogens and agents of spoilage, and microbiologists have helped control or limit the activities of these harmful microorganisms. The discovery and use of beneficial microbial products, such as antibiotics, have contributed to a doubling of the human life span in the last century.

A microbiologist who works in any of these areas of biotechnology should consider the longer-term impacts of possible technical decisions. Our first challenge, as microbiologists, is to understand, as much as is possible, the potential impacts of new products and processes on the broader society as well as on microbiology. An essential part of this responsibility is to be able to communicate effectively with the various "societal stakeholders" about the immediate and longer-term potential impacts of microbial-based (and other) technologies.

1. Discuss possible ethical and ecological impacts of a particular product or process discussed in this chapter. Think in terms of the broadest possible impacts in your discussion of this problem.
2. Define industrial ecology.
3. What are the biosphere and anthrosphere? Why might you think the term anthrosphere was coined?

Summary

41.1 Water Purification and Sanitary Analysis

a. Water purification can involve the use of sedimentation, coagulation, chlorination, and rapid and slow sand filtration. Chlorination may lead to the formation of organic disinfection by-products, including trihalomethane (THM) compounds, which are potential carcinogens (**figure 41.1**).

b. *Cryptosporidium, Cyclospora,* viruses, and *Giardia* are of concern, as conventional water purification and chlorination will not always assure their removal and inactivation to acceptable limits.

c. Indicator organisms are used to indicate the presence of pathogenic microorganisms. Most probable number (MPN) and membrane filtration procedures are employed to estimate the number of indicator organisms present. Presence-absence (P-A) tests for coliforms and defined substrate tests for coliforms and *E. coli* allow 100 ml water volumes to be tested with minimum time and materials (**table 41.1; figures 41.2** and **41.3**).

d. Molecular techniques based on the polymerase chain reaction (PCR) can be used to detect waterborne pathogens such as Shigalike-toxin producing *E. coli* O157:H7, when a preenrichment step is used.

41.2 Wastewater Treatment

a. The biochemical oxygen demand (BOD) test is an indirect measure of organic matter that can be oxidized by the aerobic microbial community. In this assay, oxygen should never limit the rate of reaction. The chemical oxygen demand

(COD) and total organic carbon (TOC) tests provide information on carbon that is not biodegraded in the 5-day BOD test.

b. Conventional sewage treatment is a controlled intensification of natural self-purification processes, and it can involve primary, secondary, and tertiary treatment (**figure 41.4**).

c. Constructed wetlands involve the use of aquatic plants (floating, emergent, submerged) and their associated microorganisms for the treatment of liquid wastes (**figure 41.7**).

d. Home treatment systems operate on general self-purification principles. The conventional septic tank (**figure 41.8**) provides anaerobic liquefaction and digestion whereas the aerobic leach-field allows oxidation of the soluble effluent. These systems are now being designed to provide nitrogen and phosphorus removal, to lessen impacts of on-site sewage treatment systems on vulnerable marine and freshwaters.

e. Groundwater is an important resource that can be affected by pollutants from septic tanks and other sources. This vital water source must be protected and improved.

41.3 Microorganisms Used in Industrial Microbiology

a. Industrial microbiology has been used to manufacture such products as antibiotics, amino acids, and organic acids and has had many important positive effects on animal and human health. Most work in this area has been carried out using microorganisms isolated from nature or modified by the use of classic mutation techniques. Biotechnology involves the use of molecular techniques to modify and improve microorganisms.

b. Finding new microorganisms in nature for use in biotechnology is a continuing challenge. Only about 1% of the observable microbial community has been grown (**table 41.6**), but major advances in growing "uncultured" microbes are being made.

c. Selection and mutation continue to be important approaches for identifying new microorganisms. These well-established procedures are now being complemented by molecular techniques, including metabolic engineering and combinatorial biology. With combinatorial biology (**table 41.7**), it is possible to transfer genes from one organism to another organism, and to form new products.

d. Site-directed mutagenesis and protein engineering are used to modify gene expression. These approaches are leading to new and often different products with new properties (**figure 41.11**).

e. Protein evolution is of increasing interest. This involves exploiting microbial responses to stress in adaptive mutation and forced evolution, with the hope of identifying microorganisms with new properties. Alternatively, an in vitro approach can be used.

41.4 Microorganism Growth in Controlled Environments

a. Microorganisms can be grown in controlled environments of various types using fermenters and other culture systems. If defined constituents are used, growth parameters can be chosen and varied in the course of growing a microorganism. This approach is used particularly for the production of amino acids, organic acids, and antibiotics.

41.5 Major Products of Industrial Microbiology

a. A wide variety of compounds are produced in industrial microbiology that impact our lives in many ways (**table 41.13**). These include antibiotics, amino acids, organic acids, biopolymers such as the cyclodextrins, and biosurfactants (**figures 41.16–41.19**). Microorganisms also can be used as biocatalysts to carry out specific chemical reactions (**figure 41.20**).

b. Specialty nonantibiotic compounds are an important part of industrial microbiology and biotechnology. These include widely used antitumor agents (**table 41.15**).

41.6 Biodegradation and Bioremediation by Natural Communities

a. Microorganism growth in complex natural environments such as soils and waters is used to to carry out environmental management processes, including bioremediation, plant inoculation, and other related activities. In these cases, the microbes themselves are not final products.

b. Biodegradation is a critical part of natural systems mediated largely by microorganisms. This can involve minor changes in a molecule, fragmentation, or mineralization (**figure 41.21**).

c. Biodegradation can be influenced by many factors, including oxygen presence or absence, humic acids, and the presence of readily usable organic matter. Reductive dehalogenation proceeds best under anoxic conditions, and the presence of organic matter can facilitate modification of recalcitrant compounds in the process of cometabolism.

d. The structure of organic compounds influences degradation. If constituents are in specific locations on a molecule, as in the *meta* position (**figure 41.22**), or if varied structural isomers are present degradation can be affected.

e. Degradation management can be carried out in place, whether this be large marine oil spills, soils, or the subsurface (**figure 41.25**). Such large-scale efforts usually involve the use of natural microbial communities.

f. Degradation can lead to increased toxicity in many cases. If not managed carefully, widespread pollution can occur. Iron corrosion is a particular concern with methanogens and *Desulfovibrio* playing important roles in this process.

g. Plants can be used to stimulate biodegradation processes during phytoremediation. This can involve extraction, filtering, stabilization, and volatilization of pollutants (**figure 41.26** and **table 41.16**).

41.7 Bioaugmentation

a. Microorganisms can be added to environments that contain complex microbial communities with greater success if living or inert microhabitats are used. These can include living plant surfaces (seeds, roots, leaves) or inert materials such as microporous glass. *Rhizobium* is an important example of a microorganism added to a complex environment using a living microhabitat (the plant root).

41.8 Microbes As Products

a. Microorganisms are being used in a wide range of biotechnological applications such as nanotechnology and biosensors (**figures 41.28** and **41.29**).

b. Bacteria, viruses, and fungi can be used as bioinsecticides and biopesticides (**table 41.18**). *Bacillus thuringiensis* is an important biopesticide, and the BT gene has been incorporated into several important crop plants.

41.9 Impacts of Microbial Biotechnology

a. Industrial microbiology and biotechnology can have long-term and possibly unexpected positive and negative effects on the environment, and on animals and humans impacted by these technologies. Advances in biotechnology should be considered in a broad ecological and societal context, which is the focus of industrial ecology.

Key Terms

activated sludge 1055
adaptive mutation 1062
bioaugmentation 1080
biocatalyst 1074

biochemical oxygen demand (BOD) 1055
biodegradation 1075
bioinsecticides 1083

biopesticide 1083
biopolymer 1073
bioprospecting 1060
bioremediation 1075

biosensor 1083
biotransformation 1074
bulking sludge 1055
chemical oxygen demand (COD) 1054

Critical Thinking Questions

1. You wish to develop a constructed wetland for removal of metals from a stream at one site, and at another site, you wish to treat acid mine drainage. How might you approach each of these problems?

2. What alternatives, if any, can one use for protection against microbiological infection when swimming in polluted recreational water? Assume that you are part of a water rescue team.

3. What possible alternatives could be used to eliminate N and P releases from sewage treatment systems? What suggestions could you make that might lead to new technologies?

4. What further technological approaches will be required to culture microorganisms that have not yet been grown? Consider the roles of nutrient fluxes, communication molecules, and competition.

5. *Deinococcus radiodurans* is a species of bacteria that is highly resistant to radiation. Can you think of a biotechnological application? How would you test its utility?

6. Most commercial antibiotics are produced by actinomycetes, and only a few are synthesized by fungi and other bacteria. From physiological and environmental viewpoints, how might you attempt to explain this observation?

7. The terms biosphere and anthrosphere have been used, together with the term industrial ecology. How does microbial biotechnology relate to these concerns?

8. Discuss the risks of releasing genetically modified microbes or ones that are not natural to the particular environment. What would be your concerns? What precautions, if any, would you take?

9. Why, when a microorganism is removed from a natural environment and grown in the laboratory, will it usually not be able to effectively colonize its original environment if it is grown and added back? Consider the nature of growth media used in the laboratory in comparison to growth conditions in a soil or water when attempting to understand this fundamental problem in microbial ecology.

10. Why might *Bacillus thuringiensis* bioinsecticides be of interest in other areas of biotechnology? Consider the molecular aspects of their mode of action.

Learn More

Cameotra, S. S., and Makkar, R. S. 2004. Recent applications of biosurfactants as biological and immunological molecules. *Curr. Opin. Microbiol.* 7:262–66.

Compant, S.; Duffy, B.; Nowak, J.; Clément, C.; and Barka, E. A. 2005. Use of plant growth-promoting bacteria for biocontrol of plant diseases: Principles, action, and future prospects. *Appl. Env. Microbiol.* 71:4951–59.

de Maagd, R. A.; Bravo, A.; and Crickmore, N. 2001. How *Bacillus thuringiensis* has evolved specific toxins to colonize the insect world. *Trends Genet.* 17(4):193–99.

Drum, R. W., and Gordon, R. 2003. Star Trek replicators and diatom nanotechnology. *Trends Biotechnol.* 21:325–27.

Gross, R. A., and Kalra, B. 2002. Biodegradable polymers for the environment. *Science* 297:803–807.

Hurst, C. J.; Crawford, R. L.; Knudsen, G. R.; McInerney, M. J.; and Stetzenbach, L. D. 2002. *Manual of environmental microbiology,* 2nd ed. Washington, D.C.: ASM Press.

Rittmann, B. E., and McCarty, P. L. 2001. *Environmental biotechnology: Principles and applications.* New York: McGraw-Hill.

Selifonova, O.; Valle, F.; and Schellenberger, V. 2001. Rapid evolution of novel traits in microorganisms. *Appl. Environ. Microbiol.* 67:3645–49.

Shelton, A. M.; Zhao, J.-Z.; and Roush, R. T. 2002. Economic, ecological, food safety, and social consequences of Bt transgenic plants. *Annu. Rev. Entomol.* 47:845–81.

Teusink, B., and Smid, E. J. 2006. Modelling strategies for the industrial exploitation of lactic acid bacteria. *Nature Rev. Microbiol.* 4:46–56.

Wackett, L. P., and Hershberger, C. D. 2001. Biocatalysts and biodegradation: Microbial transformations of organic compounds. Washington, D.C.: ASM Press.

Yaun, L.; Kurek, I.; English, J.; and Keenan, R. 2005. Laboratory-directed protein evolution. *Microbiol. Molec. Biol. Rev.* 69:373–92.

Zhang, Y.-X.; Perry, K.; Vinci, V. A.; Powell, K.; Stemmer, W. P. C.; and del Cardayré, S. B. 2002. Genome shuffling leads to rapid phenotypic improvement in bacteria. *Nature* 415:644–46.

Appendix I
A Review of the Chemistry of Biological Molecules

Appendix I provides a brief summary of the chemistry of organic molecules with particular emphasis on the molecules present in microbial cells. Only basic concepts and terminology are presented; introductory textbooks in biology and chemistry should be consulted for a more extensive treatment of these topics.

ATOMS AND MOLECULES

Matter is made of elements that are composed of atoms. An element contains only one kind of atom and cannot be broken down to simpler components by chemical reactions. An atom is the smallest unit characteristic of an element and can exist alone or in combination with other atoms. When atoms combine they form molecules. Molecules are the smallest particles of a substance. They have all the properties of the substance and are composed of two or more atoms.

Although atoms contain many subatomic particles, three directly influence their chemical behavior—protons, neutrons, and electrons. The atom's nucleus is located at its center and contains varying numbers of protons and neutrons (**figure AI.1**). Protons have a positive charge, and neutrons are uncharged. The mass of these particles and the atoms that they compose is given in terms of the atomic mass unit (AMU), which is 1/12 the mass of the most abundant carbon isotope. Often the term dalton (Da) is used to express the mass of molecules. It also is 1/12 the mass of an atom of ^{12}C or 1.661×10^{-24} grams. Both protons and neutrons have a mass of about one dalton. The atomic weight is the actual measured weight of an element and is almost identical to the mass number for the element, the total number of protons and neutrons in its nucleus. The mass number is indicated by a superscripted number preceding the element's symbol (e.g., ^{12}C, ^{16}O, and ^{14}N).

Negatively charged particles called electrons circle the atomic nucleus (figure AI.1). The number of electrons in a neutral atom equals the number of its protons and is given by the atomic number, the number of protons in an atomic nucleus. The atomic number is characteristic of a particular type of atom. For example, carbon has an atomic number of six, hydrogen's number is one, and oxygen's is eight (**table AI.1**).

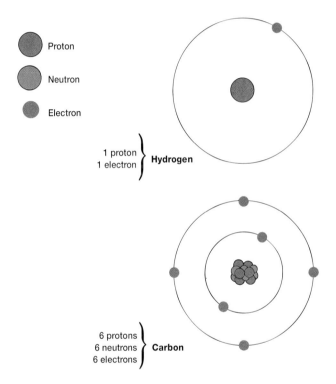

Figure AI.1 Diagrams of Hydrogen and Carbon Atoms.
The electron orbitals are represented as concentric circles.

Table AI.1	Atoms Commonly Present in Organic Molecules			
Atom	**Symbol**	**Atomic Number**	**Atomic Weight**	**Number of Chemical Bonds**
Hydrogen	H	1	1.01	1
Carbon	C	6	12.01	4
Nitrogen	N	7	14.01	3
Oxygen	O	8	16.00	2
Phosphorus	P	15	30.97	5
Sulfur	S	16	32.06	2

The electrons move constantly within a volume of space surrounding the nucleus, even though their precise location in this volume cannot be determined accurately. This volume of space in which an electron is located is called its orbital. Each orbital can contain two electrons. Orbitals are grouped into shells of different energy that surround the nucleus. The first shell is closest to the nucleus and has the lowest energy; it contains only one orbital. The second shell contains four orbitals, one circular and three shaped like dumbbells (**figure AI.2a**). It can contain up to eight electrons. The third shell has even higher energy and holds more than eight electrons. Shells are filled beginning with the innermost and moving outward. For example, carbon has six electrons, two in its first shell and four in the second (figures AI.1 and AI.2b). The electrons in the outermost shell are the ones that participate in chemical reactions. The most stable condition is achieved when the outer shell is filled with electrons. Thus the number of bonds an element can form depends on the number of electrons required to fill the outer shell. Since carbon has four electrons in its outer shell and the shell is filled when it contains eight electrons, it can form four covalent bonds (table AI.1).

CHEMICAL BONDS

Molecules are formed when two or more atoms associate through chemical bonding. Chemical bonds are attractive forces that hold together atoms, ions, or groups of atoms in a molecule or other substance. Many types of chemical bonds are present in organic molecules; three of the most important are covalent bonds, ionic bonds, and hydrogen bonds.

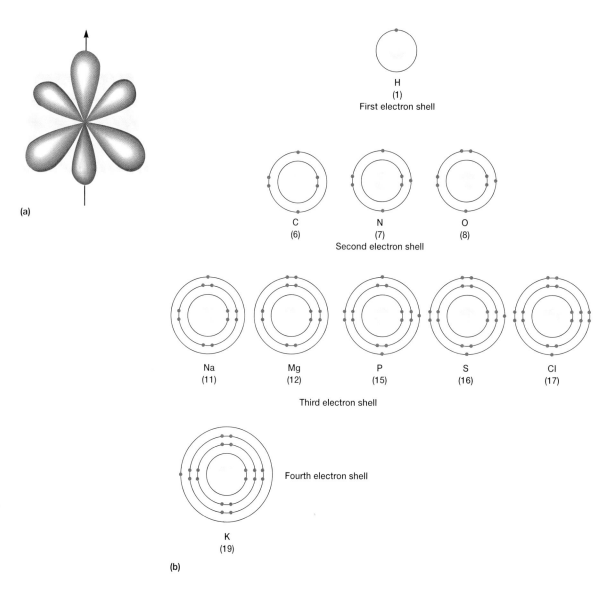

Figure AI.2   Electron Orbitals.   (a) The three dumbbell-shaped orbitals of the second shell. The orbitals lie at right angles to each other.
(b) The distribution of electrons in some common elements. Atomic numbers are given in parentheses.

Figure AI.3   The Covalent Bond.   A hydrogen molecule is formed when two hydrogen atoms share electrons.

H_2

C_6H_{14} (Hexane)

(a)

C_6H_{12} (Cyclohexane)

(b)

C_6H_6 (Benzene)

(c)

Figure AI.4   Hydrogen Bonds.   Representative examples of hydrogen bonds present in biological molecules.

Figure AI.5   Hydrocarbons.   Examples of hydrocarbons that are **(a)** linear, **(b)** cyclic, and **(c)** aromatic.

In covalent bonds, atoms are joined together by sharing pairs of electrons (**figure AI.3**). If the electrons are equally shared between identical atoms (e.g., in a carbon-carbon bond), the covalent bond is strong and nonpolar. When two different atoms such as carbon and oxygen share electrons, the covalent bond formed is polar because the electrons are pulled toward the more electronegative atom, the atom that more strongly attracts electrons (the oxygen atom). A single pair of electrons is shared in a single bond; a double bond is formed when two pairs of electrons are shared.

Atoms often contain either more or fewer electrons than the number of protons in their nuclei. When this is the case, they carry a net negative or positive charge and are called ions. Cations carry positive charges and anions have a net negative charge. When a cation and an anion approach each other, they are attracted by their opposite charges. This ionic attraction that holds two groups together is called an ionic bond. Ionic bonds are much weaker than covalent bonds and are easily disrupted by a polar solvent such as water. For example, the Na^+ cation is strongly attracted to the Cl^- anion in a sodium chloride crystal, but sodium chloride dissociates into separate ions (ionizes) when dissolved in water. Ionic bonds are important in the structure and function of proteins and other biological molecules.

When a hydrogen atom is covalently bonded to a more electronegative atom such as oxygen or nitrogen, the electrons are unequally shared and the hydrogen atom carries a partial positive charge. It will be attracted to an electronegative atom such as oxygen or nitrogen, which carries an unshared pair of electrons; this attraction is called a hydrogen bond (**figure AI.4**). Although an individual hydrogen bond is weak, there are so many hydrogen bonds in proteins and nucleic acids that they play a major role in determining protein and nucleic acid structure.

ORGANIC MOLECULES

Most molecules in cells are organic molecules, molecules that contain carbon. Since carbon has four electrons in its outer shell, it tends to form four covalent bonds in order to fill its outer shell with eight electrons. This property makes it possible to form chains and rings of carbon atoms that also can bond with hydrogen and other atoms (**figure AI.5**). Although adjacent carbons usually are connected by single bonds, they may be joined by double or triple bonds. Rings that have alternating single and double bonds, like the benzene ring, are called aromatic rings. The hydrocarbon chain or ring provides a chemically inactive skeleton to which more reactive groups of atoms may be attached. These reactive groups with specific properties are known as functional groups. They usually contain atoms of oxygen, nitrogen, phosphorus, or sulfur (**figure AI.6**) and are largely responsible for most characteristic chemical properties of organic molecules.

Organic molecules are often divided into classes based on the nature of their functional groups. Ketones have a carbonyl group within the carbon chain, whereas alcohols have a hydroxyl on the chain. Organic acids have a carboxyl group, and amines have an amino group (**figure AI.7**).

Organic molecules may have the same chemical composition and yet differ in their molecular structure and properties. Such molecules are called isomers. One important class of isomers is the stereoisomers. Stereoisomers have the same atoms arranged in the

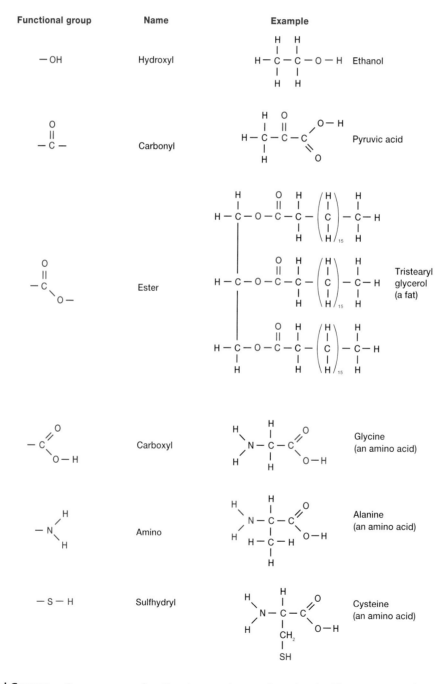

Functional group	Name	Example
— OH	Hydroxyl	Ethanol
— C — (with =O)	Carbonyl	Pyruvic acid
— C (with =O, O—)	Ester	Tristearyl glycerol (a fat)
— C (with =O, O—H)	Carboxyl	Glycine (an amino acid)
— N (with H, H)	Amino	Alanine (an amino acid)
— S — H	Sulfhydryl	Cysteine (an amino acid)

Figure AI.6 Functional Groups. Some common functional groups in organic molecules. The groups are shown in red.

Type of molecule	Example
Alcohol	$CH_3 — CH_2 — OH$
Aldehyde	$CH_3 — C \overset{\displaystyle O}{\underset{\displaystyle H}{}}$
Amine	$CH_3 — CH_2 — NH_2$
Ester	$CH_3 — \overset{\displaystyle O}{C} — O — CH_2 — CH_3$
Ether	$CH_3 — CH_2 — O — CH_2 — CH_3$
Ketone	$CH_3 — \overset{\displaystyle O}{C} — CH_3$
Organic acid	$CH_3 — C \overset{\displaystyle O}{\underset{\displaystyle OH}{}}$

Figure AI.7   Types of Organic Molecules.   These are classified on the basis of their functional groups.

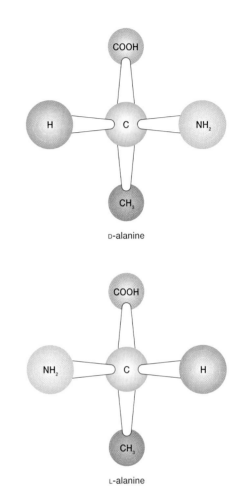

Figure AI.8   The Stereoisomers of Alanine.   The α-carbon is in gray, L-alanine is the form usually present in proteins.

same nucleus-to-nucleus sequence but differ in the spatial arrangement of their atoms. For example, an amino acid such as alanine can form stereoisomers (**figure AI.8**). L-Alanine and other L-amino acids are the stereoisomer forms normally present in proteins.

CARBOHYDRATES

Carbohydrates are aldehyde or ketone derivatives of polyhydroxy alcohols. The smallest and least complex carbohydrates are the simple sugars or monosaccharides. The most common sugars have five or six carbons (**figure AI.9**). A sugar in its ring form has two isomeric structures, the α and β forms, that differ in the orientation of the hydroxyl on the aldehyde or ketone carbon, which is called the anomeric or glycosidic carbon (**figure AI.10**). Microorganisms have many sugar derivatives in which a hydroxyl is replaced by an amino group or some other functional group (e.g., glucosamine).

Two monosaccharides can be joined by a bond between the anomeric carbon of one sugar and a hydroxyl or the anomeric

carbon of the second (**figure AI.11**). The bond joining sugars is a glycosidic bond and may be either α or β depending on the orientation of the anomeric carbon. Two sugars linked in this way constitute a disaccharide. Some common disaccharides are maltose (two glucose molecules), lactose (glucose and galactose), and sucrose (glucose and fructose). If 10 or more sugars are linked together by glycosidic bonds, a polysaccharide is formed. For example, starch and glycogen are common polymers of glucose that are used as sources of carbon and energy (**figure AI.12**).

LIPIDS

All cells contain a heterogeneous mixture of organic molecules that are relatively insoluble in water but very soluble in nonpolar solvents such as chloroform, ether, and benzene. These molecules are called lipids. Lipids vary greatly in structure and include triacylglycerols, phospholipids, steroids, carotenoids, and many other types. Among other functions, they serve as membrane

Figure AI.9   Common Monosaccharides.   Structural formulas for both the open chains and the ring forms are provided.

Glucose

Mannose

Galactose

Fructose

Ribose

Figure AI.10   The Interconversion of Monosaccharide Structures.   The open chain form of glucose and other sugars is in equilibrium with closed ring structures (depicted here with Haworth projections). Aldehyde sugars form cyclic hemiacetals, and keto sugars produce cyclic hemiketals. When the hydroxyl on carbon one of cyclic hemiacetals projects above the ring, the form is known as a β form. The α form has a hydroxyl that lies below the plane of the ring. The same convention is used in showing the α and β forms of hemiketals such as those formed by fructose.

β-D-glucose

α-D-glucose

components, storage forms for carbon and energy, precursors of other cell constituents, and protective barriers against water loss.

Most lipids contain fatty acids, which are monocarboxylic acids that often are straight chained but may be branched. Saturated fatty acids lack double bonds in their carbon chains, whereas unsaturated fatty acids have double bonds. The most common fatty acids are 16 or 18 carbons long.

Two good examples of common lipids are triacylglycerols and phospholipids. Triacylglycerols are composed of glycerol esterified to three fatty acids (**figure AI.13a**). They are used to store carbon and energy. Phospholipids are lipids that contain at least one phosphate group and often have a nitrogenous constituent as well. Phosphatidylethanolamine is an important phospholipid frequently present in bacterial membranes (figure AI.13b). It is

Figure AI.11   Common Disaccharides. (a) The formation of maltose from two molecules of an α glucose. The bond connecting the glucose extends between carbons one and four, and involves the α form of the anomeric carbon. Therefore, it is called an α (1 → 4) glycosidic bond. **(b)** Sucrose is composed of a glucose and a fructose joined to each other through their anomeric carbons, and αβ (1 → 2) bond. **(c)** The milk sugar lactose contains galactose and glucose joined by a β (1 → 4) glycosidic bond.

Figure AI.12   Glycogen and Starch Structure.   (a) An overall view of the highly branched structure characteristic of glycogen and most starch. The circles represent glucose residues. **(b)** A close-up of a small part of the chain (shown in blue in part *a*) revealing a branch point with its α (1 → 6) glycosidic bond, which is colored blue.

$$
\begin{array}{c}
\quad\quad\quad\quad O \\
\quad\quad\quad\quad \| \\
CH_2 - O - C - R \\
\; | \quad\quad\quad\; O \\
\quad\quad\quad\quad \| \\
CH \;\; - O - C - R \\
\; | \quad\quad\quad\; O \\
\quad\quad\quad\quad \| \\
CH_2 - O - C - R
\end{array}
$$

(a)

$$
\begin{array}{c}
\quad\quad\quad\quad O \\
\quad\quad\quad\quad \| \\
CH_2 - O - C - R \\
\; | \quad\quad\quad\; O \\
\quad\quad\quad\quad \| \\
CH \;\; - O - C - R \\
\; | \quad\quad\quad\; O \\
\quad\quad\quad\quad \| \\
CH_2 - O - P - O - CH_2 - CH_2 - \overset{+}{N}H_3 \\
\quad\quad\quad\quad\; | \\
\quad\quad\quad\quad O^-
\end{array}
$$

(b)

Figure AI.13   Examples of Common Lipids.   (a) A triacylglycerol or neutral fat. **(b)** The phospholipid phosphatidylethanolamine. The R groups represent fatty acid side chains.

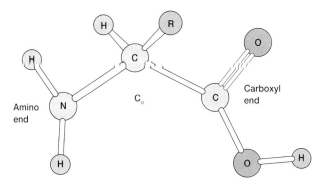

Figure AI.14   L-Amino Acid Structure.   The uncharged form is shown.

composed of two fatty acids esterified to glycerol. The third glycerol hydroxyl is joined with a phosphate group, and ethanolamine is attached to the phosphate. The resulting lipid is very asymmetric with a hydrophobic nonpolar end contributed by the fatty acids and a polar, hydrophilic end. In cell membranes the hydrophobic end is buried in the interior of the membrane, while the polar-charged end is at the membrane surface and exposed to water.

Proteins

The basic building blocks of proteins are amino acids. An amino acid contains a carboxyl group and an amino group on its alpha carbon (**figure AI.14**). About 20 amino acids are normally found in proteins; they differ from each other with respect to their side chains (**figure AI.15**). In proteins, amino acids are linked together by peptide bonds between their carboxyls and α-amino groups to form linear polymers called peptides (**figure AI.16**). If a peptide contains more than 30 amino acids, it usually is called a polypeptide. Each protein is composed of one or more polypeptide chains and has a molecular weight greater than about 6,000 to 7,000.

Proteins have three or four levels of structural organization and complexity. The primary structure of a protein is the sequence of the amino acids in its polypeptide chain or chains. The structure of the polypeptide chain backbone is also considered part of the primary structure. Each different polypeptide has its own amino acid sequence that is a reflection of the nucleotide sequence in the gene that codes for its synthesis. The polypeptide

chain can coil along one axis in space into various shapes like the α-helix (**figure AI.17**). This arrangement of the polypeptide in space around a single axis is called the secondary structure. Secondary structure is formed and stabilized by the interactions of amino acids that are fairly close to one another on the polypeptide chain. The polypeptide with its primary and secondary structure can be coiled or organized in space along three axes to form a more complex, three-dimensional shape (**figure AI.18**). This level of organization is the tertiary structure (**figure AI.19**). Amino acids more distant from one another on the polypeptide chain contribute to tertiary structure. Secondary and tertiary structures are examples of conformation, molecular shape that can be changed by bond rotation and without breaking covalent bonds. When a protein contains more than one polypeptide chain, each chain with its own primary, secondary, and tertiary structure associates with the other chains to form the final molecule. The way in which polypeptides associate with each other in space to form the final protein is called the protein's quaternary structure (**figure AI.20**).

The final conformation of a protein is ultimately determined by the amino acid sequence of its polypeptide chains. Under proper conditions a completely unfolded polypeptide will fold into its normal final shape without assistance.

Protein secondary, tertiary, and quaternary structure is largely determined and stabilized by many weak noncovalent forces such as hydrogen bonds and ionic bonds. Because of this, protein shape often is very flexible and easily changed. This flexibility is very important in protein function and in the regulation of enzyme activity. Because of their flexibility, however, proteins readily lose their proper shape and activity when exposed to harsh conditions. The only covalent bond commonly involved in the secondary and tertiary structure of proteins is the disulfide bond. The disulfide bond is formed when two cysteines are linked through their sulfhydryl groups. Disulfide bonds generally strengthen or stabilize protein structure but are not especially important in directly determining protein conformation.

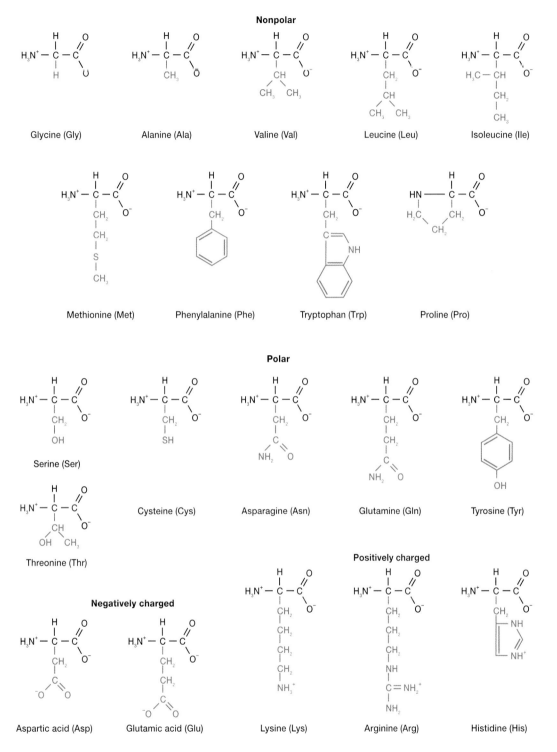

Figure AI.15   The Common Amino Acids.   The structures of the α-amino acids normally found in proteins. Their side chains are shown in blue, and they are grouped together based on the nature of their side chains—nonpolar, polar, negatively charged (acid), or positively charged (basic). Proline is actually an imino acid rather than an amino acid.

Figure AI.16   A Tetrapeptide Chain.   The end of the chain with a free α-amino group is the amino or N terminal. The end with the free α-carboxyl is the carboxyl or C terminal. One peptide bond is shaded in blue.

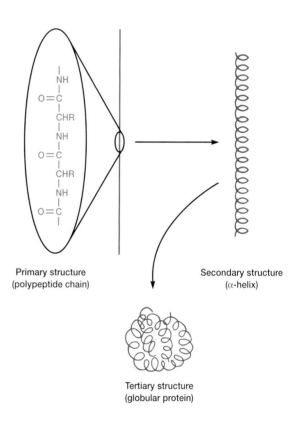

Figure AI.18   Secondary and Tertiary Protein Structures.
The formation of secondary and tertiary protein structures by folding a polypeptide chain with its primary structure.

Figure AI.17   The α-Helix.   A polypeptide twisted into one type of secondary structure, the α-helix. The helix is stabilized by hydrogen bonds joining peptide bonds that are separated by three amino acids.

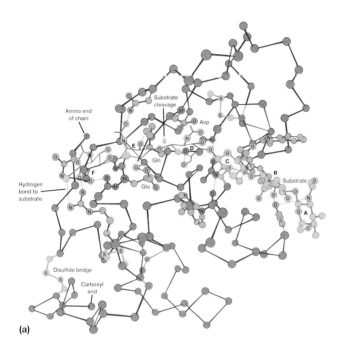

(a)

Figure AI.19   Lysozyme.   The tertiary structure of the enzyme lysozyme. **(a)** A diagram of the protein's polypeptide backbone with the substrate hexasaccharide shown in blue. The point of substrate cleavage is indicated. **(b)** A space-filling model of lysozyme. The figure on the right shows the empty active site with some of its more important amino acids indicated. On the left the enzyme has bound its substrate (in pink).

B²

B¹

Trp 62
Asp 52
Trp 63
Asp 101
Asp 103

(b)

NUCLEIC ACIDS

The nucleic acids, deoxyribonucleic acid (DNA) and ribonucleic acid (RNA), are polymers of deoxyribonucleosides and ribonucleosides joined by phosphate groups. The nucleosides in DNA contain the purines adenine and guanine, and the pyrimidine bases thymine and cytosine. In RNA the pyrimidine uracil is substituted for thymine. Because of their importance for genetics and molecular biology, the chemistry of nucleic acids is introduced earlier in the text. The structure and synthesis of purines and pyrimidines are discussed in chapter 10 (pp. 241–42). The structures of DNA and RNA are described in chapter 11 (pp. 252–53).

adenine arabinoside or **vidarabine**   An antiviral agent used especially to treat keratitis and encephalitis caused by the herpes simplex virus. (855)

adenosine diphosphate (ADP) (ah-den′o-sēn)   The nucleoside diphosphate usually formed upon the breakdown of ATP when it provides energy for work. (171)

adenosine 5′-triphosphate (ATP)   The triphosphate of the nucleoside adenosine, which is a high energy molecule or has high phosphate group transfer potential and serves as the cell's major form of energy currency. (171)

adhesin (ad-he′zin)   A molecular component on the surface of a microorganism that is involved in adhesion to a substratum or cell. Adhesion to a specific host tissue usually is a preliminary stage in pathogenesis, and adhesins are important virulence factors. (820)

adjuvant (aj′ə-vənt)   Material added to an antigen to increase its immunogenicity. Common examples are alum, killed *Bordetella pertussis*, and an oil emulsion of the antigen, either alone (Freund's incomplete adjuvant) or with killed mycobacteria (Freund's complete adjuvant). (901)

adult T-cell leukemia   A type of white blood cell cancer caused by the HTLV-1 virus. (935)

aerial mycelium   The mat of hyphae formed by actinomycetes that grows above the substrate, imparting a fuzzy appearance to colonies. (589)

aerobe (a′er-ōb)   An organism that grows in the presence of atmospheric oxygen. (139)

aerobic anoxygenic photosynthesis   Photosynthetic process in which electron donors such as organic matter or sulfide, which do not result in oxygen evolution, are used under aerobic conditions. (650)

aerobic respiration (res″pī-ra′shun)   A metabolic process in which molecules, often organic, are oxidized with oxygen as the final electron acceptor. (205)

aerotolerant anaerobes   Microbes that grow equally well whether or not oxygen is present. (139)

aflatoxin (af″lah-tok′sin)   A polyketide secondary fungal metabolite that can cause cancer. (1027)

agar (ahg′ar)   A complex sulfated polysaccharide, usually from red algae, that is used as a solidifying agent in the preparation of culture media. (111)

agglutinates   The visible aggregates or clumps formed by an agglutination reaction. (799, 876)

agglutination reaction (ah-gloo″tī-na′shun)   The formation of an insoluble immune complex by the cross-linking of cells or particles. (799)

agglutinin (ah-gloo″tī-nin)   The antibody responsible for an agglutination reaction. (799)

AIDS   *See* acquired immune deficiency syndrome.

airborne transmission   The type of infectious organism transmission in which the pathogen is truly suspended in the air and travels over a meter or more from the source to the host. (892)

akinetes   Specialized, nonmotile, dormant, thick-walled resting cells formed by some cyanobacteria. (525)

alcoholic fermentation   A fermentation process that produces ethanol and CO_2 from sugars. (208)

alga (al′gah)   A common term for a series of unrelated groups of photosynthetic eucaryotic microorganisms lacking multicellular sex organs (except for the charophytes) and conducting vessels. Most are now considered protists. (605)

algicide (al′jī-sīd)   An agent that kills algae. (151)

alkalophile   A microorganism that grows best at pHs from about 8.5 to 11.5. (134)

allele   An alternative form of a gene. (329)

allergen (al′er-jen)   An antigen that induces an allergic response. (803)

allergic contact dermatitis   An allergic reaction caused by haptens that combine with proteins in the skin to form the allergen that produces the immune response. (808)

allergy (al′er-je)   *See* hypersensitivity.

allochthonous (ăl′ək-thə-nəs)   Substances not native to a given environment. Nutrient influx into freshwater ecosystems (e.g., lakes and streams) is often of terrestrial, or allochthonous, origin. (682)

allograft (al′o-graft)   A transplant between genetically different individuals of the same species. (810)

allosteric enzyme (al″o-ster′ik)   An enzyme whose activity is altered by the noncovalent binding of a small effector or modulator molecule at a regulatory site separate from the catalytic site; effector binding causes a conformational change in the enzyme and its catalytic site, which leads to enzyme activation or inhibition. (181)

allotype   Allelic variants of antigenic determinant(s) found on antibody chains of some, but not all, members of a species, which are inherited as simple Mendelian traits. (791)

alpha hemolysis   A greenish zone of partial clearing around a bacterial colony growing on blood agar. (584, 828)

Alphaproteobacteria   One of the five classes of proteobacteria, each with distinctive 16S rRNA sequences. This group contains most of the oligotrophic proteobacteria; some have unusual metabolic modes such as methylotrophy, chemolithotrophy, and nitrogen fixing ability. Many have distinctive morphological features. (540)

alternative complement pathway   An antibody-independent pathway of complement activation that includes the C3–C9 components of the classical pathway and several other serum protein factors (e.g., factor B and properdin). (764)

alveolar macrophage   A vigorously phagocytic macrophage located on the epithelial surface of the lung alveoli where it ingests inhaled particulate matter and microorganisms. (761)

amantadine (ah-man′tah-den)   An antiviral agent used to prevent type A influenza infections. (855)

amebiasis (amebic dysentery) (am″e-bi′ah-sis)   An infection with amoebae, often resulting in dysentery; usually it refers to an infection by *Entamoeba histolytica*. (1012)

amensalism (a-men′səl-iz-əm)   A relationship in which the product of one organism has a negative effect on another organism. (732)

American trypanosomiasis   *See* trypanosomiasis.

Ames test   A test that uses a special *Salmonella* strain to test chemicals for mutagenicity and potential carcinogenicity. (325)

amino acid activation   The initial stage of protein synthesis in which amino acids are attached to transfer RNA molecules. (276)

aminoacyl or **acceptor site (A site)**   The ribosomal site that contains an aminoacyl-tRNA at the beginning of the elongation cycle during protein synthesis; the growing peptide chain is transferred to the aminoacyl-tRNA and lengthens by an amino acid. (284)

aminoglycoside antibiotics (am″ī-no-gli′ko-sīd)   A group of antibiotics synthesized by *Streptomyces* and *Micromonospora*, which contain a cyclohexane ring and amino sugars; all aminoglycoside antibiotics bind to the small ribosomal subunit and inhibit protein synthesis. (845)

amoeboid movement   Moving by means of cytoplasmic flow and the formation of pseudopodia (temporary cytoplasmic protrusions). (613)

amphibolic pathways (am″fe-bol′ik)   Metabolic pathways that function both catabolically and anabolically. (194)

amphitrichous (am-fit′rĕ-kus)   A cell with a single flagellum at each end. (67)

amphotericin B (am″fo-ter′i-sin)   An antibiotic from a strain of *Streptomyces nodosus* that is used to treat systemic fungal infections; it also is used topically to treat candidiasis. (854)

anabolism (ah-nab′o-lizm″)   The synthesis of complex molecules from simpler molecules with the input of energy. (168)

anaerobe (an-a′er-ōb)   An organism that grows in the absence of free oxygen. (139)

anaerobic digestion (an″a-er-o′bik)   The microbiological treatment of sewage wastes under anaerobic conditions to produce methane. (1056)

anaerobic respiration   An energy-yielding process in which the electron transport chain acceptor is an inorganic molecule other than oxygen. (205)

anagenesis   Changes in gene frequencies and distribution among species; the accumulation of small genetic changes within a population that introduces genetic variability but are not enough to result in either speciation or extinction. (477)

anammox reaction   The coupled use of nitrite as an electron acceptor and ammonium ion as an electron donor under anaerobic conditions to yield nitrogen gas. (649)

anamnestic response (an″am-nes′tik)   The recall, or the remembering, by the immune system of a prior response to a given antigen. (774)

anaphylaxis (an″ah-fī-lak′sis)   An immediate (type I) hypersensitivity reaction following exposure of a sensitized individual to the appropriate antigen. Mediated by reagin antibodies, chiefly IgE. (803)

anaplasia The reversion of an animal cell to a more primitive, undifferentiated state. (460)

anaplerotic reactions (an'ah-plĕ-rot'ik) Reactions that replenish depleted tricarboxylic acid cycle intermediates. (239)

anergy (an'ər-je) A state of unresponsiveness to antigens. Absence of the ability to generate a sensitivity reaction to substances that are expected to be antigenic. (803)

anisogamy (ăn-ī-sŏg'ə-mē) In the sexual reproduction of certain protists, the union of gametes that are different in morphology or physiology. (609)

annotation The process of determining the location and potential function of specific genes and genetic elements in a genome sequence. (388)

anogenital condylomata (venereal warts) (kon"dī-lo" mah-tah) Warts that are sexually transmitted and caused by types 6, 11, and 42 human papillomavirus. Usually occur around the cervix, vulva, perineum, anus, anal canal, urethra, or glans penis. (938)

anoxic (ə-nok'sik) Without oxygen present. (668)

anoxygenic photosynthesis Photosynthesis that does not oxidize water to produce oxygen; a form of photosynthesis characteristic of purple and green photosynthetic bacteria and heliobacteria. (218, 521)

antheridium (an"ther-id'e-um; pl., **antheridia**) A male gamete-producing organ, which may be unicellular or multicellular. (638)

anthrax (an'thraks) An infectious disease of warm-blooded animals, especially of cattle and sheep, caused by *Bacillus anthracis*. The disease can be transmitted to humans through contact with bacteria or spore-contaminated animal substances, such as hair, feces, or hides; and ingestion or inhalation of spores. Cutaneous anthrax is characterized by skin lesions that become necrotic. Inhalation anthrax is almost always fatal. (987)

antibiotic (an"tī-bi-ot'ik) A microbial product or its derivative that kills susceptible microorganisms or inhibits their growth. (164, 835)

antibody (immunoglobulin) (an'tī-bod"e) A glycoprotein made by plasma cells (mature B cells) in response to the introduction of an antigen. The antibody-binding region of an antibody molecule assumes a final configuration that is the three-dimensional mirror image of the antigen that stimulated its synthesis. Thus, the antibody can bind to the antigen with exact specificity. (744, 789)

antibody affinity The strength of binding between an antigen and an antibody. (774)

antibody-dependent cell-mediated cytotoxicity (ADCC) The killing of antibody-coated target cells by cells with Fc receptors that recognize the Fc region of the bound antibody. Most ADCC is mediated by NK cells that have the Fc receptor or CD16 on their surface. (748)

antibody-mediated immunity *See* humoral immunity.

antibody titer An approximation of the antibody concentration required to react with an antigen. (795)

anticodon triplet The base triplet on a tRNA that is complementary to the triplet codon on mRNA. (277)

antigen (an'tī-jen) A foreign (nonself) substance (such as a protein, nucleoprotein, polysaccharide, or sometimes a glycolipid) to which lymphocytes respond; also known as an immunogen because it induces the immune response. (744, 774)

antigen-binding fragment (Fab) "Fragment antigen binding." A monovalent antigen-binding fragment of an immunoglobulin molecule that consists of one light chain and part of one heavy chain, linked by interchain disulfide bonds. (790)

antigen processing The hydrolytic digestion of antigens to produce antigen fragments. Antigen fragments are often collected by the class I or class II MHC molecules and presented on the surface of a cell. Antigen processing can occur by proteasome action on antigens that entered a cell by means other than phagocytosis. This is known as endogenous antigen processing. Antigen processing can also occur during phagocytosis as antigens are degraded within the phagolysosome. This is known as exogenous antigen processing. (780)

antigenic determinant site (epitope) The molecular configuration of the variable region of an antibody molecule that interacts with the epitope of an antigen. (774)

antigenic drift A small change in the antigenic character of an organism that allows it to avoid attack by the immune system. (890, 916)

antigenic shift A major change in the antigenic character of an organism that makes it unrecognized by host immune mechanisms. (890, 916)

antigen-presenting cells Antigen-presenting cells (APCs) are cells that take in protein antigens, process them, and present antigen fragments to B cells and T cells in conjunction with class II MHC molecules so that the cells are activated. Macrophages, B cells, dendritic cells, and Langerhans cells may act as APCs. (780)

antimetabolite (an"tī-mĕ-tab'o-līt) A compound that blocks metabolic pathway function by competitively inhibiting a key enzyme's use of a metabolite because it closely resembles the normal enzyme substrate. (846)

antimicrobial agent An agent that kills microorganisms or inhibits their growth. (152)

antisense RNA A single-stranded RNA with a base sequence complementary to a segment of another RNA molecule that can specifically bind to the target RNA and alter its activity. (305)

antisepsis (an"tī-sep'sis) The prevention of infection or sepsis. (151)

antiseptic (an"tī-sep'tik) Chemical agents applied to tissue to prevent infection by killing or inhibiting pathogens. (151)

antiserum (an"tī-se'rum) Serum containing induced antibodies. (795)

antitoxin (an"tī-tok'sin) An antibody to a microbial toxin, usually a bacterial exotoxin, that combines specifically with the toxin, in vivo and in vitro, neutralizing the toxin. (799, 824)

apical complex (ap'ī-kal) A set of organelles characteristic of members of the protist subdivision Apicomplexa: polar rings, subpellicular microtubules, conoid, rhoptries, and micronemes. (619)

apicomplexan (a'pī-kom-plek'san) A protist that lacks special locomotor organelles but has an apical complex and a spore-forming stage. It is either an intra- or extracellular parasite of animals; a member of the subdivision Apicomplexa. (619)

apoenzyme (ap"o-en'zīm) The protein part of an enzyme that also has a nonprotein component. (176)

apoptosis (ap"o-to'sis) Programmed cell death. The fragmentation of a cell into membrane-bound particles that are eliminated by phagocytosis. Apoptosis is a physiological suicide mechanism that preserves homeostasis and occurs during normal tissue turnover. It causes cell death in pathological circumstances, such as exposure to low concentrations of xenobiotics and infections by HIV and various other viruses. (819)

aporepressor An inactive form of the repressor protein, which becomes the active repressor when the corepressor binds to it. (295)

appressorium A flattened region of hypha found in some plant-infecting fungi that aids in penetrating the host plant cell wall. Occurs in both pathogenic fungi and nonpathogenic mycorrhizal fungi. (640)

arbuscular mycorrhizal (AM) fungi The mycorrhizal fungi in a symbiotic fungus-root association that penetrate the outer layer of the root, grow intracellularly, and form characteristic much-branched hyphal structures called arbuscules. (698)

arbuscules Branched, treelike structures formed in cells of plant roots colonized by endotrophic mycorrhizal fungi. (699)

Archaea The domain that contains procaryotes with isoprenoid glycerol diether or diglycerol tetraether lipids in their membranes and archaeal rRNA (among many differences). (3, 503)

artificially acquired active immunity The type of immunity that results from immunizing an animal with a vaccine. The immunized animal now produces its own antibodies and activated lymphocytes. (778)

artificially acquired passive immunity The type of immunity that results from introducing into an animal antibodies that have been produced either in another animal or by in vitro methods. Immunity is only temporary. (778)

ascocarp (as'ko-karp) A multicellular structure in ascomycetes lined with specialized cells called asci in which nuclear fusion and meiosis produce ascospores. An ascocarp can be open or closed and may be referred to as a fruiting body. (638)

ascogenous hypha A specialized hypha that gives rise to one or more asci. (638)

bronchial asthma   An example of an atopic allergy involving the lower respiratory tract. (804)

brucellosis (brōō′sə-lō′sĭs)   A disease caused by the α-proteobacterium *Brucella*. The disease is characterized by cycles of (undulating) fever, sweating, weakness, and headache. It is transmitted to humans by direct contact with diseased animals or through ingestion of infected meat, milk, or cheese. It is also known as Bang's disease, Malta fever, Mediterranean fever, Rock fever, and undulant fever. (990)

bubo (bu′bo)   A tender, inflamed, enlarged lymph node that results from a variety of infections. (962)

bubonic plague   *See* plague.

budding   A vegetative outgrowth of yeast and some bacteria as a means of asexual reproduction; the daughter cell is smaller than the parent. (543)

bulking sludge   Sludges produced in sewage treatment that do not settle properly, usually due to the development of filamentous microorganisms. (1055)

bursa of Fabricius (bər′sə fə-bris′e-əs)   Found in birds; the blind saclike structure located on the posterior wall of the cloaca; it performs a thymuslike function. A primary lymphoid organ where B-cell maturation occurs. Bone marrow is the equivalent in mammals. (750)

burst   *See* rise period.

burst size   The number of phages released by a host cell during the lytic life cycle. (430)

butanediol fermentation   A type of fermentation most often found in the family *Enterobacteriaceae* in which 2,3-butanediol is a major product; acetoin is an intermediate in the pathway and may be detected by the Voges-Proskauer test. (209)

C

Calvin cycle   The main pathway for the fixation (or reduction and incorporation) of CO_2 into organic material by photoautotrophs; it also is found in chemolithoautotrophs. (228)

campylobacteriosis   A disease caused by species of the bacterium *Campylobacter,* primarily *Campylobacter jejuni,* characterized by an inflammatory, sometimes bloody, diarrhea and may present as a dysentery syndrome. (979)

cancer (kan′ser)   A malignant tumor that expands locally by invasion of surrounding tissues and systemically by metastasis. (461)

candidal vaginitis   Infection of the vagina caused by the fungus *Candida* sp. (1017)

candidiasis (kan″dĭ-di′ah-sis)   An infection caused by *Candida* species of dimorphic fungi, commonly involving the skin. (1017)

capsid (kap′sid)   The protein coat or shell that surrounds a virion's nucleic acid. (409)

capsomer (kap′so-mer)   The ring-shaped morphological unit of which icosahedral capsids are constructed. (410)

capsule   A layer of well-organized material, not easily washed off, lying outside the bacterial cell wall. (65)

carbonate equilibrium system   The interchange among CO_2, HCO_3^-, and CO_3^{2-} that keeps oceans buffered between pH 7.6 to 8.2. (668)

carboxysomes   Polyhedral inclusion bodies that contain the CO_2 fixation enzyme ribulose 1,5-bisphosphate carboxylase; found in cyanobacteria, nitrifying bacteria, and thiobacilli. (49, 229, 524)

caries (kar′e-ēz)   Tooth decay. (993)

carotenoids (kah-rot′e-noids)   Pigment molecules, usually yellowish in color, that are often used to aid chlorophyll in trapping light energy during photosynthesis. (217)

carrier   An infected individual who is a potential source of infection for others and plays an important role in the epidemiology of a disease. (891)

caseous lesion (ka′se-us)   A lesion resembling cheese or curd; cheesy. Most caseous lesions are caused by *Mycobacterium tuberculosis.* (954)

casual carrier   An individual who harbors an infectious organism for only a short period. (892)

catabolism (kah-tab′o-lizm)   That part of metabolism in which larger, more complex molecules are broken down into smaller, simpler molecules with the release of energy. (168)

catabolite repression (kah-tab′o-līt)   Inhibition of the synthesis of several catabolic enzymes by a metabolite such as glucose. (308)

catalyst (kat′ah-list)   A substance that accelerates a reaction without being permanently changed itself. (176)

catalytic site   *See* active site.

catenanes (kăt′ə-nāns′)   Circular, covalently closed nucleic acid molecules that are locked together like the links of a chain. (263)

cathelicidins   Antimicrobial cationic peptides that are produced by a variety of cells (e.g., neutrophils, respiratory epithelial cells, and alveolar macrophages). They are linear, alpha-helical peptides that arise from precursor proteins having an N-terminal cathepsin L inhibitor domain and a C-terminus, which gives rise to the mature peptide of 12 to 80 amino acids. (736, 762)

catheter (kath′e-ter)   A tubular instrument for withdrawing fluids from a cavity of the body, especially one for introduction into the bladder through the urethra for the withdrawal of urine. (862)

caveola (ka-ve-o′lə)   A small flask-shaped invagination of the plasma membrane formed during one type of pinocytosis. (86)

CD4⁺ cell   *See* T-helper cell.

CD8⁺ cell   *See* cytotoxic lymphyocyte.

CD95   Cell surface receptor that initiates apoptosis when ligand binds. *See also* Fas-FasL. (782)

cell cycle   The sequence of events in a cell's growth-division cycle between the end of one division and the end of the next. In eucaryotic cells, it is composed of the G_1 period, the S period in which chromosomes are replicated, the G_2 period, and the M period (mitosis). (92, 119)

cell-mediated immunity   The type of immunity that results from T cells coming into close contact with foreign cells or infected cells to destroy them; it can be transferred to a nonimmune individual by the transfer of cells. (814)

cellular slime molds   Protists with a vegetative phase consisting of amoeboid cells that aggregate to form a multicellular pseudoplasmodium. They belong to the subdivision *Dictyostelia;* they were formerly considered fungi. (614)

cellulitis (sel″u-li′tis)   A diffuse spreading infection of subcutaneous skin tissue caused by streptococci, staphylococci, or other organisms. The tissue is inflamed with edema, redness, pain, and interference with function. (957)

cellulose   The major structural carbohydrate of plants cell walls; a linear (β1 → 4) glucan. (688)

cell wall   The strong layer or structure that lies outside the plasma membrane; it supports and protects the membrane and gives the cell shape. (94)

central tolerance   The process by which immune cells are rendered inactive. (803)

cephalosporin (sef″ah-lo-spōr′in)   A group of β-lactam antibiotics derived from the fungus *Cephalosporium,* which share the 7-aminocephalosporanic acid nucleus. (844)

Chagas' disease   *See* trypanosomiasis.

chancre (shang′ker)   The primary lesion of syphilis occurring at the site of entry of the infection. (976)

chancroid (shang′kroid)   A sexually transmitted disease caused by the γ-proteobacterium *Haemophilus ducreyi.* Chancroid is an important cofactor in the transmission of the AIDS virus. Also known as genital ulcer disease due to the painful circumscribed ulcers that form on the penis or entrance to the vagina. (971)

chaperone proteins   Proteins that assist in the folding and stabilization of other proteins. Some are also involved in directing newly synthesized proteins to protein secretion systems, or to other locations in the cell. (284)

chemical oxygen demand (COD)   The amount of chemical oxidation required to convert organic matter in water and wastewater to CO_2. (1054)

chemiosmotic hypothesis (kem″e-o-os-mot′ik)   The hypothesis that a proton gradient and an electrochemical gradient are generated by electron transport and then used to drive ATP synthesis by oxidative phosphorylation. (202)

chemoheterotroph   (ke″mo-het′er-o-trŏf″)   *See* chemoorganotrophic heterotrophs.

chemolithoautotroph   A microorganism that oxidizes reduces inorganic compounds to derive both energy and electrons; CO_2 is the carbon source. Also called chemolithotrophic autotroph. (212)

chemolithoheterotroph   A microorganism that uses reduced inorganic compounds to drive both energy and electrons; organic molecules are used as the carbon source. Also called mixotroph. (103)

chemoorganoheterotroph   A microorganism that uses organic compounds as sources of energy, electrons, and carbon for biosynthesis. Also called chemoheterotroph and chemoorganotrophic heterotroph. (103)

chemoreceptors Special protein receptors in the plasma membrane or periplasmic space that bind chemicals and trigger the appropriate chemotaxic response. (71)

chemostat (ke'mo-stat) A continuous culture apparatus that feeds medium into the culture vessel at the same rate as medium containing microorganisms is removed; the medium in a chemostat contains one essential nutrient in a limiting quantity. (131)

chemotaxis (ke"mo-tak'sis) The pattern of microbial behavior in which the microorganism moves toward chemical attractants and/or away from repellents. (71)

chemotherapeutic agents (ke"mo-ther-ah-pu'tik) Compounds used in the treatment of disease that destroy pathogens or inhibit their growth at concentrations low enough to avoid doing undesirable damage to the host. (164, 835)

chemotrophs (ke'mo-trōfs) Organisms that obtain energy from the oxidation of chemical compounds. (103)

chickenpox (varicella) (chik'en-poks) A highly contagious skin disease, usually affecting 2- to 7-year-old children; it is caused by the varicella-zoster virus, which is acquired by droplet inhalation into the respiratory system. (914)

chiral (ki'rəl) Having handedness: consisting of one or another stereochemical form. (1076)

chitin (ki'tin) A tough, resistant, nitrogen-containing polysaccharide forming the walls of certain fungi, the exoskeleton of arthropods, and the epidermal cuticle of other surface structures of certain protists and animals. (631)

chlamydiae (klə-mid'e-e) Members of the genera *Chlamydia* and *Chlamydiophila*: gram-negative, coccoid cells that reproduce only within the cytoplasmic vesicles of host cells using a life cycle that alternates between elementary bodies and reticulate bodies. (531)

chlamydial pneumonia (klə-mid'e-əl noo-mo'ne-ə) A pneumonia caused by *Chlamydiophila pneumoniae*. Clinically, infections are mild and 50% of adults have antibodies to the chlamydiae. (948)

chloramphenicol (klo"ram-fen'ĭ-kol) A broad-spectrum antibiotic that is produced by *Streptomyces venzuelae* or synthetically; it binds to the large ribosomal subunit and inhibits the peptidyl transferase reaction. (846)

chlorophyll (klor'o-fil) The green photosynthetic pigment that consists of a large tetrapyrrole ring with a magnesium atom in the center. (216)

chloroplast (klo'ra-plast) A eucaryotic plastid that contains chlorophyll and is the site of photosynthesis. (90)

chlorosomes Elongated, intramembranous vesicles found in the green sulfur and nonsulfur bacteria that contain light-harvesting pigments. Sometimes called chlorobium vescicles. (523)

cholera (kol'er-ah) An acute infectious enteritis, endemic and epidemic in Asia, which periodically spreads to the Middle East, Africa, Southern Europe, and South America; caused by *Vibrio cholerae*. (983)

choleragen (kol'er-ah-gen) The cholera toxin; an extremely potent protein molecule made by strains of

Vibrio cholerae in the small intestine after ingestion of feces-contaminated water or food. It acts on epithelial cells to cause hypersecretion of chloride and bicarbonate and an outpouring of large quantities of fluid from the mucosal surface. (983)

chromatic adaptation The capacity of cyanobacteria to alter the ratio of their light-harvesting or accessory pigments in response to changes in the spectral quality (wavelengths) of light. (525)

chromatin (kro'mah-tin) The DNA-containing portion of the eucaryotic nucleus; the DNA is almost always complexed with histones. It can be very condensed (heterochromatin) or more loosely organized and genetically active (euchromatin). (91)

chromoblastomycosis (kro"mo-blas"to-mi-ko'sis) A chronic fungal skin infection, producing wartlike nodules that may ulcerate. It is caused by the black molds *Phialophora verrucosa* or *Fonsecaea pedrosoi*. (1010)

chromogen (kro'me-jen) A colorless substrate that is acted on by an enzyme to produce a colored end product. (879)

chromophore group (kro"mo-fōr) A chemical group with double bonds that absorbs visible light and gives a dye its color. (26)

chromosomal nuclei Nuclei found in some protists in which the chromosomes remain condensed throughout the cell cycle. (609)

chromosomes (kro'mo-somz) The bodies that have most or all of the cell's DNA and contain most of its genetic information (mitochondria and chloroplasts also contain DNA and genes). (91)

chronic carrier An individual who harbors a pathogen for a long time. (892)

chytrids A term used to describe the *Chytridiomycota*, which are simple terrestrial and aquatic fungi that produce motile zoospores with single, posterior, whiplash flagella. (635)

cilia (sil'e-ah) Threadlike appendages extending from the surface of some protozoa that beat rhythmically to propel them; cilia are membrane-bound cylinders with a complex internal array of microtubules, usually in a 9 + 2 pattern. (95)

citric acid cycle *See* tricarboxylic acid (TCA) cycle.

class I MHC molecule *See* major histocompatibility complex.

class II MHC molecule *See* major histocompatibility complex.

class switching The change in immunoglobulin isotype (or class) secretion that results during B-cell and then plasma cell differentiation. (795)

classical complement pathway The antibody-dependent pathway of complement activation; it leads to the lysis of pathogens and stimulates phagocytosis and other host defenses. (766)

classification The arrangement of organisms into groups based on mutual similarity or evolutionary relatedness. (478)

clonal selection The process by which an antigen selects the best-fitting B-cell receptor, activating that B cell, resulting in the synthesis of antibody and clonal expansion. (798)

clone (klōn) A group of genetically identical cells or organisms derived by asexual reproduction from a single parent. (798)

clostridial myonecrosis (klo-strid'e-al mi"o-ne-kro'sis) Death of individual muscle cells caused by clostridia. Also called gas gangrene. (965)

clue cells Vaginal epithelial cells covered with bacteria. The name comes from the fact that identification of these cells offered a clue to the disease etiology. (971)

cluster of differentiation molecules (CDs) Functional cell surface proteins or receptors that can be measured in situ from peripheral blood, biopsy samples, or other body fluids. They can be used to identify leukocyte subpopulations. Some examples include interleukin-2 receptor (IL-2R), CD4, CD8, CD25, and intercellular adhesion molecule-1 (ICAM-1). (776)

coaggregation The collection of bacteria on a substrate such as a tooth surface because of cell-to-cell recognition of genetically distinct bacterial types. Many of these interactions appear to be mediated by a lectin on one bacterium that interacts with a complementary carbohydrate receptor on another bacterium. (991)

coagulase (ko-ag'u-las) An enzyme that induces blood clotting; it is characteristically produced by pathogenic staphylococci. (582)

coated vesicles The clathrin coated vesicles formed by receptor-mediated endocytosis. (87)

coccidioidomycosis (kok-sid"e-oi"do-mi-ko'sis) A fungal disease caused by *Coccidioides immitis* that exists in dry, highly alkaline soils. Also known as valley fever, San Joaquin fever, or desert rheumatism. (1000)

coccolithophore Photosynthetic protists belonging to the phylum Stramenopila. They are characterized by coccoliths—intricate cell walls made of calcite. (624)

coccus (kok'us, pl. **cocci**, kok'si) A roughly spherical bacterial cell. (39)

code degeneracy The genetic code is organized in such a way that often there is more than one codon for each amino acid. (275)

codon (ko'don) A sequence of three nucleotides in mRNA that directs the incorporation of an amino acid during protein synthesis or signals the stop of translation. (264)

coenocytic (se"no-sit'ik) Refers to a multinucleate cell or hypha formed by repeated nuclear divisions not accompanied by cell divisions. (119, 631)

coenzyme (ko-en'zīm) A loosley bound cofactor that often dissociates from the enzyme active site after product has been formed. (176)

cofactor The nonprotein component of an enzyme; it is required for catalytic activity. (176)

cold sore A lesion caused by the herpes simplex virus; usually occurs on the border of the lips or nares. Also known as a fever blister or herpes labialis. (931)

colicin (kol'ĭ-sin) A plasmid-encoded protein that is produced by enteric bacteria and binds to specific receptors on the cell envelope of sensitive target bacteria, where it may cause lysis or attack specific intracellular sites such as ribosomes. (763)

coliform (ko′lĭ-form) A gram-negative, non-sporing, facultative rod that ferments lactose with gas formation within 48 hours at 35°C. (1052)

colonization (kol″ə-nĭ-za′shən) The establishment of a site of microbial reproduction on an inanimate surface or organism without necessarily resulting in tissue invasion or damge. (820)

colony An assemblage of microorganisms growing on a solid surface such as the surface of an agar culture medium; the assemblage often is directly visible, but also may be seen only microscopically. (113)

colony forming units (CFU) The number of microorganisms that form colonies when cultured using spread plates or pour plates, an indication of the number of viable microorganisms in a sample. (130)

colony stimulating factor (CSF) A protein that stimulates the growth and development of specific cell populations (e.g., granulocyte-CSF stimulates granulocytes to be made from their precursor stem cells). (768)

colorless sulfur bacteria A diverse group of non-photosynthetic proteobacteria that can oxidize reduced sulfur compounds such as hydrogen sulfide. Many are lithotrophs and derive energy from sulfur oxidation. Some are unicellular, whereas others are filamentous gliding bacteria. (550)

comedo (kom′ĕ-do; pl., **comedones**) A plug of dried sebum in an excretory duct of the skin. (737)

cometabolism The modification of a compound not used for growth by a microorganism, which occurs in the presence of another organic material that serves as a carbon and energy source. (1077)

commensal (kŏ-men′sal) Living on or within another organism without injuring or benefiting the other organism. (729)

commensalism (kŏ-men′sal-izm″) A type of symbiosis in which one individual gains from the association and the other is neither harmed nor benefited. (729)

common cold An acute, self-limiting, and highly contagious virus infection of the upper respiratory tract that produces inflammation, profuse discharge, and other symptoms. (932)

common-source epidemic An epidemic that is characterized by a sharp rise to a peak and then a rapid, but not as pronounced, decline in the number of individuals infected; it usually involves a single contaminated source from which individuals are infected. (889)

common vehicle transmission The transmission of a pathogen to a host by means of an inanimate medium or vehicle. (894)

communicable disease A disease associated with a pathogen that can be transmitted from one host to another. (888)

community An assemblage of different types of organisms or a mixture of different microbial populations. (643)

compatible solute A low-molecular-weight molecule used to protect cells against changes in solute concentrations (osmolarity) in their habitat; it can exist at high concentrations within the cell and still be compatible with metabolism and growth. (132)

competent A procaryotic cell that can take up free DNA fragments and incorporate them into its genome during transformation. (343)

competition An interaction between two organisms attempting to use the same resource (nutrients, space, etc.). (732)

competitive exclusion principle Two competing organisms overlap in resource use, which leads to the exclusion of one of the organisms. (732)

complementarity determining regions (CDRs) Hypervariable regions in an immunoglobulin protein that form the three-dimensional binding sites for epitope binding. (791)

complementary DNA (cDNA) A DNA copy of an RNA molecule (e.g., a DNA copy of an mRNA). (358)

complement system A group of plasma proteins that plays a major role in an animal's defensive immune response. (763)

complex medium Culture medium that contains some ingredients of unknown chemical composition. (111)

complex viruses Viruses with capsids having a complex symmetry that is neither icosahedral nor helical. (428)

composting The microbial processing of fresh organic matter under moist, aerobic conditions, resulting in the accumulation of a stable humified product, which is suitable for soil improvement and stimulation of plant growth. (1075)

compromised host A host with lowered resistance to infection and disease for any of several reasons. The host may be seriously debilitated (due to malnutrition, cancer, diabetes, leukemia, or another infectious disease), traumatized (from surgery or injury), immunosuppressed, or have an altered microbiota due to prolonged use of antibiotics. (740, 1016)

concatemer A long DNA molecule consisting of several genomes linked together in a row. (446)

conditional mutations Mutations with phenotypes that are expressed only under certain environmental conditions. (323)

confocal scanning laser microscope (CSLM) A light microscope in which monochromatic laser-derived light scans across the specimen at a specific level and illuminates one area at a time to form an image. Stray light from other parts of the specimen is blocked out to give an image with excellent contrast and resolution. (34)

congenital (neonatal) herpes An infection of a newborn caused by transmission of the herpesvirus during vaginal delivery. (934)

congenital rubella syndrome A wide array of congenital defects affecting the heart, eyes, and ears of a fetus during the first trimester of pregnancy, and caused by the rubella virus. (920)

congenital syphilis Syphilis that is acquired in utero from the mother. (976)

conidiospore (ko-nid′e-o-spōr) An asexual, thin-walled spore borne on hyphae and not contained within a sporangium; it may be produced singly or in chains. (633)

conidium (ko-nid′e-um; pl., **conidia**) See conidiospore.

conjugants (kon′joo-gants) Complementary mating types among protists that participate in a form sexual reproduction called conjugation. (620)

conjugation (kon″ju-ga′shun) 1. The form of gene transfer and recombination in procaryotes that requires direct cell-to-cell contact. 2. A complex form of sexual reproduction commonly employed by protists. (337, 620, 609)

conjugative plasmid A plasmid that carries the genes that enable its transfer to other bacteria during conjugation (e.g., F plasmid). (53, 334)

conjunctivitis of the newborn See ophthalmia neonatorum. (975)

consensus sequence A commonly occurring sequence of nucleotides within a genetic element such as the Pribnow box of bacterial promoters. (269)

consortium A physical association of two different organisms, usually beneficial to both organisms. (717)

constant region (C$_L$ and C$_H$) The part of an antibody molecule that does not vary greatly in amino acid sequence among molecules of the same class, subclass, or type. (790)

constitutive mutant A strain that produces an inducible enzyme continually, regardless of need, because of a mutation in either the operator or regulator gene. (293)

constructed wetlands Intentional creation of marshland plant communities and their associated microorganisms for environmental restoration or to purify water by the removal of bacteria, organic matter, and chemicals as the water passes through the aquatic plant communities. (1058)

consumer An organism that feeds directly on living or dead animals, by ingestion or by phagocytosis. (652)

contact transmission Transmission of the pathogen by contact of the source or reservoir of the pathogen with the host. (892)

continuous culture system A culture system with constant environmental conditions maintained through continual provision of nutrients and removal of wastes. See also chemostat. (131)

contractile vacuole (vak′u-ōl) In protists and some animals, a clear fluid-filled cell vacuole that takes up water from within the cell and then contracts, releasing it to the outside through a pore in a cyclical manner. Contractile vacuoles function primarily in osmoregulation and excretion. (607)

convalescent carrier (kon″vah-les′ent) An individual who has recovered from an infectious disease but continues to harbor large numbers of the pathogen. (891)

cooperation A positive but not obligatory interaction between two different organisms. (726)

coral bleaching The loss of photosynthetic pigments by either physiological inhibition or expulsion of the coral photosynthetic endosymbiont, zooxanthellae, a dinoflagellate. Bleached corals may be temporarily or permanently damaged. (719)

corepressor (ko″re-pre′sor) A small molecule that inhibits the synthesis of a repressible enzyme. (295)

cortex The layer of the bacterial endospore that is particularly important in conferring heat resistance to the endospore. (73)

cosmid (koz′mid) A plasmid vector with lambda phage *cos* sites that can be packaged in a phage capsid; it is useful for cloning large DNA fragments. (370)

cristae (kris′te) Infoldings of the inner mitochondrial membrane. (89)

crossing-over A process in which segments of two adjacent DNA strands are exchanged; breaks occur in both strands, and the exposed ends of each strand join to those of the opposite segment on the other strand. (330)

crown gall disease A plant tumor, or gall, caused by the certain species of the α-proteobacterium *Agrobacterium,* most commonly *A. tumefaciens.* (546)

cryptin Antimicrobial peptides produced by Paneth cells in the intestines. (761)

cryptococcosis (krip″to-kok-o′sis) An infection caused by the basidiomycete, *Cryptococcus neoformans,* which may involve the skin, lungs, brain, or meninges. (1001)

cryptosporidiosis (krip″to-spo-rid″e-o′sis) Infection with protozoa of the genus *Cryptosporidium.* The most common symptoms are prolonged diarrhea, weight loss, fever, and abdominal pain. (1014)

crystallizable fragment (Fc) The stem of the Y portion of an antibody molecule. Cells such as macrophages bind to the Fc region; it also is involved in complement activation. (790)

cutaneous anthrax (ku-ta′ne-us an′thraks) A form of anthrax involving the skin. (988)

cutaneous diphtheria (ku-ta′ne-us dif-the′re-ah) A skin disease caused by *Corynebacterium diphtheriae* that infects wound or skin lesions, causing a slow-healing ulceration. (949)

cyanobacteria (si″ah-no-bak-te′re-ah) A large group of gram-negative bacteria that carry out oxygenic photosynthesis using a system like that present in photosynthetic eucaryotes. (524)

cyclic photophosphorylation (fo″to-fos″for-ĭ-la′shun) The formation of ATP when light energy is used to move electrons cyclically through an electron transport chain during photosynthesis; only photosystem I participates. (217)

cyclosporiasis A disease caused by infection with the protozoan parasite *Cyclospora cayetanensis.* Infection has been associated with ingestion of contaminated produce. *Cyclospora* oocysts contain two sporocysts, each containing two sporozoites. Un-

sporulated oocysts are passed in the stool, and sporulation occurs outside the host. (1014)

cyst (sist) A general term used for a specialized microbial cell enclosed in a wall. Cysts are formed by protists and a few bacteria. They may be dormant, resistant structures formed in response to adverse conditions or reproductive cysts that are a normal stage in the life cycle. (608)

cytochromes (si′to-krōms) Heme proteins that carry electrons, usually as members of electron transport chains. (174)

cytokine (si′to-kīn) A general term for proteins released by a cell in response to inducing stimuli, which are mediators that influence other cells. Produced by lymphocytes, monocytes, macrophages, and other cells. (748, 766)

cytokinesis Processes that apportion the cytoplasm and organelles, synthesize a septum, and divide a cell into two daughter cells during cell division. (121)

cytomegalovirus inclusion disease (si″to-meg″ah-lo-vi′rus) An infection caused by the cytomegalovirus and marked by nuclear inclusion bodies in enlarged infected cells. (933)

cytopathic effect (si″to-path′ik) The observable change that occurs in cells as a result of viral replication. Examples include ballooning, binding together, clustering, or even death of the cultured cells. (418, 470, 866)

cytoplasmic matrix (si″to-plaz′mik) The protoplasm of a cell that lies within the plasma membrane and outside any other organelles. In bacteria it is the substance between the cell membrane and the nucleoid. (48, 83)

cytoproct A specific site in certain protists (e.g., ciliates) where digested material is expelled. (608)

cytosine (si′to-sēn) A pyrimidine 2-oxy-4-aminopyrimidine found in nucleosides, nucleotides, and nucleic acids. (241)

cytoskeleton (si″to-skel′ĕ-ton) A network of microfilaments, microtubules, intermediate filaments, and other components in the cytoplasm of eucaryotic cells that helps give them shape. (83)

cytostome (si′to-stōm) A permanent site in a ciliate protist at which food is ingested. (608)

cytotoxic T lymphoryte (CTL) (si″to-tok′sik) A type of T cell that recognizes antigen in class I MHC molecules, destroying the cell on which the antigen is displayed. Also called CD8$^+$ cell. (782)

cytotoxin (si′to-tok′sin) A toxin or antibody that has a specific toxic action upon cells; cytotoxins are named according to the cell for which they are specific (e.g., nephrotoxin). (825)

D

Dane particle A 42 nm spherical particle that is one of three that are seen in hepatitis B virus infections. The Dane particle is the complete virion. (936)

dark-field microscopy Microscopy in which the specimen is brightly illuminated while the background is dark. (21)

dark reactivation The excision and replacement of thymine dimers in DNA that occurs in the absence of light. (215)

deamination (de-am″i-na′shun) The removal of amino groups from amino acids. (212)

decimal reduction time (*D* or *D* value) The time required to kill 90% of the microorganisms or spores in a sample at a specified temperature. (154)

decomposer An organism that breaks down complex materials into simpler ones, including the release of simple inorganic products. Often a decomposer such as an insect or earthworm physically reduces the size of substrate particles. (656)

defensin (de-fens′sin) Specific peptides produced by neutrophils that permeabilize the outer and inner membranes of certain microorganisms, thus killing them. (762)

defined medium Culture medium made with components of known composition. (111)

Deltaproteobacteria One of the five classes of proteobacteria. Chemoorganotrophic bacteria that usually are either predators on other bacteria or anaerobes that generate sulfide from sulfate and sulfite. (562)

denaturation (de-na″chur-a′shun) A change in protein shape that destroys its activity; the term is also sometimes applied to changes in nucleic-acid shape. (179)

denaturing gradient gel electrophoresis (DGGE) A technique by which DNA is rendered single-stranded (denatured) while undergoing electrophoresis so that DNA fragments of the same size can be separated according to nucleotide sequence rather than molecular weight. This is accomplished by preparing gels with a gradient of a chemical that denatures DNA. As fragments migrate from the negative to positive pole of the gel, they stop when they become single-stranded. (661)

dendritic cell (den-drit′ ik) An antigen-presenting cell that has long membrane extensions resembling the dendrites of neurons. These cells are found in the lymph nodes, spleen, and thymus (interdigitating dendritic cells); skin (Langerhans cells); and other tissues (interstitial dendritic cells). They express MHC class II and B7 costimulatory molecules and present antigens to T-helper cells. (747)

dendrogram A treelike diagram that is used to graphically summarize mutual similarities and relationships between organisms. (479)

denitrification (de-ni″trī-fī-ka′shən) The reduction of nitrate to gaseous products, primarily nitrogen gas, during anaerobic respiration. (205, 649)

dental plaque (plak) A thin film on the surface of teeth consisting of bacteria embedded in a matrix of bacterial polysaccharides, salivary glycoproteins, and other substances. (991)

deoxyribonucleic acid (DNA) (de-ok″se-ri″bo-nu-kle′ik) The nucleic acid that constitutes the genetic

intermediate host   The host that serves as a temporary but essential environment for development of a parasite and completion of its life cycle. (816)

interspecies hydrogen transfer   The linkage of hydrogen production from organic matter by anaerobic heterotrophic microorganisms to the use of hydrogen by other anaerobes in the reduction of carbon dioxide to methane. This avoids possible hydrogen toxicity. (729)

intertriginous candidiasis   A skin infection caused by *Candida* species. Involves those areas of the body, usually opposed skin surfaces, that are warm and moist (axillae, groin, skin folds). (1017)

intoxication (in-tok"si-ka'shun)   A disease that results from the entrance of a specific toxin into the body of a host. The toxin can induce the disease in the absence of the toxin-producing organism. (824)

intraepidermal lymphocytes   T cells found in the epidermis of the skin that express the γδ T-cell receptor. (759)

intranuclear inclusion body (in"trə-noo'kle-ər)   A structure found within cells infected with the cytomegalovirus. (933)

intrinsic factors   Food-related factors such as moisture, pH, and available nutrients that influence microbial growth. (1024)

intron (in'tron)   A noncoding intervening sequence in a split or interrupted gene, which codes for RNA that is missing from the final RNA product. (273)

invasiveness (in-va'siv-nes)   The ability of a microorganism to enter a host, grow and reproduce within the host, and spread throughout its body. (816)

ionizing radiation   Radiation of very short wavelength and high energy that causes atoms to lose electrons or ionize. (141, 156)

isogamy   The fusion of two morphologically and physiologically similar gametes during sexual reproduction in protists. (609)

isotype (i'so-tīp)   A variant form of an immunoglobulin (e.g., an immunoglobulin class, subclass, or type) that occurs in every normal individual of a particular species. Usually the characteristic antigenic determinant is in the constant region of H and L chains. (791)

J

Jaccard coefficient (S_J)   An association coefficient used in numerical taxonomy; it is the proportion of characters that match, excluding those that both organisms lack. (479)

J chain   A polypeptide present in polymeric IgM and IgA that links the subunits together. (792)

jock itch   *See* tinea cruris. (1009)

K

kallikrein   An enzyme that acts on kininogen, releasing the active bradykinin protein. (757)

keratitis (ker"ah-ti'tis)   Inflammation of the cornea of the eye. (1013)

kinetoplast (ki-ne'to-plast)   A special structure in the mitochondrion of certain protists. It contains the mitochondrial DNA. (90)

kinetosome   Intracellular microtubular structure that serves as the base of cilia in ciliated protists. Similar in structure to a centriole. Also referred to as a basal body. (608)

Kirby-Bauer method   A disk diffusion test to determine the susceptibility of a microorganism to chemotherapeutic agents. (840)

Koch's postulates (koks pos'tu-lāts)   A set of rules for proving that a microorganism causes a particular disease. (9)

Koplik's spots (kop'liks)   Lesions of the oral cavity caused by the measles (rubeola) virus that are characterized by a bluish white speck in the center of each. (918)

Korarchaeota   A proposed phylum in the *Archaea* domain. To date, it is based entirely on uncultured microbes that have been identified through 16S rRNA nucleotide sequences cloned directly from the environment. (511)

Korean hemorrhagic fever   An acute infection caused by a virus that produces varying degrees of hemorrhage, shock, and sometimes death. (923)

Krebs cycle   *See* tricarboxylic acid (TCA) cycle.

L

lactic acid fermentation (lak'tik)   A fermentation that produces lactic acid as the sole or primary product. (208)

lactoferrin   An iron-sequestering protein released from macrophages and neutrophils into plasma. (759)

lager   Pertaining to the process of aging beers to allow flavor development. (1044)

lag phase   A period following the introduction of microorganisms into fresh culture medium when there is no increase in cell numbers or mass during batch culture. (123)

Lancefield system (group) (lans'feld)   One of the serologically distinguishable groups (as group A, group B) into which streptococci can be divided. (584, 876)

Langerhans cell   Cell found in the skin that internalizes antigen and moves in the lymph to lymph nodes where it differentiates into a dendritic cell. (758)

Lassa fever   An acute, contagious, viral disease of central western Africa. The disease is characterized by fever, inflammation, muscular pains, and difficulty swallowing. (942)

late mRNA   Messenger RNA produced later in a virus infection, which codes for proteins needed in capsid construction and virus release. (431)

latent period (la'tent)   The initial phase in the one-step growth experiment in which no phages are released. (429)

latent virus infections   Virus infections in which the virus stops reproducing and remains dormant for a period before becoming active again. (461)

lateral gene transfer   *See* horizontal gene transfer.

leader sequence   A nontranslated sequence at the 5′ end of mRNA that lies between the operator and the initiation codon; it aids in the initiation and regulation of transcription. (265)

lectin complement pathway (lek'tin)   An antibody-independent pathway of complement activation that is initiated by microbial lectins (proteins that bind carbohydrates) and includes the C3–C9 components of the classical pathway. (765)

leghemaglobin   A heme-containing pigment produced in leguminous plants. It is similar in structure to vertebrate hemoglobin; however, it has a higher affinity for oxygen. It functions to protect nodule-forming, nitrogen-fixing bacteria from oxygen, which would poison their nitrogenase. (70)

legionellosis (le"jə-nel-o'sis)   *See* Legionnaires' disease.

Legionnaires' disease (legionellosis)   A pulmonary infection, caused by *Legionella pneumophila*. (949)

leishmanias (lēsh"ma'ne-ăs)   Trypanosomal protists of the genus *Leishmania,* that cause the disease leishmaniasis. (1004)

leishmaniasis (lēsh"mah-ni'ah-sis)   A group of human diseases caused by the protists called leishmanias. (612, 1004)

lepromatous (progressive) leprosy (lep-ro'mah-tus lep'ro-se)   A relentless, progressive form of leprosy in which large numbers of *Mycobacterium leprae* develop in skin cells, killing the skin cells and resulting in the loss of features. Disfiguring nodules form all over the body. (966)

leprosy (lep'ro-se) or **Hansen's disease**   A severe disfiguring skin disease caused by *Mycobacterium leprae*. (966)

lethal dose 50 (LD$_{50}$)   Refers to the dose or number of organisms that will kill 50% of an experimental group of hosts within a specified time period. (423, 817)

leukemia (loo-ke'me-ah)   A progressive, malignant disease of blood-forming organs, marked by distorted proliferation and development of leukocytes and their precursors in the blood and bone marrow. Certain leukemias are caused by viruses (HTLV-1, HTLV-2). (935)

leukocidin (loo"ko-si'din)   A microbial toxin that can damage or kill leukocytes. (828)

leukocyte (loo′ko-sīt) Any colorless white blood cell. Can be classified into granular and agranular lymphocytes. (744)

lichen (li′ken) An organism composed of a fungus and either photosynthetic protists or cyanobacteria in a symbiotic association. (731)

lipid raft A microdomain in the plasma membrane that is enriched for particular lipids and proteins. (81)

lipopolysaccharide (LPSs) (lip″o-pol″e-sak′ah-rīd) A molecule containing both lipid and polysaccharide, which is important in the outer membrane of the gram-negative cell wall. (58)

listeriosis (lis-ter″e-o′sis) A sporadic disease of animals and humans, particularly those who are immunocompromised or pregnant, caused by the bacterium *Listeria monocytogenes*. (984)

lithotroph (lith′o-trōf) An organism that uses reduced inorganic compounds as its electron source. (103)

lobopodia Rounded pseudopodia found in some amoeboid protists. (613)

log phase *See* exponential phase.

lophotrichous (lo-fot′ri-kus) A cell with a cluster of flagella at one or both ends. (67)

Lyme disease (LD, Lyme borreliosis) (līm) A tick-borne disease caused by the spirochete *Borrelia burgdorferi*. (961)

lymph node A small secondary lymphoid organ that contains lymphocytes, macrophages, and dendritic cells. It serves as a site for (1) filtration and removal of foreign antigens and (2) the activation and proliferation of lymphocytes. (750)

lymphocyte (lim′fo-sīt) A nonphagocytic, mononuclear leukocyte (white blood cell) that is an immunologically competent cell, or its precursor. Lymphocytes are present in the blood, lymph, and lymphoid tissues. *See* B cell and T cell. (748)

lymphocytic choriomeningitis (LCM) An enveloped, single-stranded RNA virus that causes a nonbacterial meningitis in mice and, occasionally, in humans. (942)

lymphogranuloma venereum (LGV) (lim″fo-gran″u-lo′mah) A sexually transmitted disease caused by *Chlamydia trachomatis* serotypes L$_1$–L$_3$, which affect the lymph organs in the genital area. (975)

lymphokine (lim′fo-kin) A biologically active glycoprotein (e.g., IL-1) secreted by activated lymphocytes, especially sensitized T cells. It acts as an intercellular mediator of the immune response and transmits growth, differentiation, and behavioral signals. (767)

lyophilization Freezing and dehydrating samples as a means of preservation. Many microorganisms and natural products can be lyophilized for long-term storage. Commonly referred to as freeze-drying. (1063)

lysis (li′sis) The rupture or physical disintegration of a cell. (61)

lysogenic (li-so-jen′ik) *See* lysogens.

lysogens (li′so-jens) Bacterial and archaeal cells that are carrying a provirus and can produce viruses under the proper conditions. (345, 438)

lysogeny (li-soj′e-ne) The state in which a viral genome remains within the bacterial or achaeal cell after infection and reproduces along with it rather than taking control of the host cell and destroying it. (345, 438)

lysosome (li′so-sōm) A spherical membranous eucaryotic organelle that contains hydrolytic enzymes and is responsible for the intracellular digestion of substances. (86)

lysozyme (li′so-zīm) An enzyme that degrades peptidoglycan by hydrolyzing the $\beta(1 \rightarrow 4)$ bond that joins *N*-acetylmuramic acid and *N*-acetylglucosamine. (61, 759)

lytic cycle (lit′ik) A virus life cycle that results in the lysis of the host cell. (345, 428)

M

macroevolution Major evolutionary change leading to either speciation or extinction. (477)

macrolide antibiotic (mak′ro-līd) An antibiotic containing a macrolide ring, a large lactone ring with multiple keto and hydroxyl groups, linked to one or more sugars. (846)

macromolecule (mak″ro-mol′ĕ-kūl) A large molecule that is a polymer of smaller units joined together. (226)

macromolecule vaccine A vaccine made of specific, purified macromolecules derived from pathogenic microorganisms. (778)

macronucleus (mak″ro-nu′kle-us) The larger of the two nuclei in ciliate protists. It is normally polyploid and directs the routine activities of the cell. (609)

macrophage (mak′ro-fāj) The name for a large mononuclear phagocytic cell, present in blood, lymph, and other tissues. Macrophages are derived from monocytes. They phagocytose and destroy pathogens; some macrophages also activate B cells and T cells. (746)

maduromycosis (mah-du′ro-mi-ko′sis) A subcutaneous fungal infection caused by *Madurella mycetoma*; also termed an eumycotic mycetoma. (1010)

madurose The sugar derivative 3-O-methyl-D-galactose, which is characteristic of several actinomycete genera that are collectively called maduromycetes. (601)

magnetosomes Magnetite particles in magnetotactic bacteria that are tiny magnets and allow the bacteria to orient themselves in magnetic fields. (50)

maintenance energy The energy a cell requires simply to maintain itself or remain alive and functioning properly. It does not include the energy needed for either growth or reproduction. (177)

major histocompatibility complex (MHC) A chromosome locus encoding the histocompatibility antigens. Class I MHC molecules are cell surface glycoproteins present on all nucleated cells; class II MHC glycoproteins are on antigen-presenting cells. Class I MHC glycoproteins present endogenous antigens to CD8$^+$ T cells. Class II MHC glycoproteins present exogenous antigens to CD4$^+$ T cells. (778)

malaria (mah-la′re-ah) A serious infectious illness caused by the parasitic protozoan *Plasmodium*. Malaria is characterized by bouts of high chills and fever that occur at regular intervals. (1001)

malt (mawlt) Grain soaked in water to soften it, induce germination, and activate its enzymes. The malt is then used in brewing and distilling. (1043)

Marburg viral hemorrhagic fever An acute, infectious disease caused by the Marburg hemorrhagic fever virus. Symptoms include varying degrees of bleeding and shock. (942)

mash The soluble materials released from germinated grains and prepared as a microbial growth medium. (1043)

mashing The process in which cereals are mixed with water and incubated in order to degrade their complex carbohydrates (e.g., starch) to more readily usable forms such as simple sugars. (1041)

mast cell A white blood cell that produces vasoactive molecules (e.g., histamine) and stores them in vacuoles near the cell membrane where they are released upon cell stimulation by external triggers. Mast cells can bind IgE proteins by their Fc region; IgE capture (by the Fab region) of antigen acts as a trigger for the release of vasoactive molecules. (747)

mating type A strain of a eucaryotic organism that can mate sexually with another strain of the same species. Commonly refers to strains of fungal mating types (MAT) (e.g., MAT α and MAT **a** of *Saccharomyces cerevisiae*. (634)

M cell Specialized cell of the intestinal mucosa and other sites, such as the urogenital tract, that delivers the antigen from the apical face of the cell to lymphocytes clustered within the pocket in its basolateral face. (759)

mean growth rate constant (k) The rate of microbial population growth expressed in terms of the number of generations per unit time. (126)

measles (rubeola) (me′zelz) A highly contagious skin disease that is endemic throughout the world. It is caused by a morbilli virus in the family *Paramyxoviridae*, which enters the body through the respiratory tract or through the conjunctiva. (917)

medical mycology (mi-kol′o-je) The discipline that deals with the fungi that cause human disease. (997)

meiosis (mi-o′sis) The sexual process in which a diploid cell divides and forms two haploid cells. (94)

melting temperature (T_m) The temperature at which double-stranded DNA separates into individual strands; it is dependent on the G + C content of the DNA and is used to compare genetic material in microbial taxonomy. (483)

membrane attack complex (MAC) The complex complement components (C5b–C9) that create a pore in the plasma membrane of a target cell and leads to cell lysis. C9 probably forms most of the actual pore. (764)

membrane-disrupting exotoxin A type of exotoxin that lyses host cells by disrupting the integrity of the plasma membrane. (828)

membrane filter technique The use of a thin porous filter made from cellulose acetate or some other polymer to collect microorganisms from water, air, and food. (156, 1052)

memory cell An inactive lymphocyte clone derived from a sensitized B or T cell capable of an accentuated response to a subsequent antigen exposure. (786)

meningitis (men″in-ji′tis) A condition that refers to inflammation of the brain or spinal cord meninges (membranes). The disease can be divided into bacterial (septic) meningitis and aseptic meningitis syndrome (caused by nonbacterial sources). (950)

merozygote A partially diploid procaryotic cell produced by horizontal gene transfer; in most cases some of the extra genetic material is destroyed and some is incorporated into the recipient's chromosome by homologous recombination, restoring the haploid state. (330)

mesophile (mes′o-fīl) A microorganism with a growth optimum around 20 to 45°C, a minimum of 15 to 20°C, and a maximum about 45°C or lower. (138)

messenger RNA (mRNA) Single-stranded RNA synthesized from a nucleic acid template (DNA in cellular organisms, RNA in some viruses) during transcription; mRNA binds to ribosomes and directs the synthesis of protein. (251)

metabolic channeling (mĕt″ah-bol′ik) The localization of metabolites and enzymes in different parts of a cell. (180)

metabolic control engineering Modification of the controls for biosynthetic pathways without altering the pathways themselves in order to improve process efficiency. (1062)

metabolic pathway engineering (MPE) The use of molecular techniques to improve the efficiency of pathways that synthesize industrially important products. (1062)

metabolism (me-tab′o-lizm) The total of all chemical reactions in the cell; almost all are enzyme catalyzed. (167)

metachromatic granules (met″ah-kro-mat′ik) Granules of polyphosphate in the cytoplasm of some bacteria that appear a different color when stained with a blue basic dye. They are storage reservoirs for phosphate. Sometimes called volutin granules. (50)

metagenomics Also called environmental or community genomics, metagenomics is the study of genomes recovered from environmental samples without first isolating members of the microbial community and growing them in pure cultures. (402)

metastasis (mĕ-tas′tah-sis) The transfer of a disease like cancer from one organ to another not directly connected with it. (461)

methanogens (meth′ə-no-jens″) Strictly anaerobic archaea that derive energy by converting CO_2, H_2, formate, acetate, and other compounds to either methane or methane and CO_2. (510)

methanotrophy The ability to grow on methane as the sole carbon source; such a microorganism is called a methanotroph. (510)

methylotroph A bacterium that uses reduced one-carbon compounds such as methane and methanol as its sole source of carbon and energy. (544)

Michaelis constant (K_m) (mī-ka′lis) A kinetic constant for an enzyme reaction that equals the substrate concentration required for the enzyme to operate at half maximal velocity. (178)

microaerophile (mi″kro-a′er-o-fīl) A microorganism that requires low levels of oxygen for growth, around 2 to 10%, but is damaged by normal atmospheric oxygen levels. (139)

microbial ecology The study of microorganisms in their natural environments, with a major emphasis on physical conditions, processes, and interactions that occur on the scale of individual microbial cells. (643)

microbial loop The mineralization of organic matter synthesized by photosynthetic microorganisms through the activity of other microbes, such as bacteria and protozoa. This process "loops" minerals and carbon dioxide back for reuse by the primary producers and makes the organic matter unavailable to higher consumers. (656. 670)

microbial mat A firm structure of layered microorganisms with complementary physiological activities that can develop on surfaces in aquatic environments. (655)

microbial transformation (mi-kro′be-al) *See* bioconversion. (1074)

microbiology (mi″kro-bi-ol′o-je) The study of organisms that are usually too small to be seen with the naked eye. Special techniques are required to isolate and grow them. (1)

microbivory The use of microorganisms as a food source by organisms (e.g., protists) that can ingest or phagocytose them. (657)

microenvironment (mi″kro-en-vi′ron-ment) The immediate environment surrounding a microbial cell or other structure, such as a root. (653)

microevolution *See* anagenesis. (477)

microfilaments (mi″kro-fil′ah-ments) Protein filaments, about 4 to 7 nm in diameter, that are present in the cytoplasmic matrix of eucaryotic cells and play a role in cell structure and motion. (83)

micronucleus (mi″kro-nu′kle-us) The smaller of the two nuclei in ciliate protists. Micronuclei are diploid and involved only in genetic recombination and the regeneration of macronuclei. (609)

micronutrients Nutrients such as zinc, manganese, and copper that are required in very small quantities for growth and reproduction. Also called trace elements. (101)

microorganism (mi″kro-or′gan-izm) An organism that is too small to be seen clearly with the naked eye. (1)

microsporidia A primitive, fungal, obligate intracellular parasite of animals, primarily vertebrates. The infectious spore (0.5 to 2.0 μm) contains a coiled polar tubule that is used for injecting the spore contents into host cells where the sporoplasm undergoes mitotic division, producing more spores. Microsporidia are known to cause infections of the conjunctiva, cornea, and intestine. Chronic intractable diarrhea, fever, malaise, and weight loss symptoms are similar to those of cryptosporidiosis. (1018)

microtubules (mi″kro-tu′buls) Small cylinders, about 25 nm in diameter, made of tubulin proteins and present in the cytoplasmic matrix and flagella of eucaryotic cells; they are involved in cell structure and movement. (83)

miliary tuberculosis (mil′e-a-re) An acute form of tuberculosis in which small tubercles are formed in a number of organs of the body because *M. tuberculosis* is disseminated throughout the body by the bloodstream. Also known as reactivation tuberculosis. (954)

mineralization The conversion of organic nutrients into inorganic material during microbial growth and metabolism. (569, 646)

mineral soil Soil that contains less than 20% organic carbon. (688)

minimal inhibitory concentration (MIC) The lowest concentration of a drug that will prevent the growth of a particular microorganism. (840)

minimal lethal concentration (MLC) The lowest concentration of a drug that will kill a particular microorganism. (840)

minus, or **negative strand** The virus nucleic acid strand that is complementary in base sequence to the viral mRNA. (416)

mismatch repair A type of DNA repair in which a portion of a newly synthesized strand of DNA containing mismatched base pairs is removed and replaced, using the parental strand as a template. (326)

missense mutation A single base substitution in DNA that changes a codon for one amino acid into a codon for another. (320)

mitochondrion (mi″to-kon′dre-on) The eucaryotic organelle that is the site of electron transport, oxidative phosphorylation, and pathways such as the Krebs cycle; it provides most of a nonphotosynthetic cell's energy under aerobic conditions. It is constructed of an outer membrane and an inner membrane, which contains the electron transport chain. (88)

mitosis (mi-to′sis) A process that takes place in the nucleus of a eucaryotic cell and results in the formation of two new nuclei, each with the same number of chromosomes as the parent. (92)

mixed acid fermentation A type of fermentation carried out by members of the family *Enterobacteriaceae* in which ethanol and a complex mixture of organic acids are produced. (209)

mixotrophy A mode of metabolism in which oxidation of an inorganic substrate provides energy while an organic carbon is used, although this is sometimes supplemented by carbon fixation. Most commonly seen in certain protists. (103)

modified atmosphere packaging (MAP) Addition of gases such as nitrogen and carbon dioxide to packaged foods in order to inhibit the growth of spoilage organisms. (1026)

mold Any of a large group of fungi that cause mold or moldiness and that exist as multicellular filamentous colonies; also the deposit or growth caused by such fungi. Molds typically do not produce macroscopic fruiting bodies. (631)

molecular chaperones *See* chaperone proteins.

molecular chronometers Nucleic acid and protein sequences thought to gradually change over time in a random fashion and at a steady rate, and which might be used to determine phylogenetic relationships. (488)

monoclonal antibody (mAb) (mon″o-klōn′al) An antibody of a single type that is produced by a population of genetically identical plasma cells (a clone); a monoclonal antibody is typically produced from a cell culture derived from the fusion product of a cancer cell and an antibody-producing cell (a hybridoma). (799, 864)

monocyte (mon′o-sīt) A mononuclear phagocytic leukocyte that circulates briefly in the bloodstream before migrating to the tissues where it becomes a macrophage. (746)

monocyte-macrophage system The collection of fixed phagocytic cells (including macrophages, monocytes, and specialized endothelial cells) located in the liver, spleen, lymph nodes, and bone marrow. This system is an important component of the host's general nonspecific (innate) defense against pathogens. (746)

monokine (mon′o-kīn) A generic term for a cytokine produced by mononuclear phagocytes (macrophages or monocytes). (767)

monotrichous (mon-ot′rī-kus) Having a single flagellum. (67)

morbidity rate (mor-bid′i-te) Measures the number of individuals who become ill as a result of a particular disease within a susceptible population during a specific time period. (887)

mordant (mor′dant) A substance that helps fix dye on or in a cell. (26)

mortality rate (mor-tal′i-te) The ratio of the number of deaths from a given disease to the total number of cases of the disease. (887)

most probable number (MPN) The statistical estimation of the probable population in a liquid by diluting and determining end points for microbial growth. (1052)

mucociliary blanket The layer of cilia and mucus that lines certain portions of the respiratory system; it traps microorganisms up to 10 μm in diameter and then transports them by ciliary action away from the lungs. (761)

mucociliary escalator The mechanism by which respiratory ciliated cells move material and microorganisms, trapped in mucus, out of the pharynx, where it is spit out or swallowed. (761)

mucosal-associated lymphoid tissue (MALT) Organized and diffuse immune tissues found as part of the mucosal epithelium. It can be specialized to the gut (GALT), the bronchial system (BALT), or skin (SALT). (759)

multicloning site (MCS) A region of DNA on a cloning vector that has a number of restriction enzyme recognition sequences to facilitate the introduction, or cloning, of a gene. (367)

multi-drug-resistant strains of tuberculosis (MDR-TB) A multi-drug-resistant strain is defined as *Mycobacterium tuberculosis* resistant to isoniazid and rifampin, with or without resistance to other drugs. (954)

multilocus sequence typing (MLST) A method for genotypic classification of procaryotes within a single genus using nucleotide differences among five to seven housekeeping genes. (486)

mumps An acute generalized disease that occurs primarily in school-age children and is caused by a paramyxovirus that is transmitted in saliva and respiratory droplets. The principal manifestation is swelling of the parotid salivary glands. (919)

murein *See* peptidoglycan. (55)

must The juices of fruits, including grapes, that can be fermented for the production of alcohol. (1041)

mutagen (mu′tah-jen) A chemical or physical agent that causes mutations. (318)

mutation (mu-ta′shun) A permanent, heritable change in the genetic material. (317)

mutualism (mu′tu-al-izm″) A type of symbiosis in which both partners gain from the association and are unable to survive without it. The mutualist and the host are metabolically dependent on each other. (718)

mutualist (mu′tu-al-ist) An organism associated with another in an obligatory relationship that is beneficial to both. (718)

mycelium (mi-se′le-um) A mass of branching hyphae found in fungi and some bacteria. (40, 631)

mycobiont The fungal partner in a lichen. (731)

mycolic acids Complex 60 to 90 carbon fatty acids with a hydroxyl on the β-carbon and an aliphatic chain on the α-carbon; found in the cell walls of mycobacteria. (596)

mycologist (mi-kol′o-jist) A person specializing in mycology; a student of mycology. (629)

mycology (mi-kol′o-je) The science and study of fungi. (629)

mycoplasma (mi″ko-plaz′mah) Bacteria that are members of the class *Mollicutes* and order *Mycoplasmatales;* they lack cell walls and cannot synthesize peptidoglycan precursors; most require sterols for growth. (572)

Mycoplasma pneumoniae (mi″ko-plaz′mal nu-mo′ne-ah) A type of pneumonia caused by *Mycoplasma pneumoniae*. Spread involves airborne droplets and close contact. (896)

mycorrhizal fungi Fungi that form stable, mutualistic relationships on (ectomycorrhizal) or in (endomycorrhizal) the root cells of vascular plants. The plants provide carbohydrate for the fungi, while the fungal hyphae extend into the soil and bring nutrients (e.g., nitrogen and phosphorus) to the plants, thereby enhancing plant nutrient uptake. (700)

mycorrhizosphere The region around ectomycorrhizal mantles and hyphae in which nutrients released from the fungus increase the microbial population and its activities. (700)

mycosis (mi-ko′sis; pl., **mycoses**) Any disease caused by a fungus. (629, 997)

mycotoxicology (mi-ko′tok″si-kol′o-je) The study of fungal toxins and their effects on various organisms. (629)

myeloma cell (mi″e-lo′mah) A tumor cell that is similar to the cell type found in bone marrow. Also, a malignant, neoplastic plasma cell that produces large quantities of antibodies and can be readily cultivated. (800)

myositis (mi″o-si′tis) Inflammation of a striated or voluntary muscle. (957)

myxobacteria A group of gram-negative, aerobic soil bacteria characterized by gliding motility, a complex life cycle with the production of fruiting bodies, and the formation of myxospores. (564)

Myxogastria *See* acellular slime molds.

myxospores (mik′so-spōrs) Special dormant spores formed by the myxobacteria. (565)

N

narrow-spectrum drugs Chemotherapeutic agents that are effective only against a limited variety of microorganisms. (837)

natural attenuation The decrease in the level of an enviromental contaminant that results from natural chemical, physical, and biological processes. (1081)

natural classification A classification system that arranges organisms into groups whose members share many characteristics and reflect as much as possible the biological nature of organisms. (478)

natural killer (NK) cell A type of white blood cell that has a lineage independent of the granulocyte, B-cell, and T-cell lineages. It is often called a non-B, non-T lymphocyte because it does not make antibody; nor does it make cytokines associated with T cells. NK cells are part of the innate immune system, exhibiting MHC-independent cytolytic activity against virus-infected and tumor cells. (748)

naturally acquired active immunity The type of active immunity that develops when an individual's immunologic system comes into contact with an appropriate antigenic stimulus during the course of normal activities; it usually arises as the result of recovering from an infection and lasts a long time. (776)

naturally acquired passive immunity The type of temporary immunity that involves the transfer of antibodies from one individual to another. (777)

necrotizing fasciitis (nek′ro-tīz″ing fas″e-i′tis) A disease that results from a severe invasive group A streptococcus infection. Necrotizing fasciitis is an infection of the subcutaneous soft tissues, particularly of fibrous tissue, and is most common on the extremities. It begins with skin reddening, swelling, pain, and cellulitis and proceeds to skin breakdown and gangrene after 3 to 5 days. (957)

negative selection   The process by which lymphocytes that recognize host (self) antigens undergo apoptosis or become anergic (inactive). (803)

negative staining   A staining procedure in which a dye is used to make the background dark while the specimen is unstained. (26)

Negri bodies (na'gre)   Masses of viruses or unassembled viral subunits found within the neurons of rabies-infected animals. (944)

neoplasia   Abnormal cell growth and reproduction due to a loss of regulation of the cell cycle; produces a tumor in solid tissues. (461)

neurotoxin (nu''ro-tok'sin)   A toxin that is poisonous to or destroys nerve tissue; especially the toxins secreted by *C. tetani, Corynebacterium diphtheriae,* and *Shigella dysenteriae.* (825)

neutrophil (noo'tro-fil)   A mature white blood cell in the granulocyte lineage formed in bone marrow. It has a nucleus with three to five lobes and is very phagocytic. (747)

neutrophile (nu'ston''ik)   Microorganisms that grow best at a neutral pH range between pH 5.5 and 8.0. (134)

niche   The function of an organism in a complex system, including place of the organism, the resources used in a given location, and the time of use. (653)

nicotinamide adenine dinucleotide (NAD⁺) (nik''o-tin'ah-mīd)   An electron-carrying coenzyme; it is particularly important in catabolic processes and usually donates its electrons to the electron transport chain under aerobic conditions. (173)

nicotinamide adenine dinucleotide phosphate (NADP⁺)   An electron-carrying coenzyme that most often participates as an electron carrier in biosynthetic metabolism. (173)

Nipah virus   A single-stranded RNA virus named for the Nipah village in Malaysia where it was first associated with disease. (943)

nitrification (ni''trī-fī-ka'shun)   The oxidation of ammonia to nitrate. (213, 546, 648)

nitrifying bacteria (ni'trī-fi''ing)   Chemolithotrophic, gram-negative bacteria that are members of several families within the phyllum *Proteobacteria* that oxidize ammonia to nitrite and nitrite to nitrate. (213, 546)

nitrogenase (ni'tro-jen-ās)   The enzyme that catalyzes biological nitrogen fixation. (237)

nitrogen fixation   The metabolic process in which atmospheric molecular nitrogen (N_2) is reduced to ammonia; carried out by cyanobacteria, *Rhizobium,* and other nitrogen-fixing procaryotes. (236, 648)

nitrogen oxygen demand (NOD)   The demand for oxygen in sewage treatment, caused by nitrifying microorganisms. (1055)

nitrogen saturation point   The point at which mineral nitrogen (e.g., NO_3^-, NH_4^+), when added to an ecosystem, can no longer be incorporated into organic matter through biological processes. (690)

nocardioforms   Bacteria that resemble members of the genus *Nocardia;* they develop a substrate mycelium that readily breaks up into rods and coccoid elements (a quality sometimes called fugacity). (596)

nomenclature (no'men-kla''tūr)   The branch of taxonomy concerned with the assignment of names to taxonomic groups in agreement with published rules. (478)

noncyclic photophosphorylation (fo''to-fos''for-i-la'shun)   The process in which light energy is used to make ATP when electrons are moved from water to $NADP^+$ during photosynthesis; both photosystem I and photosystem II are involved. (218)

noncytopathic virus   A virus that does not kill its host cell by viral release-induced lysis. (819)

nondiscrete microorganism   A microorganism, best exemplified by a filamentous fungus, that does not have a defined and predictable cell structure or distinct edges and boundaries. The organism can be defined in terms of the cell structure and its cytoplasmic contents. (663)

nongonococcal urethritis (NGU) (u''rə-thri'tis)   Any inflammation of the urethra not caused by *Neisseria gonorrhoeae.* (976)

nonsense codon   A codon that does not code for an amino acid but is a signal to terminate protein synthesis. (275)

nonsense mutation   A mutation that converts a sense codon to a nonsense or stop codon. (321)

nonspecific immune response (innate or **natural immunity)**   *See* nonspecific resistance.

nonspecific resistance   Refers to those general defense mechanisms that are inherited as part of the innate structure and function of each animal; also known as nonspecific, innate or natural immunity. (743)

normal microbiota (also indigenous microbial population, microflora, microbial flora) (mmi''kro-bi-o'tah)   The microorganisms normally associated with a particular tissue or structure. (734)

nosocomial infection (nos''o-ko'me-al)   An infection that develops within a hospital (or other type of clinical care facility) and is acquired during the stay of the patient. (908)

nuclear envelope (nu'kle-ar)   The complex double-membrane structure forming the outer boundary of the eucaryotic nucleus. It is covered by pores through which substances enter and leave the nucleus. (91)

nucleic acid hybridization (nu-kle'ik)   The process of forming a hybrid double-stranded DNA molecule using a heated mixture of single-stranded DNAs from two different sources; if the sequences are fairly complementary, stable hybrids will form. (483)

nucleocapsid (nu''kle-o-kap'sid)   The nucleic acid and its surrounding protein coat or capsid; the basic unit of virion structure. (409)

nucleoid (nu'kle-oid)   An irregularly shaped region in the procaryotic cell that contains its genetic material. (52)

nucleolus (nu-kle'o-lus)   The organelle, located within the eucaryotic nucleus and not bounded by a membrane, that is the location of ribosomal RNA synthesis and the assembly of ribosomal subunits. (91)

nucleoside (nu'kle-o-sīd'')   A combination of ribose or deoxyribose with a purine or pyrimidine base. (241)

nucleosome (nu'kle-o-sōm'')   A complex of histones and DNA found in eucaryotic chromatin and some archaea; the DNA is wrapped around the surface of the beadlike histone complex. (253)

nucleotide (nu'kle-o-tīd)   A combination of ribose or deoxyribose with phosphate and a purine or pyrimidine base; a nucleoside plus one or more phosphates. (241)

nucleus (nu'kle-us)   The eucaryotic organelle enclosed by a double-membrane envelope that contains the cell's chromosomes. (91)

numerical aperture   The property of a microscope lens that determines how much light can enter and how great a resolution the lens can provide. (19)

numerical taxonomy   The grouping by numerical methods of taxonomic units into taxa based on their character states. (479)

nutrient (nu'tre-ent)   A substance that supports growth and reproduction. (101)

nystatin (nis'tah-tin)   A polyene antibiotic from *Streptomyces noursei* that is used in the treatment of *Candida* infections of the skin, vagina, and alimentary tract. (854)

O

O antigen   A polysaccharide antigen extending from the outer membrane of some gram-negative bacterial cell walls; it is part of the lipopolysaccharide. (60)

obligate aerobes   Organisms that grow only in the presence of oxygen. (139)

obligate anaerobes   Microorganisms that cannot tolerate the presence of oxygen and die when exposed to it. (139)

odontopathogens   Dental pathogens. (991)

Okazaki fragments   Short stretches of polynucleotides produced during discontinuous DNA replication. (260)

oligonucleotide   A short fragment of DNA or RNA, usually artificially synthesized, used in a number of molecular genetic techniques such as DNA sequencing, polymerase chain reaction, and Southern blotting. (361)

oligonucleotide signature sequence   Short, conserved nucleotide sequences that are specific for a phylogenetically defined group of organisms. The signature sequences found in small subunit rRNA molecules are most commonly used. (485)

oligotrophic environment (ol″ĭ-go-trof′ik) An environment containing low levels of nutrients, particularly nutrients that support microbial growth. (142, 676)

oncogene (ong″ko-jēn) A gene whose activity is associated with the conversion of normal cells to cancer cells. (461)

oncovirus A virus known to be associated with the development of cancer. (463)

one-step growth experiment An experiment used to study the reproduction of lytic phages in which one round of phage reproduction occurs and ends with the lysis of the host bacterial population. (428)

onychomycosis (on″i-ko-mi-ko′sis) A fungal infection of the nail plate producing nails that are opaque, white, thickened, friable, and brittle. Also called ringworm of the nails and tinea unguium. Caused by *Trichophyton* and other fungi such as *C. albicans*. (1017)

oocyst (o′o-sist) Cyst formed around a zygote of malaria and related protozoa. (1014)

öomycetes (o″o-mi-se′tēz) A collective name for protists also known as the water molds. Formerly thought to be fungi. (622)

open reading frame (ORF) A sequence of DNA not interrupted by a stop codon and with an apparent promoter and ribosome binding site at the 5′ end and a terminator at the 3′ end. It is usually determined by nucleic acid sequencing studies. (388)

operator The segment of DNA to which the repressor protein binds; it controls the expression of the genes adjacent to it. (295)

operon (op′er-on) The sequence of bases in DNA that contains one or more structural genes together with the operator controlling their expression. (295)

ophthalmia neonatorum (of-thal′me-ah ne″o-nat-or-um) A gonorrheal eye infection in a newborn, which may lead to blindness. Also called conjunctivitis of the newborn. (975)

opportunistic microorganism or **pathogen** A microorganism that is usually free-living or a part of the host's normal microbiota, but which may become pathogenic under certain circumstances, such as when the immune system is compromised. (740, 816, 1016)

opsonization (op″so-ni-za′shun) The coating of foreign substances by antibody, complement proteins, or fibronectin to make the substances more readily recognized by phagocytic cells. (763, 792)

optical tweezer The use of a focused laser beam to drag and isolate a specific microorganism from a complex microbial mixture. (664)

oral candidiasis *See* thrush.

orchitis (or-ki′tis) Inflammation of the testes. (919)

organelle (or″gah-nel′) A structure within or on a cell that performs specific functions and is related to the cell in a way similar to that of an organ to the body. (79)

organotrophs Organisms that use reduced organic compounds as their electron source. (103)

origin of replication (*ori*) A site on a chromosome or plasmid where DNA replication is initiated. (120)

ornithosis *See* psittacosis.

ortholog A gene found in the genomes of two or more different organisms that share a common ancestry. The products of orthologous genes are presumed to have similar functions. (388)

osmophilic microorganisms (oz″mo-fil′ik) Microorganisms that grow best in or on media of high solute concentration. (1024)

osmotolerant Organisms that grow over a fairly wide range of water activity or solute concentration. (134)

osmotrophy A form of nutrition in which soluble nutrients are absorbed through the cytoplasmic membrane; found in procaryotes, fungi, and some protists. (632)

Ouchterlony technique *See* double diffusion agar assay.

outbreak The sudden, unexpected occurrence of a disease in a given population. (886)

outer membrane A special membrane located outside the peptidoglycan layer in the cell walls of gram-negative bacteria. (55)

ovular nuclei The morphology of nuclei found in some protists that is characterized by a large nucleus (up to 100 μm in diameter) with many peripheral nuclei. (609)

oxidation-reduction (redox) reactions Reactions involving electron transfers; the electron donor (reductant) gives electrons to an electron acceptor (oxidant). (172)

oxidative burst The generation of reactive oxygen species, primarily superoxide anion (O_2^-) and hydrogen peroxide (H_2O_2) by a plant or an animal, in response to challenge by a potential bacterial, fungal, or viral pathogen. *See* respiratory burst. (701)

oxidative phosphorylation (fos″for-ĭ-la′-shun) The synthesis of ATP from ADP using energy made available during electron transport initiated by the oxidation of a chemical energy source. (202)

oxygenic photosynthesis Photosynthesis that oxidizes water to form oxygen; the form of photosynthesis characteristic of plants, protists and cyanobacteria. (216, 520)

P

pacemaker enzyme The enzyme in a metabolic pathway that catalyzes the slowest or rate-limiting reaction; if its rate changes, the pathway's activity changes. (183)

pandemic (pan-dem′ik) An increase in the occurrence of a disease within a large and geographically widespread population (often refers to a worldwide epidemic). (887)

Paneth cell (pah′ net) A specialized epithelial cell of the intestine. Paneth cells secrete hydrolytic enzymes and antimicrobial proteins and peptides. (734, 761)

pannus (pan′us) A superficial vascularization of the cornea with infiltration of granulation tissue. (979)

paralog Two or more genes in the genome of a single organism that arose through duplication of a common ancestral gene. The products of paralogous genes generally have different, although frequently related, functions. (388)

parasite (par′ah-sīt) An organism that lives on or within another organism (the host) and benefits from the association while harming its host. Often the parasite obtains nutrients from the host. (816)

parasitism (par′ah-si″tizm) A type of symbiosis in which one organism benefits from the other and the host is usually harmed. (730)

parasporal body An intracellular, solid protein crystal made by the bacterium *Bacillus thuringiensis*. Upon ingestion by one of over 100 different insect species, the protein becomes extremely toxic. This toxin is the basis of the bacterial insecticide Bt. (580)

parenteral route (pah-ren′ter-al) A route of drug administration that is nonoral (e.g., by injection). (849)

parfocal (par-fo′kal) A microscope that retains proper focus when the objectives are changed. (18)

paronychia (par″o-nik′e-ah) Inflammation involving the folds of tissue surrounding the nail; usually caused by *Candida albicans*. (1017)

parsimony analysis A method for developing phylogenetic trees based on the estimation of the minimum number of nucleotide or amino acid sequence changes needed to give the sequences being compared. (489)

particulate organic matter (POM) Nutrients that are not dissolved or soluble, generally referring to freshwater and marine ecosystems. This includes microorganisms or their cellular debris after senscence or viral lysis. (671)

passive diffusion The process in which molecules move from a region of higher concentration to one of lower concentration as a result of random thermal agitation. (106)

passive immunization The induction of temporary immunity by the transfer of immune products, such as antibodies or sensitized T cells, from an immune vertebrate to a nonimmune one. (777)

Pasteur effect (pas-tur′) The decrease in the rate of sugar catabolism and change to aerobic respiration that occurs when microorganisms are switched from anaerobic to aerobic conditions. (207)

pasteurization (pas″ter-ĭ-za′-shun) The process of heating milk and other liquids to destroy microorganisms that can cause spoilage or disease. (153, 1030)

pathogen (path'o-jən) Any virus, bacterium, or other agent that causes disease. (734, 743, 816)

pathogen-associated molecular pattern (PAMP) Conserved molecular structures that occur in patterns on microbial surfaces. The structures and their patterns are unique to particular microorganisms and invariant among members of a given microbial group. (753)

pathogenicity (path″o-je-nis′ĭ-te) The condition or quality of being pathogenic, or the ability to cause disease. (734, 816)

pathogenicity island A 10 to 200 Kb segment of DNA in some pathogens that contains the genes responsible for virulence; often it codes for the type III secretion system that allows the pathogen to secrete virulence proteins and damage host cells. A pathogen may have more than one pathogenicity island. (822)

pathogenic potential The degree that a pathogen causes morbid signs and symptoms. (816)

pathway architecture The analysis, design, and modification of biochemical pathways to increase process efficiency. (1062)

pattern recognition receptor A receptor found on macrophages and other phagocytic cells that binds to pathogen-associated molecular patterns on microbial surfaces. (753)

ped A natural soil aggregate, formed partly through bacterial and fungal growth in the soil. (692)

pellicle (pel′ĭ-k′l) A relatively rigid layer of proteinaceous elements just beneath the plasma membrane in many protists. The plasma membrane is sometimes considered part of the pellicle. (94, 607)

pelvic inflammatory disease (PID) A severe infection of the female reproductive organs. The disease results when gonococci or chlamydiae infect the uterine tubes and surrounding tissue. (975)

penicillins (pen″ĭ-sil′ ins) A group of antibiotics containing a β-lactam ring, which are active against gram-positive bacteria. (61, 841)

penton or **pentamer** A capsomer composed of five protomers. (411)

pentose phosphate pathway (pen′tōs) The pathway that oxidizes glucose 6-phosphate to ribulose 5-phosphate and then converts it to a variety of three to seven carbon sugars; it forms several important products (NADPH for biosynthesis, pentoses, and other sugars) and also can be used to degrade glucose to CO_2. (196)

peplomer or spike (pep′lo-mer) A protein or protein complex that extends from the virus envelope and often is important in virion attachment to the host cell surface. (412)

peptic ulcer disease A gastritis caused by *Helicobacter pylori*. (967)

peptide interbridge (pep′tīd) A short peptide chain that connects the tetrapeptide chains in some peptidoglycans. (56)

peptidoglycan (pep″tĭ-do-gli′kan) A large polymer composed of long chains of alternating *N*-acetylglucosamine and *N*-acetylmuramic acid residues. The polysaccharide chains are linked to each other through connections between tetrapeptide chains attached to the *N*-acetylmuramic acids. It provides much of the strength and rigidity possessed by bacterial cell walls. (55)

peptidyl or **donor site (P site)** The site on the ribosome that contains the peptidyl-tRNA at the beginning of the elongation cycle during protein synthesis. (284)

peptidyl transferase The 23S rRNA ribozyme that catalyzes the transpeptidation reaction in protein synthesis; in this reaction, an amino acid is added to the growing peptide chain. (284)

peptones (pep′tōns) Water-soluble digests or hydrolysates of proteins that are used in the preparation of culture media. (111)

perforin pathway The cytotoxic pathway that uses perforin protein, which polymerizes to form membrane pores that help destroy cells during cell-mediated cytotoxicity. Perforin is produced by cytotoxic T cells and NK cells and stored in granules that are released when a target cell is contacted. (782)

period of infectivity Refers to the time during which the source of an infectious disease is infectious or is disseminating the pathogen. (891)

periodontal disease (per″e-o-don′tal) A disease located around the teeth or in the periodontium. (993)

periodontitis (per″e-o-don-ti′tis) An inflammation of the periodontium. (993)

periodontium (per″e-o-don′she-um) The tissue investing and supporting the teeth, including the cementum, periodontal ligament, alveolar bone, and gingiva. (993)

periodontosis (per″e-o-don-to′sis) A degenerative, noninflammatory condition of the periodontium, which is characterized by destruction of tissue. (994)

peripheral tolerance The inhibition of effector lymphocyte activity resulting in the lack of a specific response. (803)

periplasm (per′ĭ-plaz-əm) The substance that fills the periplasmic space. (55)

periplasmic flagella The flagella that lie under the outer sheath and extend from both ends of the spirochete cell to overlap in the middle and form the axial filament. Also called axial fibrils and endoflagella. (70, 532)

periplasmic space (per″i-plas′mik) or **periplasm** (per′ĭ-plazm) The space between the plasma membrane and the outer membrane in gram-negative bacteria, and between the plasma membrane and the cell wall in gram-positive bacteria. (55)

peristalsis The muscular contractions of the gut that propel digested foods and waste through the intestinal tract. (761)

peritrichous (pĕ-rit′rĭ-kus) A cell with flagella distributed over its surface. (67)

permease (per′me-ās) A membrane-bound carrier protein or a system of two or more proteins that transports a substance across the membrane. (106)

pertussis (pər-tus′is) An acute, highly contagious infection of the respiratory tract, most frequently affecting young children, usually caused by *Bordetella pertussis* or *B. parapertussis*. Consists of peculiar paroxysms of coughing, ending in a prolonged crowing or whooping respiration; hence the name whooping cough. (955)

petri dish (pe′tre) A shallow dish consisting of two round, overlapping halves that is used to grow microorganisms on solid culture medium; the top is larger than the bottom of the dish to prevent contamination of the culture. (117)

phage (fāj) *See* bacteriophage.

phagocytic vacuole (fag″o-sit′ik vak′u-ol) A membrane-delimited vacuole produced by cells carrying out phagocytosis. It is formed by the invagination of the plasma membrane and contains solid material. (608)

phagocytosis (fag″o-si-to′sis) The endocytotic process in which a cell encloses large particles in a membrane-delimited phagocytic vacuole or phagosome and engulfs them. (86, 752)

phagolysosome (fag″o-li′so-sōm) The vacuole that results from the fusion of a phagosome with a lysosome. (755)

phagosome A membrane-enclosed vacuole formed by the invagination of the cell membrane during endocytosis. (86)

phagovar (fag′o-var) A specific phage type. (873)

pharyngitis (far″in-ji′tis) Inflammation of the pharynx, often due to a *S. pyogenes* infection. (958)

phase-contrast microscope A microscope that converts slight differences in refractive index and cell density into easily observed differences in light intensity. (21)

phenetic system A classification system that groups organisms together based on the similarity of their observable characteristics. (478)

phenol coefficient test A test to measure the effectiveness of disinfectants by comparing their activity against test bacteria with that of phenol. (165)

phosphatase (fos′fah-tās″) An enzyme that catalyzes the hydrolytic removal of phosphate from molecules. (241)

phosphate group transfer potential A measure of the ability of a phosphorylated molecule such as ATP to transfer its phosphate to water and other acceptors. It is the negative of the $\Delta G^{o′}$ for the hydrolytic removal of phosphate. (171)

phospholipase An enzyme that hydrolyzes a specific ester bond in phospholipids. There are four types: A, B, C, and D. (828)

phosphorelay system A mechanism for regulating either transcription or enzyme activity that involves the transfer of a phosphate group from one molecule to another. Covalent addition of the phosphate group to enzymes or other proteins can either activate or inhibit their activity. Examples include the phosphotransferase system (PTS) of group translocation and two-component regulatory systems. (109, 300, 302)

photolithoautotroph An organism that uses light energy, an inorganic electron source (e.g., H_2O, H_2, H_2S), and CO_2 as its carbon source. Also called photolithotrophic autotroph. (103)

photoorganoheterotroph A microorganism that uses light energy, organic electron sources, and organic molecules as a carbon source. Also called photoorganotrophic heterotroph. (103)

photoreactivation (fo″to-re-ak″tī-va′shun) The process in which blue light is used by a photoreactivating enzyme to repair thymine dimers in DNA by splitting them apart. (326)

photosynthate Nutrient material that leaks from phototrophic organisms. Photosynthate contributes to the pool of dissolved organic matter that is available to microorganisms. (671)

photosynthesis (fo″to-sin′thĕ-sis) The trapping of light energy and its conversion to chemical energy, which is then used to reduce CO_2 and incorporate it into organic form. (214)

photosystem I The photosystem in eucaryotic cells and cyanobacteria that absorbs longer wavelength light, usually greater than about 680 nm, and transfers the energy to chlorophyll P700 during photosynthesis; it is involved in both cyclic photophosphorylation and noncyclic photophosphorylation. (217)

photosystem II The photosystem in eucaryotic cells and cyanobacteria that absorbs shorter wavelength light, usually less than 680 nm, and transfers the energy to chlorophyll P680 during photosynthesis; it participates in noncyclic photophosphorylation. (217)

phototaxis The ability of certain phototrophic bacteria to move, either by gliding or swimming motility, in response to a light source. In positive phototaxis, the microbe moves toward the light; in negative, it moves away. (525)

phototrophs Organisms that use light as their energy source. (103)

phycobiliproteins Photosynthetic pigments that are composed of proteins with attached tetrapyrroles; they are found in cyanobacteria. (217)

phycobilisomes Special particles on the membranes of cyanobacteria that contain photosynthetic pigments and electron transport chains. (524)

phycobiont (fi″ko-bi′ont) The photosynthetic protist or cyanobacterial partner in a lichen. (731)

phycocyanin (fi″ko-si′an-in) A blue phycobiliprotein pigment used to trap light energy during photosynthesis. (217)

phycoerythrin (fi″ko-er′i-thrin) A red photosynthetic phycobiliprotein pigment used to trap light energy. (217)

phycology (fi-kol′o-je) The study of algae. (605)

phyllosphere The surface of plant leaves. (696)

phylogenetic or **phyletic classification system** (fi″lo-jĕ-net′ik, fi-let′ik) A classification system based on evolutionary relationships rather than the general similarity of characteristics. (478)

phylogenetic tree A graph made of nodes and branches, much like a tree in shape, that shows phylogenetic relationships between groups of organisms and sometimes also indicates the evolutionary development of groups. (489)

phylotype A taxon that is characterized only by its nucleic acid sequence; generally discovered during metagenomic analysis. (402)

phytoplankton (fi″to-plank′ton) A community of floating photosynthetic organisms, largely composed of photosynthetic protists and cyanobacteria. (670)

phytoremediation The use of plants and their associated microorganisms to remove, contain, or degrade environmental contaminants. (1079)

picoplankton Planktonic microbes between 0.2 and 2.0 μm in size. This includes the cyanobacterial genera *Prochlorococcus* and *Synechococcus,* which together can account for over half the carbon fixation in some open ocean ecosystems. (670)

piedra (pe-a′drah) A fungal disease of the hair in which white or black nodules of fungi form on the shafts. (1008)

pitching Pertaining to inoculation of a nutrient medium with yeast, for example, in beer brewing. (1043)

plague ((plāg) An acute febrile, infectious disease, caused by the γ-proteobacterium *Yersinia pestis,* which has a high mortality rate; the two major types are bubonic plague and pneumonic plague. (962)

plankton (plank′ton) Free-floating, mostly microscopic microorganisms that can be found in almost all waters; a collective name. (606)

planktonic (adj.) *See* plankton.

plantar warts Viral infections of the epithelia comprising the sole of the foot. (938)

plaque (plak) 1. A clear area in a lawn of bacteria or a localized area of cell destruction in a layer of animal cells that results from the lysis of the bacteria by bacteriophages or the destruction of the animal cells by animal viruses. 2. The term also refers to dental plaque, a film of food debris, polysaccharides, and dead cells that cover the teeth. (418)

plasma cell A mature, differentiated B lymphocyte that synthesizes and secretes antibody; a plasma cell lives for only 5 to 7 days. (748, 786)

plasmalemma The plasma membrane in protists. (607)

plasma membrane The selectively permeable membrane surrounding the cell's cytoplasm; also called the cell membrane, plasmalemma, or cytoplasmic membrane. (42)

plasmid (plaz′mid) A double-stranded DNA molecule that can exist and replicate independently of the chromosome or may be integrated with it. A plasmid is stably inherited, but is not required for the host cell's growth and reproduction. (53, 334)

plasmid fingerprinting A technique used to identify microbial isolates as belonging to the same strain because they contain the same number of plasmids with the identical molecular weights and similar phenotypes. (875)

plasmodial (acellular) slime mold (plaz-mo′de-al) A member of the protist division *Amoebozoa (Myxogastria)* that exists as a thin, streaming, multinucleate mass of protoplasm, which creeps along in an amoeboid fashion. (614)

plasmodium (plaz-mo′de-um; pl., **plasmodia**) A stage in the life cycle of myxogastria protists; a

multinucleate mass of protoplasm surrounded by a membrane. Also, a parasite of the genus *Plasmodium.* (614)

plasmolysis (plaz-mol′ĭ-sis) The process in which water osmotically leaves a cell, which causes the cytoplasm to shrivel up and pull the plasma membrane away from the cell wall. (61)

plastid (plas′tid) A cytoplasmic organelle of algae and higher plants that contains pigments such as chlorophyll, stores food reserves, and often carries out processes such as photosynthesis. (90)

pleomorphic (ple″o-mor′fik) Refers to bacteria that are variable in shape and lack a single, characteristic form. (41)

plus strand or **positive strand** The virus nucleicacid strand that is equivalent in base sequence to the viral mRNA. (416)

***Pneumocystis carinii* pneumonia (PCP);** (noo″mo-sis-tis) A type of pneumonia caused by the protist *Pneumocystis jiroveci.* (1019)

pneumonic plague *See* plague.

point mutation A mutation that affects only a single base pair in a specific location. (318)

polar flagellum A flagellum located at one end of an elongated cell. (67)

poliomyelitis (polio or **infantile paralysis)** (po″le-o-mi″e-li′tis) An acute, contagious viral disease that attacks the central nervous system, injuring or destroying the nerve cells that control the muscles and sometimes causing paralysis. (940)

poly-β-hydroxybutyrate (PHB) (hi-drok″se-bu′tī-rāt) A linear polymer of β-hydroxybutyrate used as a reserve of carbon and energy by many bacteria. (49)

polymerase chain reaction (PCR) An in vitro technique used to synthesize large quantities of specific nucleotide sequences from small amounts of DNA. It employs oligonucleotide primers complementary to specific sequences in the target gene and special heat-stable DNA polymerases. (362)

polymorphonuclear leukocyte (pol″e-mor″fo-noo′kle-ər) A leukocyte that has a variety of nuclear forms. (746)

polyphasic taxonomy An approach in which taxonomic schemes are developed using a wide range of phenotypic and genotypic information. (478)

polyribosome (pol″e-ri′bo-sōm) A complex of several ribosomes with a messenger RNA; each ribosome is translating the same message. (276)

Pontiac fever A bacterial disease caused by *Legionella pneumophila* that resembles an allergic disease more than an infection. First described from Pontiac, Michigan. *See* Legionnaires' disease. (950)

population An assemblage of organisms of the same type. (643)

porin proteins Proteins that form channels across the outer membrane of gram-negative bacterial cell walls. Small molecules are transported through these channels. (60)

postherpetic neuralgia The severe pain after a herpes infection. (915)

posttranscriptional modification The processing of the initial RNA transcript, pre-mRNA, to form mRNA. (272)

potable (po'tah-b'l) Refers to water suitable for drinking. (1052)

pour plate A petri dish of solid culture medium with isolated microbial colonies growing both on its surface and within the medium, which has been prepared by mixing microorganisms with cooled, still liquid medium and then allowing the medium to harden. (115)

precipitation (or **precipitin**) **reaction** (pre-sip″ĭ-ta'shun) The reaction of an antibody with a soluble antigen to form an insoluble precipitate. (799)

precipitin (pre-sip'ĭ-tin) The antibody responsible for a precipitation reaction. (799)

pre-mRNA In eucaryotes, the RNA transcript of DNA made by RNA polymerase II; it is processed to form mRNA. (272)

prevalence rate Refers to the total number of individuals infected at any one time in a given population regardless of when the disease began. (887)

Pribnow box A special base sequence in the promoter that is recognized by the RNA polymerase and is the site of initial polymerase binding. (269)

primary amebic meningoencephalitis An infection of the meninges of the brain by the free-living amoebae *Naegleria* or *Acanthamoeba*. (1013)

primary (frank) pathogen Any organism that causes a disease in the host by direct interaction with or infection of the host. (816)

primary metabolites Microbial metabolites produced during active growth of an organism. (1068)

primary producer Photoautotrophic and chemoautotrophic organisms that incorporate carbon dioxide into organic carbon and thus provide new biomass for the ecosystem. (656, 669)

primary production The incorporation of carbon dioxide into organic matter by photosynthetic organisms and chemoautotrophic bacteria. (656)

primary treatment The first step of sewage treatment, in which physical settling and screening are used to remove particulate materials. (1055)

primosome A complex of proteins that includes the enzyme primase, which is responsible for synthesizing the RNA primers needed for DNA replication. (260)

prion (pri'on) An infectious agent consisting only of protein; prions cause a variety of spongiform encephalopathics such as scrapie in sheep and goats. (468)

probe (prōb) A short, labeled nucleic acid segment complementary in base sequence to part of another nucleic acid, which is used to identify or isolate the particular nucleic acid from a mixture through its ability to bind specifically with the target nucleic acid. (358, 389)

probiotic A living organism that may provide health benefits beyond its nutritional value when ingested. (739, 1039)

procaryotic cells (pro″kar-e-ot'ik) Cells that lack a true, membrane-enclosed nucleus; bacteria are procaryotic and have their genetic material located in a nucleoid. (2, 97)

procaryotic species A collection of bacterial or archaeal strains that share many stable properties and differ significantly from other groups of strains. (480)

prodromal stage (pro-dro'mal) The period during the course of a disease in which there is the appearance of signs and symptoms, but they are not yet distinctive and characteristic enough to make an accurate diagnosis. (888)

promoter The region on DNA at the start of a gene that the RNA polymerase binds to before beginning transcription. (265)

propagated epidemic An epidemic that is characterized by a relatively slow and prolonged rise and then a gradual decline in the number of individuals infected. It usually results from the introduction of an infected individual into a susceptible population, and the pathogen is transmitted from person to person. (889)

prophage (pro'fāj) The latent form of a temperate phage that remains within the lysogen, usually integrated into the host chromosome. (345, 438)

prostheca (pros-the'kah) An extension of a bacterial cell, including the plasma membrane and cell wall, that is narrower than the mature cell. Found in the genus *Caulobacter*, an important model bacterium. (543)

prosthetic group (pros-thet'ik) A tightly bound cofactor that remains at the active site of an enzyme during its catalytic activity. (176)

protease (pro'te-ās) An enzyme that hydrolyzes proteins to their constituent amino acids. Also called a proteinase. (212)

proteasome A large, cylindrical protein complex that degrades ubiquitin-labeled proteins to peptides in an ATP-dependent process. (86)

protein engineering (pro'tēn) The rational design of proteins by constructing specific amino acid sequences through molecular techniques, with the objective of modifying protein characteristics. (1061)

protein modeling The process by which the amino acid sequence of a protein is analyzed using software that is designed to predict its three-dimensional structure. (394)

protein splicing The post-translational process in which part of a precursor polypeptide is removed before the mature polypeptide folds into its final shape; it is carried out by self-splicing proteins that remove inteins and join the remaining exteins. (288)

Proteobacteria (pro″te-o-bak-tēr'e-ah) A large phylum of gram-negative bacteria, that 16S rRNA sequence comparisons show to be phylogenetically related; *Proteobacteria* contain the purple photosynthetic bacteria and their relatives and are composed of the α, β, γ, δ, and ε subgroups. (539)

proteome The complete collection of proteins that an organism produces. (393)

proteomics The study of the structure and function of cellular proteins. (393)

protist (pro'tist) Unicellular (and rarely acellular) eucaryotic organisms that lack cellular differentiation into tissues. Vegetative cell differentiation is limited to cells involved in sexual reproduction, alternate vegetative morphology, or resting states such as cysts. Protists vary in morphology and metabolism, including phototrophy, heterotrophy, and mixotrophy. Many phototrophic and mixotrophic forms (e.g., diatoms and dinoflagellates) are frequently referred to as algae. (3, 491)

protistology The study of protists. (605)

protomer An individual subunit of a viral capsid; a capsomer is made of protomers. (409)

proton motive force (PMF) The force arising from a gradient of protons and a membrane potential that is thought to power ATP synthesis and other processes. (202)

proto-oncogene The normal cellular form of a gene that when mutated or overexpressed results in or contributes to malignant transformation of a cell. (461)

protoplast (pro'to-plast) A bacterial or fungal cell with its cell wall completely removed. It is spherical in shape and osmotically sensitive. (48)

protoplast fusion The joining of cells that have had their walls weakened or completely removed. (1061)

prototroph (pro'to-trōf) A microorganism that requires the same nutrients as the majority of naturally occurring members of its species. (323)

protozoan or **protozoon** (pro″to-zo'an, pl. **protozoa**) A unicellular or acellular eucaryotic; chemoorganotrophic protist whose organelles have the functional role of organs and tissues in more complex forms. Protozoa vary greatly in size, morphology, nutrition, and life cycle. (3, 605)

protozoology (pro″to-zo-ol'o-je) The study of protozoa. (605)

proviral DNA Viral DNA that has been integrated into host cell DNA. In retroviruses it is the double-stranded DNA copy of the RNA genome. (457)

pseudopodium or **pseudopod** (soo″do-po'de-um) A nonpermanent cytoplasmic extension of the cell body by which amoeboid protists move and feed. (613)

psittacosis (**ornithosis**; sit″ah-ko'sis) A disease due to a strain of *Chlamydia psittaci*, first seen in parrots and later found in other birds and domestic fowl (in which it is called ornithosis). It is transmissible to humans. (991)

psychrophile (si'kro-fīl) A microorganism that grows well at 0°C and has an optimum growth temperature of 15°C or lower and a temperature maximum around 20°C. (137)

psychrotroph A microorganism that grows at 0°C, but has a growth optimum between 20 and 30°C, and a maximum of about 35°C. (138)

pulmonary anthrax (pul'mo-ner″e) A form of anthrax involving the lungs. Also known as woolsorter's disease. (988)

pulmonary syndrome hantavirus *See* hantavirus pulmonary syndrome. (923)

puncutated equilibria The observation based on the fossil record that evolution does not proceed at a slow and linear pace, but rather is periodically interrupted by rapid bursts of speciation and extinction driven by abrupt changes in environmental conditions. (477)

pure culture A population of cells that are identical because they arise from a single cell. (113)

purine (pu′rin) A basic, heterocyclic, nitrogen-containing molecule with two joined rings that occurs in nucleic acids and other cell constituents; most purines are oxy or amino derivatives of the purine skeleton. The most important purines are adenine and guanine. (241)

purple membrane An area of the plasma membrane of *Halobacterium* that contains bacteriorhodopsin and is active in photosynthetic light energy trapping. (515)

putrefaction (pu″trĕ-fak′shun) The microbial decomposition of organic matter, especially the anaerobic breakdown of proteins, with the production of foul-smelling compounds such as hydrogen sulfide and amines. (1024)

pyrenoid (pi′rĕ-noid) The differentiated region of the chloroplast that is a center of starch formation in some photosynthetic protists. (90, 608)

pyrimidine (pi-rim′i-dēn) A basic, heterocyclic, nitrogen-containing molecule with one ring that occurs in nucleic acids and other cell constituents; pyrimidines are oxy or amino derivatives of the pyrimidine skeleton. The most important pyrimidines are cytosine, thymine, and uracil. (241)

Q

Q fever An acute zoonotic disease caused by the rickettsia *Coxiella burnetii*. (964)

Quellung reaction The increase in visibility or the swelling of the capsule of a microorganism in the presence of antibodies against capsular antigens. (876)

quinolones A class of broad-spectrum antibiotics, derived from nalidixic acid, that bind to bacterial DNA gyrase, inhibiting DNA replication and transcription. This group of antibiotics is bacteriocidal. (847)

quorum sensing The process in which bacteria monitor their own population density by sensing the levels of signal molecules that are released by the microorganisms. When these signal molecules reach a threshold concentration, quorum-dependent genes are expressed. (144, 309)

R

rabies (ra′bēz) An acute infectious disease of the central nervous system, which affects all warm-blooded animals (including humans). It is caused by an ssRNA virus belonging to the genus *Lyssavirus* in the family *Rhabdoviridae*. (943)

racking The removal of sediments from wine bottles. (1043)

radappertization The use of gamma rays from a cobalt source for control of microorganisms in foods. (1031)

radioimmunoassay (RIA) (ra″de-o-im″u no ac′a) A very sensitive assay technique that uses a purified radioisotope-labeled antigen or antibody to compete for antibody or antigen with unlabeled standard and samples to determine the concentration of a substance in the samples. (882)

rational drug design An approach to new drug development based a specific cellular macromolecule or target. (398)

reactivation tuberculosis *See* miliary tuberculosis. (954)

reactive nitrogen intermediate (RNI) Charged nitrogen radicals intermediate between various stable nitrogen molecules. (755)

reactive oxygen intermediate (ROI) Charged oxygen radicals intermediate between various stable oxygen molecules; also called reactive oxygen species (ROS). (755)

reading frame The way in which nucleotides in DNA and mRNA are grouped into codons or groups of three for reading the message contained in the nucleotide sequence. (264)

reagin (re′ah-jin) Antibody that mediates immediate hypersensitivity reactions. IgE is the major reagin in humans. (803)

real-time PCR A type of polymerase chain reaction (PCR) that quantitatively measures the amount of template in a sample as the amount of fluorescently labeled amplified product. (363)

receptor-mediated endocytosis A type of endocytosis that involves the specific binding of molecules to membrane receptors followed by the formation of coated vesicles. The substances being taken in are concentrated during the process. (86)

recombinant DNA technology The techniques used in carrying out genetic engineering; they involve the identification and isolation of a specific gene, the insertion of the gene into a vector such as a plasmid to form a recombinant molecule, and the production of large quantities of the gene and its products. (357)

recombinant-vector vaccine The type of vaccine that is produced by the introduction of one or more of a pathogen's genes into attenuated viruses or bacteria. The attenuated virus or bacterium serves as a vector, replicating within the vertebrate host and expressing the gene(s) of the pathogen. The pathogen's antigens induce an immune response. (904)

recombination (re″kom-bi-na′shun) The process in which a new recombinant chromosome is formed by combining genetic material from two organisms. (329)

recombinational repair A DNA repair process that repairs damaged DNA when there is no remaining template; a piece of DNA from a sister molecule is used. (329)

Redfield ratio The carbon-nitrogen-phosphorus ratio of marine microorganisms. This ratio is important for predicting limiting factors for microbial growth. (670)

red tides Red tides occur frequently in coastal areas and often are associated with population blooms of dinoflagellates. Dinoflagellate pigments are responsible for the red color of the water. Under these conditions, the dinoflagellates often produce saxitoxin, which can lead to paralytic shellfish poisoning. More commonly called harmful algal blooms (HABs). (621)

reducing power Molecules such as NADH and NADPH that temporarily store electrons. The stored electrons are used in anabolic reactions such as CO_2 fixation and the synthesis of monomers (e.g., amino acids). (168)

reductive dehalogenation The cleavage of carbon-halogen bonds by anaerobic bacteria that creates a strong electron-donating environment. (1075)

refraction (re-frak′shun) The deflection of a light ray from a straight path as it passes from one medium (e.g., glass) to another (e.g., air). (17)

refractive index (re-frak′tiv) The ratio of the velocity of light in the first of two media to that in the second as it passes from the first to the second. (17)

regulator T cell Regulator T cells control the development of effector T cells. Two types exist: T-helper cells (CD4$^+$ cells) and T-suppressor cells. There are three subsets of T-helper cells: T_H1, T_H2, and T_H0. T_H1 cells produce IL-2, IFN-γ, and TNF-β. They effect cell-mediated immunity and are responsible for delayed-type hypersensitivity reactions and macrophage activation. T_H2 cells produce IL-4, IL-5, IL-6, IL-10, IL-13. They are helpers for B-cell antibody responses and humoral immunity; they also support IgE responses and eosinophilia. T_H0 cells exhibit an unrestricted cytokine profile. (781)

regulatory mutants Mutant organisms that have lost the ability to limit synthesis of a product, which normally occurs by regulation of activity of an earlier step in the biosynthetic pathway. (1071)

regulon A collection of genes or operons that is controlled by a common regulatory protein. (307)

replica plating A technique for isolating mutants from a population by plating cells from each colony growing on a nonselective agar medium onto plates with selective media or environmental conditions, such as the lack of a nutrient or the presence of an antibiotic or a phage; the location of mutants on the original plate can be determined from growth patterns on the replica plates. (324)

replicase An RNA-dependent RNA polymerase used to replicate the genome of an RNA virus. (455)

replication (rep″li-ka′shun) The process in which an exact copy of parental DNA (or viral RNA) is made with the parental molecule serving as a template. (251)

replication fork The Y-shaped structure where DNA is replicated. The arms of the Y contain template strand and a newly synthesized DNA copy. (256)

replicative form (RF) A double-stranded form of nucleic acid that is formed from a single-stranded virus genome and used to synthesize new copies of the genome. (436, 455) exposure). It results from the binding of antigen to activated T lymphocytes, which then release cytokines and trigger inflammation and macrophage attacks that damage tissue. Type IV hypersensitivity is seen in contact dermititis from poison ivy, leprosy, and tertiary syphilis. (807)

type I protein secretion pathway *See* ABC protein secretion pathway.

type II protein secretion pathway A system that transports proteins from the periplasm across the outer membrane of gram-negative bacteria. (65)

type III protein secretion pathway A system in gram-negative bacteria that secretes virulence factors and injects them into host cells. (65)

type strain The microbial strain that is the nomenclatural type or holder of the species name. A type strain will remain within that species should nomenclature changes occur. (480)

typhoid fever (ti-foid) A bacterial infection transmitted by contaminated food, water, milk, or shellfish. The causative organism is *Salmonella enterica* serovar Typhi, which is present in human feces. (984)

U

ultramicrobacteria Bacteria that can exist normally in a miniaturized form or which are capable of miniaturization under low-nutrient conditions. They may be 0.2 µm or smaller in diameter. (671)

ultraviolet (UV) radiation Radiation of fairly short wavelength, about 10 to 400 nm, and high energy. (142, 156)

universal phylogenetic tree A phylogenetic tree that considers the evolutionary relationships among organism from all three domains of life: *Bacteria, Archaea,* and *Eucarya.* (475)

uracil (u'rah-sil) The pyrimidine 2,4-dioxypyrimidine, which is found in nucleosides, nucleotides, and RNA. (241)

V

vaccine A preparation of either killed microorganisms; living, weakened (attenuated) microorganisms; or inactivated bacterial toxins (toxoids). It is administered to induce development of the immune response and protect the individual against a pathogen or a toxin. (901)

vaccinomics The application of genomics and bioinformatics to vaccine development. (901)

valence (va'lens) The number of antigenic determinant sites on the surface of an antigen or the number of antigen-binding sites possessed by an antibody molecule. (774)

vancomycin A glycopeptide antibiotic obtained from *Nocardia orientalis* effective only against gram-positive bacteria. Vancomycin is bactericidal because it binds to the D-alanine amino acids of peptidogylcan precursor units, inhibiting peptidoglycan

synthesis and altering cell wall permeability. While the final cidal events are similar to those of penicillin, the mechanism of action is different. (845)

variable region (V_L and V_H) The region at the N-terminal end of immunoglobulin heavy and light chains whose amino acid sequence varies between antibodies of different specificity. Variable regions form the antigen binding site. (790)

vasculitis (vas"ku-li'tis) Inflammation of a blood vessel. (960)

vector (vek'tor) 1. In genetic engineering, another name for a cloning vector. A DNA molecule that can replicate (a replicon) and transports a piece of inserted foreign DNA, such as a gene, into a recipient cell. It may be a plasmid, phage, cosmid or artificial chromosome. 2. In epidemiology, it is a living organism, usually an arthropod or other animal, that transfers an infective agent between hosts. (358, 818, 892)

vector-borne transmission The transmission of an infectious pathogen between hosts by means of a vector. (896)

vehicle (ve'ĭ-k'l) An inanimate substance or medium that transmits a pathogen. (894)

venereal syphilis (ve-ne're-al sif'ĭ-lis) A contagious, sexually transmitted disease caused by the spirochete *Treponema pallidum.* (976)

venereal warts *See* anogenital condylomata.

verrucae vulgaris (v'ĕ-roo'se vul-ga'ris; s. **verruca vulgaris**) The common wart; a raised, epidermal lesion with horny surface caused by an infection with a human papillomavirus. (938)

vesicular nucleus The most common nuclear morphology seen in protists, characterized by a nucleus 1 to 10 µm in diameter, spherical, with a distinct nucleolus and uncondensed chromosomes. (609)

viable but nonculturable (VBNC) microorganisms Microbes that have been determined to be living in a specific environment (either in nature or the laboratory) but are not actively growing and cannot be cultured under standard laboratory conditions. (125)

vibrio (vib're-o) A rod-shaped bacterial cell that is curved to form a comma or an incomplete spiral. (40)

viral hemagglutination (vi'ral hem"ah-gloo"tĭna'shun) The clumping or agglutination of red blood cells caused by some viruses. (876)

viral hemorrhagic fevers (VHF) A group of illnesses caused by several distinct viruses, all of which cause symptoms of fever and bleeding in infected humans. (941)

viral neutralization An antibody-mediated process in which IgG, IgM, and IgA antibodies bind to some viruses during their extracellular phase and inactivate or neutralize them. (799)

viremia (vi-re'me-ə) The presence of viruses in the blood stream. (819)

viricide (vir'ĭ-sīd) An agent that inactivates viruses so that they cannot reproduce within host cells. (151)

virion (vi're-on) A complete virus particle that represents the extracellular phase of the virus life cycle; at the simplest, it consists of a protein capsid surrounding a single nucleic acid molecule. (409)

virioplankton Viruses that occur in waters; high levels are found in marine and freshwater environments. (679)

viroid (vi'roid) An infectious agent that is a single-stranded RNA not associated with any protein; the RNA does not code for any proteins and is not translated. (467)

virology (vi-rol'o-je) The branch of microbiology that is concerned with viruses and viral diseases. (407)

virulence (vir'u-lens) The degree or intensity of pathogenicity of an organism as indicated by case fatality rates and/or ability to invade host tissues and cause disease. (816)

virulence factor A bacterial product, usually a protein or carbohydrate, that contributes to virulence or pathogenicity. (816)

virulent viruses Viruses that lyse their host cells during the reproductive cycle. (345)

virus An infectious agent having a simple acellular organization with a protein coat and a nucleic acid genome, lacking independent metabolism, and reproducing only within living host cells. (3, 409)

virusoid An infectious agent that is composed only of single-stranded RNA that encodes some but not all proteins required for its replication; can be replicated and transmitted to a new host only if it infects a cell also infected by a virus that serves as a helper virus. (468)

vitamin An organic compound required by organisms in minute quantities for growth and reproduction because it cannot be synthesized by the organism; vitamins often serve as enzyme cofactors or parts of cofactors. (105)

volutin granules (vo-lu'tin) *See* metachromatic granules.

W

wart An epidermal tumor of viral origin. (938)

wastewater treatment The use of physical and biological processes to remove particulate and dissolved material from sewage and to control pathogens. (1055)

water activity (a_w) A quantitative measure of water availability in the habitat; the water activity of a solution is one-hundredth its relative humidity. (134)

water mold A common term for an öomycete. (3)

West Nile fever (encephalitis) A neurological viral disease that is spread from birds to humans by mosquitoes. It first appeared in the United States in 1999 and subsequently has been reported in almost every state. (924)

white blood cell (WBC) Blood cells having innate or acquired immune function. They are named for the white or buffy layer in which they are found when blood is centrifuged. (744)

white piedra   A fungal infection caused by the yeast *Trichosporon beigelii* that forms light-colored nodules on the beard and mustache. (1008)

whole-cell vaccine   A vaccine made from complete pathogens, which can be of four types: inactivated viruses; attenuated viruses; killed microorganisms; and live, attenuated microbes. (901)

whole-genome shotgun sequencing   An approach to genome sequencing in which the complete genome is broken into random fragments, which are then individually sequenced. Finally the fragments are placed in the proper order using sophisticated computer programs. (384)

Widal test (ve-dahl′)   A test involving agglutination of typhoid bacilli when they are mixed with serum containing typhoid antibodies from an individual having typhoid fever; used to detect the presence of *Salmonella typhi* and *S. paratyphi*. (876)

Winogradsky column   A glass column with an anaerobic lower zone and an aerobic upper zone, which allows growth of microorganisms under conditions similar to those found in a nutrient-rich lake. (675)

wort   The filtrate of malted grains used as the substrate for the production of beer and ale by fermentation. (1041)

X

xenograft (zen″o-graft)   A tissue graft between animals of different species. (810)

xerophilic microorganisms (ze″ro-fil′ik)   Microorganisms that grow best under low a_w conditions, and may not be able to grow at high a_w values. (1024)

Y

yeast (yēst)   A unicellular, uninuclear fungus that reproduces either asexually by budding or fission, or sexually through spore formation. (631)

yeast artificial chromosome (YAC)   Engineered DNA that contains all the elements required to propagate a chromosome in yeast and which is used to clone foreign DNA fragments in yeast cells. (370)

yellow fever   An acute infectious disease caused by a flavivirus, which is transmitted to humans by mosquitoes. The liver is affected and the skin turns yellow in this disease. (924)

YM shift   The change in shape by dimorphic fungi when they shift from the yeast (Y) form in the animal body to the mold or mycelial form (M) in the environment. (632)

Z

zidovudine   A drug that inhibits nucleoside reverse transcriptase (also known as AZT, ZDV, or retrovir) used as an anti-HIV treatment. (856)

zoonosis (zo″o-no′sis; pl. *zoonoses*)   A disease of animals that can be transmitted to humans. (892)

zooxanthella (zo″o-zan-thel′ah)   A dinoflagellate found living symbiotically within cnidarians and other invertebrates. (620, 719)

Z ring   A ring-shaped structure that forms on the cytoplasmic side of the plasma membrane during the bacterial cell cycle. Formation of the Z ring is the first step in formation of the septum, which will eventually divide the parent cell into two daughter cells. (122)

z **value**   The increase in temperature required to reduce the decimal reduction time to one-tenth of its initial value. (154)

zygomycetes (zi″go-mi-se′tez)   A division of fungi that usually has a coenocytic mycelium with chitinous cell walls. Sexual reproduction normally involves the formation of zygospores. The group lacks motile spores. (635)

zygospore (zi′go-spōr)   A thick-walled, sexual, resting spore characteristic of the zygomycetous fungi. (634)

zygote (zi′gōt)   The diploid (2n) cell resulting from the fusion of male and female gametes. (614, 620)